Human Anatomy and Physiology Laboratory Manual

Elaine N. Marieb, R.N., Ph.D.

Holyoke Community College

FOURTH EDITION

The Benjamin/Cummings Publishing Company, Inc.

Menlo Park, California • Reading, Massachusetts
New York • Don Mills, Ontario • Wokingham, U.K. • Amsterdam
Bonn • Paris • Milan • Madrid • Sydney • Singapore • Tokyo
Seoul • Taipei • Mexico City • San Juan, Puerto Rico

Executive Editor: Johanna Schmid
Sponsoring Editor: Lisa Moller
Assistant Editors: Natasha Banta, Thor Ekstrom
Production Editor: Lisa Weber
Art Supervisors: Kathleen Cameron, Carol Ann Smallwood
Photo Editor: Kathleen Cameron
Design Manager: Don Kesner
Composition and Film Manager: Lillian Hom
Composition and Film Buyer: Vivian McDougal
Permissions Editor: Marty Granahan
Manufacturing Supervisor: Merry Free Osborn
Marketing Manager: Nina Horne
Artists: Val Felts, Karl Miyajima, Kristin Otwell, Denise Schmidt
Copyeditor: Anita Wagner
Proofreader: Christine Sabooni
Indexer: William J. Richardson Associates, Inc.
Text Designer: Gary Head
Page Layout: Brenn Lea Pearson
Compositor/Film Preparation: GTS Graphics
Printer: Von Hoffmann Press
Cover Designer: Yvo Riezebos
Cover Illustrator: Joseph Maas/Paragon[3]

The Author and Publisher believe that the lab experiments described in this publication, when conducted in conformity with the safety precautions described herein and according to the school's laboratory safety procedures, are reasonably safe for the students to whom this manual is directed. Nonetheless, many of the described experiments are accompanied by some degree of risk, including human error, the failure or misuse of laboratory or electrical equipment, mismeasurement, spills of chemicals, and exposure to sharp objects, heat, bodily fluids, blood, or other biologics. The Author and Publisher disclaim any liability arising from such risks in connection with any of the experiments contained in this manual. If students have any questions or problems with materials, procedures, or instructions on any experiment, they should *always* ask their instructor for help before proceeding.

The Benjamin/Cummings Series in Human Anatomy and Physiology

Textbooks

By R. A. Chase
The Bassett Atlas of Human Anatomy (1989)

By S. W. Langjahr and R. D. Brister
Coloring Atlas of Human Anatomy, second edition (1992)

By E. N. Marieb
Human Anatomy and Physiology, third edition (1995)
Human Anatomy and Physiology, Study Guide, third edition (1995)
Human Anatomy and Physiology Laboratory Manual, Cat Version, fifth edition (1996)
Human Anatomy and Physiology Laboratory Manual, Fetal Pig Version, fifth edition (1996)
Human Anatomy Laboratory Manual, first edition (1992)
Essentials of Human Anatomy and Physiology, fourth edition (1994)
The A & P Coloring Workbook: A Complete Study Guide, fourth edition (1994)

By E. N. Marieb and J. Mallatt
Human Anatomy, first edition (1992)

Videotapes by California State University, Sacramento

By R. L. Vines and A. Hinderstein
Human Musculature (1989)

By R. L. Vines and University Media Services
Human Nervous System (1992)
The Cardiovascular System: The Heart (1995)
The Cardiovascular System: The Blood Vessels (1995)

By R. L. Vines and J. Barrena
Selected Actions of Hormones and Other Chemical Messengers (1994)

ISBN 0-8053-4813-1

3 4 5 6 7 8 9 10—VH—99 98 97 96

Credits appear following Appendix E.

Cover printed with soy-based inks.

The Benjamin/Cummings Publishing Company, Inc.
2725 Sand Hill Road
Menlo Park, CA 94025

Table of Contents

Exercises with a red asterisk (*) contain physiology experiments.

The Circulatory System

The Respiratory System

The Digestive System

The Urinary System

The Reproductive System, Development, and Heredity

Surface Anatomy

Preface to the Instructor

Each new edition of this book brings a matured sensibility for the way teachers teach and students learn. This sensibility is achieved through years of teaching the subject and by listening to the suggestions of other instructors as well as those of students enrolled in multifaceted health-care programs. The fourth edition of *Human Anatomy and Physiology Laboratory Manual* (in previous editions called *Human Anatomy and Physiology: Brief Version*) has been developed to facilitate the A&P laboratory experience for both teachers and students.

As with previous editions of this manual, this edition is intended for students in the introductory human anatomy and physiology courses. It presents a wide range of laboratory experiences for students concentrating in nursing, physical therapy, occupational therapy, respiratory therapy, dental hygiene, pharmacology, health, and physical education, as well as biology and premedical programs. It differs from the Cat and Fetal Pig Versions (Fifth Editions, 1996) in that it does not include detailed dissection guidelines for any particular laboratory animal. The manual's coverage is intentionally broad, allowing it to serve both one- and two-semester courses.

ORGANIZATION

The generous variety of experiments in this manual provides flexibility that enables instructors to gear their courses to specific academic programs, or to their own teaching preferences. The manual is still independent of any textbook, so it contains the background discussions and terminology necessary to perform all experiments. Such a self-contained learning aid eliminates the need for students to bring a textbook into the laboratory.

Each of the 46 exercises leads students toward a coherent understanding of the structure and function of the human body. The manual begins with anatomical terminology and an orientation to the body, which together provide the necessary tools for studying the various body systems. The exercises that follow reflect the dual focus of the manual—both anatomical and physiological aspects receive considerable attention. As the various organ systems of the body are introduced, the initial exercises focus on organization, from the cellular to the organ system level. As indicated by the Table of Contents, the anatomical exercises are followed by physiological experiments that familiarize students with various aspects of body functioning and promote an appreciation for the critical understanding that function follows structure. Homeostasis is continually emphasized as a requirement for optimal health. Pathological conditions are viewed as a loss of homeostasis; these discussions can be recognized by the homeostasis imbalance logo within the descriptive material of each exercise. This holistic approach encourages an integrated understanding of the human body.

NEW FEATURES

In this revision, I really tried to respond to reviewers' and users' feedback concerning trends that are having an impact on the anatomy and physiology laboratory experience, most importantly:

- the growing reluctance of students to perform experiments using living laboratory animals and the declining popularity of animal dissection exercises
- the increased use of multimedia (e.g., videodiscs, computers, etc.) in the laboratory and, hence, the subsequent desire for more computer simulation exercises and advanced instrumentation

Among the specific changes newly implemented to address these trends are the following:

1. Appendix C with four new physiological exercises using Intelitool equipment and students as subjects (instead of laboratory animals) has been added to this edition. The new exercises are identified by the number of the exercise they may be used to replace plus the letter ***i*** to indicate the Intelitool instrumentation. These include:

   - Exercise 16i Muscle Physiology
   - Exercise 22i Human Reflex Physiology
   - Exercise 31i Conduction System of the Heart and Electrocardiography
   - Exercise 37i Respiratory System Physiology

2. A new videotape is available to qualified adopters. *Selected Actions of Hormones and Other Chemical Messengers* by Rose Leigh Vines and Juanita Barrena of California State University, Sacramento, replicates the preparation and observations required for the last three experiments in Exercise 28 on hormonal actions. For instructors who decide not to have their students conduct live animal experiments, this videotape provides an educationally sound and time-saving alternative.

3. Appendix D, also new to this edition, correlates some of the required anatomical laboratory observations with the corresponding sections of the A.D.A.M. Standard Program (available for purchase from The Benjamin/Cummings Publishing Company). Using A.D.A.M. to complement the printed manual descriptions of anatomical structures provides an extremely useful study method for visually oriented students.

4. In response to reviewers' feedback on Exercise 39, Chemical and Physical Processes of Digestion, that the preparation was unwieldy and the time required before results were available was excessive, that exercise has been completely rewritten. The current exercise uses microdrop technique and/or tests microquantities. This approach is much easier to implement and results are clearer and quickly available for evaluation.

5. A listing of barcodes corresponding to images on the *Anatomy and PhysioShow: The Videodisc* (available to qualified adopters) is now provided in Appendix E. These barcodes are useful pedagogically to both instructors (for demonstration) and students (for self-study and review) because they provide immediate access to images related to topics being studied in the lab.

6. The Human Cardiovascular System videotapes, prepared by Rose Leigh Vines and University Media Service, California State University, Sacramento, provide a comprehensive human cadaver study of heart anatomy, enhanced by an investigation of valve function, and a beautifully executed survey of the major blood vessels of the body. This pair of videotapes will be an excellent addition to the multimedia library of any A&P laboratory.

7. The two new icons (visual mnemonics) that appear in this edition alert instructors and students to the subject areas whose study can be embellished by using multimedia products available from The Benjamin/Cummings Publishing Company.

The A.D.A.M. "walking man" icon indicates where use of the A.D.A.M. software would enhance the study and comprehension of the laboratory topics.

The video icon signifies that specific images on *Anatomy and PhysioShow: The Videodisc* can be used as an added resource for a given laboratory exercise.

8. In continuing recognition of the value of the printed page as a source of information, more than 40 new pieces of art and 45 new photographs have been added to this edition as appropriate and as space allowed. Because art plays such a critical and immediate role in helping students visualize anatomical and physiological concepts, every effort has been made to offer a figure, either line art or photo, for essentially every structure that is examined in the lab.

9. The anatomical terminology in this edition has been updated to match that in *Human Anatomy and Physiology,* Third Edition (by this author), to avoid student confusion over conflicting terms.

10. Finally, the laboratory review sheets have been expanded and now include more labeling exercises, particularly for the muscular system anatomy exercise.

SPECIAL FEATURES

Virtually all the special features appreciated by the adopters of the third edition are retained.

- The prologue, "Getting Started—What to Expect, The Scientific Method, Scientific Notation, and Metrics," explains the scientific method, the logical, practical, and reliable way of approaching and solving problems in the laboratory. It also reviews the use of exponents, metric units, and interconversions.
- Each exercise begins with learning objectives.
- Key terms appear in boldface print, and each term is defined when introduced.
- Illustrations are large and of exceptional quality. Full-color photographs and drawings highlight, differentiate, and focus student attention on important structures.
- Body structures are studied from simple-to-complex levels, and physiological experiments allow ample opportunity for student observation, manipulation, and experimentation.
- There are numerous physiological experiments for each organ system, ranging from simple experiments that can be performed without specialized tools to more complex experiments using laboratory equipment and instrumentation techniques.
- Space is provided for recording and interpreting experimental results, and tear-out laboratory review sheets, located toward the end of the manual, are designed to accompany each lab exercise. The review sheets require students to label diagrams and answer multiple-choice, short-answer, and review questions.
- In addition to the figures and suggested models, isolated animal organs such as the sheep heart and pig kidney are employed because of their exceptional similarity to human organs. If a major dissection animal is to be used, either the Cat or Fetal Pig Version of the *Human Anatomy and Physiology Manual,* Fifth Edition, is recommended.
- The ten-page Histology Atlas has 60 color photomicrographs, and the edges of this atlas are colored blue to allow it to be quickly located. The histology plates formerly located on the inside front and back covers have been moved to the Histology Atlas proper. The photomicrographs selected are those deemed most helpful to students because they correspond closely with slides typically viewed in the lab. While most such tissues are stained with hematoxylin and eosin (H & E), a few depicted in the Histology Atlas are stained with differential stains to allow selected cell populations to be identified in a given tissue. For ex-

ample, the glucagon-producing alpha cells can be differentiated from the insulin-secreting beta cells in the pancreatic islets (Plate 20). Line drawings, corresponding exactly to the plates in the Histology Atlas, appear in appropriate places in the text and add to the utility and effectiveness of the atlas. Since these diagrams can be colored by the student to replicate the stains of the slides they are viewing, they provide a valuable learning aid.

- The seven-plate Human Anatomy Atlas, providing full-color photographs of many organ systems, is easily identified by its pink-edged pages. The two color photographs of the Beauchene skull, located between the Histology Atlas and the Human Anatomy Atlas, highlight the three-dimensional quality of the bones of the skull and their relationships to each other.
- All exercises involving body fluids (blood, urine, saliva) incorporate current Centers for Disease Control (CDC) guidelines for handling human body fluids. Since it is important that nursing students, in particular, learn how to safely handle bloodstained articles, the human focus has been retained. However, the decision to allow testing of human (student) blood or to use animal blood in the laboratory is left to the discretion of the instructor in accordance with institutional guidelines. The CDC guidelines for handling body fluids are reinforced by the laboratory safety procedures described on the inside front cover of this text and the *Instructor's Guide.* The inside cover can be photocopied and posted in the lab to help students become well versed in laboratory safety.
- In addition to the two new icons that highlight the A.D.A.M. and videodisc correlation appendixes, the five logos used previously to alert students to special features or instructions have been retained. These include:

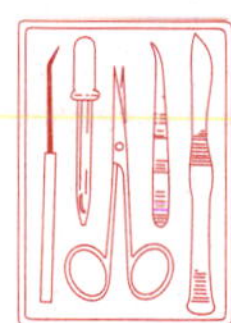

The dissection tray icon appears at the beginning of lab activities to be conducted by the students.

The homeostasis imbalance icon directs the student's attention to conditions representing a loss of homeostasis.

A blood and body fluid icon appears when blood or other body fluids such as saliva or urine must be handled and signifies where special self-protective measures are to be taken.

A safety icon notifies students that specific safety precautions are to be observed when using certain equipment or conducting particular lab procedures (for example, a hood is to be used when working with ether, and so on).

The cooperative learning icon is seen before experiment heads or procedures where learning would be enhanced and/or time saved by having students work together (in pairs or teams).

SUPPLEMENTS

The *Instructor's Guide* that accompanies all versions of the *Human Anatomy and Physiology Laboratory Manual* contains a wealth of information for those teaching this course. Instructors can find help in planning the experiments, ordering equipment and supplies, anticipating pitfalls and problem areas, and locating audiovisual material. The probable in-class time required for each lab is indicated by a clock icon. Other useful resources are the Trends in Instrumentation section that describes the latest laboratory equipment and technological teaching tools available. Additional supplements, all mentioned above, include these available free of charge to qualified adopters:

- *Anatomy and PhysioShow: The Videodisc*
- *Selected Actions of Hormones and Other Chemical Messengers* videotape by Rose Leigh Vines and Juanita Barrena
- *Human Musculature* videotape by Rose Leigh Vines and Allan Hinderstein
- *The Human Cardiovascular System: The Heart* videotape by Rose Leigh Vines and University Media Services, California State University, Sacramento
- *The Human Cardiovascular System: The Blood Vessels* videotape by Rose Leigh Vines and University Media Services, California State University, Sacramento
- *Human Nervous System* videotape by Rose Leigh Vines and University Media Services, California State University, Sacramento

A.D.A.M. SOFTWARE

Available for purchase from The Benjamin/Cummings Publishing Company to enhance student learning are:

- A.D.A.M. Standard
- A.D.A.M.-Benjamin/Cummings Cardiovascular Physiology Module by Elaine Marieb and Marvin Branstrom
- A.D.A.M.-Benjamin/Cummings Muscle Physiology Module by Marvin Branstrom

ACKNOWLEDGMENTS

I wish to thank the following reviewers for their contributions to this edition: Merlyn Anderberg (Spokane Falls Community College), John Beneski (West Chester University), Moges Bizuneh (Ivy Tech, Indianapolis), Linda Burroughs (Bucks County Community College), Jim Constantine (Bristol Community College), Steven Gehnrich (Salisbury State University), Anne Golubisky (St. Joseph's College), Lois Harlston (Tennessee State University), Henry Knizesky (Mercy College), Jerri Lindsey (Tarrant County Junior College), Ron McCune (Crafton Hills College), Michael McMahon (Union University), David Mense (Southeastern Community College), Clyde Miller (Columbus State Community

College), Robert Palma (Community College of Philadelphia), Don Puder (College of Southern Idaho), Laura Ritt (Burlington County College), Annette Schaefer (New York City Technical College), Joyce B. Stone (Memphis State University), and Don Yeltman (Davis and Elkins College). Special thanks to Intelitool's president Steven J. Flora and Steve Schact, director of Intelitool's Academic Product division, who reviewed Appendix C to assure the accuracy of our presentation.

My continued thanks to my colleagues at Benjamin/Cummings who helped me in the production of this edition, especially Lisa Moller, my new sponsoring editor, and Johanna Schmid, my "old" one who "guardian angeled" this work until Lisa was on board. Kudos also to Lisa Weber, associate production editor for this edition. Her precise approach to her work and her easygoing nature have made my job easy. Also not to be forgotten are Natasha Banta, assistant editor, for her diligent work; Thor Ekstrom, the assistant editor who handled hundreds of details (and "flogged" me daily to get this preface in); Thomas Viano, assistant editor, who helped with videodisc and A.D.A.M. correlations; and Kathleen Cameron, who is responsible for the photo research.

"Off campus" but equally important are Yvo Riezebos who conjured up the cover idea; Valerie Felts, Karl Miyajima, Kristin Otwell, and Denise Schmidt, who drew the new art pieces for this edition; and Anita Wagner, the very competent copy editor who is now an old hand at working on my books. A special debt of gratitude is owed Peter Zao (North Idaho College), who produced the much more efficient enzyme experiments for Exercise 39 and wrote the Intelitool exercises for Appendix C. Thanks also to Carlene Tonini, who catalogued the new barcodes in Appendix D and the A.D.A.M. correlations in Appendix E. Rose Leigh Vines deserves a hand for the significant roles she and the team from Sacramento State played in producing the excellent new videotapes on the cardiovascular system and on hormonal actions (details on p. vi).

Finally, since preparations for the next edition begin well in advance of its publication, I invite users of this edition to send me their comments and suggestions for subsequent editions.

Elaine N. Marieb
Department of Science, Engineering, and Math
Holyoke Community College
303 Homestead Avenue
Holyoke, MA 01040

Preface to the Student

Hopefully, your laboratory experiences will be exciting times for you. Like any unfamiliar experience, it really helps if you know in advance what to expect and what will be expected of you.

LABORATORY ACTIVITIES

The A&P laboratory exercises in this manual are designed to help you gain a broad understanding of both anatomy and physiology. So, you can anticipate that you will be examining models, dissecting isolated animal organs, and using a microscope to look at tissue slides (anatomical approaches). You will also investigate chemical conditions or observe changes in both living and nonliving systems, and conduct experiments that examine responses of living organisms, such as frogs or yourself, to various stimuli (physiological approaches).

Because some students question the use of animals in the laboratory setting, their concerns need to be addressed. Be assured that the commercially available major dissection animals and preserved organ specimens used in the A&P labs are *not* harvested from animals raised specifically for dissection purposes. Instead, cats brought to the SPCA or other animal shelters for humane extermination are sent to qualified biological supply houses, where they are prepared for laboratory use according to USDA guidelines. Organs that are of no use to the meat packing industry (such as the brain, heart, or lungs) are sent from slaughterhouses to biological supply houses for preparation.

Relative to using live animals for experimentation, every effort is being made to find alternative methods that do not use living animals to study physiological concepts. For example, new to this edition are the computerized Intelitool exercises provided in Appendix C. These exercises employ the easy-to-use Intelitool apparatus and use you and your classmates as subjects. Such Intelitool-based approaches can provide alternatives to the traditional exercises on muscle activity and electrocardiography that employ frogs as experimental subjects. Exercise 16B, a computer simulation that studies the properties of contracting muscles, provides still another alternative to using live animal specimens. Without a doubt, computer simulations offer certain advantages: (1) they allow you to experiment at length without time constraints of traditional animal experiments, in which fragile living tissues need to be kept alive and viable for the duration of the experiment; and (2) they make it possible to investigate certain concepts (such as isometric versus isotonic contraction in this case) that would be difficult or impossible to explore in traditional exercises. Yet, the main disadvantage of computer simulations is that the real-life aspects of experimentation are sacrificed. An animated frog muscle or heart on a computer screen is not really a substitute for observing the responses of actual muscle tissue.

Although it might appear that the choice is simply to decide which approach offers fewer disadvantages, unfortunately adequate software applications are still unavailable. Consequently, living animal experiments remain an important part of the approach of this manual to the study of human A&P. However, wherever possible, the minimum number of animals needed to demonstrate a particular point are used. Furthermore, more instructor-delivered and videotaped demonstrations of live animal experiments are suggested.

If you use living animals for experiments, you will be expected to handle them humanely. Inconsiderate treatment of laboratory animals will not be tolerated in your A&P laboratory.

BARCODES

The barcodes found in Appendix E will allow you to access topic-related images from the new *Anatomy and PhysioShow: The Videodisc* as you work in the laboratory (provided your classroom has the appropriate equipment).

A.D.A.M. CORRELATIONS

If the A.D.A.M. CD-ROM software is available for your use, Appendix D correlating the various laboratory topics with specific frames of the A.D.A.M. software will provide an invaluable resource to help you in your studies.

ICONS/VISUAL MNEMONICS

I have tried to make this manual very easy for you to use, and to this end, seven different icons (visual mnemonics) are described below:

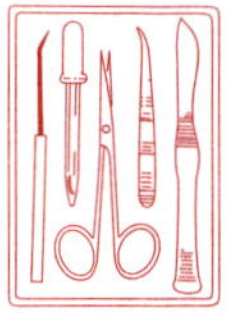

The **dissection tray icon** appears at the beginning of lab activities. Since most exercises have some explanatory background provided before the experiment(s), this visual cue alerts you that your lab involvement is imminent.

The **blood/body fluid icon** appears where blood or other body fluids (saliva, urine) must be handled, and it signifies that you should take special measures to protect yourself.

The **safety icon** alerts you to special precautions in handling lab equipment or conducting certain procedures, e.g., use a ventilating hood when using volatile chemicals.

The **cooperative learning icon** is seen before procedures where your learning will be enhanced (or time saved) if you work with a partner (or team).

The **homeostasis imbalance icon** appears where a clinical disorder is described to indicate what happens when there is a structural abnormality or physiological malfunction, i.e., a loss of homeostasis.

The **A.D.A.M. walking man icon** alerts you when the use of A.D.A.M. would enhance your laboratory experience.

The **video icon** indicates that there are images on *Anatomy and PhysioShow: The Videodisc* that correlate with laboratory observations you are being asked to make.

HINTS FOR SUCCESS IN THE LABORATORY

With the possible exception of those who have photographic memories, most students can use helpful hints and guidelines to ensure that they have successful lab experiences.

1. Perhaps the best bit of advice is to attend all your scheduled labs and to participate in all the assigned exercises. Learning is an *active* process.
2. Scan the scheduled lab exercise and the questions in the review section in the back of the manual that pertain to it *before* going to lab.
3. Be on time. Most instructors explain what the lab is about, pitfalls to avoid, and the sequence or format to be followed at the beginning of the lab session. If you are late, not only will you miss this information, you will not endear yourself to the instructor.
4. Review your lab notes after completing the lab session to help you focus on and remember the important concepts.
5. Keep your work area clean and neat. This reduces confusion and accidents.
6. Assume that all lab chemicals and equipment are sources of potential danger to you. Follow directions for equipment use and observe the laboratory safety guidelines provided inside the front cover of this manual.
7. Keep in mind the real value of the laboratory experience—a place for you to observe, manipulate, and experience "hands on" activities that will dramatically enhance your understanding of the lecture presentations.

I really hope that you enjoy your A&P laboratories and that this lab manual makes learning about intricate structures and functions of the human body a fun and rewarding process. I'm always open to constructive criticism and suggestions for improvement in future editions. If you have any, please write to me.

Elaine N. Marieb
Department of Science, Engineering and Math
Holyoke Community College
303 Homestead Avenue
Holyoke, MA 01040

Getting Started—What to Expect, The Scientific Method, Scientific Notation, and Metrics

Two hundred years ago science was largely a plaything of wealthy patrons, but today's world is dominated by science and its technology. Whether or not we believe that such domination is desirable, we all have a responsibility to try to understand the goals and methods of science that have seeded this knowledge and technological explosion.

The biosciences are very special and exciting because they open the doors to an understanding of all the wondrous workings of living things. A course in human anatomy and physiology (a minute subdivision of bioscience) provides such insights in relation to your own body. Although some experience in scientific studies is helpful when beginning a study of anatomy and physiology, perhaps the single most important prerequisite is curiosity.

Gaining an understanding of science is a little like becoming acquainted with another person. Even though a written description can provide a good deal of information about the person, you can never really know another unless there is personal contact. And so it is with science—if you are to know it well, you must deal with it intimately.

The laboratory is the setting for "intimate contact" with science. It is where scientists test their ideas (do research), the essential purpose of which is to provide a basis from which predictions about scientific phenomena can be made. Likewise, it will be the site of your "intimate contact" with the subject of human anatomy and physiology as you are introduced to the methods and instruments used in biological research.

For many students, human anatomy and physiology is taken as an introductory-level course; and their scientific background exists, at best, as a dim memory. If this is your predicament, this prologue may be just what you need to fill in a few gaps and to get you started on the right track before your actual laboratory experiences begin. So—let's get to it!

THE SCIENTIFIC METHOD

Science would quickly stagnate if new knowledge were not continually derived from and added to it. The approach commonly used by scientists when they investigate various aspects of their respective disciplines is called the **scientific method.** This method is *not* a single rigorous technique that must be followed in a lockstep manner. It is nothing more or less than a logical, practical, and reliable way of approaching and solving problems of every kind—scientific or otherwise—to gain knowledge. It comprises five major steps.

Step 1: Observation of Phenomena

The crucial first step involves observation of some phenomenon of interest. In other words, before a scientist can investigate anything, he or she must decide on a *problem* or focus for the investigation. In most college laboratory experiments, the problem or focus has been decided for you. However, to illustrate this important step, we will assume that you want to investigate the true nature of apples, particularly green apples. In such a case you would begin your studies by making a number of different observations concerning apples.

Step 2: Statement of the Hypothesis

Once you have decided on a focus of concern, the next step is to design a significant question to be answered. Such a question is usually posed in the form of a **hypothesis,** an unproven conclusion that attempts to explain some phenomenon. (At its crudest level, a hypothesis can be considered to be a "guess" or an intuitive hunch that tentatively explains some observation.) Generally, scientists do not restrict themselves to a single hypothesis; instead, they usually pose several and then test each one systematically.

We will assume that, to accomplish step 1, you go to the supermarket and randomly select apples from several bins. When you later eat the apples, you find that the green apples are sour, but the red and yellow apples are sweet. From this observation, you might conclude (*hypothesize*) that "green apples are sour." This statement would represent your current understanding of green apples. You might also reasonably predict that, if you were to buy more apples, any green ones you buy will be sour. Thus, you would have gone beyond your initial observation that "these" green apples are sour to the prediction that "all" green apples are sour.

Any good hypothesis must meet several criteria. First, *it must be testable.* This characteristic is far more important than its being correct. The tests may prove the hypothesis incorrect; or new information may require that the hypothesis be modified. Clearly the accuracy of a prediction in the green apple example or in any scientific study depends on the accuracy of the initial information on which it is based.

In our example, no great harm will come from an inaccurate prediction—that is, were we to find that some green apples are sweet. However, in some cases human life may depend on the accuracy of the prediction. Take the case of testing drugs for their effectiveness in treating disease. If one set of observations erroneously indicates that the drugs are risky but very effective, such a conclusion could lead to the death of the subsequent drug recipient(s). This illustrates two points: (1) Repeated testing of scientific ideas is important, particularly because scientists working on the same problem do not always agree in their conclusions. The studies on the use of saccharin and amino acid sweeteners are only two examples. (2) Conclusions drawn from scientific tests are only as accurate as the information on which they are based; therefore, careful observation is essential, even at the very outset of a study.

A second criterion is that, even though hypotheses are guesses of a sort, *they must be based on measurable, describable facts. No mysticism can be theorized.* We cannot conjure up, to support our hypothesis, forces that have not been shown to exist. For example, as scientists, we cannot say that the tooth fairy took Johnny's tooth unless we can prove that the tooth fairy exists!

Third, a hypothesis *must not be anthropomorphic.* Human beings tend to anthropomorphize—that is, to relate all experiences to human experience. Because man is a social animal influenced by culture, these two characteristics tend to promote biased thinking. Whereas we could state that bears instinctively protect their young, it would be anthropomorphic to say that bears love their young, because love is a human emotional response. Thus, the initial hypothesis must be stated without interpretation.

Step 3: Data Collection

Once the initial hypothesis has been stated, scientists plan experiments that will provide data (or evidence) to verify or disprove their hypotheses—that is, they *test* their hypotheses. Data are accumulated by making qualitative or quantitative observations of some sort. The observations are often aided by the use of various types of equipment such as cameras, microscopes, stimulators, or various electronic devices that allow chemical and physiologic measurements to be made.

Observations referred to as **qualitative** are those we can make with our senses—that is, by using our vision, hearing, or sense of taste, smell, or touch. The color of an object, its texture, the relationship of one part to another, and its relative size (large versus small) may all be part of a qualitative description. For some quick practice in qualitative observation, compare and contrast* an orange and an apple.

Whereas the differences between an apple and an orange are obvious, this is not always the case in biological observations. Quite often a scientist tries to detect very subtle differences that cannot be determined by qualitative observations; and data must be derived from measurements made using a variety of scientific equipment. Such observations based on precise measurements of one type or another are **quantitative observations.** Examples of quantitative observations include careful measurements of body or organ dimensions such as mass, size, and volume; measurement of volumes of oxygen consumed during metabolic studies; determination of the concentration of glucose and other chemicals in urine; and determination of the differences in blood pressure and pulse under conditions of rest and exercise. An apple and an orange could be compared quantitatively by performing chemical measurements of the relative amounts of sugar and water in a given volume of fruit flesh, by analyzing the pigments and vitamins present in the apple skin and orange peel, and so on.

A valuable part of data gathering is the use of experiments to verify or disprove a hypothesis. An **experiment** is a procedure designed to describe the factors in a given situation that affect one another (that is, to discover cause and effect) under certain conditions.

Two general rules govern experimentation. The first of these rules is that the experiment(s) should be conducted in such a manner that every **variable** (any factor that might affect the outcome of the experiment) is under the control of the experimenter. The experimenter manipulates the **independent variables** and observes the effects of this manipulation on the **dependent** (or **response**) **variable.** For example, if the goal is to determine the effect of body temperature on breathing rate, the value measured (breathing rate) is called the dependent variable because it "depends on" the value chosen for the independent variable (body temperature). The ideal way to perform such an experiment is to set up and run a series of tests that are all identical, except for one specific factor that is varied.

One specimen (or group of specimens) is used as the **control** against which all other experimental samples are compared. The importance of the control sample cannot be overemphasized. It is essential to know how the system you are investigating works under normal circumstances before you can be sure that the results obtained from experimentation are due solely to the manipulation of the independent variable(s). Taking our example one step further, if we wanted to investigate the effects of body temperature (the independent variable) on breathing rate (the dependent variable), we could collect data on the breathing rate of individuals with "normal" body temperature (the implicit control group), and compare these data to breathing-rate measurements obtained from groups of individuals with higher and lower body temperatures. The control group

* *Compare* means to emphasize the similarities between two things, whereas *contrast* means that the differences are to be emphasized.

would provide the "normal standard" against which all other samples would be compared relative to the dependent variable.

The second rule governing experimentation is that valid results require that testing be done on large numbers of subjects. It is essential to understand that it is nearly impossible to control all possible variables in biological tests. Indeed, there is a bit of scientific wisdom that mirrors this truth—that is, that laboratory animals, even in the most rigidly controlled and carefully designed experiments, "will do as they damn well please." Thus, stating that the testing of a drug for its pain-killing effects was successful after having tested it on only one postoperative patient would be scientific suicide. Large numbers of patients would have to receive the drug and be monitored for a decrease in postoperative pain before such a statement could have any scientific validity. Then, other researchers would have to be able to uphold those conclusions by running similar experiments. *Repeatability* is an important part of the scientific method, and is the primary basis for acceptance or rejection of many hypotheses.

During experimentation and observation, data must be carefully recorded. Usually, such initial, or raw, data are recorded in tabular (table) form. The table should be labeled to show the variables investigated and the results for each sample. At this point, *accurate recording* of observations is the primary concern. Later, these raw data will be reorganized and manipulated to show more explicitly the outcome of the experimentation.

Some of the observations that you will be asked to make in the anatomy and physiology laboratory will require that a drawing be made. Don't panic! The purpose of making drawings (in addition to providing a record) is to force you to observe things very closely. You need not be an artist (most biological drawings are simple outline drawings), but you do need to be neat and as accurate as possible. It is advisable to use a 4H pencil to do your drawings, because it is easily erased and doesn't smudge. Before beginning to draw, you should examine your specimen closely, studying it as though you were going to have to draw it from memory. For example, when looking at cells you should ask yourself questions such as "What is their shape—the relationship of length and width? How are they joined together?" Then decide precisely what you are going to show and how large the drawing must be to show the necessary detail. After making the drawing, add labels in the margins; and connect them, by straight lines (leader lines), to the structures being named.

Step 4: Manipulation and Analysis of Data

The form of the final data varies, depending on the nature of the data collected. Usually, the final data represent information converted from the original measured values (raw data) to some other form. This may mean that averaging or some other statistical treatment must be applied, or it may require conversions from one kind of units to another. In other cases, graphs may be needed to display the data.

ELEMENTARY TREATMENT OF DATA Only very elementary statistical treatment of data is required in this manual. For example, you will be expected to understand and/or compute an average (mean), percentages, and a range.

Two of these statistics, the average and the range, are useful in describing the *typical* case among a large number of samples evaluated. Let us use a simple example. We will assume that the following heart rates (in beats/min) were recorded during an experiment: 64, 70, 82, 94, 85, 75, 72, 78. If you put these numbers in numerical order, the **range** is easily computed, because the range is the difference between the highest and lowest numbers obtained (highest number minus lowest number). What is the range of the set of numbers just provided?

1. ______________________________*

The **average,** or **mean,** is obtained by summing the items and dividing the sum by the number of items. Compute the average for the set of numbers just provided:

2. ______________________________

The word *percent* comes from the Latin meaning "for 100"; thus *percent,* indicated by the percent sign, %, means parts per 100 parts. Thus, if we say that 45% of Americans have type O blood, what we are really saying is that among each group of 100 Americans, 45 (45/100) can be expected to have type O blood.

It is very easy to convert any number (including decimals) to a percent. The rule is to move the decimal point two places to the right and add the percent sign. If no decimal point appears, it is *assumed* to be at the end of the number; and zeros are added to fill any empty spaces. Two examples follow:

0.25 = 0.2 5 = 25%
5 = 5 = 500%

Change the following numbers to percents:

3. 38.2 = __________ 5. 1.6 = __________

4. 402 = __________

Note that although you are being asked here to convert numbers to percents, percents by themselves are meaningless. We always speak in terms of a percentage *of* something.

To change a percent to a whole number (or decimal), remove the percent sign, and move the decimal point two places to the left. Change the following percents to whole numbers or decimals:

6. 36% = __________ 8. 25777% = __________

7. 800% = __________ 9. 0.05% = __________

* Answers are given on page xviii.

MAKING AND READING LINE GRAPHS For some laboratory experiments you will be required to show your data (or part of them) graphically. Simple line graphs allow relationships within the data to be shown interestingly and allow trends (or patterns) in the data to be demonstrated. An advantage of properly drawn graphs is that they save the reader's time because the essential meaning of large numbers of statistical data can be visualized at a glance.

To aid in making accurate graphs, graph paper (or a printed grid in the manual) is used. Line graphs have both horizontal and vertical scales. Each scale should have uniform intervals—that is, each unit measured on the scale should require the same distance along the scale as any other. Variations from this rule may be misleading and result in false interpretations of the data. By convention, the condition that is manipulated (the independent variable) in the experimental series is plotted on the X-axis (the horizontal axis); and the value that we then measure (the dependent variable) is plotted on the Y-axis (the vertical axis). To plot the data, a dot or a small *x* is placed at the precise point where the two variables (measured for each sample) meet; and then a line (this is called the **curve**) is drawn to connect the plotted points.

Sometimes, you will see the curve on a line graph extended beyond the last plotted point. This is (supposedly) done to predict "what comes next." When you see this done, be skeptical. The information provided by such a technique is only slightly more accurate than that provided by a crystal ball!

To read a line graph, pick any point on the line, and match it with the information directly below on the horizontal scale and with that directly to the left of it on the vertical scale. Figure G.1 is a graph that illustrates the relationship between breaths per minute (respiratory rate) and body temperature. Answer the following questions about this graph:

10. What was the respiratory rate at a body temperature of 96°F? _______

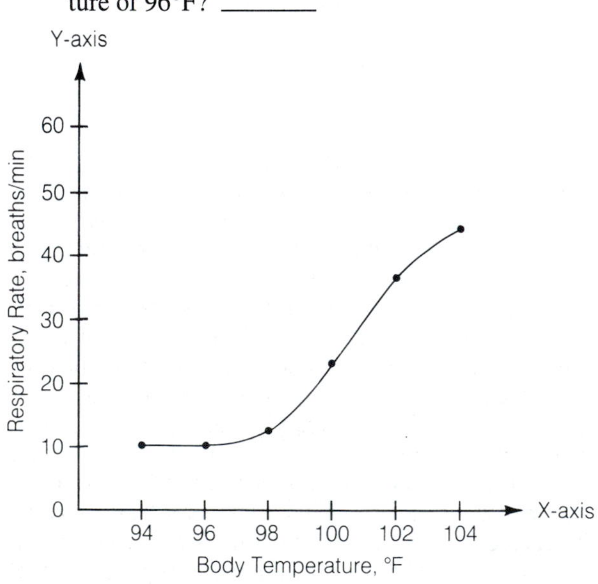

FG.1

Example of graphically presented data.

11. Between 98° and 102°F, the respiratory rate increased from

_______ to _______ breaths per minute.

12. Between which two body temperature readings was the increase in breaths per minute greatest?

____

13. Are the intervals on each scale uniform?

____

Step 5: Reporting Conclusions of the Study

Drawings, tables, and graphs alone do not suffice as the final presentation of scientific results. The final step requires that you provide a straightforward description of the conclusions drawn from your results. If possible, your findings should be compared to those of other investigators working on the same problem. (For laboratory investigations conducted by students, these comparative figures are provided by classmates.)

It is important to realize that scientific investigations do not always yield the anticipated results. If there are discrepancies between your results and those of others, or what you expected to find based on your class notes or textbook readings, this is the place to try to explain those discrepancies.

Results are often only as good as the observation techniques used. Depending on the type of experiment conducted, several questions may need to be answered. Did you weigh the specimen carefully enough? Did you balance the scale first? Was the subject's blood pressure actually as high as you recorded it, or did you record it hastily (and inaccurately)? If you did record it accurately, is it possible that the subject was emotionally upset about something, which (even though the matter of concern had nothing to do with the experiment) might have given falsely high data for the variable being investigated? Attempting to explain an unexpected result will often teach you more than you would have learned from anticipated results.

When the experiment produces results that are consistent with the hypothesis, then the hypothesis can be said to have reached a higher level of certainty. There is now a greater probability that the hypothesis is correct.

A hypothesis that has been validated by many different investigators is called a **theory.** Theories are useful in two important ways. First, they link sets of data; and second, they make predictions that may lead to additional avenues of investigation. (Okay, we know this with a high degree of certainty; what's next?)

When a theory has been repeatedly verified and appears to have wide applicability in biology, it may assume the status of a **biological principle.** A principle is a statement that applies with a high degree of probability to a range of events. For example, "Living matter is made of cells or cell products" is a principle stated in many biology texts. It is a sound and useful principle, and will continue to be used as such—unless new findings prove it wrong.

We have been through quite a bit of background concerning the scientific method and what its use entails. Because it is important that you remember the phases of the scientific method, they are summarized here:

1. Observation of some phenomenon
2. Statement of a hypothesis (based on the observations)
3. Collection of data (testing the hypothesis with controlled experiments)
4. Manipulation and analysis of the data
5. Reporting of the conclusions of the study

SCIENTIFIC NOTATION AND METRICS

No matter how highly developed our ability to observe, observations have scientific value only if they can be communicated to others. This necessitates the use of scientific notation and the widely accepted system of metric measurements.

Scientific Notation

Because quantitative measurements often yield very large or very small numbers, you are quite likely to encounter numbers such as 3.5×10^{12} or 10^{-3}. It is important that you understand what this **scientific notation** means.

Scientific notation is dependent on the properties of exponents and on the movement of the decimal point when multiplying or dividing by 10. When you multiply 10 by itself, you get a product that is one followed by zeros. The number of zeros (two, in this case) in the product is equal to the number of times you have used 10 as a factor and is shown as an exponent. Thus, the notation 10^2 has these parts:

$$\text{base} \rightarrow 10^2 \leftarrow \text{exponent}$$

and translates to "the base 10 multiplied by itself (10 × 10)."

The powers of 10 are represented as follows:

$10^0 = 1$ (Any number followed by a zero exponent is one.)

$10^1 = 10$ $(10 \times 1 = 10)$

$10^2 = 100$ $(10 \times 10 = 100)$

$10^3 = 1000$ $(10 \times 10 \times 10 = 1000)$

$10^4 = 10{,}000$ $(10 \times 10 \times 10 \times 10 = 10{,}000)$

As you can see, each time the exponent is increased by one, another zero (× 10) is added to the answer.

When you multiply any number by a power of 10 written with exponents, the decimal point is moved to the right the number of times shown in the exponent. Thus:

$$3.25 \times 10^1 = 3.25 = 32.5$$

$$3.25 \times 10^3 = 3.25 = 3250$$

$$3.25 \times 10^5 = 3.25 = 325{,}000$$

By using such exponential notation, very large numbers may be written in far simpler form.

Write the following numbers using the proper scientific notation:

14. $140{,}000 = 1.4 \times$ _______

15. $9{,}650{,}000 = 9.65 \times$ _______

16. $852 = 8.52 \times$ _______

17. $10 = 1.0 \times$ _______

Notice that proper scientific notation entails only one number to the left of the decimal point. Thus 1.03×10^3 is correct, but 10.3×10^2 is not.

In the above examples, all of the numbers used were greater than one. Scientific notation can also be used to report numbers less than one. To do this, negative exponents are used. For example, in

$$3.25 \times 10^2$$

the positive exponent means that the decimal point is to be moved two places to the right, and the number designated is 325 (3.25 × 10 × 10). However, in

$$3.25 \times 10^{-2}$$

the negative exponent means that the number is to be divided by the power of 10 indicated by the exponent and the decimal point is to be moved two places to the left. The number so designated is 0.0325 [3.25 ÷ (10 × 10)].

Thus, the rule for converting scientific notation (using powers of 10) to decimal notation is to move the decimal point the number of places indicated by the exponent. When the exponent is positive (with or without a plus sign), the decimal point is moved to the right. When the exponent is negative (always provided with a minus sign), the decimal point is moved to the left.

For a little practice, write the following numbers in scientific notation: (18–23)

$140{,}000 = 1.4 \times$ _______ $45{,}000 = 4.5 \times$ _______

$0.0000063 = 6.3 \times$ _______ $0.265 = 2.65 \times$ _______

$0.00054 = 5.4 \times$ _______ $0.10 = 1.0 \times$ _______

Metrics

Without measurement, we would be limited to qualitative description. However, with a system of measurement, quantitative description becomes possible.

Anyone can establish a system of measurement. All that is required is a reference point; and, historically, much of our common (the British) system of measurement evolved from units based on objects everyone knew. For example, horses were measured in "hands," and a "fathom" was the distance between outstretched arms. However, the variability in such measurements is immediately apparent—for example, an infant's hand is substantially smaller than that of an adult. Therefore, for precise and repeatable communication of information, the agreed-upon system of measurement used by scientists is the **metric system,** a nonvarying standard of reference.

TABLE G.1 Commonly Used Units of the Metric System, and Their Fractions and Multiples

Measurement	Unit	Fraction or multiple		Prefix	Symbol
Length	Meter (m)	10^6	one million	mega	M
Volume	Liter (L; l with prefix)	10^3	one thousand	kilo	k
Mass	Gram (g)	10^{-1}	one tenth	deci	d
Time*	Second (s)	10^{-2}	one hundredth	centi	c
Temperature	Degree Celsius (°C)	10^{-3}	one thousandth	milli	m
		10^{-6}	one millionth	micro	μ
		10^{-9}	one billionth	nano	n

* The accepted standard for time is the second; and thus hours and minutes are used in scientific, as well as everyday, measurement of time. The prefixes used in the designation of units of length, mass, and volume are also used in specifying units of time. However, because minutes and hours are terms that indicate *multiples* of seconds, the only prefixes generally used are those indicating *fractional portions* of seconds—for example, millisecond and microsecond.

A major advantage of the metric system is that it is based on units of 10. This allows rapid conversion to workable numbers so that neither very large nor very small figures need be used in calculations. Fractions or multiples of the standard units of length, volume, mass, time, and temperature have been assigned specific names. Table G.1 shows the commonly used units of the metric system, along with the prefixes used to designate fractions and multiples thereof.

To change from smaller units to larger units, you must *divide* by the appropriate factor of 10 (because there are fewer of the larger units). For example, a milliunit (milli = one thousandth), such as a milliliter or millimeter, is one step smaller than a centiunit (centi = one hundredth), such as a centiliter or centimeter. Thus to change milliunits to centiunits, you must divide by 10. On the other hand, when converting from larger units to smaller ones, you must *multiply* by the appropriate factor of 10 (because there will be more of the smaller units). A partial scheme for conversions between the metric units is shown below.

Students studying a science or preparing for a profession in the health-related fields find that certain of the metric units are encountered and dealt with more frequently than others. Thus, the objectives of the sections that follow are to provide a brief overview of these most-used measurements and to help you gain some measure of confidence in dealing with them. (A listing of the most frequently used conversion factors, for conversions between British and metric system units, is provided in Appendix A.)

LENGTH MEASUREMENTS The metric unit of length is the **meter (m)**. In addition to measuring things in meters, you will be expected to measure smaller objects in centimeters or millimeters. Subcellular structures are measured in micrometers.

To help you picture these units of length, some equivalents follow:

One meter (m) is slightly longer than one yard (1 m = 39.37 in.).
One centimeter (cm) is approximately the width of a piece of chalk. (Note: there are 2.54 cm in 1 in.)
One millimeter (mm) is approximately the thickness of the wire of a paper clip or of a mark made by a No. 2 pencil lead.
One micrometer (μm) is extremely tiny and can be measured only microscopically.

Make the following conversions between metric units of length: (24–28)

352 cm = ________ mm   12 cm = ________ mm

150 km = ________ m   1 mm = ________ m

2000 μm = ________ mm

Now, circle the answer that would make the most sense in each of the following statements:

29. A match (in a matchbook) is (0.3, 3, 30) cm long.
30. A standard-size American car is about 4 (mm, cm, m, km) long.
31. John pole-vaults a height of 5 meters, whereas Gerry vaults a height of 5 yards. Does John or Gerry make the more difficult vault?

____

microunit ⇄ (÷1000 / ×1000) milliunit ⇄ (÷10 / ×10) centiunit ⇄ (÷100 / ×100) unit ⇄ (÷1000 / ×1000) kilounit

smallest ⇄ largest

VOLUME MEASUREMENTS The metric unit of volume is the liter. A **liter (l,** or sometimes **L,** especially without a prefix) is slightly more than a quart (1 L = 1.057 quarts). Liquid products measured in liters are becoming more common, and laboratory solutions often come in 1-liter quantities. Liquid volumes measured out for lab experiments are usually measured in milliliters (ml). (The terms *ml* and *cc,* cubic centimeter, are used interchangeably in laboratory and medical settings.)

To help you visualize metric volumes, the equivalents of some common substances follow:

A 12-oz can of soda is just slightly more than 360 ml.
A cup of coffee is approximately 180 ml.
A fluid ounce is 30 ml (cc).
A teaspoon of vanilla is about 5 ml (cc), and many drug injections are given in 5-ml volumes.

Compute the following:

32. How many 5-ml injections can be prepared from 1 liter of a medicine?

____

33. A 450-ml volume of alcohol is _______ L.

34. The volume of one grape is approximately 0.004 L. What is the volume of the grape in milliliters?

____

MASS MEASUREMENTS Although many people use the terms *mass* and *weight* interchangeably, this usage is inaccurate. **Mass** is the amount of matter in an object; and an object has a constant mass, regardless of where it is—that is, at sea level, on a mountaintop, or in outer space. However, weight varies with gravitational pull; the greater the gravitational pull, the greater the weight. Thus, our astronauts are said to be weightless* when in outer space, but they still have the same mass as they do on earth.

The metric unit of mass is the **gram (g)**, and most objects weighed in the laboratory will be measured in terms of this unit or fractions thereof. Medical dosages are usually prescribed in milligrams (mg) or micrograms (μg); and, in the clinical agency, body weight (particularly of infants) is typically specified in kilograms (kg) (1 kg = 2.2 lb).

The following examples are provided to help you become familiar with the masses of some common objects:

Two aspirin tablets have a mass of approximately 1 g.
A nickel has a mass of 5 g.
The mass of an average woman (132 lb) is 60 kg.

* Astronauts are not *really* weightless. It is just that they and their surroundings are being pulled toward the earth at the same speed; and so, in reference to their environment, they appear to float.

Make the following conversions:

35. 300 g = _______ mg = _______ μg

36. 4000 μg = _______ mg = _______ g

37. A nurse must administer to her patient, Mrs. Smith, 5 mg of a drug per kg of body mass. Mrs. Smith weighs 140 lb. How many grams of the drug should the nurse administer to her patient?

_______ g

TEMPERATURE MEASUREMENTS In the laboratory and in the clinical agency, temperature is measured both in metric units (degrees Celsius, °C) and in British units (degrees Fahrenheit, °F). Thus it helps to be familiar with both temperature scales.

The temperatures of boiling and freezing water can be used to compare the two scales:

The boiling point of water is 100°C and 212°F.
The freezing point of water is 0°C and 32°F.

As you can see, the range from the freezing point to the boiling point of water on the Celsius scale is 100 degrees, whereas the comparable range on the Fahrenheit scale is 180 degrees. Hence, one degree on the Celsius scale represents a greater change in temperature. Normal body temperature is approximately 98.6°F and 37°C.

To convert from the Fahrenheit scale to the Celsius scale (or vice versa), the following equation is used:

$$°C = 5(°F - 32)/9$$

For example, to convert 180°C to °F:

$$\begin{aligned} °C &= 5(°F - 32)/9 \\ 180 &= 5(°F - 32)/9 \\ 1620 &= 5(°F - 32) \\ 1620 &= 5(°F) - 160 \\ 1780 &= 5(°F) \\ 356 &= °F \end{aligned}$$

and to convert 72°F to °C:

$$\begin{aligned} °C &= 5(°F - 32)/9 \\ °C &= 5(72 - 32)/9 \\ °C &= 5(40)/9 \\ °C &= 200/9 \\ °C &= 22.2 \end{aligned}$$

Perform the following temperature conversions:

38. Convert 38°C to °F: _______

39. Convert 158°F to °C: _______

Answers

1. range of 94–64; 30 beats/min
2. average 77.5
3. 3820%
4. 40200%
5. 160%
6. 0.36
7. 8
8. 257.77
9. 0.0005
10. 10 breaths/min
11. 12 to 36
12. interval between 100–102° (went from 22 to 36 breaths/min)
13. yes
14. 10^5
15. 10^6
16. 10^2
17. 10^1
18. 1.4×10^5
19. 6.3×10^{-6}
20. 5.4×10^{-4}
21. 4.5×10^4
22. 2.65×10^{-1}
23. 1.0×10^{-1}
24. cm = 3520 mm
25. km = 150,000 m
26. μm = 2 mm
27. cm = 120 mm
28. mm = 0.001 m
29. 3 cm
30. m long
31. John
32. 200
33. 0.45 L
34. 4 ml
35. 300 g = 3×10^5 mg = 3×10^8 μg
36. 4000 μg = 4 mg = 4×10^{-3} g (0.004)
37. 0.32 g
38. 100.4°F
39. 70°C

The Language of Anatomy

OBJECTIVES

1. To describe the anatomical position verbally or by demonstration.
2. To use proper anatomical terminology to describe body directions, planes, and surfaces.
3. To name the body cavities and indicate the important organs in each.

MATERIALS

Human torso model (dissectible)
Human skeleton
Demonstration: sectioned and labeled kidneys (three separate kidneys uncut or cut so that (a) entire, (b) transverse section, and (c) longitudinal sectional views are visible)

See Appendix D, Exercise 1 for links to A.D.A.M. Standard.

See Appendix E, Exercise 1 for links to *Anatomy and PhysioShow: The Videodisc.*

Most of us have a natural curiosity about our bodies. This fact is amply demonstrated by infants, who early in life become fascinated with their own waving hands or their mother's nose. The study of the gross anatomy of the human body elaborates on this fascination. Unlike the infant, however, the student of anatomy must learn to identify and observe the dissectible body structures formally. The purpose of any gross-anatomy experience is to examine the three-dimensional relationships of body structures—a goal that can never completely be achieved by using illustrations and models, regardless of their excellence.

When beginning the study of any science, the student is often initially overcome by jargon unique to the subject. The study of anatomy is no exception. But without this specialized terminology, confusion is inevitable. For example, what do *over, on top of, superficial to, above,* and *behind* mean in reference to the human body? Anatomists have an accepted set of reference terms that are universally understood. These allow body structures to be located and identified with a minimum of words and a high degree of clarity. Thus it is not surprising that physicians' orders and progress notes, therapists' records, and nurses' notes use *anatomical terminology* to describe body parts, regions, positions, and activities. The ability to understand and use correct anatomical terminology is a skill that distinguishes health care personnel who are successful and comfortable in their chosen profession from those perpetually unsure of just what is expected of them.

This exercise presents some of the most important anatomical terminology used to describe the body and introduces you to basic concepts of **gross anatomy,** the study of body structures visible to the naked eye.

ANATOMICAL POSITION

When anatomists or doctors refer to specific areas of the human body, they do so in accordance with a universally accepted standard position called the **anatomical position.** It is essential to understand this position, because much of the body terminology employed in this book refers to this body positioning, regardless of the position the body happens to be in. In the anatomical position the human body is erect, with feet together, head and toes pointed forward, and arms hanging at the sides with palms facing forward (Figure 1.1).

- Assume the anatomical position, and note that it is not particularly comfortable. The hands are held unnaturally forward rather than hanging partially cupped toward the thighs.

SURFACE ANATOMY

Body surfaces provide visible landmarks for study of the body.

Anterior Body Landmarks

Note the following regions in Figure 1.2a:

Abdominal: pertaining to the anterior body trunk region inferior to the ribs
Antebrachial: pertaining to the forearm
Antecubital: pertaining to the anterior surface of the elbow
Axillary: pertaining to the armpit

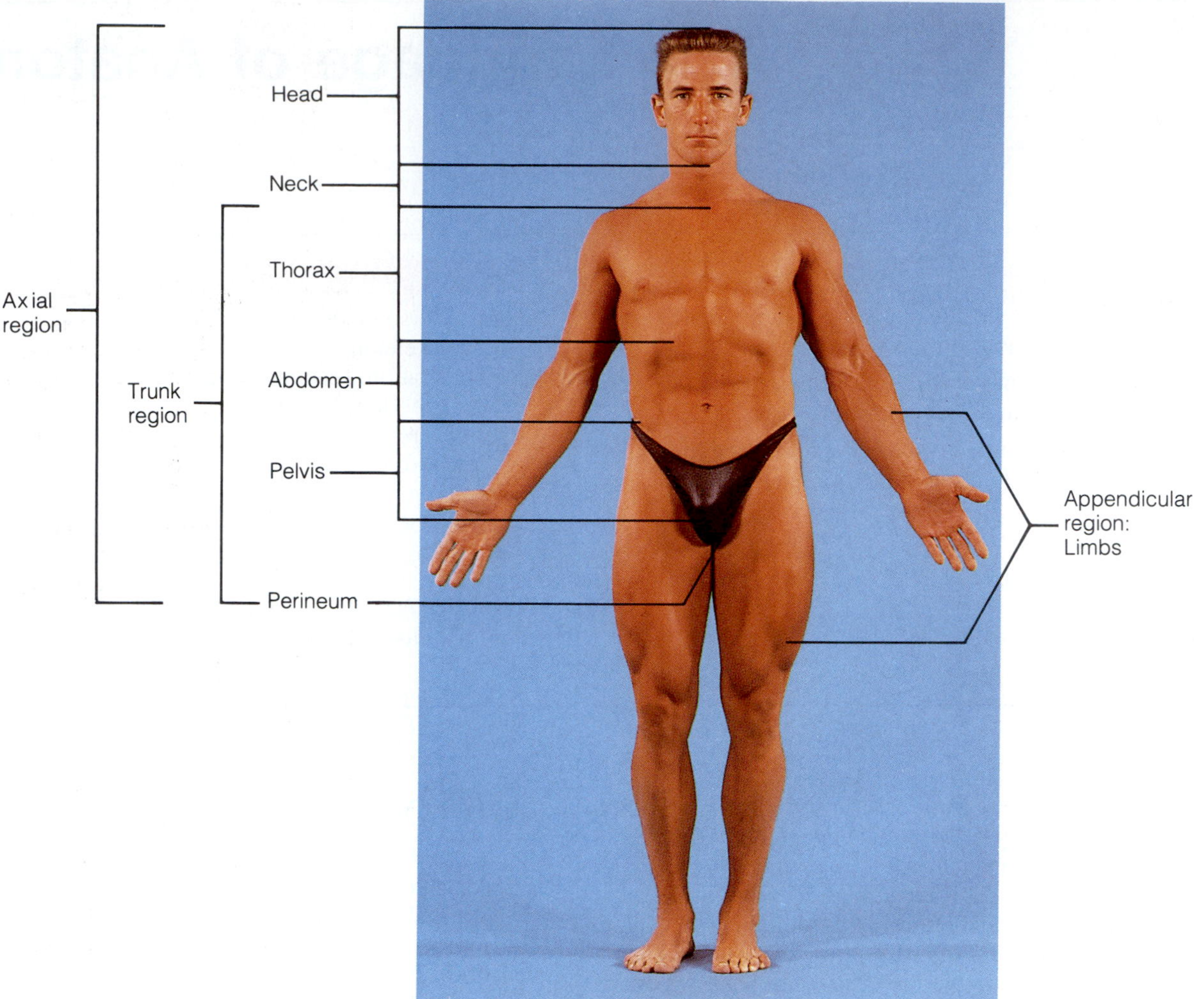

F1.1

Anatomical position.

Brachial: pertaining to the arm
Buccal: pertaining to the cheek
Carpal: pertaining to the wrist
Cervical: pertaining to the neck region
Coxal: pertaining to the hip
Crural: pertaining to the leg
Deltoid: pertaining to the roundness of the shoulder caused by the underlying deltoid muscle
Digital: pertaining to the fingers or toes
Femoral: pertaining to the thigh
Frontal: pertaining to the forehead
Hallux: pertaining to the great toe
Inguinal: pertaining to the groin
Mammary: pertaining to the breast
Manus: pertaining to the hand
Mental: pertaining to the chin
Nasal: pertaining to the nose
Oral: pertaining to the mouth
Orbital: pertaining to the bony eye socket (orbit)
Palmar: pertaining to the palm of the hand
Patellar: pertaining to the anterior knee (kneecap) region
Pedal: pertaining to the foot
Pelvic: pertaining to the pelvis region
Peroneal: pertaining to the side of the leg
Pollex: pertaining to the thumb
Pubic: pertaining to the genital region
Sternal: pertaining to the region of the breastbone
Tarsal: pertaining to the ankle
Thoracic: pertaining to the chest
Umbilical: pertaining to the navel

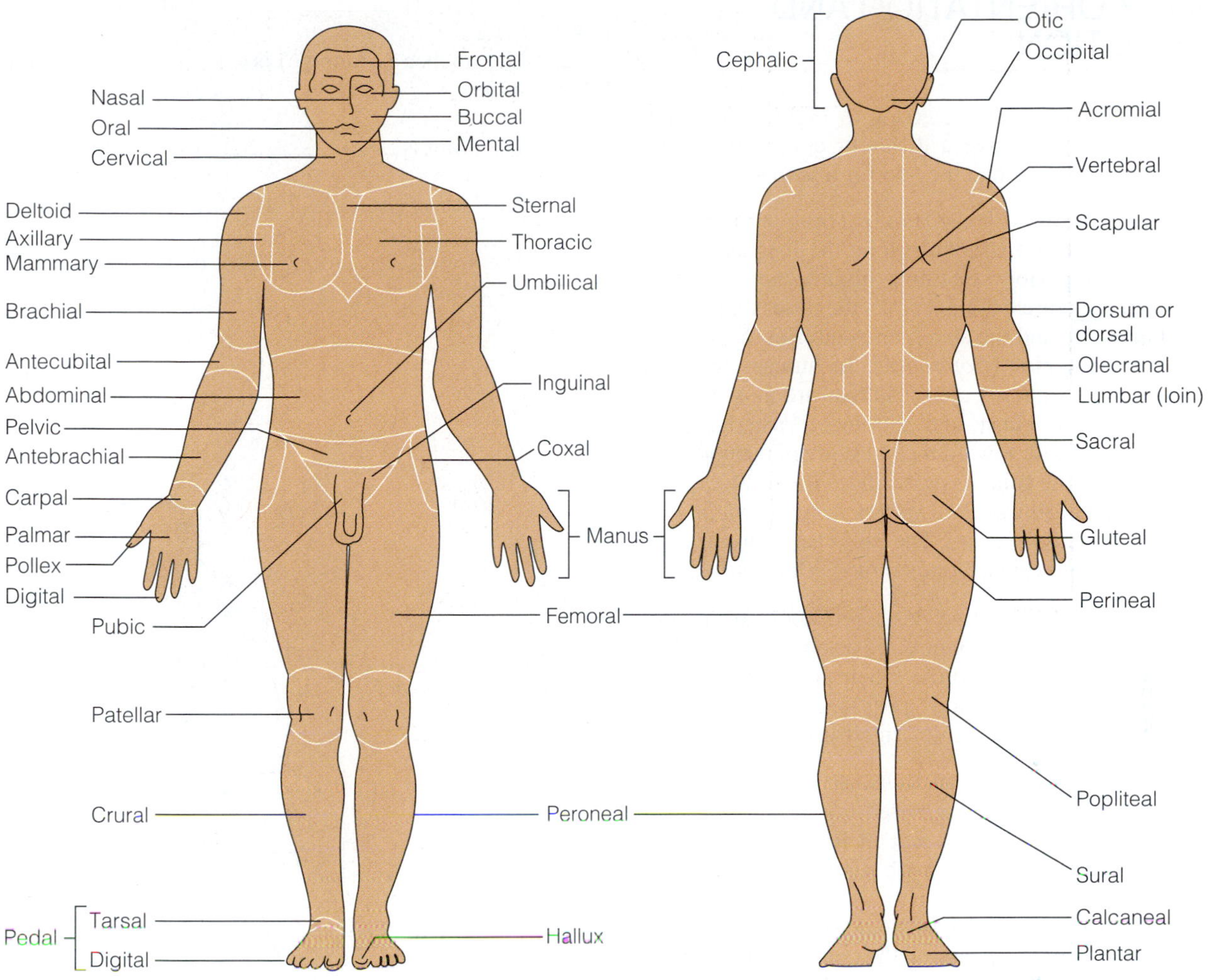

F1.2

Surface anatomy. **(a)** Anterior body landmarks. **(b)** Posterior body landmarks.

Posterior Body Landmarks

Note the following body surface regions in Figure 1.2b:

Acromial: pertaining to the point of the shoulder
Calcaneal: pertaining to the heel of the foot
Cephalic: pertaining to the head
Dorsum: pertaining to the back
Gluteal: pertaining to the buttocks or rump
Lumbar: pertaining to the area of the back between the ribs and hips; the loin
Occipital: pertaining to the posterior aspect of the head or base of the skull
Olecranal: pertaining to the posterior aspect of the elbow
Otic: pertaining to the ear
Perineal: pertaining to the region between the anus and external genitalia
Plantar: pertaining to the sole of the foot
Popliteal: pertaining to the back of the knee
Sacral: pertaining to the region between the hips (overlying the sacrum)
Scapular: pertaining to the scapula or shoulder blade area
Sural: pertaining to the calf or posterior surface of the leg
Vertebral: pertaining to the area of the spinal column

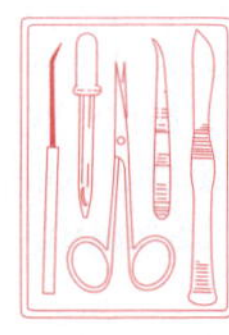

Locate the anterior and posterior body landmarks on yourself, your lab partner, and a torso model before continuing.

BODY ORIENTATION AND DIRECTION

Study the terms below, referring to Figure 1.3. Notice that certain terms have a different connotation for a four-legged animal than they do for a human.

Superior/inferior (*above/below*): These terms refer to placement of a structure along the long axis of the body. Superior structures always appear above other structures. For example, the nose is superior to the mouth, and the abdomen is inferior to the chest.

Anterior/posterior (*front/back*): In humans the most anterior structures are those that are most forward — the face, chest, and abdomen. Posterior structures are those toward the backside of the body. For instance, the spine is posterior to the heart.

Medial/lateral (*toward the midline/away from the midline or median plane*): The ear is lateral to the nose; the sternum (breastbone) is medial to the ribs.

The terms of position described above depend on an assumption of anatomical position. The next four term pairs are more absolute. Their applicability is not relative to a particular body position, and they consistently have the same meaning in all vertebrate animals.

Cephalad/caudal (*toward the head/toward the tail*): In humans these terms are used interchangeably with *superior* and *inferior.* But in four-legged animals they are synonymous with *anterior* and *posterior* respectively.

Dorsal/ventral (*backside/belly side*): These terms are used chiefly in discussing the comparative anatomy of animals, assuming the animal is standing. *Dorsum* is a Latin word meaning "back." Thus, *dorsal* refers to the backside of the animal's body or of any other structures; e.g., the posterior surface of the leg is its dorsal surface. The term *ventral* derives from the Latin term *venter,* meaning "belly," and always refers to the belly side of animals. In humans the terms *ventral* and *dorsal* are used interchangeably with the terms *anterior* and *posterior,* but in four-legged animals *ventral* and *dorsal* are synonymous with *inferior* and *superior* respectively.

Proximal/distal (*nearer the trunk or attached end/farther from the trunk or point of attachment*): These terms are used primarily to locate various areas of the body limbs. For example, the fingers are distal to the elbow; the knee is proximal to the toes.

Superficial/deep (*toward or at the body surface/away from the body surface or more internal*): These terms locate body organs according to their relative closeness to the body surface. For example, the lungs are deep to the rib cage, and the skin is superficial to the skeletal muscles.

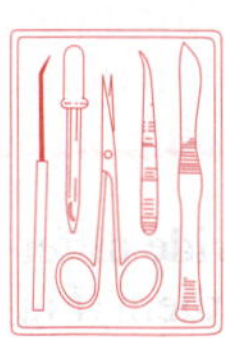

Before continuing, use a human torso model, a skeleton, or your own body to specify the relationship between the following structures. Use the correct anatomical terminology:

1. The wrist is _______________ to the hand.
2. The trachea (windpipe) is _______________ to the spine.
3. The brain is _______________ to the spinal cord.
4. The kidneys are _______________ to the liver.
5. The nose is _______________ to the cheekbones.

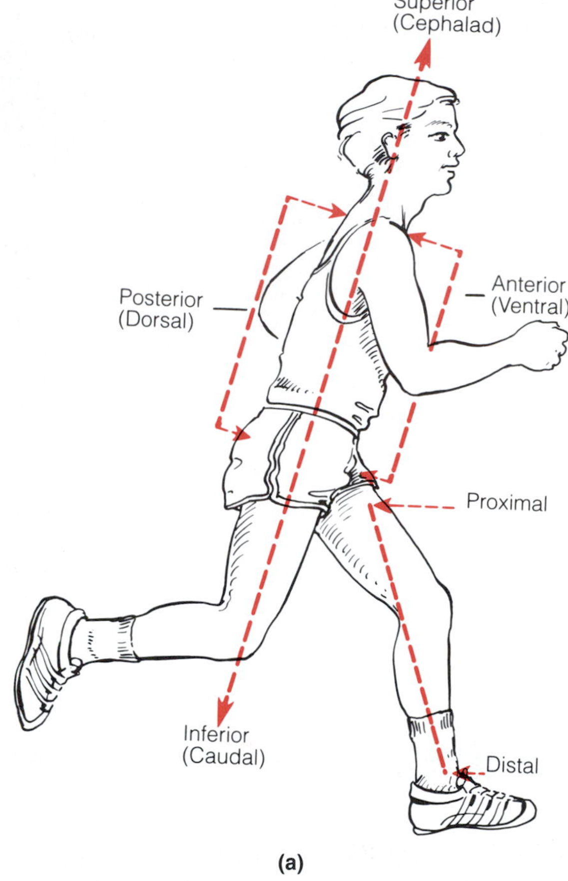

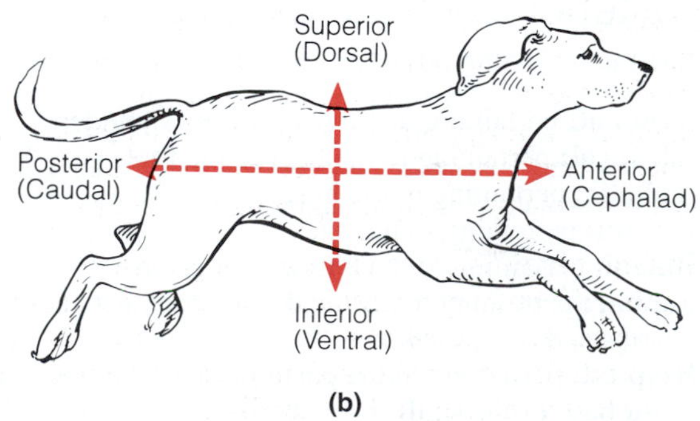

F1.3

Anatomical terminology describing body orientation and direction. (a) With reference to a human. **(b)** With reference to a four-legged animal.

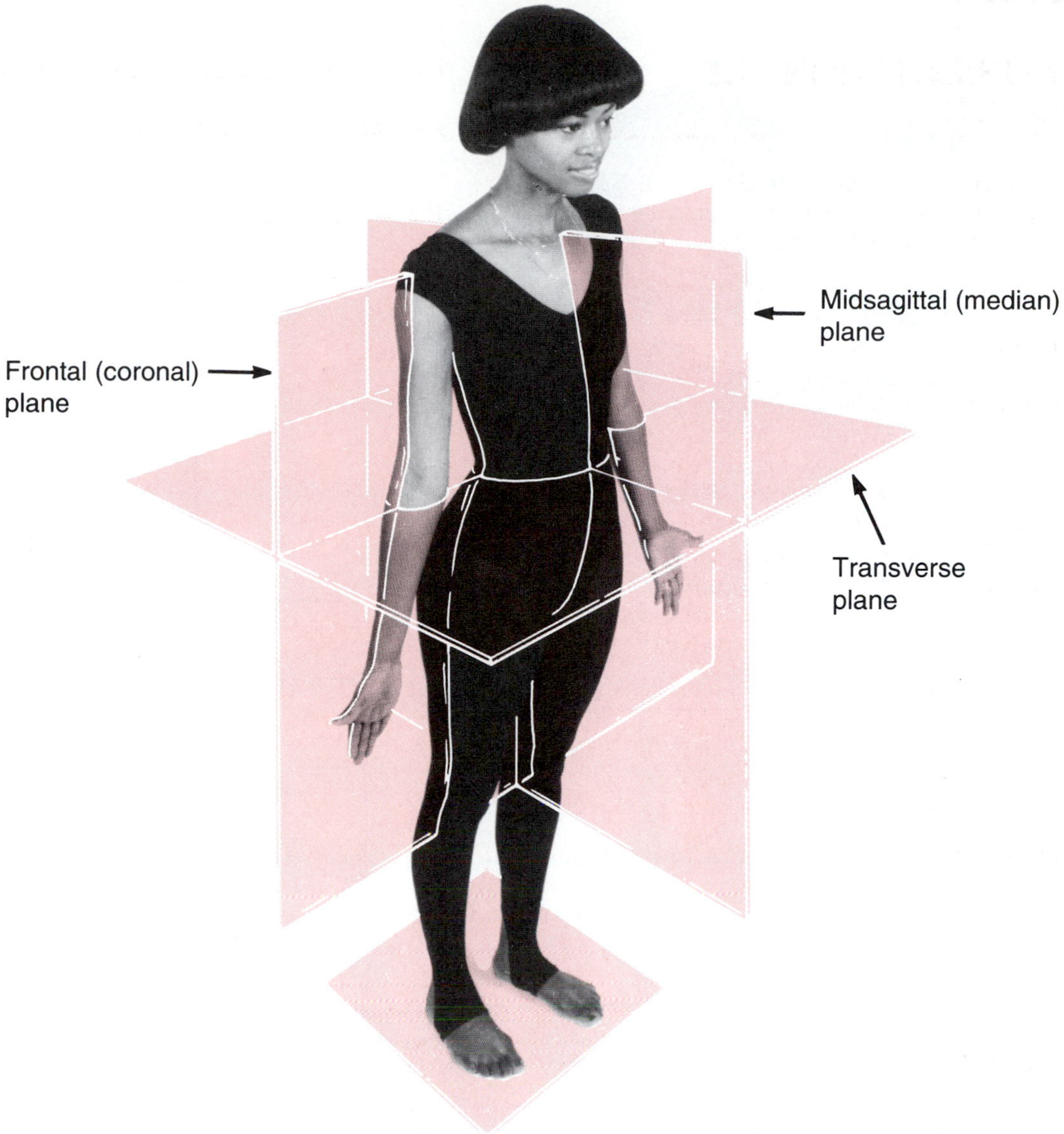

F1.4

Planes of the body.

BODY PLANES AND SECTIONS

The body is three-dimensional and, in order to observe its internal structures, it is often helpful and necessary to make use of a **section,** or cut. When the section is made through the body wall or through an organ, it is made along an imaginary surface or line called a **plane.** Anatomists commonly refer to three planes (Figure 1.4) or sections which lie at right angles to one another.

Sagittal plane: A plane that runs longitudinally, dividing the body into right and left parts, is referred to as a sagittal plane. If it divides the body into equal parts, right down the median plane of the body, it is called a **midsagittal,** or **median, plane.** All other planes are referred to as **parasagittal planes.**

Frontal plane: Sometimes called a **coronal plane,** the frontal plane is a longitudinal plane that divides the body (or an organ) into anterior and posterior parts.

Transverse plane: A transverse plane runs horizontally, dividing the body into superior and inferior parts. When organs are sectioned along the transverse plane, the sections are commonly called **cross sections.**

As shown in Figure 1.5, a sagittal or frontal plane section of an organ provides quite a different view than a transverse section.

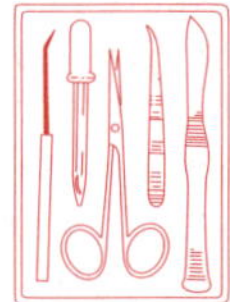

Go to the demonstration area and observe the transversely and longitudinally cut organ specimens. Pay close attention to the different details of structure in the samples.

BODY CAVITIES

The body has two sets of cavities, which provide different degrees of protection to the organs within them (Figure 1.6).

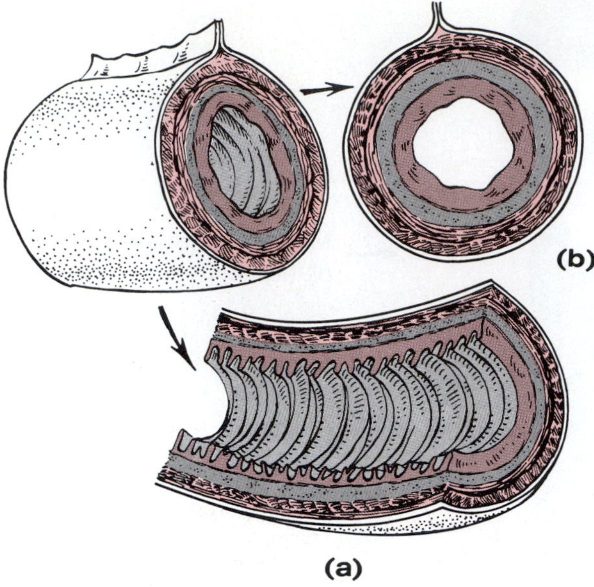

F1.5

Segment of the small intestine. **(a)** Cut longitudinally. **(b)** Cut transversely.

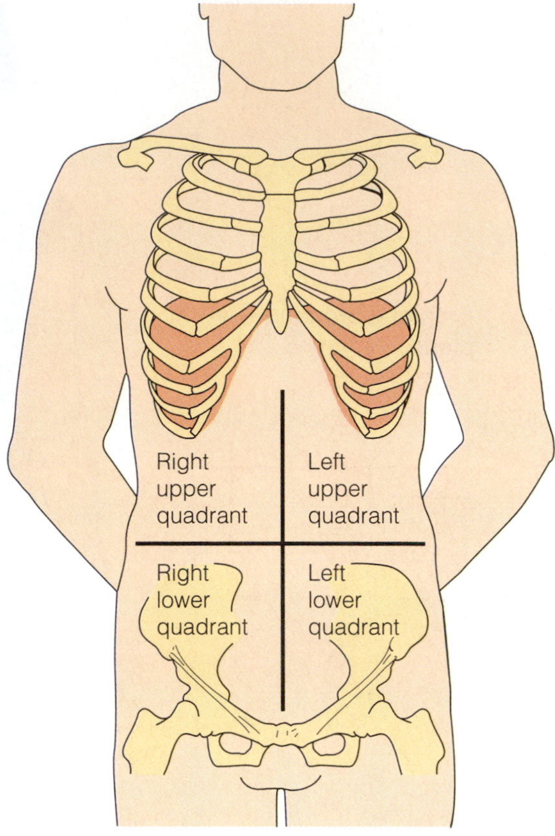

F1.7

Abdominopelvic surface and cavity. **(a)** The four quadrants. **(b)** Nine regions delineated by four planes. The superior horizontal plane is just inferior to the ribs; the inferior horizontal plane is at the superior aspect of the hip bones. The vertical planes are just medial to the nipples. **(c)** Anterior view of the abdominopelvic cavity showing superficial organs.

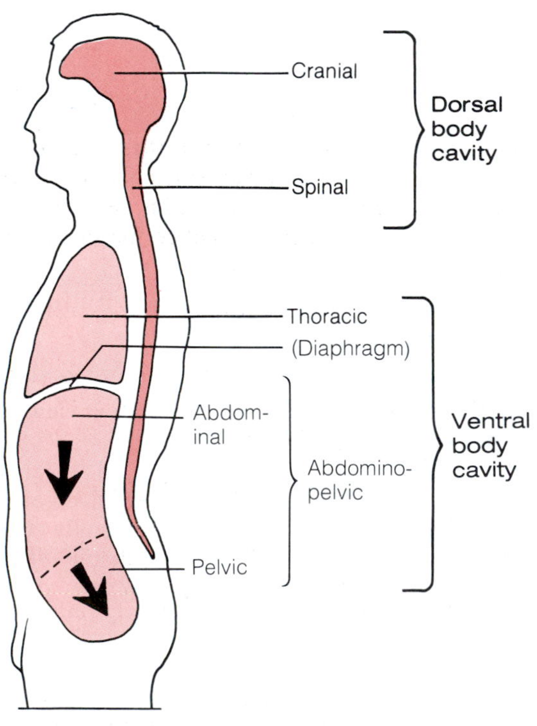

F1.6

Body cavities. Arrows indicate the angle of the relationship between the abdominal and pelvic cavities.

Dorsal Body Cavity

The dorsal body cavity can be subdivided into the **cranial cavity,** in which the brain is enclosed within the rigid skull, and the **spinal cavity,** within which the delicate spinal cord is protected by the bony vertebral column. Because the cord is a continuation of the brain, these cavities are continuous with each other.

Ventral Body Cavity

Like the dorsal cavity, the ventral body cavity is subdivided. The superior **thoracic cavity** is separated from the rest of the ventral cavity by the dome-shaped diaphragm. The heart and lungs, located in the thoracic cavity, are afforded some measure of protection by the bony rib cage. The cavity inferior to the diaphragm is often referred to as the **abdominopelvic cavity,** since there is no further physical separation of the ventral cav-

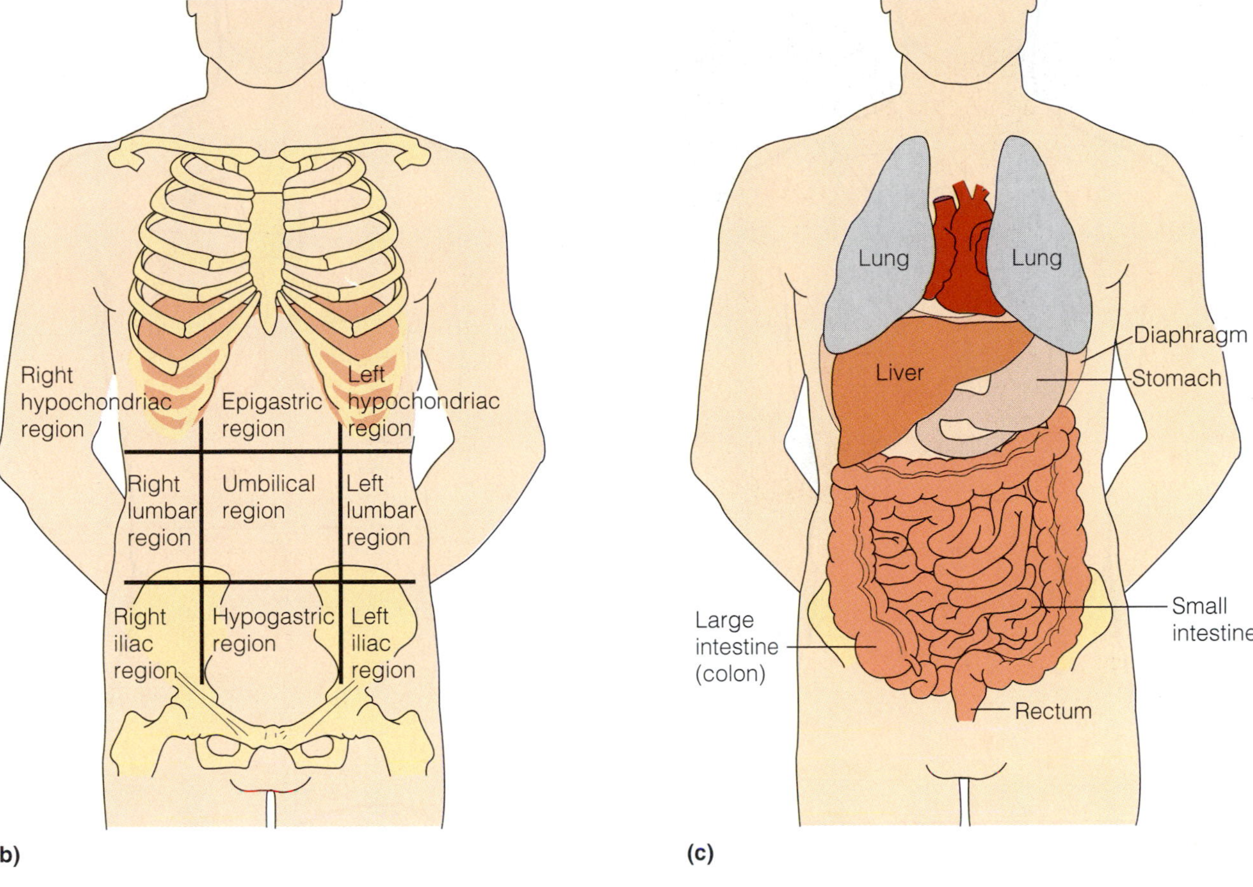

F1.7 *(continued)*

ity. Some prefer to subdivide the abdominopelvic cavity into a superior **abdominal cavity,** which houses the stomach, intestines, liver, and other organs, and an inferior **pelvic cavity,** partially enclosed by the bony pelvis and containing the reproductive organs, bladder, and rectum. Notice in Figure 1.6 that the abdominal and pelvic cavities are not continuous with each other in a straight plane but that the pelvic cavity is tipped away from the perpendicular.

ABDOMINOPELVIC QUADRANTS AND REGIONS Because the abdominopelvic cavity is quite large and contains many organs, it is helpful to divide it up into smaller areas for discussion or study. The scheme used by most physicians and nurses divides the abdominal surface (and the abdominopelvic cavity deep to it) into four approximately equal regions called **quadrants.** These quadrants are named according to their relative position—that is, *right upper quadrant, right lower quadrant, left upper quadrant,* and *left lower quadrant* (see Figure 1.7a).

Another scheme, commonly used by anatomists, divides the abdominal surface and abdominopelvic cavity into nine separate regions by four planes, as shown in Figure 1.7b. Although the names of these nine regions are unfamiliar to you now, with a little patience and study they will become easier to remember. As you read through the descriptions of these nine regions below and locate them in the figure, notice the organs they contain by referring to Figure 1.7c.

Umbilical region: the centermost region, which includes the umbilicus

Epigastric region: immediately above the umbilical region; overlies most of the stomach

Hypogastric region: immediately below the umbilical region; encompasses the pubic area

Iliac regions: lateral to the hypogastric region and overlying the hip bones

Lumbar regions: between the ribs and the flaring portions of the hip bones

Hypochondriac regions: flanking the epigastric region and overlying the lower ribs

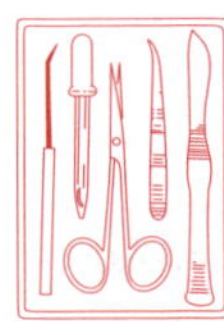

Locate the regions of the abdominal surface on a torso model and on yourself before continuing.

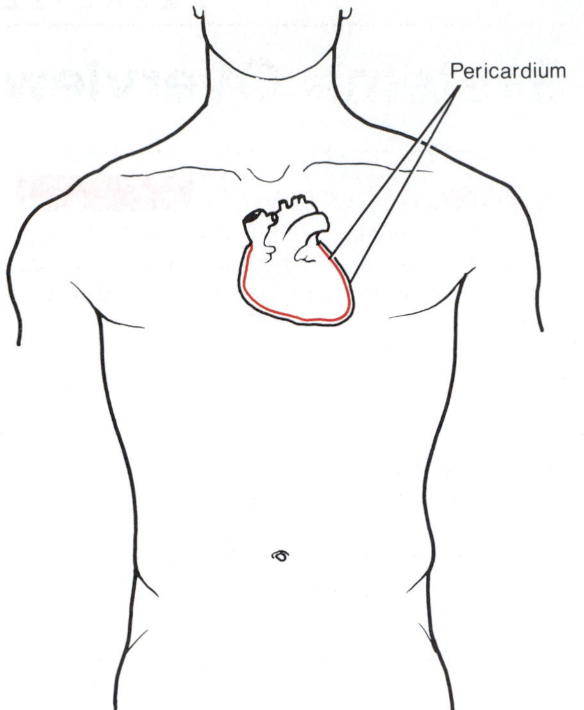

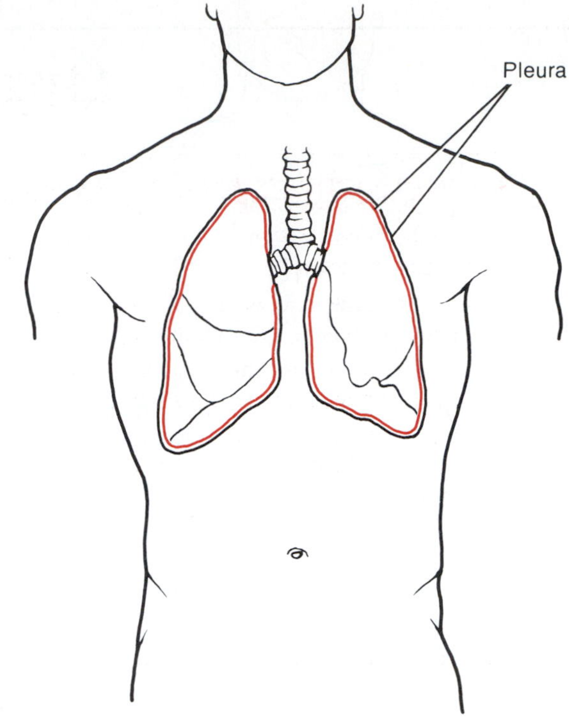

F1.8

Serous membranes. The visceral layer is shown in color; the parietal layer is shown in black.

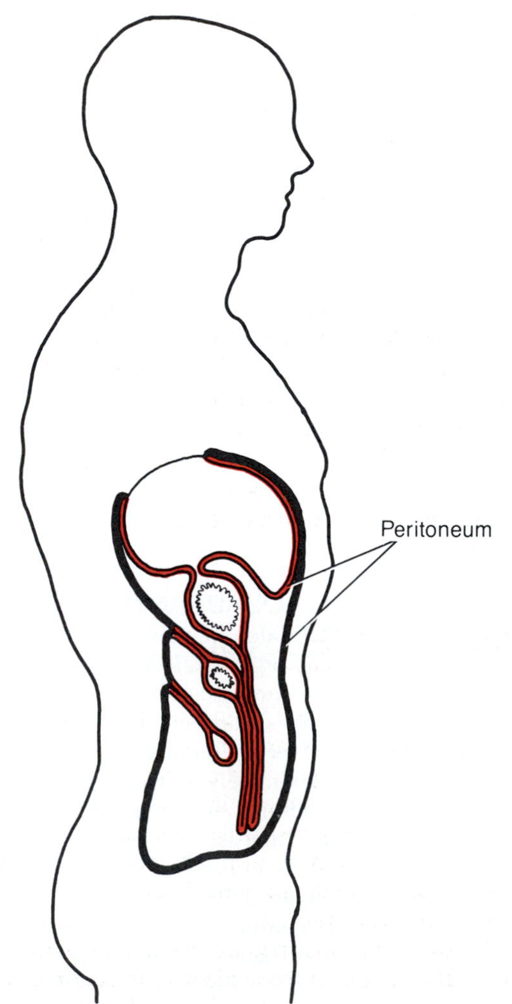

SEROUS MEMBRANES OF THE VENTRAL BODY CAVITY The walls of the ventral body cavity and the outer surfaces of the organs it contains are covered with an exceedingly thin, double-layered membrane called the **serosa,** or **serous membrane.** The part of the membrane lining the cavity walls is referred to as the **parietal serosa,** and it is continuous with a similar membrane, the **visceral serosa,** covering the external surface of the organs within the cavity. These membranes produce a thin lubricating fluid that allows the visceral organs to slide over one another or to rub against the body wall without friction. Serous membranes also compartmentalize the various organs so that infection of one organ is prevented from spreading to others.

The specific names of the serous membranes depend on the structures they envelop. Thus the serosa lining the abdominal cavity and covering its organs is the **peritoneum,** that enclosing the lungs is the **pleura,** and that around the heart is the **pericardium** (Figure 1.8).

Organ Systems Overview

OBJECTIVES

1. To name the human organ systems and indicate the major functions of each.
2. To list two or three organs of each system, and categorize the various organs by organ system.
3. To identify these organs in a dissected rat or on a dissectible human torso model or human cadaver.
4. To identify the correct organ system for each organ when presented with a list of organs (as studied in the laboratory).

MATERIALS

Freshly killed or preserved rat predissected by instructor as a demonstration or for student dissection (one for every two to four students), or dissected human cadaver
Dissecting pans and pins
Scissors
Forceps
Twine
Disposable plastic gloves
Human torso model (dissectible)

See Appendix D, Exercise 2 for links to A.D.A.M. Standard.

The basic unit or building block of all living things is the **cell.** Cells fall into four different categories according to their structures and functions. Each of these corresponds to one of the four **tissue** types: epithelial, muscular, nervous, and connective. An **organ** is a structure composed of two or more tissue types that performs a specific function for the body. For example, the small intestine, which digests and absorbs nutrients, is composed of all four tissue types. An **organ system** is a group of organs that act together to perform a particular body function. For example, the organs of the digestive system work together to ensure that food moving through the digestive system is properly broken down and that the end products are absorbed into the bloodstream to provide nutrients and fuel for all the body's cells. In all, there are eleven organ systems, which are described in Table 2.1. The lymphatic system also encompasses a *functional system* called the immune system, which is composed of an army of mobile *cells* that act to protect the body from foreign substances. Read through this summary before beginning the rat dissection.

RAT DISSECTION

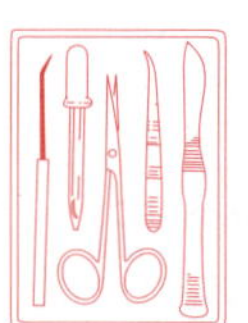

Now you will have a chance to observe the size, shape, location, and distribution of the organs and organ systems. Many of the external and internal structures of the rat are quite similar in structure and function to those of the human, so a study of the gross anatomy of the rat should help you understand your own physical structure.

The following instructions have been written to complement and direct the student's dissection and observation of a rat, but the descriptions for organ observations from procedure 4 (p. 12) apply as well to superficial observations of a previously dissected human cadaver. In addition, the general instructions for observing external structures can easily be extrapolated to serve human cadaver observations.

Note that four of the organ systems listed in Table 2.1 will not be studied at this time (integumentary, skeletal, muscular, and nervous), as they require microscopic study or more detailed dissection.

External Structures

1. If your instructor has provided a predissected rat, go to the demonstration area to make your observations. Alternatively, if you and/or members of your group will be dissecting the specimen, obtain a preserved or freshly killed rat (one for every two to four students), a dissecting pan, dissecting pins, scissors, forceps, and disposable gloves and bring them to your laboratory bench.

2. Don the gloves before beginning your observations. This precaution is particularly important when handling freshly killed animals, which may harbor internal parasites.

3. Observe the major divisions of the animal's body—head, trunk, and extremities. Compare these divisions to those of humans.

Oral Cavity

Examine the structures of the oral cavity. Identify the teeth and tongue. Observe the extent of the hard palate (the portion underlain by bone) and the soft palate (immediately posterior to the hard palate, with no bony support). Notice that the posterior end of the oral cavity leads into the throat, or pharynx. The pharynx is a passageway used by both the digestive and respiratory systems.

TABLE 2.1 Overview of Organ Systems of the Body

Organ system	Major component organs	Function
Integumentary (Skin)	Epidermal and dermal regions; cutaneous sense organs and glands	· Protects deeper organs from mechanical, chemical, and bacterial injury, and desiccation (drying out) · Excretes salts and urea · Aids in regulation of body temperature · Produces vitamin D
Skeletal	Bones, cartilages, tendons, ligaments, and joints	· Body support and protection of internal organs · Provides levers for muscular action · Cavities provide a site for blood cell formation
Muscular	Muscles attached to the skeleton	· Primary function is to contract or shorten; in doing so, skeletal muscles allow locomotion (running, walking, etc.), grasping and manipulation of the environment, and facial expression · Generates heat
Nervous	Brain, spinal cord, nerves, and sensory receptors	· Allows body to detect changes in its internal and external environment and to respond to such information by activating appropriate muscles or glands · Helps maintain homeostasis of the body via rapid transmission of electrical signals
Endocrine	Pituitary, thyroid, parathyroid, adrenal, and pineal glands; ovaries, testes, and pancreas	· Helps maintain body homeostasis, promotes growth and development; produces chemical "messengers" (hormones) that travel in the blood to exert their effect(s) on various "target organs" of the body
Cardiovascular	Heart, blood vessels, and blood	· Primarily a transport system that carries blood containing oxygen, carbon dioxide, nutrients, wastes, ions, hormones, and other substances to and from the tissue cells where exchanges are made; blood is propelled through the blood vessels by the pumping action of the heart · Antibodies and other protein molecules in the blood act to protect the body
Lymphatic	Lymphatic vessels, lymph nodes, spleen, thymus, tonsils, and scattered collections of lymphoid tissue	· Picks up fluid leaked from the blood vessels and returns it to the blood · Cleanses blood of pathogens and other debris · Houses lymphocytes that act via the immune response to protect the body from foreign substances (antigens)
Respiratory	Nasal passages, pharynx, larynx, trachea, bronchi, and lungs	· Keeps the blood continuously supplied with oxygen while removing carbon dioxide
Digestive	Oral cavity, esophagus, stomach, small and large intestines, and accessory structures (teeth, salivary glands, liver, and pancreas)	· Breaks down ingested foods to minute particles, which can be absorbed into the blood for delivery to the body cells · Undigested residue removed from the body as feces
Urinary	Kidneys, ureters, bladder, and urethra	· Rids the body of nitrogen-containing wastes (urea, uric acid, and ammonia), which result from the breakdown of proteins and nucleic acids by body cells · Maintains water, electrolyte, and acid-base balance of blood
Reproductive	Male: testes, scrotum, penis, and duct system, which carries sperm to the body exterior	· Provides germ cells (sperm) for perpetuation of the species
	Female: ovaries, uterine tubes, uterus, and vagina	· Provides germ cells (eggs) and the female uterus houses the developing fetus until birth

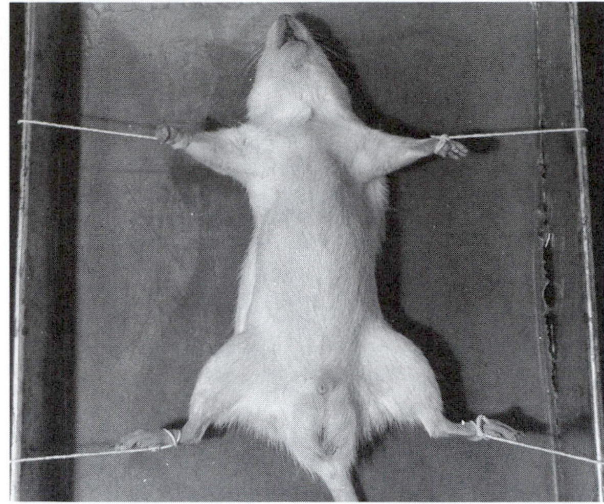

(a)

(b)

Ventral Body Cavity

1. Pin the animal to the wax of the dissecting pan by placing its dorsal side down and securing its extremities to wax. If the dissecting pan is not waxed, you will need to secure the animal with twine as shown in Figure 2.1a. (Some may prefer this method in any case.) Obtain the roll of twine. Make a loop knot around one upper limb, pass the twine under the pan, and secure the opposing limb. Repeat for the lower extremities.

2. Lift the abdominal skin with a forceps, and cut through it with the scissors (Figure 2.1b). Close the scissor blades and insert them under the cut skin. Moving in a cephalad direction, open and close the blades to loosen the skin from the underlying connective tissue and muscle. Once this skin-freeing procedure has been completed, cut the skin along the body midline, from the pubic region to the lower jaw (Figure 2.1c). Make a lateral cut about halfway down the ventral surface of each limb. Complete the job of freeing the skin with the scissor tips, and pin the flaps to the tray (Figure 2.1d). The underlying tissue that is now exposed is the skeletal musculature of the body wall and limbs. It allows voluntary body movement. Notice that the muscles are packaged in sheets of pearly white connective tissue (fascia), which protect the muscles and bind them together.

3. Carefully cut through the muscles of the abdominal wall in the pubic region, avoiding the underlying organs. Remember, *to dissect* means "to separate"—not mutilate! Now, hold and lift the muscle layer with a forceps and cut through the muscle layer from the pubic

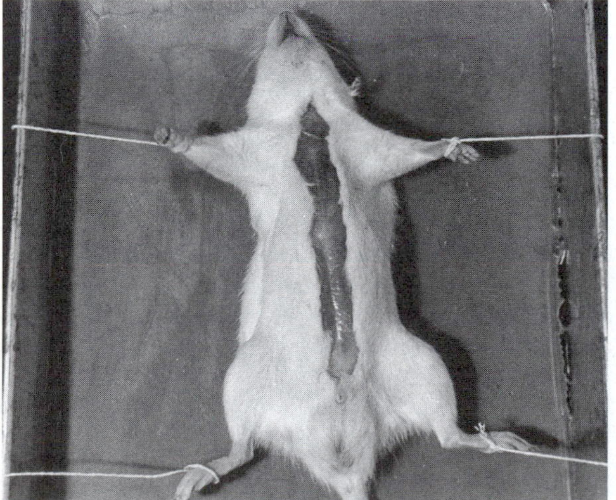

(c)

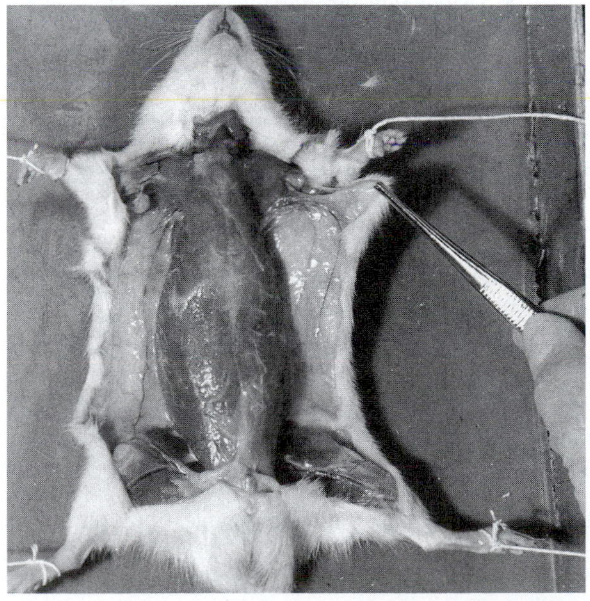

(d)

F2.1

Rat dissection: Securing for dissection and the initial incision. **(a)** Securing the rat to the dissection tray with twine. **(b)** Using scissors to make the incision on the median line of the abdominal region. **(c)** Completed incision from the pelvic region to the lower jaw. **(d)** Reflection (folding back) of the skin to expose the underlying muscles.

region to the bottom of the rib cage. Make two lateral cuts through the rib cage (Figure 2.2). A thin membrane attached to the inferior boundary of the rib cage should be obvious; this is the **diaphragm,** which separates the thoracic and abdominal cavities. Cut the diaphragm away to loosen the rib cage. You can now lift the ribs to view the contents of the thoracic cavity.

4. Examine the structures of the thoracic cavity, starting with the most superficial structures and working deeper. As you work, refer to Figure 2.3 which shows the superficial organs.

Thymus: an irregular mass of glandular tissue overlying the heart.

Push the thymus to the side to view the heart.

Heart: median oval structure enclosed within the pericardium (serous membrane sac).
Lungs: flanking the heart on either side.

Now observe the throat region.

Trachea: tubelike "windpipe" running medially down the throat; part of the respiratory system.

Follow the trachea into the thoracic cavity; note where it divides into two branches. These are the bronchi.

Bronchi: two passageways that plunge laterally into the tissue of the two lungs.

To expose the esophagus, push the trachea to one side.

Esophagus: a food chute; the part of the digestive system that transports food from the pharynx (throat) to the stomach.

Follow the esophagus through the diaphragm to its junction with the stomach.

Stomach: a C-shaped organ important in food digestion and temporary food storage.

5. Examine the superficial structures of the abdominopelvic cavity. Beginning with the stomach, trace the rest of the digestive tract.

Small intestine: connected to the stomach and ending just before a large saclike cecum.
Cecum: the initial portion of the large intestine.
Large intestine: a large muscular tube coiled within the abdomen.

Follow the course of the large intestine to the rectum, which is partially covered by the urinary bladder.

Rectum: terminal part of the large intestine; continuous with the anal canal.
Anus: the opening of the digestive tract (anal canal) to the exterior.

Now lift the small intestine with the forceps to view the mesentery.

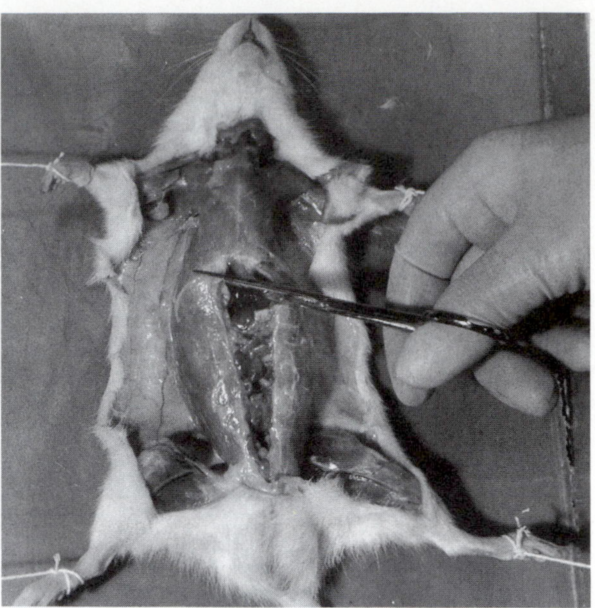

F2.2

Rat dissection: Making lateral cuts at the base of the rib cage.

Mesentery: an apronlike serous membrane; suspends many of the digestive organs in the abdominal cavity. Notice that it is heavily invested with blood vessels and, more likely than not, riddled with large fat deposits.

Locate the remaining abdominal structures.

Pancreas: a diffuse gland; rests dorsal to and in the mesentery between the first portion of the small intestine and the stomach.
Spleen: a dark red organ curving around the left lateral side of the stomach; considered part of the lymphatic system and often called the red blood cell graveyard.
Liver: large and brownish red; the most superior organ in the abdominal cavity, directly beneath the diaphragm.

6. To locate the deeper structures of the abdominopelvic cavity, cut through the superior margin of the stomach and the distal end of the large intestine and lay them aside. (Refer to Figure 2.4 (p. 14) as you work.)

Examine the posterior wall of the abdominal cavity to locate the two kidneys.

Kidneys: bean-shaped organs; retroperitoneal (behind the peritoneum).
Adrenal glands: large glands that sit astride the superior margin of each kidney; considered part of the endocrine system.

Carefully strip away part of the peritoneum and attempt to follow the course of one of the ureters to the bladder.

Ureter: tube running from the indented region of a kidney to the urinary bladder.

Urinary bladder: the sac that serves as a reservoir for urine.

7. In the midline of the body cavity lying between the kidneys are the two principal abdominal blood vessels. Identify each.

Inferior vena cava: the large vein that returns blood to the heart from the lower regions of the body.

Descending aorta: deep to the inferior vena cava; the largest artery of the body; carries blood away from the heart down the midline of the body.

8. Only a cursory examination of reproductive organs will be done. First determine if the animal is a male or female. Observe the ventral body surface beneath the tail. If a saclike scrotum and a single body opening are visible, the animal is a male. If three body openings are present, it is a female. (See Figure 2.4.)

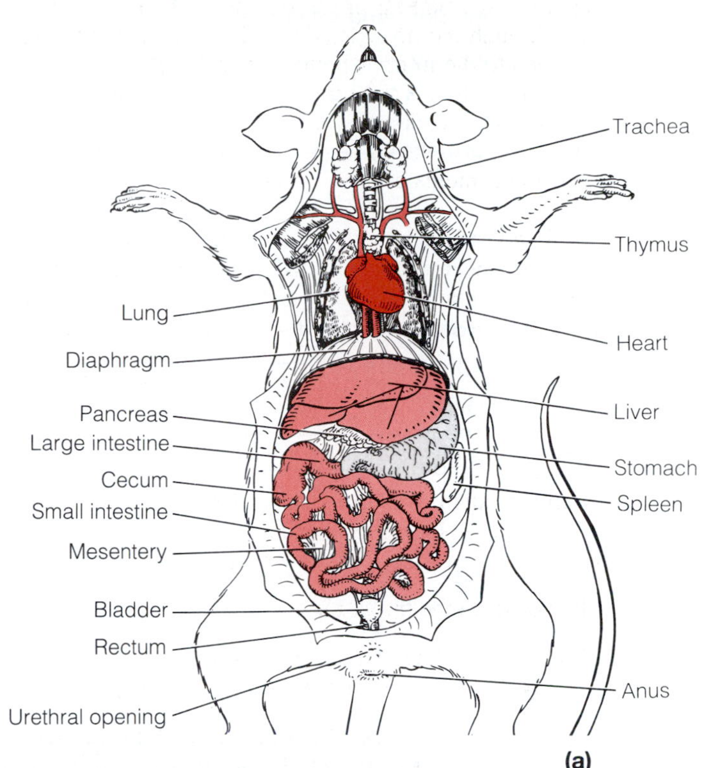

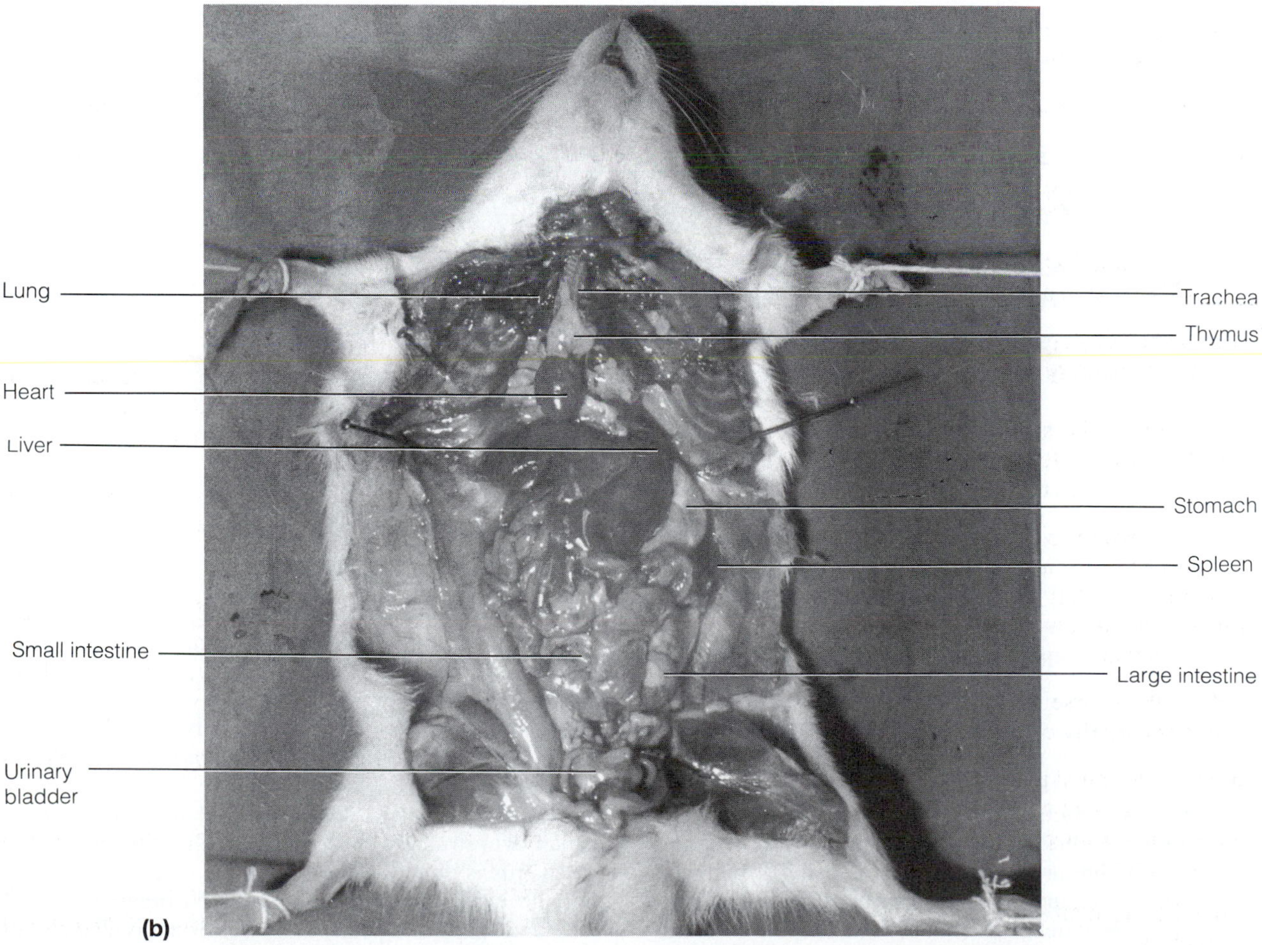

F2.3

Rat dissection: superficial organs of the thoracic and abdominal cavities. **(a)** Diagrammatic view. **(b)** Photograph.

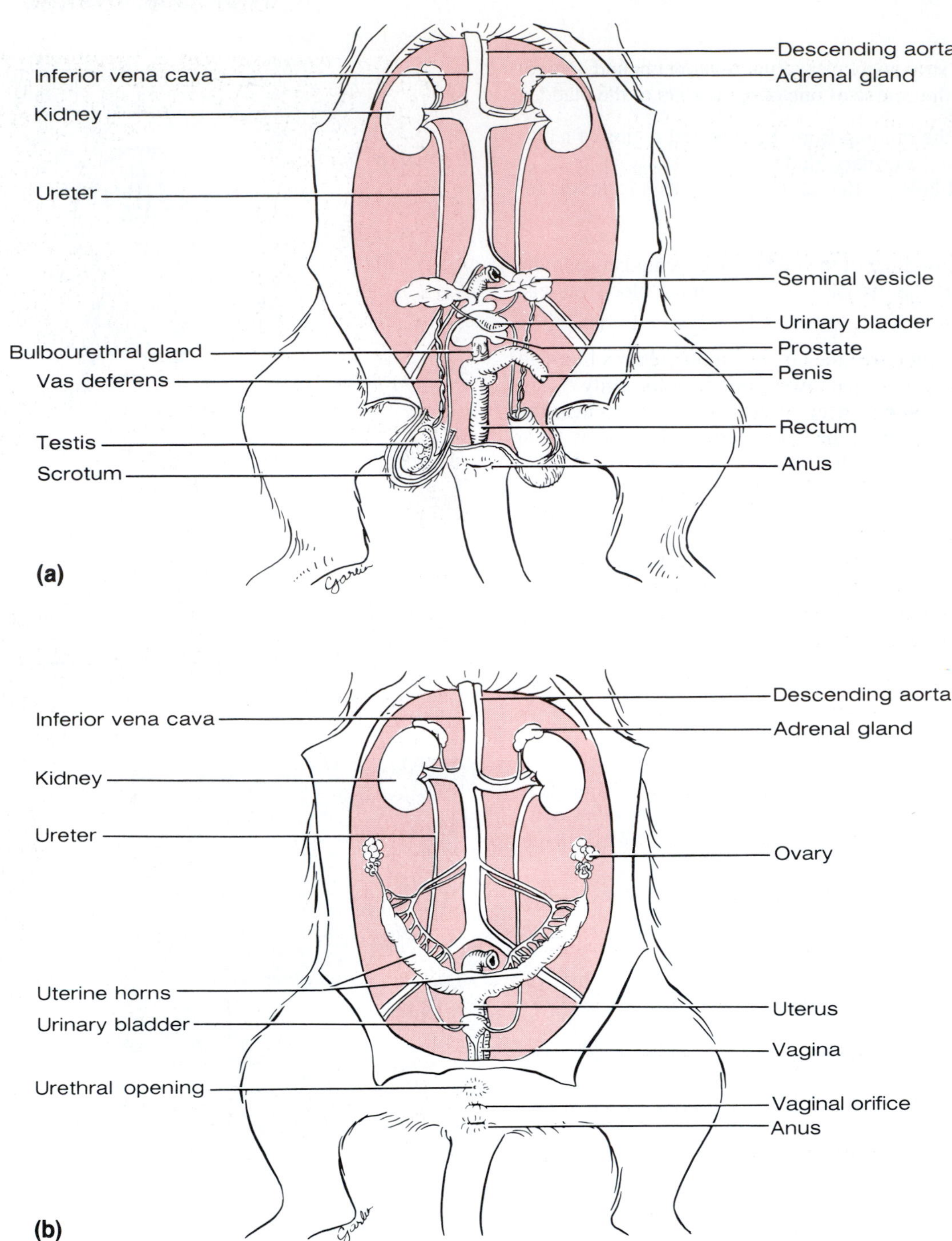

F2.4

Rat dissection: deeper organs of the abdominal cavity and the reproductive structures. **(a)** Male. **(b)** Female.

MALE ANIMAL Make a shallow incision into the **scrotum.** Loosen and lift out the oval **testis.** Exert a gentle pull on the testis to identify the slender **vas deferens,** or sperm duct, which carries sperm from the testis superiorly into the abdominal cavity and joins with the urethra. The urethra runs through the penis of the male and carries both urine and sperm out of the body. Identify the **penis,** extending from the bladder to the ventral body wall. Figure 2.4a indicates other glands of the male reproductive system, but they need not be identified at this time.

FEMALE ANIMAL Inspect the pelvic cavity to identify the Y-shaped **uterus** lying against the dorsal body wall and beneath the bladder (Figure 2.4b). Follow one of the uterine horns superiorly to identify an **ovary,** a small oval structure at the end of the uterine horn. (The rat uterus is quite different from the uterus of a human female, which is a single-chambered organ about the size and shape of a pear.) The inferior undivided part of the rat uterus is continuous with the vagina, which leads to the body exterior. Identify the **vaginal orifice** (external vaginal opening).

9. When you have finished your observations, store or dispose of the rat according to your instructor's directions. Wash the dissecting pan and tools with laboratory detergent. Dispose of the gloves. Then wash and dry your hands before continuing with the examination of the torso model.

EXAMINING THE HUMAN TORSO MODEL

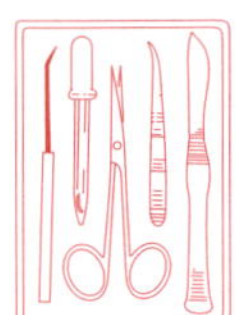

Examine a human torso model to identify the organs listed below. (Note: If a torso model is not available, Figure 2.5 may be used for this part of the exercise.)

Dorsal body cavity: brain, spinal cord

Thoracic cavity: heart, lungs, bronchi, trachea, esophagus, diaphragm

Abdominopelvic cavity: liver, stomach, pancreas, spleen, small intestine, large intestine, rectum, kidneys, ureters, bladder, adrenal gland, descending aorta, inferior vena cava

As you observe these structures, locate the nine abdominopelvic areas studied earlier and determine which organs would be found in each area.

1. Umbilical region: ______________________
2. Epigastric region: ______________________
3. Hypogastric region: ______________________
4. Right iliac region: ______________________
5. Left iliac region: ______________________
6. Right lumbar region: ______________________
7. Left lumbar region: ______________________
8. Right hypochondriac region: ______________________
9. Left hypochondriac region: ______________________

Would you say that the shape and location of the human organs are similar or dissimilar to those of the rat?

____

Assign each of the organs just identified to one of the organ system categories below.

Digestive: ______________________

Urinary: ______________________

Cardiovascular: ______________________

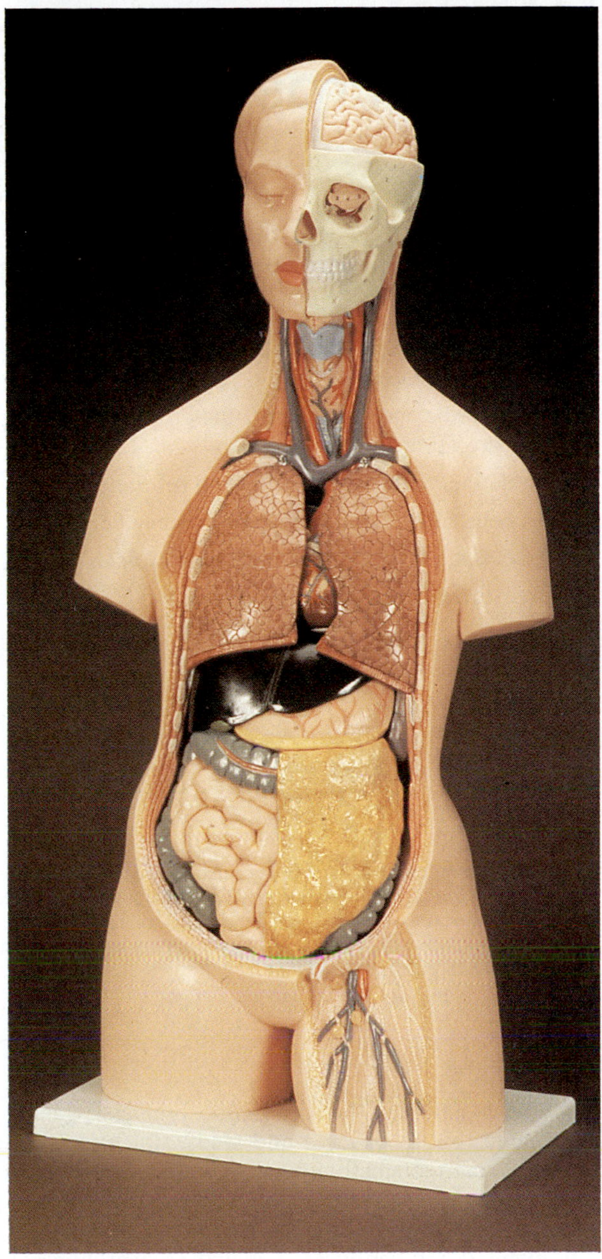

F2.5

Human torso model.

Reproductive: ______________________

Respiratory: ______________________

Lymphatic: ______________________

Nervous: ______________________

3

EXERCISE

The Microscope

OBJECTIVES

1. To identify the parts of the microscope and list the function of each.
2. To describe and demonstrate the proper techniques for care of the microscope.
3. To define *total magnification* and *resolution.*
4. To demonstrate proper focusing technique.
5. To define *parfocal, field,* and *depth of field.*
6. To estimate the size of objects in a field.

MATERIALS

Compound microscope
Millimeter ruler
Prepared slides of the letter *e* or newsprint
Immersion oil
Lens paper
Prepared slide of grid ruled in millimeters (grid slide)
Prepared slide of 3 crossed colored threads
Clean microscope slide and coverslip
Toothpicks (flat-tipped)
Physiologic saline in a dropper bottle
Methylene blue stain (dilute) in a dropper bottle
Filter paper
Forceps
Beaker containing fresh 10% household bleach solution for wet mount disposal
Disposable autoclave bag

Note to the Instructor: The slides and coverslips used for viewing cheek cells are to be soaked for 2 hours (or longer) in 10% bleach solution and then drained. The slides, coverslips, and disposable autoclave bag (containing used toothpicks) are to be autoclaved for 15 min at 121°C and 15 pounds pressure to insure sterility. After autoclaving, the disposable autoclave bag may be discarded in any disposal facility and the glassware washed with laboratory detergent and reprepared for use. These instructions apply as well to any blood-stained glassware or disposable items used in other experimental procedures.

With the invention of the microscope, biologists gained a valuable tool to observe and study structures (like cells) that are too small to be seen by the unaided eye. As a result, many of the theories basic to the understanding of biologic sciences have been established. This exercise will familiarize you with the workhorse of microscopes—the compound microscope—and provide you with the necessary instructions for its proper use.

CARE AND STRUCTURE OF THE COMPOUND MICROSCOPE

The **compound microscope** is a precision instrument and should always be handled with care. At all times you must observe the following rules for its transport, cleaning, use, and storage:

- When transporting the microscope, hold it in an upright position with one hand on its arm and the other supporting its base. Avoid jarring the instrument when setting it down.
- Use only special grit-free lens paper to clean the lenses. Clean all lenses before and after use.
- Always begin the focusing process with the lowest-power objective lens in position, changing to the higher-power lenses as necessary.
- Never use the coarse adjustment knob with the high-power or oil immersion lenses.
- A coverslip must always be used with temporary (wet mount) preparations.
- Before putting the microscope in the storage cabinet, remove the slide from the stage, rotate the lowest-power objective lens into position, and replace the dust cover.
- Never remove any parts from the microscope; inform your instructor of any mechanical problems that arise.

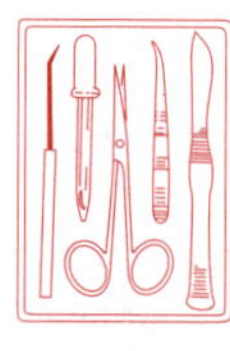

1. Obtain a microscope and bring it to the laboratory bench. (Use the proper carrying technique!) Compare your microscope with the illustration in Figure 3.1 and identify the following microscope parts:

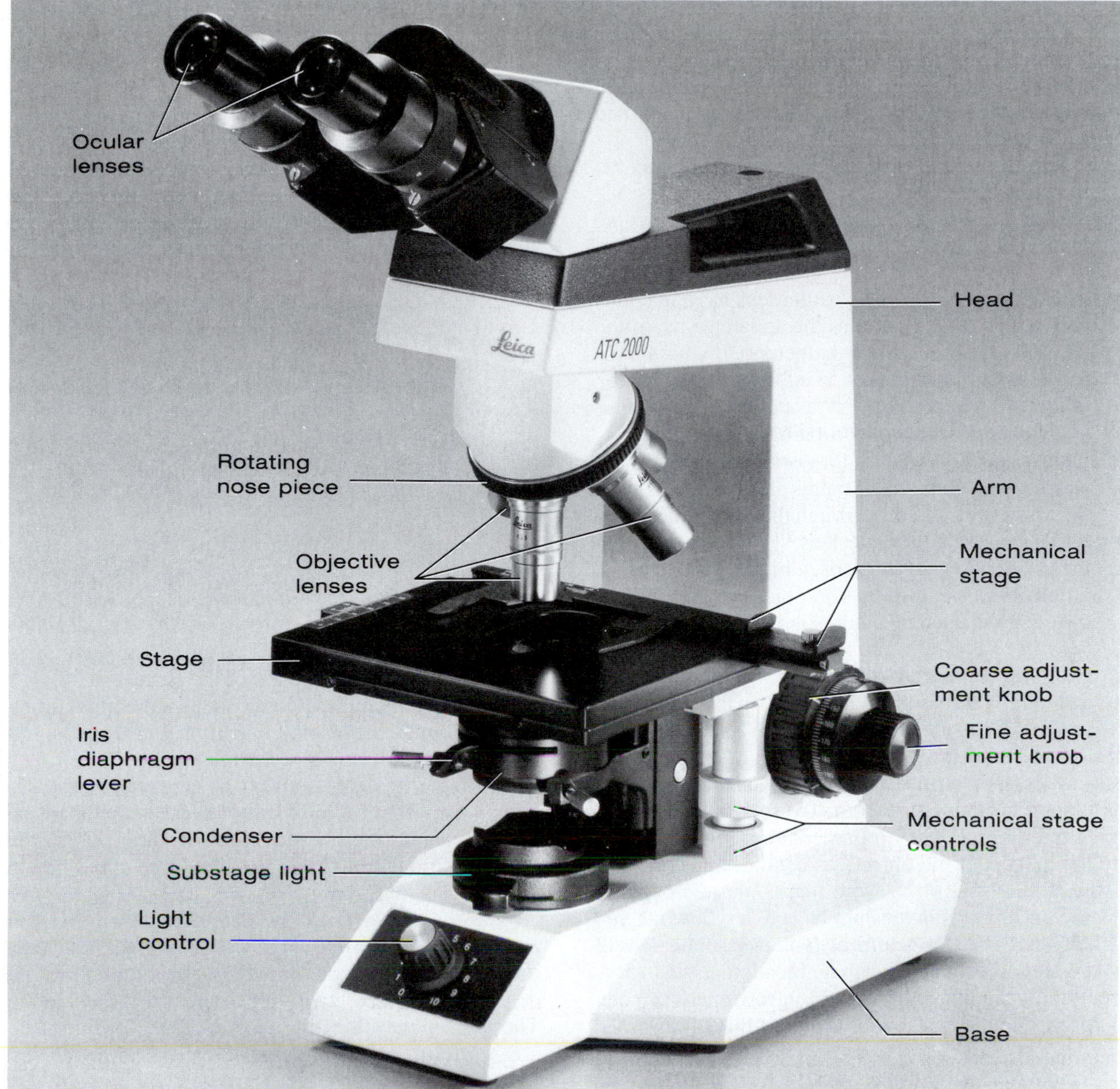

F3.1

Compound microscope and its parts.

Base: supports the microscope. (Note: Some microscopes are provided with an inclination joint, which allows the instrument to be tilted backward for viewing dry preparations.)

Substage light (or *mirror*): located in the base. In microscopes with a substage light source, the light passes directly upward through the microscope. If a mirror is used, light must be reflected from a separate free-standing lamp.

Stage: the platform the slide rests on while being viewed. The stage always has a hole in it to permit light to pass through both it and the specimen. Some microscopes have a stage equipped with *spring clips;* others have a clamp-type *mechanical stage* as shown in Figure 3.1. Both hold the slide in position for viewing; in addition, the mechanical stage permits precise movement of the specimen.

Condenser: concentrates the light on the specimen. The condenser may have a height-adjustment knob that raises and lowers the condenser to vary light delivery. Generally, the best position for the condenser is close to the inferior surface of the stage.

Iris diaphragm lever: arm attached to the condenser that regulates the amount of light passing through the condenser. The iris diaphragm permits the best possible contrast when viewing the specimen.

Coarse adjustment knob: used to focus the specimen.

Fine adjustment knob: used for precise focusing once coarse focusing has been completed.

Head or **body tube:** supports the objective lens system (which is mounted on a movable nosepiece), and the ocular lens or lenses.

Arm: vertical portion of the microscope connecting the base and head.

Ocular (or *eyepiece*): depending on the microscope, there will be one or two lenses at the superior end of the head or body tube. Observations are made through the ocular(s). An ocular lens has a magnification of 10× (it increases the apparent size of the object by ten times or ten diameters). If your microscope has a **pointer** (used to indicate a specific area of the viewed specimen), it is attached to the ocular and can be positioned by rotating the ocular lens.

Nosepiece: generally carries three objective lenses and permits sequential positioning of these lenses over the light beam passing through the hole in the stage.

Objective lenses: adjustable lens system that permits the use of a **low-power lens,** a **high-power lens,** or an **oil immersion lens.** The objective lenses have different magnifying and resolving powers.

2. Examine the objectives carefully, noting their relative lengths and the numbers inscribed on their sides. On most microscopes, the low-power (l.p.) objective is the shortest and generally has a magnification of 10×. The high-power (h.p.) objective is of intermediate length and has a magnification range from 40× to 50×, depending on the microscope. The oil immersion objective is usually the longest of the objectives and has a magnifying power of 95× to 100×. Note that some microscopes lack the oil immersion lens but have a very low magnification lens called the **scanning lens.** A scanning lens is a very short objective with a magnification of 4× to 5×. Record the magnification of each objective lens of your microscope in the first row of the chart that follows.

3. Rotate the low-power objective until it clicks into position, and turn the coarse adjustment knob about 180 degrees. Notice how far the stage (or objective) travels during this adjustment. Move the fine adjustment knob 180 degrees, noting again the distance that the stage (or the objective) moves.

Magnification and Resolution

The microscope is an instrument of magnification. In the compound microscope, magnification is achieved through the interplay of two lenses—the ocular lens and the objective lens. The objective lens magnifies the specimen to produce a **real image** that is projected to the ocular. This real image is magnified by the ocular lens to produce the **virtual image** seen by your eye (Figure 3.2).

The **total magnification** of any specimen being viewed is equal to the power of the ocular lens multiplied by the power of the objective lens used. For example, if the ocular lens magnifies 10× and the objective lens being used magnifies 45×, the total magnification is 450×.

Determine the total magnification you may achieve with each of the objectives on your microscope and record the figures on the second row of the chart. At this time, also cross out the column relating to a lens that your microscope does not have, and record the number of your microscope at the top of the chart.

The compound light microscope has certain limitations. Although the level of magnification is almost limitless, the **resolution** (or resolving power), the ability to discriminate two close objects as separate, is not. The human eye can resolve objects about 100 μm apart, but the compound microscope has a resolution of 0.2 μm under ideal conditions. Objects closer than 0.2 μm are seen as a single fused image.

Resolving power (*RP*) is determined by the amount and physical properties of the visible light that enters the microscope. In general, the greater the amount of light delivered to the objective lens, the greater the resolution. The size of the objective lens aperture (opening) decreases with increasing magnification, allowing less light to enter the objective. Thus, you will probably find it necessary to increase the light intensity at the higher magnifications.

Summary Chart for Microscope # ____________

	Scanning	Low power	High power	Oil immersion
Magnification of objective lens	×	×	×	×
Total magnification	×	×	×	×
Detail observed				
Field size (diameter)	mm μm	mm μm	mm μm	mm μm
Working distance		mm	mm	mm

VIEWING OBJECTS THROUGH THE MICROSCOPE

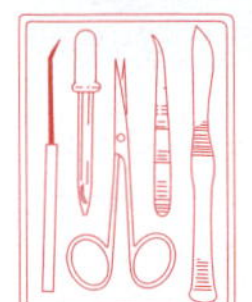

1. Obtain a millimeter ruler, a prepared slide of the letter e or newsprint, a dropper bottle of immersion oil, and some lens paper. Adjust the condenser to its highest position and switch on the light source of your microscope. (If the light source is not built into the base, use the curved surface of the mirror to reflect the light up into the microscope.)

2. Secure the slide on the stage so that the letter *e* is centered over the light beam passing through the stage. If you are using a microscope with spring clips, make sure the slide is secured at both ends. If your microscope has a mechanical stage, open the jaws of its slide retainer (holder) by using the control lever (typically) located at the rear left corner of the mechanical stage. Insert the slide squarely within the confines of the slide retainer. Check to see that the slide is resting on the stage (and not on the mechanical stage frame) before releasing the control lever.

3. With your lowest power (scanning or low-power) objective in position over the stage, use the coarse adjustment knob to bring the objective and stage as close together as possible.

4. Look through the ocular and adjust the light for comfort. Now use the coarse adjustment knob to focus slowly away from the *e* until it is as clearly focused as possible. Complete the focusing with the fine adjustment knob.

5. Sketch the letter in the circle just as it appears in the **field** (the area you see through the microscope).

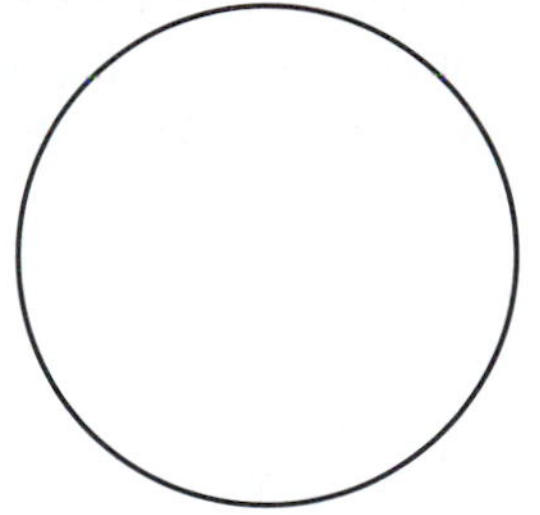

What is the total magnification? ________ ×

How far is the bottom of the objective from the specimen? In other words, what is the **working distance?**

________ mm

Use a millimeter ruler to make this measurement and record it in the chart on page 18. How has the apparent orientation of the *e* changed top to bottom, right to left, and so on?

____

____

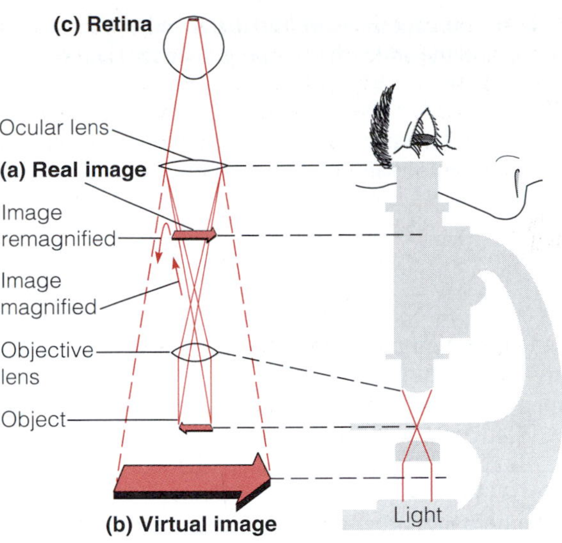

F3.2

Image formation in light microscopy. Light passing through the objective lens forms a real image (a). The real image serves as the object for the ocular lens, which remagnifies the image and forms the virtual image (b). The virtual image passes through the lens of the eye and is focused on the retina (c).

6. Move the slide slowly away from you on the stage as you view it through the ocular. In what direction does the image move?

____

Move the slide to the left. In what direction does the image move?

____

At first this change in orientation will confuse you, but with practice you will learn to move the slide in the desired direction with no problem.

7. Without touching the focusing knobs, increase the magnification by rotating the next higher magnification lens (low-power or high-power) into position over the stage. Using the fine adjustment only, sharpen the focus.* What new details become clear?

____

____

What is the total magnification now? ______________ ×

* Today most good laboratory microscopes are parfocal; that is, the slide should be in focus (or nearly so) at the higher magnifications once you have properly focused in l.p. If you are unable to swing the objective into position without raising the objective, your microscope is not parfocal. Consult your instructor.

As best you can, measure the distance between the objective and the slide (the working distance) and record it on the chart.

Why should the coarse focusing knob *not* be used when focusing with the higher-powered objective lenses?

____

____

Is the image larger or smaller? ______________

Approximately how much of the letter *e* is visible now?

____

Is the field larger or smaller? ______________

Why is it necessary to center your object (or the portion of the slide you wish to view) before changing to a higher power?

____

Move the iris diaphragm lever while observing the field. What happens?

____

Is it more desirable to increase *or* decrease the light when changing to a higher magnification?

__________ Why? ______________

____

8. If you have just been using the low-power objective, repeat the steps given in direction 7 using the high-power objective lens. Record the total magnification, approximate working distance, and information on detail observed on the chart on p. 18.

9. Without touching the focusing knob, rotate the high-power lens out of position so that the area of the slide over the opening in the stage is unobstructed. Place a drop of immersion oil over the *e* on the slide and rotate the oil immersion lens into position. Set the condenser at its highest point (closest to the stage), and open the diaphragm fully. Adjust the fine focus and fine-tune the light for the best possible resolution.

Is the field again decreased in size? ______________

What is the total magnification with the immersion lens?

__________ ×

Is the working distance less *or* greater than it was when the high-power lens was focused?

____

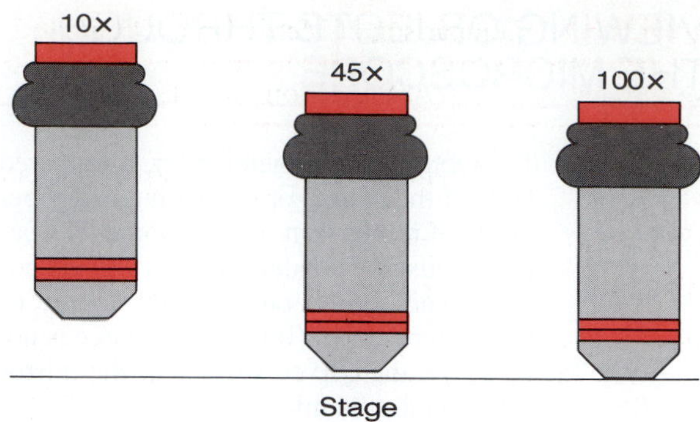

F3.3

Relative working distances of the 10×, 45×, and 100× objectives.

Compare your observations on the relative working distances of the objective lenses with the illustration in Figure 3.3. Explain why it is desirable to begin the focusing process in low power.

____

____

____

10. Rotate the oil immersion lens slightly to the side and remove the slide. Clean the oil immersion lens carefully with lens paper and then clean the slide in the same manner with a fresh piece of lens paper.

DETERMINING THE SIZE OF THE MICROSCOPE FIELD

By this time you should know that the size of the microscope field decreases with increasing magnification. For future microscope work, it will be useful to determine the diameter of each of the microscope fields. This information will allow you to make a fairly accurate estimate of the size of the objects you view in any field. For example, if you have calculated the field diameter to be 4 mm and the object being observed extends across half this diameter, you can estimate the length of the object to be approximately 2 mm.

Microscopic specimens are usually measured in micrometers and millimeters, both units of the metric system. You can get an idea of the relationship and meaning of these units from Table 3.1. A more detailed treatment appears in Appendix A.

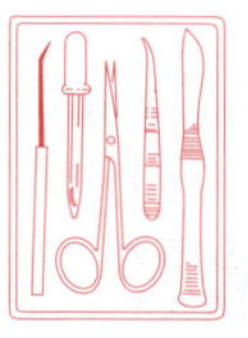

1. Return the letter *e* slide and obtain a grid slide, a slide prepared with graph paper ruled in millimeters. Each of the squares in the grid is 1 mm on each side. Use your lowest power objective to bring the grid lines into focus.

TABLE 3.1 Comparison of Metric Units of Length

Metric unit	Abbreviation	Equivalent
Meter	m	(about 39.3 in.)
Centimeter	cm	10^{-2}m
Millimeter	mm	10^{-3}m
Micrometer (or micron)	μm (μ)	10^{-6}m
Nanometer (or millimicrometer, or millimicron	nm (mμ)	10^{-9}m
Angstrom	Å	10^{-10}m

2. Move the slide so that one grid line touches the edge of the field on one side, and then count the number of squares you can see across the diameter of the field. If you can see only part of a square, as in the accompanying diagram, estimate the part of a millimeter that the partial square represents.

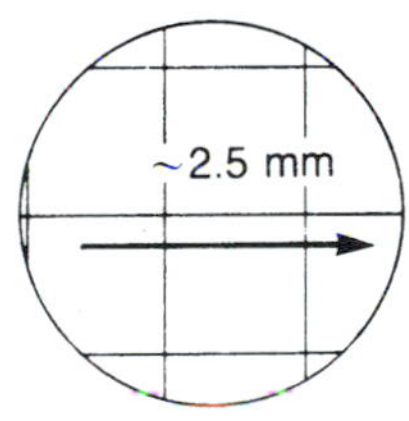

For future reference, record this figure in the appropriate space marked "field size" on the summary chart on page 18. (If you have been using the scanning lens, repeat the procedure with the low-power objective lens.) Complete the chart by computing the approximate diameter of the high-power and immersion fields. Say the diameter of the low-power field (total magnification of 50×) is 2 mm. You would compute the diameter of a high-power field with a total magnification of 100× as follows:

$$2 \text{ mm} \times 50 = Y \text{ (diameter of h.p. field)} \times 100$$
$$100 \text{ mm} = 100Y$$
$$1 \text{ mm} = Y \text{ (diameter of the h.p. field)}$$

The formula is:
Diameter of the l.p. field (mm) × Total magnification of the l.p. field = Diameter of field Y × Total magnification of field Y

3. Estimate the length (longest dimension) of the following microscopic objects. *Base your calculations on the field sizes you have determined for your microscope.*

a. Object seen in low-power field:

approximate length:

_______ mm.

b. Object seen in high-power field:

approximate length:

_______ mm,

or _______ μm.

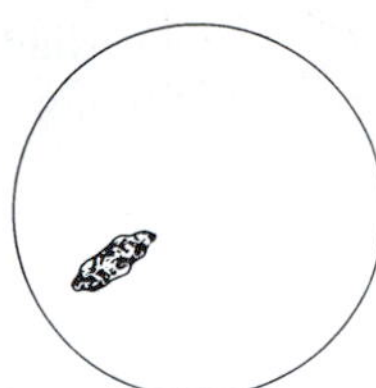

c. Object seen in oil immersion field:

approximate length:

_______ μm.

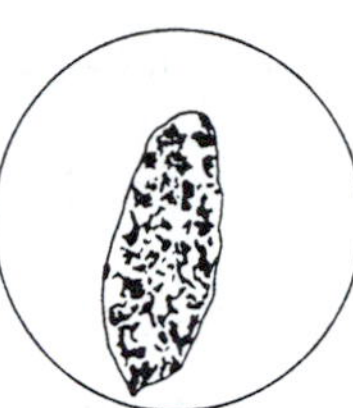

4. If an object viewed with the oil immersion lens looked like the field depicted just below, could you determine its approximate size from this view?

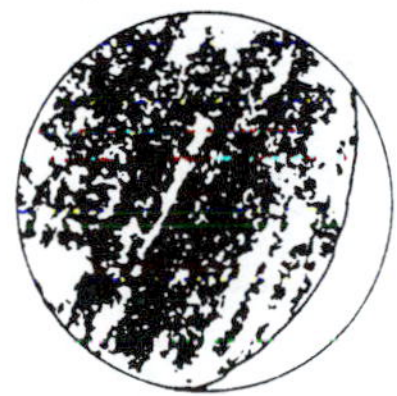

If not, then how could you determine it? _______________

____

____

PERCEIVING DEPTH

Any microscopic specimen has depth as well as length and width; it is rare indeed to view a tissue slide with just one layer of cells. Normally you can see two or three cell thicknesses. Therefore, it is important to learn how to determine relative depth with your microscope.*

* In microscope work the **depth of field** (the depth of the specimen clearly in focus) is greater at lower magnifications.

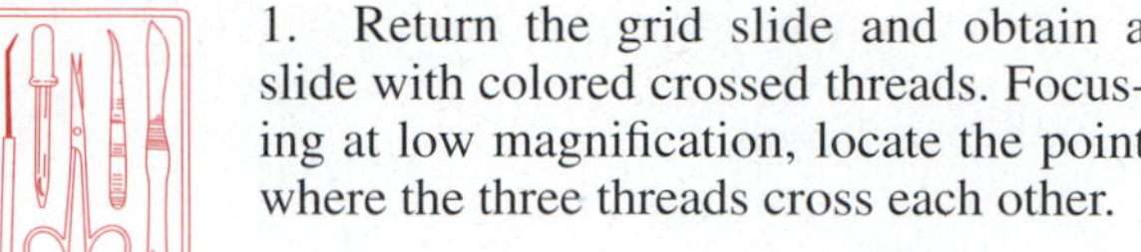

1. Return the grid slide and obtain a slide with colored crossed threads. Focusing at low magnification, locate the point where the three threads cross each other.

2. Use the iris diaphragm lever to greatly reduce the light, thus increasing the contrast. Focus down with the coarse adjustment until the threads are out of focus, then slowly focus upward again, noting which thread comes into clear focus first. This one is the lowest or most inferior thread. (You will see two or even all three threads, so you must be very careful in determining which one comes into clear focus first.) Record your observations:

_______________ thread over _____________________

Continue to focus upward until the uppermost thread is clearly focused. Again record your observation.

_______________ thread over _____________________

Which thread is uppermost? ____________________

Lowest? ______________________________________

PREPARING AND OBSERVING A WET MOUNT

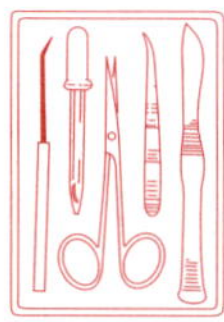

1. Obtain the following: a clean microscope slide and coverslip, a flat-tipped toothpick, a dropper bottle of physiologic saline, a dropper bottle of methylene blue stain, forceps, and filter paper.

2. Place a drop of physiologic saline in the center of the slide. Using the flat end of the toothpick, *gently* scrape the inner lining of your cheek. Agitate the end of the toothpick containing the cheek scrapings in the drop of saline (Figure 3.4a).

Immediately discard the used toothpick in the disposable autoclave bag provided at the supplies area.

3. Add a tiny drop of the methylene blue stain to the preparation. (These epithelial cells are nearly transparent and thus difficult to see without the stain, which colors the nuclei of the cells and makes them look much darker than the cytoplasm.) Stir again and then dispose of the toothpick as described above.

4. Hold the coverslip with the forceps so that its bottom edge touches one side of the fluid drop (Figure 3.4b), then *carefully* lower the coverslip onto the preparation (Figure 3.4c). *Do not just drop the coverslip,* or you will trap large air bubbles under it, which will obscure the cells. *A coverslip should always be used with a wet mount* to prevent soiling the lens if you should misfocus.

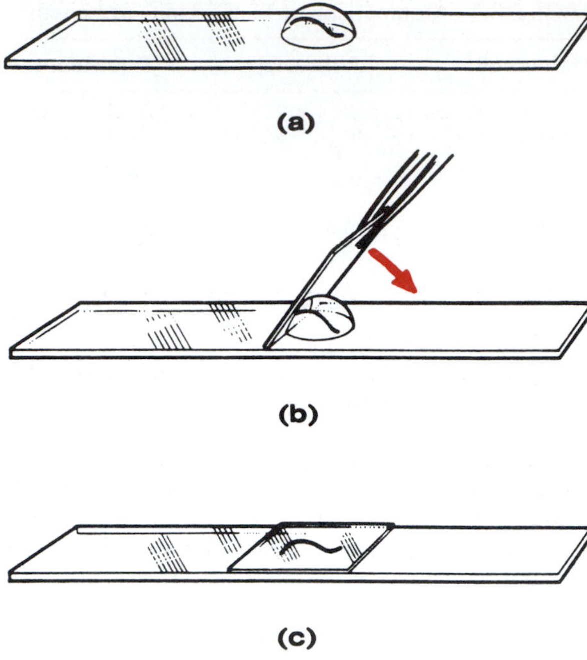

F3.4

Procedure for preparation of a wet mount. (a) The object is placed in a drop of water (or saline) on a clean slide, **(b)** a coverslip is held at a 45° angle with forceps, and **(c)** it is lowered carefully over the water and the object.

5. Examine your preparation carefully. The coverslip should be closely apposed to the slide. If there is excess fluid around its edges, you will need to remove it. Obtain a piece of filter paper, fold it in half, and use the folded edge to absorb the excess fluid.

Before continuing, discard the filter paper in the disposable autoclave bag.

6. Place the slide on the stage and locate the cells in low power. You will probably want to dim the light with the iris diaphragm to provide more contrast for viewing the lightly stained cells. Furthermore, a wet mount will dry out quickly in bright light, because a bright light source is hot.

7. Cheek epithelial cells are very thin, six-sided cells. In the cheek, they provide a smooth, tilelike lining, as shown in Figure 3.5.

8. Make a sketch of the epithelial cells that you observe.

Approximately how wide are the cheek epithelial cells?

_______ mm

Why do *your* cheek cells look different than those illustrated in Figure 3.5? (Hint: what did you have to *do* to your cheek to obtain them?)

____

____

9. When you have completed your observations, dispose of your wet mount preparation in the beaker of bleach solution.

10. Before leaving the laboratory, make sure all other materials are properly discarded or returned to the appropriate laboratory station. Clean the microscope lenses and put the dust cover on the microscope before you return it to the storage cabinet.

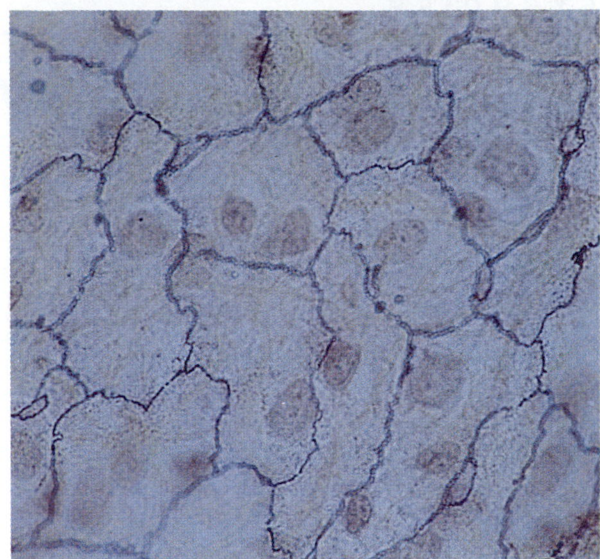

F3.5

Epithelial cells of the cheek cavity (surface view, 488×).

4 EXERCISE

The Cell—Anatomy and Division

OBJECTIVES

1. To define *cell, organelle,* and *inclusion.*
2. To identify on a cell model or diagram the following cellular regions and to list the major function of each: nucleus, cytoplasm, and plasma membrane.
3. To identify and list the major functions of the various organelles studied.
4. To compare and contrast specialized cells with the concept of the "generalized cell."
5. To define *interphase, mitosis,* and *cytokinesis.*
6. To list the stages of mitosis and describe the events of each stage.
7. To identify the mitotic phases on projected slides or appropriate diagrams.
8. To explain the importance of mitotic cell division and its product.

MATERIALS

Three-dimensional model of the "composite" animal cell or laboratory chart of cell anatomy
Prepared slides of simple squamous epithelium ($AgNO_3$ stain), teased smooth muscle, human blood cell smear, and sperm
Compound microscope
Prepared slides of whitefish blastulae
Three-dimensional models of mitotic states

See Appendix E, Exercise 4 for links to *Anatomy and PhysioShow: The Videodisc.*

Note to the Instructor: See directions for handling of toothpicks and wet mount preparations on p. 16.

The **cell,** defined as the structural and functional unit of all living things, is a very complex entity. The cells of the human body are highly diverse; and their differences in size, shape, and internal composition reflect their specific roles in the body. Nonetheless cells do have many common anatomical features, and there are some functions that all must perform to sustain life. For example, all cells have the ability to maintain their boundaries, to metabolize, to digest nutrients and dispose of wastes, to grow and reproduce, to move, and to respond to a stimulus. Most of these functions are considered in detail in later exercises. This exercise focuses on structural similarities that typify the "composite," or "generalized," cell and considers only the function of cell reproduction (cell division). Transport mechanisms (the means by which substances cross the plasma membrane) are dealt with separately in Exercise 5.

ANATOMY OF THE COMPOSITE CELL

In general, all cells have three major regions, or parts, that can readily be identified with a light microscope: the **nucleus,** the **plasma membrane,** and the **cytoplasm.** The nucleus is usually seen as a round or oval structure near the center of the cell. It is surrounded by cytoplasm, which in turn is enclosed by the plasma membrane. Since the advent of the electron microscope, even smaller cell structures—organelles—have been identified. Figure 4.1a is a diagrammatic representation of the fine structure of the composite cell; Figure 4.1b depicts cellular structure as revealed by the electron microscope.

Nucleus

The nucleus is often described as the control center of the cell and is necessary for cell reproduction. A cell that has lost or ejected its nucleus (for whatever reason) is literally programmed to die because the nucleus is the site of the "genes," or genetic material—DNA.

When the cell is not dividing, the genetic material is loosely dispersed throughout the nucleus in a threadlike form called **chromatin.** When the cell is in the process of dividing to form daughter cells, the chromatin coils and condenses to form dense, darkly staining rodlike bodies called **chromosomes**—much in the way a stretched spring becomes shorter and thicker when it is released. (Cell division is discussed later in this exercise.) Notice the appearance of the nucleus carefully—it is somewhat nondescript when a cell is healthy. When the nucleus appears dark and the chromatin becomes clumped, this is an indication that the cell is dying and undergoing degeneration.

The nucleus also contains one or more small round bodies, called **nucleoli,** composed primarily of proteins and ribonucleic acid (RNA). The nucleoli are assembly sites for ribosomal particles (particularly abundant in

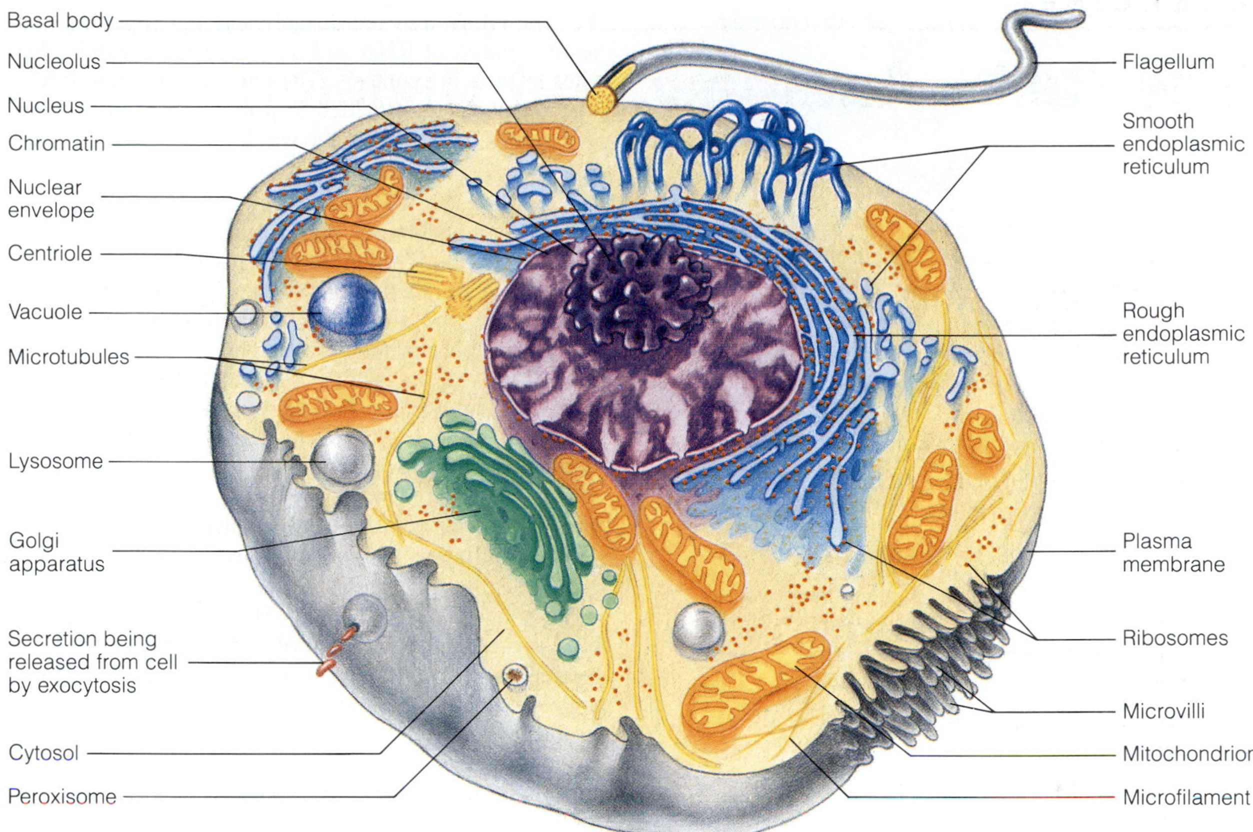

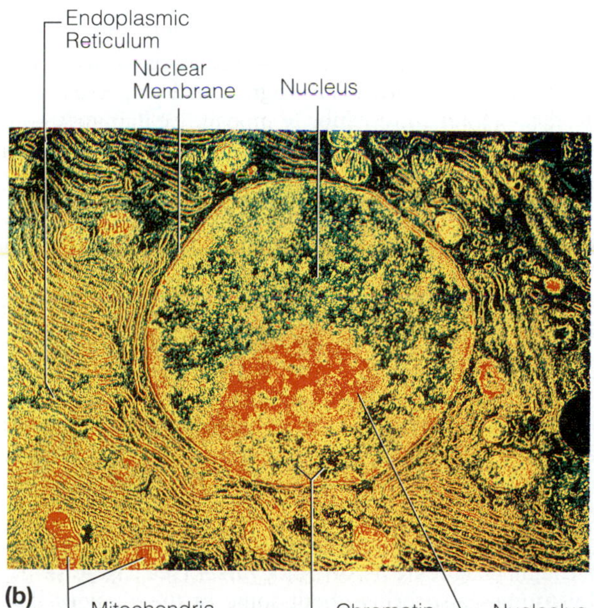

F4.1

Anatomy of the composite animal cell. (a) Diagrammatic view. **(b)** Transmission electron micrograph (10,000×).

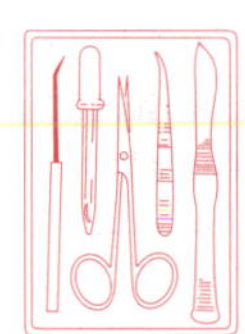

Identify the nuclear membrane, chromatin, nucleoli, and the nuclear pores in Figure 4.1a and b.

the cytoplasm), which are the actual protein synthesizing "factories."

The nucleus is bound by a double-layered porous membrane, the **nuclear membrane** (or nuclear envelope). The nuclear membrane is similar in composition to other cellular membranes, but it is distinguished by its large *nuclear pores*, which permit easy passage of protein and RNA molecules.

Plasma Membrane

The **plasma membrane** separates cell contents from the surrounding environment. Its main structural building blocks are phospholipids (fats) and globular protein molecules, but some of the externally facing proteins and lipids have sugar (carbohydrate) side chains attached to them which are important in cellular interactions (Figure 4.2). Described by the fluid-mosaic model, the membrane appears to have a bimolecular lipid core that the protein molecules float in. Occasional cholesterol molecules dispersed in the fluid phospholipid bilayer help stabilize it.

Besides providing a protective barrier for the cell, the plasma membrane plays an active role in determining which substances may enter or leave the cell and in what quantity, maintains a resting potential that is essential to normal functioning of excitable cells, and

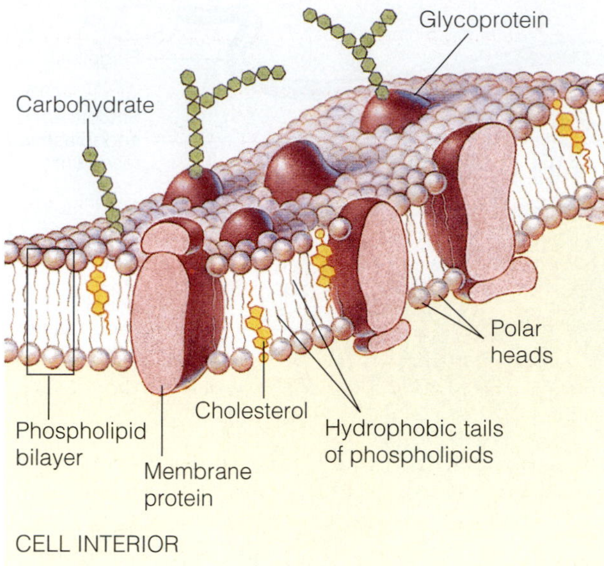

F4.2

Structural details of the plasma membrane.

plays a vital role in cell signaling and cell-to-cell interactions. In some cells the membrane is thrown into minute fingerlike projections or folds called **microvilli**, which greatly increase the surface area of the cell available for absorption or passage of materials, and binding of signaling molecules.

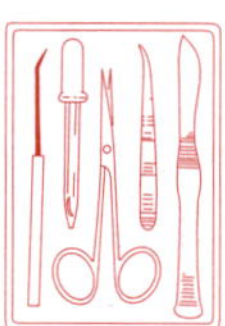

Identify the phospholipid and protein portions of the plasma membrane in Figure 4.2. Also locate the sugar side chains and cholesterol molecules. Identify the microvilli in Figure 4.1.

Cytoplasm and Organelles

The cytoplasm consists of the cell contents outside the nucleus. It is the major site of most activities carried out by the cell. Suspended in the **cytosol,** the fluid cytoplasmic material, are many small structures called **organelles** (literally, small organs). The organelles are the metabolic machinery of the cell, and they are highly organized to carry out specific functions for the cell as a whole. The organelles include the ribosomes, endoplasmic reticulum, Golgi apparatus, lysosomes, peroxisomes, mitochondria, cytoskeletal elements, and centrioles.

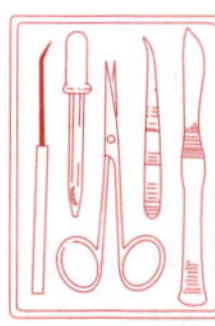

Each organelle type is summarized in Table 4.1 and described briefly next. Read through this material and then, as best you can, locate the organelles in both Figure 4.1a and b.

- The **ribosomes** are densely staining spherical bodies composed of RNA and protein. They are the actual sites of protein synthesis. They are seen floating free in the cytoplasm or attached to a membranous structure. When they are attached, the whole ribosome-membrane complex is called the *rough endoplasmic reticulum.*

- The **endoplasmic reticulum** (**ER**) is a highly folded system of membranous tubules and cisternae (sacs) that extends throughout the cytoplasm. The ER is continuous with the Golgi apparatus, nuclear membrane, and plasma membrane. Thus it is assumed that the ER provides a system of channels for the transport of cellular substances (primarily proteins) from one part of the cell to another or to the cell exterior. The ER exists in two forms; a particular cell may have both or only one, depending on its specific functions. The **rough ER,** as noted earlier, is studded with ribosomes. Its cisternae modify and store the newly formed proteins and dispatch them to other areas of the cell. The external face of the rough ER is involved in phospholipid and cholesterol synthesis. The amount of rough ER is closely correlated with the amount of protein a cell manufactures and is especially abundant in cells that make protein products for export—for example, pancreas cells, which produce digestive enzymes destined for the small intestine. The **smooth ER** has no protein synthesis–related function but is present in conspicuous amounts in cells that produce steroid-based hormones—for example, the interstitial cells of the testes, which produce testosterone, and in cells that are highly active in lipid metabolism and drug detoxification activities—liver cells, for instance.

- The **Golgi apparatus** is a stack of flattened sacs with bulbous ends that is generally found close to the nucleus. Within its cisterns, the proteins delivered to it from the rough ER are modified (by attachment of sugar groups), segregated, and packaged into membranous vesicles that are ultimately incorporated into the plasma membrane, or become secretory vesicles that release their contents from the cell, or become lysosomes.

- The **lysosomes,** which appear in various sizes, are membrane-bound sacs containing an array of powerful digestive enzymes. A product of the packaging activities of the Golgi apparatus, the lysosomes contain *acid hydrolase* enzymes capable of digesting worn-out cell structures and foreign substances that enter the cell through phagocytosis or pinocytosis (see Exercise 5). Lysosomes also bring about some of the changes that occur during menstruation, when the uterine lining is sloughed off. Since they have the capacity of total cell destruction, the lysosomes are often referred to as the "suicide sacs" of the cell.

- **Peroxisomes,** like lysosomes, are enzyme-containing sacs. However, their *oxidase* enzymes have a different task. Using oxygen, they detoxify a number of harmful substances, most importantly free radicals. Per-

TABLE 4.1 Cytoplasmic Organelles

Organelle	Location and function
Ribosomes	Tiny spherical bodies composed of RNA and protein; actual sites of protein synthesis; floating free or attached to a membranous structure (the rough ER) in the cytoplasm
Endoplasmic reticulum (ER)	Membranous system of tubules that extends throughout the cytoplasm; two varieties: rough or granular ER—studded with ribosomes (tubules of the rough ER provide an area for storage and transport of the proteins made on the ribosomes to other cell areas; external face synthesizes phospholipids and cholesterol); smooth or agranular ER—no protein synthesis–related function (a site of steroid and lipid synthesis, lipid metabolism, and drug detoxification)
Golgi apparatus	Stack of flattened sacs with bulbous ends and associated small vesicles; found close to the nucleus; role in packaging proteins or other substances for export from the cell or incorporation into the plasma membrane and in packaging lysosomal enzymes
Lysosomes	Various-sized membranous sacs containing powerful digestive enzymes; function to digest worn-out cell organelles and foreign substances that enter the cell; since they have the capacity of total cell destruction if ruptured, referred to as "suicide sacs of the cell"
Peroxisomes	Small lysosome-like membranous sacs containing oxidase enzymes that detoxify alcohol, hydrogen peroxide, and other harmful chemicals
Mitochondria	Generally rod-shaped bodies with a double membrane wall; inner membrane is thrown into folds, or cristae; contain enzymes that oxidize foodstuffs to produce cellular energy (ATP); often referred to as "powerhouses of the cell"
Centrioles	Paired, cylindrical bodies lie at right angles to each other, close to the nucleus; direct the formation of the mitotic spindle during cell division; form the bases of cilia and flagella
Cytoskeletal elements: microtubules, intermediate filaments, and microfilaments	Provide cellular support; function in intracellular transport; microtubules form the internal structure of the centrioles and help determine cell shape; intermediate filaments, stable elements composed of a variety of proteins, resist mechanical forces acting on cells; microfilaments are formed largely of actin, a contractile protein, and thus are important in cell mobility (particularly in muscle cells)

oxisomes are particularly abundant in kidney and liver cells, cells that are actively involved in detoxification.

- The **mitochondria** are generally rod-shaped bodies with a double-membrane wall; the inner membrane is thrown into folds, or *cristae*. Oxidative enzymes on or within the mitochondria catalyze the reactions of the Krebs cycle and the electron transport chain (collectively called oxidative respiration), in which foods are broken down to produce energy. The released energy is captured in the bonds of ATP (adenosine triphosphate) molecules, which are then transported out of the mitochondria to provide a ready energy supply to power the cell. Every living cell requires a constant supply of ATP for its many activities. Since the mitochondria provide the bulk of this ATP, they are referred to as the powerhouses of the cell.

- The **cytoskeletal elements** ramify throughout the cytoplasm forming an internal scaffolding called the *cytoskeleton* that supports and moves substances within the cell. The **microtubules** are basically slender tubules formed of proteins called *tubulins* which have the ability to aggregate and then disaggregate spontaneously. Microtubules organize the cytoskeleton and direct formation of the spindle formed by the centrioles during cell division. They also act in the transport of substances down the length of elongated cells (such as neurons), suspend organelles, and help maintain cell shape by providing rigidity to the soft cellular substance. **Intermediate filaments** are *stable* proteinaceous cytoskeletal elements that act as internal guy wires to resist mechanical (pulling) forces acting on cells. **Microfilaments,** ribbon or cordlike elements, are formed of contractile proteins. Because of their ability to shorten and then relax to assume a more elongated form, these are important in cell mobility and are very conspicuous in cells that are highly specialized to contract (such as muscle cells). A cross-linked network of microfilaments braces and strengthens the internal face of the plasma membrane.

The cytoskeletal structures are labile and minute. With the exception of the microtubules of the spindle, which are very obvious during cell division (see pp. 30–31), and the microfilaments of skeletal muscle cells (see p. 108), they are rarely seen, even in electron micrographs, and are not depicted in Figure 4.1b. However, special stains can reveal the plentiful supply of these very important organelles (see Plate 5 in the Histology Atlas).

- The paired **centrioles** lie close to the nucleus in all animal cells capable of reproducing themselves. They are rod-shaped bodies that lie at right angles to each other. Internally each centriole is composed of nine triplets of microtubules. During cell division, the centrioles direct the formation of the mitotic spindle. Centrioles also form the basis for cell projections called cilia and flagella (described below).

In addition to these cell structures, some cells have projections called **flagella,** which propel the cells, or **cilia,** which allow cells to sweep substances along a tract. Identify the cilium in Figure 4.1a.

The cell cytoplasm contains various other substances and structures, including stored foods (glycogen granules and lipid droplets), pigment granules, crystals of various types, water vacuoles, and ingested foreign materials. But these are not part of the active metabolic machinery of the cell and are therefore called **inclusions.**

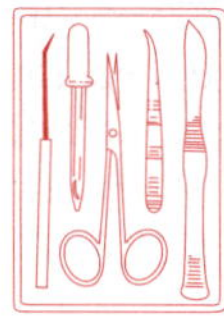

Once you have located all of these structures in Figure 4.1a, examine the cell model (or cell chart) to repeat and reinforce your identifications.

OBSERVING DIFFERENCES AND SIMILARITIES IN CELL STRUCTURE

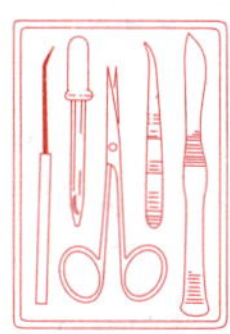

1. Obtain a compound microscope and prepared slides of simple squamous epithelium, sperm, smooth muscle cells (teased), and human blood.

2. Observe each slide under the microscope, carefully noting similarities and differences in the cells. (The oil immersion lens will be needed to observe blood and sperm.) Distinguish the limits of the individual cells, and notice the shape and position of the nucleus in each case. When you look at the human blood smear, direct your attention to the red blood cells, the pink-stained cells that are most numerous. The color photomicrographs illustrating a blood smear (Plate 55) and sperm (Plate 50) that appear in the Histology Atlas may be helpful in this cell structure study. Sketch your observations in the circles provided.

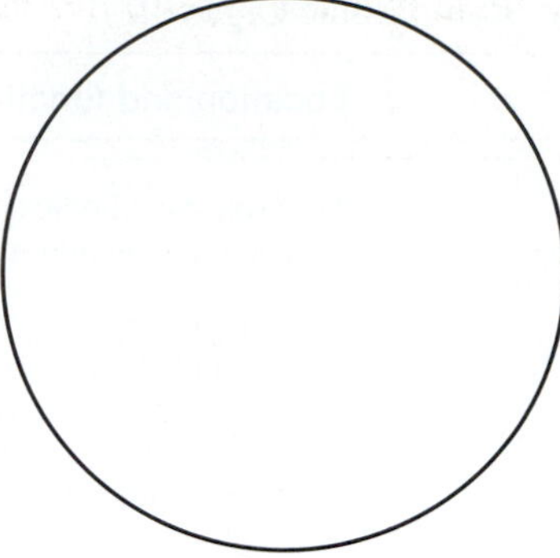

Simple squamous epithelium

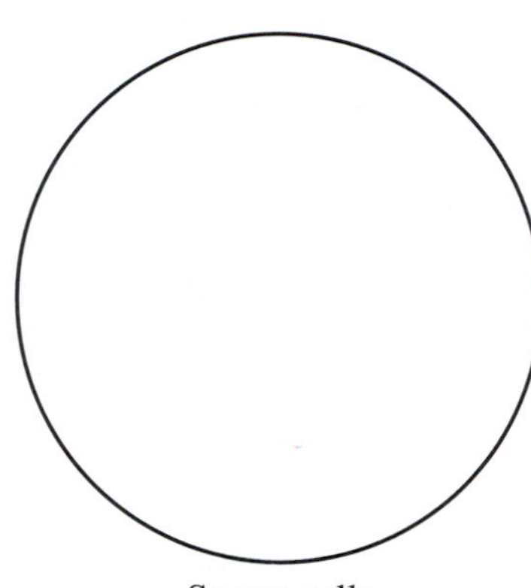

Sperm cells

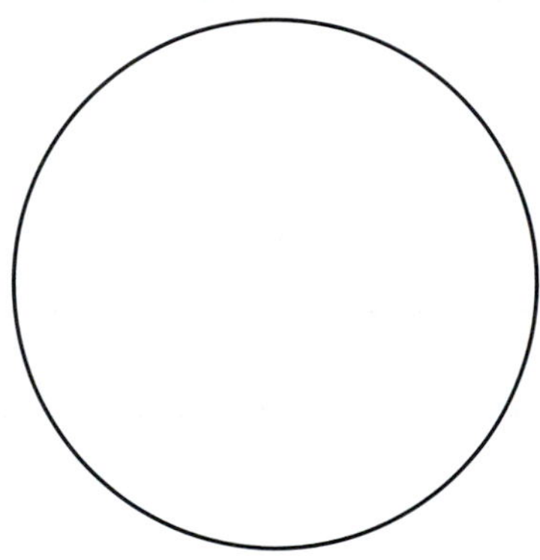

Human red blood cells

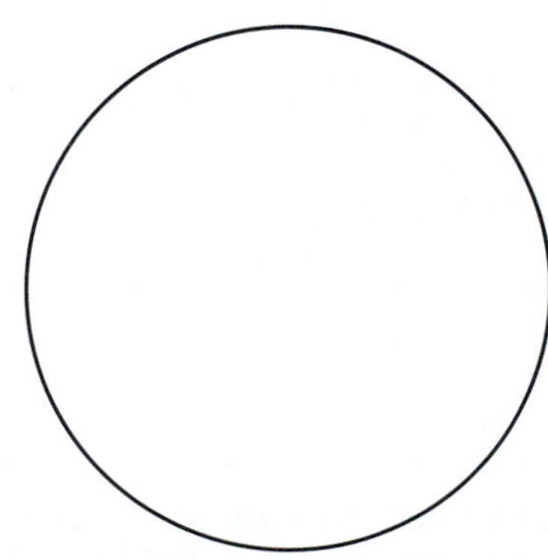

Teased smooth muscle cells

3. How do these four cell types differ in shape and size?

____

____

____

____

____

How might cell shape affect cell function?

____

____

____

____

____

Which cells have visible projections?

____

How do these projections relate to the function of this cell?

____

____

____

Do any of these cells lack a cell membrane? __________

A nucleus? ______________________________________

In the cells with a nucleus, can you discern nucleoli?

____

Were you able to observe any of the organelles in these cells? ____________________ Why or why not?

____

____

____

CELL DIVISION: MITOSIS AND CYTOKINESIS

A cell's *life cycle* is the series of changes it goes through from the time it is formed until it reproduces itself. It encompasses two stages—**interphase,** the longer period during which the cell grows and carries out its usual activities, and **cell division,** when the cell reproduces itself by dividing. In an interphase cell about to divide, the genetic material (DNA) is replicated (duplicated exactly). Once this important event has occurred, cell division ensues.

Cell division in all cells other than bacteria consists of a series of events collectively called mitosis and cytokinesis. **Mitosis** is nuclear division; **cytokinesis** is the division of the cytoplasm, which begins after mitosis is nearly complete. Although mitosis is usually accompanied by cytokinesis, in some instances cytoplasmic division does not occur, leading to the formation of binucleate (or multinucleate) cells. This is relatively common in the human liver and during embryonic development of skeletal muscle cells.

The process of **mitosis** results in the formation of two daughter nuclei that are genetically identical to the mother nucleus. This distinguishes mitosis from **meiosis,** a specialized type of nuclear division that occurs only in the reproductive organs (testes or ovaries). Meiosis, which yields four daughter nuclei that differ genetically and in composition from the mother nucleus, is used only for the production of eggs and sperm (gametes) for sexual reproduction. The function of cell division, including mitosis and cytokinesis in the body, is to increase the number of cells for growth and repair while maintaining their genetic heritage.

The stages of mitosis illustrated in Figure 4.3 include the following events:

Prophase (Figure 4.3b and c): At the onset of cell division, the chromatin threads coil and shorten to form densely staining, short, barlike **chromosomes.** By the middle of prophase the chromosomes appear as double-stranded structures (each strand is a **chromatid**) connected by a small median body called a **centromere.** The centrioles separate from one another and act as focal points for the assembly of two systems of microtubules, the **mitotic spindle** which forms between the centrioles and the **asters** which radiate outward from the ends of the spindle and anchor it to the plasma membrane. Some of the spindle fibers, the *kinetochore fibers*, attach to special protein complexes on each chromosome's centromere. Spindle fibers that do not attach to the chromosomes are called *polar fibers*. The spindle acts as a scaffolding for the attachment and movement of the chromosomes during later mitotic stages. Meanwhile, the nuclear membrane and the nucleolus break down and disappear.

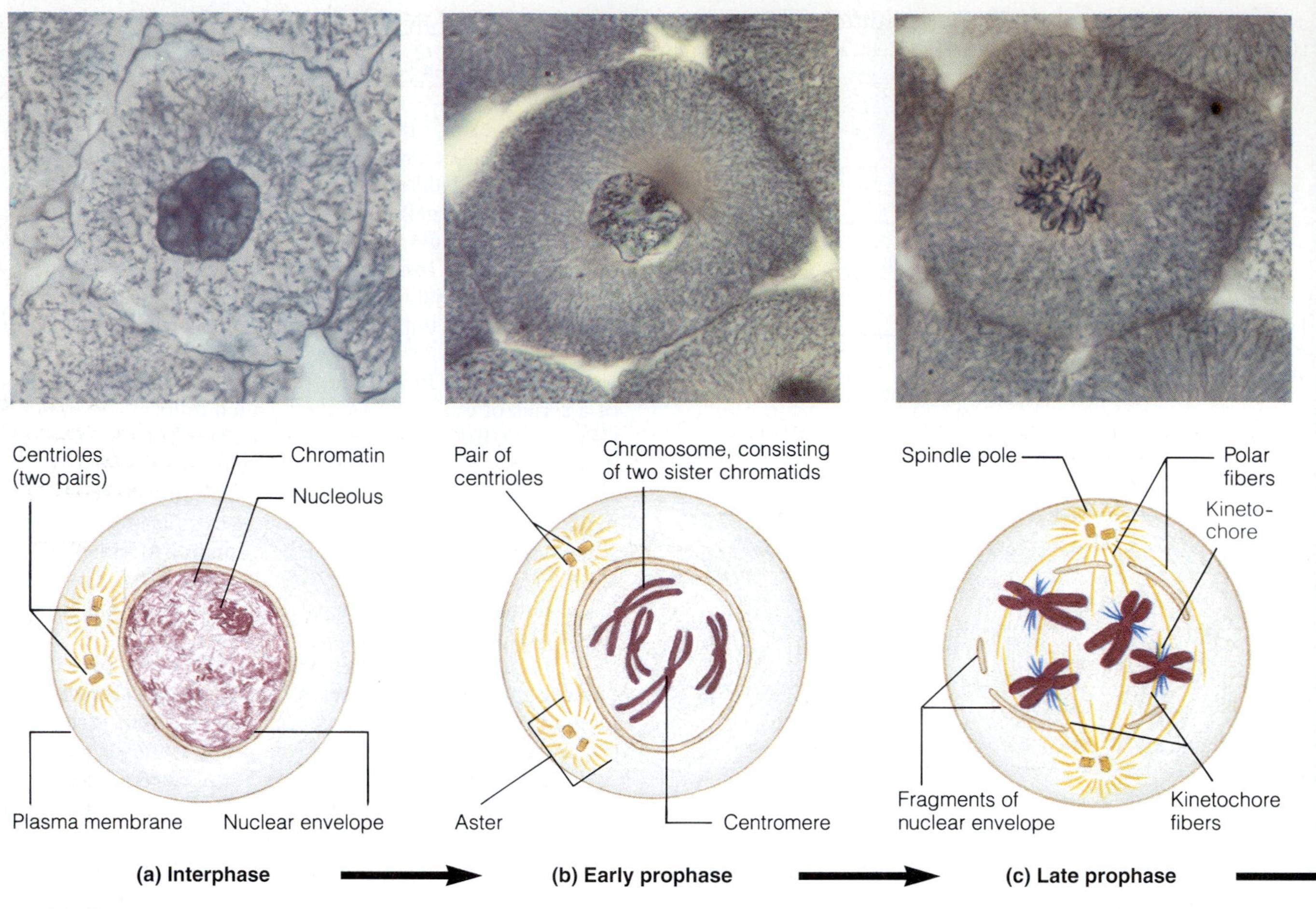

F4.3

The interphase cell and the stages of mitosis. The cells shown are from an early embryo of a whitefish. Photomicrographs are above; corresponding diagrams are below. (Micrographs approximately 600×).

Metaphase (Figure 4.3d): A brief stage, during which the chromosomes migrate to the central plane or equator of the spindle and align along that plane in a straight line (the so-called *metaphase plate*) from the superior to the inferior region of the spindle (lateral view). Viewed from the poles of the cell (end view), the chromosomes appear to be arranged in a "rosette," or circle, around the widest dimension of the spindle.

Anaphase (Figure 4.3e): During anaphase, the centromeres split, and the chromatids (now called chromosomes again) separate from one another and then progress slowly toward opposite ends of the cell. The chromosomes are pulled by the kinetochore fibers attached to their centromeres, their "arms" dangling behind them. Anaphase is complete when poleward movement ceases.

Telophase (Figure 4.3f): During telophase, the events of prophase are essentially reversed. The chromosomes clustered at the poles begin to uncoil and resume the chromatin form, the spindle breaks down and disappears, a nuclear membrane forms around each chromatin mass, and nucleoli appear in each of the daughter nuclei.

Mitosis is essentially the same in all animal cells, but depending on the type of tissue, it takes from 5 minutes to several hours to complete. In most cells, centriole replication is deferred until interphase of the next cell cycle.

Cytokinesis, or the division of the cytoplasmic mass, begins during telophase (Figure 4.3f), and provides a good guideline for where to look for the mitotic figures of telophase. In animal cells, a *cleavage furrow* begins to form approximately over the equator of the spindle, and eventually splits or pinches the original cytoplasmic mass into two portions. Thus at the end of cell division two daughter cells exist, each smaller in cyto-

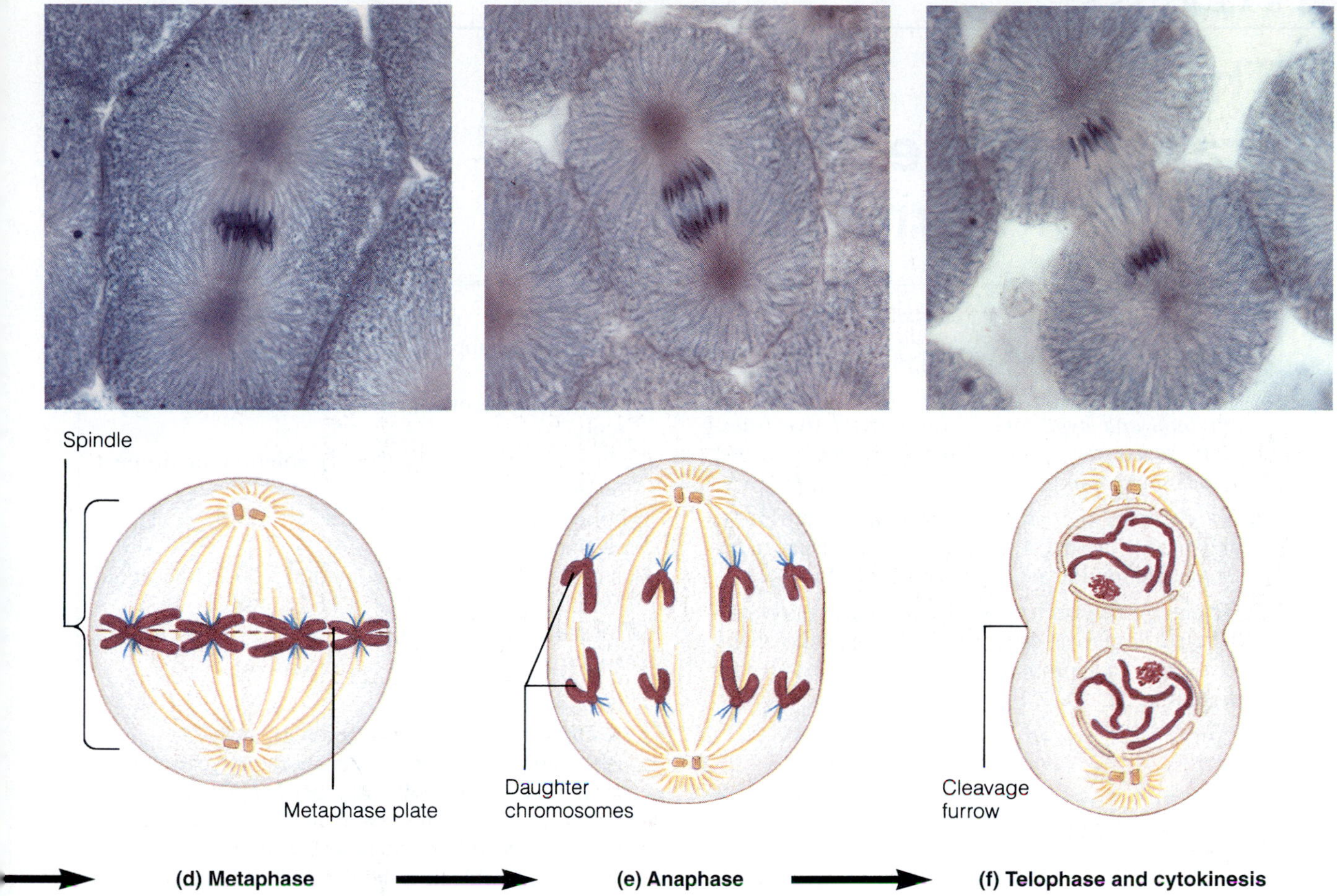

Spindle

Metaphase plate

Daughter chromosomes

Cleavage furrow

(d) Metaphase

(e) Anaphase

(f) Telophase and cytokinesis

F4.3 *(continued)*

plasmic mass than the mother cell but genetically identical to it. The daughter cells grow and carry out the normal spectrum of metabolic processes until it is their turn to divide.

Cell division is extremely important during the body's growth period. Most cells (excluding nerve cells) undergo mitosis until puberty, when normal body size is achieved and overall body growth ceases. After this time in life, only certain cells routinely carry out cell division—for example, cells subjected to abrasion (epithelium of the skin and lining of the gut). Other cell populations—such as liver cells—stop dividing but retain this ability should some of them be removed or damaged. Skeletal muscle, cardiac muscle, and nervous tissue completely lose this ability to divide and thus are severely handicapped by injury. Throughout life, the body retains its ability to repair cuts and wounds and to replace some of its aged cells.

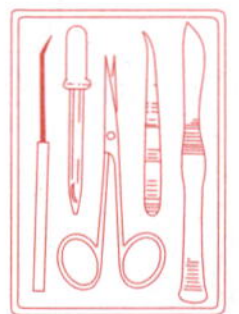

Obtain a prepared slide of whitefish blastulae to study the stages of mitosis. The cells of each *blastula* (a stage of embryonic development consisting of a hollow ball of cells) are at approximately the same mitotic stage, so it may be necessary to observe more than one blastula to view all the mitotic stages. The exceptionally high rate of mitosis observed in this tissue is typical of embryos, but if occurring in specialized tissues, it can be an indication of cancerous cells, which also have an extraordinarily high mitotic rate. Examine the slide carefully, identifying the four mitotic stages and the process of cytokinesis. Compare your observations with Figure 4.3, and verify your identifications with your instructor.

5

EXERCISE

The Cell— Transport Mechanisms and Cell Permeability

OBJECTIVES

1. To define *differential permeability; diffusion* (*dialysis* and *osmosis*); *Brownian motion; isotonic, hypotonic, and hypertonic solutions; passive transport; active transport; pinocytosis; phagocytosis;* and *solute pump.*
2. To describe the processes that account for the movement of substances across the plasma membrane and to indicate the driving force for each.
3. To determine which way substances will move passively through a differentially permeable membrane (given appropriate information on concentration differences).

MATERIALS

For passive transport experiments:
Clean slides and coverslips
Forceps
Glass stirring rods
15-ml graduated cylinders
Compound microscopes
Hot plate and large beaker for hot water bath

Brownian motion:
Milk in dropper bottles

Diffusion:
Petri plate containing 12 ml of 1.5% agar-agar
Methylene blue dye crystals
Potassium permanganate dye crystals
Millimeter rulers

Four dialysis sacs or small Hefty "alligator" sandwich bags
Beakers (250 ml)
40% glucose solution
Fine twine or dialysis tubing clamps
10% NaCl solution, boiled starch solution
Laboratory balance
Benedict's solution in dropper bottle
Test tubes in racks, test tube holder
Wax marker
Small funnel
Silver nitrate ($AgNO_3$) in dropper bottle
Lugol's iodine solution in dropper bottle

Vials of animal (mammalian) blood obtained from a biological supply house or veterinarian—at option of instructor
Physiologic (mammalian) saline solution in dropper bottle
1.5% sodium chloride (NaCl) solution in dropper bottle
Distilled water in dropper bottle
Medicine dropper
Basin and wash bottles containing 10% household bleach solution
Disposable autoclave bag
Disposable plastic gloves

Diffusion demonstrations:

1: Diffusion of a dye through water
Prepared the morning of the laboratory session with setup time noted. Potassium permanganate crystals placed in a 1000-ml graduated cylinder, and distilled water added slowly and with as little turbulence as possible to fill to the 1000-ml mark.

2: Osmometer
Just before the laboratory begins, the broad end of a thistle tube is closed with a differentially permeable dialysis membrane, and the tube is secured to a ring stand. Molasses is added to approximately 5 cm above the thistle tube bulb and the bulb is immersed in a beaker of distilled water. At the beginning of the lab session, the level of the molasses in the tube is marked with a wax pencil.

Filtration:
Ring stand, ring, clamp
Filter paper, funnel
Solution containing a mixture of uncooked starch, powdered charcoal, and copper sulfate ($CuSO_4$)

Active transport:
Culture of starved amoeba (*Amoeba proteus*)
Depression slide
Coverslip (glass)
Tetrahymena pyriformis culture
Compound microscope

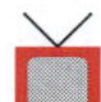

See Appendix E, Exercise 5 for links to *Anatomy and PhysioShow: The Videodisc.*

Note to the Instructor: See directions for handling wet mount preparations and disposable supplies on page 16, Exercise 3.

Because of its molecular composition, the plasma membrane is selective about what passes through it. It allows nutrients to enter the cell but keeps out undesirable substances. By the same token, valuable cell proteins and other substances are kept within the cell, and excreta or wastes pass to the exterior. This property is known as **differential,** or **selective, permeability.** Transport through the plasma membrane occurs in two basic ways. In **active transport,** the cell provides energy (ATP) to power the transport process. In the other, **passive transport,** the transport process is driven by concentration or pressure differences.

PASSIVE TRANSPORT

All molecules vibrate randomly (because of their intrinsic kinetic energy) at all temperatures above absolute zero (about −460°F). In general, the smaller the particle, the more kinetic energy it possesses and the faster it moves. This random movement (Figure 5.1) may be detected indirectly by observing a suspension like milk. The larger particles can be seen moving randomly as they are hit and deflected by the smaller, more rapidly moving particles. The zigzag movement of the larger particles is known as **Brownian motion.**

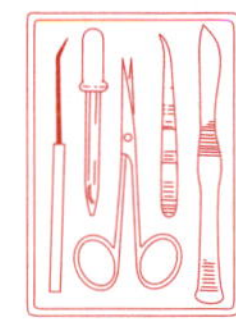

1. Obtain a clean slide and coverslip, forceps, compound microscope, dropper bottle of milk, and a hot plate and bring them to your laboratory bench.

2. Make a wet mount of milk; that is, place a small drop of milk on a slide and cover carefully with a coverslip. Allow the slide to stand on the microscope stage for about 10 minutes before observing. While waiting, put the hot plate on and set to a low temperature.

3. Observe the slide with high power and then with the oil immersion lens. Keep the light as dim as possible to increase the contrast. As the minute solvent (water) molecules collide with the fat globules of the milk, you can see the much larger fat globules ricochet in an erratic manner (Brownian motion).

4. Place the preparation on the warm hot plate for a few seconds. Observe again. How has the *rate* of Brownian motion changed?

____

____

What can you conclude about the effect of increased temperature on the kinetic energy of molecules?

____

____

____

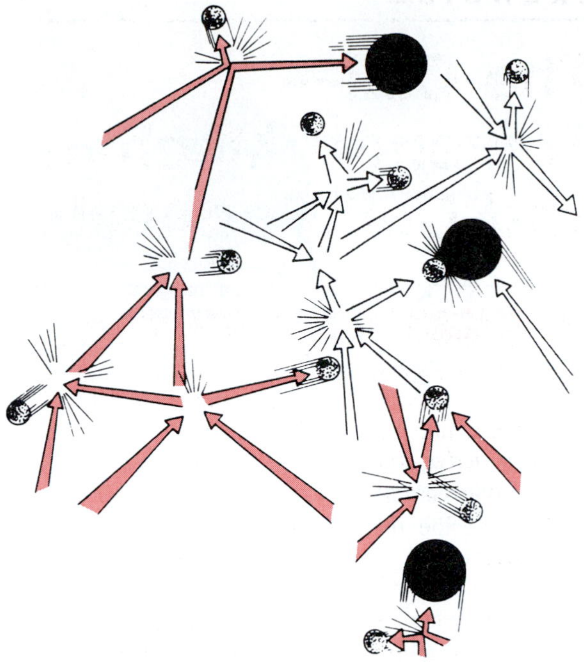

F5.1

Random movement and numerous collisions cause molecules to become evenly distributed. The small spheres represent water molecules; the large spheres represent glucose molecules.

Diffusion

If a **concentration gradient** (difference in concentration) exists, molecules eventually become evenly distributed through random molecular motion (Figure 5.1). **Diffusion** is the movement of molecules from a region of their higher concentration to a region of their lower concentration. The driving force is the kinetic energy of the molecules themselves.

There are many examples of diffusion in nonliving systems. For example, if a bottle of ether was uncorked at the front of the laboratory, very shortly thereafter you would be nodding, as the ether molecules become distributed throughout the room. The ability to smell a friend's cologne shortly after he or she has entered the room is another such example.

The diffusion of particles into and out of cells is modified by the plasma membrane, which constitutes a physical barrier. In general, molecules diffuse passively through the plasma membrane if they are small enough to pass through its pores (and are aided by an electrical gradient), or if they can dissolve in the lipid portion of the membrane (as in the case of CO_2 and O_2). The diffusion of solutes (particles dissolved in water) through a semipermeable membrane is called **simple diffusion.** The diffusion of water through a semipermeable membrane is called **osmosis.** Both simple diffusion and osmosis involve the movement of a substance from an area of its higher concentration to one of its lower concentration, i.e., down its concentration gradient.

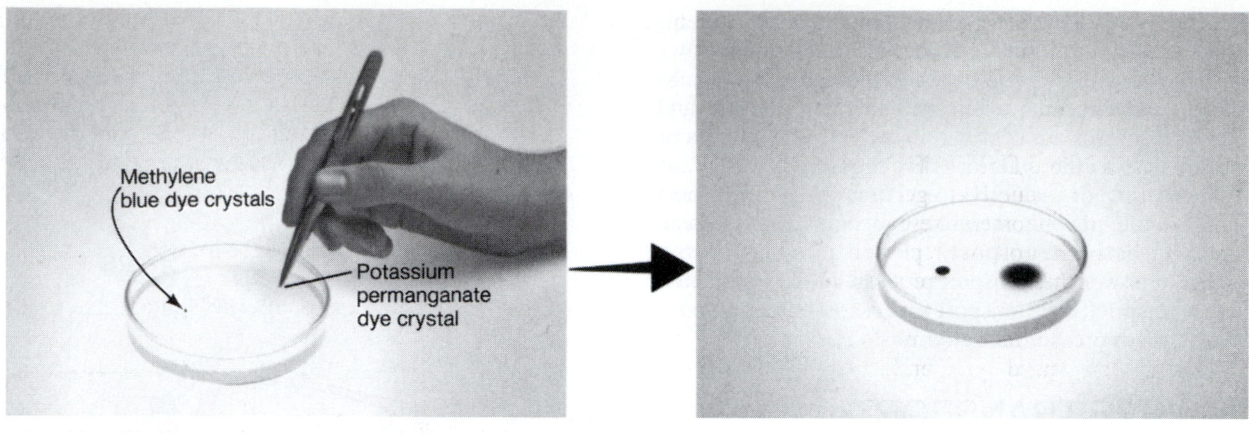

Equal amounts of methylene blue and potassium permanganate crystals are placed on agar medium with forceps.

Diffusion distances measured at 15 minute intervals for one and ½ hrs. Measure the radius of the diffusion ring.

F5.2

Setup for comparing the diffusion rates of molecules of methylene blue and potassium permanganate through an agar gel.

DIFFUSION OF A DYE THROUGH AN AGAR GEL The relationship between molecular weight and the rate of diffusion can be examined easily by observing the diffusion of the molecules of two different types of dye through an agar gel. The dyes used in this experiment are methylene blue, which has a molecular weight of 320 and is deep blue in color, and potassium permanganate, a purple dye with a molecular weight of 158. Although the agar gel appears quite solid, it is primarily (98.5%) water and allows free movement of the diffusing dye molecules through it.

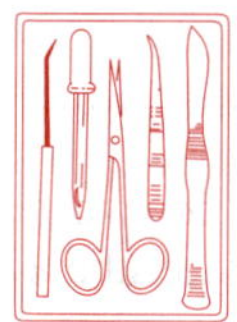

1. Obtain a petri dish containing agar gel, a forceps, a millimeter ruler, and containers of methylene blue crystals and potassium permanganate crystals, and bring them to your bench.

⚠ *Avoid contact between your skin and the dye crystals by using the forceps to pick up the crystals.*

2. Select a crystal of potassium permanganate dye, and place it gently on the agar gel surface. Place an equal amount of the smaller methylene blue crystals on the agar surface approximately 10 centimeters away from the potassium permanganate crystal (Figure 5.2). Record the time.

____

3. At 15-minute intervals, use the millimeter ruler to measure the distance the dye has diffused from each crystal source. These observations should be continued for 1½ hours, and the results recorded in the chart to the right.

Time (min)	Diffusion of methylene blue (mm)	Diffusion of potassium permanganate (mm)
15		
30		
45		
60		
75		
90		

Which dye diffused more rapidly? ________________

What is the relationship between molecular weight and rate of molecular movement (diffusion)?

____

Why did the dye molecules move? ________________

____

Compute the rate of diffusion of the potassium permanganate molecules in millimeters per minute (mm/min) and record.

_______ mm/min

Compute the rate of diffusion of the methylene blue molecules in mm/min and record.

_______ mm/min

DIFFUSION OF A DYE THROUGH WATER Make a mental note to yourself to go to demonstration area 1 at the end of the laboratory session to observe the extent of diffusion of the potassium permanganate dye through water. At that time, follow the directions given next.

1. Measure the number of millimeters from the bottom of the graduated cylinder the dye has diffused and record.

_______ mm

2. Make note of the time the demonstration was set up and the time of your observation. Then compute the rate of the dye's diffusion through water and record below.

Time of setup _______

Time of observation _______

Rate of diffusion _______ mm/min

3. Does the potassium permanganate dye move (diffuse) more rapidly through water or the agar gel? (Explain your answer.)

____

____

____

____

DIFFUSION THROUGH NONLIVING MEMBRANES A diffusion experiment providing information on the passage of water and solutes through semipermeable membranes, which may be applied to the study of transport mechanisms in living membrane-bound cells, is outlined next.

1. Obtain four dialysis sacs,* a small funnel, a graduated cylinder, a wax marker, and four beakers (250 ml). Number the beakers 1 to 4 with the wax marker, and half fill all of them with distilled water except beaker 2, to which you should add 40% glucose solution.

2. Prepare the dialysis sacs one at a time. Using the funnel, half fill each with 20 ml of the specified liquid (see below). Press out the air, fold over the open end of the sac, and tie it securely with fine twine. Before proceeding to the next sac, quickly and carefully blot the sac dry by rolling it on a paper towel, and weigh it with a laboratory balance. Record the weight, and then drop the sac into the corresponding beaker. Be sure the sac is completely covered by the beaker solution, adding more solution if necessary.

- Sac 1: 40% glucose solution. Weight: __________ g
- Sac 2: 40% glucose solution. Weight: __________ g
- Sac 3: 10% NaCl solution. Weight: __________ g
- Sac 4: boiled starch solution. Weight: __________ g

Allow sacs to remain undisturbed in the beakers for 1 hour. (Use this time to continue with other experiments.)

3. After an hour, get a beaker of water boiling on the hot plate. Obtain the supplies you will need to determine your experimental results: dropper bottles of Benedict's solution, silver nitrate solution, and Lugol's iodine, a test tube rack, 4 test tubes, and a test tube holder.

4. Quickly and gently blot sac 1 dry and weigh it. (Note: Do not squeeze the sac during the blotting process.)

Weight of sac 1: _______ g

Has there been any change in weight? ______________

Conclusions? ______________________________

____

____

____

Place 5 ml of Benedict's solution in each of two test tubes. Put 4 ml of the beaker fluid into one test tube and 4 ml of the sac fluid into the other. Mark the tubes for identification and then place them in a beaker containing boiling water. Boil 2 minutes. Cool slowly. If a green, yellow, or rusty red precipitate forms, the test is positive, meaning that glucose is present. If the solution remains the original blue color, the test is negative.

Was glucose still present in the sac? ______________

Was glucose present in the beaker? ______________

Conclusions? ______________________________

____

____

5. Blot gently and weigh sac 2: ______________ g

Was there an *increase* or *decrease* in weight? ________

* Dialysis sacs are selectively permeable membranes with pores of a particular size. The selectivity of living membranes depends on more than just pore size, but using the dialysis sacs will allow you to examine selectivity due to this factor.

With 40% glucose in the sac and 40% glucose in the beaker, would you expect to see any net movements of water (osmosis) or of glucose molecules (simple diffusion)?

_________ Why or why not? _________

____

____

6. Blot gently and weigh sac 3: _____________ g

Was there any change in weight? _____________

Conclusions? _____________

____

Take a 5 ml sample of beaker 3 solution and put it in a clean test tube. Add a drop of silver nitrate. The appearance of a white precipitate or cloudiness indicates the presence of AgCl, which is formed by the reaction of $AgNO_3$ with NaCl (sodium chloride).

Results? _____________

Conclusions? _____________

____

7. Blot gently and weigh sac 4: _____________ g

Was there any change in weight? _____________

Conclusions? _____________

____

Take a 5 ml sample of beaker 4 solution and add a couple of drops of Lugol's iodine solution. The appearance of a black color is a positive test for the presence of starch. Did any starch diffuse from the sac into the beaker?

_______ Explain: _____________

____

____

8. In which of the test situations did net osmosis occur?

____

In which of the test situations did net simple diffusion occur?

____

What conclusions can you make about the relative size of glucose, starch, NaCl, and water molecules?

____

____

____

____

With what cell structure can the dialysis sac be compared?

____

____

9. Before leaving the laboratory, observe demonstration 2, the *osmometer demonstration* set up before the laboratory session to follow the movement of water through a membrane (osmosis). Measure the distance the water column has moved during the laboratory period and record below. (The position of the meniscus in the thistle tube at the beginning of the laboratory period is marked with wax pencil.)

Distance the meniscus has moved: _______ mm

DIFFUSION THROUGH LIVING MEMBRANES

To examine permeability properties of cell membranes, conduct the following two experiments.

Experiment 1: 1. The following supplies should be available at your laboratory bench to conduct this experimental series: a clean slide and coverslip, vial of animal blood, medicine dropper, physiologic saline, 1.5% sodium chloride solution, 3 test tubes, test tube rack, glass stirring rod, 15 ml graduated cylinder, filter paper, and plastic gloves.

2. Label 3 test tubes A, B, and C, and prepare them as follows:

- A: add 2 ml physiologic saline
- B: add 2 ml 1.5% sodium chloride solution
- C: add 2 ml distilled water

3. Don the gloves, and use a medicine dropper to add 5 drops of animal blood to each test tube. Stir each test tube with the glass rod, rinsing between each sample.

4. Hold each test tube in front of this printed page. **Record** the clarity of print seen through the fluid in each tube.

Test tube A _____________

Test tube B _____________

Test tube C _____________

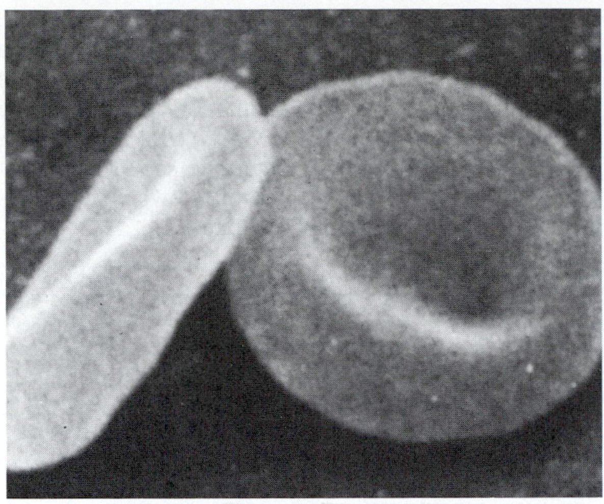

(a)

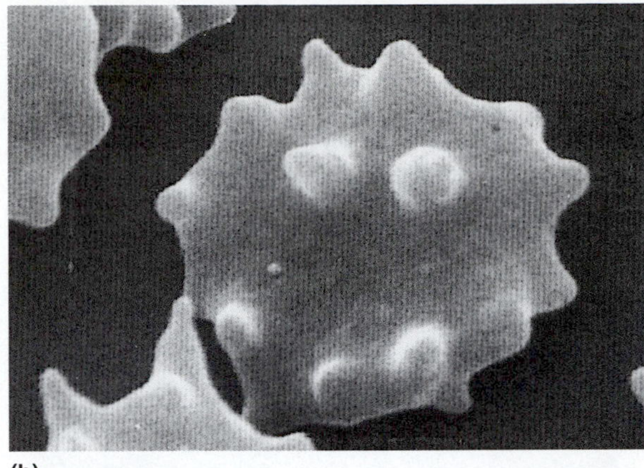

(b)

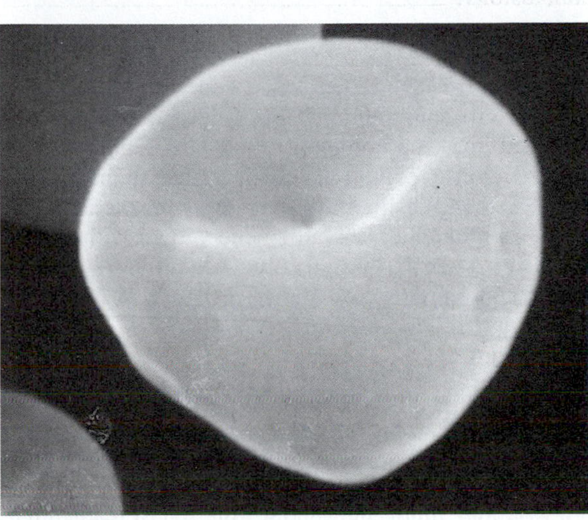

(c)

F5.3

Influence of hypertonic and hypotonic solutions on red blood cells. (a) Red blood cells suspended in an isotonic solution, where the cells retain their normal size and shape. **(b)** Red blood cells suspended in hypertonic solution. As the cells lose water to the external environment, they shrink and become prickly, a phenomenon called crenation. **(c)** Red blood cells suspended in a hypotonic solution. Notice their spherical bloated shape, a result of excessive water intake.

Experiment 2: Now you will conduct a microscope study of red blood cells suspended in the same three solutions. The objective is to determine if these solutions have any effect on cell shape by promoting net osmosis.

1. Place a very small drop of physiologic saline on a slide. Using the medicine dropper, add a small drop of animal blood to the saline on the slide. Tilt the slide to mix, cover with a coverslip, and immediately examine the preparation under the high-power lens. Notice that the red blood cells retain their normal smooth disclike shape (see Figure 5.3a). This is because the physiologic saline is **isotonic** to the cells. That is, it contains a concentration of nonpenetrating solutes equal to that in the cells (same solute-solvent ratio). Consequently, the cells neither gain nor lose water by osmosis.

2. Prepare another wet mount of animal blood, but this time use 1.5% saline solution as the suspending medium. After 5 minutes, carefully observe the red blood cells under high power. What is happening to the normally smooth disc shape of the red blood cells?

____

____

____

This crinkling-up process, called **crenation,** is due to the fact that the 1.5% sodium chloride solution is slightly hypertonic to the cell sap of the red blood cell. A **hypertonic** solution contains more nonpenetrating solutes (thus less water) than are present in the cell. Under these circumstances, water tends to leave the cells by osmosis. Compare your observations to Figure 5.3b.

3. Add a drop of distilled water to the edge of the coverslip. Fold a piece of filter paper in half and place its folded edge at the opposite edge of the coverslip; it will absorb the saline solution and draw the distilled water across the cells. Watch the red blood cells as they float across the field. After about 5 minutes have passed, describe the change in their appearance.

____

____

____

____

Distilled water contains *no* solutes (it is 100% water). Distilled water and *very* dilute solutions (that is, those containing less than 0.9% nonpenetrating solutes) are **hypotonic** to the cell. In a hypotonic solution, the red blood cells first "plump up" (Figure 5.3c) but then they suddenly start to disappear. The red blood cells burst as the water floods into them, leaving "ghosts" in their wake. This phenomenon is called **hemolysis.**

How do your observations of test tube C in Experiment 1 correlate with what you have just observed under the microscope?

____

____

____

4. Place the blood-soiled slides and test tube in the bleach-containing basin. Put the gloves you used into the disposable autoclave bag. Obtain a wash (squirt) bottle containing 10% bleach solution and squirt the bleach liberally over the bench area where blood was handled. Wipe the bench down with a paper towel wet with the bleach solution and allow it to dry before continuing.

Filtration

Filtration is the process by which water and solutes are forced through a membrane from an area of higher hydrostatic (fluid) pressure into an area of lower hydrostatic pressure. Like diffusion, it is a passive process. For example, fluids and solutes filter out of the capillaries in the kidneys into the kidney tubules because the blood pressure in the capillaries is greater than the fluid pressure in the tubules. Filtration is not a selective process. The amount of filtrate (fluids and solutes) formed depends almost entirely on the pressure gradient (difference in pressure on the two sides of the membrane) and on the size of the membrane pores.

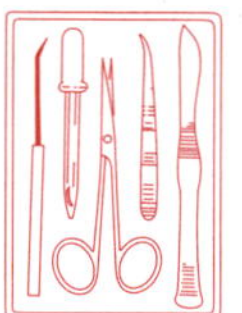

1. Obtain the following equipment: a ring stand, ring, and ring clamp; a piece of filter paper; a beaker; a solution containing uncooked starch, powdered charcoal, and copper sulfate; and a dropper bottle of Lugol's iodine. Attach the ring to the ring stand with the clamp.

2. Fold the filter paper in half twice, open it into a cone, and place it in a funnel. Place the funnel in the ring of the ring stand and place a beaker under the funnel. Shake the starch solution, and fill the funnel with it to just below the top of the filter paper. When the steady stream of filtrate changes to countable filtrate drops, count the number of drops formed in 10 seconds and record.

______________ drops

When the funnel is half empty, again count the number of drops formed in 10 seconds and record the count.

______________ drops

3. After all the fluid has passed through the filter, check the filtrate and paper to see which materials were retained by the paper. (Note: If the filtrate is blue, the copper sulfate passed. Check both the paper and filtrate for black particles to see if the charcoal passed. Finally, add Lugol's iodine to a 2 ml filtrate sample in a test tube. If the sample turns blue/black when iodine is added, starch is present in the filtrate.)

Passed: ______________________________

Retained: ______________________________

What does the filter paper represent? ______________

During which counting interval was the filtration rate greatest? ______________________________

Explain: ______________________________

____

____

____

What characteristic of the three solutes determined whether or not they passed through the filter paper?

____

ACTIVE TRANSPORT

Whenever a cell expends cellular energy (ATP) to move substances across its boundaries, the process is referred to as an *active transport process.* Substances moved by active means are generally unable to pass by diffusion. They may be too large to pass through the pores; they may not be lipid-soluble; or they may have to move against rather than with a concentration gradient.

In one type of active transport, substances move across the membrane by combining with a protein carrier molecule; the process resembles an enzyme-substrate interaction. ATP provides the driving force, and in many cases the substances move against concentration or electrochemical gradients or both. Some of the substances that are moved into the cells by such carriers, commonly called **solute pumps,** are amino acids and some sugars. Both solutes are lipid-insoluble and too large to pass through the pores, but are necessary for cell life. On the other hand, sodium ions (Na^+) are ejected from cells by active transport. There is more Na^+ outside the cell than there is inside, so the Na^+ tends to remain in the cell unless actively transported out.

Pinocytosis and phagocytosis also require ATP. In **pinocytosis** (cell drinking), the cell membrane sinks beneath the material to form a small vesicle, which then pinches off into the cell interior (see Figure 5.4). Pinocytosis is most common for taking in liquids containing protein or fat.

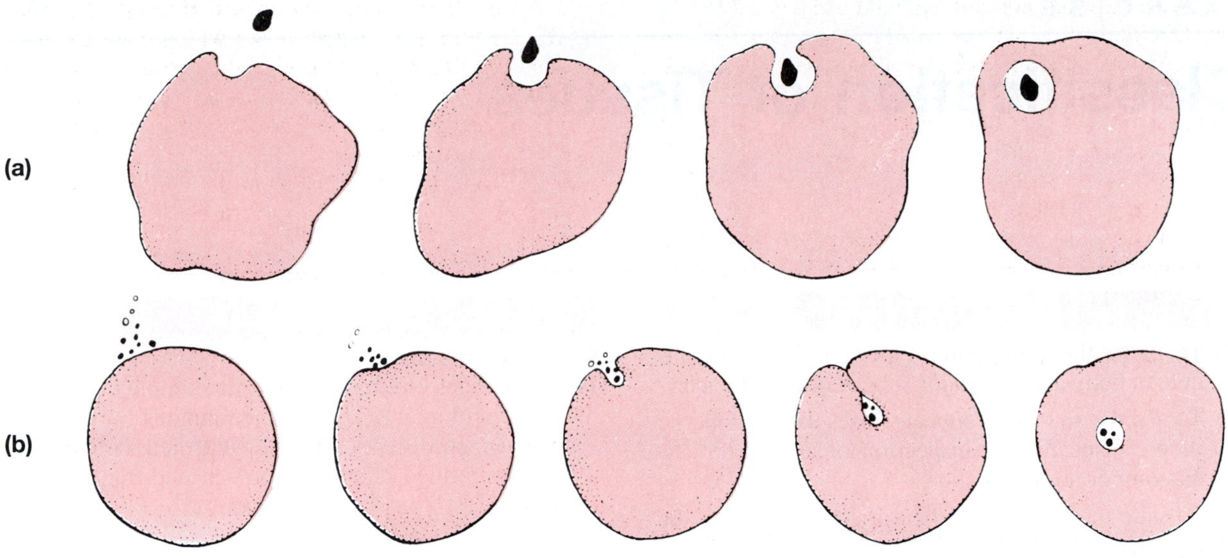

F5.4

Phagocytosis and pinocytosis. **(a)** In phagocytosis, cellular extensions (pseudopodia) flow around the external particle and enclose it within a vacuole. **(b)** In pinocytosis, dissolved proteins gather on the external surface of the plasma membrane, causing the membrane to invaginate and to incorporate a droplet of the fluid.

In **phagocytosis** (cell eating), parts of the plasma membrane and cytoplasm expand and flow around a relatively large or solid material (for example, bacteria or cell debris) and engulf it (Figure 5.4). The membranous sac thus formed, called a *phagosome,* is then fused with a lysosome and its contents are digested. In the human body, phagocytic cells are mainly found among the white blood cells and macrophages that act as scavengers and help protect the body from disease-causing microorganisms and cancer cells.

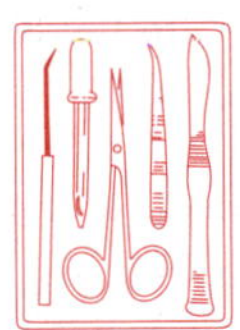

1. Obtain a drop of starved *Amoeba proteus* culture and place it on a coverslip. Add a drop of *Tetrahymena pyriformis* culture (an amoeba "meal") to the amoeba-containing drop, and then quickly but gently invert the coverslip over the well of a depression slide.

2. Locate an amoeba under low power. Keep the light as dim as possible; otherwise the amoeba will "ball up" and begin to disintegrate.

3. Watch as the amoeba phagocytizes the *Tetrahymena* by forming pseudopods that engulf it. In unicellular organisms like the amoeba, phagocytosis is an important food-getting mechanism, but in higher organisms, it is more important as a protective device, as mentioned above.

4. Return all equipment to the appropriate supply areas and rinse glassware used.

6 EXERCISE

Classification of Tissues

OBJECTIVES

1. To name the four major types of tissues in the human body and the major subcategories of each.
2. To identify the tissue subcategories through microscopic inspection or inspection of an appropriate diagram or projected slide.
3. To state the location of the various tissue types in the body.
4. To state the general functions and structural characteristics of each of the four major tissue types.

MATERIALS

Compound microscope

Prepared slides of simple squamous, simple cuboidal, simple columnar, stratified squamous (nonkeratinized), stratified cuboidal, stratified columnar, pseudostratified ciliated columnar, and transitional epithelium

Prepared slides of mesenchyme; of adipose, areolar, reticular, and dense (both regular [tendon] and irregular [dermis]) connective tissues; of hyaline and elastic cartilage; of fibrocartilage; of bone (cross section); and of blood

Prepared slides of skeletal, cardiac, and smooth muscle (longitudinal sections)

Prepared slide of nervous tissue (spinal cord smear)

See Appendix E, Exercise 6 for links to *Anatomy and PhysioShow: The Videodisc.*

Exercise 4 describes cells as the building blocks of life and the all-inclusive functional units of unicellular organisms. But in higher organisms cells do not usually operate as isolated, independent entities. In humans and other multicellular organisms, cells depend on one another and cooperate to maintain homeostasis in the body.

With a few exceptions (parthenogenic organisms), even the most complex animal starts out as a single cell, the fertilized egg, which divides almost endlessly. The trillions of cells that result become specialized for a particular function; some become supportive bone, others the transparent lens of the eye, still others skin cells, and so on. Thus a division of labor exists, with certain groups of cells highly specialized to perform functions that benefit the organism as a whole. Cell specialization brings about great sophistication of achievement but carries with it certain hazards, because when a small specific group of cells is indispensable, any inability to function on its part can paralyze or destroy the entire body.

Groups of cells that are similar in structure and function are called **tissues.** The four primary tissue types—epithelium, connective tissue, nervous tissue, and muscle—have distinctive structures, patterns, and functions. The four primary tissues are further divided into subcategories, as described shortly.

To perform specific body functions, the tissues are organized into such **organs** as the heart, kidneys, and lungs. Most organs contain several representatives of the primary tissues, and the arrangement of these tissues determines the organ's structure and function. Thus **histology,** the study of tissues, complements a study of gross anatomy and provides the structural basis for a study of organ physiology.

The main objective of this exercise is to familiarize you with the major similarities and dissimilarities of the primary tissues, so that when the tissue makeup of an organ is described, you will be able to more easily understand (and perhaps even predict) the organ's major function. Because epithelium and some types of connective tissue will not be considered again, they are emphasized more than muscle, nervous tissue, and bone (a connective tissue), which are covered in more depth in later exercises.

EPITHELIAL TISSUE

Epithelial tissue, or **epithelium,** covers surfaces. For example, epithelium covers the external body surface (as the epidermis), lines its cavities and tubules, and generally marks off our "insides" from our outsides.

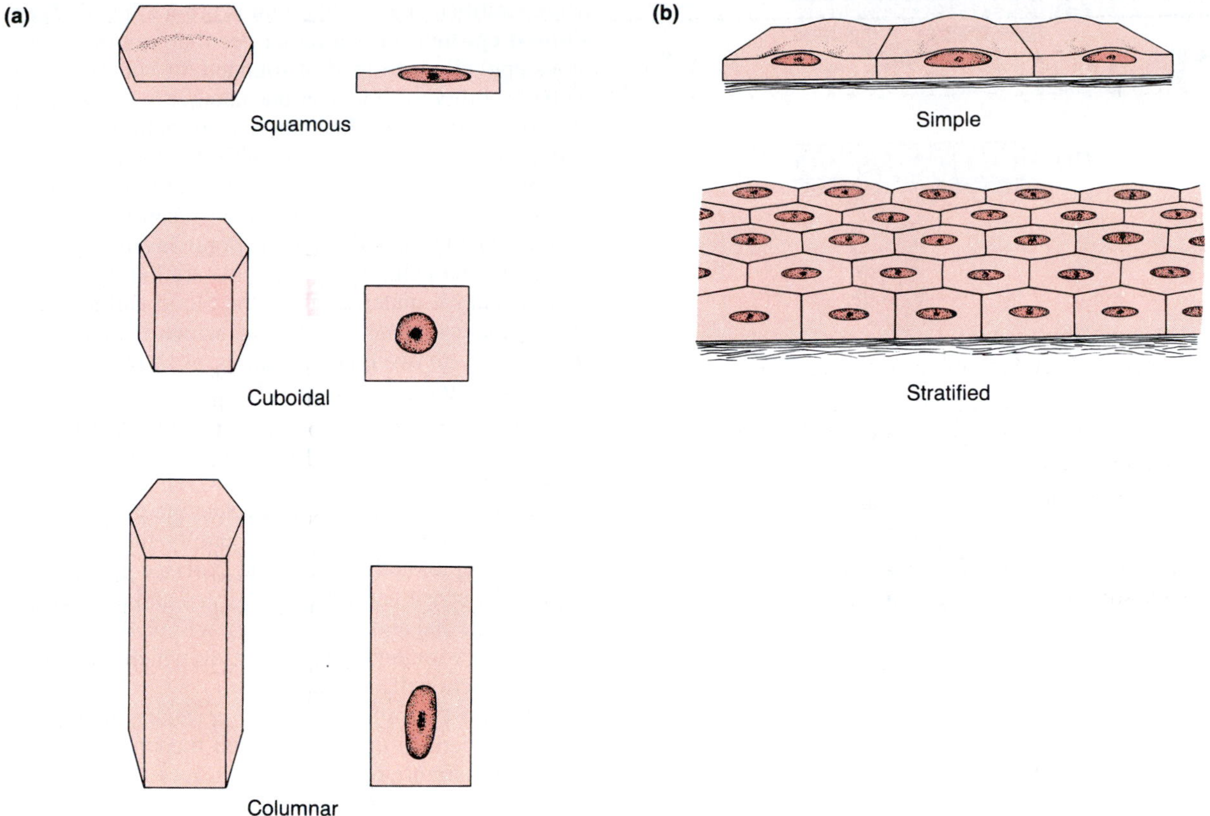

F6.1

Classification of epithelia. (a) Classification on the basis of cell shape. For each category, a whole cell is shown on the left and a longitudinal section is shown on the right. **(b)** Classification on the basis of arrangement (relative number of layers).

Since the various endocrine (hormone-producing) and exocrine glands of the body almost invariably develop from epithelial membranes, glands too are logically classed as epithelium.

Epithelial functions include protection, absorption, filtration, excretion, secretion, and sensory reception. For example, the epithelium covering the body protects against bacterial invasion and chemical damage; that lining the respiratory tract is ciliated to sweep dust and other foreign particles away from the lungs. Epithelium specialized to absorb substances lines the stomach and small intestine. In the kidney tubules, the epithelium absorbs, secretes, and filters. Secretion is a specialty of the glands.

Epithelium generally exhibits the following characteristics:

- Cells fit closely together to form membranes, or sheets of cells, and are bound together by specialized junctions.
- The membranes always have one free surface, called the *apical surface.*
- The cells are attached to an adhesive **basement membrane,** an amorphous material secreted partly by the epithelial cells (*basal lamina*) and connective tissue cells (*reticular lamina*) that lie adjacent to each other.
- Epithelial tissues have no blood supply of their own (are avascular), but depend on diffusion of nutrients from the underlying connective tissue.
- If well nourished, epithelial cells can easily regenerate themselves. This is an important characteristic because many epithelia are subjected to a good deal of friction.

The covering and lining epithelia are classified according to two criteria—cell shape and arrangement or relative number of layers (Figure 6.1). **Squamous** (scalelike), **cuboidal** (cubelike), and **columnar** (column-shaped) epithelial cells are the general types based on shape. On the basis of arrangement, there are **simple** epithelia, consisting of one layer of cells attached to the basement membrane, and **stratified** epithelia, consisting of more than one layer of cells. The terms denoting shape and arrangement of the epithelial cells are combined to describe the epithelium fully. *Stratified epithelia are named according to the cells at the apical surface of the epithelial membrane,* not those resting on the basement membrane.

There are, in addition, two less easily categorized types of epithelia. **Pseudostratified epithelium** is actually a simple columnar epithelium (one layer of cells), but because its cells extend varied distances from the basement membrane, it gives the false appearance of

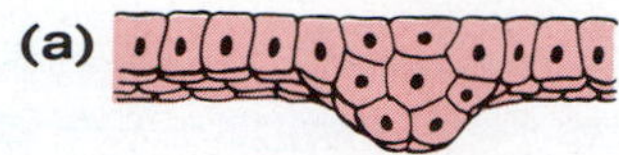

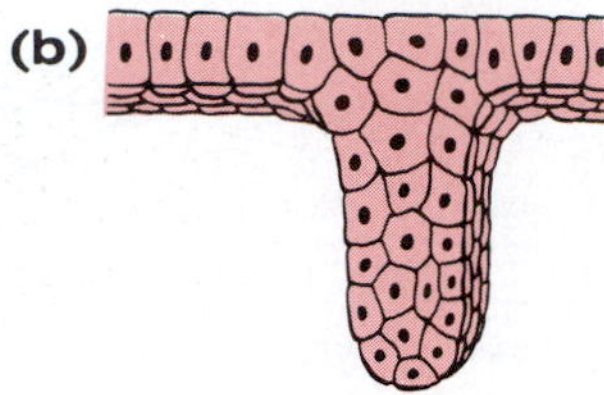

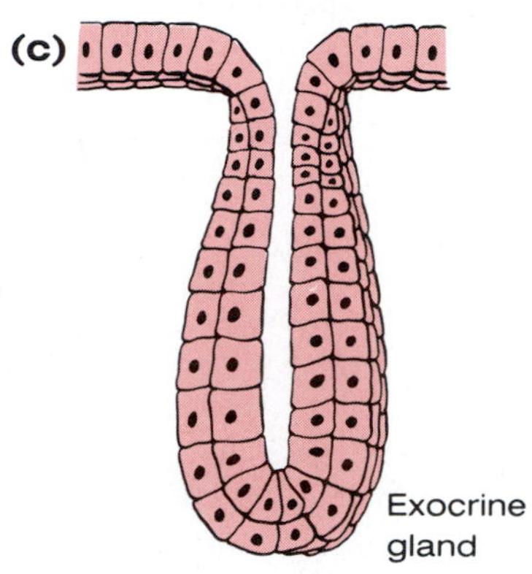

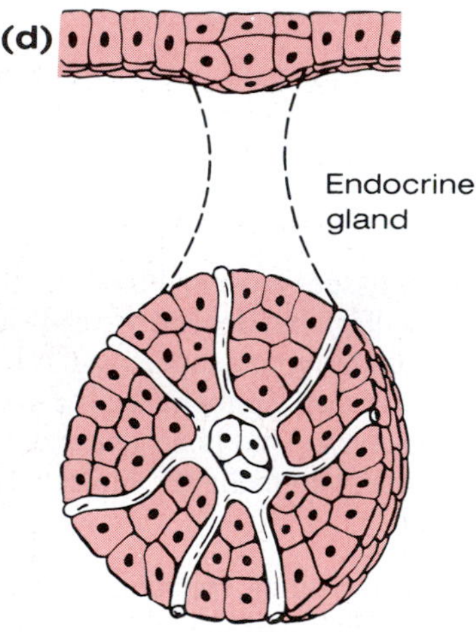

F6.2

Formation of endocrine and exocrine glands from epithelial sheets. (a) Epithelial cells grow and push into the underlying tissue. **(b)** A cord of epithelial cells forms. **(c)** In an exocrine gland, a lumen (cavity) forms. The inner cells form the duct, the outer cells produce the secretion. **(d)** In a forming endocrine gland, the connecting duct cells atrophy, leaving the secretory cells with no connection to the epithelial surface. However, they do become heavily invested with blood and lymphatic vessels that receive the secretions.

being stratified. This epithelium is often ciliated. **Transitional epithelium** is a rather peculiar stratified squamous epithelium formed of rounded, or "plump," cells with the ability to slide over one another to allow the organ to be stretched. Transitional epithelium is found only in urinary system organs subjected to periodic distension, such as the bladder. The superficial cells are flattened (like true squamous cells) when the organ is distended and rounded when the organ is empty.

Epithelial cells forming glands are highly specialized to remove materials from the blood and to manufacture them into new materials, which they then secrete. There are two types of glands, as shown in Figure 6.2. **Endocrine glands** lose their surface connection (duct) as they develop; thus they are referred to as ductless glands. Their secretions (all hormones) are extruded directly into the blood or the lymphatic vessels that weave through the glands. **Exocrine glands** retain their ducts, and their secretions empty through these ducts to an epithelial surface. The exocrine glands—including the sweat and oil glands, liver, and pancreas—are both external and internal; they will be discussed in conjunction with the organ systems to which their products are functionally related.

The most common types of epithelia, their most common locations in the body, and their functions are described in Figure 6.3.

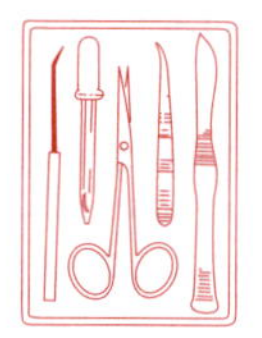

Obtain slides of simple squamous, simple cuboidal, simple columnar, stratified squamous (nonkeratinized), pseudostratified ciliated columnar, stratified cuboidal, stratified columnar, and transitional epithelia. Examine each carefully, and notice how the epithelial cells fit closely together to form intact sheets of cells, a necessity for a tissue that forms linings or covering membranes. Scan each epithelial type for modifications for specific functions, such as cilia (motile cell projections that help to move substances along the cell surface), and microvilli, which increase the surface area for absorption. Also be alert for goblet cells, which secrete lubricating mucus (see Plate 1 of the Histology Atlas). Compare your observations with the photomicrographs in Figure 6.3.

While working, check the questions in the laboratory review section for this exercise. A number of the questions there refer to some of the observations you are asked to make during your microscopic study.

CONNECTIVE TISSUE

Connective tissue is found in all parts of the body as discrete structures or as part of various body organs. It is the most abundant and widely distributed of the tissue types.

The connective tissues perform a variety of functions, but they primarily protect, support, and bind together other tissues of the body. For example, bones are composed of connective tissue (**bone** or **osseous tissue**),

(Text continues on p. 47)

(a) Simple squamous epithelium

Description: Single layer of flattened cells with disc-shaped central nuclei and sparse cytoplasm; the simplest of the epithelia.

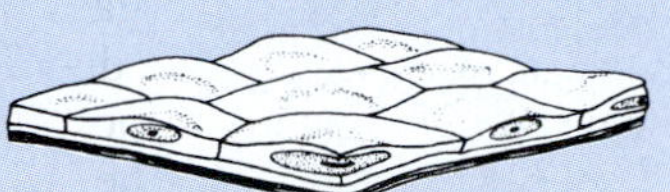

Location: Air sacs of lungs; kidney glomeruli; lining of heart, blood vessels, and lymphatic vessels; lining of ventral body cavity (serosae).

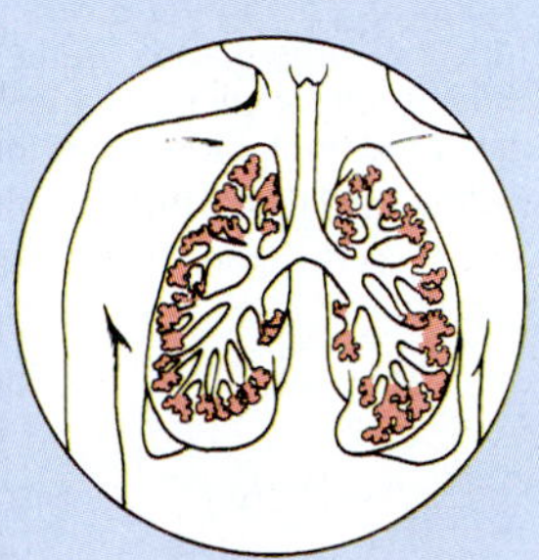

Function: Allows passage of materials by diffusion and filtration in sites where protection is not important; secretes lubricating substances in serosae.

Photomicrograph: Simple squamous epithelium forming walls of alveoli (air sacs) of the lung (576×).

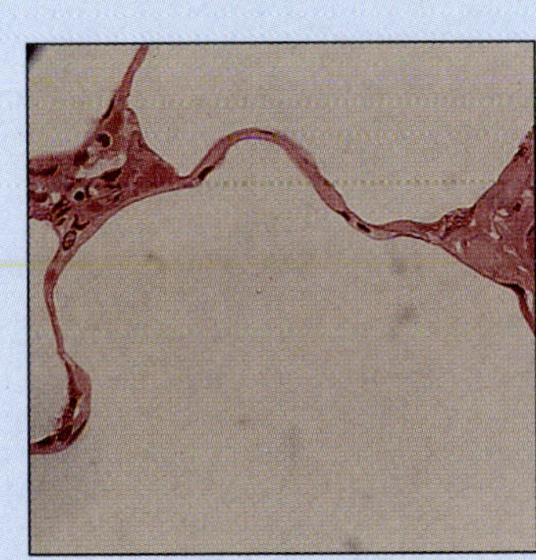

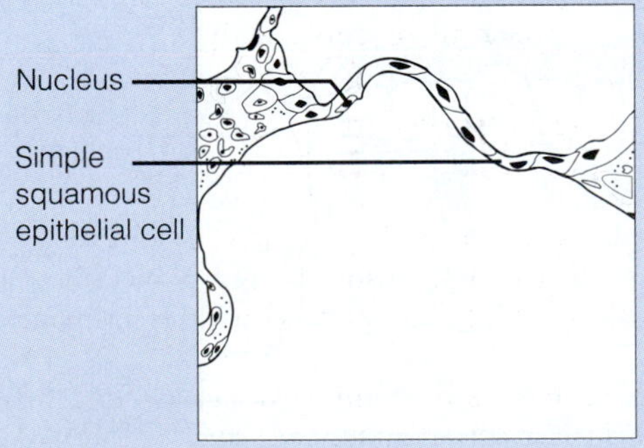

(b) Simple cuboidal epithelium

Description: Single layer of cubelike cells with large, spherical central nuclei.

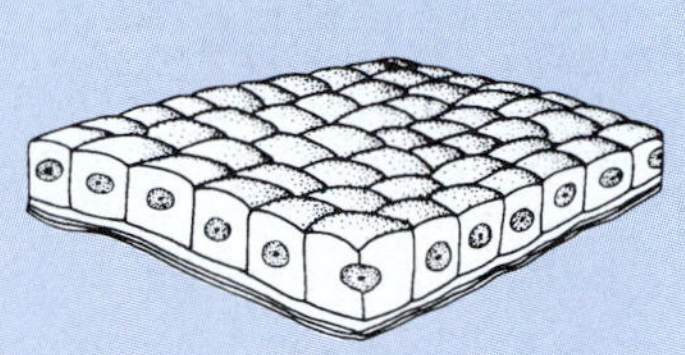

Location: Kidney tubules; ducts and secretory portions of small glands; ovary surface.

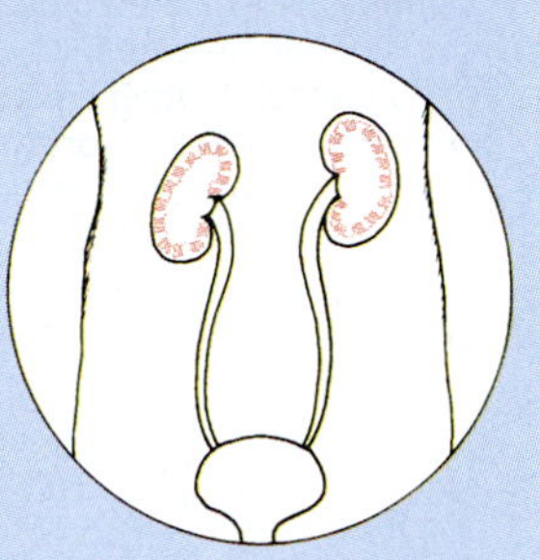

Function: Secretion and absorption.

Photomicrograph: Simple cuboidal epithelium in kidney tubules (576×).

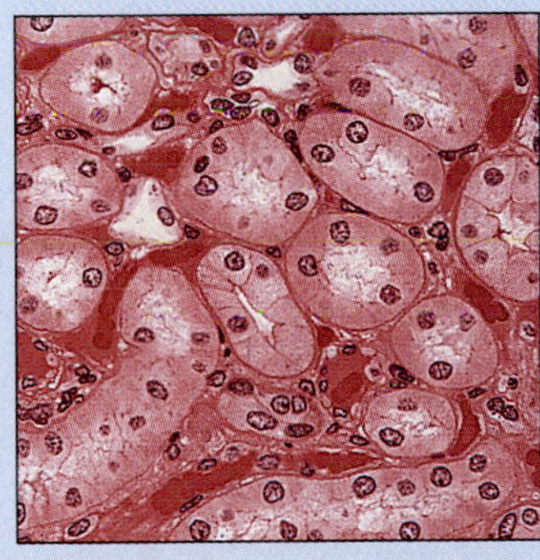

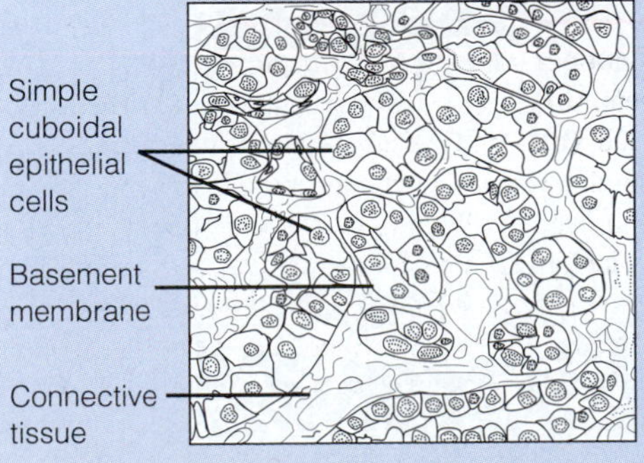

F6.3

Epithelial tissues. Simple epithelia (a–b).

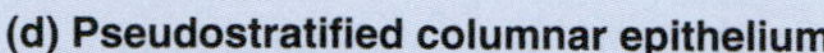

(c) Simple columnar epithelium

Description: Single layer of tall cells with *oval* nuclei; some cells bear cilia; layer may contain mucus-secreting glands (goblet cells).

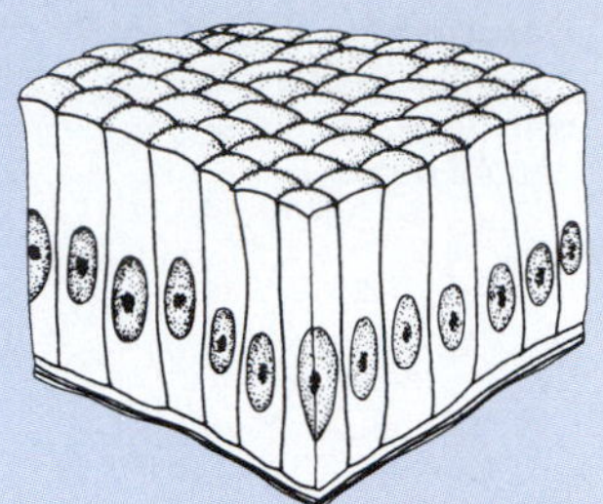

Location: Nonciliated type lines most of the digestive tract (stomach to anal canal), gallbladder and excretory ducts of some glands; ciliated variety lines small bronchi, uterine tubes, and some regions of the uterus.

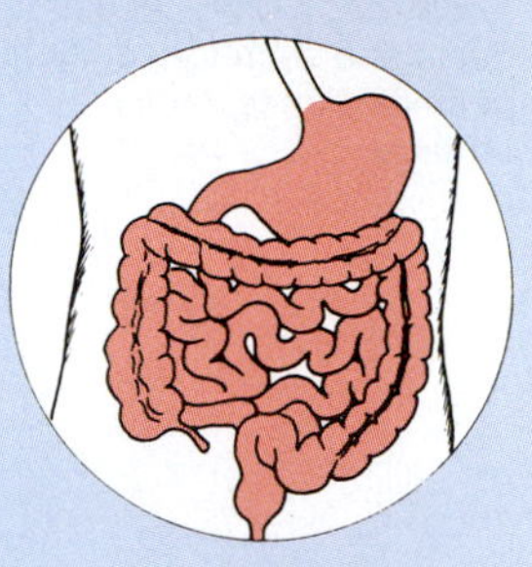

Function: Absorption; secretion of mucus, enzymes, and other substances; ciliated type propels mucus (or reproductive cells) by ciliary action.

Photomicrograph: Simple columnar epithelium of the gallbladder mucosa (576×).

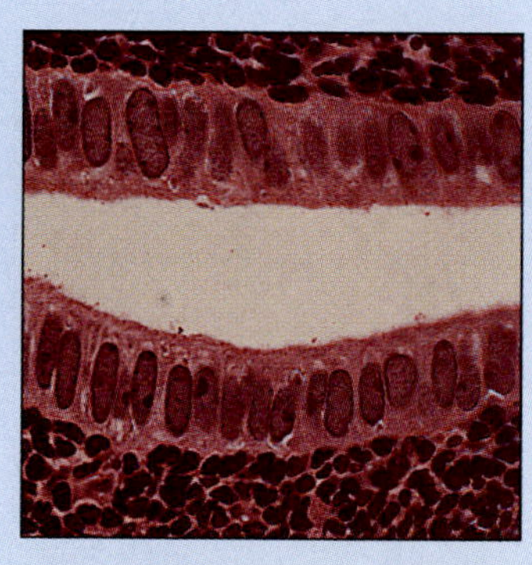

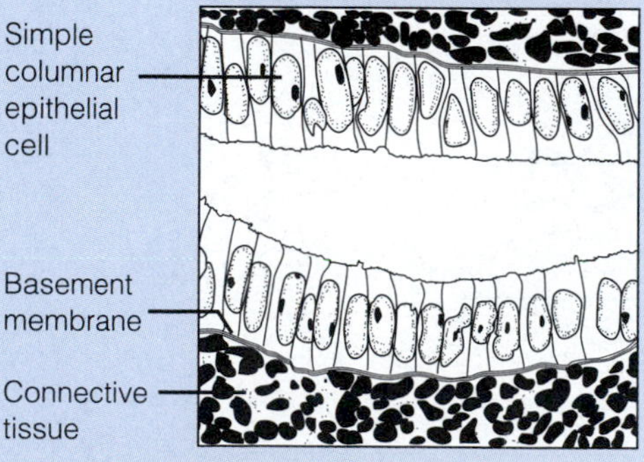

(d) Pseudostratified columnar epithelium

Description: Single layer of cells of differing heights, some not reaching the free surface; nuclei seen at different levels; may contain goblet cells and bear cilia.

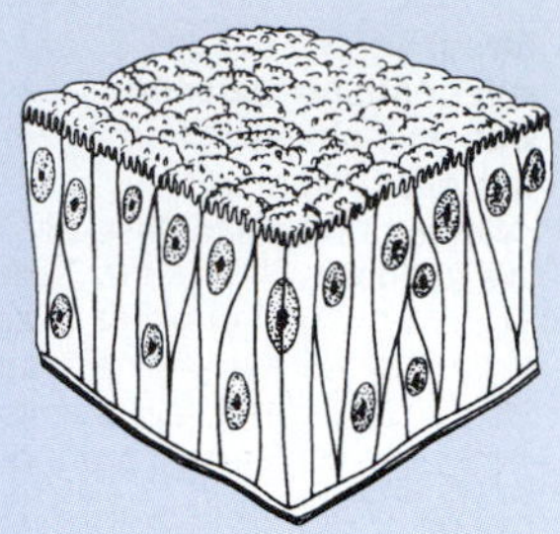

Location: Nonciliated type in ducts of large glands, parts of male urethra; ciliated variety lines the trachea, most of the upper respiratory tract.

Function: Secretion, particularly of mucus; propulsion of mucus by ciliary action.

Photomicrograph: Pseudostratified ciliated columnar epithelium lining the human trachea (612×).

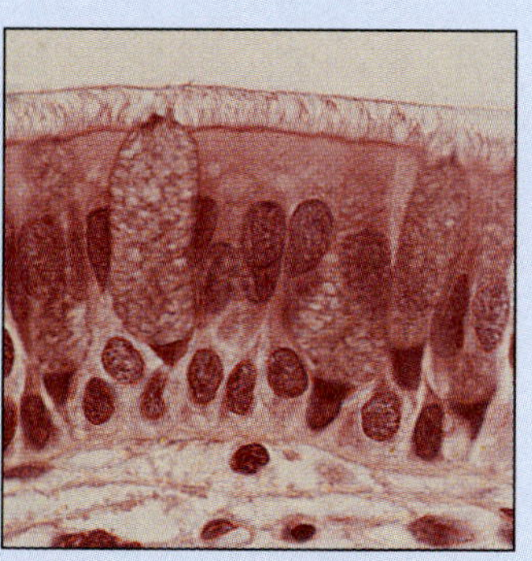

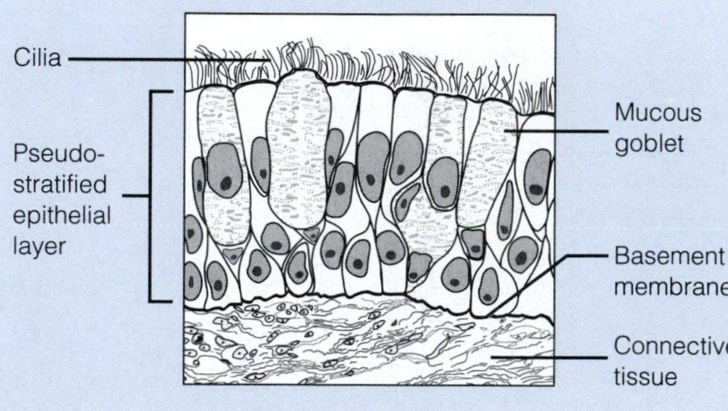

F6.3 (*continued*)

Simple epithelia (c and d).

(e) Stratified squamous epithelium

Description: Thick membrane composed of several cell layers; basal cells are cuboidal or columnar and metabolically active; surface cells are flattened (squamous); in the keratinized type, the surface cells are full of keratin and dead; basal cells are active in mitosis and produce the cells of the more superficial layers.

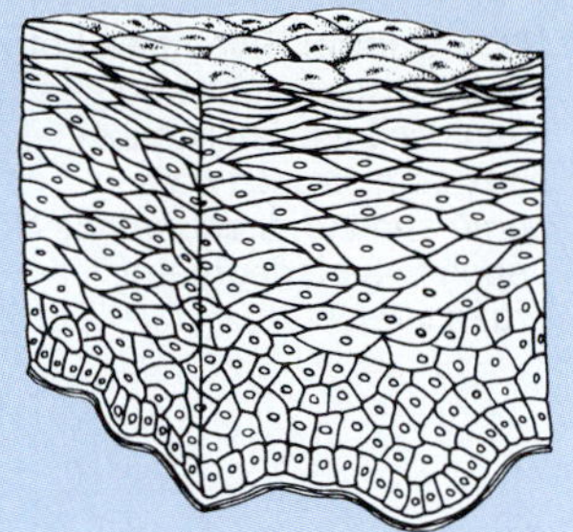

Location: Nonkeratinized type forms the moist linings of the esophagus, mouth, and vagina; keratinized variety forms the epidermis of the skin, a dry membrane.

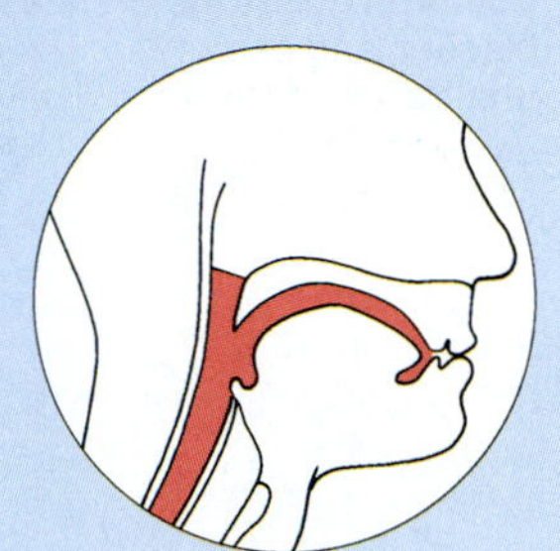

Function: Protects underlying tissues in areas subjected to abrasion.

Photomicrograph: Stratified squamous epithelium lining of the esophagus (144×).

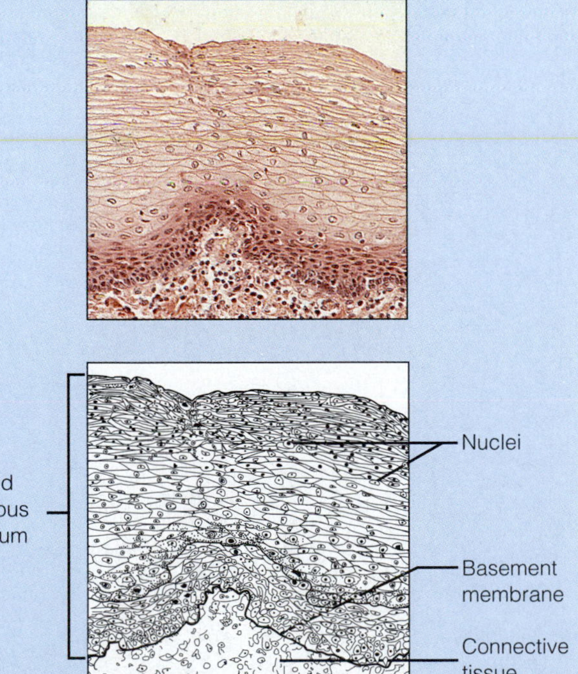

(f) Stratified cuboidal epithelium

Description: Generally two layers of cubelike cells.

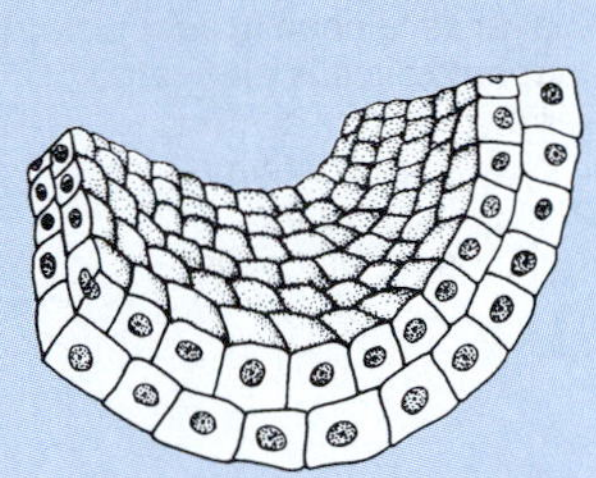

Location: Largest ducts of sweat glands, mammary glands, and salivary glands.

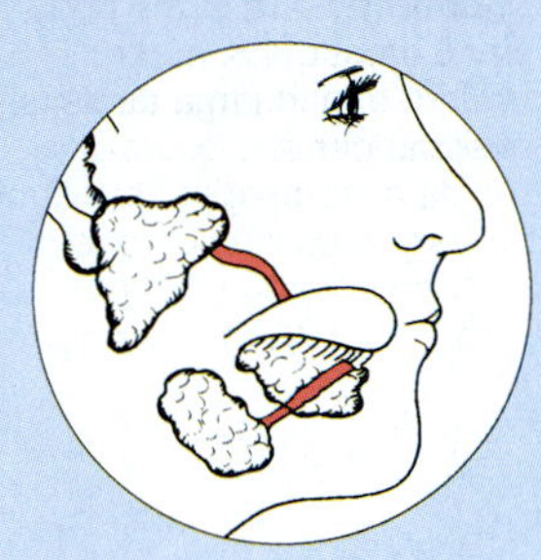

Function: Protection.

Photomicrograph: Stratified cuboidal epithelium forming a salivary gland duct (86×).

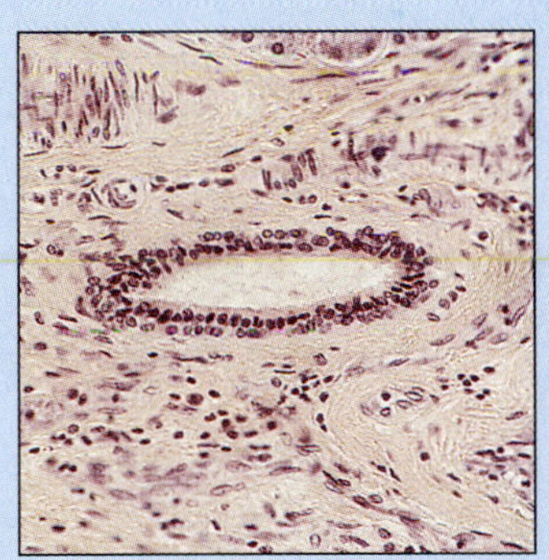

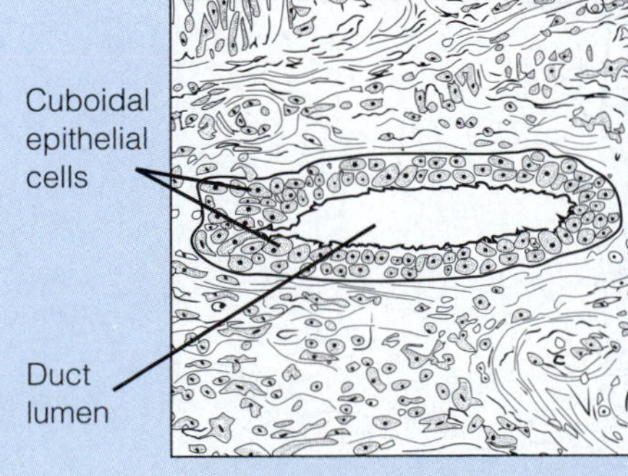

F6.3 *(continued)*

Stratified epithelia (e and f).

(g) Stratified columnar epithelium

Description: Several cell layers; basal cells usually cuboidal; superficial cells elongated and columnar.

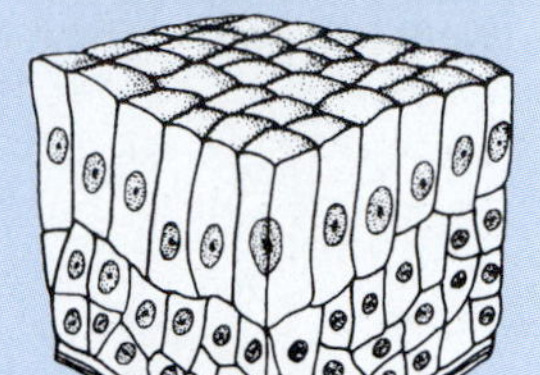

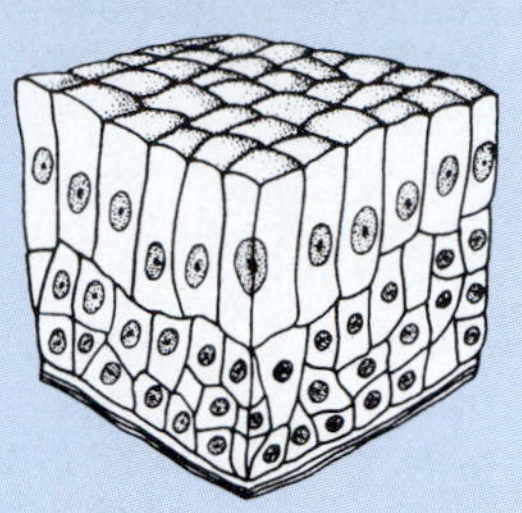

Location: Rare in the body; small amounts in male urethra and in large ducts of some glands.

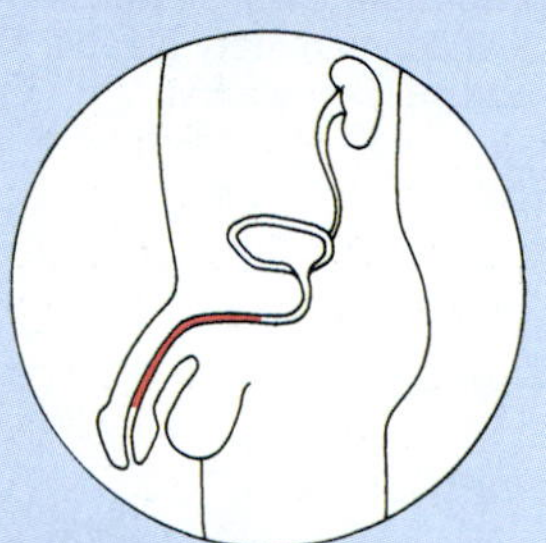

Function: Protection; secretion.

Photomicrograph: Stratified columnar epithelium lining of the male urethra (429×).

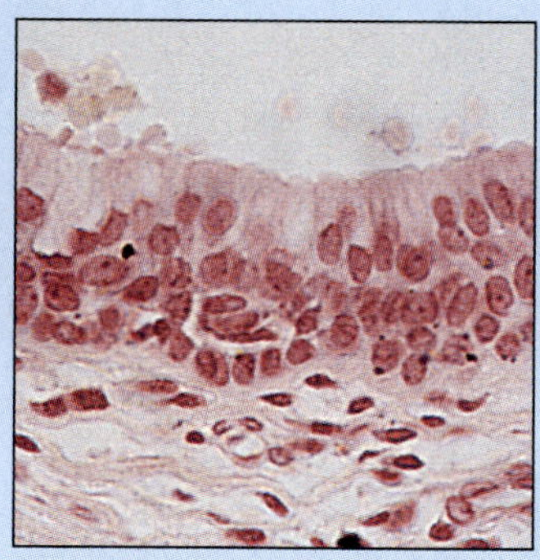

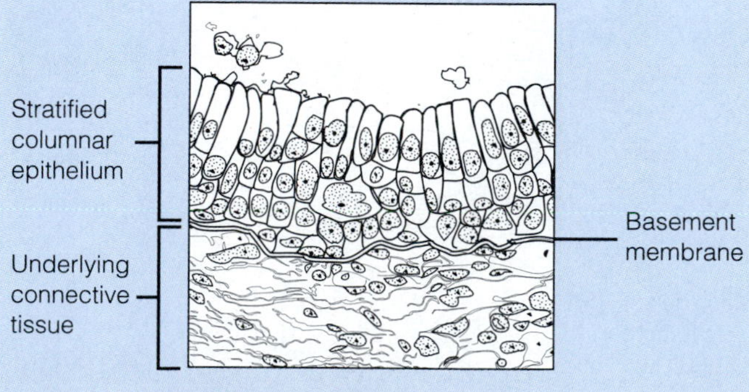

(h) Transitional epithelium

Description: Resembles both stratified squamous and stratified cuboidal; basal cells cuboidal or columnar; surface cells dome-shaped or squamous-like, depending on degree of organ stretch.

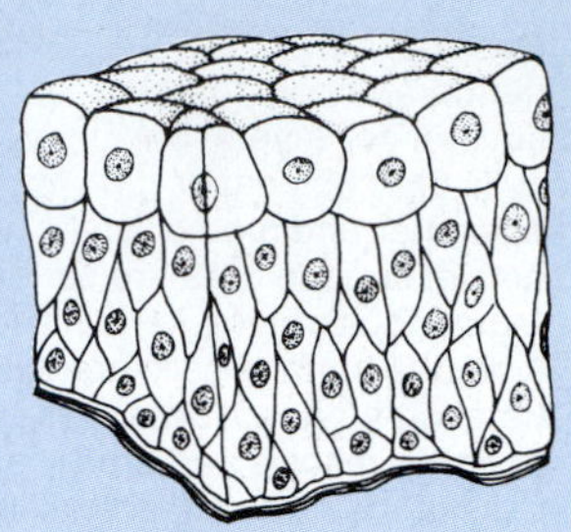

Location: Lines the ureters, bladder, and part of the urethra.

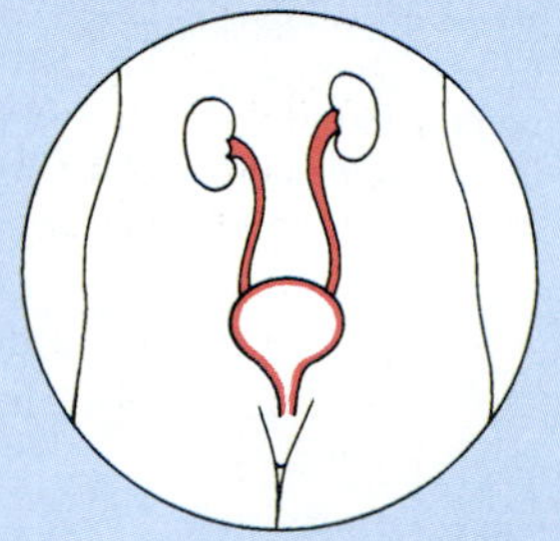

Function: Stretches readily and permits distension of urinary organ by contained urine.

Photomicrograph: Transitional epithelium lining of the bladder, relaxed state (252×); note the bulbous, or rounded, appearance of the cells at the surface; these cells flatten and become elongated when the bladder is filled with urine.

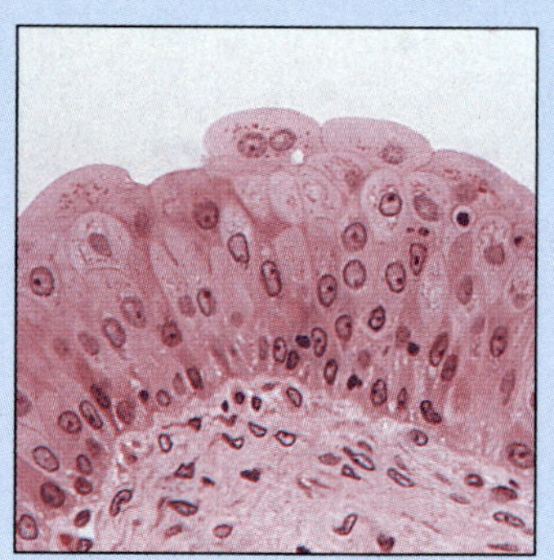

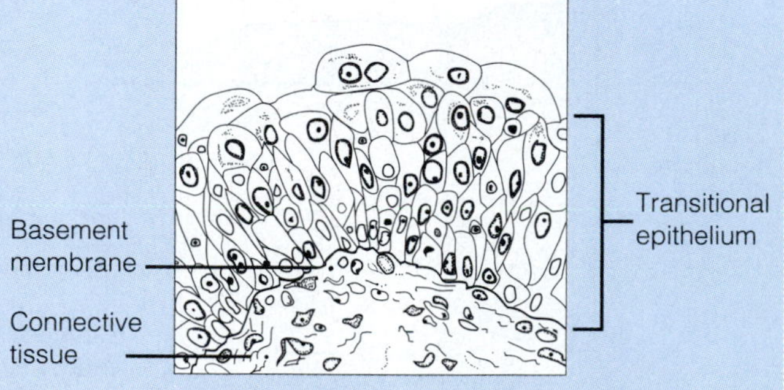

F6.3 *(continued)*

Stratified epithelia (g and h).

and they protect and support other body tissues and organs. The ligaments and tendons (**dense connective tissue**) bind the bones together or bind skeletal muscles to bones.

Areolar connective tissue is a soft packaging material that cushions and protects body organs. Fat (**adipose**) tissue provides insulation for the body tissues and a source of stored food. Blood-forming (**hematopoietic**) tissue replenishes the body's supply of red blood cells. In addition, connective tissue also serves a vital function in the repair of all body tissues since many wounds are repaired by connective tissue in the form of scar tissue.

The characteristics of connective tissue include the following:

- With a few exceptions (cartilages, tendons, and ligaments), connective tissues are well vascularized.
- Connective tissues are composed of many types of cells.
- There is a great deal of noncellular, nonliving material (matrix) between the cells of connective tissue.

The nonliving material between the cells—the **extracellular matrix**—deserves a bit more explanation because it distinguishes connective tissue from all other tissues. It is produced by the cells and then extruded. The matrix is primarily responsible for the strength associated with connective tissue, but there is variation. At one extreme, adipose tissue is composed mostly of cells. At the opposite extreme, bone and cartilage have very few cells and large amounts of matrix.

The matrix has two components—ground substance and fibers. The **ground substance** is composed chiefly of glycoproteins and large charged polysaccharide molecules. Depending on its specific composition, the ground substance may be liquid, semisolid, gel-like, or very hard. When the matrix is firm, as in cartilage and bone, the connective tissue cells reside in cavities in the matrix called *lacunae.* The fibers, which provide support, include **collagenic** (white) **fibers, elastic** (yellow) **fibers,** and **reticular** (fine collagenic) **fibers.** Of these, the collagen fibers are most abundant.

Generally speaking, the ground substance functions as a molecular sieve, or medium through which nutrients and other dissolved substances can diffuse between the blood capillaries and the cells. The fibers in the matrix hinder diffusion somewhat and make the ground substance less pliable. The properties of the connective tissue cells and the makeup and arrangement of their matrix elements vary tremendously, accounting for the amazing diversity of this tissue type. Nonetheless, the connective tissues have a common structural plan seen best in *areolar connective tissue* (Figure 6.4), a soft packing tissue that occurs throughout the body. Since all other connective tissues are variations of areolar, it is considered the model or prototype of the connective tissues. Notice in Figure 6.4 that areolar tissue has all three varieties of fibers, but they are sparsely arranged in its transparent gel-like ground substance. The cell type that

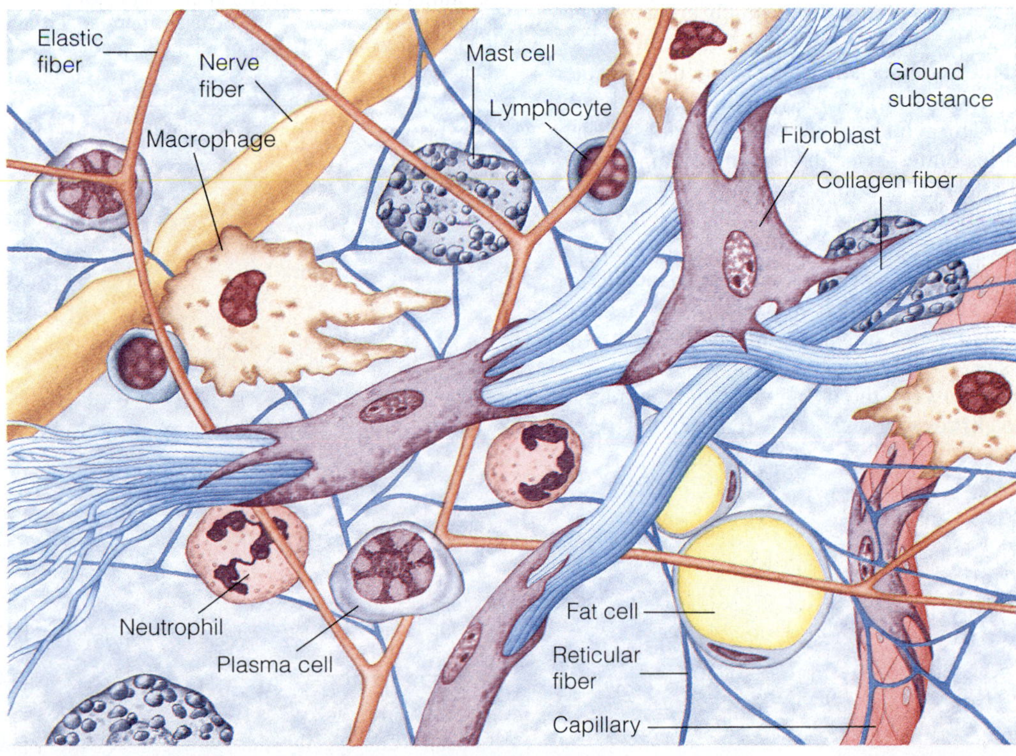

F6.4

Areolar connective tissue: A prototype (model) connective tissue. Note the various cell types and the three classes of fibers (collagen, reticular, elastic) embedded in the ground substance.

secretes its matrix is the *fibroblast,* but a wide variety of other cells including phagocytic cells like macrophages and certain white blood cells and mast cells which act in the inflammatory response are present as well. The more durable connective tissues, such as bone, cartilage, and the dense fibrous varieties, characteristically have a firm ground substance and many more fibers.

There are four main types of adult connective tissue, all of which typically have large amounts of matrix. These are connective tissue proper (which includes areolar, adipose, reticular, and dense [fibrous] connective tissues), cartilage, bone, and blood. All of these derive from an embryonic tissue called *mesenchyme.* Figure 6.5 lists the general characteristics, location, and function of some of the connective tissues found in the body.

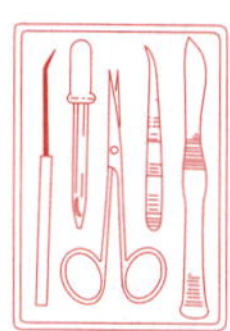

Obtain prepared slides of mesenchyme; of adipose, areolar, reticular, dense regular and irregular connective tissue; of elastic connective tissue; of hyaline and elastic cartilage and fibrocartilage; and of osseous connective tissue (bone). Compare your observations with the views in Figure 6.5.

Distinguish between the living cells and the matrix and pay particular attention to the denseness and arrangement of the matrix. For example, notice how the matrix of the dense fibrous connective tissues, making up tendons and the dermis of the skin, is chock-full of collagenic fibers, and that in the regular variety (tendon), the fibers are all running in the same direction, whereas in the dermis they appear to be running in many directions.

While examining the areolar connective tissue, a soft "packing tissue," notice how much empty space there appears to be, and distinguish between the collagen fibers and the coiled elastic fibers. Also, try to locate a **mast cell,** which has large darkly staining granules in its cytoplasm. This cell type releases histamine that makes capillaries quite permeable during inflammatory reactions and allergies and thus is partially responsible for that "runny nose" of allergies.

In adipose tissue, locate a cell (signet ring cell) in which the nucleus can be seen pushed to one side by the large fat-filled vacuole which appears to be a large empty space. Also notice how little matrix there is in fat or adipose tissue.

Distinguish between the living cells and the matrix in the dense fibrous, bone, and hyaline cartilage preparations.

Embryonic connective tissue

(a) Mesenchyme

Description: Embryonic connective tissue; gel-like ground substance containing fine fibers; star-shaped mesenchymal cells.

Location: Primarily in embryo.

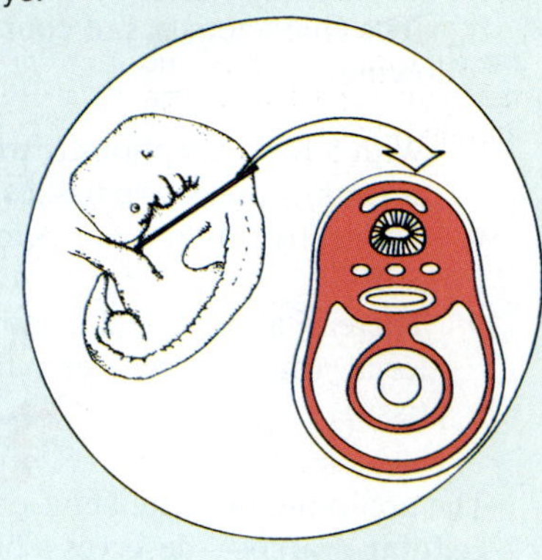

Function: Gives rise to all other connective tissue types.

Photomicrograph: Mesenchymal tissue, an embryonic connective tissue (1072×); the clear-appearing background is the fluid ground substance of the matrix; notice the fine, sparse fibers.

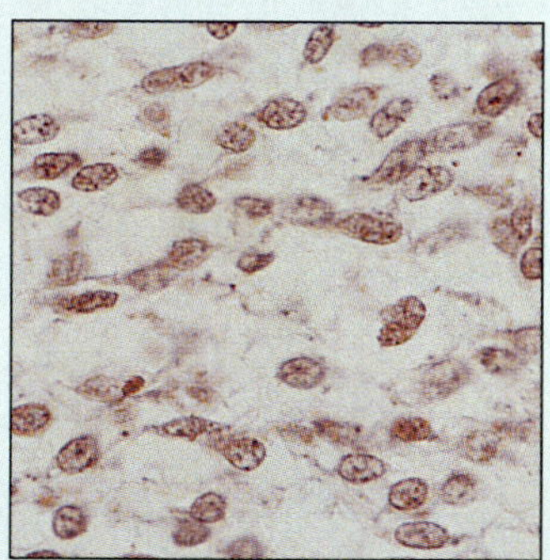

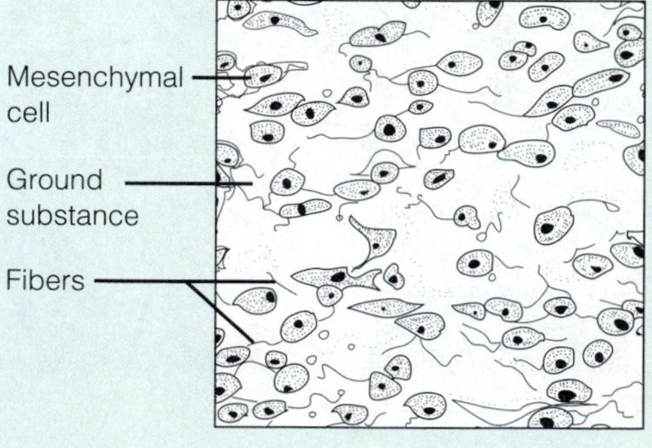

F6.5

Connective tissues.

Connective tissue proper: Loose connective tissue (b to d)

(b) Areolar connective tissue

Description: Gel-like matrix with all three fiber types; cells: fibroblasts, macrophages, mast cells, and some white blood cells.

Location: Widely distributed under epithelia of body, e.g., forms lamina propria of mucous membranes; packages organs, surrounds capillaries.

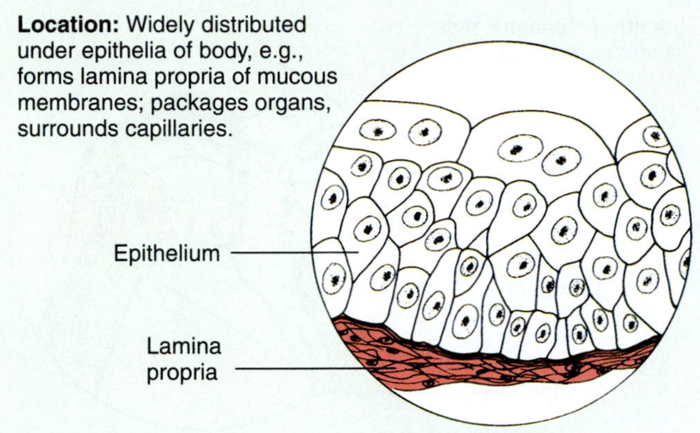

Function: Wraps and cushions organs; its macrophages phagocytize bacteria; plays important role in inflammation; holds and conveys tissue fluid.

Photomicrograph: Areolar connective tissue, a soft packaging tissue of the body (170×).

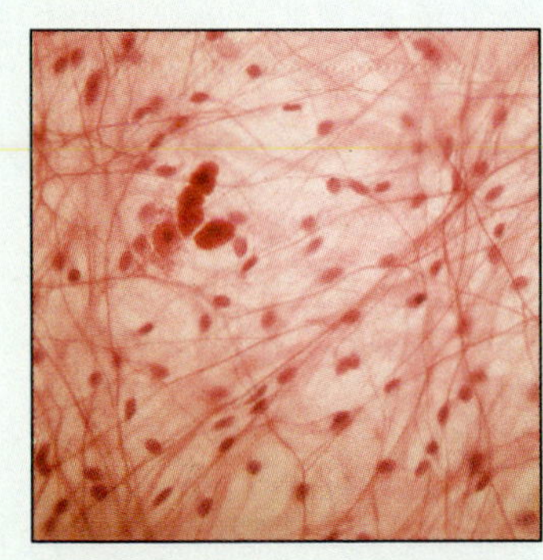

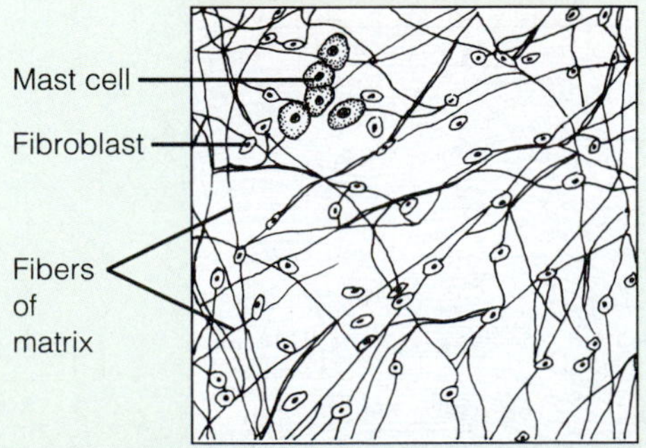

(c) Adipose tissue

Description: Matrix as in areolar, but very sparse; closely packed adipocytes, or fat cells, have nucleus pushed to the side by large fat droplet.

Location: Under skin; around kidneys and eyeballs; in bones and within abdomen; in breasts.

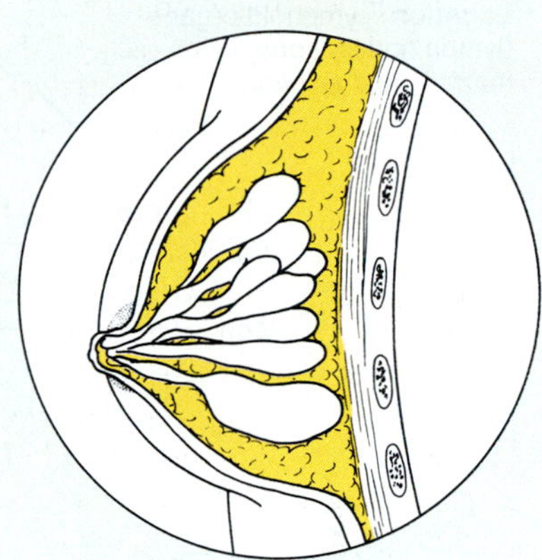

Function: Provides reserve food fuel; insulates against heat loss; supports and protects organs.

Photomicrograph: Adipose tissue from the subcutaneous layer under the skin (864×).

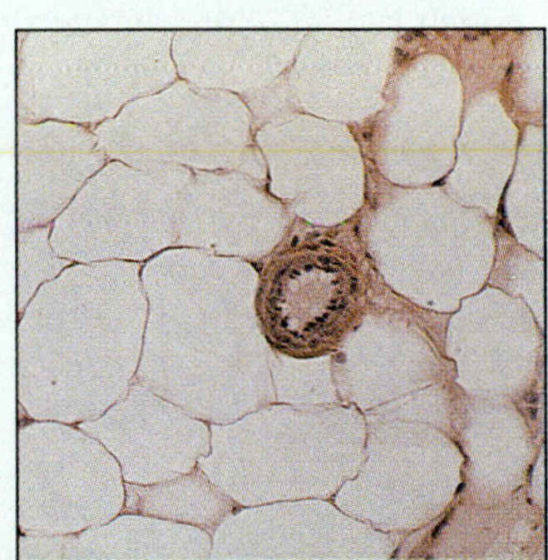

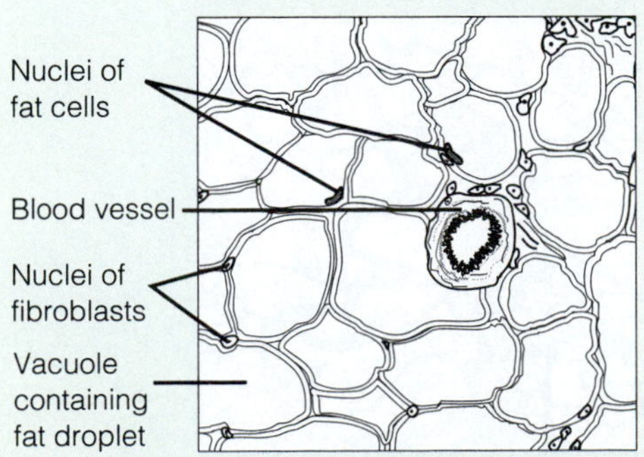

F6.5 *(continued)*

Connective tissue proper: Dense connective tissue (e and f)

(d) Reticular connective tissue

Description: Network of reticular fibers in a typical loose ground substance; reticular cells predominate.

Location: Lymphoid organs (lymph nodes, bone marrow, and spleen).

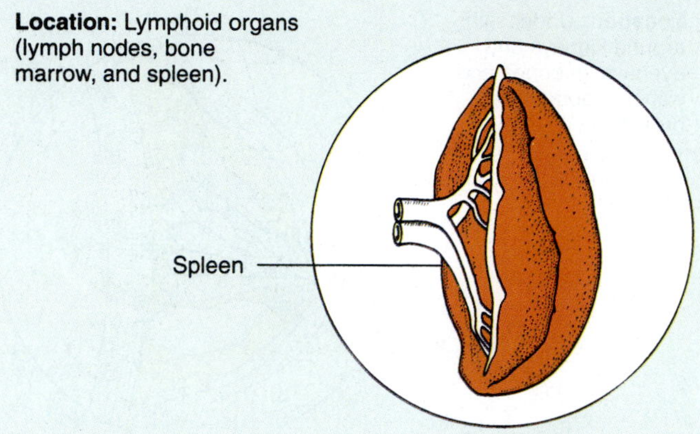

Function: Fibers form a soft internal skeleton that supports other cell types.

Photomicrograph: Dark-staining network of reticular connective tissue fibers forming the internal skeleton of the spleen (1125×).

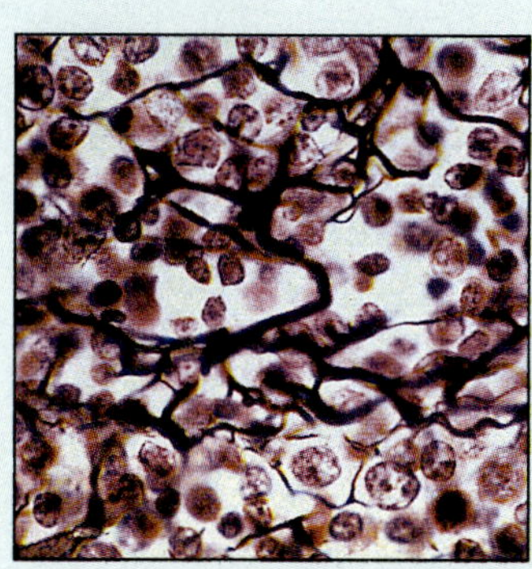

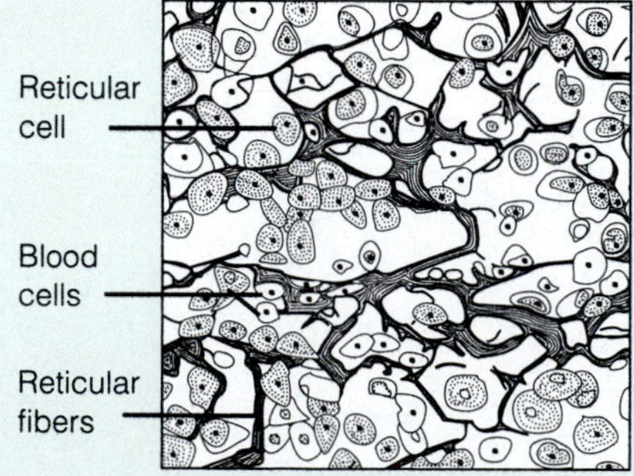

(e) Dense regular connective tissue

Description: Primarily parallel collagen fibers; a few elastin fibers; major cell type is the fibroblast.

Location: Tendons, most ligaments, aponeuroses.

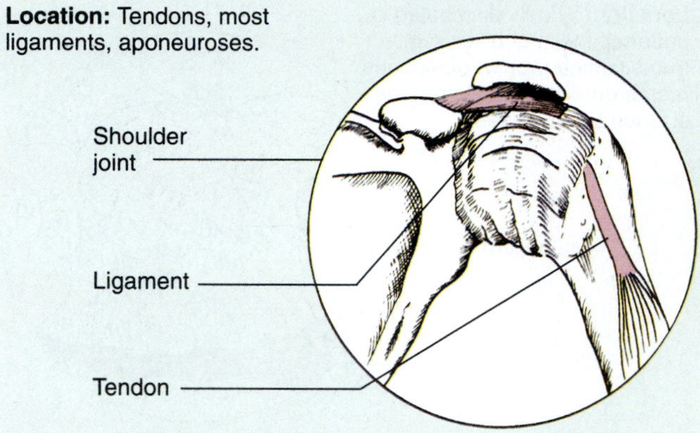

Function: Attaches muscles to bones or to muscles; attaches bones to bones; withstands great tensile stress when pulling force is applied in one direction.

Photomicrograph: Dense regular connective tissue from a tendon (576×).

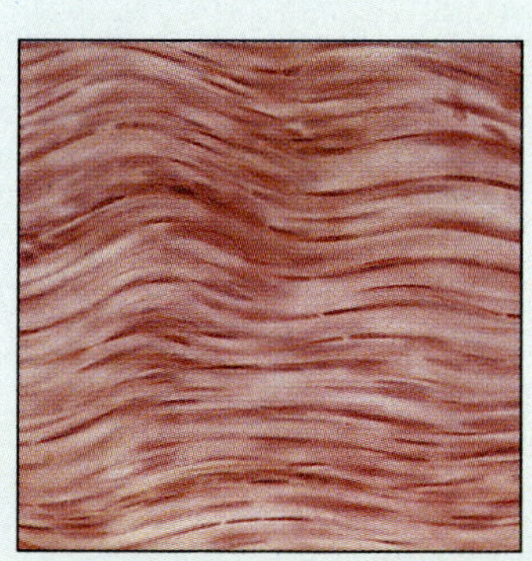

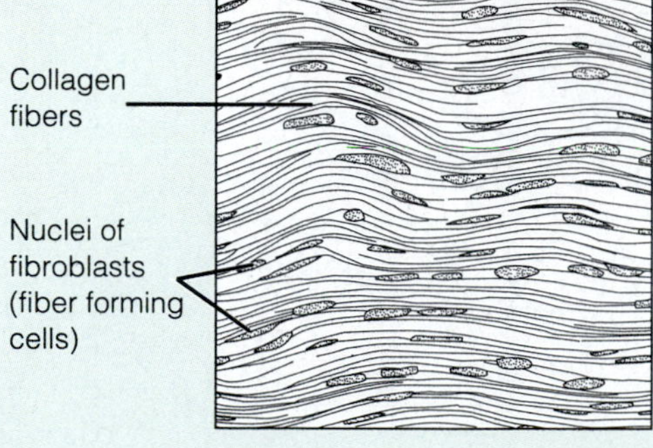

F6.5 (*continued*)

Cartilage: (g to i)

(f) Dense irregular connective tissue

Description: Primarily irregularly arranged collagen fibers; some elastic fibers; major cell type is the fibroblast.

Location: Dermis of the skin; submucosa of digestive tract; fibrous capsules of organs and of joints.

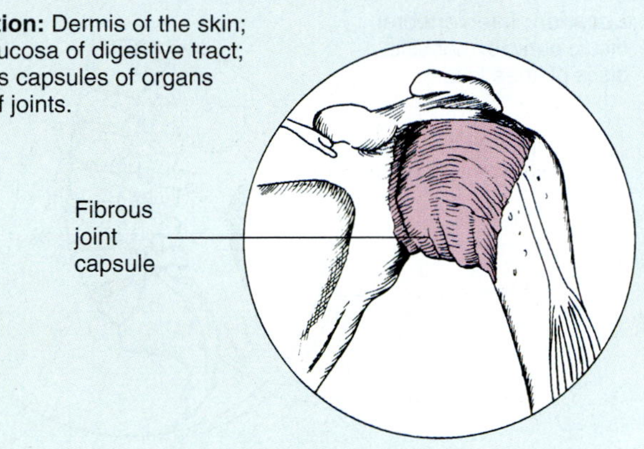

Function: Able to withstand tension exerted in many directions; provides structural strength.

Photomicrograph: Dense irregular connective tissue from the dermis of the skin (268×).

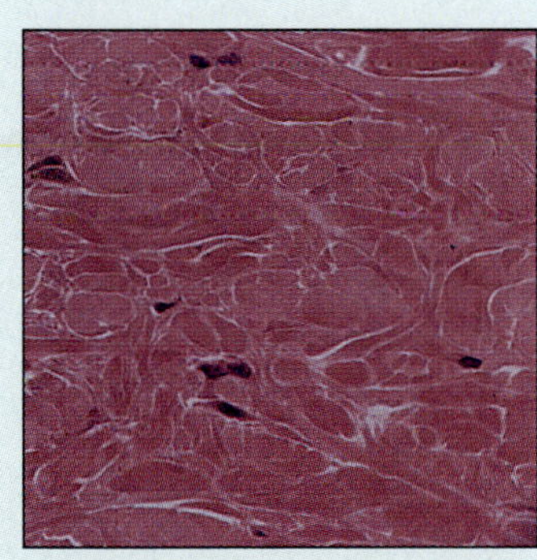

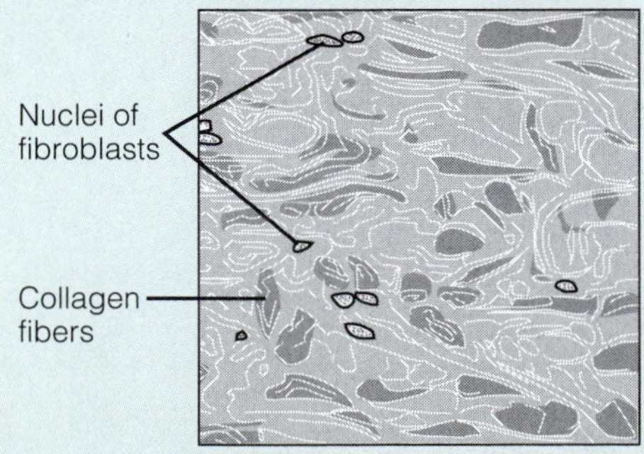

(g) Hyaline cartilage

Description: Amorphous but firm matrix; collagen fibers form an imperceptible network; chondroblasts produce the matrix and when mature (chondrocytes) lie in lacunae.

Location: Forms most of the embryonic skeleton; covers the ends of long bones in joint cavities; forms costal cartilages of the ribs; cartilages of the nose, trachea, and larynx.

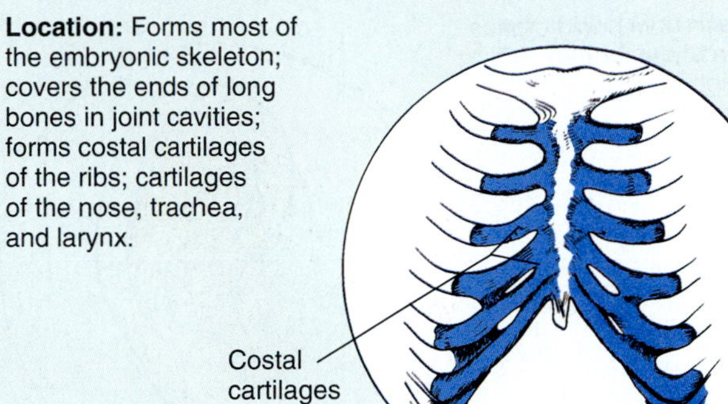

Function: Supports and reinforces; has resilient cushioning properties; resists compressive stress.

Photomicrograph: Hyaline cartilage from the trachea (469×).

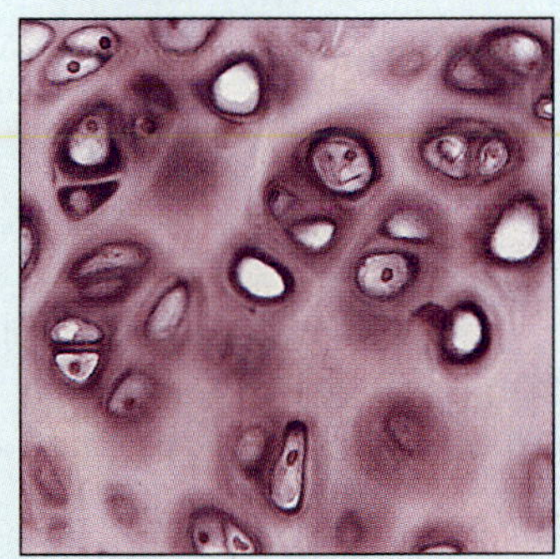

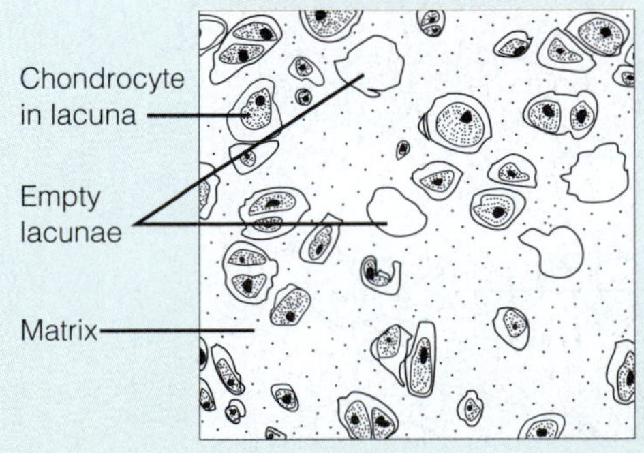

F6.5 *(continued)*

(h) Elastic cartilage

Description: Similar to hyaline cartilage, but more elastic fibers in matrix.

Location: Supports the external ear (pinna); epiglottis.

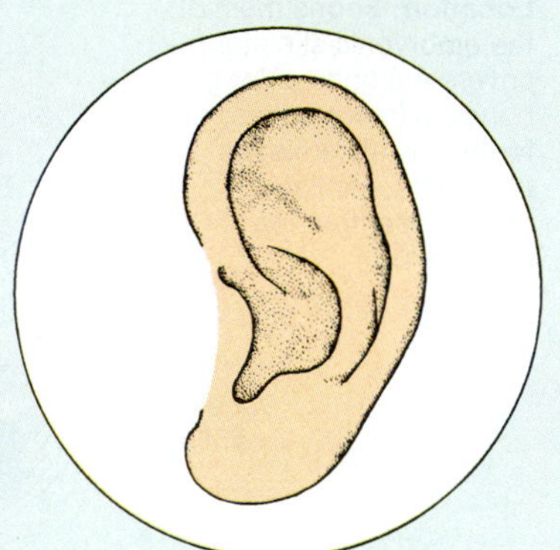

Function: Maintains the shape of a structure while allowing great flexibility.

Photomicrograph: Elastic cartilage from the human ear pinna; forms the flexible skeleton of the ear (144×).

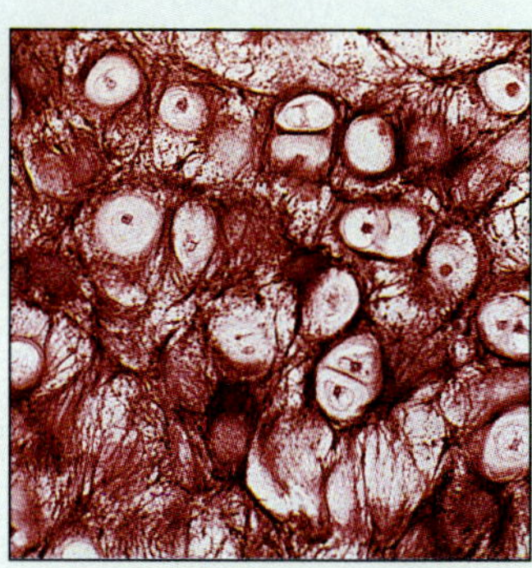

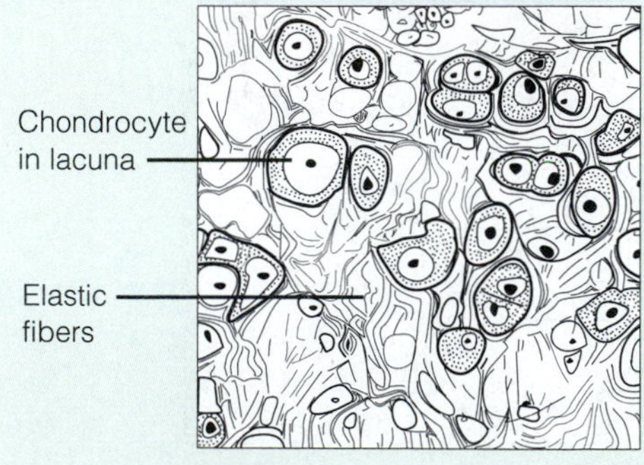

(i) Fibrocartilage

Description: Matrix similar but less firm than in hyaline cartilage; thick collagen fibers predominate.

Location: Intervertebral discs; pubic symphysis; discs of knee joint.

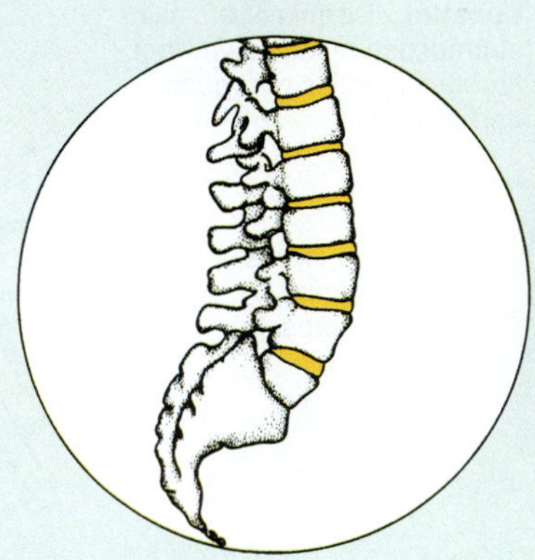

Function: Tensile strength with the ability to absorb compressive shock.

Photomicrograph: Fibrocartilage of an intervertebral disc (823×).

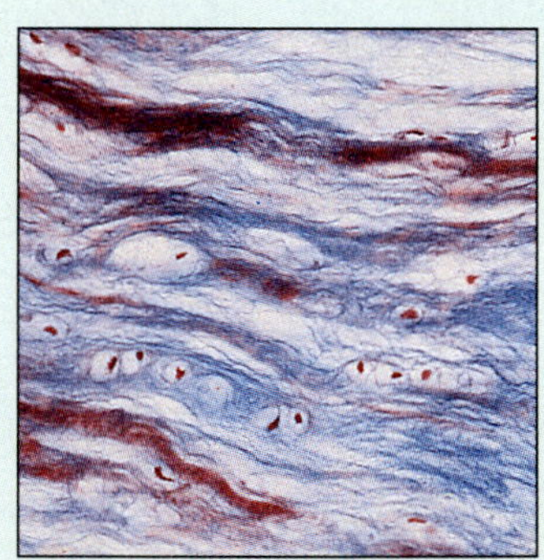

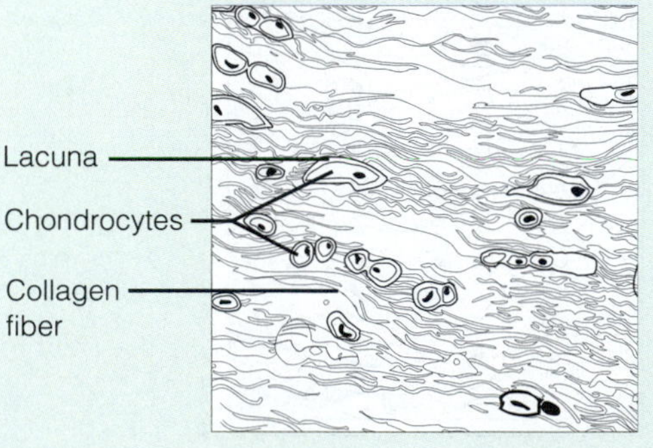

F6.5 (*continued*)

Others: (j and k)

(j) Bone (osseous tissue)

Description: Hard, calcified matrix containing many collagen fibers; osteocytes lie in lacunae. Very well vascularized.

Location: Bones

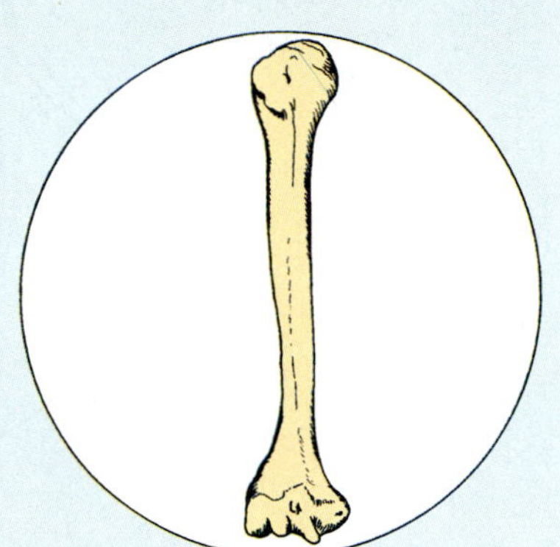

Function: Bone supports and protects (by enclosing); provides levers for the muscles to act on; stores calcium and other minerals and fat; marrow inside bones is the site for blood cell formation (hematopoiesis).

Photomicrograph: Cross-sectional view of bone (144×).

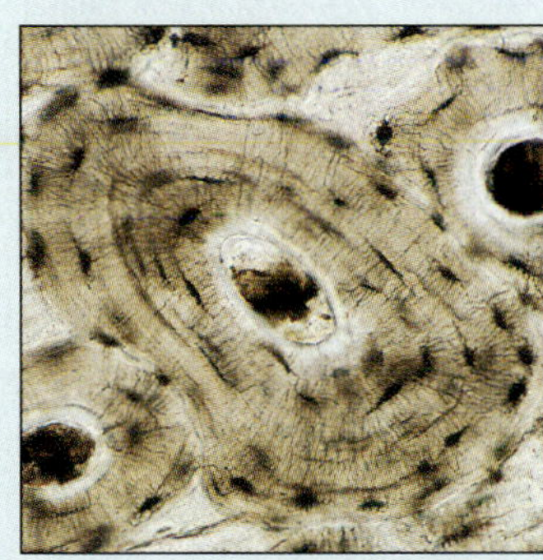

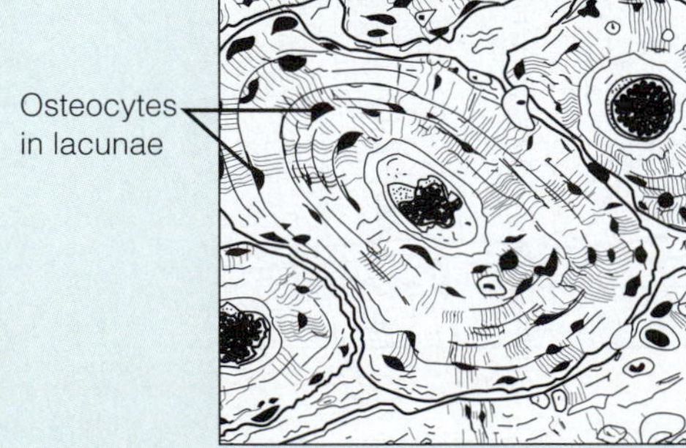

(k) Blood

Description: Red and white blood cells in a fluid matrix (plasma).

Location: Contained within blood vessels.

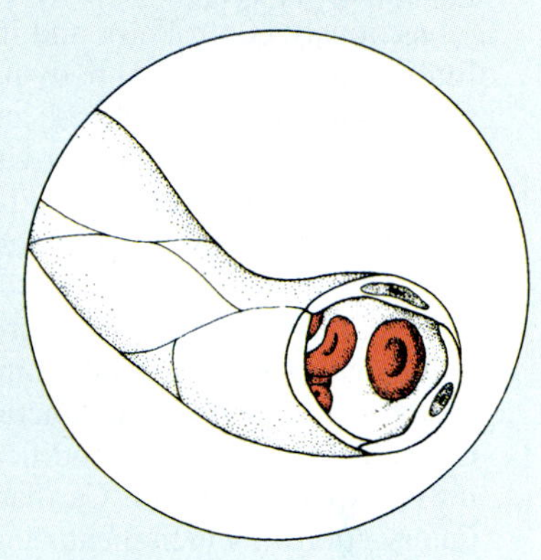

Function: Transport of respiratory gases, nutrients, wastes, and other substances.

Photomicrograph: Smear of human blood (1076×); two white blood cells (neutrophils) are seen surrounded by red blood cells.

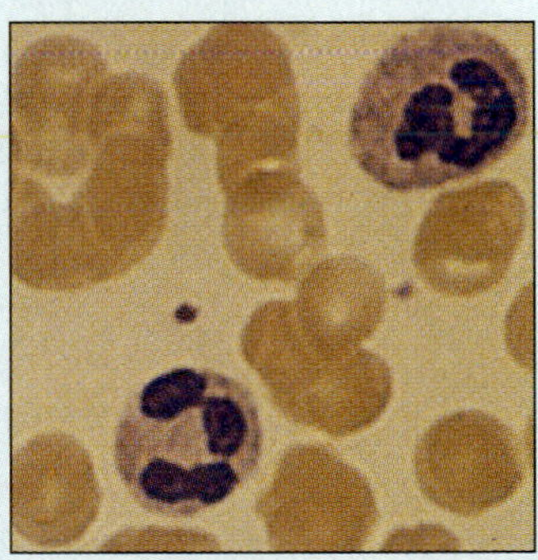

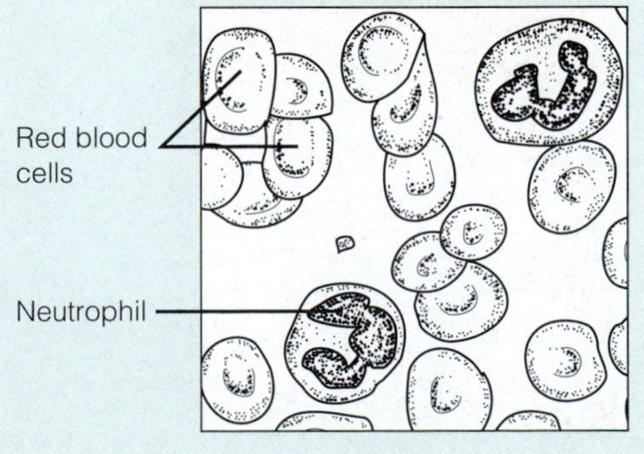

F6.5 *(continued)*

MUSCLE TISSUE

Muscle tissue is highly specialized to contract (shorten) in order to produce movement of some body parts. As you might expect, muscle cells tend to be quite elongated, providing a long axis for contraction. The three basic types of muscle tissue are described briefly here; cardiac muscle and skeletal muscle are treated more completely in later exercises.

Skeletal muscle, the "meat," or flesh, of the body, is attached to the skeleton. It is under voluntary control (consciously controlled), and its contraction moves the limbs and other external body parts. The cells of skeletal muscles are long, cylindrical, and multinucleate (several nuclei per cell); they have obvious *striations* (stripes).

Cardiac muscle is found only in the heart. As it contracts, the heart acts as a pump, propelling the blood through the blood vessels. Cardiac muscle, like skeletal muscle, has striations. But cardiac cells are branching uninucleate (or occasionally binucleate) cells that interdigitate (fit together) at junctions called **intercalated discs.** These structural modifications allow the cardiac muscle to act as a unit. Cardiac muscle is under involuntary control, which means that we cannot voluntarily or consciously control the operation of the heart.

Smooth muscle, or *visceral muscle,* is found mainly in the walls of hollow organs (digestive and urinary tract organs, uterus, blood vessels). Typically there are two layers that run at right angles to each other; consequently its contraction can constrict or dilate the lumen (cavity) of an organ and propel substances along predetermined pathways. Smooth muscle cells are quite different in appearance from those of skeletal or cardiac muscle. No striations are visible, and the uninucleate smooth muscle cells are spindle-shaped.

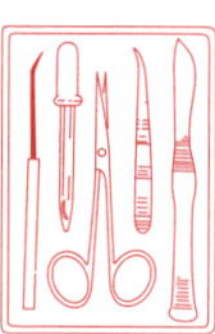

Obtain and examine prepared slides of skeletal, cardiac, and smooth muscle. Notice their similarities and dissimilarities in both your observations and in the illustrations in Figure 6.6.

(a) Skeletal muscle

Description: Long, cylindrical, multinucleate cells; obvious striations.

Location: In skeletal muscles attached to bones or occasionally to skin.

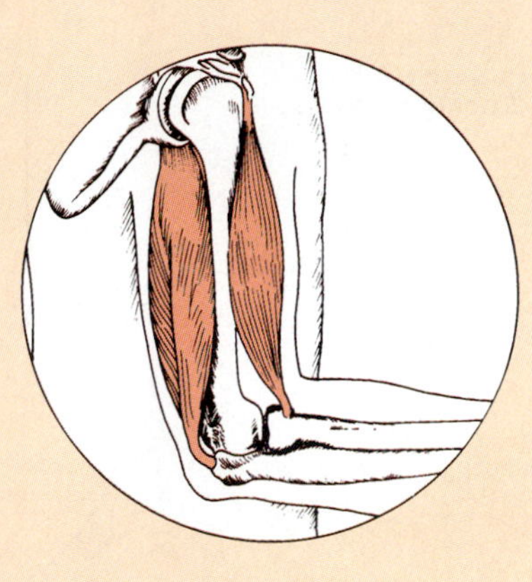

Function: Voluntary movement; locomotion; manipulation of the environment; facial expression. Voluntary control.

Photomicrograph: Skeletal muscle (approx. 576×). Notice the obvious banding pattern and the fact that these large cells are multinucleate.

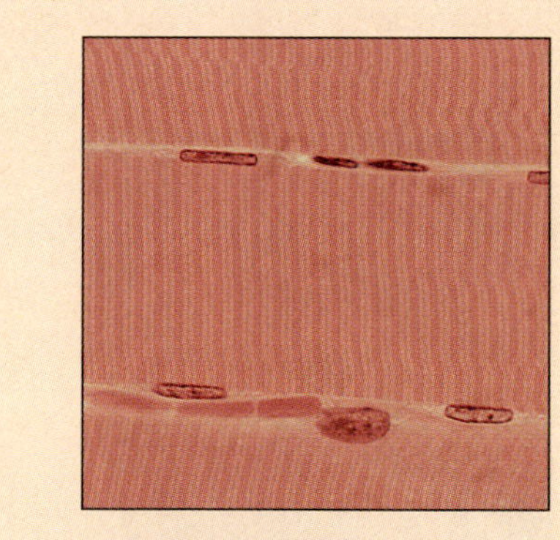

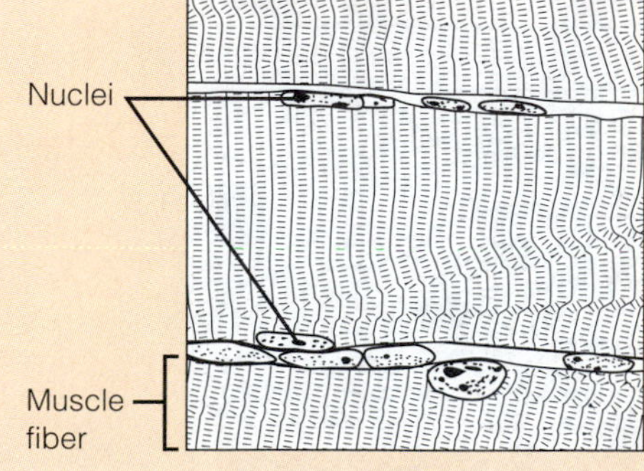

F6.6

Muscle tissues.

(b) Cardiac muscle

Description: Branching, striated, generally uninucleate cells that interdigitate at specialized junctions (intercalated discs).

Location: The walls of the heart.

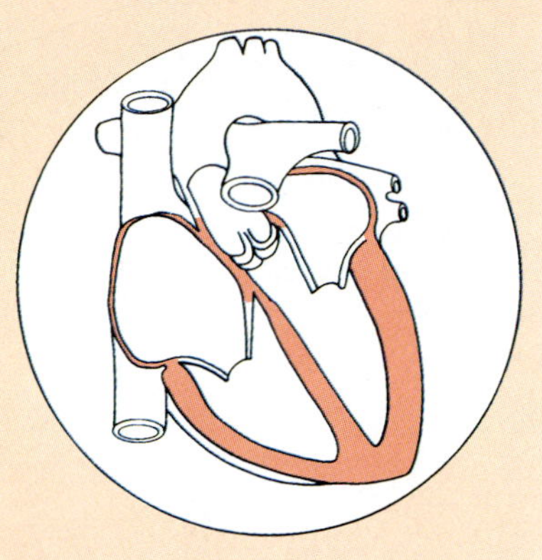

Function: As it contracts, it propels blood into the circulation; involuntary control.

Photomicrograph: Cardiac muscle (576×); notice the striations, branching of fibers, and the intercalated discs.

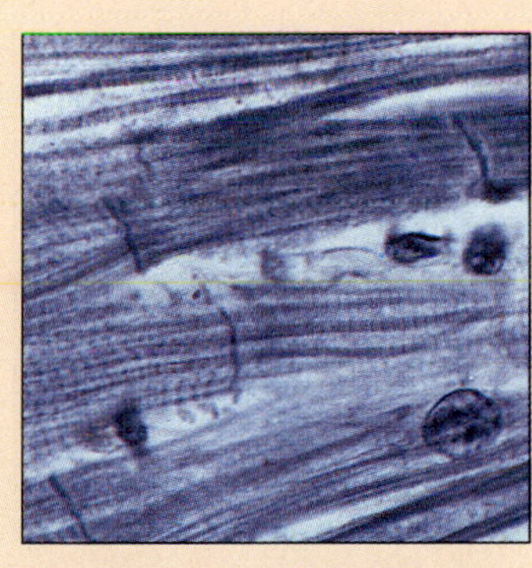

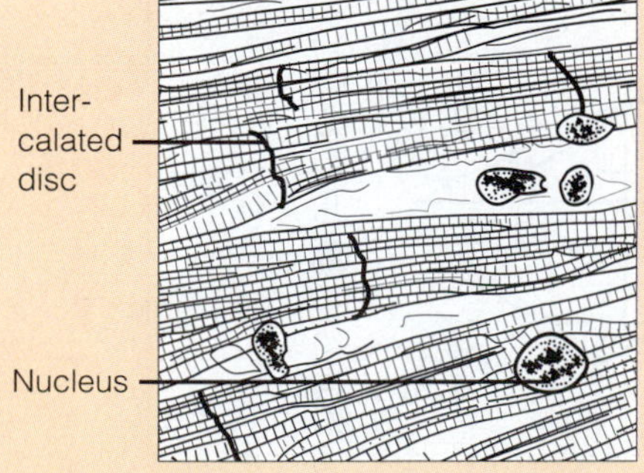

(c) Smooth muscle

Description: Spindle-shaped cells with central nuclei; cells arranged closely to form sheets; no striations.

Location: Mostly in the walls of hollow organs.

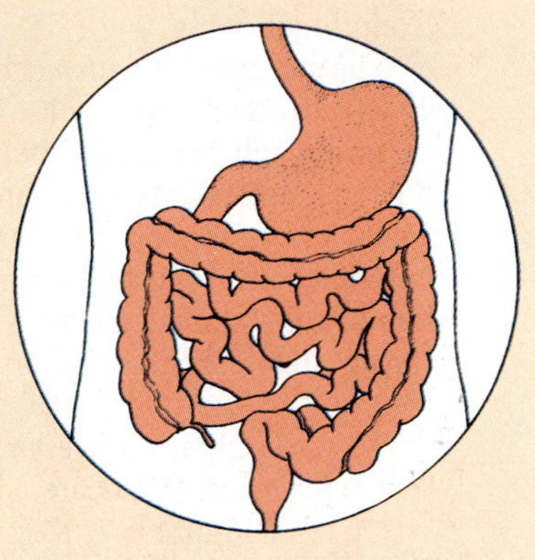

Function: Propels substances or objects (foodstuffs, urine, a baby) along internal passageways; involuntary control.

Photomicrograph: Sheet of smooth muscle (324×).

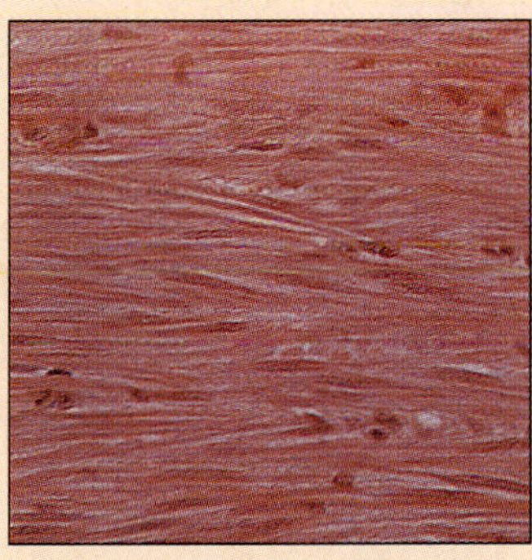

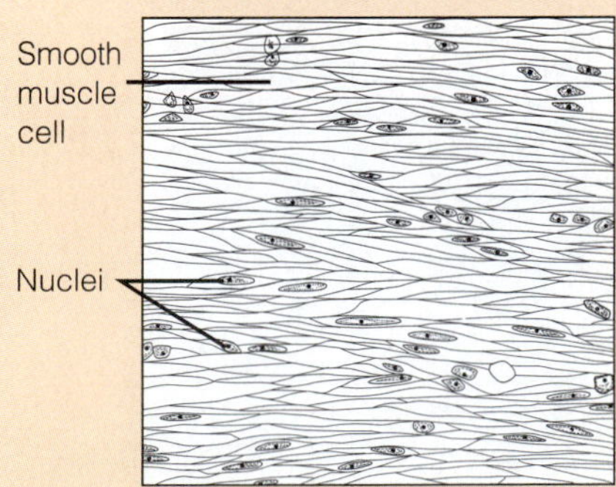

F6.6 (*continued*)

NERVOUS TISSUE

Nervous tissue is composed of two major cell populations. The **neuroglia** are special supporting cells that protect, support, and insulate the more delicate neurons. The **neurons** are highly specialized to receive stimuli (irritability) and to conduct waves of excitation, or impulses, to all parts of the body (conductivity). They are the cells that are most often associated with nervous system functioning.

The structure of neurons is markedly different from that of all other body cells. They all have a nucleus-containing cell body, and their cytoplasm is drawn out into long extensions (cell processes)—sometimes as long as 3 feet (about 1 m), which allows a single neuron to conduct an impulse over relatively long distances. More detail about the anatomy of the different classes of neurons and neuroglia appears in Exercise 17.

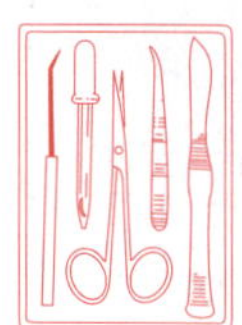

Obtain a prepared slide of a spinal cord smear. Locate a neuron and compare it to Figure 6.7. Keep the light dim—this will help you see the cellular extensions of the neurons. Also see Plates 4 and 5 in the Histology Atlas.

Description: Neurons are branching cells; cell processes that may be quite long extend from the nucleus-containing cell body; nonirritable supporting cells (not illustrated) also contribute to nervous tissue.

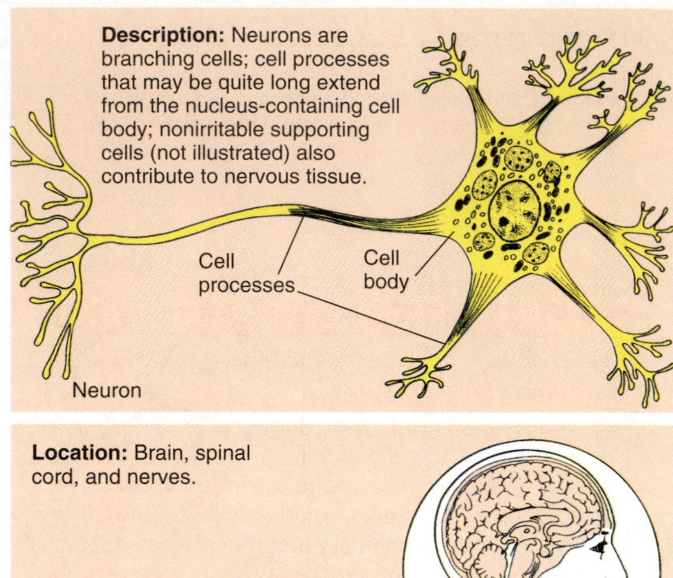

Location: Brain, spinal cord, and nerves.

Function: Transmit electrical signals from sensory receptors and to effectors (muscles and glands) which control their activity.

Photomicrograph: Neuron (170×).

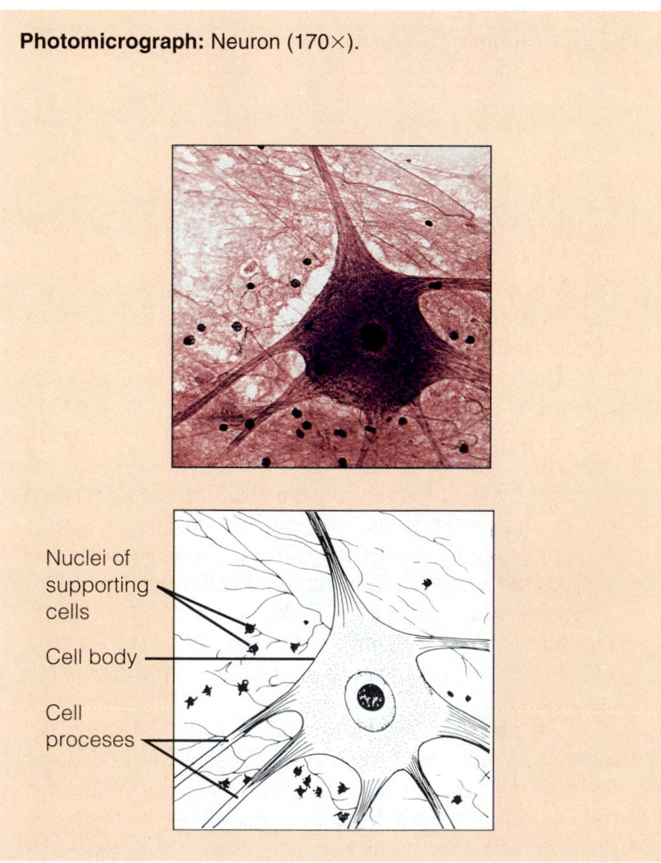

F6.7

Nervous tissue.

The Integumentary System

OBJECTIVES

1. To recount several important functions of the skin, or integumentary system.
2. To recognize and name during observation of an appropriate model, diagram, projected slide, or microscopic specimen the following skin structures: epidermis (and note relative positioning of its strata), dermis (papillary and reticular layers), hair follicles and hair, sebaceous glands, and sweat glands.
3. To name the layers of the epidermis and describe the characteristics of each.
4. To compare the properties of the epidermis to those of the dermis.
5. To describe the distribution and function of the skin derivatives—sebaceous glands, sweat glands, and hairs.
6. To differentiate between eccrine and apocrine sweat glands.
7. To enumerate the factors determining skin color.
8. To describe the function of melanin.
9. To identify the major regions of nails.

MATERIALS

Skin model (three-dimensional, if available)
Compound microscope
Prepared slide of human skin with hair follicles
Sheet of #20 bond paper ruled to mark off cm^2 areas
Scissors
Betadine swabs, or Lugol's iodine and cotton swabs
Adhesive tape

See Appendix E, Exercise 7 for links to *Anatomy and PhysioShow: The Videodisc.*

The **skin,** or **integument,** is often considered an organ system because of its extent and complexity. It is much more than an external body covering; architecturally, the skin is a marvel. It is tough yet pliable, a characteristic that enables it to withstand constant insult from outside agents.

The skin has many functions, most (but not all) concerned with protection. It insulates and cushions the underlying body tissues and protects the entire body from mechanical damage (bumps and cuts), chemical damage (acids, alkalis, and the like), thermal damage (heat), and bacterial invasion (by virtue of its acid mantle and continuous surface). The hardened uppermost layer of the skin (the cornified layer) prevents water loss from the body surface. The skin's abundant capillary network (under the control of the nervous system) plays an important role in regulating heat loss from the body surface.

The skin has other functions as well. For example, it acts as a mini-excretory system; urea, salts, and water are lost through the skin pores in sweat. The skin also has important metabolic duties. For example, like liver cells, it carries out some chemical conversions that activate or inactivate certain drugs and hormones, and it is the site of vitamin D synthesis for the body. Finally, the cutaneous sense organs are located in the dermis.

BASIC STRUCTURE OF THE SKIN

The skin has two distinct regions—the superficial *epidermis* composed of epithelium and an underlying connective tissue *dermis.* These layers are firmly "cemented" together along an undulating border. But friction, such as the rubbing of a poorly fitting shoe, may cause them to separate, resulting in a blister. Immediately deep to the dermis is the **hypodermis** or **superficial fascia** (primarily adipose tissue), which is not considered part of the skin. The main skin areas and structures are described below.

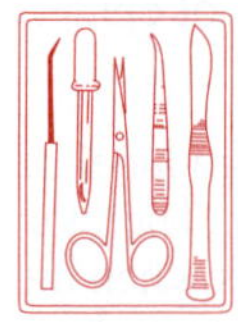

As you read, locate the following structures on Figure 7.1 and on a skin model.

Epidermis

Structurally, the avascular epidermis is a keratinized stratified squamous epithelium consisting of four distinct cell types and four or five distinct layers.

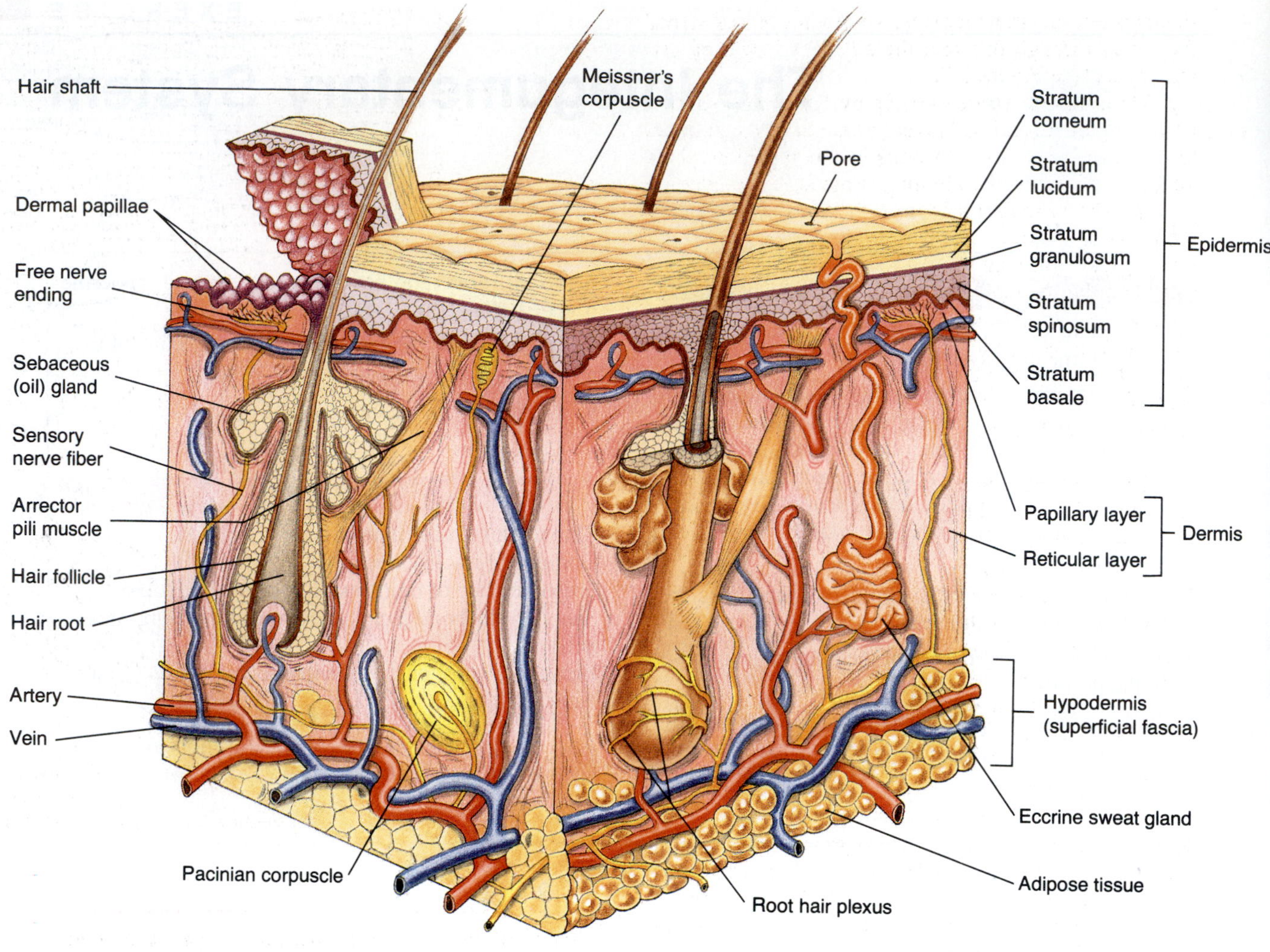

F7.1

Skin structure. Three-dimensional view of the skin and the underlying hypodermis. The epidermis and dermis have been pulled apart at the left corner to reveal the dermal papillae.

CELLS OF THE EPIDERMIS Most epidermal cells are **keratinocytes** (literally, keratin cells), epithelial cells that function mainly to produce keratin fibrils. **Keratin** is a fibrous protein that gives the epidermis its durability and protective capabilities.

Far less abundant are the following types of epidermal cells (Figure 7.2):

- **Melanocytes**—spidery black cells that produce the brown-to-black pigment called **melanin.** The skin tans because melanin production increases when the skin is exposed to sunlight. The melanin provides a protective pigment umbrella over the nuclei of the cells in the deeper epidermal layers, thus shielding their genetic material (DNA) from the damaging effects of ultraviolet radiation. A concentration of melanin in one spot is called a *freckle*.
- **Langerhans' cells**—phagocytic cells (macrophages) that play a role in immunity.
- **Merkel cells**—in conjunction with sensory nerve endings, Merkel cells form sensitive touch receptors called *Merkel discs* located at the epidermal-dermal junction.

LAYERS OF THE EPIDERMIS From deep to superficial, the layers of the epidermis are the stratum basale, stratum spinosum, stratum granulosum, stratum lucidum, and stratum corneum (Figure 7.1).

The **stratum basale** (basal layer) is a single row of cells immediately adjacent to the dermis. Its cells are constantly undergoing mitotic cell division to produce millions of new cells daily, hence its alternate name *stratum germinativum.* About a quarter of the cells in this stratum are melanocytes, whose processes thread their way through this and the adjacent layers of keratinocytes (see Figure 7.2).

The **stratum spinosum** (spiny layer) is a stratum consisting of several cell layers immediately superficial to the basal layer. Its cells contain *tonofilaments*, thick bundles of intermediate filaments made of a prekeratin protein. The stratum spinosum cells appear spiky (hence their name) because, as the skin tissue is prepared for histological examination, they shrink but their desmosomes hold tight. Cells divide fairly rapidly in this stratum, but less so than in the stratum basale. Cells in the basal and spiny layers are the only ones to receive adequate nourishment (via diffusion of nutrients from the dermis). So as their daughter cells are pushed upward and away from the source of nutrition, they gradually die.

The **stratum granulosum** (granular layer) is a thin layer named for the abundant granules its cells contain. These granules are of two types: (1) *laminated granules*, which contain a waterproofing lipid that is secreted into the extracellular space; and (2) *keratohyalin granules*, which combine with the tonofilaments in the more superficial layers to form the keratin fibrils. At the upper border of this layer, the cells are beginning to die.

The **stratum lucidum** (clear layer) is a very thin translucent band of flattened dead keratinocytes with indistinct boundaries. It is not present in thin skin.

The outermost epidermal layer, the **stratum corneum** (horny layer), consists of some 20 to 30 cell layers, and accounts for the bulk of the epidermal thickness. Cells in this layer, like those in the stratum lucidum (where it exists), are dead and their flattened scalelike remnants are fully keratinized. They are constantly rubbing off and being replaced by division of the deeper cells.

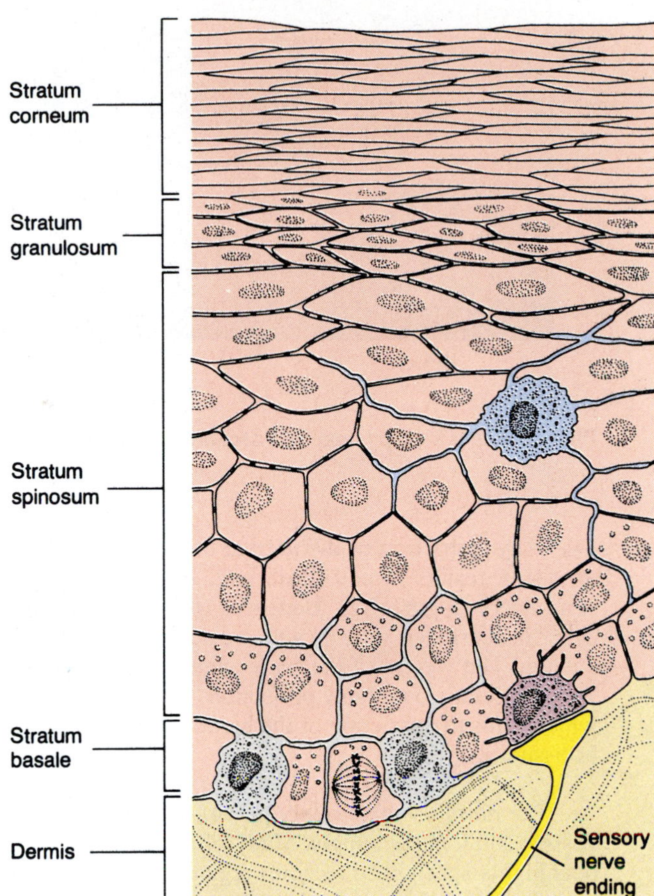

F7.2

Diagram showing the main features—layers and relative numbers of the different cell types—in epidermis of thin skin. The keratinocytes (pink) form the bulk of the epidermis. Less numerous are the melanocytes (gray), which produce the pigment melanin; Langerhans' cells (blue), which function as macrophages; and Merkel cells (purple). A sensory nerve ending (yellow), extending from the dermis, is depicted in association with the Merkel cell forming a Merkel disc (touch receptor). Notice that the keratinocytes, but not the other cell types, are joined by numerous desmosomes.

Dermis

The dense irregular connective tissue making up the dermis consists of two principal regions—the papillary and reticular areas. Like the epidermis, the dermis varies in thickness. For example, the skin is particularly thick on the palms of the hands and soles of the feet and is quite thin on the eyelids.

The **papillary layer** is the more superficial dermal region. It is very uneven and has fingerlike projections from its superior surface, the **dermal papillae,** which attach it to the epidermis above. These projections produce unique patterns of ridges that remain unchanged throughout life, which are reflected in fingerprints. Abundant capillary networks in the papillary layer furnish nutrients for the epidermal layers and allow heat to radiate to the skin surface. The pain and touch receptors (Meissner's corpuscles) are also found here.

The **reticular layer** is the deepest skin layer. It contains many arteries and veins, sweat and sebaceous glands, and pressure receptors.

Both the papillary and reticular layers are heavily invested with collagenic and elastic fibers. The elastic fibers give skin its exceptional elasticity in youth. In old age, the number of elastic fibers decreases and the subcutaneous layer loses fat, which leads to wrinkling and inelasticity of the skin. Fibroblasts, adipose cells, various types of macrophages (which are important in the body's defense), and other cell types are found throughout the dermis.

The abundant dermal blood supply allows the skin to play a role in the regulation of body temperature. When body temperature is high, the arterioles serving the skin dilate, and the capillary network of the dermis becomes engorged with the heated blood. Thus body heat is allowed to radiate from the skin surface. If the environment is cool and body heat must be conserved, the arterioles constrict so that blood bypasses the dermal capillary networks.

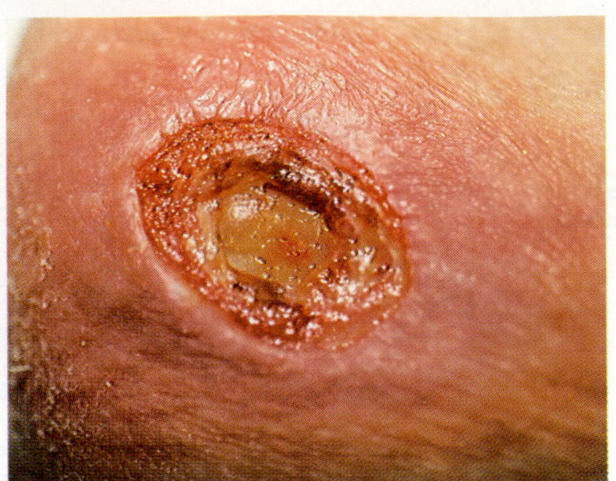

F7.3

Photograph of a deep (stage III) decubitus ulcer.

Any restriction of the normal blood supply to the skin results in cell death and, if severe enough, skin ulcers (Figure 7.3). **Bedsores (decubitus ulcers)** occur in bedridden patients who are not turned regularly enough. The weight of the body exerts pressure on the skin, especially over bony projections (hips, heels, etc.), which leads to restriction of the blood supply and death of tissue. ■

The dermis is also richly provided with lymphatic vessels and a nerve supply. Many of the nerve endings bear highly specialized receptor organs that, when stimulated by environmental changes, transmit messages to the central nervous system for interpretation. Some of these receptors—bare nerve endings (pain receptors), a Meissner's corpuscle, a Pacinian corpuscle, and a root hair plexus—are shown in Figure 7.1. (These receptors are discussed in depth in Exercise 23.)

Skin Color

Skin color is a result of three factors—the relative amount of two pigments (melanin and carotene) in skin and the degree of oxygenation of the blood. People who produce large amounts of melanin have brown-toned skin. In light-skinned people, who have less melanin, the dermal blood supply flushes through the rather transparent cell layers above, giving the skin a rosy glow. *Carotene* is a yellow-orange pigment present primarily in the stratum corneum and in the adipose tissue of the hypodermis. Its presence is most noticeable when large amounts of carotene-rich foods (carrots, for instance) are eaten.

Skin color may be an important diagnostic tool. For example, flushed skin may indicate hypertension, fever, or embarrassment, whereas pale skin is typically seen in anemic individuals. When the blood is inadequately oxygenated, as during asphyxiation and serious lung disease, both the blood and the skin take on a bluish or cyanotic cast. **Jaundice,** in which the tissues become yellowed, is almost always diagnostic for liver disease, whereas a bronzing of the skin hints that a person's adrenal cortex is hypoactive **(Addison's disease).** ■

APPENDAGES OF THE SKIN

The appendages of the skin—hair, nails, and cutaneous glands—are all derivatives of the epidermis, but they reside in the dermis. They originate from the stratum basale and grow downward into the deeper skin regions.

Cutaneous Glands

The cutaneous glands fall primarily into two categories: the sebaceous glands and the sweat glands (Figure 7.1). The **sebaceous glands** are found nearly all over the skin, except for the palms of the hands and the soles of the feet. Their ducts usually empty into a hair follicle, but some open directly onto the skin surface.

The product of the sebaceous glands, called **sebum,** is a mixture of oily substances and fragmented cells. The sebum is a lubricant that keeps the skin soft and moist (a natural skin cream) and keeps the hair from becoming brittle. The sebaceous glands become particularly active during puberty when more male hormones (androgens) begin to be produced; thus, the skin tends to become oilier during this period of life. *Blackheads* are accumulations of dried sebum and bacteria; *acne* is due to active infection of the sebaceous glands.

Epithelial openings, called *pores*, are the outlets for the **sweat (sudoriferous) glands.** These exocrine glands are widely distributed in the skin. Sweat glands are subcategorized by the composition of their secretions. The **eccrine glands,** which are distributed all over the body, produce clear perspiration, consisting primarily of water, salts (NaCl), and urea. The **apocrine glands,** found predominantly in the axillary and genital areas, secrete a milky protein- and fat-rich substance (also containing water, salts, and urea) that is an excellent nutrient medium for the microorganisms typically found on the skin.

The sweat glands, under the control of the nervous system, are an important part of the body's heat-regulating apparatus. They secrete perspiration when the external temperature or body temperature is high. When this water-based substance evaporates, it carries excess body heat with it. Thus evaporation of greater amounts of perspiration provides an efficient means of dissipating body heat when the capillary cooling system is not sufficient or is unable to maintain body temperature homeostasis.

Hair

Hairs are found over the entire body surface, except for thick-skinned areas (the palms of the hands, the soles of the feet), parts of the external genitalia, the nipples, and the lips. A hair, enclosed in a hair **follicle,** is also an epithelial structure (Figure 7.4). The portion of the hair enclosed within the follicle is called the **root;** the portion projecting from the scalp surface is called the **shaft.** The hair is formed by mitosis of the well-nourished germinal epithelial cells at the basal end of the follicle (the **hair bulb**). As the daughter cells are pushed farther away from the growing region, they die and become keratinized; thus the bulk of the hair shaft, like the bulk of the epidermis, is dead material.

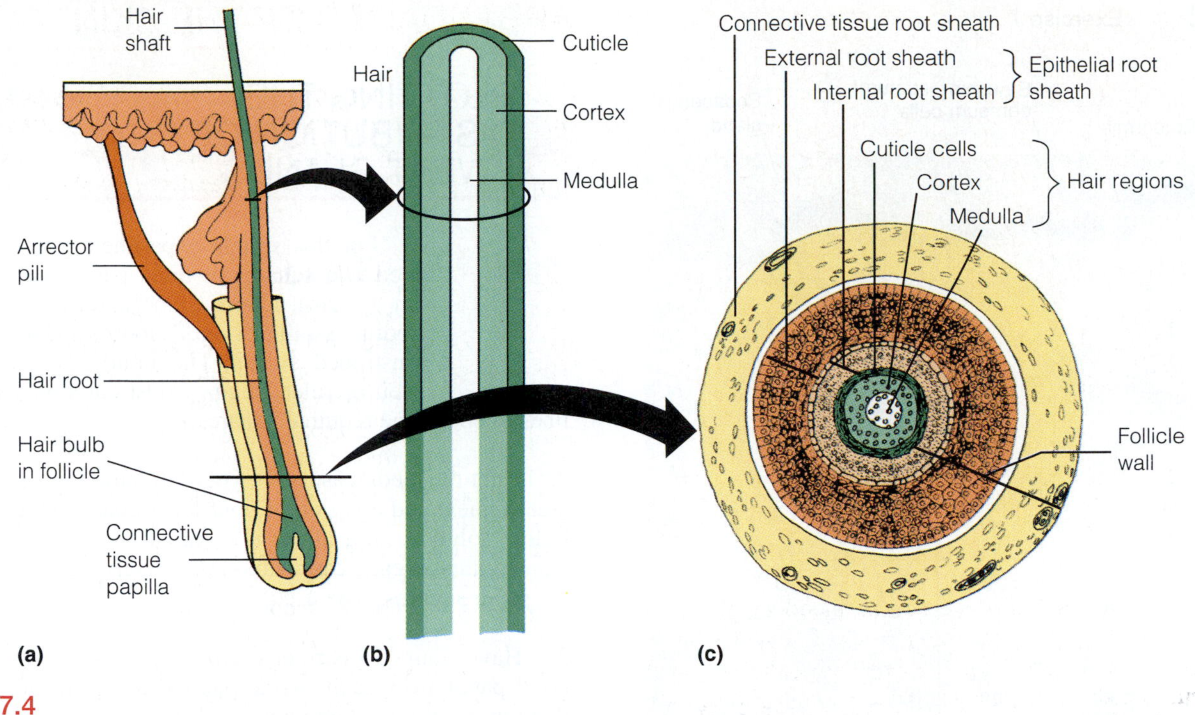

F7.4

Structure of a hair and hair follicle. **(a)** Longitudinal section of a hair within its follicle. **(b)** Enlarged longitudinal section of a hair. **(c)** Cross section of a hair and hair follicle.

A hair (Figure 7.4b) consists of a central region (medulla) surrounded first by the cortex and then by a protective cuticle. Abrasion of the cuticle results in "split ends." Hair color is a manifestation of the amount and kind of melanin pigment within the hair cortex.

The hair follicle is structured from both epidermal and dermal cells (Figure 7.4c). Its inner *epithelial root sheath*, with two parts (internal and external), is enclosed by the *connective tissue root sheath*, which is essentially dermal tissue. A small nipple of dermal tissue, the *connective tissue papilla*, protrudes into the hair bulb from the connective tissue sheath and provides nutrition to the growing hair. If you look carefully at the structure of the hair follicle (see Figure 7.1), you will see that it generally is in a slanted position. Small bands of smooth muscle cells—**arrector pili**—connect each hair follicle to the papillary layer of the dermis. When these muscles contract (during cold or fright), the hair follicle is pulled upright, dimpling the skin surface with "goose bumps." This phenomenon is especially dramatic in a scared cat, whose fur actually stands on end to increase its apparent size. The activity of the arrector pili muscles also exerts pressure on the sebaceous glands surrounding the follicle, causing a small amount of sebum to be released.

Nails

Nails, the hornlike derivatives of the epidermis, consist of a *free edge*, a *body* (visible attached portion), and a *root* (embedded in the skin and adhering to an epithelial **nail bed**).The borders of the nail are overlapped by skin folds called **nail folds;** the thick proximal nail fold is the **eponychium,** commonly called the cuticle (Figure 7.5).

The germinal cells in the **nail matrix,** the thickened proximal part of the nail bed, are responsible for nail growth. As the nail cells are produced by the matrix, they become heavily keratinized and die. Thus, nails, like hairs, are mostly nonliving material.

Nails are transparent and nearly colorless, but they appear pink because of the blood supply in the underlying dermis. The exception to this is the proximal region of the thickened nail matrix which appears as a white crescent called the *lunula.* When someone is cyanotic due to a lack of oxygen in the blood, the nail beds take on a blue cast.

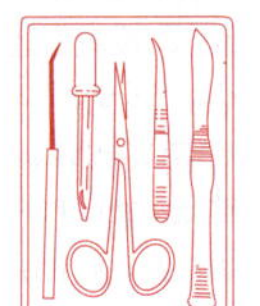

Identify the nail structures shown in Figure 7.5 on yourself or your lab partner.

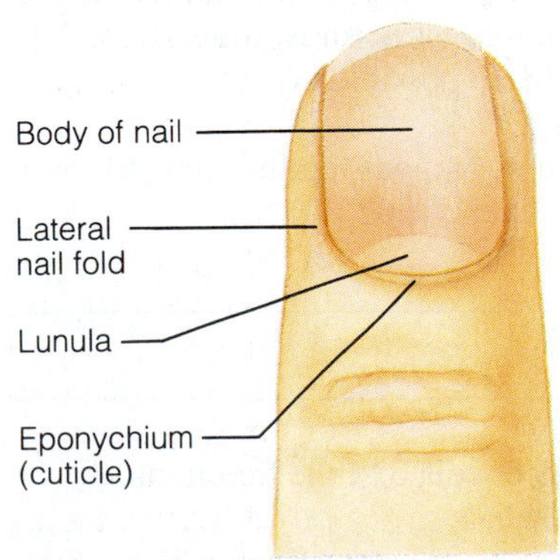

F7.5

Structure of a nail.

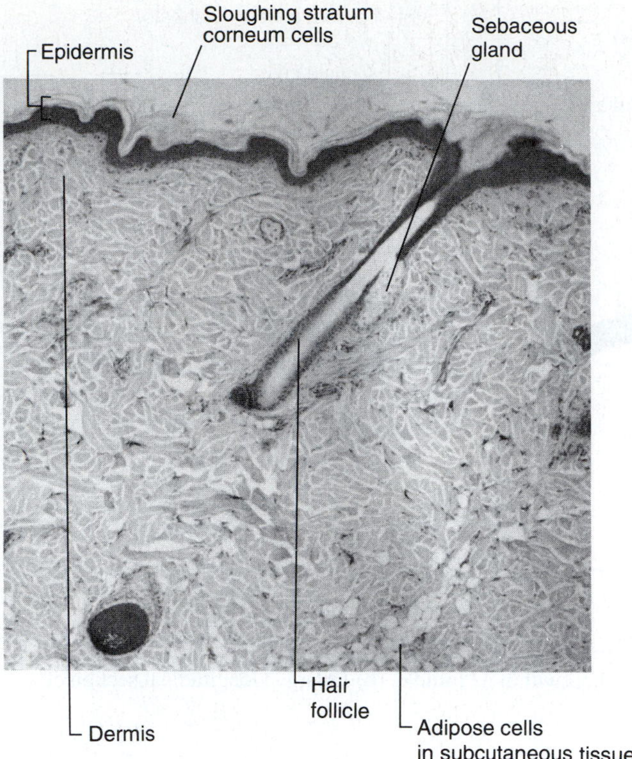

F7.6

Photomicrograph of skin (20×).

EXAMINATION OF THE MICROSCOPIC STRUCTURE OF THE SKIN

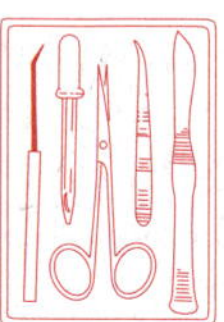

Obtain a prepared slide of human skin, and study it carefully under the microscope. Compare your tissue slide to the view shown in Figure 7.6, and identify as many of the structures diagrammed in Figure 7.1 as possible.

How is this stratified squamous epithelium different from that observed in Exercise 6?

___

___

How do these differences relate to the functions of these two similar epithelia?

___

___

PLOTTING THE DISTRIBUTION OF SWEAT GLANDS

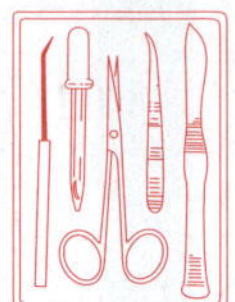

1. For this simple experiment you will need two squares of bond paper (each 1 cm × 1 cm), adhesive tape, and a betadine (iodine) swab *or* Lugol's iodine and a cotton-tipped swab. (The bond paper has been preruled in cm^2—just cut along the lines to obtain the required squares.)

2. Paint the medial aspect of your left palm (avoid the crease lines) and a region of your left forearm with the iodine solution, and allow it to dry thoroughly. The painted area in each case should be slightly larger than the paper squares to be used.

3. Have your lab partner *securely* tape a square of bond paper over each iodine-painted area, and leave them in place for 20 minutes. (If it is very warm in the laboratory while this test is being conducted, good results may be obtained within 10 to 15 minutes.)

4. After 20 minutes, remove the paper squares, and count the number of blue-black dots on each square. The presence of a blue-black dot on the paper indicates an active sweat gland. (The iodine in the pore is dissolved in the sweat and reacts chemically with the starch in the bond paper to produce the blue-black color.) Thus "sweat maps" have been produced for the two skin areas.

5. Which skin area tested has the greater density of sweat glands?

___

___

___

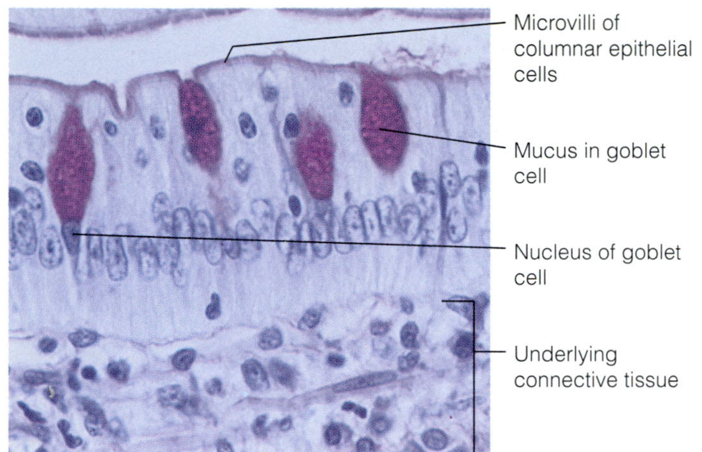

PLATE 1 Simple columnar epithelium. Mucus in goblet cells stains pink in this view (543X)

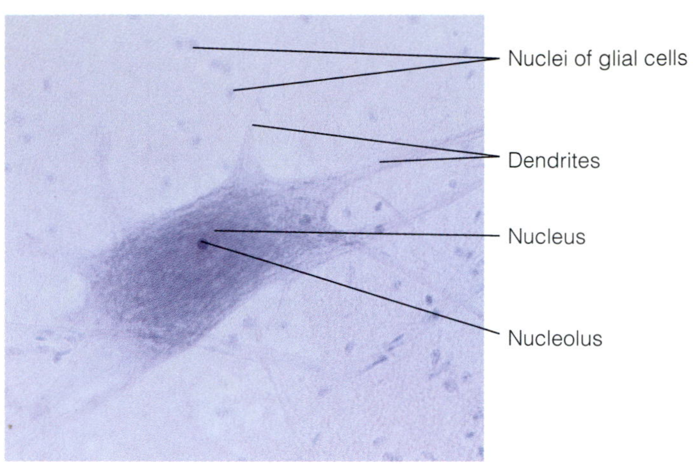

PLATE 4 Multipolar neuron in spinal cord smear (212X)

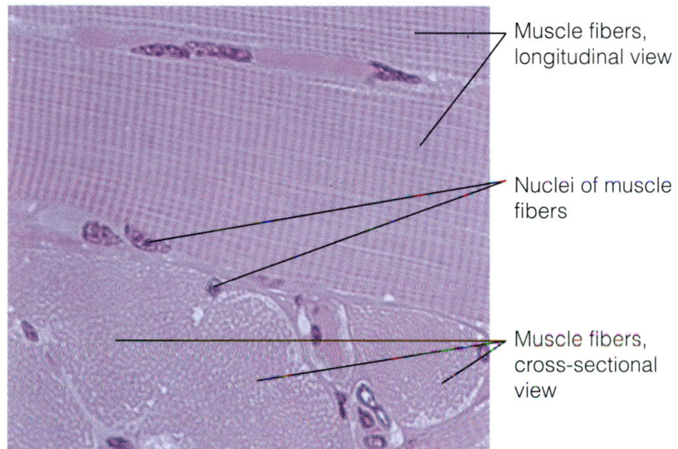

PLATE 2 Skeletal muscle, transverse and longitudinal views shown (543X)

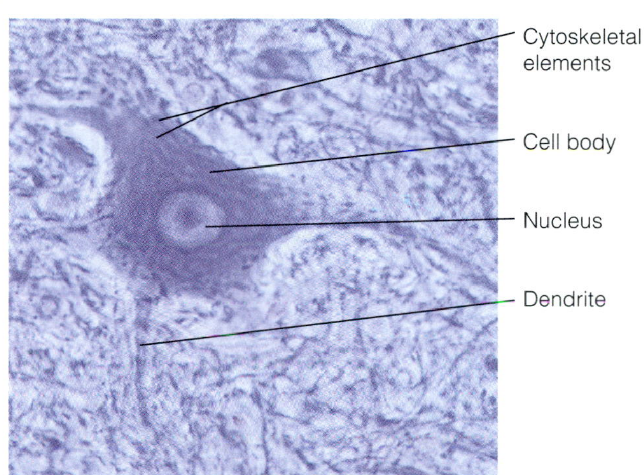

PLATE 5 Neuron stained to allow cytoskeletal elements in their processes to be seen (265X)

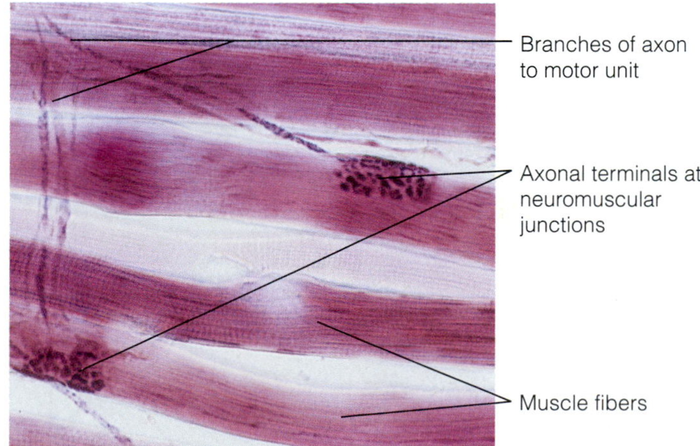

PLATE 3 Part of a motor unit (265X)

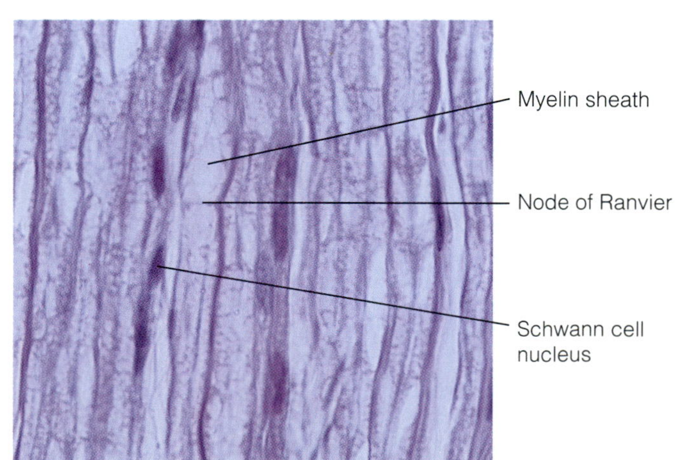

PLATE 6 Longitudinal view of myelinated axons (543X). Myelin sheaths appear "bubbly" because most of the fatty myelin is dissolved during slide preparation

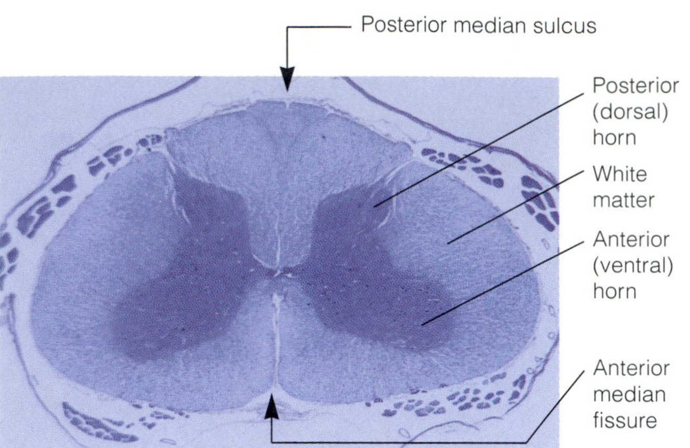

PLATE 7 Adult spinal cord, cross-sectional view (12X)

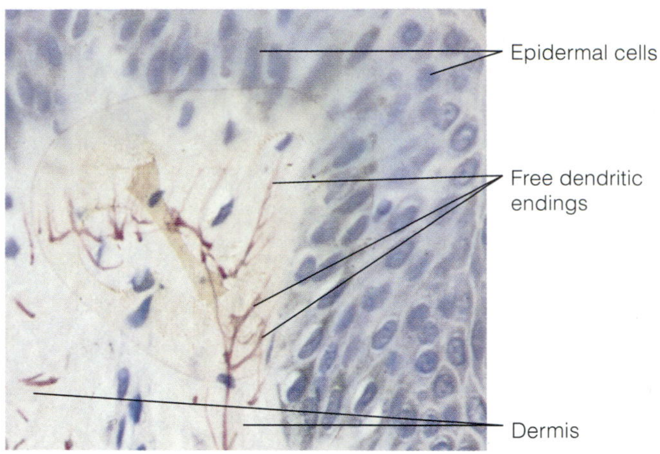

PLATE 10 Free dendritic endings at the dermal–epidermal junction (480X)

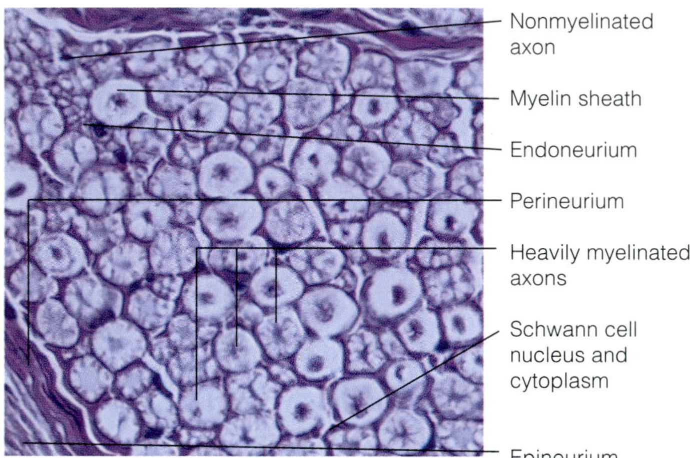

PLATE 8 Cross section of a portion of a peripheral nerve (543X). Heavily myelinated fibers are identified by a centrally-located axon, surrounded by an unstained ring of myelin, and then a peripheral rim of pink-staining Schwann cell

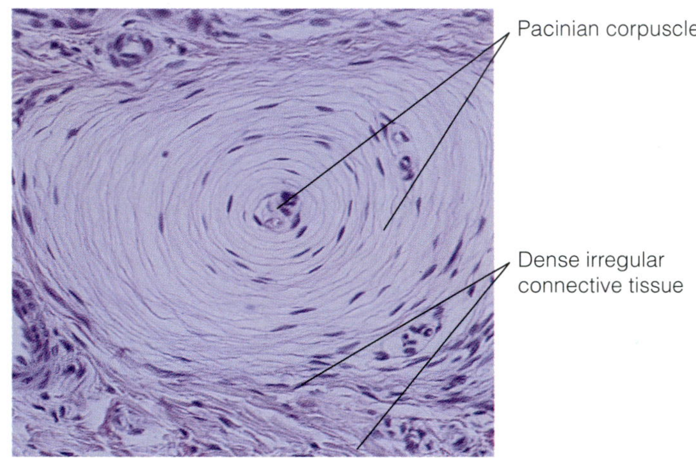

PLATE 11 Pacinian corpuscle in the hypodermis (212X)

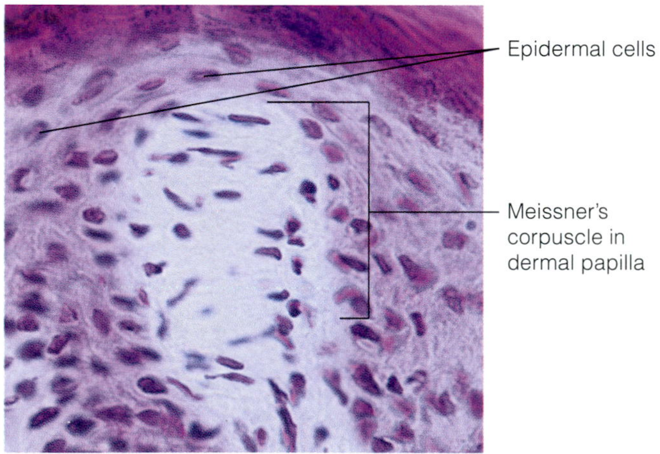

PLATE 9 Meissner's corpuscle in a dermal papilla (559X)

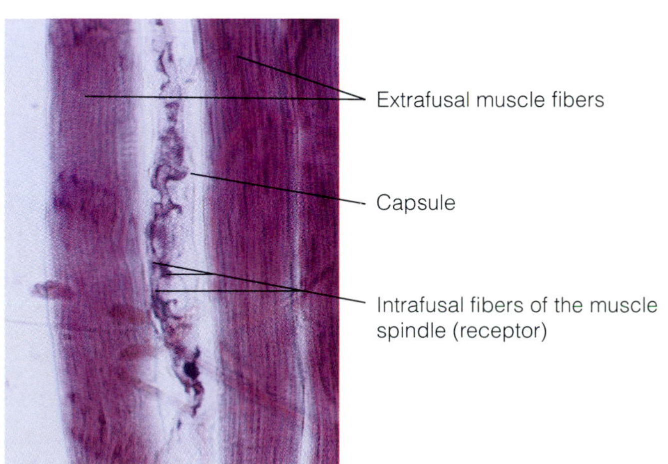

PLATE 12 Longitudinal section of a muscle spindle (305X)

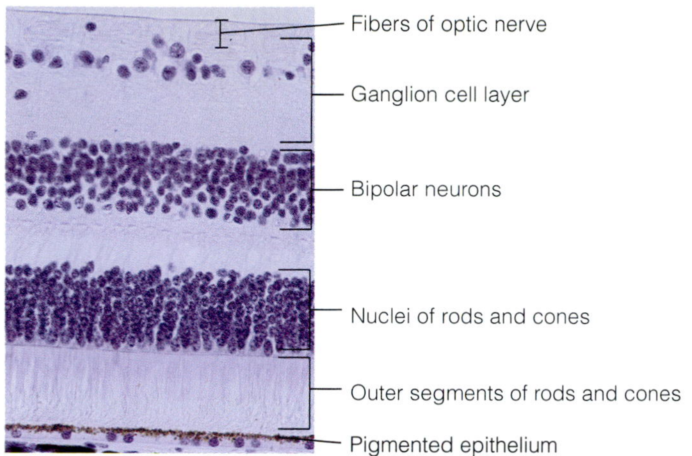

PLATE 13 Structure of the retina of the eye (353X)

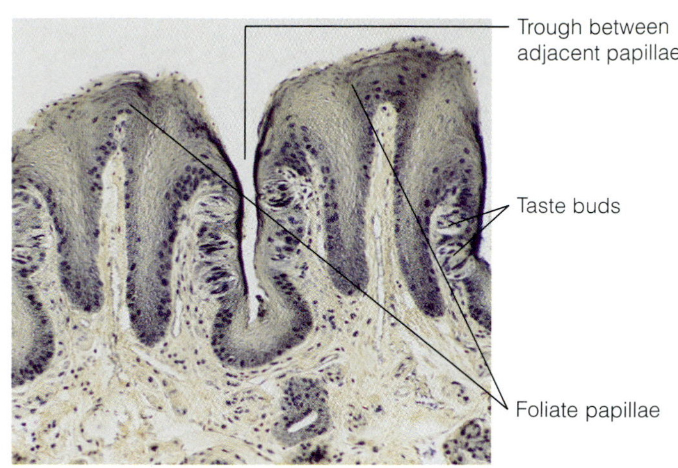

PLATE 16 Location of the taste buds on the lateral aspects of foliate papillae of the tongue (133X)

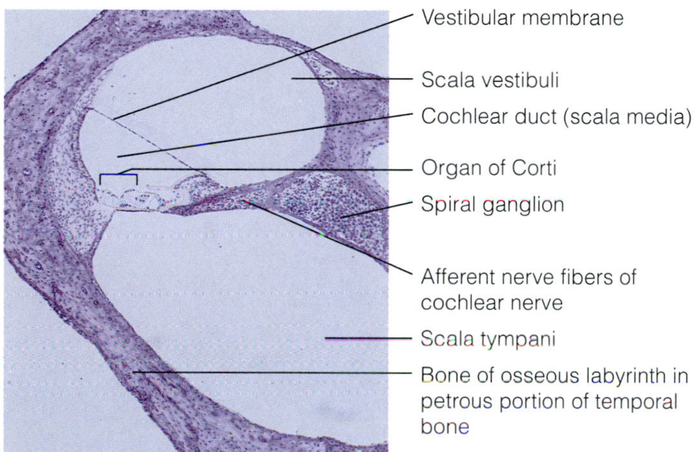

PLATE 14 Cross-sectional view of one turn of the cochlea showing the three scalas, and the location of the organ of Corti (36X)

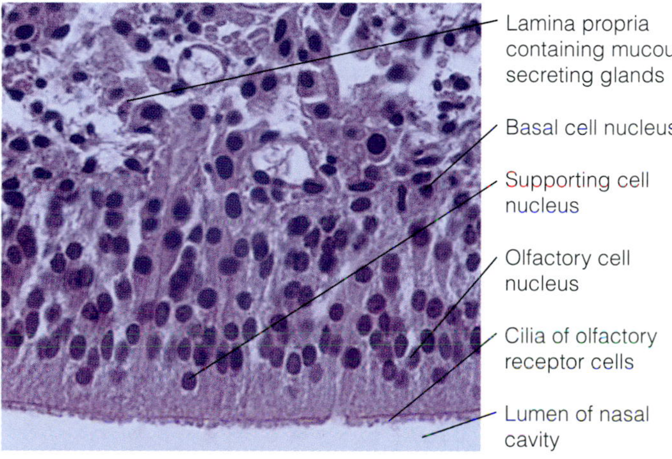

PLATE 17 Olfactory epithelium. From the lamina propria to the nasal cavity, the general arrangement of cells in this pseudostratified epithelium is basal cells, olfactory receptor cells, and supporting cells (543X)

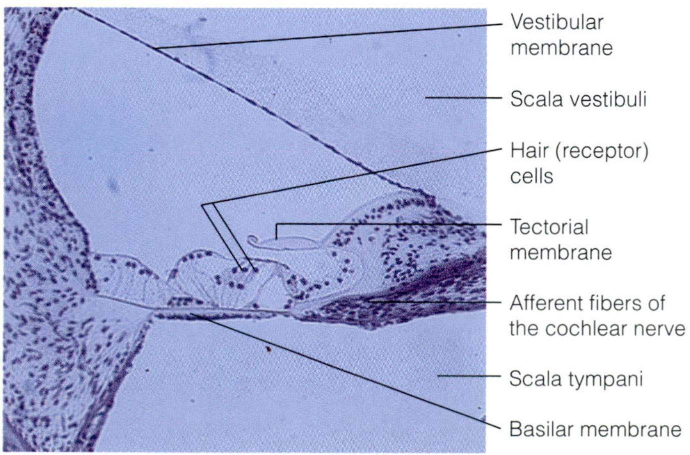

PLATE 15 The organ of Corti (106X)

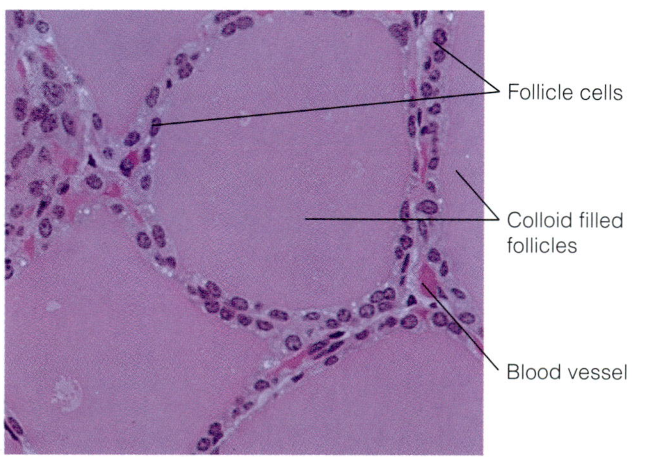

PLATE 18 The thyroid gland (345X)

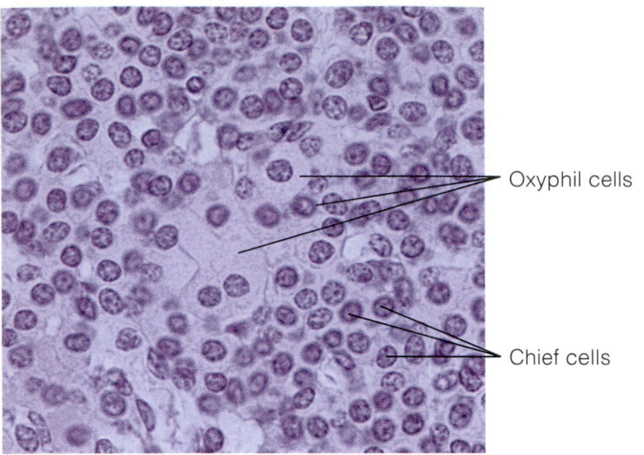

PLATE 19 Parathyroid gland tissue (543X)

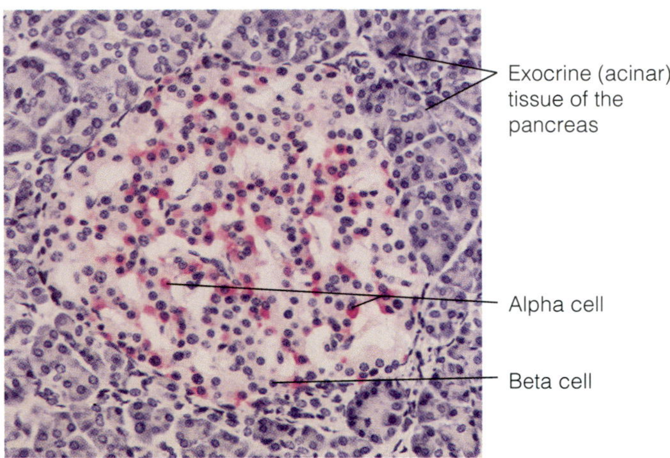

PLATE 20 Pancreatic islet stained differentially to allow identification of the glucagon-secreting alpha cells and the insulin-secreting beta cells (199X)

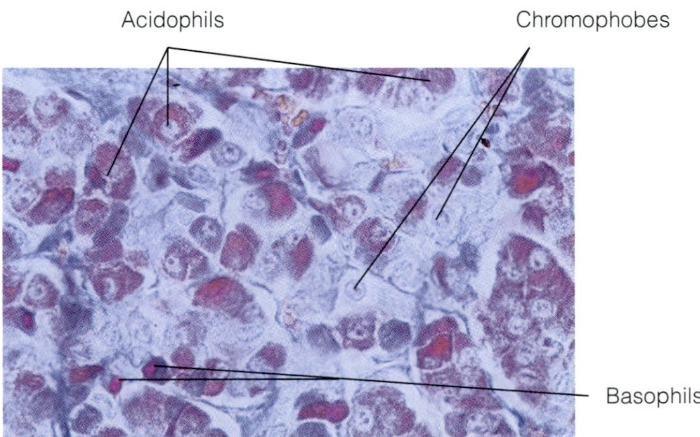

PLATE 21 Anterior pituitary gland. Differentially staining allows the acidophils, basophils, and chromophobes to be distinguished (434X)

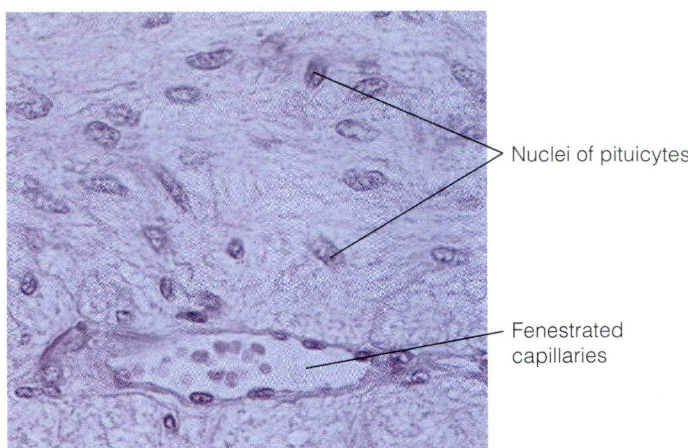

PLATE 22 Posterior pituitary (543X). The axons of the neurosecretory cells are indistinguishable from the cytoplasm of the pituicytes.

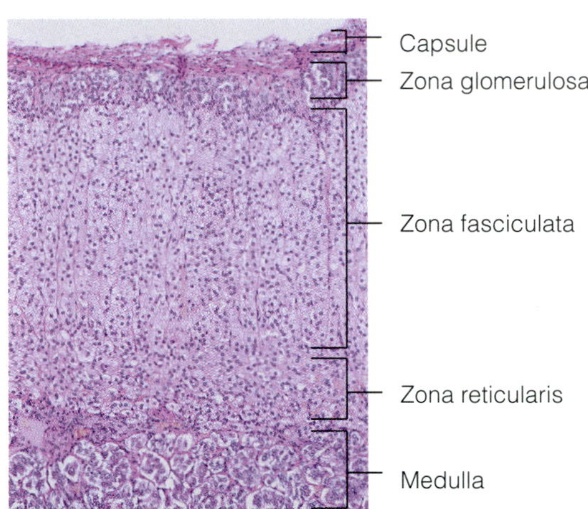

PLATE 23 Histologically distinct regions of the adrenal gland (56X)

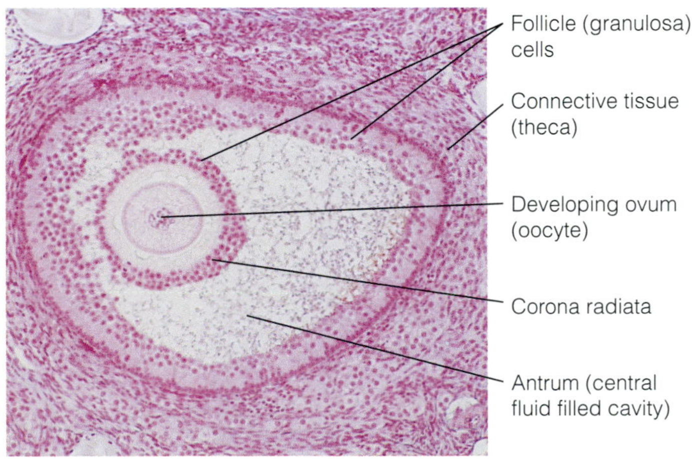

PLATE 24 A vesicular follicle of the ovary (106X)

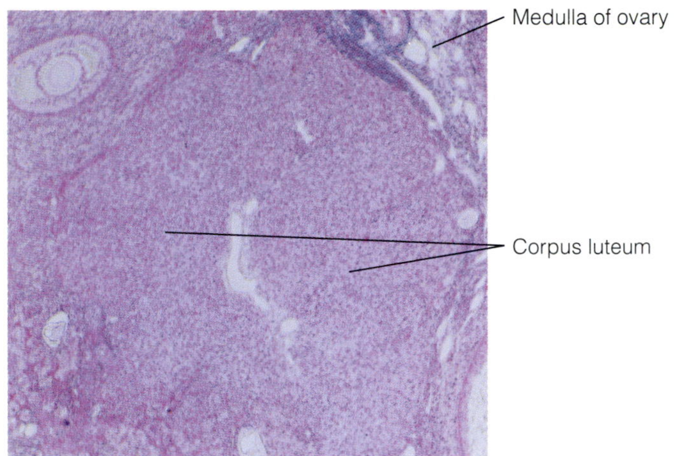

PLATE 25 The glandular corpus luteum of an ovary (45X)

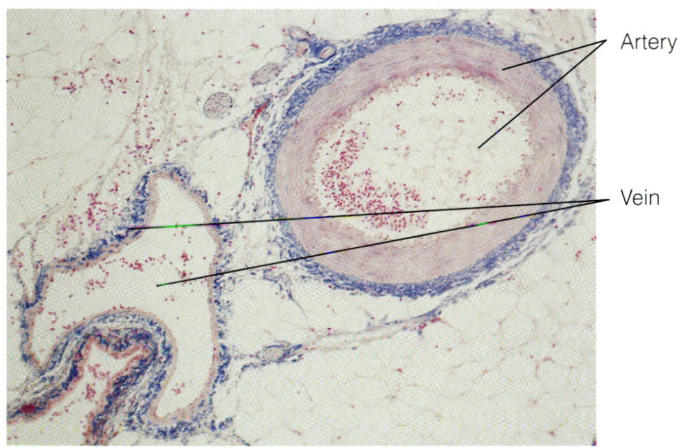

PLATE 26 Cross-sectional view of a small artery and vein (158X)

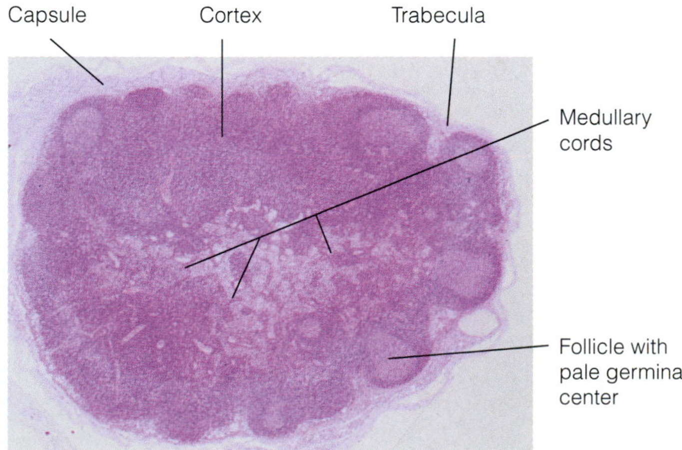

PLATE 27 Main structural features of a lymph node (18X)

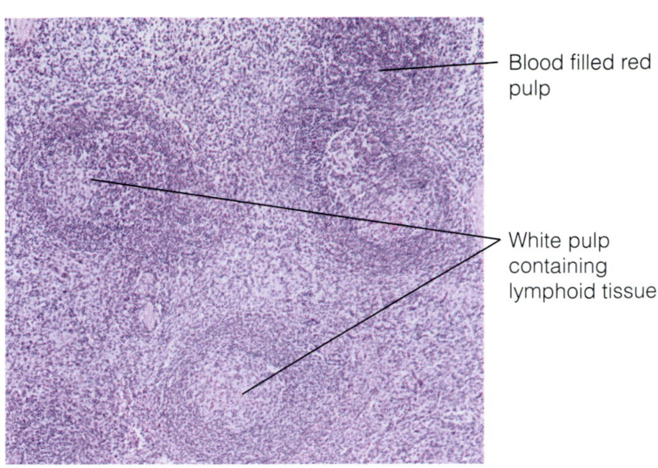

PLATE 28 Microscopic anatomy of a portion of the spleen showing the red and white pulp regions (56X)

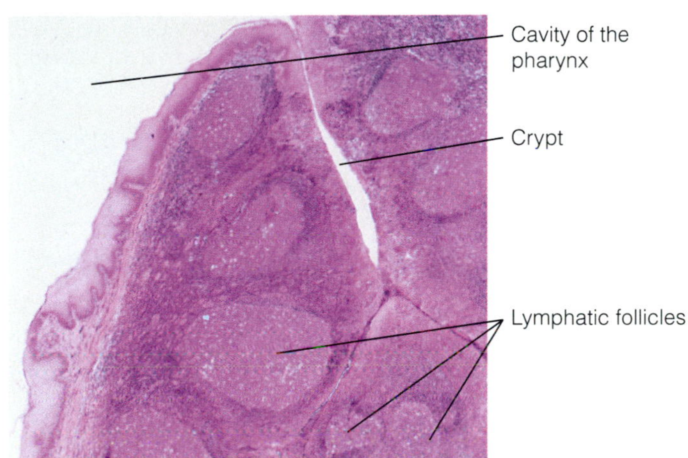

PLATE 29 Histology of a palatine tonsil. The luminal surface is covered with epithelium which invaginates deeply to form crypts (27X)

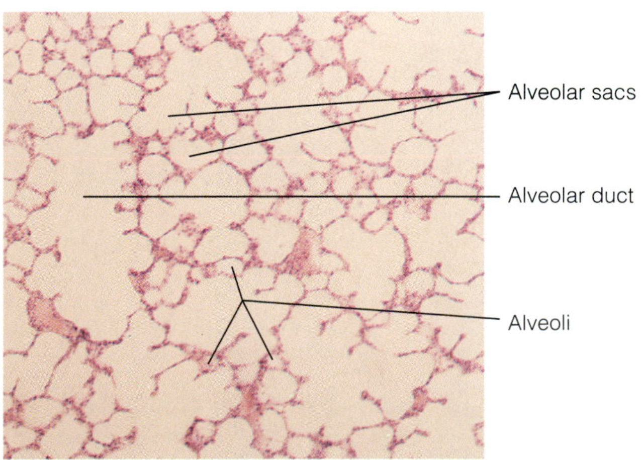

PLATE 30 Photomicrograph of part of the lung showing alveoli and alveolar ducts and sacs (34X)

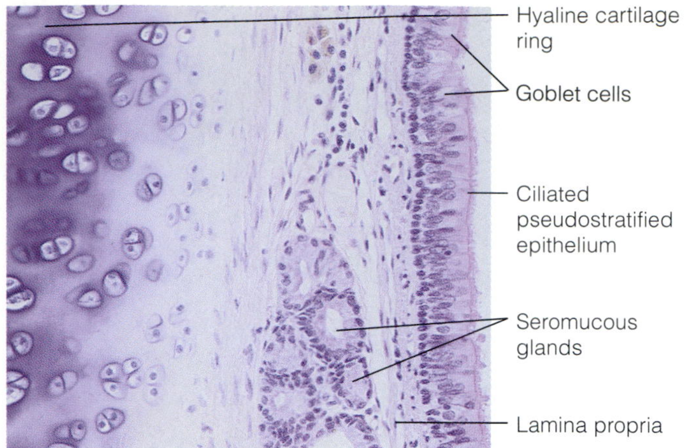

PLATE 31 Cross-section through the trachea showing the pseudostratified ciliated epithelium, glands, and part of the supporting ring of hyaline cartilage (159X)

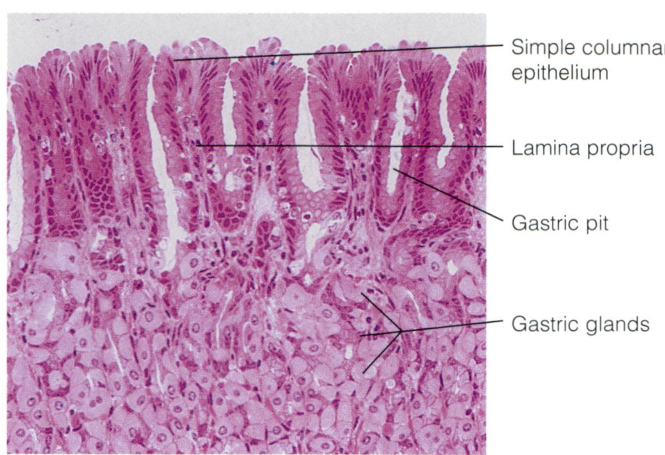

PLATE 34 Detailed structure of the gastric glands and pits (212X)

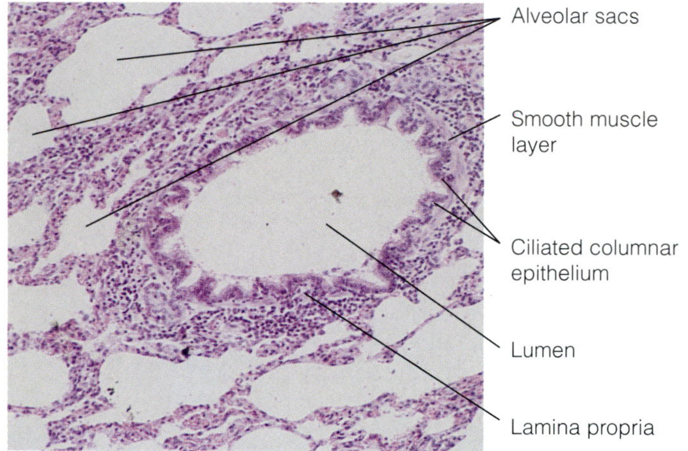

PLATE 32 Bronchiole, cross-sectional view (106X)

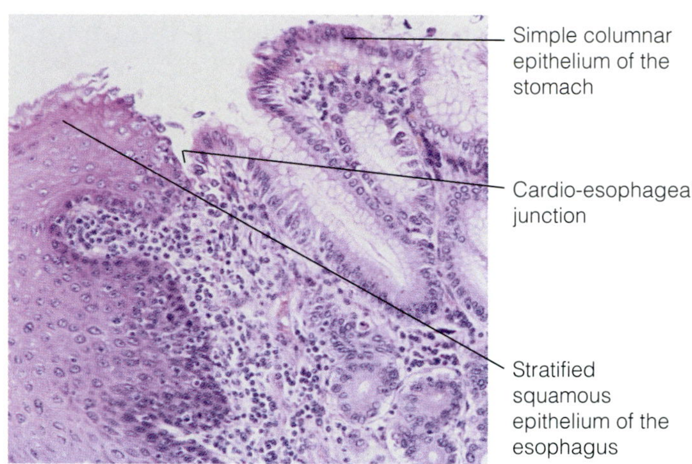

PLATE 35 Gastroesophageal junction showing the meeting of the simple columnar epithelium of the stomach and the stratified squamous epithelium of the esophagus (133X)

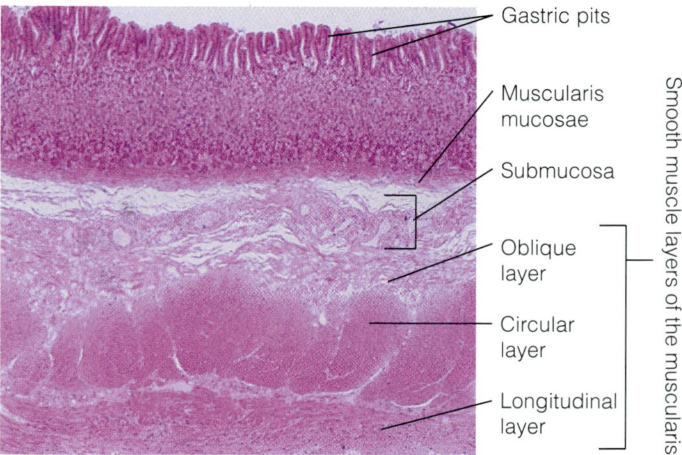

PLATE 33 Stomach. Low-power cross-sectional view through its wall showing three tunics (34X)

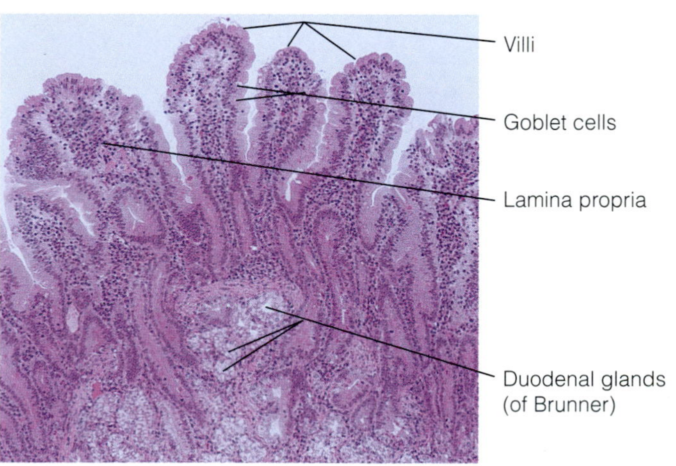

PLATE 36 Cross-sectional view of the duodenum showing villi and duodenal glands (66X)

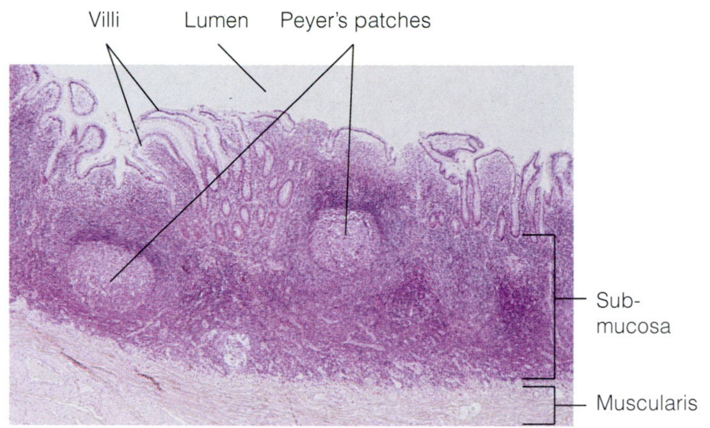

PLATE 37 Ileum, showing Peyer's patches (36X)

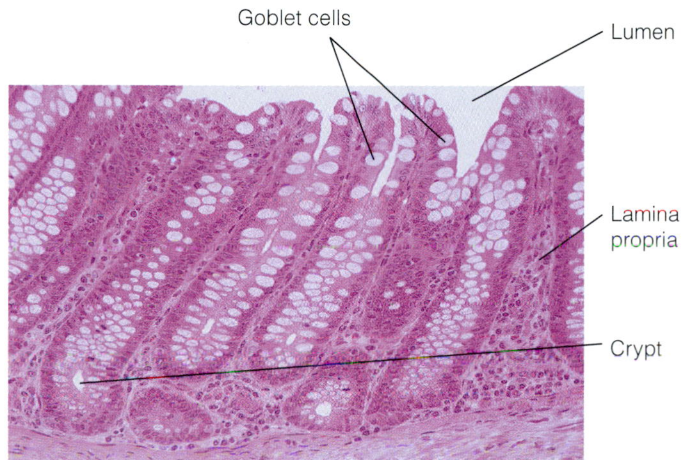

PLATE 38 Large intestine. Cross-sectional view showing the abundant goblet cells of the mucosa (127X)

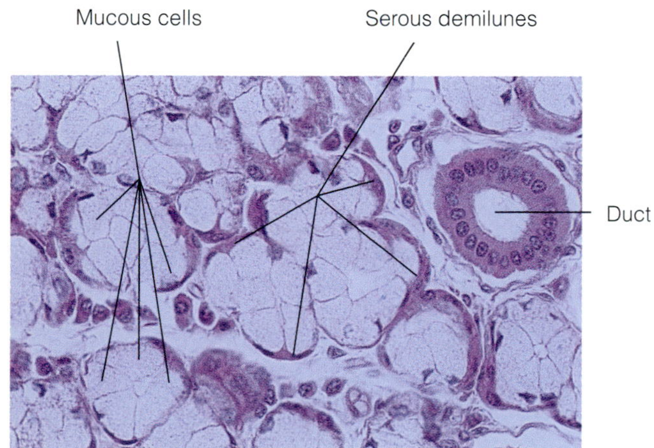

PLATE 39 Sublingual salivary glands (350X)

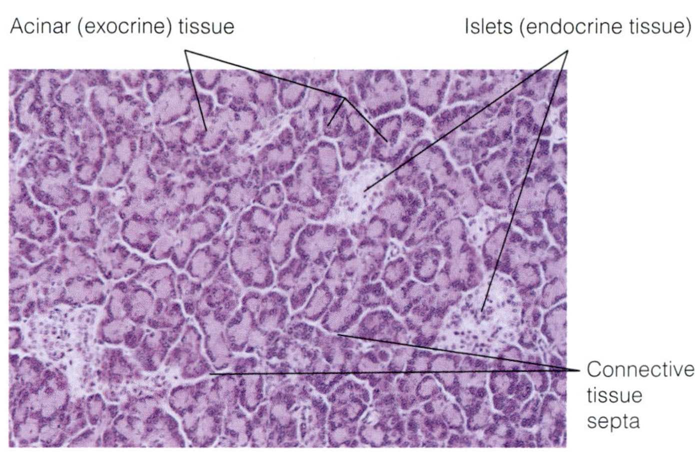

PLATE 40 Pancreas tissue. Exocrine and endocrine (islets) areas clearly visible (106X)

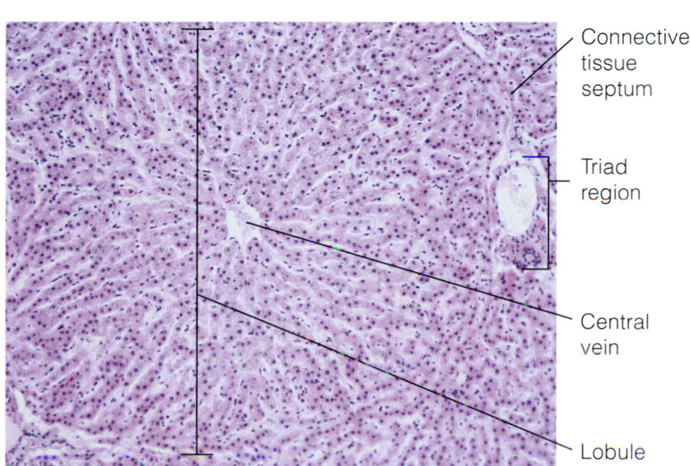

PLATE 41 Pig liver. Structure of the liver lobules (66X)

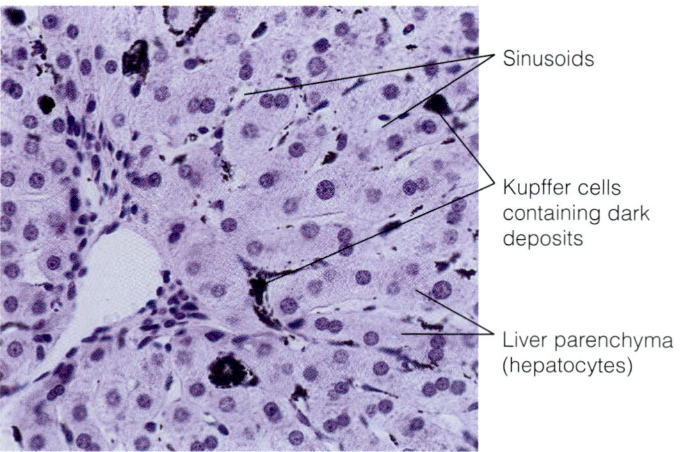

PLATE 42 Liver stained to show the location of the phagocytic cells (Kupffer cells) lining the sinusoids (265X)

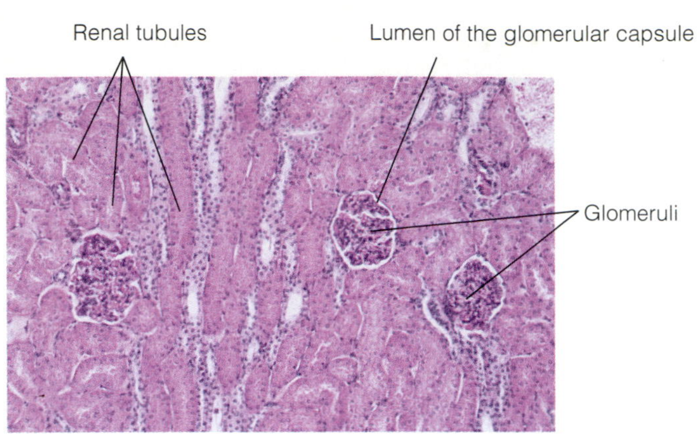

PLATE 43 Renal cortex of the kidney (85X)

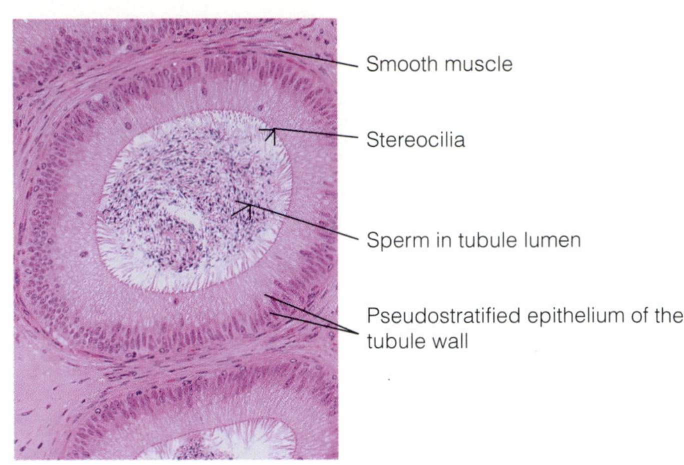

PLATE 46 Epididymis (148X)

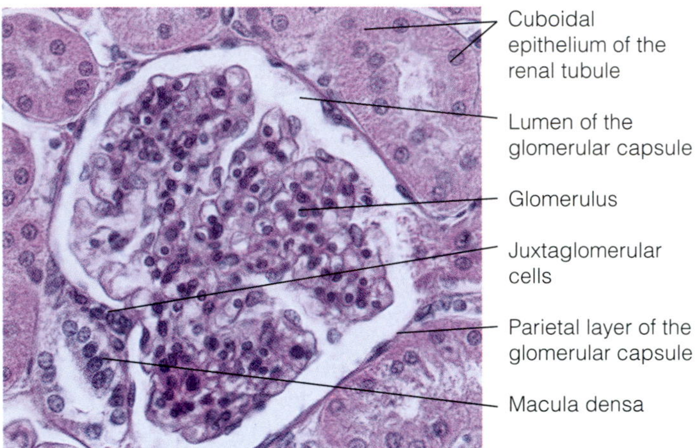

PLATE 44 Detailed structure of a glomerulus (345X)

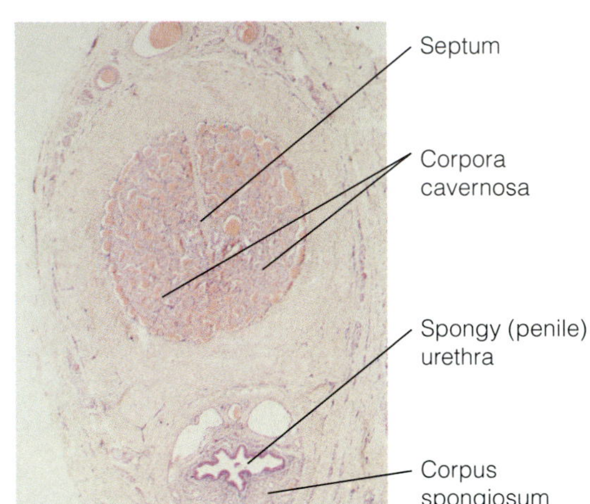

PLATE 47 Penis, transverse section (13X)

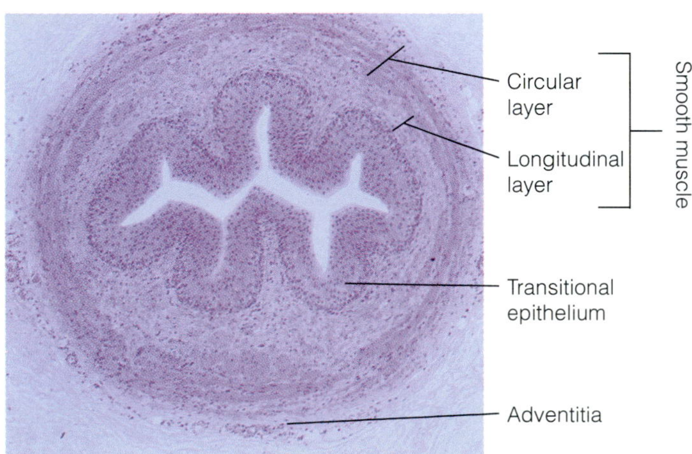

PLATE 45 Cross-section of the ureter (45X)

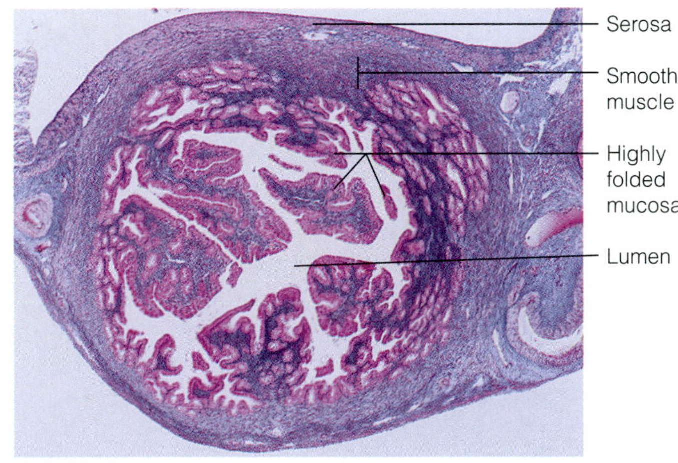

PLATE 48 Cross-sectional view of the uterine tube (34X)

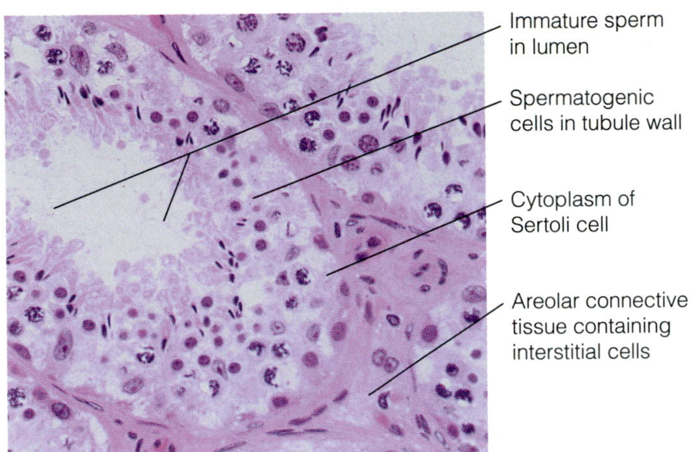

PLATE 49 Parts of two seminiferous tubules and the intervening connective tissue (345X)

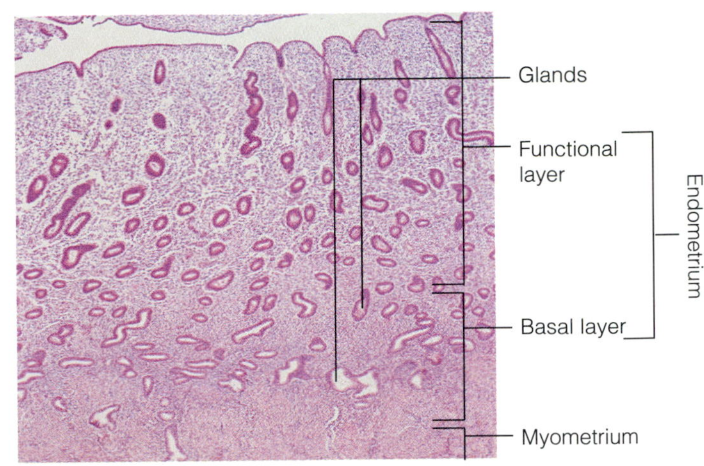

PLATE 52 Uterus: proliferative phase (17X)

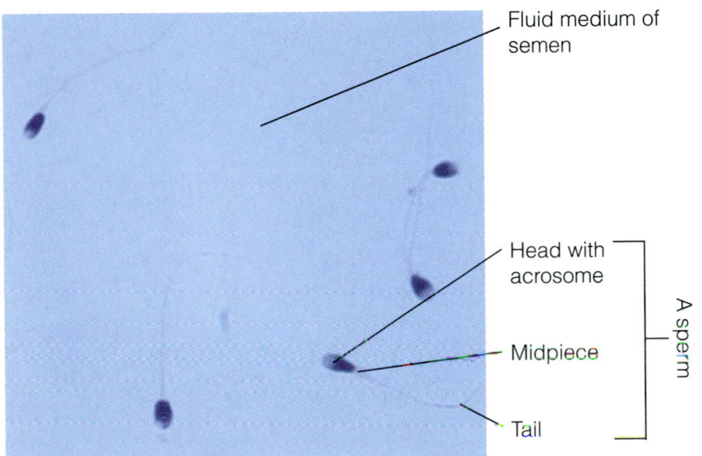

PLATE 50 Semen, the product of ejaculation, consisting of sperm and fluids secreted by the accessory glands (particularly the prostate and seminal vesicles) (489X)

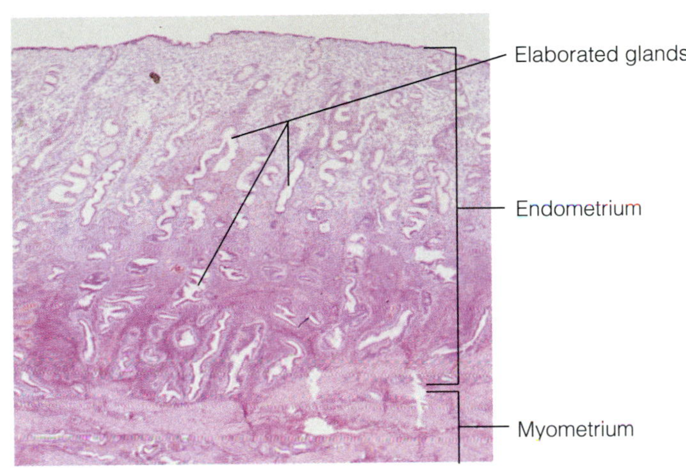

PLATE 53 Uterus: secretory phase (17X)

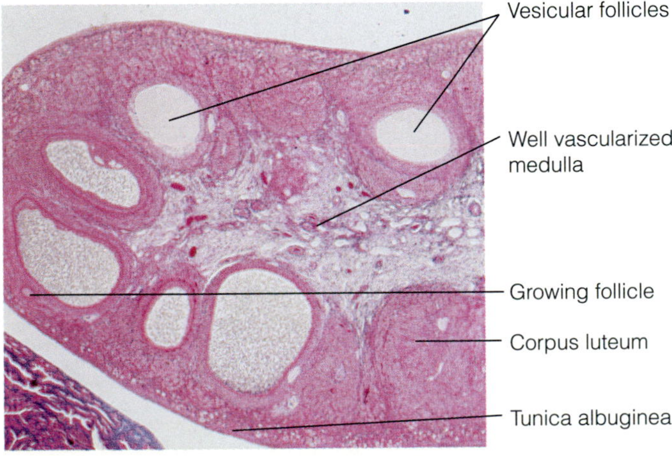

PLATE 51 The ovary, showing its follicles in various stages of development (17X)

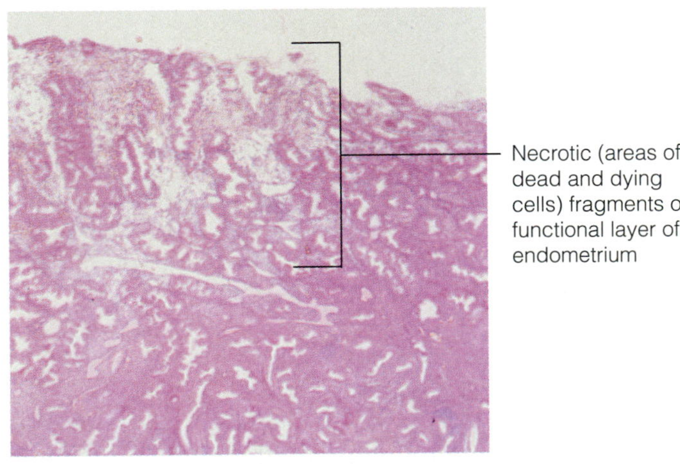

PLATE 54 Uterus: Menstrual phase (17X)

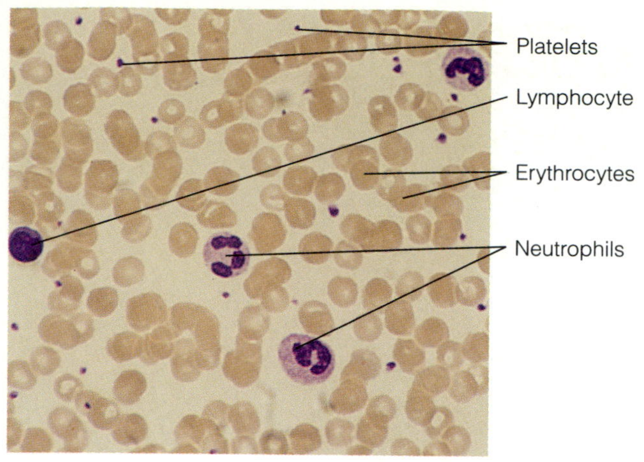

PLATE 55 Human blood smear (543X)

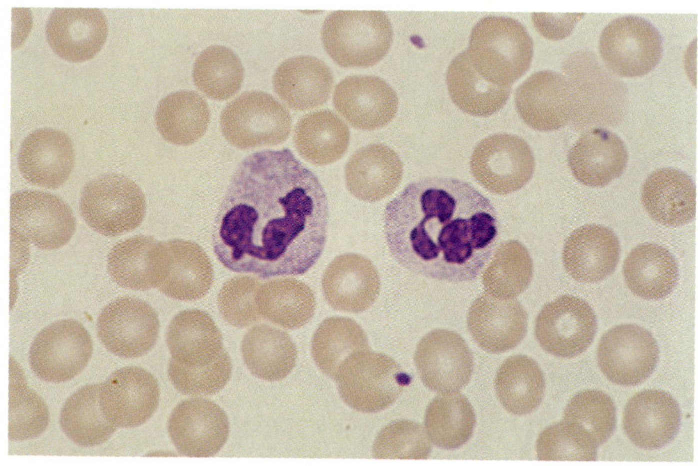

PLATE 56 Two neutrophils surrounded by erythrocytes (848X)

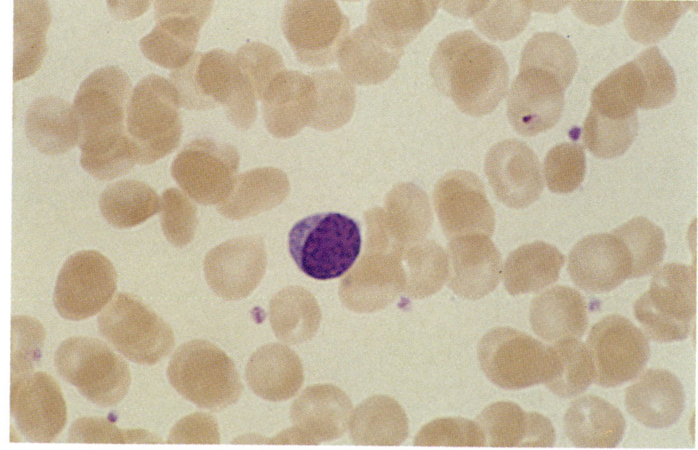

PLATE 57 A lymphocyte surrounded by erythrocytes (848X)

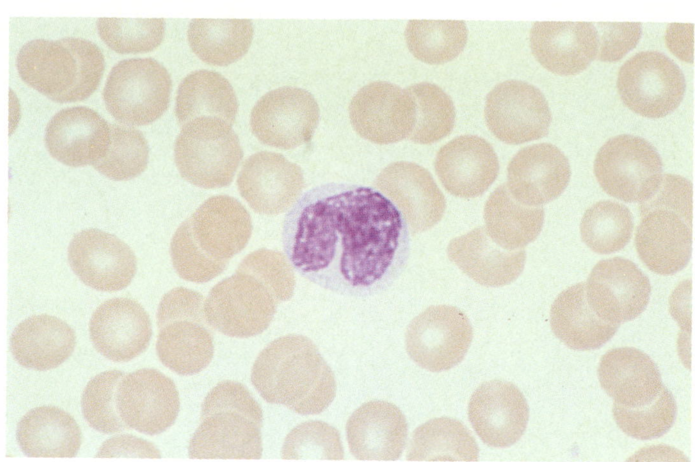

PLATE 58 A monocyte surrounded by erythrocytes (848X)

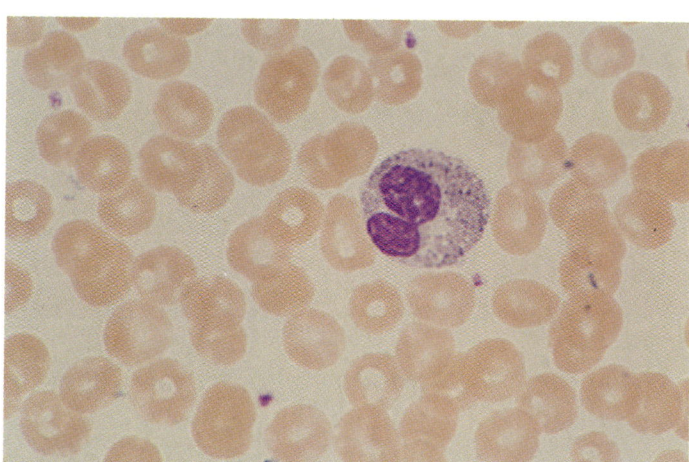

PLATE 59 An eosinophil surrounded by erythrocytes (848X)

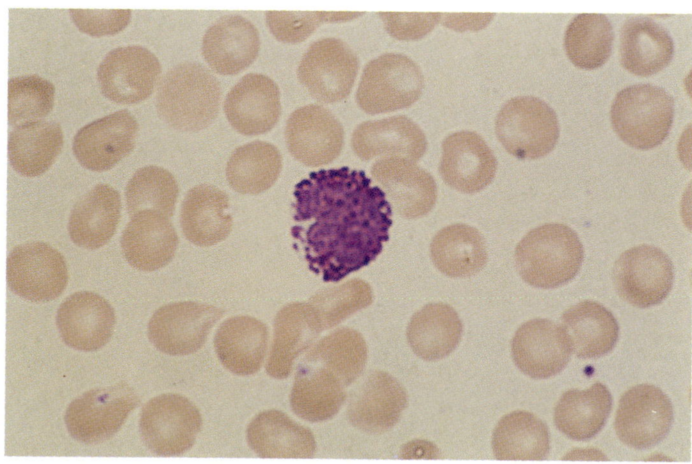

PLATE 60 A basophil surrounded by erythrocytes (848X)

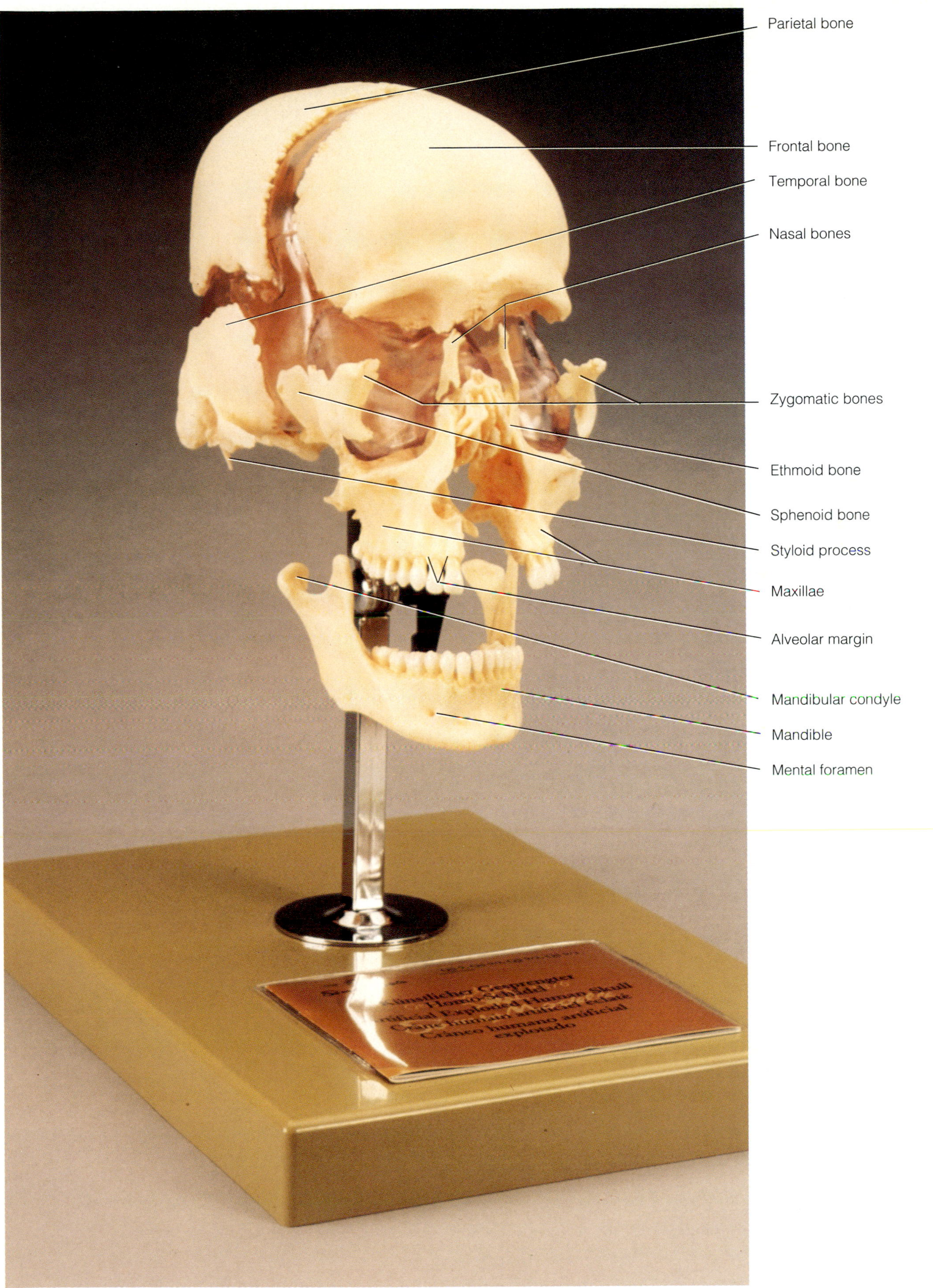

Beauchene skull, frontal view. See pages 73–78. (Somso model)

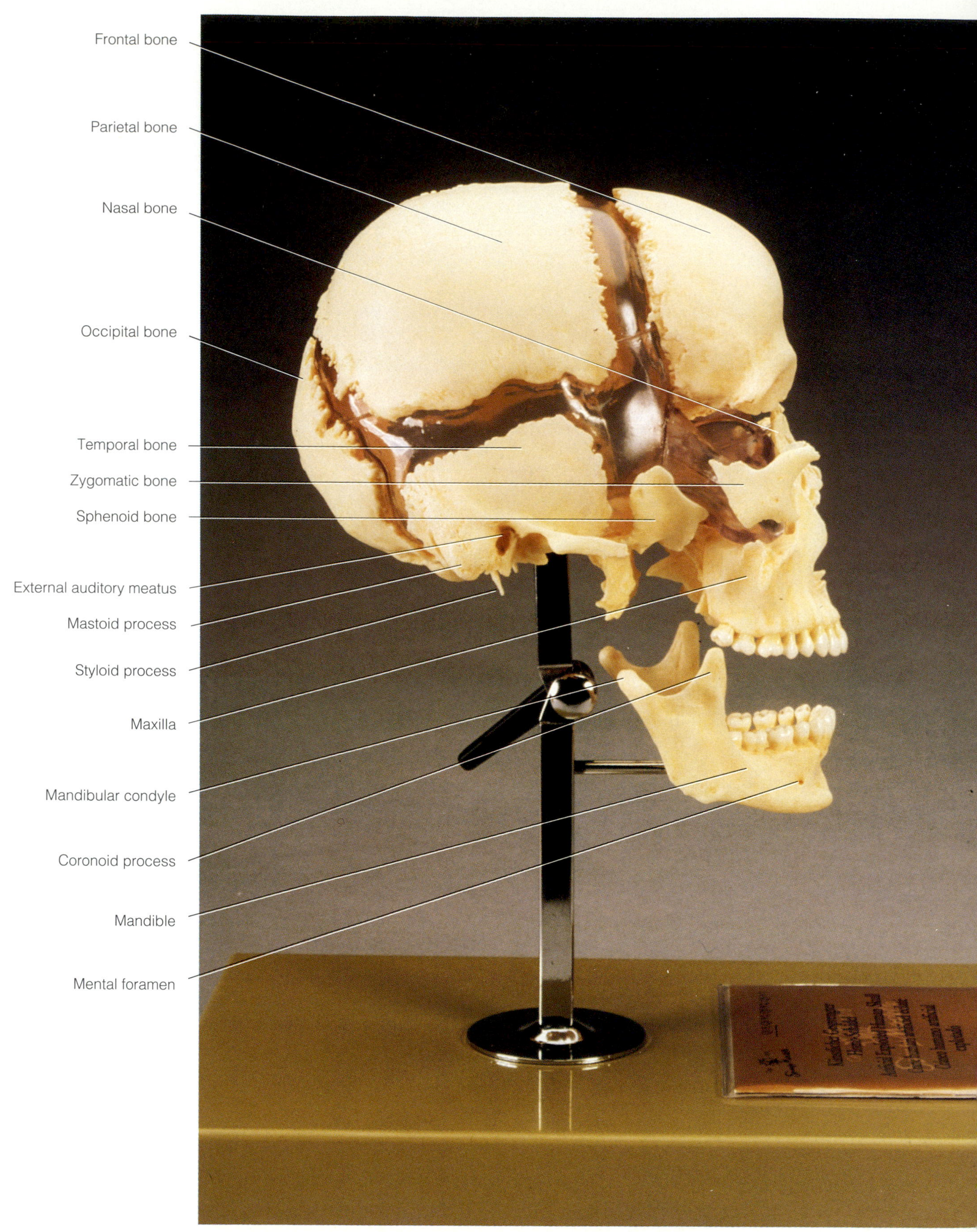

Beauchene skull, side view. See pages 73–78. (Somso model)

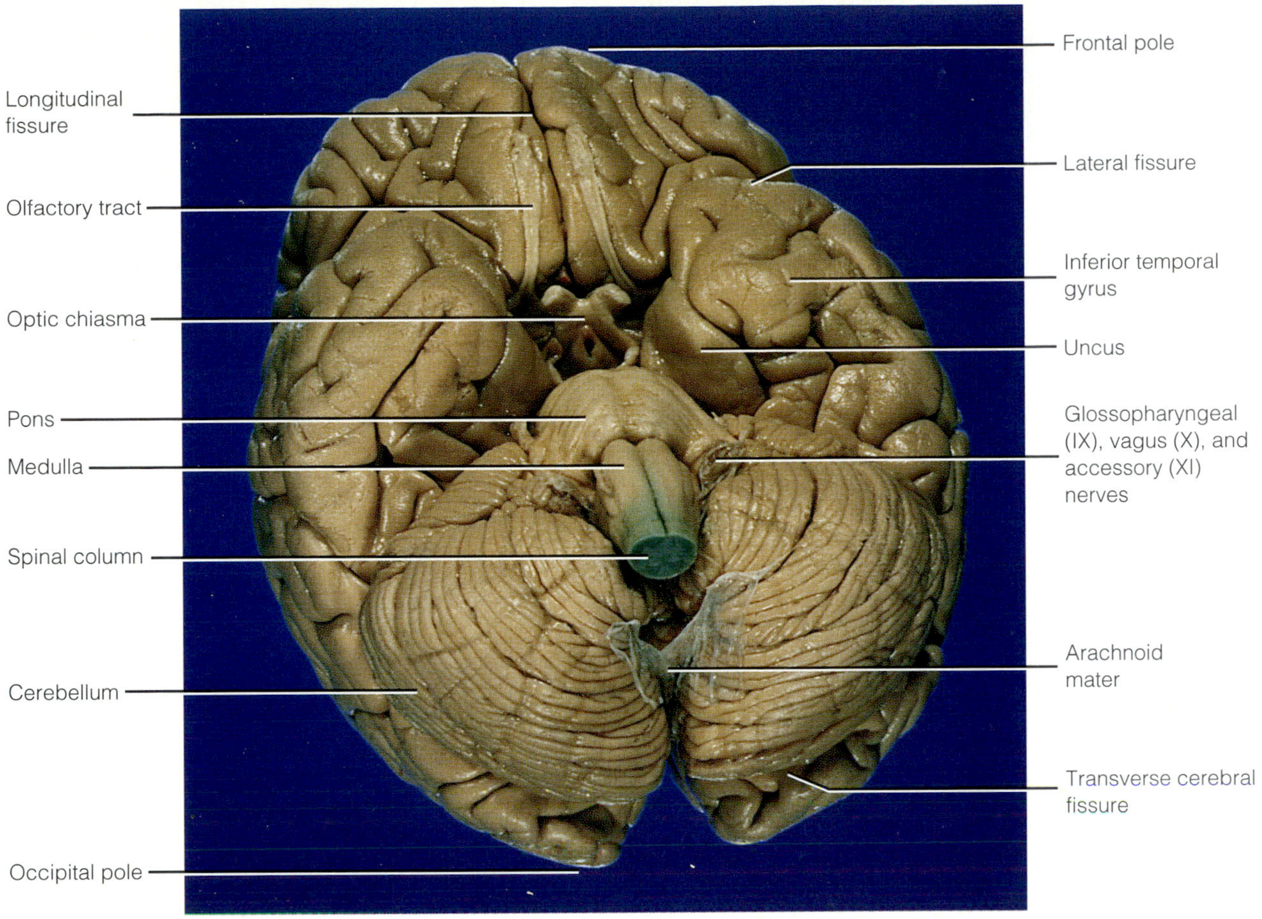

Frontal pole
Longitudinal fissure
Lateral fissure
Olfactory tract
Inferior temporal gyrus
Optic chiasma
Uncus
Pons
Glossopharyngeal (IX), vagus (X), and accessory (XI) nerves
Medulla
Spinal column
Arachnoid mater
Cerebellum
Transverse cerebral fissure
Occipital pole

PLATE A Base of the brain.

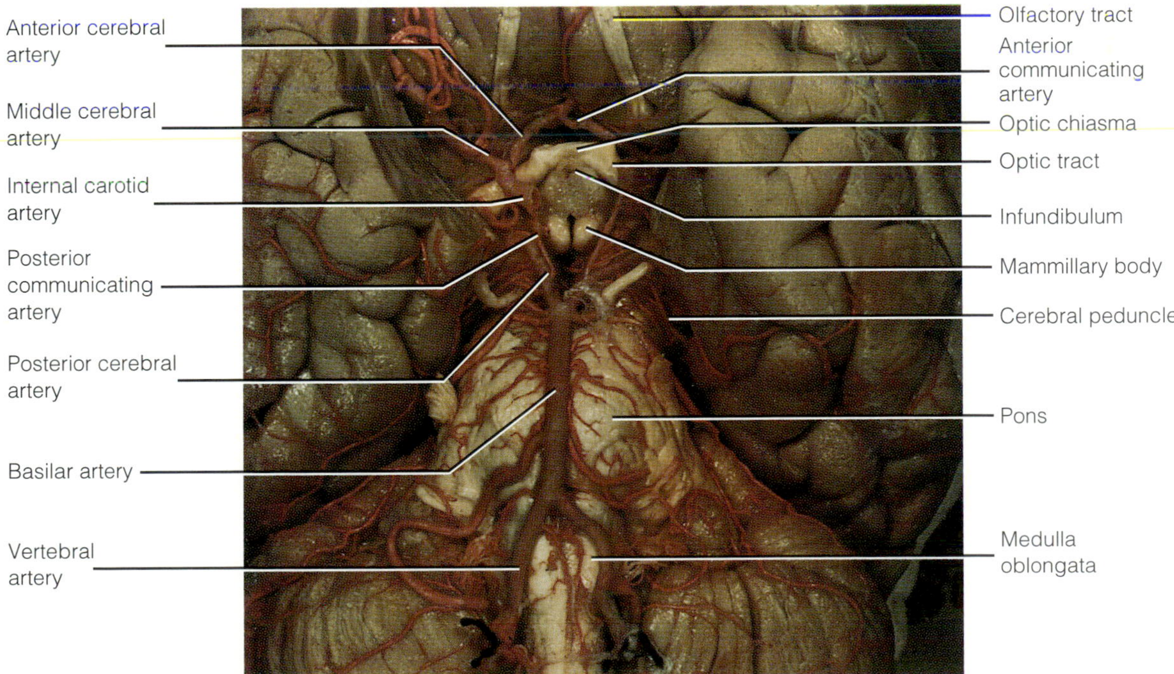

PLATE B Circle of Willis.

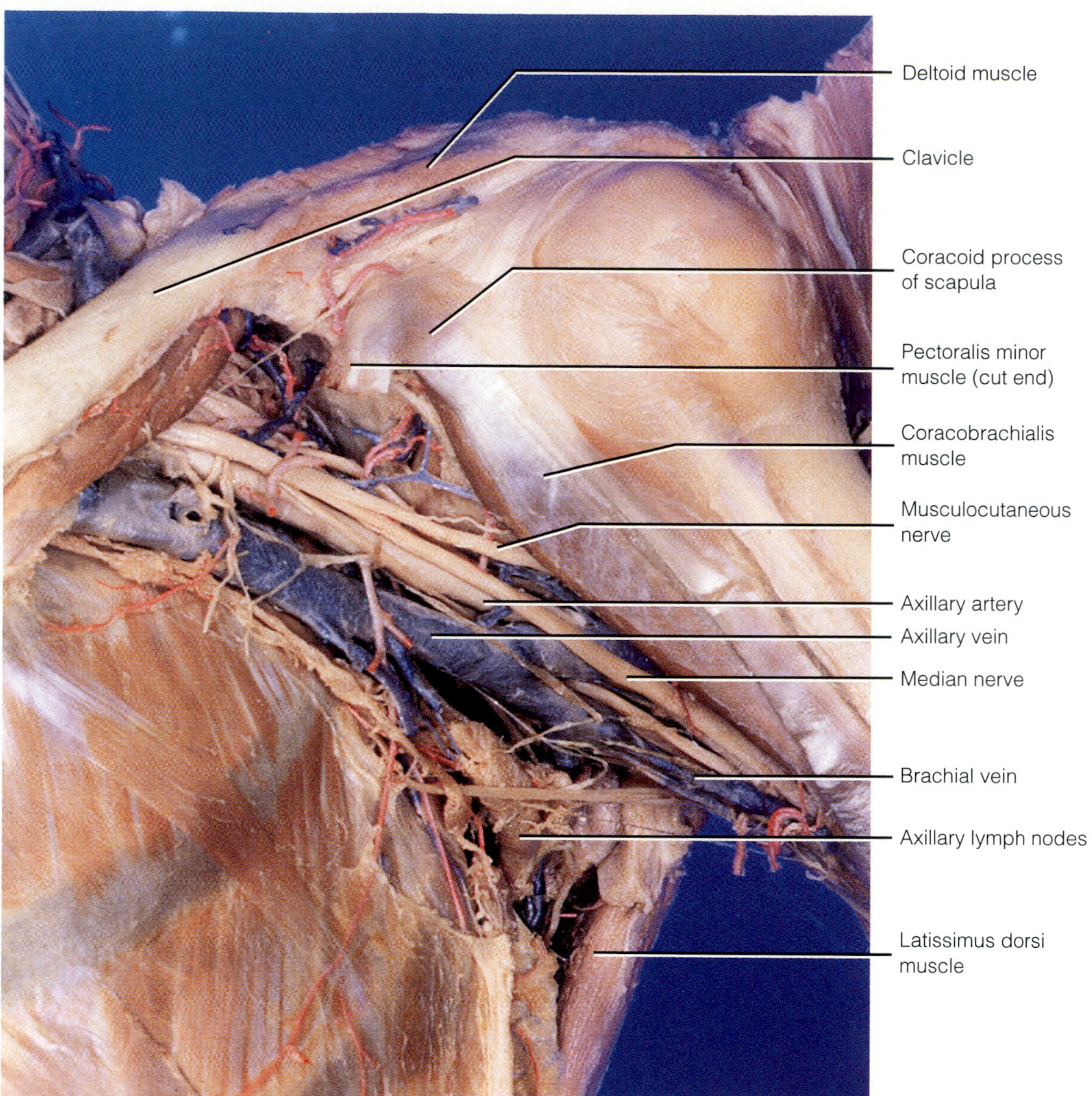

PLATE C Brachial plexus and axilla.

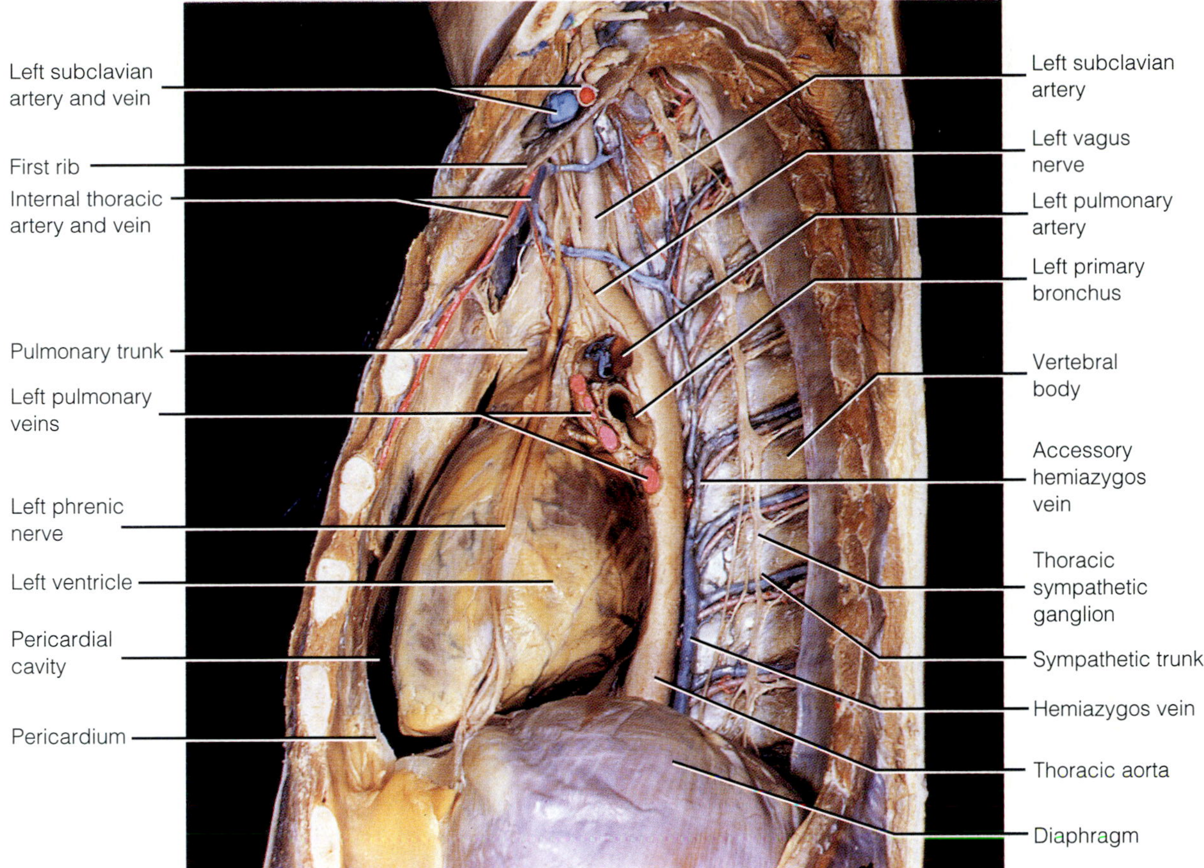

PLATE D Lateral view of mediastinum.

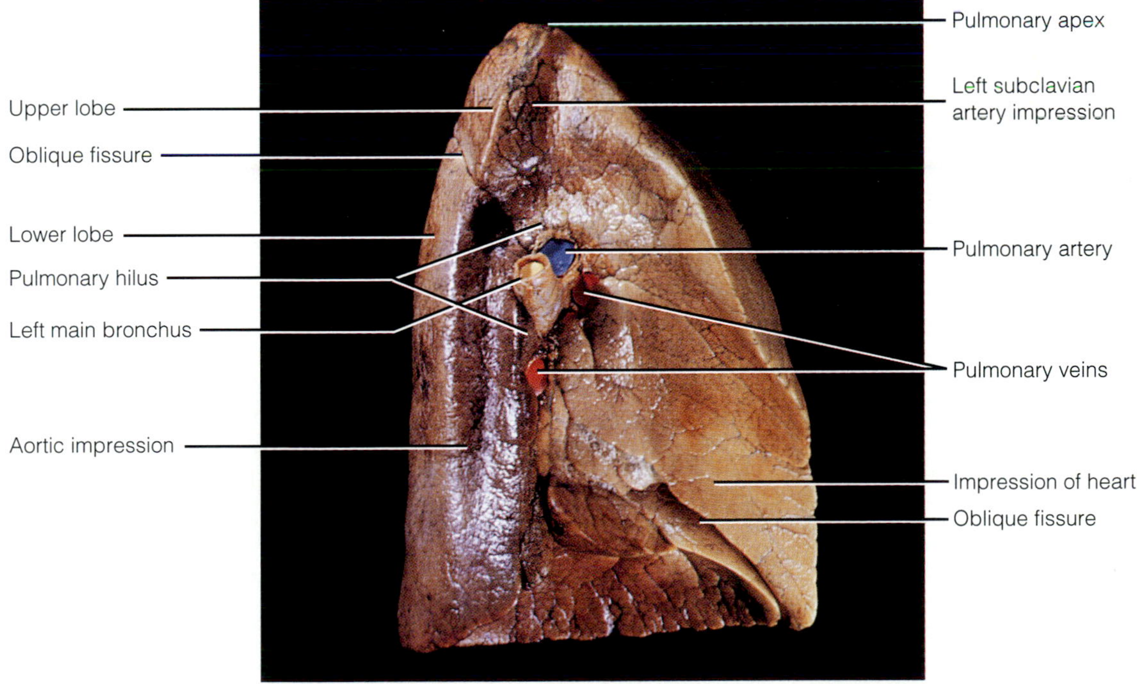

PLATE E Left lung, mediastinal surface.

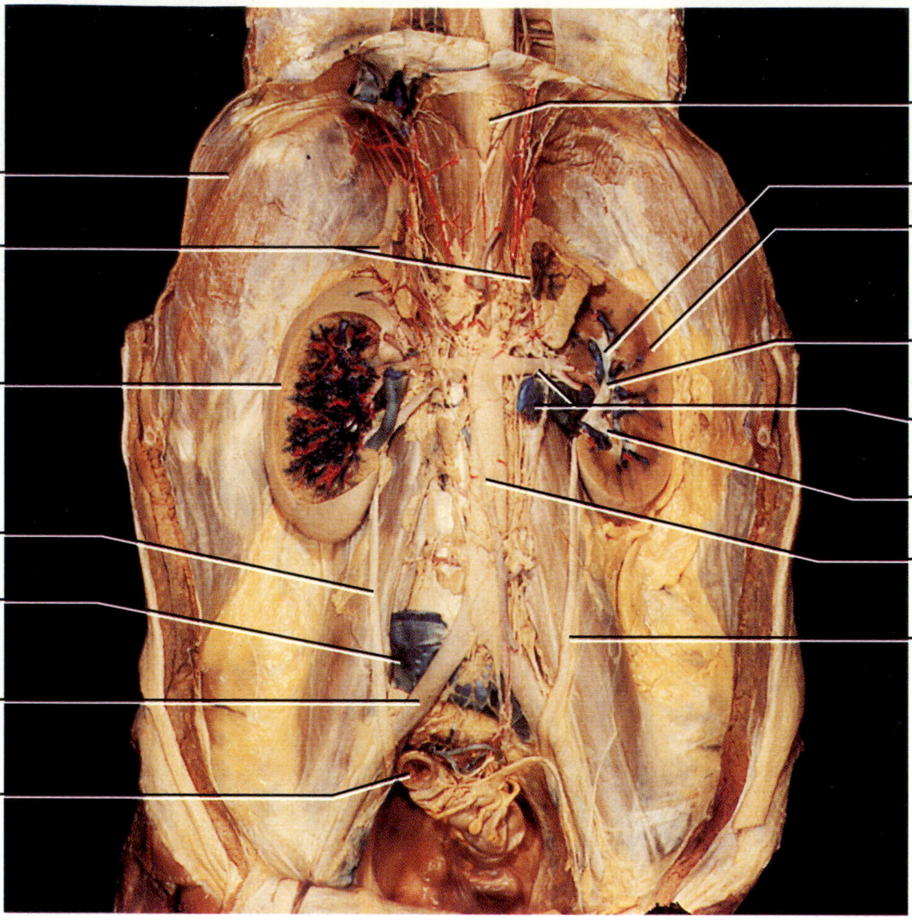

Esophagus
Diaphragm
Minor calyx
Suprarenal glands
Left kidney
Major calyx
Right kidney
Left renal artery and vein
Renal pelvis
Right ureter
Abdominal aorta
Inferior vena cava
Left ureter
Right common iliac artery
Sigmoid colon

PLATE F Retroperitoneum, kidneys dissected.

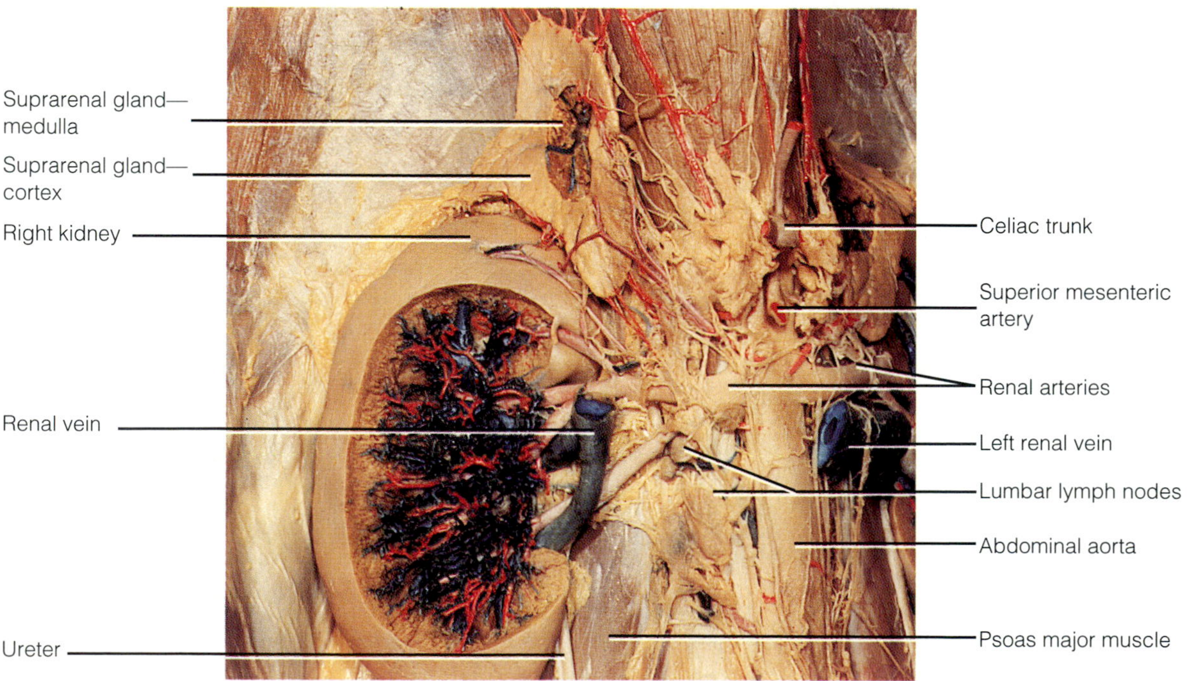

PLATE G Right kidney dissected, detail.

Classification of Body Membranes

OBJECTIVES

1. To compare the structure and function of the major membrane types.
2. To list the general functions of each membrane type and note its location in the body.
3. To recognize by microscopic examination cutaneous, mucous, and serous membranes.

MATERIALS

Compound microscope
Prepared slides of trachea (cross section) and small intestine (cross section)
Prepared slide of serous membrane (e.g., mesentery)
Longitudinally cut fresh beef joint (if available)

See Appendix D, Exercise 8 for links to A.D.A.M. Standard.

See Appendix E, Exercise 8 for links to *Anatomy and PhysioShow: The Videodisc.*

The body membranes, which cover surfaces, line body cavities, and form protective (and often lubricating) sheets around organs, fall into two major categories. These are the so-called *epithelial membranes* and the *synovial membranes.*

EPITHELIAL MEMBRANES

The term "epithelial membrane" is used in various ways. Here we will define an **epithelial membrane** as a simple organ consisting of an epithelial sheet bound to an underlying layer of connective tissue. Most of the covering and lining epithelia take part in forming one of the three common varieties of epithelial membranes: cutaneous, mucous, or serous.

The **cutaneous membrane** (Figure 8.1a) is the skin, a dry membrane with a keratinizing epithelium (the epidermis). Since the skin is discussed in some detail in Exercise 7, the mucous and serous membranes will receive our attention here.

Mucous Membranes

The **mucous membranes** (**mucosae**) are composed of epithelial cells resting on a layer of loose connective tissue called the **lamina propria.** They line all body cavities that open to the body exterior—the respiratory (Figure 8.1b), digestive, and urinary tracts. All mucosae are "wet" membranes because they are continuously bathed by secretions (or, in the case of urinary tract mucosae, urine). Although mucous membranes often secrete mucus, this is not a requirement. The mucous membranes of both the digestive and respiratory tracts secrete mucus, but that of the urinary tract does not.

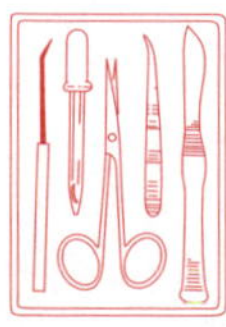

Using Plates 31 and 36 of the Histology Atlas as guides, examine a slide made from a cross section of the trachea and another of the small intestine. Draw the mucosa of each in the appropriate circle, and fully identify each epithelial type. Remember to look for the epithelial cells at the free surface. Also search the epithelial sheets for **goblet cells**—columnar epithelial cells with a large mucus-containing vacuole (goblet) in their apical cytoplasm. Which mucosa type contains goblet cells?

____

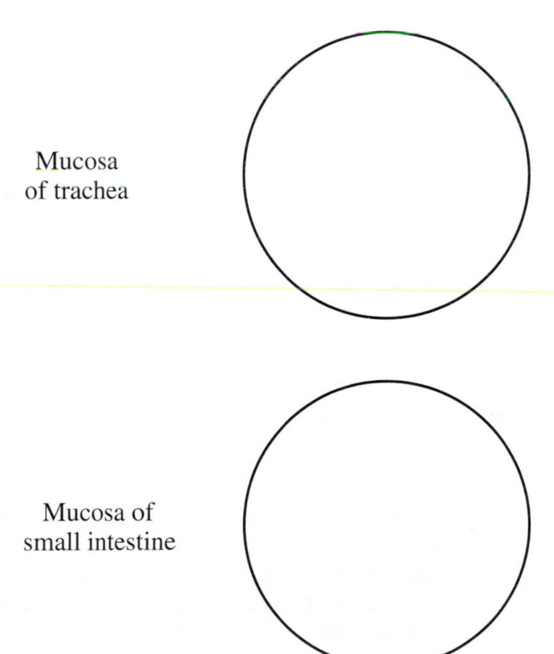

Compare and contrast the roles of these two mucous membranes.

____

____

____

____

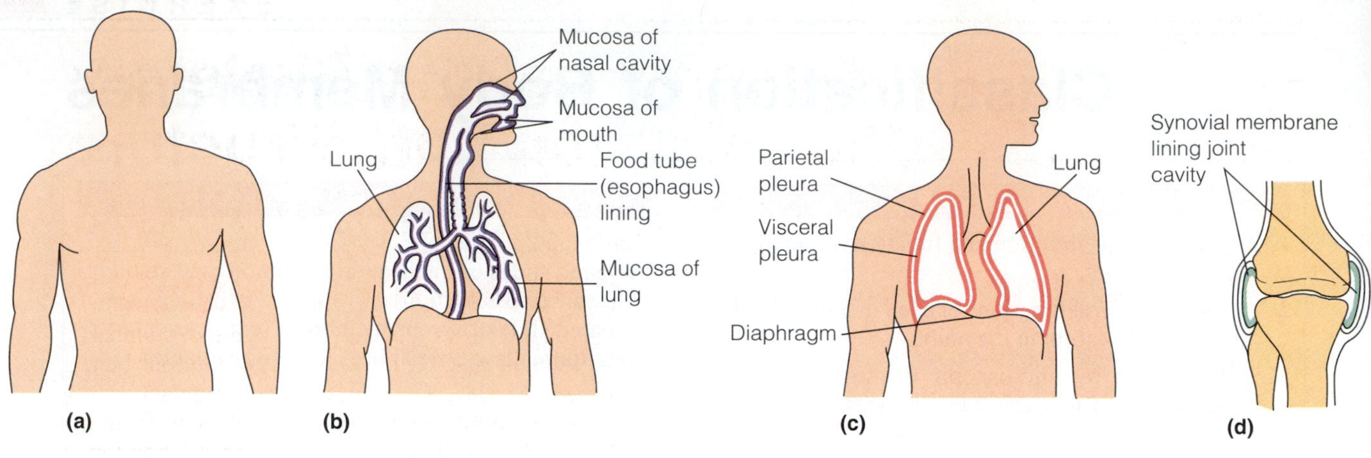

F8.1

Body membranes. The epithelial membranes (a–c) are composite membranes with epithelial and connective tissue elements. **(a)** The cutaneous membrane, or skin, covers and protects the body surface. **(b)** Mucous membranes line body cavities (hollow organs) that open to the exterior. **(c)** Serous membranes line the closed ventral cavity of the body. One example, the pleura, is illustrated here. **(d)** Synovial membranes, derived solely from connective tissue, form smooth linings in joint cavities.

Serous Membranes

The **serous membranes** (**serosae**) are also epithelial membranes (Figure 8.1c). They are composed of a layer of simple squamous epithelium on a scant amount of loose connective tissue. The serous membranes generally occur in twos. The parietal layer lines a body cavity, and the visceral layer covers the outside of the organs in that cavity (see also Figure 1.8, p. 8). In contrast to the mucous membranes, which line open body cavities, the serous membranes line body cavities that are closed to the exterior (with the exception of the female peritoneal cavity and the dorsal body cavity). The serosae secrete a thin fluid (serous fluid) that lubricates the organs and body walls and thus reduces friction as the organs slide across one another and against the body cavity walls. A serous membrane also lines the interior of blood vessels (endothelium) and the heart (endocardium). In capillaries, the entire wall is composed of serosa that serves as a selectively permeable membrane between the blood and the tissue fluid of the body.

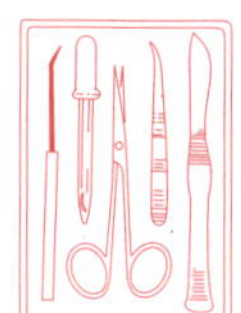

Examine a prepared slide of a serous membrane and diagram it in the circle provided here.

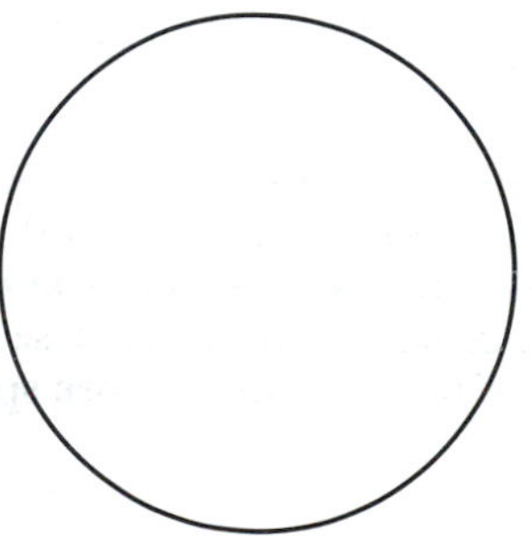

What are the specific names of the serous membranes covering the heart and lining the cavity in which it resides (respectively)?

____

The abdominal viscera and visceral cavity (respectively)?

____

SYNOVIAL MEMBRANES

Synovial membranes, unlike the mucous and serous membranes, are composed entirely of connective tissue; they contain no epithelial cells. These membranes line the cavities surrounding the joints, providing a smooth surface and secreting a lubricating fluid. They also line smaller sacs of connective tissue (bursae and tendon sheaths), which cushion structures moving against each other, as during muscle activity. Figure 8.1d illustrates the positioning of a synovial membrane in the joint cavity.

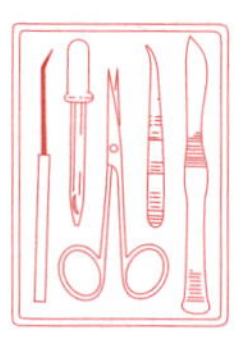

If a freshly sawed beef joint is available, visually examine the interior surface of the joint capsule to observe the smooth texture of the synovial membrane.

Overview of the Skeleton: Classification and Structure of Bones and Cartilages

OBJECTIVES

1. To list at least three functions of the skeletal system.
2. To identify the four main kinds of bones.
3. To identify surface bone markings and their function.
4. To identify the major anatomical areas on a longitudinally cut long bone (or diagram of one).
5. To identify the major regions and structures of an osteon in a histologic specimen of compact bone (or diagram of one).
6. To explain the role of the inorganic salts and organic matrix in providing flexibility and hardness to bone.
7. To locate and identify the three major types of skeletal cartilages.

MATERIALS

Disarticulated bones (identified by name or number) that demonstrate classic examples of the four bone classifications (long, short, flat, and irregular)
Long bone sawed longitudinally (beef bone from a slaughterhouse, if possible, or prepared laboratory specimen)
Compound microscope
Prepared slides of ground bone (cross section), hyaline cartilage, elastic cartilage, and fibrocartilage
3-D model of microscopic structure of compact bone
Long bone soaked in 10% nitric acid (or vinegar) until flexible
Long bone baked at 250°F for more than 2 hours
Disposable plastic gloves
Articulated skeleton

See Appendix D, Exercise 9 for links to A.D.A.M. Standard.

See Appendix E, Exercise 9 for links to *Anatomy and PhysioShow: The Videodisc.*

The **skeleton** is constructed of two of the most supportive tissues found in the human body—cartilage and bone. In embryos, the skeleton is predominantly composed of hyaline cartilage, but in the adult, most of the cartilage is replaced by more rigid bone. Cartilage persists only in such isolated areas as the bridge of the nose, the larynx, the trachea, joints, and parts of the rib cage.

Besides supporting and protecting the body as an internal framework, the skeleton provides a system of levers with which the skeletal muscles work to move the body. In addition, the bones store lipids and many minerals (most importantly calcium). Finally, the red marrow cavities of bones provide a site for hematopoiesis (blood cell formation).

The skeleton is made up of bones that are connected at joints, or articulations. The skeleton is subdivided into two divisions: the **axial skeleton** (those bones that lie around the body's center of gravity) and the **appendicular skeleton** (bones of the limbs, or appendages) (Figure 9.1).

Before beginning your study of the skeleton, imagine for a moment that your bones have turned to putty. What if you were running when this metamorphosis took place? Now imagine your bones forming a continuous metal framework within your body, somewhat like a network of plumbing pipes. What problems could you envision with this arrangement? These images should help you understand how well the skeletal system provides support and protection, as well as facilitating movement.

BONE MARKINGS

Even a casual observation of the bones will reveal that bone surfaces are not featureless smooth areas but are scarred with an array of bumps, holes, and ridges. These **bone markings** reveal where bones form joints with other bones, where muscles, tendons, and ligaments were attached, and where blood vessels and nerves passed. Bone markings fall into two categories: projections, or processes which grow out from the bone and serve as sites of muscle attachment or help form joints; and depressions or cavities, indentations or openings in the bone that often serve as conduits for nerves and blood vessels. The bone markings are summarized in Table 9.1.

CLASSIFICATION OF BONES

The 206 bones of the adult skeleton are composed of two basic kinds of osseous tissue that differ in their texture. **Compact** bone looks smooth and homogeneous; **spongy** (or *cancellous*) bone is composed of small trabeculae (bars) of bone and lots of open space.

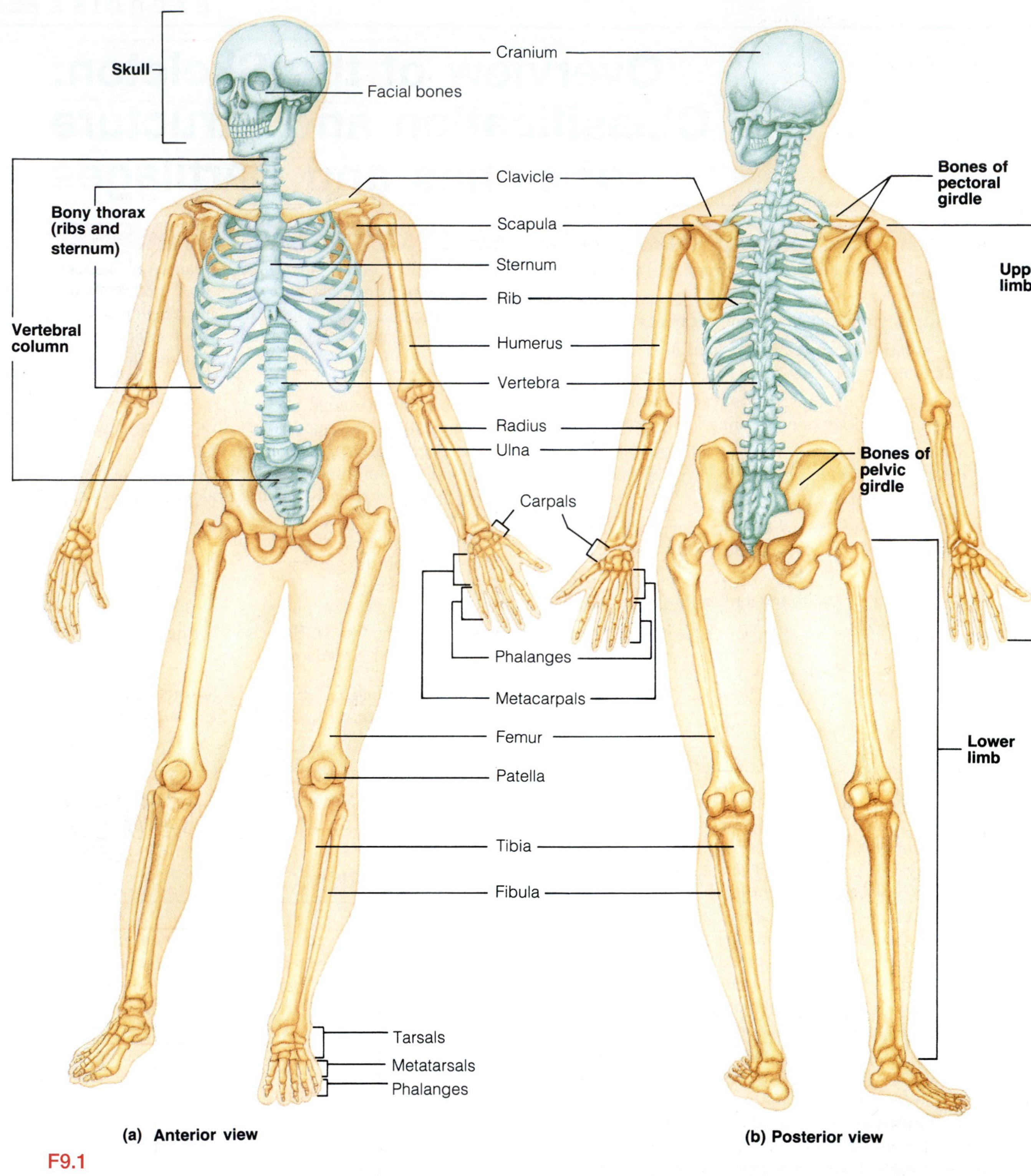

F9.1

The human skeleton. The bones of the axial skeleton are colored green to distinguish them from the bones of the appendicular skeleton.

Bones may be classified further on the basis of their relative gross anatomy into four groups: long, short, flat, and irregular bones.

Long bones, such as the femur (Figure 9.1), are much longer than they are wide, generally consisting of a shaft with heads at either end. Long bones are composed predominantly of compact bone. **Short bones** are typically cube-shaped, and they contain more spongy bone than compact bone. See the tarsals and carpals in Figure 9.1.

Flat bones are generally thin, with two thin layers of compact bone sandwiching a layer of spongy bone between them. Although the name "flat bone" implies a structure that is level or horizontal, many flat bones are curved (for example, the bones of the skull). Bones that do not fall into one of the preceding categories are clas-

TABLE 9.1 Bone Markings

Name of bone marking	Description	Illustration
Projections that are sites of muscle and ligament attachment		
Tuberosity	Large rounded projection; may be roughened	
Crest	Narrow ridge of bone; usually prominent	
Trochanter	Very large, blunt, irregularly shaped process. (The only examples are on the femur.)	
Line	Narrow ridge of bone; less prominent than a crest	
Tubercle	Small rounded projection or process	
Epicondyle	Raised area on or above a condyle	
Spine	Sharp, slender, often pointed projection	
Projections that help to form joints		
Head	Bony expansion carried on a narrow neck	
Facet	Smooth, nearly flat articular surface	
Condyle	Rounded articular projection	
Ramus	Armlike bar of bone	
Depressions and openings allowing blood vessels and nerves to pass		
Meatus	Canal-like passageway	
Sinus	Cavity within a bone, filled with air and lined with mucous membrane	
Fossa	Shallow, basinlike depression in a bone, often serving as an articular surface	
Groove	Furrow	
Fissure	Narrow, slitlike opening	
Foramen	Round or oval opening through a bone	

sified as **irregular bones.** The vertebrae are irregular bones (see Figure 9.1).

Some anatomists also recognize two other subcategories of bones. **Sesamoid bones** are small bones formed in tendons. The patellas (kneecaps) are sesamoid bones. **Wormian bones** are tiny bones between cranial bones. Except for the patellas, the sesamoid and Wormian bones are not included in the bone count given above because they vary in number and location in different individuals.

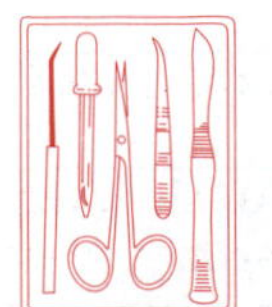

Examine the isolated (disarticulated) bones (numbered) on display. See if you can find specific examples of the bone markings described in Table 9.1. Then classify each of the bones into one of the four anatomical groups by recording its name or number in the chart at right. Verify your identifications with your instructor before leaving the laboratory.

Long	Short	Flat	Irregular

GROSS ANATOMY OF THE TYPICAL LONG BONE

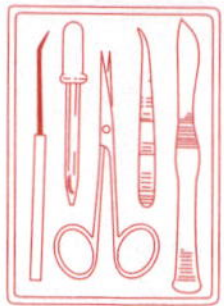

1. Obtain a long bone that has been sawed along its longitudinal axis. If a cleaned dry bone is provided, no special preparations need be made.

If the bone supplied is a fresh beef bone, don plastic gloves before beginning your observations.

With the help of Figure 9.2, identify the shaft, or **diaphysis.** Observe its smooth surface, which is composed of compact bone. If you are using a fresh specimen, carefully pull away the **periosteum,** or fibrous membrane covering, to view the bone surface. Notice that many fibers of the periosteum penetrate into the bone. These fibers are called **Sharpey's fibers.** The periosteum is the source of the blood vessels and nerves that invade the bone. **Osteoblasts** (bone-forming cells) on its inner face secrete the bony matrix that increases the girth of the long bone.

2. Now inspect the **epiphysis,** the end of the long bone. Notice that it is composed of a thin layer of compact bone that encloses spongy bone.

3. Identify the **articular cartilage,** which covers the epiphyseal surface in place of the periosteum. Since it is composed of glassy hyaline cartilage, it provides a smooth surface to prevent friction at joint surfaces.

4. If the animal was still young and growing, you will be able to see the **epiphyseal plate,** a thin area of hyaline cartilage that provides for longitudinal growth of the bone during youth. Once the long bone has stopped growing, these areas are replaced with bone and appear as thin, barely discernible remnants—the **epiphyseal lines.**

5. In an adult animal, the central cavity of the shaft (*medullary cavity*) is essentially a storage region for adipose tissue, or **yellow marrow.** In the infant, this area is involved in forming blood cells, and so **red marrow** is found in the marrow cavities. In adult bones, the red marrow is confined to the interior of the epiphyses, where it occupies the spaces between the trabeculae of spongy bone.

6. If you are examining a fresh bone, look carefully to see if you can distinguish the delicate **endosteum** lining the shaft. In a living bone, **osteoclasts** (bone-destroying cells) are found on the inner surface of the endosteum, against the compact bone of the diaphysis. As the bone grows in diameter on its external surface, it is constantly being broken down on its inner surface. Thus the thickness of the compact bone layer composing the shaft remains relatively constant.

7. If you have been working with a fresh bone specimen, return it to the appropriate area and properly dispose of your gloves, as designated by your instructor. Wash your hands before continuing on to the microscope study.

Longitudinal bone growth at epiphyseal discs follows a predictable sequence and provides a reliable indicator of the age of children exhibiting normal growth. In cases in which problems of long-bone growth are suspected (for example, pituitary dwarfism), X rays are taken to view the width of the growth plates. An abnormally thin epiphyseal plate indicates growth retardation. ■

CHEMICAL COMPOSITION OF BONE

Bone is one of the hardest materials in the body. Although relatively light, bone has a remarkable ability to resist tension and shear forces that continually act on it. An engineer would tell you that a cylinder (like a long bone) is one of the strongest structures for its mass. Thus nature has given us an extremely strong, exceptionally simple (almost crude), and flexible supporting system without sacrificing mobility.

The hardness of bone is due to the inorganic calcium salts deposited in its ground substance. Its flexibility comes from the organic elements of the matrix, particularly the collagenic fibers.

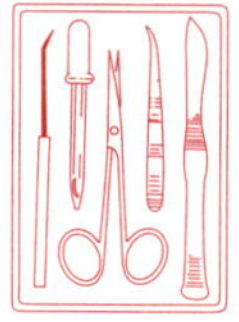

Obtain a bone sample that has been soaked in nitric acid (or vinegar) and one that has been baked. Heating removes the organic part of bone, while acid dissolves out the minerals. Do the treated bones retain the structure of untreated specimens?

___

Gently apply pressure to each bone sample. What happens to the heated bone?

___

___

The bone treated with acid?

___

___

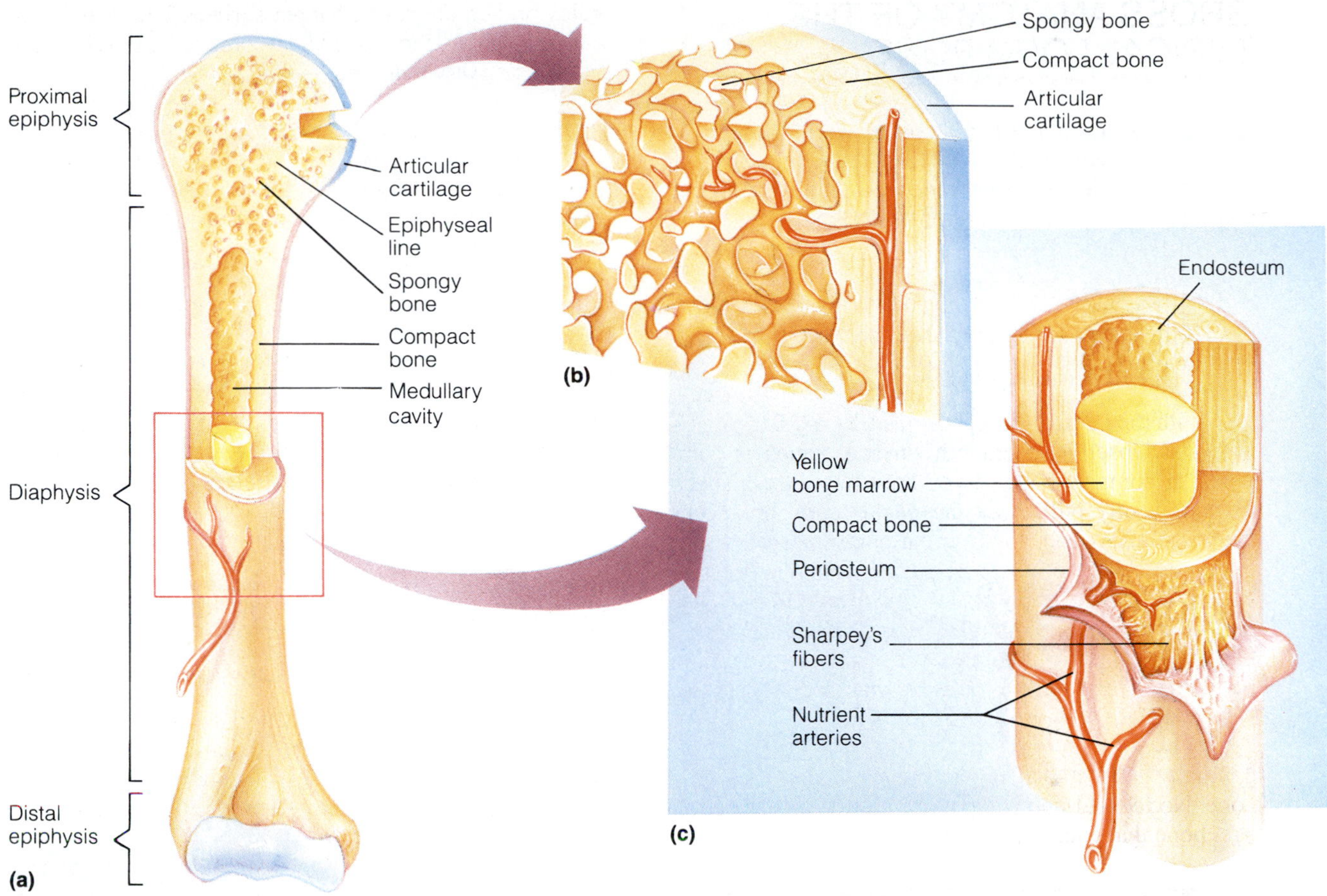

F9.2

The structure of a long bone (humerus of the arm). (a) Anterior view with longitudinal section cut away at the proximal end. **(b)** Pie-shaped, three-dimensional view of spongy bone and compact bone of the epiphysis. **(c)** Cross section of shaft (diaphysis). Note that the external surface of the diaphysis is covered by a periosteum, but the articular surface of the epiphysis is covered with hyaline cartilage.

What does the acid appear to remove from the bone?

___

What does baking appear to do to the bone?

___

___

In rickets, the bones are not properly calcified. Which of the demonstration specimens would more closely resemble the bones of a child with rickets?

___

MICROSCOPIC STRUCTURE OF COMPACT BONE

As you have seen, spongy bone has a spiky, open-work appearance, resulting from the arrangement of the **trabeculae** that compose it, while compact bone appears to be dense and homogeneous. Microscopic examination of compact bone, however, reveals that it is riddled with passageways carrying blood vessels, nerves, and lymphatic vessels that provide the living bone cells with needed substances and a way to eliminate wastes. Indeed, bone histology is much easier to understand when you recognize that bone tissue is organized around its blood supply.

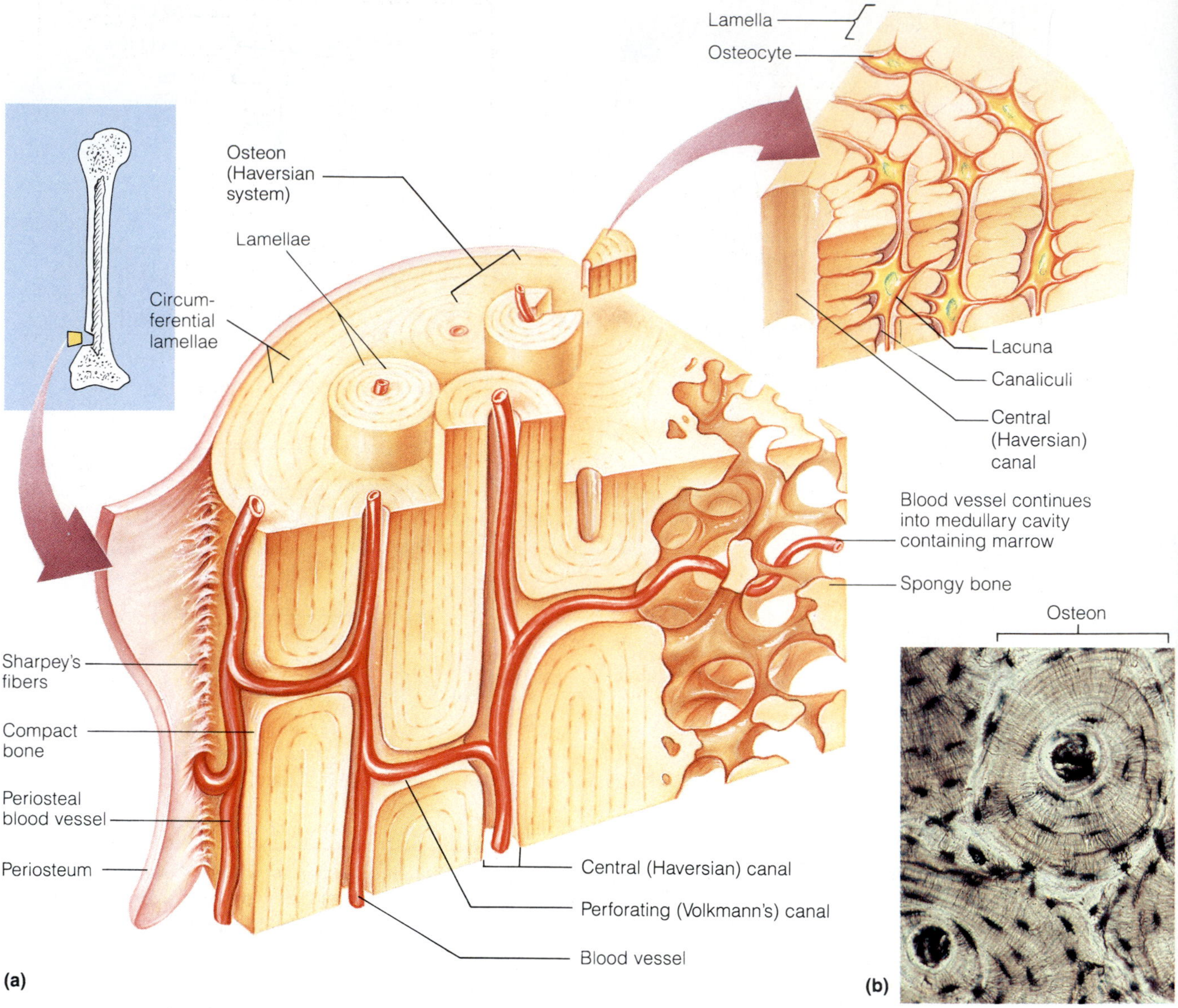

F9.3

Microscopic structure of compact bone. (a) Diagrammatic view of a pie-shaped segment of compact bone, illustrating its structural units (osteons). The inset shows a more highly magnified view of a portion of one osteon. Notice the position of osteocytes in lacunae (cavities of the matrix). **(b)** Photomicrograph of a cross-sectional view of one osteon (90×).

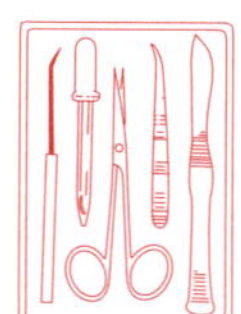

1. Obtain a prepared slide of ground bone and examine it under low power. Using Figure 9.3 as a guide, focus on a **central (Haversian) canal.** The central canal runs parallel to the long axis of the bone and carries blood vessels, nerves, and lymph vessels through the bony matrix. Identify the **osteocytes** (mature bone cells) in **lacunae** (chambers), which are arranged in concentric circles (concentric **lamellae**) around the central canal. A central canal and all the concentric lamellae surrounding it are referred to as an **osteon** or **Haversian system.** Also identify **canaliculi,** tiny canals radiating outward from a central canal to the lacunae of the first lamella and then from lamella to lamella. The canaliculi form a dense transportation network through the hard bone matrix, connecting all the living cells of the osteon to the nutrient supply. The canaliculi allow each cell to take what it needs for nourishment and to pass along the excess to the next osteocyte. You may need a higher-power magnification to see the fine canaliculi.

2. Also note the **perforating (Volkmann's) canals** in Figure 9.3. These canals run into the compact bone and marrow cavity from the periosteum, at right angles to the shaft. With the central canals, the perforating canals complete the communication pathway between the bone interior and its external surface.

3. If a model of bone histology is available, identify the same structures on the model.

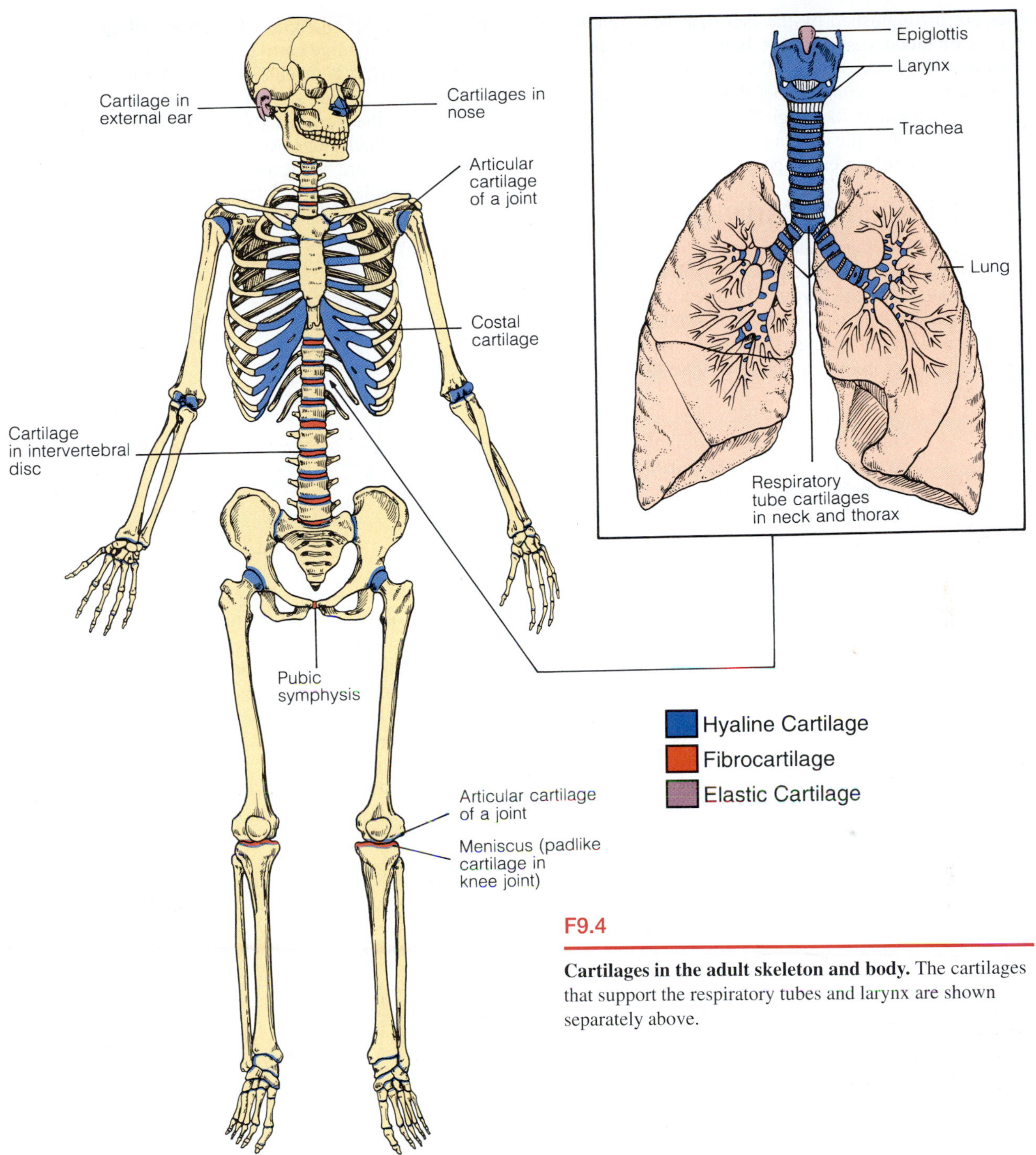

F9.4

Cartilages in the adult skeleton and body. The cartilages that support the respiratory tubes and larynx are shown separately above.

CARTILAGES OF THE SKELETON

Location and Basic Structure

As mentioned earlier, cartilaginous regions of the skeleton have a fairly limited distribution in adults (Figure 9.4). The most important of these skeletal cartilages are (1) **articular cartilages,** which cover the bone ends at movable joints; (2) **costal cartilages,** found connecting the ribs to the sternum (breastbone); (3) **laryngeal cartilages,** which largely construct the larynx (voicebox); (4) **tracheal** and **bronchial cartilages,** which reinforce other passageways of the respiratory system; (5) **nasal cartilages,** which support the external nose, (6) **intervertebral discs,** which separate and cushion bones of the spine (vertebrae); and (7) the cartilage supporting the external ear.

The skeletal cartilages consist of some variety of *cartilage tissue,* which typically consists primarily of water and is fairly resilient. Additionally, cartilage tissues are distinguished by the fact that they contain no nerves or blood vessels. Like bones, each cartilage is surrounded by a covering of dense connective tissue, called a *perichondrium* (rather than a periosteum). The perichondrium acts like a girdle to resist distortion of the cartilage when the cartilage is subjected to pressure. It also plays a role in cartilage growth and repair.

Classification of Cartilage

The skeletal cartilages have representatives from each of the three cartilage tissue types—hyaline, elastic, and fibrocartilage. Although you have already studied cartilage tissues (Exercise 6), some of that information will be recapped briefly here and you will have a chance to review the microscope structure unique to each cartilage type.

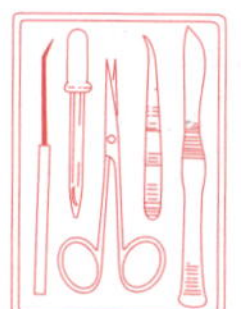

Obtain prepared slides of hyaline cartilage, elastic cartilage, and fibrocartilage and bring them to your laboratory bench for viewing.

As you read through the descriptions of these cartilage types, keep in mind that the bulk of cartilage tissue consists of a nonliving *matrix* (containing a jellylike ground substance and fibers) secreted by chondrocytes.

HYALINE CARTILAGE **Hyaline cartilage** looks like frosted glass when viewed by the unaided eye. As easily seen in Figure 9.4, most skeletal cartilages are composed of hyaline cartilage. Its chondrocytes, snugly housed in lacunae, appear spherical and collagen fibers are the only fiber type in its matrix. Hyaline cartilage provides sturdy support with some resilience or "give." Draw a small section of hyaline cartilage in the circle below. Label the chrondrocytes, lacunae, and the cartilage matrix. Compare your drawing to Figure 6.5g, p. 51.

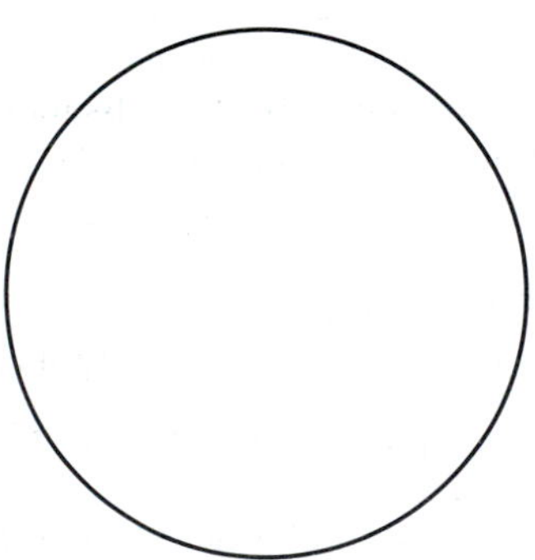

ELASTIC CARTILAGE **Elastic cartilage** can be envisioned as "hyaline cartilage with more elastic fibers." Consequently, it is much more flexible than hyaline cartilage and it tolerates repeated bending better. Essentially, only the cartilages of the external ear and the epiglottis (which flops over and covers the larynx when we swallow) are made of elastic cartilage. Focus on how this cartilage differs from hyaline cartilage as you diagram it in the circle below. Compare your illustration to Figure 6.5h, p. 52.

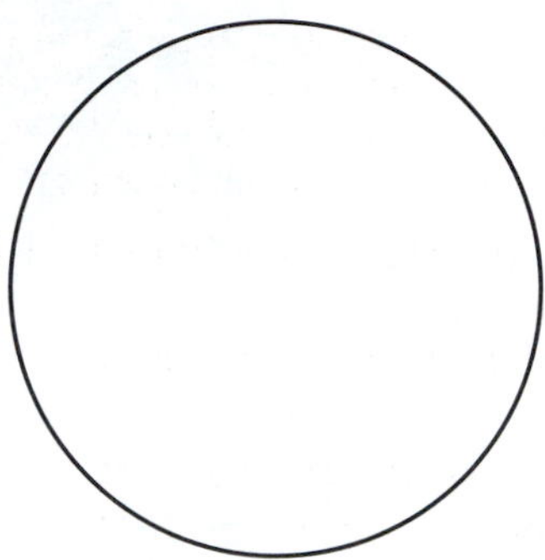

FIBROCARTILAGE **Fibrocartilage** consists of rows of chondrocytes alternating with rows of thick collagen fibers. This tissue looks like a cartilage-dense regular connective tissue hybrid, and it is always found where hyaline cartilage joins a tendon or ligament. Fibrocartilage has great tensile strength and can withstand heavy compression. Hence, its use to construct the intervertebral discs and the cartilages within the knee joint makes a lot of sense (see Figure 9.4). Sketch a section of this tissue in the circle provided and then compare your sketch to the view seen in Figure 6.5i, p. 52.

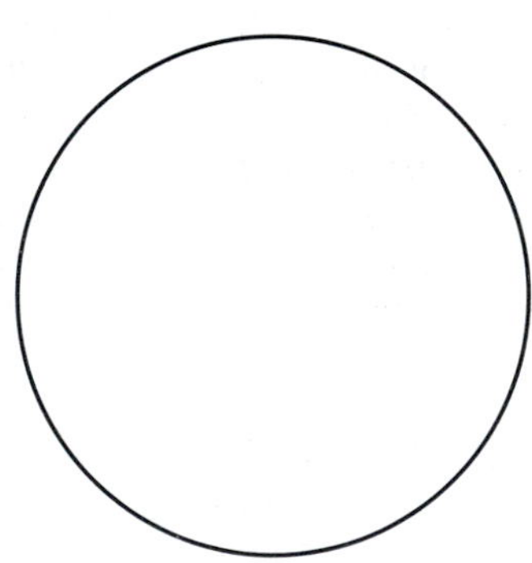

The Axial Skeleton

OBJECTIVES

1. To identify and name the three bone groups composing the axial skeleton (skull, bony thorax, and vertebral column).
2. To identify the bones composing the axial skeleton, either by examining the isolated bones or by pointing them out on an articulated skeleton or a skull, and to name the important bone markings on each.
3. To distinguish by examination the different types of vertebrae.
4. To discuss the importance of the intervertebral fibrous discs and spinal curvatures.
5. To distinguish the three abnormal spinal curvatures (lordosis, kyphosis, and scoliosis).

MATERIALS

Intact skull and Beauchene skull
X rays of individuals with scoliosis, lordosis, and kyphosis (if available)
Articulated skeleton, articulated vertebral column
Isolated cervical, thoracic, and lumbar vertebrae, sacrum, and coccyx

See Appendix D, Exercise 10 for links to A.D.A.M. Standard.

See Appendix E, Exercise 10 for links to *Anatomy and PhysioShow: The Videodisc.*

The **axial skeleton** (the green portion of Figure 9.1 on p. 66) can be divided into three parts: the skull, the vertebral column, and the bony thorax.

THE SKULL

The **skull** is composed of two sets of bones. Those of the **cranium** enclose and protect the fragile brain tissue. The **facial bones** present the eyes in an anterior position and form the base for the facial muscles, which make it possible for us to present our feelings to the world. All but one of the bones of the skull are joined by interlocking joints called *sutures;* the mandible, or lower jawbone, is attached to the rest of the skull by a freely movable joint.

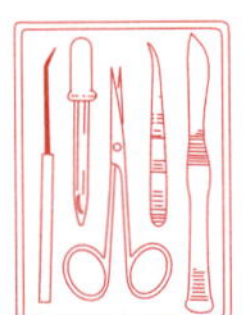

The bones of the skull, shown in Figures 10.1 through 10.4, are described below. As you read through this material, identify each bone on an intact (and/or Beauchene) skull. Note that important bone markings are listed beneath the bones on which they appear and that a color-coding dot before each bone name indicates its color in the figures.

The Cranium

The cranium may be divided into two major areas for study—the **cranial vault** or **calvaria,** forming the superior, lateral, and posterior walls of the skull, and the **cranial floor** or **base,** forming the skull bottom. Internally, the cranial floor has three distinct concavities, the **anterior, middle,** and **posterior cranial fossae** (see Figure 10.4). The brain sits in these fossae, completely enclosed by the cranial vault.

Eight large flat bones construct the cranium. *With the exception of two paired bones (the parietals and the temporals), all are single bones.* Sometimes the six ossicles of the middle ear are also considered part of the cranium. Because the ossicles are functionally part of the hearing apparatus, their consideration is deferred to Exercise 25, Special Senses: Hearing and Equilibrium.

○ FRONTAL See Figures 10.1, 10.2, and 10.4. Anterior portion of cranium; forms the forehead, superior part of the orbit, and floor of anterior cranial fossa.

Supraorbital foramen (notch): opening above each orbit allowing blood vessels and nerves to pass.
Glabella: smooth area between the eyes.

○ PARIETAL See Figures 10.1 and 10.2. Posterolateral to the frontal bone, forming sides of cranium.

Sagittal suture: midline articulation point of the two parietal bones.
Coronal suture: point of articulation of parietals with frontal bone.

○ TEMPORAL See Figures 10.1 through 10.4. Inferior to parietal bone on lateral skull. The temporals can be divided into four major parts: the **squamous region** abuts the parietals; the **tympanic region** surrounds the external ear opening; the **mastoid region** is the area posterior to the ear; and the **petrous region** forms the lateral region of the skull base.

Important markings associated with the flaring squamous region (Figures 10.2 and 10.3) include:

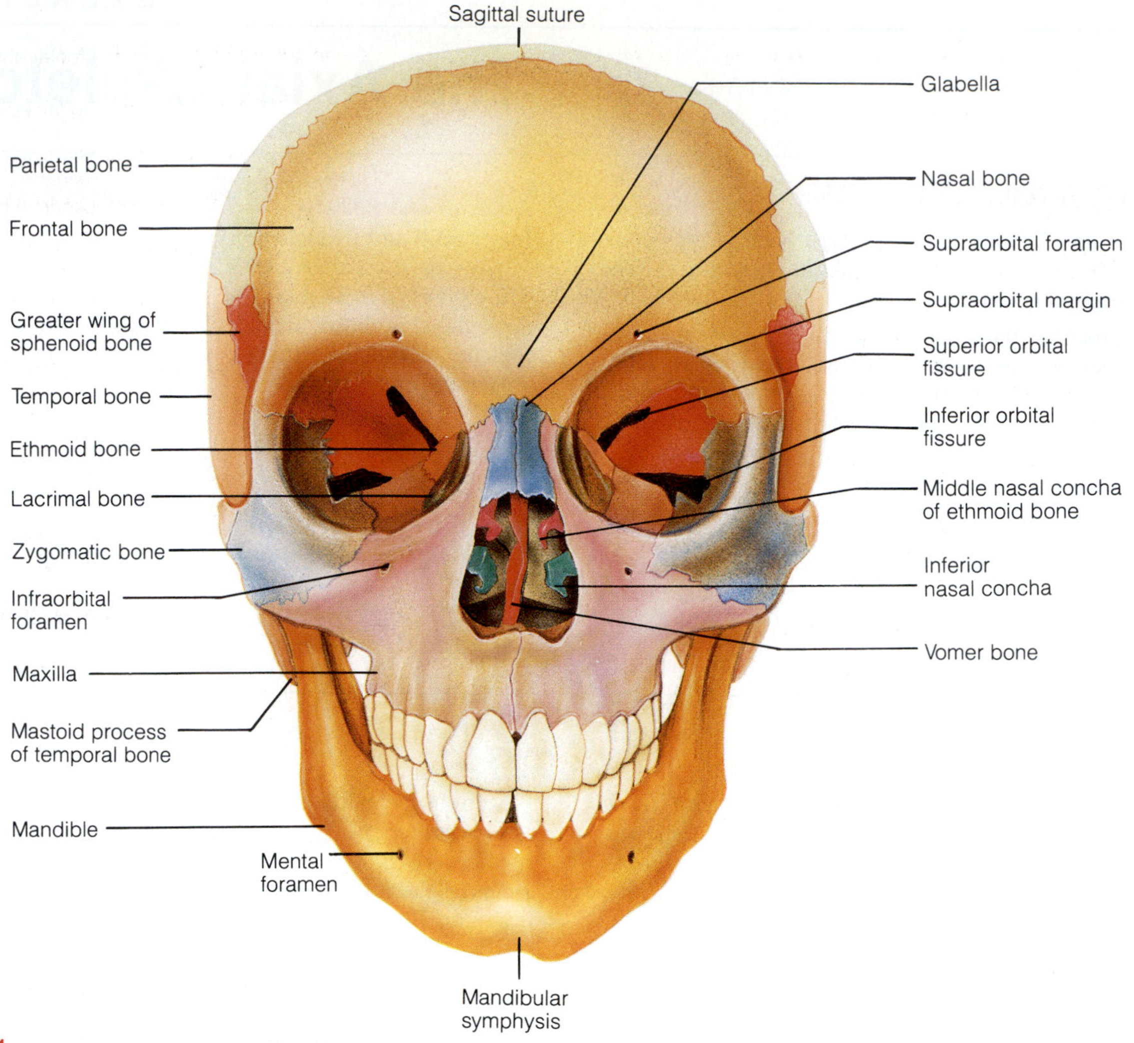

F10.1

Anatomy of the anterior aspect of the skull.

Squamosal suture: point of articulation of the temporal bone with parietal bone.

Zygomatic process: a bridgelike projection joining the zygomatic bone (cheekbone) anteriorly. Together these two bones form the *zygomatic arch.*

Mandibular fossa: rounded depression on the inferior surface of the zygomatic process (anterior to the ear); forms the socket for the mandibular condyle, the point where the mandible (lower jaw) joins the cranium.

The markings of the tympanic region (Figures 10.2 and 10.3) include:

External auditory meatus: canal leading to eardrum and middle ear.

Styloid (*stylo*=stake, pointed object) **process:** needlelike projection inferior to external auditory meatus; attachment point for muscles and ligaments of the neck. This process is often missing from (broken off) demonstration skulls.

Prominent structures in the mastoid region (Figures 10.2 and 10.3) are:

Mastoid process: rough projection inferior and posterior to external auditory meatus; attachment site for muscles.

The mastoid process is full of air cavities and is so close to the middle ear, a trouble spot for infections, that it often becomes infected too, a condition referred to as **mastoiditis.** Because the mastoid area is separated from the brain by only a very thin layer of bone, an ear infection that has spread to the mastoid process can inflame the brain coverings or the meninges. The latter condition is known as **meningitis.** ■

Stylomastoid foramen: tiny opening between the mastoid and styloid processes through which cranial nerve VII leaves the cranium.

The petrous region (Figure 10.3), which helps form the middle and posterior cranial fossae, exhibits several obvious foramina with important functions:

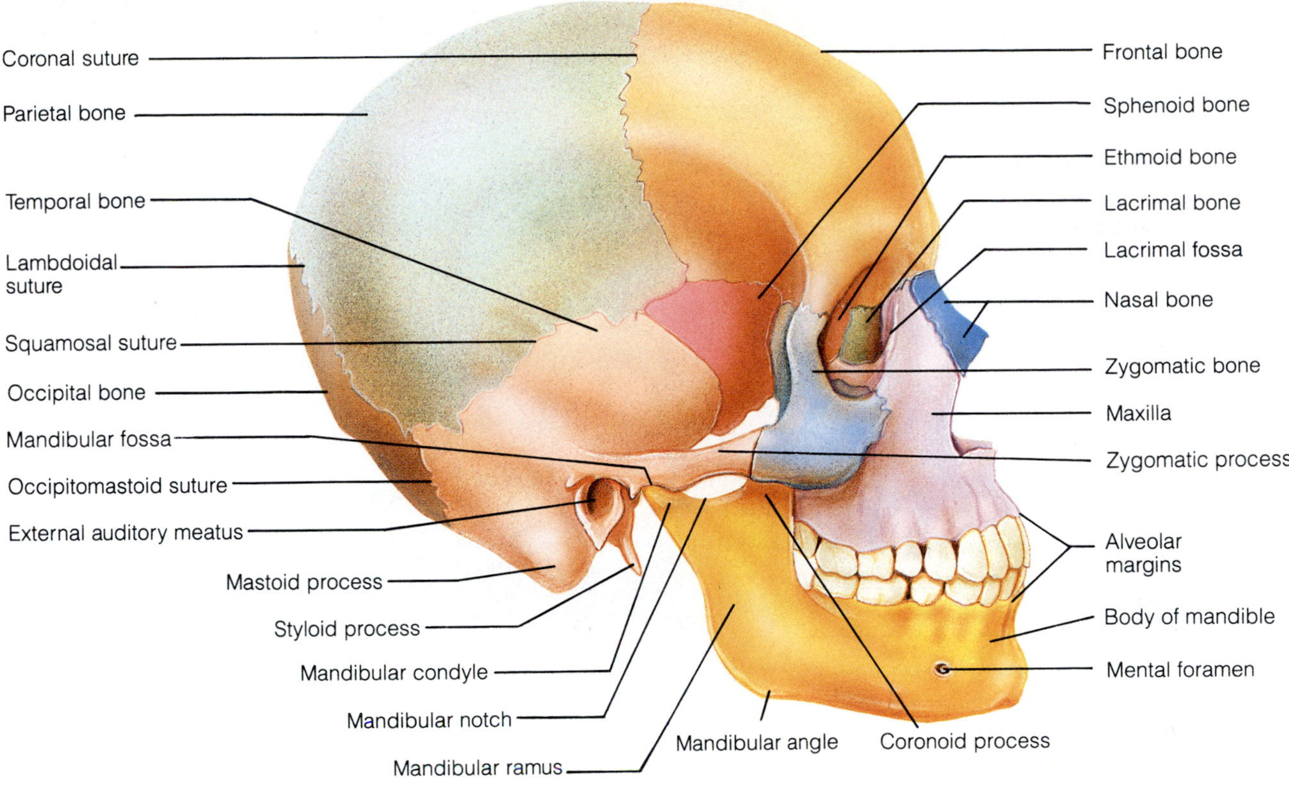

F10.2

External anatomy of the right lateral aspect of the skull.

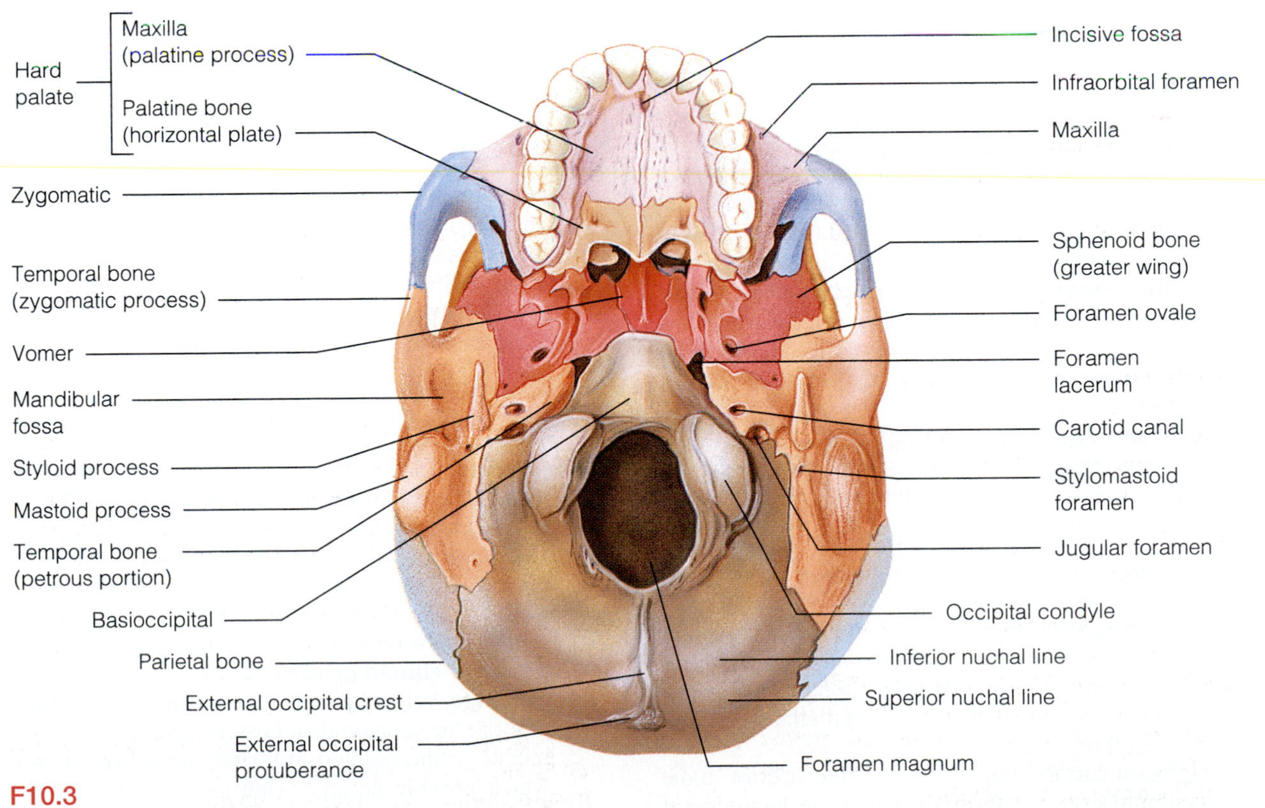

F10.3

Inferior superficial view of the skull, mandible removed.

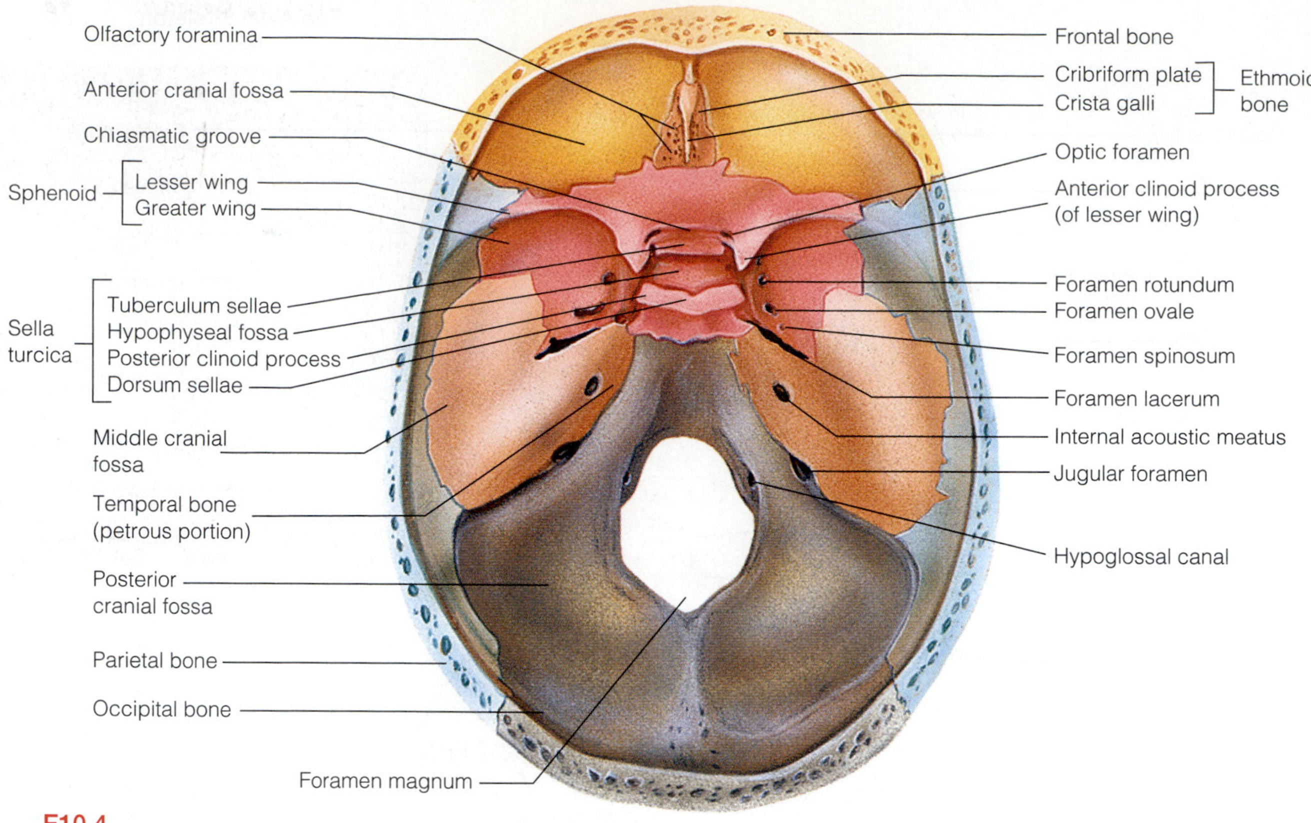

F10.4

Superior view of the floor of the cranial cavity, calvaria removed.

Jugular foramen: opening medial to styloid process through which the internal jugular vein and cranial nerves IX, X, and XI pass.

Carotid canal: opening, medial to the styloid process, through which the internal carotid artery passes into the cranial cavity.

Internal acoustic meatus: opening on posterior aspect (petrous portion) of temporal bone allowing passage of cranial nerves VII and VIII (Figure 10.4).

Foramen lacerum: a jagged opening between the petrous temporal bone and the sphenoid providing passage for a number of small nerves, and for the internal carotid artery to enter the middle cranial fossa (after it passes through part of the temporal bone).

● OCCIPITAL See Figures 10.2, 10.3, 10.4. Most posterior bone of cranium—forms floor and back wall. Joins sphenoid bone anteriorly via its narrow basioccipital region.

Lambdoidal suture: site of articulation of occipital bone and parietal bones.

Foramen magnum: large opening in base of occipital, which allows the spinal cord to join with the brain.

Occipital condyles: rounded projections lateral to the foramen magnum that articulate with the first cervical vertebra (atlas).

Hypoglossal canal: opening medial and superior to the occipital condyle through which the hypoglossal nerve (cranial nerve XII) passes.

External occipital crest and protuberance: midline prominences posterior to the foramen magnum.

● SPHENOID See Figures 10.1 through 10.4. Bat-shaped bone forming the anterior plateau of the middle cranial fossa across the width of the skull.

Greater wings: portions of the sphenoid seen exteriorly anterior to the temporal and forming a portion of the orbits of the eyes.

Superior orbital fissures: jagged openings in orbits providing passage for cranial nerves III, IV, V, and VI to enter the orbit where they serve the eye.

The sphenoid bone can be seen in its entire width if the top of the cranium (calvaria) is removed (Figure 10.4).

Sella turcica (Turk's saddle): a saddle-shaped region in the sphenoid midline which nearly encloses the pituitary gland in a living person. The pituitary gland sits in the **hypophyseal fossa** portion of the sella turcica. This fossa is abutted before and aft respectively by the **tuberculum sellae** and the **dorsum sellae.** The dorsum sellae terminates laterally in the **posterior clinoid processes.**

Lesser wings: bat-shaped portions of the sphenoid anterior to the sella turcica. Posteromedially these terminate in the pointed **anterior clinoid processes,** which provide an anchoring site for securing the brain within the skull.

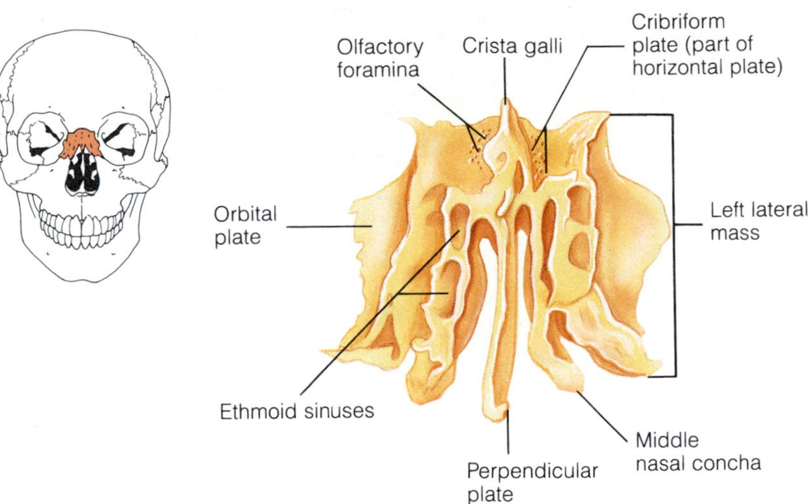

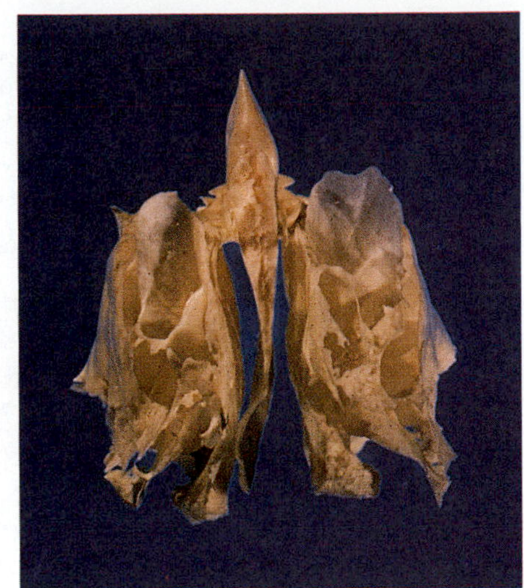

F10.5

The ethmoid bone. Anterior view.

Optic foramina: openings in the bases of the lesser wings through which the optic nerves enter the orbits to serve the eyes; these foramina are connected by the *chiasmatic groove*.

Foramen rotundum: opening lateral to sella turcica providing passage for a branch of the fifth cranial nerve. (This foramen is not visible on an inferior view of the skull.)

Foramen ovale: opening posterior to the sella turcica that allows passage of a branch of the fifth cranial nerve.

ETHMOID See Figures 10.1, 10.2, 10.4, and 10.5. Irregularly shaped bone anterior to the sphenoid. Forms the roof of the nasal cavity, upper nasal septum, and part of the medial orbit walls.

Crista galli (cock's comb): vertical projection providing a point of attachment for the dura mater (outermost membrane covering of the brain).

Cribriform plates: bony plates lateral to the crista galli through which olfactory fibers pass to the brain from the nasal mucosa. Together the cribriform plates and the midline crista galli form the *horizontal plate* of the ethmoid bone.

Perpendicular plate: Inferior projection of the ethmoid that forms the superior part of the nasal septum.

Lateral masses: Irregularly shaped thin-walled bony regions flanking the perpendicular plate laterally. Their lateral surfaces (*orbital plates*) shape part of the medial orbit wall.

Superior and middle nasal conchae (turbinates): thin, delicately coiled plates of bone extending medially from the lateral masses of the ethmoid into the nasal cavity. The conchae make air flow through the nasal cavity more efficient and greatly increase the surface area of the mucosa that covers them, thus increasing the mucosa's ability to warm and humidify incoming air.

Facial Bones

Of the 14 bones composing the face, 12 are paired. *Only the mandible and vomer are single bones.* An additional bone, the hyoid bone, although not a facial bone, is considered here because of its location. Refer to Figures 10.1 through 10.4 to find the structures described below.

MANDIBLE See Figures 10.1, 10.2, and 10.6. The lower jawbone, which articulates with the temporal bones providing the only freely movable joints of the skull.

Body: horizontal portion; forms the chin.

Ramus: vertical extension of the body on either side.

Mandibular condyle: articulation point of the mandible with the mandibular fossa of the temporal bone.

Coronoid process: jutting anterior portion of the ramus; site of muscle attachment.

Angle: posterior point at which ramus meets the body.

Mental foramen: prominent opening on the body (lateral to the midline) that transmits the mental blood vessels and nerve to the lower jaw.

Mandibular foramen: open the lower jaw of the skull to identify this prominent foramen on the medial aspect of the mandibular ramus. This foramen permits passage of the nerve involved with tooth sensation (mandibular branch of cranial nerve V) and is the site where the dentist injects Novocain to prevent pain while working on the lower teeth.

Alveolar margin: superior margin of mandible, contains sockets in which the teeth lie.

Mandibular symphysis: anterior median depression indicating point of mandibular fusion.

MAXILLAE See Figures 10.1, 10.2, and 10.3. Two bones fused in a median suture; form the upper jawbone and part of the orbits. All facial bones, except the mandible, join the maxillae. Thus they are the main, or keystone, bones of the face.

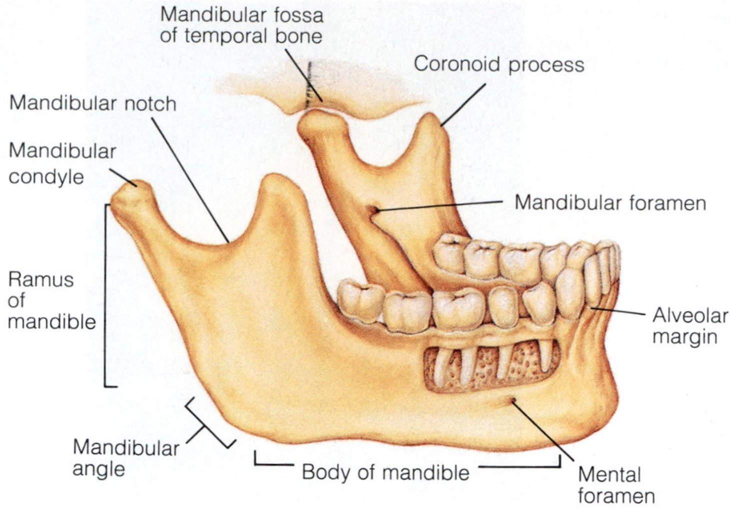

F10.6

The mandible. Isolated anterolateral view.

Alveolar margin: inferior margin containing sockets (alveoli) in which teeth lie.

Palatine processes: form the anterior hard palate.

Infraorbital foramen: opening under the orbit carrying the infraorbital nerves and blood vessels to the nasal region.

Incisive fossa: large bilateral opening located posterior to the central incisor tooth of the maxilla and piercing the hard palate; transmits the nasopalatine arteries and blood vessels.

PALATINE See Figure 10.3. Paired bones posterior to the palatine processes; form posterior hard palate and part of the orbit.

ZYGOMATIC See Figures 10.1, 10.2, and 10.3. Lateral to the maxilla; forms the portion of the face commonly called the cheekbone, and forms part of the lateral orbit. Its three processes are named for the bones with which they articulate.

LACRIMAL See Figures 10.1 and 10.2. Fingernail-sized bones forming a part of the medial orbit walls between the maxilla and the ethmoid. Each lacrimal bone is pierced by an opening, the **lacrimal fossa,** which serves as a passageway for tears (*lacrima* means "tear").

NASAL See Figures 10.1 and 10.2. Small rectangular bones forming the bridge of the nose.

VOMER (*vomer* = plow) See Figures 10.1 and 10.3. Blade-shaped bone in median plane of nasal cavity that forms the posterior and inferior nasal septum.

INFERIOR NASAL CONCHAE (turbinates) See Figure 10.1. Thin curved bones protruding medially from the lateral walls of the nasal cavity; serve the same purpose as the turbinate portions of the ethmoid bone (described earlier).

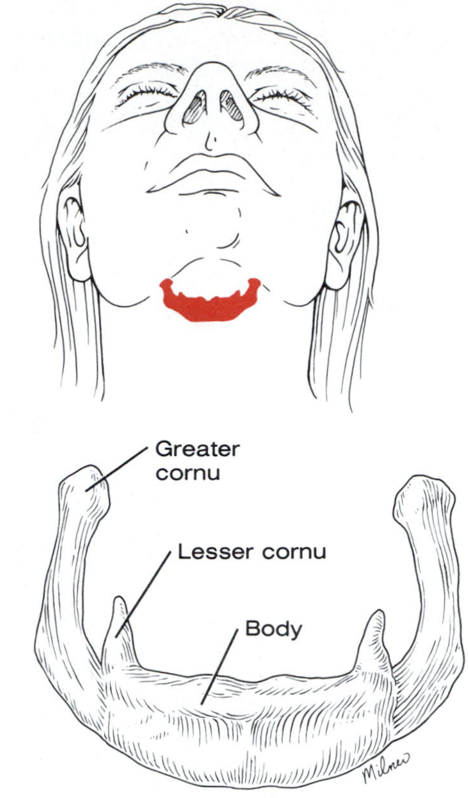

F10.7

Hyoid bone.

Hyoid Bone

Not really considered or counted as a skull bone. Located in the throat above the larynx (Figure 10.7); serves as a point of attachment for many tongue and neck muscles. Does not articulate with any other bone, and is thus unique. Horseshoe-shaped with a body and two pairs of horns, or **cornua.**

Paranasal Sinuses

Four skull bones—maxillary, sphenoid, ethmoid, and frontal—contain sinuses (mucosa-lined air cavities), which lead into the nasal passages (see Figure 10.8). These paranasal sinuses lighten the facial bones and may act as resonance chambers for speech. The maxillary sinus is the largest of the sinuses found in the skull.

Sinusitis, or inflammation of the sinuses, sometimes occurs as a result of an allergy or bacterial invasion of the sinus cavities. In such cases, some of the connecting passageways between the sinuses and nasal passages may become blocked with thick mucus or infectious material. Then, as the air in the sinus cavities is absorbed, a partial vacuum forms. The result is a sinus headache localized over the inflamed sinus area. Severe sinus infections may require surgical drainage to relieve this painful condition. ■

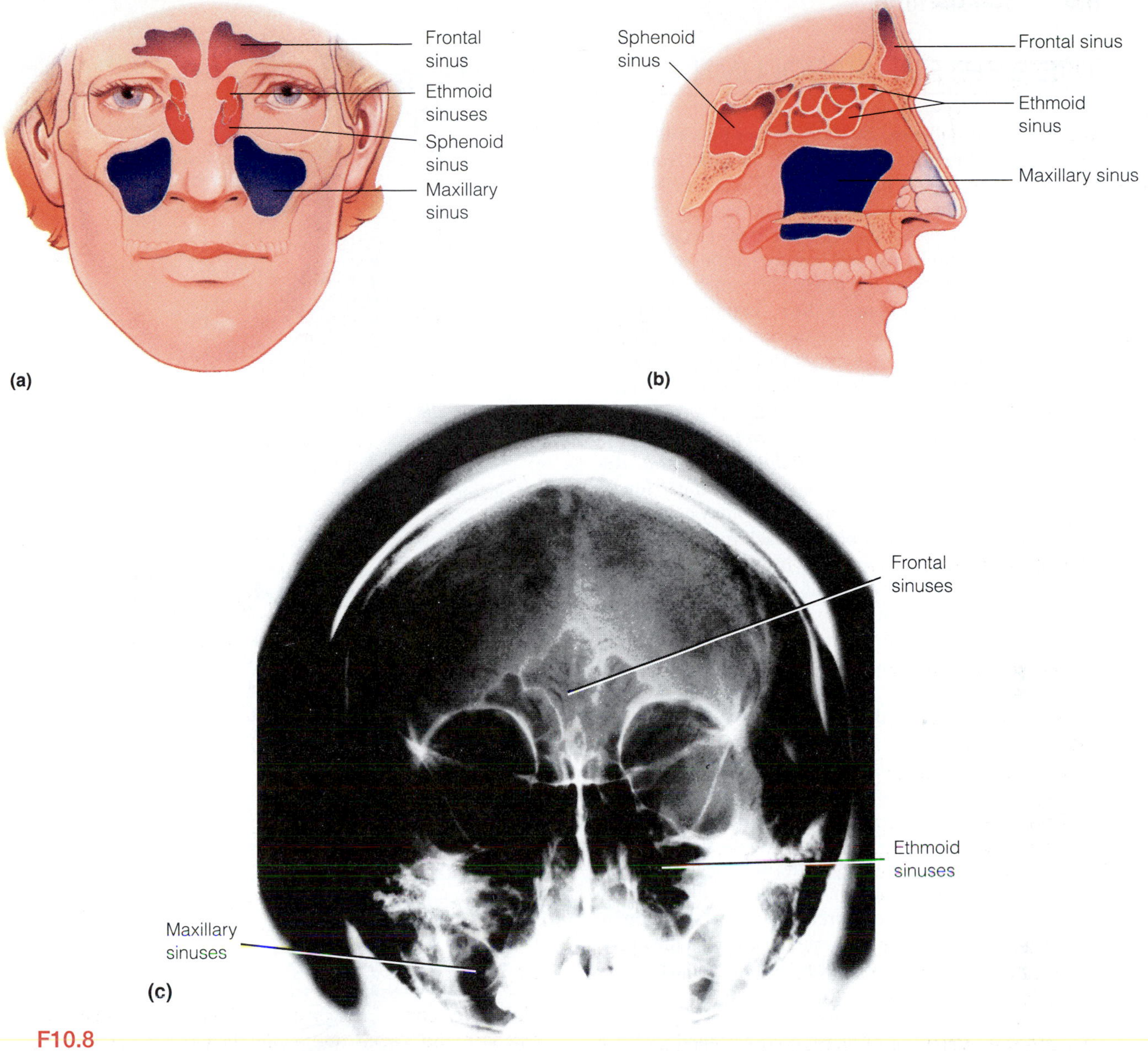

F10.8

Paranasal sinuses. **(a)** Anterior "see-through" view. **(b)** As seen in a sagittal section of the head. **(c)** Skull X ray showing three of the paranasal sinuses, anterior view.

Palpation of Selected Skull Markings

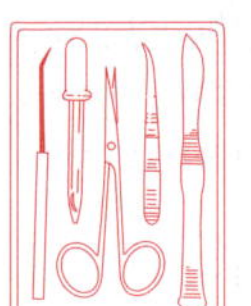

Palpate the following areas on yourself:

- Zygomatic bone and arch. (The most prominent part of your cheek is your zygomatic bone. Follow the posterior course of the zygomatic arch to its junction with your temporal bone.)
- Mastoid process (the rough area behind your ear).
- Temporomandibular joints. (Open and close your jaws to locate these.)
- Greater wing of sphenoid. (Find the indentation posterior to the orbit and superior to the zygomatic arch on your lateral skull.)
- Superior orbital foramen. (Apply firm pressure along the superior orbital margin to find the indentation resulting from this foramen.)
- Inferior orbital foramen. (Apply firm pressure along the inferomedial border of the orbit to locate this large foramen.)
- Mandibular angle (most inferior and posterior aspect of the mandible).
- Mandibular symphysis (midline of chin).
- Nasal bones. (Run your index finger and thumb along opposite sides of the bridge of your nose until they "slip" medially at the inferior end of the nasal bones.)
- External occipital protuberance. (This midline projection is easily felt by running your fingers up the furrow at the back of your neck to the skull.)
- Hyoid bone. (Place a thumb high behind the lateral edge of the mandible and squeeze medially.)

THE VERTEBRAL COLUMN

The **vertebral column,** extending from the skull to the pelvis, forms the body's major axial support. Additionally, it surrounds and protects the delicate spinal cord while allowing the spinal nerves to issue from the cord via openings between adjacent vertebrae. The term *vertebral column* might suggest a rather rigid supporting rod, but this is far from the truth. The vertebral column consists of 24 single bones called **vertebrae** and two composite, or fused, bones (the sacrum and coccyx) that are connected in such a way as to provide a flexible curved structure (Figure 10.9). Of the 24 single vertebrae, the seven bones of the neck are called *cervical vertebrae;* the next 12 are *thoracic vertebrae;* and the 5 supporting the lower back are *lumbar vertebrae.* Remembering common mealtimes for breakfast, lunch, and dinner (7 A.M., 12 noon, and 5 P.M.) may help you to remember the number of bones in each region.

The vertebrae are separated by pads of fibrocartilage, **intervertebral discs,** that cushion the vertebrae and absorb shocks. Each disc is composed of two major regions, a central gelatinous *nucleus pulposus* that behaves like a fluid, and an outer ring of encircling collagen fibers called the *annulus fibrosus* that stabilizes the disc and contains the pulposus.

As a person ages, the water content of the discs decreases (as it does in other tissues throughout the body), and the discs become thinner and less compressible. This situation, along with other degenerative changes such as weakening of the ligaments and tendons of the vertebral column, predisposes older people to **ruptured discs.** A ruptured disc is a situation in which the nucleus pulposus herniates through the annulus portion and typically compresses adjacent nerves. ■

The presence of the discs and the S-shaped or springlike construction of the vertebral column prevent shock to the head in walking and running and provide flexibility to the body trunk. The thoracic and sacral curvatures of the spine are referred to as *primary curvatures,* since they are present and well developed at birth. Later the *secondary curvatures* are formed. The cervical curvature becomes prominent when the baby begins to hold its head up independently, and the lumbar curvature develops when the baby begins to walk.

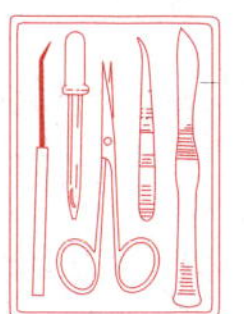

1. Observe the normal curvature of the vertebral column in your laboratory specimen, and compare it to Figure 10.9. Then examine Figure 10.10, which depicts three abnormal spinal curvatures—*scoliosis, kyphosis,* and *lordosis.* These abnormalities may result from disease or poor posture. Also examine X rays, if they are available, showing these same conditions in a living patient.

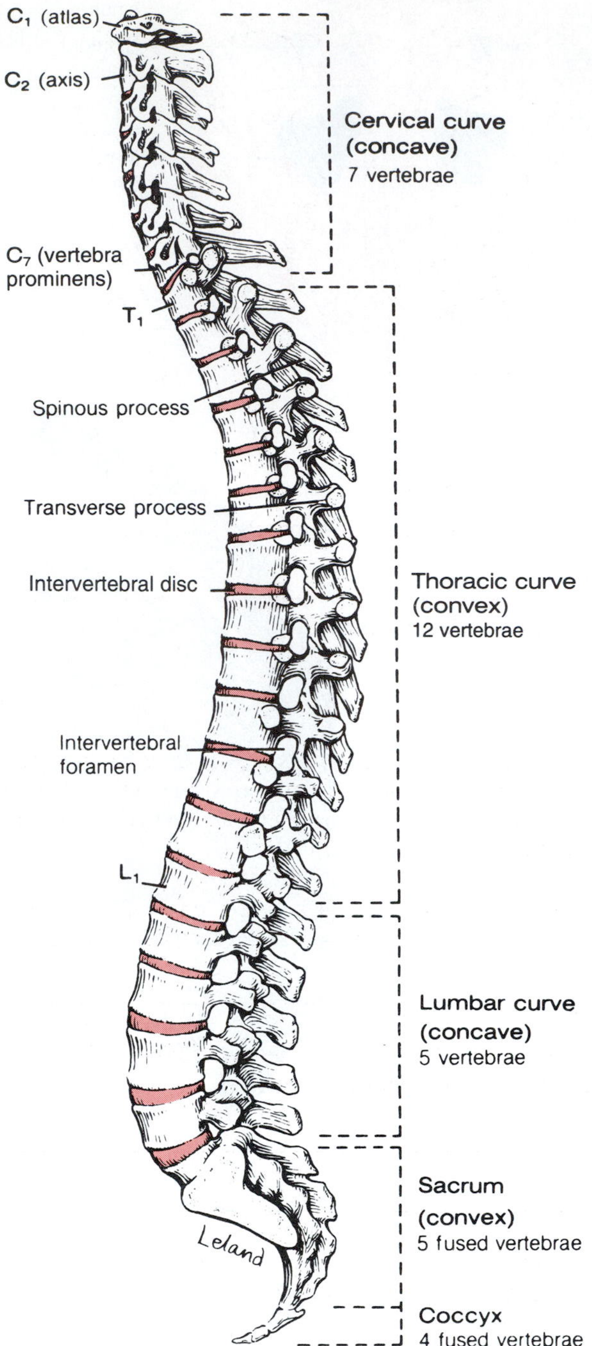

F10.9

The vertebral column. Notice the curvatures in the lateral view. (The terms *convex* and *concave* refer to the curvature of the posterior aspect of the vertebral column.)

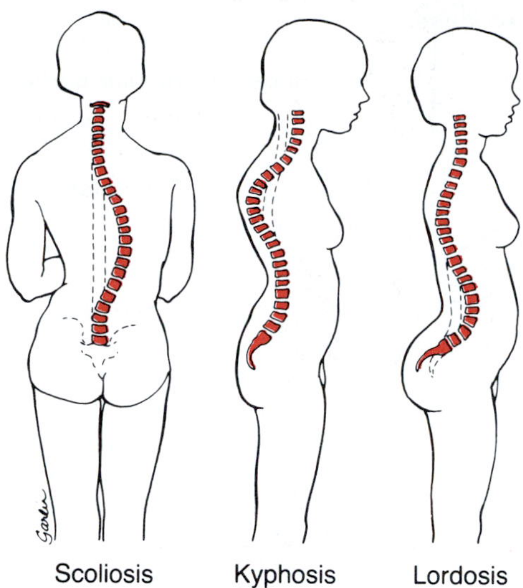

F10.10

Abnormal spinal curvatures.

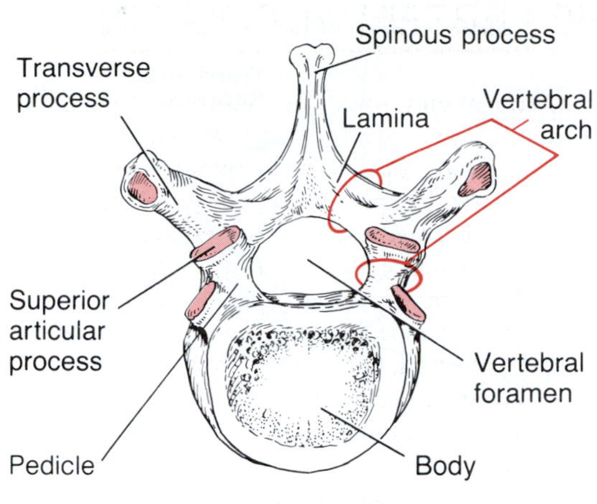

F10.11

A typical vertebra, superior view. Inferior articulating surfaces not shown.

2. Then using an articulated vertebral column (or an articulated skeleton), examine the freedom of movement between two lumbar vertebrae separated by an intervertebral disc.

When the fibrous disc is properly positioned, are the spinal cord or peripheral nerves impaired in any way?

___

Remove the disc and put the two vertebrae back together. What happens to the nerve?

___

___

What would happen to the spinal nerves in areas of malpositioned or "slipped" discs?

___

___

___

Structure of a Typical Vertebra

Although they differ in size and specific features, all vertebrae have some features in common (Figure 10.11).

Body (or centrum): rounded central portion of the vertebra, which faces anteriorly in the human vertebral column.

Vertebral arch: composed of pedicles, laminae, and a spinous process, it represents the junction of all posterior extensions from the vertebral body.

Vertebral foramen: opening enclosed by the body and vertebral arch; a conduit for the spinal cord.

Transverse processes: two lateral projections from the vertebral arch.

Spinous process: single medial and posterior projection from the vertebral arch.

Superior and inferior articular processes: paired projections lateral to the vertebral foramen that enable articulation with adjacent vertebrae. The superior articular processes typically face toward the spinous process, whereas the inferior articular processes face away from the spinous process.

Intervertebral foramina: the right and left pedicles have notches on their inferior and superior surfaces that create openings, the intervertebral foramina, for spinal nerves to leave the spinal cord between adjacent vertebrae.

Figures 10.12 and 10.13 show how specific vertebrae differ; refer to them as you read the following sections.

Cervical Vertebrae

The seven cervical vertebrae (referred to as C_1 through C_7) form the neck portion of the vertebral column. The first two cervical vertebrae (atlas and axis) are highly modified to perform special functions (see Figure 10.12). The **atlas** (C_1) lacks a body, and its lateral processes contain large concave depressions on their superior surfaces that receive the occipital condyles of the skull. This joint enables you to nod "yes." The **axis** (C_2) acts as a pivot for the rotation of the atlas (and skull) above. It bears a large vertical process, the **odontoid process,** or **dens,** which serves as the pivot point. The articulation between C_1 and C_2 allows you to rotate your head from side to side to indicate "no."

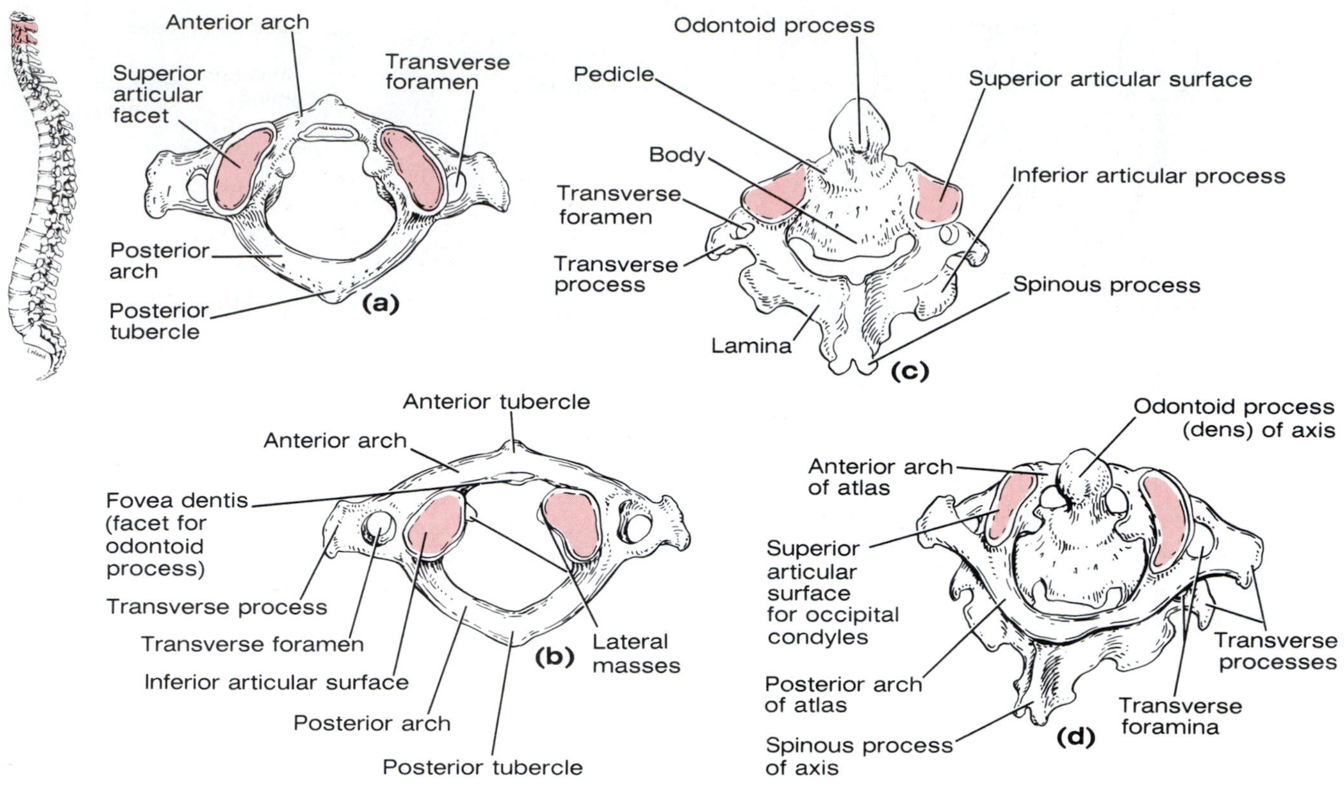

F10.12

Cervical vertebrae C_1 and C_2. **(a)** Superior view of the atlas (C_1). **(b)** Inferior view of the atlas (C_1). **(c)** Superior view of the axis (C_2). **(d)** Superior view of the articulated atlas and axis.

The more typical cervical vertebrae (C_3 through C_7) are distinguished from the thoracic and lumbar vertebrae by several features (see Figure 10.13a). They are the smallest, lightest vertebrae and the vertebral foramen is triangular. The spinous process is short and often bifurcated, or divided into two branches. The spinous process of C_7 is not branched, however, and is substantially longer than that of the other cervical vertebrae. Because the spinous process of C_7 is visible through the skin, it is called the *vertebra prominens* and is used as a landmark for counting the vertebrae. Transverse processes of the cervical vertebrae are wide, and they contain foramina through which the vertebral arteries pass superiorly on their way to the brain. Any time you see these foramina in a vertebra, you should know immediately that it is a cervical vertebra.

- Palpate your vertebra prominens.

Thoracic Vertebrae

The 12 thoracic vertebrae (referred to as T_1 through T_{12}) may be recognized by the following structural characteristics. As shown in Figure 10.13b, they have a larger body than the cervical vertebrae. The body is somewhat heart shaped, with two small articulating surfaces, or *costal demifacets* on each side (one superior, the other inferior) close to the origin of the vertebral arch. These demifacets articulate with the heads of the corresponding ribs. The vertebral foramen is oval or round, and the spinous process is long, with a sharp downward hook. The closer the thoracic vertebra is to the lumbar region, the less sharp and shorter the spinous process. Articular facets on the transverse processes articulate with the tubercles of the ribs. Besides forming the thoracic part of the spine, these vertebrae form the posterior aspect of the bony thoracic cage (rib cage). Indeed, they are the only vertebrae that articulate with the ribs.

Lumbar Vertebrae

The five lumbar vertebrae (L_1 through L_5) have massive blocklike bodies and short, thick, hatchet-shaped spinous processes extending backward horizontally (see Figure 10.13c). The superior articular facets are directed posteromedially; the inferior ones are directed anterolaterally. These structural features reduce the mobility of the lumbar region of the spine. Since most stress on the vertebral column occurs in the lumbar region, these are also the sturdiest of the vertebrae.

The spinal cord ends at the superior edge of L_2, but the outer covering of the cord, filled with cerebrospinal fluid, extends an appreciable distance beyond. Thus a *lumbar puncture* (for examination of the cerebrospinal fluid) or the administration of "saddle block" anesthesia for childbirth is normally done between L_3 and L_4 or L_4 and L_5, where there is little or no chance of injuring the delicate spinal cord.

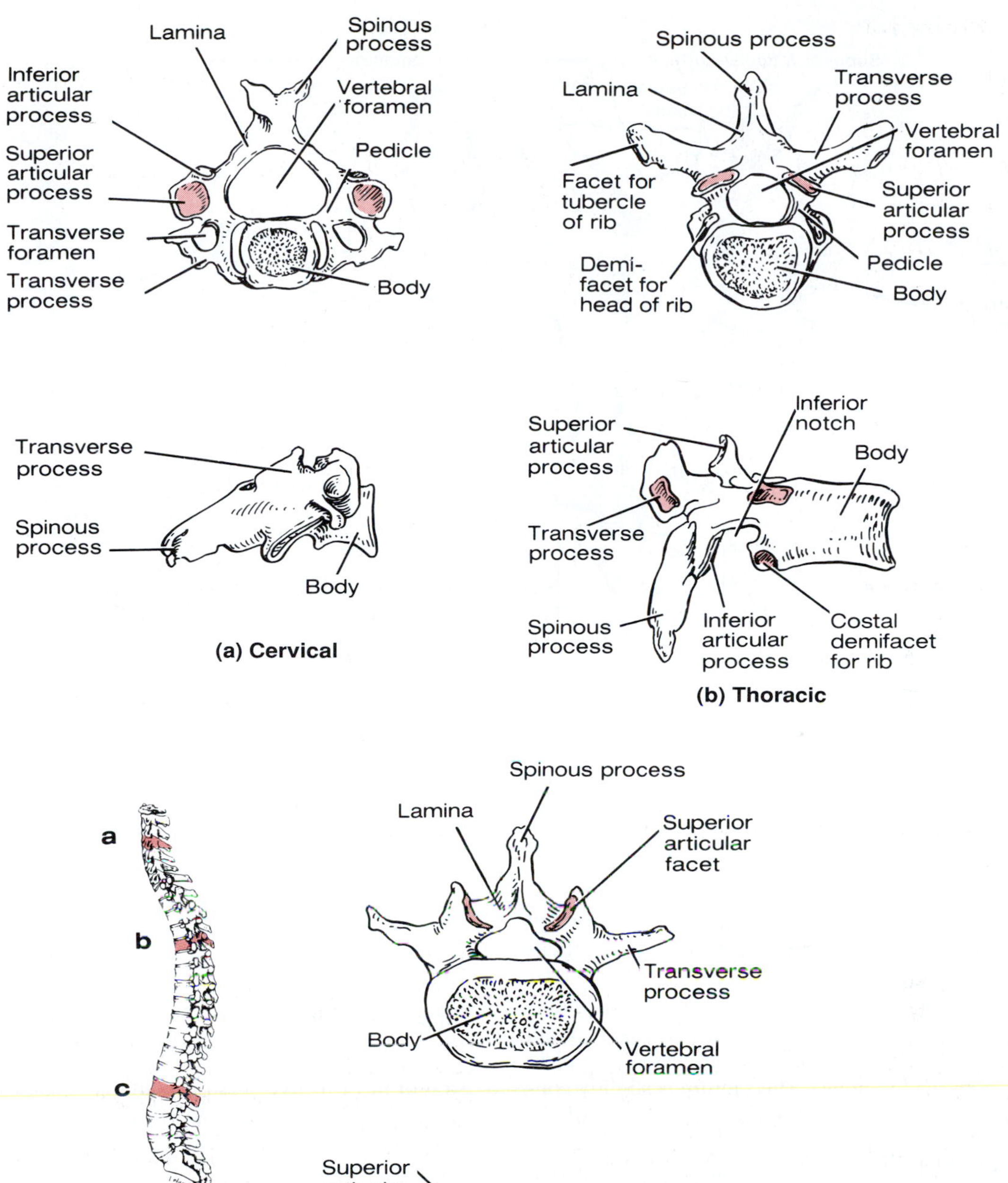

F10.13

Comparison of typical cervical, thoracic, and lumbar vertebrae (superior view above, lateral view below). (a) Cervical vertebra. **(b)** Thoracic vertebra. **(c)** Lumbar vertebra.

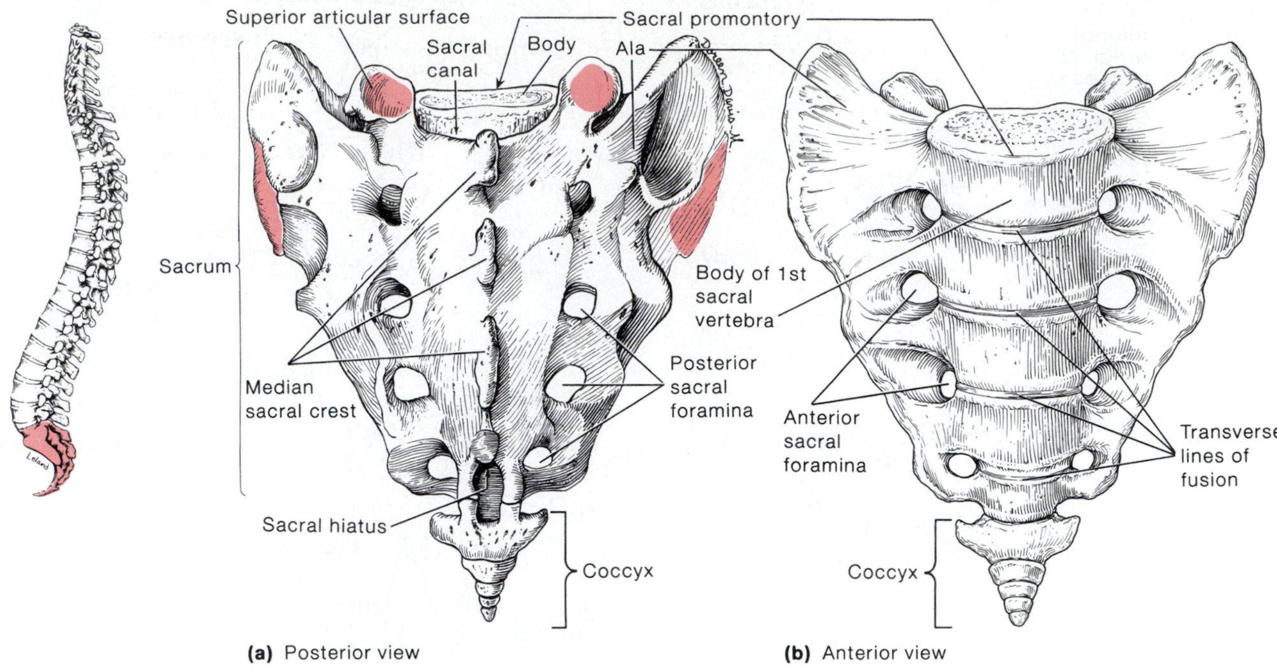

F10.14

Sacrum and coccyx. **(a)** Posterior view. **(b)** Anterior view.

The Sacrum

The **sacrum** (Figure 10.14) is a composite bone formed from the fusion of five vertebrae. Superiorly it articulates with L_5, and inferiorly it connects with the coccyx. The **median sacral crest** is a remnant of the spinous processes of the fused vertebrae. The winglike **alae,** formed by fusion of the transverse processes, articulate laterally with the hip bones. The sacrum is slightly concave anteriorly and forms the posterior border of the pelvis. Four ridges (lines of fusion) cross the anterior part of the sacrum, and **sacral foramina** are located at either end of these ridges. These foramina allow blood vessels and nerves to pass. The vertebral canal continues inside the sacrum as the **sacral canal** and terminates near the coccyx via an enlarged opening called the **sacral hiatus.** The **sacral promontory** (anterior border of the body of S_1) is an important anatomical landmark for obstetricians.

- Attempt to palpate the median sacral crest of your sacrum. (This is more easily done by thin people and [obviously] in privacy.)

The Coccyx

The **coccyx** (see Figure 10.14) is formed from the fusion of three to five small irregularly shaped vertebrae. It is literally the human tailbone, a vestige of the tail that other vertebrates have. The coccyx is attached to the sacrum by ligaments.

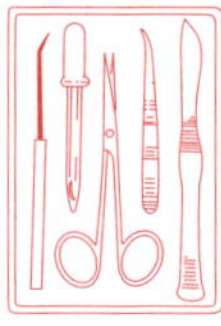

Obtain examples of each type of vertebra and examine them carefully, comparing them to Figures 10.12, 10.13, and 10.14 and to each other.

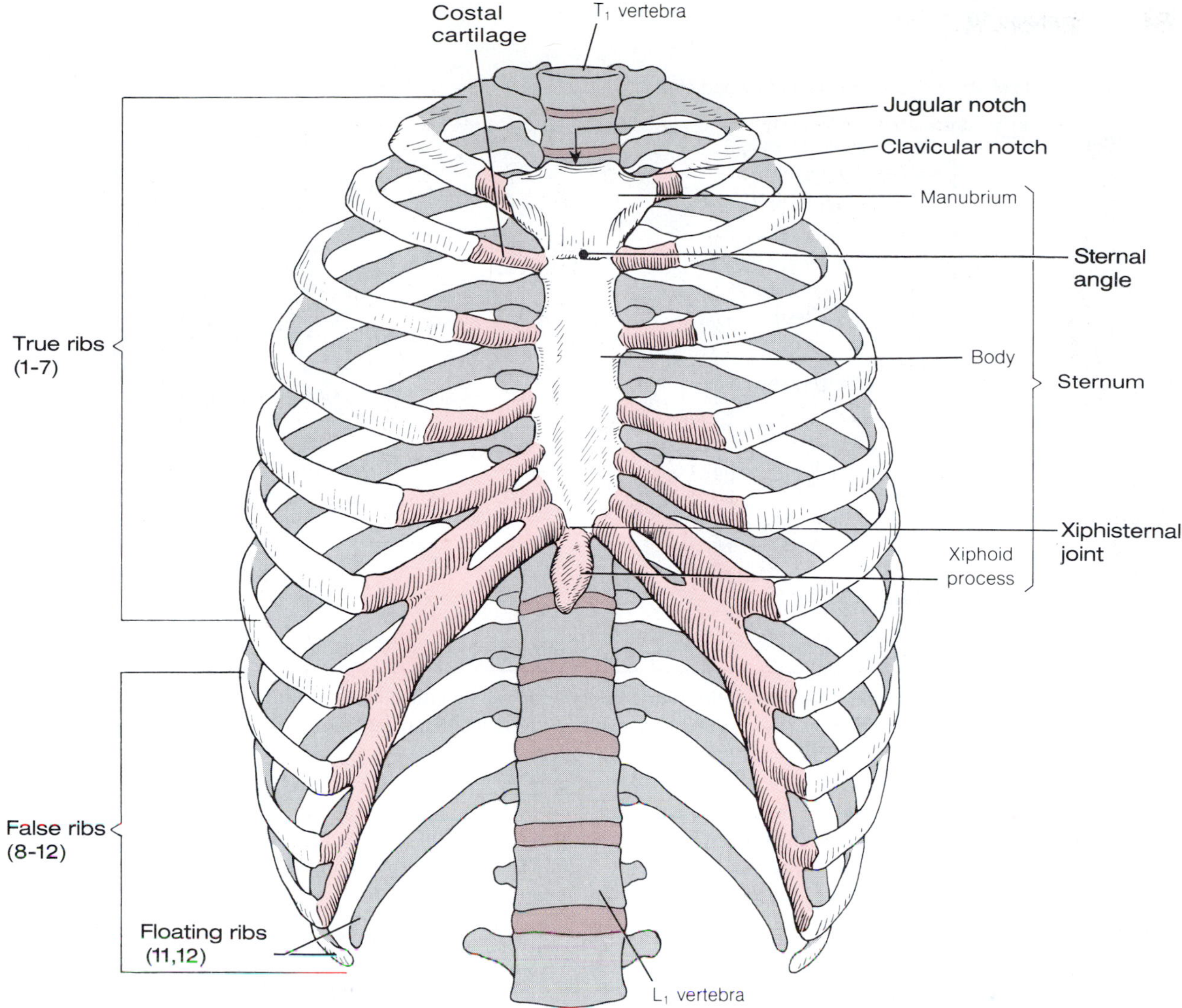

F10.15

Bony thorax, anterior view. (The pink areas are cartilages.)

THE BONY THORAX

The **bony thorax** is composed of the sternum, ribs, and thoracic vertebrae (Figure 10.15). It is also referred to as the **thoracic cage** because of its appearance and because it forms a protective cone-shaped enclosure around the organs of the thoracic cavity (heart and lungs, for example).

The Sternum

The **sternum** (breastbone), a typical flat bone, is a result of the fusion of three bones—the manubrium, body, and xiphoid process. It is attached to the first seven pairs of ribs. The superiormost **manubrium** looks like the knot of a tie; it articulates with the clavicle (collarbone) laterally. The **body (gladiolus)** forms the bulk of the sternum. The **xiphoid process** constructs the inferior end of the sternum and lies at the level of the fifth intercostal space. Although it is made of hyaline cartilage in children, it is usually ossified in adults.

In some people, the xiphoid process projects dorsally. This may present a problem because physical trauma to the chest can push such a xiphoid into the heart or liver (both immediately deep to the process), causing massive hemorrhage. ■

The sternum has three important bony landmarks—the jugular notch and the sternal angle. The **jugular notch** (concave upper border of the manubrium) can be palpated easily; generally it is at the level of the third thoracic vertebra. The **sternal angle** is a result of the manubrium and body meeting at a slight angle to each other, so that a transverse ridge is formed at the level of the second ribs. It provides a handy reference point for counting ribs to locate the second intercostal space for

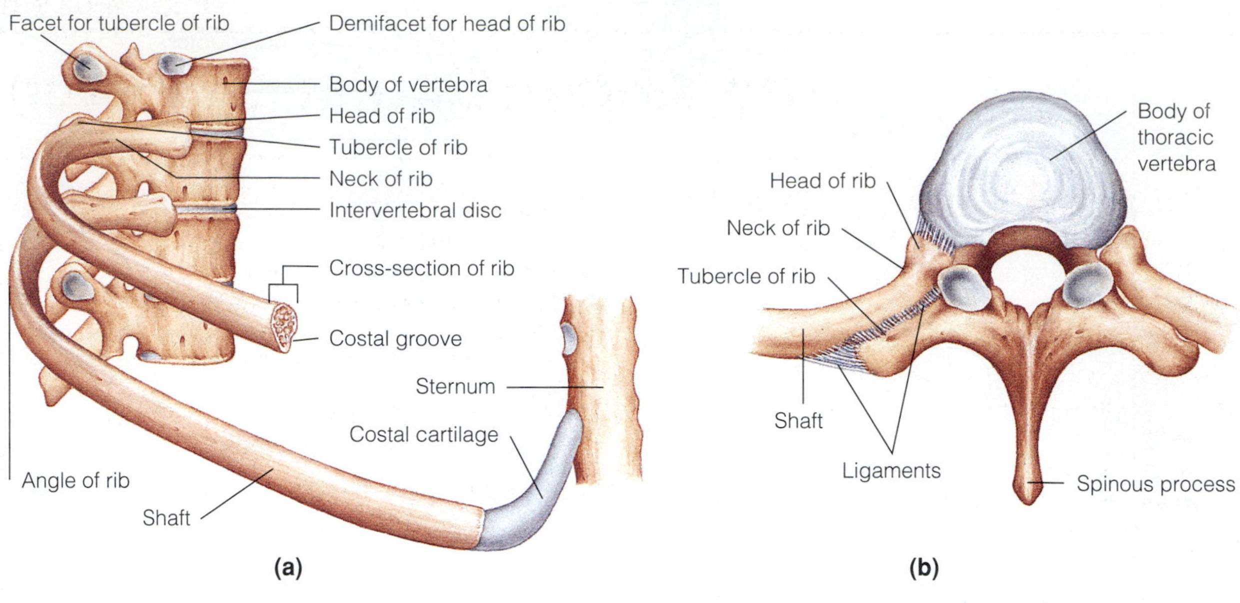

F10.16

Structure of a "typical" true rib and its articulations. (a) Vertebral and sternal articulations of a typical true rib. **(b)** Superior view of the articulation between a rib and a thoracic vertebra, with costovertebral ligaments shown on left side only.

listening to certain heart valves, and is an important anatomical landmark for thoracic surgery. The **xiphisternal joint,** the point where the sternal body and xiphoid process fuse, lies at the level of the ninth thoracic vertebra.

- Palpate your sternal angle and jugular notch.

Because of its accessibility, the sternum is a favored site for obtaining samples of blood-forming (hematopoietic) tissue for the diagnosis of suspected blood diseases. A needle is inserted into the marrow of the sternum and the sample withdrawn (sternal puncture).

The Ribs

The 12 pairs of **ribs** form the walls of the thoracic cage (see Figures 10.15 and 10.16). All of the ribs articulate posteriorly with the vertebral column via their heads and tubercles and then curve downward and toward the anterior body surface. The first seven pairs, called the *true,* or *vertebrosternal, ribs,* attach directly to the sternum by their "own" costal cartilages. The next three pairs, called *false,* or *vertebrochondral, ribs,* have indirect cartilage attachments to the sternum. The last two pairs, called *floating,* or *vertebral, ribs*, have no sternal attachment at all.

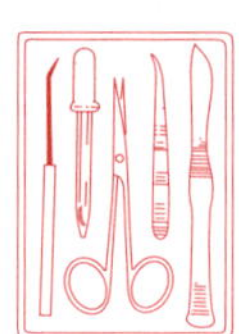

First take a deep breath to expand your chest. Notice how your ribs seem to move outward and how your sternum rises. Then examine an articulated skeleton to observe the relationship between the ribs and the vertebrae.

The Appendicular Skeleton

OBJECTIVES

1. To identify on an articulated skeleton the bones of the pectoral and pelvic girdles and their attached limbs.
2. To arrange unmarked, disarticulated bones in their proper relative position to form the entire skeleton.
3. To differentiate between a male and a female pelvis.
4. To discuss the common features of the human appendicular girdles (pectoral and pelvic), and to note how their structure relates to their specialized functions.
5. To identify specific bone markings in the appendicular skeleton.

MATERIALS

Articulated skeletons
Disarticulated skeletons (complete)
Articulated pelves (male and female for comparative study)
X rays of bones of the appendicular skeleton

See Appendix D, Exercise 11 for links to A.D.A.M. Standard.

See Appendix E, Exercise 11 for links to *Anatomy and PhysioShow: The Videodisc.*

The **appendicular skeleton** (the gold-colored portion of Figure 9.1) is composed of the 126 bones of the appendages and the pectoral and pelvic girdles, which attach the limbs to the axial skeleton. Although the bones of the upper and lower limbs are quite different in their functions and mobility, they have the same fundamental plan, with each limb composed of three major segments connected together by freely movable joints.

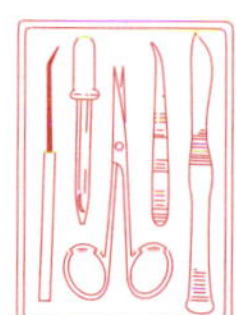

Carefully examine each of the bones described and identify the characteristic bone markings of each. The markings aid in determining whether a bone is the right or left member of its pair. *This is a very important instruction because, before completing this laboratory exercise, you will be constructing your own skeleton.* Additionally, when corresponding X rays are available, compare the actual bone specimen to its X-ray image.

BONES OF THE PECTORAL GIRDLE AND UPPER EXTREMITY

The Pectoral Girdle

The paired **pectoral,** or **shoulder, girdles** (Figure 11.1) each consist of two bones—the anterior clavicle and the posterior scapula. The shoulder girdles function to at-

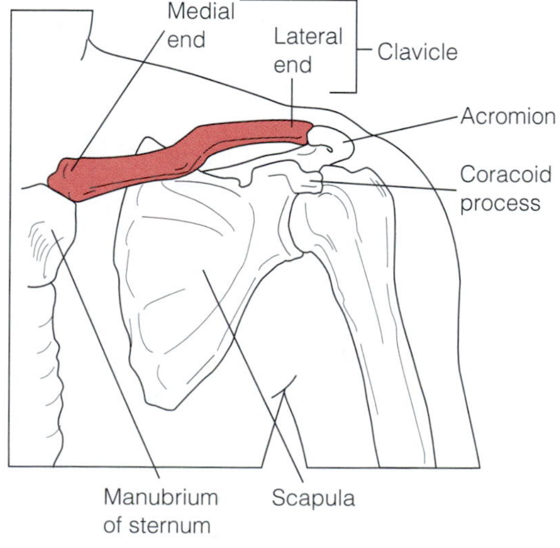

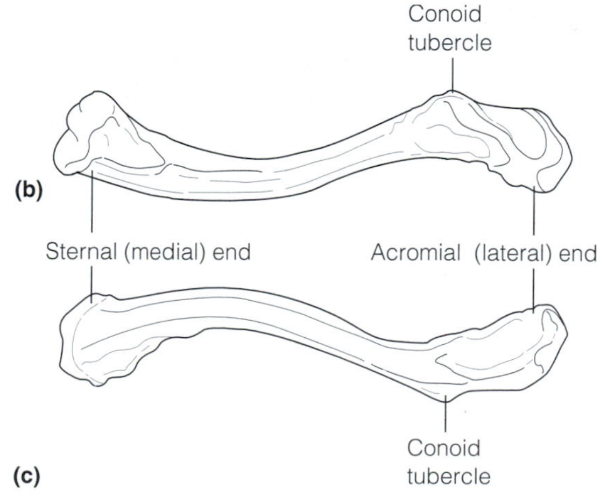

F11.1

Bones of the pectoral (shoulder) girdle. (a) Left pectoral girdle articulated to show the relationship of the girdle to the bones of the thorax and arm. **(b)** Left clavicle, superior view. **(c)** Left clavicle, inferior view.

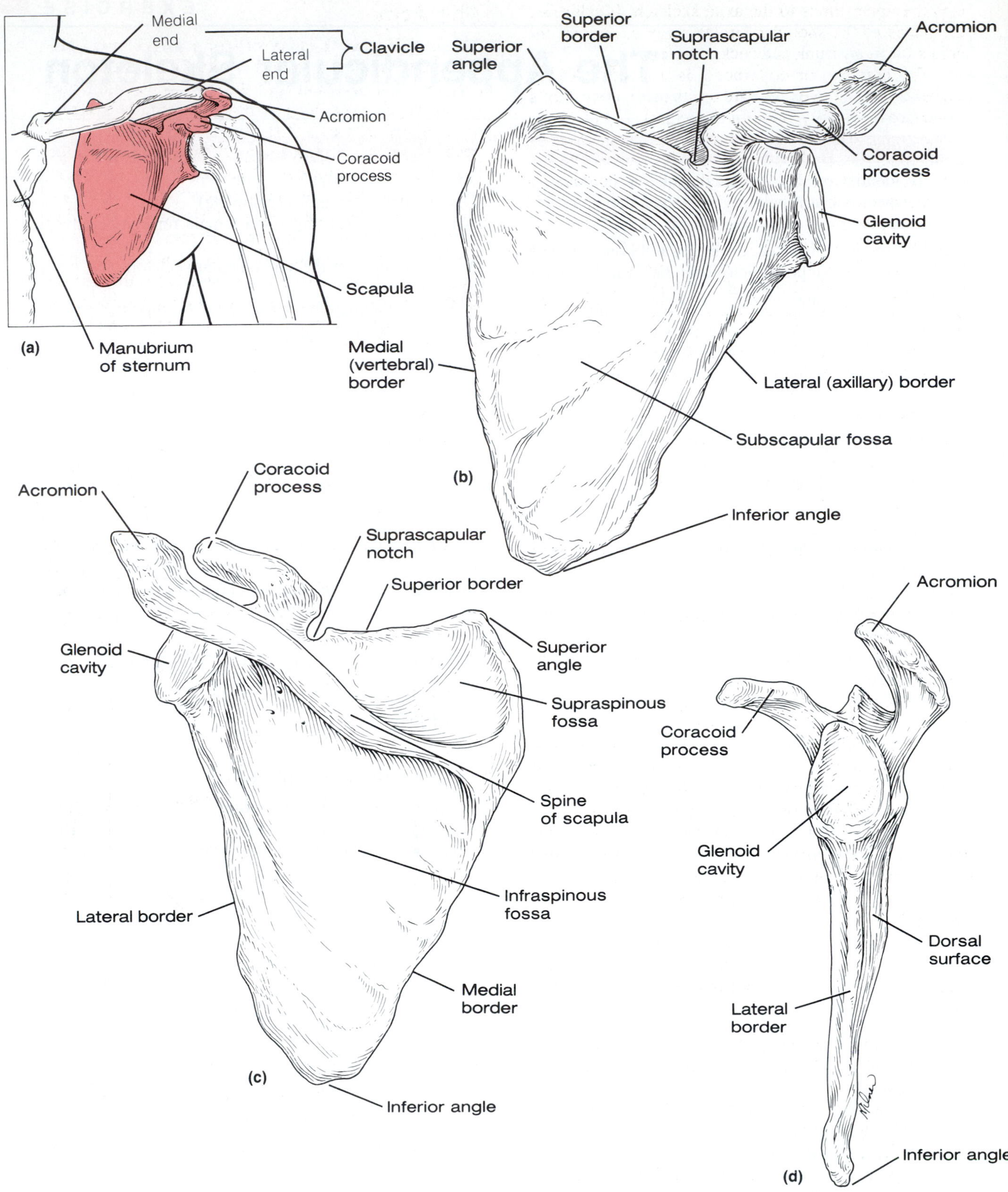

F11.2

Bones of the pectoral (shoulder) girdle. (a) Left pectoral girdle articulated to show the relationship of the girdle to the bones of the thorax and arm. **(b)** Left scapula, anterior view. **(c)** Left scapula, posterior view. **(d)** Left scapula, lateral view.

tach the upper limbs to the axial skeleton. In addition, the bones of the shoulder girdles serve as attachment points for many trunk and neck muscles.

The **clavicle,** or collarbone, is a slender doubly curved bone—convex forward on its medial two-thirds and concave laterally. Its *sternal* (medial) *end,* which attaches to the sternal manubrium, is rounded or triangular in cross section. The sternal end projects above the manubrium and can be felt and (usually) seen forming the lateral walls of the *jugular notch* (see Figure 10.15, p. 85). The *acromial* (lateral) *end* of the clavicle is flattened where it articulates with the scapula to form part of the shoulder joint. On its posteroinferior surface is the prominent **conoid tubercle.** This projection serves to attach a ligament and provides a handy landmark for determining whether a given clavicle is from the right or left side of the body. The clavicle serves as an anterior brace, or strut, to hold the arm away from the top of the thorax.

The **scapulae** (Figure 11.2), or shoulder blades, are generally triangular and are commonly called the "wings" of humans. Each scapula has a flattened body and two important processes—the **acromion** (the enlarged end of the spine of the scapula) and the beaklike **coracoid process** (*corac* = crow, raven). The acromion connects with the clavicle; the coracoid process points anteriorly over the tip of the shoulder joint and serves as a point of attachment for some of the muscles of the upper limb. The **suprascapular notch** at the base of the coracoid process allows nerves to pass. The scapula has no direct attachment to the axial skeleton but is loosely held in place by trunk muscles.

The scapula has three angles: superior, inferior, and lateral. The inferior angle provides a landmark for auscultating (listening to) lung sounds. The scapula also has three named borders: superior, medial (vertebral), and lateral (axillary). Several shallow depressions (fossae) appear on both sides of the scapula and are named according to location; i.e., there are the *anterior subscapular fossa* and the *posterior infraspinous* and *supraspinous fossae.* The **glenoid cavity,** a shallow socket that receives the head of the arm bone, is located in the lateral angle.

The shoulder girdle is exceptionally light and allows the upper limb a degree of mobility not seen anywhere else in the body. This is due to the following factors:

- The sternoclavicular joints are the *only* site of attachment of the shoulder girdles to the axial skeleton.
- The relative looseness of the scapular attachment allows it to slide back and forth against the thorax with muscular activity.
- The glenoid cavity is shallow, and does little to stabilize the shoulder joint.

However, this exceptional flexibility exacts a price: the arm bone (humerus) is very susceptible to dislocation, and fracture of the clavicle disables the entire upper limb.

The Arm

The arm (Figure 11.3) consists of a single bone—the **humerus,** a typical long bone. At its proximal end is the rounded *head,* which fits into the shallow glenoid cavity of the scapula. The head is separated from the shaft by the *anatomical neck* and the more constricted *surgical neck,* which is a common site of fracture. Opposite the head are two prominences, the **greater** and **lesser tubercles** (from lateral to medial aspect), separated by a groove (the **intertubercular** or **bicipital groove**) that guides the tendon of the biceps muscle to its point of attachment (the superior rim of the glenoid cavity). In the midpoint of the shaft is a roughened area, the **deltoid tuberosity,** where the large fleshy shoulder muscle, the deltoid, attaches. Just inferior to the deltoid tuberosity is the **radial groove,** which indicates the pathway of the radial nerve.

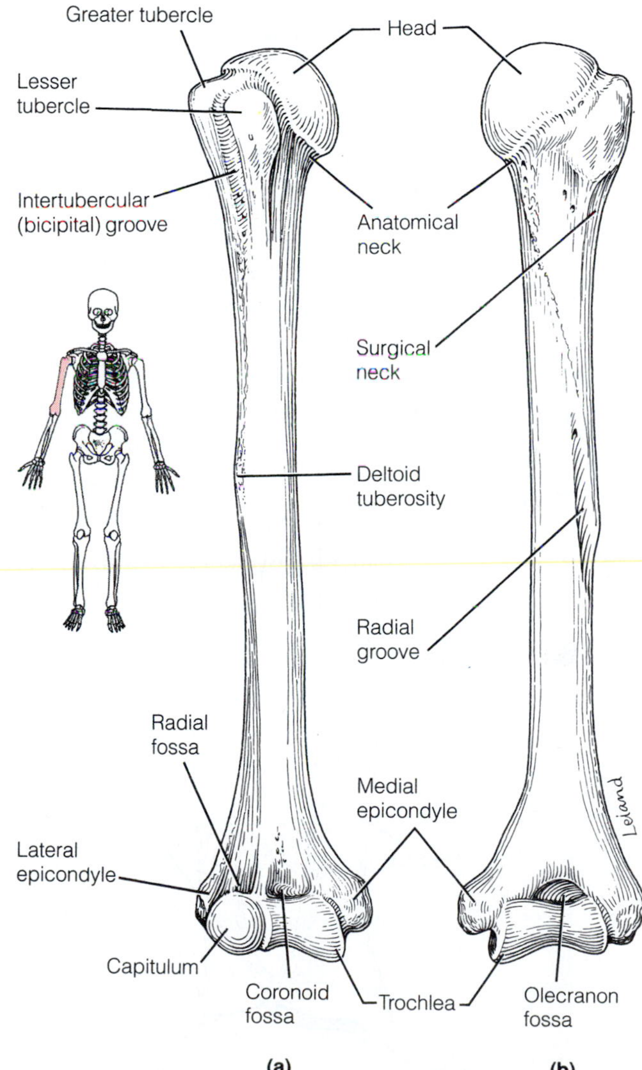

F11.3

Bones of the right arm. **(a)** Humerus, anterior view. **(b)** Humerus, posterior view.

At the distal end of the humerus are two condyles—the medial **trochlea** (looking rather like a spool), which articulates with the ulna, and the lateral **capitulum,** which articulates with the radius of the forearm. This condyle pair is flanked medially by the **medial epicondyle** and laterally by the **lateral epicondyle.**

The medial epicondyle is commonly referred to as the "funny bone." The large ulnar nerve runs in a groove beneath the medial epicondyle, and when this region is sharply bumped, we are quite likely to experience a temporary, but excruciatingly painful, tingling sensation. This event is called "hitting the funny bone," a strange expression, because it is certainly *not* funny!

Above the trochlea on the anterior surface is a depression, the **coronoid fossa;** on the posterior surface is the **olecranon fossa.** These two depressions allow the corresponding processes of the ulna to move freely when the elbow is flexed and extended. A small **radial fossa,** lateral to the coronoid fossa, receives the head of the radius when the elbow is flexed.

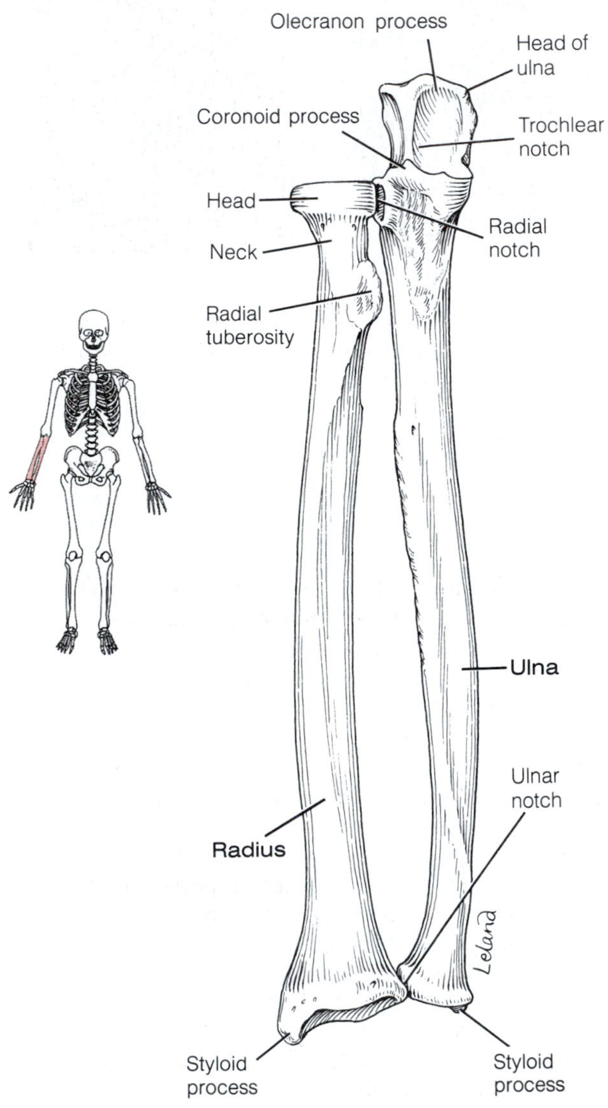

F11.4

Bones of the right forearm. Radius and ulna, anterior view.

The Forearm

Two bones, the radius and the ulna, compose the skeleton of the forearm, or antebrachium (see Figure 11.4). When the body is in the anatomical position, the **radius** is in the lateral position in the forearm and the radius and ulna are parallel. Proximally, the disc-shaped head of the radius articulates with the capitulum of the humerus. Just below the head, on the medial aspect of the shaft, is a prominence called the **radial tuberosity,** the point of attachment for the tendon of the biceps muscle of the arm. Distally, the small **ulnar notch** reveals where it articulates with the end of the ulna.

The **ulna** is the medial bone of the forearm. Its proximal head bears the anterior **coronoid process** and the posterior **olecranon process,** which are separated by the **trochlear notch.** Together these processes grip the trochlea of the humerus in a plierslike joint. The small **radial notch** on the lateral side of the coronoid process articulates with the head of the radius. The slimmer distal end of the ulna bears a small medial **styloid process,** which serves as a point of attachment for the ligaments of the wrist.

The Wrist

The wrist is referred to anatomically as the **carpus,** and the eight bones composing it are the **carpals.** The carpals are arranged in two irregular rows of four bones each, which are illustrated in Figure 11.5. In the proximal row (lateral to medial) are the scaphoid, lunate, triangular, and pisiform bones; the scaphoid and lunate articulate with the distal end of the radius. In the distal row are the trapezium, trapezoid, capitate, and hamate. The carpals are bound closely together by ligaments, which restrict movements between them.

The Hand

The hand, or manus (see Figure 11.5), consists of two groups of bones: the **metacarpals** (bones of the palm) and the **phalanges** (bones of the fingers). The metacarpals are numbered 1 to 5 from the thumb side of the hand toward the little finger. When the fist is clenched, the heads of the metacarpals become prominent as the knuckles. Each hand contains 14 phalanges. There are 3 phalanges in each finger except the thumb, which has only proximal and distal phalanges.

Surface Anatomy of the Pectoral Girdle and the Upper Limb

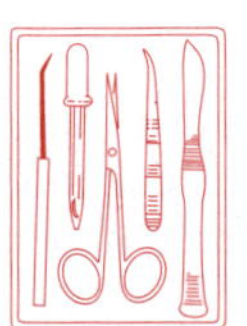

Before continuing on to study the bones of the pelvic girdle, take the time to identify the following bone markings related to the upper appendage on the skin surface. It is usually preferable to observe and palpate the bone markings on your lab partner, particularly since many of these markings can only be seen from the dorsal aspect.

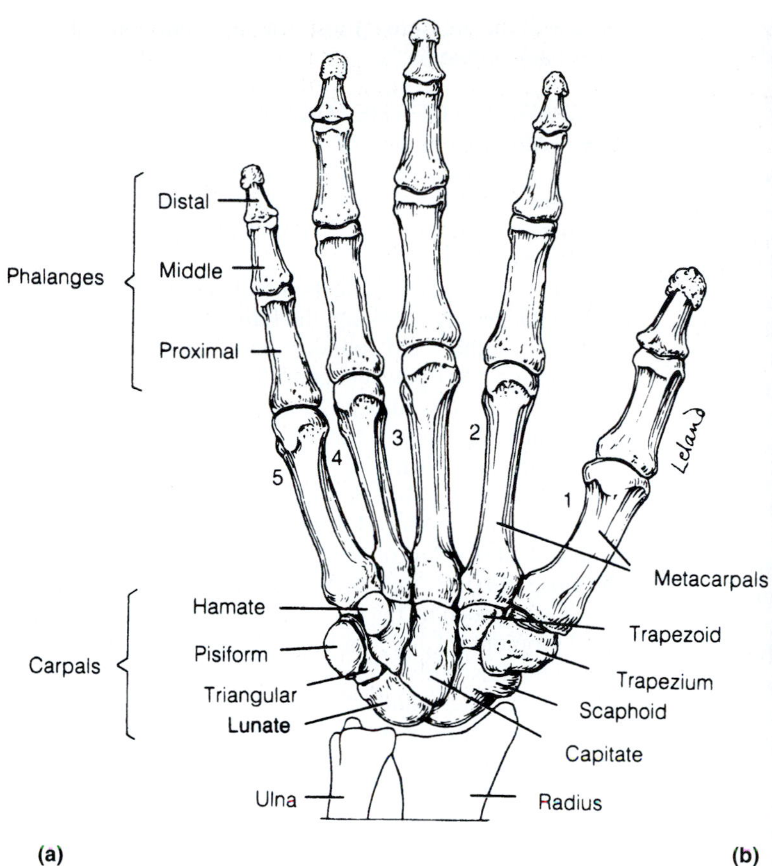

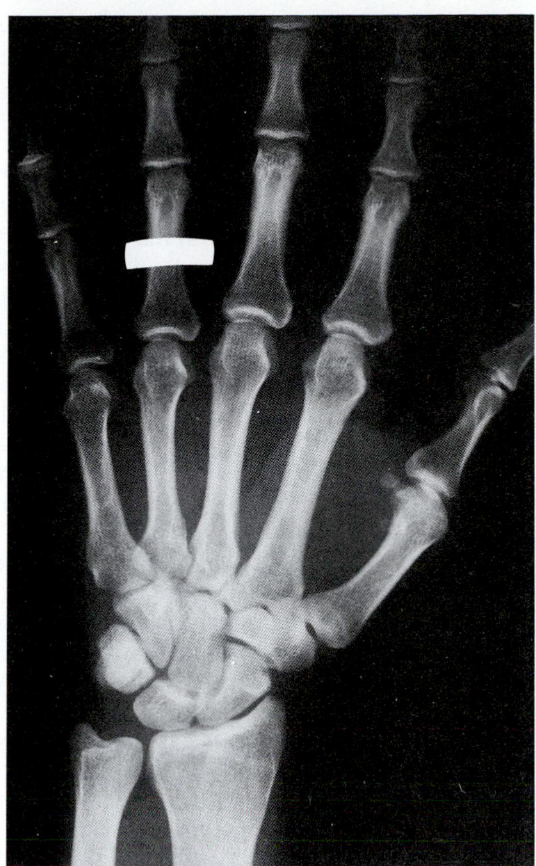

F11.5

Bones of the right wrist and hand. **(a)** Anterior view showing the relationships of the carpals, metacarpals, and phalanges. **(b)** X ray. White bar on phalanx 1 of the ring finger shows the position at which a ring would be worn.

- Clavicle: Palpate the clavicle along its entire length from sternum to shoulder.
- Acromioclavicular joint: The high point of the shoulder, which represents the junction point between the clavicle and the acromion of the scapular spine.
- Spine of the scapula: Extend your arm at the shoulder so that your scapula is moved posteriorly. As you do this, your scapular spine will be seen as a winglike protrusion on your dorsal thorax and can be easily palpated by your lab partner.
- Lateral epicondyle of the humerus: The inferiormost projection at the lateral aspect of the distal humerus. After you have located the epicondyle, run your finger posteriorly into the hollow immediately dorsal to the epicondyle. This is the site where the extensor muscles of the hand are attached and is a common site of the often excruciating pain of tennis elbow, a condition in which those muscles and their tendons are abused physically.
- Medial epicondyle of the humerus: Feel this medial projection at the distal end of the humerus.
- Olecranon process of the ulna: Work your elbow—flexing and extending—as you palpate its dorsal aspect to feel the olecranon process of the ulna moving into and out of the olecranon fossa on the dorsal aspect of the humerus.
- Styloid process of the ulna: With the hand in the anatomical position, feel out this small inferior projection on the medial aspect of the distal end of the ulna.
- Styloid process of the radius: Find this projection at the distal end of the radius (lateral aspect). It is most easily located by moving the hand medially at the wrist. Once you have palpated the styloid process, move your fingers just medially to the process (onto the anterior wrist). Press firmly and then let up slightly on the pressure. You should be able to feel your pulse at this pressure point, which lies over the radial artery (radial pulse).
- Metacarpophalangeal joints (knuckles): Clench your fist and find the first set of flexed-joint protrusions beyond the wrist—these are your metacarpophalangeal joints.

BONES OF THE PELVIC GIRDLE AND LOWER LIMB

The Pelvic Girdle

The **pelvic girdle,** or **hip girdle,** (Figure 11.6) is formed by the two **coxal bones** (**ossa coxae** or hip bones). The two coxal bones together with the sacrum and coccyx form the **bony pelvis.** In contrast to the bones of the shoulder girdle, those of the pelvic girdle are heavy and massive, and they are attached securely to the axial skeleton. The sockets for the heads of the femurs (thigh bones) are deep and heavily reinforced by ligaments to ensure a stable, strong limb attachment. The ability to bear weight is more important here than exceptional mobility and flexibility. The combined weight of the upper body rests on the pelvis (specifically, where the hip bones meet the sacrum).

Each coxal bone is a result of the fusion of three bones—the ilium, ischium, and pubis—which are distinguishable in the young child. The **ilium** is a large flaring bone forming the major portion of the coxal bone. It connects posteriorly, via its **auricular surface,** with the sacrum at the **sacroiliac joint.** The superior margin of the iliac bone, the **iliac crest,** is rough; when you rest your hands on your hips, you are palpating your iliac crests. The iliac crest terminates anteriorly in the **anterior superior spine** and posteriorly in the **posterior superior spine.** Two inferior spines are located below these. The shallow **iliac fossa** marks its internal surface, and a shallow ridge, the **arcuate line,** outlines the pelvic inlet, or pelvic brim.

The **ischium** is the "sit-down" bone, forming the most inferior and posterior portion of the coxal bone. The most outstanding marking on the ischium is the **ischial tuberosity,** which receives the weight of the body when sitting. The **ischial spine,** superior to the ischial tuberosity, is an important anatomical landmark of the pelvic cavity. (See Comparison of the Male and Female Pelves, below.) The obvious **lesser** and **greater sciatic notches** allow nerves and blood vessels to pass to and from the thigh. The sciatic nerve passes through the latter.

The **pubis** is the most anterior portion of the coxal bone. Fusion of the **rami** of the pubic bone anteriorly and the ischium posteriorly forms a bar of bone enclosing the **obturator foramen,** through which blood vessels and nerves run from the pelvic cavity into the thigh. The pubic bones of each hip bone meet anteriorly at the **pubic crest** to form a cartilaginous joint called the **pubic symphysis.** At the lateral end of the pubic crest is the *pubic tubercle* (not obvious in Figure 11.6) to which the important *inguinal ligament* attaches.

The ilium, ischium, and pubis fuse at the deep hemispherical socket called the **acetabulum** (literally, "vinegar cup"), which receives the head of the thigh bone.

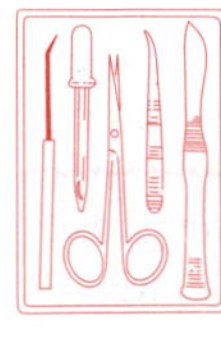

Before continuing with the bones of the lower limbs, take the time to examine an articulated pelvis. Note how each coxal bone articulates with the sacrum posteriorly and how the two coxal bones join at the pubic symphysis. The sacroiliac joint, because of the pressure it must bear, is a common site of lower back problems.

COMPARISON OF THE MALE AND FEMALE PELVES Although bones of males are usually larger, heavier, and have more prominent bone markings, the male and female skeletons are very similar. The outstanding exception to this generalization is pelvic structure (Figure 11.7).

The female pelvis reflects modifications for childbearing. Generally speaking, the female pelvis is wider, shallower, lighter, and rounder than that of the male. Not only must her pelvis support the increasing size of a fetus, but it must also be large enough to allow the infant's head (its largest dimension) to descend through the birth canal at birth.

To describe pelvic sex differences, a few more terms must be introduced. Anatomically, the pelvis can be described in terms of a false pelvis and a true pelvis. The **false pelvis** is that portion superior to the arcuate line; it is bounded by the alae of the ilia laterally and the sacral promontory and lumbar vertebrae posteriorly. Although the false pelvis supports the abdominal viscera, it does not restrict childbirth in any way. The **true pelvis** is the region inferior to the arcuate line that is almost entirely surrounded by bone. Its posterior boundary is formed by the sacrum. The ilia, ischia, and pubic bones define its limits laterally and anteriorly.

The dimensions of the true pelvis, particularly its inlet and outlet, are critical if delivery of a baby is to be uncomplicated; and they are carefully measured by the obstetrician. The **pelvic inlet,** or **pelvic brim,** is the opening delineated by the sacral promontory posteriorly and the arcuate lines of the ilia anterolaterally. It is the superiormost margin of the true pelvis. Its widest dimension is from left to right, that is, along the frontal plane. The **pelvic outlet** is the inferior margin of the true pelvis. It is bounded anteriorly by the pelvic arch, laterally by the ischia, and posteriorly by the sacrum and coccyx. Since both the coccyx and the ischial spines protrude into the outlet opening, a sharply angled coccyx or large, sharp ischial spines can dramatically narrow the outlet. The largest dimension of the outlet is the anterior-posterior diameter.

The major differences between the male and female pelves are summarized below (see Figure 11.7).

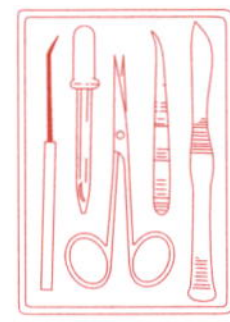

Examine male and female pelves for the following differences:

- The female inlet is larger and more circular.
- The female pelvis as a whole is shallower, and the bones are lighter and thinner.
- The female sacrum is broader and less curved, and the pubic arch is more rounded.
- The female acetabula are smaller and farther apart, and the ilia flare more laterally.
- The female ischial spines are shorter, farther apart, and everted, thus enlarging the pelvic outlet.

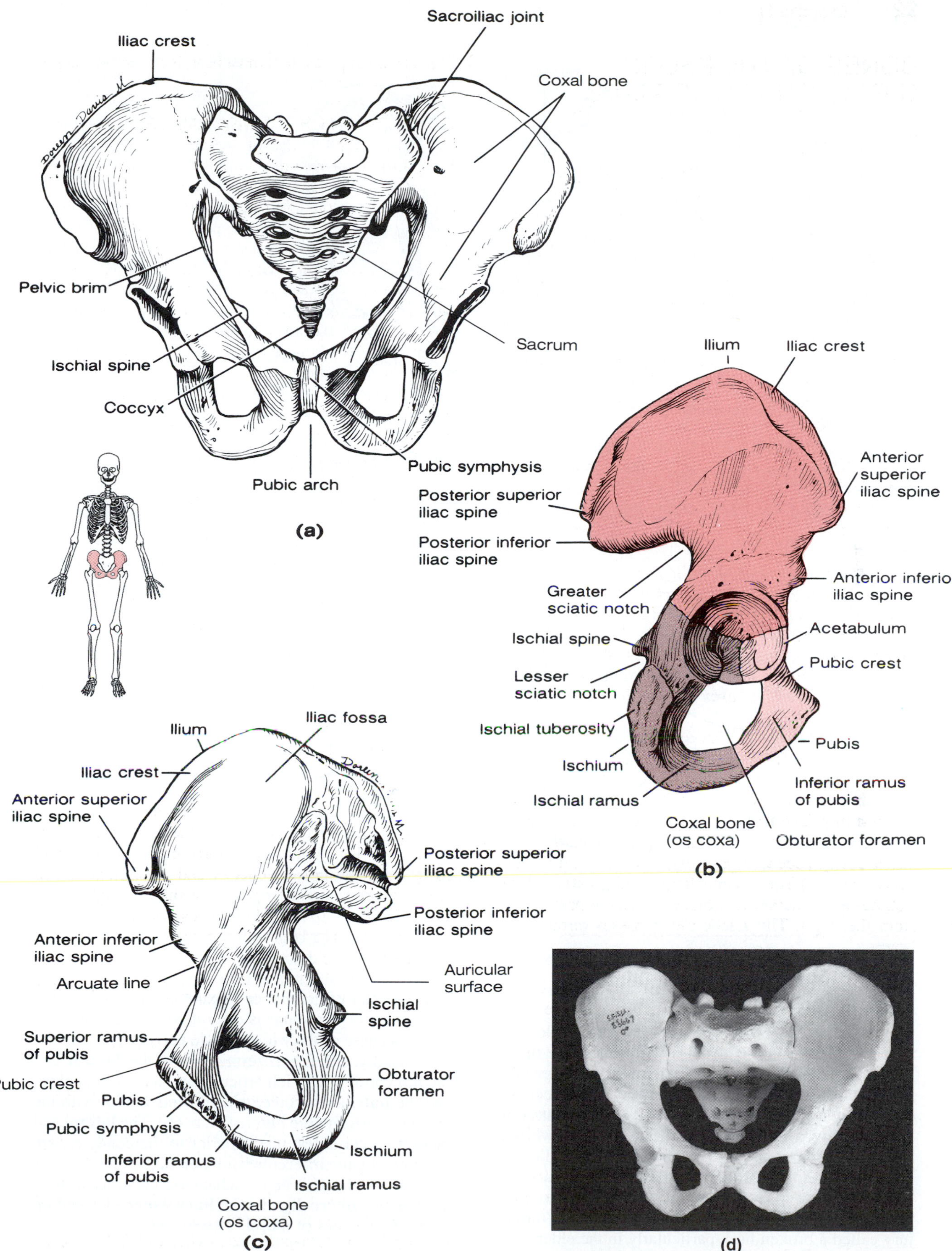

F11.6

Bones of the pelvic girdle. **(a)** Articulated bony pelvis, showing the two coxal bones, which together comprise the pelvic girdle, and the sacrum. **(b)** Right coxal bone, lateral view, showing the point of fusion of the ilium, ischium, and pubic bones. **(c)** Right coxal bone, medial view. **(d)** Photograph of male pelvis.

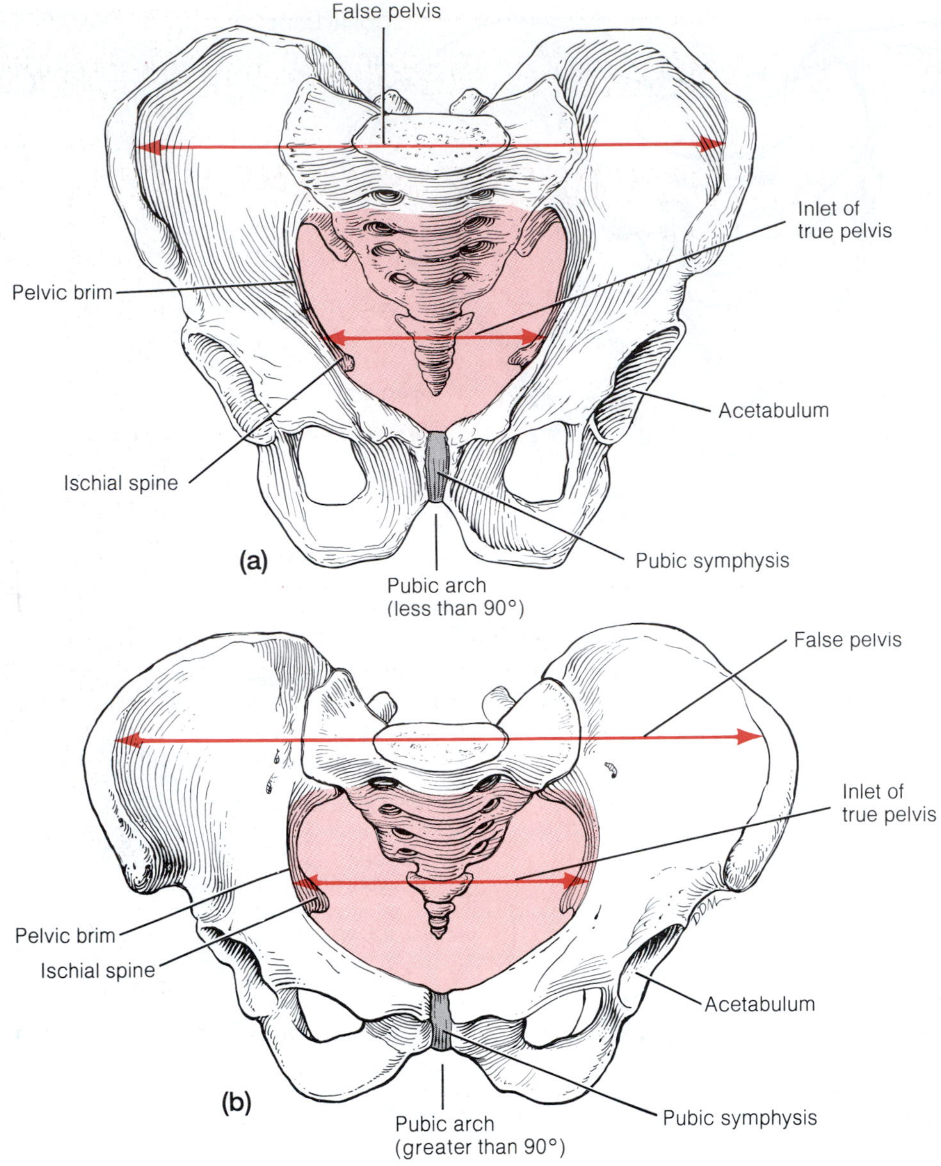

F11.7

Comparison of male and female pelves, anterior views. (a) Male. **(b)** Female.

The Thigh

The **femur,** or thigh bone (Figure 11.8), is the sole bone of the thigh and is the heaviest, strongest bone in the body. The ball-like head of the femur articulates with the hip bone via the deep, secure socket of the acetabulum. Obvious in the femur's head is a small central pit called the **fovea capitis** ("pit of the head") from which a small ligament runs to the acetabulum. The head of the femur is carried on a short, constricted *neck,* which angles laterally to join the shaft. The neck is the weakest part of the femur and is a common fracture site (an injury called a broken hip), particularly in the elderly. At the junction of the shaft and neck are the **greater** and **lesser trochanters** (separated posteriorly by the **intertrochanteric crest** and anteriorly by the **intertrochanteric line**).

The femur inclines medially as it runs downward to the leg bones; this brings the knees in line with the body's center of gravity, or maximum weight. The medial course of the femur is even more noticeable in females because of the wider female pelvis.

Distally, the femur terminates in the **lateral** and **medial condyles,** which articulate with the tibia below, and the **patellar surface,** which forms a joint with the patella anteriorly (see Figure 9.1, p. 66). The **lateral** and **medial epicondyles,** just superior to the condyles, are separated by the **intercondylar notch.**

The trochanters and trochanteric crest, as well as the **gluteal tuberosity** and the **linea aspera** located on the shaft, are sites of muscle attachment.

The Leg

Two bones, the tibia and the fibula, form the skeleton of the leg (see Figure 11.9). The **tibia,** or *shinbone,* is the larger and more medial of the two leg bones. At the

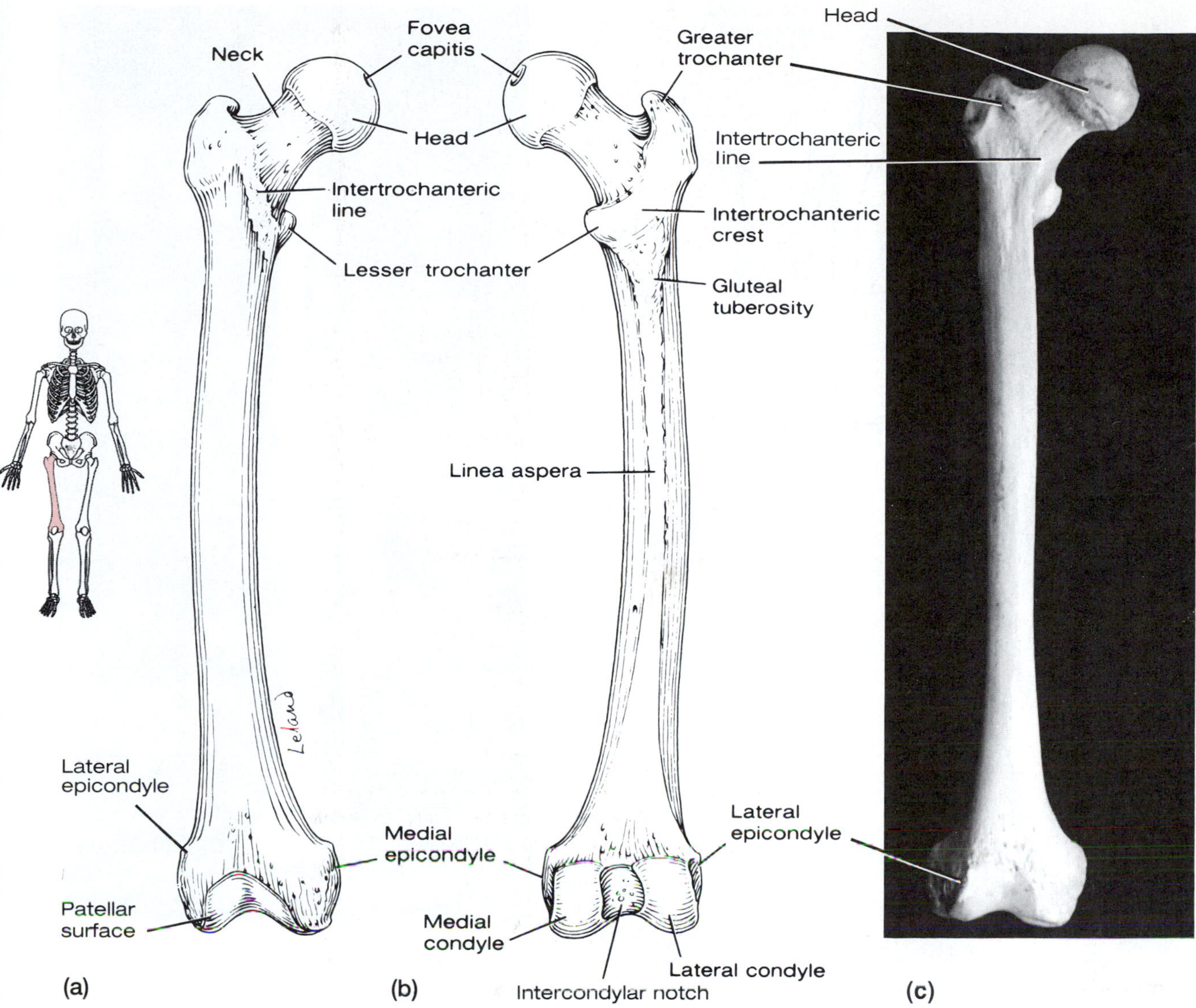

F11.8

Bone of the right thigh. (**a**) Femur, anterior view. (**b**) Femur, posterior view. (**c**) Photo of femur, anterior view.

proximal end, the **medial** and **lateral condyles** (separated by the **intercondylar eminence**) receive the distal end of the femur to form the knee joint. The **tibial tuberosity,** a roughened protrusion on the anterior tibial surface (just below the condyles), is the site of attachment of the patellar (kneecap) ligament. Small facets on its superior and inferior lateral surface articulate with the fibula. Distally, a process called the **medial malleolus** forms the inner (medial) bulge of the ankle, and the smaller distal end articulates with the talus bone of the foot. The anterior surface of the tibia is a sharpened ridge (anterior crest) that is relatively unprotected by muscles. It is easily felt beneath the skin.

The **fibula,** which lies parallel to the tibia, takes no part in forming the knee joint. Its proximal head articulates with the lateral condyle of the tibia. The fibula is thin and sticklike with a sharp anterior crest. It terminates distally in the **lateral malleolus,** which forms the outer part, or lateral bulge, of the ankle.

The Foot

The bones of the foot include the 7 **tarsal** bones, 5 **metatarsals,** which form the instep, and 14 **phalanges,** which form the toes (see Figure 11.10). Body weight is concentrated on the two largest tarsals which form the posterior aspect of the foot, the *calcaneus* (heel bone) and the *talus,* which lies between the tibia and the calcaneus. The other tarsals are named and identified in Figure 11.10. Like the fingers of the hand, each toe has 3 phalanges except the great toe, which has 2.

The bones in the foot are arranged to produce three strong arches—two longitudinal arches (medial and lateral) and one transverse arch (Figure 11.11). Ligaments, binding the foot bones together, and tendons of the foot muscles hold the bones firmly in the arched position but still allow a certain degree of give. Weakened arches are referred to as fallen arches or flat feet.

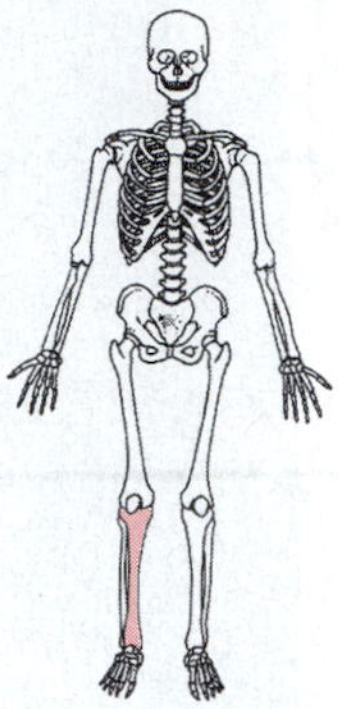

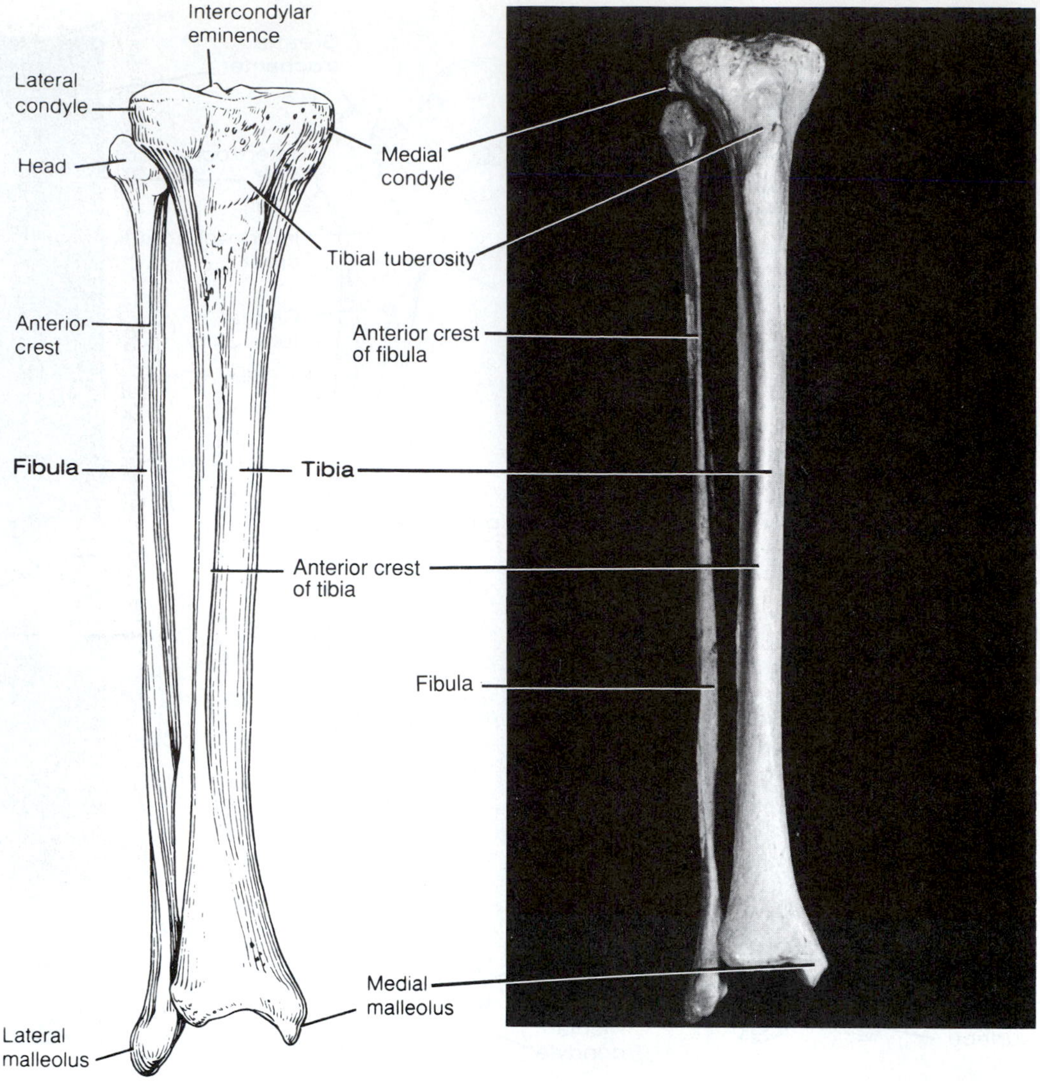

Intercondylar eminence
Lateral condyle
Head
Medial condyle
Tibial tuberosity
Anterior crest
Anterior crest of fibula
Fibula
Tibia
Anterior crest of tibia
Fibula
Medial malleolus
Lateral malleolus

F11.9

Bones of the right leg. Tibia and fibula, anterior view.

Surface Anatomy of the Pelvic Girdle and Lower Limb

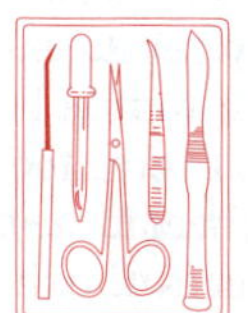

Locate and palpate the following bone markings on yourself and/or your lab partner.

- Iliac crest and anterior superior iliac spine: Rest your hands on your hips—they will be overlying the iliac crests. Trace the crest as far posteriorly as you can and then follow it anteriorly to the anterior superior iliac spine. This latter bone marking is easily felt in almost everyone, and is clearly visible through the skin (and perhaps the clothing) of very slim people. (The posterior superior iliac spine is much less obvious and is usually indicated only by a dimple in the overlying skin. Check it out in the mirror tonight.)
- Greater trochanter of the femur: This is easier to locate in females than in males because of the wider female pelvis; also it is more likely to be clothed by bulky muscles in males. Try to locate it on yourself as the most lateral point of the proximal femur. It typically lies about 6–8 inches below the iliac crest.
- Patella and tibial tuberosity: Feel your kneecap and palpate the ligaments attached to its borders. Follow the inferior patellar ligament to the tibial tuberosity.
- Medial and lateral condyles of the femur and tibia: As you move from the patella inferiorly on the medial (and then the lateral) knee surface, you will feel first the femoral and then the tibial condyle.
- Medial malleolus: Feel the medial protrusion of your ankle, the medial malleolus of the distal tibia.
- Lateral malleolus: Feel the bulge of the lateral aspect of your ankle, the lateral malleolus of the fibula.
- Calcaneus: Attempt to follow the extent of your calcaneus or heel bone.

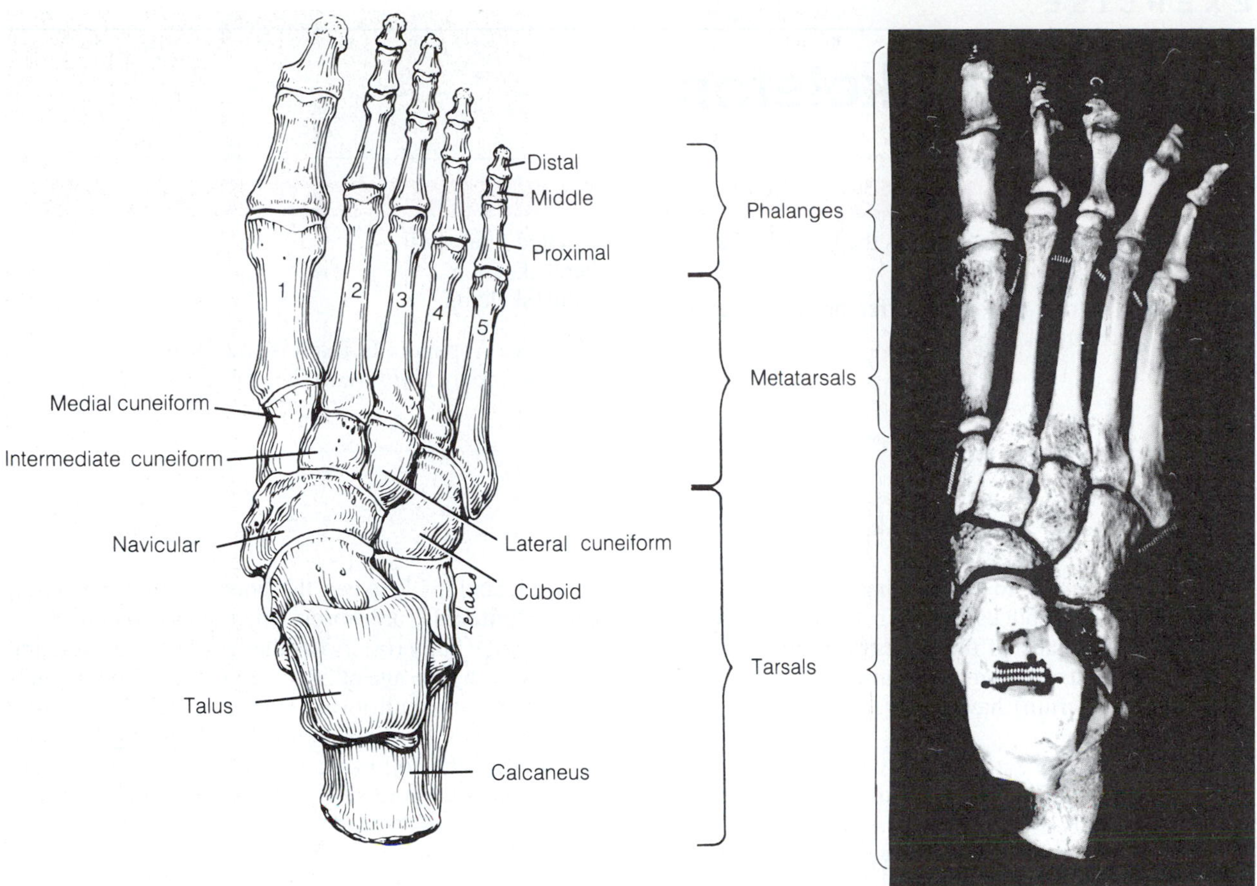

F11.10

Bones of the right ankle and foot, superior view.

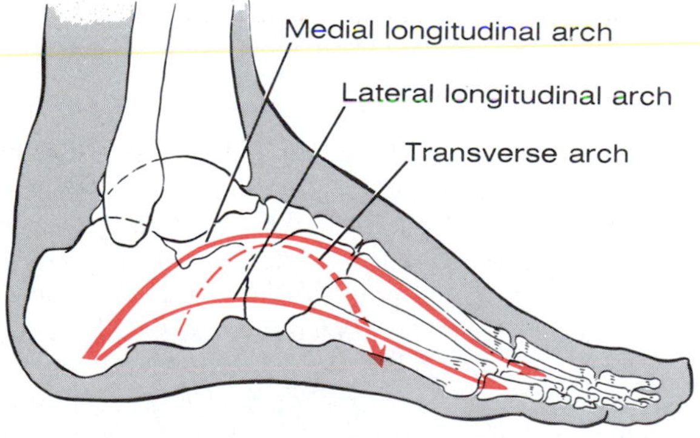

F11.11

Arches of the foot.

APPLYING KNOWLEDGE: CONSTRUCTING A SKELETON

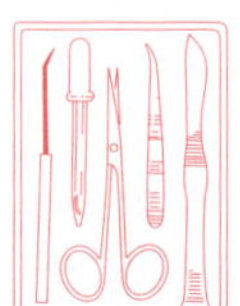

1. When you finish examining the disarticulated bones of the appendicular skeleton and yourself, work with your lab partner to arrange the disarticulated bones on the laboratory bench in their proper relative positions to form an entire skeleton. Careful observations of the bone markings should help you distinguish between right and left members of bone pairs.

2. When you believe that you have accomplished this task correctly, ask the instructor to check your arrangement to ensure that it is correct. If it is not, go to the articulated skeleton and check your bone arrangements. Also review the descriptions of the bone markings as necessary to correct your bone arrangement.

12

EXERCISE

The Fetal Skeleton

OBJECTIVES

1. To define *fontanel* and discuss the function and fate of fontanels in the fetus.
2. To demonstrate important differences between the fetal and adult skeletons.

MATERIALS

Isolated fetal skull
Fetal skeleton
Adult skeleton

See Appendix E, Exercise 12 for links to *Anatomy and PhysioShow: The Videodisc.*

A human fetus about to be born has 275 bones, many more than the 206 bones found in the adult skeleton. This is because many of the bones described as single bones in the adult skeleton (for example, the coxal bone, sternum, and sacrum) have not yet fully ossified and fused in the fetus.

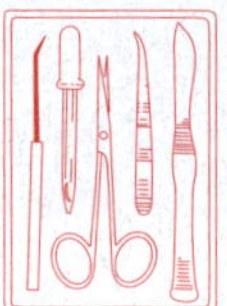

1. Obtain a fetal skull and study it carefully. Make observations as needed to answer the following questions. Does it have the same bones as the adult skull? How does the size of the fetal face relate to the cranium? How does this compare to what is seen in the adult?

2. Indentations between the bones of the fetal skull, called **fontanels,** are fibrous membranes. These areas will become bony (ossify) as the fetus ages, completing the process by the age of 20 to 22 months. The fontanels allow the fetal skull to be compressed slightly during birth and also allow for brain growth during late fetal life. Locate the following fontanels on the fetal skull with the aid of Figure 12.1: anterior (or frontal) fontanel, mastoid fontanel, sphenoidal fontanel, and posterior (or occipital) fontanel.

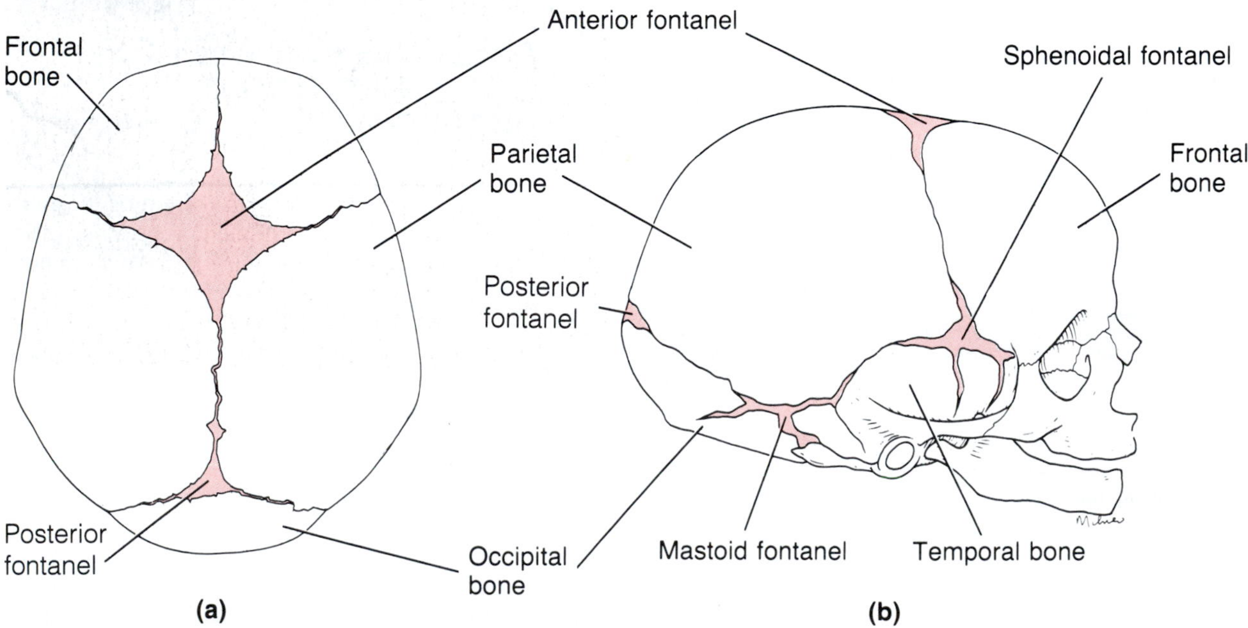

F12.1

The fetal skull. (**a**) Superior view. (**b**) Lateral view.

3. Notice that some of the cranial bones have conical protrusions. These are growth centers. Notice also that the frontal bone is still bipartite, and the temporal bone is incompletely ossified, little more than a ring of bone in the fetus.

4. Obtain a fetal skeleton (Figure 12.2) and examine it carefully, noting differences between it and an adult skeleton. Pay particular attention to the vertebrae, sternum, frontal bone of the cranium, patellae (kneecaps), coxal bones, carpals and tarsals, and rib cage.

5. Check the questions in the review section before completing this study to ensure that you have made all of the necessary observations.

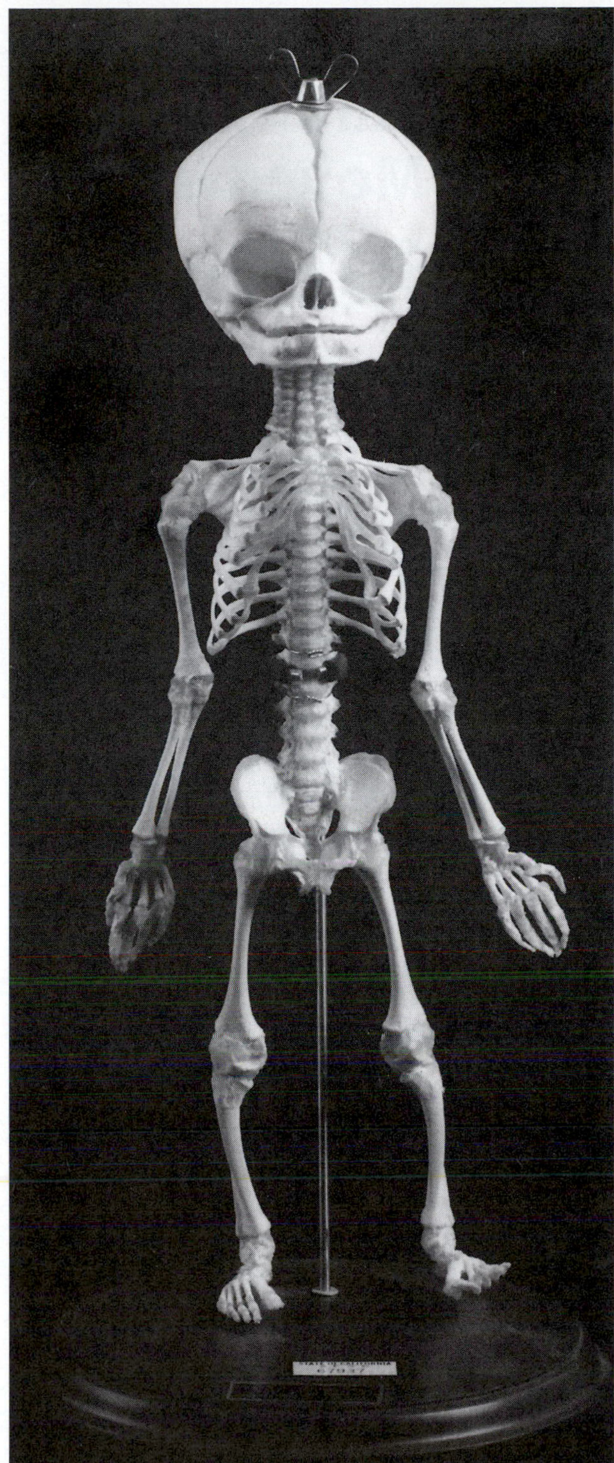

F12.2

The fetal skeleton.

13

EXERCISE

Articulations and Body Movements

OBJECTIVES

1. To name the three structural categories of joints, and to compare their structure and mobility.
2. To identify the types of synovial joints.
3. To define *origin* and *insertion* of muscles.
4. To demonstrate or identify the various body movements.

MATERIALS

Articulated skeleton
Skull
Diarthrotic beef joint (fresh or preserved), preferably a knee joint
Disposable gloves
Anatomical chart of joint types (if available)
X rays of normal and arthritic joints (if available)

See Appendix D, Exercise 13 for links to A.D.A.M. Standard.

See Appendix E, Exercise 13 for links to *Anatomy and PhysioShow: The Videodisc.*

With rare exceptions, every bone in the body is connected to, or forms a joint with, at least one other bone. **Articulations,** or joints, perform two functions for the body. They (1) hold the bones together and (2) allow the rigid skeletal system some flexibility so that gross body movements can occur.

TYPES OF JOINTS

Joints may be classified structurally or functionally. The *structural classification* is based on whether there is connective tissue fiber, cartilage, or a joint cavity between the articulating bones. Structurally, there are *fibrous, cartilaginous,* and *synovial joints.*

The functional classification focuses on the amount of movement allowed at the joint. On this basis, there are **synarthroses,** or immovable joints; **amphiarthroses,** or slightly movable joints; and **diarthroses,** or freely movable joints. Freely movable joints predominate in the limbs, whereas immovable and slightly movable joints are largely restricted to the axial skeleton, where firm bony attachments and protection of enclosed organs are a priority.

As a general rule, fibrous joints are immovable, and synovial joints are freely movable. Cartilaginous joints offer both rigid and slightly movable examples. Since the structural categories are more clear-cut, we will use the structural classification here and indicate functional properties as appropriate.

Fibrous Joints

In **fibrous joints,** the bones are joined by fibrous tissue. No joint cavity is present. The amount of movement allowed depends on the length of the fibers uniting the bones. Although some fibrous joints are slightly movable, most are synarthrotic and permit virtually no movement.

The two major types of fibrous joints are sutures and syndesmoses. In **sutures** (Figure 13.1d) the irregular edges of the bones interlock and are united by very short connective tissue fibers, as in most joints of the skull. In **syndesmoses** the articulating bones are connected by short ligaments of dense fibrous tissue; the bones do not interlock. The joint at the distal end of the tibia and fibula is an example of a syndesmosis (Figure 13.1e). Although this syndesmosis allows some give, it is classed functionally as a synarthrosis.

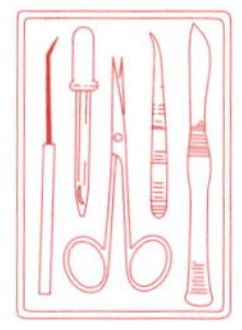

Examine a human skull again. Notice that adjacent bone surfaces do not actually touch but are separated by fibrous connective tissue. Also examine a skeleton and anatomical chart of joint types for examples of fibrous joints.

Cartilaginous Joints

In **cartilaginous joints,** the articulating bone ends are connected by a plate or pad of cartilage. No joint cavity is present. The two major types of cartilaginous joints are synchondroses and symphyses. Although there is variation, most cartilaginous joints are *slightly movable* (amphiarthroses) functionally. In **symphyses** (*symphysis* means "a growth together") the bones are connected by a broad, flat disc of **fibrocartilage.** The intervertebral joints and the pubic symphysis of the pelvis are symphyses (see Figure 13.1b and c). In **synchondroses** the bony portions are united by hyaline cartilage. The articulation of the costal cartilage of the first rib with the sternum (Figure 13.1a) is a synchondrosis, but perhaps the best examples of synchondroses are the epiphyseal

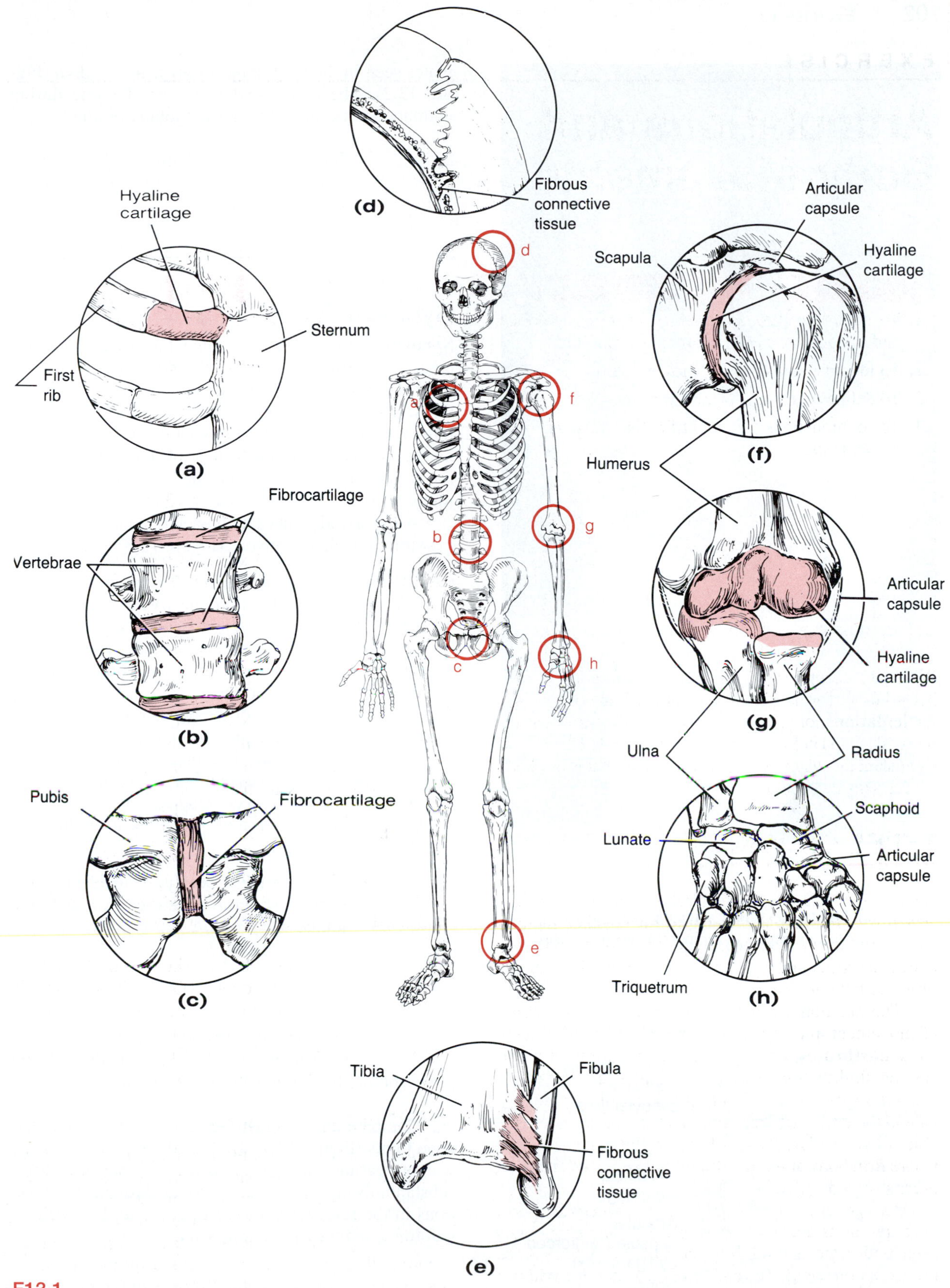

F13.1

Types of joints. Joints to the left of the skeleton are cartilaginous joints; joints above and below the skeleton are fibrous joints; joints to the right of the skeleton are synovial joints. **(a)** Synchondrosis (joint between costal cartilage of rib 1 and the sternum). **(b)** Symphyses (intervertebral discs of fibrocartilage connecting adjacent vertebrae). **(c)** Symphysis (fibrocartilaginous pubic symphysis connecting the pubic bones anteriorly). **(d)** Suture (fibrous connective tissue connecting interlocking skull bones). **(e)** Syndesmosis (fibrous connective tissue connecting the distal ends of the tibia and fibula). **(f)** Synovial joint (multiaxial shoulder joint). **(g)** Synovial joint (uniaxial elbow joint). **(h)** Synovial joints (biaxial intercarpal joints of the hand).

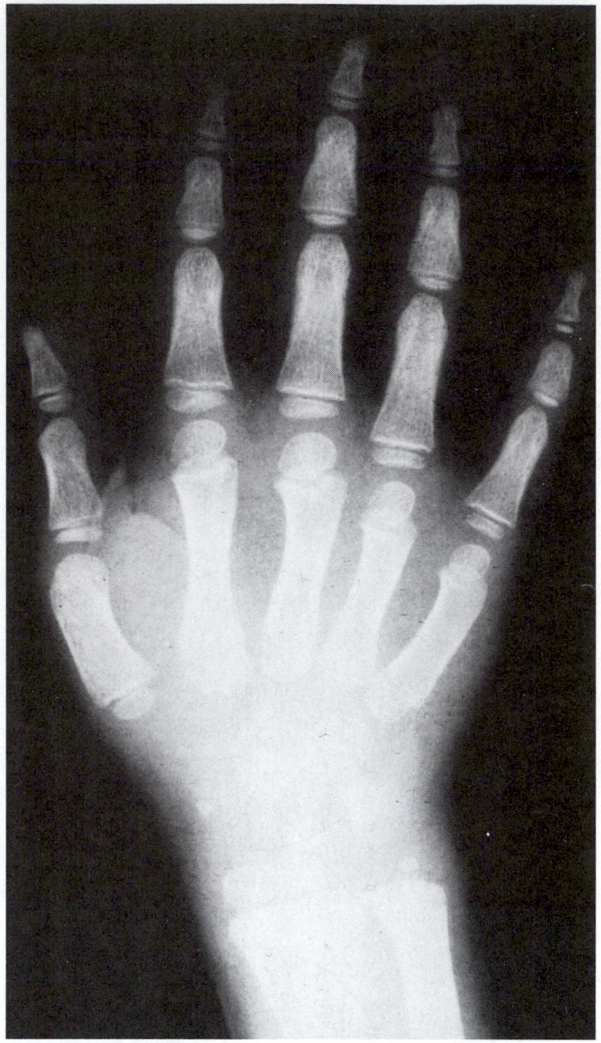

F13.2

X ray of the hand of a child. Notice the cartilaginous epiphyseal plates, examples of temporary synchondroses.

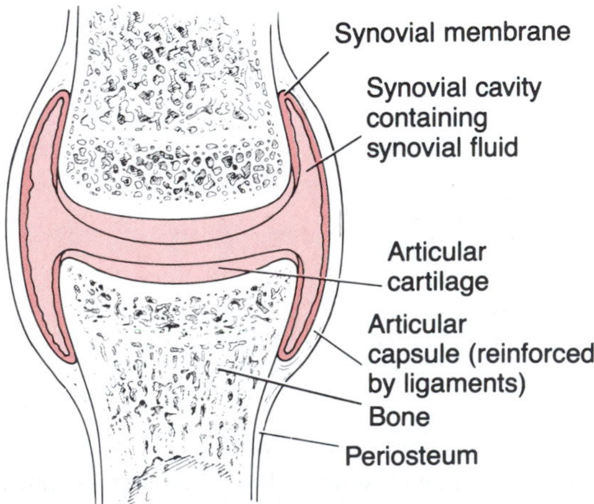

F13.3

Major structural features of a synovial joint.

plates seen in the long bones of growing children (Figure 13.2). The epiphyseal plates are flexible during childhood but eventually they are totally ossified.

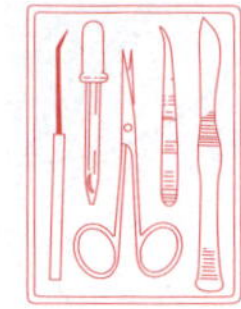

Identify the cartilaginous joints on a human skeleton and on an anatomical chart of joint types.

Synovial Joints

Synovial joints are those in which the articulating bone ends are separated by a joint cavity containing synovial fluid (see Figure 13.1f–h). All synovial joints are diarthroses, or freely movable joints. Their mobility varies, however; some synovial joints can move in only one plane, and others can move in several directions (multiaxial movement). Most joints in the body are synovial joints.

All synovial joints are characterized by the following structural characteristics (Figure 13.3):

- The joint surfaces are enclosed by an *articular capsule* (a sleeve of fibrous connective tissue).
- The interior of this capsule is lined with a smooth connective tissue membrane, called *synovial membrane,* which produces a lubricating fluid (synovial fluid) that reduces friction.
- Articulating surfaces of the bones forming the joint are covered with hyaline (*articular*) cartilage.
- The articular capsule is typically reinforced with ligaments and may contain bursae (fluid-filled sacs that reduce friction where tendons cross bone).
- Fibrocartilage pads may be present within the capsule.

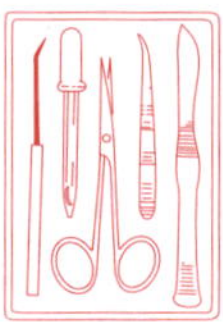

1. Examine a beef joint to identify the general structural features of diarthrotic joints.

⚠ If the joint is freshly obtained from the slaughterhouse and you will be handling it, don plastic gloves before beginning your observations.

2. Compare and contrast the structure of the hip and knee joints (Figure 13.4). Both of these joints are large weight-bearing joints of the lower limb but they differ substantially in their security. Read through the questions in the review section that pertain to this exercise before beginning your comparison.

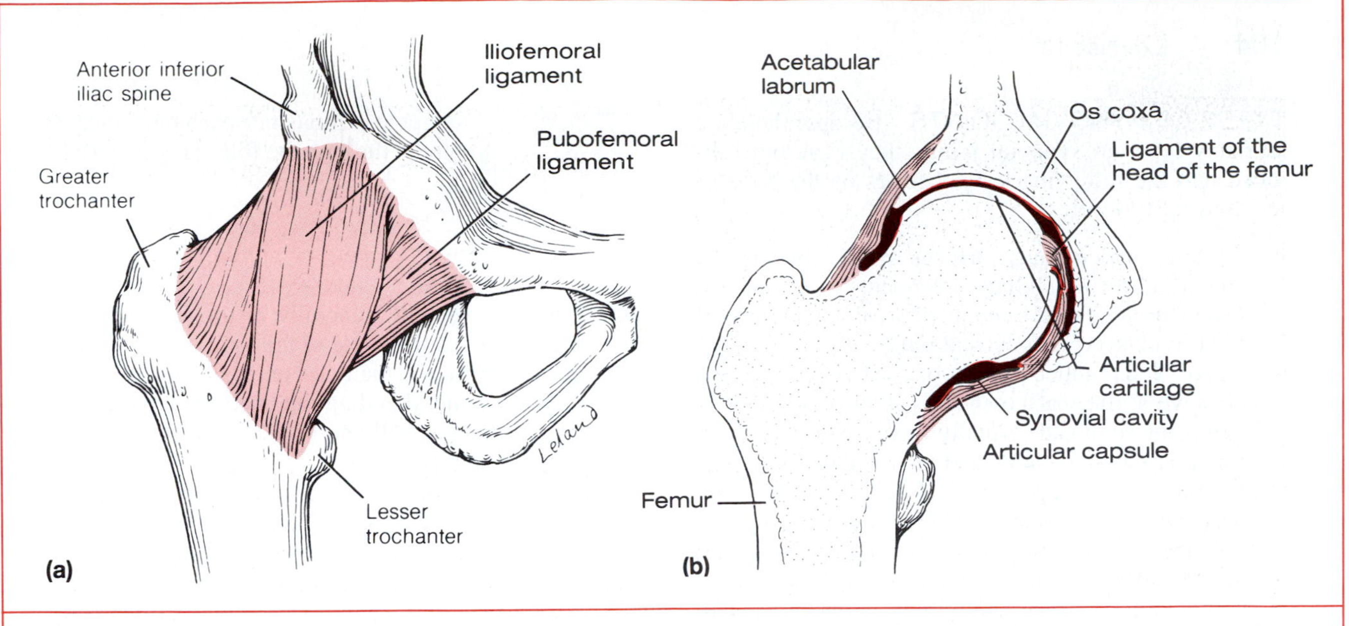

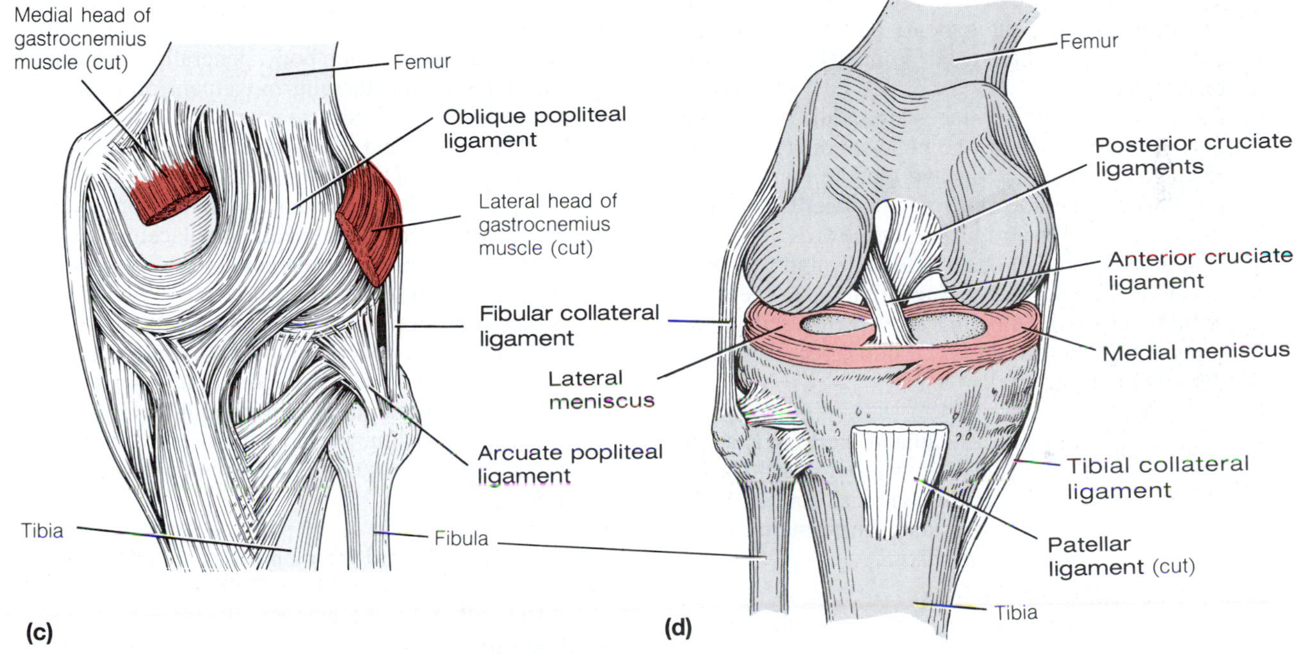

F13.4

Comparative anatomy of the hip and knee joints. **(a)** Ligaments of the right hip joint, anterior view. **(b)** Right hip joint, frontal section view. **(c)** Posterior ligaments of right knee joint. **(d)** Anterior view of the flexed right knee. Patella and articular capsule removed to allow ligaments and menisci to be seen. **(e)** Midsagittal section of right knee joint.

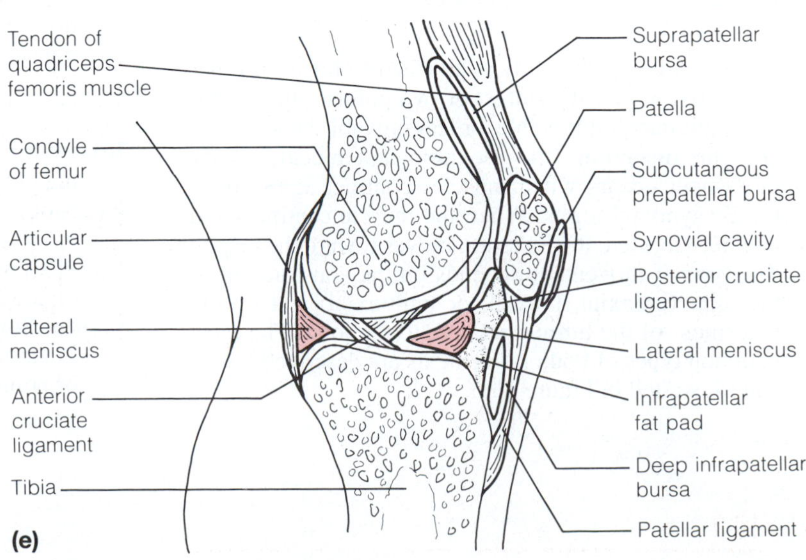

TYPES OF SYNOVIAL JOINTS Because there are so many types of synovial joints, they have been divided into the following subcategories on the basis of movements allowed:

- Gliding: Articulating surfaces are flat or slightly curved, allowing sliding movements in one or two planes. Examples are the intercarpal and intertarsal joints and the vertebrocostal joints.
- Hinge: The rounded process of one bone fits into the concave surface of another to allow movement in one plane (uniaxial), usually flexion and extension. Examples are the elbow and interphalangeal joints.
- Pivot: The rounded or conical surface of one bone articulates with a shallow depression or foramen in another bone to allow uniaxial rotation, as in the joint between the atlas and axis (C_1 and C_2).
- Condyloid: The oval condyle of one bone fits into an ellipsoidal depression in another bone, allowing biaxial (two-way) movement. The wrist joint and the metacarpal-phalangeal joints (knuckles) are examples.
- Saddle: Articulating surfaces are saddle shaped; the articulating surface of one bone is convex, and the reciprocal surface is concave. Saddle joints, which are biaxial, include the joint between the thumb metacarpal and the trapezium of the wrist.
- Ball and socket: The ball-shaped head of one bone fits into a cuplike depression of another. These are multiaxial joints, allowing movement in all directions and pivotal rotation. Examples are the shoulder and hip joints.

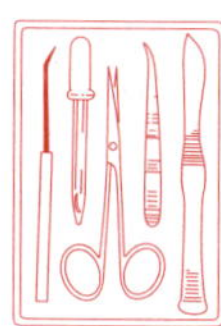

Examine the articulated skeleton, anatomical charts, and yourself to identify the subcategories of synovial joints. Make sure you understand the terms *uniaxial, biaxial,* and *multiaxial.*

BODY MOVEMENTS

Every muscle of the body is attached to bone (or other connective tissue structures) at two points—the **origin** (the stationary, immovable, or less movable attachment) and the **insertion** (the movable attachment). Body movement occurs when muscles contract across diarthrotic synovial joints. When the muscle contracts and its fibers shorten, the insertion moves toward the origin. The type of movement depends on the construction of the joint (uniaxial, biaxial, or multiaxial) and on the placement of the muscle relative to the joint. The most common types of body movements are described below and illustrated in Figure 13.5.

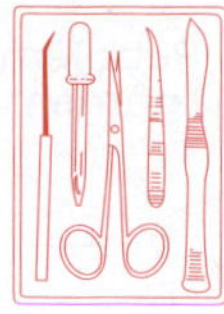

Attempt to demonstrate each movement as you read through the following material:

Flexion: a movement, generally in the sagittal plane, that decreases the angle of the joint and lessens the distance between the two bones. Flexion is typical of hinge joints (bending the knee or elbow), but is also common at ball-and-socket joints (bending forward at the hip).

Extension: a movement that increases the angle of a joint and the distance between two bones or parts of the body (straightening the knee or elbow). Extension is the opposite of flexion. If extension is greater than 180 degrees (bending the trunk backward), it is termed *hyperextension.*

Abduction: movement of a limb away from the midline or median plane of the body, generally on the frontal plane, or the fanning movement of fingers or toes when they are spread apart.

Adduction: movement of a limb toward the midline of the body. Adduction is the opposite of abduction.

Rotation: movement of a bone around its longitudinal axis without lateral or medial displacement. Rotation, a common movement of ball-and-socket joints, also describes the movement of the atlas around the odontoid process of the axis.

Circumduction: a combination of flexion, extension, abduction, and adduction commonly observed in ball-and-socket joints like the shoulder. The proximal end of the limb remains stationary, and the distal end moves in a circle. The limb as a whole outlines a cone.

Pronation: movement of the palm of the hand from an anterior or upward-facing position to a posterior or downward-facing position. This action moves the distal end of the radius across the ulna.

Supination: movement of the palm from a posterior position to an anterior position (the anatomical position). Supination is the opposite of pronation. During supination, the radius and ulna are parallel.

The last four terms refer to movements of the foot:

Inversion: a movement that results in the medial turning of the sole of the foot.

Eversion: a movement that results in the lateral turning of the sole of the foot; the opposite of inversion.

Dorsiflexion: a movement of the ankle joint in a dorsal direction (standing on one's heels).

Plantar flexion: a movement of the ankle joint in which the foot is flexed downward (standing on one's toes or pointing the toes).

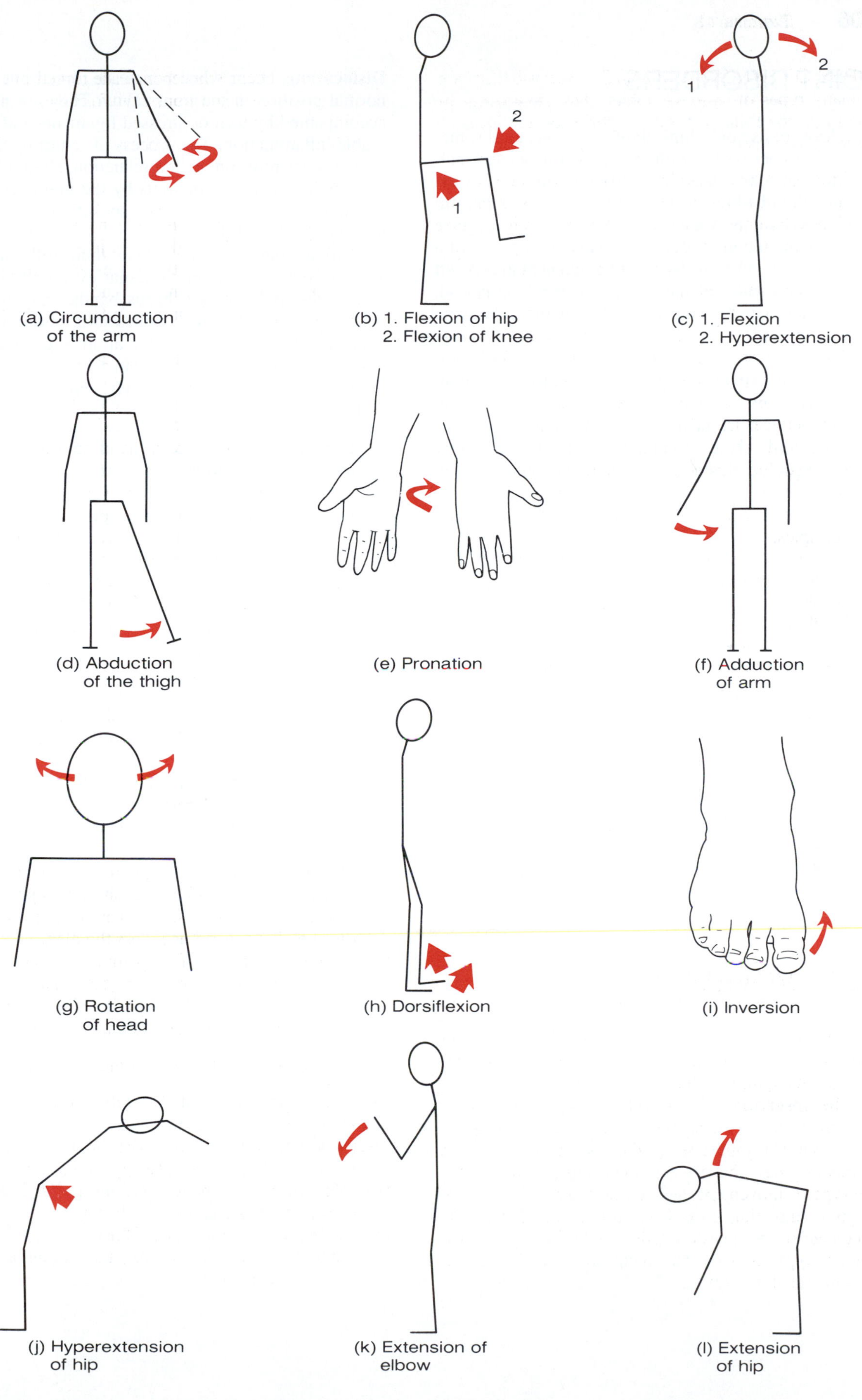

F13.5

Movements occurring at synovial joints of the body.

JOINT DISORDERS

Most of us don't think about our joints until something goes wrong with them. Joint pains and malfunctions may be caused by a variety of things. For example, a hard blow to the knee can cause a painful bursitis, known as "water on the knee," due to damage to, or inflammation of, the patellar bursa. Slippage of a fibrocartilage pad or the tearing of a ligament may result in a painful condition that persists over a long period, since these poorly vascularized structures heal so slowly.

Sprains and dislocations are other types of joint problems. In a **sprain,** the ligaments reinforcing a joint are damaged by excessive stretching or are torn away from the bony attachment. Since both ligaments and tendons are cords of dense connective tissue with a poor blood supply, sprains heal slowly and are quite painful. **Dislocations** occur when bones are forced out of their normal position in the joint cavity. They are normally accompanied by torn or stressed ligaments and considerable inflammation. The process of returning the bone to its proper position, called reduction, should be done only by a physician. Attempts by the untrained person to"snap the bone back into its socket" are often more harmful than helpful.

Advancing years also take their toll on joints. Weight-bearing joints in particular eventually begin to degenerate. *Adhesions* (fibrous bands) may form between the surfaces where bones join, and extraneous bone tissue (*spurs*) may grow along the joint edges. Such degenerative changes lead to the complaint so often heard from the elderly: "My joints are getting so stiff. . . ."

- If possible compare an X ray of an arthritic joint to one of a normal joint. ■

EXERCISE 14

Microscopic Anatomy, Organization, and Classification of Skeletal Muscle

OBJECTIVES

1. To describe the structure of skeletal muscle from gross to microscopic levels.
2. To define and explain the role of the following:

actin	*myofilament*	*tendon*
myosin	*perimysium*	*endomysium*
fiber	*aponeurosis*	*epimysium*
myofibril		

3. To describe the structure of a neuromuscular junction and to explain its role in muscle function.
4. To define: *agonist* (prime mover), *antagonist, synergist, fixator, origin,* and *insertion.*
5. To cite criteria used in naming skeletal muscles.

MATERIALS

Three-dimensional model of skeletal muscle cells (if available)
Forceps
Dissecting needles
Microscope slides and coverslips
0.9% saline solution in dropper bottles
Chicken breast or thigh muscle (freshly obtained from the meat market)
Compound microscope
Histologic slides of skeletal muscle (longitudinal and cross-sectional) and skeletal muscle showing neuromuscular junctions
Three-dimensional model of skeletal muscle showing neuromuscular junction (if available)

See Appendix E, Exercise 14 for links to *Anatomy and PhysioShow: The Videodisc.*

The bulk of the body's muscle is called **skeletal muscle** because it is attached to the skeleton (or associated connective tissue structures). Skeletal muscle influences body contours and shape, allows you to grin and frown, provides a means of locomotion, and enables you to manipulate the environment. The balance of the body's muscle—smooth and cardiac muscle—as the major component of the walls of hollow organs and the heart is involved with the transport of materials within the body.

Each of the three muscle types has a structure and function uniquely suited to its task in the body. However, because the term *muscular system* applies specifically to skeletal muscle, the primary objective of this unit is to investigate the structure and function of skeletal muscle.

Skeletal muscle is also known as *voluntary muscle* (because it can be consciously controlled) and as *striated muscle* (because it appears to be striped). As you might guess from both of these alternative names, skeletal muscle has some very special characteristics. Thus an investigation of skeletal muscle should begin at the cellular level.

THE CELLS OF SKELETAL MUSCLE

Skeletal muscle is composed of relatively large, long cylindrical cells ranging from 10 to 100 μm in diameter and up to 6 cm in length. However, the cells of large, hard-working muscles like the antigravity muscles of the hip are extremely coarse, ranging up to 25 cm in length, and can be seen with the naked eye.

Skeletal muscle cells (Figure 14.1a) are multinucleate: Multiple oval nuclei can be seen just beneath the plasma membrane (called the *sarcolemma* in these cells). The nuclei are pushed peripherally by the longitudinally arranged **myofibrils,** which nearly fill the sarcoplasm (Figure 14.1b). Alternating light (I) and dark (A) bands along the length of the perfectly aligned myofibrils give the muscle fiber as a whole its striped appearance.

Electron microscope studies have revealed that the myofibrils are made up of even smaller threadlike structures called **myofilaments** (Figure 14.1b and d). The myofilaments are composed largely of two varieties of contractile proteins—**actin** and **myosin**—which slide past each other during muscle activity to bring about shortening or contraction of the muscle cells. It is the highly specific arrangement of the myofilaments within the myofibrils that is responsible for the banding pattern in skeletal muscle. The actual contractile units of muscle, called **sarcomeres,** extend from the middle of one I band (its Z line) to the middle of the next along the length of the myofibrils. (See Figure 14.1c and d.)

1. Look at the three-dimensional model of skeletal muscle cells, noting the relative shape and size of the cells. Identify the nuclei, myofibrils, and light and dark bands.

2. Obtain forceps, two dissecting needles, slide and coverslip, and a dropper bottle of saline solution. With forceps, remove a very small piece of muscle from the chicken breast (or thigh). Place the tis-

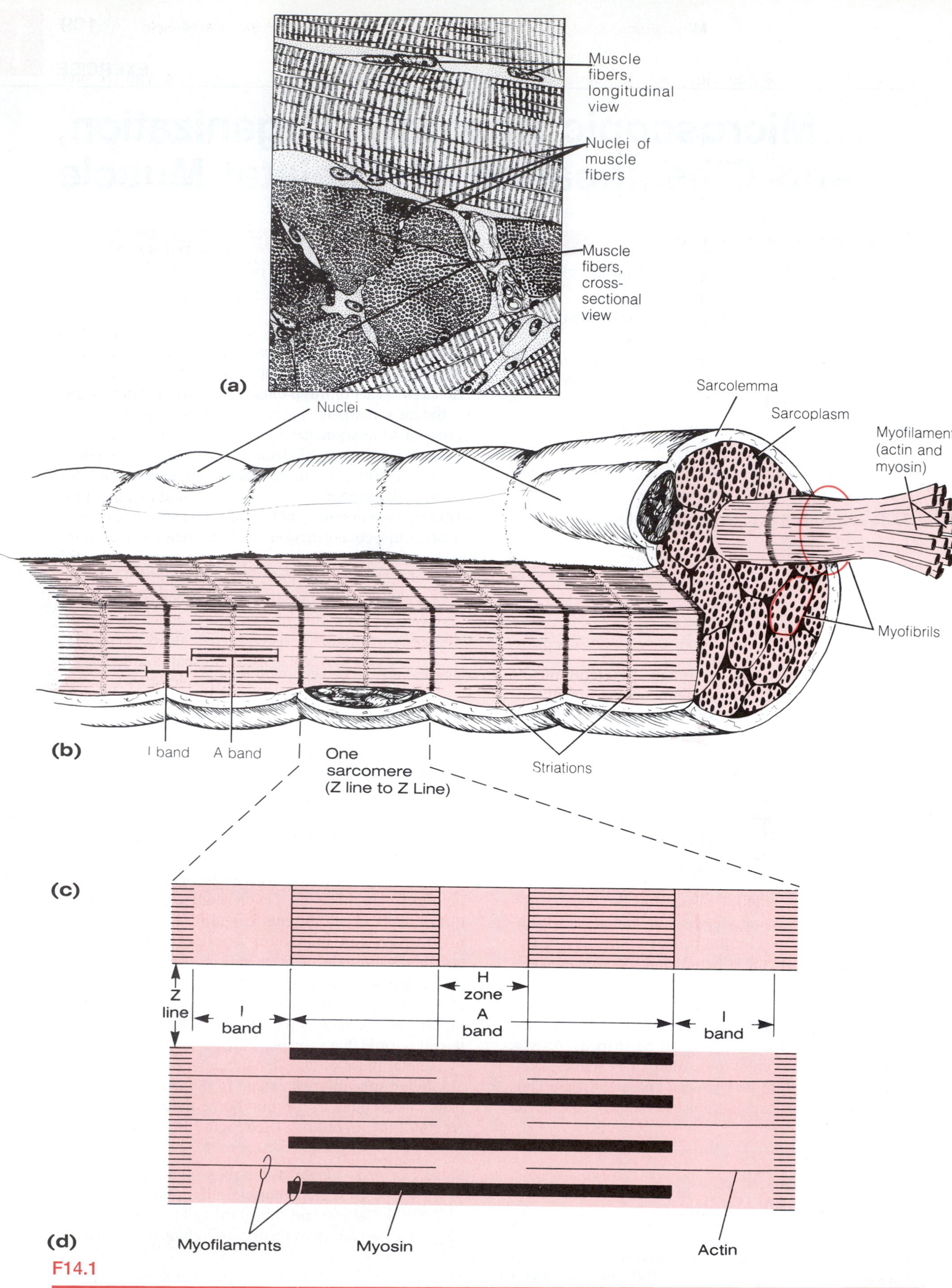

F14.1

Structure of skeletal muscle cells. **(a)** Muscle fibers, longitudinal and transverse views. (See corresponding photomicrograph in Plate 2 of the Histology Atlas.) **(b)** A portion of a skeletal muscle cell; one myofibril has been extended and disrupted to indicate its myofilament composition. **(c)** One sarcomere of the myofibril. **(d)** Banding pattern in the sarcomere. (From H. E. Huxley, "The Contraction of Muscle." © November 1958 by Scientific American, Inc. All rights reserved.)

sue on a clean microscope slide, and add a drop of the saline solution.

3. Pull the muscle fibers apart with the dissecting needles (tease them) until you have a fluffy-looking mass of tissue. Cover the teased tissue with a coverslip, and observe under the high-power lens of a microscope. Look for the banding pattern. Regulate the light carefully to obtain the highest possible contrast.

4. Now compare your observations with Figure 14.1a and with what can be seen with professionally prepared muscle tissue. Obtain a slide of skeletal muscle (longitudinal section), and view it under high power. From your observations, draw a small section of a muscle fiber in the space provided here. Label the nuclei, sarcolemma, and A and I bands.

What structural details become apparent with the prepared slide?

____

____

ORGANIZATION OF SKELETAL MUSCLE CELLS INTO MUSCLES

Muscle fibers are soft and surprisingly fragile. Thus thousands of muscle fibers are bundled together with connective tissue to form the organs we refer to as skeletal muscles (Figure 14.2). Each muscle fiber is enclosed in a delicate, areolar connective tissue sheath called **endomysium.** Several sheathed muscle fibers are wrapped by a collagenic membrane called **perimysium,** forming a bundle of fibers called a **fascicle,** or **fasciculus.** A large number of fascicles are bound together by a substantially coarser "overcoat" of dense connective tissue called an **epimysium,** which sheathes the entire muscle. These epimysia blend into the **deep fascia,** still coarser sheets of dense connective tissue that bind muscles into functional groups, and into strong cordlike **tendons** or sheetlike **aponeuroses,** which attach muscles to each other or indirectly to bones. As noted in Exercise 13, a muscle's more movable attachment is called its *insertion* whereas its fixed (or immovable) attachment is the *origin.*

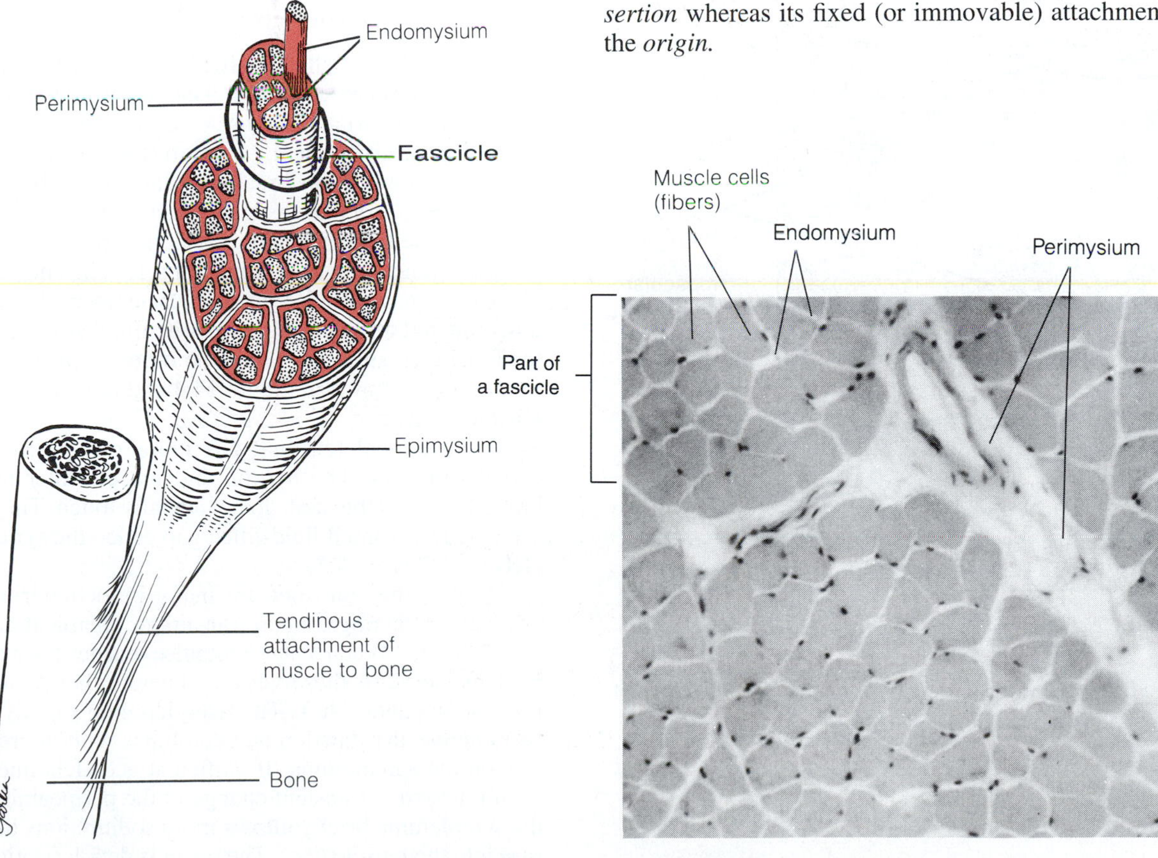

F14.2

Connective tissue coverings of skeletal muscle (64×).

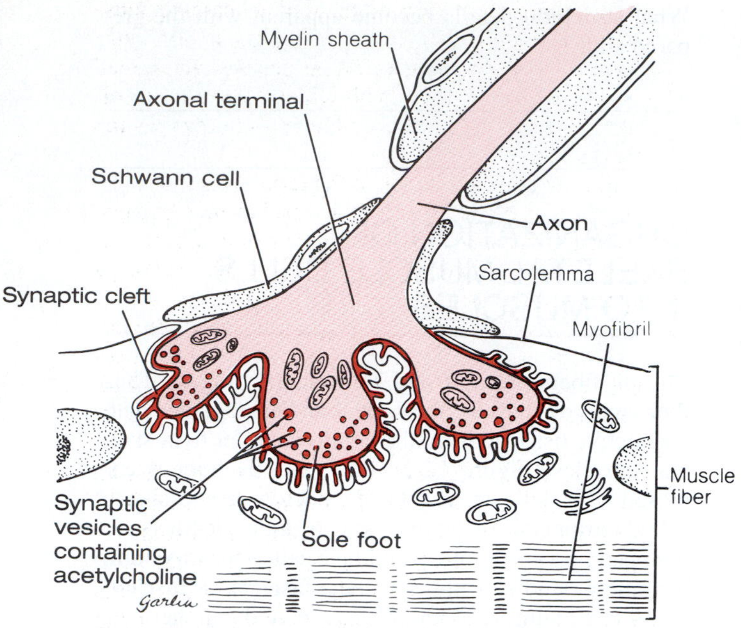

F14.3

The neuromuscular junction.

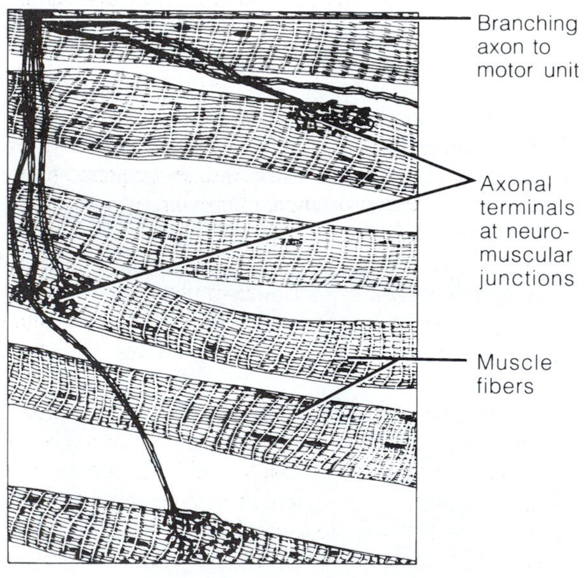

F14.4

A portion of a motor unit. (Corresponding photomicrograph is Plate 3 in the Histology Atlas.)

Tendons perform several functions, two of the most important being to provide durability and to conserve space. Because tendons are tough collagenic connective tissue, they can span rough bony prominences that would destroy the more delicate muscle tissues. Because of their relatively small size, more tendons than fleshy muscles can pass over a joint.

In addition to supporting and binding the muscle fibers, and providing strength to the muscle as a whole, the connective tissue wrappings provide a route for the entry and exit of nerves and blood vessels that serve the muscle fibers. The larger, more powerful muscles have relatively more connective tissue than muscles involved in fine or delicate movements.

As we age, the mass of the muscle fibers decreases, and the amount of connective tissue increases; thus the skeletal muscles gradually become more sinewy, or "stringier." ■

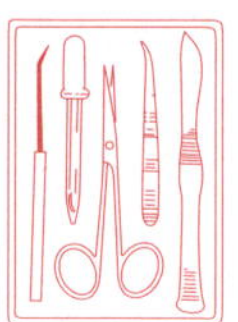

Obtain a slide showing a cross section of skeletal muscle tissue. Using Figure 14.2 as a reference, identify the muscle fibers, endomysium, perimysium, and epimysium (if visible).

THE NEUROMUSCULAR JUNCTION

Voluntary muscle cells are always stimulated by motor neurons via nerve impulses. The junction between a nerve fiber (axon) and a muscle cell is called a **neuromuscular,** or **myoneural, junction** (Figure 14.3).

Each motor axon breaks up into many branches called *axonal terminals* as it approaches the muscle, and each of these branches participates in forming a neuromuscular junction with a single muscle cell. Thus a single neuron may stimulate many muscle fibers. Together, a neuron and all the muscle cells it stimulates make up the functional structure called the **motor unit.** Part of a motor unit is shown in Figure 14.4 and in Plate 3 of the Histology Atlas.

Each axonal terminal has numerous projections called **sole feet.** The neuron and muscle fiber membranes, close as they are, do not actually touch. They are separated by a small fluid-filled gap called the **synaptic cleft** (see Figure 14.3).

Within the sole foot are many mitochondria and vesicles containing a neurotransmitter chemical called acetylcholine. When a nerve impulse reaches the axonal endings, some of these vesicles liberate their contents into the synaptic cleft. The acetylcholine rapidly diffuses across the junction and combines with the receptors on the sarcolemma. If sufficient acetylcholine has been released, a transient change in the permeability of the sarcolemma briefly allows more sodium ions to diffuse into the muscle fiber. The result is depolarization of the sarcolemma and subsequent contraction of the muscle fiber.

1. If possible, examine a three-dimensional model of skeletal muscle cells that illustrates the neuromuscular junction. Identify the structures just described.

2. Obtain a slide of skeletal muscle stained to show a portion of a motor unit. Examine the slide under high power to identify the axonal fibers extending leashlike to the muscle cells. Follow one of the axonal fibers to its terminus to identify the oval-shaped axonal terminal. Compare your observations to Figure 14.4. Sketch a small section in the space provided, labeling the motor axon, its terminal branches, sole feet, and muscle fibers.

CLASSIFICATION OF SKELETAL MUSCLES

Naming Skeletal Muscles

Remembering the names of the skeletal muscles is a monumental task, but certain clues help. Muscles are named on the basis of the following criteria:

- **Direction of muscle fibers:** Some muscles are named in reference to some imaginary line, usually the midline of the body or the longitudinal axis of a limb bone. A muscle with fibers (and fascicles) running parallel to that imaginary line will have the term *rectus* (straight) in its name. For example, the rectus abdominis is the straight muscle of the abdomen. Likewise, the terms *transverse* and *oblique* indicate that the muscle fibers run at right angles and obliquely (respectively) to the imaginary line.
- **Relative size of the muscle:** Terms such as *maximus* (largest), *minimus* (smallest), *longus* (long), and *brevis* (short) are often used in naming muscles—as in gluteus maximus and gluteus minimus.
- **Location of the muscle:** Some muscles are named according to the bone with which they are associated. For example, the frontalis muscle overlies the frontal bone.
- **Number of origins:** When the term *biceps, triceps,* or *quadriceps* forms part of a muscle name, you can generally assume that the muscle has two, three, or four origins (respectively). For example, the biceps muscle of the arm has two heads, or origins.
- **Location of the muscle's origin and insertion:** For example, the sternocleidomastoid muscle has its origin on the sternum (*sterno*) and clavicle (*cleido*), and inserts on the mastoid process of the temporal bone.
- **Shape of the muscle:** For example, the deltoid muscle is roughly triangular (*deltoid* = "triangle"), and the trapezius muscle resembles a trapezoid.
- **Action of the muscle:** For example, all the adductor muscles of the anterior thigh bring about its adduction, and all the extensor muscles of the wrist extend the wrist.

Types of Muscles

Most often, body movements are not a result of the contraction of a single muscle but instead reflect the coordinated action of several muscles acting together. Muscles that are primarily responsible for producing a particular movement are called **prime movers,** or **agonists.**

Muscles that oppose or reverse a movement are called **antagonists.** When a prime mover is active, the fibers of the antagonist are stretched and in the relaxed state. The antagonist can also regulate the prime mover by providing some resistance, to prevent overshoot or to stop its action.

It should be noted that antagonists can be prime movers in their own right. For example, the biceps muscle of the arm (a prime mover of elbow flexion) is antagonized by the triceps (a prime mover of elbow extension).

Synergists contribute substantially to the action of agonists by reducing undesirable or unnecessary movement. Contraction of a muscle crossing two or more joints would cause movement at all joints spanned if the synergists were not there to stabilize them. For example, you can make a fist without bending your wrist only because synergist muscles stabilize the wrist joint and allow the prime mover to exert its force at the finger joints.

Fixators, or fixation muscles, are specialized synergists. They immobilize the origin of a prime mover so that all the tension is exerted at the insertion. Muscles that help maintain posture are fixators; so too are muscles of the back that stabilize or "fix" the scapula during arm movements.

15

EXERCISE

Gross Anatomy of the Muscular System

OBJECTIVES

1. To name and locate the major muscles of the human body (on a torso model, a human cadaver, laboratory chart, or diagram) and state the action of each.
2. To explain how muscle actions are related to their location.
3. To name muscle origins and insertions as required by the instructor.
4. To identify antagonists of the major prime movers.
5. To name and locate muscles on a dissection animal.
6. To recognize similarities and differences between human and cat musculature.

MATERIALS

Disposable gloves or protective skin cream
Preserved and injected cat (one for every two to four students)
Dissecting trays and instruments
Name tag and large plastic bag
Paper towels
Embalming fluid
Human torso model or large anatomical chart showing human musculature
Human cadaver for demonstration (if available)
Human Musculature videotape*

See Appendix D, Exercise 15 for links to A.D.A.M. Standard.

See Appendix E, Exercise 15 for links to *Anatomy and PhysioShow: The Videodisc.*

*Available to qualified adopters from Benjamin/Cummings.

IDENTIFICATION OF HUMAN MUSCLES

Muscles of the Head and Neck

The muscles of the head serve many specific functions. For instance, the muscles of facial expression differ from most skeletal muscles because they insert into the skin (or other muscles) rather than into bone. As a result, they move the facial skin, allowing a wide range of emotions to be shown on the face. Other muscles of the head are the muscles of mastication, which manipulate the mandible during chewing, and the six extrinsic eye muscles located within the orbit, which aim the eye. (Orbital muscles are studied in conjunction with the anatomy of the eye in Exercise 24.) Neck muscles are primarily concerned with the movement of the head and shoulder girdle. Figures 15.1 and 15.2 are summary figures illustrating the superficial musculature of the body as a whole. Head and neck muscles are discussed in Tables 15.1 and 15.2 and shown in Figures 15.3 and 15.4.

Carefully read the description of each muscle and visualize what happens when the muscle contracts. After reading the tables and identifying the head and neck muscles in Figures 15.3 and 15.4, use a torso model or an anatomical chart to again identify as many of these muscles as possible. (If a human cadaver is available for observation, specific instructions for muscle examination will be provided by your instructor.) Then carry out the following palpations on yourself:

- To demonstrate how the temporalis works, clench your teeth. The masseter can also be palpated at this time at the angle of the jaw.

Muscles of the Trunk

The trunk musculature includes muscles that move the vertebral column; anterior thorax muscles that act to move ribs, head, and arms; and muscles of the abdominal wall that play a role in the movement of the vertebral column but more importantly form the "natural girdle," or the major portion of the abdominal body wall.

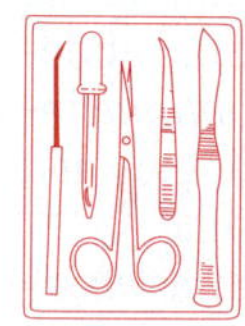

The trunk muscles are described in Tables 15.3 and 15.4 and shown in Figures 15.5 and 15.6. As before, identify the muscles in the figure as you read the tabular descriptions and then identify them on the torso or laboratory chart.

(*Text continues on p. 125*)

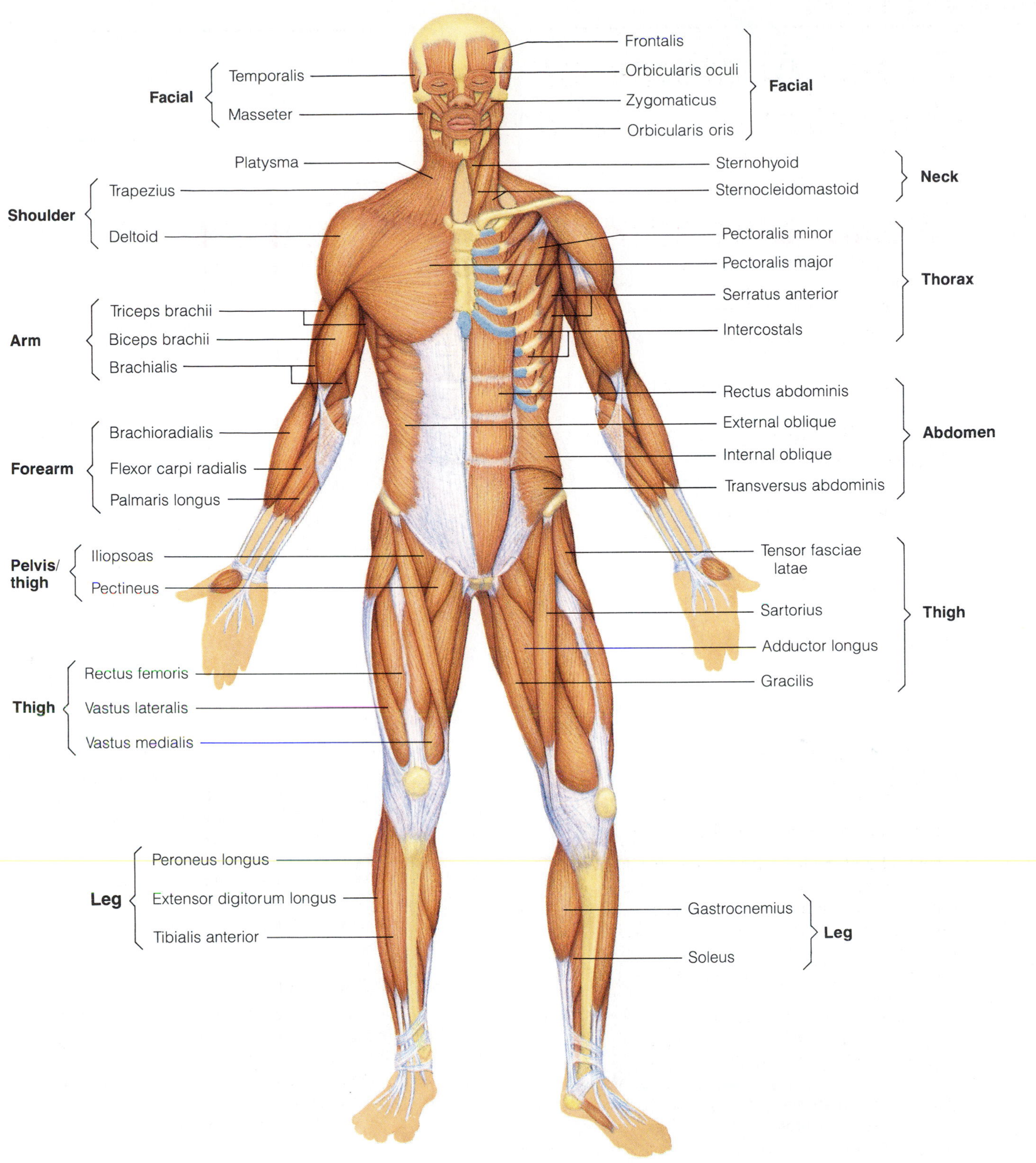

F15.1

Anterior view of superficial muscles of the body. The abdominal surface has been partially dissected on the left side of the body to show somewhat deeper muscles.

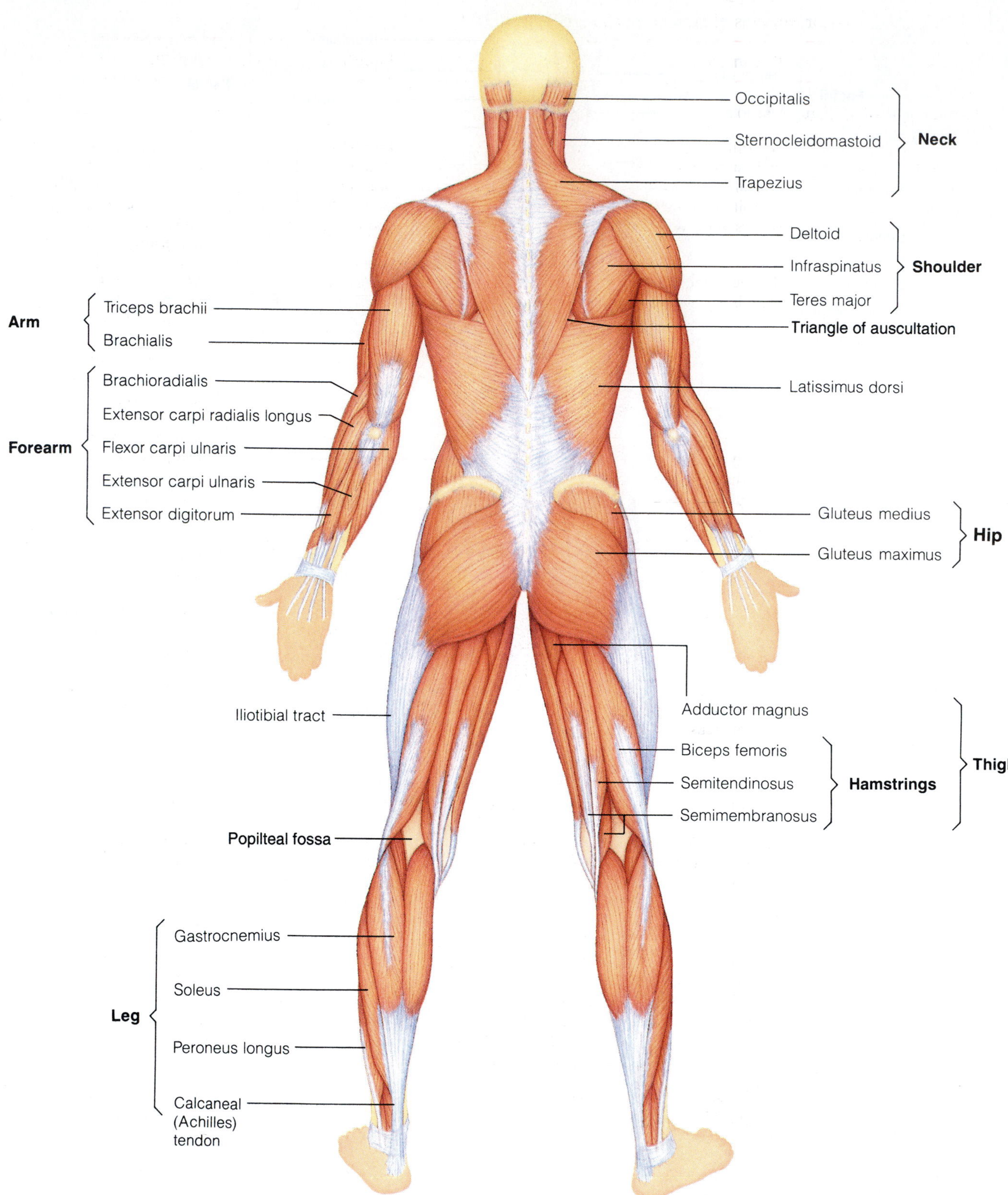

F15.2

Posterior view of superficial muscles of the body.

TABLE 15.1 Major Muscles of Human Head (see Figure 15.3)

Muscle	Comments	Origin	Insertion	Action
Facial Expression (Figure 15.3a)				
Epicranius—frontalis and occipitalis	Bipartite muscle consisting of frontalis and occipitalis, which covers dome of skull	Frontalis: cranial aponeurosis (galea aponeurotica); occipitalis: occipital bone	Frontalis: skin of eyebrows and root of nose; occipitalis: cranial aponeurosis	With aponeurosis fixed, frontalis raises eyebrows; occipitalis fixes aponeurosis and pulls scalp posteriorly
Orbicularis oculi	Sphincter muscle of eyelids	Frontal and maxillary bones and ligaments around orbit	Encircles orbit and inserts in tissue of eyelid	Various parts can be activated individually; closes eyes, produces blinking, squinting, and draws eyebrows downward
Corrugator supercilii	Small muscle; activity associated with that of orbicularis oculi	Arch of frontal bone above nasal bone	Skin of eyebrow	Draws eyebrows medially; wrinkles skin of forehead vertically
Levator labii superioris	Thin muscle between orbicularis oris and inferior eye margin	Zygomatic bone and infraorbital margin of maxilla	Skin and muscle of upper lip and border of nostril	Raises and furrows upper lip; flares nostril (as in disgust)
Zygomaticus—major and minor	Extends diagonally from corner of mouth to cheekbone	Zygomatic bone	Skin and muscle at corner of mouth	Raises lateral corners of mouth upward (smiling muscle)
Risorius	Slender muscle; runs laterally to zygomaticus	Fascia of masseter muscle	Skin at corner of mouth	Draws corner of lip laterally; tenses lip; zygomaticus synergist
Depressor labii inferioris	Small muscle from lower lip to jawbone	Body of mandible lateral to its midline	Skin and muscle of lower lip	Draws lower lip downward
Depressor anguli oris	Small muscle lateral to depressor labii inferioris	Body of mandible below incisors	Skin and muscle at angle of mouth below insertion of zygomaticus	Zygomaticus antagonist; draws corners of mouth downward and laterally
Orbicularis oris	Multilayered sphincter muscle of lips with fibers that run in many different directions	Arises indirectly from maxilla and mandible; fibers blended with fibers of other muscles associated with lips	Encircles mouth; inserts into muscle and skin at angles of mouth	Closes mouth; purses and protrudes lips (kissing muscle)
Mentalis	One of muscle pair forming V-shaped muscle mass on chin	Mandible below incisors	Skin of chin	Protrudes lower lip; wrinkles chin
Buccinator	Principal muscle of cheek; runs horizontally, deep to the masseter	Molar region of maxilla and mandible	Orbicularis oris	Draws corner of mouth laterally; compresses cheek (as in whistling); holds food between teeth during chewing

(*continued*)

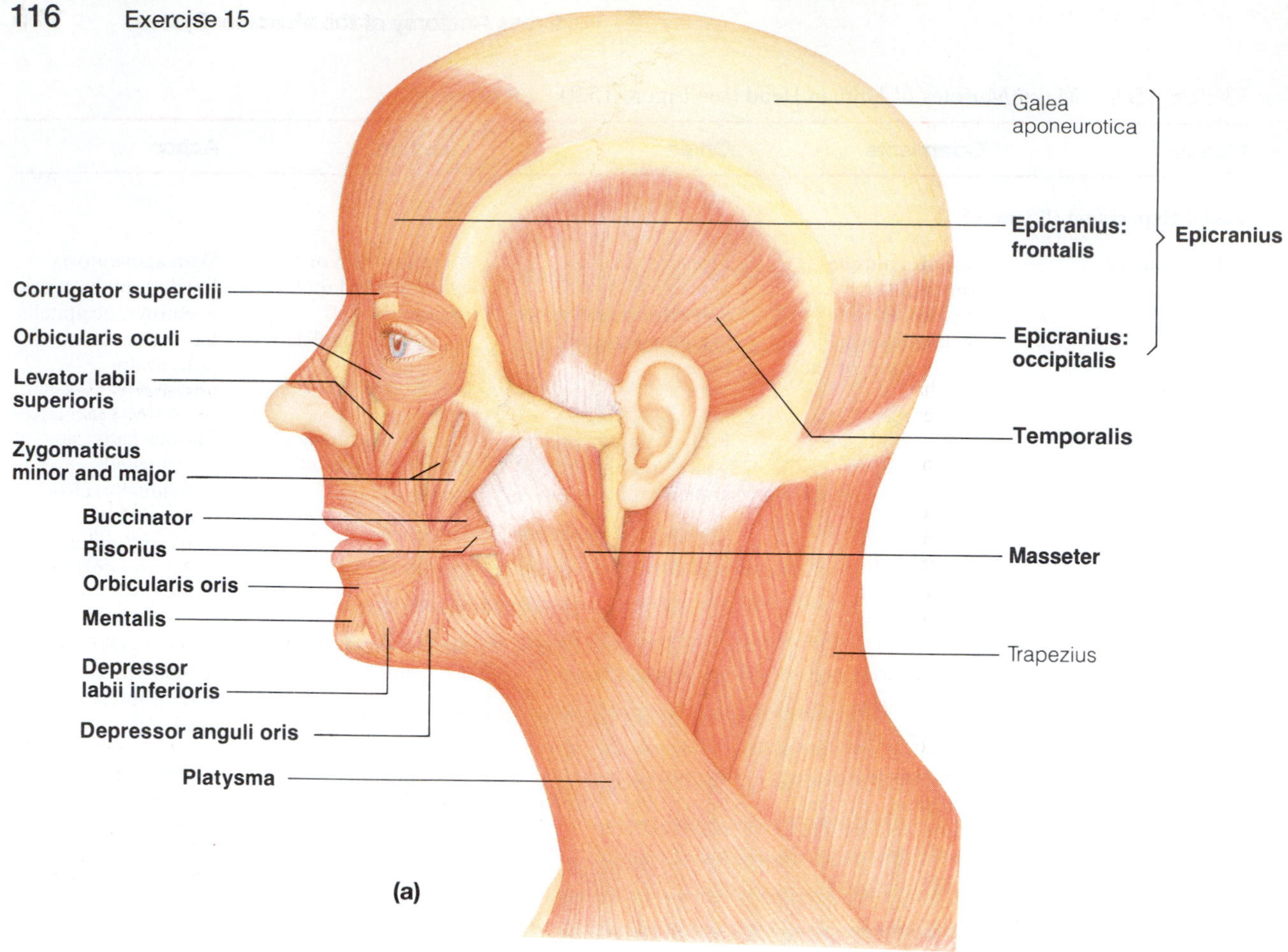

Galea aponeurotica
Epicranius: frontalis
Epicranius
Epicranius: occipitalis
Temporalis
Masseter
Trapezius
Corrugator supercilii
Orbicularis oculi
Levator labii superioris
Zygomaticus minor and major
Buccinator
Risorius
Orbicularis oris
Mentalis
Depressor labii inferioris
Depressor anguli oris
Platysma

(a)

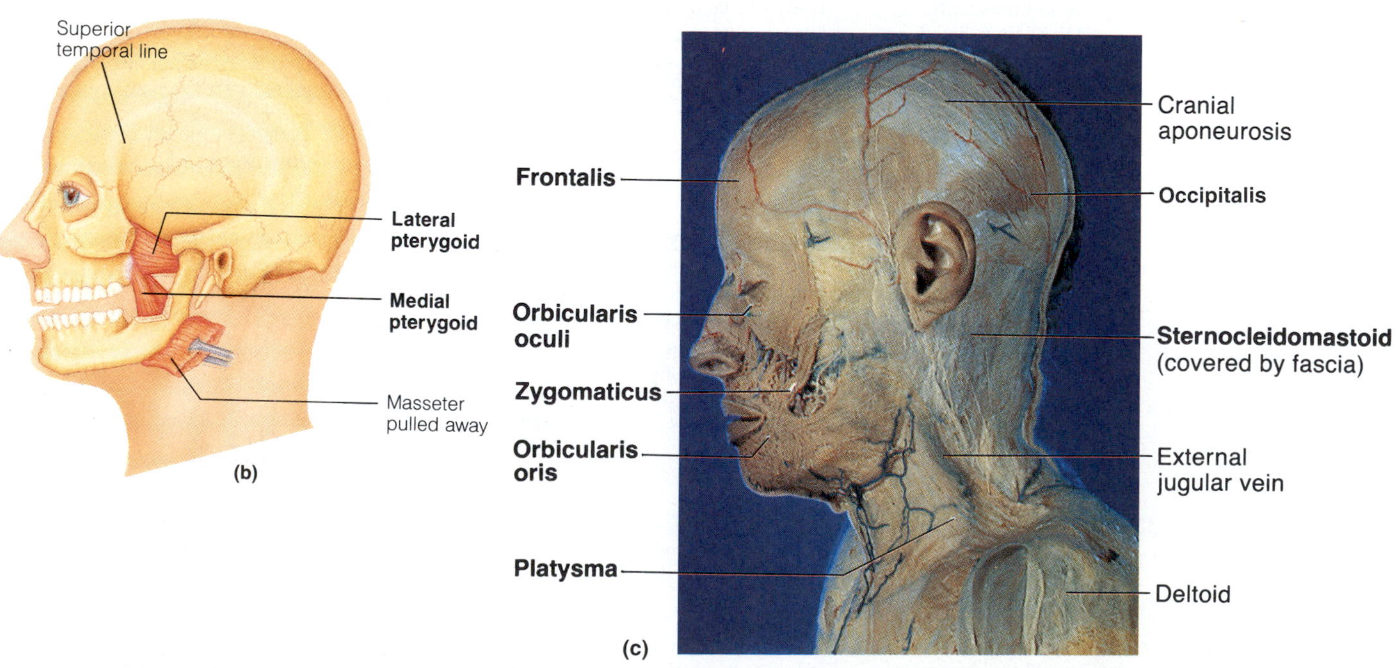

F15.3

Muscles of the scalp, face, and neck; left lateral view. (**a**) Superficial muscles. (**b**) The deep chewing muscles, the medial and lateral pterygoid muscles. (**c**) Photo of superficial structures of head and neck.

TABLE 15.1 *(Continued)*

Muscle	Comments	Origin	Insertion	Action
Mastication (Figure 15.3a,b)				
Masseter	Extends across jawbone; can be palpated on forcible closure of jaws	Zygomatic process and arch	Angle and ramus of mandible	Closes jaw and elevates mandible
Temporalis	Fan-shaped muscle over temporal bone	Temporal fossa	Coronoid process of mandible	Closes jaw; elevates and retracts mandible
Buccinator	(See muscles of facial expression.)			
Pterygoid—medial	Runs along internal (medial) surface of mandible (thus largely concealed by that bone)	Sphenoid, palatine, and maxillary bones	Medial surface of mandibular ramus and angle	Synergist of temporalis and masseter; closes and elevates mandible; in conjunction with lateral pterygoid, aids in grinding movements
Pterygoid—lateral	Superior to medial pterygoid	Greater wing of sphenoid bone	Mandibular condyle	Protracts jaw (moves it anteriorly); in conjunction with medial pterygoid, aids in grinding movements of teeth

TABLE 15.2 Anterolateral Muscles of Human Neck (see Figure 15.4)

Muscle	Comments	Origin	Insertion	Action
Superficial				
Platysma	Unpaired muscle: thin, sheetlike superficial neck muscle, not strictly a head muscle but plays role in facial expression (see Fig. 15.3a)	Fascia of chest (over pectoral muscles) and deltoid	Lower margin of mandible, skin, and muscle at corner of mouth	Depresses mandible; pulls lower lip back and down; i.e., produces downward sag of the mouth
Sternocleidomastoid	Two-headed muscle located deep to platysma on anterolateral surface of neck; fleshy parts on either side indicate limits of anterior and posterior triangles of neck	Manubrium of sternum and medial portion of clavicle	Mastoid process of temporal bone	Simultaneous contraction of both muscles of pair causes flexion of neck forward, generally against resistance (as when lying on the back); acting independently, rotate head toward shoulder on opposite side
Scalenes—anterior, middle, and posterior	Located more on lateral than anterior neck; deep to platysma (see Fig. 15.4b)	Transverse processes of cervical vertebrae	Anterolaterally on first two ribs	Flex and slightly rotate neck; elevate first two ribs (aid in inspiration)

(continued)

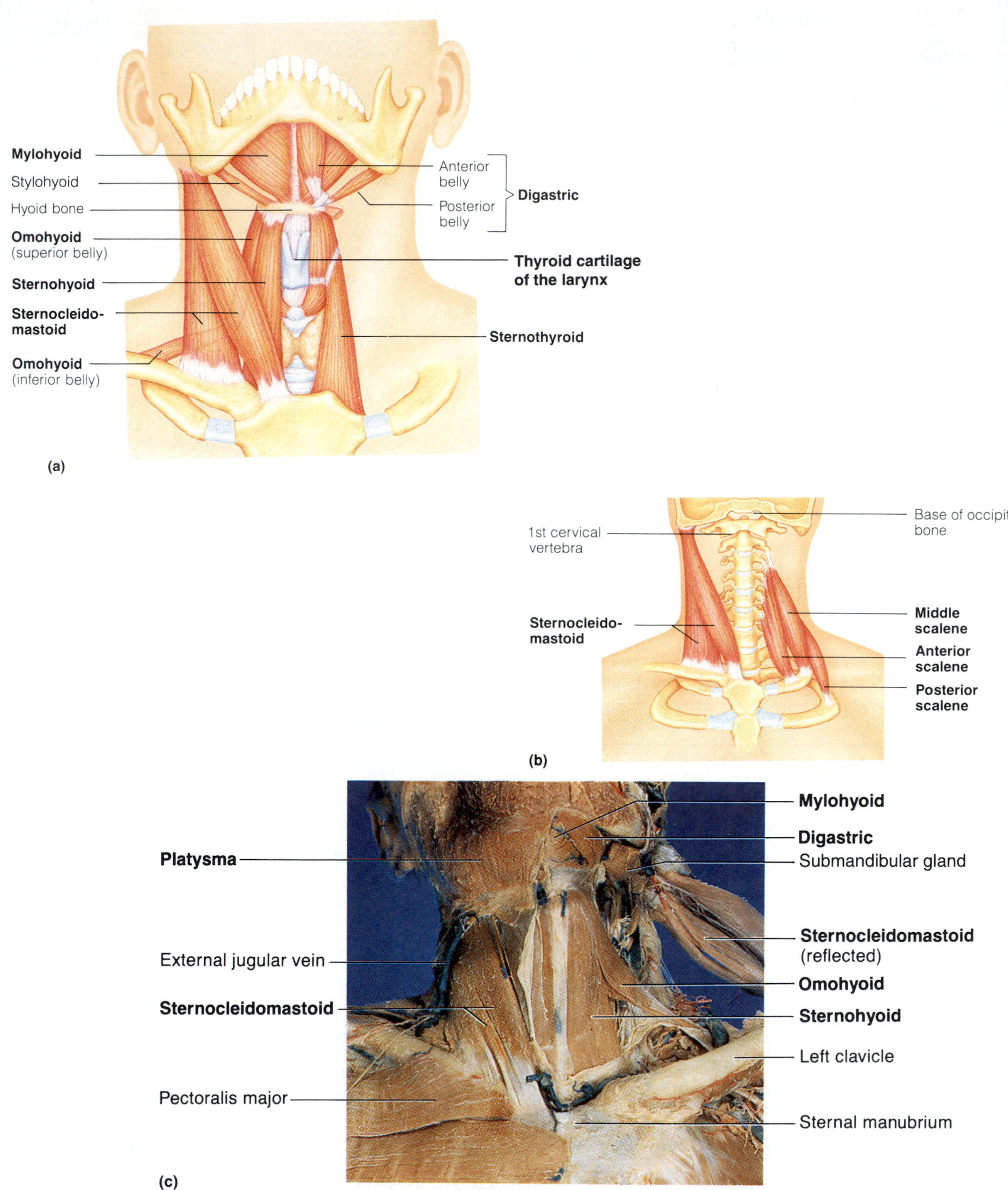

F15.4

Muscles of the neck and throat. **(a)** Anterior view of deep neck (suprahyoid and infrahyoid) muscles. **(b)** Muscles of the anterolateral neck. The superficial platysma muscle and deeper neck muscles have been removed to show the origins and insertions of the sternocleidomastoid and scalene muscles clearly. **(c)** Photo of the anterior and lateral regions of the neck. The fascia has been partially removed (left side of photo) to expose the sternocleidomastoid muscle. On the right side of the photo, the sternocleidomastoid muscle is reflected to expose the sternohyoid and omohyoid muscles.

TABLE 15.2 *(Continued)*

Muscle	Comments	Origin	Insertion	Action
Deep (Figure 15.4a,c)				
Digastric	Consists of two bellies united by an intermediate tendon; assumes a V-shaped configuration under chin	Lower margin of mandible (anterior belly) and mastoid process (posterior belly)	By a connective tissue loop to hyoid bone	Acting in concert, elevate hyoid bone; open mouth and depress mandible
Mylohyoid	Just deep to digastric; forms floor of mouth	Medial surface of mandible	Hyoid bone	Elevates hyoid bone and base of tongue during swallowing
Sternohyoid	Runs most medially along neck; straplike	Posterior surface of manubrium	Lower margin of body of hyoid bone	Acting with sternothyroid and omohyoid (all inferior to hyoid bone), depresses larynx and hyoid bone if mandible is fixed; may also flex skull
Sternothyroid	Lateral to sternohyoid; straplike	Manubrium and medial end of clavicle	Thyroid cartilage of larynx	(See Sternohyoid, above)
Omohyoid	Straplike with two bellies; lateral to sternohyoid	Superior surface of scapula	Hyoid bone	(See Sternohyoid above)

TABLE 15.3 Anterior Muscles of Human Thorax, Shoulder, and Abdominal Wall (see Figure 15.5)

Muscle	Comments	Origin	Insertion	Action
Thorax and Shoulder (Figure 15.5a)				
Pectoralis major	Large fan-shaped muscle covering upper portion of chest	Clavicle, sternum, cartilage of first six ribs, and aponeurosis of external oblique muscle	Fibers converge to insert by short tendon into greater tubercle of humerus	Prime mover of arm flexion; adducts, medially rotates arm; with arm fixed, pulls chest upward (thus also acts in forced inspiration)
Serratus anterior	Deep and superficial portions; beneath and inferior to pectoral muscles on lateral rib cage	Lateral aspect of first to eighth (or ninth) ribs	Vertebral border of anterior surface of scapula	Moves scapula forward toward chest wall; rotates scapula causing inferior angle to move laterally and upward

(continued)

TABLE 15.3 *(Continued)*

Muscle	Comments	Origin	Insertion	Action
Deltoid	Fleshy triangular muscle forming shoulder muscle mass	Lateral third of clavicle; acromion and spine of scapula	Deltoid tuberosity of humerus	Acting as a whole, prime mover of arm abduction; when only specific fibers are active, can aid in flexion, extension, and rotation of humerus
Pectoralis minor	Flat, thin muscle directly beneath and obscured by pectoralis major	Anterior surface of third, fourth, and fifth ribs, near their costal cartilages	Coracoid process of scapula	With ribs fixed, draws scapula forward and inferiorly; with scapula fixed, draws rib cage superiorly
Intercostals—external	11 pairs lie between ribs; fibers run obliquely downward and forward toward sternum	Inferior border of rib above (not shown in figure)	Superior border of rib below	Pulls ribs toward one another to elevate rib cage; aids in inspiration
Intercostals—internal	11 pairs lie between ribs; fibers run deep and at right angles to those of external intercostals	Superior border of rib below	Inferior border of rib above (not shown in figure)	Draws ribs together to depress rib cage; aids in forced expiration; antagonistic to external intercostals
Abdominal Wall (Figure 15.5b and c)				
Rectus abdominis	Medial superficial muscle, extends from pubis to rib cage; ensheathed by aponeuroses of oblique muscles; segmented	Pubic crest and symphysis	Xiphoid process and costal cartilages of fifth through seventh ribs	Flexes vertebral column; increases abdominal pressure; fixes and depresses ribs; stabilizes pelvis during walking
External oblique	Most superficial lateral muscle; fibers run downward and medially; ensheathed by an aponeurosis	Anterior surface of last eight ribs	Linea alba,* pubic tubercles, and iliac crest	See Rectus abdominis, above; also aids muscles of back in trunk rotation and lateral flexion
Internal oblique	Fibers run at right angles to those of external oblique, which it underlies	Lumbodorsal fascia, iliac crest, and inguinal ligament	Linea alba, pubic crest, and costal cartilages of last three ribs	As for External oblique
Transversus abdominis	Deepest muscle of abdominal wall; fibers run horizontally	Inguinal ligament, iliac crest, and cartilages of last five or six ribs	Linea alba and pubic crest	Compresses abdominal contents

*The linea alba ("white line") is a narrow, tendinous sheath that runs along the middle of the abdomen from the sternum to the pubic symphysis. It is formed by the fusion of the aponeurosis of the external oblique and transversus muscles.

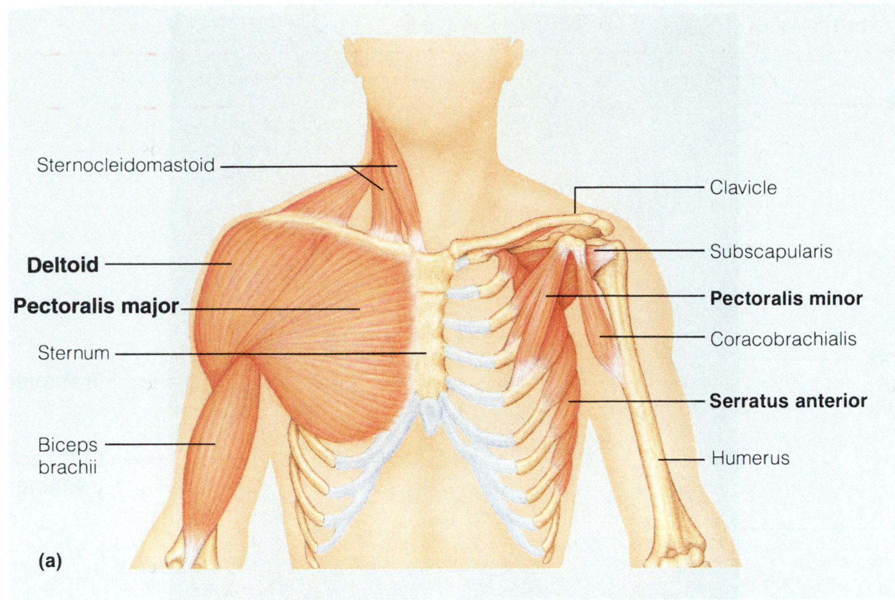

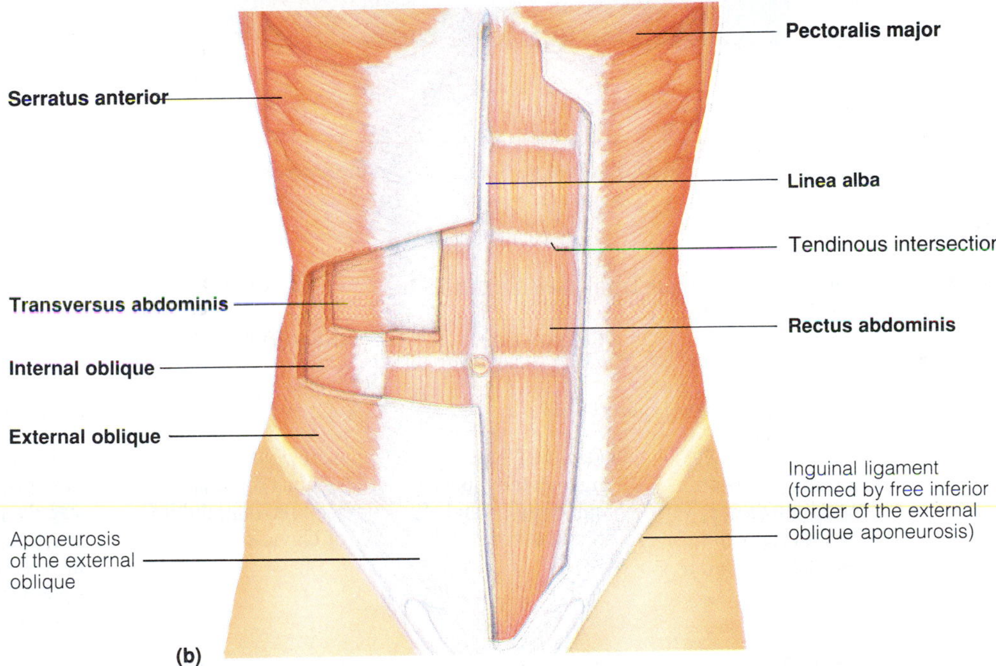

F15.5

Anterior muscles of the thorax, shoulder, and abdominal wall. (a) Anterior thorax. The superficial pectoralis major and deltoid muscles that effect arm movements are illustrated on the left. These muscles have been removed on the right side of the figure to illustrate the pectoralis minor, serratus anterior, and subscapularis muscles. **(b)** Anterior view of the muscles forming the anterolateral abdominal wall. The superficial muscles have been partially cut away on the left side of the diagram to reveal the deeper internal oblique and transversus abdominis muscles.

(*continued*)

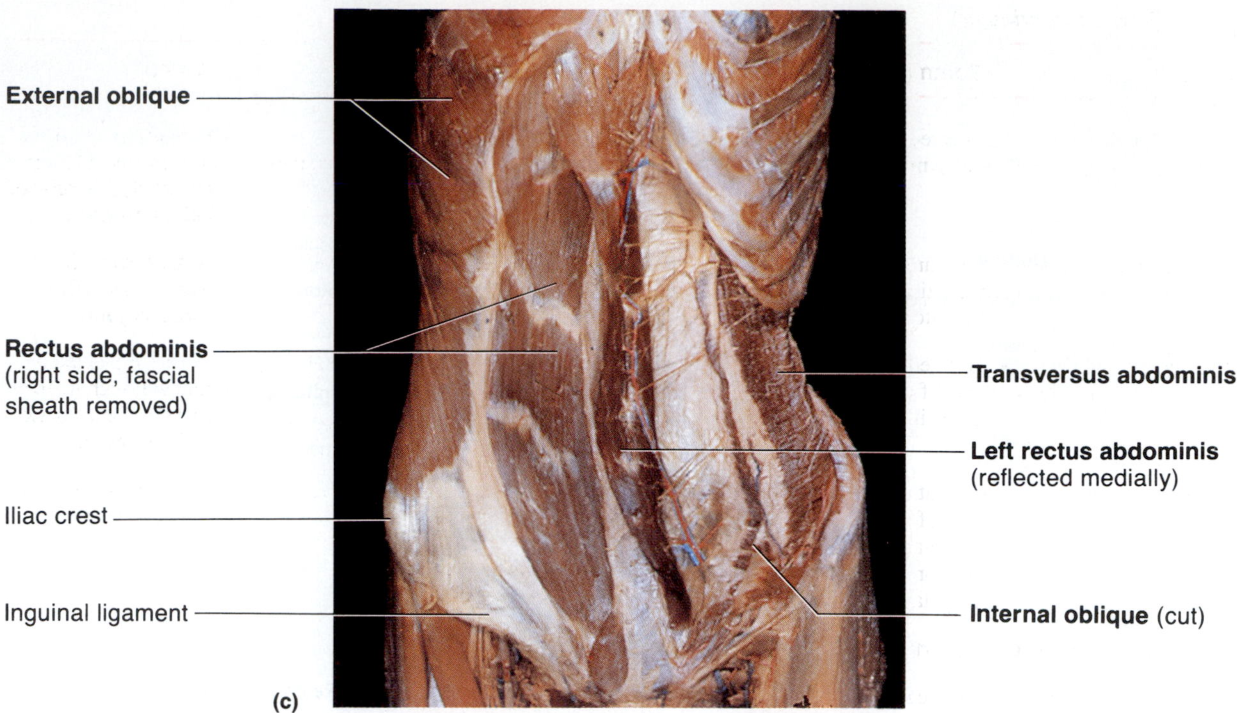

F15.5 (*continued*)

Anterior muscles of the thorax, shoulder, and abdominal wall. (c) Photo of the anterolateral abdominal wall.

TABLE 15.4 Posterior Muscles of Human Trunk (see Figure 15.6)

Muscle	Comments	Origin	Insertion	Action
Muscles of the Neck, Shoulder, and Thorax (Figure 15.6a)				
Trapezius	Most superficial muscle of posterior neck and thorax; very broad origin and insertion	Occipital bone; ligamentum nuchae; spines of C_7 and all thoracic vertebrae	Acromion and spinous process of scapula; lateral third of clavicle	Extends head; retracts (adducts) scapula and stabilizes it; upper fibers elevate scapula; lower fibers depress it
Latissimus dorsi	Broad flat muscle of lower back (lumbar region); extensive superficial origins	Indirect attachment to spinous processes of lower six thoracic vertebrae, lumbar vertebrae, lower 3 to 4 ribs, and iliac crest	Floor of intertubercular groove of humerus	Prime mover of arm extension; adducts and medially rotates arm; depresses scapula; brings arm down in power stroke, as in striking a blow
Infraspinatus	Partially covered by deltoid and trapezius; a rotator cuff muscle	Infraspinous fossa of scapula	Greater tubercle of humerus	Lateral rotation of humerus; helps hold head of humerus in glenoid cavity
Teres minor	Small muscle inferior to infraspinatus; a rotator cuff muscle	Lateral margin of scapula	Greater tuberosity of humerus	As for infraspinatus

(*continued*)

TABLE 15.4 (*Continued*)

Muscle	Comments	Origin	Insertion	Action
Teres major	Located inferiorly to teres minor	Posterior surface at inferior angle of scapula	Crest of lesser tubercle of humerus	Extends, medially rotates, and adducts humerus; synergist of latissimus dorsi
Supraspinatus	Obscured by trapezius and deltoid; a rotator cuff muscle	Supraspinous fossa of scapula	Greater tubercle of humerus	Assists abduction of humerus; stabilizes shoulder joint
Levator scapulae	Located at back and side of neck, deep to trapezius	Transverse processes of C_1 through C_4	Superior vertebral border of scapula	Raises and adducts scapula; with fixed scapula, flexes neck to the same side
Rhomboids—major and minor	Beneath trapezius and inferior to levator scapulae; run from vertebral column to scapula	Spinous processes of C_7 and T_1 through T_5	Vertebral border of scapula	Pull scapula medially (retraction) and elevate it
Muscles Associated with the Vertebral Column (Figure 15.6b)				
Semispinalis	Deep composite muscle of the back—thoracis, cervicis, and capitis portions	Transverse processes of C_7–T_{12}	Occipital bone and spinous processes of cervical vertebrae and T_1–T_4	Acting together, extend head and vertebral column; acting independently (right vs. left) causes rotation toward the opposite side

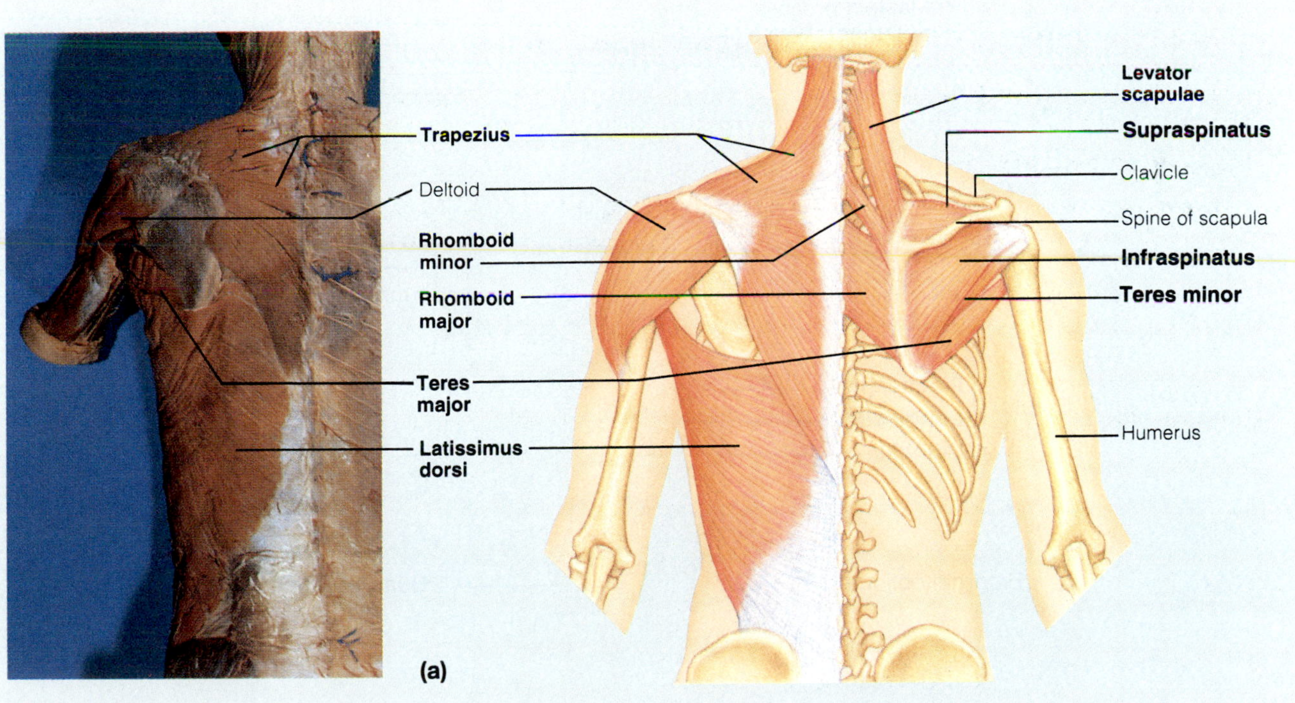

F15.6

Muscles of the neck, shoulder, and thorax, posterior view. **(a)** The superficial muscles of the back are shown for the left side of the body, with a corresponding photograph. The superficial muscles are removed on the right side of the illustration to reveal the deeper muscles acting on the scapula and the rotator cuff muscles that help to stabilize the shoulder joint.

(continued)

TABLE 15.4 (*Continued*)

Muscle	Comments	Origin	Insertion	Action
Erector spinae	A long tripartite muscle composed of iliocostalis (lateral), longissimus, and and spinalis (medial) muscle columns; superficial to semispinalis muscles; extends from pelvis to head	Sacrum, iliac crest, transverse processes of lumbar, thoracic and cervical vertebrae, and/or ribs 3–12 depending on specific part	Ribs and transverse processes of vertebrae about six segments above origin. Longissimus also inserts into mastoid process	All act to extend and abduct the vertebral column; fibers of the longissimus also extend head
Splenius (see Figure 15.6c)	Superficial muscle (capitis and cervicis parts) just deep to levator scapulae and superficial to erector spinae	Ligamentum nuchae and spinous processes of C_7–T_6	Mastoid process, occipital bone, and transverse processes of C_2–C_4	As a group, extend or hyperextend head; when only one side is active, head is rotated and bent toward the same side
Quadratus lumborum	Forms greater portion of posterior abdominal wall	Iliac crest and iliolumbar fascia	Inferior border twelfth rib; transverse processes of lumbar vertebrae	Each flexes vertebral column laterally; together extend the lumbar spine and fix the twelfth rib

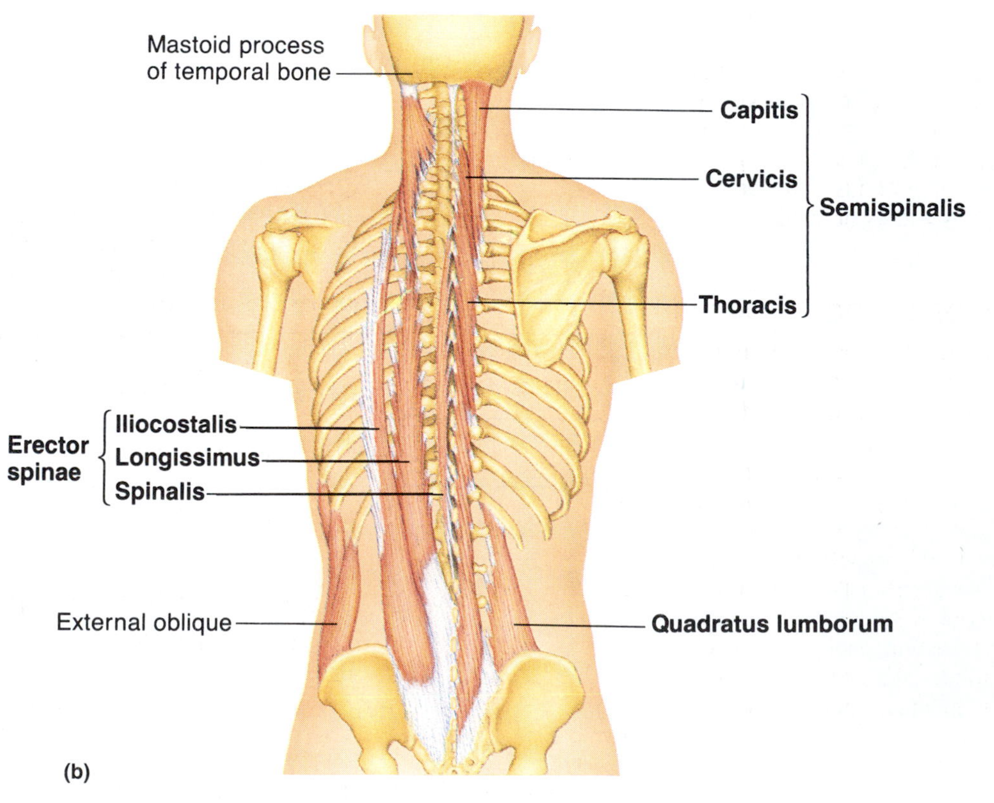

F15.6 (*continued*)

Muscles of the neck, shoulder, and thorax, posterior view. **(b)** The erector spinae and semispinalis muscles which respectively form the intermediate and deep muscle layers of the back associated with the vertebral column.

F15.6 (*continued*)

Muscles of the neck, shoulder, and thorax, posterior view. **(c)** Deep (splenius) muscles of the posterior neck. Superficial muscles have been removed.

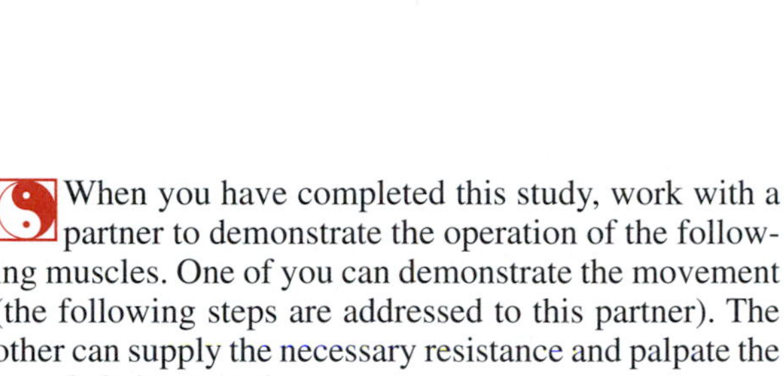

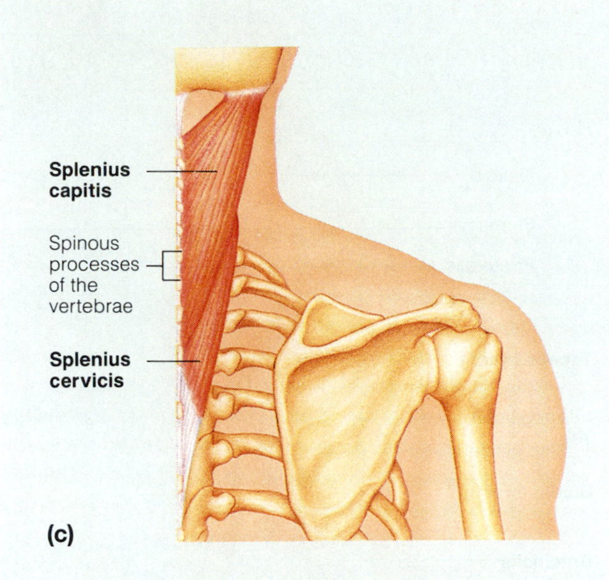

When you have completed this study, work with a partner to demonstrate the operation of the following muscles. One of you can demonstrate the movement (the following steps are addressed to this partner). The other can supply the necessary resistance and palpate the muscle being tested.

1. Start by fully abducting the arm and extending the elbow. Now try to adduct the arm against resistance. You are exercising the *latissimus dorsi.*
2. To observe the *deltoid,* attempt to abduct your arm against resistance. Now attempt to elevate your shoulder against resistance; you are contracting the upper portion of the *trapezius.*
3. The *pectoralis major* comes into play when you press your hands together at chest level with your elbows widely abducted.

Muscles of the Upper Limb

The muscles that act on the upper limb fall into three groups: those that move the arm, those causing movement at the elbow, and those effecting movements of the wrist and hand.

The muscles that cross the shoulder joint to insert on the humerus and move the arm (subscapularis, supraspinatus and infraspinatus, deltoid, and so on) are primarily trunk muscles that originate on the axial skeleton or shoulder girdle. These muscles are included with the trunk muscles.

The second group of muscles, which cross the elbow joint and move the forearm, consists of muscles forming the musculature of the humerus. These muscles arise primarily from the humerus and insert in forearm bones. They are responsible for flexion, extension, pronation, and supination. The origins, insertions, and actions of these muscles are summarized in Table 15.5 and the muscles are shown in Figure 15.7.

The third group composes the musculature of the forearm. For the most part, these muscles insert on the digits and produce movements at the wrist and fingers. In general, muscles acting on the wrist and hand are more easily identified if their insertion tendons are located first. These muscles are described in Table 15.6 and illustrated in Figure 15.8.

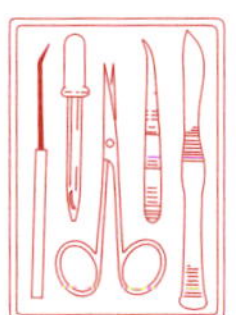

First study the tables and figures, then see if you can identify these muscles on a torso model, anatomical chart, or cadaver. Complete this portion of the exercise with palpation demonstrations as outlined next.

- To observe the *biceps brachii,* attempt to flex your forearm (hand supinated) against resistance. The insertion tendon of this biceps muscle can also be felt in the lateral aspect of the antecubital fossa (where it runs toward the radius to attach).
- If you acutely flex your elbow and then try to extend it against resistance, you can demonstrate the action of your *triceps brachii.*
- Strongly flex your wrist and make a fist. Palpate your contracting wrist flexor muscles (which originate from the medial epicondyle of the humerus) and their insertion tendons, which can be easily felt at the anterior aspect of the wrist.
- Flare your fingers to identify the tendons of the *extensor digitorum* muscle on the dorsum of your hand.

Muscles of the Lower Limb

Muscles that act on the lower limb cause movement at the hip, knee, and foot joints. Since the human pelvic girdle is composed of heavy fused bones that allow very little movement, no special group of muscles is necessary to stabilize it. This is unlike the shoulder girdle, where several muscles (mainly trunk muscles) are needed to stabilize the scapulae.

Muscles acting on the thigh (femur) cause various movements at the multiaxial hip joint (flexion, extension, rotation, abduction, and adduction). These include the iliopsoas, the adductor group, and other muscles

(*Text continues on p. 130*)

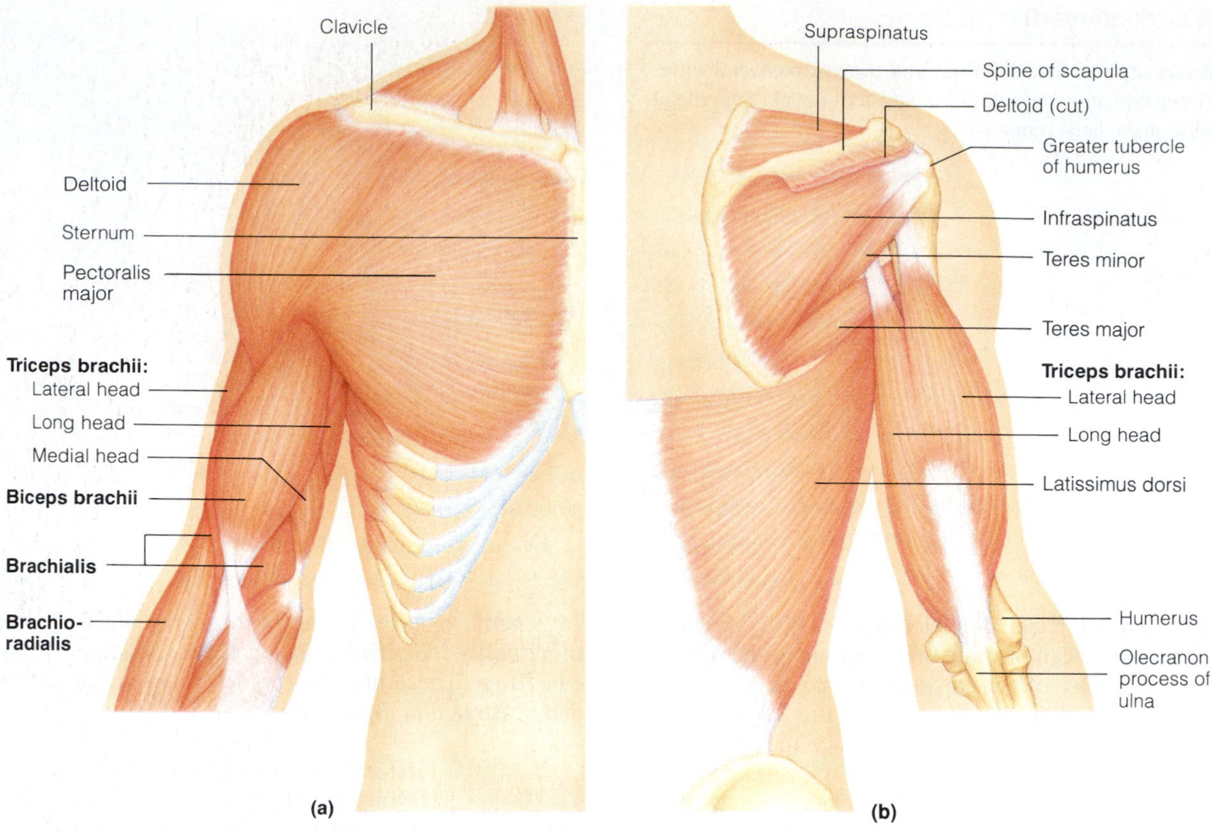

F15.7

Muscles causing movements of the forearm. (a) Superficial muscles of the anterior thorax, shoulder, and arm, anterior view. **(b)** Posterior aspect of the arm showing the lateral and long heads of the triceps brachii muscle.

TABLE 15.5 Muscles of Human Humerus That Act on the Forearm (see Figure 15.7)

Muscle	Comments	Origin	Insertion	Action
Triceps brachii	Sole, large fleshy muscle of posterior humerus; three-headed origin	Long head: inferior margin of glenoid cavity; lateral head: posterior humerus; medial head: distal radial groove on posterior humerus	Olecranon process of ulna	Powerful forearm extensor; antagonist of forearm flexors (brachialis and biceps brachii)
Biceps brachii	Most familiar muscle of anterior humerus because this two-headed muscle bulges when forearm is flexed	Short head: coracoid process; tendon of long head runs in intertubercular groove and within capsule of shoulder joint	Radial tuberosity	Flexion (powerful) of elbow and supination of forearm; "it turns the corkscrew and pulls the cork"; weak arm flexor
Brachioradialis	Superficial muscle of lateral forearm; forms lateral boundary of antecubital fossa	Lateral ridge at distal end of humerus	Base of styloid process of radius	Forearm flexor (weak)
Brachialis	Immediately deep to biceps brachii	Distal portion of anterior humerus	Coronoid process of ulna	A major flexor of forearm

TABLE 15.6 Muscles of Human Forearm That Act on Hand and Fingers (see Figure 15.8)

Muscle	Comments	Origin	Insertion	Action
Anterior Compartment (Figure 15.8a,b,c)				
Superficial				
Pronator teres	Seen in a superficial view between proximal margins of brachioradialis and flexor carpi ulnaris	Medial epicondyle of humerus and coronoid process of ulna	Midshaft of radius	Acts synergistically with pronator quadratus to pronate forearm; weak forearm flexor
Flexor carpi radialis	Superficial; runs diagonally across forearm	Medial epicondyle of humerus	Base of second and third metacarpals	Powerful flexor of wrist; abducts hand
Palmaris longus	Small fleshy muscle with a long tendon; medial to flexor carpi radialis	Medial epicondyle of humerus	Palmar aponeurosis	Flexes wrist (weak)
Flexor carpi ulnaris	Superficial; medial to palmaris longus	Medial epicondyle of humerus and olecranon process of ulna	Base of fifth metacarpal	Powerful flexor of wrist; adducts hand

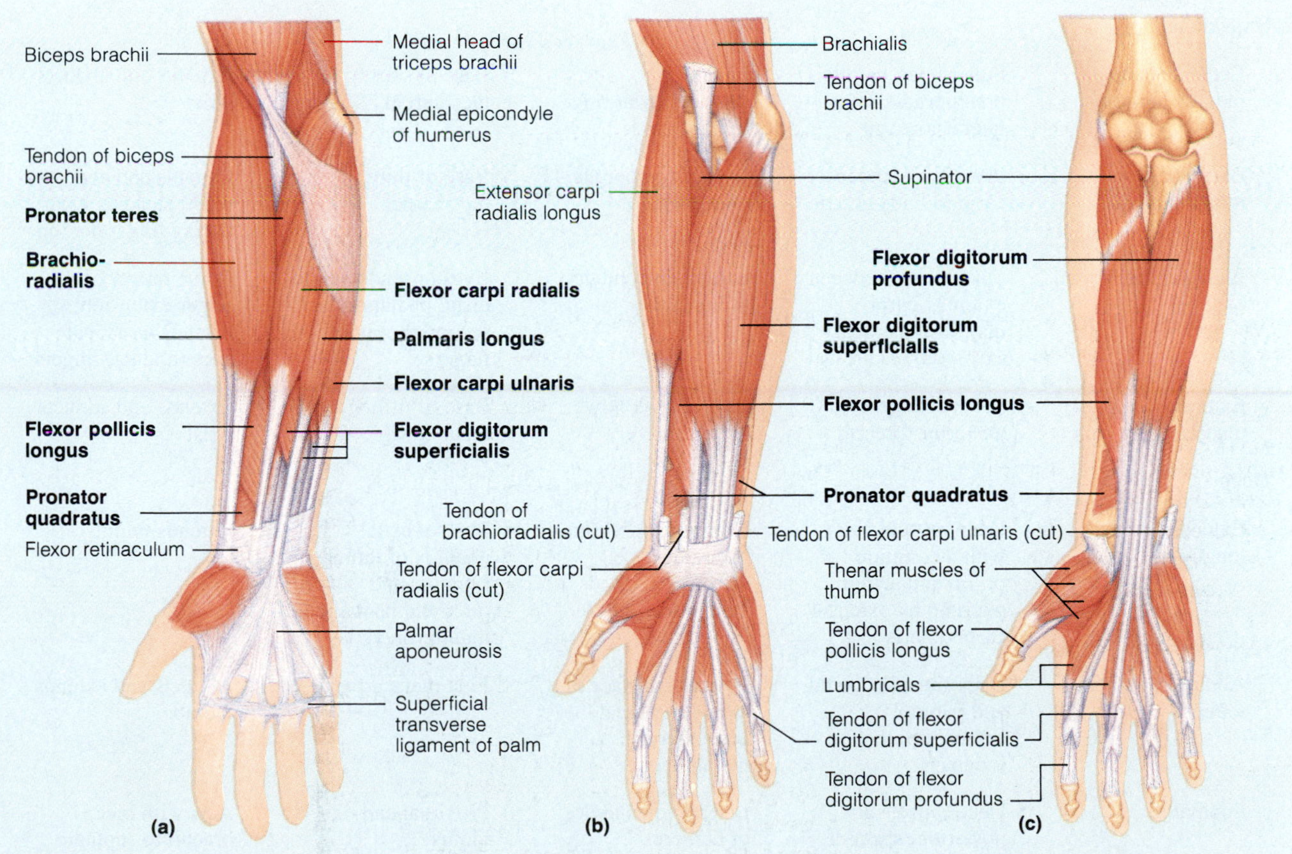

F15.8

Muscles of the forearm and wrist. **(a)** Superficial anterior view of right forearm and hand. **(b)** The brachioradialis, flexors carpi radialis and ulnaris, and palmaris longus muscles have been removed to reveal the position of the somewhat deeper flexor digitorum superficialis. **(c)** Deep muscles of the anterior compartment. Superficial muscles have been removed. Note: The thenar muscles of the thumb and the lumbricals that help move the fingers are illustrated here but are not described in Table 15.6.

(continued)

TABLE 15.6 (*Continued*)

Muscle	Comments	Origin	Insertion	Action
Flexor digitorum superficialis	Deeper muscle; overlain by muscles named above; visible at distal end of forearm	Medial epicondyle of humerus, medial surface of ulna, and anterior border of radius	Middle phalanges of second through fifth fingers	Flexes wrist and middle phalanges of second through fifth fingers
Deep				
Flexor pollicis longus	Deep muscle of anterior forearm; distal to and paralleling lower margin of flexor digitorum superficialis	Anterior surface of radius, and interosseous membrane	Distal phalanx of thumb	Flexes thumb (*pollix* is Latin for "thumb"); weak flexor of wrist
Flexor digitorum profundus	Deep muscle; overlain entirely by flexor digitorum superficialis	Anteromedial surface of ulna and interosseous membrane	Distal phalanges of second through fifth fingers	Sole muscle that flexes distal phalanges; assists in wrist flexion
Pronator quadratus	Deepest muscle of distal forearm	Distal portion of anterior ulnar surface	Anterior surface of radius, distal end	Pronates forearm
Posterior Compartment (Figure 15.8d,e,f)				
Superficial				
Extensor carpi radialis longus	Superficial; parallels brachioradialis on lateral forearm	Lateral supracondylar ridge of humerus	Base of second metacarpal	Extends and abducts wrist
Extensor carpi radialis brevis	Posterior to extensor carpi radialis longus	Lateral epicondyle of humerus	Base of third metacarpal	Extends and abducts wrist; steadies wrist during finger flexion
Extensor digitorum	Superficial; between extensor carpi ulnaris and extensor carpi radialis brevis	Lateral epicondyle of humerus	By four tendons into distal phalanges of second through fifth fingers	Prime mover of finger extension; extends wrist; can flare (abduct) fingers
Extensor carpi ulnaris	Superficial; medial posterior forearm	Lateral epicondyle of humerus	Base of fifth metacarpal	Extends and adducts wrist
Deep				
Extensor pollicis longus and brevis	Deep muscle pair with a common origin and action; overlain by extensor carpi ulnaris	Dorsal shaft of ulna and radius, interosseous membrane	Base of distal phalanx of thumb (longus) and proximal phalanx of thumb (brevis)	Extends thumb
Abductor pollicis longus	Deep muscle; lateral and parallel to extensor pollicis longus	Posterior surface of radius and ulna; interosseous membrane	First metacarpal	Abducts and extends thumb
Supinator	Deep muscle at posterior aspect of elbow	Lateral epicondyle of humerus	Proximal end of radius	Acts with biceps brachii to supinate forearm; antagonist of pronator muscles

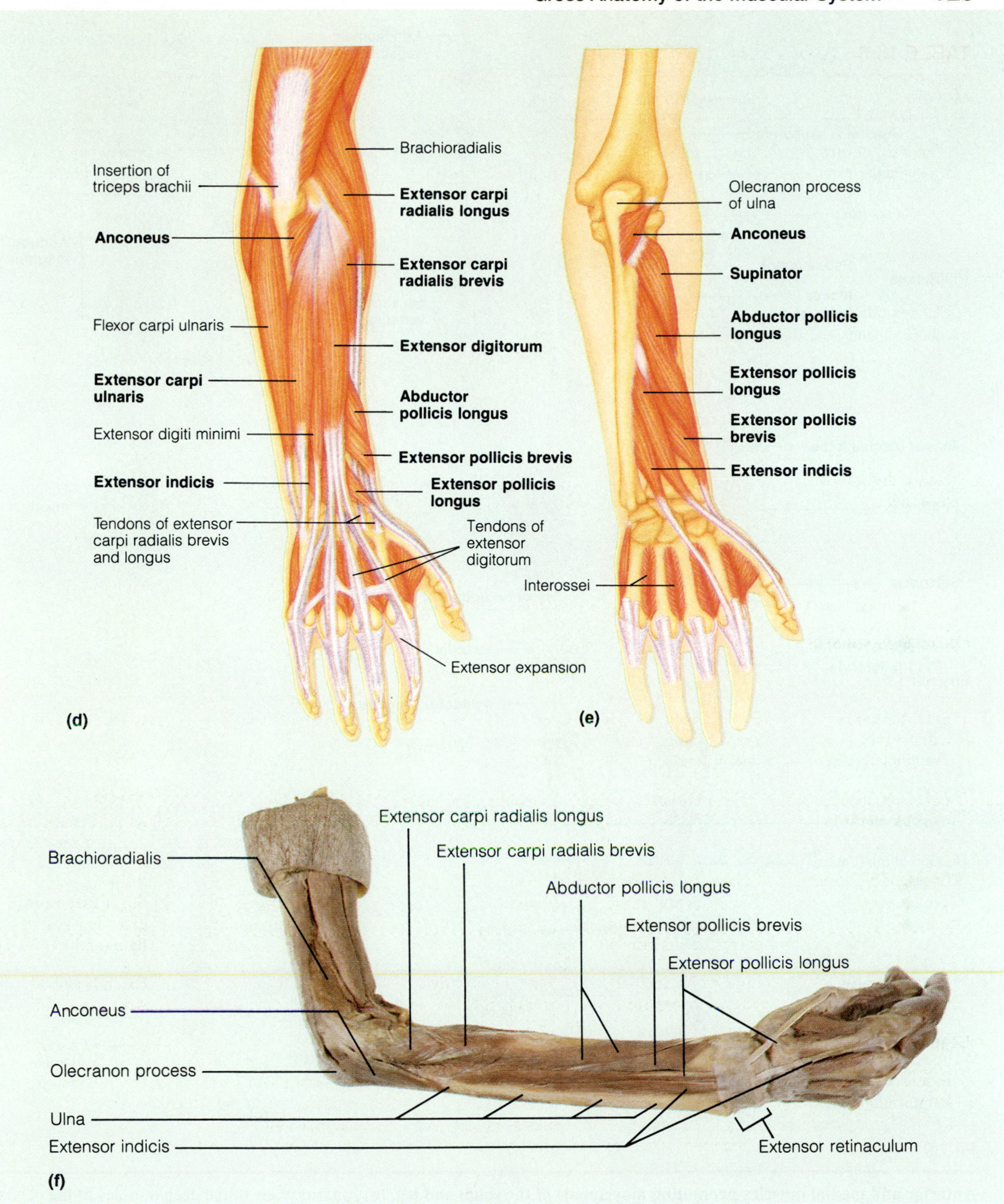

F15.8 *(continued)*

(d) Superficial muscles, posterior view. **(e)** Deep posterior muscles; superficial muscles have been removed. The interossei, the deepest layer of intrinsic hand muscles, are also illustrated. **(f)** Photo of deep posterior muscles of the right forearm. The superficial muscles have been removed.

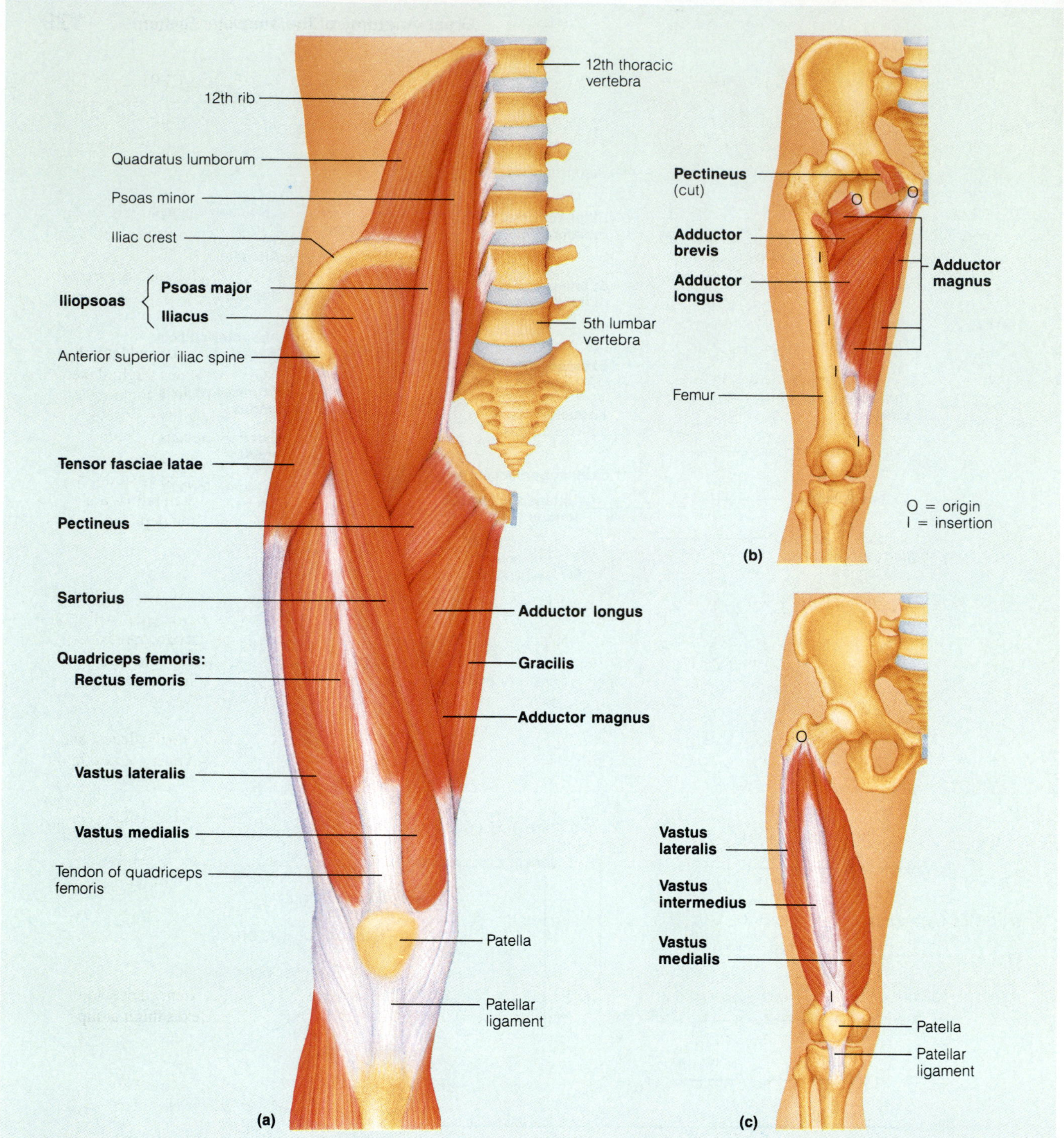

F15.9

Anterior and medial muscles promoting movements of the thigh and leg. **(a)** Anterior view of the deep muscles of the pelvis and superficial muscles of the right thigh. **(b)** Adductor muscles of the medial compartment of the thigh. **(c)** The vastus muscles (isolated) of the quadriceps group.

summarized in Tables 15.7 and 15.8 and illustrated in Figures 15.9 and 15.10.

Muscles acting on the leg form the major musculature of the thigh. (Anatomically the term *leg* refers only to that portion between the knee and the ankle.) The thigh muscles cross the knee to allow its flexion and extension. They include the hamstrings and the quadriceps and, along with the muscles acting on the thigh, are described in Tables 15.7 and 15.8 and illustrated in Figures 15.9 and 15.10. Since some of these muscles also have attachments on the pelvic girdle, they can cause movement at the hip joint.

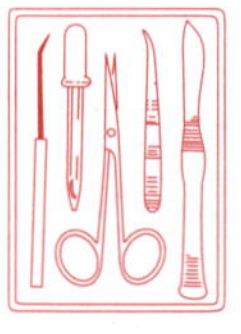

The muscles originating on the leg and acting on the foot and toes are described in Table 15.9 and shown in Figures 15.11 and 15.12. Identify the muscles as instructed previously.

(*Text continues on p. 134*)

TABLE 15.7 Muscles Acting on Human Thigh and Leg, Anterior and Medial Aspects (see Figure 15.9)

Muscle	Comments	Origin	Insertion	Action
Origin on the Pelvis				
Iliopsoas—iliacus and psoas major	Two closely related muscles; fibers pass under inguinal ligament to insert into femur via a common tendon	Iliacus: iliac fossa; psoas major: transverse processes, bodies, and discs of T_{12} and lumbar vertebrae	Lesser trochanter of femur	Flex trunk on thigh; major flexor of hip (or thigh on pelvis when pelvis is fixed)
Sartorius	Straplike superficial muscle running obliquely across anterior surface of thigh to knee	Anterior superior iliac spine	By an aponeurosis into medial aspect of proximal tibia	Flexes and laterally rotates thigh; flexes knee; known as "tailor's muscle" because it helps bring about cross-legged position in which tailors are often depicted
Medial Compartment				
Adductors—magnus, longus, and brevis	Large muscle mass forming medial aspect of thigh; arise from front of pelvis and insert at various levels on femur	Magnus: ischial and pubic rami; longus: pubis near pubic symphysis; brevis: body and inferior ramus of pubis	Magnus: linea aspera and adductor tubercle of femur; longus and brevis: linea aspera	Adduct and laterally rotate and flex thigh; posterior part of magnus is also a synergist in thigh extension
Pectineus	Overlies adductor brevis on proximal thigh	Pectineal line of pubis	Inferior to lesser trochanter of femur	Adducts, flexes, and laterally rotates thigh
Gracilis	Straplike superficial muscle of medial thigh	Inferior ramus and body of pubis	Medial surface of head of tibia	Adducts thigh; flexes and medially rotates leg, especially during walking
Anterior Compartment				
Quadriceps*				
Rectus femoris	Superficial muscle of thigh; runs straight down thigh; only muscle of group to cross hip joint; arises from two heads	Anterior inferior iliac spine and superior margin of acetabulum	Tibial tuberosity	Extends knee and flexes thigh at hip
Vastus lateralis	Forms lateral aspect of thigh	Greater trochanter and linea aspera	Tibial tuberosity	Extends knee
Vastus medialis	Forms medial aspect of thigh	Linea aspera	Tibial tuberosity	Extends knee
Vastus intermedius	Obscured by rectus femoris; lies between vastus lateralis and vastus medialis on anterior thigh	Anterior and lateral surface of femur (not shown in figure)	Tibial tuberosity	Extends knee
Tensor fasciae latae	Enclosed between fascia layers of thigh	Anterior aspect of iliac crest and anterior superior iliac spine	Iliotibial band of fascia lata	Flexes, abducts, and medially rotates thigh

*The quadriceps form the flesh of the anterior thigh and have a common insertion in the tibial tuberosity via the patellar tendon. They are powerful leg extensors, enabling humans to kick a football, for example.

TABLE 15.8 Muscles Acting on Human Thigh and Leg, Posterior Aspect (see Figure 15.10)

Muscle	Comments	Origin	Insertion	Action
Origin on Pelvis				
Gluteus maximus	Largest and most superficial of gluteal muscles (which form buttock mass)	Dorsal ilium, sacrum, and coccyx	Gluteal tuberosity of femur and iliotibial tract*	Complex, powerful hip extensor (most effective when hip is flexed, as in climbing stairs—but not as in walking); antagonist of iliopsoas; laterally rotates thigh
Gluteus medius	Partially covered by gluteus maximus	Upper lateral surface of ilium	Greater trochanter of femur	Abducts and medially rotates thigh; steadies pelvis during walking
Gluteus minimus	Smallest and deepest gluteal muscle	Inferior surface of ilium (not shown in figure)	Greater trochanter of femur	Abducts and medially rotates thigh
Posterior Compartment				
Hamstrings†				
Biceps femoris	Most lateral muscle of group; arises from two heads	Ischial tuberosity (long head); linea aspera and distal femur (short head)	Tendon passes laterally to insert into head of fibula and lateral condyle of tibia	Extends thigh; laterally rotates leg on thigh; flexes knee
Semitendinosus	Medial to biceps femoris	Ischial tuberosity	Medial aspect of upper tibial shaft	Extends thigh; flexes knee; medially rotates leg
Semimembranosus	Deep to semitendinosus	Ischial tuberosity	Medial condyle of tibia	Extends thigh; flexes knee; medially rotates leg

*The iliotibial tract, a thickened lateral portion of the fascia lata, ensheathes all the muscles of the thigh. It extends as a tendinous band from the iliac crest to the knee.

†The hamstrings are the fleshy muscles of the posterior thigh. The name comes from the butchers' practice of using the tendons of these muscles to hang hams for smoking. As a group, they are strong extensors of the hip; they counteract the powerful quadriceps by stabilizing the knee joint when standing.

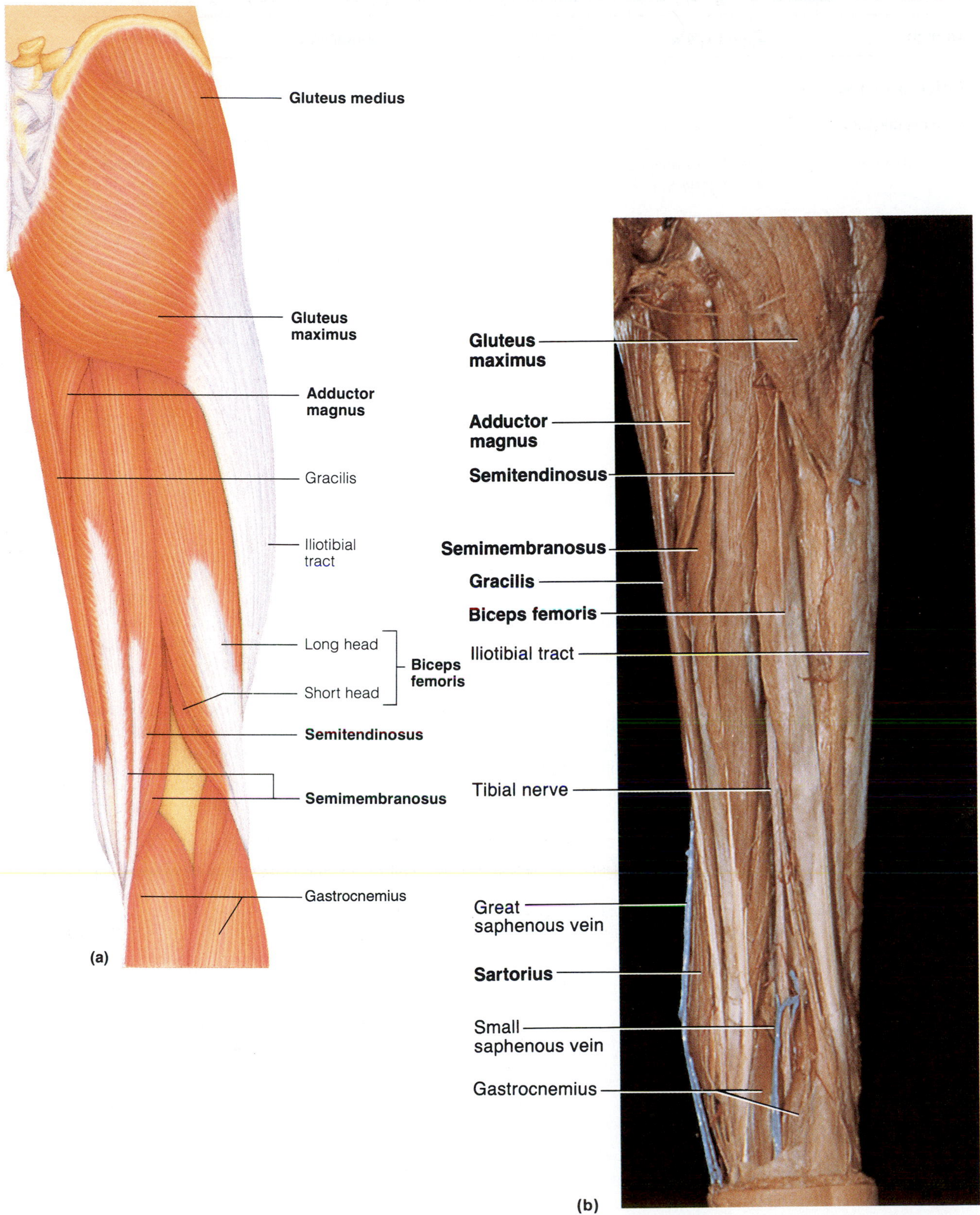

F15.10

Muscles of the posterior aspect of the right hip and thigh. **(a)** Superficial view showing the gluteus muscles of the buttock and hamstring muscles of the thigh. **(b)** Photo of muscles of the posterior thigh.

TABLE 15.9 Muscles Acting on Human Foot and Ankle (see Figures 15.11 and 15.12)

Muscle	Comments	Origin	Insertion	Action
Posterior Compartment				
Superficial (Figure 15.11a,b)				
Triceps surae	Muscle pair that shapes posterior calf		Via common tendon (calcaneal or Achilles) into heel	Plantar flex foot
Gastrocnemius	Superficial muscle of pair; two prominent bellies	By two heads from medial and lateral condyles of femur	Calcaneus via calcaneal tendon	Crosses knee joint; thus also can flex knee (when foot is dorsiflexed)
Soleus	Deep to gastrocnemius	Proximal portion of tibia and fibula	Calcaneus via calcaneal tendon	Plantar flexion; is an important muscle for locomotion

(*continued on p. 136*)

Complete this exercise by performing the following palpation demonstrations with your lab partner.

- Go into a deep knee bend and palpate your own *gluteus maximus* muscle as you extend your hip to resume the upright posture.
- Demonstrate the contraction of the anterior *quadriceps femoris* by trying to extend your knee against resistance. Do this while seated and note how the patellar tendon reacts. The *biceps femoris* of the posterior thigh comes into play when you flex your knee against resistance.
- Now stand on your toes. Have your partner palpate the lateral and medial heads of the *gastrocnemius* and follow it to its insertion in the calcaneal tendon.
- Dorsiflex and invert your foot while palpating your *tibialis anterior* muscle (which parallels the sharp anterior crest of the tibia laterally).

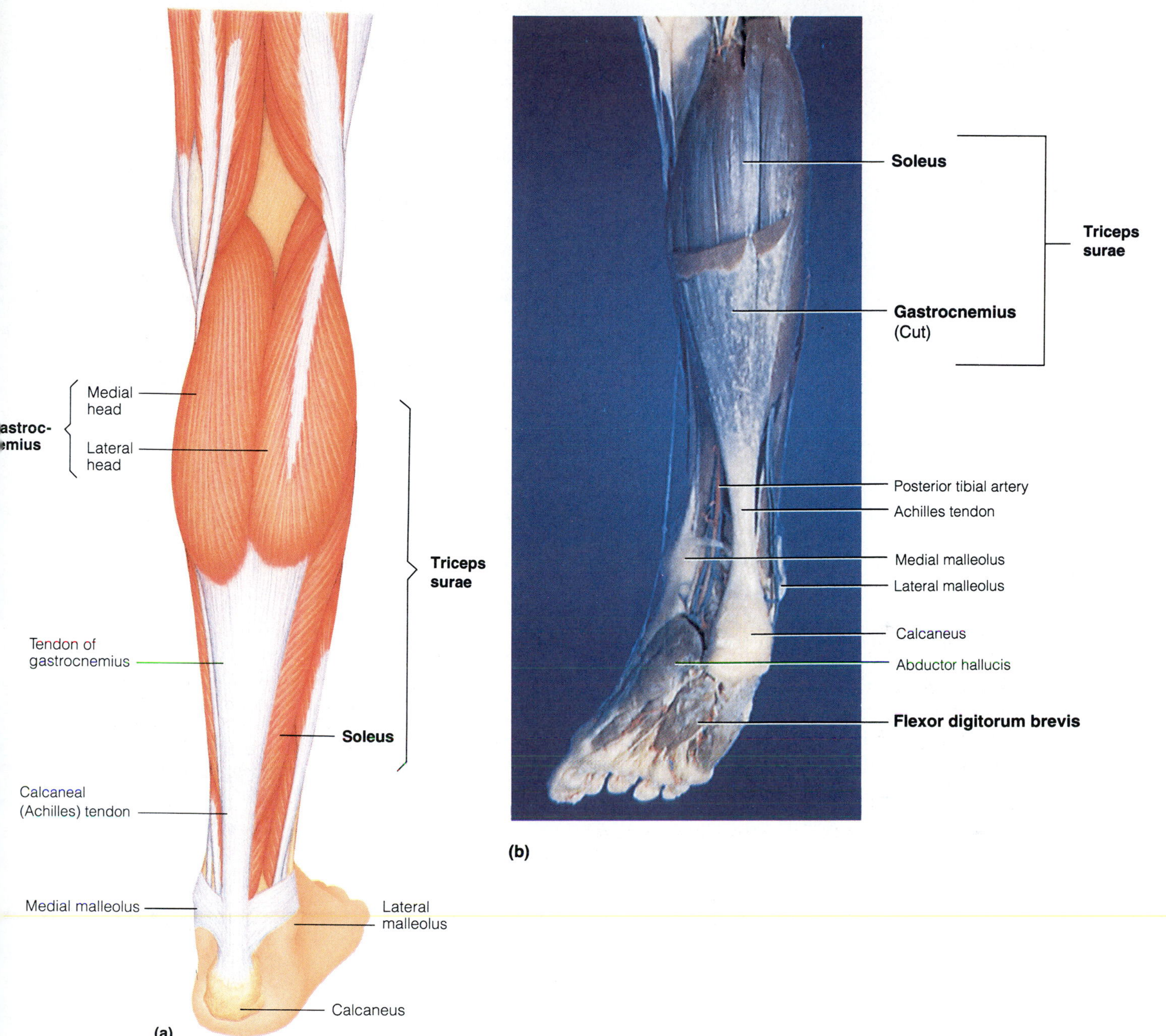

F15.11

Muscles of the posterior aspect of the right leg. **(a)** Superficial view of the posterior leg. **(b)** Photo of posterior aspect of right leg. The gastrocnemius has been transected and its superior part removed.

(continued)

TABLE 15.9 *(Continued)*

Muscle	Comments	Origin	Insertion	Action
Deep (Figure 15.11c,d)				
Popliteus	Thin muscle at posterior aspect of knee	Lateral condyle of femur	Proximal tibia	Flexes and rotates leg medially to "unlock" extended knee when knee flexion begins
Tibialis posterior	Thick muscle deep to soleus	Superior portion of tibia and fibula and interosseous membrane	Tendon passes obliquely behind medial malleolus and under arch of foot; inserts into several tarsals and metatarsals 2–4	Prime mover of foot inversion; plantar flexes foot
Flexor digitorum longus	Runs medial to and partially overlies tibialis posterior	Posterior surface of tibia	Distal phalanges of second through fifth toes	Flexes toes; plantar flexes and inverts foot
Flexor hallucis longus (see also Figure 15.12)	Lies lateral to inferior aspect of tibialis posterior	Middle portion of fibula shaft	Tendon runs under foot to insert on distal phalanx of great toe	Flexes great toe; plantar flexes and inverts foot; the "push-off muscle" during walking
Lateral Compartment (Figure 15.11c and Figure 15.12a,b)				
Peroneus longus	Superficial lateral muscle; overlies fibula	Head and upper portion of fibula	By long tendon under foot to first metatarsal and medial cuneiform	Plantar flexes and everts foot; helps keep foot flat on ground
Peroneus brevis	Smaller muscle; deep to peroneus longus	Distal portion of fibula shaft	By tendon running behind lateral malleolus to insert on proximal end of fifth metatarsal	Plantar flexes and everts foot, as part of peronei group

(continued on p. 138)

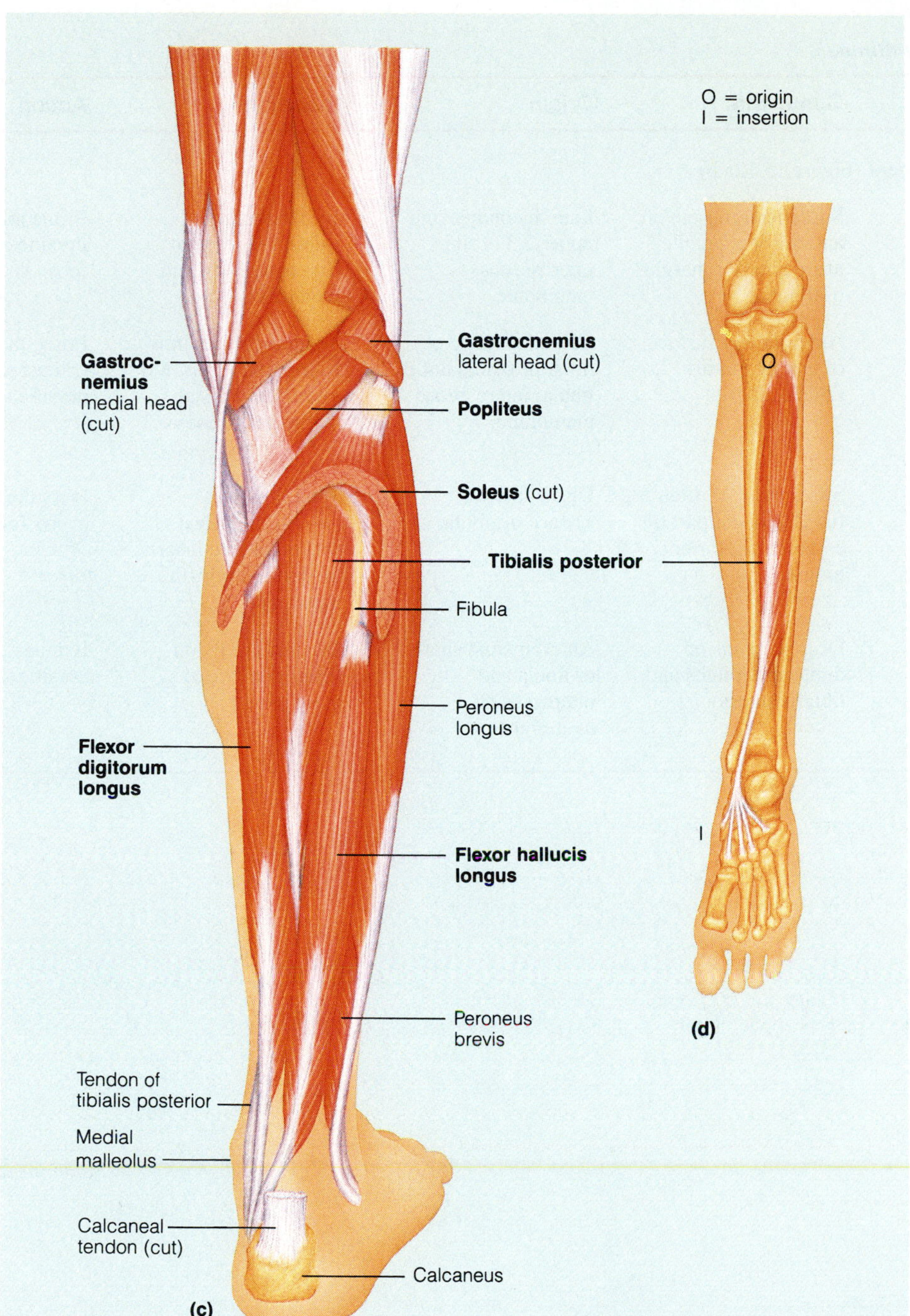

F15.11 *(continued)*

Muscles of the posterior aspect of the right leg. (c) The triceps surae has been removed to show the deep muscles of the posterior compartment. **(d)** Tibialis posterior shown in isolation so that its origin and insertion may be visualized.

TABLE 15.9 *(Continued)*

Muscle	Comments	Origin	Insertion	Action
Anterior Compartment (Figure 15.12a,b)				
Tibialis anterior	Superficial muscle of anterior leg; parallels sharp anterior margin of tibia	Lateral condyle and upper 2/3 of tibia; interosseous membrane	By tendon into inferior surface of first cuneiform and metatarsal 1	Prime mover of dorsiflexion; inverts foot
Extensor digitorum longus	Anterolateral surface of leg; lateral to tibialis anterior	Lateral condyle of tibia; proximal 3/4 of fibula; interosseous membrane	Tendon divides into four parts; insert into middle and distal phalanges of toes 2–5	Prime mover of toe extension; dorsiflexes foot
Peroneus tertius	Small muscle; often fused to distal part of extensor digitorum longus	Distal anterior surface of fibula	Tendon passes anterior to lateral malleolus and inserts on dorsum of fifth metatarsal	Dorsiflexes and everts foot
Extensor hallucis longus	Deep to extensor digitorum longus and tibialis anterior	Anteromedial shaft of fibula and interosseous membrane	Tendon inserts on distal phalanx of great toe	Extends great toe; dorsiflexes foot

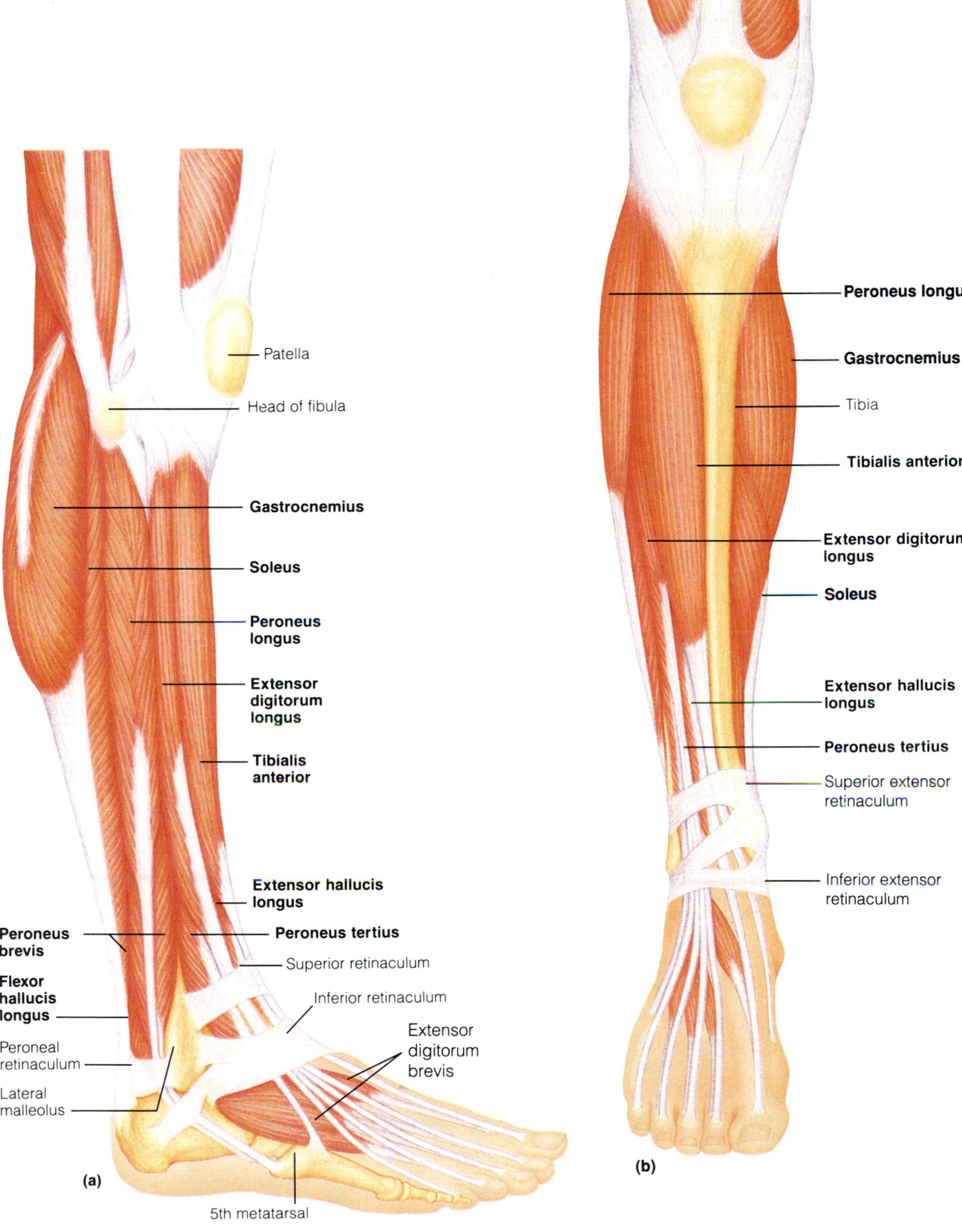

F15.12

Muscles of the anterolateral aspect of the right leg. (a) Superficial view of lateral aspect of the leg, illustrating the positioning of the lateral compartment muscles (peroneus longus and brevis) relative to anterior and posterior leg muscles. **(b)** Superficial view of anterior leg muscles.

16A

EXERCISE

Muscle Physiology (Frog Experimentation)

OBJECTIVES

1. To observe muscle contraction on the microscopic level and describe the role of ATP and various ions in muscle contraction.
2. To define and explain the physiologic basis of the following:
   action potential
   subthreshold stimulus
   threshold stimulus
   maximal stimulus
   treppe
   wave summation
   multiple motor unit summation
   tetanus
   muscle fatigue
   absolute and *relative refractory periods*
   depolarization
   repolarization
3. To trace the events that result from the electrical stimulation of a muscle.
4. To explain why the "all-or-none" law is demonstrated by the activity of a single muscle cell but not an intact skeletal muscle.
5. To recognize that a graded response of skeletal muscle is a function of the number of muscle fibers stimulated and the frequency of the stimulus.
6. To name and describe the phases of a muscle twitch.
7. To distinguish between a muscle twitch and a sustained (tetanic) contraction and to describe their importance in normal muscle activity.
8. To demonstrate how the kymograph or polygraph can be used to obtain pertinent and representative recordings of various physiologic events of skeletal muscle activity.
9. To explain the significance of muscle tracings obtained during experimentation.

MATERIALS

ATP muscle kits (glycerinated rabbit psoas muscle;* ATP and salt solutions obtainable from Carolina Biological Supply, Item #20-3525)
Petri dishes
Microscope slides
Cover glasses
Millimeter ruler
Compound microscope
Dissecting microscope
Small beaker (50 ml)
Frog Ringer's solution
Scissors
Metal needle probes
Pointed glass probes (teasing needles)
Medicine dropper
Cotton thread
Forceps
Disposable gloves
Glass or porcelain plate
Pithed bullfrog†
Apparatus A or B
A: physiograph (polygraph), polygraph paper and ink, myograph, pin and clip electrodes, stimulator output extension cable, transducer cable, straight pins, frog board, laboratory stand, clamp
B: kymograph, kymograph paper (smoking stand, burner, and glazing fluid if using a smoke-writing apparatus), skeletal muscle lever, signal magnet, laboratory stand, clamp, electronic stimulators

See Appendix E, Exercise 16A for links to *Anatomy and PhysioShow: The Videodisc.*

Notes to the Instructor:

* At the beginning of the lab, the muscle bundle should be removed from the test tube and cut into ~ 2-cm lengths. Both the cut muscle segments and the entubed glycerol should be put into a petri dish. One muscle *segment* is sufficient for each two to four students making observations.

†Bullfrogs to be pithed by lab instructor as needed for student experimentation. (If instructor prefers that students pith their own specimens, an instructional sheet on that procedure suitable for copying for student handouts is provided in the Instructor's Guide.)

Some instructors may prefer students to study muscle physiology with a computer simulation exercise. Exercise 16B should satisfy this preference. Alternatively, Exercise 16i in Appendix C uses student subjects and the Intelitool electromyography apparatus.

MUSCLE ACTIVITY

The contraction of skeletal and cardiac muscle fibers can be considered in terms of three events—electrical excitation of the muscle cell, excitation-contraction coupling, and shortening of the muscle cell due to sliding of the myofilaments within it.

At rest, all cells maintain a potential difference, or voltage, across their plasma membrane; the inner face of the membrane is approximately –60 to –90 millivolts (mV) compared with the cell exterior. This potential difference is a result of differences in membrane permeability to cations, most importantly sodium (Na^+) and potassium (K^+) ions. Intracellular potassium concentration is much greater than its extracellular concentration, and intracellular sodium concentration is considerably less than its extracellular concentration. Hence, steep concentration gradients across the membrane exist for both cations. However, because the plasma membrane is slightly more permeable to K^+ than to Na^+, Na^+ influx into the cell is inadequate to balance K^+ outflow. The result of this unequal Na^+-K^+ diffusion across the membrane establishes the cell's **resting membrane potential.** The resting membrane potential is of particular interest in excitable cells, like muscle cells and neurons, because changes in that voltage underlie their ability to do work (to contract or to signal respectively in muscle cells and neurons).

Action Potential

When a muscle cell is stimulated, the sarcolemma becomes temporarily permeable to sodium, which rushes into the cell. This sudden influx of sodium ions alters the membrane potential. That is, the cell interior becomes less negatively charged at that point, an event called **depolarization.** When depolarization reaches a certain level and the sarcolemma momentarily changes its polarity, a depolarization wave travels along the sarcolemma. Even as the influx of sodium ions occurs, the sarcolemma becomes impermeable to sodium and permeable to potassium ions. Consequently, potassium ions leak out of the cell, restoring the resting membrane potential (but not the original ionic conditions), an event called **repolarization.** The repolarization wave follows the depolarization wave across the sarcolemma. This rapid depolarization and repolarization of the membrane that is propagated along the entire membrane from the point of stimulation is called the **action potential.**

While the sodium gates are still open, there is no possibility of another response and the muscle cell is said to be in the **absolute refractory period.** The **relative refractory period** is the period after the sodium gates have closed when potassium gates are open and repolarization is ongoing. Especially strong stimuli may provoke a contraction during this part of the refractory period. Repolarization restores the muscle cell's irritability. Temporarily, the sodium-potassium pump, which actively transports K^+ into the cell and Na^+ out of the cell, need not be "revved up." But, if the cell is stimulated to contract again and again in rapid-fire order, the loss of potassium and gain of sodium occurring during action potential generation begins to hamper its ability to respond. And so, eventually the sodium-potassium pump must be activated to reestablish the ionic concentrations of the resting state.

Contraction

The propagation of the action potential along the sarcolemma causes the release of calcium ions (Ca^{2+}) from storage depots (tubules of the sarcoplasmic reticulum) within the muscle cell. When the calcium ions bind to regulatory proteins on the actin myofilaments, they act as an ionic trigger that initiates contraction, and the actin and myosin filaments slide past each other. Once the action potential has ceased, the calcium ions are almost immediately transported back into the tubules of the sarcoplasmic reticulum. Instantly the muscle cell relaxes.

The events of the contraction process can most simply be summarized as follows: muscle cell contraction is initiated by generation and transmission of an action potential along the sarcolemma. This electrical event is coupled with the sliding of the myofilaments—contraction—by the release of Ca^{2+}. Keep in mind this sequence of events as you conduct the experiments.

OBSERVATION OF MUSCLE FIBER CONTRACTION

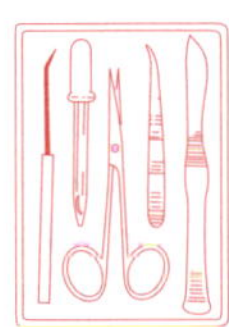

In this simple observational experiment, you will have the opportunity to review your understanding of muscle cell anatomy and to watch fibers contracting (or not contracting) in response to the presence of certain chemicals (ATP, and potassium and magnesium ions).

1. Obtain the following materials from the supply area: 2 glass teasing needles, 3 glass microscope slides and cover glasses, millimeter ruler, dropper vials containing the following solutions: (a) 0.25% ATP in triply distilled water; (b) 0.25% ATP plus 0.05*M* KCl plus 0.001*M* $MgCl_2$ in distilled water; and (c) 0.05*M* KCl plus 0.001*M* $MgCl_2$ in distilled water; a petri dish, and a small portion of a previously cut muscle bundle segment. While you are at the supply area, place the muscle fibers in the petri dish and pour a small amount of glycerol (the fluid in the supply petri dish) over your muscle cells. Also obtain both a compound and a dissecting microscope and bring them to your laboratory bench.

2. Using the fine glass needles, tease the muscle segment to separate its fibers. The objective is to isolate *single* muscle cells or fibers for observation. Be patient and work carefully so that the fibers do not get torn during this isolation procedure.

3. Transfer one or more of the fibers (or the thinnest strands you have obtained) onto a clean microscope slide with a glass needle, and cover it with a cover glass. Examine the fiber under low and then high power magnifications to observe the striations and the smoothness of the fibers when they are in the relaxed state.

4. Transfer three or four fibers to a second clean microscope slide with a glass needle. Using the needle as a prod, carefully position the fibers so that they are parallel to one another and as straight as possible. Place this slide under a dissecting microscope and measure the length of each fiber by holding a millimeter ruler adjacent to it. Alternatively, you can rest the microscope slide *on* the millimeter ruler to make your length determinations. Record the fiber lengths on the chart at the bottom of this page.

5. Flood the fibers (situated under the dissecting microscope) with several drops of the solution containing ATP, potassium ions, and magnesium ions. Watch the reaction of the fibers after adding the solution. After 30 seconds (or slightly longer), remeasure each fiber and record the observed lengths on the chart. Also, observe the fibers to see if any width changes have occurred. Calculate the degree (or percentage) of contraction by using the following simple formula, and record this information on the chart also.

$$\text{initial length (mm)} - \text{contracted length (mm)} = \text{degree of contraction (mm)}$$

then:

$$\frac{\text{degree of contraction (mm)}}{\text{initial length (mm)}} \times 100 = \_\_\_\_\_\_ \text{ \% contraction}$$

6. Carefully transfer one of the contracted fibers to a clean microscope slide, cover with a cover glass, and observe with the compound microscope. Mentally compare your initial observations with the view you are observing now. What differences do you see? (Be specific.)

____

____

What zones (or bands) have disappeared?

____

7. Repeat steps 3 to 6 twice more, using clean slides and fresh muscle cells. First use the solution of ATP in distilled water (no salts). Then, use the solution containing only salts (no ATP) for the third series.

What degree of contraction was observed when ATP was applied in the absence of potassium and magnesium ions?

____

What degree of contraction was observed when the muscle fibers were flooded with a K^+- and Mg^{2+}-containing solution that lacked ATP?

____

What conclusions can you draw about the importance of ATP, and potassium and magnesium ions to the contractile process?

____

____

INDUCTION OF CONTRACTION IN THE FROG GASTROCNEMIUS MUSCLE

Physiologists have learned a great deal about the way muscles function by isolating muscles from laboratory animals and then stimulating these muscles to observe their responses. Various stimuli—electrical shock, temperature changes, extremes of pH, certain chemicals—elicit muscle activity, but laboratory experiments of this type typically use electrical shock. This is because it is easier to control the onset and cessation of electrical shock, as well as the strength of the stimulus.

Preparing a Muscle for Experimentation

The preparatory work that precedes the recording of muscle activity tends to be quite time consuming. If you work in teams of two or three, the work can be divided. While one of you is setting up the recording apparatus (kymograph or physiograph), one or two students can dissect the frog leg. Experimentation should begin as soon as the dissection is completed.

Various types of apparatus are used to record muscle contraction. All include a way to mark time intervals, a way to indicate exactly when the stimulus was applied, and a way to measure the magnitude of the contractile response. Instructions are provided here for setting up physiograph (Figure 16A.1) and kymograph (Figure 16A.2) apparatus. Specific instructions for use of recording apparatus during recording will be provided by your instructor.

	Muscle fiber 1	Muscle fiber 2	Muscle fiber 3
Initial length (mm)			
Contracted length (mm)			
% contraction			

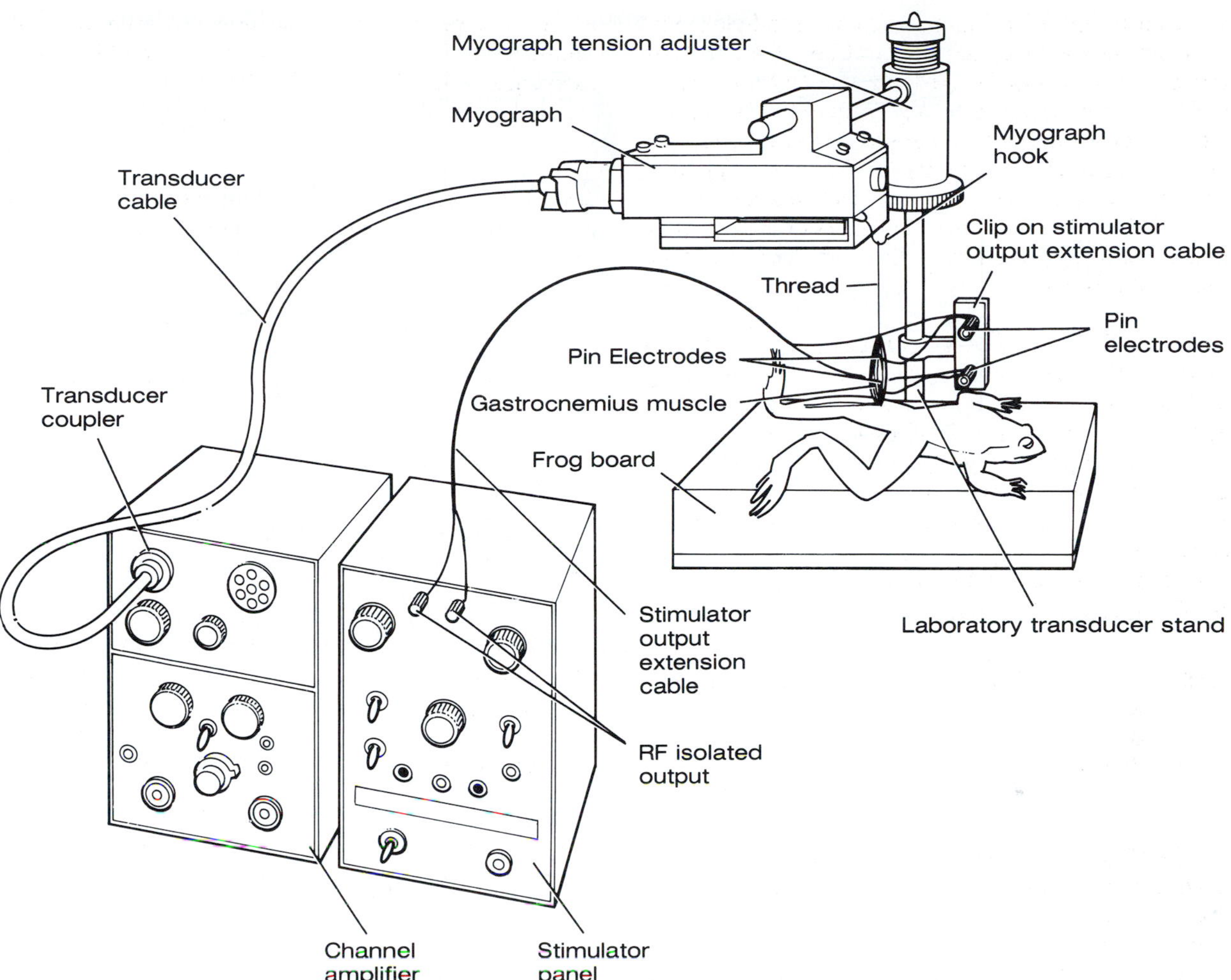

Materials:
Channel amplifier and stimulator transducer cable
Stimulator panel and stimulator output extension cable
Myograph
Myograph tension adjuster
Transducer stand
Two pin electrodes
Frog board and straight pins
Prepared frog (gastrocnemius muscle freed and calcaneal tendon ligated with thread)
Frog Ringer's solution

1. Connect myograph to transducer stand and attach frog board to stand.

2. Attach transducer cable to myograph and to input connection on amplifier channel.

3. Attach stimulator output extension cable to output on stimulator panel (red to red, black to black).

4. Using clip at opposite end of extension cable, attach cable to bottom of transducer stand adjacent to frog board.

5. Attach two pin electrodes securely to electrodes on clip.

6. Place knee of prepared frog in clip on frog board and secure by inserting a straight pin through tissues of frog. Keep frog muscle moistened with Ringer's solution.

7. Attach thread ligating the calcaneal tendon of frog to myograph leaf-spring hook.

8. Adjust position of myograph on stand to produce a constant tension on thread attached to muscle (taut but not tight). Gastrocnemius muscle should hang vertically directly below myograph hook.

9. Insert free ends of pin electrodes into muscle, one at proximal end and other at distal end.

F16A.1

Physiograph setup for frog gastrocnemius experiments.

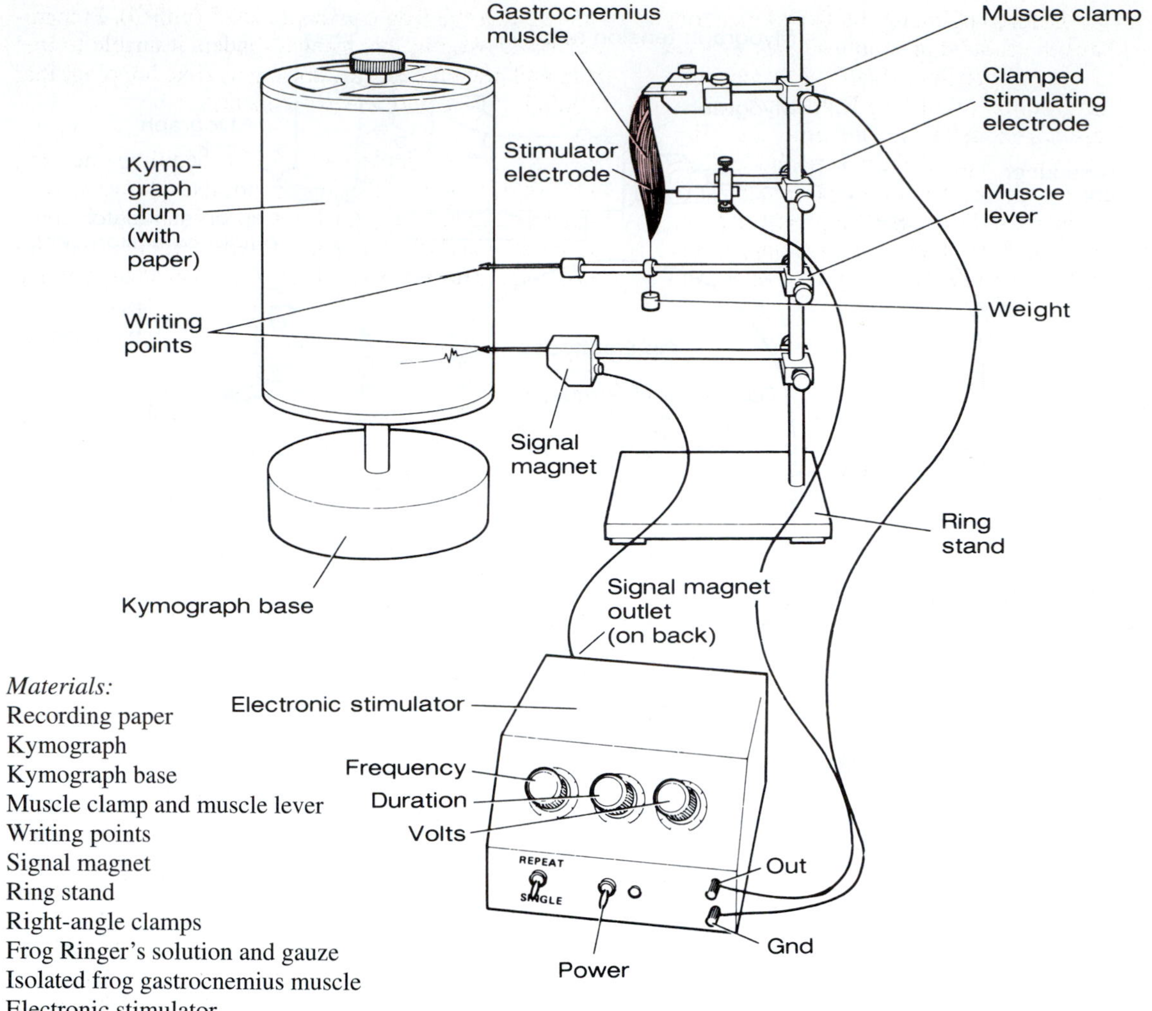

Materials:
Recording paper
Kymograph
Kymograph base
Muscle clamp and muscle lever
Writing points
Signal magnet
Ring stand
Right-angle clamps
Frog Ringer's solution and gauze
Isolated frog gastrocnemius muscle
Electronic stimulator

1. Attach muscle clamp to supporting rod and secure cut femur in clamp.

2. Position and clamp muscle lever to supporting rod beneath muscle tissue. Tie thread ligating the calcaneal tendon to hook on the muscle lever so that muscle hangs vertically. Keep muscle moist with Ringer's solution from this point until completion of preparations and experimentation.

3. Hang a 10-g weight on muscle lever just below thread attachment to provide proper tension on lever. Muscle lever should rest horizontally. Writing tip should lightly touch paper surface on drum. Test its position by pulling lightly and directly upward on thread. Writing tip should remain in contact with drum surface.

4. Clamp signal magnet to supporting rod. Tip of its writing stylus should line up directly beneath that of muscle lever. *Precise alignment is important* so that a correlation can be seen between the point of stimulus and onset of muscle contractions.

5. Position stimulator to one side so that area immediately in front of recording apparatus is free for manipulations during and after experimentation. Turn all its switches to OFF and all dials to left. Plug the stimulator into the power outlet.

6. Connect two gray wires from signal magnet outlet (back of stimulator) to binding posts on signal magnet. To operate signal magnet, toggle switch (upper left on front of stimulator) must be ON (flipped up). Signal magnet will indicate up to 30 pulses per second when it is operating. If you use frequencies greater than 30 pulses per second during experimentation, turn signal magnet toggle switch to OFF (down) position.

7. Wash platinum tips of stimulator electrode in distilled water and dry. Connect electrode to white and black output terminals of stimulator and clamp electrode to supporting rod so that its tips rest on muscle tissue.

F16A.2

Kymograph setup for frog gastrocnemius experiments.

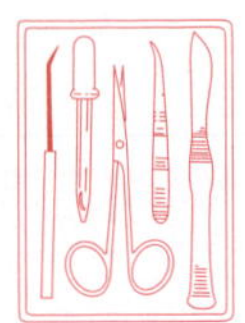

1. Before beginning the frog dissection, have the following supplies ready at your laboratory bench: a small beaker containing 20 to 30 ml of frog Ringer's solution, scissors, a metal needle probe, a glass probe with a pointed tip, a medicine dropper, cotton thread, forceps, a glass or porcelain plate, and disposable gloves. While these supplies are being accumulated, one member of your team should notify the instructor that you are ready to begin experimentation, so that the frog can be prepared (pithed). Preparation of the frog in this manner renders it unable to feel pain and prevents reflex movements (like hopping) that would interfere with the experiments.

2. All students that will be handling the frog should obtain and don disposable plastic gloves. Obtain a pithed frog and place it ventral surface down on the glass plate. Make an incision into the skin approximately midthigh (Figure 16A.3), and then continue

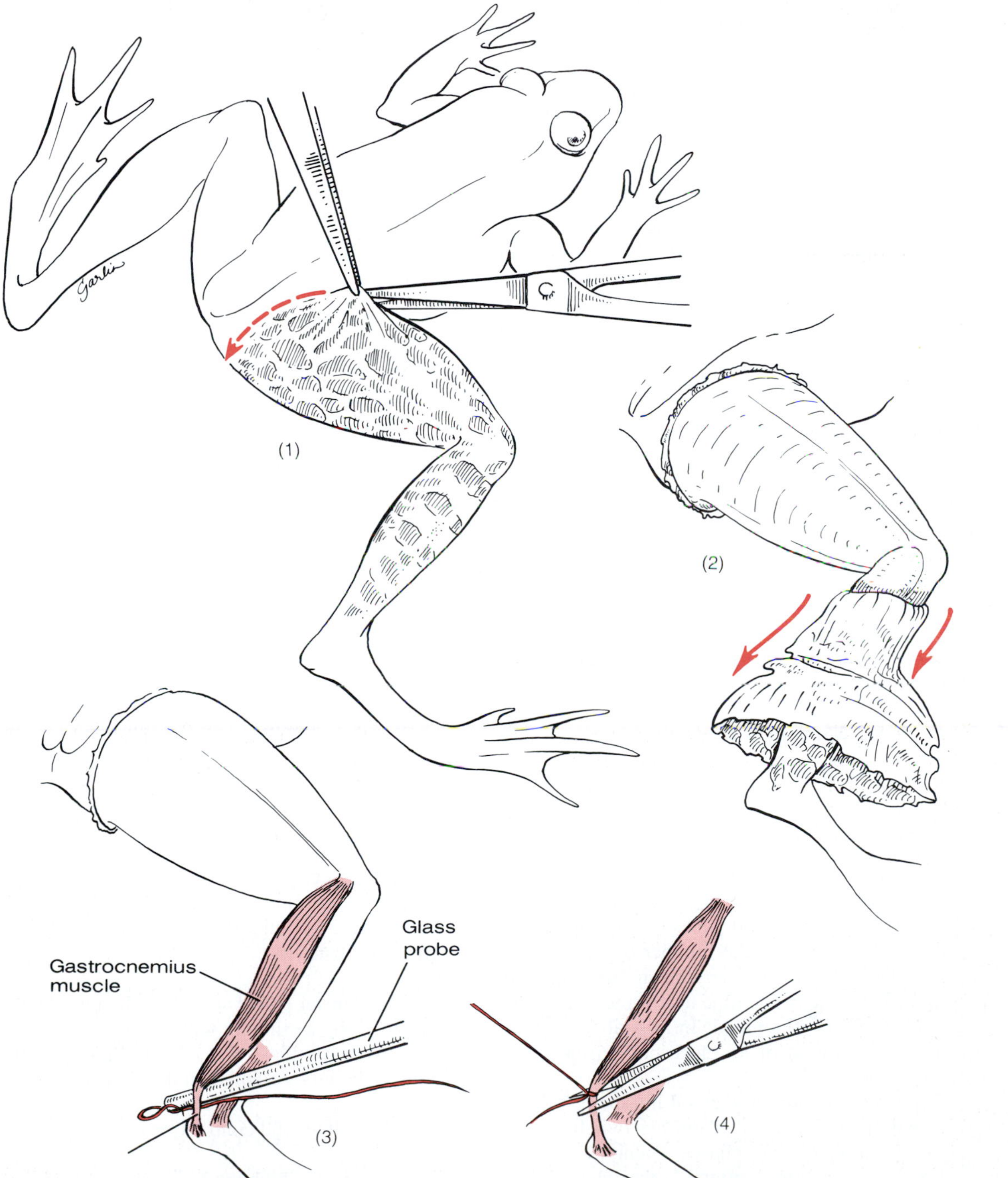

F16A.3

Preparation of the frog gastrocnemius muscle. Numbers indicate the sequence of manipulation.

the cut completely around the thigh. Grasp the skin with the forceps and strip it from the leg and hindfoot. The skin tends to adhere more at the joints, but a careful, persistent pulling motion—somewhat like pulling off a nylon stocking—will enable you to remove it in one piece. *From this point on, the exposed muscle tissue should be kept moistened* with the Ringer's solution to prevent spontaneous twitches.

3. Identify the gastrocnemius muscle (the fleshy muscle of the posterior calf) and the calcaneal (Achilles) tendon that secures it to the heel.

4. Slip a glass probe under the gastrocnemius muscle and run it along the entire length and under the calcaneal tendon to free them from the underlying tissues.

5. Cut a piece of thread about 10 in. long and use the glass probe to slide the thread under the calcaneal tendon. Knot the thread firmly around the tendon and then sever the tendon distal to the thread. Alternatively, you can bend a common pin into a Z-shape and insert the pin securely into the tendon. The thread is then attached to the opposite end of the pin. If you are using a physiograph, once the tendon has been tied or pinned, the frog is ready for experimentation (see Figure 16A.1). If you are using a kymograph, the gastrocnemius muscle must be completely isolated, as described in step 6.

6. Cut away the fibulotibial bone just distal to the knee. Expose the femur of the thigh and cut it completely through at midthigh. Remove as much of the thigh muscle tissue as possible by carefully cutting it away with the scissors. The isolated gastrocnemius muscle can now be mounted on the muscle bar, and the stimulating electrodes of the kymograph can be attached (see Figure 16A.2). About halfway through the laboratory period, dissect the second leg for use.

Recording Muscle Activity

The **"all-or-none"** law of muscle physiology states that a muscle cell will contract maximally when stimulated adequately. Skeletal muscles, however, consisting of thousands of muscle cells, react to stimuli with graded responses. Thus muscle contractions can be weak or vigorous, depending on the requirements of the task. Graded responses (different degrees of shortening) of a skeletal muscle depend on the number of muscle cells being stimulated. In the intact organism, the number of motor units firing at any one time determines how many muscle cells will be stimulated. In this laboratory, the frequency and strength of an electrical current determines the response.

A single contraction of skeletal muscle is called a **muscle twitch.** A tracing of a muscle twitch (Figure 16A.4) shows three distinct phases: latent, contraction, and relaxation. The **latent phase** is the interval from stimulus application until the muscle begins to shorten. Although no activity is indicated on the tracing during this phase, important electrical and chemical changes are occurring within the muscle. During the **contraction phase,** the muscle fibers shorten; the tracing shows an increasingly higher needle deflection and the tracing peaks. During the **relaxation phase,** represented by a downward curve of the tracing, the muscle fibers relax and lengthen. On a slowly moving recording surface, the single muscle twitch appears as a spike (rather than a bell-shaped curve, as in Figure 16A.4), but on a rapidly moving recording surface, the three distinct phases just described become recognizable.

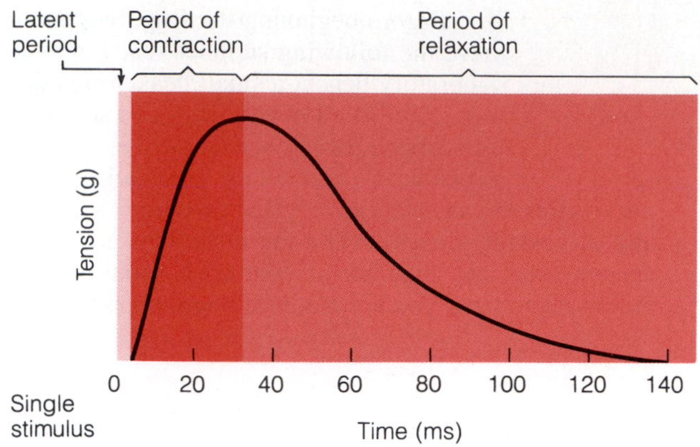

F16A.4

Tracing of a muscle twitch.

DETERMINING THE THRESHOLD STIMULUS

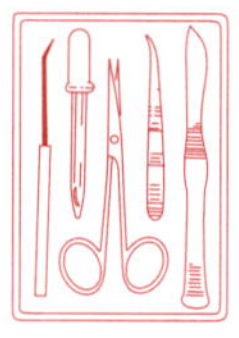

1. Assuming that you have already set up the recording apparatus, set the time marker to deliver one pulse per second and set the paper speed at a slow rate, approximately 0.1 cm per second.

2. Set the duration control on the stimulator between 7 and 15 msec, multiplier × one and the voltage control at zero V, multiplier × one. Turn the sensitivity control knob of the stimulator fully clockwise (lowest value, greatest sensitivity).

3. Administer single stimuli to the muscle at intervals of 1 to 2 sec, beginning with 0.1 V and increasing each successive stimulus by 0.1 V until a contraction is obtained (shown by a spike on the paper).

At what voltage did contraction occur? ____________V

The voltage at which the first perceptible contractile response is obtained is called the **threshold stimulus.** All stimuli applied prior to this point are termed **subthreshold stimuli,** because at those voltages no response was elicited.

4. Stop the recording and mark the record to indicate the threshold stimulus, voltage, and time. *Do not remove the record from the recording surface;* continue with the next experiment. Remember: keep the muscle preparation moistened with Ringer's solution at all times.

GRADED MUSCLE RESPONSE TO INCREASED STIMULUS INTENSITY

1. Follow the previous setup instructions, but set the voltage control at the threshold voltage (as determined in the first experiment).

2. Deliver single stimuli at 1- or 2-sec intervals. Initially increase the voltage between shocks by 0.5 V; then increase the voltage by 1 to 2 V between shocks as the experiment continues, until contraction height increases no further. Stop the recording apparatus.

What voltage produced the highest spike (thus the maximal strength of contraction)?

______________ V

This voltage, called the **maximal stimulus** (for *your* muscle specimen), is the weakest stimulus at which all muscle cells are being stimulated. As the voltage was increased to this point, more and more muscle cells (motor units) were activated, resulting in a stronger and stronger contraction. Past this point, an increase in the intensity of the stimulus will not produce any greater contractile response. This phenomenon, called **multiple motor unit summation,** or **recruitment,** is the process by which the increased contractile strength reflects the relative number of muscle cells stimulated.

3. Mark the record *multiple motor unit summation.* Record the maximal stimulus voltage and the time you completed the experiment. Continue on to the next experiment.

TIMING THE MUSCLE TWITCH

1. Follow the previous setup directions, but set the voltage for the maximal stimulus (as determined in the preceding experiment) and set the paper advance or recording speed at maximum. Record the paper speed setting:

______________ mm/sec

2. Determine the time required for the paper to advance 1 mm by using the formula:

$$\frac{1 \text{ mm}}{\text{mm/sec (paper speed)}}$$

(Thus, if your paper speed is 25 mm/sec, each mm on the chart equals 0.04 sec.) Record the computed value:

1 mm = ______________ sec

3. Deliver single stimuli at 2- to 3-sec intervals to obtain several "twitch" curves. Stop the recording.

4. Determine the duration of the latent, contraction, and relaxation phases of the twitches and record here:

Duration of latent period: ______________ sec

Duration of contraction period: ______________ sec

Duration of relaxation period: ______________ sec

5. Label the record to indicate the point of stimulus, the beginning of contraction, the end of contraction, and the end of relaxation.

6. Allow the muscle to rest (but keep it moistened) before continuing with the next experiment.

THE TREPPE, OR STAIRCASE, PHENOMENON

As a muscle is stimulated to contract, a curious phenomenon is observed in the tracing pattern of the first few twitches. Even though the stimulus intensity is unchanged, the height of the individual spikes increases in a stepwise manner—producing a sort of staircase pattern called **treppe** (Figure 16A.5). This phenomenon is not well understood, but the following explanation has been offered: in the muscle cell's resting state, there is much less Ca^{2+} in the sarcoplasm and the enzyme systems in the muscle cell are less efficient than after it has contracted a few times. As the muscle cell begins to contract, the intracellular concentration of Ca^{2+} rises dramatically, and the heat generated by muscle activity increases the efficiency of the enzyme systems. As a result, the muscle becomes more efficient and contracts more vigorously. This is the physiologic basis of the warm-up period prior to competition in sports events.

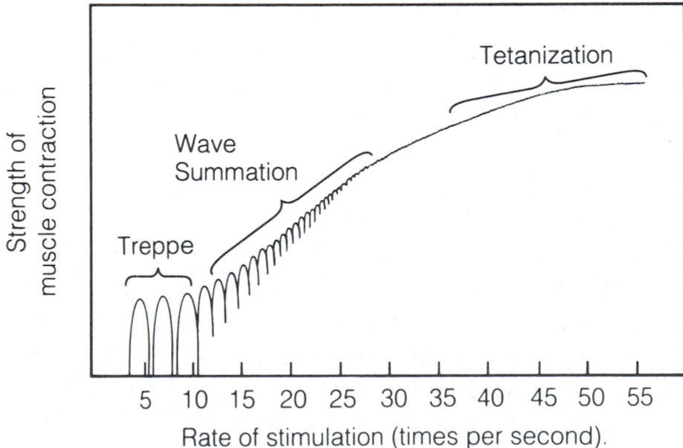

F16A.5

Treppe, wave summation, and tetanization. Progressive summation of successive contractions occurs as the rate of stimulation is increased. Tetanization occurs when the rate of stimulation reaches approximately 35 per second, and maximum contraction force occurs at a stimulation rate of approximately 50 per second.

1. Set up the apparatus as in the previous experiment, again setting the voltage to the maximal stimulus.

2. Deliver single stimuli at 1-sec intervals until the strength of contraction does not increase further.

3. Stop the recording apparatus and mark the record *treppe.* Note also the number of contractions (and seconds) required to reach the constant contraction magnitude. Record the voltage used and the time when you completed this experiment. Continue on to the next experiment.

GRADED MUSCLE RESPONSE TO INCREASED STIMULUS FREQUENCY Muscles subjected to frequent stimulation, without a chance to relax, exhibit two kinds of responses—wave summation and tetanus—depending on the level of stimulus frequency (Figure 16A.5).

Wave Summation: If a muscle is stimulated with a rapid series of stimuli of the same intensity before it has had a chance to relax completely, the response to the second and subsequent stimuli will be greater than to the first stimulus. This phenomenon, called **wave,** or **temporal, summation,** occurs because the muscle is already in a partially contracted state when subsequent stimuli are delivered.

1. With the recorder running at slow speed and the stimulus intensity set to the maximal stimulus, stimulate the muscle at a rate of 15 to 25 stimuli per second.

2. Shut off the recorder and label the record as *wave summation.* Note also the time, the voltage, and the frequency.

Tetanus: Stimulation of a muscle at an even higher frequency will produce a "fusion" (tetanization) of the summated twitches. In effect, a single sustained contraction is achieved in which no evidence of relaxation can be seen (Figure 16A.5). **Tetanus** is a feature of normal skeletal muscle functioning; the single muscle twitch is primarily a laboratory phenomenon.

1. To demonstrate tetanus, maintain the conditions used for wave summation except for the frequency of stimulation. Set the stimulator to deliver 60 stimuli per second.

2. As soon as you obtain a single smooth, sustained contraction (with no evidence of relaxation), discontinue stimulation and shut off the recorder.

3. Label the tracing with the conditions of experimentation, the time, and the area of tetanus.

MUSCLE FATIGUE **Muscle fatigue,** the loss of the ability to contract, is believed to be a result of the oxygen debt that occurs in the tissue after prolonged activity (through the accumulation of such waste products as lactic acid as well as the depletion of ATP). True muscle fatigue rarely occurs in the body, because it is most often preceded by a subjective feeling of fatigue. Furthermore, fatigue of the neuromuscular junctions typically precedes fatigue of the muscle.

1. To demonstrate muscle fatigue, set up an experiment like the tetanus experiment, but continue stimulation until the muscle completely relaxes and the contraction curve returns to the base line.

2. Measure the time interval between the beginning of complete tetanus and the beginning of fatigue (when the tracing begins its downward curve). Mark the record appropriately.

3. Determine the time required for complete fatigue to occur (the time interval from the beginning of fatigue until the return of the curve to the base line). Mark the record appropriately.

4. Allow the muscle to rest (keeping it moistened with Ringer's solution) for 10 min, and then repeat the experiment.

What was the effect of the rest period on the fatigued muscle?

___

___

What is the physiologic basis for this reaction?

___

___

THE EFFECT OF LOAD ON SKELETAL MUSCLE When the fibers of a skeletal muscle are slightly stretched by a weight or tension, the muscle responds by contracting more forcibly and thus is capable of doing more work. If the load is increased beyond the optimum, the latent period becomes longer, contractile force decreases, and relaxation (fatigue) occurs more quickly. With excessive stretching, the muscle is unable to develop any tension and no contraction occurs. Since the filaments no longer overlap at all with this degree of stretching, the sliding force cannot be generated.

If your equipment allows you to add more weights to the muscle specimen or to increase the tension on the muscle, perform the following experiment to determine the effect of loading on skeletal muscle, and to develop a work curve for the frog's gastrocnemius muscle.

1. Set the stimulator to deliver the maximal voltage as previously determined.

2. Stimulate the unweighted muscle with single shocks at 1- to 2-sec intervals to achieve three or four muscle twitches.

3. Stop the recording apparatus and add 10 g of weight or tension to the muscle. Restart and advance the recording about 1 cm, and then stimulate again to obtain three or four spikes.

4. Repeat the previous step seven more times, increasing the weight by 10 g each time until the total load on the muscle is 80 g or the muscle fails to respond. If the calcaneal tendon tears, the weight will drop, thus ending the trial. In such cases, you will need to prepare another frog's leg to continue the experiments and the maximal stimulus will have to be determined for the new muscle preparation.

5. When these "loading" experiments are completed, discontinue recording and remove the tracing. Mark the curves on the record to indicate the load (in grams).

6. Measure the height of contraction (in millimeters) for each sequence of twitches obtained with each load, and insert this information into the chart below.

7. Compute the work done by the muscle for each twitch (load) sequence.

Weight of load (g) $\times$ *distance load lifted (mm)* = *work done*

Enter these calculations into the chart in the columns labeled Trial 1.

Load (g)	Distance load lifted (mm)		Work done	
	Trial 1	Trial 2	Trial 1	Trial 2
0				
10				
20				
30				
40				
50				
60				
70				
80				

8. Allow the muscle to rest for 5 minutes. Then conduct a second trial in the same manner (i.e., repeat steps 2 through 7). Record this second set of calculations in the columns labeled Trial 2. Be sure to keep the muscle well moistened with Ringer's solution during the resting interval.

9. Using two different colors, plot a line graph of work done against the weight on the accompanying grid for each trial. Label each plot appropriately.

10. Dismantle all apparatus and prepare the equipment for storage. Dispose of the frog remains in the appropriate container. Discard the gloves as instructed and wash and dry your hands.

11. Inspect your records of the experiments and make sure each is fully labeled with the experimental conditions, the date, and the names of those who conducted the experiments. (If you used a smoked-drum kymograph recording system, "fix" the records before leaving the laboratory.) For future reference, attach a tracing (or a photocopy of the tracing) for each experiment to this page.

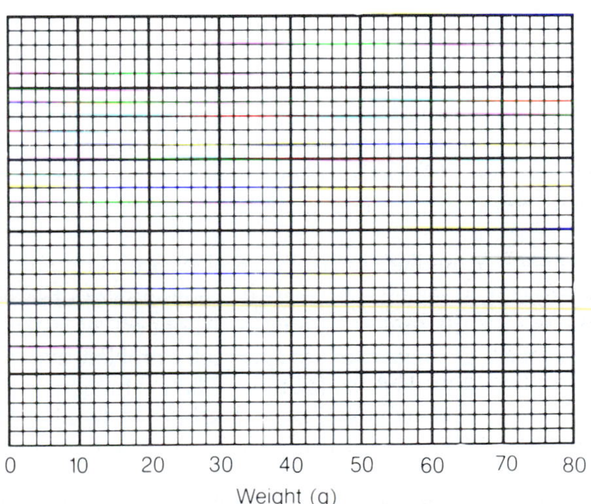

16B

EXERCISE

Muscle Physiology (Computerized Simulations)

OBJECTIVES

1. To define important terms used in describing muscle physiology.
2. To identify the two ways that the mode of stimulation can influence muscle force.
3. To draw a graph relating stimulus strength and twitch force to illustrate graded muscle response.
4. To explain how slow, smooth, sustained contraction is possible in a skeletal muscle.
5. To understand the relationships between passive, active, and total forces.
6. To identify the conditions under which muscle contraction is isometric or isotonic.
7. To describe in terms of length and force the transitions between isometric and isotonic conditions during a single muscle twitch.
8. To describe the effects of afterload and starting length on the initial velocity of shortening.
9. To explain why muscle force remains constant during isotonic shortening.
10. To explain experimental results in terms of muscle structure.

MATERIALS

Minimum equipment required:
IBM PC/XT/AT or compatible
256K RAM (or better)
Color graphics adapter (CGA) and compatible graphics monitor
Software:
Mechanical Properties of Active Muscle (available in 3.5-inch and 5.25-inch versions from Benjamin/Cummings)

Many important physiological concepts of skeletal muscle contraction can be demonstrated using computer simulations. The set of simulations you will be using here investigates the mechanical properties of active skeletal muscle. The programs graphically present all the equipment and materials necessary for you, the investigator, to set up experimental conditions and observe the results. In student-conducted laboratory investigations there are many ways to approach a problem and the same is true of these simulations. The instructions provided will guide your investigations, but you should also attempt alternate approaches to gain insight into the logical methods employed in scientific experimentation.

Try the following approach: The first time through the programs, follow the instructions closely and answer the questions they pose as you go along. Then try your own ideas, asking "What if ... ?" questions to test the validity of your theories. Major advantages of these computer simulations are that the muscle cannot be damaged accidentally, lab equipment will not break down at the worst possible time, and you will have ample time to think critically about the processes in question.

Various types of apparatus are used in research settings, and you will find minor differences in the operation of the equipment used in these simulations. Think about what is happening in each situation, because it is important to understand how you are experimentally manipulating the muscle for a complete understanding of the results. To aid you in this endeavor, each of the exercises has the following elements:

1. Introduction to the System (explains equipment used)
2. Some Definitions (briefly defines terms)
3. Experiments (designed to manipulate skeletal muscle)

Work your way through each section in turn so that you become familiar with the simulated equipment you will be using and the terminology involved.

GETTING STARTED

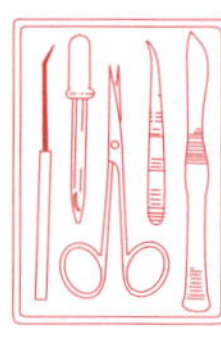

Begin by making sure you have the computer equipment listed in the materials section. If your computer is already running, proceed to step 3.

1. Turn the computer on. If you see a C prompt (such as C:\ >), proceed. If you are in Windows, exit Windows and the C prompt should appear.
2. Insert the muscle program diskette (Mechanical Properties of Active Muscle) into the proper drive and type the letter of that drive followed by a colon (for example, **A:**). When the prompt for that drive appears, type **GO** and press (ENTER). You must keep the program diskette in the drive for the entire time you are using the simulation.

3. The title screen will appear after a few seconds. Press any key to move on to the next information screen and again to advance to the main Index of Programs. This is a list of all the choices available to you. Select option 1 to see an overview of all programs available. This will display the opening screen of each program in turn and let you examine it in more detail if you wish. Instructions will appear in boxes on the screen. If you get lost, pressing ESC gets you back to the Index of Programs.

ELECTRICAL STIMULATION

When stimulated by the nervous system (or electric shock), skeletal muscle will contract and exert force. Unlike single cells or motor units, which follow the all-or-none law of muscle physiology, a whole muscle responds to stimuli with a graded response. A *motor unit* consists of a motor neuron and all the muscle cells it innervates. Hence, activation of the neuron innervating a single motor unit will cause all muscle cells in that unit to fire simultaneously in an all-or-none fashion. The graded contractile response of a whole muscle reflects the number of motor units firing at a given time. Strong muscle contraction implies many motor units are activated (and each unit has maximally contracted); weak contraction means few motor units are active (however, the activated units are maximally contracted). By increasing the number of motor units firing, a process called *recruitment,* we can produce a slow, steady increase in muscle force.

Regardless of the number of motor units activated, a single contraction of skeletal muscle is called a *muscle twitch.* A tracing of a muscle twitch shows three distinct phases: latent, contraction, and relaxation (Figure 16B.1). The **latent phase** is the short period between the time of stimulation and the beginning of contraction. Although no force is generated during this interval, important electrical and chemical changes are taking place within the muscle in preparation for contraction. During **contraction,** the myofilaments are sliding and the muscle shortens. **Relaxation** takes place when contraction has ended and the muscle returns to its normal resting state (and length).

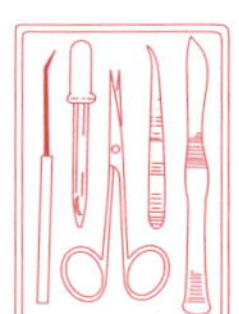

Select option 2: Electrical Stimulation. This first selection simulates an **isometric** (fixed-length) **contraction** of an isolated skeletal muscle and allows you to investigate how the strength and frequency of an electrical stimulus affect muscle activity. Note that these simulations involve *indirect* stimulation by an electrode placed on the surface of the muscle. This differs from the situation *in vivo,* where each fiber in the muscle is *directly* stimulated via a nerve ending. Now spend a few moments on the introduction to the system and definitions of terms used.

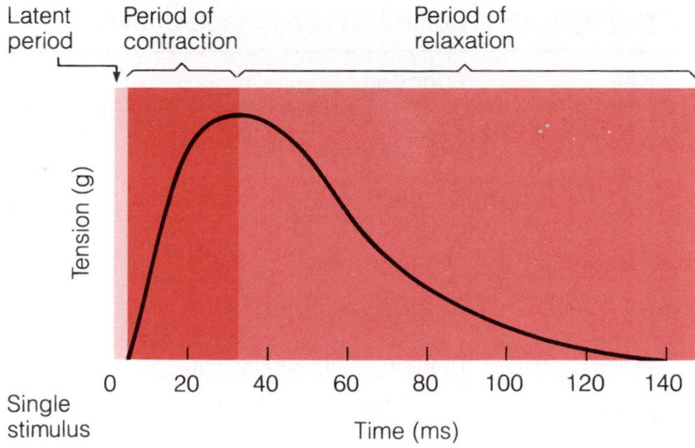

F16B.1

Tracing of a single muscle twitch.

Single Stimulus

Select option 3: Single Stimulus. The opening screen will appear in a few seconds (Figure 16B.2).

The oscilloscope display is the most important part of the screen, because it is where all contraction data are graphically presented for analysis. *Time* is displayed on the horizontal axis, where a full sweep is 1 second. It is marked off in 0.1-second intervals. *Force* is displayed on the vertical axis on an arbitrary scale from 0 to 5. Familiarize yourself with the control keys (Table 16B.1) before proceeding.

1. Press **S** once. Since the *voltage* is set to zero, no muscle activity will result. However, notice that a yellow line moves across the bottom of the screen—this

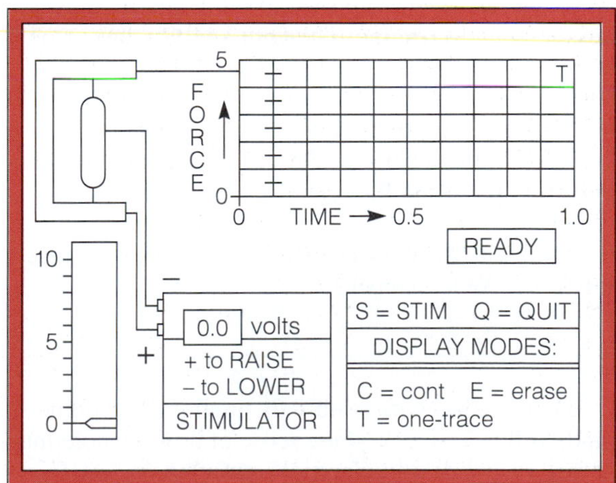

F16B.2

Opening screen of Electrical Stimulation.

Table 16B.1 Control Keys for Single Stimulus

Key	Function
+	Increase voltage
–	Decrease voltage
S	Stimulate muscle
P	Pause display
H	Halt (abort) display
C	Continuous trace mode
T	One-trace mode
E	Erase all tracings
Q	Quit (return to menu)

line will indicate the muscle force later. The red line moving across the top of the screen indicates time. A small vertical line marks the exact time of stimulus.

2. Press **+** until the voltmeter reads 5. Then press **S** once and release it. You will see the muscle react, and a *twitch force* will appear on the screen. Notice that the latent phase is not represented on the screen, providing a cleaner tracing for experimental purposes. The Erase key **E** is used as needed to clear the screen. The Continuous mode key **C** allows the tracing to wrap around the screen continuously until force = 0. Its companion, the One-trace key **T**, will stop the tracing when it has reached the end of a single sweep. The mode is indicated by a *C* or *T* in the upper righthand corner of the screen. The Halt key **H** aborts the tracing, and the Pause key **P** causes the tracing to stop momentarily until you press it again to resume. The Quit key **Q** returns you to the program menu.

3. Try changing the voltage and notice how force of contraction also changes. Observe the contraction and relaxation phases in the tracings.

4. Stimulate the muscle again before it has a chance to fully relax. What happens?

____

5. Feel free to experiment with anything that comes to mind, as this will give you a sense of how a whole muscle will respond to an electrical stimulus.

DETERMINING THE THRESHOLD STIMULUS

1. Erase the oscilloscope display and set the voltage control to zero. Set the display mode to single trace.

2. Using single stimuli, increase the voltage in 0.1-volt increments until the first trace of contraction is seen. This is the **threshold** voltage, below which no contraction occurs. Record the threshold voltage.

______________ V

GRADED MUSCLE RESPONSE TO INCREASED STIMULUS INTENSITY Examine the tracings just recorded. Notice that as voltage is increased, contractile force also increases. Also pay attention to the relative amount of the muscle being stimulated (it turns red as stimulus is applied).

1. Try using higher voltages. As more voltage is delivered to the whole muscle, greater numbers of muscle fibers are activated, thereby increasing the total force produced by the muscle. This result is similar to that occurring *in vivo,* where the recruitment of additional motor units increases the force of muscle contraction. This phenomenon is called **multiple motor unit summation.**

Is there a stimulus voltage beyond which there appears to be no further increase in muscle contraction?

____

Why is this so? ______________________________

____

This voltage is the lowest stimulus intensity (strength) necessary to activate all cells within the muscle and is called the **maximal stimulus.** The maximum contraction thus produced is called the *maximal response.*

2. Do a systematic study of force produced over the entire range of stimulus voltages in 1.0-volt increments and record your data in the chart at the top of p. 153. Then plot the results in the accompanying grid.

From the graph, estimate the following:

Threshold voltage ______________ V

Maximal stimulus ______________ V

Is a muscle twitch an all-or-none response? __________

Justify your answer.

____

____

GRADED MUSCLE RESPONSE TO INCREASED STIMULUS FREQUENCY In addition to multiple motor unit summation, another way to increase the amount of force produced by muscle is wave (temporal) summation. Unlike multiple motor unit summation which relies on increased stimulus intensity, **wave summation** is achieved by increasing the stimulus frequency (rate of stimulus delivery to the muscle). Wave summation occurs because the muscle is already in a

Stimulus Voltage	Twitch Force

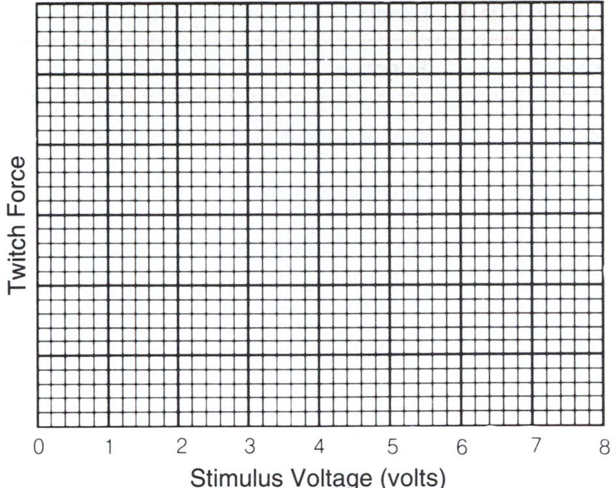

partially contracted state when subsequent stimuli are delivered.

Tetanus can be considered an extreme form of wave summation that results in a steady, sustained contraction. In effect, the muscle does not have any chance to relax due to the very high frequency of stimulation. This "fuses" the force peaks so that we observe a smooth tracing.

WAVE SUMMATION

1. Erase all tracings, set voltage at maximal stimulus, and use the continuous tracing mode. Stimulate once and then again when the muscle has relaxed about halfway. Is the peak force produced greater than that produced by the first stimulus?

2. Try stimulating again at a greater frequency. Is the contraction more forceful?

3. Try manipulating the stimulus frequency to produce a constant sustained contraction (tetanus) so that Force = 2. Is the force produced steady, or does it rise and fall periodically?

4. Try again at Force = 3. Is the tracing smooth this time?

5. So far, you have been using maximal stimulus voltage. What do you think would happen if you used a lower voltage?

Try it to prove (or disprove) your hypothesis.

6. Try adjusting both stimulus voltage and frequency to achieve sustained contraction at Force = 2 and Force = 3. Is it possible?

7. Use the concepts of stimulus intensity and frequency to explain how human skeletal muscles work to produce smooth, steady contractions at all desired levels of force.

8. When finished experimenting in this area, press **Q** to quit and return to the menu.

Multiple Stimulus

Select option 4: Multiple Stimulus. The screen will appear much like that for the Single Stimulus equipment. However, notice that this equipment works somewhat differently from that in Single Stimulus. You can encounter many different types of physiological equipment in the laboratory, and this set of exercises is typical of that situation. The equipment in this section functions in the same way as that of Single Stimulus, but you now have the additional ability to select exact stimulus rates. The basic function keys are the same as

Table 16B.2 Control Keys for Multiple Stimulus

Key	Function
K	Stimulate on/off
R	Rate selection box
I	Increase stimulus rate
D	Decrease stimulus rate
V	Return to experiment
C	Continuous trace mode
T	One-trace mode
E	Erase all tracings
Q	Quit (return to menu)

those for Single Stimulus with a few additions (Table 16B.2). As before, familiarize yourself with all control keys before beginning the experiment.

Press **R** and notice that a chart comes up that allows you to select stimulus rate. **I** increases the stimulation rate, **D** decreases it. As before, + and − increase or decrease voltage. Pressing **K** will start stimulation, pressing it again will stop it, and pressing it once more will stimulate again. Pressing **C** puts the oscilloscope in the continuous mode: the tracing will continue running until you stop it or the force is decreased to zero.

FUSION FREQUENCY

1. Set the stimulator at the maximal stimulus.
2. Stimulate at this intensity for all available rates, keeping all tracings on the screen.

At what stimulation rate is tetany produced?

______________ stimuli per second

This rate is the *fusion frequency.* As before, try producing smooth, sustained contraction (tetany) at Force = 2 and Force = 3 by adjusting the stimulus rate. Do the results support your earlier ideas?

____

MUSCLE FATIGUE A prolonged period of sustained contraction will result in **muscle fatigue,** a condition in which the tissue loses its ability to contract. Fatigue is easily demonstrated using the Multiple Stimulus program.

1. Set the voltage control at the maximal stimulus, stimulus rate at 20/sec, and screen on continuous sweep. Stimulate once and observe the results.

   Does the force eventually begin to fall (muscle fatigue)?

   ____

2. Erase the screen. Stimulate as before, but when fatigue has developed, turn off the stimulator (press **K** again) for a second or two, then turn back on.

   Do you see any evidence of recovery?__________

3. When finished with this section, return to the Index of Programs by pressing (ESC).

ISOMETRIC CONTRACTION

Isometric contraction is the condition in which muscle length does not change regardless of the amount of force generated by the muscle (*iso* = same, *metric* = length). This is accomplished experimentally by holding both ends of the muscle in a fixed position while stimulating it electrically. *Resting length* (length of the muscle before contraction) is an important factor in determining the amount of force that a muscle can develop. *Passive force* is generated by stretching the muscle and is due to the elastic properties of the tissue itself. *Active force* is generated by the physiological contraction of the muscle. *Total force* is the sum of passive and active forces.

This program allows you to set the resting length of the experimental muscle and stimulate it with a single maximal stimulus shock. You can then construct a graph relating the forces generated to muscle length. These principles can be applied to human muscles in order to understand how optimum resting length enables a muscle to produce maximum force. To understand why muscle tissue behaves as it does, it is necessary to comprehend *how* contraction works at the cellular level. If you have difficulty understanding the results of this exercise, review the sliding filament model of muscle contraction.

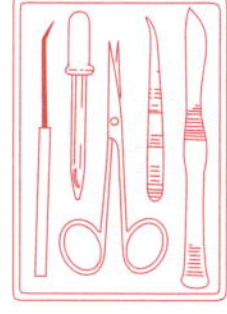

From the Index of Programs, select option 3: Isometric Contraction. Spend a few moments on the program's Introduction to the System and Some Definitions sections to gain an understanding of the essential elements of the system.

From the next menu, again select option 3: Isometric Contraction program. The opening screen (Figure 16B.3) will appear in a few seconds.

The tip of the length-changing device indicates muscle length on the length axis below and is set at approximately 21 units. The Force-Time graph displays the single muscle twitch. Notice a small horizontal line next to the force scale; this will display resting force of the muscle. Familiarize yourself with the control keys (Table 16B.3) before proceeding.

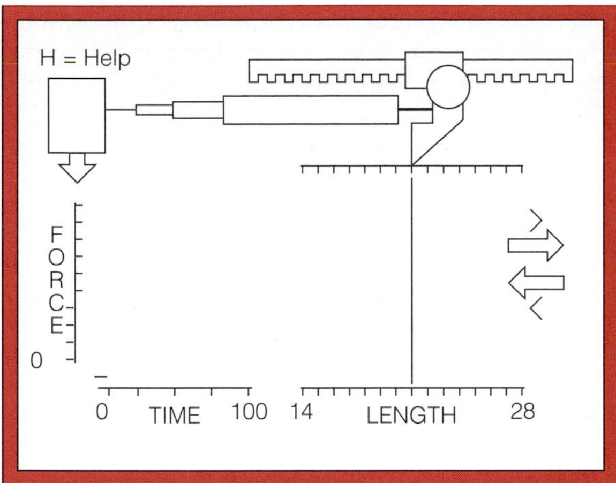

F16B.3

Opening screen of Isometric Contraction.

1. Shorten the muscle to a length of 14 units.

2. Press **M** once to mark *passive* (resting) *force* (a small dot should appear just above the length axis on the Force-Length graph).

3. Stimulate once by pressing **S**. You should see a single force tracing on the Force-Time graph.

4. Press **M** once again to mark *peak twitch* (active) *force*.

5. Increasing the muscle length by 1 unit each time and repeating the Mark-Stimulate-Mark pattern (steps 2–4), make tracings for the entire length range.

What is happening to the resting and peak twitch forces as muscle length is increased?

___

6. When you have finished plotting all the data, connect points by pressing **L**.

Can you explain the dip in the upper curve? (Keep in mind you are measuring total muscle force.)

___

___

7. Display the ideal curves by pressing **D**. Now is the reason for the dip evident?

8. Press **X** to label the graph. Notice that what we have been calling peak force is actually total force and is the sum of the resting and active forces.

Table 16B.3 Control Keys for Isometric Contraction

Key	Function
+ or >	Lengthen muscle
− or <	Shorten muscle
S	Stimulate muscle
M	Mark data point
F	Show data values
L	Draw line graph
D	Display ideal curves
C	Clear screen
E	Erase tracings only
R	Resume work
H	Help screen
Q	Quit (return to menu)

9. When you understand these concepts, return to the Index of Programs (ESC).

ISOTONIC CONTRACTION

During **isotonic contraction,** muscle length changes, but the force produced does not (*iso* = same, *tonic* = force). Unlike the isometric exercise, where both ends of the muscle were held in a fixed position, the setup for isotonic contraction requires one end of the muscle to be free. Variable weights can then be attached to the free end, while the other end of the muscle is held fixed. If the weight is not too great, the muscle will be able to lift it with a certain velocity. You can think of lifting an object from the floor as an example: If the object is light, it can be lifted quickly (high velocity), whereas a heavier weight will be lifted more slowly (with a lower velocity). Try to transfer the idea of what is happening in the simulation to the muscles of your arm when you are attempting to lift a weight. The two important variables in this exercise are *starting length* of the muscle and the *afterload* (weight) applied. As before, some background in the sliding filament model of contraction will be helpful.

This program allows you to change muscle length and afterload so that you can investigate the effects of changes in these variables on the speed of skeletal muscle shortening. Both variables can be independently altered and results are graphically presented on the screen.

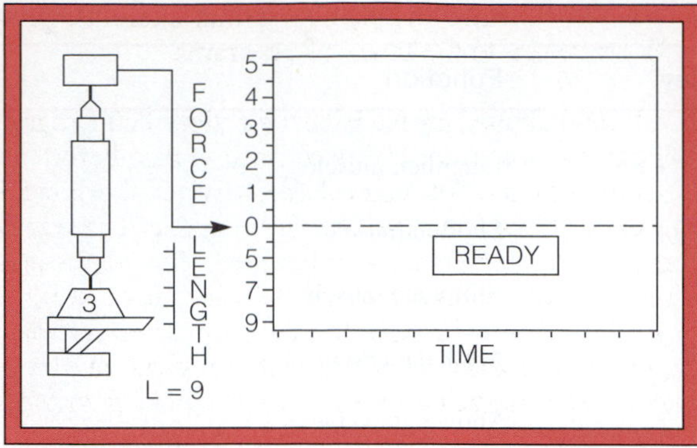

F16B.4

Opening screen of Isotonic Contraction.

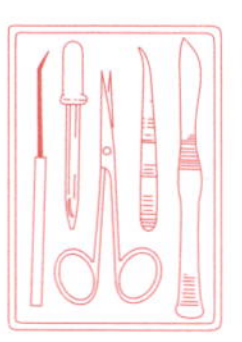

Choose option 4: Isotonic Contraction from the Index of Programs and, as before, spend some time on the introductory sections before proceeding. Select option 3: Contraction and Relaxation. The opening screen will appear in a few seconds (Figure 16B.4).

The length scale displays 5 to 9 arbitrary units and the force scale indicates a range from 0 to 5 units. Initially, muscle length is set at 9 units and afterload weight at 3 units. Familiarize yourself with the equipment layout and control keys (Table 16B.4) before proceeding.

1. Stimulate the muscle once and observe the tracings. Notice the Ready box while the tracing is being produced. It will indicate *isometric* conditions (length is constant but force may be changing) and *isotonic* conditions (force is constant but length is changing).

2. To change weight or length, press the appropriate key (**W** or **L**) and select the variable you wish. The upper tracing displays force as a function of time. The lower tracing indicates length as a function of time.

   Does the flat part of the tracings correspond to isotonic conditions or isometric conditions?

   ____

3. Analyze length tracings carefully because they indicate initial velocity of shortening. Think about what the curve means and the following concept should become clear: Using a given weight, a specific change in the length of the muscle will occur in a certain amount of time (heavier afterloads require more time). This means that the *steepness* of the tracing is an indication of initial velocity or speed of shortening; the steeper the tracing, the higher the initial velocity.

Table 16B.4 Control Keys for Isotonic Contraction

Key		Function
S		Stimulate muscle
V		View screen
W		Change afterload
L		Change length
G		Graph data
Graph options:		
	P	Plot current point
	D	Draw curve
	R	Return to program
	X	Erase graph
E		Erase screen
H		Help screen
Q		Quit (return to menu)

THE EFFECT OF LOAD ON SKELETAL MUSCLE

1. Keeping the resting length constant at 9 units, run through the entire range of afterload weights. After each weight tracing has been completed, press **G** to activate the graph, **P** to plot that data point, and **R** to return to the program. When the entire set of afterload weights has been plotted, press **D** while the graph is open to draw the plot.

What is the relationship between afterload weight (Fa) and the initial velocity of shortening?

____

____

What about afterload and maximum length change?

____

2. Now keep afterload weight constant (W = 2) and run through the entire range of starting lengths.

What is the relationship between starting length and initial velocity of shortening?

____

____

____

3. Try the same experiment with a different weight.

Is there a similar pattern? ______________________

Can you set up a contraction that is entirely isometric?

____

One that is entirely isotonic? ____________________

Explain why or why not.

____

____

____

____

4. When you are finished with this exercise, press (ESC) to return to the Index of Programs.

5. If time allows, try the Biker selection. This program simulates a bike rider trying to finish a race before exhaustion sets in. You control the gears of the bicycle, and the object of the exercise is to shift effectively so that the biker achieves his or her goal. Bar graphs indicating conditions and a force-velocity curve are provided to aid you in your task. Apply what you have learned in this exercise along with a logical interpretation of the force-velocity curve and it will be possible to finish the race.

6. When finished with all exercises, escape to the Index of Programs and select option 8: Quit. Remove the simulation program diskette and return the computer to its starting condition.

17 EXERCISE

Histology of Nervous Tissue

OBJECTIVES

1. To differentiate between the functions of neurons and neuroglia.
2. To list four types of neuroglia cells.
3. To identify the important anatomical characteristics of a neuron on an appropriate diagram or projected slide.
4. To state the functions of axons, dendrites, axonal terminals, neurofibrils, and myelin sheaths.
5. To explain how a nerve impulse is transmitted from one neuron to another.
6. To explain the role of Schwann cells in the formation of the myelin sheath.
7. To classify neurons according to structure and function.
8. To distinguish between a nerve and a tract and between a ganglion and a nucleus.
9. To describe the structure of a nerve, identifying the connective tissue coverings (endoneurium, perineurium, and epineurium) and citing their functions.

MATERIALS

Model of a "typical" neuron (if available)
Compound microscope
Histologic slides of an ox spinal cord smear and teased myelinated nerve fibers
Prepared slides of Purkinje cells (cerebellum), pyramidal cells (cerebrum), and a dorsal root ganglion
Prepared slide of a nerve (cross section)

See Appendix E, Exercise 17 for links to *Anatomy and PhysioShow: The Videodisc.*

The nervous system is the master integrating and coordinating system, continuously monitoring and processing sensory information both from the external environment and from within the body. Every thought, action, and sensation is a reflection of its activity. Like a computer, it processes and integrates new "inputs" with information previously fed into it ("programmed") to produce an appropriate response ("readout"). However, no computer can possibly compare in complexity and scope to the human nervous system.

Despite its complexity, nervous tissue is made up of just two principal cell populations: **neurons** and the **neuroglia** (glial cells). The *neuroglia,* literally "nerve glue," include *astrocytes, oligodendrocytes, microglia,* and *ependymal cells* (Figure 17.1). These cells serve the needs of the neurons by acting as supportive and protective cells, myelinating cells, and phagocytes. In addition, they probably serve some nutritive function by acting as a selective barrier between the capillary blood supply and the neurons. Although neuroglia resemble neurons in some ways (they have fibrous cellular extensions), they are not capable of generating and transmitting nerve impulses, a capability that is highly developed in neurons. Our focus in this exercise is the highly irritable neurons.

NEURON ANATOMY

The delicate **neurons** are the structural units of nervous tissue. They are highly specialized to transmit messages (nerve impulses) from one part of the body to another. Although neurons differ structurally, they have many identifiable features in common (Figure 17.2). All have a **cell body** from which slender processes or fibers extend. Although neuron cell bodies are typically found in the CNS (central nervous system: brain and spinal cord) in clusters called **nuclei,** occasionally they reside in **ganglia** (collections of neuron cell bodies outside the CNS). They make up the gray matter of the nervous system. Neuron processes running through the CNS form **tracts** of white matter; outside the CNS they form the peripheral **nerves.**

The neuron cell body contains a large round nucleus surrounded by cytoplasm (*neuroplasm*). The cytoplasm is riddled with neurofibrils and with darkly staining structures called Nissl bodies. **Neurofibrils,** the cytoskeletal elements of the neuron, have a support and intracellular transport function. **Nissl bodies,** an elaborate type of rough endoplasmic reticulum, are involved in the metabolic activities of the cell.

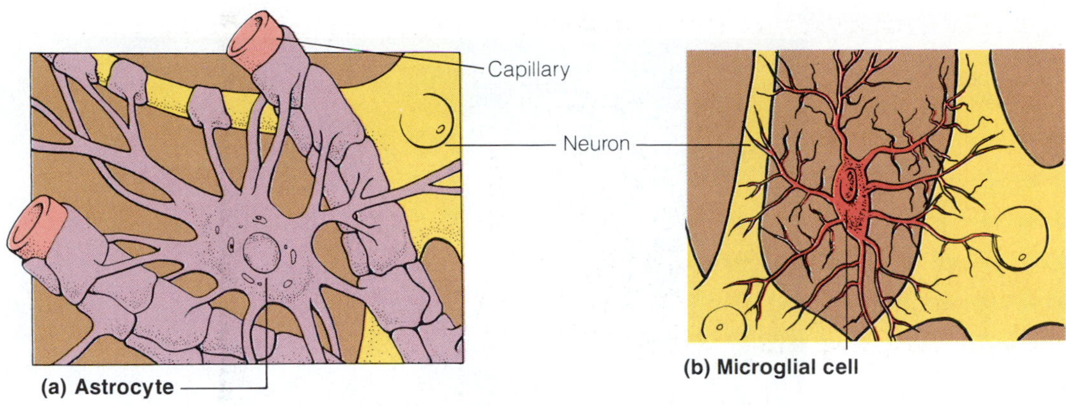

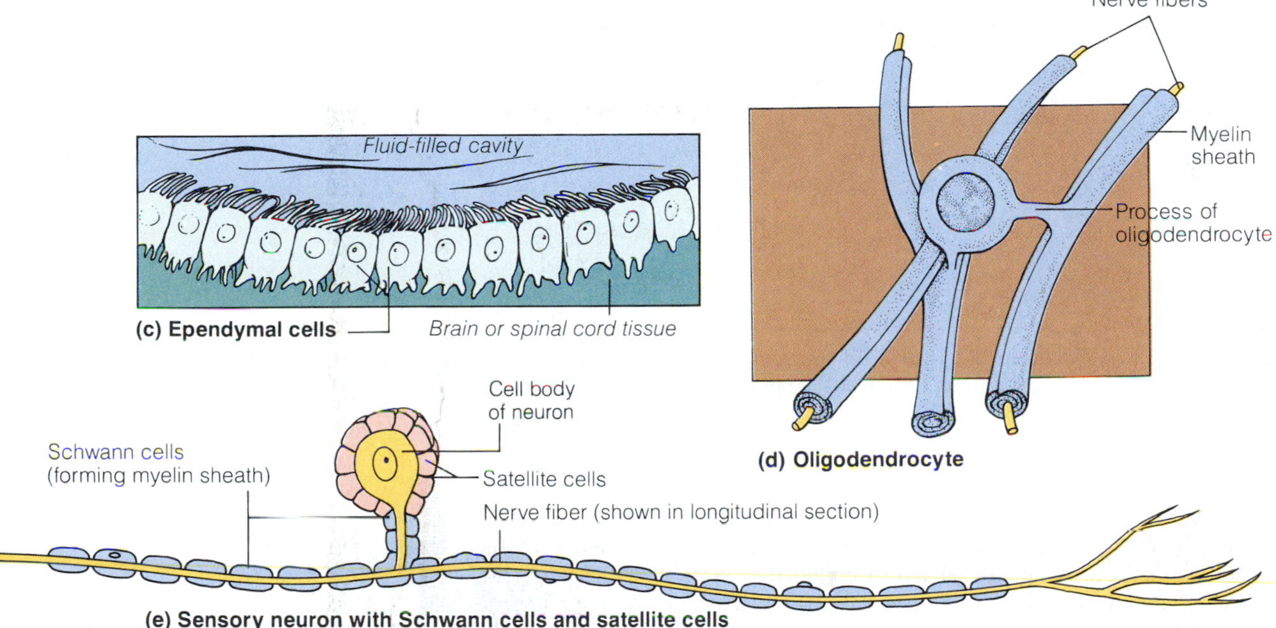

F17.1

Supporting cells of nervous tissue. (**a**) Astrocyte. (**b**) Microglial cell. (**c**) Ependymal cells. (**d**) Oligodendrocyte. (**e**) Neuron with Schwann cells and satellite cells.

According to the older, traditional scheme, neuron processes that conduct electrical currents *toward* the cell body are called **dendrites;** and those that carry impulses away *from* the nerve cell body are called **axons.** When it was discovered that this functional scheme had pitfalls (some axons carry impulses *both* toward and away from the cell body), a newer functional definition of neuron processes was adopted. According to this scheme, dendrites are *receptive regions* (they bear receptors for neurotransmitters released by other neurons), whereas axons are *nerve impulse generators* and *transmitters.* Neurons have only one axon (which may branch into **collaterals**) but may have many dendrites, depending on the neuron type. Notice that the term *nerve fiber* is a synonym for axon and is, thus, quite specific.

In general, a neuron is excited by other neurons when their axons release neurotransmitters close to its dendrites or cell body. The electrical current generated travels across the cell body and down the axon. As Fig-

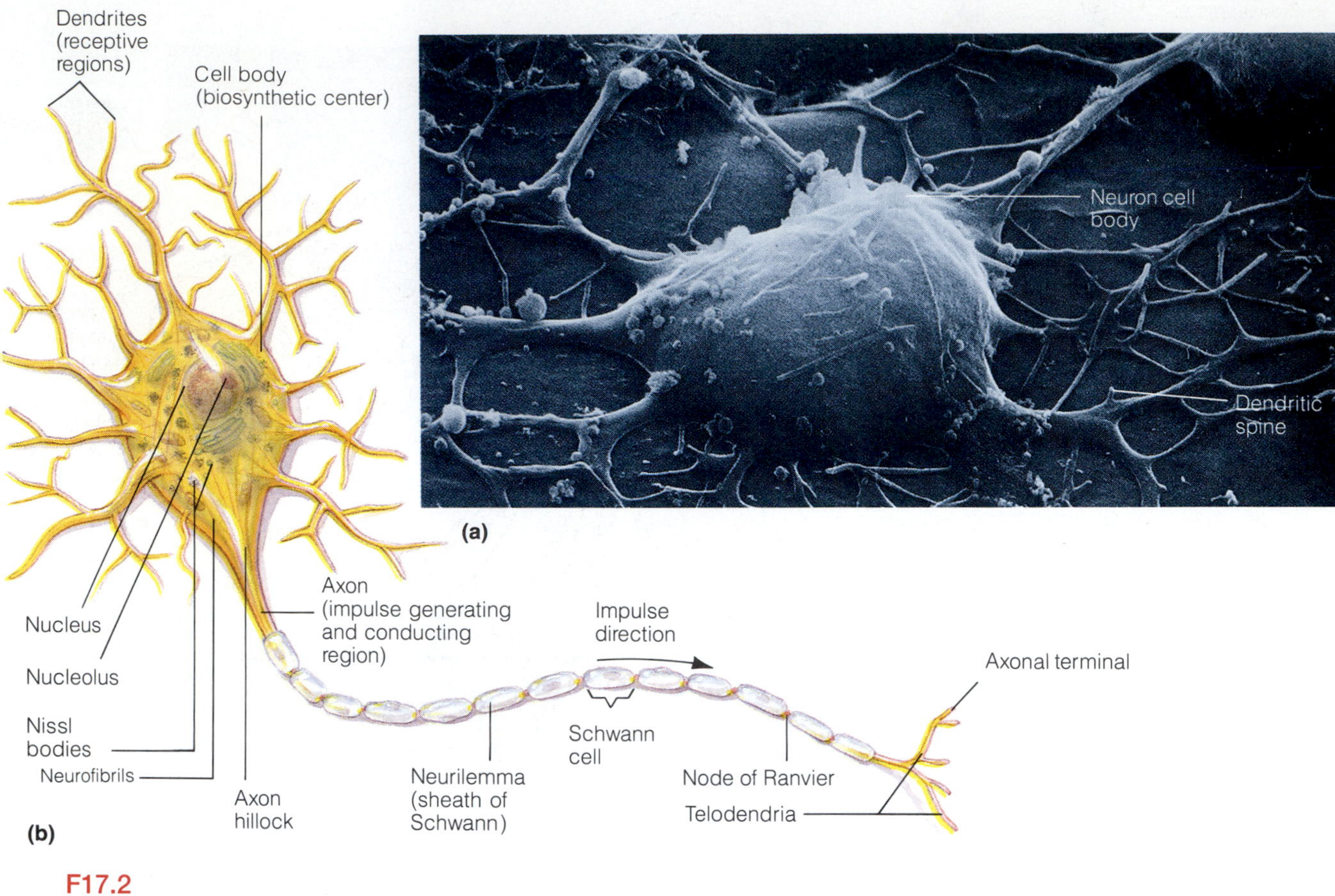

F17.2

Structure of a typical motor neuron. **(a)** Photomicrograph showing the neuron cell body and dendrites with obvious dendritic spines, which are synapse sites (5000×). **(b)** Diagrammatic view.

ure 17.2 shows, the axon (in motor neurons) begins at a slightly enlarged cell body structure called the **axon hillock** and ends in many small structures called **axonal terminals,** or synaptic knobs. These terminals store the neurotransmitter chemical in tiny vesicles. Each axonal terminal is separated from the cell body or dendrites of the next (postsynaptic) neuron by a tiny gap called the **synaptic cleft.** Thus, although they are close, there is no actual physical contact between neurons. When an impulse reaches the axonal terminals, some of the synaptic vesicles rupture and release neurotransmitter into the synaptic cleft. The neurotransmitter then diffuses across the synaptic cleft to bind to membrane receptors on the next neuron, initiating the action potential.*

Many drugs can influence the transmission of impulses at synapses. Some, like caffeine, are stimulants which decrease the receptor neuron's threshold and make it more irritable. Others block transmission by binding competitively with the receptor sites or by interfering with the release of neurotransmitter by the axonal terminals. As might be anticipated, some of these drugs are used as painkillers or tranquilizers. ■

Most long nerve fibers are covered with a fatty material called *myelin,* and such fibers are referred to as **myelinated fibers.** Axons in the peripheral nervous system are typically heavily myelinated by special cells called **Schwann cells,** which wrap themselves tightly around the axon jelly-roll fashion (Figure 17.3). During the wrapping process, the cytoplasm is squeezed from between adjacent layers of the Schwann cell membranes, so that when the process is completed a tight core of plasma membrane material (protein-lipoid material) encompasses the axon. This wrapping is the **myelin sheath.** The Schwann cell nucleus and the bulk of its cytoplasm ends up just beneath the outermost portion of its plasma membrane. This peripheral part of the Schwann cell and its plasma membrane is referred to as the **neurilemma.** Since the myelin sheath is formed by many individual Schwann cells, it is a discontinuous sheath; the gaps or indentations in the sheath are called **nodes of Ranvier** (see Figure 17.2).

* Specialized synapses in skeletal muscle are called neuromuscular junctions. They are discussed in Exercise 14.

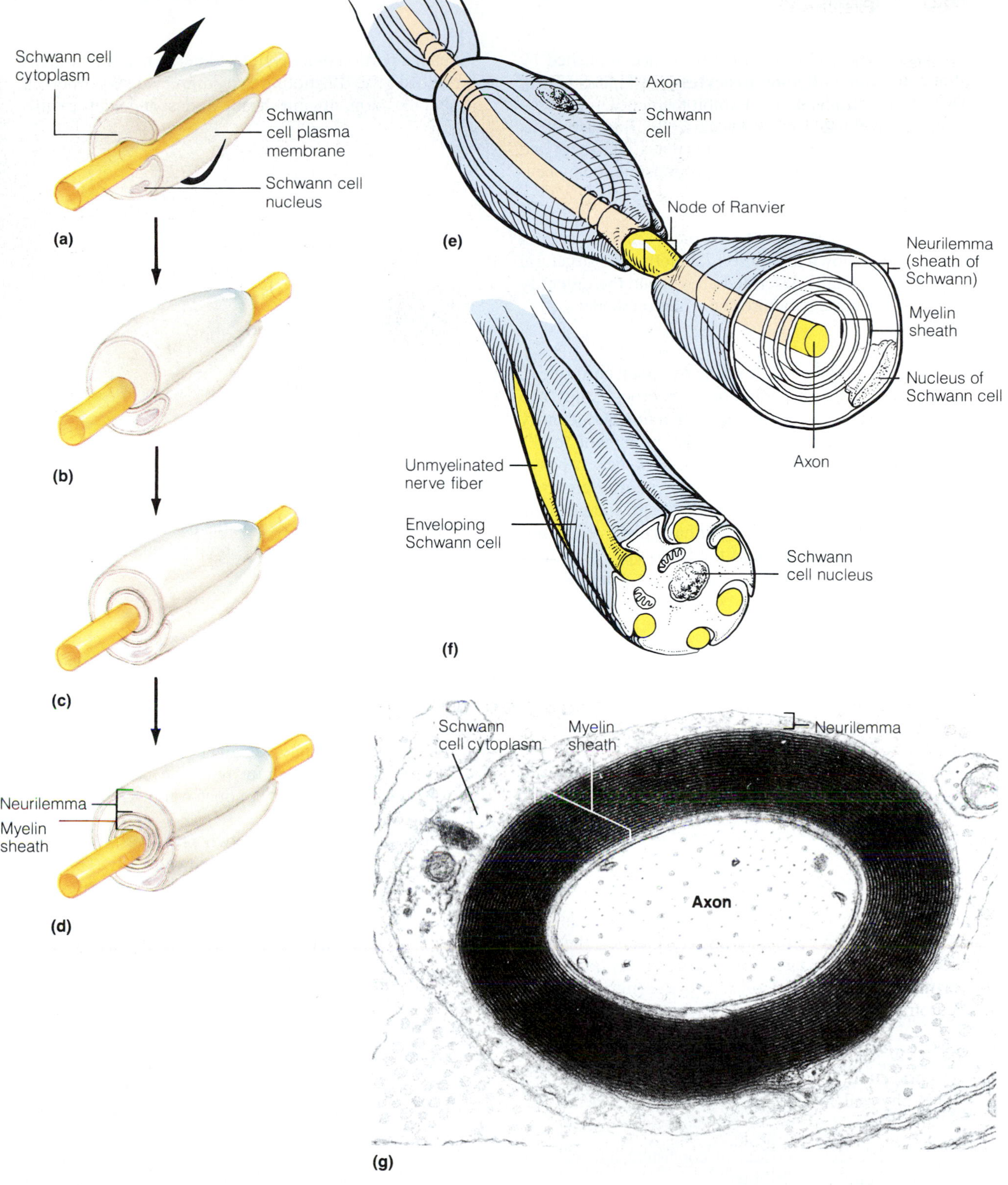

F17.3

Myelination of neuron processes by individual Schwann cells. (a–d) A Schwann cell becomes apposed to an axon and envelops it in a trough. It then begins to rotate around the axon, wrapping it loosely in successive layers of its plasma membrane. Eventually, the Schwann cell cytoplasm is forced from between the membranes and comes to lie peripherally just beneath the exposed portion of the Schwann cell membrane. The tight membrane wrappings surrounding the axon form the myelin sheath. The area of Schwann cell cytoplasm and its exposed membrane are referred to as the neurilemma or sheath of Schwann. **(e)** Longitudinal view of myelinated axon showing portions of adjacent Schwann cells and the node of Ranvier between them. **(f)** Unmyelinated fibers. Schwann cells may associate loosely with several axons, which they partially invest. In such cases, Schwann cell coiling around the axons does not occur. **(g)** Photomicrograph of a myelinated axon, cross-sectional view (20000×).

Within the CNS, myelination is accomplished by glial cells called **oligodendrocytes** (see Figure 17.1d). These CNS sheaths do not exhibit the neurilemma seen in fibers myelinated by Schwann cells. Because of its chemical composition, myelin insulates the fibers and greatly increases the speed of neurotransmission by neuron fibers.

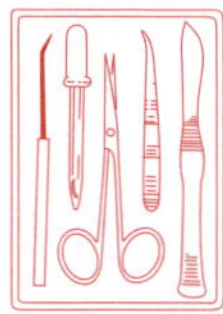

1. Study the typical motor neuron shown in Figure 17.2, noting the structural details described above, and then identify these structures on a neuron model.

2. Obtain a prepared slide of the ox spinal cord smear, which has large, easily identifiable neurons. Study one representative neuron under oil immersion and identify the cell body; the nucleus; the large, prominent "owl's eye" nucleolus; and the granular Nissl bodies. If possible, distinguish the axon from the many dendrites. Sketch the cell in the space provided here, and label the important anatomical details you have observed. Compare your sketch to Plate 4 of the Histology Atlas. Also examine Plate 5 which differentiates the neuronal processes more clearly.

3. Obtain a prepared slide of teased myelinated nerve fibers. Using Plate 6 of the Histology Atlas as a guide, identify the following: nodes of Ranvier, neurilemma, axis cylinder (the axon itself), Schwann cell nuclei, and myelin sheath.

Do the nodes seem to occur at consistent intervals, or are they irregularly distributed?

___

Explain the significance of this finding: ___

___

___

___

___

Sketch a portion of a myelinated nerve fiber in the space provided here, illustrating two or three nodes of Ranvier. Label the axon, myelin sheath, nodes, and neurilemma.

NEURON CLASSIFICATION

Neurons may be classified on the basis of structure or of function. Figure 17.4 depicts both schemes.

Basis of Structure

Structurally, neurons may be differentiated according to the number of processes attached to the cell body. In **unipolar neurons,** one very short process, which divides into *peripheral* and *central processes,* extends from the cell body. Functionally, only the most distal portions of the peripheral process act as dendrites; the rest acts as an axon along with the central process. Nearly all neurons that conduct impulses toward the CNS are unipolar.

Bipolar neurons have two processes—one axon and one dendrite—attached to the cell body. This neuron type is quite rare, typically found only as part of the receptor apparatus of the eye, ear, and olfactory mucosa.

Many processes issue from the cell body of **multipolar** neurons, all classified as dendrites except for a single axon. Most neurons in the brain and spinal cord (CNS neurons) and those whose axons carry impulses away from the CNS fall into this last category.

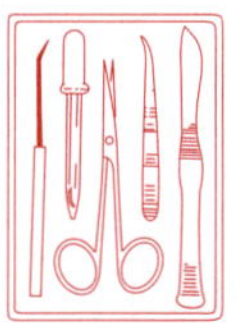

Obtain prepared slides of Purkinje cells of the cerebellar cortex, pyramidal cells of the cerebral cortex, and a dorsal root ganglion. As you observe them under the microscope, try to pick out the anatomical details depicted in Figure 17.5. Notice that the neurons of the cerebral and cerebellar tissues (both brain tissues) are extensively branched; in contrast, the neurons of the dorsal root ganglion are more rounded. You may also be able to identify astrocytes (a type of neuroglia) in the brain tissue slides if you examine them closely.

Which of these neuron types would be classified as multipolar neurons?

___

Which as unipolar? ___

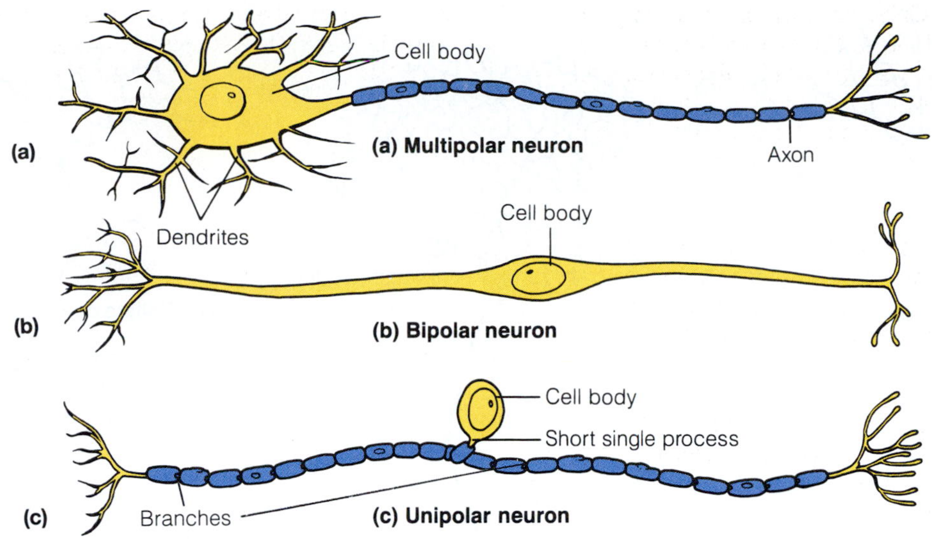

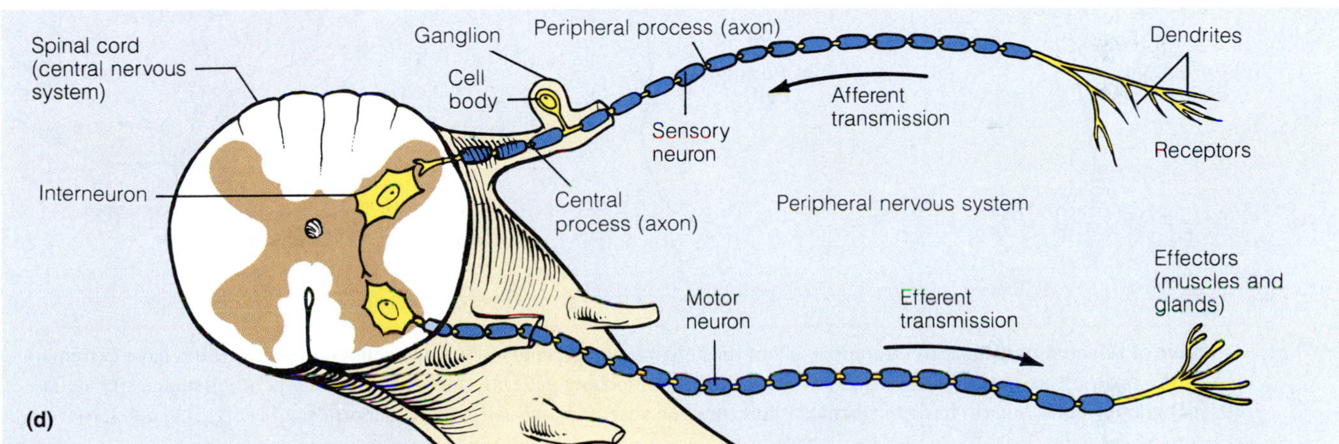

F17.4

Classification of neurons. **(a–c)** On the basis of structure: **(a)** multipolar; **(b)** bipolar; **(c)** unipolar. **(d)** On the basis of function, there are sensory, motor, and association neurons. Sensory (afferent) neurons conduct impulses from the body's sensory receptors to the central nervous system; most are unipolar neurons with their nerve cell bodies in ganglia in the peripheral nervous system (PNS). Motor (efferent) neurons transmit impulses from the CNS to effectors such as muscles and glands. Association neurons (interneurons) complete the communication line between sensory and motor neurons. They are typically multipolar and their cell bodies reside in the CNS.

Basis of Function

In general, neurons carrying impulses from the sensory receptors in the internal organs (viscera) or in the skin are termed **sensory,** or **afferent, neurons** (see Figure 17.4d). The dendritic endings of sensory neurons are often equipped with specialized receptors that are stimulated by specific changes in their immediate environment. The structure and function of these receptors is considered separately in Exercise 23 (General Sensation). The cell bodies of sensory neurons are always found in a ganglion outside the CNS, and these neurons are typically unipolar.

Neurons carrying activating impulses from the CNS to the viscera and/or body muscles and glands are termed **motor,** or **efferent, neurons.** Motor neurons are most often multipolar and their cell bodies are almost always located in the CNS.

The third functional category of neurons is the **association neurons,** or **interneurons,** which are situated between and contribute to pathways that connect sensory and motor neurons. Their cell bodies are always

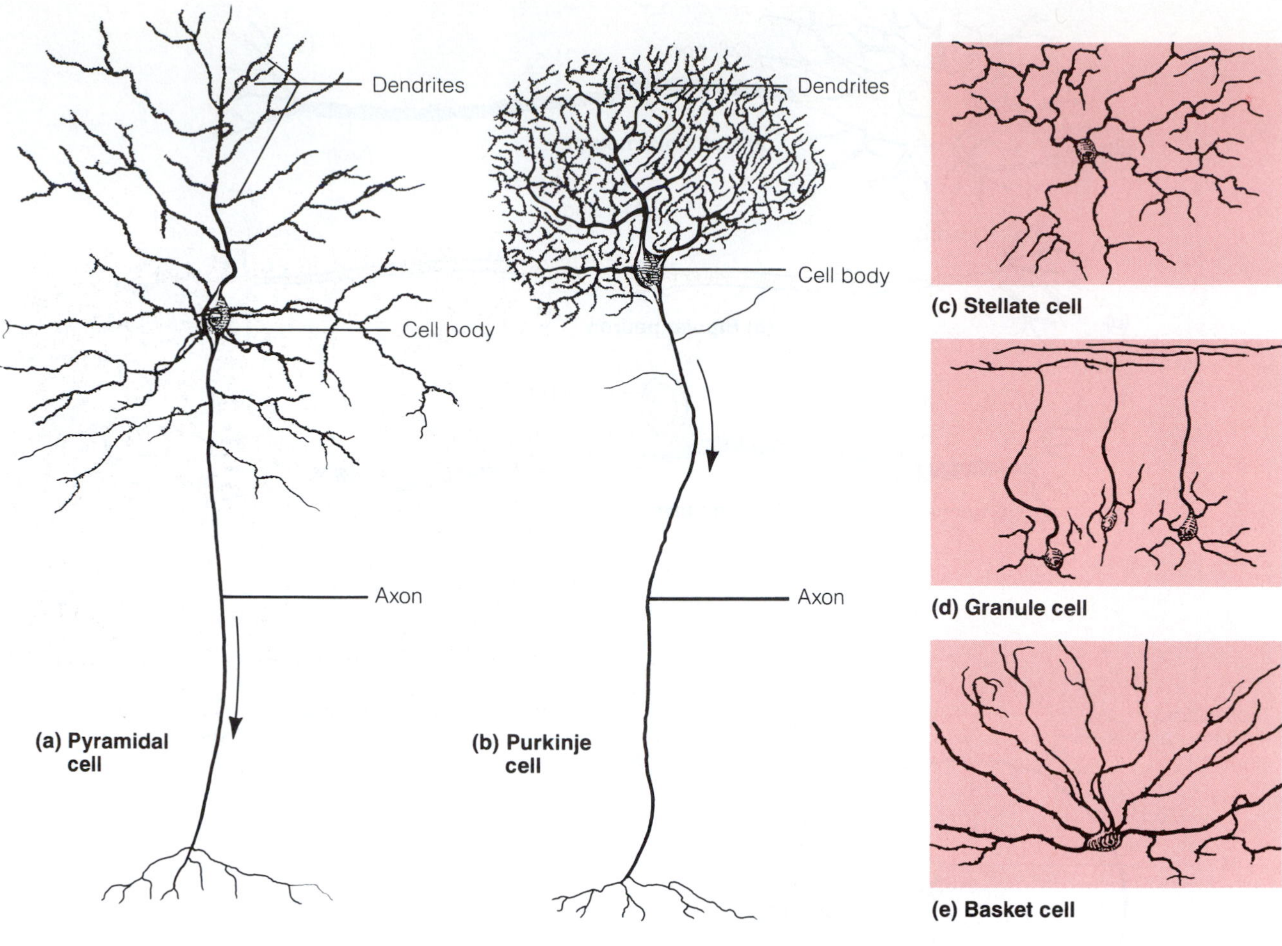

F17.5

Structure of selected neurons. (**a**) Pyramidal cell of the cerebral cortex and (**b**) Purkinje cells of the cerebellum have extremely long axons, but they are easily distinguished by their dendrite-branching patterns, whereas cerebellar neurons such as (**c**) stellate cells, (**d**) granule cells, and (**e**) basket cells have short (or even absent) axons and profuse dendrites.

located within the CNS and they are multipolar neurons structurally.

STRUCTURE OF A NERVE

A nerve is a bundle of neuron fibers or processes wrapped in connective tissue coverings that extends to and/or from the CNS and visceral organs or structures of the body periphery (such as skeletal muscles, glands, and skin).

Within a nerve, each fiber is surrounded by a delicate connective tissue sheath called an **endoneurium,** which insulates it from the other neuron processes adjacent to it. (The endoneurium is often mistaken for the myelin sheath; it is instead an additional sheath that surrounds the myelin sheath.) Groups of fibers are bound by a coarser connective tissue, called the **perineurium,** to form bundles of fibers called **fascicles.** Finally, all the fascicles are bound together by a tough, white, fibrous connective tissue sheath called the **epineurium,** forming the cordlike nerve (Figure 17.6). In addition to the connective tissue wrappings, blood vessels and lymphatic vessels serving the fibers also travel within a nerve.

Like neurons, nerves are classified according to the direction in which they transmit impulses. Nerves carrying both sensory (afferent) and motor (efferent) fibers are called **mixed nerves;** all spinal nerves are mixed nerves. Nerves that carry only sensory processes and

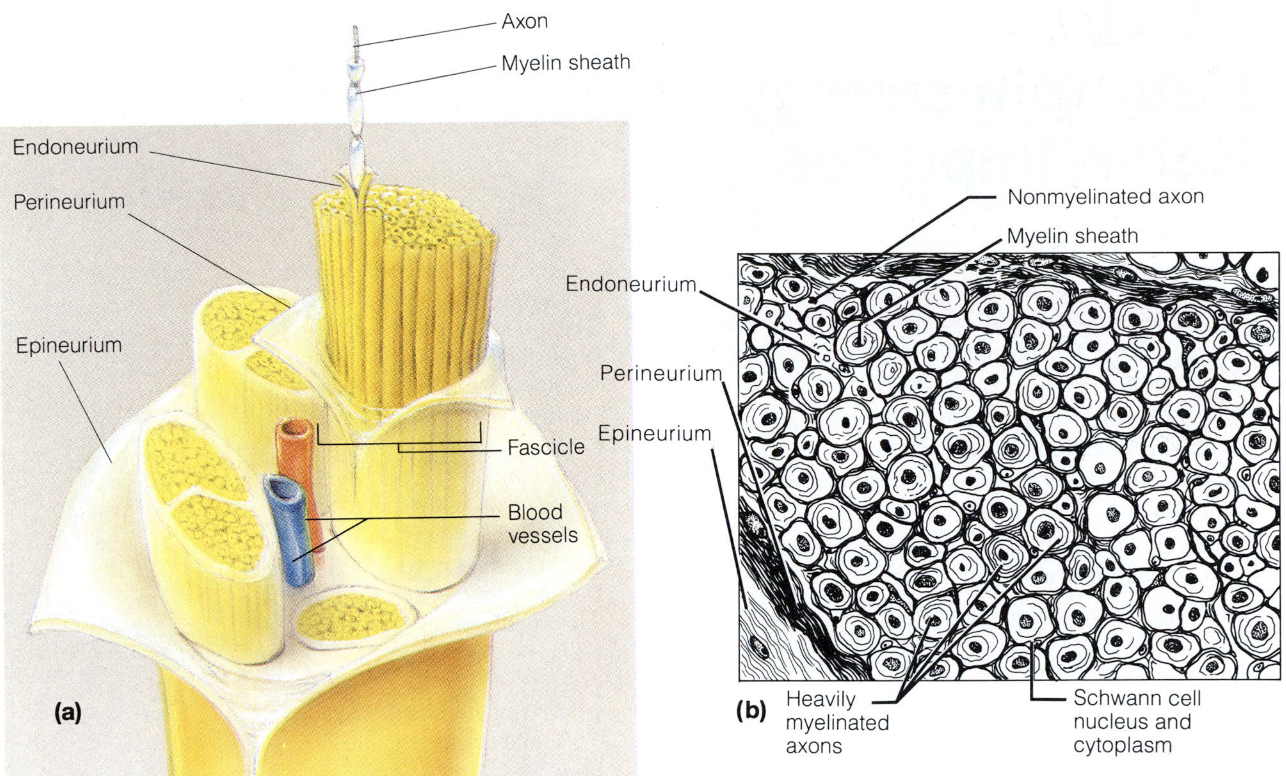

F17.6

Structure of a nerve showing connective tissue wrappings. (a) Three-dimensional view of a portion of a nerve. **(b)** Cross-sectional view corresponding to Plate 8 of the Histology Atlas.

conduct impulses only toward the CNS are referred to as **sensory,** or **afferent, nerves.** A few of the cranial nerves are pure sensory nerves, but the majority are mixed nerves. The ventral roots of the spinal cord, which carry only motor fibers, can be considered **motor,** or **efferent, nerves.**

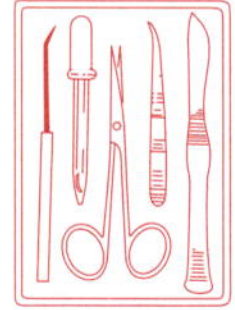

Examine under the compound microscope a prepared cross section of a peripheral nerve. Identify nerve fibers, myelin sheaths, fascicles, and endoneurium, perineurium, and epineurium sheaths. If desired, sketch the nerve in the space to the right.

18

EXERCISE

Neurophysiology of Nerve Impulses

OBJECTIVES

1. To list the two major physiological properties of neurons.
2. To describe the polarized and depolarized states of the nerve cell membrane and to describe the events that lead to generation and conduction of a nerve impulse.
3. To explain how a nerve impulse is transmitted from one neuron to another.
4. To define *action potential, depolarization, repolarization, relative refractory period,* and *absolute refractory period.*
5. To list various substances and factors that can stimulate neurons.
6. To recognize that neurotransmitters may be either stimulatory or inhibitory in nature.
7. To state the site of action of the blocking agents ether and curare.

MATERIALS

*Rana pipiens**
Dissecting instruments and tray
Ringer's solution (frog) in dropper bottles, some at room temperature and some in an ice bath
Thread
Glass rods or probes
Glass plates or slides
Ring stand and clamp
Stimulator; platinum electrodes
Oscilloscope
Nerve chamber
Filter paper
0.01% hydrochloric acid (HCl) solution
Sodium chloride (NaCl) crystals
Heat-resistant mitts
Bunsen burner
Absorbent cotton
Ether
Pipettes
1-cc syringe with small-gauge needle
0.5% tubocurarine solution
Frog board
Disposable plastic gloves
Safety goggles

See Appendix E, Exercise 18 for links to *Anatomy and PhysioShow: The Videodisc.*

*Instructor to provide freshly pithed frogs (Rana pipiens) for student experimentation.

THE NERVE IMPULSE

Neurons have two major physiologic properties: **irritability**, or the ability to respond to stimuli and convert them into nerve impulses, and **conductivity,** the ability to transmit the impulse to other neurons, muscles, or glands. In a resting neuron (as in resting muscle cells), the exterior surface of the membrane is slightly more positively charged than the inner surface, as shown in Figure 18.1a. This difference in electrical charge on the two sides of the membrane results in a voltage across the plasma membrane referred to as the **resting membrane potential,** and a neuron in this state is said to be **polarized.** In the resting state, the predominant intracellular ion is potassium (K^+), and sodium ions (Na^+) are found in greater concentration in the extracellular fluids. The resting potential is maintained by a very active sodium-potassium pump, which transports Na^+ out of the cell and K^+ into the cell.

When the neuron is activated by a stimulus of adequate intensity—a **threshold stimulus**—the membrane at its *trigger zone,* typically the axon hillock (or the most peripheral part of a sensory neuron's axon), briefly becomes more permeable to sodium (sodium gates are opened). Sodium ions rush into the cell, increasing the number of positive ions inside the cell and reversing the polarity (Figure 18.1b). Thus the interior of the membrane becomes less negative at that point and the exterior surface becomes less positive—a phenomenon called **depolarization.** When depolarization reaches a certain point such that the local membrane polarity changes (momentarily the external face becomes negative and the internal face becomes positive), it initiates an **action potential*** (Figure 18.1c).

* If the stimulus is of less than threshold intensity, depolarization is limited to a small area of the membrane, and no action potential is generated.

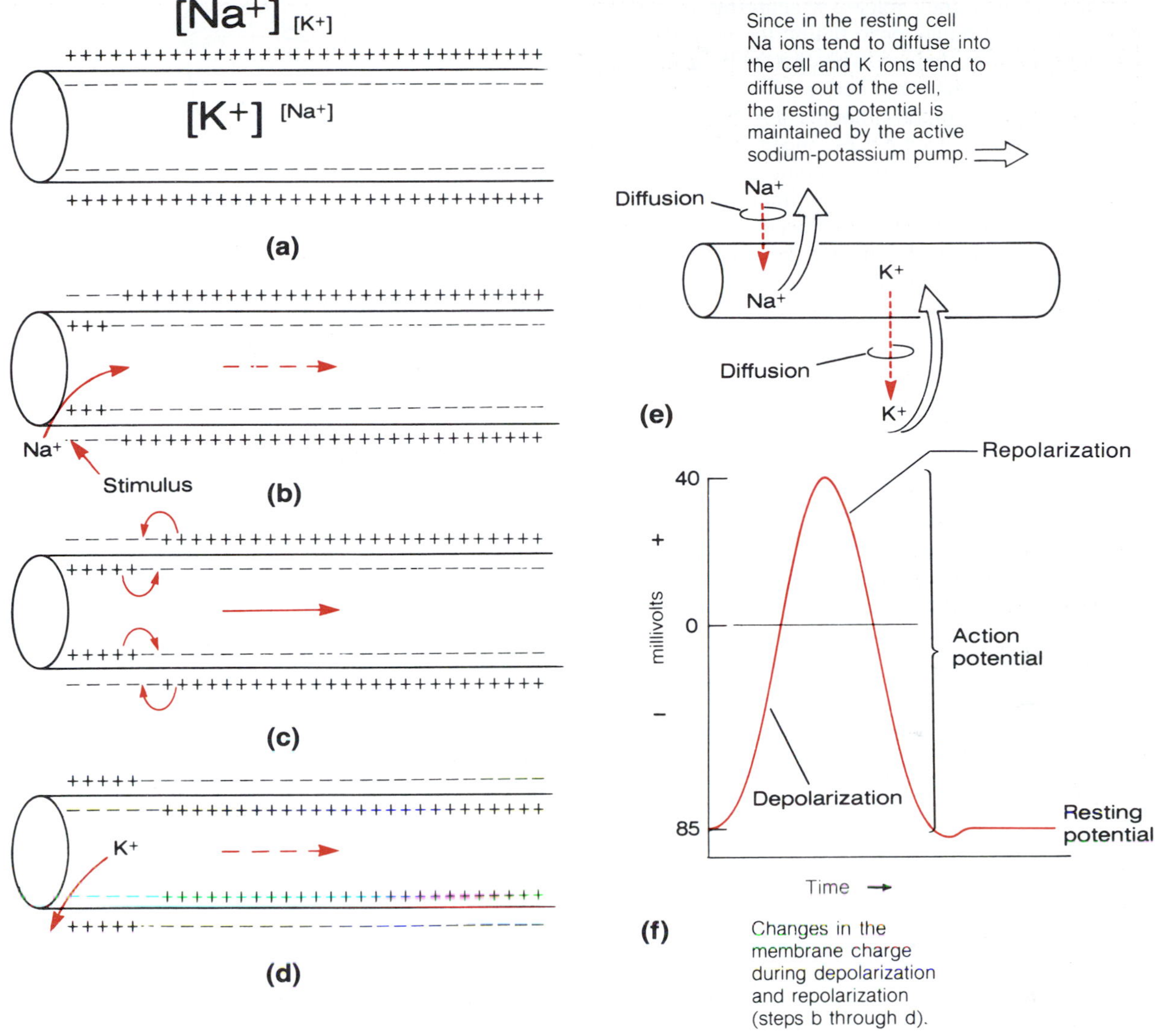

F18.1

The nerve impulse. **(a)** Resting membrane potential (−85 mV). There is an excess of positive ions outside the cell, with Na^+ the predominant extracellular fluid ion and K^+ the predominant intracellular ion. The plasma membrane has a low permeability to Na^+. **(b)** Depolarization—reversal of the resting potential. Application of a stimulus changes the membrane permeability, and Na^+ ions are allowed to diffuse rapidly into the cell. **(c)** Generation of the action potential or nerve impulse. If the stimulus is of adequate intensity, the depolarization wave spreads rapidly along the entire length of the membrane. **(d)** Repolarization—reestablishment of the resting potential. The negative charge on the internal plasma membrane surface and the positive charge on its external surface are reestablished by diffusion of K^+ ions out of the cell, proceeding in the same direction as in depolarization. **(e)** The original ionic concentrations of the resting state are restored by the sodium-potassium pump. **(f)** A tracing of an action potential.

Within a millisecond after the inward influx of sodium, the membrane permeability is again altered. As a result, Na^+ permeability decreases, K^+ permeability increases, and K^+ rushes out of the cell. Since K^+ ions are positively charged, their movement out of the cell reverses the membrane potential again, so that the external membrane surface is again positive relative to the internal membrane face (Figure 18.1d). This event, called **repolarization,** reestablishes the resting membrane potential. When the sodium gates are open, the neuron is totally insensitive to additional stimuli and is said to be in an **absolute refractory period.** During the time of repolarization, the neuron is nearly insensitive to further stimulation. A very strong stimulus may reactivate it, however, thus this period is referred to as the **relative refractory period.**

Once generated, the action potential is a self-propagating phenomenon that spreads rapidly along the entire length of the neuron. It is never partially transmitted; that is, it is an all-or-none response. This propagation of the action potential in neurons is also called the **nerve impulse.** When the nerve impulse reaches the axonal terminals, they release a neurotransmitter that acts either to stimulate or to inhibit the next neuron in the transmission chain. (Note that only stimulatory transmitters are considered here.)

Because only minute amounts of sodium and potassium ions have changed places, once repolarization has been completed, the neuron can quickly respond again to a stimulus. In fact, thousands of impulses can be generated before ionic imbalances prevent the neuron from transmitting impulses. Eventually, however, it is necessary to restore the original ionic concentrations on the two sides of the membrane; this is accomplished by enhanced activity of the Na^{+}-K^{+} pump (Figure 18.1e).

PHYSIOLOGY OF NERVE FIBERS

In this laboratory session, you will investigate the functioning of nerve fibers by subjecting the sciatic nerve of a frog to various types of stimuli and blocking agents. Work in groups of two to four to lighten the work load.

Eliciting a Nerve Impulse

In the first set of experiments, stimulation of the nerve and generation of the action potential will be indicated by the contraction of the gastrocnemius muscle. Because you will make no mechanical recording (unless your instructor asks you to), you must keep complete and accurate records of all experimental procedures and results.

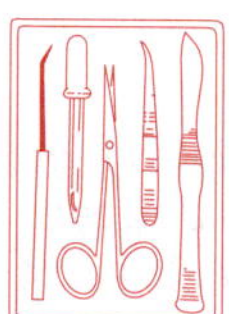

1. Don gloves to protect yourself from any parasites the frogs might have. Request and obtain a pithed frog from your instructor and bring it to your laboratory bench. Also obtain dissecting instruments and a tray from the supply area.

2. Prepare the sciatic nerve as illustrated in Figure 18.2. Place the pithed frog on the dissecting tray, dorsal side down. Make a cut through the skin around the circumference of the frog approximately halfway down the trunk, and then pull the skin down over the muscles of the legs. Open the abdominal cavity and push the abdominal organs to one side to expose the origin of the glistening white sciatic nerve, which arises from the last three spinal nerves. Once the sciatic nerve has been exposed, it should be kept continually moist with room temperature Ringer's solution.

3. Using a glass probe, slip a piece of thread moistened with Ringer's solution under the sciatic nerve close to its origin at the vertebral column. Make a single ligature (tie it firmly with the thread), and then cut through the nerve roots to free the proximal end of the sciatic nerve from its attachments. Using a glass rod or probe, carefully separate the posterior thigh muscles to locate and then free the sciatic nerve, which runs down the posterior aspect of the thigh.

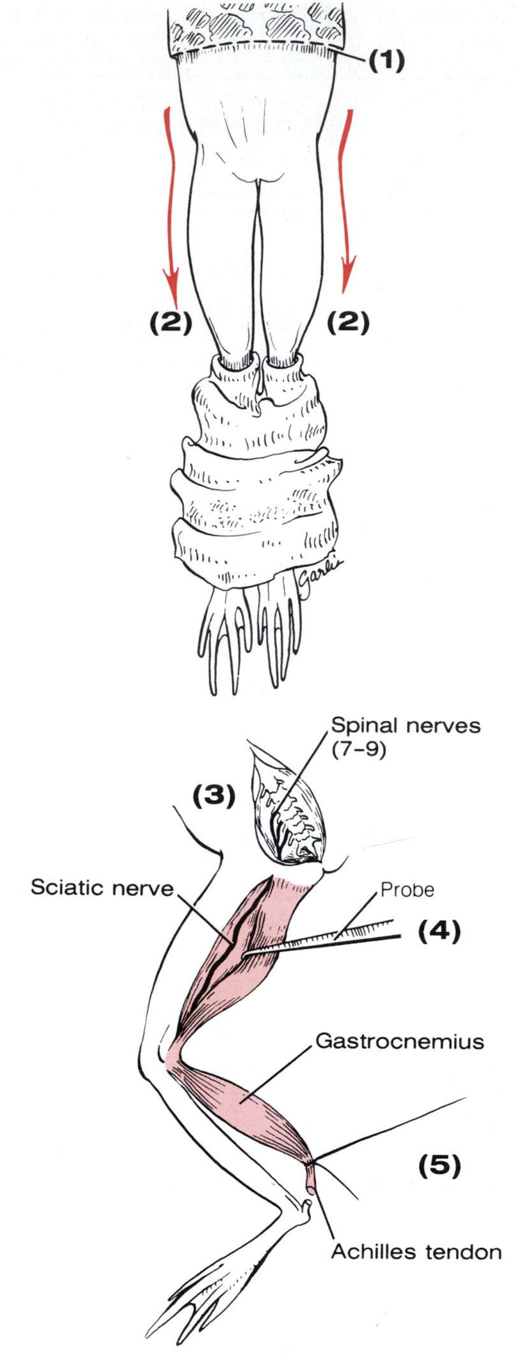

F18.2

Removal of the sciatic nerve and gastrocnemius muscle. **(1)** Cut through the frog's skin around the circumference of the trunk. **(2)** Pull the skin down over the trunk and legs. **(3)** Make a longitudinal cut through the abdominal musculature and expose the roots of the sciatic nerve (arising from spinal nerves 7–9). Ligate the nerve and cut the roots proximal to the ligature. **(4)** Use a glass probe to expose the sciatic nerve beneath the posterior thigh muscles. **(5)** Ligate the Achilles tendon and cut it free distal to the ligature. **(6)** Release the gastrocnemius muscle from the connective tissue of the knee region.

4. Tie a piece of thread around the Achilles tendon of the gastrocnemius muscle, and then cut through the tendon distal to the ligature to free the gastrocnemius muscle from the heel. Using a scalpel, very carefully release the gastrocnemius muscle from the connective tissue in the knee region. At this point you should have completely freed both the gastrocnemius muscle and the sciatic nerve, which innervates it.

5. With glass rods, transfer the muscle to a glass plate or slide, and then attach the slide to a ring stand with a clamp. Allow the end of the sciatic nerve to hang over the free edge of the glass slide, so that it is easily accessible for stimulation. Remember to keep the nerve moist at all times.

6. You are now ready to investigate the response of the sciatic nerve to various stimuli, beginning with electrical stimulation. Using the stimulator and platinum electrodes, stimulate the sciatic nerve with single shocks, gradually increasing the intensity of the stimulus until the threshold stimulus is determined.

(The muscle as a whole will just barely contract at the threshold stimulus.) Record the voltage of this stimulus:

______________ V

Continue to increase the voltage until you find the point beyond which no further increase occurs in the strength of muscle contraction—that is, the point at which the maximal contraction of the muscle is obtained. Record this voltage below.

______________ V

Delivering multiple or repeated shocks to the sciatic nerve causes volleys of impulses in the nerve. Shock the nerve with multiple stimuli. Observe the response of the muscle. How does this response compare with the response to the single electrical shocks?

____

____

____

7. To investigate mechanical stimulation, pinch the free end of the nerve by firmly pressing it between two glass rods or by pinching it with forceps. What is the result?

____

8. Chemical stimulation can be tested by applying a small piece of filter paper saturated with 0.01% hydrochloric acid (HCl) solution to the free end of the nerve. What is the result?

____

Drop a few grains of salt (NaCl) on the free end of the nerve. What is the result?

____

9. Now test thermal stimulation. Wearing the heat-resistant mitts, heat a glass rod for a few moments over a Bunsen burner. Then touch the rod to the free end of the nerve. What is the result?

____

What do these muscle reactions say about the irritability and conductivity of neurons?

____

____

____

Although most neurons within the body are stimulated to the greatest degree by a particular stimulus (in many cases, a chemical neurotransmitter), a variety of other stimuli may trigger nerve impulses, as illustrated by the experimental series just conducted. Generally, no matter what type of stimulus is present, if the affected part responds by becoming activated, it will always react in the same way. Familiar examples are the well-known phenomenon of "seeing stars" when you receive a blow to the head or press on your eyeball (try it), both of which trigger impulses in your optic nerves.

Inhibiting the Nerve Impulse

Numerous physical factors and chemical agents can impair the ability of nerve fibers to function. For example, deep pressure and cold temperature both block nerve impulse transmission by preventing the local blood supply from reaching the nerve fibers. Local anesthetics, alcohol, and numerous other chemicals are also very effective at blocking nerve transmission. Ether, one such chemical blocking agent, will be investigated first.

⚠ Since ether is extremely volatile and explosive, be sure that laboratory fans are on and that *all Bunsen burners are off* during this procedure. *Don safety glasses before beginning this procedure.*

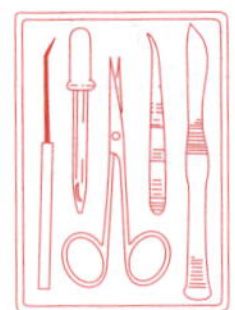

1. Clamp a second glass slide to the ring stand slightly below the first slide of the apparatus setup for the previous experiment. With glass rods, gently position the sciatic nerve on this second slide, allowing a small portion of the nerve's distal end to extend over the edge. Place a piece of absorbent cotton soaked with ether under the midsection of the nerve on the slide, prodding it into position with a glass rod. Using a voltage slightly above the threshold stimulus, stimulate the distal end of the nerve at 2-min intervals until the muscle fails to respond. (If the cotton dries be-

fore this, rewet it with ether using a pipette.) How long did it take for anesthesia to occur?

______________ sec

2. Once anesthesia has occurred, stimulate the nerve beyond the anesthetized area, between the ether-soaked pad and the muscle. What is the result?

____

3. Remove the ether-soaked pad and flush the nerve fibers with saline. Again stimulate the nerve at its distal end at 2-min intervals. How long does it take for recovery?

____

Does ether exert its blocking effect on the nerve fibers *or* on the muscle cells?

________________________ Explain your reasoning.

____

____

If sufficient frogs are available and time allows, you may do the following classic experiment. In the 1800s Claude Bernard described an investigation into the effect of curare on nerve-muscle interaction. *Curare* was used by some South American Indian tribes to tip their arrows. Victims struck with these arrows were paralyzed, but the paralysis was not accompanied by loss of sensation.

1. Prepare another frog as described in steps 1 through 3 of Eliciting a Nerve Impulse. However, in this case position the frog ventral side down on a frog board. In exposing the sciatic nerve, take care not to damage the blood vessels in the thigh region, as the success of the experiment depends on maintaining the blood supply to the muscles of the leg.

2. Expose and gently tie the left sciatic nerve so that it can be lifted away from the muscles of the leg for stimulation. Slip another length of thread under the nerve, and then tie the thread tightly around the thigh muscles to cut off circulation to the leg. The sciatic nerve should be above the thread and *not in* the ligatured tissue. Expose and ligature the sciatic nerve of the *right* leg in the same manner, but this time do *not* ligate the thigh muscles.

⚠ 3. Using a syringe and needle, slowly and carefully inject 1 cc of 0.5% tubocurarine into the dorsal lymph sac of the frog.* The dorsal lymph sacs are located dorsally at the level of the scapulae, so introduce the needle of the syringe just beneath the skin between the scapulae and toward one side of the spinal column. *Handle the tubocurarine very carefully, because it is extremely poisonous.* Do not get any on your skin.

4. Wait 15 min after injection of the tubocurarine to allow it to be distributed throughout the body in the blood and lymphatic stream. Then stimulate electrically the left sciatic nerve. Be careful not to touch any of the other tissues with the electrode. Gradually increasing the voltage, deliver single shocks until the threshold stimulus is determined for this specimen.

Threshold stimulus: ______________ V

Now stimulate the right sciatic nerve with the same voltage intensity. Is there any difference in the reaction of the two muscles?

_______ If so, explain. ______________________

____

____

If you did not find any difference, wait an additional 10 to 15 min and restimulate both sciatic nerves.

What is the result? __________________________

5. To determine the site at which tubocurarine acts, directly stimulate each gastrocnemius muscle. What is the result?

____

____

____

Explain the difference between the responses of the right and left sciatic nerves.

____

____

____

Explain the results when the muscles were stimulated directly.

____

____

____

At what site does tubocurarine (or curare) act?

____

* To obtain 1 cc of the tubocurarine, inject 1 cc of air into the vial through the rubber membrane, and then draw up 1 cc of the chemical into the syringe.

Visualizing the Action Potential with an Oscilloscope

The *oscilloscope* is an instrument that visually displays the rapid but extremely minute changes in voltage that occur during an action potential. The oscilloscope is similar to a TV set in that the screen display is produced by a stream of electrons generated by an electron gun (cathode) at the rear of a tube. The electrons pass through the tube and between two sets of plates that lie alongside the beam pathway. When the electrons reach the fluorescent screen, they create a tiny glowing spot. The plates determine the placement of the glowing spot by controlling the vertical or horizontal sweep of the beam. Vertical movement represents the voltage of the input signal, and horizontal movement indicates the time base. When there is no electrical output signal to the oscilloscope, the electron beam sweeps horizontally (left to right) across the screen, but when the plates are electrically stimulated, the path of electrons is deflected vertically.

In this exercise, a frog's sciatic nerve will be electrically stimulated, and the action potentials generated will be observed on the oscilloscope. The dissected nerve will be placed in contact with two pairs of electrodes—*stimulating* and *recording*. The stimulating electrodes will be used to deliver a pulse of electricity to a point on the sciatic nerve. At another point on the nerve, a pair of recording electrodes connected to the oscilloscope will deliver the current to the plates inside the tube, and the electrical pulse will be recorded on the screen as a vertical deflection, or a *stimulus artifact* (Figure 18.3a). As the nerve is stimulated with increasingly higher voltage, the stimulus artifact increases in amplitude as well. When the stimulus voltage reaches a high enough level (threshold), an action potential will be generated by the nerve, and a *second* vertical deflection will appear on the screen, approximately 2 milliseconds after the stimulus artifact (Figure 18.3b–d). This second deflection reports the potential difference between the two recording electrodes—that is, between the first recording electrode which has already depolarized (and is in the process of repolarizing as the action potential travels along the nerve) and the second recording electrode.

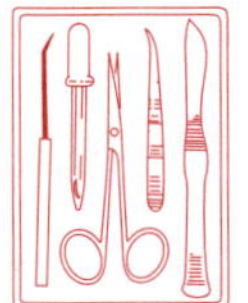

1. Obtain a nerve chamber, an oscilloscope, a stimulator, frog Ringer's solution (room temperature), a dissecting needle, and glass probes. Set up the experimental apparatus as illustrated in Figure 18.4. Connect the two stimulating electrodes to the output terminals of the stimulator and the two recording electrodes to the preamplifier of the oscilloscope.

2. Obtain another pithed frog, and prepare one of its sciatic nerves for experimentation as indicated on page 182 in steps 1 through 3 under Eliciting a Nerve Impulse. While working, be careful not to touch the nerve with your fingers, and do not allow the nerve to touch the frog's skin.

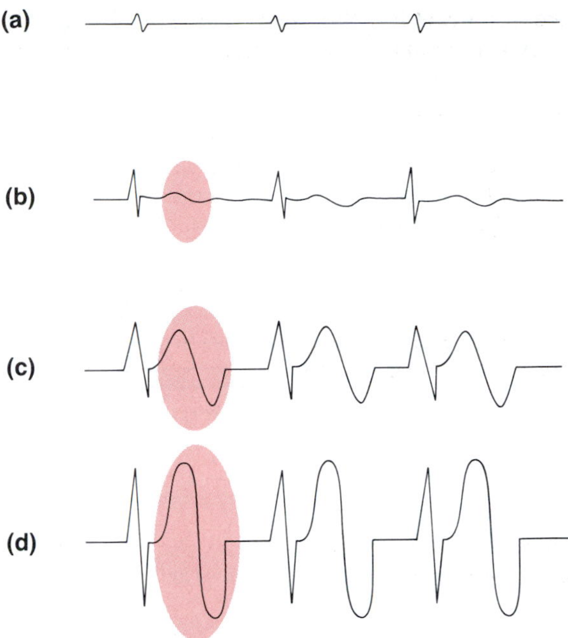

F18.3

Oscilloscope scans of nerve stimulation using stimuli with increasing intensities. The first action potential in each scan is circled. **(a)** Stimulus artifacts only; no action potential produced. Subthreshold stimulation. **(b)** Threshold stimulation. **(c)** Submaximal stimulation. **(d)** Maximal stimulus.

3. When you have freed the sciatic nerve to the knee region with the glass probe, slip another thread length beneath that end of the nerve and make a ligature. Cut the nerve distal to this tied thread and then carefully lift the cut nerve away from the thigh of the frog by holding the threads at the nerve's proximal and distal ends. Place the nerve in the nerve chamber so that it rests across all four electrodes (the two stimulating and two recording electrodes) as shown in Figure 18.4. Flush the nerve with room temperature frog Ringer's solution.

4. Adjust the horizontal sweep according to the instructions given in the manual or by your instructor, and set the stimulator duration, frequency, and amplitude to their lowest settings.

5. Begin to stimulate the nerve with single stimuli, slowly increasing the voltage until a threshold stimulus is achieved. The action potential will appear as a small rounded "hump" immediately following the stimulus artifact. Record the voltage of the threshold stimulus:

 V

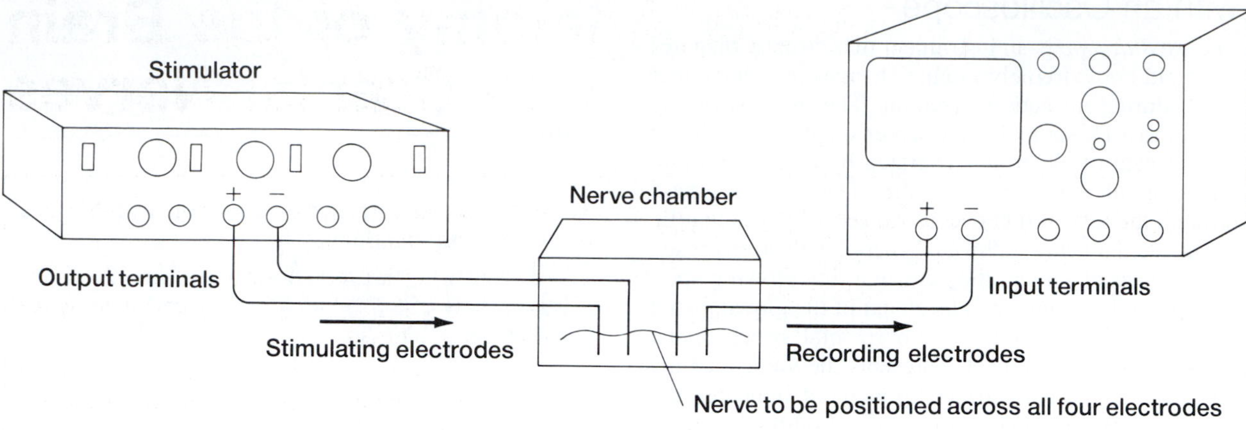

F18.4

Setup for oscilloscope visualization of action potentials in a nerve.

6. Flush the nerve with the Ringer's solution and continue to increase the voltage, watching as the vertical deflections produced by the action potentials become diphasic (show both upward and downward vertical deflections). Record the voltage at which the action potential reaches its maximal amplitude; this is the maximal stimulus:

______________ V

7. Set the stimulus voltage at a level just slightly lower than the maximal stimulus and gradually increase the frequency of stimulation. What is the effect on the size (amplitude) of the action potential?

____

8. Flush the nerve with saline once again, and allow it to sit for a few minutes while you obtain a bottle of Ringer's solution from the ice bath. Repeat steps 5 and 6 while your partner continues to flush the nerve preparation with the cold saline. Record the threshold and maximal stimuli and watch the oscilloscope pattern carefully to detect any differences in the velocity or speed of conduction from what was seen previously.

Threshold stimulus ______________ V

Maximal stimulus ______________ V

9. Flush the nerve preparation with room temperature Ringer's solution again and then gently lift the nerve by its attached threads and turn it around so that the end formerly resting on the stimulating electrodes now rests on the recording electrodes and vice versa. Stimulate the nerve. Is the impulse conducted in the opposite direction?

____

10. Dispose of the frog remains and gloves in the appropriate containers, and return your equipment to the proper supply area.

EXERCISE 19

Gross Anatomy of the Brain and Cranial Nerves

OBJECTIVES

1. To identify the following brain structures on a dissected specimen, human brain model (or slices), or appropriate diagram, and to state their functions:
   - *Cerebral hemisphere structures:* lobes, important fissures, lateral ventricles, basal nuclei, corpus callosum, fornix, septum pellucidum
   - *Diencephalon structures:* thalamus, intermediate mass, hypothalamus, optic chiasma, pituitary gland, mammillary bodies, pineal body, choroid plexus of the third ventricle, interventricular foramen
   - *Brain stem structures:* corpora quadrigemina, cerebral aqueduct, cerebral peduncles of the midbrain, pons, medulla, fourth ventricle
   - *Cerebellum structures:* cerebellar hemispheres, vermis, arbor vitae
2. To describe the composition of gray and white matter.
3. To locate the well-recognized functional areas of the human cerebral hemispheres.
4. To define *gyri* and *fissures* (*sulci*).
5. To identify the three meningeal layers and state their function, and to locate the falx cerebri, falx cerebelli, and tentorium cerebelli.
6. To state the function of the arachnoid villi and dural sinuses.
7. To discuss the formation, circulation, and drainage of cerebrospinal fluid.
8. To identify at least four pertinent anatomical differences between the human brain and that of the sheep (or other mammal).
9. To identify the cranial nerves by number and name on an appropriate model or diagram, stating the origin and function of each.

MATERIALS

Human brain model (dissectible)
3-D model of ventricles
Preserved human brain (if available)
Coronally sectioned human brain slice (if available)
Preserved sheep brain (meninges and cranial nerves intact)
Dissecting tray and instruments
Protective skin cream or disposable gloves
Materials as needed for cranial nerve testing
The Human Nervous System: The Brain and Cranial Nerves videotape*

See Appendix D, Exercise 19 for links to A.D.A.M. Standard.

See Appendix E, Exercise 19 for links to *Anatomy and PhysioShow: The Videodisc.*

*Available to qualified adopters from Benjamin/Cummings.

When viewed alongside all nature's animals, humans are indeed unique, and the key to their uniqueness is found in the brain. Only in humans has the brain region called the cerebrum become so elaborated and grown out of proportion that it overshadows other brain areas. Other animals are primarily concerned with informational input and response for the sake of survival and preservation of the species, but human beings devote considerable time to nonsurvival ends. They are the only animals who manipulate abstract ideas and search for knowledge for its own sake, who are capable of emotional response and artistic creativity, or who can anticipate the future and guide their lives according to ethical and moral values. For all this, humans can thank their overgrown cerebrum (cerebral hemispheres).

We can be considered composite reflections of our brain's experience. If all past sensory input could mysteriously and suddenly be "erased," we would be unable to walk, talk, or communicate in any manner. Spontaneous movement would occur, as in a fetus, but no voluntary integrated function of any type would be possible. Clearly we would cease to be the same individuals.

Because of the complexity of the nervous system, its anatomical structures are usually considered in terms of two principal divisions: the **central nervous system (CNS)** and the **peripheral nervous system (PNS).** The central nervous system consists of the brain and spinal cord, which primarily interpret incoming sensory information and issue instructions based on past experience. The peripheral nervous system consists of the cranial and spinal nerves, ganglia, and sensory receptors. These structures serve as communication lines as they carry impulses—from the sensory receptors to the CNS and from the CNS to the appropriate glands or muscles.

In this exercise both CNS (brain) and PNS (cranial nerves) structures will be studied because of their close anatomical relationship.

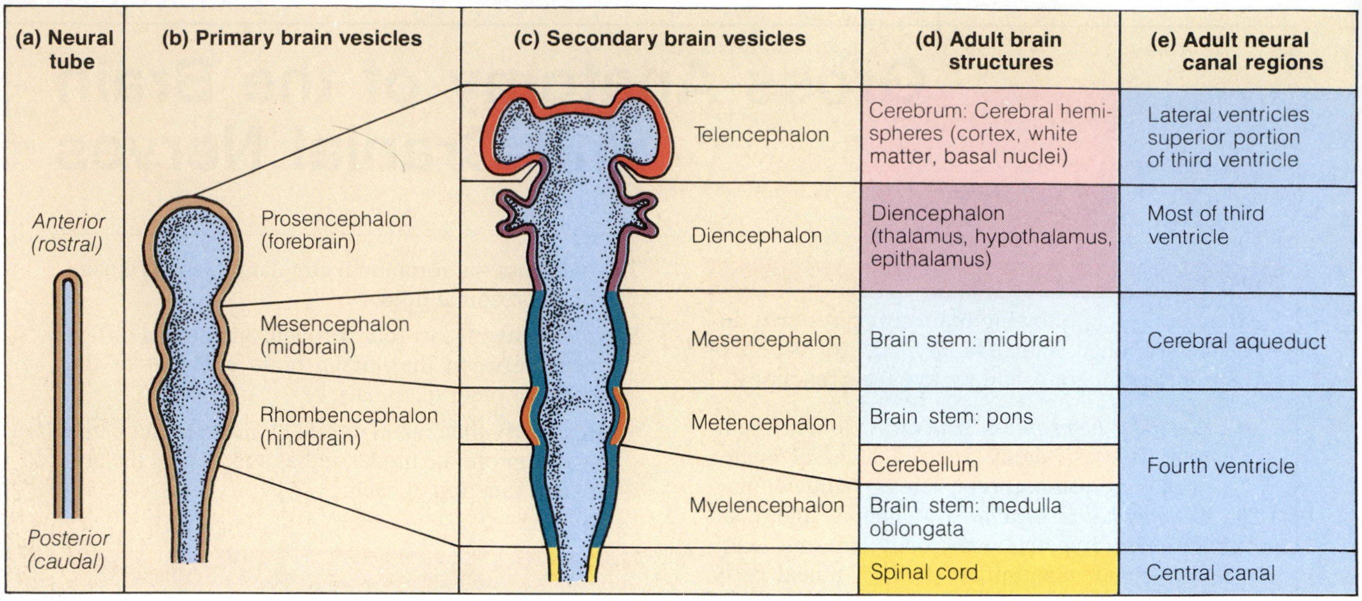

(a) Neural tube	(b) Primary brain vesicles	(c) Secondary brain vesicles	(d) Adult brain structures	(e) Adult neural canal regions
Anterior (rostral)	Prosencephalon (forebrain)	Telencephalon	Cerebrum: Cerebral hemispheres (cortex, white matter, basal nuclei)	Lateral ventricles superior portion of third ventricle
		Diencephalon	Diencephalon (thalamus, hypothalamus, epithalamus)	Most of third ventricle
	Mesencephalon (midbrain)	Mesencephalon	Brain stem: midbrain	Cerebral aqueduct
	Rhombencephalon (hindbrain)	Metencephalon	Brain stem: pons	Fourth ventricle
			Cerebellum	
		Myelencephalon	Brain stem: medulla oblongata	
Posterior (caudal)			Spinal cord	Central canal

F19.1

Embryonic development of the human brain. (a) The neural tube becomes subdivided into **(b)** the primary brain vesicles, which subsequently form **(c)** the secondary brain vesicles, which differentiate into **(d)** the adult brain structures. **(e)** The adult structures derived from the neural canal.

THE HUMAN BRAIN

During embryonic development of all vertebrates, the CNS first makes its appearance as a simple tubelike structure, the **neural tube,** that extends down the dorsal median plane. By the fourth week, the human brain begins to form as an expansion of the anterior or rostral end of the neural tube (the end toward the head). Shortly thereafter, constrictions appear, dividing the developing brain into three major regions—**forebrain, midbrain,** and **hindbrain** (Figure 19.1). The remainder of the neural tube becomes the spinal cord.

During fetal development, two anterior outpocketings extend from the forebrain and grow rapidly to form the cerebral hemispheres. Because of space restrictions imposed by the skull, the cerebral hemispheres are forced to grow posteriorly and inferiorly, and finally end up enveloping and obscuring the rest of the forebrain and most midbrain structures. Somewhat later in development, the dorsal portion of the hindbrain also enlarges to produce the cerebellum. The central canal of the neural tube, which remains continuous throughout the brain and cord, becomes enlarged in four regions of the brain, forming chambers called **ventricles** (see Figure 19.8a, p. 182).

External Anatomy

Generally, the brain is considered in terms of four major regions: the cerebral hemispheres, diencephalon, brain stem, and cerebellum. The relationship between these four anatomical regions and the structures of the forebrain, midbrain, and hindbrain is also outlined in Figure 19.1.

CEREBRAL HEMISPHERES The **cerebral hemispheres** are the most superior portion of the brain (Figure 19.2). Their entire surface is thrown into elevated ridges of tissue called **gyri** that are separated by depressed areas called **fissures** or **sulci.** Of the two types of depressions, the fissures are deeper. Many of the fissures and gyri are important anatomical landmarks.

The cerebral hemispheres are divided by a single deep fissure, the **longitudinal fissure.** The **central sulcus** divides the **frontal lobe** from the **parietal lobe,** and the **lateral sulcus** separates the **temporal lobe** from the parietal lobe. The **parieto-occipital sulcus,** which divides the **occipital lobe** from the parietal lobe, is not visible externally. Notice that the cerebral hemisphere lobes are named for the cranial bones that lie over them.

Some important functional areas of the cerebral hemispheres have also been located (Figure 19.2d). The **primary somatosensory area** is located in the **postcentral gyrus** of the parietal lobe. Impulses traveling from the body's sensory receptors (such as those for pressure, pain, and temperature) are localized in this area of the brain. ("This information is from my big toe.") Immediately posterior to the primary somatosensory area is the **somatosensory association area,** in which the meaning of incoming stimuli is analyzed. ("Ouch! I have a *pain* there.") Thus, the somatosensory association area allows you to become aware of pain, coldness, a light touch, and the like.

Impulses from the special sense organs are interpreted in other specific areas also noted in Figure 19.2d. For example, the visual areas are in the posterior portion of the occipital lobe and the auditory area is located in the temporal lobe in the gyrus bordering the lateral sulcus. The olfactory area is deep within the temporal lobe

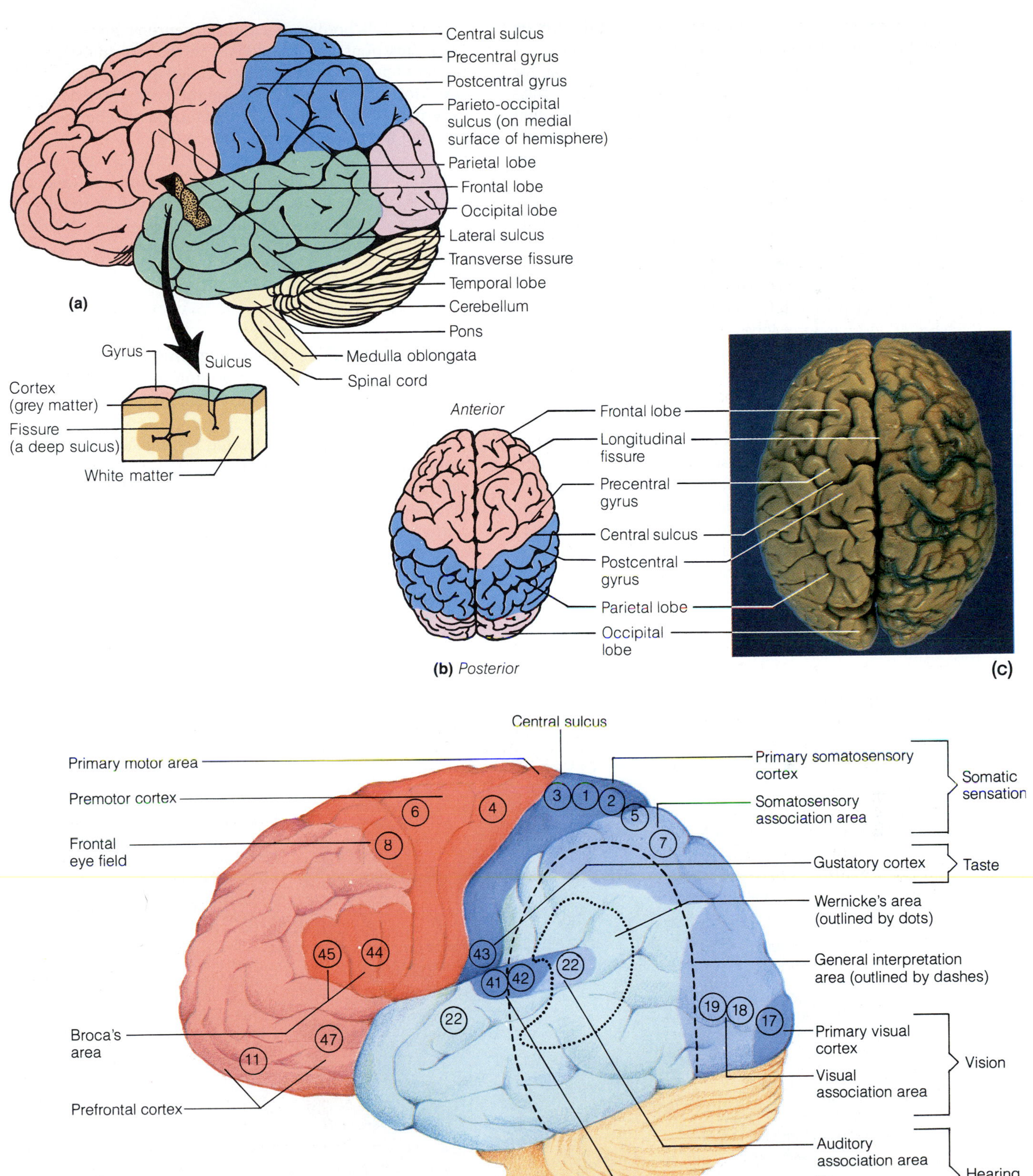

F19.2

External structure (lobes and fissures) of the cerebral hemispheres. (a) Left lateral view of the brain. **(b)** Superior view. **(c)** Photograph of the superior aspect of the human brain. **(d)** Functional areas of the left cerebral cortex. The olfactory area, which is deep to the temporal lobe on the medial hemispheric surface, is not identified. Numbers indicate brain regions plotted by the Brodman system.

along its medial surface, in a region called the **uncus** (see Figure 19.4b).

The **primary motor area,** which is responsible for conscious or voluntary movement of the skeletal muscles, is located in the **precentral gyrus** of the frontal lobe. A specialized motor speech area called **Broca's area** is found at the base of the precentral gyrus just above the lateral sulcus. Damage to this area (which is located only in one cerebral hemisphere, usually the left) reduces or eliminates the ability to articulate words. Areas involved in intellect, complex reasoning, and personality lie in the anterior portions of the frontal lobes, in a region called the **prefrontal cortex.**

A rather poorly defined region at the junction of the parietal and temporal lobes is Wernicke's area, an area in which unfamiliar words are sounded out. Like Broca's area, Wernicke's area is located in one cerebral hemisphere only, typically the left.

Although there are many similar functional areas in both cerebral hemispheres, such as motor and sensory areas, each hemisphere is also a "specialist" in certain ways. For example, the left hemisphere is the "language brain" in most of us, because it houses centers associated with language skills and speech. The right hemisphere is more specifically concerned with abstract, conceptual, or spatial processes—skills associated with artistic or creative pursuits.

The cell bodies of cerebral neurons involved in these functions are found only in the outermost gray matter of the cerebrum, the area called the **cerebral cortex.** Most of the balance of cerebral tissue—the deeper **cerebral white matter**—is composed of fiber tracts carrying impulses to or from the cortex.

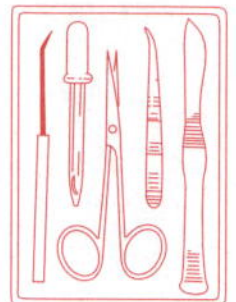

Using a model of the human brain (and a preserved human brain, if available), identify the areas and structures of the cerebral hemispheres described above.

DIENCEPHALON The **diencephalon,** sometimes considered the most superior portion of the brain stem, is embryologically part of the forebrain, along with the cerebral hemispheres.

Turn the brain model so the ventral surface of the brain can be viewed. Using Figure 19.3 as a guide, start superiorly and identify the externally visible structures that mark the position of the floor of the diencephalon. These are the **olfactory bulbs** and **tracts, optic nerves, optic chiasma** (where the fibers of the optic nerves partially cross over), **optic tracts, pituitary gland,** and **mammillary bodies.**

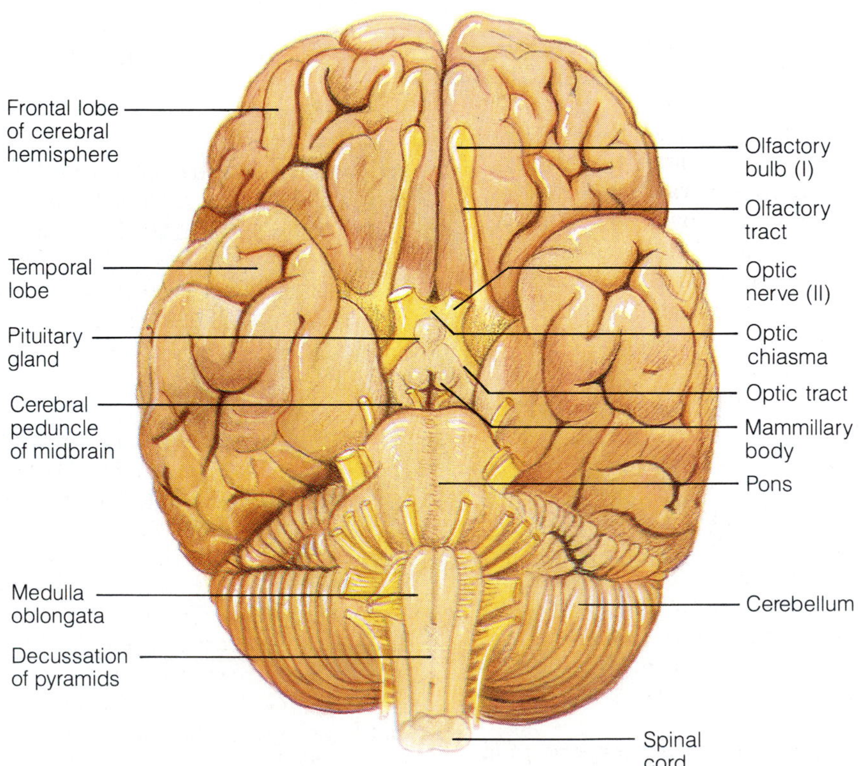

F19.3

Ventral aspect of the human brain, showing the three regions of the brain stem. Only a small portion of the midbrain can be seen; the rest is surrounded by other brain regions. (See corresponding Plate A in the Human Anatomy Atlas.)

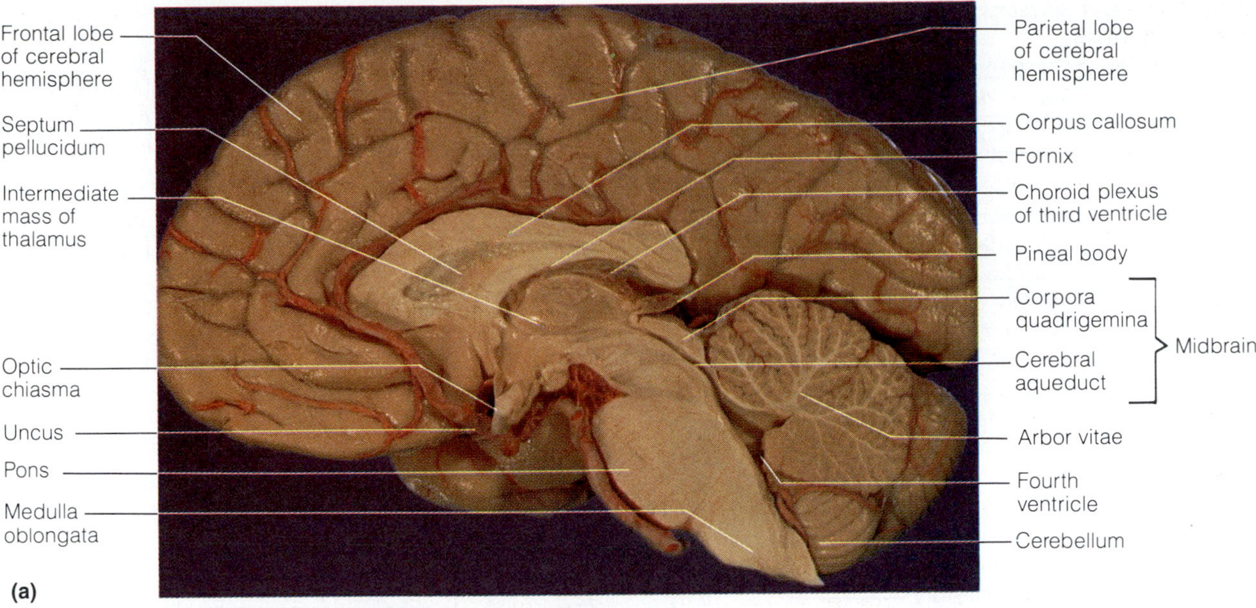

F19.4

Diencephalon and brain stem structures as seen in a midsagittal section of the brain. (a) Photograph.

BRAIN STEM Continue inferiorly to identify the **brain stem** structures—the **cerebral peduncles** (fiber tracts in the **midbrain** connecting the pons below with cerebrum above), the pons, and the medulla oblongata. *Pons* means "bridge," and the **pons** consists primarily of motor and sensory fiber tracts connecting the brain with lower CNS centers. The lowest brain stem region, the **medulla,** is also composed primarily of fiber tracts. You can see the **decussation of pyramids,** a crossover point for the major motor tract (pyramidal tract) descending from the motor areas of the cerebrum to the cord, on the medulla's anterior surface. The medulla also houses many vital autonomic centers involved in the control of heart rate, respiratory rhythm, and blood pressure as well as involuntary centers involved in the initiation of vomiting, swallowing, and so on.

CEREBELLUM

1. Turn the brain model so you can see the dorsal aspect. Identify the large cauliflowerlike **cerebellum,** which projects dorsally from under the occipital lobe of the cerebrum. Note that, like the cerebrum, the cerebellum has two major hemispheres and a convoluted surface (see Figure 19.6). It also has an outer cortex made up of gray matter with an inner region of white matter.

2. Remove the cerebellum to view the **corpora quadrigemina,** located on the posterior aspect of the midbrain, a brain stem structure. The two superior prominences are the **superior colliculi** (visual reflex centers); the two smaller inferior prominences are the **inferior colliculi** (auditory reflex centers).

Internal Anatomy

The deeper structures of the brain have also been well mapped. Like the external structures, these can be studied in terms of the four major regions.

CEREBRAL HEMISPHERES

1. Take the brain model apart so you can see a median sagittal view of the internal brain structures (Figure 19.4). Observe the model closely to see the extent of the outer cortex (gray matter), which contains the cell bodies of cerebral neurons. [The pyramidal cells of the cerebral motor cortex (studied in Exercise 17, pp. 162 and 164) are representative of the neurons seen in the precentral gyrus.]

2. Observe the deeper area of white matter, which is composed of fiber tracts. The fiber tracts found in the cerebral hemisphere white matter are called *association tracts* if they connect two portions of the same hemisphere, *projection tracts* if they run between the cerebral cortex and the lower brain or spinal cord, and *commissures* if they run from one hemisphere to another. Observe the large **corpus callosum,** the major commissure connecting the cerebral hemispheres. The corpus callosum arches above the structures of the diencephalon and roofs over the lateral ventricles. Note also the **fornix,** a bandlike fiber tract concerned with olfaction as well as limbic system functions, and the membranous **septum pellucidum,** which separates the lateral ventricles of the cerebral hemispheres.

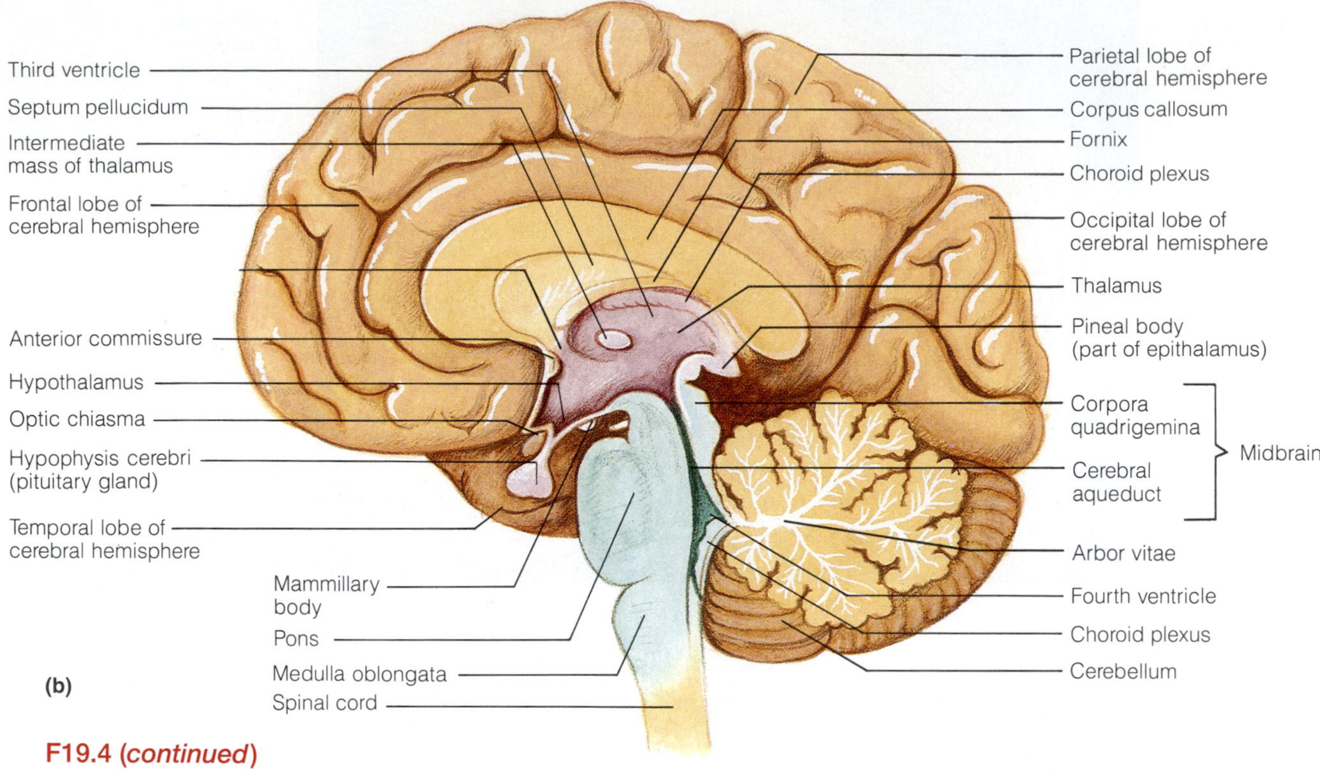

F19.4 ***(continued)***

Diencephalon and brain stem structures in a midsagittal section of the brain. (b) Diagrammatic view.

3. In addition to the gray matter of the cerebral cortex, there are several "islands" of gray matter (clusters of neuron cell bodies) called **nuclei** buried deep within the white matter of the cerebral hemispheres. One important group of cerebral nuclei, called the **basal nuclei,** flank the lateral and third ventricles. You can see the basal nuclei if you have an appropriate dissectible model or a coronally or cross-sectioned human brain slice. Otherwise, Figure 19.5 will suffice.

The basal nuclei, which are important subcortical motor nuclei (and part of the so-called *extrapyramidal system*), are involved in regulating voluntary motor activities. The most important of them are the arching, comma-shaped **caudate nucleus,** the **claustrum,** the **amygdaloid nucleus** (located at the tip of the caudate nucleus), and the **lentiform nucleus,** which is composed of the **putamen** and **globus pallidus nuclei.** The **corona radiata,** a spray of projection fibers coursing down from the precentral (motor) gyrus, combines with sensory fibers traveling to the sensory cortex to form a broad band of fibrous material called the **internal capsule.** The internal capsule passes between the diencephalon and the basal nuclei, and gives these basal nuclei a striped appearance. This is why the caudate nucleus and the lentiform nucleus are sometimes referred to collectively as the **corpus striatum,** or "striped body."

4. Examine the relationship of the lateral ventricles and corpus callosum to the diencephalon structures; that is, hypothalamus, thalamus, and third ventricle—from the cross-sectional viewpoint (see Figure 19.5b).

DIENCEPHALON

1. The major internal structures of the diencephalon are the thalamus, hypothalamus, and epithalamus (Figure 19.4). The **thalamus** consists of two large lobes of gray matter that laterally enclose the shallow third ventricle of the brain. A slender stalk of thalamic tissue, the **intermediate mass,** or **massa intermedia,** connects the two thalamic lobes and bridges the ventricle. The thalamus is a major integrating and relay station for sensory impulses passing upward to the cortical sensory areas for localization and interpretation. Locate also the **interventricular foramen** *(foramen of Monro),* a tiny orifice connecting the third ventricle with the lateral ventricle on the same side.

2. The **hypothalamus** makes up the floor and the inferolateral walls of the third ventricle. It is an important autonomic center involved in regulation of body temperature, water balance, and fat and carbohydrate metabolism as well as in many other activities and drives (sex, hunger, thirst). Locate again the pituitary gland, or **hypophysis,** which hangs from the anterior floor of the hypothalamus by a slender stalk, the **infundibulum.** (The pituitary gland is usually not present in preserved brain specimens.) In life, the pituitary rests in the hypophyseal fossa of the sella turcica of the sphenoid bone. Its function is discussed in Exercise 27.

Anterior to the pituitary, identify the optic chiasma portion of the optic pathway to the brain. The **mammillary bodies,** relay stations for olfaction, bulge exteriorly from the floor of the hypothalamus just posterior to the pituitary gland.

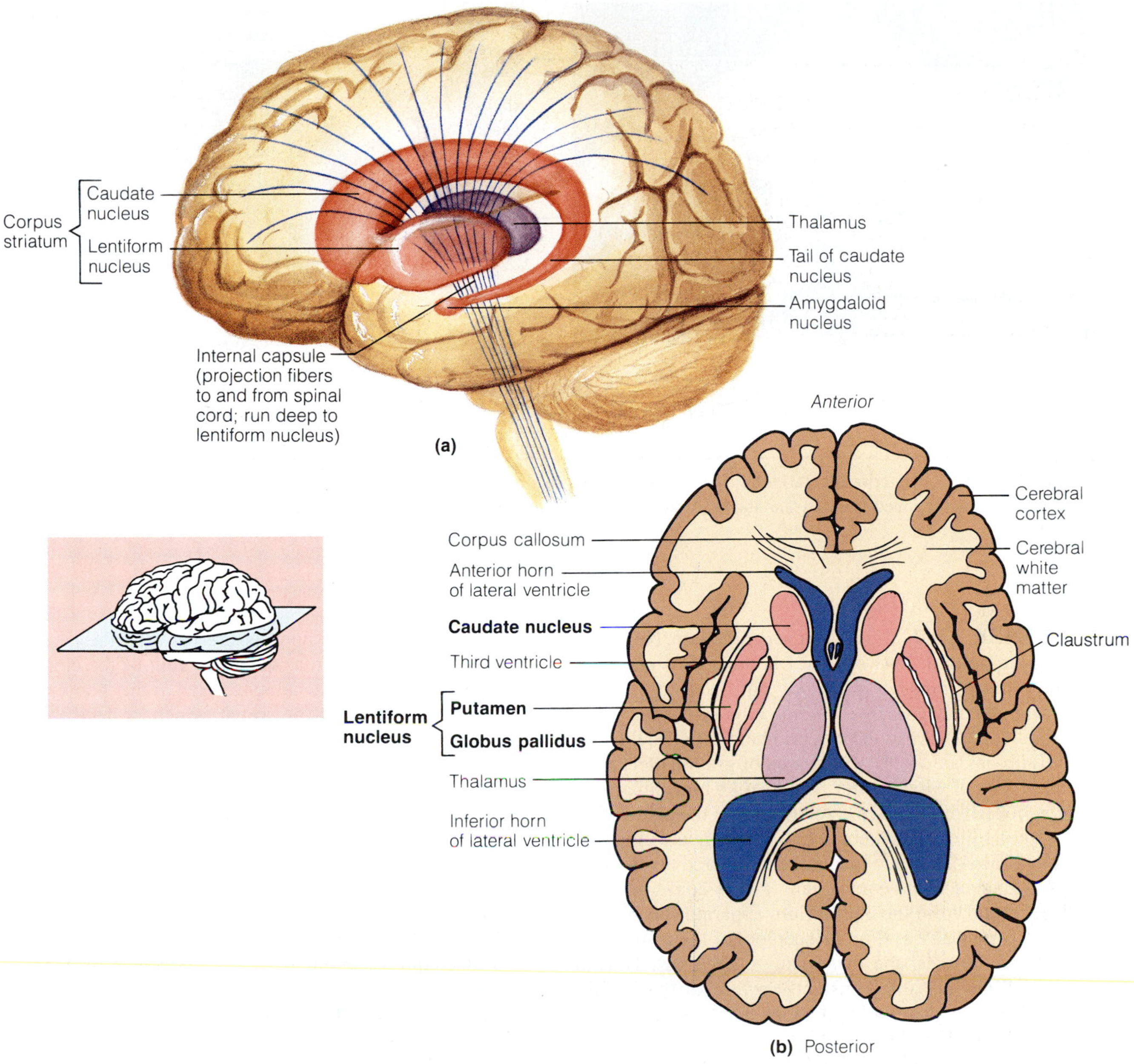

F19.5

Basal nuclei. (a) Three-dimensional view of the basal nuclei showing their positions within the cerebrum. **(b)** A transverse section of the cerebrum and diencephalon showing the relationship of the basal nuclei to the thalamus and the lateral and third ventricles.

3. The **epithalamus** forms the roof of the third ventricle and is the most dorsal portion of the diencephalon. Important structures in the epithalamus are the **pineal body,** or **gland** (a neuroendocrine structure), and the **choroid plexus** of the third ventricle. The choroid plexuses, knotlike collections of capillaries within each ventricle, form the cerebrospinal fluid.

BRAIN STEM

1. Now trace the short midbrain from the mammillary bodies to the rounded pons below. Continue to refer to Figure 19.4. The **cerebral aqueduct** is a slender canal traveling through the midbrain; it connects the third ventricle to the fourth ventricle in the hindbrain below. The cerebral peduncles and the rounded corpora quadrigemina make up the midbrain tissue anterior and posterior (respectively) to the cerebral aqueduct.

2. Locate the hindbrain structures. Trace the rounded pons to the medulla oblongata below, and identify the fourth ventricle posterior to these structures. Attempt to identify the single median aperture and the two lateral apertures, three orifices found in the walls of the fourth ventricle. These apertures serve as conduits for cerebrospinal fluid to circulate into the subarachnoid space from the fourth ventricle.

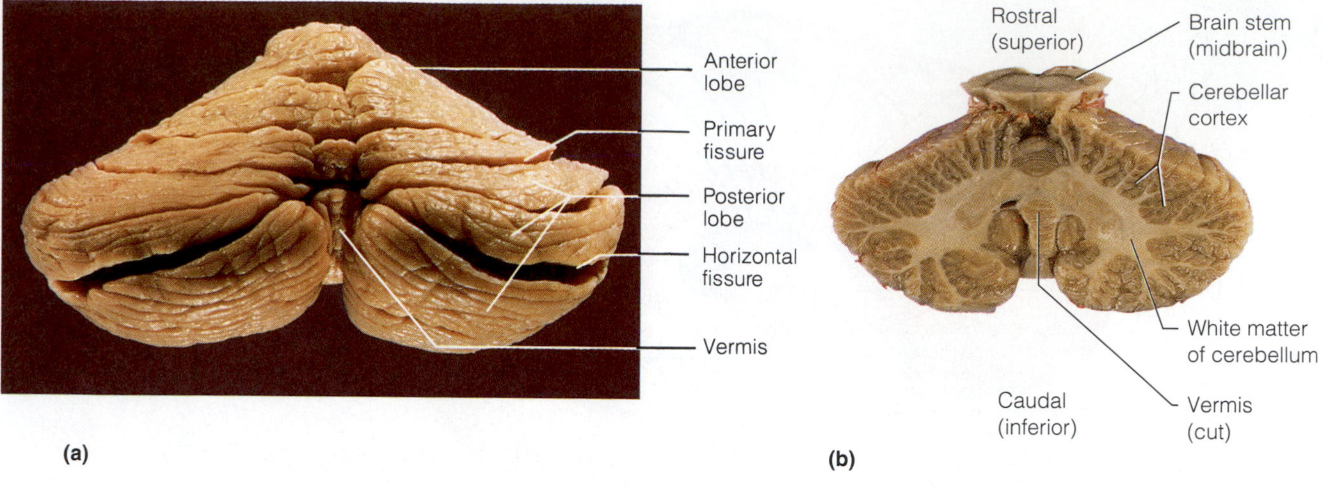

F19.6

Cerebellum. **(a)** Posterior (dorsal) view. **(b)** The cerebellum, sectioned to reveal its cortex and medullary regions. (Note that the cerebellum is sectioned frontally and the brain stem is sectioned horizontally in this posterior view.)

CEREBELLUM Examine the cerebellum. Notice that it is composed of two lateral hemispheres each with three lobes [*anterior, posterior,* and (a deep) *flocculonodular*] connected by a midline lobe called the **vermis** (Figure 19.6). As in the cerebral hemispheres, the cerebellum has an outer cortical area of gray matter and an inner area of white matter. The treelike branching of the cerebellar white matter is referred to as the **arbor vitae,** or tree of life. The cerebellum is concerned with unconscious coordination of skeletal muscle activity and control of balance and equilibrium. Fibers converge on the cerebellum from the equilibrium apparatus of the inner ear, visual pathways, proprioceptors of the tendons and skeletal muscles, and from many other areas. Thus the cerebellum remains constantly aware of the position and state of tension of the various body parts.

Meninges of the Brain

The brain (and spinal cord) are covered and protected by three connective tissue membranes called **meninges** (Figure 19.7). The outermost meninx is the leathery **dura mater,** a double-layered membrane. One of its layers (the *periosteal layer*) is attached to the inner surface of the skull, forming the periosteum. The other (the *meningeal layer*) forms the outermost brain covering and is continuous with the dura mater of the spinal cord.

The dural layers are fused together except in three places where the inner membrane extends inward to form a septum that secures the brain to structures inside the cranial cavity. One such extension, the **falx cerebri,** dips into the longitudinal fissure between the cerebral hemispheres to attach to the crista galli of the ethmoid bone of the skull. The cavity created at this point is the large **superior sagittal sinus**, which collects blood draining from the brain tissue. The **falx cerebelli,** separating the two cerebellar hemispheres, and the **tentorium cerebelli,** separating the cerebrum from the cerebellum below, are two other important inward folds of the inner dural membrane.

The middle meninx, the weblike **arachnoid mater,** or simply the **arachnoid,** underlies the dura mater and is partially separated from it by the **subdural space.** Threadlike projections bridge the **subarachnoid space** to attach the arachnoid to the innermost meninx, the **pia mater.** The delicate pia mater is highly vascular and clings tenaciously to the surface of the brain, following its convolutions.

In life, the subarachnoid space is filled with cerebrospinal fluid. Specialized projections of the arachnoid tissue called **arachnoid villi** protrude through the dura mater to allow the cerebrospinal fluid to drain back into the venous circulation via the superior sagittal sinus and other dural sinuses.

Meningitis, inflammation of the meninges, is a serious threat to the brain because of the intimate association between the brain and meninges. Should infection spread to the neural tissue of the brain itself, life-threatening **encephalitis** may occur. Meningitis is often diagnosed by taking a sample of cerebrospinal fluid from the subarachnoid space. ■

(*Text continues on p. 183*)

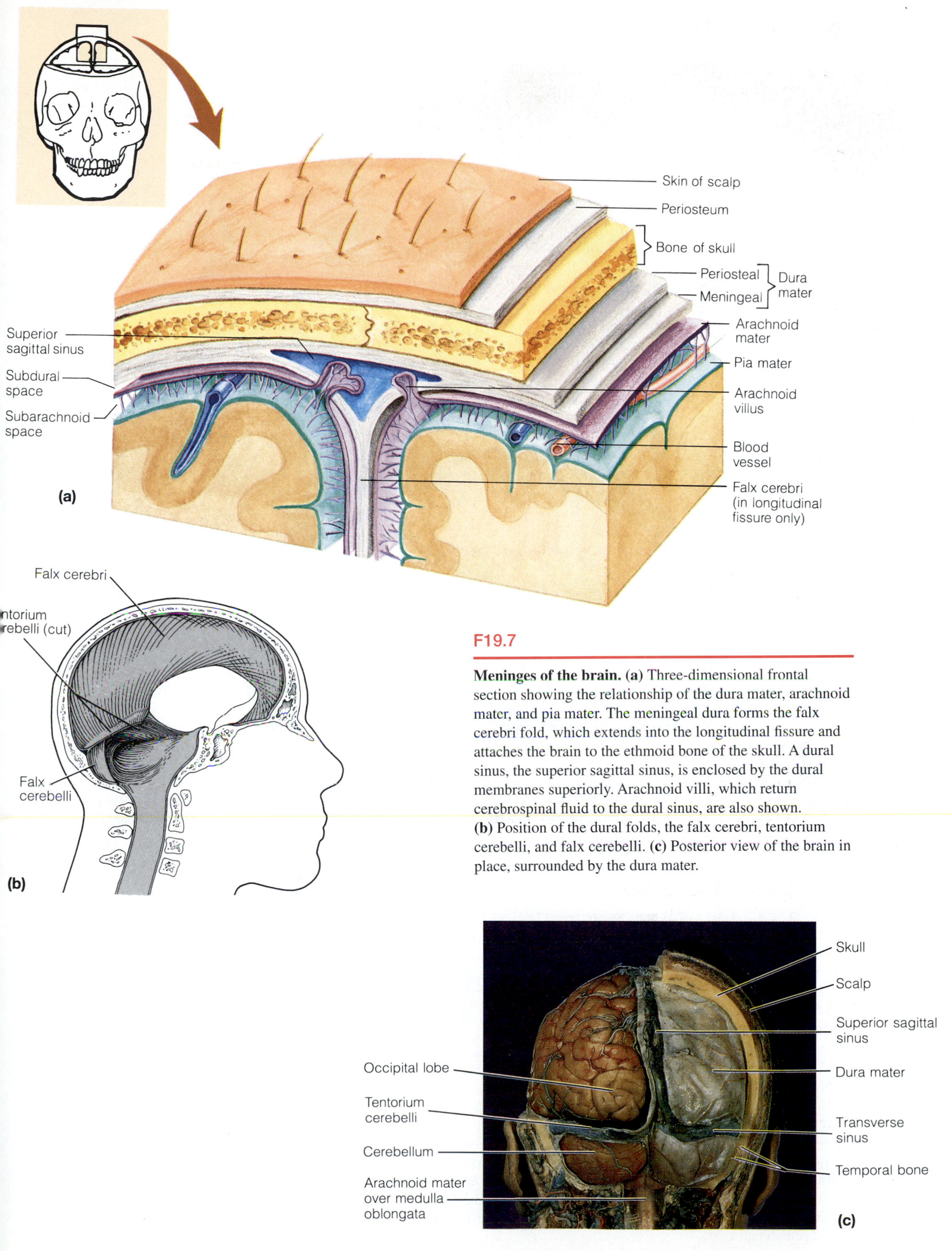

F19.7

Meninges of the brain. (a) Three-dimensional frontal section showing the relationship of the dura mater, arachnoid mater, and pia mater. The meningeal dura forms the falx cerebri fold, which extends into the longitudinal fissure and attaches the brain to the ethmoid bone of the skull. A dural sinus, the superior sagittal sinus, is enclosed by the dural membranes superiorly. Arachnoid villi, which return cerebrospinal fluid to the dural sinus, are also shown. **(b)** Position of the dural folds, the falx cerebri, tentorium cerebelli, and falx cerebelli. **(c)** Posterior view of the brain in place, surrounded by the dura mater.

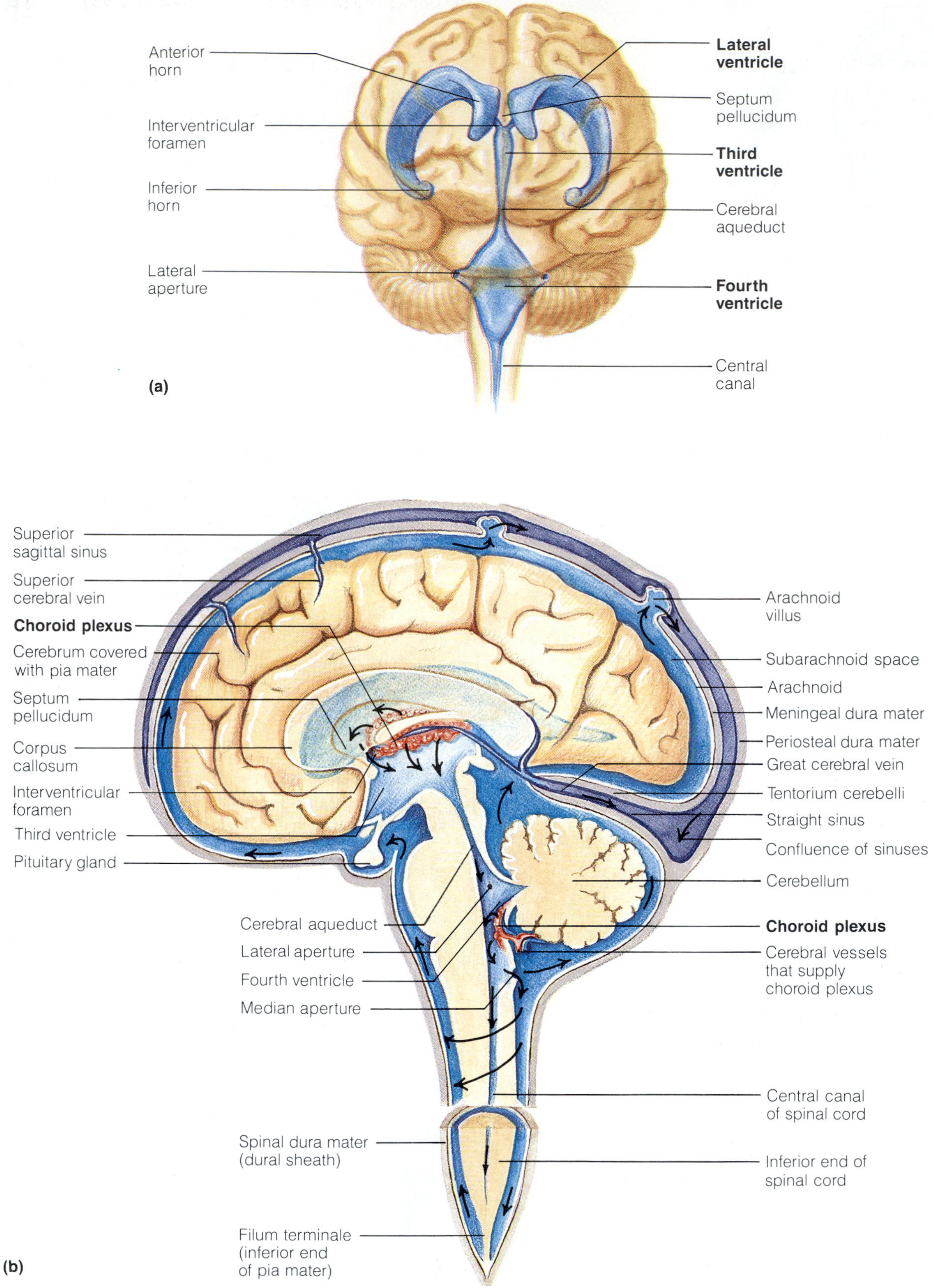

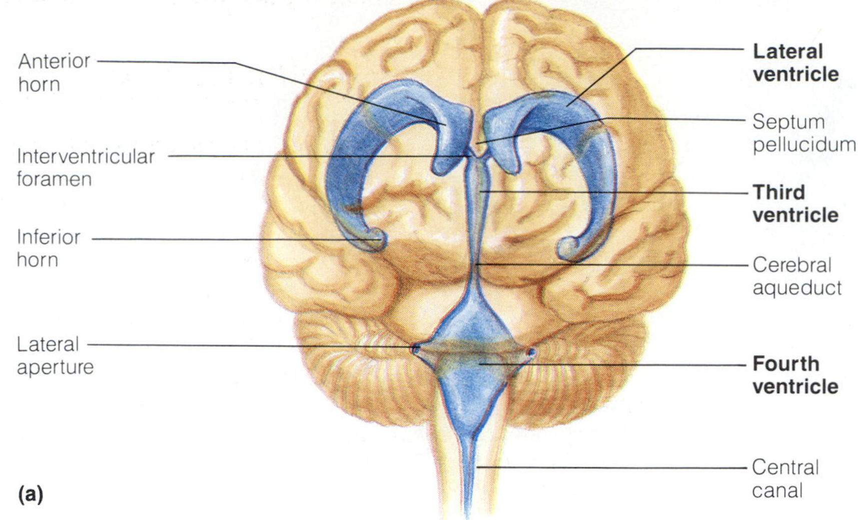

F19.8

Location and circulatory pattern of cerebrospinal fluid. (a) Anterior view; note that different regions of the large lateral ventricles are indicated by the terms *anterior horn, posterior horn,* and *inferior horn.* **(b)** The cerebrospinal fluid flows from the lateral ventricles, through the interventricular foramina, into the third ventricle, and then into the fourth ventricle via the cerebral aqueduct. (The relative position of the right lateral ventricle is indicated by the pale blue area deep to the corpus callosum and septum pellucidum.)

Cerebrospinal Fluid

The cerebrospinal fluid, much like plasma in composition, is continually formed by the **choroid plexuses**, small capillary knots hanging from the roof of the ventricles of the brain. The cerebrospinal fluid in and around the brain forms a watery cushion that protects the delicate brain tissue against blows to the head.

Within the brain, the cerebrospinal fluid circulates from the two lateral ventricles (in the cerebral hemispheres) into the third ventricle via the **interventricular foramina,** and then through the cerebral aqueduct of the midbrain into the fourth ventricle in the hindbrain (Figure 19.8). Some of the fluid reaching the fourth ventricle continues down the central canal of the spinal cord, but the bulk of it circulates into the subarachnoid space, exiting through the three foramina in the walls of the fourth ventricle (the two lateral and the single median apertures). The fluid returns to the blood in the dural sinuses via the arachnoid villi.

Ordinarily, cerebrospinal fluid forms and drains at a constant rate. However, under certain conditions—for example, obstructed drainage or circulation resulting from tumors or anatomical deviations—the cerebrospinal fluid accumulates and exerts increasing pressure on the brain which, uncorrected, causes neurological damage in adults. In infants, **hydrocephalus** (literally, water on the brain) is indicated by a gradually enlarging head. Since the infant's skull is still flexible and contains fontanels, it can expand to accommodate the increasing size of the brain. ■

CRANIAL NERVES

The **cranial nerves** are part of the peripheral nervous system and not part of the brain proper, but they are most appropriately identified in conjunction with the study of brain anatomy. The 12 pairs of cranial nerves primarily serve the head and neck. Only one pair, the vagus nerves, extends into the thoracic and abdominal cavities. All but the first two pairs (olfactory and optic nerves) arise from the brain stem and pass through foramina in the base of the skull to reach their destination.

The cranial nerves are numbered consecutively, and in most cases their names reflect the major structures they control. The cranial nerves are described by name, number (Roman numeral), origin, course, and function in Table 19.1. This information should be committed to memory. A mnemonic device that might be helpful for remembering the cranial nerves in order is "*O*n *o*ccasion, *o*ur *t*rusty *t*ruck *a*cts *f*unny—*v*ery *g*ood *v*ehicle *a*ny*h*ow." The first letter of each word and the "a" and "h" of the final word "anyhow" will remind you of the first letter of the cranial nerve name.

Most cranial nerves are mixed nerves (containing both motor and sensory fibers). However, close scrutiny of Table 19.1 will reveal that three pairs of cranial nerves (optic, olfactory, and vestibulocochlear) are purely sensory in function.

You may recall that the cell bodies of neurons are always located within the central nervous system (cortex or nuclei) or in specialized collections of cell bodies (ganglia) outside the CNS. Neuron cell bodies of the sensory cranial nerves are located in ganglia; those of the mixed cranial nerves are found both within the brain and in peripheral ganglia.

(Text continues on p. 186)

TABLE 19.1 The Cranial Nerves (see Figure 19.9)

Number and name	Origin and course	Function	Testing
I. Olfactory	Fibers arise from olfactory mucosa and run through cribriform plate of ethmoid bone to synapse with olfactory bulbs.	Purely sensory—carries impulses associated with sense of smell.	Person is asked to sniff aromatic substances, such as oil of cloves and vanilla, and to identify each.
II. Optic	Fibers arise from retina of eye and pass through optic foramen in sphenoid bone. Fibers of the two optic nerves then take part in forming optic chiasma (with partial crossover of fibers) after which they continue on to thalamus as the optic tracts. Final fibers of this pathway travel from the thalamus to the optic cortex as the optic radiation.	Purely sensory—carries impulses associated with vision.	Vision and visual field are determined with eye chart and by testing the point at which the person first sees an object (finger) moving into the visual field. Fundus of eye viewed with ophthalmoscope to detect papilledema (swelling of optic disc, or point at which optic nerve leaves the eye) and to observe blood vessels.
III. Oculomotor	Fibers emerge from midbrain and exit from skull via superior orbital fissure to run to eye.	Mixed—somatic motor fibers to inferior oblique and superior, inferior, and medial rectus muscles, which direct eyeball, and to levator palpebrae muscles of eyelid; parasympathetic fibers to iris and smooth muscle controlling lens shape (reflex responses to varying light intensity and focusing of eye for near vision); contains proprioceptive sensory fibers carrying impulses from extrinsic eye muscles.	Pupils are examined for size, shape, and equality. Pupillary reflex is tested with penlight (pupils should constrict when illuminated). Convergence for near vision is tested, as is subject's ability to follow objects up, down, side to side, and diagonally.
IV. Trochlear	Fibers emerge from midbrain and exit from skull via superior orbital fissure to run to eye.	Mixed—provides somatic motor fibers to superior oblique muscle (an extrinsic eye muscle); conveys proprioceptive impulses from same muscle to brain.	Tested in common with cranial nerve III.
V. Trigeminal	Fibers emerge from pons and form three divisions, which exit separately from skull: mandibular division through foramen ovale in sphenoid bone, maxillary division via foramen rotundum in sphenoid bone, and ophthalmic division through superior orbital fissure of eye socket.	Mixed—major sensory nerve of face; conducts sensory impulses from skin of face and anterior scalp, from mucosae of mouth and nose, and from surface of eyes; mandibular division also contains motor fibers that innervate muscles of mastication and muscles of floor of mouth.	Sensations of pain, touch, and temperature are tested with safety pin and hot and cold objects. Corneal reflex tested with wisp of cotton. Motor branch assessed by asking person to clench his teeth, open mouth against resistance, and move jaw side to side.
VI. Abducens	Fibers leave inferior region of pons and exit from skull via superior orbital fissure to run to eye.	Carries motor fibers to lateral rectus muscle of eye and proprioceptive fibers from same muscle to brain.	Tested in common with cranial nerve III.

(*continued*)

TABLE 19.1 *(continued)*

Number and name	Origin and course	Function	Testing
VII. Facial	Fibers leave pons and travel through temporal bone via internal acoustic meatus, exiting via stylomastoid foramen to reach the face.	Mixed—supplies somatic motor fibers to muscles of facial expression and parasympathetic motor fibers to lacrimal and salivary glands; carries sensory fibers from taste receptors of anterior portion of tongue.	Anterior two-thirds of tongue is tested for ability to taste sweet (sugar), salty, sour (vinegar), and bitter (quinine) substances. Symmetry of face is checked. Subject is asked to close eyes, smile, whistle, and so on. Tearing is assessed with ammonia fumes.
VIII. Vestibulocochlear	Fibers run from inner-ear equilibrium and hearing apparatus, housed in temporal bone, through internal acoustic meatus to enter pons.	Purely sensory—vestibular branch transmits impulses associated with sense of equilibrium from vestibular apparatus and semicircular canals; cochlear branch transmits impulses associated with hearing from cochlea.	Hearing is checked by air and bone conduction using tuning fork.
IX. Glossopharyngeal	Fibers emerge from medulla and leave skull via jugular foramen to run to throat.	Mixed—somatic motor fibers serve pharyngeal muscles, and parasympathetic motor fibers serve salivary glands; sensory fibers carry impulses from pharynx, tonsils, posterior tongue (taste buds), and pressure receptors of carotid artery.	Position of the uvula is checked. Gag and swallowing reflexes are checked. Subject is asked to speak and cough. Posterior third of tongue may be tested for taste.
X. Vagus	Fibers emerge from medulla and pass through jugular foramen and descend through neck region into thorax and abdomen.	Mixed—fibers carry somatic motor impulses to pharynx and larynx and sensory fibers from same structures; very large portion is composed of parasympathetic motor fibers, which supply heart and smooth muscles of abdominal visceral organs; transmits sensory impulses from viscera.	As for cranial nerve IX (IX and X are tested in common, since they both innervate muscles of throat and mouth).
XI. Accessory	Fibers arise from medulla and superior aspect of spinal cord and travel through jugular foramen to reach muscles of neck and back.	Mixed—provides somatic motor fibers to sternocleidomastoid and trapezius muscles and to muscles of soft palate, pharynx, and larynx (spinal and medullary fibers respectively); proprioceptive impulses are conducted from these muscles to brain.	Sternocleidomastoid and trapezius muscles are checked for strength by asking person to rotate head and shoulders against resistance.
XII. Hypoglossal	Fibers arise from medulla and exit from skull via hypoglossal canal to travel to tongue.	Mixed—carries somatic motor fibers to muscles of tongue and proprioceptive impulses from tongue to brain.	Person is asked to protrude and retract tongue. Any deviations in position are noted.

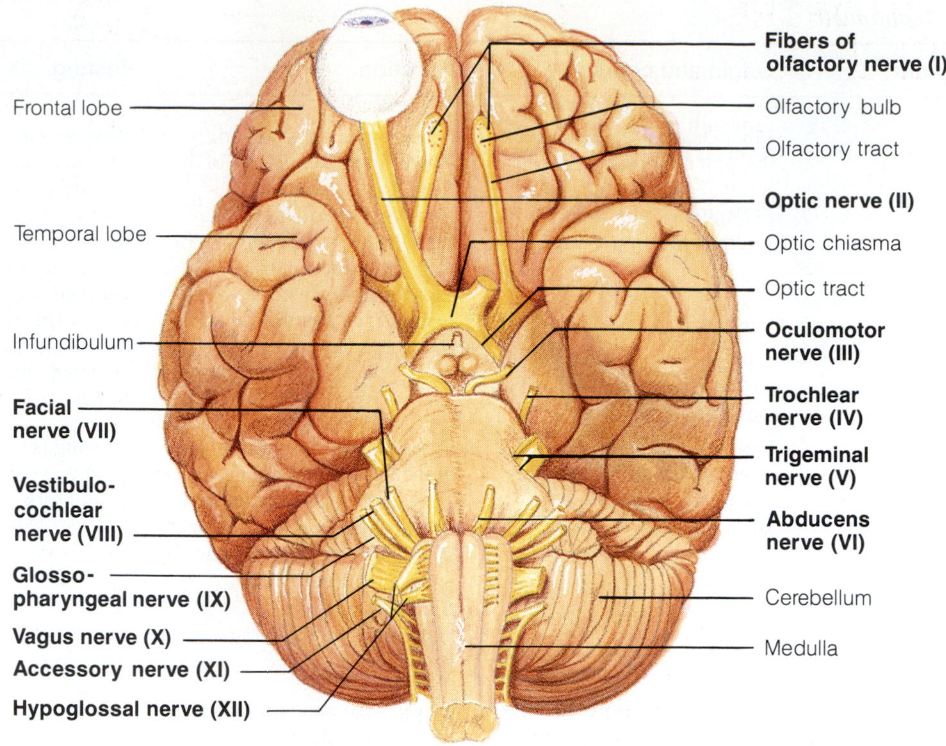

F19.9

Ventral aspect of the human brain, showing the cranial nerves. (See corresponding Plate A in the Human Anatomy Atlas.)

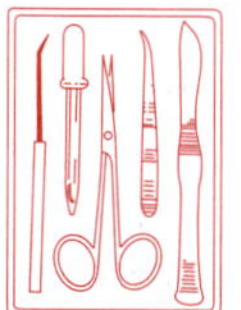

1. Observe the anterior surface of the brain model to identify the cranial nerves. Figure 19.9 may also aid you in this study. Notice that the first (olfactory) cranial nerves are not visible on the model because they consist only of those short axons that run from the nasal mucosa through the cribriform plate of the ethmoid bone. (However, the synapse points of the first cranial nerves, the *olfactory bulbs,* are visible on the model.)

2. The last column of Table 19.1 describes techniques for testing cranial nerves, which is an important part of any neurologic examination. This information may help you understand cranial nerve function, especially as it pertains to some aspects of brain function. If materials are provided for cranial nerve testing, conduct tests of cranial nerve function following directions given in the "testing" column of the table.

3. Several cranial nerve ganglia are named here. *Using your textbook or an appropriate reference,* name the cranial nerve the ganglion is associated with and state its location.

Cranial nerve ganglion	Cranial nerve	Site of ganglion
trigeminal		
geniculate		
inferior		
superior		
spiral		
vestibular		

DISSECTION OF THE SHEEP BRAIN

The brain of any mammal is enough like the human brain to warrant comparison. Obtain a sheep brain, protective skin cream or disposable gloves, dissecting pan, and instruments, and bring them to your laboratory bench.

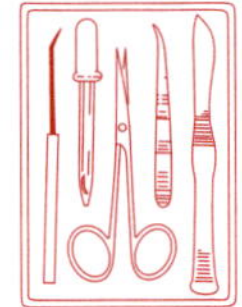

1. Place the intact sheep brain ventral surface down on the dissecting pan and observe the dura mater. Feel its consistency and note its toughness. Cut through the dura mater along the line of the longitudinal fissure (which separates the cerebral hemispheres) to enter the superior sagittal sinus. Gently force the cerebral hemispheres apart laterally to expose the corpus callosum deep to the longitudinal fissure.

2. Carefully remove the dura mater and examine the superior surface of the brain. Notice that, like the human brain, its surface is thrown into convolutions (fissures and gyri). Locate the arachnoid mater, which appears on the brain surface as a delicate "cottony" material spanning the fissures. In contrast, the innermost meninx, the pia mater, closely follows the cerebral contours.

Dorsal Structures

1. Refer to Figures 19.10a and c as a guide in identifying the following structures. The cerebral hemispheres should be easy to locate. How do the size of the sheep's cerebral hemispheres and the depth of the fissures compare to those in the human brain?

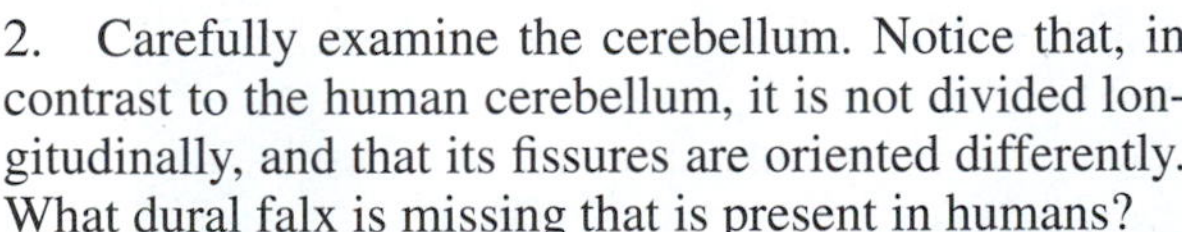

2. Carefully examine the cerebellum. Notice that, in contrast to the human cerebellum, it is not divided longitudinally, and that its fissures are oriented differently. What dural falx is missing that is present in humans?

____

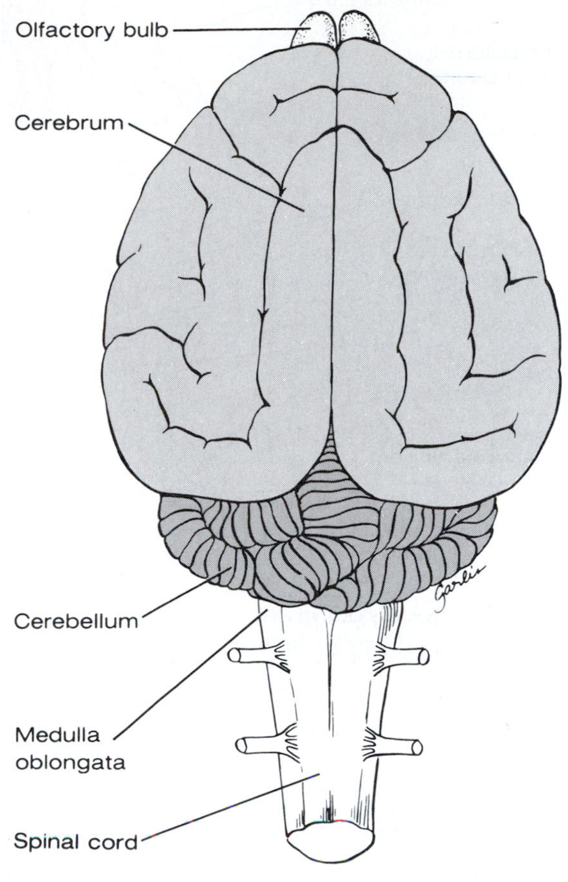

F19.10

Intact sheep brain. **(a)** Dorsal view.

3. Locate the three pairs of cerebellar peduncles, fiber tracts that connect the cerebellum to other brain structures, by lifting the cerebellum dorsally away from the brain stem. The most posterior pair, the inferior cerebellar peduncles, connect the cerebellum to the medulla. The middle cerebellar peduncles attach the cerebellum to the pons, and the superior cerebellar peduncles run from the cerebellum to the midbrain.

(Text continues on p. 189)

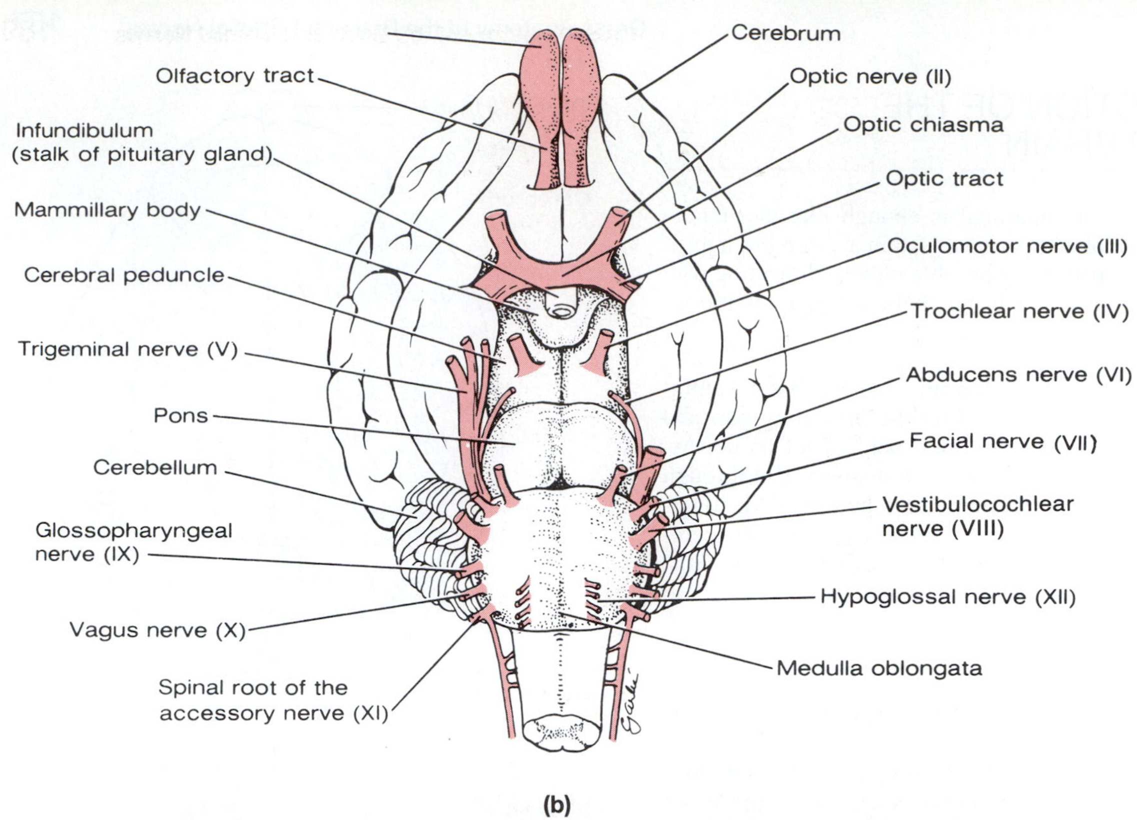

(b)

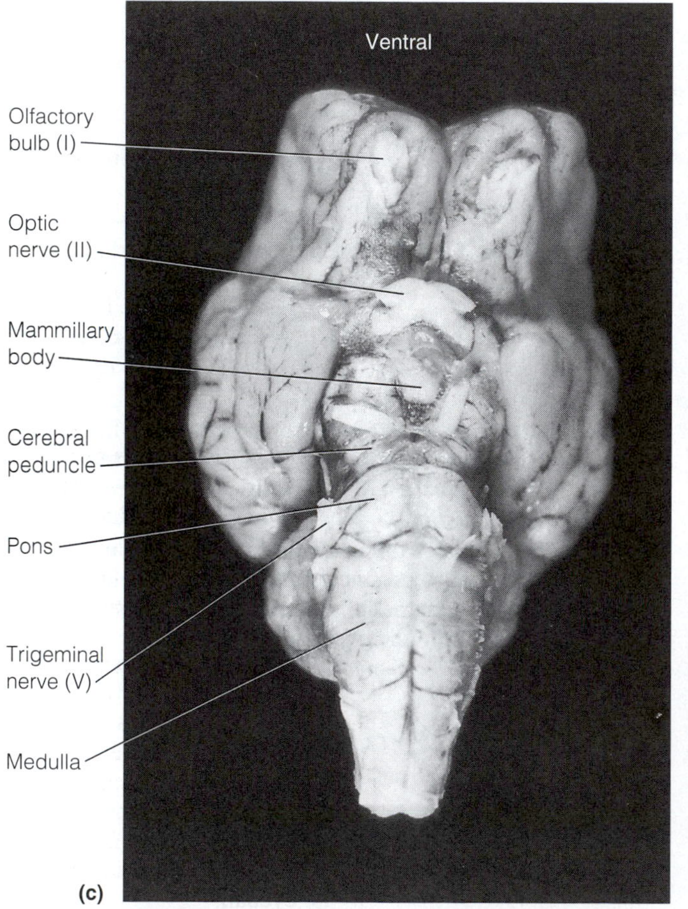

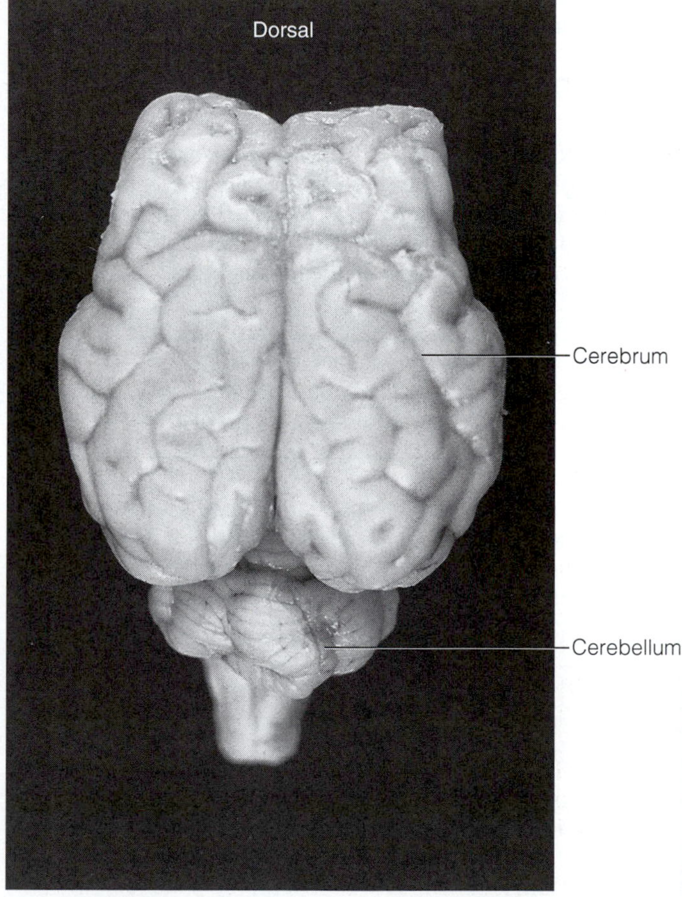

(c)

F19.10 (*continued*)

Intact sheep brain. **(b)** Ventral view. **(c)** Photographs showing ventral and dorsal views.

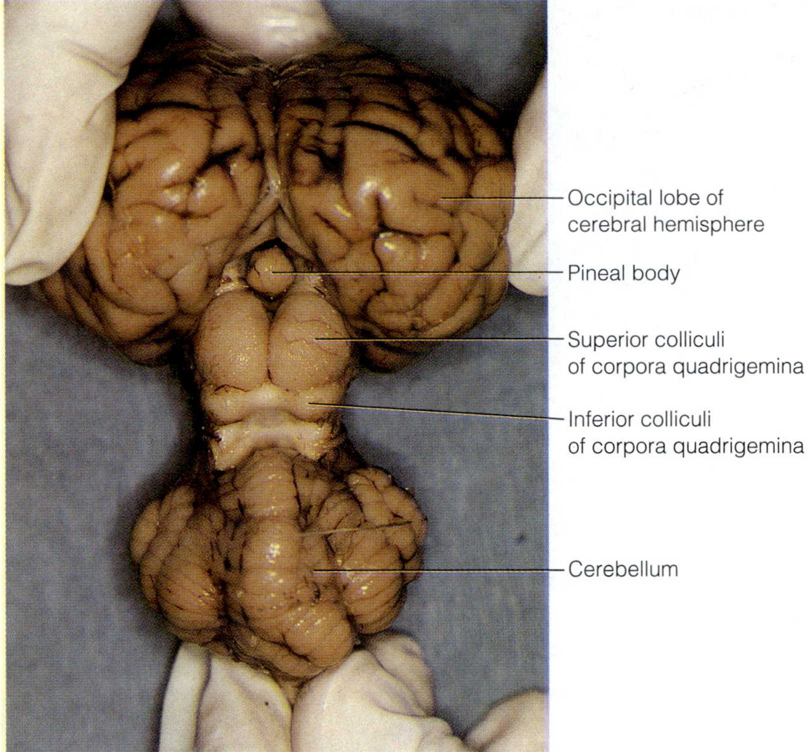

F19.11

Means of exposing the dorsal midbrain structures of the sheep brain.

4. To expose the dorsal surface of the midbrain, gently spread the cerebrum and cerebellum apart, as shown in Figure 19.11. Identify the corpora quadrigemina, which appear as four rounded prominences on the dorsal midbrain surface. What is the function of the corpora quadrigemina?

____

____

Also locate the pineal body, which appears as a small oval protrusion in the midline just anterior to the corpora quadrigemina.

Ventral Structures

Figure 19.10b and c shows the important features of the ventral surface of the brain.

1. Look for the clublike olfactory bulbs anteriorly, on the inferior surface of the frontal lobes of the cerebral hemispheres. Axons of olfactory neurons run from the nasal mucosa through the perforated cribriform plate of the ethmoid bone to synapse with the olfactory bulbs.

How does the size of these olfactory bulbs compare with those of humans?

____

Is the sense of smell more important as a protective and a food-getting sense in sheep or in humans?

____

2. The optic nerve (II) carries sensory impulses from the retina of the eye. Thus this cranial nerve is involved in the sense of vision. Identify the optic nerves, optic chiasma, and optic tracts.

3. Posterior to the optic chiasma, two structures protrude from the ventral aspect of the hypothalamus—the infundibulum (stalk of the pituitary gland) immediately posterior to the optic chiasma and the mammillary body. Notice that the sheep's mammillary body is a single rounded eminence; in humans it is a double structure.

4. Identify the cerebral peduncles on the ventral aspect of the midbrain, just posterior to the mammillary body of the hypothalamus. The cerebral peduncles are fiber tracts connecting the cerebrum and medulla. Identify the large oculomotor nerves (III), which arise from the ventral midbrain surface, and the tiny trochlear nerves (IV), which can be seen at the junction of the midbrain and pons. Both of these cranial nerves provide motor fibers to extrinsic muscles of the eyeball.

5. Move posteriorly from the midbrain to identify first the pons and then the medulla oblongata, both hindbrain structures composed primarily of ascending and descending fiber tracts.

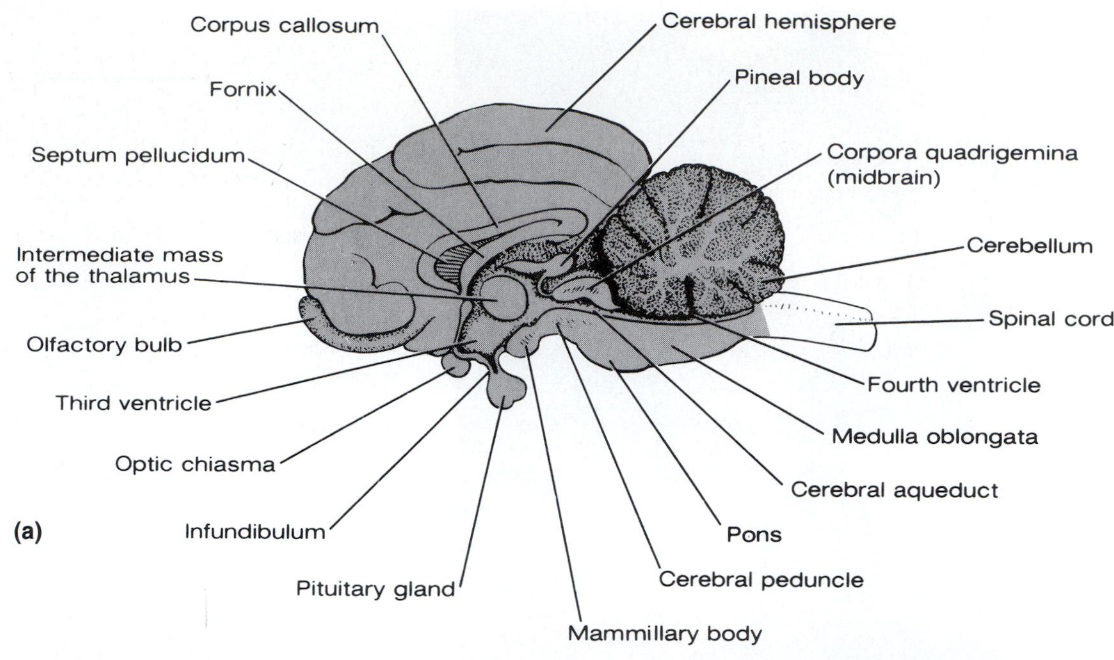

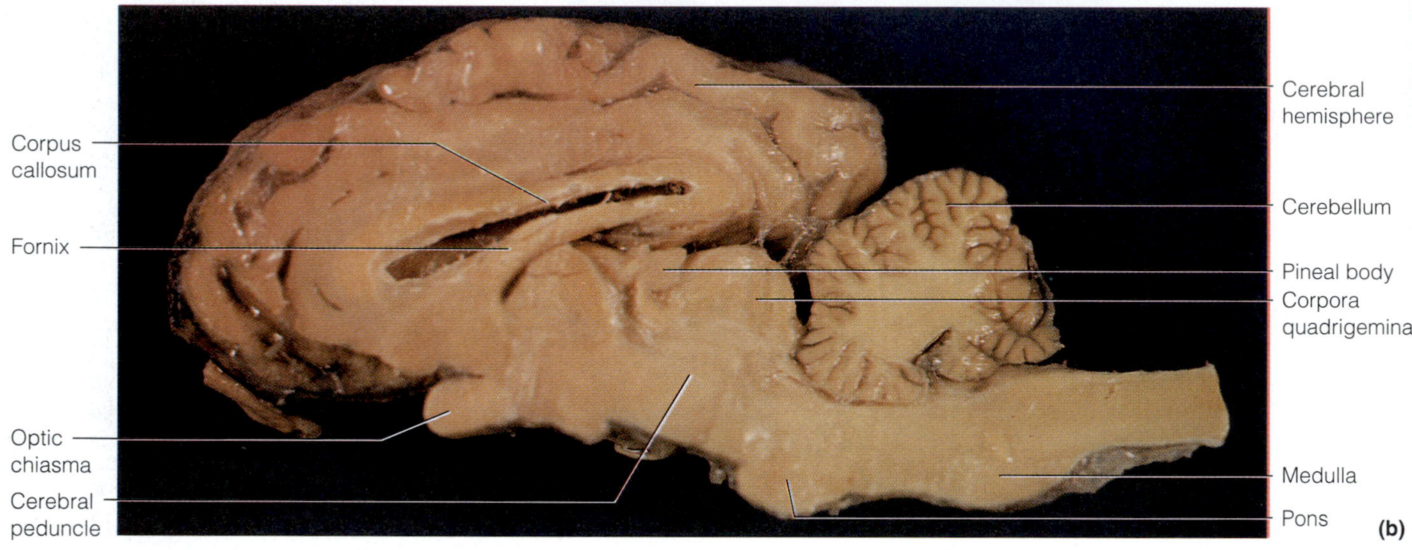

F19.12

Sagittal section of the sheep brain showing internal structures. **(a)** Diagrammatic view. **(b)** Photograph.

6. Return to the junction of the pons and midbrain and proceed posteriorly to identify the following cranial nerves, all arising from the pons:

- Trigeminal nerves (V), which are involved in chewing and sensations of the head and face
- Abducens nerves (VI), which abduct the eye (and thus work in conjunction with cranial nerves III and IV)
- Facial nerves (VII), large nerves are involved in taste sensation, gland function (salivary and lacrimal glands), and facial expression

7. Continue posteriorly to identify

- Vestibulocochlear nerves (VIII), purely sensory nerves which are involved with hearing and equilibrium
- Glossopharyngeal nerves (IX), which contain motor fibers innervating throat structures and sensory fibers transmitting taste stimuli (in conjunction with cranial nerve VII)
- Vagus nerves (X), often called "wanderers," which serve many organs of the head, thorax, and abdominal cavity
- Accessory nerves (XI), which serve muscles of the neck, larynx, and shoulder; notice that the accessory nerves arise from both the medulla and the spinal cord
- Hypoglossal nerves (XII), which stimulate tongue and neck muscles

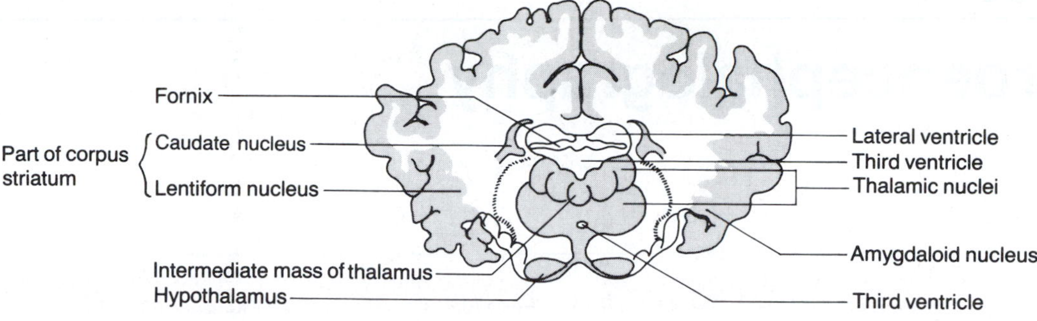

F19.13

Frontal section of a sheep brain. Major structures revealed are the location of major basal nuclei deep in the interior, the thalamus, hypothalamus, and lateral and third ventricles.

Internal Structures

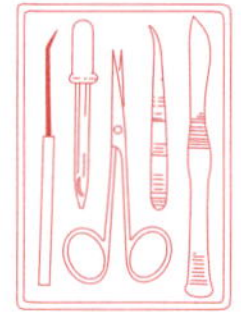

1. The internal structure of the brain can only be examined after further dissection. Place the brain ventral side down on the dissecting pan and make a cut completely through it in a superior to inferior direction. Cut through the longitudinal fissure, corpus callosum, and midline of the cerebellum. Refer to Figure 19.12 as you work.

2. The thin nervous tissue membrane immediately ventral to the corpus callosum that separates the lateral ventricles is the septum pellucidum. Pierce this membrane and probe the lateral ventricle cavity. The fiber tract ventral to the septum pellucidum and anterior to the third ventricle is the fornix.

How does the size of the fornix in this brain compare with the human fornix?

____

____

Why do you suppose this is so? (Hint: What is the function of this band of fibers?)

____

____

____

3. Identify the thalamus, which forms the walls of the third ventricle and is located posterior and ventral to the fornix. The intermediate mass spanning the ventricular cavity appears as an oval protrusion of the thalamic wall. Anterior to the intermediate mass, locate the interventricular foramen, a canal connecting the lateral ventricle on the same side with the third ventricle.

4. The hypothalamus forms the floor of the third ventricle. Identify the optic chiasma, infundibulum, and mammillary body on its exterior surface. You can see the pineal body at the superoposterior end of the third ventricle, just beneath the junction of the corpus callosum and fornix.

5. Locate the midbrain by identifying the corpora quadrigemina that form its dorsal roof. Follow the cerebral aqueduct (the narrow canal connecting the third and fourth ventricles) through the midbrain tissue to the fourth ventricle. Identify the cerebral peduncles, which form its anterior walls.

6. Identify the pons and medulla, which lie anterior to the fourth ventricle. The medulla continues into the spinal cord without any obvious anatomical change, but the point at which the fourth ventricle narrows to a small canal is generally accepted as the beginning of the spinal cord.

7. Identify the cerebellum posterior to the fourth ventricle. Note its internal treelike arrangement of white matter, the arbor vitae.

8. If time allows, obtain another sheep brain and section it along the frontal plane so that the cut passes through the infundibulum. Compare your specimen to the diagrammatic view in Figure 19.13, and attempt to identify all the structures shown in the figure.

9. Check with your instructor to determine if cow spinal cord sections (preserved) are available for the spinal cord studies in Exercise 21. If not, save the small portion of the spinal cord from your brain specimen. Otherwise, dispose of all the organic debris in the appropriate laboratory containers and clean the dissecting instruments and tray before leaving the laboratory.

20

EXERCISE

Electroencephalography

OBJECTIVES

1. To define *electroencephalogram* and to discuss its clinical significance.
2. To describe or recognize typical tracings of the most common brain wave patterns (alpha, beta, theta, and delta waves) and to indicate the conditions under which each is most likely to be predominant.
3. To state the source of brain waves.
4. To define *alpha block.*
5. To monitor electroencephalography and recognize alpha rhythm.
6. To describe the effect of a sudden sound, mental concentration, and alkalosis on brain wave patterns.

MATERIALS

Oscilloscope and EEG lead-selector box or polygraph and high-gain preamplifier
Cot
Electrode gel
EEG electrodes and leads
Collodion gel or long elastic EEG straps

BRAIN WAVE PATTERNS AND THE ELECTROENCEPHALOGRAM

Any physiologic investigation of the brain can emphasize and expose only a very minute portion of its activity. Higher brain functions, such as consciousness and logical reasoning, are extremely difficult to investigate. It is obviously much easier to do experiments on the brain's input-output functions, some of which can be detected with appropriate recording equipment. Still, the ability to record brain activity does not necessarily guarantee an understanding of the brain.

The **electroencephalogram (EEG),** a record of the electrical activity of the brain, can be obtained through electrodes placed at various points on the skin or scalp of the head. This electrical activity, which is recorded as waves (Figure 20.1), is not completely understood at present but may be regarded as action potentials generated by brain neurons.

Certain characteristics of brain waves are known. They have a frequency of 1 to 30 hertz (Hz) or cycles per second, a dominant rhythm of 10 Hz, and an average amplitude (voltage) of 20 to 100 microvolts (μV). They vary in frequency in different brain areas, occipital waves having a lower frequency than those associated with the frontal and parietal lobes.

The first of the brain waves to be described by scientists were the **alpha waves** (or alpha rhythm). Alpha waves have an average frequency range of 8 to 13 Hz and are produced when the individual is in a relaxed state with the eyes closed. **Alpha block,** suppression of the alpha rhythm, occurs if the eyes are opened or if the individual begins to concentrate on some mental problem or visual stimulus. Under these conditions, the waves decrease in amplitude but increase in frequency. Under conditions of fright or excitement, the frequency increases still more.

Beta waves, closely related to alpha waves, are faster (14 to 25 Hz) and have a lower amplitude. They are typical of the attentive or alert state.

Very large (high-amplitude) waves with a frequency of 4 Hz or less that are seen in deep sleep are **delta waves. Theta waves** are large, abnormally contoured waves with a frequency of 4 to 7 Hz. Although theta waves are normal in children, they represent emotional problems or some sort of neural imbalance in adults.

Brain waves change with age, sensory stimuli, brain pathology or disease, and the chemical state of the body. (Glucose deprivation, oxygen poisoning, and sedatives all interfere with the rhythmic activity of brain output by disturbing the metabolism of the neurons.) Sleeping individuals and patients in a coma have EEGs that are slower (or lower frequency) than the alpha rhythm of normal adults. Fright, epileptic seizures, and various types of drug intoxication are associated with comparatively faster cortical activity. Thus impairment of cortical function is indicated by neuronal activity that is either too fast or too slow; unconsciousness occurs at both extremes of the frequency range.

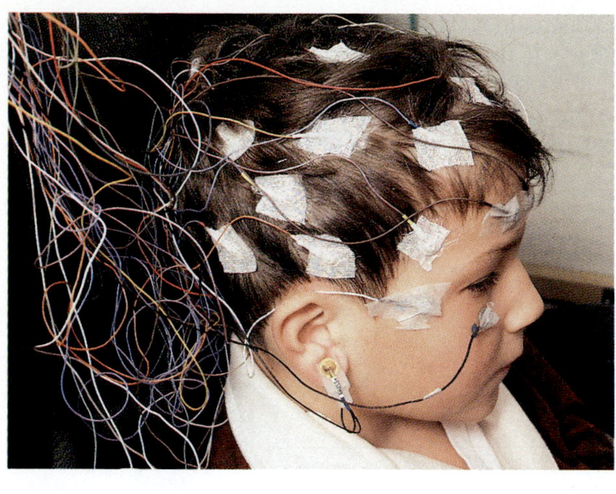

(a)

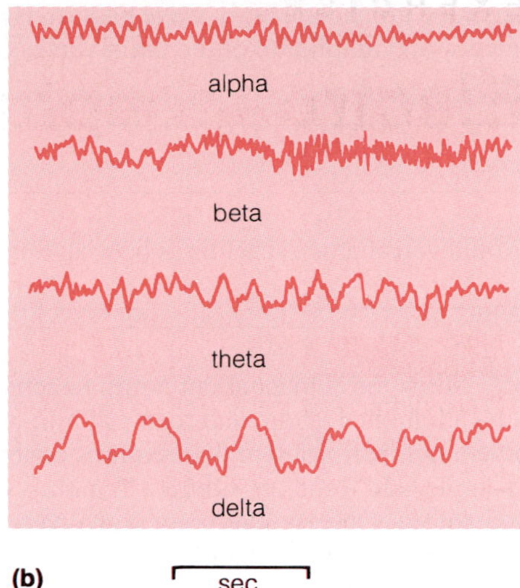

(b)

F20.1

Electroencephalography and brain waves. (a) To obtain a recording of brain wave activity (an EEG), electrodes are positioned on the patient's scalp and attached to a recording device called an electroencephalograph. **(b)** Typical EEGs. Alpha waves are typical of the awake but relaxed state; beta waves occur in the awake, alert state; theta waves are common in children but not in normal awake adults; delta waves occur during deep sleep.

Since spontaneous brain waves are always present, even during unconsciousness and coma, the absence of brain waves (a "flat" EEG) is taken as clinical evidence of death. The EEG is used clinically to diagnose and localize many types of brain lesions, including epileptic lesions, infections, abscesses, and tumors. ■

OBSERVING BRAIN WAVE PATTERNS

If one electrode (the *active electrode*) is placed over a particular cortical area and another (the *indifferent electrode*) is placed over an inactive part of the head, such as the earlobe, all of the activity of the cortex underlying the active electrode will, theoretically, be recorded. The inactive area provides a zero reference point or a base line, and the EEG represents the difference between "activities" occurring under the two electrodes.

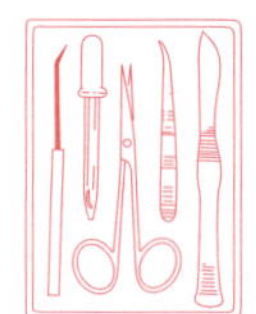

1. Connect the EEG selector box to the oscilloscope preamplifier, or connect the high-gain preamplifier to the polygraph channel amplifier. Adjust the horizontal sweep and sensitivity according to the directions given in the manual or by your instructor.

2. Prepare the subject. The subject should lie undisturbed with eyes closed in a quiet, dimly lit area. (Someone who is able to relax easily makes a good subject.) Apply a small amount of electrode gel to the subject's forehead above the left eye and on the left earlobe. Press an electrode to each prepared area and secure them (1) by applying a film of collodion gel to the electrode surface and the adjacent skin, or (2) with a long elastic EEG strap (knot tied at the back of the head). If collodion gel is used, allow it to dry before continuing.

3. Connect the active frontal lead (forehead) to the EEG selector box outlet marked "L Frontal." Connect the lead from the indifferent electrode (earlobe) to the ground outlet (or to the appropriate input terminal on the high-gain preamplifier).

4. Turn the oscilloscope or polygraph on, and observe the EEG pattern of the relaxed subject for a period of 5 min. If the subject is truly relaxed, you should see a typical alpha-wave pattern. (If the subject is unable to relax and the alpha-wave pattern does not appear in this time interval, test another subject.) Discourage all muscle movement during the monitoring period.*

5. Abruptly and loudly clap your hands. The subject's eyes should open and alpha block should occur. Ob-

* Note that 60-cycle "noise" (appearing as fast, regular, low amplitude waves superimposed on the more irregular brain waves) may interfere with the tracings being made, particularly if the laboratory has a lot of electrical equipment.

serve the immediate brain wave pattern. How do the frequency and amplitude of the brain waves change?

___

___

Would you characterize this as beta rhythm? ___

Why? ___

6. Allow the subject about 5 min to achieve complete relaxation once again, and then ask him or her to compute a number problem that requires concentration (for example, add 3 and 36, subtract 7, multiply by 2, add 50 and so on). Observe the brain wave pattern during the period of mental computation.

7. Once again allow the subject to relax until alpha rhythm resumes. Then, instruct him or her to hyperventilate for 3 min. *Be sure to tell the subject when to stop hyperventilating.* Hyperventilation rapidly flushes carbon dioxide out of the lungs, decreasing carbon dioxide levels in the blood and producing respiratory alkalosis.

Observe the changes in the rhythm and amplitude of the brain waves occurring during the period of hyperventilation. Record your observations:

___

___

___

___

___

___

___

___

EXERCISE 21

Spinal Cord, Spinal Nerves, and the Autonomic Nervous System

OBJECTIVES

1. To identify important anatomical areas on a spinal cord model or appropriate diagram of the spinal cord, and to cite the neuron type found in these areas (where applicable).
2. To indicate two major areas where the spinal cord is enlarged, and to explain the reasons for this anatomical characteristic.
3. To define *conus medullaris, cauda equina,* and *filum terminale.*
4. To locate on a diagram the fiber tracts in the spinal cord and to state their functional importance.
5. To list two major functions of the spinal cord.
6. To name the meningeal coverings of the spinal cord and state their function.
7. To describe the origin, fiber composition, and distribution of the spinal nerves, differentiating between roots, the spinal nerve proper, and rami, and to discuss the result of transecting these structures.
8. To discuss the distribution of the dorsal rami and ventral rami of the spinal nerves.
9. To identify the four major nerve plexuses, the major nerves of each, and their distribution.
10. To identify the site of origin and the function of the sympathetic and parasympathetic divisions of the autonomic nervous system, and to state how the autonomic nervous system differs from the somatic nervous system.

MATERIALS

Spinal cord model (cross section)
Laboratory charts of the spinal cord and spinal nerves and sympathetic chain
Red and blue pencils
Preserved cow spinal cord sections with meninges and nerve roots intact (or spinal cord segment saved from the brain dissection in Exercise 19)
Dissecting tray and instruments
Dissecting microscope
Histologic slide of spinal cord (cross section)
Protective skin cream or disposable gloves
Compound microscope
The Human Nervous System: The Spinal Cord and Spinal Nerves videotape*

See Appendix D, Exercise 21 for links to A.D.A.M. Standard.

See Appendix E, Exercise 21 for links to *Anatomy and PhysioShow: The Videodisc.*

*Available to qualified adopters from Benjamin/Cummings

ANATOMY OF THE SPINAL CORD

The cylindrical **spinal cord,** a continuation of the brain stem, is an association and communication center. It plays a major role in spinal reflex activity and provides neural pathways to and from higher nervous centers. Enclosed within the vertebral canal of the spinal column, the spinal cord extends from the foramen magnum of the skull to the first or second lumbar vertebra, where it terminates in the cone-shaped **conus medullaris** (Figure 21.1). Like the brain, it is cushioned and protected by meninges. The dura mater and arachnoid meningeal coverings extend beyond the conus medullaris, approximately to the level of S_2, and a fibrous extension of the pia mater extends even farther (into the coccygeal canal) as the **filum terminale.**

The fact that the meninges, filled with cerebrospinal fluid, extend well beyond the end of the spinal cord provides an excellent site for removing cerebrospinal fluid for analysis (as when bacterial or viral infections of the spinal cord or its meningeal coverings are suspected) without endangering the delicate spinal cord. This procedure, called a *lumbar tap,* is usually performed below L_3. Additionally, "saddle block" or caudal anesthesia for childbirth is normally administered (injected) between L_3 and L_5.

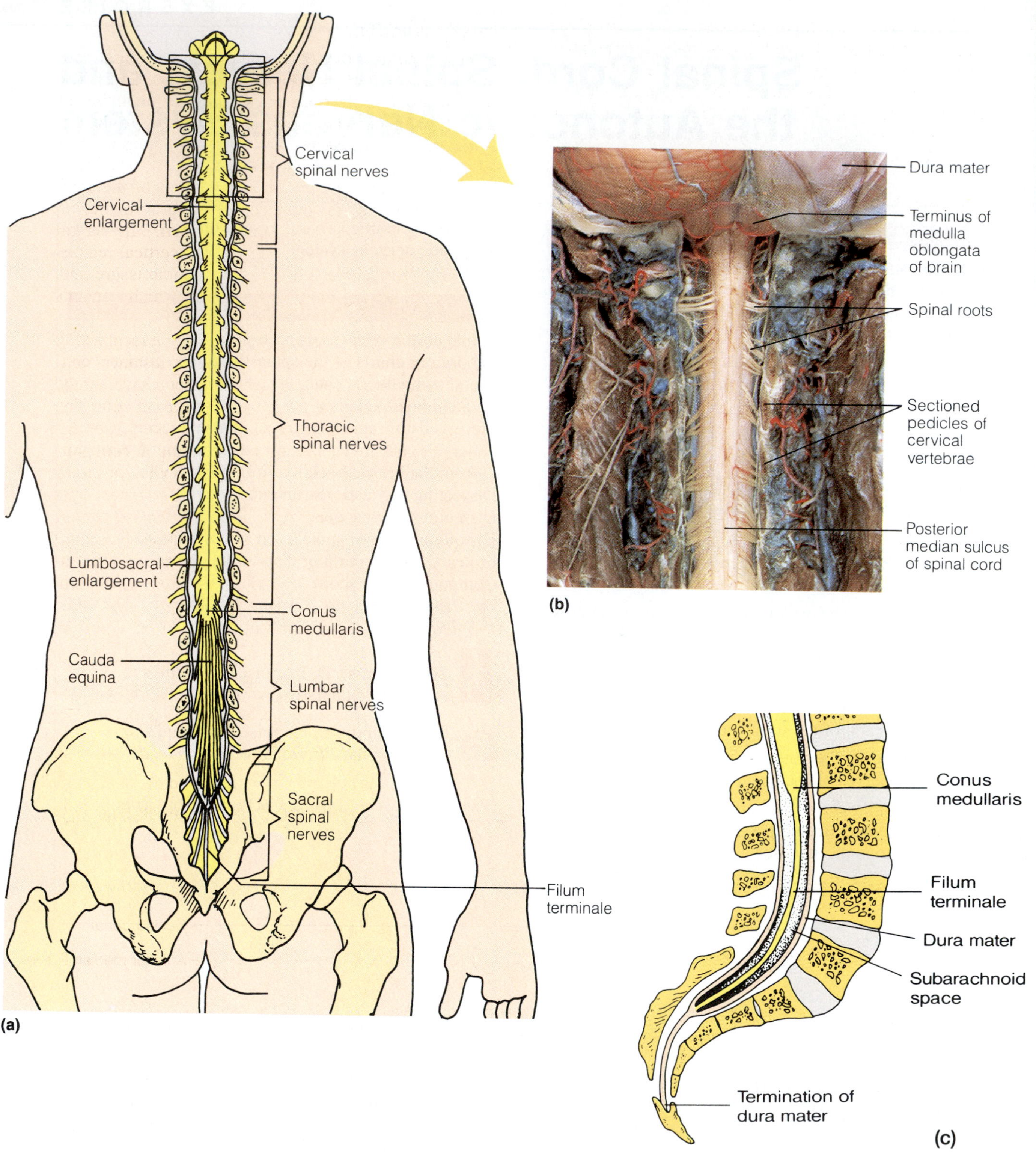

F21.1

Structure of the spinal cord. (a) The vertebral arches have been removed to show the dorsal aspect of the spinal cord (and its nerve roots). The dura mater is cut and reflected laterally. The various regions of the spinal cord are indicated in relation to the vertebral column as cervical, thoracic, lumbar, and sacral. **(b)** Photograph of the cervical region of the spinal cord; the meningeal coverings have been removed to show the spinal roots that give rise to the spinal nerves. **(c)** Lateral view, showing the extent of the filum terminale.

In humans, 31 pairs of spinal nerves arise from the spinal cord and pass through intervertebral foramina to serve the body area at their approximate level of emergence. The cord is about the size of a thumb in circumference for most of its length, but there are obvious enlargements in the cervical and lumbar areas where the nerves serving the upper and lower limbs issue from the cord.

Because the spinal cord does not extend to the end of the vertebral column, the spinal roots emerging from the inferior end of the cord must travel through the vertebral canal for some distance before exiting at the appropriate intervertebral foramina. This collection of spinal roots traversing the inferior end of the vertebral canal is called the **cauda equina** because of its similarity to a horse's tail (the literal translation of *cauda equina*).

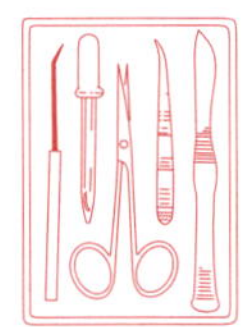

Obtain a model of a cross section of a spinal cord and identify its structures as they are described next.

Gray Matter

In cross section, the **gray matter** of the spinal cord looks like a butterfly or the letter H (Figure 21.2). The two posterior projections are called the **posterior,** or **dorsal, horns;** the two anterior projections are the **anterior,** or **ventral, horns.** The tips of the anterior horns are broader and less tapered than those of the posterior horns. In the thoracic and lumbar regions of the cord, there is also a lateral outpocketing of gray matter on each side referred to as the **lateral horn.** The central area of gray matter connecting the two vertical regions is the **gray commissure.** The gray commissure surrounds the **central canal** of the cord, which contains cerebrospinal fluid.

Neurons with specific functions can be localized in the gray matter. The posterior horns, for instance, contain association neurons and sensory fibers that enter the cord from the body periphery via the **dorsal root.** The cell bodies of these sensory neurons are found in an enlarged area of the dorsal root called the **dorsal root ganglion.** The anterior horns contain cell bodies of motor neurons of the somatic nervous system (voluntary system), which send their axons out via the **ventral root** of the cord to enter the adjacent spinal nerve. The **spinal nerves** are formed from the fusion of the dorsal and ventral roots. The lateral horns, where present, contain

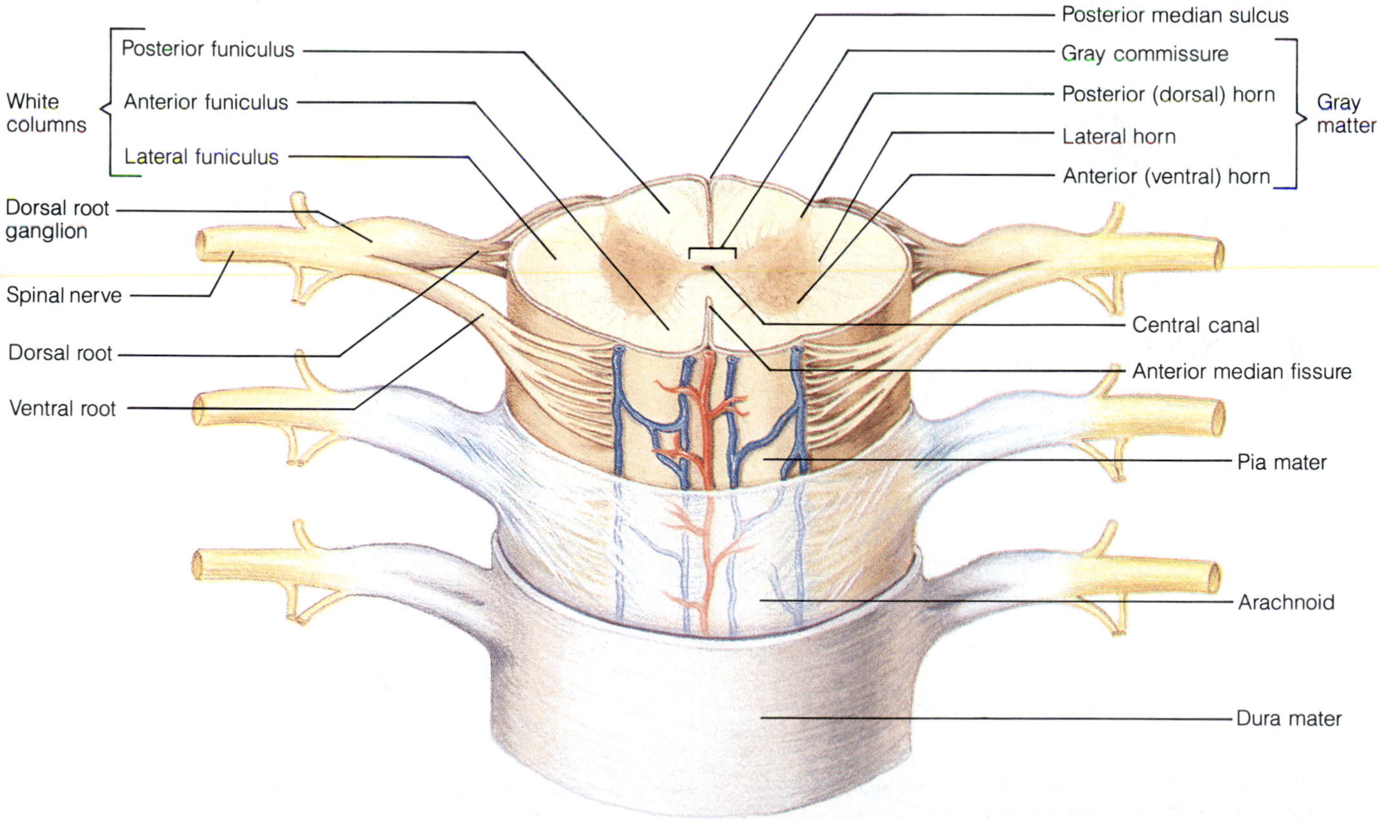

F21.2

Anatomy of the human spinal cord (three-dimensional view).

cell bodies of motor neurons of the autonomic nervous system (sympathetic division). Their axons also leave the cord via the ventral roots, along with those of the motor neurons of the anterior horns.

White Matter

The **white matter** of the spinal cord is nearly bisected by fissures (see Figure 21.2). The more open anterior fissure is the **anterior median fissure,** and the posterior one is the **posterior median sulcus.** The white matter is composed of myelinated fibers—some running to higher centers, some traveling from the brain to the cord, and some conducting impulses from one side of the cord to the other.

Because of the irregular shape of the gray matter, the white matter on each side of the cord can be divided into three primary regions or *white columns:* the **posterior, lateral,** and **anterior funiculi.** Each funiculus contains a number of fiber **tracts** composed of axons with the same origin, terminus, and function. Tracts conducting sensory impulses to the brain are called *ascending,* or *sensory, tracts;* those carrying impulses from the brain to the skeletal muscles are *descending,* or *motor, tracts.*

Because it serves as the transmission pathway between the brain and the body periphery, the spinal cord is an extremely important functional area. Even though it is protected by meninges and cerebrospinal fluid in the vertebral canal, it is highly vulnerable to traumatic injuries, such as might occur in an automobile accident.

When the cord is transected (or severely traumatized), both motor and sensory functions are lost in body areas normally served by that (and lower) regions of the spinal cord. Injury to certain spinal cord areas may even result in a permanent flaccid paralysis of both legs (paraplegia) or of all four limbs (quadriplegia). ■

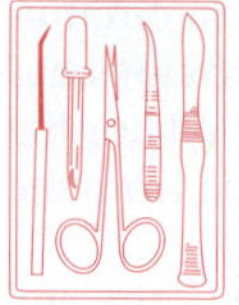

With the help of your textbook or a laboratory chart showing the tracts of the spinal cord, label Figure 21.3 with the tract names that follow. Since each tract is represented on both sides of the cord, for clarity you can label the motor tracts on the right side of the diagram and the sensory tracts on the left side of the diagram. *Color ascending tracts red and descending tracts blue.* Then fill in the functional importance of each tract beside its name below. As you work, try to be aware of how the naming of the tracts is related to their anatomical distribution.

Fasciculus gracilis ______________________

Fasciculus cuneatus ______________________

Dorsal spinocerebellar ______________________

Ventral spinocerebellar ______________________

Lateral spinothalamic ______________________

Ventral spinothalamic ______________________

Lateral corticospinal ______________________

Ventral corticospinal ______________________

Rubrospinal ______________________

Tectospinal ______________________

Vestibulospinal ______________________

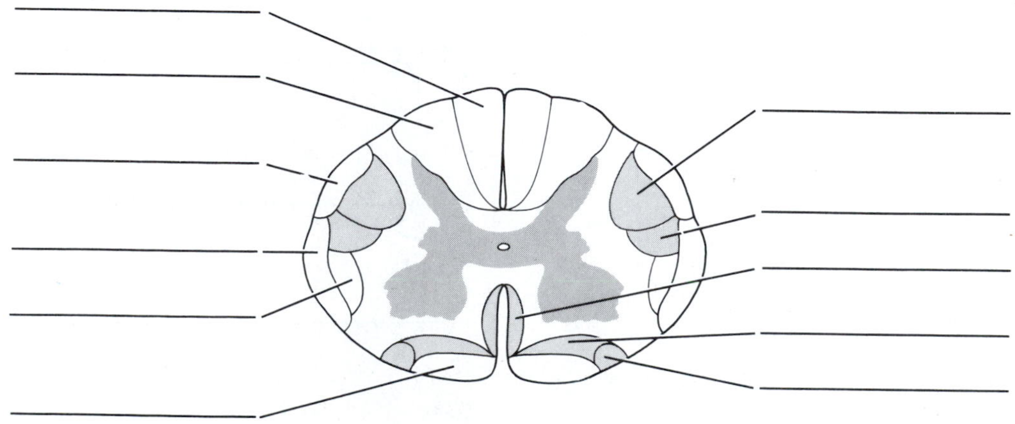

F21.3

Cross section of the spinal cord showing the relative positioning of its major tracts.

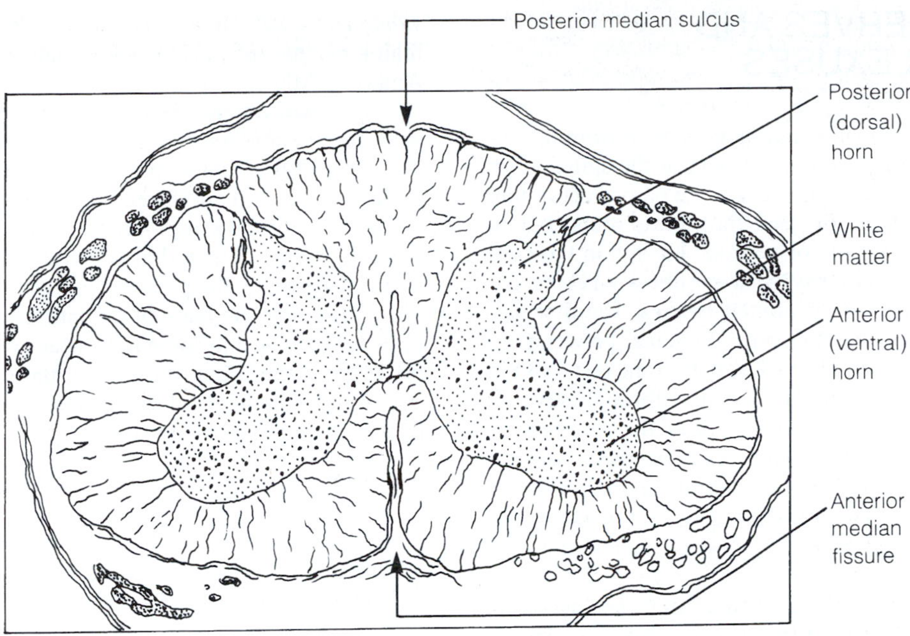

F21.4

Cross section of the spinal cord. See corresponding Plate 7 in the Histology Atlas.

Spinal Cord Dissection

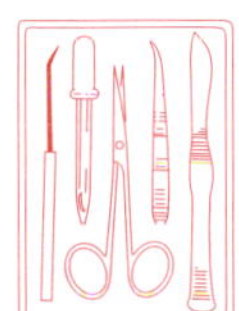

1. Obtain a dissecting tray and instruments and a segment of preserved spinal cord (from a cow or saved from the brain specimen used in Exercise 19). Identify the tough outer meninx (dura mater) and the weblike arachnoid mater.

What name is given to the third meninx, and where is it found?

____

____

Peel back the dura mater and observe the fibers making up the dorsal and ventral roots. If possible, identify a dorsal root ganglion.

2. Cut a thin cross section of the cord and identify the anterior and posterior horns of the gray matter with the naked eye or with the aid of a dissecting microscope.

How can you be certain that you are correctly identifying the anterior and posterior horns?

____

____

Also identify the central canal, white matter, anterior median fissure, posterior median sulcus, and posterior, anterior, and lateral funiculi.

3. Obtain a prepared slide of the spinal cord (cross section) and a compound microscope. Refer to Figure 21.4 as you examine the slide carefully under low power. Observe the shape of the central canal.

Is it basically circular or oval? ______________________________

Name the glial cell type that lines this canal. ______________

What would you expect to find in this canal in the living animal?

____

Can any neuron cell bodies be seen? ______________________

Where? ______________________________

____

What type of neurons would these most likely be—motor, association, or sensory?

____

____

SPINAL NERVES AND NERVE PLEXUSES

The 31 pairs of human spinal nerves arise from the fusions of the ventral and dorsal roots of the spinal cord. Figure 21.5 shows how the nerves are named according to their point of issue. Because the ventral roots contain myelinated axons of motor neurons located in the cord and the dorsal roots carry sensory fibers entering the cord, all spinal nerves are **mixed nerves.** The first pair of spinal nerves leaves the vertebral canal between the base of the occiput and the atlas, but all the rest exit via the intervertebral foramina. The first through seventh pairs of cervical nerves emerge *above* the vertebra for which they are named. C_8 emerges between C_7 and T_1. (Notice that there are 7 cervical vertebrae, but 8 pairs of cervical nerves.) The remaining spinal nerve pairs emerge from the spinal cord *below* the same-numbered vertebra.

Almost immediately after emerging, each nerve divides into **dorsal** and **ventral rami.** (Thus each spinal nerve is only about 1 or 2 cm long.) The rami, like the spinal nerves, contain both motor and sensory fibers. The smaller dorsal rami serve the skin and musculature of the posterior body trunk at their approximate level of emergence. The ventral rami of spinal nerves T_2–T_{12} pass anteriorly as the **intercostal nerves** to supply the muscles of intercostal spaces, and the skin and muscles of the anterior and lateral trunk. The ventral rami of all other spinal nerves form complex networks of nerves called **plexuses.** These plexuses serve the motor and sensory needs of the muscles and skin of the limbs. The fibers of the ventral rami unite in the plexuses (with a few rami supplying fibers to more than one plexus). From the plexuses the fibers diverge again to form peripheral nerves, each of which contains fibers from more than one spinal nerve. The four major nerve plexuses and their chief peripheral nerves are illustrated in Figures 21.5 and 21.6 and are described below. Their names and site of origin should be committed to memory.

Cervical Plexus and the Neck

The **cervical plexus** arises from the ventral rami of C_1 through C_5 to supply muscles of the shoulder and neck. The major motor branch of this plexus is the **phrenic nerve,** which arises from C_3–C_4 (plus some fibers from C_5) and passes into the thoracic cavity in front of the first rib to innervate the diaphragm. The primary danger of a broken neck is that the phrenic nerve may be severed, leading to paralysis of the diaphragm and cessation of breathing. A jingle to help you remember the rami (roots) forming the phrenic nerves is "C_3, C_4, C_5 keep the diaphragm alive."

Brachial Plexus and the Upper Limb

The **brachial plexus** is large and complex, arising from the ventral rami of C_5 through C_8 and T_1. The plexus, after being rearranged consecutively into *trunks, divisions,* and *cords,* finally becomes subdivided into five major *peripheral nerves.* (See Plate C in the Human Anatomy Atlas.)

The **axillary nerve,** which serves the muscles and skin of the shoulder, has the most limited distribution. The large **radial nerve** passes down the posterolateral surface of the arm and forearm, supplying all the extensor muscles of the arm, forearm, and hand and the skin along its course. The radial nerve is often injured in the axillary region by the pressure of a crutch or by hanging one's arm over the back of a chair. The **median nerve** passes down the anteromedial surface of the arm to supply most of the flexor muscles in the forearm and several muscles in the hand (plus the skin of the lateral surface of the palm of the hand).

- Hyperextend your wrist to identify the long, obvious tendon of your palmaris longus muscle, which crosses the exact midline of the anterior wrist. Your median nerve lies immediately deep to that tendon, and the radial nerve lies just *lateral* to it.

The **musculocutaneous nerve** supplies the arm muscles that flex the forearm and the skin of the lateral surface of the forearm. The **ulnar nerve** travels down the posteromedial surface of the arm. It courses around the medial epicondyle of the humerus to supply the flexor carpi ulnaris, the ulnar head of the flexor digitorum profundus of the forearm, and all intrinsic muscles of the hand not served by the median nerve. It supplies the skin of the medial third of the hand, both the anterior and posterior surfaces. Trauma to the ulnar nerve, which often occurs when the elbow is hit, produces a smarting sensation commonly referred to as "hitting the funny bone."

Lumbosacral Plexus and the Lower Limb

The **lumbosacral plexus,** which serves the pelvic region of the trunk and the lower limbs, is actually a complex of two plexuses, the lumbar plexus and the sacral plexus (see Figure 21.6). The **lumbar plexus** arises from ventral rami of L_1 through L_4 (and sometimes T_{12}). Its nerves serve the lower abdominopelvic region and the anterior thigh. The largest nerve of this plexus is the **femoral nerve,** which passes beneath the inguinal ligament to innervate the anterior thigh muscles. The cutaneous branches of the femoral nerve (median and anterior femoral cutaneous and the saphenous nerves) supply the skin of the anteromedial surface of the entire lower limb.

Arising from L_4 through S_4, the nerves of the **sacral plexus** supply the buttock, the posterior surface of the thigh, and virtually all sensory and motor fibers of the leg and foot. The major peripheral nerve of this plexus is the **sciatic nerve,** the largest nerve in the body. The sciatic nerve leaves the pelvis through the greater sciatic notch and travels down the posterior thigh, serving its flexor muscles and skin. In the popliteal region, the sciatic nerve divides into the **common peroneal nerve** and the **tibial nerve,** which together supply the balance of the leg muscles and skin, both directly and via several branches.

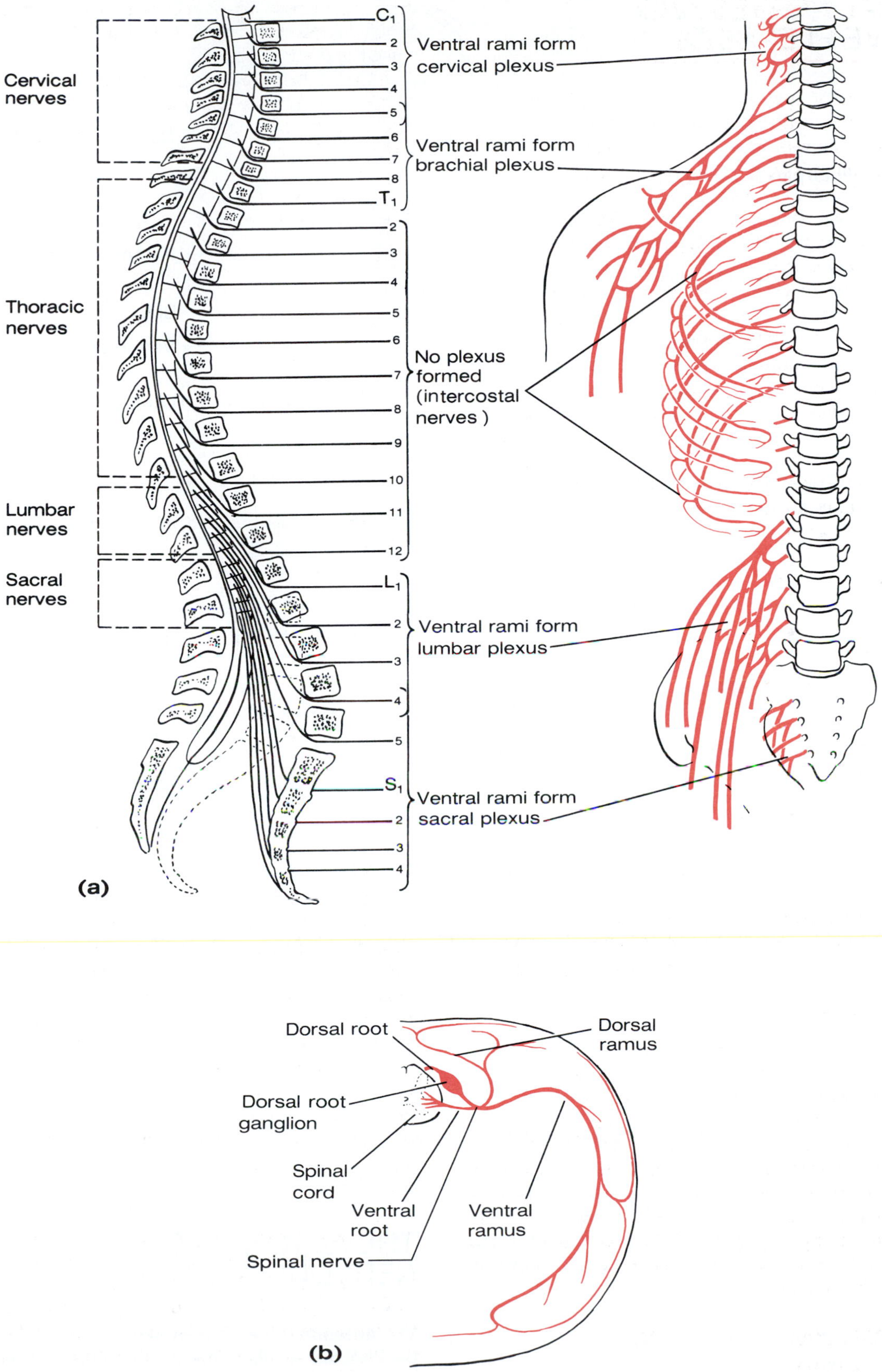

Cervical nerves
Thoracic nerves
Lumbar nerves
Sacral nerves
C_1 2 3 4 5 6 7 8 T_1 2 3 4 5 6 7 8 9 10 11 12 L_1 2 3 4 5 S_1 2 3 4
Ventral rami form cervical plexus
Ventral rami form brachial plexus
No plexus formed (intercostal nerves)
Ventral rami form lumbar plexus
Ventral rami form sacral plexus
(a)

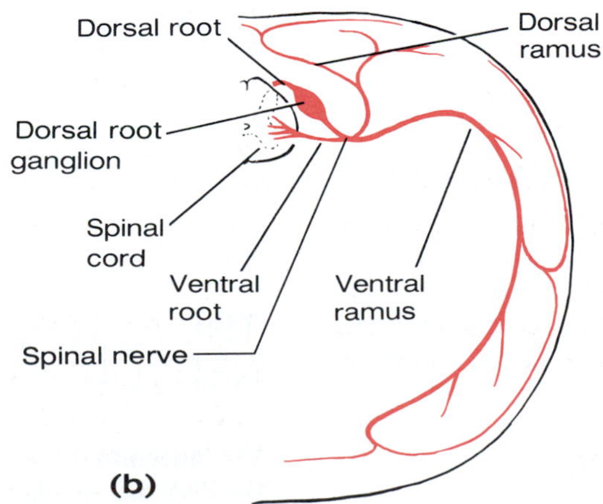

F21.5

Human spinal nerves. **(a)** Relationship of spinal nerves to vertebrae (areas of plexuses formed by the ventral rami are indicated). **(b)** Relative distribution of the ventral and dorsal rami of a spinal nerve (cross section of left trunk).

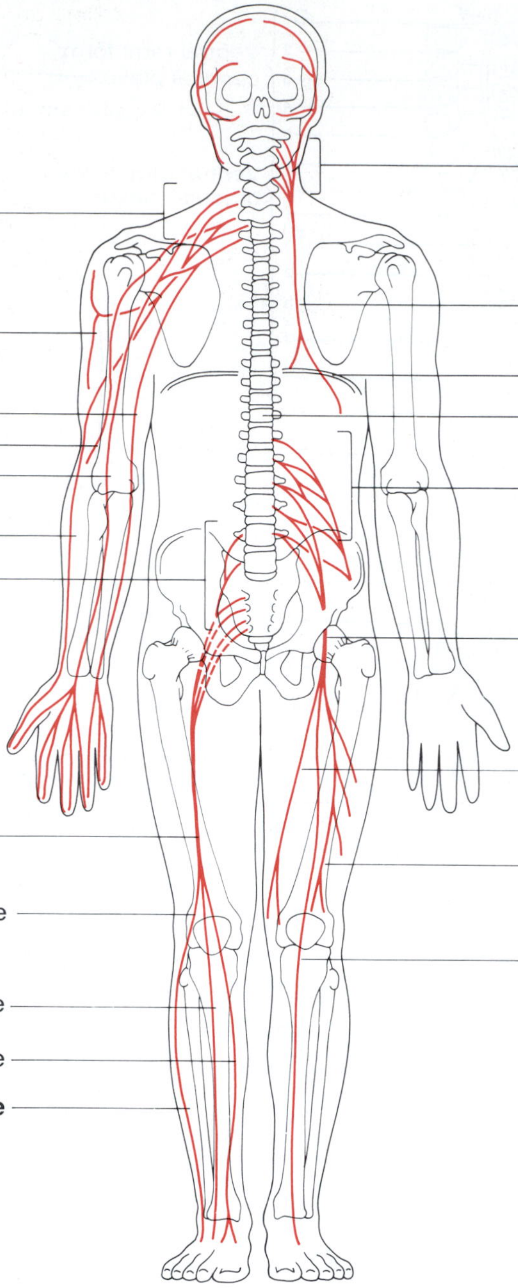

F21.6

Nerve plexuses and major nerves arising from each. For clarity, each plexus is illustrated only on one side of the body.

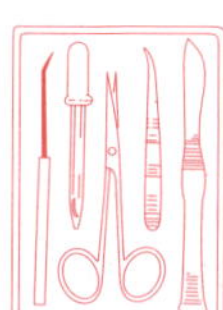

Identify each of the four major nerve plexuses (and its major nerves) shown in Figure 21.6 on a large laboratory chart. Trace the course of the nerves.

THE AUTONOMIC NERVOUS SYSTEM

The **autonomic nervous system** is the subdivision of the PNS that regulates body activities that are generally not under conscious control. It is composed of a special group of motor neurons serving cardiac muscle (the heart), smooth muscle (found in the walls of the visceral organs and blood vessels), and internal glands. Because these structures typically function without conscious

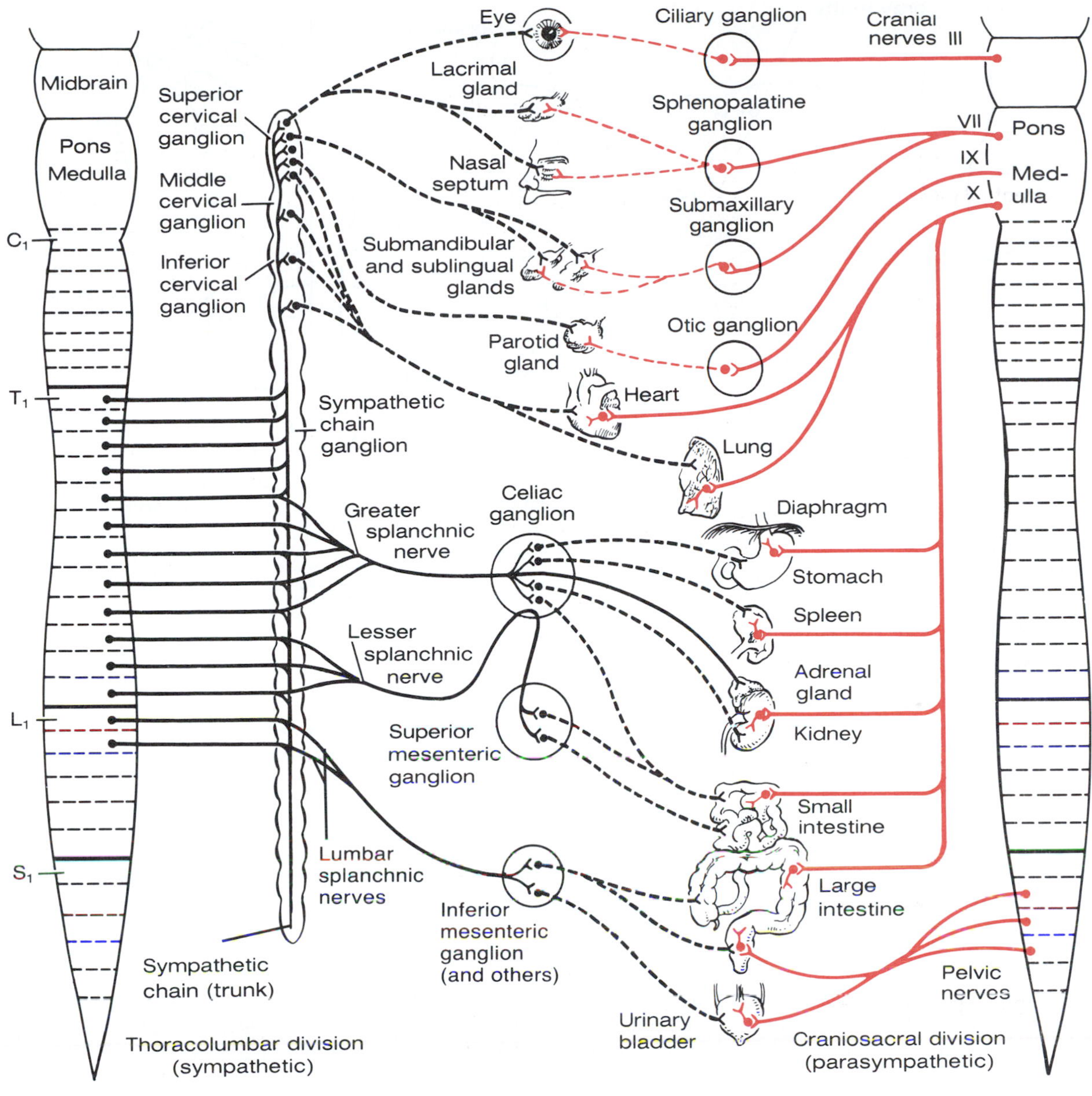

F21.7

The autonomic nervous system. Solid lines indicate preganglionic nerve fibers; dashed lines indicate postganglionic nerve fibers.

control, this system is often referred to as the *involuntary nervous system.*

There is a basic anatomical difference between the motor pathways of the **somatic** (voluntary) **nervous system,** which innervates the skeletal muscles, and those of the autonomic nervous system. In the somatic division, the cell bodies of the motor neurons reside in the CNS (spinal cord or brain), and their axons, sheathed in spinal nerves, extend all the way to the skeletal muscles they serve. However, the autonomic nervous system consists of chains of two motor neurons. The first motor neuron of each pair, called the *preganglionic neuron,* resides in the brain or cord. Its axon leaves the CNS to synapse with the second motor neuron (*postganglionic neuron*), whose cell body is located in a ganglion outside the CNS. The axon of the postganglionic neuron then extends to the organ it serves.

The autonomic nervous system has two major functional subdivisions (Figure 21.7). These, the sympathetic and parasympathetic divisions, serve most of the same organs, but generally cause opposing or antagonistic effects.

Parasympathetic Division

The preganglionic neurons of the **parasympathetic,** or **craniosacral,** division are located in brain nuclei of cranial nerves III, VII, IX, X and in the S_2 through S_4 level of the spinal cord. The axons of the preganglionic neurons of the cranial region travel in their respective cranial nerves to the *immediate area* of the head and neck organs to be stimulated. There they synapse with the postganglionic neuron in a **terminal,** or **intramural** (literally, "within the walls"), **ganglion.** The postgangli-

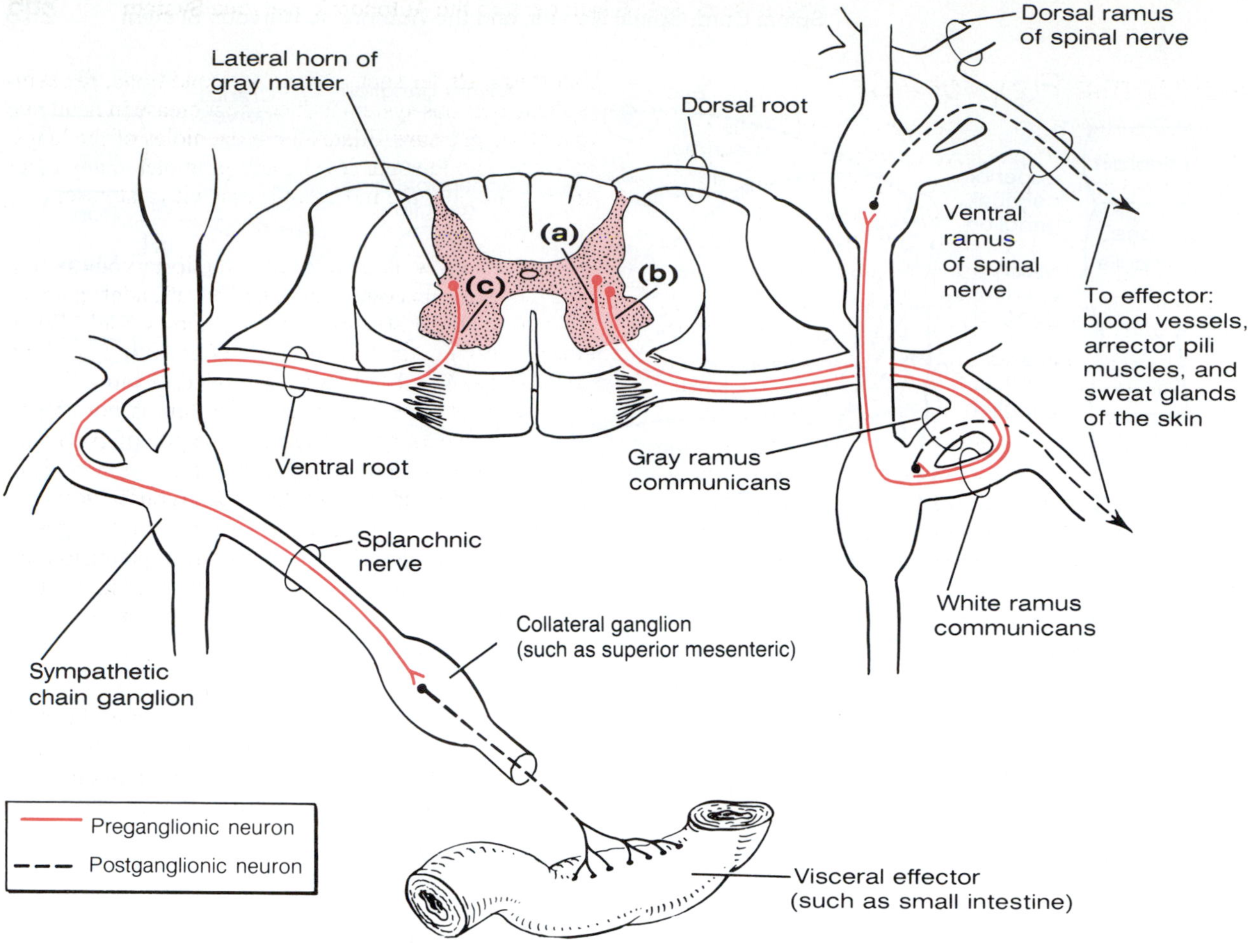

Lateral horn of gray matter
Dorsal root
Dorsal ramus of spinal nerve
(a)
(b)
(c)
Ventral ramus of spinal nerve
To effector: blood vessels, arrector pili muscles, and sweat glands of the skin
Ventral root
Gray ramus communicans
Splanchnic nerve
White ramus communicans
Collateral ganglion (such as superior mesenteric)
Sympathetic chain ganglion
Preganglionic neuron
Postganglionic neuron
Visceral effector (such as small intestine)

F21.8

Sympathetic pathways. (a) Synapse in a sympathetic chain (paravertebral) ganglion at the same level. **(b)** Synapse in a sympathetic chain ganglion at a different level. **(c)** Synapse in a collateral (prevertebral) ganglion.

onic neuron then sends out a very short axon to the organ it serves. In the sacral region, the preganglionic axons leave the ventral roots of the spinal cord and collectively form the **pelvic splanchnic nerves,** which travel to the pelvic cavity. In the pelvic cavity, the preganglionic axons synapse with the postganglionic neurons in ganglia located on or close to the organs served.

Sympathetic Division

The preganglionic neurons of the **sympathetic,** or **thoracolumbar,** division are in the lateral horns of the gray matter of the spinal cord from T_1 through L_2. The preganglionic axons leave the cord via the ventral root (in conjunction with the axons of the somatic motor neurons), enter the spinal nerve, and then travel briefly in the ventral ramus (Figure 21.8). From the ventral ramus, they pass through a small branch called the **white ramus communicans** to enter a **paravertebral ganglion** in the **sympathetic chain,** or **trunk** (see Plate D in the Human Anatomy Atlas), which lies alongside the vertebral column (the literal meaning of *paravertebral*).

Having reached the ganglion, a preganglionic axon may take one of three main courses (see Figure 21.8). First, it may synapse with a postganglionic neuron in the sympathetic chain at that level. Second, the axon may travel upward or downward through the sympathetic chain to synapse with a postganglionic neuron in a paravertebral ganglion at another level. In either of these two instances, the postganglionic axons then reenter the ventral or dorsal ramus of a spinal nerve via a **gray ramus communicans** and travel in the ramus to innervate skin structures (sweat glands, arrector pili muscles attached to hair follicles, and the smooth muscles of blood vessel walls). Third, the axon may pass through the ganglion without synapsing and form part of the **thoracic, lumbar,** and **sacral splanchnic nerves,** which travel to the viscera to synapse with a postganglionic neuron in a **prevertebral** or **collateral ganglion.** The major prevertebral ganglia—the *celiac, superior mesenteric, inferior mesenteric,* and *hypogastric ganglia*—supply the abdominal and pelvic visceral organs. The postganglionic axon then leaves the ganglion and travels to a nearby visceral organ which it innervates.

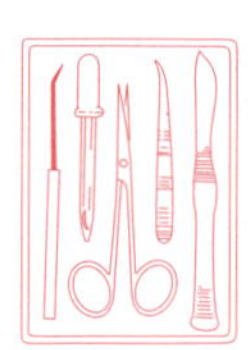

Locate the sympathetic chain on the spinal nerve chart.

Autonomic Functioning

As noted earlier, most body organs served by the autonomic nervous system receive fibers from both the sympathetic and parasympathetic divisions. The only exceptions are the structures of the skin (sweat glands and arrector pili muscles attached to the hair follicles), the pancreas and liver, the adrenal medulla, and essentially all blood vessels except those of the external genitalia, all of which receive sympathetic innervation only. When both divisions serve an organ, they have antagonistic effects. This is because their postganglionic axons release different neurotransmitters. The parasympathetic fibers, called **cholinergic fibers,** release acetylcholine; the sympathetic postganglionic fibers, called **adrenergic fibers,** release norepinephrine. (However, there are isolated examples of postganglionic sympathetic fibers, such as those serving blood vessels in the skeletal muscles, that release acetylcholine.) The preganglionic fibers of both divisions release acetylcholine.

The parasympathetic division is often referred to as the housekeeping, or "resting and digesting," system because it maintains the visceral organs in a state most suitable for normal functions and internal homeostasis; that is, it promotes normal digestion and elimination. In contrast, activation of the sympathetic division is referred to as the "fight or flight" response because it readies the body to cope with situations that threaten homeostasis. Under such emergency conditions, the sympathetic nervous system induces an increase in heart rate and blood pressure, dilates the bronchioles of the lungs, increases blood sugar levels, and promotes many other effects that help the individual cope with a stressor.

As we grow older, our sympathetic nervous system gradually becomes less and less efficient, particularly in causing vasoconstriction of blood vessels. When elderly people stand up quickly after sitting or lying down, they often become light-headed or faint. This is because the sympathetic nervous system is not able to react quickly enough to counteract the pull of gravity by activating the vasoconstrictor fibers. So, blood pools in the feet. This condition, **orthostatic hypotension,** is a type of low blood pressure resulting from changes in body position as described. Orthostatic hypotension can be prevented to some degree if *slow* changes in position are made. This gives the sympathetic nervous system a little more time to react and adjust. ■

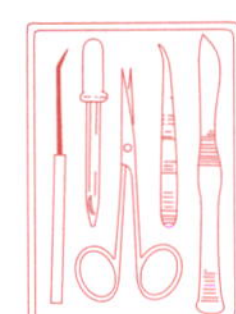

Several body organs are listed in the chart below. *Using your textbook as a reference,* list the effect of the sympathetic and parasympathetic divisions on each.

Organ or function	Parasympathetic effect	Sympathetic effect
Heart		
Bronchioles of lungs		
Digestive tract activity		
Urinary bladder		
Iris of the eye		
Blood vessels (most)		
Penis/clitoris		
Sweat glands		
Adrenal medulla		
Pancreas		

EXERCISE 22

Human Reflex Physiology

OBJECTIVES

1. To define *reflex* and reflex arc.
2. To name, identify, and describe the function of each element of a reflex arc.
3. To indicate why reflex testing is an important part of every physical examination.
4. To describe and discuss several types of reflex activities as observed in the laboratory; to indicate the functional or clinical importance of each; and to categorize each as a somatic or autonomic reflex action.
5. To explain why cord-mediated reflexes are generally much faster than those involving input from the higher brain centers.
6. To investigate differences in reaction time of reflexes and unlearned responses.

MATERIALS

Reflex hammer
Sharp pencils
Cot (if available)
Absorbent cotton (sterile)
Tongue depressor
Metric and 12-in. ruler
Flashlight
100- or 250-ml beaker
10- or 25-ml graduated cylinder
Lemon juice in dropper bottle
Wide-range pH paper
Large laboratory bucket containing freshly prepared 10% household bleach solution (for saliva-soiled glassware)
Disposable autoclave bag
Wash bottle containing 10% bleach solution

See Appendix E, Exercise 22 for links to *Anatomy and PhysioShow: The Videodisc.*

Note to Instructor:

Exercise 22i in Appendix C uses Intelitool equipment to examine human reflex activity.

THE REFLEX ARC

Reflexes are rapid, predictable, involuntary motor responses to stimuli; they are mediated over neural pathways called **reflex arcs**.

All reflex arcs have five essential components (Figure 22.1a):

1. The *receptor* reacts to a stimulus.
2. The *sensory neuron* conducts the afferent impulses to the CNS.
3. The *integration center* consists of one or more synapses in the CNS.
4. The *motor neuron* conducts the efferent impulses from the integration center to an effector.
5. The *effector,* muscle fibers or glands, respond to the efferent impulses by contracting or secreting a product, respectively.

The simple patellar or knee-jerk reflex shown in Figure 22.1b is an example of a simple, two-neuron, *monosynaptic* (literally, "one synapse") reflex arc. It will be demonstrated in the laboratory. However, most reflexes are more complex and *polysynaptic,* involving the participation of one or more association neurons in the reflex arc pathway. A three-neuron reflex arc (flexor reflex) is diagrammed in Figure 22.1c. Since delay or inhibition of the reflex may occur at the synapses, the more synapses encountered in a reflex pathway, the more time is required to effect the reflex.

Reflexes of many types may be considered programmed into the neural anatomy. For example, many *spinal reflexes* (reflexes that are initiated and completed at the spinal cord level such as the flexor reflex) occur without the involvement of higher brain centers. These reflexes work equally well in decerebrate animals (those in which the brain has been destroyed), as long as the spinal cord is functional. Conversely, other reflexes require the involvement of functional brain tissue, since many different inputs must be evaluated before the appropriate reflex is determined. Superficial cord reflexes and pupillary responses to light are in this category. In addition, although many spinal reflexes do not require the involvement of higher centers, the brain is "advised" of spinal cord reflex activity and may alter it by facilitating or inhibiting the reflexes.

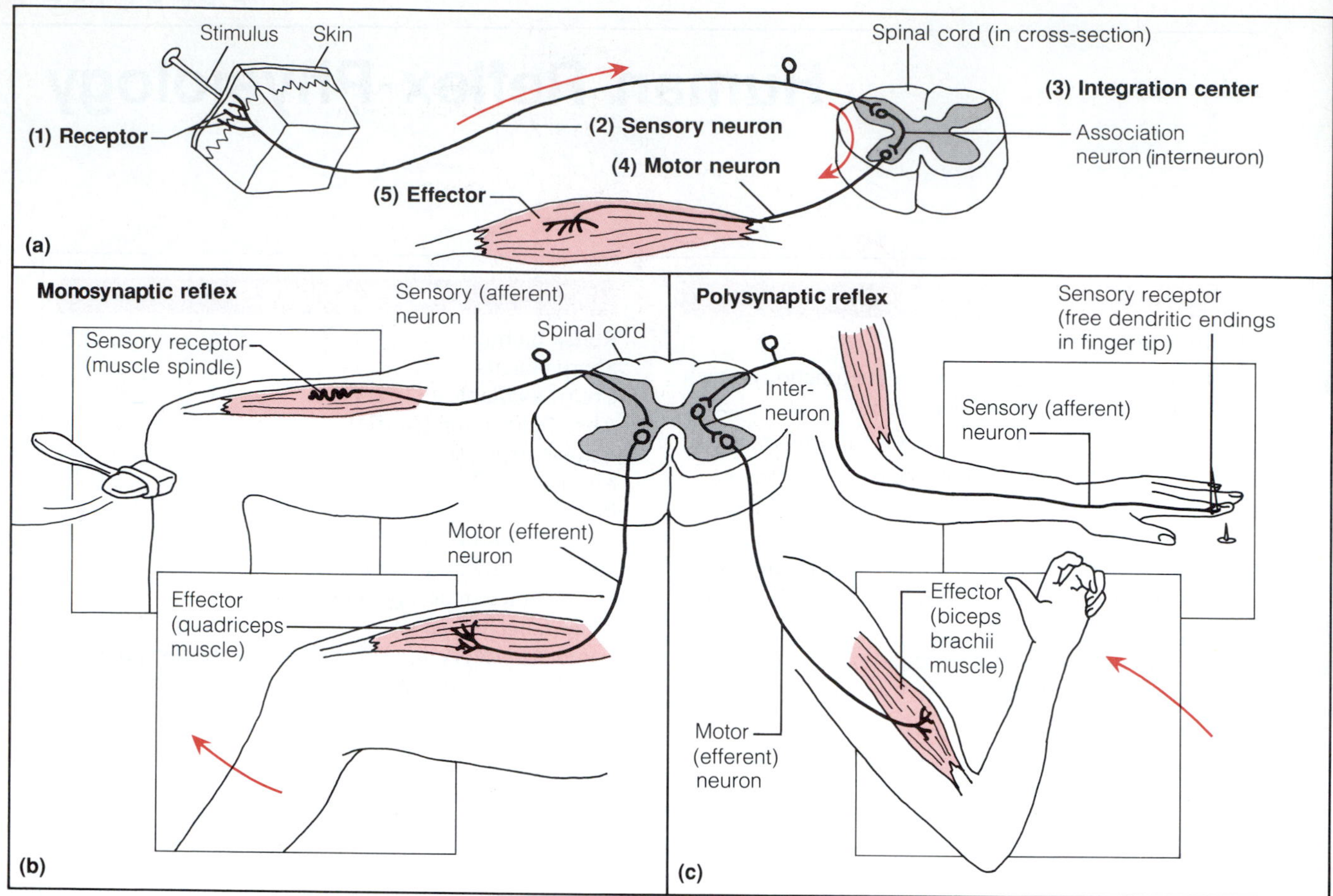

F22.1

Simple reflex arcs. (a) Components of all human reflex arcs: receptor, sensory neuron, integration center (one or more synapses in the CNS), motor neuron, and effector. **(b)** Monosynaptic reflex arc. **(c)** Polysynaptic reflex arc. The integration center is in the spinal cord, and in each example the receptor and effector are in the same limb.

Reflex testing is an important diagnostic tool for assessing the condition of the nervous system. Distorted, exaggerated, or absent reflex responses may indicate degeneration or pathology of portions of the nervous system, often before other signs are apparent.

If the spinal cord is damaged, the easily performed reflex tests can help pinpoint the area (level) of spinal cord injury. Motor nerves above the injured area may be unaffected, whereas those at or below the lesion site may be unable to participate in normal reflex activity. ■

Reflexes can be categorized into one of two large groups: somatic reflexes and autonomic reflexes. **Autonomic** (or visceral) **reflexes** are mediated through the autonomic nervous system and are not subject to conscious control. These reflexes activate smooth muscles, cardiac muscle, and the glands of the body and they regulate body functions such as digestion, elimination, blood pressure, salivation, and sweating. **Somatic reflexes** include all those reflexes that involve stimulation of skeletal muscles by the somatic division of the nervous system. An example of such a reflex is the rapid withdrawal of a hand from a hot object.

SOMATIC REFLEXES

There are several types of somatic reflexes, including several that you will be eliciting during this laboratory session—the stretch, crossed extensor, superficial cord, corneal, and gag reflexes. Some require only spinal cord activity; others require brain involvement as well.

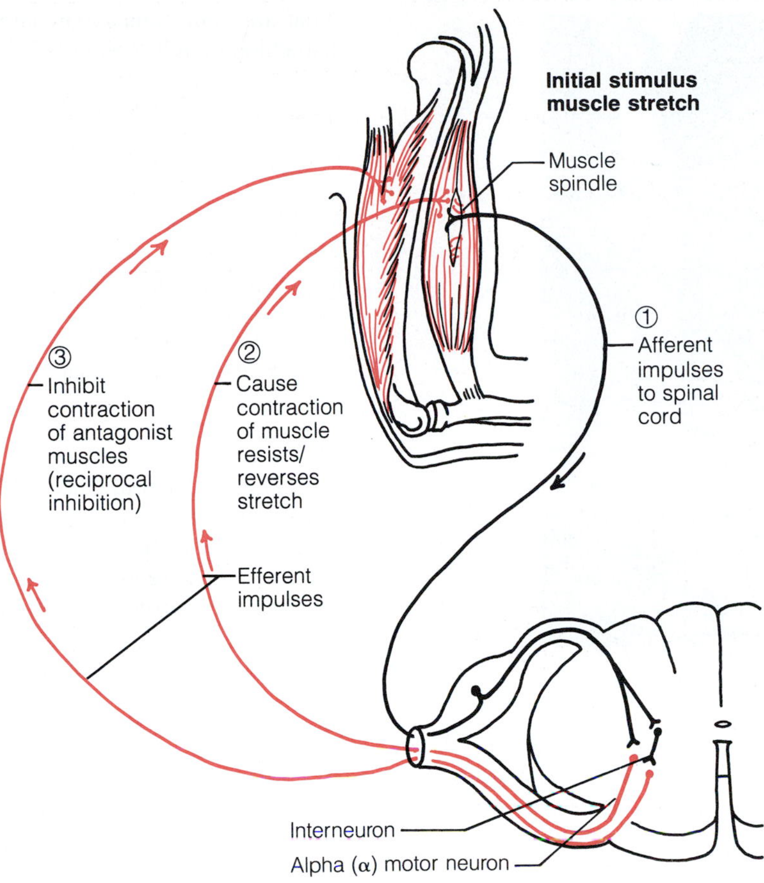

F22.2

Events of the stretch reflex by which muscle stretch is damped. The events are shown in circular fashion. **(1)** Stretching of the muscle activates a muscle spindle. **(2)** Impulses transmitted by afferent fibers from muscle spindle to alpha motor neurons in the spinal cord result in activation of the stretched muscle, causing it to contract. **(3)** Impulses transmitted by afferent fibers from muscle spindle to interneurons in the spinal cord result in reciprocal inhibition of the antagonist muscle.

Spinal Reflexes

STRETCH REFLEXES **Stretch reflexes** are important postural reflexes, normally acting to maintain posture, balance, and locomotion. Stretch reflexes are initiated by tapping a tendon, which stretches the muscle the tendon is attached to. This stimulates the muscle spindles and causes reflex contraction of the stretched muscle or muscles, which resists further stretching. Even as the primary stretch reflex is occurring, impulses are being sent to other destinations as well. For example, branches of the afferent fibers (from the muscle spindles) also synapse with interneurons (association neurons) controlling the antagonist muscles (Figure 22.2). The inhibition of the antagonist muscles that follows, called *reciprocal inhibition,* causes them to relax and prevents them from resisting (or reversing) the contraction of the stretched muscle caused by the main reflex arc. Additionally, impulses are relayed to higher brain centers (largely via the dorsal white columns) to advise of muscle length, speed of shortening, and the like—information needed to maintain muscle tone and posture. Stretch reflexes tend to be hypoactive or absent in cases of peripheral nerve damage or ventral horn disease, and hyperactive in corticospinal tract lesions. They are absent in deep sedation and coma.

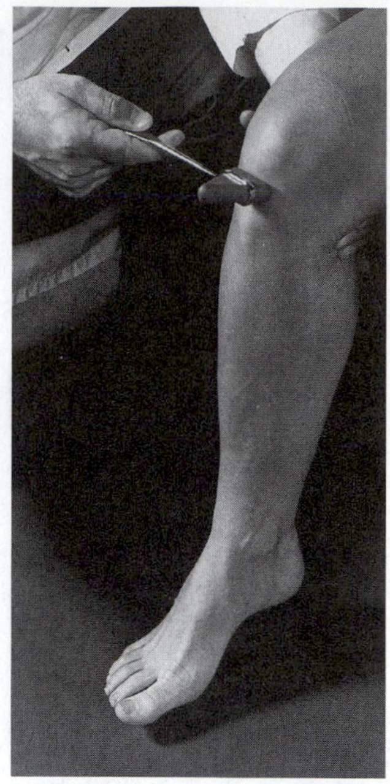

F22.3

Testing the patellar reflex. The examiner supports the subject's knee so that the subject's muscles are relaxed, and then strikes the patellar ligament with the reflex hammer. The proper location may be ascertained by palpation of the patella.

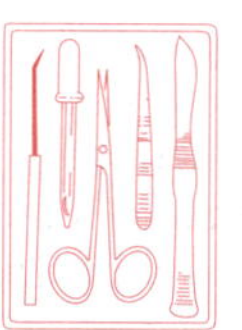

1. Test the **patellar,** or knee-jerk, **reflex** by seating a subject on the laboratory bench with legs hanging free (or with knees crossed). Tap the patellar ligament sharply with the reflex hammer just below the knee to elicit the knee-jerk response, which assesses the L_2–L_4 level of the spinal cord (Figure 22.3). Test both knees and record your observations.

___

___

Which muscles contracted? ___

___

What nerve is carrying the afferent and efferent impulses?

___

2. Test the effect of mental distraction on the patellar reflex by having the subject add a column of three-digit numbers while you test the reflex again. Is the response greater than or less than the first response?

___

What are your conclusions about the effect of mental distraction on reflex activity?

___

___

3. Now test the effect of muscular activity occurring simultaneously in other areas of the body. Have the subject clasp the edge of the laboratory bench and vigorously attempt to pull it upward with both hands. At the same time, test the patellar reflex again. Is the response more or less vigorous than the first response?

___

4. Fatigue also influences the reflex response. The subject should jog in position until she or he is very fatigued (**really fatigued**—no slackers). Test the patellar reflex again and record whether it is more or less vigorous than the first response.

___

Would you say that nervous system activity *or* muscle function is responsible for the changes you have just observed?

___

Explain your reasoning. ___

___

___

___

5. The **Achilles,** or ankle-jerk, **reflex** assesses the first two sacral segments of the spinal cord. With your shoe removed and your foot dorsiflexed slightly to increase the tension of the gastrocnemius muscle, have your partner sharply tap your calcaneal (Achilles) tendon with the reflex hammer (Figure 22.4).

What is the result? ___

___

Does the contraction of the gastrocnemius normally result in the activity you have observed?

___

CROSSED EXTENSOR REFLEX The **crossed extensor reflex** is more complex than the stretch reflex. It consists of a flexor, or withdrawal, reflex followed by extension of the opposite limb.

This reflex is quite obvious when, for example, a stranger suddenly and strongly grips one's arm. The immediate response is to withdraw the clutched arm and

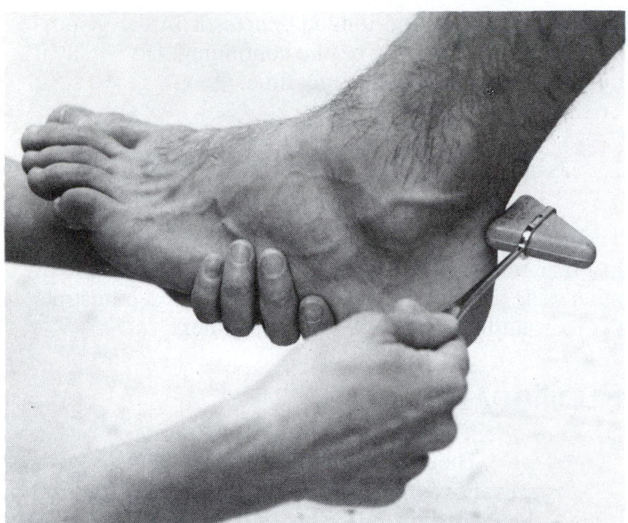

F22.4

Testing the Achilles reflex. The examiner slightly dorsiflexes the subject's ankle by supporting the foot lightly in the hand, and then taps the Achilles tendon just above the ankle.

push the intruder away with the other arm. The reflex is more difficult to demonstrate in a laboratory because it is anticipated, and under these conditions the extensor part of the reflex may be inhibited.

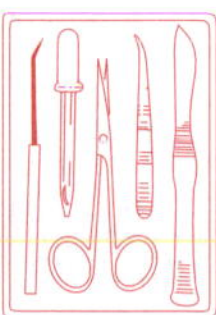

The subject should sit with eyes closed and with the dorsum of one hand resting on the laboratory bench. Obtain a sharp pencil and suddenly prick the subject's index finger. What are the results?

___

___

___

Did the extensor part of this reflex seem to be slow compared to the other reflexes you have observed?

___

What are the reasons for this? ___

___

___

The reflexes that have been demonstrated so far—the stretch and crossed extensor reflexes—are examples of reflexes in which the reflex pathway is initiated and completed at the spinal cord level.

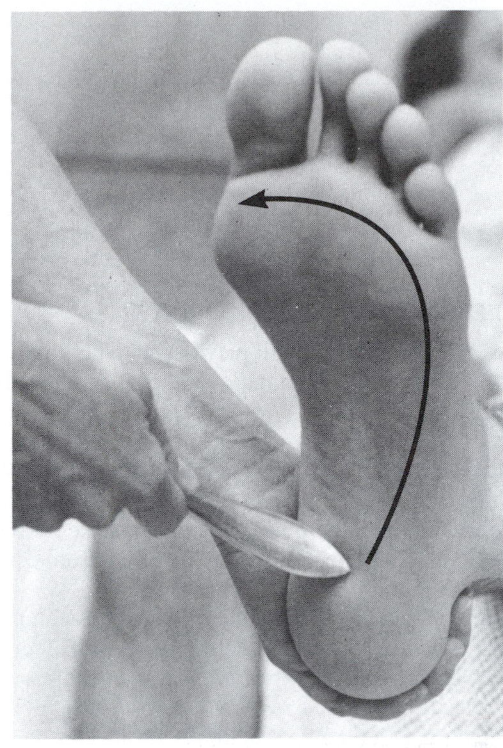

F22.5

Testing the plantar reflex. Using a moderately sharp object, the examiner strokes the lateral border of the subject's sole, starting at the heel and continuing toward the big toe across the ball of the foot.

SUPERFICIAL CORD REFLEXES The **superficial cord reflexes** (abdominal, cremaster, and plantar reflexes) result from pain and temperature changes. They are initiated by stimulation of receptors in the skin and mucosae. The superficial cord reflexes depend *both* on functional upper-motor pathways and on the cord-level reflex arc. Since only the plantar reflex can be tested conveniently in a laboratory setting, we will use this as our example.

The **plantar reflex,** an important neurological test, is elicited by stimulating the cutaneous receptors in the sole of the foot. In adults, stimulation of these receptors causes the toes to flex and move closer together. Damage to the pyramidal (or corticospinal) tract, however, produces *Babinski's sign,* an abnormal response in which the toes flare and the great toe moves in an upward direction. (In newborn infants, Babinski's sign is seen due to incomplete myelination of the nervous system.)

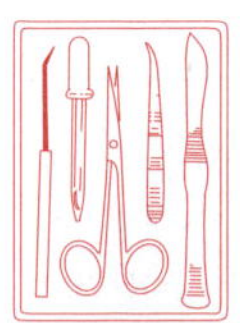

Have the subject remove a shoe and lie on the cot or laboratory bench with knees slightly bent and thighs rotated so that the lateral side of the foot rests on the cot. Alternatively, the subject may sit up and rest the lateral surface of the foot on a chair. Draw the handle of the reflex hammer firmly down the lateral side of the exposed sole from the heel to the base of the great toe (Figure 22.5).

What is the response? ______________________________

____

____

Is this a normal plantar reflex or Babinski's sign?

____

Cranial Nerve Reflex Tests

In these experiments, you will be working with your lab partner to illustrate two somatic reflexes mediated by cranial nerves.

CORNEAL REFLEX The **corneal reflex** is mediated through the trigeminal nerve (cranial nerve V). The absence of this reflex is an ominous sign, because it often indicates damage to the brain stem, resulting from compression of the brain or other trauma.

Stand to one side of the subject; the subject should look away from you toward the opposite wall. Wait a few seconds and then quickly, *but gently,* touch the subject's cornea (on the side toward you) with a wisp of absorbent cotton. What is the reaction?

____

What is the function of this reflex?

____

____

____

Was the sensation that of touch *or* of pain?

____

Why? ______________________________

____

GAG REFLEX The **gag reflex** tests the somatic motor responses of cranial nerves IX and X. When the oral mucosa on the side of the uvula is stroked, each side of the mucosa should rise, and the amount of elevation should be equal.*

For this experiment, select a subject who does not have a queasy stomach, because regurgitation is a possibility. Stroke the oral mucosa on each side of the subject's uvula with a tongue depressor. What happens?

____

____

* The uvula is the fleshy tab hanging from the roof of the mouth just above the root of the tongue.

Discard the used tongue depressor in the disposable autoclave bag before continuing. Do *not* lay it on the laboratory bench at any time.

AUTONOMIC REFLEXES

The autonomic reflexes include the pupillary, ciliospinal, and salivary reflexes, as well as a multitude of other reflexes. Work with your partner to demonstrate the four autonomic reflexes described next.

Pupillary Reflexes

There are several types of pupillary reflexes. The **pupillary light reflex** and the **consensual reflex** will be examined here. In both of these pupillary reflexes, the retina of the eye is the receptor, the optic nerve (cranial nerve II) contains the afferent fibers, the oculomotor nerve (cranial nerve III) is responsible for conducting efferent impulses to the eye, and the smooth muscle of the iris is the effector. Many central nervous system centers are involved in the integration of these responses. Absence of the normal pupillary reflexes is generally a late indication of severe trauma or deterioration of the vital brain stem tissue due to metabolic imbalance.

1. Conduct the reflex testing in an area where the lighting is relatively dim. Before beginning, obtain a metric ruler to measure and record the size of the subject's pupils as best you can.

Right pupil: ________ mm Left pupil: ________ mm

2. Stand to the left of the subject to conduct the testing. The subject should shield his or her right eye by holding a hand vertically between the eye and the right side of the nose.

3. Shine a flashlight into the subject's left eye. What is the pupillary response?

____

Measure the size of the left pupil: ____________ mm

4. Observe the right pupil. Has the same type of change (called a *consensual response*) occurred in the right eye?

____

Measure the size of the right pupil: ____________ mm

The consensual response, or any reflex observed on one side of the body when the other side has been stimulated, is called a **contralateral response.** The pupillary light response, or any reflex occurring on the same side stimulated, is referred to as an **ipsilateral response.**

When a contralateral response occurs, what does this indicate about the pathways involved?

___

___

Was the sympathetic *or* the parasympathetic division of the autonomic nervous system active during the testing of these reflexes?

___

What is the function of these pupillary responses?

___

___

___

Ciliospinal Reflex

The **ciliospinal reflex** is another example of reflex activity in which pupillary responses can be observed. This response may initially seem a little bizarre, especially in view of the consensual reflex just demonstrated. While observing the subject's eyes, gently stroke the skin (or just the hairs) on the left side of the back of the subject's neck, close to the hairline.

What is the reaction of the left pupil? ___

The reaction of the right pupil? ___

If you see no reaction, repeat the test using a gentle pinch in the same area.

The response you should have noted—pupillary dilation—is consistent with the pupillary changes occurring when the sympathetic nervous system is stimulated. Such a response may also be elicited in a single pupil when more impulses from the sympathetic nervous system reach it for any reason. For example, when the left side of the subject's neck was stimulated, sympathetic impulses to the left iris increased, resulting in the ipsilateral reaction of the left pupil.

On the basis of your observations, would you say that the sympathetic innervation of the two irises is closely integrated?

___ Why or why not? ___

___

___

Salivary Reflex

Unlike the other reflexes, in which the effectors were smooth or skeletal muscles, the effectors of the **salivary reflex** are glands. The salivary glands secrete varying amounts of saliva in response to reflex activation.

1. Obtain a small beaker, a graduated cylinder, lemon juice, and wide-range pH paper. After refraining from swallowing for 2 minutes, the subject is to expectorate (spit) the accumulated saliva into a small beaker. Using the graduated cylinder, measure the volume of the expectorated saliva and determine its pH.

Volume: ___ cc pH: ___

2. Now place 2 or 3 drops of lemon juice on the subject's tongue. Allow the lemon juice to mix with the saliva for 5 to 10 seconds, and then determine the pH of the subject's saliva by touching a piece of pH paper to the tip of his tongue.

pH: ___

As before, the subject is to refrain from swallowing for 2 minutes. After the 2 minutes is up, again collect and measure the volume of the saliva and determine its pH.

Volume: ___ cc pH: ___

3. How does the volume of saliva collected after the application of the lemon juice compare with the volume of the first saliva sample?

___

How does the final saliva pH reading compare to the initial reading?

___

To that obtained 10 seconds after the application of lemon juice?

___

What division of the autonomic nervous system mediates the reflex release of saliva?

___

Dispose of the saliva-containing beakers and the graduated cylinders in the laboratory bucket that contains bleach and put the used pH paper into the disposable autoclave bag. Wash the bench down with 10% bleach solution before continuing.

REACTION TIME OF UNLEARNED RESPONSES

The time required for reaction to a stimulus depends on many factors—sensitivity of the receptors, velocity of nerve conduction, the number of neurons and synapses involved, and the speed of effector activation, to name just a few. The type of response to be elicited is also important. If the response involves a reflex arc, the synapses are facilitated and response time will be short. If, on the other hand, the response can be categorized as an unlearned response, then a far larger number of neural pathways and many types of higher intellectual activities—including choice and decision making—will be involved, and the time for response will be considerably lengthened.

There are various ways of testing reaction time of unlearned responses. The tests range in difficulty from simple to ultrasophisticated. Since the objective here is to demonstrate the major time difference between reflexes and unlearned responses, the simple approach will suffice.

1. Using a reflex hammer, elicit the patellar reflex in your partner. Note the relative reaction time needed for this reflex to occur.

2. Now test the reaction time for unlearned responses. The subject should hold his hand out, with the thumb and index finger extended. Hold a 12-in. ruler so that its end is exactly 1 in. above the subject's outstretched hand. The ruler should be in the vertical position with the numbers reading from the bottom up. When the ruler is dropped, the subject should be able to grasp it between thumb and index finger as it passes, without having to change position. Have the subject catch the ruler five times, varying the time between trials. The relative speed of reaction can be determined by reading the number on the ruler at the point of the subject's fingertips. (Thus if the number at the fingertips is 6 in., the subject was unable to catch the ruler until 7 in. of length had passed through his fingers; 6 in. of ruler length plus 1 in. to account for the distance of the ruler above the hand.) Record the number of inches that pass through the subject's fingertips for each trial:

Trial 1: ____________ in. Trial 4: ____________ in.

Trial 2: ____________ in. Trial 5: ____________ in.

Trial 3: ____________ in.

3. Perform the test again, but this time say a simple word each time you release the ruler. Designate a specific word as a signal for the subject to catch the ruler. On all other words, the subject is to allow the ruler to pass through his fingers. Trials in which the subject erroneously catches the ruler are to be disregarded. Record the distance the ruler travels for five *successful* trials:

Trial 1: ____________ in. Trial 4: ____________ in.

Trial 2: ____________ in. Trial 5: ____________ in.

Trial 3: ____________ in.

Did the addition of a specific word to the stimulus increase or decrease the reaction time?

____

4. Perform the testing once again to investigate the subject's reaction to word association. As you drop the ruler, say a word—for example, *hot.* The subject is to respond with a word he associates with the stimulus word—for example, *cold*—catching the ruler as he responds. If he is unable to make a word association, he must allow the ruler to pass through his fingers. Record the distance the ruler travels for five successful trials, as well as the number of times the ruler is not caught by the subject.

Trial 1: ____________ in. Trial 4: ____________ in.

Trial 2: ____________ in. Trial 5: ____________ in.

Trial 3: ____________ in.

The number of times the subject was unable to catch the ruler:

____

You should have noticed quite a large variation in reaction time in this series of trials. Why is this so?

____

____

____

____

General Sensation

OBJECTIVES

1. To recognize various types of general sensory receptors as studied in the laboratory, and to describe the function and location of each type.
2. To define *exteroceptor, interoceptor,* and *proprioceptor.*
3. To demonstrate and relate differences in relative density and distribution of tactile and thermoreceptors in the skin.
4. To define *tactile localization* and describe how this ability varies in different areas of the body.
5. To explain the tactile two-point discrimination test, and to state its anatomical basis.
6. To define *referred pain.*
7. To define *adaptation, negative afterimage,* and *projection.*

MATERIALS

Compound microscope
Histologic slides of Pacinian corpuscles (longitudinal section), Meissner's corpuscles, Golgi tendon organs, and muscle spindles
Mall probes (obtainable from Carolina Biological Supply, code #62-7400), or pointed metal rods of approximately 1-in. diameter and 2- to 3-in. length, with insulated blunt end
Large beaker of ice water; chipped ice
Hot water bath set at 45°C
Laboratory thermometer
Fine-point, felt-tipped markers (black, red, and blue)
Small metric ruler
Von Frey's hairs (or 1-in. lengths of horsehair glued to a matchstick) or sharp pencil
Towel
Caliper or esthesiometer
Four coins (nickels or quarters)
Three large finger bowls or 1000-ml beakers

See Appendix E, Exercise 23 for links to *Anatomy and PhysioShow: The Videodisc.*

It can be said without reservation that people are irritable creatures. Hold a sizzling steak before them and their mouths water. Flash your high beams in their eyes on the highway and they cuss. Tickle them and they giggle. These "irritants" (the steak, the light, and the tickle) and many others are stimuli that continually assault us.

The body's **sensory receptors** react to stimuli or changes within the body and in the external environment. The tiny sensory receptors of the *general senses* react to touch, pressure, pain, heat, cold, and changes in position and are distributed throughout the body. In contrast to these widely distributed *general sensory receptors,* the receptors of the special senses are large, complex *sense organs* or small, localized groups of receptors. The *special senses* include sight, hearing, equilibrium, smell, and taste.

Sensory receptors may be classified according to the source of their stimulus. **Exteroceptors** react to stimuli in the external environment, and typically they are found close to the body surface. Exteroceptors include the simple cutaneous receptors in the skin and the highly specialized receptor structures of the special senses (the vision apparatus of the eye, and the hearing and equilibrium receptors of the ear, for example). **Interoceptors** respond to stimuli arising within the body. Interoceptors are found in the internal visceral organs, and include stretch receptors (in walls of hollow organs), chemoreceptors, and others. The **proprioceptors,** located in joint capsules and in skeletal muscles and their tendons, also respond to internal stimuli. They provide information on the position and degree of stretch of those structures.

The receptors of the special sense organs are complex and deserve considerable study. Thus the special senses (vision, hearing, equilibrium, taste, and smell) are covered separately in Exercises 24 through 26. Only the anatomically simpler general sensory receptors—cutaneous receptors and proprioceptors—will be studied in this exercise.

STRUCTURE OF GENERAL SENSORY RECEPTORS

You cannot become aware of changes in the environment unless your sensory neurons and their receptors are operating properly. Sensory receptors are modified dendritic endings (or specialized cells associated with the dendrites) that are sensitive to certain environmental

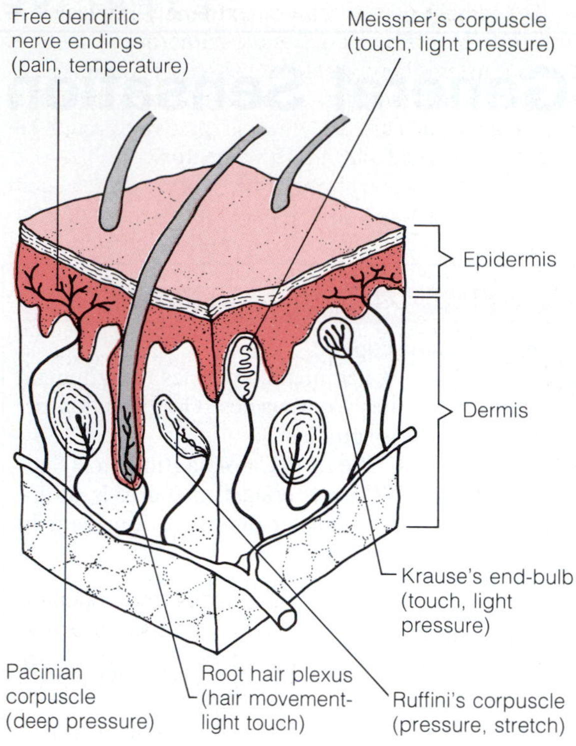

F23.1

Cutaneous receptors. Free dendritic nerve endings, root hair plexus, Meissner's corpuscle, Pacinian corpuscle, Krause's end bulb, and Ruffini's corpuscle. See also Plates 9, 10, and 11 of the Histology Atlas.

stimuli. They react to such stimuli by initiating a nerve impulse. Several histologically distinct types of general sensory receptors have been identified in the skin. Their structures are depicted in Figure 23.1.

Many references link each type of receptor to specific stimuli, but there is still considerable controversy about the precise qualitative function of each receptor. It may be that the responses of all these receptors overlap considerably. Certainly, intense stimulation of any of them is always interpreted as pain.

The least specialized of the cutaneous receptors are the **free dendritic endings** of sensory neurons (Figure 23.1 and Plate 10 of the Histology Atlas), which respond chiefly to pain and temperature. (The pain receptors are widespread in the skin and make up a sizable portion of the visceral interoceptors.) Certain free dendritic endings associate with specific epidermal cells to form **Merkel discs,** or entwine in hair follicles to form **root hair plexuses.** Both Merkel discs and root hair plexuses function as light touch receptors.

The other cutaneous receptors are a bit more complex, with the dendritic endings *encapsulated* by connective tissue cells. **Meissner's corpuscles,** commonly referred to as *tactile receptors* because they respond to light touch, are located in the dermal papillae of the skin. Although **Krause's end bulbs** and **Ruffini's corpuscles** were presumed to be thermoreceptors, now most authorities classify Krause's end bulbs with Meissner's corpuscles as a special kind of touch, or light pressure, receptor. Ruffini's corpuscles appear to respond to deep pressure and stretch stimuli. As you inspect Figure 23.1, note that all of the encapsulated receptors are quite similar, with the possible exception of the **Pacinian corpuscles** (deep pressure receptors), which are anatomically more distinctive and lie deepest in the dermis.

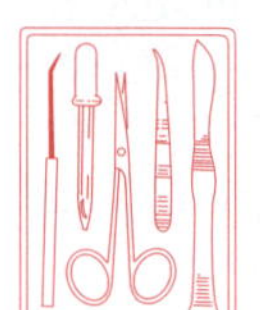

1. Obtain histologic slides of Pacinian and Meissner's corpuscles. Locate, under low power, a Meissner's corpuscle in the dermal layer of the skin. As mentioned above, these are usually found in the dermal papillae. Then, switch to the oil immersion lens for a detailed study. Notice that the naked dendritic fibers within the capsule are aligned parallel to the skin surface. Compare your observations to Figure 23.1 and Plate 9 of the Histology Atlas.

2. Next observe a Pacinian corpuscle located much deeper in the dermis. Try to identify the slender naked dendrite ending in the center of the receptor and the heavy capsule of connective tissue surrounding it (which looks rather like an onion cut lengthwise). Also, notice how much larger the Pacinian corpuscles are than the Meissner's corpuscles. Compare your observations to the views shown in Figure 23.1 and Plate 11 of the Histology Atlas.

3. Obtain slides of muscle spindles and Golgi tendon organs, the two major types of proprioceptors (see Figure 23.2). In the slide of **muscle spindles,** note that minute extensions of the dendrites of the sensory neurons coil around specialized slender skeletal muscle cells called **intrafusal cells,** or **fibers.** The **Golgi tendon organs** are composed of dendrites that ramify through the tendon tissue close to the muscle tendon attachment. Stretching of the muscle or tendon excites both types of receptors, which then transmit impulses that ultimately reach the cerebellum for interpretation. Compare your observations to Figure 23.2.

RECEPTOR PHYSIOLOGY

Sensory receptors act as *transducers,* changing environmental stimuli into afferent nerve impulses. Since the action potential generated in all nerve fibers is essentially identical, the stimulus is identified entirely by the area of the brain's sensory cortex that is stimulated (which, of course, differs for the various afferent nerves).

Three qualities of cutaneous sensations have traditionally been recognized: tactile (touch), temperature, and pain. Mapping these sensations on the skin has revealed that the sensory receptors for these qualities are not distributed uniformly. Instead, they have discrete locations and are characterized by clustering at certain points—**punctate distribution.**

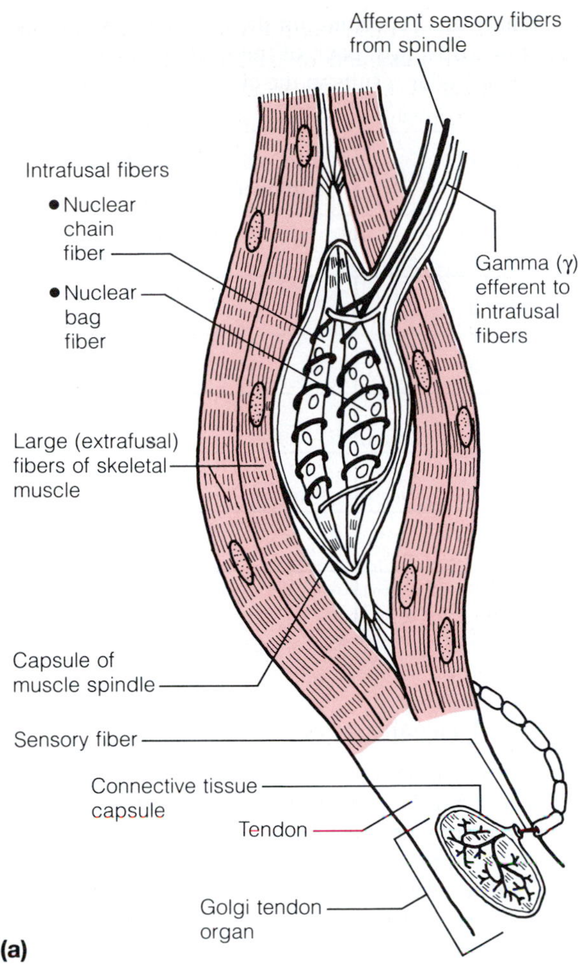

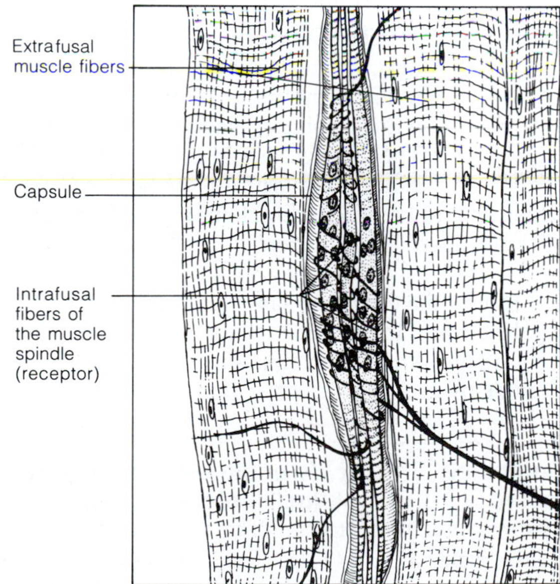

F23.2

Proprioceptors. **(a)** Diagrammatic view of a muscle spindle and Golgi tendon organ. **(b)** Drawing of a photomicrograph of a muscle spindle. See corresponding Plate 12 in the Histology Atlas.

The simple pain receptors, extremely important in protecting the body, are the most numerous. Touch receptors cluster where greater sensitivity is desirable, as on the hands and face. It is surprising to learn that rather large areas of the skin are quite insensitive to touch because of a relative lack of touch receptors.

There are several simple experiments you can conduct to investigate the location and physiology of cutaneous receptors. In each of the following activities, work in pairs with one person as the subject and the other as the experimenter. After you have completed an experiment, switch roles and go through the procedures again so that all class members obtain individual results. Keep an accurate account of each test that you perform.

Density and Location of Temperature and Touch Receptors

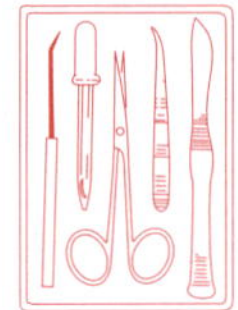

1. Obtain 2 Mall probes (or temperature rods), Von Frey hairs (or a sharp pencil), and black, red, and blue felt-tipped markers and bring them to your laboratory bench.

2. Place one Mall probe (or temperature rod) in a beaker of ice water and the other in a water bath controlled at 45°C. With a felt marker, draw a square (2 cm on each side) on the ventral surface of the subject's forearm. During the following tests, the subject's eyes are to remain closed. The subject should tell the examiner when a stimulus is detected.

3. Working in a systematic manner from one side of the marked square to the other, gently touch the Von Frey's hairs or a sharp pencil tip to different points within the square. The *hairs* should be applied with a pressure that just causes them to bend; use the same pressure for each contact. Do not apply deep pressure. The goal is to stimulate only the more superficially located Meissner's corpuscles (as opposed to the Pacinian corpuscles located in the subcutaneous tissue). Mark with a *black dot* all points at which the touch is perceived.

4. Remove the Mall probe (or temperature rod) from the ice water and quickly wipe it dry. Repeat the procedure outlined above, noting all points of cold perception with a *blue dot.** Perception should be for temperature, not simply touch.

5. Remove the second Mall probe from the 45°C water bath and repeat the procedure once again, marking all points of heat perception with a *red dot.**

* The Mall probes will have to be returned to the water baths approximately every 2 minutes to maintain the desired testing temperatures.

6. After each student has acted as the subject, produce a "map" of your own receptor areas in the squares provided here.

2 cm

Student 1 ______________

2 cm

Student 2 ______________

How does the density of the heat receptors correspond to that of the touch receptors?

____

To that of the cold receptors? ______________

____

On the basis of your observations, which of these three types of receptors appears to be most abundant (at least in the area tested)?

____

Two-Point Discrimination Test

The density of the touch receptors varies significantly in different areas of the body. In general, areas that have the greatest density of tactile receptors have a heightened ability to "feel." These areas correspond to areas that receive the greatest motor innervation; thus they are also typically areas of fine motor control.

On the basis of this information, which areas of the body do you *predict* will have the greatest density of touch receptors?

____

____

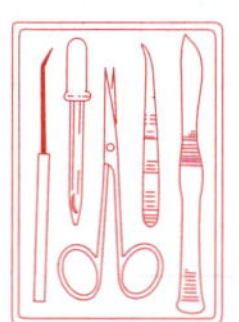

1. Using a caliper or esthesiometer and a metric ruler, test the ability of the subject to differentiate two distinct sensations when the skin is touched simultaneously at two points. Beginning with the face, start with the caliper arms completely together. Gradually increase the distance between the arms, testing the subject's skin after each adjustment. Continue with this testing procedure until the subject reports that *two points* of contact can be felt. This measurement, the smallest distance at which two points of contact can be felt, is the **two-point threshold.**

2. Repeat this procedure on the back and palm of the hand, fingertips, lips, back of the neck, and back of the calf. Record your results in the chart below.

Body area tested	Two-point threshold (millimeters)
Face	
Back of hand	
Palm of hand	
Fingertips	
Lips	
Back of neck	
Back of calf	

Tactile Localization

Tactile localization is the ability to determine which portion of the skin has been touched. The tactile receptor field of the body periphery has a corresponding "touch" field in the brain's somatosensory association area. Some body areas are well represented with touch receptors, allowing tactile stimuli to be localized with great accuracy, but the density of the touch receptors in other body areas allows only a crude discrimination.

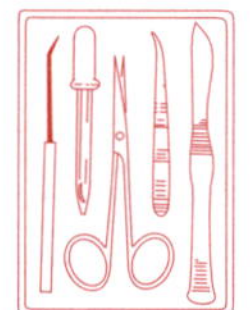

1. The subject's eyes should be closed during the testing. The experimenter touches the palm of the subject's hand with a pointed black felt-tipped marker. The subject should then try to touch the exact point with his or her own marker, which should be of a different color. Measure the error of localization in millimeters.

2. Repeat the test in the same spot twice more, recording the error of localization for each test. Average the results of the three determinations and record it in the chart below.

Body area tested	Average error of localization (millimeters)
Palm of hand	
Fingertip	
Ventral forearm	
Ventral surface of arm	
Upper back	

Does the ability to localize the stimulus improve the second time?

__________ The third time? __________ Explain. _____

____

____

3. Repeat the above procedure on a fingertip, the ventral forearm, the ventral surface of the upper arm, and the upper back between the shoulder blades. Record the averaged results in the chart.

Adaptation of Touch Receptors

The number of impulses transmitted by sensory receptors often changes both with the intensity of the stimulus and with the length of time the stimulus is applied. In many cases, when a stimulus is applied for a prolonged period, the rate of receptor discharge slows and conscious awareness of the stimulus declines or is lost until some type of stimulus change occurs. This phenomenon is referred to as **adaptation.** The touch receptors adapt particularly rapidly, which is highly desirable. Who, for instance, would want to be continually aware of the pressure of clothing on their skin? The simple experiments to be conducted next allow you to investigate this phenomenon of adaptation.

1. The subject's eyes should be closed. Place a coin on the anterior surface of the subject's forearm, and determine how long the sensation persists for the subject. Duration of the sensation:

______________ sec

2. Repeat the test, placing the coin at a different forearm location. How long does the sensation persist at the second location?

______________ sec

3. After awareness of the sensation has been lost at the second site, stack three more coins atop the first one.

Does the pressure sensation return? ______________

If so, for how long is the subject aware of the pressure in this instance?

______________ sec

Are the same receptors being stimulated when the four coins, rather than the one coin, are used?

__________ Explain. ______________________________

____

4. To further illustrate the adaptation of touch receptors—in this case, the root hair plexuses of the hair follicles—gently and slowly bend one hair shaft with a pen or pencil until it springs back (away from the pencil) to its original position. Is the tactile sensation greater when the hair is being slowly bent or when it springs back?

____

Why is the adaptation of the touch receptors in the hair follicles particularly important to a woman who wears her hair in a ponytail? If the answer is not immediately apparent, consider the opposite phenomenon: what would happen, in terms of sensory input from her hair follicles, if these receptors did not exhibit adaptation?

____

____

Adaptation of Temperature Receptors

Adaptation of the temperature receptors can be tested using some very unsophisticated methods.

1. Obtain three large finger bowls or 1000-ml beakers and fill the first with 45°C water. Have the subject immerse her or his left hand in the water and report the sensation. Keep the left hand immersed for 1 min and then also immerse the right hand in the same bowl.

What is the sensation of the left hand when it is first immersed?

____

What is the sensation of the left hand after 1 min as compared to the sensation in the right hand just immersed?

____

Had adaptation occurred in the left hand? ___________

2. Rinse both hands in tap water, dry them, and wait 5 min before conducting the next test. Just before beginning the test, refill the finger bowl with fresh 45°C water, fill a second with ice water, and fill a third with water at room temperature.

3. Place the *left* hand in the ice water and the *right* hand in the 45°C water. What is the sensation in each hand after 2 min as compared to the sensation perceived when the hands were first immersed?

____

____

Which hand seemed to adapt more quickly?

____

4. After reporting these observations, the subject should then place both hands simultaneously into the finger bowl containing the water at room temperature. Record the sensation in the left hand:

___

The right hand: ___

The sensations that the subject experiences when both hands were put into room-temperature water are called **negative afterimages.** They are explained by the fact that sensations of heat and cold depend on the rapidity of heat loss or gain by the skin and differences in the temperature gradient.

Referred Pain

Experiments on pain receptor localization and adaptation are commonly conducted in the laboratory. However, there are certain problems with such experiments. Pain receptors are densely distributed in the skin, and they adapt very little, if at all. (This lack of adaptability is due to the protective function of the receptors. The sensation of pain often indicates tissue damage or trauma to body structures.) Thus no attempt will be made in this exercise to localize the pain receptors or to prove their nonadaptability, since both would cause needless discomfort to those of you acting as subjects and would not add any additional insight.

However, the phenomenon of referred pain is easily demonstrated in the laboratory, and such experiments provide information that may be useful in explaining common examples of this phenomenon. **Referred pain** is a sensory experience in which pain is perceived as arising in one area of the body when in fact another, often quite remote area is receiving the painful stimulus. Thus the pain is said to be "referred" to a different area. The phenomenon of **projection,** the process by which the brain refers sensations to their *usual* point of stimulation, provides the most simple explanation of such experiences. Many of us have experienced referred pain as a radiating pain in the forehead after quickly swallowing an ice-cold drink. Referred pain is important in many types of clinical diagnosis, since damage to many visceral organs results in this phenomenon. For example, inadequate oxygenation of the heart muscle often results in pain being referred to the chest wall and left shoulder (*angina pectoris*), and the reflux of gastric juice into the esophagus causes a sensation of intense discomfort in the thorax referred to as *heartburn.* In addition, amputees often report *phantom limb* pain—feelings of pain that appear to be coming from a part of the body that is no longer there.

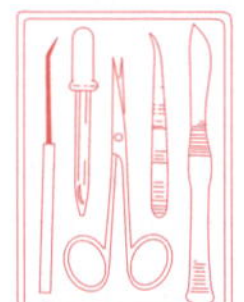

Immerse the subject's elbow in a finger bowl containing ice water. In the chart below, record the quality (such as discomfort, tingling, or pain) and the quality progression of the sensations he or she reports for 2 min. Also record the location of the perceived sensations. The ulnar nerve, which serves the medial third of the hand, is involved in the phenomenon of referred pain experienced during this test. How does the localization of this referred pain correspond to the areas served by the ulnar nerve?

___

___

___

___

Time of observation	Quality of sensation	Localization of sensation
On immersion		
After 1 min		
After 2 min		

Special Senses: Vision

OBJECTIVES

1. To describe the structure and function of the accessory visual structures.
2. To identify the structural components of the eye when provided with a model, an appropriate diagram, or a preserved sheep or cow eye, and list the function(s) of each.
3. To describe the cellular makeup of the retina.
4. To discuss the mechanism of image formation on the retina.
5. To trace the visual pathway to the optic cortex and note the effects of damage to various parts of this pathway.
6. To define the following terms:

   refraction | *myopia*
   accommodation | *hyperopia*
   convergence | *cataract*
   astigmatism | *glaucoma*
   emmetropia | *conjunctivitis*

7. To discuss the importance of the pupillary and convergence reflexes.
8. To explain the difference between rods and cones with respect to visual perception and retinal localization.
9. To state the importance of an ophthalmoscopic examination.

MATERIALS

Dissectible eye model
Chart of eye anatomy
Preserved cow or sheep eye
Dissecting pan and instruments
Protective skin cream or disposable gloves

Histologic section of an eye showing retinal layers
Compound microscope

Snellen eye chart (floor marked with chalk to indicate 20-ft distance from posted Snellen chart)
Ishihara's color-blindness plates
Laboratory lamp or penlight
Ophthalmoscope
1-inch-diameter discs of colored paper (white, red, blue, green)
White, red, blue, and green chalk
Metric ruler; meter stick
Test tubes
Common straight pins

See Appendix D, Exercise 24 for links to A.D.A.M. Standard.

See Appendix E, Exercise 24 for links to *Anatomy and PhysioShow: The Videodisc.*

ANATOMY OF THE EYE

External Anatomy and Accessory Structures

The adult human eye is a sphere measuring about 2.5 cm (1 inch) in diameter. Only about one-sixth of the eye's anterior surface is observable; the remainder is enclosed and protected by a cushion of fat and the walls of the bony orbit.

Six **extrinsic eye muscles** attached to the exterior surface of each eyeball control eye movement and make it possible for the eye to follow a moving object. The names and positioning of these extrinsic muscles are noted in Figure 24.1. Their actions are given in the chart accompanying that figure.

The anterior surface of each eye is protected by the **eyelids,** or **palpebrae.** (See Figure 24.2.) The medial and lateral junctions of the upper and lower eyelids are referred to as the **medial** and **lateral canthus** (respectively). The **caruncle,** a fleshy elevation at the medial canthus, produces a whitish oily secretion. A mucous membrane, the **conjunctiva,** lines the internal surface of the eyelids (as the *palpebral conjunctiva*) and continues over the anterior surface of the eyeball to its junction with the corneal epithelium (as the *ocular,* or *bulbar, conjunctiva*). The conjunctiva secretes mucus, which aids in lubricating the eyeball. Inflammation of the conjunctiva, often accompanied by redness of the eye, is called **conjunctivitis.**

Name	Innervation (cranial nerve)	Action
Lateral rectus	VI	Moves eye horizontally (laterally)
Medial rectus	III	Moves eye horizontally (medially)
Superior rectus	III	Elevates eye
Inferior oblique	III	Elevates eye and turns it laterally
Inferior rectus	III	Depresses eye
Superior oblique	IV	Depresses eye and turns it laterally

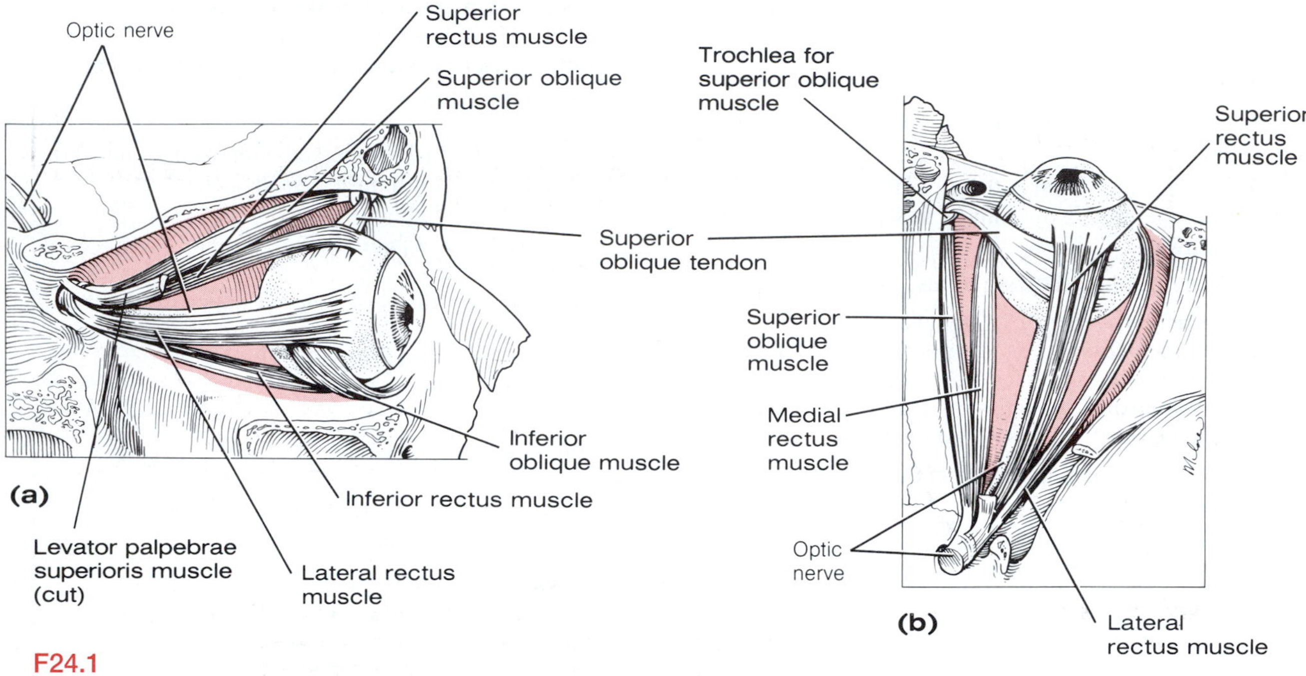

F24.1

Extrinsic muscles of the eye. (a) Lateral view of right eye. **(b)** Superior view of right eye.

Projecting from the border of each eyelid is a row of short hairs, the **eyelashes.** The **ciliary glands,** a type of sweat gland, lie between the eyelash hair follicles and help lubricate the eyeball. An inflammation of one of these glands is called a **sty.** Small sebaceous glands associated with the hair follicles and the larger **meibomian glands,** located posterior to the eyelashes, secrete an oily substance.

The **lacrimal apparatus** consists of the **lacrimal gland, lacrimal canals, lacrimal sac,** and the **nasolacrimal duct.** The lacrimal glands are situated superior to the lateral aspect of each eye. They continually liberate a dilute salt solution (tears) that flows onto the anterior surface of the eyeball through several small ducts. The tears flush across the eyeball into the lacrimal canals medially, then into the lacrimal sac, and finally into the nasolacrimal duct, which empties into the nasal cavity. The lacrimal secretion also contains **lysozyme,** an antibacterial enzyme. Because it constantly flushes the eyeball, the lacrimal fluid cleanses and protects the eye surface as it moistens and lubricates it. As we age, our eyes tend to become dry due to decreased lacrimation, and thus are more vulnerable to bacterial invasion and irritation.

Observe the eyes of another student and identify as many of the accessory structures as possible. Ask the student to look to the left. What extrinsic eye muscles are responsible for this action?

Right eye ______________________________

Left eye ______________________________

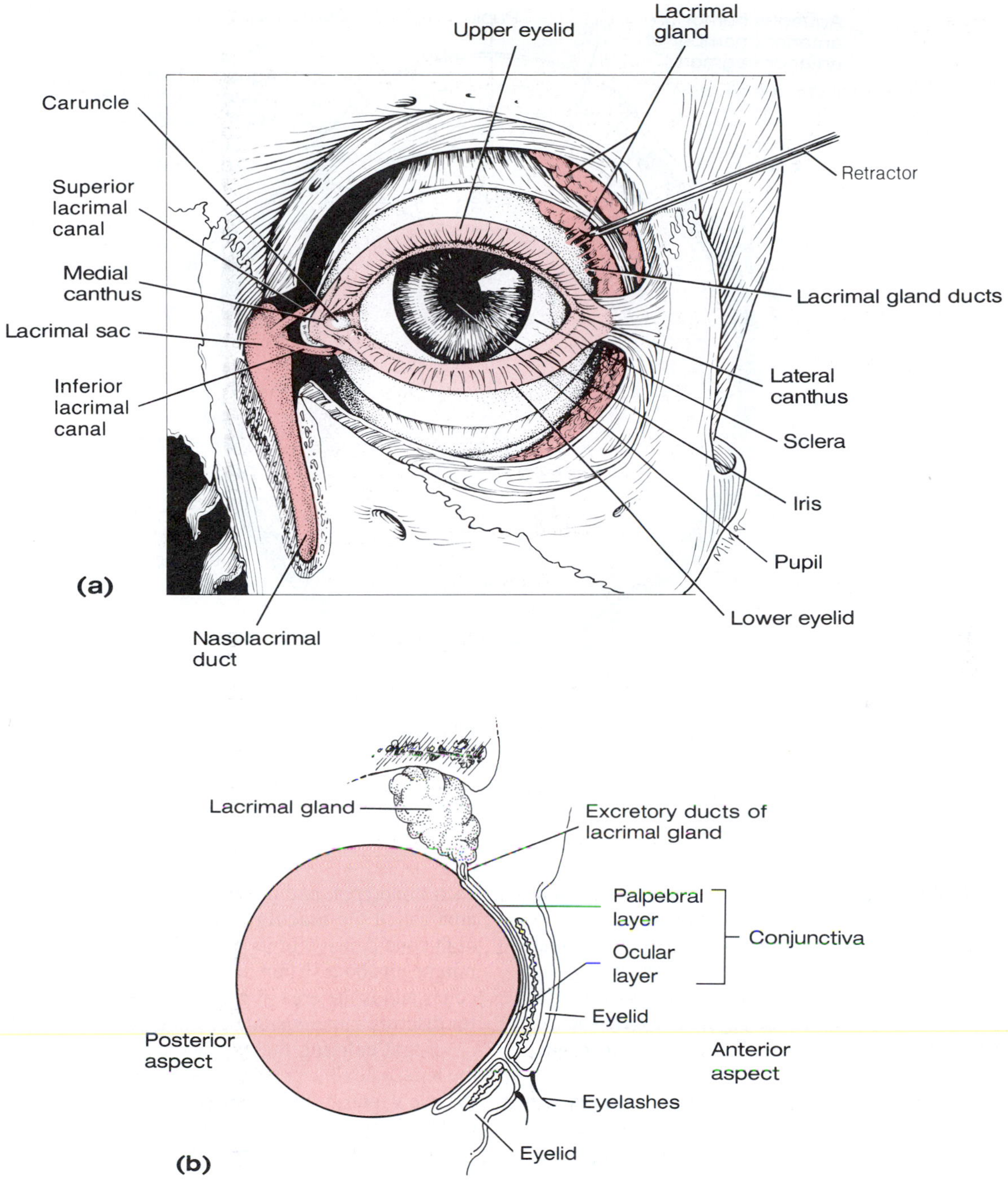

Upper eyelid
Lacrimal gland
Caruncle
Retractor
Superior lacrimal canal
Medial canthus
Lacrimal gland ducts
Lacrimal sac
Inferior lacrimal canal
Lateral canthus
Sclera
Iris
Pupil
(a)
Lower eyelid
Nasolacrimal duct

Lacrimal gland
Excretory ducts of lacrimal gland
Palpebral layer
Ocular layer
Conjunctiva
Eyelid
Posterior aspect
Anterior aspect
Eyelashes
Eyelid
(b)

F24.2

External anatomy of the eye and accessory structures. (a) Anterior view. **(b)** Sagittal section.

Internal Anatomy of the Eye

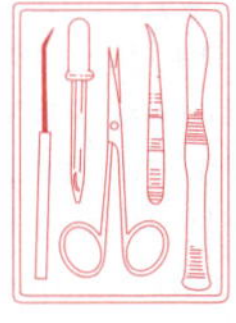

Obtain a dissectible eye model and identify its internal structures as they are described below. As you work, also refer to Figure 24.3.

Anatomically, the wall of the eye is constructed of three tunics, or coats. The outermost **fibrous tunic** is a protective layer composed of dense avascular connective tissue. It has two obviously different regions: The opaque white **sclera** forms the bulk of the fibrous tunic and is observable anteriorly as the "white of the eye." Its anteriormost portion is modified structurally to form the transparent **cornea,** through which light enters the eye.

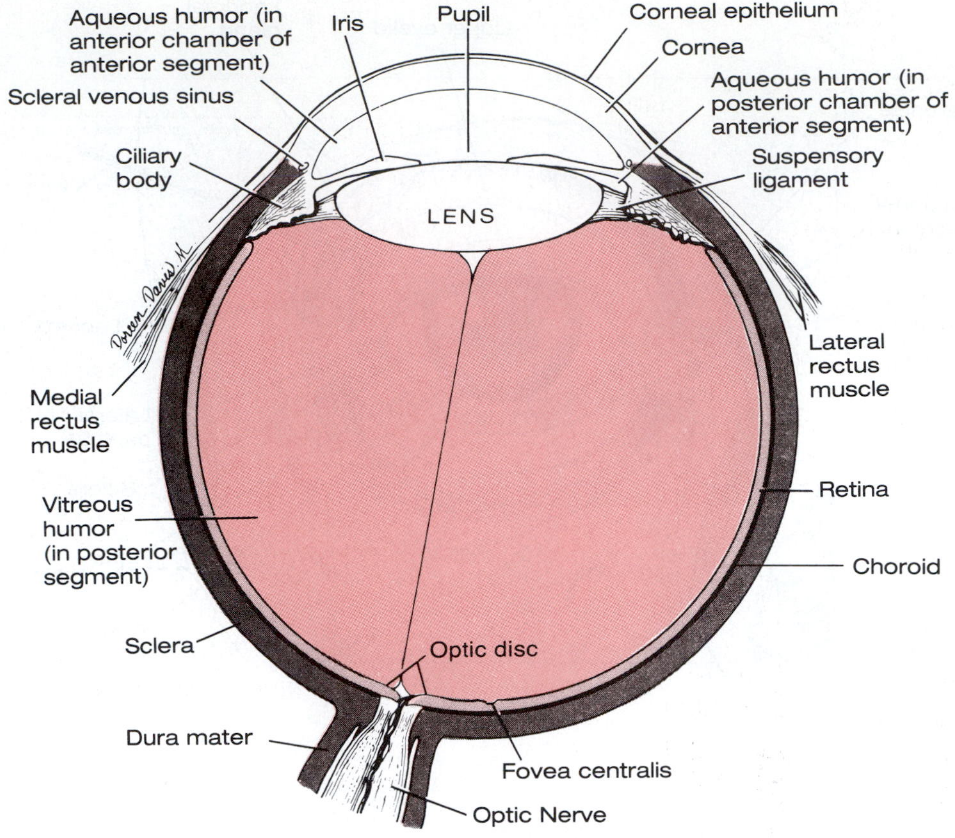

F24.3

Internal anatomy of the eye (transverse section).

The middle tunic, called the **uvea,** is the **vascular tunic.** Its posteriormost part, the **choroid,** is a richly vascular nutritive layer that contains a dark pigment that prevents light scattering within the eye. Anteriorly, the choroid is modified to form the **ciliary body,** to which the lens is attached, and then the pigmented **iris.** The iris is incomplete, resulting in a rounded opening, the **pupil,** through which light passes.

The iris is composed of circularly and radially arranged smooth muscle fibers and acts as a reflexively activated diaphragm to regulate the amount of light entering the eye. In close vision and bright light, the circular muscles of the iris contract, and the pupil constricts. In distant vision and in dim light, the radial fibers contract, enlarging (dilating) the pupil and allowing more light to enter the eye.

The innermost **sensory tunic** of the eye is the delicate, two-layered **retina.** The outer **pigmented epithelial layer** abuts and lines the entire uvea. The transparent inner **neural (nervous) layer** extends anteriorly only to the ciliary body. It contains the photoreceptors, **rods** and **cones,** which begin the chain of electrical events that ultimately result in the transduction of light energy into nerve impulses that are transmitted to the optic cortex of the brain. Vision is the result. The photoreceptor cells are distributed over the entire neural retina, except where the optic nerve leaves the eyeball. This site is called the **optic disc,** or blind spot. Lateral to each blind spot, and directly posterior to the lens, is an area called the **macula lutea** (yellow spot), an area of high cone density. In its center is the **fovea centralis,** a minute pit about ½ mm in diameter, which contains only cones and is the area of greatest visual acuity. Focusing for discriminative vision occurs in the fovea centralis.

Light entering the eye is focused on the retina by the **lens,** a flexible crystalline structure held vertically in the eye's interior by the **suspensory ligament** attached to the ciliary body. Activity of the ciliary muscle, which accounts for the bulk of ciliary body tissue, changes lens thickness to allow light to be properly focused on the retina.

In the elderly the lens becomes increasingly hard and opaque. **Cataracts,** which often result from this process, cause vision to become hazy or entirely obstructed. ■

The lens divides the eye into two segments: the **anterior segment** anterior to the lens, which contains a clear watery fluid called the **aqueous humor,** and the **posterior segment** behind the lens, filled with a gel-like substance, the **vitreous humor,** or **vitreous body.** The anterior segment is further divided into **anterior** and **posterior chambers,** located before and after the iris, respectively. The aqueous humor is continually formed by the capillaries of the **ciliary processes** of the ciliary

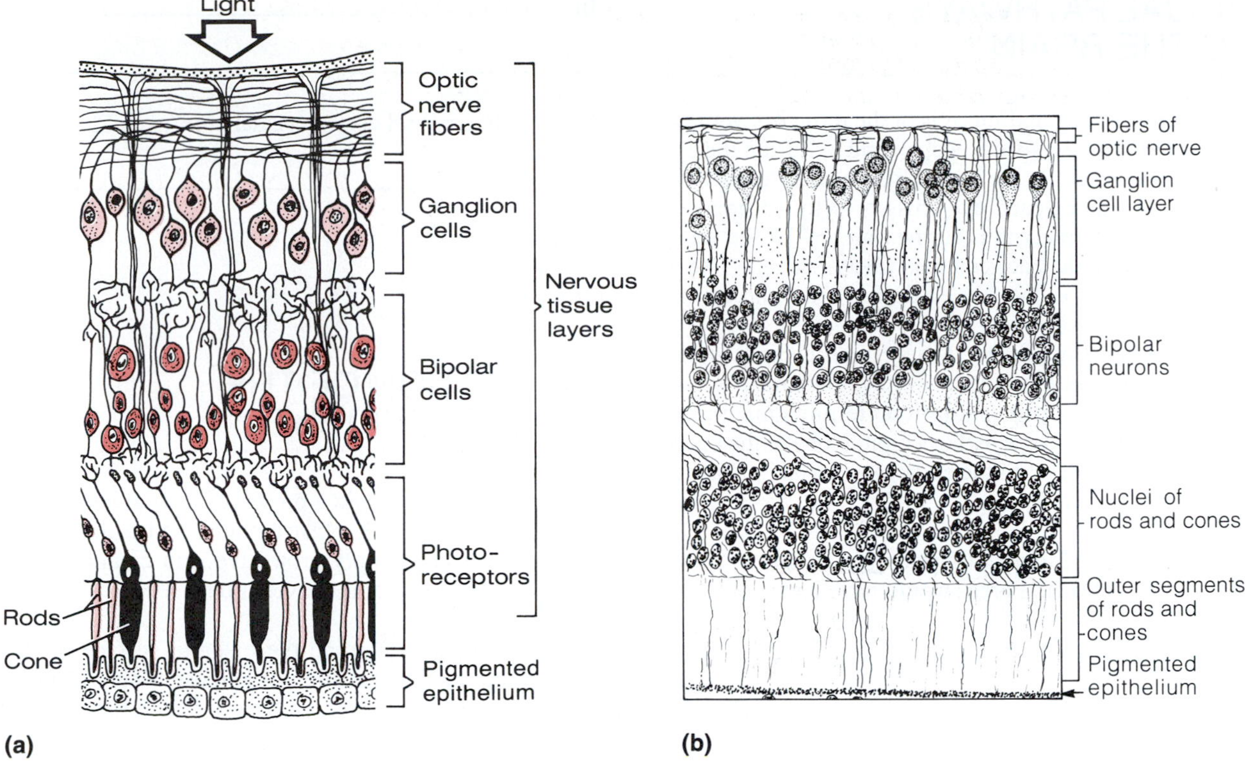

F24.4

Microscopic anatomy of the cellular layers of the retina. **(a)** Diagrammatic view. **(b)** Line drawing of the photomicrograph provided in Plate 13 of the Histology Atlas.

body. It helps to maintain the intraocular pressure of the eye and provides nutrients for the avascular lens and cornea. The aqueous humor is reabsorbed into the **scleral venous sinus (canal of Schlemm).** The vitreous humor provides the major internal reinforcement of the posterior part of the eyeball, and helps to keep the neural layer of the retina pressed firmly against the wall of the eyeball. It is formed *only* before birth.

Anything that interferes with drainage of the aqueous fluid increases intraocular pressure. When intraocular pressure reaches dangerously high levels, the retina and optic nerve are compressed, resulting in pain and possible blindness, a condition called **glaucoma.** ■

MICROSCOPIC ANATOMY OF THE RETINA

As described above, the retina consists of two main types of cells: a pigmented *epithelial* layer, which abuts the choroid, and an inner cell layer composed of *neurons,* which is in contact with the vitreous humor (Figure 24.4). The inner nervous layer is composed of three major neuronal populations. These are, from outer to inner aspect, the **photoreceptors** (rods and cones), the **bipolar cells,** and the **ganglion cells.**

The **rods** are the specialized receptors for dim light. Visual interpretation of their activity is in gray tones. The **cones** are color receptors that permit high levels of visual acuity, but they function only under conditions of high light intensity; thus, for example, no color vision is possible in moonlight. Only cones are found in the fovea centralis, and their number decreases as the retinal periphery is approached. By contrast, rods are most numerous in the periphery, and their density decreases as the macula is approached.

Light must pass through the ganglion cell layer and the bipolar neuron layer to reach and excite the rods and cones. As a result of a light stimulus, the photoreceptors undergo changes in their membrane potential that ultimately influence the bipolar neurons. These in turn stimulate the ganglion cells, whose axons leave the retina in the tight bundle of fibers known as the optic nerve. The retinal layer is thickest where the optic nerve attaches to the eyeball because an increasing number of ganglion cell axons converge at this point. It thins as it approaches the ciliary body.

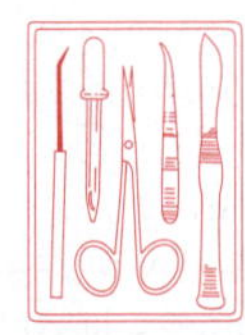

Obtain a histologic slide of a longitudinal section of the eye. Identify the retinal layers by comparing it to Figure 24.4.

VISUAL PATHWAYS TO THE BRAIN

The axons of the ganglion cells of the retina converge at the posterior aspect of the eyeball and exit from the eye as the optic nerve. At the **optic chiasma,** the fibers from the medial side of each eye cross over to the opposite side. The fiber tracts thus formed are called the **optic tracts.** Each optic tract contains fibers from the lateral side of the eye on the same side and from the medial side of the opposite eye.

The optic tract fibers synapse with neurons in the **lateral geniculate nucleus** of the thalamus, whose axons form the **optic radiation,** terminating in the **optic,** or **visual, cortex** in the occipital lobe of the brain. Here they synapse with the cortical cells, and visual interpretation occurs.

After examining Figure 24.5, determine what effects lesions in the following areas would have on vision:

In the right optic nerve ______________________________

____

Through the optic chiasma ______________________________

____

In the left optic tract ______________________________

____

In the right cerebral cortex (visual area) ______________________________

____

DISSECTION OF THE COW (SHEEP) EYE

1. Obtain a preserved cow or sheep eye, dissecting instruments, and a dissecting pan. Apply protective skin cream or don disposable gloves if desired.

2. Examine the external surface of the eye, noting the thick cushion of adipose tissue. Identify the optic nerve (cranial nerve II) as it leaves the eyeball, the remnants of the extrinsic eye muscles, the conjunctiva, the sclera, and the cornea. The normally transparent cornea is opalescent or opaque if the eye has been preserved. Refer to Figure 24.6 as you work.

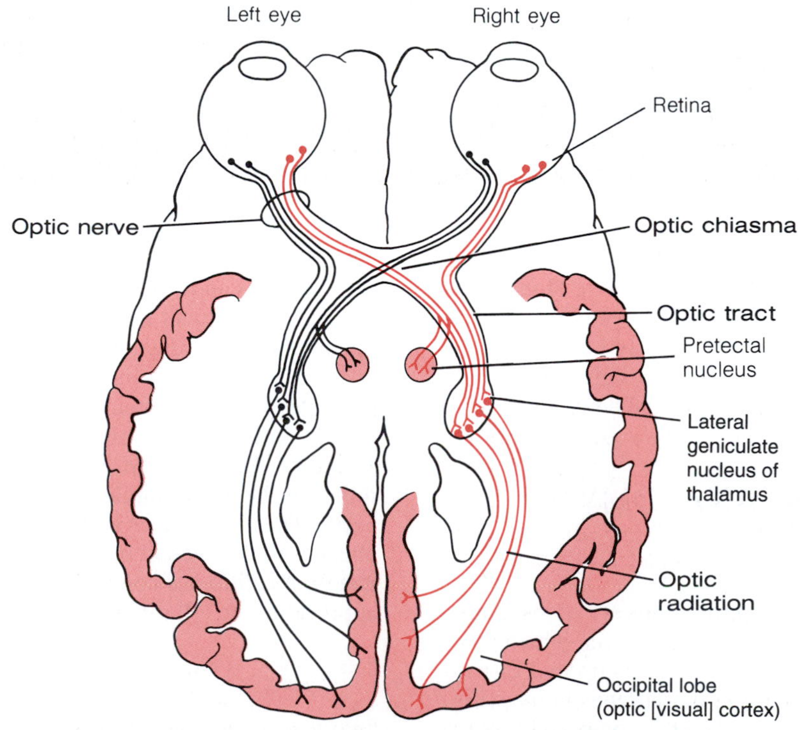

F24.5

Visual pathway to the brain. (Note that fibers from the lateral portion of each retinal field do not cross at the optic chiasma.)

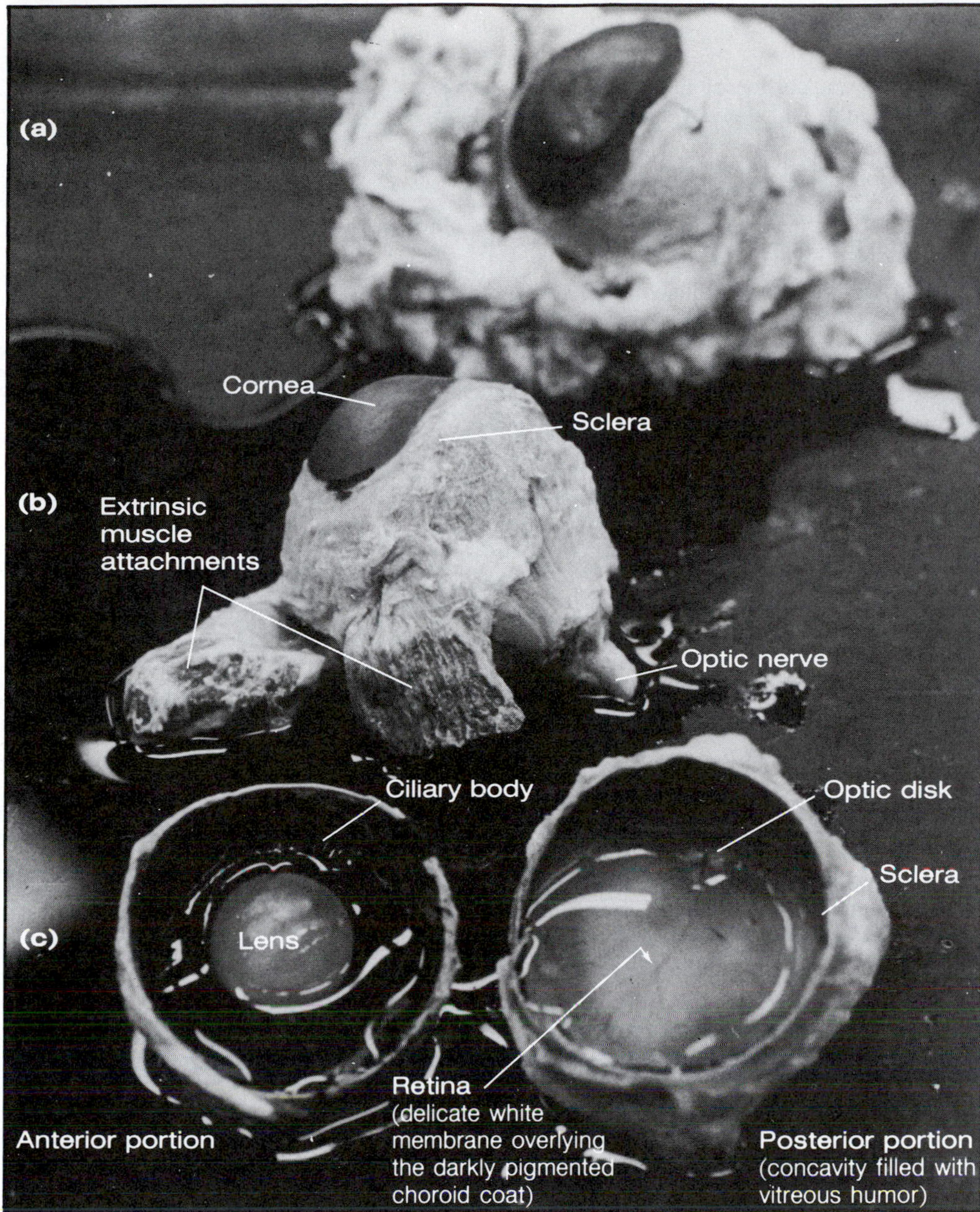

F24.6

Anatomy of the cow eye. (a) Cow eye (entire) removed from orbit (notice the large amont of fat cushioning the eyeball). **(b)** Cow eye (entire) with fat removed to show the extrinsic muscle attachments and optic nerve. **(c)** Cow eye cut along the coronal plane to reveal internal structures.

3. Trim away most of the fat and connective tissue, but leave the optic nerve intact. Holding the eye with the cornea facing downward, carefully make an incision with a sharp scalpel into the sclera about 1/4 inch above the cornea. (The sclera of the preserved eyeball is *very* tough so you will have to apply substantial pressure to penetrate it.) Using scissors, complete the incision around the circumference of the eyeball paralleling the corneal edge.

4. Carefully lift the anterior part of the eyeball away from the posterior portion. Conditions being proper, the vitreous body should remain with the posterior part of the eyeball.

5. Examine the anterior part of the eye and identify the following structures:

Ciliary body: black pigmented body that appears to be a halo encircling the lens.

Lens: biconvex structure that is opaque in preserved specimens.

Suspensory ligament: a halo of delicate fibers attaching the lens to the ciliary body.

Carefully remove the lens and identify the adjacent structures:

Iris: anterior continuation of the ciliary body penetrated by the pupil.

Cornea: more convex anteriormost portion of the sclera; normally transparent but cloudy in preserved specimens.

6. Examine the posterior portion of the eyeball. Remove the vitreous humor, and identify the following structures:

Retina: the neural layer of the retina appears as a delicate white, probably crumpled membrane that separates easily from the pigmented choroid.

Note its point of attachment. What is this point called?

____

Pigmented choroid coat: appears iridescent in the cow or sheep eye owing to a special reflecting surface called the **tapetum lucidum.** This specialized surface reflects the light within the eye and is found in the eyes of animals that live under conditions of low-intensity light. It is not found in humans.

VISUAL TESTS AND EXPERIMENTS

Demonstration of the Blind Spot

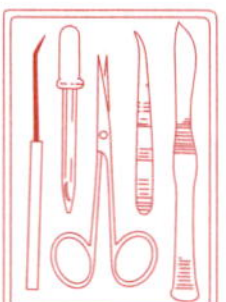

1. Hold Figure 24.7 about 18 inches from your eyes. Close your left eye, and focus your right eye on the X, which should be positioned so that it is directly in line with your right eye. Move the figure slowly toward your face, keeping your right eye focused on the X. When the dot focuses on the blind spot, which lacks photoreceptors, it will disappear.

2. Have your laboratory partner record in metric units the distance at which this occurs. The dot will reappear as the figure is moved closer. Distance at which the dot disappears:

Right eye ______________________________

Repeat the test for the left eye, this time closing the right eye and focusing the left eye on the dot. Record the distance at which the X disappears:

Left eye ______________________________

Afterimages

When light from an object strikes **rhodopsin,** the purple pigment contained in the rods of the retina, it triggers a photochemical reaction that splits rhodopsin into its colorless precursor molecules (vitamin A and a protein called opsin). This event, called *bleaching of the pigment,* initiates a chain of events leading to impulse transmission along fibers of the optic nerve. Once bleaching has occurred in a rod, the photoreceptor pigment must be resynthesized before the rod can be restimulated. This takes a certain period of time. Both phenomena—that is, the stimulation of the photoreceptor cells and their subsequent inactive period—can be demonstrated indirectly in terms of positive and negative afterimages.

X ●

F24.7

Blind spot test figure.

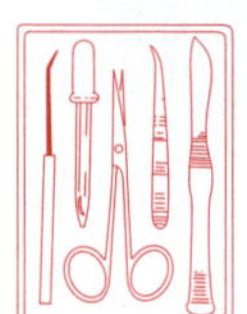

1. Stare at a bright lightbulb for a few seconds, and then gently close your eyes for approximately one minute.

2. Record, in sequence of occurrence, what you "saw" after closing your eyes:

____

____

____

The bright image of the lightbulb initially seen was a **positive afterimage** caused by the continued firing of the rods. The dark image of the lightbulb that subsequently appeared against a lighter background was the **negative afterimage,** an indication that the rhodopsin in the affected photoreceptor cells had been bleached.

Refraction, Tests for Visual Acuity, and Astigmatism

When light rays pass from one medium to another, their velocity, or speed of transmission, changes, and the rays are bent or refracted. Thus the light rays in the visual field are refracted as they encounter the cornea, lens, and vitreous humor of the eye.

The refractive index (bending power) of the cornea and vitreous humor are constant. But the lens's refractive index, or strength, can be varied by changing the lens's shape—that is, by making it more or less convex so that the light is properly converged and focused on the retina. The greater the lens convexity, or bulge, the more the light will be bent and the stronger the lens. Conversely, the less the lens convexity (the flatter it is), the less it bends the light.

In general, light from a distant source (over 20 feet) approaches the eye as parallel rays, and no change in lens convexity is necessary for it to focus properly on the retina. However, light from a close source tends to diverge, and the convexity of the lens must increase to make close vision possible. To achieve this, the ciliary muscle contracts, decreasing the tension on the suspensory ligament attached to the lens and allowing the elastic lens to "round up." Thus, a lens capable of bringing a *close* object into sharp focus is stronger (more convex) than a lens focusing on a more distant object. The ability of the eye to focus differentially for objects of near vi-

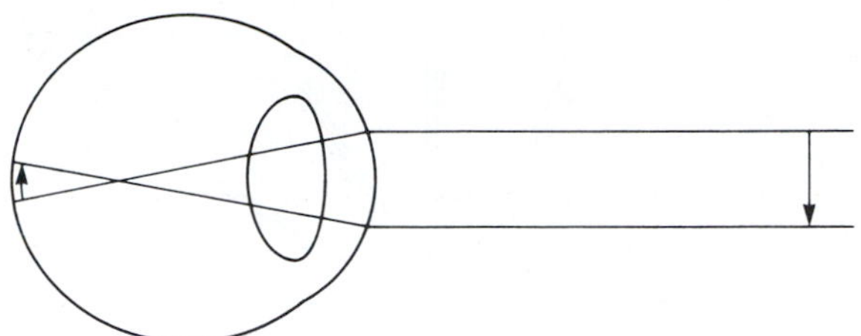

F24.8

Refraction of light in the eye, resulting in the production of a real image on the retina.

sion (less than 20 feet) is called **accommodation.** It should be noted that the image formed on the retina as a result of the refractory activity of the lens (see Figure 24.8) is a **real image** (reversed from left to right, inverted, and smaller than the object).

The normal or **emmetropic eye** is able to accommodate properly. However, visual problems may result from (1) lenses that are too strong or too "lazy" (overconverging and underconverging, respectively), (2) from structural problems such as an eyeball that is too long or too short to provide for proper focusing by the lens, or (3) a cornea or lens with improper curvatures.

Individuals in whom the image normally focuses in front of the retina are said to have **myopia,** or "nearsightedness" (Figure 24.9a); they can see close objects without difficulty, but distant objects are blurred or seen indistinctly. Correction requires a concave lens, which causes the light reaching the eye to diverge (Figure 24.9b).

If the image focuses behind the retina, the individual is said to have **hyperopia** or farsightedness. Such persons have no problems with distant vision but need glasses with convex lenses to augment the converging power of the lens for close vision (Figure 24.9c and d).

Irregularities in the curvatures of the lens and/or the cornea lead to a blurred vision problem called **astigmatism.** Cylindrically ground lenses, which compensate for inequalities in the curvatures of the refracting surfaces, are prescribed to correct the condition. ■

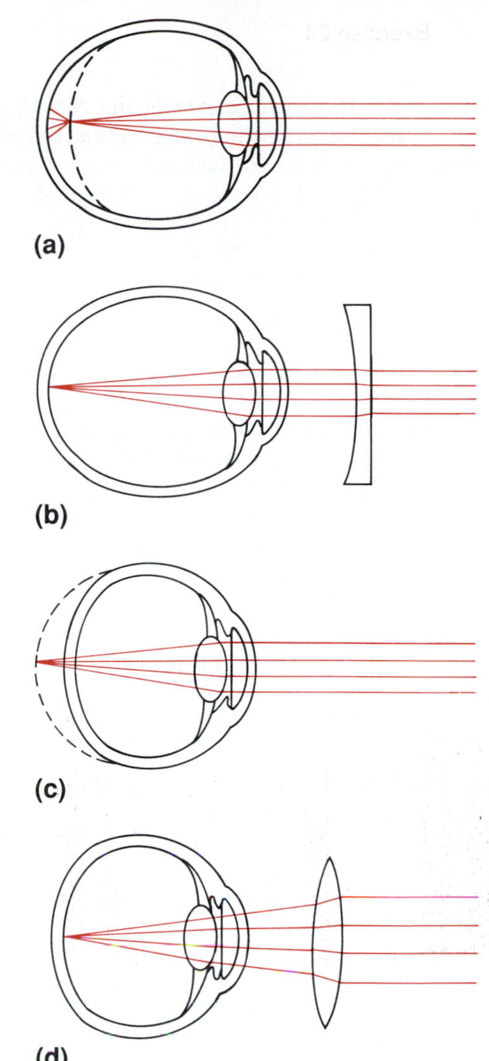

F24.9

Common refraction problems and their correction. In myopia: **(a)** light from a distant object focuses in front of the retina, and **(b)** correction involves the use of a concave lens that diverges the light before it enters the eye. In hyperopia: **(c)** light from a distant object focuses behind the retina, and **(d)** correction requires a convex lens that converges the light rays before they enter the eye.

TEST FOR NEAR-POINT ACCOMMODATION

The elasticity of the lens decreases dramatically with age, resulting in difficulty in focusing for near or close vision. This condition is called **presbyopia**—literally, old vision. Lens elasticity can be tested by measuring the **near point of accommodation.** The near point of vision is about 10 cm from the eye in young adults. It is closer in children and farther in old age.

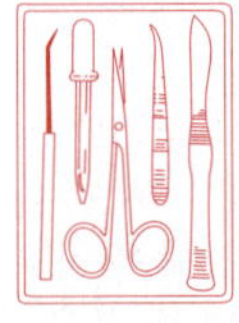

To determine your near point of accommodation, hold a common straight pin at arm's length in front of one eye. Slowly move the pin toward that eye until the pin image becomes distorted. Have your lab partner measure the distance from your eye to the pin at this point, and record the distance below. Repeat the procedure for the other eye.

Near point for right eye ______________________________

Near point for left eye ______________________________

TEST FOR VISUAL ACUITY **Visual acuity,** or sharpness of vision, is generally tested with a Snellen eye chart, which consists of letters of various sizes printed on a white card. This test is based on the fact that letters of a certain size can be seen clearly by eyes with normal vision at a specific distance. The distance at which the normal, or emmetropic, eye can read a line of letters is printed at the end of that line.

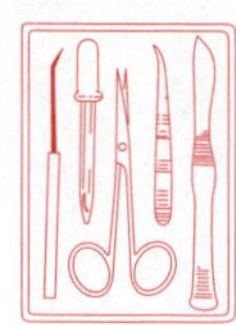

1. Have your partner stand 20 feet from the posted Snellen eye chart and cover one eye with a card or hand. As your partner reads each consecutive line aloud, check for accuracy. If this individual wears glasses, give the test twice—first with glasses off and then with glasses on.

2. Record the number of the line with the smallest-sized letters read. If it is 20/20, the person's vision for that eye is normal. If it is 20/40, or any ratio with a value less than one, he or she has less than the normal visual acuity. (Such an individual is myopic.) If the visual acuity is 20/15, vision is better than normal, because this person can stand at 20 feet from the chart and read letters that are only discernible by the normal eye at 15 feet. Give your partner the number of the line corresponding to the smallest letters read, to record in step 4.

3. Repeat the process for the other eye.

4. Have your partner test and record your visual acuity. If you wear glasses, the test results *without* glasses should be recorded first.

Visual acuity, right eye ____________________

Visual acuity, left eye ____________________

TEST FOR ASTIGMATISM The astigmatism chart (Figure 24.10) is designed to test for defects in the refracting surface of the lens and/or cornea.

View the chart first with one eye and then with the other, focusing on the center of the chart. If all the radiating lines appear equally dark and distinct, there is no distortion of your refracting surfaces. If some of the lines are blurred or appear less dark than others, at least some degree of astigmatism is present.

Is astigmatism present in your left eye? ____________

Right eye? ____________________

Test for Color Blindness

Ishihara's color plates are designed to test for deficiencies in the cones or color photoreceptor cells. There are three cone types, each containing a different light-absorbing pigment. One type primarily absorbs the red wavelengths of the visible light spectrum, another the blue wavelengths, and a third the green wavelengths. Nerve impulses reaching the brain from these different photoreceptor types are then interpreted (seen) as red, blue, and green, respectively.

The interpretation of the intermediate colors of the visible light spectrum is a result of overlapping input from more than one cone type.

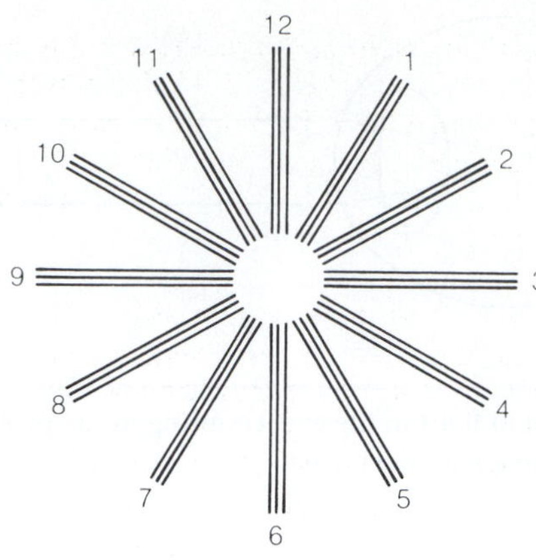

F24.10

Astigmatism testing chart.

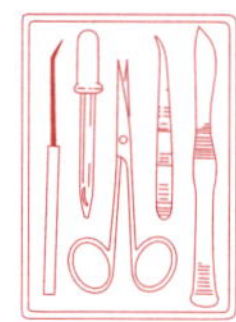

1. View the various color plates in bright light or sunlight while holding them about 30 inches away and at right angles to your line of vision. Report to your laboratory partner what you see in each plate. (Take no more than 3 seconds for each decision.)

2. Your partner is to write down your responses and then check their accuracy with the correct answers provided in the color plate book. Is there any indication that you have some degree of color blindness?

__________ If so, what type? ____________________

____

Repeat the procedure to test your partner's color vision.

Relative Positioning of Rods and Cones on the Retina

The test subject for this demonstration should have shown no color vision problems during the test for color blindness. Students may work in pairs, or two students may perform the test for the entire class. White, red, blue, and green paper discs and chalk will be needed for this demonstration.

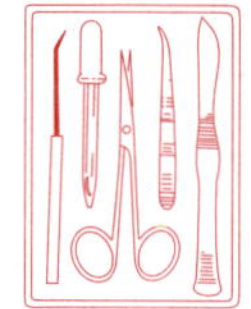

1. Position the subject about 1 foot away from the blackboard.

2. Make a small white chalk circle on the board immediately in front of the subject's right eye. Have the subject close the left eye and stare fixedly at the circle with the right eye throughout the test.

3. To map the extent of the rod field, begin to move a white paper disc into the field of vision from various sides of the visual field (beginning at least 2 feet away from the white chalk circle) and plot with white chalk dots the points at which the disc first becomes visible to the test subject.

4. Repeat the procedure, using red, green, and blue paper discs and like-colored chalk to map the cone fields. Color (not object) identification is required.

5. After the test has been completed for all four discs, connect all dots of the same color. It should become apparent that each of the color fields has a different radius and that the rod and cone distribution on the retina is not uniform. (Normal fields have white outermost, followed by blue, red, and green as the fovea is approached.)

6. Record the rod and cone color field distribution observed in the laboratory review section using appropriately colored pencils.

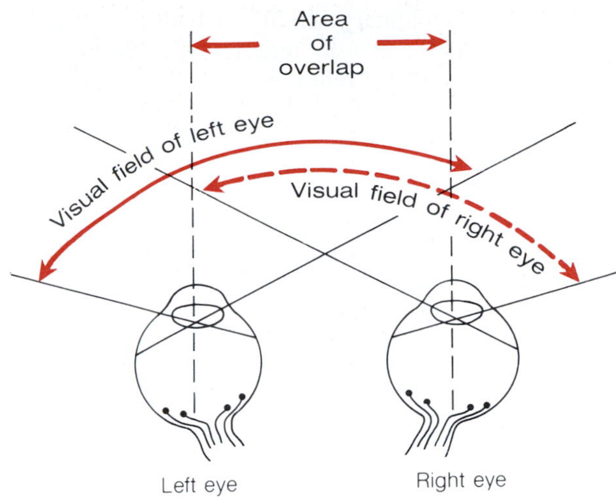

F24.11

Overlapping of the visual fields.

Tests for Binocular Vision

Humans, cats, predatory birds, and most primates are endowed with **binocular** (two-eyed) **vision.** Although both eyes look in approximately the same direction, they see slightly different views. Their visual fields, each about 170 degrees, overlap to a considerable extent; thus there is two-eyed vision at the overlap area (Figure 24.11).

In contrast, the eyes of many animals (rabbits, pigeons, and others) are more on the sides of their head. Such animals see in two different directions and thus have a panoramic field of view and **panoramic vision.**

Although both types of vision have their good points, binocular vision provides three-dimensional vision and an accurate means of locating objects in space. The slight differences between the views seen by the two eyes are fused by the higher centers of the visual cortex to give us *depth perception.* Because of the manner in which the visual cortex resolves these two different views into a single image, it is sometimes referred to as the "cyclopean eye of the binocular animal."

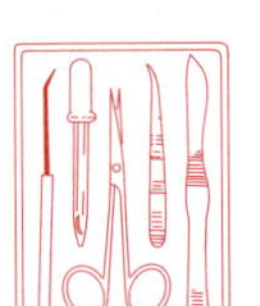

1. To demonstrate that a slightly different view is seen by each eye, perform the following simple experiment.

Close your left eye. Hold a pencil at arm's length directly in front of your right eye. Position another pencil directly beneath it and then move the lower pencil about half the distance toward you. As you move the lower pencil, make sure it remains in the *same plane* as the stationary pencil, so that the two pencils continually form a straight line. Then, without moving the pencils, close your right eye and open your left eye. Notice that with only the right eye open, the moving pencil stays in the same plane as the fixed pencil, but that when viewed with the left eye, the moving pencil is displaced laterally away from the plane of the fixed pencil.

2. To demonstrate the importance of two-eyed binocular vision for depth perception, perform this second simple experiment.

Have your laboratory partner hold a test tube erect about arm's length in front of you. With both eyes open, quickly insert a pencil into the test tube. Remove the pencil, bring it back close to your body, close one eye, and quickly and without hesitation insert the pencil into the test tube. (Do not feel for the test tube with the pencil!) Repeat with the other eye closed.

Was it as easy to dunk the pencil with one eye closed as with both eyes open?

____

Tests of Eye Reflexes

Both intrinsic (internal) and extrinsic (external) muscles are necessary for proper eye functioning. The *intrinsic muscles,* controlled by the autonomic nervous system, are those of the ciliary body (which alters the lens curvature in focusing) and the radial and circular muscles of the iris (which control pupillary size and thus regulate the amount of light entering the eye). The *extrinsic muscles* are the rectus and oblique muscles, which are attached to the eyeball exterior (see Figure 24.1). These muscles control eye movement and make it possible to keep moving objects focused on the fovea centralis. They are also responsible for **convergence,** or medial eye movements, which is essential for near vision. When convergence occurs, both eyes are directed toward the near object viewed. The extrinsic eye muscles are controlled by the somatic nervous system.

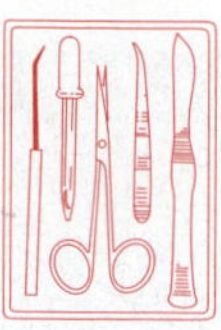

Involuntary activity of both the intrinsic and extrinsic muscle types is brought about by reflex actions that can be observed in the following experiments.

PHOTOPUPILLARY REFLEX Sudden illumination of the retina by a bright light causes the pupil to constrict reflexively in direct proportion to the light intensity. This protective response prevents damage to the delicate photoreceptor cells.

Obtain a laboratory lamp or penlight. Have your laboratory partner sit with eyes closed and hands over his or her eyes. Turn on the light and position it so that it shines on the subject's right hand. After 1 minute, ask your partner to uncover and open the right eye. Quickly observe the pupil of that eye. What happens to the pupil?

____

Shut off the light and ask your partner to uncover and open the opposite eye. What are your observations?

____

____

ACCOMMODATION PUPILLARY REFLEX Have your partner gaze for approximately 1 minute at a distant object in the lab—*not* toward the windows or another light source. Observe your partner's pupils. Then hold some printed material 6 to 10 inches from his or her face, and direct him or her to focus on it.

How does pupil size change as your partner focuses on the printed material?

____

Explain the value of this reflex. ______________________

____

____

____

CONVERGENCE REFLEX Repeat the previous experiment, this time using a pen or pencil as the close object to be focused on. Note the position of your partner's eyeballs both while he or she is gazing at the distant and at the close object. Do they change position as the object of focus is changed?

__________ In what way? ______________________

____

____

____

Explain the importance of the convergence reflex.

____

____

____

Ophthalmoscopic Examination of the Eye (Optional)

The ophthalmoscope is an instrument used to examine the *fundus,* or eyeball interior, to determine visually the condition of the retina, optic disc, and internal blood vessels. Certain pathologic conditions such as diabetes mellitus, arteriosclerosis, and degenerative changes of the optic nerve and retina can be detected by such an examination. The ophthalmoscope consists of a set of lenses mounted on a rotating disc (the **lens selection disc**), a light source regulated by a **rheostat control,** and a mirror that reflects the light so that the eye interior can be illuminated (Figure 24.12a).

The lens selection disc is positioned in a small slit in the mirror, and the examiner views the eye interior through this slit, appropriately called the **viewing window.** The focal length of each lens is indicated in diopters preceded by a + sign if the lens is convex and by a − sign if the lens is concave. When the zero (0) is seen in the **diopter window,** there is no lens positioned in the slit. The depth of focus for viewing the eye interior is changed by changing the lens.

The light is turned on by depressing the red **rheostat lock button** and then rotating the rheostat control in the clockwise direction. The aperture selection disc on the front of the instrument allows the nature of the light beam to be altered. Generally, green light allows for clearest viewing of the blood vessels in the eye interior and is most comfortable for the subject.

Once you have examined the ophthalmoscope and have become familiar with it, you are ready to conduct an eye examination.

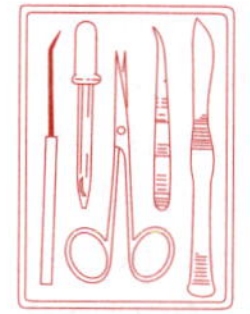

1. Conduct the examination in a dimly lit or darkened room with the subject comfortably seated and gazing straight ahead. To examine the right eye, sit face-to-face with the subject, and steady yourself by resting your left hand on the subject's head as shown in Figure 24.12b. Hold the instrument in your right hand, and use your right eye to view the eye interior. To view the left eye, use your left eye, hold the instrument in your left hand, and steady yourself with your right hand.

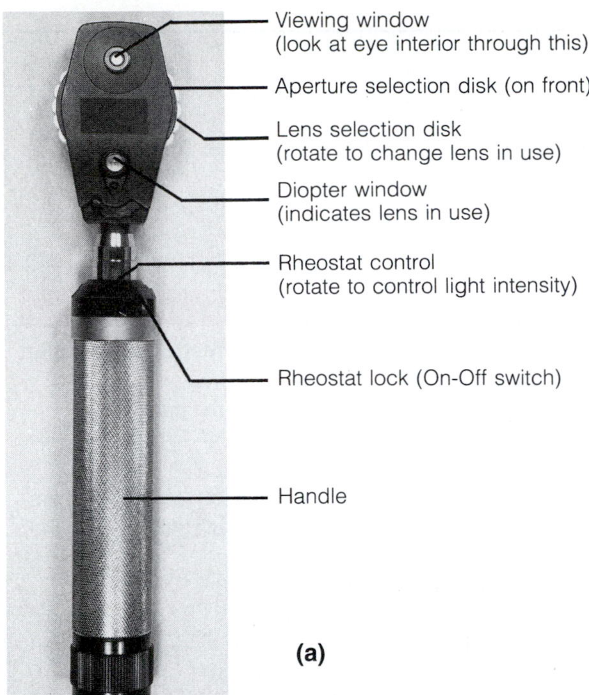

F24.12

Structure and use of an ophthalmoscope. **(a)** Structure of an ophthalmoscope. **(b)** Proper position for examining the right eye with an ophthalmoscope.

2. Begin the examination with the 0 (no lens) in position. Grasp the instrument so that the lens disc may be rotated with the index finger. Holding the ophthalmoscope about 6 inches from the subject's eye, direct the light into the pupil at a slight angle—through the pupil edge rather than directly through its center. You will see a red circular area that is the illuminated eye interior.

3. Move in as close as possible to the subject's cornea (to within 2 in.) as you continue to observe the area. Steady your instrument-holding hand on the subject's cheek if necessary (see Figure 24.12b). If both your eye and that of the subject are normal, the fundus can be viewed clearly without further adjustment of the ophthalmoscope. If the fundus cannot be focused, slowly rotate the lens disc counterclockwise until the fundus can be clearly seen. When the ophthalmoscope is correctly set, the fundus of the right eye should appear as shown in Figure 24.13. (Note: If a positive [convex] lens is required and your eyes are normal, the subject has hyperopia. If a negative [concave] lens is necessary to view the fundus and your eyes are normal, the subject is myopic.)

When the examination is proceeding correctly, the subject can often see images of retinal vessels in his own eye that appear rather like cracked glass. If you are unable to achieve a sharp focus or to see the optic disc, move medially or laterally and begin again.

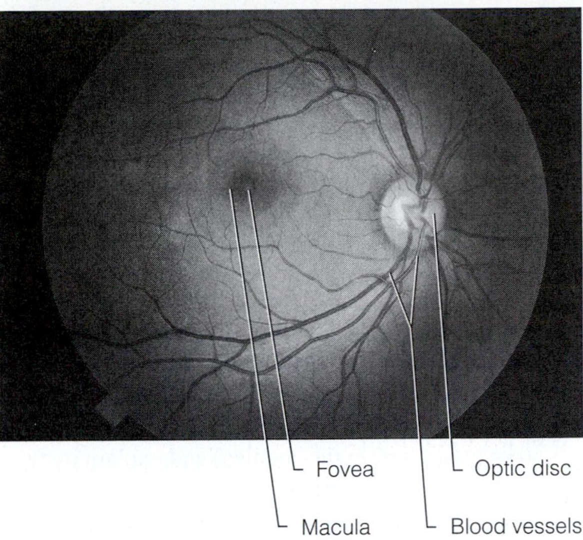

F24.13

Posterior portion of right retina. Photograph taken with slit-lamp camera.

4. Examine the optic disc for color, elevation, and sharpness of outline, and observe the blood vessels radiating from near its center. Locate the macula, lateral to the optic disc. It is a darker area in which blood vessels are absent, and the fovea appears to be a slightly lighter area in its center. The macula is most easily seen when the subject looks directly into the light of the ophthalmoscope.

⚠ Do not examine the macula for longer than 1 second at a time.

5. When you have finished examining your partner's retina, shut off the ophthalmoscope. Change places with your partner (become the subject) and repeat steps 1–4.

25 EXERCISE

Special Senses: Hearing and Equilibrium

OBJECTIVES

1. To identify the anatomical structures of the outer, middle, and inner ear by appropriately labeling a diagram.
2. To describe the anatomy of the organ of hearing (organ of Corti in the cochlea) and explain its function in sound reception.
3. To describe the anatomy of the equilibrium organs of the inner ear (cristae ampullares and maculae), and to explain their relative function in maintaining equilibrium.
4. To define or explain *central deafness, conduction deafness,* and *nystagmus.*
5. To state the purpose of the Weber, Rinne, Barany, and Romberg tests.
6. To explain how one is able to localize the source of sounds.
7. To describe the effects of acceleration on the semicircular canals.
8. To explain the role of vision in maintaining equilibrium.

MATERIALS

Three-dimensional dissectible ear model and/or chart of ear anatomy
Histologic slides of the cochlea of the ear
Compound microscope

Demonstration: Microscope focused on a crista ampullaris receptor of a semicircular canal

Pocket watch or clock that ticks
Tuning forks (range of frequencies)
Rubber mallet
Absorbent cotton
Otoscope (if available)
Alcohol swabs
12-inch ruler
Audiometer
Red and blue pencils

See Appendix E, Exercise 25 for links to *Anatomy and PhysioShow: The Videodisc.*

ANATOMY OF THE EAR

Gross Anatomy

The ear is a complex structure containing sensory receptors for hearing and equilibrium. The ear is divided into three major areas: the *outer ear,* the *middle ear,* and the *inner ear* (Figure 25.1). The outer and middle ear structures serve the needs of the sense of hearing *only,* while inner ear structures function both in equilibrium and hearing reception.

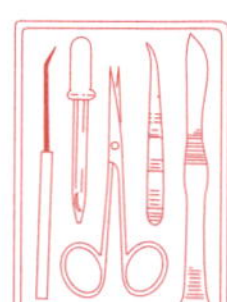

Obtain a dissectible ear model and identify the structures described below. Refer to Figure 25.1 as you work.

The **outer, external, ear** is composed primarily of the **pinna,** or **auricle,** and the **external auditory canal.** The pinna is the skin-covered cartilaginous structure encircling the auditory canal opening. In many animals, it collects and directs sound waves into the auditory canal. In humans this function of the pinna is largely lost.

The external auditory canal is a short, narrow (about 1 inch long by ¼ inch wide) chamber carved into the temporal bone. In its skin-lined walls are wax-secreting glands called **ceruminous glands.** The sound waves that enter the external auditory canal eventually encounter the **tympanic membrane,** or **eardrum,** which vibrates at exactly the same frequency as the sound wave(s) hitting it. The membranous eardrum separates the outer from the middle ear.

The **middle ear** is essentially a small chamber—the **tympanic cavity**—found within the temporal bone. The cavity is spanned by three small bones, collectively called the **ossicles** (hammer, anvil, and stirrup),* which articulate to form a lever system that transmits the vibratory motion of the eardrum to the fluids of the inner ear via the **oval window.**

Connecting the middle ear chamber with the nasopharynx is the **pharyngotympanic,** or **auditory, tube.**

* The ossicles are often referred to by their Latin names, that is, **malleus, incus,** and **stapes,** respectively.

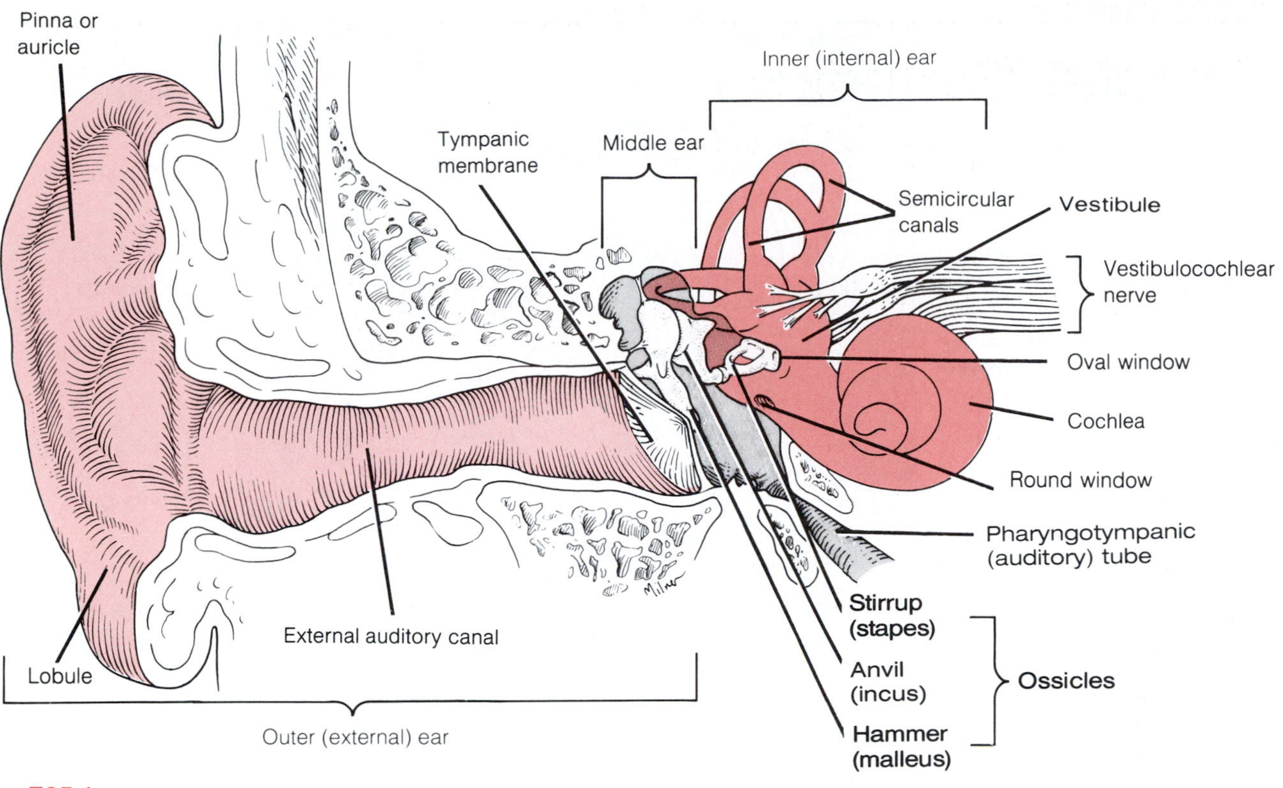

F25.1

Anatomy of the ear.

Normally this tube is flattened and closed, but swallowing or yawning can cause it to open temporarily to equalize the pressure of the middle ear cavity with external air pressure. This is an important function. The eardrum does not vibrate properly unless the pressure on both of its surfaces is the same.

Because the mucosal membranes of the middle ear cavity and nasopharynx are continuous through the pharyngotympanic tube, **otitis media,** or inflammation of the middle ear, is a fairly common condition, especially among youngsters prone to sore throats. In cases where large amounts of fluid or pus accumulate in the middle ear cavity, an emergency myringotomy (lancing of the eardrum) may be necessary to relieve the pressure. Frequently, tiny ventilating tubes are put in during the procedure. ■

The **inner,** or **internal, ear** consists of a system of bony and rather tortuous chambers called the **osseous,** or **bony, labyrinth,** which is filled with an aqueous fluid called **perilymph** (Figure 25.2). Suspended in the perilymph is the **membranous labyrinth,** a system that mostly follows the contours of the osseous labyrinth. The membranous labyrinth is filled with a more viscous fluid called **endolymph.** The three subdivisions of the bony labyrinth are the **cochlea,** the **vestibule,** and the **semicircular canals,** with the vestibule situated between the cochlea and semicircular canals.

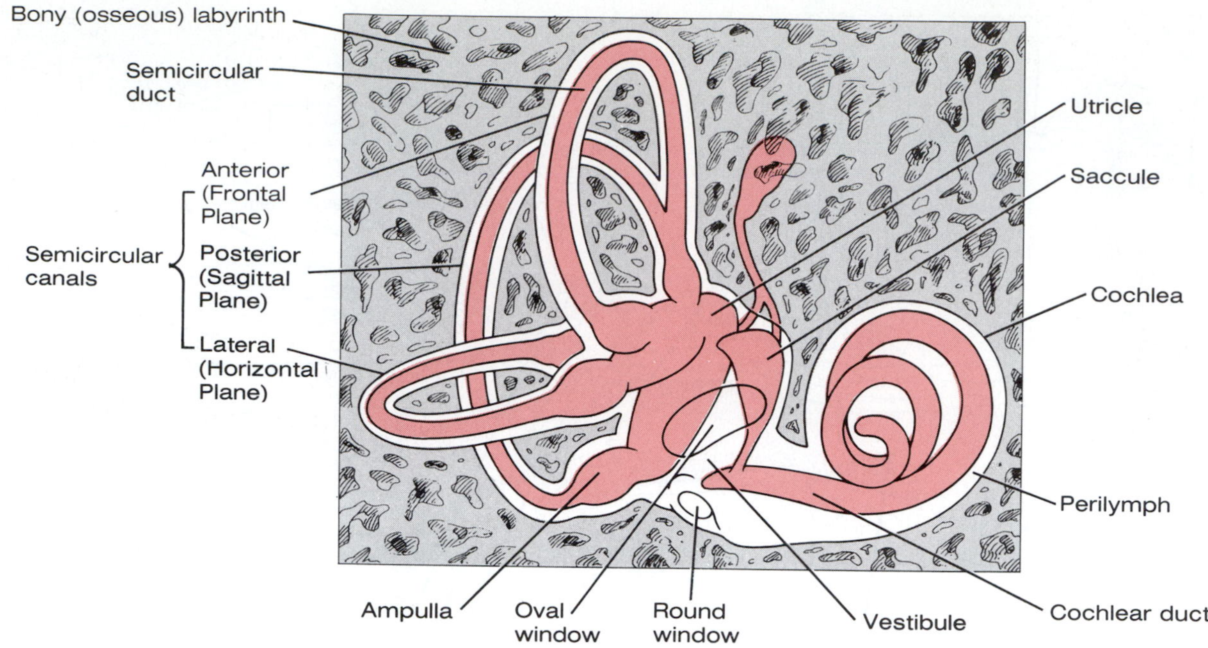

F25.2

Inner ear. Right membranous labyrinth shown within the bony labyrinth.

The snail-like cochlea (see Figures 25.2 and 25.3) contains the sensory receptors for hearing. The cochlear membranous labyrinth, the **cochlear duct,** is a soft wormlike tube about 1½ inches long. It winds through the full two and three-quarter turns of the cochlea and separates the perilymph-containing cochlear cavity into upper and lower chambers, the **scala vestibuli** and **scala tympani,** respectively. The scala vestibuli terminates at the oval window, which "seats" the foot plate of the stirrup located laterally in the tympanic cavity. The scala tympani is bounded by a membranous area called the **round window.** The cochlear duct, itself filled with endolymph, supports the **organ of Corti,** which contains the receptors for hearing—the sensory hair cells and nerve endings of the cochlear division of the vestibulocochlear nerve (VIII).

Otoscopic Examination of the Ear (Optional)

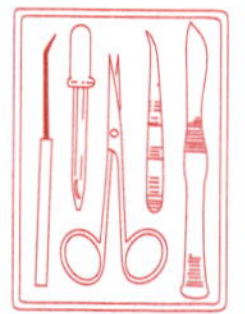

1. Obtain an otoscope and two alcohol swabs. Inspect your partner's ear canal and then select the largest-*diameter* (not length!) speculum that will fit comfortably into his or her ear to permit full visibility. Clean the speculum thoroughly with an alcohol swab, and then attach it to the battery-containing otoscope handle. Before beginning, check that the otoscope light beam is strong. (If not, obtain another otoscope or new batteries.)

2. Hold the lighted otoscope securely between your thumb and forefinger (like a pencil), and rest the little finger of the otoscope-holding hand against your partner's head when you are ready to begin the examination. This maneuver forms a brace that allows the speculum to move as your partner moves and prevents the speculum from penetrating too deeply into the ear canal during unexpected movements.

3. Grasp the ear pinna firmly and pull it up, back, and slightly laterally. If your partner experiences pain or discomfort when the pinna is manipulated, an inflammation or infection of the external ear may be present. If this occurs, do not attempt to examine the ear canal.

4. Carefully insert the speculum of the otoscope into the external auditory canal in a downward and forward direction only far enough to permit examination of the tympanic membrane or eardrum. Note its shape, color, and vascular network. The healthy tympanic membrane is pearly white. During the examination, notice if there is any discharge or redness in the canal and identify earwax.

5. After the examination, thoroughly clean the speculum with the second alcohol swab before returning the otoscope to the supply area.

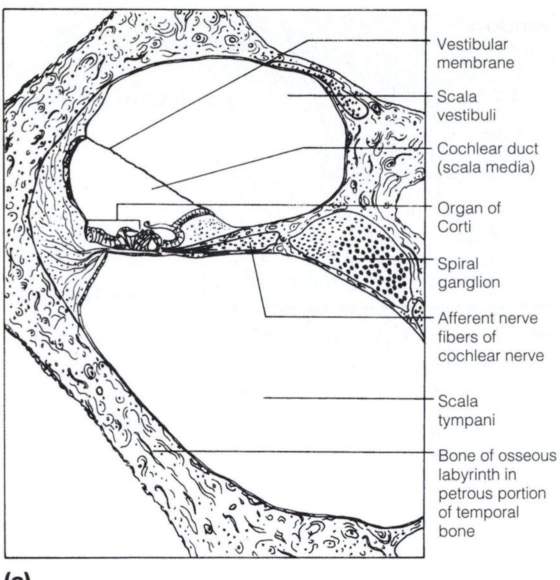

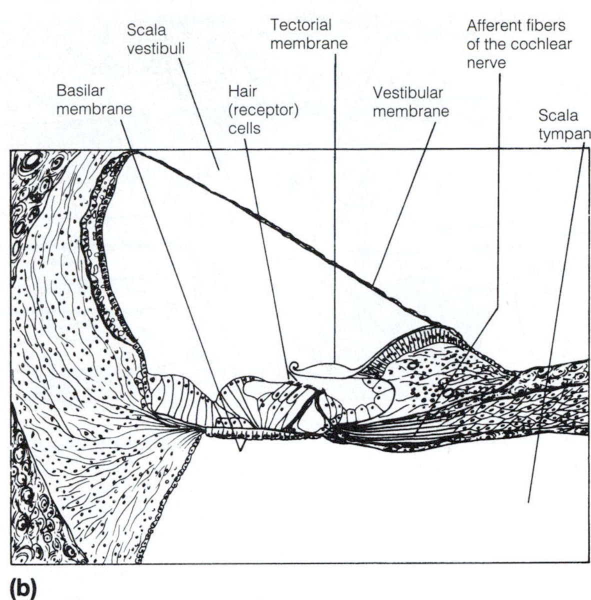

F25.3

Organ of Corti. (a) Cross section through one turn of the cochlea showing the position of the organ of Corti. (b) Enlarged view of the organ of Corti. Corresponding photomicrographs are Plates 14 and 15 respectively in the Histology Atlas.

Microscopic Anatomy of the Organ of Corti and the Mechanism of Hearing

The anatomical details of the organ of Corti are shown in Figure 25.3. The hair (auditory receptor) cells rest on the **basilar membrane,** which forms the floor of the cochlear duct, and their "hairs" (stereocilia) project into a gelatinous membrane, the **tectorial membrane,** that overlies them. The roof of the cochlear duct is called the **vestibular membrane.** The endolymph-filled chamber of the cochlear duct is the **scala media.**

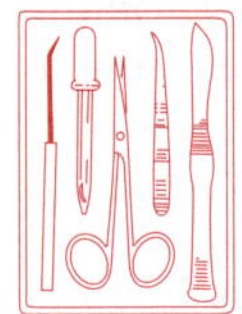

Obtain a compound microscope and a prepared microscope slide of the cochlea and identify the areas shown in Figure 25.3a and b.

The mechanism of hearing begins as sound waves pass through the external auditory canal and through the middle ear into the inner ear, where the vibration eventually reaches the organ of Corti, which contains the receptors for hearing. Many theories have attempted to explain how the organ of Corti actually responds to sound.

The popular "traveling wave" hypothesis of Von Békésy suggests that vibration of the stirrup at the oval window initiates traveling waves that cause maximal displacements of the basilar membrane where they peak and stimulate the hair cells of the organ of Corti in that region. Since the area at which the traveling waves peak is a high-pressure area, the vestibular membrane is compressed at this point and, in turn, compresses the endolymph and the basilar membrane of the cochlear duct. The resulting pressure on the perilymph in the scala tympani causes the membrane of the round window to bulge outward into the middle ear chamber, thus acting as a relief valve for the compressional wave (Figure 25.4). Von Békésy found that high-frequency waves

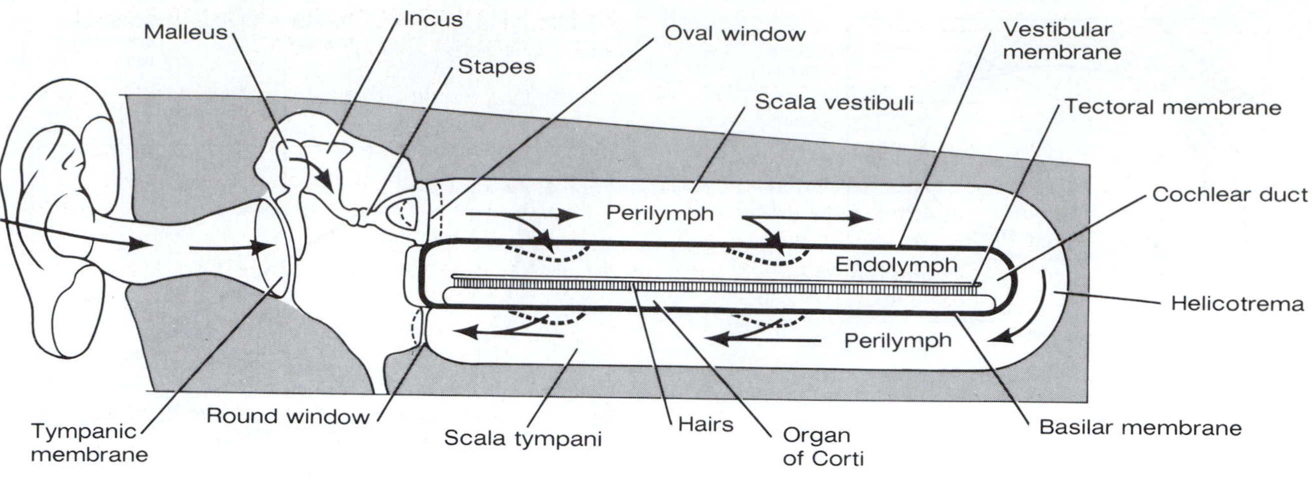

F25.4

Fluid movement in the cochlea following the stirrup thrust at the oval window. Compressional wave created causes the round window to bulge into the middle ear.

(high-pitched sounds) peaked close to the oval window and that low-frequency waves (low-pitched sounds) peaked farther up the basilar membrane near the apex of the cochlea. Although the mechanism of sound reception by the organ of Corti is not completely understood, we do know that hair cells on the basilar membrane are uniquely stimulated by sounds of various frequencies and amplitude and that once stimulated they depolarize and begin the chain of nervous impulses to the auditory centers of the temporal lobe cortex. This series of events results in the phenomenon we call hearing.

By the time most people are in their 60s, a gradual deterioration and atrophy of the organ of Corti begins, and leads to a loss in the ability to hear high tones and speech sounds. This condition, **presbycusis,** is a type of sensorineural deafness. Because many elderly people refuse to accept their hearing loss and resist using hearing aids, they begin to rely more and more on their vision for clues as to what is going on around them, and may be accused of ignoring people.

Although presbycusis is considered to be a disability of old age, it is becoming much more common in younger people as our world grows noisier. The damage (breakage of the "hairs" of the hair cells) caused by excessively loud sounds is progressive and cumulative. Each assault causes a bit more damage. Rock music played and listened to at deafening levels is definitely a contributing factor to the deterioration of hearing receptors. ■

Laboratory Tests: Hearing

Perform the following hearing tests in a quiet area.

ACUITY TEST Have your lab partner pack one ear with cotton and sit quietly with eyes closed. Obtain a ticking clock or pocket watch and hold it very close to his or her *unpacked* ear. Then slowly move it away from the ear until your partner signals that the ticking is no longer audible. Record the distance in inches at which ticking is inaudible.

Right ear ______________ Left ear ______________

Is the threshold of audibility sharp or indefinite?

____

SOUND LOCALIZATION Ask your partner to close both eyes. Hold the pocket watch at an audible distance (about 6 inches) from his or her ear, and move it to various locations (front, back, sides, and above his or her head). Have your partner locate the position by pointing in each instance. Can the sound be localized equally well at all positions?

______________ If not, at what position(s) was the sound less easily located?

____

____

The ability to localize the source of a sound depends on two factors—the difference in the loudness of the sound reaching each ear and the time of arrival of the sound at each ear. How does this information help to explain your findings?

____

____

FREQUENCY RANGE OF HEARING Obtain three tuning forks: one with a low frequency (75 to 100 Hz [cps]), one with a frequency of approximately 1000 Hz, and one with a frequency of 4000 to 5000 Hz. Strike the lowest-frequency fork on the heel of your hand or with a rubber mallet, and hold it close to your partner's ear. Repeat with the other two forks.

Which fork was heard most clearly and comfortably?

______________ Hz

Which was heard least well? ______________ Hz

WEBER TEST TO DETERMINE CONDUCTIVE AND SENSORINEURAL DEAFNESS Strike a tuning fork and place the handle of the tuning fork medially on your forehead (see Figure 25.5a). Is the tone equally loud in both ears, or is it louder in one ear?

____

If it is equally loud in both ears, you have equal hearing or equal loss of hearing in both ears. If sensorineural deafness is present in one ear, the tone will be heard in the unaffected ear, but not in the ear with sensorineural deafness. If conduction deafness is present, the sound will be heard more strongly in the ear in which there is a hearing loss. Conduction deafness can be simulated by plugging one ear with cotton to interfere with the conduction of sound to the inner ear.

RINNE TEST FOR COMPARING BONE- AND AIR-CONDUCTION HEARING

1. Strike the tuning fork, and place its handle on your partner's mastoid process (Figure 25.5b).
2. When your partner indicates that the sound is no longer audible, hold the still-vibrating prongs close to his auditory canal (Figure 25.5c). If your partner hears the fork again (by air conduction) when it is moved to that position, hearing is not impaired and the test result is to be recorded as positive (+). (Record below.)
3. Repeat the test, but this time test air conduction hearing first.
4. After the tone is no longer heard by air conduction, hold the handle of the tuning fork on the bony mastoid process. If the subject hears the tone again by bone conduction after hearing by air conduction is lost, there is some conductive deafness and the result is recorded as negative (−).
5. Repeat the sequence for the opposite ear.

Right ear ______________ Left ear ______________

Does the subject hear better by bone or by air conduction?

____

Audiometry

When the simple tuning fork tests reveal a problem in hearing, audiometer testing is usually prescribed to determine the precise nature of the hearing deficit. An *audiometer* is an instrument (specifically, an electronic oscillator with earphones) used to determine hearing acuity by exposing each ear to sound stimuli of differing

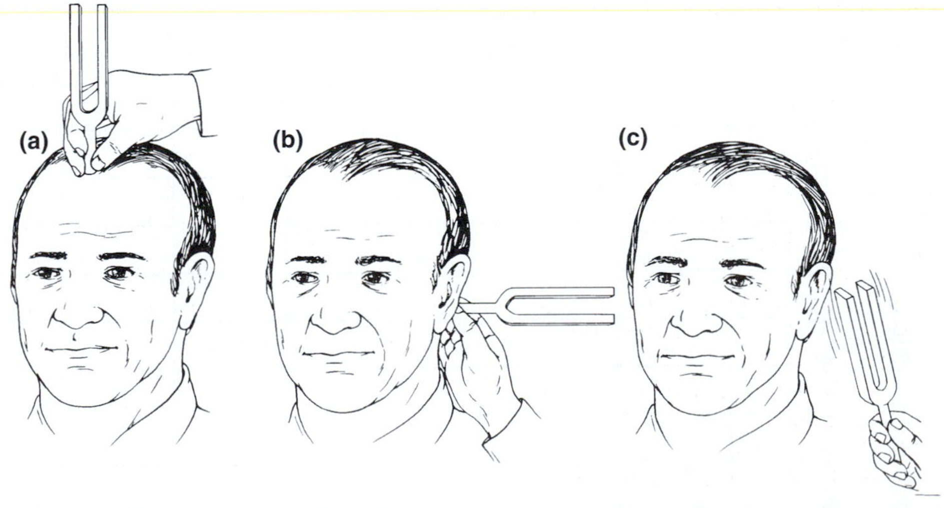

F25.5

The Weber and Rinne tuning fork tests. **(a)** The Weber test to evaluate whether the sound remains centralized (normal) or lateralizes to one side or the other (indicative of some degree of conductive or sensorineural deafness). **(b and c)** The Rinne test to compare bone conduction and air conduction.

frequencies and *intensities.* The hearing range of human beings during youth is from 30 to 22,000 Hz, but hearing acuity declines with age, with reception for the high-frequency sounds lost first. Though this loss represents a major problem for some people, such as musicians, most of us tend to be fairly unconcerned until we begin to have problems hearing sounds in the range of 125 to 8000 Hz, the normal frequency range of speech.

The basic procedure of audiometry is to initially deliver tones of different frequencies to one ear of the subject at an intensity of 0 decibels (dB). (Zero decibels is not the complete absence of sound, but rather the softest sound intensity that can be heard by a person of normal hearing at each frequency.) If the subject cannot hear a particular frequency stimulus of 0 dB, the hearing threshold level control is adjusted until the subject reports that he or she can hear the tone. The number of decibels of intensity required above 0 dB is recorded as the hearing loss. For example, if the subject cannot hear a particular frequency tone until it is delivered at 30 dB intensity, then he or she has a hearing loss of 30 dB for that frequency.

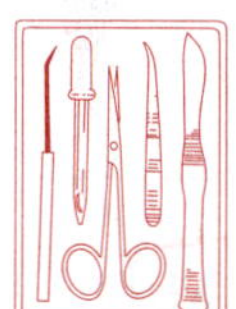

1. Before beginning the tests, examine the audiometer to identify the two tone controls: one to regulate frequency and a second to regulate the intensity (loudness) of the sound stimulus. Also identify the two output control switches that regulate the delivery of sound to one ear or the other (*red* to the right ear, *blue* to the left ear). Also find the *hearing threshold level control,* which is calibrated to deliver a basal tone of 0 dB to the subject's ears.

2. Place the earphones on the subject's head so that the red cord or ear-cushion is over the right ear and the blue cord or ear-cushion is over the left ear. Instruct the subject to raise one hand when he or she hears a tone.

3. Set the frequency control at 125 Hz and the intensity control at 0 dB. Press the red output switch to deliver a tone to the subject's right ear. If the subject does not respond, raise the sound intensity slowly by rotating the hearing level control counterclockwise until the subject reports (by raising a hand) that a tone is heard. Repeat this procedure for frequencies of 250, 500, 1000, 2000, 4000, and 8000.

4. Record the results in the chart below by marking a small red circle on the chart at each frequency-dB junction indicated. Then connect the circles with a red line to produce a hearing acuity graph for the right ear.

5. Repeat steps 3 and 4 for the left (blue) ear and record the results with blue circles and connecting lines on the chart.

Microscopic Anatomy of the Equilibrium Apparatus and Mechanisms of Equilibrium

The equilibrium apparatus of the inner ear is in the vestibular and semicircular canal portions of the bony labyrinth. Their chambers are filled with perilymph, in which membranous labyrinth structures are suspended. The vestibule contains the saclike **utricle** and **saccule,** and the semicircular chambers contain **membranous semicircular ducts.** Like the cochlear duct, these membranes are filled with endolymph and contain receptor cells that are activated by the disturbance of their cilia.

The semicircular canals are centrally involved in the **mechanism of dynamic equilibrium.** They are about ½ inch in circumference and are oriented in three planes—horizontal, frontal, and sagittal. At the base of each semicircular duct is an enlarged region, the **ampulla,** which communicates with the utricle of the vestibule. Within each ampulla is a receptor region called a **crista ampullaris,** which consists of a tuft of hair cells covered with a gelatinous cap, or **cupula** (Figure 25.6). When your head position changes in an angular direc-

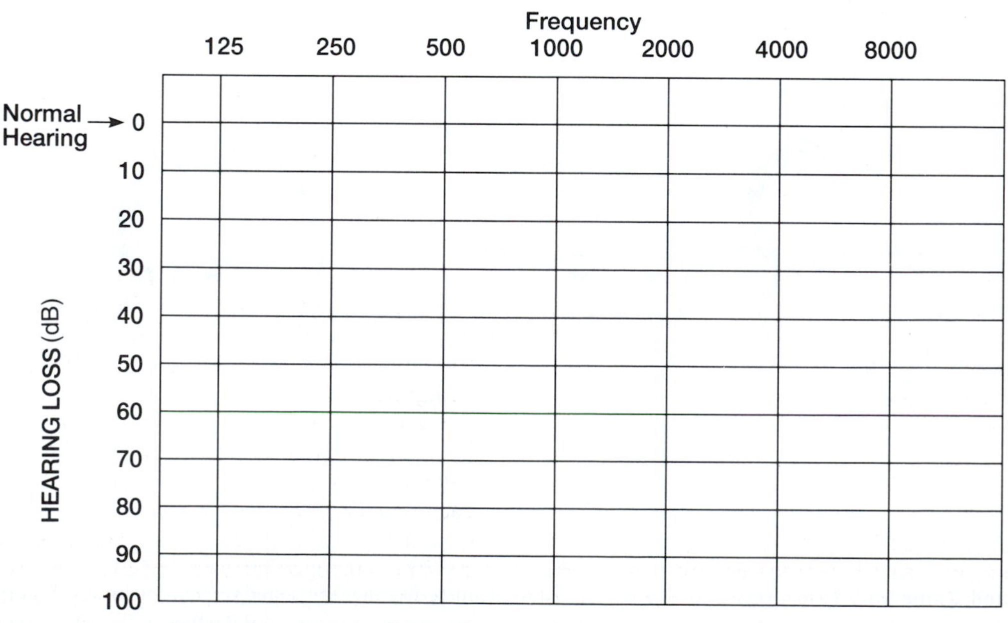

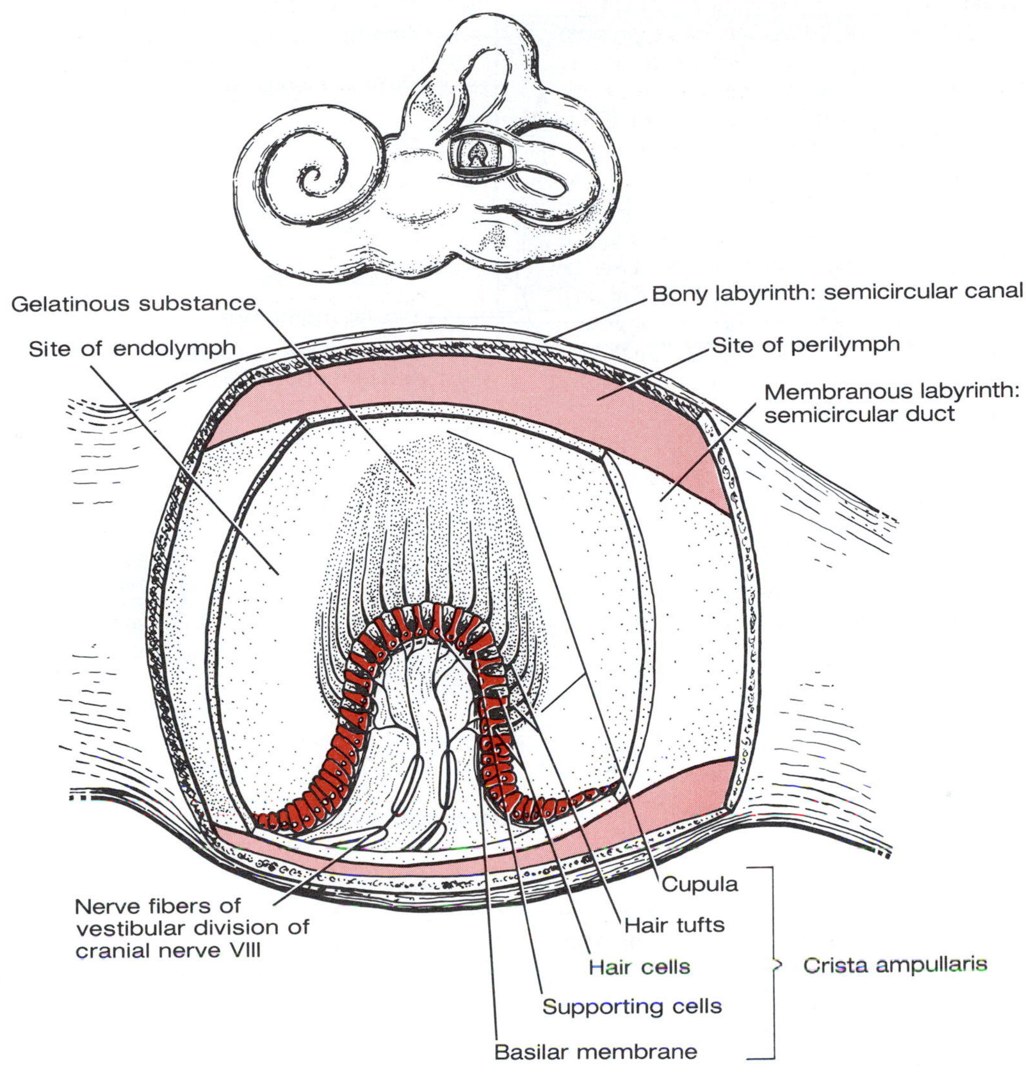

F25.6

Crista ampullaris in the semicircular canal.

tion, as when twirling on the dance floor or when taking a rough boat ride, the endolymph in the canal lags behind, pushing the cupula—like a swinging door—in a direction opposite to that of the angular motion. Depending on the ear, this movement depolarizes or hyperpolarizes the hair cells resulting in enhanced or reduced impulse transmission up the vestibular division of the eighth cranial nerve to the brain. Likewise, when the angular motion stops suddenly, the inertia of the endolymph causes it to continue to move, pushing the cupula in the same direction as the previous motion. This movement again initiates electrical changes in the hair cells. (This phenomenon accounts for the reversed motion sensation you feel when you stop suddenly after twirling.) If you begin to move at a constant rate of motion, the cupula gradually returns to its original position. The hair cells, no longer bent, send no new signals, and you lose the sensation of spinning. Thus the response of these dynamic equilibrium receptors is a reaction to *changes* in angular motion rather than to motion itself.

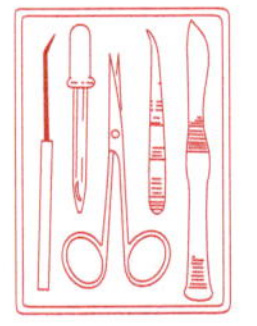

Go to the demonstration area and examine the slide of a crista ampullaris. Identify the areas depicted in Figure 25.6.

The vestibule contains **maculae,** receptors that are essential to the **mechanism of static equilibrium.** The maculae respond to gravitational pull, thus providing information on which way is up or down, and to linear or straightforward changes in speed. They are located on the walls of the saccule and utricle. The **otolithic membrane,** a gelatinous material containing small grains of calcium carbonate (**otoliths**), overrides the hair cells in each macula. As the head moves, the otoliths move in response to variations in gravitational pull (Figure 25.7). As they deflect different hair cells, they trigger hyperpolarization or depolarization of the hair cells and modify the rate of impulse transmission along the vestibular nerve.

(a)

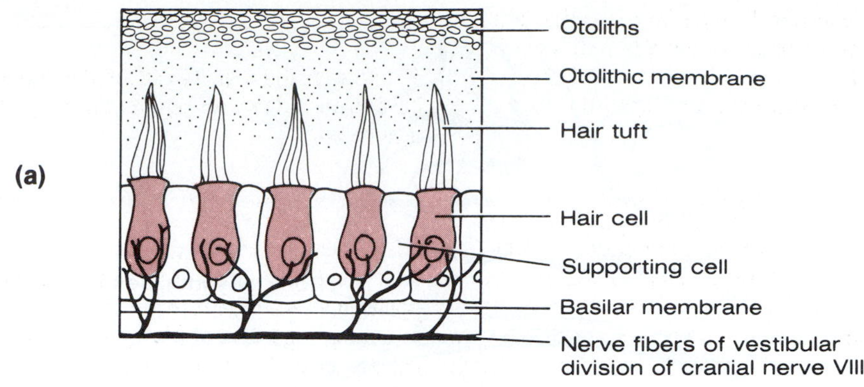

(b)

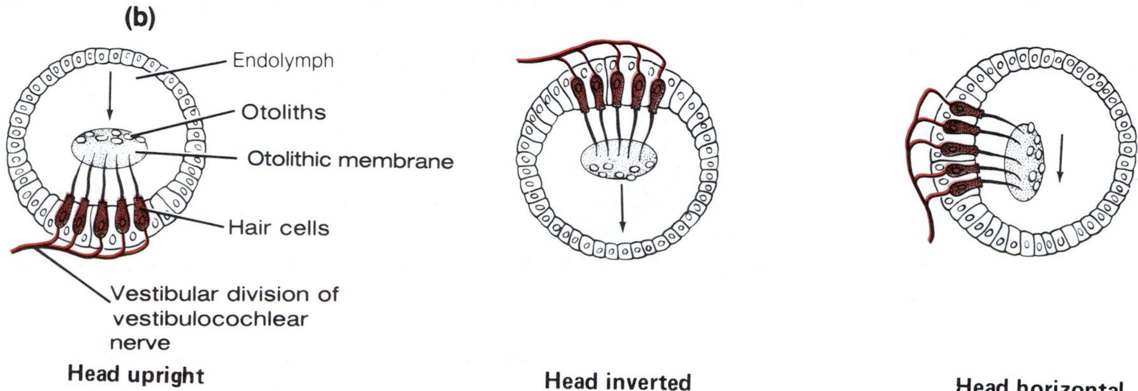

F25.7

Structure and function of static equilibrium receptors (maculae). (a) Diagrammatic view of a portion of a macula. **(b)** Stimulation of the maculae by movement of otoliths in the gelatinous otolithic membrane creates a pull on the hair cells. Arrows indicate direction of gravitational pull.

Although the receptors of the semicircular canals and the vestibule are responsible for dynamic and static equilibrium respectively, they rarely act independently. Complex interaction of many of the receptors is the rule. Also, the information these equilibrium, or balance, senses provide is significantly augmented by the proprioceptors and sight, as some of the following laboratory experiments demonstrate.

Laboratory Tests: Equilibrium

The function of the semicircular canals and vestibule are not routinely tested in the laboratory, but the following simple tests should serve to illustrate normal equilibrium apparatus functioning.

BALANCE TEST Have your partner walk a straight line, placing one foot directly in front of the other.

Is he or she able to walk without undue wobbling from side to side?

____

Did he or she experience any dizziness? ____________

The ability to walk with balance and without dizziness, unless subject to rotational forces, indicates normal function of the equilibrium apparatus.

Was nystagmus* present? ______________________

BARANY TEST (INDUCTION OF NYSTAGMUS AND VERTIGO†) This experiment evaluates the semicircular canals and should be conducted as a group effort to protect test subject(s) from possible injury.

⚠ The following precautionary notes should be read before beginning:

- The subject(s) chosen should not be easily inclined to dizziness during rotational or turning movements.
- Rotation should be stopped immediately if the subject complains of feeling nauseous.

* **Nystagmus** is the involuntary rolling of the eyes in any direction or the trailing of the eyes slowly in one direction, followed by their rapid movement in the opposite direction. It is normal after rotation; abnormal otherwise. The direction of nystagmus is that of its quick phase on acceleration.

† **Vertigo** is a sensation of dizziness and rotational movement when such movement is not occurring or has ceased.

- Because the subject(s) will experience vertigo and loss of balance as a result of the rotation, several classmates should be prepared to catch, hold, or support the subject(s) as necessary until the symptoms pass.

1. Instruct the subject to sit on a rotating chair or stool, and to hold on to the arms or seat of the chair, feet on stool rungs. The subject's head should be tilted forward approximately 30 degrees (almost touching the chest). The horizontal (lateral) semicircular canal will be stimulated when the head is in this position. The subject's eyes are to remain *open* during the test.

2. Four classmates should position themselves so that the subject is surrounded on all sides. The classmate posterior to the subject will rotate the chair.

3. Rotate the chair to the subject's right approximately 10 revolutions in 10 seconds, and then suddenly stop the rotation.

4. Immediately note the direction of the subject's resultant nystagmus; and ask him or her to describe the feelings of movement, indicating speed and direction sensation. Record below.

____

____

____

____

If the semicircular canals are operating normally, the subject will experience a sensation that the stool is still rotating immediately after it has stopped and *will* demonstrate nystagmus.

When the subject is rotated to the right, the cupula will be bent to the left, causing nystagmus during rotation in which the eyes initially move slowly to the left and then quickly to the right. Nystagmus will continue until the cupula has returned to its initial position. Then, when rotation is stopped abruptly, the cupula will be bent to the right, producing nystagmus with its slow phase to the right and its rapid phase to the left. In many subjects, this will be accompanied by a feeling of vertigo and a tendency to fall to the right.

ROMBERG TEST The Romberg test determines the integrity of the dorsal white column of the spinal cord, which transmits impulses to the brain from the proprioceptors involved with posture.

1. Have your partner stand with his or her back to the blackboard.

2. Draw one line parallel to each side of your partner's body. He or she should stand erect, with eyes open and staring straight ahead for 2 minutes while you observe any movements. Did you see any gross swaying movements?

____

3. Repeat the test. This time the subject's eyes should be closed. Note and record the degree of side-to-side movement.

____

4. Repeat the test with the subject's eyes first open and then closed. This time, however, the subject should be positioned with his or her left shoulder toward, but not touching, the board so that you may observe and record the degree of front-to-back swaying.

____

Do you think the equilibrium apparatus of the inner ear was operating equally well in all these tests?

____

The proprioceptors? ______________________________

Why was the observed degree of swaying greater when the eyes were closed?

____

____

What conclusions can you draw regarding the factors necessary for maintaining body equilibrium and balance?

____

____

____

____

ROLE OF VISION IN MAINTAINING EQUILIBRIUM To further demonstrate the role of vision in maintaining equilibrium, perform the following experiment. (Ask your lab partner to record observations and act as a "spotter.") Stand erect, with your eyes open. Raise your left foot approximately 1 foot off the floor, and hold it there for 1 minute.

Record the observations: ______________________________

____

____

Rest for 1 or 2 minutes; and then repeat the experiment with the same foot raised, but with your eyes closed. Record the observations:

____

____

26

EXERCISE

Special Senses: Taste and Olfaction

OBJECTIVES

1. To describe the structure and function of the taste receptors.
2. To describe the location and cellular composition of the olfactory epithelium.
3. To name the four basic qualities of taste sensation and list the chemical substances that elicit these sensations.
4. To point out on a diagram of the tongue the predominant location of the basic types of taste receptors (salty, sweet, sour, bitter).
5. To explain the interdependence between the senses of smell and taste.
6. To name two factors other than olfaction that influence taste appreciation of foods.
7. To define *olfactory adaptation.*

MATERIALS

Prepared histologic slides: the tongue showing taste buds; nasal olfactory epithelium (longitudinal section)
Compound microscope
Paper towels
Small mirror
Granulated sugar
Cotton-tipped swabs
Disposable autoclave bag
Paper cups; paper plates
Prepared vials of 10% NaCl, 0.1% quinine or Epsom salt solution, 5% sucrose solution, and 1% acetic acid
Beaker containing 10% bleach solution
Prepared dropper bottles of oil of cloves, oil of peppermint, and oil of wintergreen or corresponding flavors found in the condiment section of a supermarket
Equal-size food cubes of cheese, apple, raw potato, dried prunes, banana, raw carrot, and hard-cooked egg white (These prepared foods should be in an opaque container; a foil-lined egg carton would work well.)
Chipped ice

See Appendix E, Exercise 26 for links to *Anatomy and PhysioShow: The Videodisc.*

The receptors for taste and olfaction are classified as **chemoreceptors** because they respond to chemicals or volatile substances in solution. Although four relatively specific types of taste receptors have been identified, the olfactory receptors are considered sensitive to a much wider range of chemical sensations. The sense of smell is the least understood of the special senses.

LOCALIZATION AND ANATOMY OF TASTE BUDS

The **taste buds,** specific receptors for the sense of taste, are widely but not uniformly distributed in the oral cavity. Most are located on the dorsal surface of the tongue (as described next). A few are found on the soft palate, epiglottis, and inner surface of the cheeks.

The dorsal tongue surface is covered with small projections, or **papillae,** of three major types: sharp *filiform papillae* and the rounded *fungiform* and *circumvallate papillae.* The taste buds are located primarily on the sides of the circumvallate papillae (arranged in a V-formation on the posterior surface of the tongue) and on the more numerous fungiform papillae. The latter look rather like minute mushrooms and are widely distributed on the tongue. (See Figure 26.1.)

- Use a mirror to examine your tongue. Can you pick out the various papillae types?

  _______ If so, which? ______________________

  ____

Each taste bud consists largely of a globular arrangement of two types of modified epithelial cells: the **gustatory,** or **taste cells,** which are the actual receptor cells, and **supporting cells.** Several nerve fibers enter each taste bud and supply sensory nerve endings to each of the taste cells. The long microvilli of the receptor cells penetrate the epithelial surface through an opening called the **taste pore.** When these microvilli, called **gustatory hairs,** contact specific chemicals in the solution, the taste cells depolarize. The afferent fibers from the taste buds to the sensory cortex in the postcentral gyrus

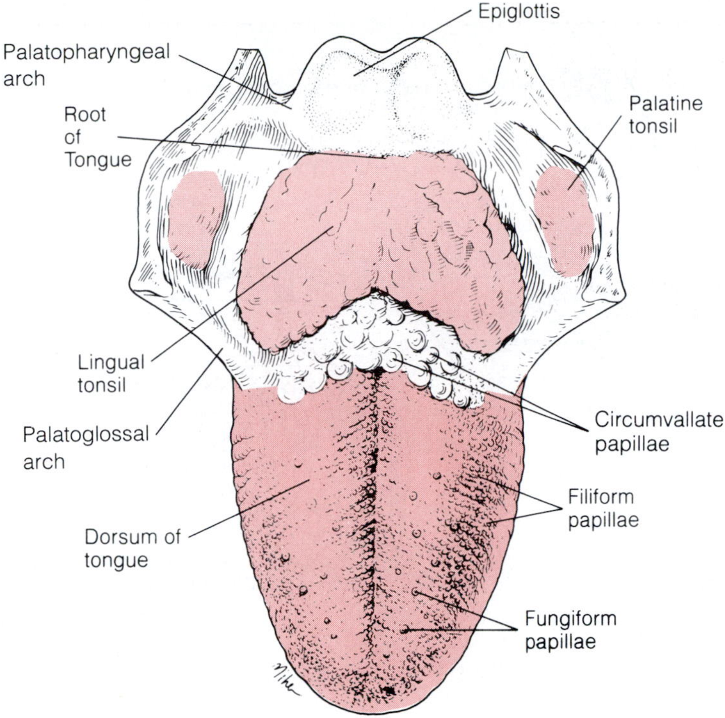

F26.1

Dorsal surface of the tongue, showing major structures.

of the brain are carried in three cranial nerves: the *facial nerve* (*VII*) serves the anterior two-thirds of the tongue; the *glossopharyngeal nerve* (*IX*) serves the posterior third of the tongue; and the *vagus nerve* (*X*) carries a few fibers from the pharyngeal region.

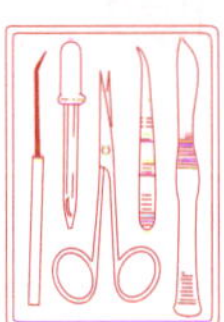

Obtain a microscope and a prepared slide of a tongue cross section. Use Figure 26.2 as a guide to aid you in locating the taste buds on the tongue papillae. Make a detailed study of one taste bud. Identify the taste pore and gustatory hairs if observed.

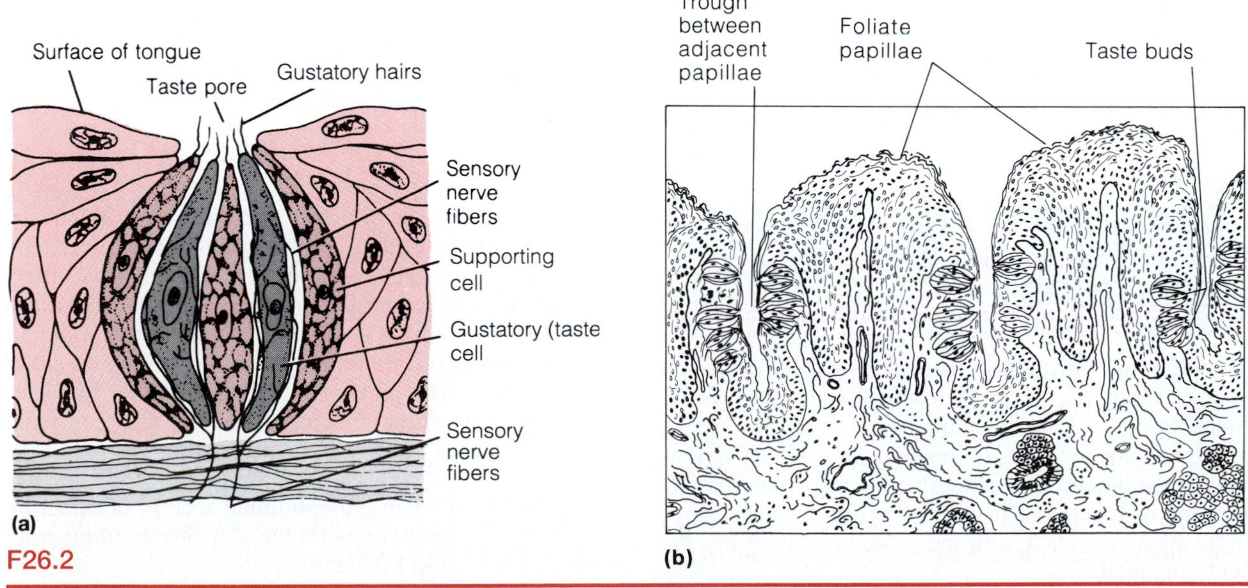

F26.2

Taste bud anatomy and localization. **(a)** Diagrammatic view of taste bud. **(b)** View of tongue papillae, showing position of taste buds. See Plate 16 in the Histology Atlas for corresponding photomicrograph.

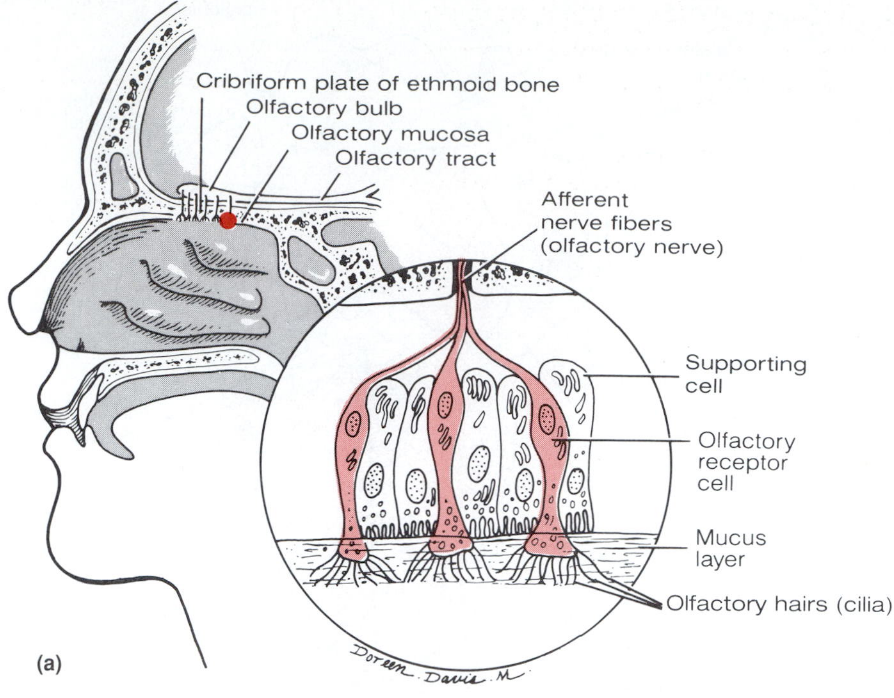

(b)

Lumen of nasal cavity

Supporting cell nucleus

Basal cell nucleus

Lamina propria containing mucous secreting glands

Cilia of olfactory receptor cells

Olfactory cell nucleus

F26.3

Cellular composition of olfactory epithelium. (a) Diagrammatic representation. **(b)** Line drawing corresponding to the photomicrograph in Plate 17 in the Histology Atlas.

LOCALIZATION AND ANATOMY OF THE OLFACTORY RECEPTORS

The **olfactory epithelium** (organ of smell) occupies an area of about 2.5 cm in the roof of each nasal cavity. Since the air entering the human nasal cavity must make a hairpin turn to enter the respiratory passages below, the nasal epithelium is in a rather poor position for performing its function. This is why sniffing, which brings more air into contact with the receptors, intensifies the sense of smell.

The specialized receptor cells in the olfactory epithelium are surrounded by **supporting cells,** nonsensory epithelial cells. The **olfactory receptor cells** are bipolar neurons whose **olfactory hairs** (actually cilia) extend outward from the epithelium. Axonal nerve fibers emerging from their basal ends penetrate the cribriform plate of the ethmoid bone and proceed as the *olfactory nerves* to synapse in the olfactory bulbs lying on either side of the crista galli of the ethmoid bone. Impulses from neurons of the olfactory bulbs are then conveyed to the olfactory portion of the cortex (uncus).

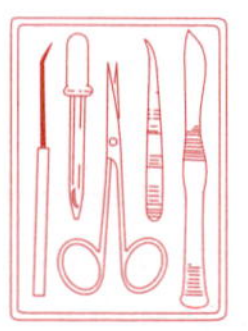

Obtain a longitudinal section of olfactory epithelium. Examine it closely, comparing it to Figure 26.3.

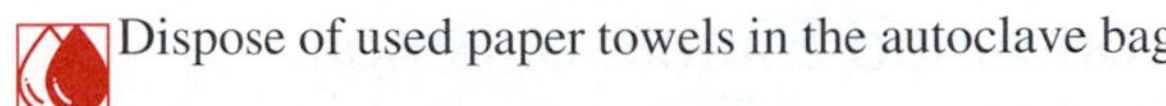

LABORATORY EXPERIMENTS

Stimulation of Taste Buds

1. Obtain several paper towels and a disposable autoclave bag and bring them to your bench.

2. With a paper towel, dry the dorsal surface of your tongue.

Immediately dispose of the paper towel in the autoclave bag.

3. Place a few sugar crystals on your dried tongue. Do *not* close your mouth. Time how long it takes to taste the sugar.

______________ sec

Why couldn't you taste the sugar immediately?

____

____

Plotting Taste Bud Distribution

When taste is tested with pure chemical compounds, taste sensations can all be grouped into one of four basic qualities—sweet, sour, bitter, or salty. Although all taste buds are believed to respond in some degree to all four classes of chemical stimuli, each type responds optimally to only one. This characteristic makes it possible to map the tongue to show the relative density of each type of taste bud.

The *sweet receptors* respond to a number of seemingly unrelated compounds such as sugars (fructose, sucrose, glucose), saccharine, and some amino acids. Some believe the common factor is the hydroxyl (OH^-) group. *Sour receptors* respond to hydrogen ions (H^+) or the acidity of the solution, *bitter receptors* to alkaloids, and *salty receptors* to metallic ions in solution.

1. Prepare to make a taste sensation map of your lab partner's tongue by obtaining the following: cotton-tipped swabs; one vial each of NaCl, quinine or Epsom salt solution, sucrose solution, and acetic acid; paper cups; and a flask of distilled or tap water.

2. Before each test, the subject should rinse his or her mouth thoroughly with water and lightly dry his or her tongue with a paper towel.

Dispose of used paper towels in the autoclave bag.

3. Generously moisten a swab with 5% sucrose solution and touch it to the center, back, tip, and sides of the dorsal surface of the subject's tongue.

4. Map, with O's on the tongue outline below, the location of the sweet receptors.

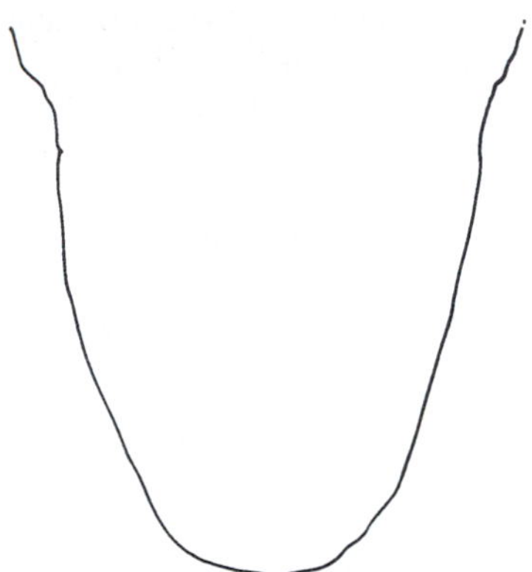

Put the used swab in the autoclave bag. Do not redip the swab into the sucrose solution.

5. Repeat the procedure with quinine (or Epsom salt solution) to map the location of the bitter receptors (use the symbol B), with NaCl to map the salt receptors (symbol +), and with acetic acid to map the sour receptors (symbol −).

Use a fresh swab for each test, and properly dispose of the swabs immediately after use.

What area of the tongue dorsum seems to lack taste receptors?

____

How closely does your localization of the different taste receptors coincide with the information in your textbook?

____

____

Combined Effects of Smell, Texture, and Temperature on Taste

EFFECTS OF SMELL AND TEXTURE

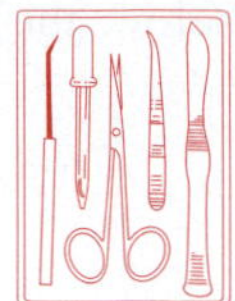

1. Ask the subject to sit with eyes closed and to pinch his or her nostrils shut.

2. Using a paper plate, obtain samples of the food items listed in the chart below. At no time should the subject be allowed to see the foods being tested.

3. Use an out-of-sequence order of food testing. For each test, place a cube of food in the subject's mouth and ask him or her to identify the food by using the following sequence of activities:

- First, manipulate the food with the tongue.
- Second, chew the food.
- Third, if a positive identification is not made with the first two techniques and the taste sense, ask the subject to release the pinched nostrils and to continue chewing with the nostrils open to determine if a positive identification can be made.

Record the results on the chart by checking the appropriate column.

Was the sense of smell equally important in all cases?

____

Where did it seem to be important and why?

____

____

EFFECT OF OLFACTORY STIMULATION There is no question that what is commonly referred to as taste depends heavily on stimulation of the olfactory receptors, particularly in the case of strongly odoriferous substances. The following experiment should illustrate this fact.

1. Obtain vials of oil of wintergreen, peppermint, and cloves and some fresh cotton-tipped swabs. Ask the subject to sit so that he or she cannot see which vial is being used, and to dry the tongue and close the nostrils.
2. Apply a drop of one of the oils to the subject's tongue. Can he or she distinguish the flavor?

   ____

3. Have the subject open the nostrils, and record the change in sensation he or she reports.

   ____

4. Have the subject rinse the mouth well and dry the tongue.
5. Prepare two swabs, each with one of the two remaining oils.
6. Hold one swab under the subject's open nostrils, while touching the second swab to the tongue.

   Record the reported sensations. _______________

   ____

   ____

   ____

Method of Identification

Food	Texture only	Chewing with nostrils pinched	Chewing with nostrils open	Identification not made
Cheese	______	______	______	______
Apple	______	______	______	______
Raw potato	______	______	______	______
Banana	______	______	______	______
Dried prunes	______	______	______	______
Raw carrot	______	______	______	______
Hard-cooked egg white	______	______	______	______

7. Dispose of the used swabs and paper towels in the autoclave bag before continuing.

Which sense, taste or smell, appears to be more important in the proper identification of a strongly flavored volatile substance?

____

EFFECT OF TEMPERATURE In addition to the effect that olfaction and food texture play in determining our taste sensations, the temperature of foods also helps determine if the food is appreciated or even tasted. To illustrate this, have your partner hold some chipped ice on the tongue for approximately a minute and then close his or her eyes. Immediately place any of the foods previously identified in his or her mouth and ask for an identification.

Results? ______________________________

____

____

Olfactory Adaptation

Obtain some absorbent cotton and two of the following oils (oil of wintergreen, peppermint, or cloves). Press one nostril shut. Hold the bottle of oil under the open nostril and exhale through the mouth. Record the time required for the odor to disappear (for olfactory adaptation to occur).

______________ sec

Repeat the procedure with the other nostril.

______________ sec

Immediately test another oil with the nostril that has just experienced olfactory adaptation. What are the results?

____

____

What conclusions can you draw? ______________________________

____

____

27

EXERCISE

Anatomy and Basic Function of the Endocrine Glands

OBJECTIVES

1. To identify and name the major endocrine glands and tissues of the body when provided with an appropriate diagram.
2. To list the hormones produced by the endocrine glands and discuss the general function of each.
3. To indicate the means by which hormones contribute to body homeostasis by giving appropriate examples of hormonal actions.
4. To cite the mechanism by which the endocrine glands are stimulated to release their hormones.
5. To describe the structural and functional relationship between the hypothalamus and the pituitary.
6. To describe a major pathologic consequence of hypersecretion and hyposecretion of each hormone considered.
7. To correctly identify the histologic structure of the anterior and posterior pituitary, thyroid, parathyroid, adrenal cortex and medulla, pancreas, testis, and ovary by microscopic inspection or when presented with an appropriate photomicrograph or diagram. (Optional)
8. To name and point out the specialized hormone-secreting cells in the above tissues as studied in the laboratory. (Optional)

MATERIALS

Human torso model
Anatomical chart of the human endocrine system
Compound microscopes
Colored pencils
Histologic slides of the anterior pituitary,* posterior pituitary, thyroid gland, parathyroid glands, adrenal gland, and pancreas,* ovary, and testis tissue

See Appendix D, Exercise 27 for links to A.D.A.M. Standard.

See Appendix E, Exercise 27 for links to *Anatomy and PhysioShow: The Videodisc.*

*Differential-staining if possible

The **endocrine system** is the second major controlling system of the body. Acting with the nervous system, it helps coordinate and integrate the activity of the body's cells. However, the nervous system employs electrochemical impulses to bring about rapid control, while the more slowly acting endocrine system employs chemical "messengers," or **hormones,** which are released into the blood to be transported throughout the body.

The term *hormone* comes from a Greek word meaning "to arouse." The body's hormones, which are steroids or amino acid-based molecules, arouse the body's tissues and cells by stimulating changes in their metabolic activity. These changes lead to growth and development and to the physiologic homeostasis of many body systems. Although all hormones are blood-borne, a given hormone affects only the biochemical activity of a specific organ or organs. Organs that respond to a particular hormone are referred to as the **target organs** of that hormone. The ability of the target tissue to respond seems to depend on the ability of the hormone to bind with specific receptors (proteins) occurring on the cells' plasma membrane or within the cells.

Although the function of some hormone-producing glands (the anterior pituitary, thyroid, adrenals, parathyroids) is purely endocrine, the function of others (the pancreas and gonads) is mixed—both endocrine and exocrine. Both types of glands are derived from epithelium, but the endocrine, or ductless, glands release their product (always hormonal) directly into the blood. The exocrine glands release their products at the body's surface or outside an epithelial membrane via ducts. In addition, there are varied numbers of hormone-producing cells within the intestine, stomach, kidney, and placenta, organs whose functions are primarily nonendocrine. Only the major endocrine organs are considered here.

GROSS ANATOMY AND BASIC FUNCTION OF THE ENDOCRINE GLANDS

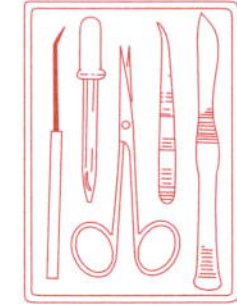

As the endocrine organs are described, *locate and identify them by name* on Figure 27.1. When you have completed the descriptive material, also locate the organs on the anatomical charts or torso.

Pituitary Gland (Hypophysis)

The *pituitary gland,* or *hypophysis,* is located in the concavity of the sella turcica of the sphenoid bone. It consists largely of two functional areas, the **adenohypophysis,** or **anterior pituitary,** and the **neurohypophysis,** or **posterior pituitary,** and is attached to the hypothalamus by a stalk called the **infundibulum.**

ADENOHYPOPHYSEAL HORMONES The adenohypophysis secretes a number of hormones. Four of these are **tropic** hormones. In each case, a tropic hor-

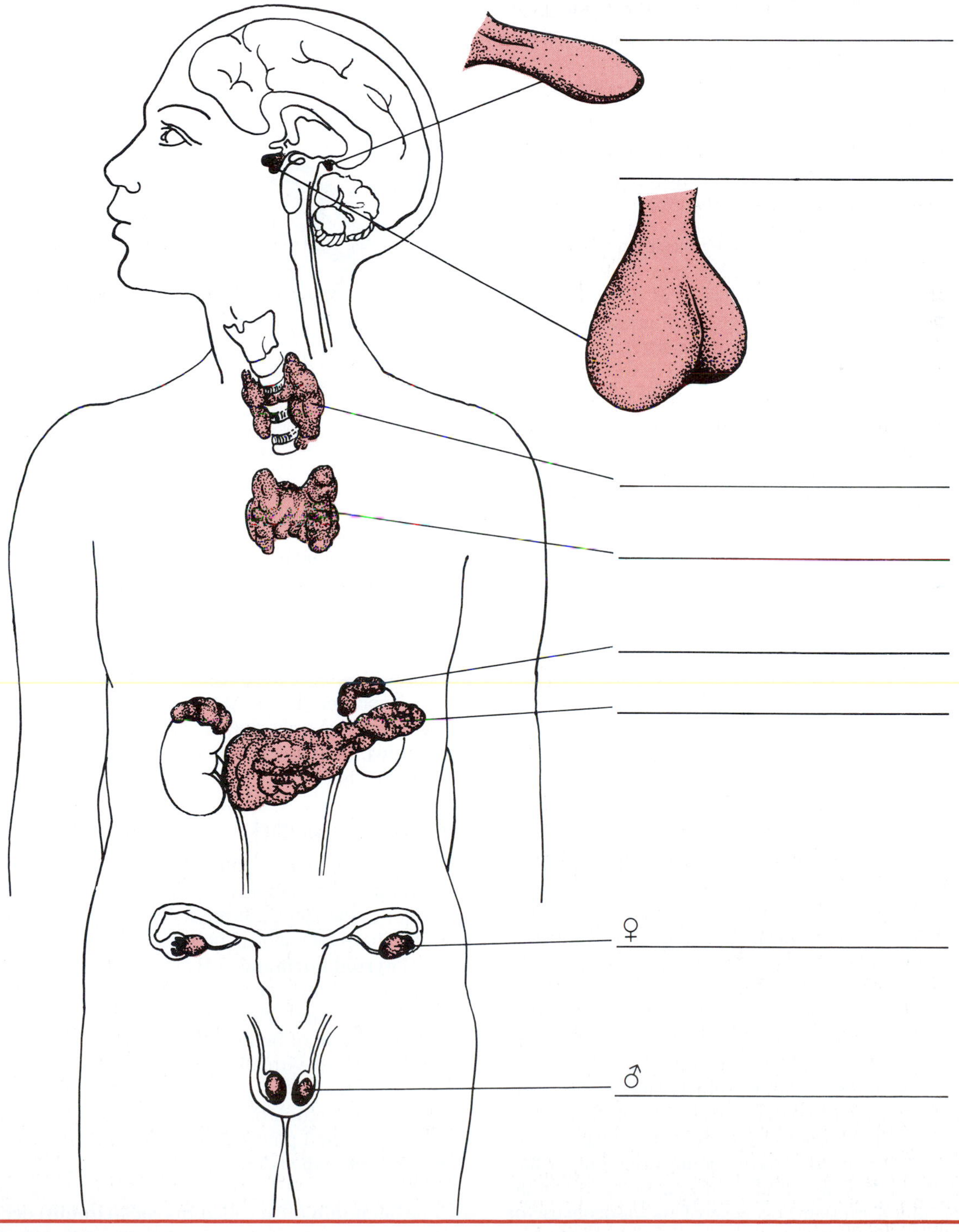

F27.1

Human endocrine organs.

mone released by the anterior pituitary stimulates its target organ, which is also an endocrine gland, to secrete its hormones. Target organ hormones then exert their effects on other body organs and tissues. The tropic hormones include:

- The **gonadotropins—follicle-stimulating hormone (FSH)** and **luteinizing hormone (LH)**—regulate gamete production and hormonal activity of the gonads (ovaries and testes). The precise roles of the gonadotropins are described in Exercise 43 along with other considerations of reproductive system physiology.
- **Adrenocorticotropic hormone (ACTH)** regulates the endocrine activity of the cortex portion of the adrenal gland.
- **Thyrotropic hormone (TSH)** influences the growth and activity of the thyroid gland.

The three other hormones produced by the anterior pituitary are not directly involved in the regulation of other endocrine glands of the body.

- **Growth hormone (GH)** is a general metabolic hormone that plays an important role in determining body size. It affects many tissues of the body; however, its major effects are exerted on the growth of muscle and the long bones of the body.
- **Prolactin (PRL)** stimulates breast development and promotes and maintains lactation by the mammary glands after childbirth. Its function in males is unknown.
- **Melanocyte-stimulating hormone (MSH),** which stimulates melanocytes to increase their synthesis of melanin pigment, does not appear to be of major significance in humans.

The anterior pituitary controls the activity of so many other endocrine glands that it has often been called the *master endocrine gland.* However, the anterior pituitary is not autonomous in its control because release of the anterior pituitary hormones is controlled by neurosecretions, *releasing or inhibiting hormones,* produced by the hypothalamus. These hypothalamic hormones are liberated into the **hypophyseal portal system,** which serves the circulatory needs of the anterior pituitary (Figure 27.2).

NEUROHYPOPHYSEAL HORMONES The neurohypophysis, or posterior pituitary, is not an endocrine gland in a strict sense, because it does not synthesize the hormones it releases. (This relationship is also indicated in Figure 27.2.) Instead, it acts as a storage area for two hormones transported to it from the paraventricular and supraoptic nuclei of the hypothalamus. The first of these hormones is **oxytocin,** which stimulates powerful uterine contractions during birth and coitus and also causes milk ejection in the lactating mother. The second, **antidiuretic hormone (ADH),** causes the distal and collecting tubules of the kidneys to reabsorb more water from the urinary filtrate, thereby reducing urine output and conserving body water. It also plays a minor role in increasing blood pressure because of its vasoconstrictor effect on the arterioles.

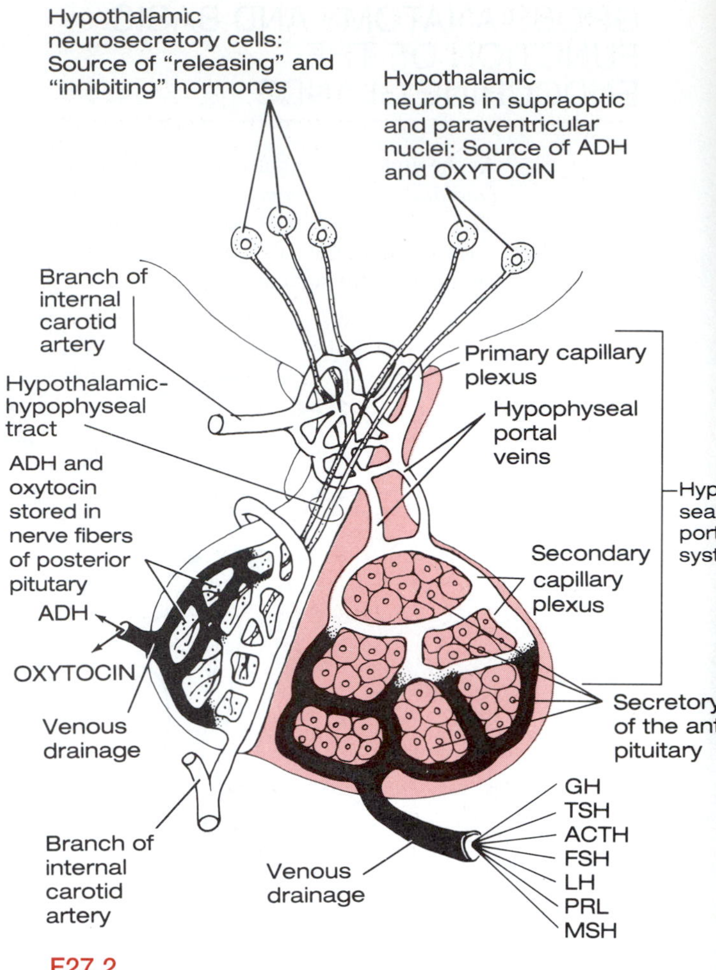

F27.2

Neural and vascular relationships between the hypothalamus and the anterior and posterior lobes of the pituitary.

Hyposecretion of ADH results in dehydration from excessive urine output, a condition called **diabetes insipidus.** Individuals with this condition experience an insatiable thirst. ■

Thyroid Gland

The *thyroid gland* is composed of two lobes joined by a central mass, or isthmus. It is located in the throat, just inferior to the larynx. It produces two major hormones, thyroid hormone and calcitonin.

Thyroid hormone (TH) is actually two physiologically active hormones known as $\mathbf{T_4}$ (thyroxine) and $\mathbf{T_3}$ (triiodothyronine). Because its primary function is to control the rate of body metabolism and cellular oxidation, TH affects virtually every cell in the body.

Hyposecretion of thyroxine leads to a condition of mental and physical sluggishness, which is called **myxedema** in the adult. ■

Calcitonin (also called **thyrocalcitonin**) decreases blood calcium levels by stimulating calcium deposit in the bones. It acts antagonistically to parathyroid hormone, the hormonal product of the parathyroid glands.

- Try to palpate your thyroid gland by placing your fingers against your windpipe. As you swallow, the thyroid gland will move up and down on the sides and front of the windpipe.

Parathyroid Glands

The *parathyroid glands* are found embedded in the posterior surface of the thyroid gland. Typically, there are two small oval glands on each lobe, but there may be more and some may be located in other regions of the neck. They secrete **parathyroid hormone (PTH),** the most important regulator of calcium-phosphate ion homeostasis of the blood. When blood calcium levels decrease below a certain critical level, the parathyroids release PTH, which causes release of calcium from bone matrix and prods the kidney to reabsorb more calcium and less phosphate from the filtrate. If blood calcium levels fall too low, **tetany** results and may be fatal.

Adrenal Glands

The two bean-shaped *adrenal,* or *suprarenal, glands* are located atop or close to the kidneys. Anatomically, the **adrenal medulla** develops from neural crest tissue and it is directly controlled by sympathetic nervous system neurons. The medullary cells respond to this stimulation by releasing **epinephrine** (80%) or **norepinephrine** (20%), which act in conjunction with the sympathetic nervous system to elicit the "flight or fight" response to stressors.

The **adrenal cortex** produces three major groups of steroid hormones, collectively called the **corticosteroids.** The **mineralocorticoids,** chiefly **aldosterone,** regulate water and electrolyte balance in the extracellular fluids, mainly by regulating sodium ion reabsorption by kidney tubules. The **glucocorticoids** (cortisone, hydrocortisone, and corticosterone) enable the body to resist long-term stressors, primarily by increasing blood glucose levels. The **gonadocorticoids,** or **sex hormones,** produced by the adrenal cortex are chiefly androgens (male sex hormones), but some estrogens (female sex hormones) are also formed. The gonadocorticoids are produced throughout life in relatively insignificant amounts; however, hypersecretion of these hormones produces abnormal hairiness (**hirsutism**), and masculinization occurs.

Pancreas

The *pancreas,* which functions as both an endocrine and exocrine gland, produces digestive enzymes as well as insulin and glucagon, important hormones concerned with the regulation of blood sugar levels. The pancreas is located partially behind the stomach in the abdomen.

Elevated blood glucose levels stimulate release of **insulin,** which decreases blood sugar levels, primarily by accelerating the transport of glucose into the body cells, where it is oxidized for energy or converted to glycogen or fat for storage.

Hyposecretion of insulin or some deficiency in the insulin receptors leads to **diabetes mellitus,** which is characterized by the inability of body cells to utilize glucose and the subsequent loss of glucose in the urine. Alterations of protein and fat metabolism also occur, but these are probably secondary to derangements in carbohydrate metabolism. ■

Glucagon acts antagonistically to insulin. Its release is stimulated by low blood glucose levels, and its action is basically hyperglycemic. It stimulates the liver, its primary target organ, to break down its glycogen stores to glucose and subsequently to release the glucose to the blood.

The Gonads

The female *gonads,* or *ovaries,* are paired, almond-sized organs located in the pelvic cavity. In addition to producing the female sex cells (ova), the ovaries produce two steroid hormone groups, the estrogens and progesterone. The endocrine and exocrine functions of the ovaries do not begin until the onset of puberty, when the anterior pituitary gonadotropic hormones prod the ovary into action that produces rhythmic ovarian cycles in which ova develop and hormonal levels rise and fall. The **estrogens** are responsible for the development of the secondary sex characteristics of the female at puberty (primarily maturation of the reproductive organs and development of the breasts) and act with progesterone to bring about cyclic changes of the uterine lining that occur during the menstrual cycle. The estrogens also help prepare the mammary glands for lactation.

Progesterone, as already noted, acts with estrogen to bring about the menstrual cycle. During pregnancy it maintains the uterine musculature in a quiescent state and helps to prepare the breast tissue for lactation.

The paired oval *testes* of the male are suspended in a pouchlike sac, the scrotum, outside the pelvic cavity. In addition to the male sex cells, sperm, the testes produce the male sex hormone, **testosterone.** Testosterone promotes the maturation of the reproductive system accessory structures, brings about the development of the secondary sex characteristics, and is responsible for the male sexual drive, or libido. Both the endocrine and exocrine functions of the testes begin at puberty under the influence of the anterior pituitary gonadotropins.

Two glands not mentioned earlier as major endocrine glands should also be briefly considered here, the thymus and the pineal gland.

Thymus

The *thymus* is a bilobed gland situated in the superior thorax, posterior to the sternum and anterior to the heart and lungs. Conspicuous in the infant, it begins to atrophy at puberty, and by old age it is relatively inconspicuous. The thymus produces a hormone called **thymosin** which helps direct the maturation and specialization of a unique population of white blood cells called T lymphocytes or T cells. T lymphocytes are responsible for the cellular immunity aspect of body defense; that is, rejection of foreign grafts, tumors, or virus-infected cells.

Pineal Body

The *pineal body,* or *epiphysis cerebri,* is a small cone-shaped gland located in the roof of the third ventricle of the brain. Its major endocrine product is **melatonin.**

The endocrine role of the pineal body in humans is still controversial, but it is known to play a role in the biological rhythms (particularly mating and migratory behavior) of other animals. In humans, melatonin appears to exert some inhibitory effect on the reproductive system that prevents precocious sexual maturation.

Once you are satisfied that you can name and locate the endocrine organs and you have labeled Figure 27.1, use an appropriate reference to define the following pathologic conditions (which have not been described here) resulting from hypersecretion or hyposecretion of the various hormones.

acromegaly ______________________________

____

____

____

____

Addison's disease ______________________________

____

____

____

____

cretinism ______________________________

____

____

____

____

Cushing's syndrome ______________________________

____

____

____

____

diabetes insipidus ______________________________

____

____

____

pituitary dwarfism ______________________________

____

____

____

eunuchism ______________________________

____

____

____

exophthalmic goiter ______________________________

____

____

____

gigantism ______________________________

____

____

____

hypercalcemia ______________________________

____

____

____

Simmonds' disease ______________________________

____

____

____

MICROSCOPIC ANATOMY OF SELECTED ENDOCRINE GLANDS (OPTIONAL)

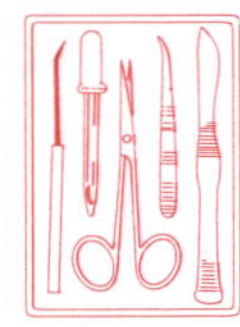

To prepare for the histologic study of the endocrine glands, obtain a microscope, one each of the slides listed in the list of materials, and colored pencils. We will study only organs in which it is possible to identify the endocrine-producing cells. Compare your observations with the line drawings in Figure 27.3a–f of the endocrine tissue photomicrographs.

Thyroid Gland

1. Scan the thyroid under low power, noting the **follicles,** spherical sacs containing a pink-stained material (*colloid*). Stored T_3 and T_4 are attached to the protein colloidal material stored in the follicles as **thyroglobulin** and are released gradually to the blood. Compare the tissue viewed to Figure 27.3a and Plate 18 in the Histology Atlas.

2. Observe the tissue under high power. Note that the walls of the follicles are formed by simple cuboidal or squamous epithelial cells that synthesize the follicular products. The **parafollicular,** or **C, cells** you see between the follicles are responsible for calcitonin production.

3. Color appropriately two or three follicles in Figure 27.3a. Label the colloid and parafollicular cells.

When the thyroid gland is actively secreting, the follicles appear small, and the colloidal material has a ruffled border. When the thyroid is hypoactive or inactive, the follicles are large and plump and the follicular epithelium appears to be squamouslike. What is the physiologic state of the tissue you have been viewing?

____

____

Parathyroid Glands

1. Observe the parathyroid tissue under low power to view its two major cell types, the **chief cells** and the **oxyphil cells.** Compare your observations to the view in Plate 19 of the Histology Atlas. The chief cells, which synthesize parathyroid hormone (PTH), are small and abundant, and arranged in thick branching cords. The function of the scattered, much larger oxyphil cells is unknown.

2. Color a small portion of the parathyroid tissue in Figure 27.3b. Label the chief cells, oxyphil cells, and the connective tissue matrix.

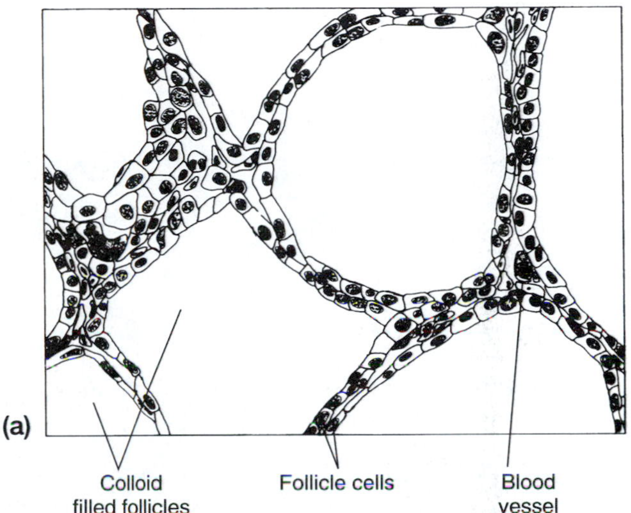

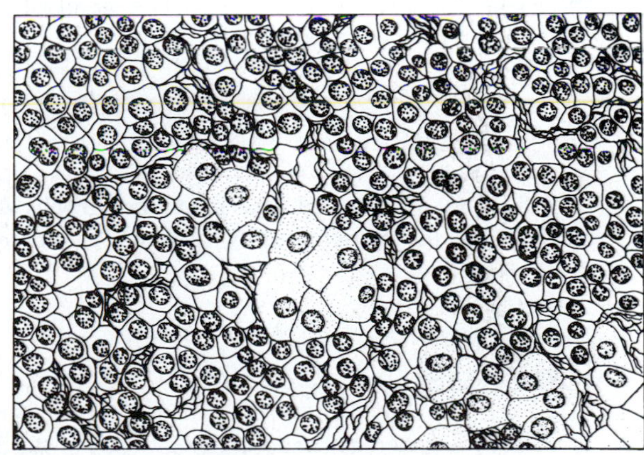

F27.3

Line drawings of Histology Atlas photomicrographs of selected endocrine organs. (a) Thyroid (see corresponding Plate 18 in the Histology Atlas); **(b)** parathyroid (see Plate 19).

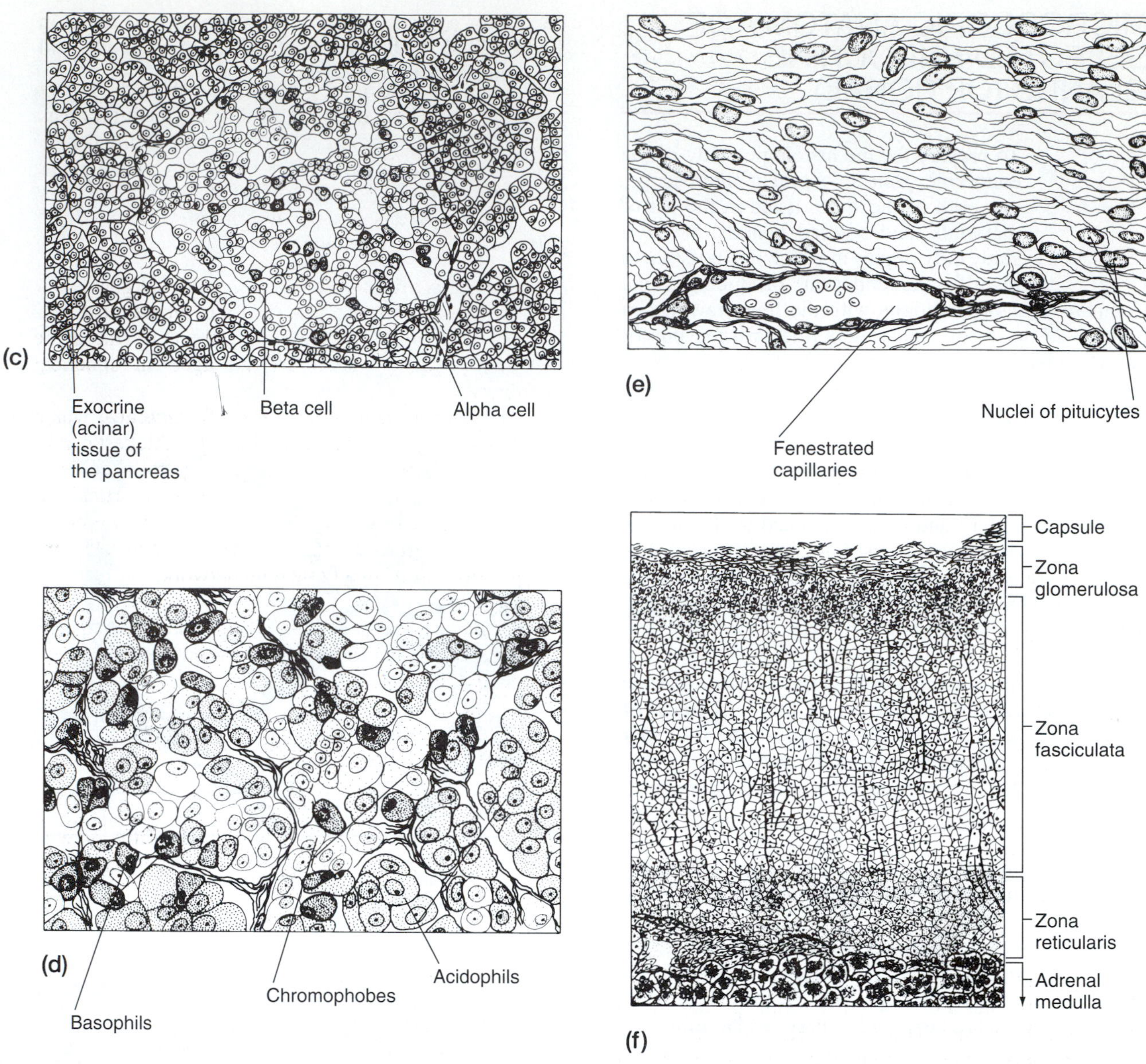

F27.3 (*continued*)

Line drawings of Histology Atlas photomicrographs of selected endocrine organs. (c) pancreas showing a pancreatic islet (see Plate 20); **(d)** anterior pituitary (see Plate 21); **(e)** posterior pituitary (see Plate 22); **(f)** adrenal gland (see Plate 23).

Pancreas

1. Observe pancreas tissue under low power to identify the roughly circular **pancreatic islets (islets of Langerhans),** the endocrine portions of the pancreas. The islets are scattered amid the more numerous acinar cells and stain differently (usually lighter), which makes their identification possible [Figure 38.15 (p. 368) and Plate 40 in the Histology Atlas].

2. Focus on an islet and examine its cells under high power. Notice that the islet cells are densely packed and have no definite arrangement. In contrast, the cuboidal acinar cells are arranged around secretory ducts. Unless special stains are used, it will not be possible to distinguish the **alpha cells,** which tend to cluster at the periphery of the islets and produce glucagon, from the **beta cells,** which synthesize insulin. With these specific stains, the beta cells are larger and stain gray-blue; and the alpha cells are smaller and appear bright pink, as hinted at in Figure 27.3c and shown in Plate 20 in the Histology Atlas. What is the product of the acinar cells?

____

3. Draw a section of the pancreas in the space below. Label the islets and the acinar cells. If possible, differentiate the alpha and beta cells of the islets by color.

Pituitary Gland

1. Observe the general structure of the pituitary gland under low power to differentiate between the glandular anterior pituitary and the neural posterior pituitary. Figure 27.3d and e should help you get started.

2. Using the high-power lens, focus on the nests of cells of the anterior pituitary. It is possible to identify the specialized cell types that secrete the specific hormones when differential stains are used. Using Plate 21 in the Histology Atlas as a guide, locate the reddish-brown stained **acidophil cells,** which produce growth hormone and prolactin, and the **basophil cells,** whose deep-blue granules are responsible for the production of the tropic hormones (TSH, ACTH, FSH, and LH). **Chromophobes,** the third cellular population, do not take up the stain and appear rather dull and colorless. The role of the chromophobes is controversial, but they apparently are not directly involved in hormone production.

3. Use appropriately colored pencils to identify acidophils, basophils, and chromophobes in Figure 27.3d.

4. Switch your focus to the posterior pituitary. Observe the nerve fibers (axons of hypophyseal neurons) that compose most of this portion of the pituitary. Also note the **pituicytes,** glial cells which are randomly distributed among the nerve fibers. Refer to Plate 22 in the Histology Atlas as you scan the slide.

What two hormones are stored here?

____

What is their source?

____

Adrenal Gland

1. Hold the slide of the adrenal gland up to the light to distinguish the outer cortex and inner medulla areas. Then scan the cortex under low power to distinguish the differences in cell appearance and arrangement in the three cortical areas. Refer to Figure 27.3f, Plate 23 in the Histology Atlas, and the descriptions below to identify the following cortical areas:

- Connective tissue capsule of the adrenal gland.
- The outermost **zona glomerulosa,** where most mineralocorticoid production occurs and where the tightly packed cells are arranged in spherical clusters.
- The deeper intermediate **zona fasciculata,** which produces glucocorticoids. This is the thickest part of the cortex. Its cells are arranged in parallel cords.
- The innermost cortical zone, the **zona reticularis** abutting the medulla, which produces sex hormones and some glucocorticoids. The cells here stain intensely and form a branching network.

2. Switch focus to view the large, lightly stained cells of the adrenal medulla under high power. Note their clumped arrangement.

What hormones are produced by the medulla?

________________ and ________________

3. Draw a representative area of each of the adrenal regions, indicating in your sketch the differences in relative cell size and arrangement.

Zona glomerulosa

Zona fasciculata

Zona reticularis

Adrenal medulla

Ovary

Because you will consider the *ovary* in greater histologic detail when you study the reproductive system, the objective in this laboratory exercise is just to identify the endocrine-producing parts of the ovary.

1. Scan an ovary slide under low power, and look for a **vesicular (Graafian) follicle,** a circular arrangement of cells enclosing a central cavity. See Figure 43.5 (p. 404) and Plate 24 in the Histology Atlas. This structure synthesizes estrogens.
2. Examine the vesicular follicle under high power, identifying the follicular cells that produce estrogens, the antrum (fluid-filled cavity), and developing ovum (if present). The ovum will be the largest cell in the follicle.
3. Draw and color a vesicular follicle below, labeling the antrum, follicle cells, and developing ovum.
4. Switch to low power, and scan the slide to find a **corpus luteum,** a large amorphous-looking area that produces progesterone (and some estrogens). A corpus luteum is shown in Plate 25 in the Histology Atlas.

Testis

1. Examine a section of a *testis* under low power. Identify the seminiferous tubules, which produce sperm, and the **interstitial cells,** which produce testosterone. The interstitial cells are scattered between the seminiferous tubules in the connective tissue matrix. The photomicrograph of seminiferous tubules, Plate 49 in the Histology Atlas, will be helpful here.
2. Draw a representative area of the testis in the space provided. Label the seminiferous tubules and area of the interstitial cells.

EXERCISE 28

Experiments on Hormonal Action

OBJECTIVES

1. To describe and explain the effect of pituitary hormones on the ovary.
2. To describe and explain the effects of hyperinsulinism.
3. To describe and explain the effect of epinephrine on the heart.
4. To understand the physiologic (and clinical) importance of metabolic rate measurement.
5. To investigate the effect of hypo-, hyper-, and euthyroid conditions on oxygen consumption and metabolic rate.
6. To assemble the necessary apparatus and properly use a manometer to obtain experimental results.
7. To calculate metabolic rate in terms of ml O_2/kg/hr.

MATERIALS

Experiment 1:

Glass desiccator; manometer, 20-ml glass syringe, two-hole cork, and T-valve (1 for every 3 to 4 students)
Soda lime (desiccant)
Hardware cloth squares
Petrolatum
Rubber tubing; tubing clamps; scissors
3-in. pieces of glass tubing
Animal balances
Heavy animal-handling gloves
Chart set up on chalkboard so that each student group can record its computed metabolic rate figures under the appropriate headings

Young rats of the same sex, obtained 2 weeks prior to the laboratory session and treated as follows for 14 days:

Group 1: control group—fed normal rat chow and water
Group 2: experimental group A—fed normal rat chow, and drinking water containing 0.02% 6-n-propylthiouracil*
Group 3: experimental group B—fed rat chow containing desiccated thyroid (2% by weight), and normal drinking water

Experiment 2[†]*:*

Female frogs (*Rana pipiens*)
Syringe (2-ml capacity)
20- to 25-gauge needle
Frog pituitary extract
Physiological saline
Battery jar
Spring or pond water
Wax marking pencils

Experiment 3[†]*:*

500- or 600-ml beakers
20% glucose solution
Commercial insulin solution (400 IU per 100 ml H_2O)
Finger bowls
Small (1½–2 in.) freshwater fish (guppy, bluegill, or sunfish—listed in order of preference)

Experiment 4[†]*:*

Frog (*Rana pipiens*)
1:1000 epinephrine (Adrenalin) solution in dropper bottles
Dissecting pan, instruments, and pins
Frog Ringer's solution in dropper bottle
Disposable gloves

*Note to the Instructor: 6-n-propylthiouracil (PTU) is degraded by light and should be stored in light-resistant containers or in the dark.

[†] The *Selected Actions of Hormones and Other Chemical Messengers* videotape (available to qualified adopters from Benjamin/Cummings) may be used in lieu of student participation in Experiments 2–4 of Exercise 28.

The endocrine system exerts many complex and interrelated effects on the body as a whole, as well as on specific organs and tissues. Most scientific knowledge about this system is contemporary, and new information is constantly being presented. Many experiments on the endocrine system require relatively large laboratory animals; are time-consuming (requiring days to weeks of observation); and often involve technically difficult surgical procedures to remove the glands or parts of them, all of which makes it difficult to conduct more general types of laboratory experiments. Nevertheless, the four technically unsophisticated experiments presented here should illustrate how dramatically hormones affect body functioning.

To conserve laboratory specimens, the experiments can be conducted by groups of four students. The use of larger working groups should not detract from benefits gained, since the major value of these experiments lies in observation.

EXPERIMENT 1: EFFECT OF THYROID HORMONE ON METABOLIC RATE

Metabolism is a broad term referring to all chemical reactions that are necessary to maintain life. It involves both *catabolism,* enzymatically controlled processes in which substances are broken down to simpler substances, and *anabolism,* processes in which larger molecules or structures are built from smaller ones. During catabolic reactions, energy is released as chemical bonds are broken. Some of the liberated energy is captured to make ATP, the energy-rich molecule used by body cells to energize all their activities; the balance is lost in the form of thermal energy or heat. Maintaining body temperature is critically related to the heat-liberating aspects of metabolism.

Various foodstuffs make different contributions to the process of metabolism. For example, carbohydrates, particularly glucose, are generally broken down or oxidized to make ATP, whereas fats are utilized to form cell membranes, myelin sheaths, and to insulate the body with a fatty cushion. (Fats are used secondarily for producing ATP, particularly when there are inadequate carbohydrates in the diet.) Proteins and amino acids tend to be carefully conserved by body cells, and understandably so, since most structural elements of the body are proteinaceous in nature.

Thyroid hormone (collectively T_3 and T_4), produced by the thyroid gland, is the single most important hormone influencing the rate of cellular metabolism and body heat production.

Under conditions of excess thyroid hormone production (hyperthyroidism), an individual's basal metabolic rate (BMR), heat production, and oxygen consumption increase, and the individual tends to lose weight and become heat-intolerant and irritable. Conversely, hypothyroid individuals become mentally and physically sluggish, obese, and are cold-intolerant because of their low BMR. ■

Many factors other than thyroid hormone levels contribute to metabolic rate (for example, body size and weight, age, and activity level), but the focus of the following experiment is to investigate how differences in thyroid hormone concentration impact metabolism.

Three groups of laboratory rats will be used. The *control group* animals are assumed to be euthyroid and to have normal metabolic rates for their relative body weights. *Experimental group A* animals have received water containing the chemical 6-n-propylthiouracil, which counteracts or antagonizes the effects of thyroid hormone in the body. *Experimental group B* animals have been fed rat chow containing dried thyroid tissue, which contains thyroid hormone. The rates of oxygen consumption (an indirect means of determining metabolic rate) in the animals of the three groups will be measured and compared to investigate the effects of hyperthyroid, hypothyroid, and euthyroid conditions.

Oxygen consumption will be measured with a simple respirometer-manometer apparatus. Each animal will be placed in a closed chamber, containing soda lime. As carbon dioxide is evolved and expired, it will be absorbed by the soda lime; therefore, the pressure changes observed will indicate the volume of oxygen consumed by the animal during the testing interval. Students will work in groups of 3 to 4 to assemble the apparatus, make preliminary weight measurements on the animals, and record the data.

Preparation of the Respirometer-Manometer Apparatus

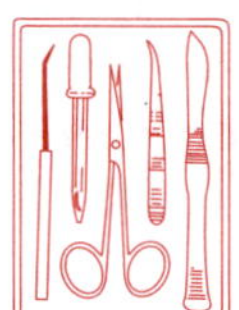

1. Obtain a desiccator, a two-hole rubber stopper, a 20-ml glass syringe, a hardware cloth square, a T-valve, scissors, rubber tubing, two short pieces of glass tubing, soda lime, a manometer, a clamp, and petrolatum, and bring them to your laboratory bench. The apparatus will be assembled as illustrated in Figure 28.1.

2. Shake soda lime into the bottom of the desiccator to thoroughly cover the glass bottom. Then place the hardware cloth on the ledge of the desiccator over the soda lime. The hardware cloth should be well above the soda lime, so that the animal will not be able to touch it. Soda lime is quite caustic and can cause chemical burns.

3. Lubricate the ends of the two pieces of glass tubing with petrolatum, and twist them into the holes in the rubber stopper until their distal ends protrude from the opposite side. *Do not plug the tubing with petrolatum.* Place the stopper into the desiccator cover, and set the cover on the desiccator temporarily.

4. Cut off a short (3-in.) piece of rubber tubing, and attach it to the top of one piece of glass tubing extending from the stopper. Cut and attach a 12- to 14-in. piece of rubber tubing to the other glass tubing. Insert the T-valve stem into the distal end of the longer-length tubing.

5. Cut another short piece of rubber tubing; attach one end to the T-valve and the other to the nib of the 20-ml syringe. Remove the plunger of the syringe and grease its sides generously with petrolatum. Insert the plunger back into the syringe barrel and work it up and down to evenly disperse the petrolatum on the inner walls of the syringe, then pull the plunger out to the 20-ml marking.

6. Cut a piece of rubber tubing long enough to reach from the third arm of the T-valve to one arm of the ma-

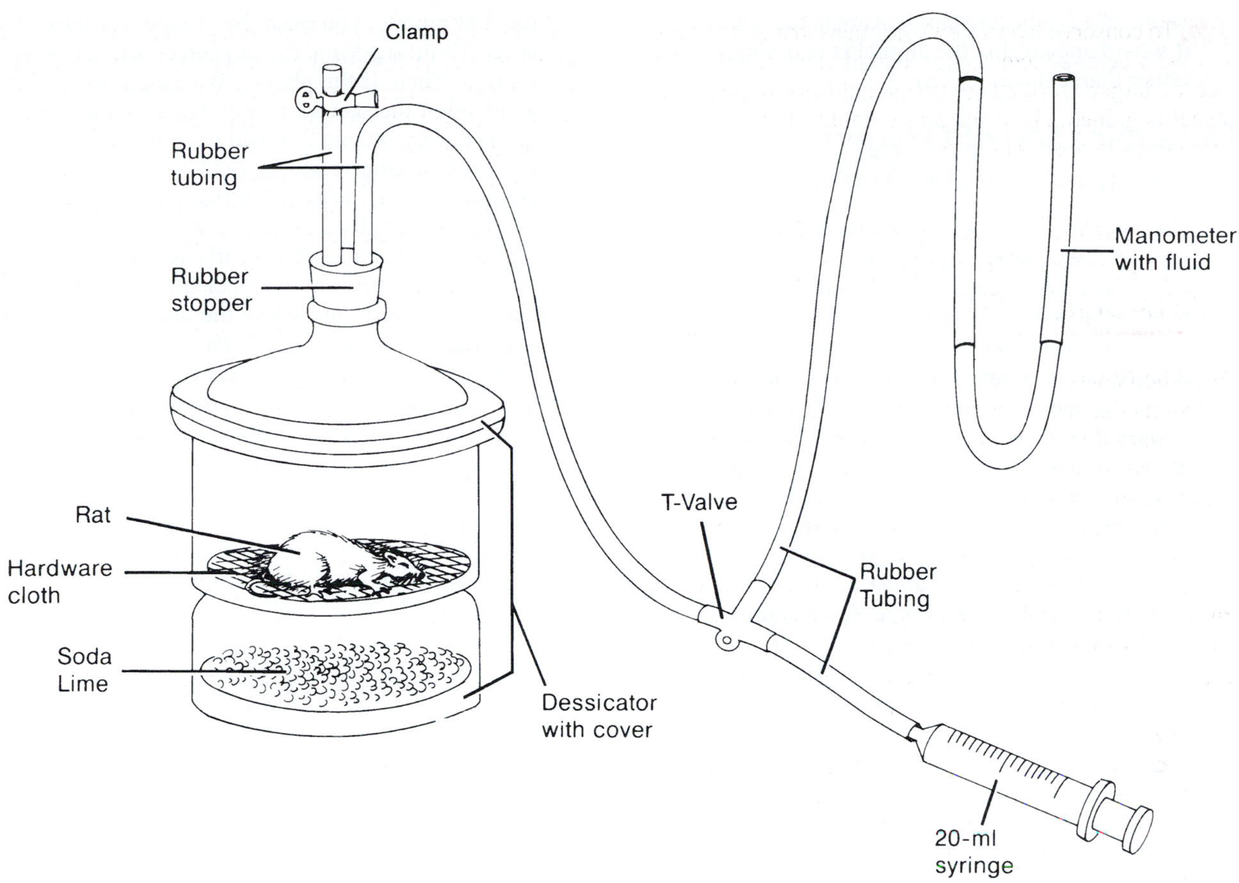

F28.1

Respirometer-manometer apparatus.

nometer. (The manometer should be partially filled with water so that a U-shaped water column is seen.) Attach the tubing to the T-valve and the manometer arm.

7. Remove the desiccator cover and generously grease its bottom edge with petrolatum. Place the cover back on the desiccator and firmly move it from side to side to spread the lubricant evenly.

8. Test the system for leaks as follows: Firmly clamp the *short* length of rubber tubing extending from the stopper. Now gently push in on the plunger of the syringe. If the system is properly sealed, the fluid in the manometer will move away from the rubber tubing attached to its arm. If there is an air leak, the manometer fluid level will not change, or it will change and then quickly return to its original level. If either of these events occurs, check all glass-to-glass or glass-to-rubber tubing connections. Smear additional petrolatum on suspect areas and test again. The apparatus must be airtight before experimentation can begin.

9. After ensuring that there are no leaks in the system, unclamp the short rubber tubing, and remove the desiccator cover.

Preparation of the Animal

1. Put on the heavy animal-handling gloves, and obtain one of the animals as directed by your instructor. Handling it gently, weigh the animal to the nearest 0.1 g on the animal balance.
2. Carefully place the animal on the hardware cloth in the desiccator. The objective is to measure oxygen usage at basal levels, so you do not want to prod the rat into high levels of activity, which would produce errors in your measurements.
3. Record the animal's group (control, experimental group A or B) and its weight in kilograms (that is, weight in grams/1000) on the data sheet on p. 262.

Equilibration of the Chamber

1. Place the lid on the desiccator, and move it slightly from side to side to seal it firmly.
2. Leave the short tubing unclamped for 7 to 10 minutes to allow for temperature equilibration in the chamber. (Since the animal's body heat will warm the air in the container, the air will expand initially. This must be allowed to occur before any measure-

ments of oxygen consumption are taken. Otherwise, it would appear that the animal is evolving oxygen rather than consuming it.)

Determination of Oxygen Consumption of the Animal

1. Once again clamp the short rubber tubing extending from the desiccator lid stopper.
2. Check the manometer to make sure that the fluid level is the same in both arms. (If not, manipulate the syringe plunger to make the fluid levels even.) Record the time and the position of the bottom of the plunger (use ml marking) in the syringe.
3. Observe the manometer fluid levels at 1-minute intervals. Each time make the necessary adjustment to bring the fluid levels even in the manometer by carefully pushing the syringe plunger further into the barrel. Determine the amount of oxygen used per minute intervals by computing the difference in air volumes within the syringe. For example, if after the first minute, you push the plunger from the 20- to the 17-ml marking, the oxygen consumption is 3 ml/min. Then, if the plunger is pushed from 17 ml to 15 ml at the second minute reading, the oxygen usage during minute 2 would be 2 ml, and so on.
4. Continue taking readings (and recording oxygen consumption per minute interval on the data sheet) for 10 consecutive minutes or until the syringe plunger has been pushed nearly to the 0-ml mark. Then unclamp the short rubber tubing, remove the desiccator cover, and allow the apparatus to stand open for 2 to 3 minutes to flush out the stale air.
5. Repeat the recording procedures for another 10-minute interval. *Make sure that you equilibrate the temperature within the chamber before beginning this second recording series.*
6. After you have recorded the animal's oxygen consumption for two 10-minute intervals, unclamp the short rubber tubing, remove the desiccator lid, and carefully return the rat to its cage.

Metabolic Rate Data Sheet

Animal used from group: ______________________

14-day prior treatment of animal: ______________________

Body weight in grams: ____________ /1000 = body weight in kg: ____________

O_2 consumption/min: Test 1		O_2 consumption/min: Test 2	
Beginning syringe reading ______ ml		Beginning syringe reading ______ ml	
min 1 ________	min 6 ________	min 1 ________	min 6 ________
min 2 ________	min 7 ________	min 2 ________	min 7 ________
min 3 ________	min 8 ________	min 3 ________	min 8 ________
min 4 ________	min 9 ________	min 4 ________	min 9 ________
min 5 ________	min 10 ________	min 5 ________	min 10 ________
________ Total O_2/10 min		________ Total O_2/10 min	

Average ml O_2 consumed/10 min: ____________

Milliliters O_2 consumed/hr: ____________

Metabolic rate: ____________ ml O_2/kg/hr

Averaged class results:

Metabolic rate of control animals: ____________ ml O_2/kg/hr

Metabolic rate of experimental group A animals (PTU-treated): ____________ ml O_2/kg/hr

Metabolic rate of experimental group B animals (desiccated thyroid-treated): ____________ ml O_2/kg/hr

Computation of Metabolic Rate

Metabolic rate calculations are generally reported in terms of Kcal/m^2/hr and require that corrections be made to present the data in terms of standardized pressure and temperature conditions. These more complex calculations will not be used here, since the object is simply to arrive at some generalized conclusions concerning the effect of thyroid hormone on metabolic rate.

1. Obtain the average figure for milliliters of oxygen consumed per 10-minute interval by adding up the minute-interval consumption figures for each 10-minute testing series and dividing the total by 2.

$$\frac{\text{Total ml } O_2 \text{ test series 1} + \text{total ml } O_2 \text{ test series 2}}{2}$$

2. Determine oxygen consumption per hour using the following formula, and record the figure on the data sheet:

$$\frac{\text{Average ml } O_2 \text{ consumed}}{\text{10 min}} \times \frac{\text{60 min}}{\text{hr}} = \text{ml } O_2\text{/hr}$$

3. To determine the metabolic rate in milliliters of oxygen consumed per kilogram of body weight per hour so that the results of all experiments can be compared, divide the figure just obtained in procedure 2 by the animal's weight in kilograms (kg = wt in lb ÷ 2.2).

$$\text{Metabolic rate} = \frac{\text{ml } O_2\text{/hr}}{\text{wt in kg}} = \_\_\_\_\_\_\_\_ \text{ml } O_2\text{/kg/hr}$$

Record the metabolic rate on the data sheet and also in the appropriate space on the chart on the chalkboard.

4. Once all groups have recorded their final metabolic rate figures on the chalkboard, average the results of each animal grouping to obtain the mean for each experimental group. Also record this information on your data sheet.

EXPERIMENT 2: EFFECT OF PITUITARY HORMONES ON THE OVARY

As indicated in Exercise 27, anterior pituitary hormones called *gonadotropins,* specifically follicle-stimulating hormone (FSH) and luteinizing hormone (LH), regulate the ovarian cycles of the female. Although amphibians normally ovulate seasonally, many can be stimulated to ovulate "on demand" by injecting an extract of pituitary hormones.

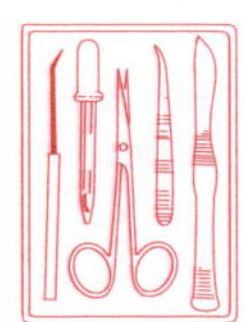

1. Don plastic gloves, and obtain two frogs. Place them in separate battery jars to bring them to your laboratory bench. Also bring back a syringe and needle, a wax marking pencil, pond or spring water, and containers of pituitary extract and physiological saline.

2. Before beginning, examine each frog for the presence of eggs. Hold the frog firmly with one hand and exert pressure on its abdomen toward the cloaca (in the direction of the legs). If ovulation has occurred, any eggs present in the oviduct will be forced out and will appear at the cloacal opening. If no eggs are present, continue with step 3. If eggs are expressed, return the animal to your instructor and obtain another frog for experimentation. Repeat the procedure for determining if eggs are present until two frogs that lack eggs have been obtained.

3. Aspirate 1 to 2 ml of the pituitary extract into a syringe. Inject the extract subcutaneously into the anterior abdominal (peritoneal) cavity of the frog you have selected to be the experimental animal. To inject into the peritoneal cavity, hold the frog with its ventral surface superiorly. Insert the needle through the skin and muscles of the abdominal wall in the lower quarter of the abdomen. Do not insert the needle far enough to damage any of the vital organs. With a wax marker, label its large battery jar "experimental," and place the frog in it. Add a small amount of pond water to the battery jar before continuing.

4. Aspirate 1 to 2 ml of normal saline into a syringe and inject it into the peritoneal cavity of the second frog —this will be the control animal. (Make sure you inject the same volume of fluid into both frogs.) Place this frog into the second battery jar, marked "control." Allow the animals to remain undisturbed for 24 hours.

5. After 24 hours,* again check each frog for the presence of eggs in the cloaca. (See step 2 above.) If no eggs are present, make arrangements with your laboratory instructor to return to the lab on the next day (at 48 hours after injection) to check your frogs for the presence of eggs.

6. Return the frogs to the terrarium before leaving or continuing with the lab.

In which of the prepared frogs was ovulation induced?

____

Specifically, what hormone in the pituitary extract causes ovulation to occur?

____

* The student will need to inject the frog the day before the lab session or return to check results the day after the scheduled lab session.

EXPERIMENT 3: EFFECTS OF HYPERINSULINISM

Many people with diabetes mellitus need injections of insulin to maintain blood sugar (glucose) homeostasis. Adequate levels of blood glucose are essential for proper functioning of the nervous system; thus, the administration of insulin must be carefully controlled. If blood glucose levels fall precipitously, the patient will go into insulin shock.

A small fish will be used to demonstrate the effects of hyperinsulinism. Since the action of insulin on the fish parallels that in the human, this experiment should provide valid information concerning its administration to humans.

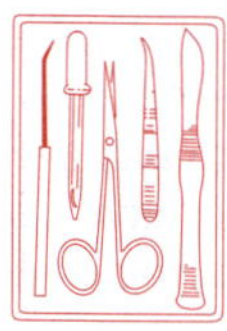

1. Prepare two finger bowls. Using a wax marker, mark one A and the other B. To finger bowl A, add 100 ml of the commercial insulin solution. To finger bowl B, add 200 ml of 20% glucose solution.

2. Place a small fish in finger bowl A and observe its actions carefully as the insulin diffuses into its bloodstream through the capillary circulation of its gills.

Approximately how long did it take for the fish to become comatose?

___

What types of activity did you observe in the fish before it became comatose?

___

___

___

___

3. When the fish is comatose, carefully transfer it to finger bowl B and observe its actions. What happens to the fish after it is transferred?

___

___

___

___

Approximately how long did it take for this recovery?

___

4. After all observations have been made and recorded, carefully return the fish to the aquarium.

EXPERIMENT 4: EFFECT OF EPINEPHRINE ON THE HEART

As noted in Exercise 27, the adrenal medulla and the sympathetic nervous system are closely interrelated, specifically because the cells of the adrenal medulla and the postganglionic axons of the sympathetic nervous system both release catecholamines. This experiment demonstrates the effects of epinephrine on the frog heart.

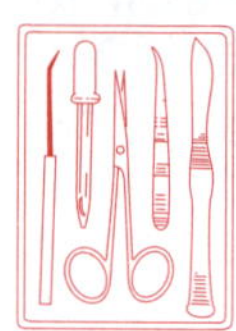

⚠ 1. Obtain a frog, dissecting instruments and pan, dropper bottle of 1:1000 epinephrine solution, and bring them to your laboratory bench. Don the gloves before beginning step 2.

2. Destroy the nervous system of the frog. (A frog used in the experiment on pituitary hormone effects may be used if your test results have already been obtained and are positive. Otherwise, obtain another frog.) Insert one blade of a scissors into its mouth as far as possible and quickly cut off the top of its head, posterior to the eyes. Then, identify the spinal cavity and insert a dissecting needle into it to destroy the spinal cord.

3. Place the frog dorsal side down on a dissecting pan, and carefully open its ventral body cavity by making a vertical incision with the scissors.

4. Identify the beating heart, and carefully cut through the saclike pericardium to expose the heart tissue.

5. Visually count the heart rate for 1 minute, and record below. Keep the heart moistened with frog Ringer's solution during this interval.

Beats per minute: ___

6. Flush the heart with epinephrine solution. Record the heartbeat rate per minute for 5 consecutive minutes.

minute 1 ___ minute 4 ___

minute 2 ___ minute 5 ___

minute 3 ___

What was the effect of epinephrine on the heart rate?

___

Was the effect long-lived? ___

7. Dispose of the frog in an appropriate container, and clean the dissecting pan and instruments before returning them to the supply area.

Blood

OBJECTIVES

1. To name the two major components of blood and state their average percentages in whole blood.
2. To describe the composition and functional importance of plasma.
3. To define *formed elements* and list the cell types composing them, cite their relative percentages, and describe their major functions.
4. To identify red blood cells, basophils, eosinophils, monocytes, lymphocytes, and neutrophils when provided with a microscopic preparation or appropriate diagram.
5. To conduct the following blood test determinations in the laboratory, and to state their norms and the importance of each.

   hematocrit
   hemoglobin determination
   clotting time
   sedimentation rate
   differential white blood cell count
   total white blood cell count
   total red blood cell count
   ABO and Rh blood typing
6. To discuss the reason for transfusion reactions resulting from the administration of mismatched blood.
7. To define *anemia, polycythemia, leukopenia, leukocytosis,* and *leukemia* and to cite a possible reason for each condition.

MATERIALS

Compound microscope
Immersion oil
Models and charts of blood cells

Demonstration station:
Microscopes set up with prepared slides demonstrating the following blood (or bone marrow) conditions:
Macrocytic, hypochromic anemia
Microcytic, hypochromic anemia
Sickle-cell anemia
Lymphocytic leukemia (chronic)
Eosinophilia

General supply area:
Plasma (obtained from an animal hospital or prepared by centrifuging animal (e.g., cattle or sheep) blood obtained from a biological supply house
Wide-range pH paper

Stained smears of human blood or, if desired by the instructor, heparinized blood obtained from an animal hospital (e.g., dog blood)
Clean microscope slides
Sterile lancets
Glass stirring rods
Alcohol swabs (wipes)
Absorbent cotton balls
Wright's stain in dropper bottle
Distilled water in dropper bottle
Test tubes
Test tube racks
Disposable gloves
Pipette cleaning solutions—(1) 10% household bleach solution, (2) distilled water, (3) 70% ethyl alcohol, (4) acetone
Bucket or large beaker containing 10% household bleach solution for slide and glassware disposal
Disposable autoclave bag
Spray bottles containing 10% bleach solution

Because many blood tests are to be conducted in this exercise, it seems advisable to set up a number of appropriately labeled supply areas for the various tests. These are designated below.

Blood cell count supply area:
Hemacytometer
Unopette reservoir system apparatus for conducting blood counts (available from Becton-Dickinson [Test #5851 for RBCs and #5855 for WBCs])
Mechanical hand counters

Hematocrit supply area:
Heparinized capillary tubes
Microhematocrit centrifuge and reading gauge (if the reading gauge is not available, millimeter ruler may be used)
Seal-ease (Clay Adams Co.) or modeling clay

Hemoglobin determination supply area:
Tallquist hemoglobin scales and test paper or a hemoglobinometer, hemolysis applicator and lens paper

(Materials list continues on next page)

Sedimentation rate supply area:
Landau Sed-rate pipettes with tubing and rack
Wide-mouthed bottle of 5% sodium citrate
Mechanical suction device
Millimeter ruler

Coagulation time supply area:
Capillary tubes (nonheparinized)
Fine triangular file

Blood typing supply area:
Blood typing sera (anti-A, anti-B, and anti-Rh [D])
Rh typing box
Wax marker
Toothpicks
Clean microscope slides

Cholesterol-measurement supply area:
Mechanical pipettor
0.1-ml pipettes (3)
10- to 15-ml graduated cylinder
Test tubes (3) and rack
Cholesterol standard solution (200 mg/100 ml)
Cholesterol reagent (Harleco)*
Water bath set at 37°C
Spectrophotometer
Cuvettes for the spectrophotometer
Wax marker

See Appendix E, Exercise 29 for links to *Anatomy and PhysioShow: The Videodisc.*

Note to the Instructor: See directions for handling of soiled glassware and disposable items on p. 16.

* Instructions for preparing Harleco reagent are provided in the Instructor's Guide.

In this exercise you will study plasma and formed elements of blood and conduct various hematological tests. These tests are extremely useful diagnostic tools for the physician because blood composition (number and types of blood cells, and chemical composition) reflects the status of many body functions and malfunctions.

ALERT: The decision to use animal blood for testing or to have students test their own blood will be made by the instructor in accordance with the educational purpose of the student group. For example, for students in the nursing or laboratory technician curricula, learning how to safely handle human blood or other human wastes is essential. If blood samples are provided and they are human blood samples, gloves should be worn while conducting the blood tests. If human blood is being tested, yours or that obtained from a clinical agency, precautions provided in the text for disposal of human waste **must be observed.** All soiled glassware is to be immersed in household bleach solution immediately after use, and disposable items (lancets, cotton balls, alcohol swabs, etc.) are to be placed in a disposable autoclave bag so that they can be sterilized before disposal. Safety glasses are to be worn throughout the laboratory session.

COMPOSITION OF BLOOD

The blood circulating to and from the body cells within the blood vessels is a rather viscous substance that varies from bright scarlet to a dull brick red, depending on the amount of oxygen it is carrying. The circulatory system of the average adult contains about 5.5 liters of blood.

Blood is classified as a type of connective tissue, because it consists of a nonliving fluid matrix (the **plasma**) in which living cells (**formed elements**) are suspended. The fibers typical of a connective tissue matrix become visible in blood only when clotting occurs. They then appear as fibrin threads, which form the structural basis for clot formation.

Over 100 different substances are dissolved or suspended in plasma (Figure 29.1), which is over 90% water. These include nutrients, gases, hormones, various wastes and metabolites, many types of proteins, and mineral salts. The composition of plasma varies continuously as cells remove or add substances to the blood.

Three types of formed elements are present in blood. The most numerous are the **erythrocytes,** or **red blood cells (RBCs),** which are literally sacs of hemoglobin molecules that transport the bulk of the oxygen carried in the blood (and a small percentage of the carbon dioxide). **Leukocytes,** or **white blood cells (WBCs),** are part of the body's nonspecific defenses and the immune system, and **platelets** function in hemostasis (blood clot formation). Formed elements normally constitute 45% of whole blood; plasma accounts for the remaining 55%.

Physical Characteristics of Plasma

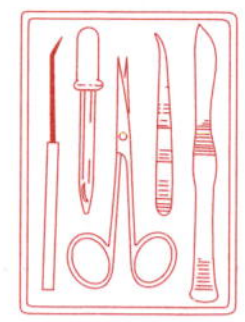

Go to the general supply area and carefully pour a few milliliters of plasma into a test tube. Also obtain some wide-range pH paper and then return to your laboratory bench to make the following simple observations.

pH OF PLASMA Test the pH of the plasma with wide-range pH paper. Record the pH observed.

____

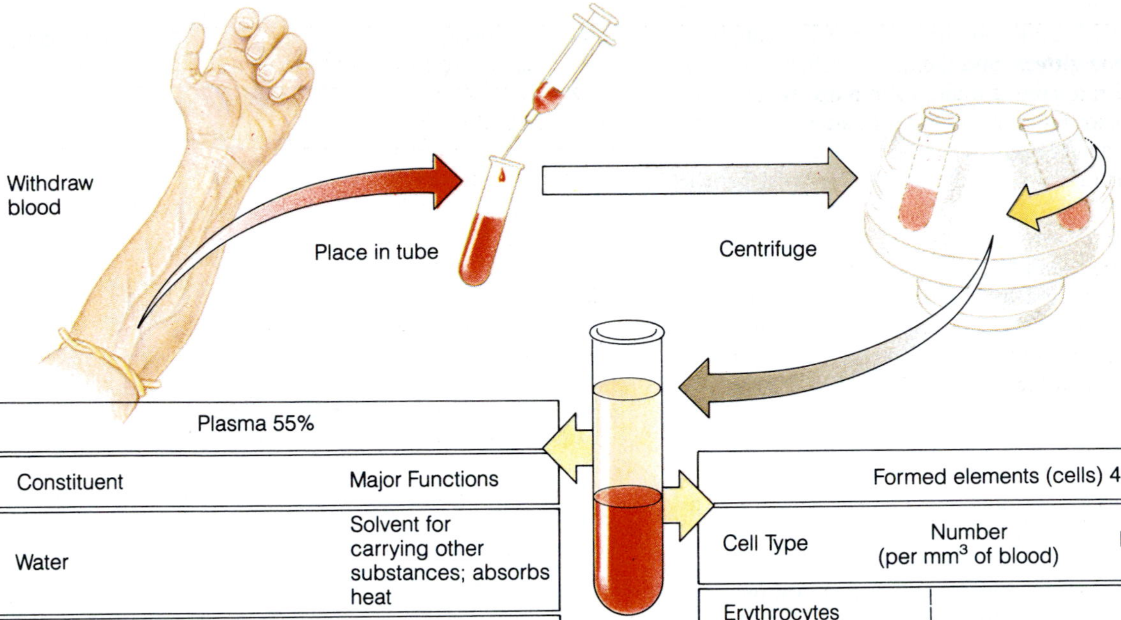

Withdraw blood

Place in tube

Centrifuge

Plasma 55%	
Constituent	**Major Functions**
Water	Solvent for carrying other substances; absorbs heat
Salts (electrolyles) Sodium Potassium Calcium Magnesium Chloride Bicarbonate	Osmotic balance, pH buffering, regulation of membrane permeability
Plasma proteins Albumin Fibrinogen Globulins	 Osmotic balance and pH buffering Clotting of blood Defense (antibodies) and lipid transport
Substances transported by blood Nutrients (glucose, fatty acids, amino acids, vitamins) Waste products of metabolism (urea, uric acid) Respiratory gases (O_2 and CO_2) Hormones	

Formed elements (cells) 45%		
Cell Type	**Number (per mm^3 of blood)**	**Functions**
Erythrocytes (red blood cells)	4–6 million	Transport oxygen and help transport carbon dioxide
Leukocytes (white blood cells) Basophil Eosinophil Neutrophil Lymphocyte Monocyte	4000–11,000	Defense and immunity
Platelets	250,000–500,000	Blood clotting

F29.1

The composition of blood.

COLOR AND CLARITY OF PLASMA Hold the test tube up to a source of natural light. Note and record its color and degree of transparency. Is it clear, translucent, or opaque?

Color ______________________________

Degree of transparency ______________________________

CONSISTENCY Dip your finger and thumb into the plasma and then press them firmly together for a few seconds. Gently pull them apart. How would you describe the consistency of plasma? Slippery, watery, sticky, or granular? Record your observations.

____

Formed Elements of Blood

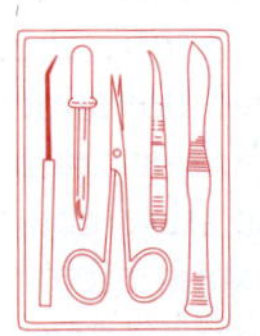

In this section, you will conduct your observations of blood cells on an already prepared (purchased) blood slide, or on a slide prepared from your own blood or blood provided by your instructor. Those using the purchased blood slide are to obtain a slide and begin their observations at step 6. Those testing blood provided by a biological supply source or an animal hospital are to obtain a tube of the supplied blood, disposable gloves, and the supplies listed in step 1, except for the lancets and alcohol swabs. After donning gloves, those students will jump to step 3b to begin their observations. If you are examining your own blood, you will perform all the steps described below *except* step 3b.

1. Obtain two glass slides, a glass stirring rod, dropper bottles of Wright's stain and distilled water, two or three lancets, cotton balls, and alcohol swabs. Bring this equipment to the laboratory bench. Clean the slides thoroughly and dry them.

2. Open the alcohol swab packet and scrub your third or fourth finger with the swab. (Because the pricked finger may be a little sore later, it is better to prepare a finger on the hand used less often.) Circumduct your hand (swing it in a cone-shaped path) for 10 to 15 seconds. This will dry the alcohol and cause your fingers to become engorged with blood. Then, open the lancet packet and grasp the lancet by its blunt end. Quickly jab the pointed end into the prepared finger to produce a free flow of blood. It is *not* a good idea to squeeze or "milk" the finger, as this forces out tissue fluid as well as blood. If the blood is not flowing freely, another puncture should be made.

Under no circumstances is a lancet to be used for more than one puncture. Dispose of the lancets in the disposable autoclave bag *immediately* after use.

3a. With a cotton ball, wipe away the first drop of blood; then allow another large drop of blood to form. Touch the blood to one of the cleaned slides approximately ½ inch from the end. Then quickly (to prevent clotting) use the second slide to form a blood smear as shown in Figure 29.2. When properly prepared, the blood smear is uniformly thin. If the blood smear appears streaked, the blood probably began to clot or coagulate before the smear was made, and another slide should be prepared. Continue at step 4.

3b. Dip a glass rod in the blood provided, and transfer a generous drop of blood to the end of a cleaned microscope slide. For the time being, lay the glass rod on a paper towel on the bench. Then, as described just above, use the second slide to make your blood smear.

4. Dry the slide by waving it in the air. When it is completely dry, it will look dull. Place it on a paper towel, and flood it with Wright's stain. Count the number of drops of stain used. Allow the stain to remain on the slide for 3 to 4 minutes and then flood the slide with an equal number of drops of distilled water. Allow the water and Wright's stain mixture to remain on the slide for 4 or 5 minutes or until a metallic green film or scum is apparent on the fluid surface. Blow on the slide gently every minute or so to keep the water and stain mixed during this interval.

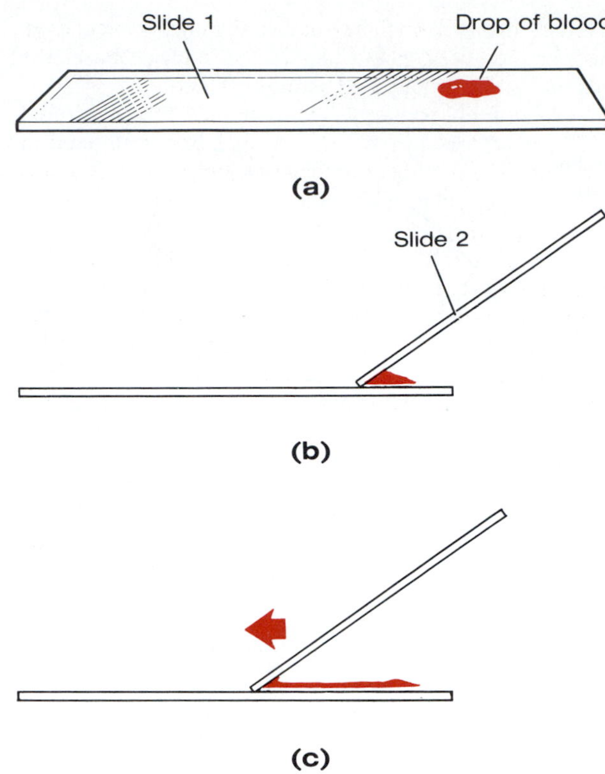

F29.2

Procedure for making a blood smear. (a) Place a drop of blood on slide 1 approximately ½ inch from one end. **(b)** Hold slide 2 at a 30° to 40° angle to slide 1 (it should touch the drop of blood) and allow blood to spread along entire bottom edge of angled slide. **(c)** Smoothly advance slide 2 to end of slide 1 (blood should run out before reaching the end of slide 1).

5. Rinse the slide with a stream of distilled water. Then flood it with distilled water, and allow it to lie flat until the slide becomes translucent and takes on a pink cast. Then stand the slide on its long edge on the paper towel, and allow it to dry completely. Once the slide is dry, you can begin your observations.

6. Obtain a microscope and scan the slide under low power to find the area where the blood smear is the thinnest. After scanning the slide in low power to find the areas with the largest numbers of nucleated WBCs, read the following descriptions of cell types, and find each one on Figure 29.1. (The formed elements are also shown in Plates 55 through 60 in the Histology Atlas.) Then, switch to the oil immersion lens and observe the slide carefully to identify each cell type.

ERYTHROCYTES Erythrocytes, or red blood cells, which average 7.5 μm in diameter, vary in color from a salmon red color to pale pink, depending on the effectiveness of the stain. They have a distinctive biconcave disk shape and appear paler in the center than at the edge (see Plate 55 in the Histology Atlas).

As you observe the slide, notice that the red blood cells are by far the most numerous blood cells seen in the field. Their number averages 4.5 million to 5.0 million cells per cubic millimeter of blood (for women and men, respectively).

Red blood cells differ from the other blood cells because they are anucleate when mature and circulating in the blood. As a result, they are unable to reproduce and have a limited life span of 100 to 120 days, after which they begin to fragment and are destroyed in the spleen and other reticuloendothelial tissues of the body.

In various anemias, the red blood cells may appear pale (an indication of decreased hemoglobin content) or may be nucleated (an indication that the bone marrow is turning out cells prematurely). ■

LEUKOCYTES Leukocytes, or white blood cells, are nucleated cells that are formed in the bone marrow from the same stem cell (*hemocytoblast*) as red blood cells. They are much less numerous than the red blood cells, averaging from 4000 to 11,000 cells per cubic millimeter. Basically, white blood cells are protective, pathogen-destroying cells that are transported to all parts of the body in the blood or lymph. Important to their protective function is their ability to move in and out of blood vessels, a process called **diapedesis,** and to wander through body tissues by **amoeboid motion** to reach sites of inflammation or tissue destruction. They are classified into two major groups, depending on whether or not they contain conspicuous granules in their cytoplasm.

Granulocytes comprise the first group. The granules in their cytoplasm stain differentially with Wright's stain, and they have peculiarly lobed nuclei, which often consist of expanded nuclear regions connected by thin strands of nucleoplasm. Additional information about the three types of granulocytes follows:

Neutrophil: The most abundant of the white blood cells (40% to 70% of the leukocyte population); nucleus consists of 3 to 7 lobes and the pale lilac cytoplasm contains fine cytoplasmic granules, which are generally indistinguishable and take up both the acidic (red) and basic (blue) dyes (*neutrophil* = neutral loving); functions as an active phagocyte. The number of neutrophils increases exponentially during acute infections. (See Plates 55 and 56.)

Eosinophil: Represents 1% to 4% of the leukocyte population; nucleus is generally figure 8 or bilobed in shape; contains large cytoplasmic granules (elaborate lysosomes) that stain red-orange with the acid dyes in Wright's stain (see Plate 59). Precise function is unknown, but they increase in number during allergies and parasite infections and may selectively phagocytize antigen-antibody complexes.

Basophil: Least abundant leukocyte type representing less than 1% of the population; large U- or S-shaped nucleus with two or more indentations. Cytoplasm contains coarse, sparse granules that are stained deep purple by the basic dyes in Wright's stain (see Plate 60). The granules contain several chemicals including histamine, a vasodilator which is discharged on exposure to antigens and helps mediate the inflammatory response.

The second group, **agranulocytes,** or **agranular leukocytes,** contains no observable cytoplasmic granules. Although found in the bloodstream, they are much more abundant in lymphoid tissues. Their nuclei tend to be closer to the norm, that is, spherical, oval, or kidney-shaped. Specific characteristics of the two types of agranulocytes are listed below.

Lymphocyte: The smallest of the leukocytes, approximately the size of a red blood cell (see Plates 55 and 57). The nucleus stains dark blue to purple, is generally spherical or slightly indented, and accounts for most of the cell mass. Sparse cytoplasm appears as a thin blue rim around the nucleus. Concerned with immunologic responses in body; one population, the B lymphocytes, oversees the production of antibodies that are released to blood. The second population, T lymphocytes, plays a regulatory role and destroys grafts, tumors, and virus-infected cells. Represents 20% to 45% of the WBC population.

Monocyte: The largest of the leukocytes; approximately twice the size of red blood cells (see Plate 58). Represents 4% to 8% of the leukocyte population. Dark blue nucleus is generally kidney-shaped; abundant cytoplasm stains gray-blue. Functions as an active phagocyte (the "long-term cleanup team"), increasing dramatically in number during chronic infections such as tuberculosis.

Students are often asked to list the leukocytes in order from the most abundant to the least abundant. The following silly phrase may help you with this task: **N**ever **l**et **m**onkeys **e**at **ba**nanas (neutrophils, lymphocytes, monocytes, eosinophils, basophils).

PLATELETS Platelets are cell fragments of large multinucleate cells (**megakaryocytes**) formed in the bone marrow. They appear as darkly staining, irregularly shaped bodies interspersed among the blood cells (see Plate 55). The normal platelet count in blood ranges from 250,000 to 500,000 per cubic millimeter. Platelets are instrumental in the clotting process that occurs in plasma when blood vessels are ruptured.

After you have identified these cell types on your slide, observe three-dimensional models of blood cells if these are available. *Do not dispose of your slide,* as it will be used later for the differential white blood cell count.

HEMATOLOGIC TESTS

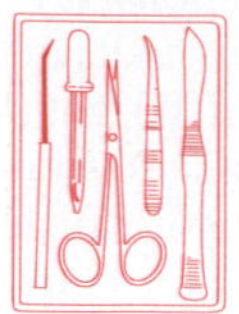

When someone enters a hospital as a patient, several hematologic tests are routinely done to determine general level of health as well as the presence of pathologic conditions. You will be conducting the most common of these tests in this exercise.

Materials such as cotton balls, lancets, and alcohol swabs are used in nearly all of the following diagnostic tests. These supplies are at the general supply area and should be properly disposed of (glassware to the "bleach bucket" and disposable items to the autoclave bag) immediately after use.

Other necessary supplies and equipment are at specific supply areas marked according to the test with which they are used. Since nearly all of the tests require a finger stab, if you will be using your own blood it might be wise to quickly read through the tests to determine in which instances more than one preparation can be done from the same finger stab. For example, the hematocrit capillary tubes and sedimentation rate samples might be prepared at the same time the chamber is being prepared for the total blood cell counts. A little preplanning will save you the discomfort of a multiply punctured finger.

An alternative to using blood obtained from the finger stab technique is using heparinized blood samples supplied by your instructor. The purpose of using heparinized tubes is to prevent the blood from clotting. Thus blood collected and stored in such tubes will be suitable for all tests except coagulation time testing.

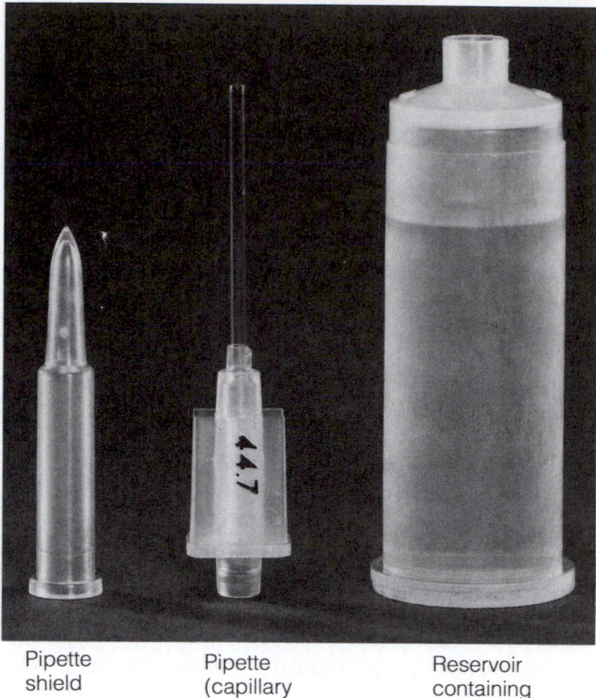

Pipette shield

Pipette (capillary tube)

Reservoir containing diluent

F29.3

The Unopette reservoir system for blood counts. The system consists of a reservoir containing a premeasured amount of RBC or WBC diluent and a plastic capillary tube pipette with a shield that is used for puncturing the reservoir top.

Total White and Red Blood Cell Counts

To conduct a **total WBC** or **RBC count,** you will dilute a known volume of blood with a fluid that prevents blood coagulation and place the mixture into a counting chamber of known volume—a **hemacytometer.** The cells are then counted by microscopic inspection, and corrections for dilution and chamber volume are made to obtain the result in cells per cubic millimeter. The white and red blood cell diluents differ. They are isotonic for the cell type to be counted and cause dissolution of the cell types not being counted. In other words, if the WBC diluent is used, the RBCs are hemolyzed and thus do not interfere with the WBC counting process. The Unopette apparatus for conducting blood counts is shown in Figure 29.3. The reservoir contains the required amount of the appropriate diluting fluid (acetic acid for WBC counts and Hayem's solution for RBC counts). Note that this hand counting technique is rather outdated, since most clinical agencies now have computerized equipment for performing blood counts.

TOTAL WHITE BLOOD CELL COUNT Since white blood cells are an important part of the body's defense system, it is essential to note any abnormalities in them. **Leukocytosis,** an abnormally high WBC count, may indicate bacterial or viral infection, metabolic disease, hemorrhage, or poisoning by drugs or chemicals. A decrease in the white cell number below 4000/mm^3 (**leukopenia**) may indicate typhoid fever, measles, infectious hepatitis or cirrhosis, tuberculosis, or excessive antibiotic or X-ray therapy. A person with leukopenia lacks the usual protective mechanisms.

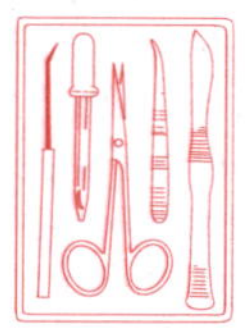

1. Obtain a hemacytometer and cover glass, cotton balls, lancets, alcohol swabs, and a hand counter. Also obtain the Unopette system for conducting WBC counts (the reservoir has a *white* bottom).

2. Puncture the top (diaphragm) of the reservoir container with the pointed end of the protective pipette shield. Then transfer a drop of animal blood to a clean slide with the glass rod. If you will be using your own blood, cleanse your finger thoroughly, and obtain a second drop of blood as described on p. 268 (step 2).

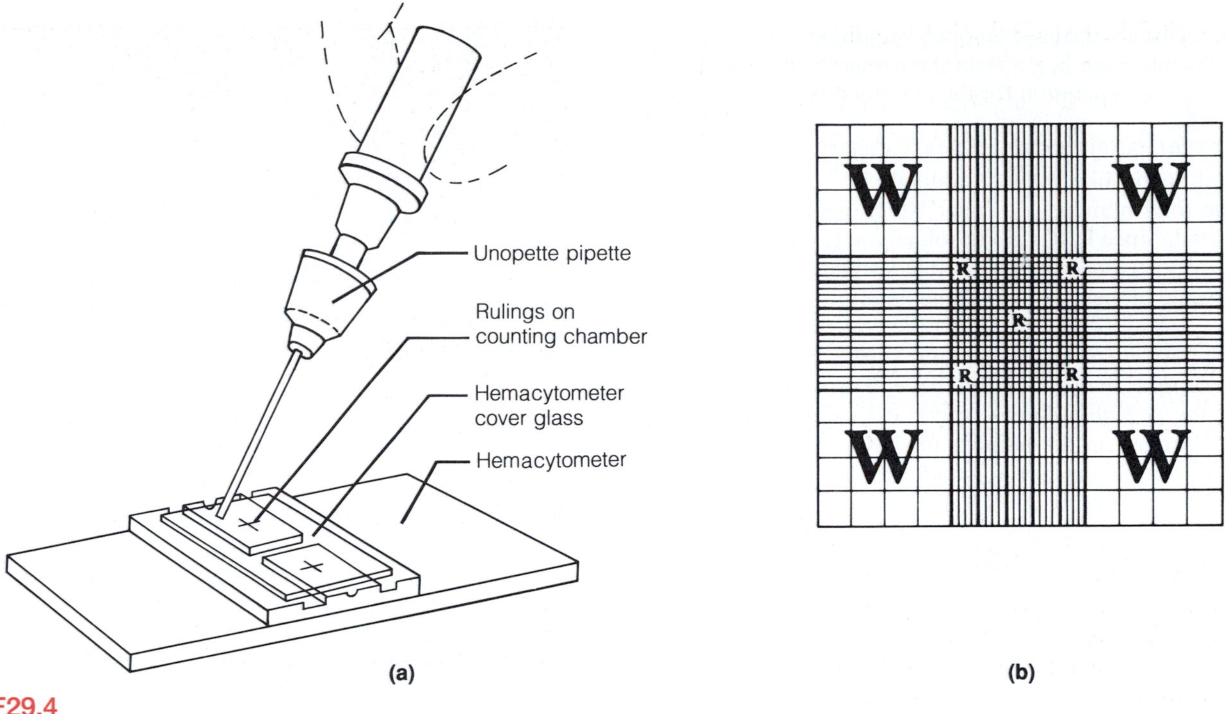

F29.4

Apparatus and preparation for blood cell counts. (a) Procedure for introducing diluted blood to the hemacytometer. **(b)** Areas of the hemacytometer used for red and white blood cell counts.

3. Now you are ready to charge the Unopette pipette with blood. Twist off the protective shield and, holding the pipette horizontally, touch the tip of the capillary tube pipette to the blood drop. When the pipette has filled by capillary action, wipe its tip with a cotton ball.

Squeeze the reservoir slightly to expel some air. While the reservoir is still compressed, insert the blood-charged pipette into the reservoir and then release the pressure on the reservoir. As you do so, the blood will be drawn into the diluent in the reservoir.

4. Squeeze the reservoir and invert it gently several times to thoroughly mix the contents. Remove the pipette from the reservoir, and then reverse it and insert its opposite end into the reservoir to convert the capillary tube to a dropper.

5. Place a coverslip on the hemacytometer, and prepare to charge it. Discard the first 2 to 3 drops—which is only diluent—from the Unopette onto a paper towel. Squeeze the reservoir to apply one drop of the diluted blood at each end of the coverslip at the junction of the hemacytometer and coverslip (Figure 29.4a), and then immediately lift the pipette away from the hemacytometer surface. *Do this quickly;* otherwise the hemacytometer will overfill, and the mixture will flow under the coverslip by capillary action. You now have two charged chambers. If the chambers are properly charged, no fluid will appear in the moat.

6. Place the hemacytometer on the microscope stage and allow it to remain undisturbed for 2 to 3 minutes to allow the cells to become evenly distributed.

7. Bring the grid lines into focus under low power (10× objective lens) and check for uniformity of white blood cell distribution. If it is uneven, switch to the other chamber for observation.

8. Move the slide into position so that one of the W areas can be seen (see Figure 29.4b), and proceed to count all WBCs observed in the four corner W areas. Use a hand counter to facilitate the counting process. *To prevent overcounts of cells at boundary lines, count only those that touch the left and upper boundary lines of the W area but not those touching the right and lower boundaries.* Record the number of WBCs counted and multiply this number by 50 to obtain the number of WBCs per cubic millimeter.*

No. WBCs: ___________ WBC/mm^3: ___________

Record this information on the data sheet on p. 280. How does your result compare with the norms for a white blood cell count?

____

* The factor of 50 is obtained by multiplying the volume dilution factor (20) by the volume correction factor (2.5): 20 × 2.5 = 50. The 2.5 volume correction factor is obtained in the following manner: Each W area on the grid is exactly 1 mm^2 × 0.1 mm deep; therefore, the volume of each W section is 0.1 mm^3. Since WBCs in four W areas are counted (a total of 0.4 mm^3), this volume must be multiplied by 2.5 (correction factor) to obtain the number of cells in 1 mm^3.

9. Discard the used Unopette equipment in the disposable autoclave bag. Clean the hemacytometer and coverslip in preparation for the red blood cell count.

TOTAL RED BLOOD CELL COUNT The red blood cell count, like the white blood cell count, determines the total number of this cell type per unit volume of blood. Since RBCs are absolutely necessary for oxygen transport, a doctor typically investigates any excessive change in their number immediately.

An increase in the number of RBCs (**polycythemia**) may result from bone marrow cancer or from living at high altitudes where less oxygen is available. A decrease in the number of RBCs results in anemia. (The term **anemia** simply indicates a decreased oxygen-carrying capacity of blood that may result from a decrease in RBC number or size or a decreased hemoglobin content of the RBCs.) A decrease in RBCs may result suddenly from hemorrhage or more gradually from conditions that destroy RBCs or hinder RBC production. ■

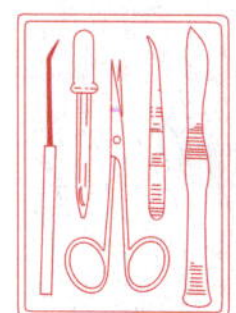

1. Obtain a Unopette apparatus for performing a RBC count (*red* bottom on the reservoir). Also obtain other supplies as required for the WBC count. Follow the same procedure for obtaining blood and charging the chamber as for the WBC count, except be very careful when reinserting the pipette into the reservoir (step 4) and when mixing the blood with the diluent (Hayem's solution, in this case).

Do not squeeze the reservoir hard because this might force Hayem's solution out the overflow opening in the reservoir. Hayem's solution contains sodium azide, an antibacterial agent that is very toxic. Hence, it is best to prevent it from splashing on your skin.

2. Allow the hemacytometer to rest on the microscope stage as before. Using the high-power lens, count all RBCs in the R-marked areas (see Figure 29.4b). Again, at the boundary lines, count only those cells touching the left and upper lines.

3. Record the number of cells counted, and multiply that number by 10,000 to compute the RBC/mm^3. (Note: the dilution factor (200) × volume correction factor (50) = 10,000.)

No. RBCs: ____________ RBCs/mm^3: ____________

How does your result compare with normal RBC values?

____

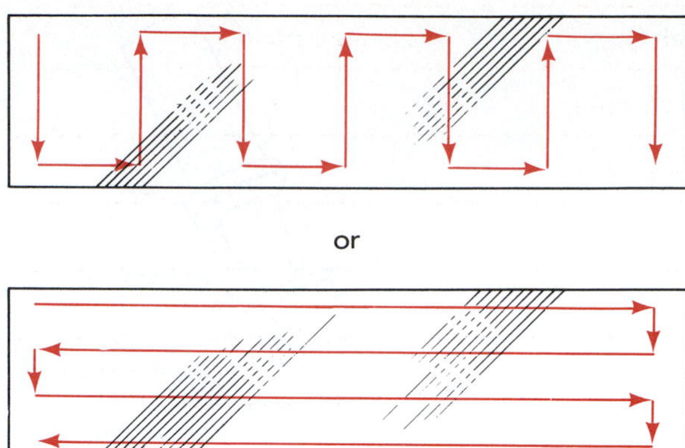

F29.5

Alternative methods of moving the slide for a differential WBC count.

4. Place all used glassware (including the hemacytometer) in the beaker containing bleach. Put disposable items, including the Unopette apparatus, in the disposable autoclave bag.

Differential White Blood Cell Count

To make a **differential white blood cell count,** 100 WBCs are counted and classified according to type. Such a count is routine in a physical examination and in diagnosing illness, since any abnormality or significant elevation in percentages of WBC types may indicate a problem or the source of pathology. Use the slide prepared for the identification of the blood cells (p. 268) for the count.

1. Begin at the edge of the smear and move the slide in a systematic manner on the microscope stage—either up and down or from side to side as indicated in Figure 29.5.

2. Record each type of white blood cell you observe by making a count on the chart on the next page (for example, 𝍸 || = 7 cells) until you have observed and recorded a total of 100 WBCs. Using the equation below, compute the percentage of each WBC type counted, and record the percentages on the data sheet at the end of this exercise.

$$\text{Percent (\%)} = \frac{\text{\# observed}}{\text{Total \# counted (100)}} \times 100$$

How does your differential white blood cell count correlate with the percentages given for each type on p. 269?

Cell type	Number observed
Neutrophils	
Eosinophils	
Basophils	
Lymphocytes	
Monocytes	

Hematocrit

The **hematocrit,** or **packed cell volume (PCV),** is routinely determined when anemia is suspected. Centrifuging whole blood spins the formed elements to the bottom of the tube, with plasma forming the top layer (see Figure 29.1). Since the blood cell population is primarily RBCs, the PCV is generally considered equivalent to the RBC volume, and this is the only value reported. However, the relative percentage of WBCs can be differentiated, and both WBC and plasma volume will be reported here. Normal hematocrit values for the male and female, respectively, are 47.0 ± 7 and 42.0 ± 5.

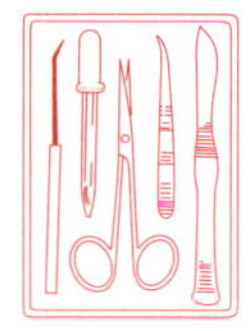

The hematocrit is determined by the micromethod, so only a drop of blood is needed. If possible, all members of the class should prepare their capillary tubes at the same time so the centrifuge can be properly balanced and run only once.

1. Obtain two heparinized capillary tubes, Seal-ease or modeling clay, a lancet, alcohol swabs, and some cotton balls.

2. Cleanse the finger, and allow the blood to flow freely. Wipe away the first few drops and, holding the red-line-marked end of the capillary tube to the blood drop, allow the tube to fill at least three-fourths full by capillary action (Figure 29.6a). If the blood is not flowing freely, the end of the capillary tube will not be completely submerged in the blood during filling, air will enter, and you will have to prepare another sample.

3. Plug the blood-containing end by pressing it into the Seal-ease or clay (Figure 29.6b). Prepare a second tube in the same manner.

4. Place the prepared tubes opposite one another in the radial grooves of the microhematocrit centrifuge with the sealed ends abutting the rubber gasket at the centrifuge periphery (Figure 29.6c). This loading procedure balances the centrifuge and prevents blood from spraying everywhere by centrifugal force. *Make a note of the numbers of the grooves your tubes are in.* When all the tubes have been loaded, make sure the centrifuge is properly balanced, and secure the centrifuge cover. Turn the centrifuge on, and set the timer for 4 or 5 minutes.

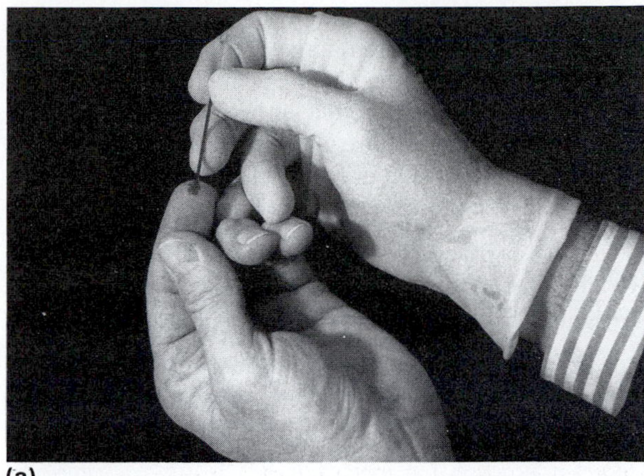

(a)

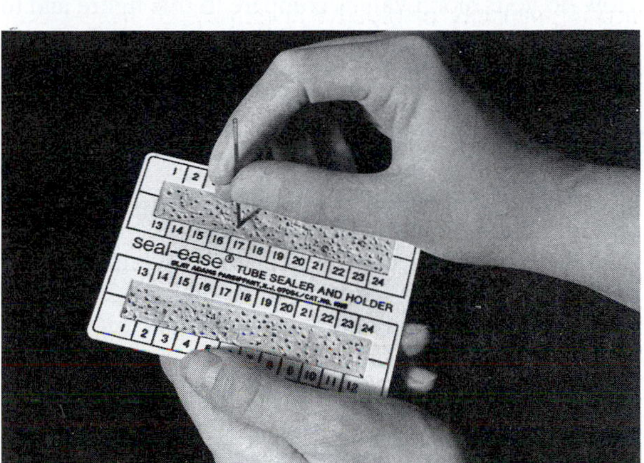

(b)

(c)

F29.6

Steps in a hematocrit determination. (a) Load a heparinized capillary tube with blood. **(b)** Plug the blood-containing end of the tube with clay. **(c)** Place the tube in a microhematocrit centrifuge. (Centrifuge must be balanced.)

5. Determine the percentage of RBCs, WBCs, and plasma by using the microhematocrit reader. The RBCs are the bottom layer, the plasma is the top layer, and the WBCs are the buff-colored layer between the two. If the reader is not available, use a millimeter ruler to measure the length of the filled capillary tube occupied by each element, and compute its percentage by using the following formula:

$$\frac{\text{Height of the column composed of the element (mm)}}{\text{Height of the original column of whole blood (mm)}} \times 100$$

Record your calculations below and on the data sheet.

% RBC ______ % WBC ______ % plasma ________

Usually WBCs constitute 1% of the total blood volume. How do your blood values compare to this figure and to the normal percentages for RBCs and plasma? (See p. 266.)

____

As a rule, a hematocrit is considered a more accurate test for determining the RBC composition of the blood than the total RBC count. A hematocrit within the normal range generally indicates a normal RBC number, whereas an abnormally high or low hematocrit is cause for concern.

Hemoglobin Concentration Determination

As noted earlier, a person can be anemic even with a normal RBC count. Since hemoglobin is the RBC protein responsible for oxygen transport, perhaps the most accurate way of measuring the oxygen-carrying capacity of the blood is to determine its hemoglobin content. Oxygen, which combines reversibly with the heme (iron-containing portion) of the hemoglobin molecule, is picked up by the blood cells in the lungs and unloaded in the tissues. Thus, the more hemoglobin molecules the RBCs contain, the more oxygen they will be able to transport. Normal blood contains 12 to 16 g hemoglobin per 100 ml blood. Hemoglobin content in men is slightly higher (14 to 18 g) than in women (12 to 16 g).

Several techniques have been developed to estimate the hemoglobin content of blood, ranging from the old, rather inaccurate Tallquist method to expensive colorimeters, which are precisely calibrated and yield highly accurate results. Directions for both the Tallquist method and a hemoglobinometer are provided here.

TALLQUIST METHOD

1. Obtain a Tallquist hemoglobin scale, lancets, alcohol swabs, and cotton balls.

2. Use instructor-provided blood or prepare the finger as previously described. (For best results, make sure the alcohol evaporates before puncturing your finger.) Place one good-sized drop of blood on the special absorbent paper provided with the color chart. The blood stain should be larger than the holes on the color chart.

3. As soon as the blood has dried and loses its glossy appearance, match its color, under natural light, with the color standards by moving the specimen under the comparison chart so that the blood stain appears at all the various apertures. (The blood should not be allowed to dry to a brown color, as this will result in an inaccurate reading.) Because the colors on the chart represent 1% variations in hemoglobin content, it may be necessary to estimate the percentage if the color of your blood sample is intermediate between two color standards.

4. On the data sheet on p. 280, record your results as the percentage of hemoglobin concentration and as grams per 100 ml of blood.

HEMOGLOBINOMETER DETERMINATION

1. Obtain a hemoglobinometer, hemolysis applicator stick, alcohol swab, and lens paper and bring them to your bench. Test the hemoglobinometer light source to make sure it is working; if not, request new batteries before proceeding and test it again.

2. Remove the blood chamber from the slot in the side of the hemoglobinometer and disassemble the blood chamber by separating the glass plates from the metal clip. Notice as you do this that the larger glass plate has an H-shaped depression cut into it that acts as a moat to hold the blood, whereas the smaller glass piece is flat and serves as a coverslip.

3. Clean the glass plates with an alcohol swab and then wipe dry with lens paper. Hold the plates by their sides to prevent smearing during the wiping process.

4. Reassemble the blood chamber (remember: larger glass piece on the bottom with the moat up), but leave the moat plate about halfway out to provide adequate exposed surface to charge it with blood.

5. Obtain a drop of blood (from the provided sample or from your fingertip as before), and place it on the depressed area of the moat plate that is closest to you (Figure 29.7a).

6. Using the wood hemolysis applicator, stir or agitate the blood to rupture (lyse) the RBCs (Figure 29.7b). This usually takes 35 to 45 seconds. Hemolysis is complete when the blood appears transparent rather than cloudy.

7. Push the blood-containing glass plate all the way into the metal clip and then firmly insert the charged blood chamber back into the slot on the side of the instrument (Figure 29.7c).

8. Hold the hemoglobinometer in your left hand with your left thumb resting on the light switch located on the underside of the instrument. Look into the eyepiece

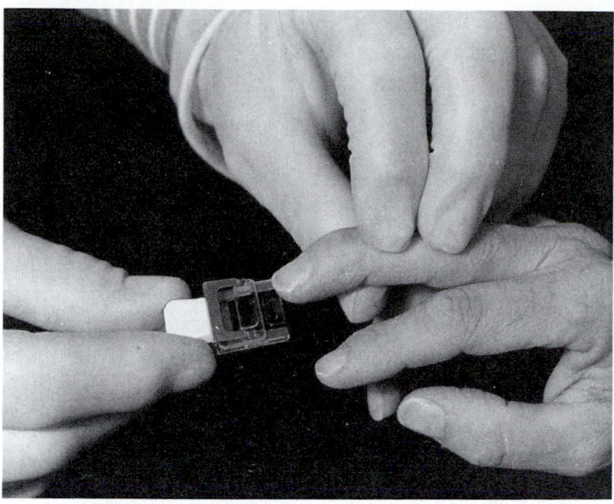

(a) A drop of blood is added to the moat plate of the blood chamber. The blood must flow freely.

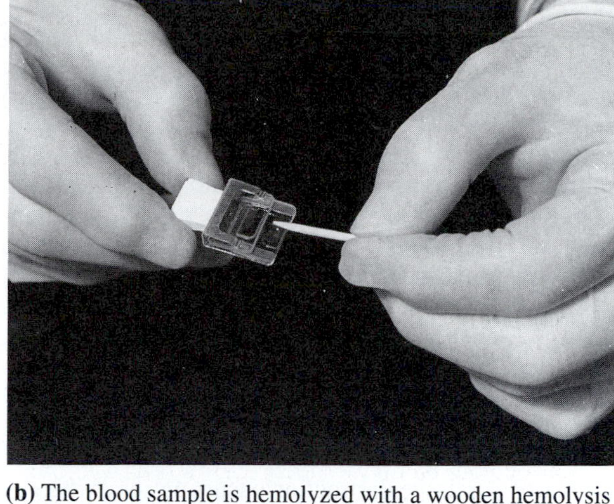

(b) The blood sample is hemolyzed with a wooden hemolysis applicator. Thirty-five to forty-five seconds are required for complete hemolysis.

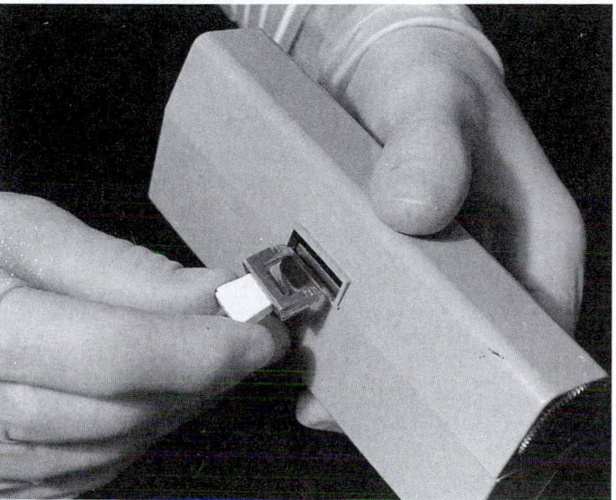

(c) The charged blood chamber is inserted into the slot on the side of the hemoglobinometer.

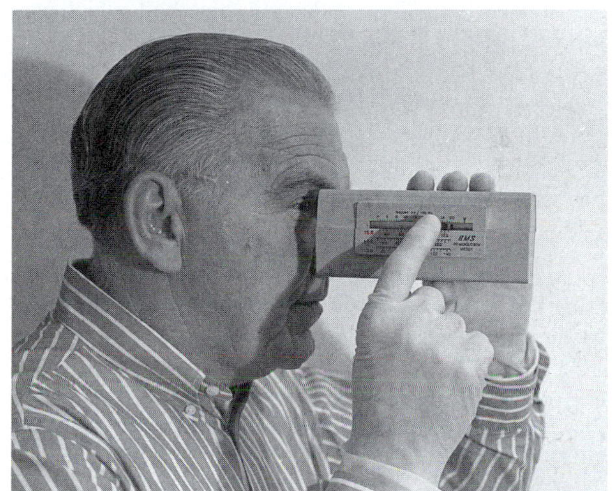

(d) The colors of the green split screen are found by moving the slide with right index finger. When the two colors match in density, the grams/100 ml and %Hb are read on the scale.

F29.7

Hemoglobin determination using a hemoglobinometer.

and notice that there is a green area divided into two halves (a split field).

9. With the index finger of your right hand, slowly move the slide on the right side of the hemoglobinometer back and forth until the two halves of the green field match (Figure 29.7d).

10. Note and record on the data sheet on p. 280 the grams Hb (hemoglobin)/100 ml blood indicated on the uppermost scale by the index mark on the slide. Also record % Hb, indicated by one of the lower scales.

11. Disassemble the blood chamber once again, and carefully place its parts (glass plates and clip) into a bleach-containing beaker.

Generally speaking, the relationship between the PCV and grams of hemoglobin per 100 ml blood is 3:1. How do your values compare?

____

Record, on the data sheet, the value obtained from your data.

Sedimentation Rate

The speed at which red blood cells settle to the bottom of a vertical tube when allowed to stand is called the **sedimentation rate.** The normal rate for adults is 0 to 6 mm/hr (averaging 3 mm/hr) and for children 0 to 8 mm/hr (averaging 4 mm/hr). Sedimentation of RBCs apparently proceeds in three stages: rouleaux formation, rapid settling, and final packing. *Rouleaux formation* (alignment of RBCs like a stack of pennies) does not occur with abnormally shaped red blood cells (as in sickle-cell anemia); therefore, the sedimentation rate is decreased. The size and number of RBCs affect the packing phase. In anemia the sedimentation rate increases; in polycythemia the rate decreases. The sedimentation rate is greater than normal during menses and pregnancy, and very high sedimentation rates may indicate infectious conditions or tissue destruction occurring somewhere in the body. Although this test is nonspecific, it alerts the diagnostician to the need for further tests to pinpoint the site of pathology. The Landau micromethod, which uses just one drop of blood, is used here.

1. Obtain lancets, cotton balls, alcohol swabs, and Landau Sed-rate pipette and tubing, the Landau rack, a wide-mouthed bottle of 5% sodium citrate, a mechanical suction device, and a millimeter ruler.

2. Use the mechanical suction device to draw up the sodium citrate to the first (most distal) marking encircling the pipette. Prepare the finger for puncture, and produce a free flow of blood.

3. Wipe off the first drop, and then draw the blood into the pipette until the mixture reaches the second encircling line. Keep the pipette tip immersed in the blood to avoid air bubbles.

4. Thoroughly mix the blood with the citrate (an anticoagulant) by drawing the mixture into the bulb and then forcing it back down into the lumen. Repeat this mixing procedure six times, and then adjust the top level of the mixture as close to the zero marking as possible. *If any air bubbles are introduced during the mixing process, discard the sample and begin again.*

5. Seal the tip of the pipette by holding it tightly against the tip of your index finger, and then carefully remove the suction device from the upper end of the pipette.

6. Stand the pipette in an exactly vertical position on the Sed-rack with its lower end resting on the base of the rack. Record the time, and allow it to stand for exactly 1 hour.

time ______________________________

7. After 1 hour, measure the number of millimeters of visible clear plasma (which indicates the amount of settling of the RBCs), and record this figure here and on the data sheet.

sedimentation rate ______________ mm/hr

8. Clean the pipette by using the mechanical suction device to draw up each of the cleaning solutions (available at the general supply area) in succession—bleach, distilled water, alcohol, and acetone. Discharge the contents each time, before drawing up the next solution. Place the cleaned pipette in the bleach-containing beaker at the general supply area, after first drawing bleach up into the pipette.

Coagulation Time

Blood clotting, or **coagulation,** is a protective mechanism that minimizes blood loss when blood vessels are ruptured. This process requires the interaction of many substances normally present in the plasma (clotting factors, or procoagulants) as well as some released by platelets and injured tissues. Basically hemostasis proceeds as follows (Figure 29.8a): The injured tissues and platelets release **thromboplastin** and **PF_3** respectively, which trigger the clotting mechanism, or cascade. Thromboplastin and PF_3 interact with other blood protein clotting factors and calcium ions to form **prothrombin activator,** which in turn converts **prothrombin** (present in plasma) to **thrombin.** Thrombin then acts enzymatically to polymerize the soluble **fibrinogen** proteins (present in plasma) into insoluble **fibrin,** which forms a meshwork of strands that traps the RBCs and forms the basis of the clot (Figure 29.8b). Normally, blood removed from the body clots within 2 to 6 minutes.

1. Obtain a *nonheparinized* capillary tube, a lancet, cotton balls, a triangular file, and alcohol swabs.

2. Use instructor-supplied blood or clean and prick the finger to produce a free flow of blood.

3. Place one end of the capillary tube in the blood drop, and hold the opposite end at a lower level to collect the sample.

4. Lay the capillary tube on a paper towel.

Record the time. ______________________________

5. At 30-sec intervals, make a small nick on the tube close to one end with the triangular file, and then carefully break the tube. Slowly separate the ends to see if a gel-like thread of fibrin spans the gap. When this occurs, record below and on the data sheet the time for coagulation to occur. Are your results within the normal time range?

____

6. Dispose of the capillary tube and used supplies in the disposable autoclave bag.

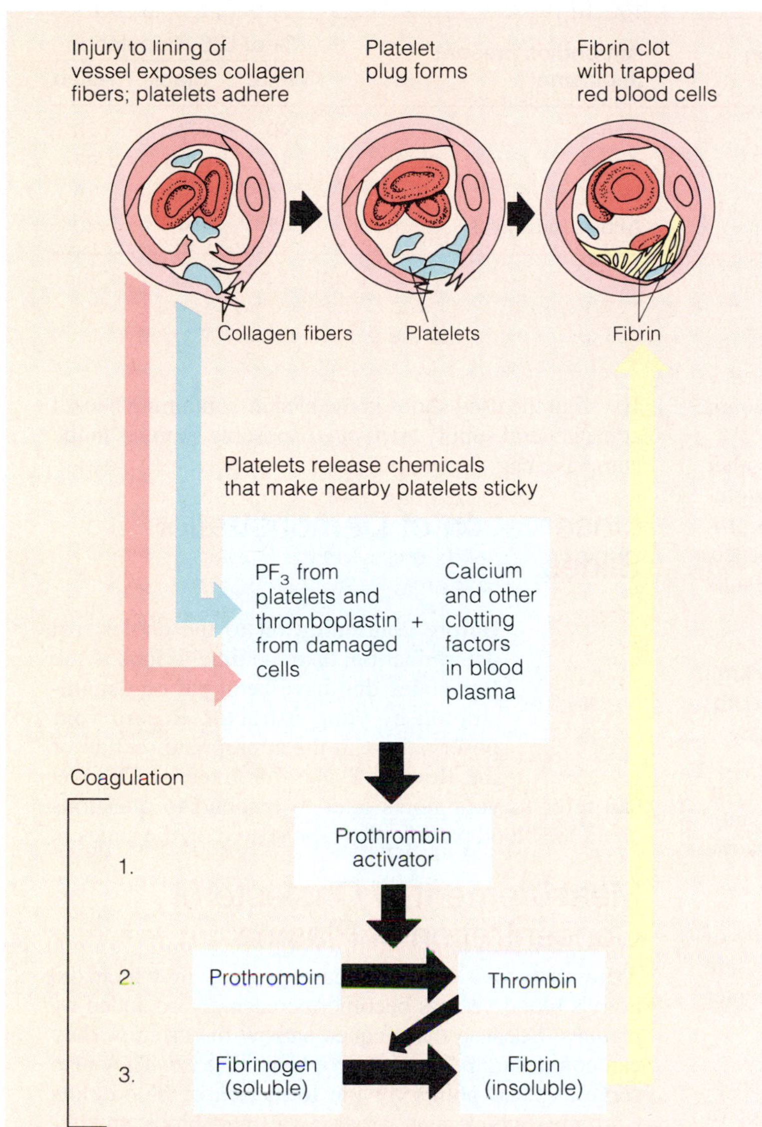

(a)

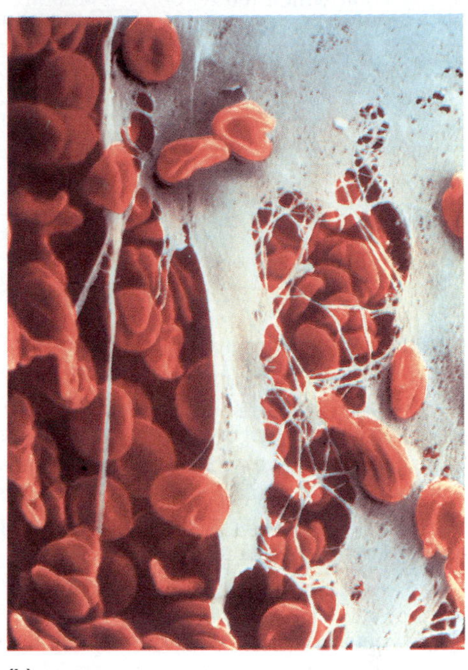

(b)

F29.8

Events of hemostasis and blood clotting. (a) Simple schematic of events. Steps numbered 1–3 represent the major events of coagulation. **(b)** Photomicrograph of RBCs trapped in a fibrin mesh (1,600×).

Blood Typing

Blood typing is a system of blood classification based on the presence of specific glycoproteins on the outer surface of the RBC plasma membrane. Such proteins are called **antigens,** or **agglutinogens,** and are genetically determined. In many cases, these antigens are accompanied by plasma proteins, **antibodies** or **agglutinins,** that react with RBCs bearing different antigens, causing them to be clumped, agglutinated, and eventually hemolyzed. It is because of this phenomenon that a person's blood must be carefully typed before a whole blood or packed cell transfusion.

Several blood typing systems exist, based on the various possible antigens, but the factors routinely typed for are antigens of the ABO and Rh blood groups which are most commonly involved in transfusion reactions. Other blood factors, such as Kell, Lewis, M, and N, are not routinely typed for unless the individual will require multiple transfusions. The basis of the ABO typing is shown in the chart at the top of p. 278.

Individuals whose red blood cells carry the Rh antigen are Rh positive (approximately 85% of the U.S. population); those lacking the antigen are Rh negative. Unlike ABO blood groups neither the blood of the Rh-positive (Rh^+) nor Rh-negative (Rh^-) individuals carries preformed anti-Rh antibodies. This is understandable in the case of the Rh-positive individual. However, Rh-negative persons who receive transfusions of Rh-positive blood become sensitized by the Rh antigens of the donor RBCs and their systems begin to produce anti-Rh antibodies. On subsequent exposures to Rh-positive blood, typical transfusion reactions occur, re-

ABO blood type	Antigens present on RBC membranes	Antibodies present in plasma	% of U.S. Population		
			White	Black	Asian
A	A	Anti-B	40	27	28
B	B	Anti-A	11	20	27
AB	A and B	None	4	4	5
O	Neither	Anti-A and anti-B	45	49	40

sulting in the clumping and hemolysis of the donor blood cells.

ALERT: Although the blood of dogs and other mammals does react with some of the human agglutinins (present in the antisera), the reaction is not as pronounced and varies with the animal blood used. Hence, the most accurate and predictable blood typing results are obtained with human blood.

1. Obtain two clean microscope slides, a wax marking pencil, anti-A, anti-B, and anti-Rh typing sera, toothpicks, lancets, alcohol swabs, and the Rh typing box.

2. Divide slide 1 into two equal halves with the wax marking pencil. Label the lower left-hand corner "anti-A" and the lower right-hand corner "anti-B." Mark the bottom of slide 2 "anti-Rh."

3. Place one drop of anti-A serum on the *left* side of slide 1. Place one drop of anti-B serum on the *right* side of slide 1. Place one drop of anti-Rh serum in the center of slide 2.

4. Cleanse your finger with an alcohol swab, pierce the finger with a lancet, and wipe away the first drop of blood. Obtain 3 drops of freely flowing blood, placing one drop on each side of slide 1 and a drop on slide 2.

5. Quickly mix each blood-antiserum sample with a *fresh* toothpick. Then dispose of the toothpicks, lancet, and used alcohol swab in the autoclave bag.

6. Place slide 2 on the Rh typing box and rock gently back and forth. (A slightly higher temperature is required for precise Rh typing than for ABO typing.)

7. After 2 minutes, observe all three blood samples for evidence of clumping. The agglutination that occurs in the positive test for the Rh factor is very fine and difficult to perceive; thus if there is any question, observe the slide under the microscope. Record your observations in the chart to the right.

8. Interpret your ABO results in light of the information in Figure 29.9. If clumping was observed on slide 2, you are Rh positive. If not, you are Rh negative.

9. Record your blood type on the data sheet.

10. Put the used slides in the bleach-containing bucket at the general supply area; put disposable supplies in the autoclave bag.

Observation of Demonstration Slides

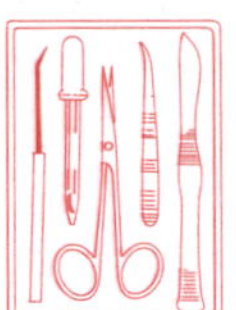

Before continuing on to the cholesterol determination, take the time to look at the five slides that have been put on demonstration by your instructor. Record your observations in the appropriate section of the Review Sheets for Exercise 29. You can refer to your notes later to respond to questions about the blood pathologies represented on the slides.

Measurement of Cholesterol Concentration in Plasma

Atherosclerosis is the disease process in which the body's blood vessels become increasingly occluded by plaques. Because the plaques narrow the arteries, they can contribute to hypertensive heart disease. They also serve as focal points for the formation of blood clots (thrombi) which may break away and block smaller vessels farther downstream in the circulatory pathway, causing heart attacks or strokes.

Ever since medical clinicians discovered that cholesterol is a major component of the smooth muscle plaques formed during atherosclerosis, it has had a bad press. Today, virtually no physical examination of an adult is considered complete until cholesterol levels are assessed along with other life-style risk factors. A nor-

	Observed (+)	Not observed (–)
Presence of clumping with anti-A		
Presence of clumping with anti-B		
Presence of clumping with anti-Rh		

mal value for plasma cholesterol in adults ranges from 130 to 200 mg/100 ml plasma; you will be making such a determination on the animal plasma provided by the instructor.

Although the total plasma cholesterol concentration is valuable information, it may be misleading, particularly if a person's high-density lipoprotein (HDL) level is high and low-density lipoprotein (LDL) level is relatively low. Cholesterol, being water insoluble, is transported in the blood complexed to lipoproteins. In general, cholesterol bound into HDLs is destined to be degraded by the liver and then eliminated from the body, whereas that forming part of the LDLs is "traveling" to the body's tissue cells. When LDL levels are excessive, cholesterol is deposited in the blood vessel walls; hence, LDLs are considered to carry the "bad" cholesterol.

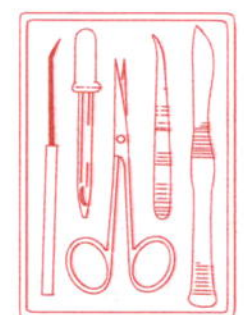

1. Go to the appropriate supply area, put three test tubes in a test tube rack, and add 5 ml of cholesterol reagent into each of the three test tubes. Mark the test tubes 1–3 with a wax marker.

2. Using the mechanical pipettor and a fresh 0.1-ml pipette each time, add 0.1 ml quantities of the following to the numbered test tubes as designated next.

- To test tube 1: Add 0.1 ml of cholesterol standard (200 mg/100 ml).
- To test tube 2: Add 0.1 ml of plasma (available at the general supply area).
- To test tube 3: Add 0.1 ml of distilled water.

3. Mix each tube by shaking it gently from side to side and then place the tubes (in the rack) into the water bath set at 37°C. (Note: if there is any question about your ability to identify your test tubes, put your initials on them with the wax marker *before* placing them into the water bath.)

Record the time: ______________________________

4. Obtain three spectrophotometer cuvettes (containers for the solutions to be "read" by the spectrophotometer). Clean them thoroughly with cuvette brushes and laboratory detergent, rinsing several times with distilled water. Invert them on a paper towel to drain.

5. After the test tubes have been in the water bath for 10 minutes, remove them and transfer their contents to similarly numbered cuvettes. Standardize the spectrophotometer at 625 nm using sample 3 (distilled water) as the *blank*.

6. Determine and record the absorbance of samples 1 and 2.

Absorbance of 1: ________ Absorbance of 2: ________

7. Calculate the cholesterol concentration (C) in sample 2 (the "unknown" plasma sample) using the following equation:

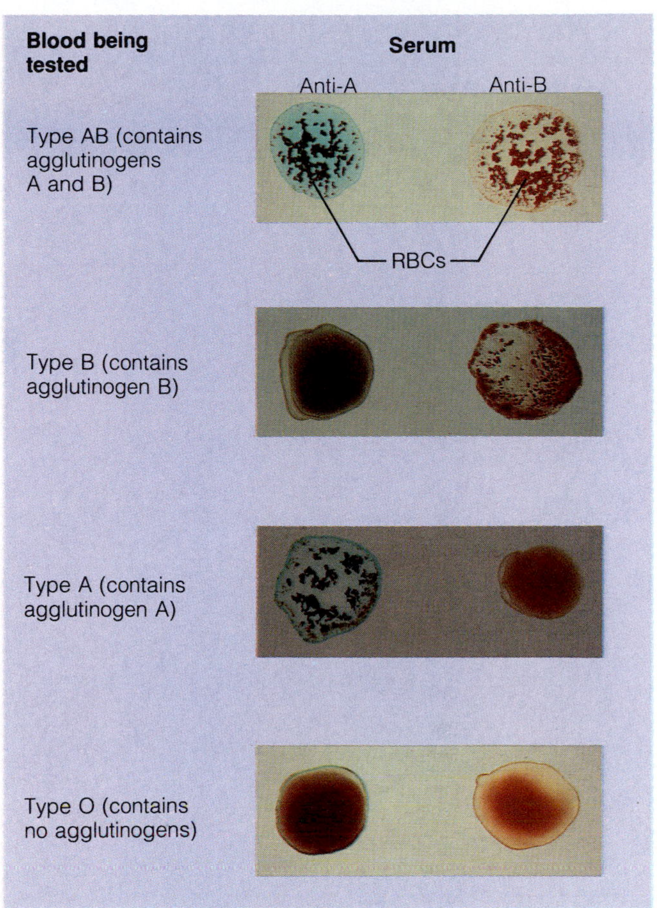

F29.9

Blood typing of ABO blood types. When serum containing anti-A or anti-B agglutinins is added to a blood sample, agglutination will occur between the agglutinin and the corresponding agglutinogen (A or B). As illustrated, agglutination occurs with both sera in blood group AB, with anti-B serum in blood group B, with anti-A serum in blood group A, and with neither serum in blood group O.

$$C_{\text{plasma}} = \frac{\text{absorbance}_{\text{plasma}}}{\text{absorbance}_{\text{standard}}} \times C_{\text{standard}}$$

Record the computed cholesterol concentration here and on the data sheet.

Cholesterol concentration: ________ mg/100 ml plasma

8. To properly dispose of cuvette solutions, liberally run cold water in the sink while carefully pouring the solutions down the drain.

9. Clean the pipettes using the mechanical pipettor to draw up each of the pipette cleaning solutions (bleach, distilled water, ethyl alcohol, and acetone) in turn, and then return them to the general supply area.

10. Before leaving the laboratory, use a paper towel saturated with bleach solution to wash down your laboratory bench.

Hematologic Test Data Sheet

Differential WBC count:

_______ % granulocytes _______ % agranulocytes

_______ % neutrophils _______ % lymphocytes

_______ % eosinophils _______ % monocytes

_______ % basophils

Total WBC count _______ WBCs/mm^3

Total RBC count _______ RBCs/mm^3

Hematocrit (PCV):

RBC _______ % of blood volume

WBC _______ % of blood volume } not generally reported

Plasma _______ % of blood volume } not generally reported

Hemoglobin content:

Tallquist method:

_______ **g/100 ml blood;** _______________ **%**

Hemoglobinometer (type: _______________ **)**

_______ **g/100 ml blood;** _______________ **%**

Ratio (PCV/grams Hb per 100 ml blood): _______________

Sedimentation rate _______ mm/hr

Coagulation time _______

Blood typing:

ABO group _______ Rh factor _______________

Cholesterol concentration _________ mg/100 ml plasma

Anatomy of the Heart

OBJECTIVES

1. To describe the location of the heart.
2. To name and locate the major anatomical areas and structures of the heart when provided with an appropriate model, diagram, or dissected sheep heart, and to explain the function of each.
3. To trace the pathway of blood through the heart.
4. To explain why the heart is called a double pump, and to compare the pulmonary and systemic circuits.
5. To explain the operation of the atrioventricular and semilunar valves.
6. To name and follow the functional blood supply of the heart.
7. To describe the histologic structure of cardiac muscle, and to note the importance of its intercalated discs and the spiral arrangement of its fibers in the heart.

MATERIALS

Torso model or laboratory chart showing heart anatomy
Red and blue pencils
Three-dimensional models of cardiac and skeletal muscle
Heart model (three-dimensional)
X ray of the human thorax for observation of the position of the heart *in situ;* X ray viewing box
Preserved sheep heart, pericardial sacs intact (if possible)
Dissecting pan and instruments
Pointed glass rods for probes
Protective skin cream or disposable gloves
Compound microscope
Histologic slides of cardiac muscle (longitudinal section)
Human Cardiovascular System: The Heart videotape*

See Appendix D, Exercise 30 for links to A.D.A.M. Standard.

See Appendix E, Exercise 30 for links to *Anatomy and PhysioShow: The Videodisc.*

* Available to qualified adopters from Benjamin/Cummings.

The major function of the **cardiovascular system** is transportation. Using blood as the transport vehicle, the system carries oxygen, digested foods, cell wastes, electrolytes, and many other substances vital to the body's homeostasis to and from the body cells. The system's propulsive force is the contracting heart, which can be compared to a muscular pump equipped with one-way valves. As the heart contracts, it forces blood into a closed system of large and small plumbing tubes (blood vessels) within which the blood is confined and circulated. This exercise deals with the structure of the heart or circulatory pump. The anatomy of the blood vessels is considered separately in Exercise 32.

GROSS ANATOMY OF THE HUMAN HEART

The **heart,** a cone-shaped organ approximately the size of a fist, is located within the mediastinum, or medial cavity, of the thorax. It is flanked laterally by the lungs, posteriorly by the vertebral column, and anteriorly by the sternum (Figure 30.1). Its more pointed **apex** extends slightly to the left and rests on the diaphragm, approximately at the level of the fifth intercostal space. Its broader **base,** from which the great vessels emerge, lies

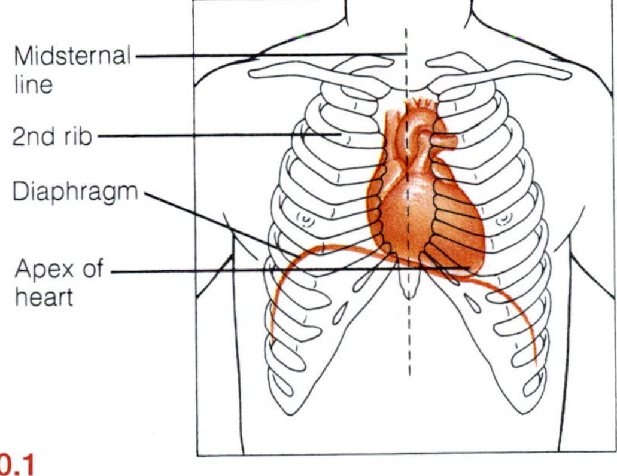

F30.1

Location of the heart in the thorax. (See also Plate D in the Human Anatomy Atlas.)

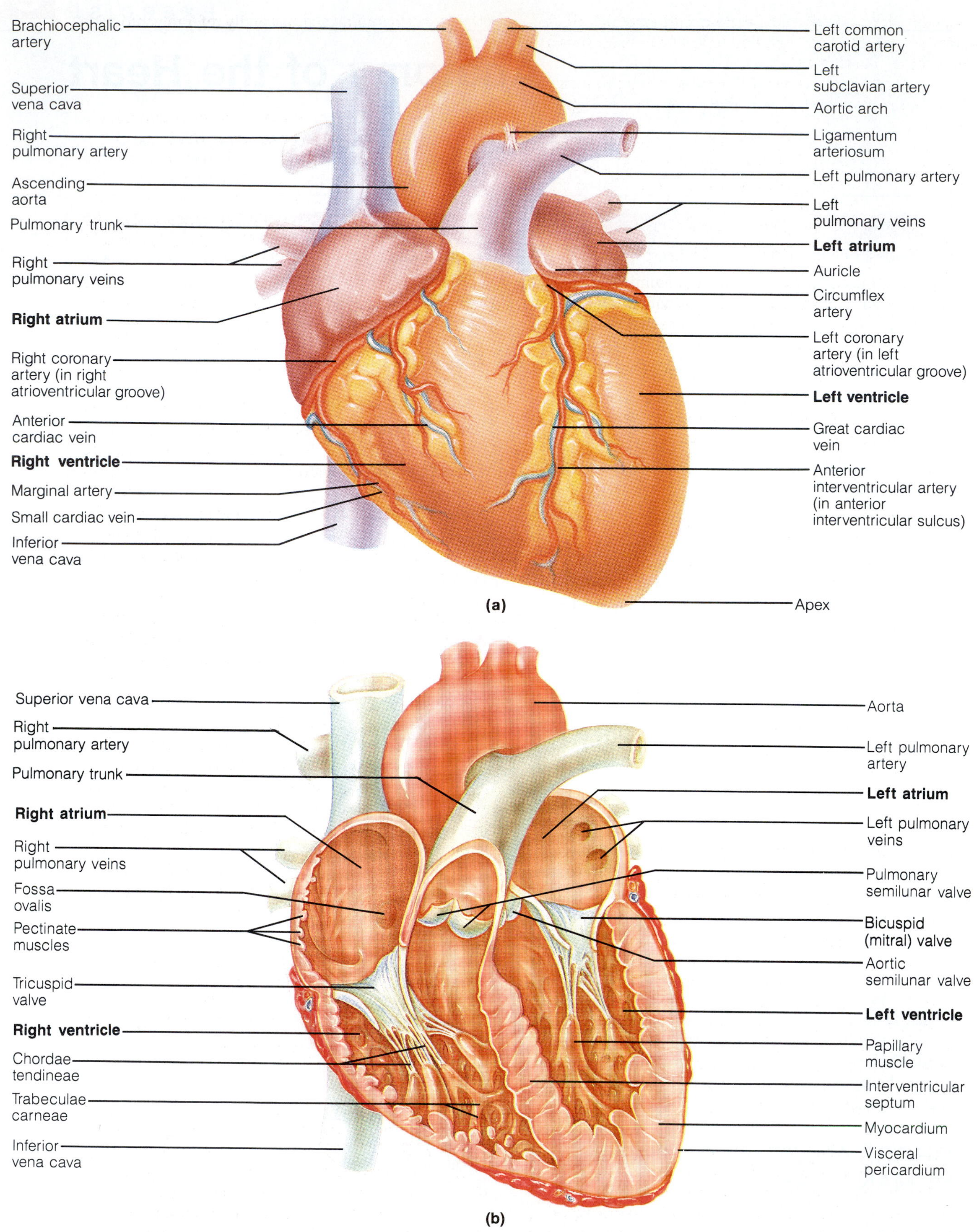

F30.2

Anatomy of the human heart. **(a)** External anterior view. **(b)** Frontal section. (See also Plate D in the Human Anatomy Atlas.)

beneath the second rib and points toward the right shoulder. *In situ,* the right ventricle of the heart forms most of its anterior surface.

If an X ray of a human thorax is available, verify the relationships described above; otherwise Figure 30.1 will suffice.

The heart is enclosed within a double-walled fibroserous sac called the pericardium. The thin **visceral pericardium,** or **epicardium,** is closely applied to the heart muscle. It reflects downward at the base of the heart to form its companion serous membrane, the outer, loosely applied **parietal pericardium,** which is attached at the heart apex to the diaphragm. Serous fluid produced by these membranes allows the heart to beat in a relatively frictionless environment. The serous parietal pericardium, in turn, lines the loosely fitting superficial **fibrous pericardium** composed of dense connective tissue.

Inflammation of the pericardium, **pericarditis,** causes painful adhesions between the serous pericardial layers. These adhesions interfere with heart movements. ■

The walls of the heart are composed primarily of cardiac muscle—the **myocardium**—which is reinforced internally by a dense fibrous connective tissue network. This network—the *fibrous skeleton of the heart*—is more elaborate and thicker in certain areas, for example, around the valves and at the base of the great vessels leaving the heart.

Figure 30.2 shows two views of the heart—an external anterior view and a frontal section. As its anatomical areas are described in the text, consult the figure. When you have pinpointed all the structures, observe the human heart model, and reidentify the same structures without reference to the figure.

Heart Chambers

The heart is divided into four chambers: two superior **atria** and two inferior **ventricles,** each lined by a thin serous membrane called the **endocardium.** The septum that divides the heart longitudinally is referred to as the **interatrial** or **interventricular septum,** depending on which chambers it partitions. Functionally, the atria are receiving chambers and are relatively ineffective as pumps. Blood flows into the atria under low pressure from the veins of the body. The right atrium receives relatively oxygen-poor blood from the body via the **superior** and **inferior venae cavae.** Four **pulmonary veins** deliver oxygen-rich blood from the lungs to the left atrium.

The inferior thick-walled ventricles, which form the bulk of the heart, are the discharging chambers. They force blood out of the heart into the large arteries that emerge from its base. The right ventricle pumps blood into the **pulmonary trunk,** which routes blood to the lungs to be oxygenated. The left ventricle discharges blood into the **aorta,** from which all systemic arteries of the body diverge to supply the body tissues. Discussions of the heart's pumping action usually refer to ventricular activity.

Heart Valves

Four valves enforce a one-way blood flow through the heart chambers. The **atrioventricular (AV) valves,** located between the atrial and ventricular chambers on each side, prevent backflow into the atria when the ventricles are contracting. The left atrioventricular valve, also called the **mitral** or **bicuspid valve,** consists of two cusps, or flaps, of endocardium. The right atrioventricular valve, the **tricuspid valve,** has three cusps (Figure 30.3). Tiny white collagenic cords called the **chordae tendineae** (literally, "heart strings") anchor the cusps to the ventricular walls. The chordae tendineae originate from small bundles of cardiac muscle, called **papillary muscles,** that project from the myocardial wall (see Figure 30.2).

When blood is flowing passively into the atria and then into the ventricles during **diastole** (the period of ventricular relaxation), the AV valve flaps hang limply into the ventricular chambers and then are carried passively toward the atria by the accumulating blood. When the ventricles contract (**systole**) and compress the blood in their chambers, the intraventricular blood pressure rises, causing the valve flaps to be reflected superiorly, which closes the AV valves. The chordae tendineae, pulled taut by the contracting papillary muscles, anchor the flaps in a closed position that prevents backflow into the atria during ventricular contraction. If unanchored, the flaps would blow upward into the atria rather like an umbrella being turned inside out by a strong wind.

The second set of valves, the **pulmonary** and **aortic semilunar valves,** each composed of three pocketlike cusps, guards the bases of the two large arteries leaving the ventricular chambers. The valve cusps are forced open and flatten against the walls of the artery as the ventricles discharge their blood into the large arteries during systole. However, when the ventricles relax, blood flows backward toward the heart and the cusps fill with blood, closing the semilunar valves and preventing arterial blood from reentering the heart.

PULMONARY, SYSTEMIC, AND CARDIAC CIRCULATIONS

Pulmonary and Systemic Circulations

The heart functions as a double pump. The right side serves as the **pulmonary circulation** pump, shunting the carbon dioxide–rich blood entering its chambers to the lungs to unload carbon dioxide and pick up oxygen, and then back to the left side of the heart. The function of this circuit is strictly to provide for gas exchange. The second circuit, which carries oxygen-rich blood from the left heart through the body tissues and back to the right heart is called the **systemic circulation.** It supplies the functional blood supply to all body tissues.

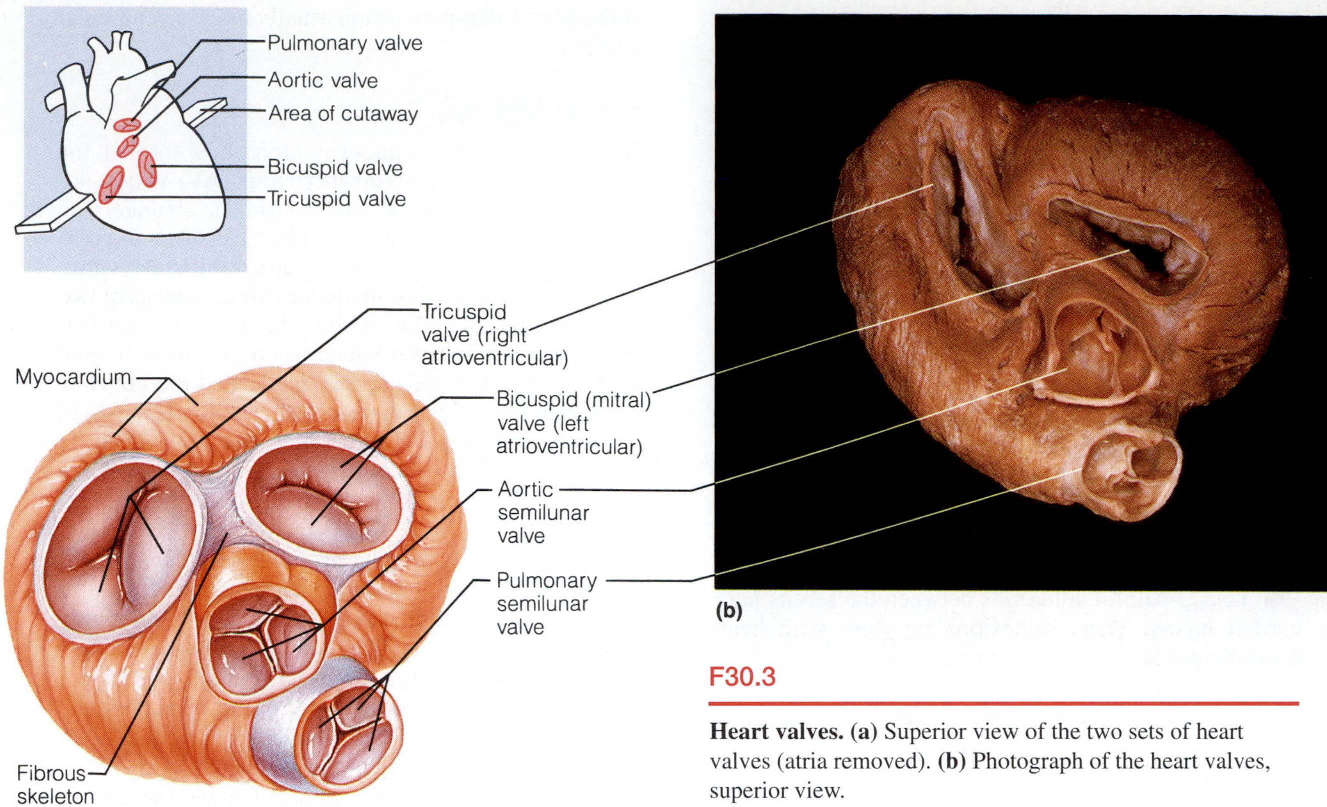

F30.3

Heart valves. **(a)** Superior view of the two sets of heart valves (atria removed). **(b)** Photograph of the heart valves, superior view.

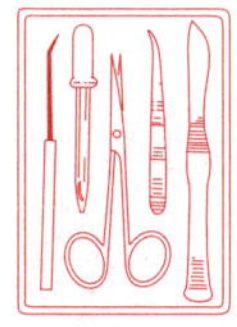

Trace the pathway of blood through the heart by adding arrows to the frontal section diagram (see Figure 30.2b). Use red arrows for the oxygen-rich blood and blue arrows for the less oxygen-rich blood.

Cardiac Circulation

Even though the heart chambers are almost continually bathed with blood, this contained blood does not nourish the myocardium. The functional blood supply of the heart is provided by the right and left coronary arteries (see Figures 30.2 and 30.4). The **coronary arteries** issue from the base of the aorta just above the aortic semilunar valve and encircle the heart in the **atrioventricular groove** at the junction of the atria and ventricles. They then ramify over the heart's surface, the right coronary artery supplying the posterior surface of the ventricles and the lateral aspect of the right side of the heart, largely through its **posterior interventricular** and **marginal artery** branches. The left coronary artery supplies the anterior ventricular walls and the laterodorsal part of the left side of the heart via its two major branches, the **anterior interventricular artery** and the **circumflex artery.** The coronary arteries and their branches are compressed during systole and fill when the heart is relaxed. The myocardium is largely drained by the **great, middle,** and **small cardiac veins,** which empty into the **coronary sinus.** The coronary sinus, in turn, empties into the right atrium. In addition, several **anterior cardiac veins** empty directly into the right atrium (Figure 30.4).

MICROSCOPIC ANATOMY OF CARDIAC MUSCLE

Cardiac muscle is found in only one place—the heart. The heart acts as a vascular pump, propelling blood to all tissues of the body; cardiac muscle is thus very important to life. Cardiac muscle is involuntary, ensuring a constant blood supply.

The cardiac cells, only sparingly invested in connective tissue, are arranged in spiral or figure-8-shaped bundles (Figure 30.5). When the heart contracts, its internal chambers become smaller (or are temporarily obliterated), forcing the blood into the large arteries leaving the heart.

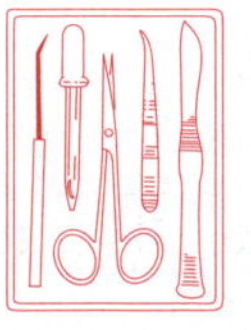

1. Observe the three-dimensional model of cardiac muscle, examining its branching cells and the areas where the cells interdigitate, the **intercalated discs.** These two structural features provide a continuity to cardiac muscle not seen in other

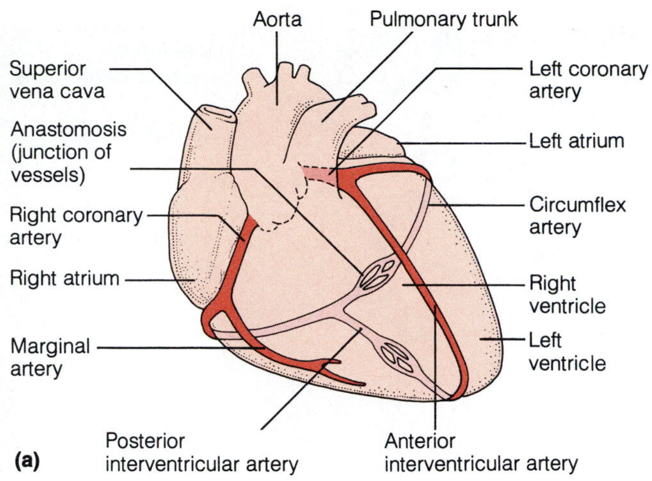

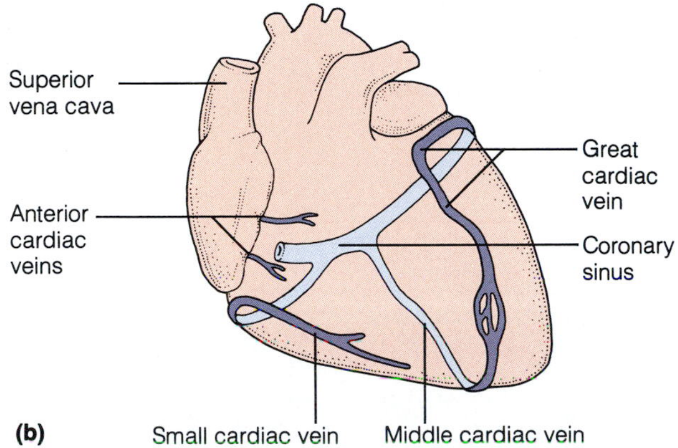

F30.4

Cardiac circulation. **(a)** Main coronary arteries. **(b)** Major cardiac veins.

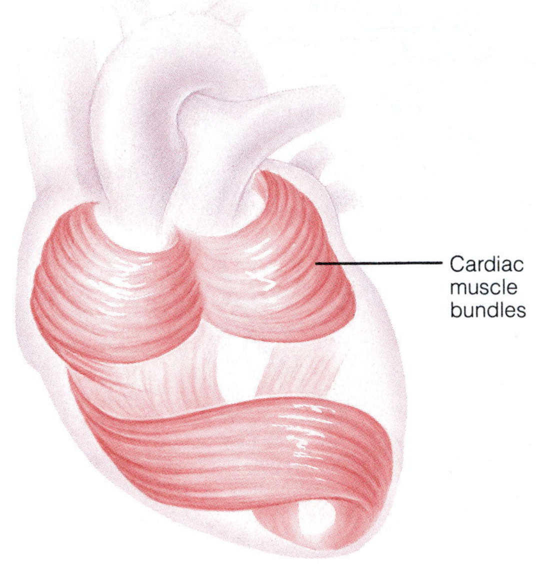

F30.5

Longitudinal view of the heart chambers showing the spiral arrangement of the cardiac muscle fibers.

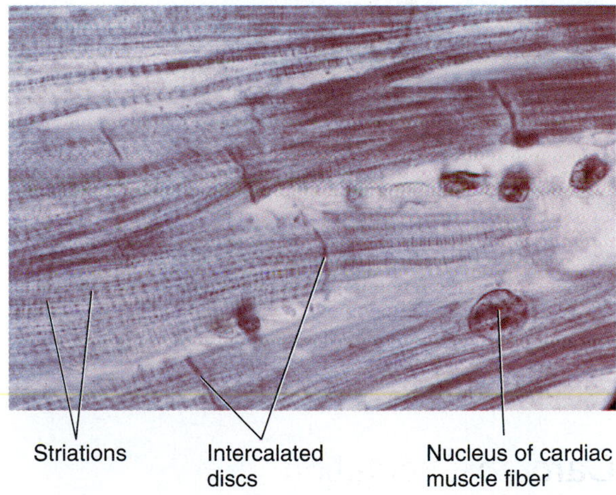

F30.6

Photomicrograph of cardiac muscle (688×).

muscle tissues and allow close coordination of heart activity.

2. Compare the model of cardiac muscle to the model of skeletal muscle. Note the similarities and differences between the two kinds of muscle tissue.

3. Obtain and observe a longitudinal section of cardiac muscle under high power. Identify the nucleus, striations, intercalated discs, and sarcolemma of the individual cells and then compare your observations to the view seen in Figure 30.6.

DISSECTION OF THE SHEEP HEART

Dissection of a sheep heart is valuable because it is similar in size and structure to the human heart. Also, a dissection experience allows you to view structures in a way not possible with models and diagrams. Refer to Figure 30.7 as you proceed with the dissection.

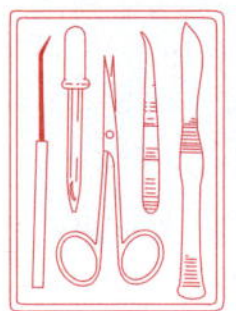

1. Obtain a preserved sheep heart, a dissection tray, dissecting instruments and, if desired, protective skin cream or gloves. Rinse the sheep heart in cold water to remove excessive preservatives and to flush out any trapped blood clots. Now you are ready to make your observations.

2. Observe the texture of the pericardium. Also, note its point of attachment to the heart. Where is it attached?

____

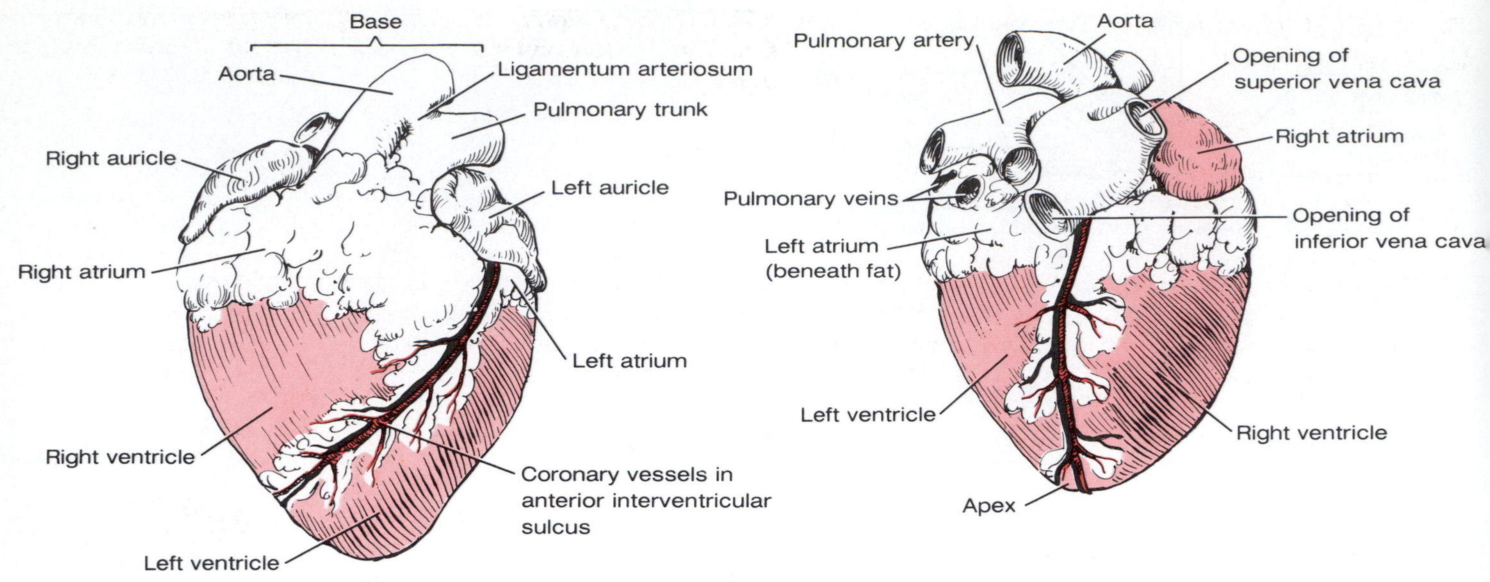

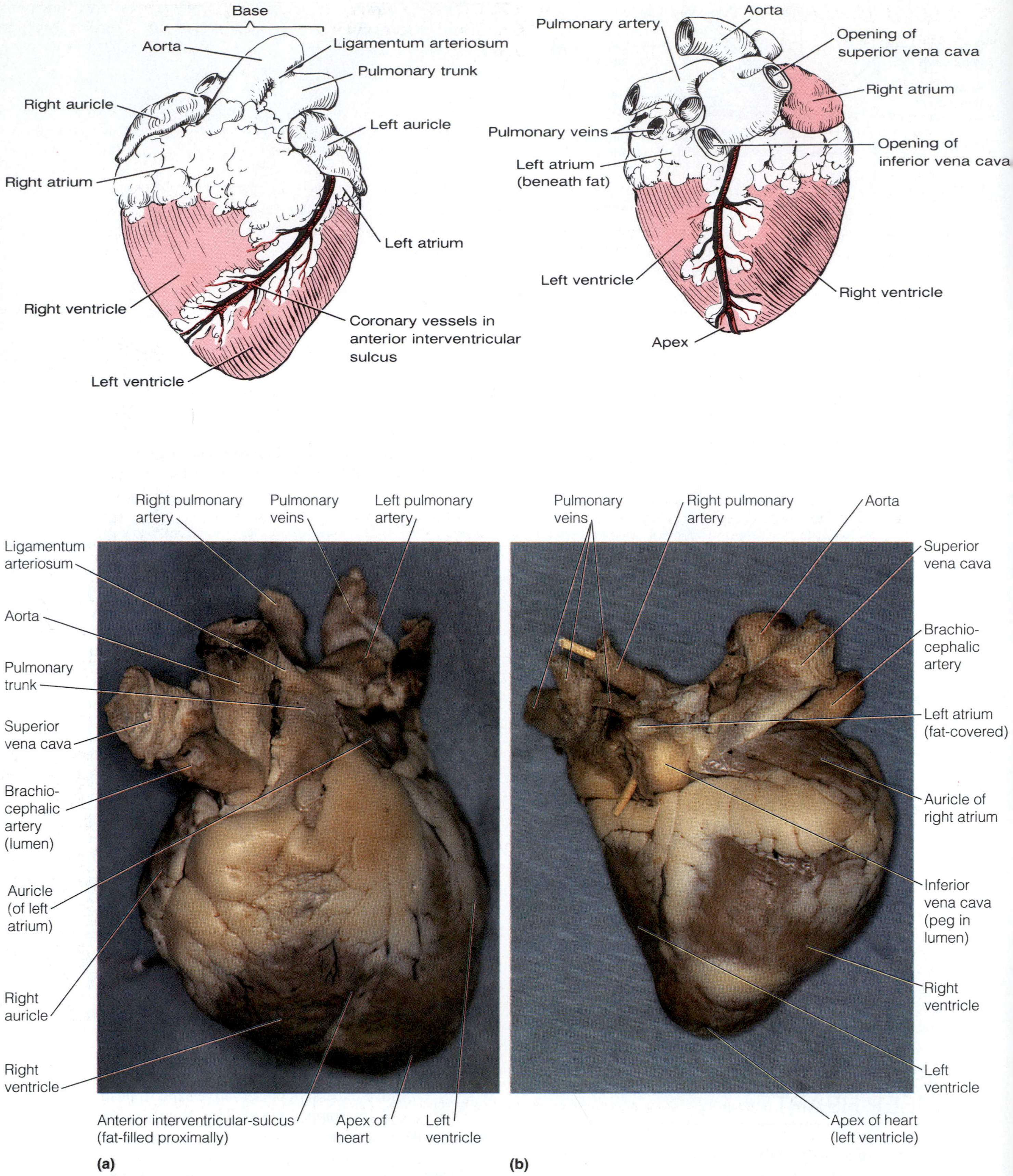

Right pulmonary artery
Pulmonary veins
Left pulmonary artery
Ligamentum arteriosum
Aorta
Pulmonary trunk
Superior vena cava
Brachio-cephalic artery (lumen)
Auricle (of left atrium)
Right auricle
Right ventricle
Anterior interventricular-sulcus (fat-filled proximally)
Apex of heart
Left ventricle

(a)

Pulmonary veins
Right pulmonary artery
Aorta
Superior vena cava
Brachio-cephalic artery
Left atrium (fat-covered)
Auricle of right atrium
Inferior vena cava (peg in lumen)
Right ventricle
Left ventricle
Apex of heart (left ventricle)

(b)

F30.7

Anatomy of the sheep heart. **(a)** Anterior view. **(b)** Posterior view. Diagrammatic views at top; photographs at bottom.

3. If the serous pericardial sac is still intact, slit open the parietal pericardium and cut it from its attachments. Observe the visceral pericardium (epicardium). Using a sharp scalpel, carefully pull a little of this serous membrane away from the myocardium. How does its position, thickness, and apposition to the heart differ from those of the parietal pericardium?

___

___

___

4. Examine the external surface of the heart. Notice the accumulation of adipose tissue, which in many cases marks the separation of the chambers and the location of the coronary arteries that nourish the myocardium. Carefully scrape away some of the fat with a scalpel to expose the coronary blood vessels.

5. Identify the base and apex of the heart, and then identify the two wrinkled **auricles,** earlike flaps of tissue projecting from the atrial chambers. The balance of the heart muscle is ventricular tissue. To identify the left ventricle, compress the ventricular chambers on each side of the longitudinal fissures carrying the coronary blood vessels. The side that feels thicker and more solid is the left ventricle. The right ventricle feels much thinner and somewhat flabby when compressed. This difference reflects the greater demand placed on the left ventricle, which must pump blood through the much longer systemic circulation, a pathway with much higher resistance than the pulmonary circulation served by the right ventricle. Hold the heart in its anatomical position (Figure 30.7a), with the anterior surface uppermost. In this position the left ventricle composes the entire apex and the left side of the heart.

6. Identify the pulmonary trunk and the aorta extending from the superior aspect of the heart. The pulmonary trunk is the more anterior, and you may see its division into the right and left pulmonary arteries if it has not been cut too closely to the heart. The thicker-walled aorta, which branches almost immediately, is located just beneath the pulmonary trunk. The first branch of the sheep aorta, the **brachiocephalic artery,** is identifiable unless the aorta has been cut immediately as it leaves the heart. The brachiocephalic artery splits to form the right carotid and subclavian arteries, which supply the right side of the head and right forelimb, respectively.

Carefully clear away some of the fat between the pulmonary trunk and the aorta to expose the **ligamentum arteriosum,** a cordlike remnant of the **ductus arteriosus.** (In the fetus, the ductus arteriosus allows blood to pass directly from the pulmonary trunk to the aorta, thus bypassing the nonfunctional fetal lungs.)

7. Cut through the wall of the aorta until you see the aortic semilunar valve. Identify the two openings to the coronary arteries just above the valve. Insert a probe into one of these holes to see if you can follow the course of a coronary artery across the heart.

8. Turn the heart to view its posterior surface. The heart will appear as shown in Figure 30.7b. Notice that the right and left ventricles appear equal-sized in this view. Identify the four thin-walled pulmonary veins entering the left atrium. (It may or may not be possible to locate the pulmonary veins from this vantage point, depending on how they were cut as the heart was removed.) Identify the superior and inferior venae cavae entering the right atrium. Compare the approximate diameter of the superior vena cava with the diameter of the aorta.

Which is larger? ___

Which has thicker walls? ___

Why do you suppose these differences exist?

___

___

___

9. Insert a probe into the superior vena cava and use scissors to cut through its wall so that you can view the interior of the right atrium. Do not extend your cut entirely through the right atrium or into the ventricle. Observe the right atrioventricular valve.

How many flaps does it have? ___

Pour some water into the right atrium and allow it to flow into the ventricle. Slowly and gently squeeze the right ventricle to watch the closing action of this valve. (If you squeeze too vigorously, you'll get a face full of water!) Drain the water from the heart before continuing.

10. Return to the pulmonary trunk and cut through its anterior wall until you can see the pulmonary semilunar valve. Pour some water into the base of the pulmonary trunk to observe the closing action of this valve. How does its action differ from that of the atrioventricular valve?

___

___

___

After observing semilunar valve action, drain the heart once again. Return to the superior vena cava, and continue the cut made in its wall through the right atrium and right atrioventricular valve into the right ventricle. Parallel the anterior border of the interventricular septum until you "round the corner" to the dorsal aspect of the heart (Figure 30.8).

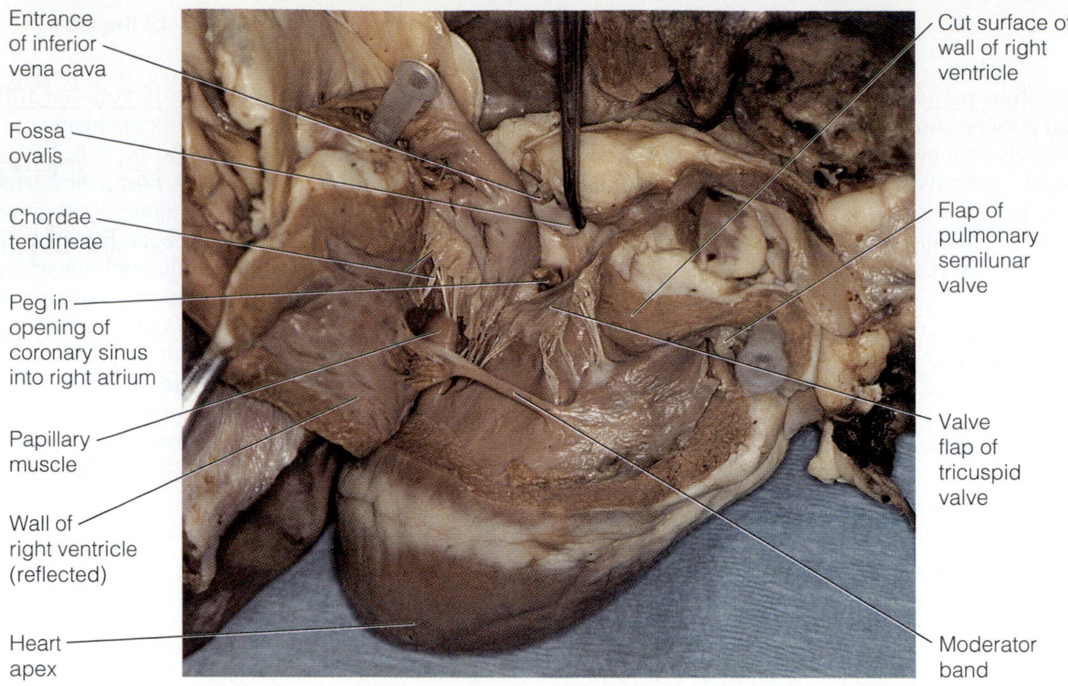

F30.8

Right side of the heart opened and reflected to reveal internal structures.

11. Reflect the cut edges of the superior vena cava, right atrium, and right ventricle to obtain the view seen in Figure 30.8. Observe the comblike ridges of muscle throughout most of the right atrium. This is called **pectinate muscle** (*pectin* means "comb"). Identify, on the ventral atrial wall, the large opening of the inferior vena cava and follow it to its external opening with a probe. Notice that the atrial walls in the vicinity of the venae cavae are smooth and lack the roughened appearance (pectinate musculature) of the other regions of the atrial walls. Just below the inferior vena caval opening, identify the opening of the **coronary sinus,** which returns venous blood of the coronary circulation to the right atrium. Nearby, locate an oval depression, the **fossa ovalis,** in the interatrial septum. This depression marks the site of an opening in the fetal heart, the **foramen ovale,** which allows blood to pass from the right to the left atrium, thus bypassing the fetal lungs.

12. Identify the papillary muscles in the right ventricle, and follow their attached chordae tendineae to the flaps of the tricuspid valve. Notice the pitted and ridged appearance (**trabeculae carneae**) of the inner ventricular muscle.

13. Make a longitudinal incision through the aorta and continue it into the left ventricle. Notice how much thicker the myocardium of the left ventricle is than that of the right ventricle. Compare the *shape* of the left ventricular cavity to the shape of the right ventricular cavity.

____

____

____

Are the papillary muscles and chordae tendineae observed in the right ventricle also present in the left ventricle?

____

Count the number of cusps in the left atrioventricular valve. How does this compare with the number seen in the right atrioventricular valve?

____

____

How do the sheep valves compare with their human counterparts?

____

14. Continue your incision from the left ventricle superiorly into the left atrium. Reflect the cut edges of the atrial wall, and attempt to locate the entry points of the pulmonary veins into the left atrium. Follow the pulmonary veins to the heart exterior with a probe. Notice how thin-walled these vessels are.

15. Properly dispose of the organic debris, and clean the dissecting tray and instruments before leaving the laboratory.

Conduction System of the Heart and Electrocardiography

OBJECTIVES

1. To list and localize the elements of the intrinsic conduction, or nodal, system of the heart; and to describe how impulses are initiated and conducted through this system and the myocardium.
2. To interpret the ECG in terms of depolarization and repolarization events occurring in the myocardium; and to identify the P, QRS, and T waves on an ECG recording.
3. To calculate the heart rate, QRS interval, P-R interval, and Q-T interval from an ECG obtained during the laboratory period.
4. To define *tachycardia, bradycardia,* and *fibrillation.*

MATERIALS

ECG recording apparatus*
Electrode paste; alcohol swabs
Cot

See Appendix E, Exercise 31 for links to *Anatomy and PhysioShow: The Videodisc.*

***For instructors using Intelitool, Exercise 31i in Appendix C uses Cardiocomp.**

THE INTRINSIC CONDUCTION SYSTEM

Heart contraction results from a series of electrical potential changes (depolarization waves) that travel through the heart preliminary to each beat. The ability of cardiac muscle to beat is intrinsic—it does not depend on impulses from the nervous system to initiate its contraction and will continue to contract rhythmically even if all nerve connections are severed. However, two types of controlling systems exert their effects on heart activity. One of these involves nerves of the autonomic nervous system, which accelerate or decrease the heartbeat rate depending on which division is activated. The second system is the **intrinsic conduction system,** or **nodal system,** of the heart, consisting of specialized noncontractile myocardial tissue. The intrinsic conduction system ensures that heart muscle depolarizes in an orderly and sequential manner (from atria to ventricles) and that the heart beats as a coordinated unit.

The components of the intrinsic conduction system include the **SA (sinoatrial) node,** located in the right atrium just inferior to the entrance to the superior vena cava; the **AV (atrioventricular) node** in the lower atrial septum at the junction of the atria and ventricles; the **AV bundle (bundle of His)** and right and left **bundle branches,** located in the interventricular septum; and the **Purkinje fibers,** which ramify within the muscle bundles of the ventricular walls. The Purkinje fiber network is much denser and more elaborate in the left ventricle because of the larger size of this chamber (Figure 31.1).

The SA node, which has the highest rate of discharge, provides the stimulus for contraction. Because it sets the rate of depolarization for the heart as a whole, the SA node is often referred to as the *pacemaker.* From the SA node, the impulse spreads throughout the atria and to the AV node. This electrical wave is immediately followed by atrial contraction. At the AV node, the impulse is momentarily delayed (approximately 0.1 sec), allowing the atria to complete their contraction. It then passes through the AV bundle, the right and left bundle

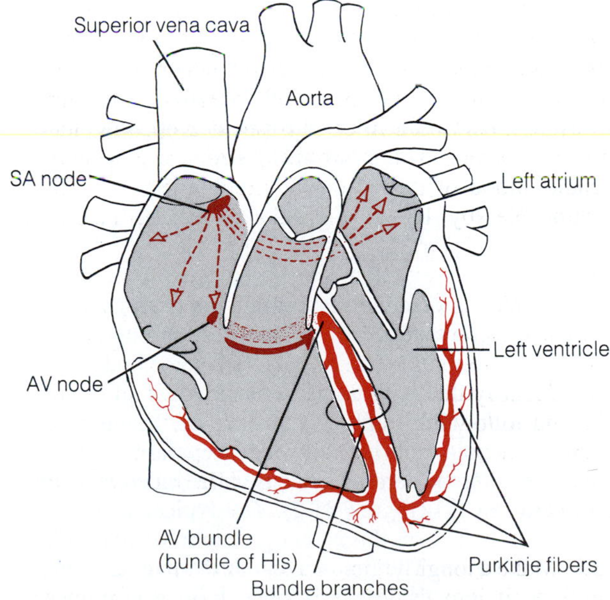

F31.1

The intrinsic conduction system of the heart. Dashed-line arrows indicate transmission of the impulse from the SA node through the atria. Solid arrow indicates transmission of the impulse from the AV node to the AV bundle.

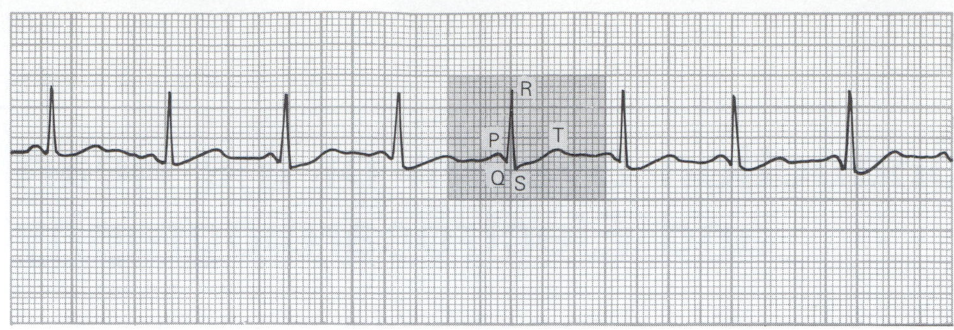

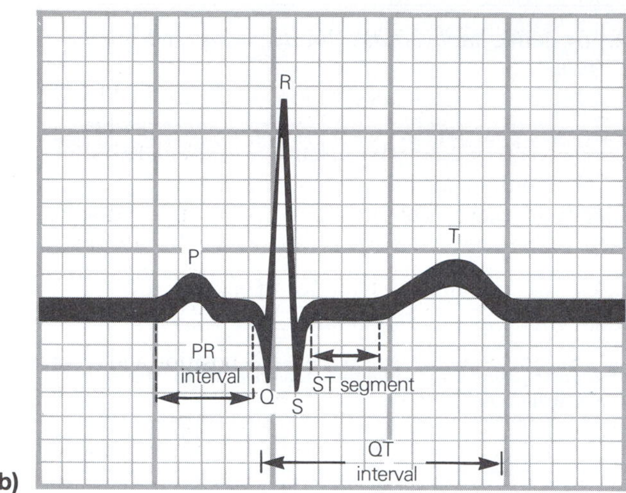

Time: small squares = 0.04 sec
1 large square = 0.20 sec
5 large squares = 1.00 sec

F31.2

The normal electrocardiogram. (a) Regular sinus rhythm. **(b)** Waves, segments, and intervals of a normal ECG.

branches, and the Purkinje fibers, finally resulting in ventricular contraction. Note that the atria and ventricles are separated from one another by a region of electrically inert connective tissue; so the depolarization wave can be transmitted to the ventricles only via the tract between the AV node and AV bundle. Thus, any damage to the AV node-bundle pathway partially or totally insulates the ventricles from the influence of the SA node. Although autorhythmic cells are found throughout the heart, their rates of spontaneous depolarization differ. The nodal system increases the rate of heart depolarization and synchronizes heart activity.

ELECTROCARDIOGRAPHY

The conduction of impulses through the heart generates electrical currents that eventually spread throughout the body. These impulses can be detected on the body's surface and recorded with an instrument called an *electrocardiograph.* The graphic recording of the electrical changes (depolarization followed by repolarization) occurring during the cardiac cycle is called an **electrocardiogram (ECG)** (Figure 31.2). The typical ECG consists of a series of three recognizable waves called *deflection waves.* The first wave, the **P wave,** is a small wave that indicates depolarization of the atria immediately before atrial contraction. The large **QRS complex,** resulting from ventricular depolarization, has a complicated shape (primarily because of the variability in size of the two ventricles and the time differences required for these chambers to depolarize). It precedes ventricular contraction. The **T wave** results from currents propagated during ventricular repolarization. The repolarization of the atria, which occurs during the QRS interval, is generally obscured by the large QRS complex.

It is important to understand what an ECG does and does not show: First, an ECG is a record of voltage and time—nothing else. Although we can and do infer that muscle contraction follows its excitation, sometimes it does not. Secondly, an ECG records electrical events occurring in relatively large amounts of muscle tissue (i.e., the bulk of the heart muscle), *not* the activity of nodal tissue which, like muscle contraction, can only be inferred. Nonetheless, abnormalities of the deflection waves and changes in the time intervals of the ECG are useful in detecting myocardial infarcts or problems with the conduction system of the heart. The P-R (P-Q) interval represents the time between the beginning of atrial depolarization and ventricular depolarization. Thus, it typically includes the period during which the depolarization wave passes to the AV node, atrial systole, and the passage of the excitation wave to the balance of the conducting system. Generally, the P-R interval is about 0.16 to 0.18 sec. A longer interval may suggest a partial AV heart block caused by damage to the AV node. In total heart block, no impulses are transmitted through the AV node, and the atria and ventricles beat independently of one another—the atria at the SA node rate and the ventricles at their intrinsic rate, which is considerably slower.

If the QRS interval (normally 0.08 sec) is prolonged, it may indicate a right or left bundle branch block in which one ventricle is contracting later than the other. The Q-T interval is the period from the beginning of ventricular depolarization through repolarization and includes the time of ventricular contraction (the S-T segment). With a heart rate of 70 beats/min, this interval is normally 0.31 to 0.41 sec. As the rate increases, this interval becomes shorter; conversely, when the heart rate drops, the interval is longer.

A heart rate over 100 beats/min is referred to as **tachycardia;** a rate below 60 beats/min is **bradycardia.** Although neither condition is pathological, prolonged tachycardia may progress to **fibrillation,** a condition of rapid uncoordinated heart contractions which makes the heart useless as a pump. Bradycardia in athletes is a positive finding; that is, it indicates an increased efficiency of cardiac functioning. Because *stroke volume* (the amount of blood ejected by a ventricle with each contraction) increases with physical conditioning, the heart can contract more slowly and still meet circulatory demands.

Twelve standard leads are used to record an ECG for diagnostic purposes. Three of these are bipolar leads that measure the voltage difference between the arms, or an arm and a leg, and nine are unipolar leads. Together the 12 leads provide a fairly comprehensive picture of the electrical activity of the heart.

For this investigation, four electrodes are used (Figure 31.3), and results are obtained from the three *standard limb leads* (also shown in Figure 31.3). Several types of polygraphs or ECG recorders are available. Your instructor will provide specific directions on how to set up and use the available apparatus.

Understanding the Standard Limb Leads

As you might expect, electrical activity recorded by any lead depends on the location and orientation of the recording electrodes. Clinically, it is assumed that the heart lies in the center of a triangle with sides of equal lengths (*Einthoven's triangle*) and that the recording connections are made at the vertices (corners) of that triangle. But in practice, the electrodes connected to each arm and to the left leg are considered to connect to the triangle vertices. The standard limb leads record the voltages generated in the extracellular fluids surrounding the heart by the ion flows occurring simultaneously in many cells between any two of the connections. A recording using lead I (RA-LA) which connects the right arm (RA) and the left arm (LA) is most sensitive to electrical activity spreading horizontally across the heart. Lead II (RA-LL) and lead III (LA-LL) record activity along the vertical axis (from the base of the heart to its apex), but from different orientations. The significance of Einthoven's triangle is that the sum of the voltages of leads I and III equals that in lead II (Einthoven's law). Hence, if the voltages of two of the standard leads are recorded, that of the third lead can be determined mathematically.

Preparing the Subject

1. Place electrode paste on four electrode plates and position each electrode as follows after scrubbing the skin at that attachment site with an alcohol swab. Attach an electrode to the anterior surface of each forearm, about 2 to 3 in. above the wrist, and secure them with rubber straps. In the same manner, attach an electrode to each leg, approximately 2 to 3 in. above the medial malleolus (inner aspect of the ankle).

2. Attach the appropriate tips of the patient cable to the electrodes. The cable leads are marked RA (right arm), LA (left arm), LL (left leg), and RL (right leg, the ground).

Recording the ECG

The ECG will be recorded first under baseline (resting) conditions and then under conditions of fairly strenuous activity. Finally, recordings will be made while the subject holds his or her breath. The activity and breath-holding recordings will be compared to the baseline recordings, and you will be asked to determine the reasons for the observed differences in the recordings.

BASELINE RECORDINGS

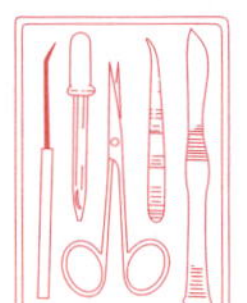

1. Position the subject comfortably in a supine position on a cot (if available), or sitting relaxed on a laboratory chair.

2. Turn on the power switch and adjust the sensitivity knob to 1. Set the paper speed to 25 mm/sec and the lead selector to the position corresponding to recording from lead I (RA-LA).

3. Set the control knob at the RUN position and record the subject's at-rest ECG from lead I for 2 to 3 minutes or until the recording stabilizes. (You will need a tracing long enough to provide each student with a representative segment.) The subject should try to relax and not move unnecessarily, because the skeletal muscle action potentials will also be picked up and recorded.

4. Stop the recording and mark it "lead I."

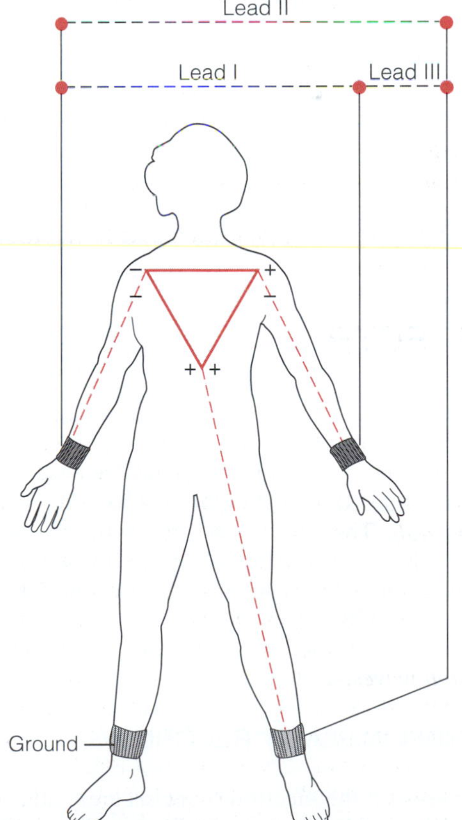

F31.3

ECG recording positions for the standard limb leads.

5. Repeat the recording procedure for leads II (RA-LL) and III (LA-LL).

6. Each student should take a representative segment of one of the lead recordings and label the record with the name of the subject and the lead used. Identify and label the P, QRS, and T waves. The calculations you perform for your recording should be based on the following information: Because the paper speed was 25 mm/sec, each millimeter of paper corresponds to a time interval of 0.04 sec. Thus, if an interval requires 4 mm of paper, its duration is 4 mm × 0.04 sec/mm = 0.16 sec.

7. Compute the heart rate. Measure the distance (mm) from the beginning of one QRS complex to the beginning of the next QRS complex, and plug this value into the equation below to find the time for one heartbeat.

_______ mm × 0.04 sec/mm = _______ sec/beat

Now find the beats per minute, or heart rate, by using the figure just computed for seconds per beat in the following equation:

$$\text{Beats/min} = \frac{1}{\_\_\_\_\_\_\ \text{sec/beat}} \times 60\ \text{sec/min}$$

Beats/min = _______

Is the obtained value within normal limits? _________

Measure the QRS interval and compute its duration.

____

Measure the Q-T interval and compute its duration.

____

Measure the P-R interval and compute its duration.

____

Are the computed values within normal limits?

____

8. At the bottom of this page, attach segments of the ECG recordings from leads I through III. Make sure you indicate the paper speed, lead, and subject's name on each tracing. To the recording on which you based your previous computations, add your calculations for the duration of the QRS, P-R intervals, and Q-T intervals above the respective area of tracing. Also record the heart rate on that tracing.

"RUNNING IN PLACE" RECORDING

1. Make sure the electrodes are securely attached to prevent electrode movement while recording the ECG.

2. Set the paper speed to 25 mm/sec, and prepare to make the recording using lead I.

3. Record the ECG while the subject is running in place for 3 min. Then have the subject sit down, but continue to record the ECG for an additional 4 minutes. *Mark the recording* at the end of the 3 minutes of running and at 1 minute after cessation of activity.

4. Stop the recording. Compute the beats/min during the third minute of running, at 1 minute after exercise, and at 4 minutes after exercise. Record below:

_______ beats/min while running in place

_______ beats/min at 1 minute after exercise

_______ beats/min at 4 minutes after exercise

5. Compare this recording with the previous recording from lead I. Which intervals are shorter in the "running" recording?

____

Does the subject's heart rate return to baseline levels by 4 minutes after exercise?

____

"BREATH-HOLDING" RECORDING

1. Position the subject comfortably in the sitting position.

2. Using lead I and a paper speed of 25 mm/sec, *begin* the recording. After approximately 10 seconds, instruct the subject to begin breath holding and mark the record to indicate the onset of the 1-min breath-holding interval.

3. Stop the recording after 1 minute and remind the subject to breathe. Compute the beats/minute during the 1-minute experimental (breath-holding) period.

Beats/min during breath holding _______________

4. Compare this recording with the lead I recording obtained under baseline conditions.

What differences are seen? _______________

____

Attempt to *explain* the physiologic reason for the differences you have seen. (Hint: a good place to start might be to check "hypoventilation" or the role of the *respiratory system* in acid-base balance of the blood.)

____

____

____

____

Anatomy of Blood Vessels

OBJECTIVES

1. To describe the tunics of blood vessel walls and state the function of each layer.
2. To correlate differences in artery, vein, and capillary structure with the functions these vessels perform.
3. To recognize a cross-sectional view of an artery and vein when provided with a microscopic view or appropriate diagram.
4. To list and/or identify the major arteries arising from the aorta, and to indicate the body region supplied by each.
5. To list and/or identify the major veins draining into the superior and inferior venae cavae, and to indicate the body regions drained.
6. To point out and/or discuss the unique features of special circulations (hepatic portal system, circle of Willis, pulmonary circulation, fetal circulation) in the body.

MATERIALS

Anatomical charts of human arteries and veins (or a three-dimensional model of the human circulatory system)

Anatomical charts of the following specialized circulations: pulmonary circulation, hepatic portal circulation, arterial supply and circle of Willis of the brain (or a brain model showing this circulation), fetal circulation

Compound microscope

Prepared microscope slides showing cross sections of an artery and vein

Human Cardiovascular System: The Blood Vessels videotape*

See Appendix D, Exercise 32 for links to A.D.A.M. Standard.

See Appendix E, Exercise 32 for links to *Anatomy and PhysioShow: The Videodisc.*

*Available to qualified adopters from Benjamin/Cummings.

The blood vessels constitute a closed transport system. As the heart contracts, blood is propelled into the large arteries leaving the heart. It moves into successively smaller arteries and then to the arterioles, which feed the capillary beds in the tissues. Capillary beds are drained by the venules, which in turn empty into veins that ultimately converge on the great veins entering the heart. Thus arteries, carrying blood away from the heart, and veins, which drain the tissues and return blood to the heart, function simply as conducting vessels or conduits. Only the tiny capillaries that connect the arterioles and venules and ramify throughout the tissues directly serve the needs of the body's cells. It is through the capillary walls that exchanges are made between tissue cells and blood.

Respiratory gases, nutrients, and wastes move along diffusion gradients. Thus, oxygen and nutrients diffuse from the blood to the tissue cells, and carbon dioxide and metabolic wastes move from the cells to the blood.

In this exercise you will examine the microscopic structure of blood vessels and will identify the major arteries and veins of the systemic circulation and other special circulations.

MICROSCOPIC STRUCTURE OF THE BLOOD VESSELS

Except for the microscopic capillaries, the walls of blood vessels are constructed of three coats, or *tunics* (Figure 32.1). The **tunica intima, or interna,** which lines the lumen of a vessel, is a single thin layer of *endothelium* (squamous cells underlain by a scant basal lamina) that is continuous with the endocardium of the heart. Its cells fit closely together, forming an extremely smooth blood vessel lining that helps decrease resistance to blood flow.

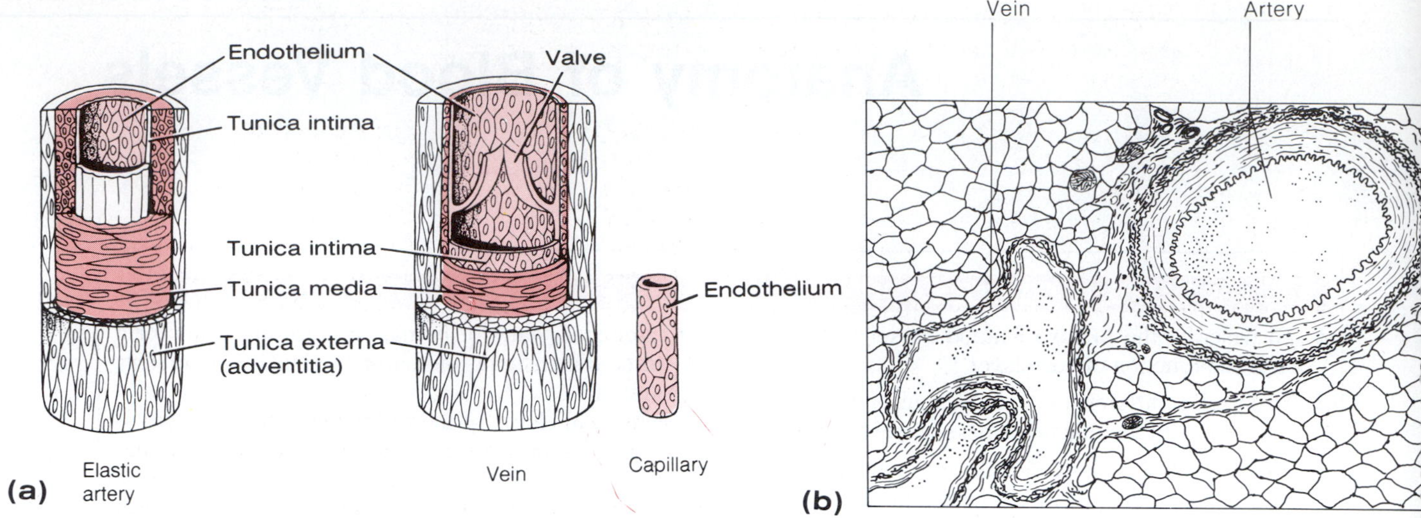

F32.1

Structure of arteries, veins, and capillaries. **(a)** Diagrammatic view. **(b)** Line drawing of a small artery (right) and vein (left), cross-sectional view, shown in Plate 26 in the Histology Atlas.

The **tunica media** is the more bulky middle coat and is composed primarily of smooth muscle and elastic tissue. The smooth muscle, under the control of the sympathetic nervous system, plays an active role in reducing or increasing the diameter of blood vessels, which in turn increases or decreases the peripheral resistance and blood pressure.

The **tunica externa,** or **adventitia,** the outermost tunic, is composed of areolar or fibrous connective tissue. Its function is basically supportive and protective.

In general, the walls of arteries are thicker than those of veins. The tunica media in particular tends to be much heavier and contains substantially more smooth muscle and elastic tissue. This anatomical difference reflects a functional difference in the two types of vessels. Arteries, which are closer to the pumping action of the heart, must be able to expand as an increased volume of blood is propelled into them and then recoil passively as the blood flows off into the circulation during diastole. Their walls must be sufficiently strong and resilient to withstand such pressure fluctuations. Since these larger arteries have such large amounts of elastic tissue in their media, they are often referred to as *elastic arteries.* Smaller arteries, further along in the circulatory pathway, are exposed to less extreme pressure fluctuations. They have less elastic tissue but still have substantial amounts of smooth muscle in their media. For this reason, they are called *muscular arteries.*

By contrast, veins, which are far removed from the heart in the circulatory pathway, are not subjected to such pressure fluctuations and are essentially low-pressure vessels. Thus, veins may be thinner-walled without jeopardy. However, the low-pressure condition itself and the fact that blood returning to the heart often flows against gravity require structural modifications to ensure that venous return equals cardiac output. Thus, the lumens of veins tend to be substantially larger than those of corresponding arteries, and valves in larger veins function to prevent backflow of blood in much the same manner as the semilunar valves of the heart. The skeletal muscle "pump" also promotes venous return; as the skeletal muscles surrounding the veins contract and relax, the blood is milked through the veins toward the heart. (Anyone who has been standing relatively still for an extended time will be happy to show you their swollen ankles, caused by blood pooling in their feet during the period of muscle inactivity!) Pressure changes that occur in the thorax during breathing also aid the return of blood to the heart.

- To demonstrate how efficiently venous valves prevent backflow of blood, perform the following simple experiment. Allow one hand to hang by your side until the blood vessels on the dorsal aspect become distended. Place two fingertips against one of the distended veins and, pressing firmly, move the superior finger proximally along the vein and then release this finger. The vein will remain flattened and collapsed despite gravity. Then remove the distal fingertip and observe the rapid filling of the vein.

The transparent walls of the tiny capillaries are only one cell layer thick, consisting of just the endothelium underlain by a basal lamina, that is, the tunica intima. Because of this exceptional thinness, exchanges are easily made between the blood and tissue cells.

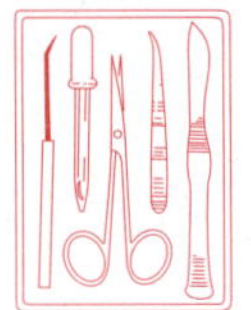

1. Obtain a slide showing a cross-sectional view of blood vessels and a microscope.

2. Using Figure 32.1b as a guide, scan the section to identify a thick-walled artery. Very often, but not always, its lumen will appear scalloped due to the constriction of its walls by the elastic tissue of the media.

3. Identify a vein. Its lumen may appear elongated or irregularly shaped and collapsed, and its walls will be considerably thinner. Notice the difference in the relative amount of elastic fibers in the media of the two vessels. Also, note the thinness of the intima layer, which is composed of flat squamous-type cells.

4. Make a drawing of your observations of the two vessel types below, and label the tunics. Try to indicate the proper size relationships relative to wall and tunic thicknesss.

Artery

Vein

MAJOR SYSTEMIC ARTERIES OF THE BODY

The **aorta** is the largest artery of the body. Extending upward as the *ascending aorta* from the left ventricle, it arches posteriorly and to the left (*aortic arch*) and then courses downward as the *descending aorta* through the thoracic cavity. It penetrates the diaphragm to enter the abdominal cavity just anterior to the vertebral column.

Figure 32.2 depicts the course of the aorta and its major branches. As you locate the arteries on the figure, be aware of ways in which you can make your memorization task easier. In many cases the name of the artery reflects the body region traversed (axillary, subclavian, brachial, popliteal), the organ served (renal, hepatic), or the bone followed (tibial, femoral, radial, ulnar). Once you have identified these arteries on the figure, attempt to locate and name them (without a reference) on a large anatomical chart or a three-dimensional model of the blood vessels.

- All arteries described here are shown in the figure, but some that are not described are also named and illustrated. Ask your instructor which arteries you are required to identify.

Ascending Aorta

The only branches of the ascending aorta are the **right** and the **left coronary arteries,** which supply the myocardium. The coronary arteries are described in Exercise 30 in conjunction with heart anatomy.

Aortic Arch

The **brachiocephalic** (literally, "arm-head") **artery** is the first branch of the aortic arch. It persists briefly before dividing into the right **common carotid artery** and the right **subclavian artery.** The common carotid divides to form the **internal carotid artery,** which serves the brain, and the **external carotid artery,** which supplies the extracranial tissues of the neck and head. The subclavian artery gives off three branches to the head and neck, the most important being the **vertebral artery,** which runs up the posterior neck to supply a portion of the brain. In the axillary region, the subclavian

(Text continues on p. 297)

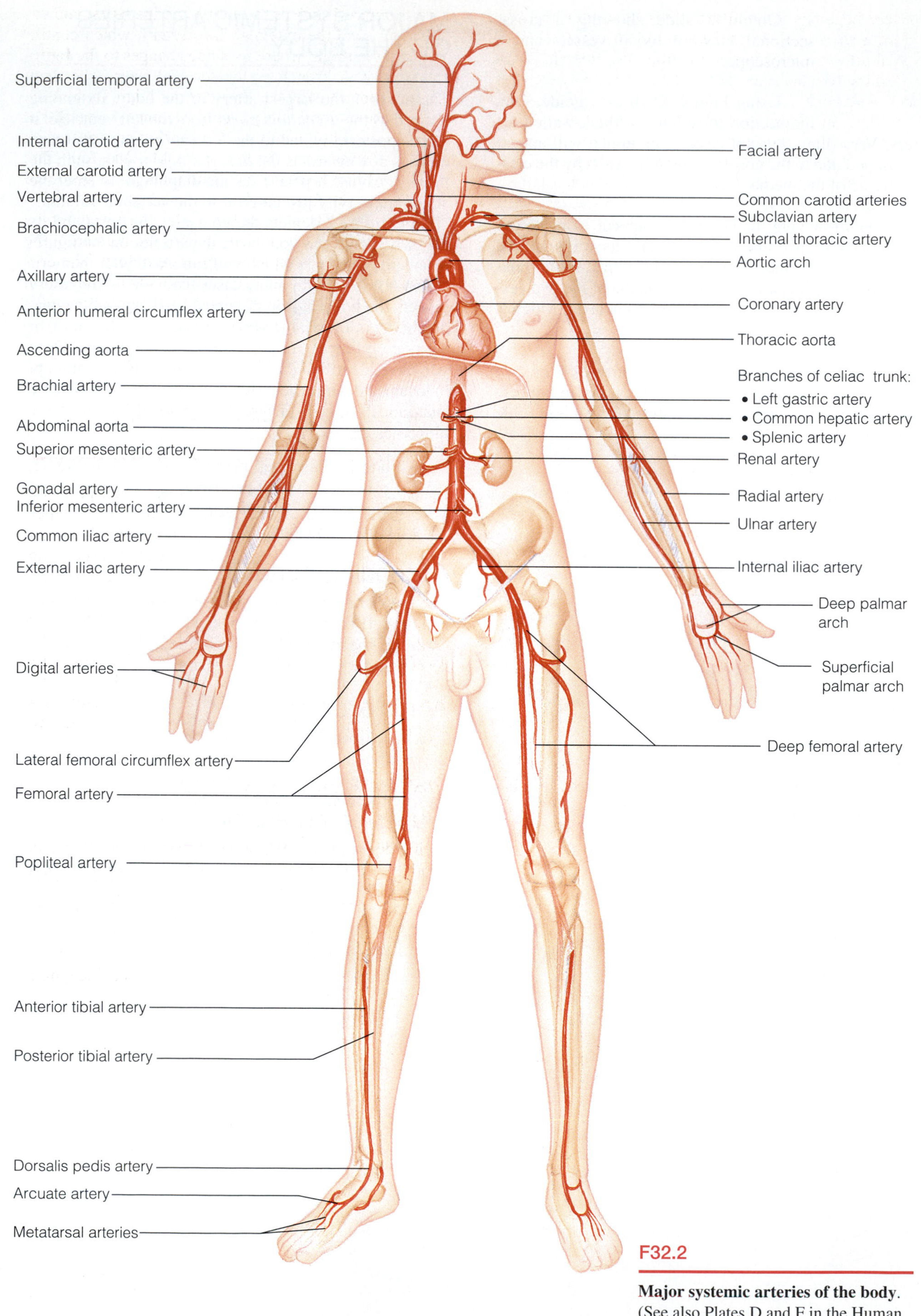

F32.2

Major systemic arteries of the body. (See also Plates D and F in the Human Anatomy Atlas.)

artery becomes the **axillary artery** and then the **brachial artery** as it enters the arm. At the elbow, the brachial artery divides into the **radial** and **ulnar arteries,** which follow the same-named bones to supply the forearm and hand.

The **left common carotid artery,** the second branch of the aortic arch, supplies the left side of the head and neck in the same manner the right common carotid serves the right side. The third branch is the **left subclavian artery,** which supplies the left upper extremity and subdivides as described for the right subclavian artery.

Descending Aorta

Not shown in Figure 32.2 are the 9 or 10 pairs of *intercostal arteries* that supply the muscles of the thoracic wall. The intercostal arteries are small branches of the descending aorta, as are the *phrenic arteries,* which supply the diaphragm. Other more major branches of the descending aorta supply the abdominal region. The **celiac trunk** is an unpaired artery that subdivides into three branches: the **left gastric** artery supplying the stomach, the **splenic artery** supplying the spleen, and the **common hepatic artery,** which provides the functional blood supply of the liver. The largest branch of the descending aorta, the **superior mesenteric artery,** supplies most of the small intestine and the first half of the large intestine. The small paired **suprarenal arteries** emerge at approximately the same level as the superior mesenteric artery and run laterally to supply the adrenal glands. (These are not shown in Figure 32.2.) The paired **renal arteries** supply the kidneys, and the **gonadal arteries,** arising from the ventral surface of the aorta slightly below the renal arteries, run inferiorly to serve the gonads. They are called **ovarian arteries** in the female and **testicular** (or **internal spermatic**) **arteries** in the male. Since these vessels must travel through the inguinal canal to supply the testes in the male, they are considerably longer in the male than in the female. The small unpaired artery supplying the second half of the large intestine is the **inferior mesenteric artery.**

In the pelvic region, the descending aorta divides into the two large **common iliac arteries.** Each of these vessels extends for about 2 inches into the pelvis before it divides into the **internal iliac artery,** which supplies the pelvic organs (bladder, rectum, and some reproductive structures, such as the uterus, uterine tubes, and vas deferens), and the **external iliac artery,** which continues into the thigh, where its name changes to the **femoral artery.** A branch of the femoral artery, the **deep femoral artery,** supplies the posterior thigh. In the knee region, the femoral artery briefly becomes the **popliteal artery;** its subdivisions—the **anterior** and **posterior tibial arteries**—supply the leg, ankle, and foot. The posterior tibial gives off one main branch, the **peroneal artery** (not shown), which serves the lateral calf (peroneal muscles). The anterior tibial artery terminates at the **dorsalis pedis** artery, which supplies the dorsum of the foot and continues on as the **arcuate artery.** The dorsalis pedis is often palpated in patients with circulation problems of the leg to determine the circulatory efficiency to the limb as a whole.

- Palpate your own dorsalis pedis artery.

MAJOR SYSTEMIC VEINS OF THE BODY

Arteries are generally located in deep, well-protected body areas. However, many veins follow a more superficial course and are often easily seen and palpated on the body surface (Figure 32.3). Most deep veins parallel the course of the major arteries, and in many cases the naming of the veins and arteries is identical except for the designation of the vessels as veins. Whereas the major systemic arteries branch off the aorta, the veins tend to converge on the venae cavae, which enter the right atrium of the heart. Veins draining the head and upper extremities empty into the **superior vena cava,** and those draining the lower body empty into the **inferior vena cava.**

Veins Draining into the Superior Vena Cava

Veins draining into the superior vena cava are named from the superior vena cava distally; *but remember that the flow of blood is in the opposite direction.*

The **right** and **left brachiocephalic veins** drain the head, neck, and upper extremities and unite to form the superior vena cava. (Note that although there is only one brachiocephalic artery, there are two brachiocephalic veins.)

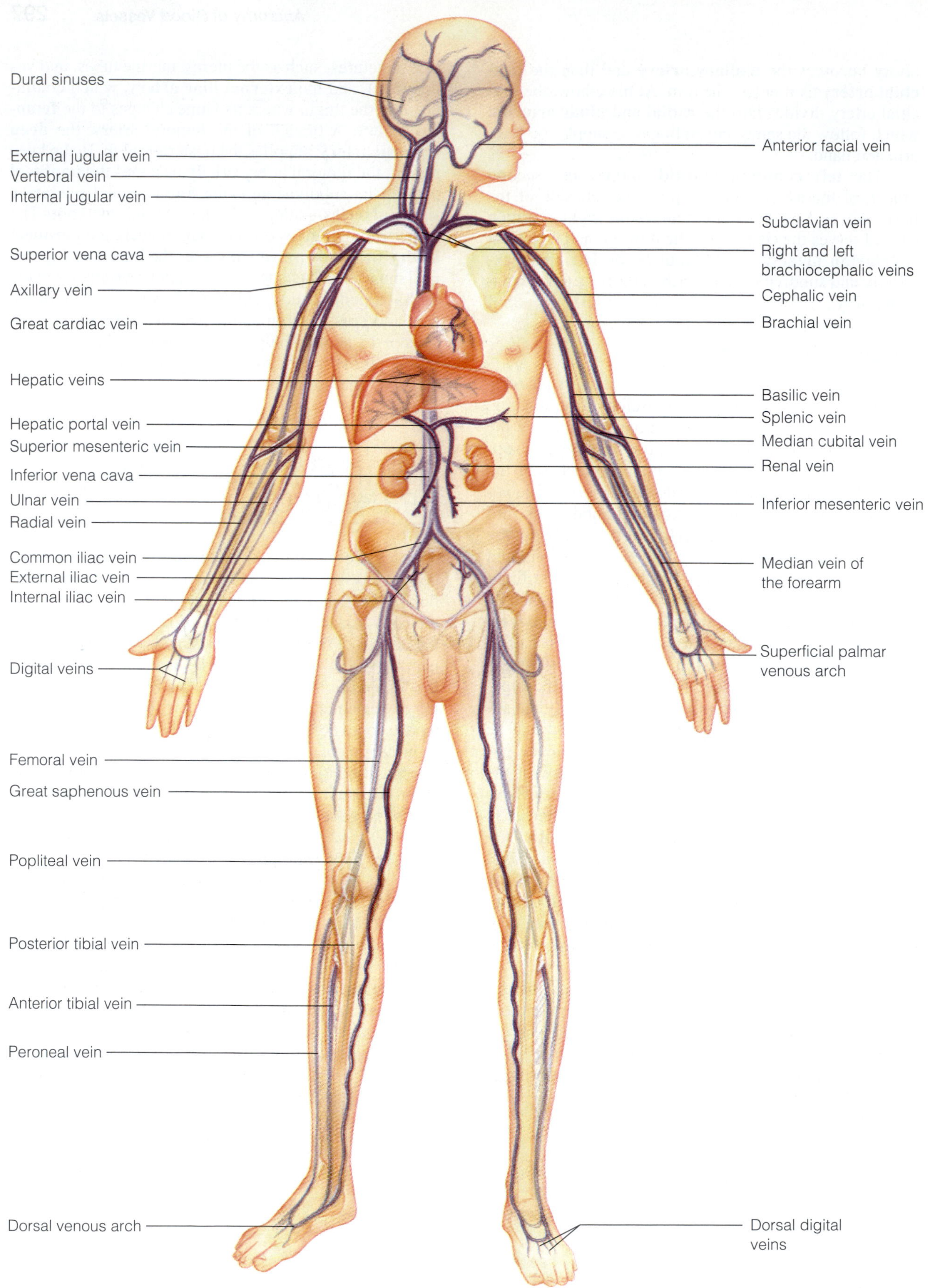

F32.3

Major systemic veins of the body. (See also Plate D in the Human Anatomy Atlas.)

Branches of the brachiocephalic veins include the **internal jugular veins,** large veins that drain the superior sagittal sinus and other dural sinuses of the brain; the **vertebral veins,** which drain the posterior aspect of the head; and the **subclavian veins,** which receive venous blood from the upper extremity. The **external jugular vein** joins the subclavian vein near its origin to return the venous drainage of the extracranial tissues of the head and neck. As the subclavian vein traverses the axilla, it becomes the **axillary vein** and then the **brachial vein** as it courses along the posterior aspect of the humerus. The brachial vein is formed by the union of the deep **radial** and **ulnar veins** of the forearm. The superficial venous drainage of the arm includes the **cephalic vein,** which courses along the lateral aspect of the arm and empties into the axillary vein; the **basilic vein,** found on the medial aspect of the arm and entering the brachial vein; and the **median cubital vein,** which runs between the cephalic and basilic veins in the anterior aspect of the elbow (this vein is often the site of choice for removing blood for testing purposes).

The **azygos system** (Figure 32.4) drains the intercostal muscles of the thorax and provides an accessory venous system to drain the abdominal wall. The **azygos vein,** which drains the right side of the thorax, enters the dorsal aspect of the superior vena cava immediately before that vessel enters the right atrium. Also part of the azygos system are the **hemiazygos** and **accessory hemiazygos veins,** which together drain the left side of the thorax and empty into the azygos vein.

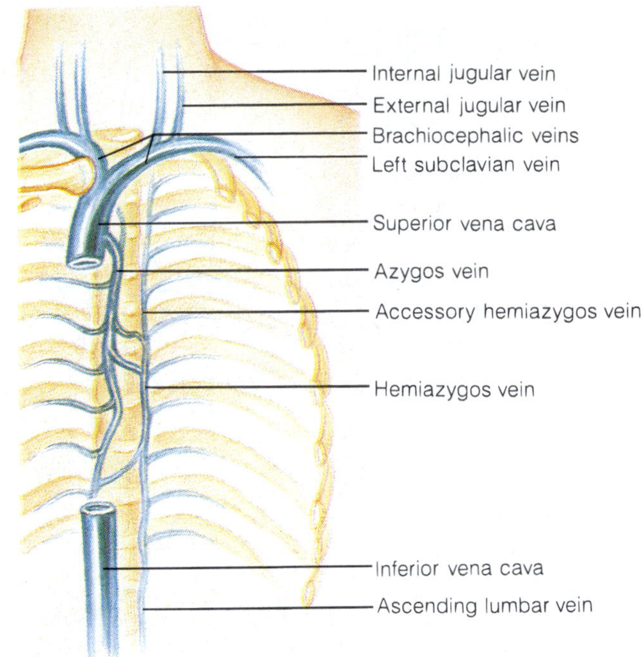

F32.4

The azygos system. (See also Plate D in the Human Anatomy Atlas.)

Veins Draining into the Inferior Vena Cava

The inferior vena cava, a much longer vessel than the superior vena cava, returns blood to the heart from all body regions below the diaphragm (see Figure 32.3). It begins in the lower abdominal region with the union of the paired **common iliac veins,** which drain venous blood from the legs and pelvis. Each common iliac vein in turn is formed by the union of the **internal iliac vein,** draining the pelvis, and the **external iliac vein,** which receives venous blood from the lower limb. Veins of the leg include the **anterior** and **posterior tibial veins,** which serve the calf and foot. The posterior tibial vein becomes the **popliteal vein** in the knee region and the **femoral vein** in the thigh. The femoral vein empties into the external iliac vein in the inguinal region. The **great saphenous vein,** a superficial vein, is the longest vein of the body. Beginning in the foot with the **dorsal venous arch,** it extends up the medial side of the leg, knee, and thigh to empty into the femoral vein.

Moving superiorly into the abdominal cavity, the inferior vena cava receives blood from the **right gonadal vein** (testicular or spermatic vein in the male; ovarian vein in the female), which drains the right gonad. (The left testicular or ovarian vein drains into the left renal vein.) The **right** and **left renal veins** drain the kidneys, and the liver is drained by the **right** and **left hepatic veins.** The unpaired veins draining the digestive tract organs empty into a special vessel, the **hepatic portal vein,** which carries this blood through the liver before it enters the systemic venous system. (The hepatic portal system is discussed separately on p. 304.)

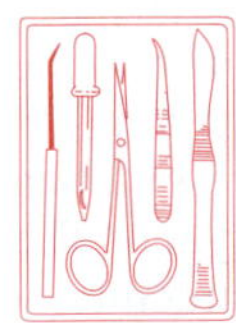

Identify the important arteries and veins on the large anatomical chart or model without referring to the figures.

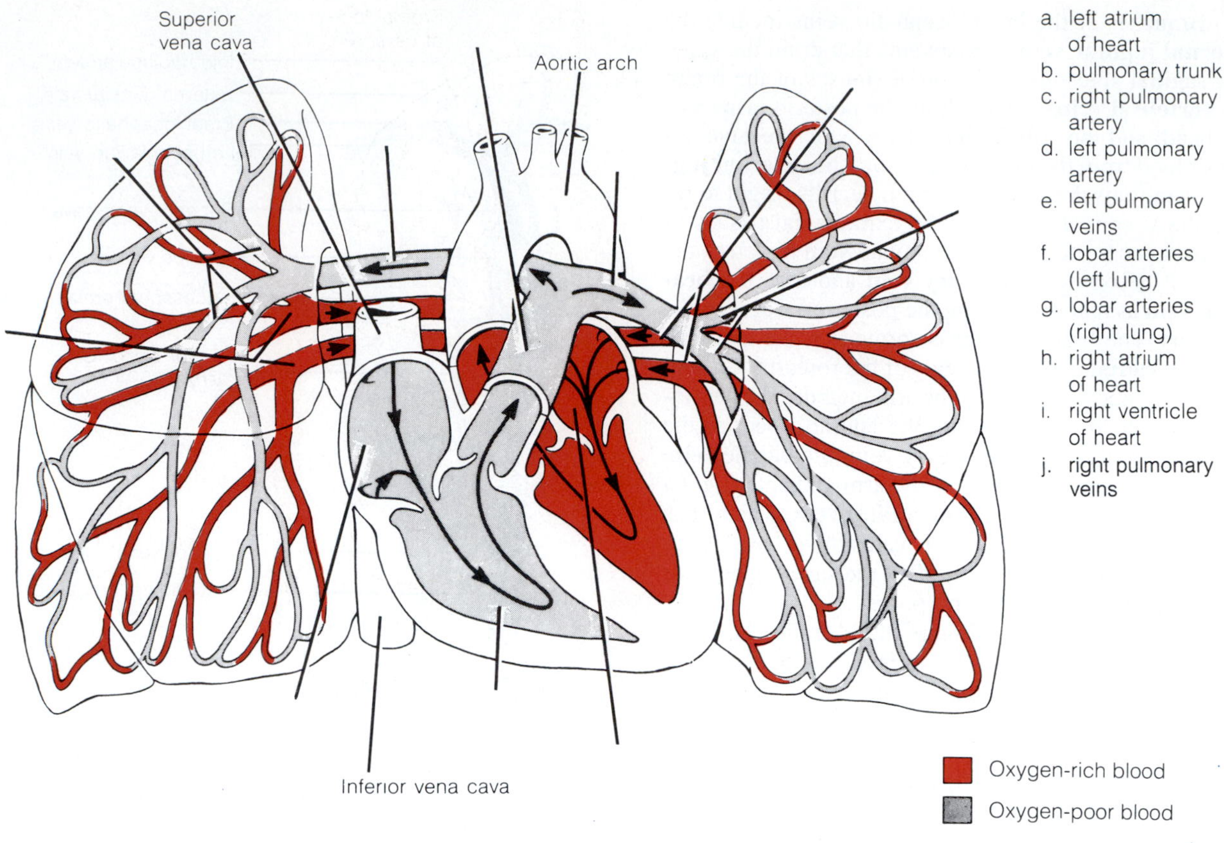

F32.5

The pulmonary circulation. (See also Plate D in the Human Anatomy Atlas.)

SPECIAL CIRCULATIONS

Pulmonary Circulation

Pulmonary circulation (discussed previously in relation to heart anatomy on p. 283) differs in many ways from Systemic circulation because it does not serve the metabolic needs of the body tissues with which it is associated (in this case, lung tissue). It functions instead to bring the blood into close contact with the alveoli of the lungs to permit gas exchanges that rid the blood of excess carbon dioxide and replenish its supply of vital oxygen. The arteries of the pulmonary circulation are structurally much like veins, and they create a low-pressure bed in the lungs. (If the arterial pressure in the systemic circulation is 120/80, the pressure in the pulmonary artery is likely to be approximately 25/10.) The functional blood supply of the lungs is provided by the **bronchial arteries,** which diverge from the thoracic portion of the descending aorta.

Pulmonary circulation begins with the large **pulmonary trunk,** which leaves the right ventricle and divides into the **right** and **left pulmonary arteries** about 2 inches above its origin (Figure 32.5). The right and left pulmonary arteries plunge into the lungs, where they subdivide into **lobar arteries** (three on the right and two on the left), which accompany the main bronchi into the lobes of the lungs. The lobar arteries branch extensively within the lungs to form arterioles, which finally terminate in the capillary networks surrounding the alveolar sacs of the lungs. Diffusion of the respiratory gases occurs across the walls of the alveoli and **pulmonary capillaries.** The pulmonary capillary beds are drained by venules, which converge to form sequentially larger veins and finally the four **pulmonary veins** (two leaving each lung), which return the blood to the left atrium of the heart.

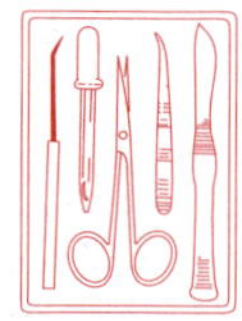

Using the terms provided in Figure 32.5, *label* all structures provided with leader lines.

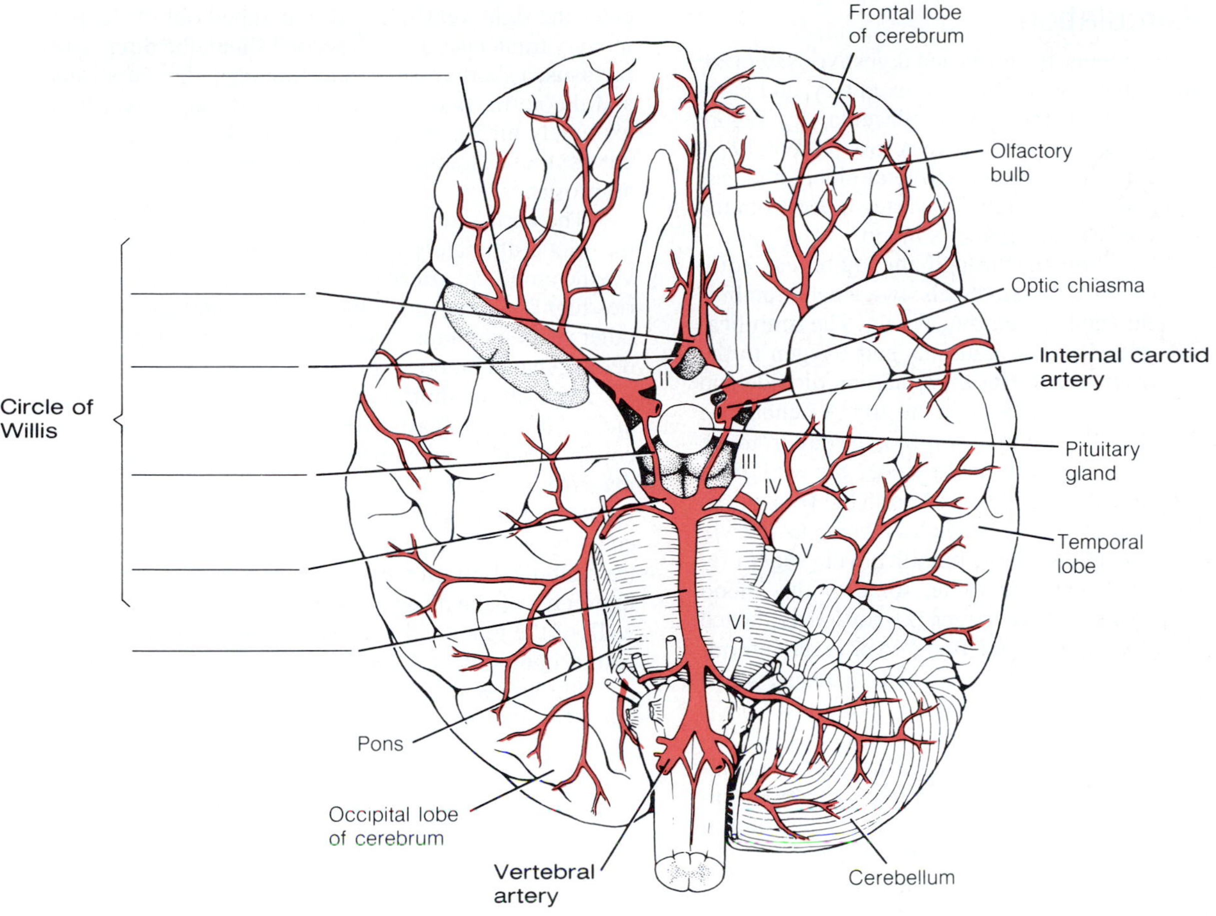

F32.6

Arterial supply of the brain. Cerebellum is not shown on the left side of the figure. (See also Plate B in the Human Anatomy Atlas.)

Arterial Supply of the Brain and the Circle of Willis

A continuous blood supply to the brain is crucial, since oxygen deprivation for even a few minutes causes irreparable damage to the delicate brain tissue. The brain is supplied by two pairs of arteries arising from the region of the aortic arch—the *internal carotid arteries* and the *vertebral arteries.* Figure 32.6 is a diagram of the brain's arterial supply.

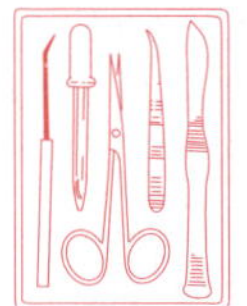

The internal carotid and vertebral arteries are labeled. As you read the description of the blood supply below, *complete the labeling of this diagram.*

The **internal carotid arteries,** branches of the common carotid arteries, follow a deep course through the neck and along the pharynx, entering the skull through the carotid canals of the temporal bone. Within the cranium, each divides into **anterior** and **middle cerebral arteries,** which supply the bulk of the cerebrum. The internal carotid arteries also contribute to the formation of the **circle of Willis,** an arterial anastomosis at the base of the brain surrounding the pituitary gland and the optic chiasma, by forming a **posterior communicating artery** on each side. The circle is completed by the **anterior communicating artery,** a short shunt connecting the right and left anterior cerebral arteries.

The paired **vertebral arteries** diverge from the subclavian arteries and pass superiorly through the foramina of the transverse process of the cervical vertebrae to enter the skull through the foramen magnum. Within the skull, the vertebral arteries unite to form a single **basilar artery,** which continues superiorly along the ventral aspect of the brain stem, giving off branches to the pons, cerebellum, and inner ear. At the base of the cerebrum, the basilar artery divides to form the **posterior cerebral arteries.** These supply portions of the temporal and occipital lobes of the cerebrum and also become part of the circle of Willis by joining with the posterior communicating arteries.

The uniting of the blood supply of the internal carotid arteries and the vertebral arteries via the circle of Willis is a protective device that theoretically provides an alternate set of pathways for blood to reach the brain tissue in the case of arterial occlusion or impaired blood flow anywhere in the system. In actuality, the communicating arteries are tiny, and in many cases the communicating system is defective.

Fetal Circulation

In a developing fetus, the lungs and digestive system are not yet functional, and all nutrient, excretory, and gaseous exchanges occur through the placenta (see Figure 32.7). Nutrients and oxygen move across placental barriers from the mother's blood into fetal blood, and carbon dioxide and other metabolic wastes move from the fetal blood supply to the mother's blood.

Fetal blood travels through the umbilical cord, which contains three blood vessels: two smaller umbilical arteries and one large umbilical vein. The **umbilical vein** carries blood rich in nutrients and oxygen to the fetus; the **umbilical arteries** carry carbon dioxide and waste-laden blood from the fetus to the placenta. The umbilical arteries, which transport blood away from the fetal heart, meet the umbilical vein at the *umbilicus* (navel, or belly button) and wrap around the vein within the cord en route to their placental attachments. Newly oxygenated blood flows in the umbilical vein superiorly toward the fetal heart. En route, some of this blood perfuses the liver, but the larger proportion is ducted through the relatively nonfunctional liver to the inferior vena cava via a vessel called the **ductus venosus,** which carries the blood to the right atrium of the heart.

Because fetal lungs are nonfunctional and collapsed, two shunting mechanisms ensure that blood almost entirely bypasses the lungs. Much of the blood entering the right atrium is shunted into the left atrium through the **foramen ovale,** a flaplike opening in the interatrial septum. The left ventricle then pumps the blood out the aorta to the systemic circulation. Blood that does enter the right ventricle and is pumped out of the pulmonary trunk encounters a second shunt, the **ductus arteriosus,** a short vessel connecting the pulmonary trunk and the aorta. Because the collapsed lungs present an extremely high-resistance pathway, blood more readily enters the systemic circulation through the ductus arteriosus.

The aorta carries blood to the tissues of the body; this blood ultimately finds its way back to the placenta via the umbilical arteries. The only fetal vessel that carries highly oxygenated blood is the umbilical vein. All other vessels contain varying degrees of oxygenated and deoxygenated blood.

At birth, or shortly after, the foramen ovale closes and becomes the **fossa ovalis,** and the ductus arteriosus collapses and is converted to the fibrous **ligamentum arteriosum.** Lack of blood flow through the umbilical vessels leads to their eventual obliteration, and the circulatory pattern becomes that of the adult. Remnants of the umbilical arteries persist as the **medial umbilical ligaments** on the inner surface of the anterior abdominal wall, of the umbilical vein as the **ligamentum teres** or **round ligament** of the liver, and of the ductus venosus as a fibrous band called the **ligamentum venosum** on the inferior surface of the liver.

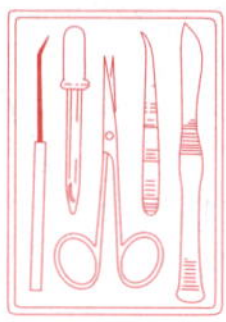

The pathway of fetal blood flow is indicated with arrows on Figure 32.7. Appropriately *label* all specialized fetal circulatory structures provided with leader lines.

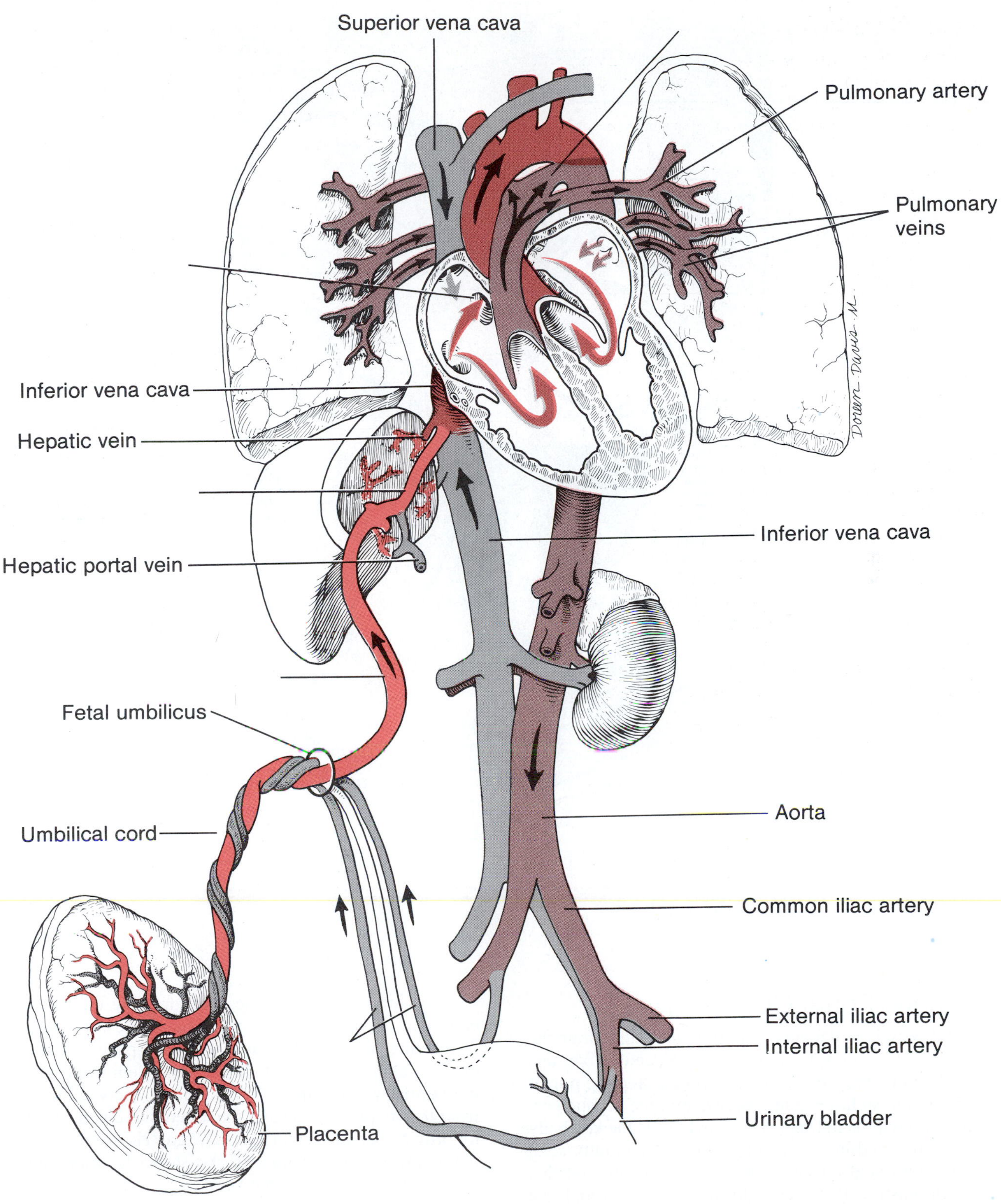

F32.7

The fetal circulation.

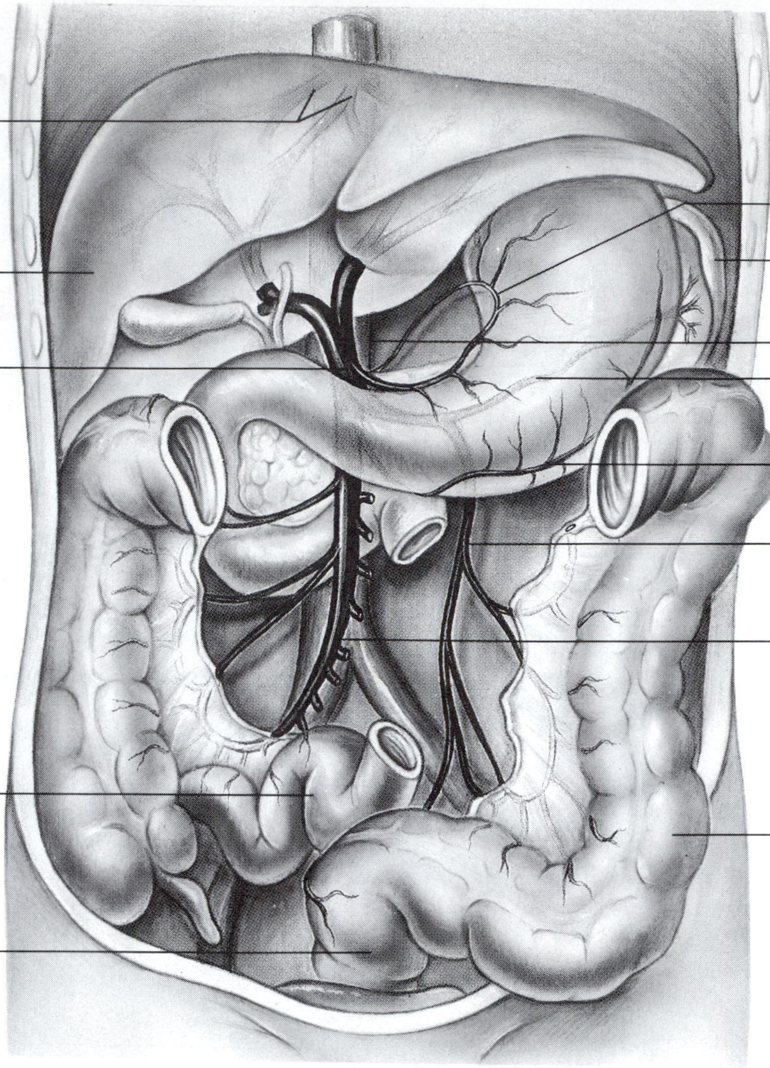

F32.8

Hepatic portal circulation of the human.

Hepatic Portal Circulation

Blood vessels of the hepatic portal circulation drain the digestive viscera, spleen, and pancreas and deliver this blood to the liver for processing via the **hepatic portal vein.** If a meal has recently been eaten, the hepatic portal blood will be nutrient rich. The liver is the key body organ involved in maintaining proper sugar, fatty acid, and amino acid concentrations in the blood, and this system ensures that these substances pass through the liver before entering the systemic circulation. As blood percolates through the liver sinusoids, some of the nutrients are removed to be stored or processed in various ways for release to the general circulation. At the same time, the hepatocytes are detoxifying alcohol and other possibly harmful chemicals present in the blood, and the liver's macrophages are removing bacteria and other debris from the passing blood. The liver in turn is drained by the hepatic veins that enter the inferior vena cava.

The **inferior mesenteric vein,** draining the transverse and terminal portions of the large intestine, joins the **splenic vein,** which drains the spleen, pancreas, and greater curvature of the stomach. The splenic vein and the **superior mesenteric vein,** which receives blood from the small intestine and the ascending colon, unite to form the hepatic portal vein. The **gastric vein,** which drains the lesser curvature of the stomach, drains directly into the hepatic portal vein.

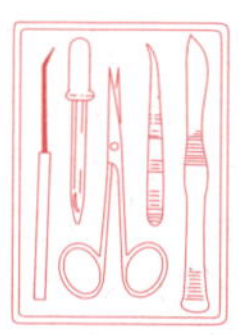

Locate the vessels named above on Figure 32.8.

Human Cardiovascular Physiology—Blood Pressure and Pulse Determinations

OBJECTIVES

1. To define *systole, diastole,* and *cardiac cycle.*
2. To indicate the normal length of the cardiac cycle, the relative pressure changes occurring within the atria and ventricles during the cycle, and the timing of valve closure.
3. To use the stethoscope to auscultate heart sounds and to relate heart sounds to cardiac cycle events.
4. To describe the clinical significance of heart sounds and heart murmurs.
5. To demonstrate the thoracic locations where the first and second heart sounds are most accurately auscultated.
6. To define *pulse, pulse deficit, blood pressure,* and *sounds of Korotkoff.*
7. To accurately determine a subject's apical and radial pulse.
8. To accurately determine a subject's blood pressure with a sphygmomanometer, and to relate systolic and diastolic pressures to events of the cardiac cycle.
9. To investigate the effects of exercise on blood pressure, pulse, and cardiovascular fitness.
10. To note factors affecting and/or determining blood flow and skin color.

MATERIALS

Stethoscope
Sphygmomanometer
Watch (or clock) with a second hand
Step stools (16 in. and 20 in. in height)
Cot (if available)
Alcohol swabs
Millimeter ruler
Ice
Small basin suitable for the immersion of one hand
Laboratory thermometer (°C)
Record or audiotape: "Interpreting Heart Sounds" (available on free loan from the local chapters of the American Heart Association)
Phonograph or tape deck
Felt marker

See Appendix D, Exercise 33 for links to A.D.A.M. Standard.

See Appendix E, Exercise 33 for links to *Anatomy and PhysioShow: The Videodisc.*

Any comprehensive study of human cardiovascular physiology takes much more time than a single laboratory period. However, it is possible to conduct investigations of a few phenomena such as pulse, heart sounds, and blood pressure, all of which reflect the heart in action and the function of blood vessels. (The electrocardiogram is studied separately in Exercise 31.) A discussion of the cardiac cycle will provide a basis for understanding and interpreting the various physiologic measurements taken.

CARDIAC CYCLE

In a healthy heart, the two atria contract simultaneously. As they begin to relax, simultaneous contraction of the ventricles occurs. According to general usage, the terms **systole** and **diastole** refer to events of ventricular contraction and relaxation, respectively. The **cardiac cycle** is equivalent to one complete heartbeat—during which both atria and ventricles contract and then relax. It is marked by a succession of changes in blood volume and pressure within the heart. Figure 33.1 is a graphic representation of the events of the cardiac cycle for the left side of the heart. Although pressure changes in the right side are lower than those in the left, the same relationships apply.

We will begin the discussion of the cardiac cycle with the heart in complete relaxation (diastole). At this point, pressure in the heart is very low, blood is flowing passively from the pulmonary and systemic circulations into the atria and on through to the ventricles, the semilunar valves are closed, and the AV valves are open. Shortly, atrial contraction occurs and atrial pressure increases, forcing residual blood into the ventricles. Then ventricular systole begins and intraventricular pressure increases rapidly, closing the AV valves. When ventricular pressure exceeds that of the large arteries leaving the heart, the semilunar valves are forced open; and the blood in the ventricular chambers is expelled through the valves. During this phase, the aortic pressure reaches approximately 120 mm Hg. During ventricular systole, the atria relax and their chambers fill with blood, which results in a gradually increasing atrial pressure. At the end of ventricular systole, the ventricles relax; the semilunar valves snap shut, preventing back-

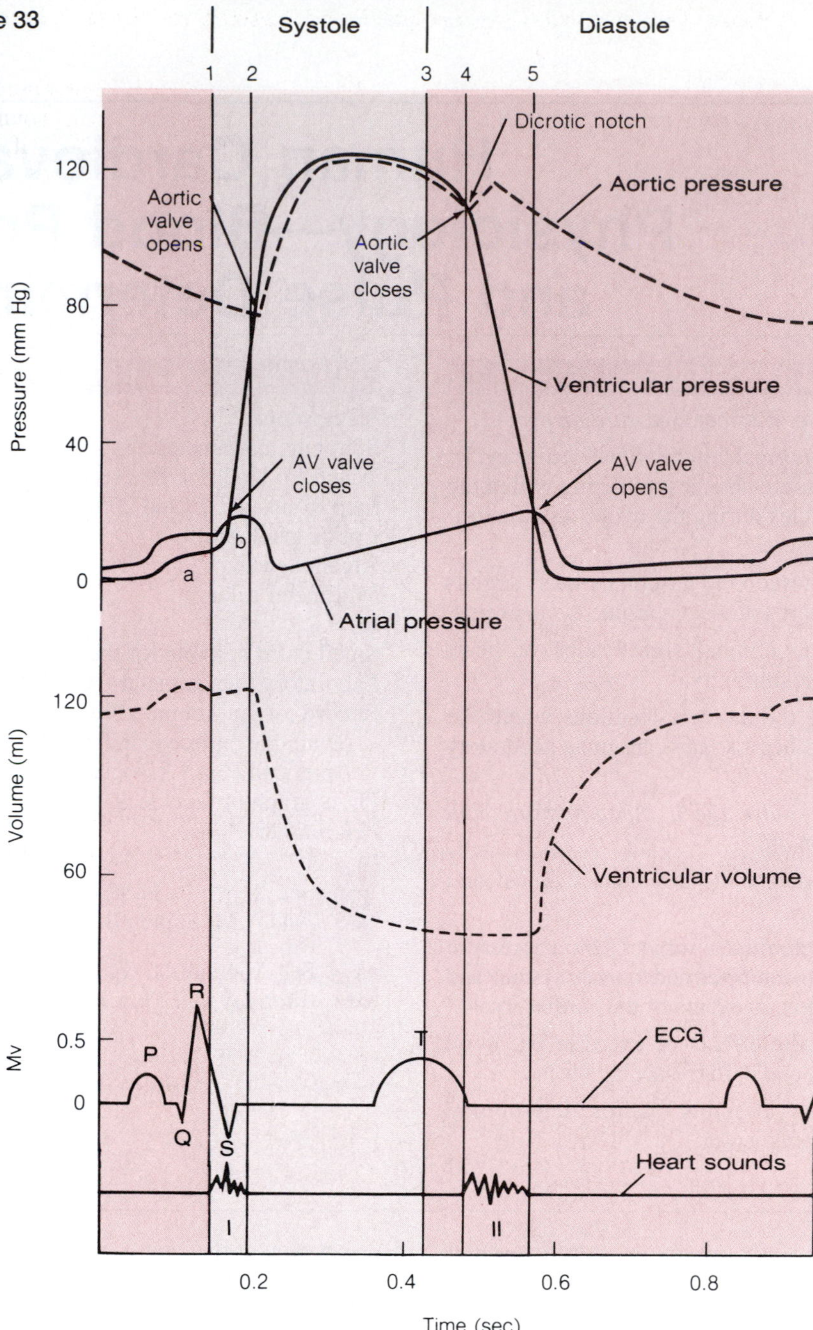

F33.1

Graphic representation of pressure and volume changes in the left ventricle during one cardiac cycle. Depicts valvular events and correlates changes to heart sounds and ECG. (ECG is considered in Exercise 31.)

flow, and momentarily, the ventricles are closed chambers. When the aortic semilunar valve snaps shut, a momentary increase in the aortic pressure results from the elastic recoil of the aorta after valve closure. This event results in the pressure fluctuation called the *dicrotic notch* (see Figure 33.1). As the ventricles relax, the pressure within them begins to drop. When intraventricular pressure is again less than atrial pressure, the AV valves are forced open, and the ventricles again begin to fill with blood. Atrial and aortic pressures decrease, and the ventricles rapidly refill, completing the cycle.

The average heart beats approximately 72 beats per minute, and so the length of the cardiac cycle is about 0.8 second. Of this time period, atrial contraction occupies the first 0.1 second, which is followed by atrial relaxation and ventricular contraction for the next 0.3 second. The remaining 0.4 second is the quiescent, or ventricular relaxation, period. When the heart beats at a more rapid pace than normal, this last period decreases.

Notice that two different types of phenomena control the movement of blood through the heart: the alternate contraction and relaxation of the myocardium, and the opening and closing of valves (which is entirely dependent on the pressure changes within the heart chambers).

Study Figure 33.1 carefully to make sure you understand what has been discussed before continuing with the next portion of the exercise.

AUSCULTATION OF HEART SOUNDS

Two distinct sounds can be heard during each cardiac cycle. These heart sounds are commonly described by the monosyllables "lub" and "dup"; and the sequence is designated lub-dup, pause, lub-dup, pause, and so on. The first heart sound (lub) is associated with closure of the AV valves at the beginning of systole. The second heart sound (dup) occurs as the semilunar valves close and corresponds with the end of systole. Figure 33.1 indicates the timing of heart sounds in the cardiac cycle.

Abnormal heart sounds are called **murmurs** and often indicate valvular problems. In valves that do not close tightly, closure is followed by a swishing sound due to the backflow of blood (regurgitation). Distinct sounds, often described as high-pitch "screeching," are associated with the tortuous flow of blood through constricted, or stenosed, valves. ■

Before auscultating your partner's heart sounds, listen to the recording "Interpreting Heart Sounds" so that you may hear both normal and abnormal heart sounds.

In the following procedure, you will auscultate heart sounds with an ordinary stethoscope. A number of more sophisticated heart-sound amplification systems are on the market, and your instructor may prefer to use one such if it is available. If so, directions for the use of this apparatus will be provided by the instructor.

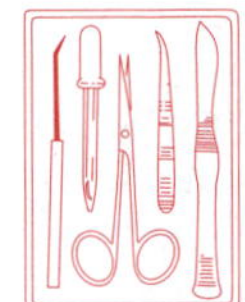

1. Obtain a stethoscope and some alcohol swabs. Heart sounds are best auscultated (listened to) if the subject's outer clothing is removed, so a male subject is preferable.

2. With an alcohol swab, clean the earpieces of the stethoscope. Allow the alcohol to dry. Notice that the earpieces are angled. For comfort and best auscultation, the earpieces should be angled in a forward direction when placed into the ears.

3. Don the stethoscope. Place the diaphragm of the stethoscope on your partner's thorax, just to the sternal side of the left nipple at the fifth intercostal space, and listen carefully for heart sounds. The first sound will be a longer, louder (more booming) sound than the second, which is short and sharp. After listening for a couple of minutes, try to time the pause between the second sound of one heartbeat and the first sound of the subsequent heartbeat.

How long is this interval? ______________________ sec

How does it compare to the interval between the first and second sounds of a single heartbeat?

____

____

4. To differentiate individual valve sounds somewhat more precisely, auscultate the heart sounds over specific thoracic regions. Refer to Figure 33.2 for the positioning of the stethoscope.

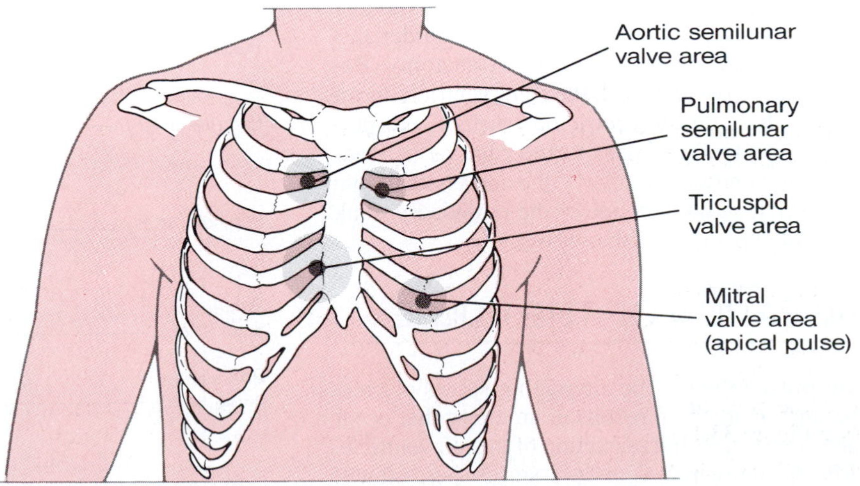

F33.2

Areas of the thorax where valvular sounds can best be detected.

Auscultation of AV Valves

As a rule, the mitral valve closes slightly before the tricuspid valve. You can hear the mitral valve more clearly if you place the stethoscope over the apex of the heart, which is at the fifth intercostal space, approximately in line with the middle region of the left clavicle. Listen to the heart sounds at this region; then move the stethoscope medially to the right margin of the sternum to auscultate the tricuspid valve. Can you detect the slight lag between the closure of the mitral and tricuspid valves?

___

There are normal variations in the site for "best" auscultation of the tricuspid valve. These range from the site depicted in Figure 33.2 (right sternal margin over fifth intercostal space) to over the sternal body in the same plane, to the left sternal margin over the fifth intercostal space. If you have difficulty hearing closure of the tricuspid valve, try one of these other locations.

Auscultation of Semilunar Valves

Again there is a slight dysynchrony of valve closure; the aortic semilunar valve normally snaps shut just ahead of the pulmonary semilunar valve. If the subject inhales deeply but gently, filling of the right ventricle will be delayed slightly (due to the compression of the thoracic blood vessels by the increased intrapulmonary pressure); and the two sounds can be heard more distinctly. Position the stethoscope over the second intercostal space, just to the *right* of the sternum. The aortic valve is best heard at this position. As you listen, have your partner take a deep breath. Then move the stethoscope to the *left* side of the sternum in the same line; and auscultate the pulmonary valve. Listen carefully; try to hear the "split" between the closure of these two valves in the second heart sound.

Although at first it may seem a bit odd that the pulmonary valve issuing from the *right* heart is heard most clearly to the *left* of the sternum and the aortic valve of the left heart is best heard at the right sternal border, this is easily explained by reviewing heart anatomy. Because the heart is twisted, with the right ventricle forming most of the anterior ventricular surface, the pulmonary trunk actually crosses to the right as it issues from the right ventricle. Similarly, the aorta issues from the left ventricle at the left side of the pulmonary trunk before arching up and over that vessel.

PALPATION OF THE PULSE

The term **pulse** refers to the alternating surges of pressure (expansion and then recoil) in an artery that occur with each contraction and relaxation of the left ventricle. Normally the pulse rate (pressure surges per minute) equals the heart rate (beats per minute), and the pulse averages 70 to 76 beats per minute in the resting state.

Parameters other than pulse rate are also useful clinically. You may also assess the regularity (or rhythmicity) of the pulse, and its amplitude and/or tension—does the blood vessel expand and recoil (sometimes visibly) with the pressure waves? Can you feel it strongly, or is it difficult to detect? Is it regular like the ticking of a clock, or does it seem to skip beats?

Superficial Pulse Points

The pulse may be felt easily on any superficial artery when the artery is compressed over a bone or firm tissue. Palpate the following pulse or pressure points on your partner by placing the fingertips of the first two or three fingers of one hand over the artery. It helps to compress the artery firmly as you begin your palpation and then immediately ease up on the pressure slightly. In each case, notice the regularity of the pulse, and assess the degree of tension or amplitude. Figure 33.3 illustrates the superficial pulse points to be palpated.

Common carotid artery: at the side of the neck

Temporal artery: anterior to the ear, in the temple region

Facial artery: clench the teeth, and palpate the pulse just anterior to the masseter muscle on the mandible (in line with the corner of the mouth)

Brachial artery: in the antecubital fossa, at the point where it bifurcates into the radial and ulnar arteries

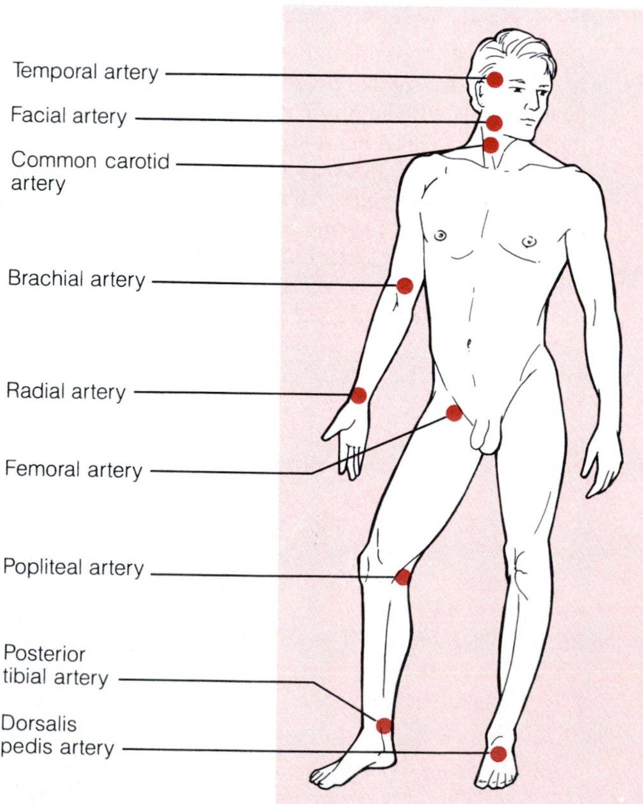

F33.3

Body sites where the pulse is most easily palpated.

Radial artery: at the lateral aspect of the wrist, above the thumb
Femoral artery: in the groin
Popliteal artery: at the back of the knee
Posterior tibial artery: just above the medial malleolus
Dorsalis pedis artery: on the dorsum of the foot

Which pulse point had the greatest amplitude?

____

Which the least? ___________________________

Can you offer any explanation for this? ____________

____

____

____

Because of its easy accessibility, the pulse is most often taken on the radial artery. With your partner sitting quietly, practice counting the radial pulse for 1 minute. Make three counts and average the results.

count 1 ____________ count 2 ____________

count 3 ____________ average ____________

Apical-Radial Pulse

The correlation between the apical and radial pulse rates can be determined by simultaneously counting them. The **apical pulse** (actually the counting of heartbeats) may be slightly faster than the radial because of a slight lag in time as the blood rushes from the heart into the large arteries where it can be palpated. However, any *large* difference between the values observed, referred to as a **pulse deficit,** may indicate cardiac impairment (a weakened heart that is unable to pump blood into the arterial tree to a normal extent—low cardiac output) or abnormal heart rhythms. In the case of atrial fibrillation or ectopic heartbeats, for instance, the second beat may follow the first so quickly that no second pulse is felt even though the apical pulse can still be heard (auscultated). Apical pulse counts are routinely ordered for those with cardiac decompensation.

With the subject sitting quietly, one student, using a stethoscope, should determine the apical pulse rate while another simultaneously counts the radial pulse rate. The stethoscope should be positioned over the fifth left intercostal space. The person taking the radial pulse should determine the starting point for the count and give the stop-count signal exactly 1 minute later. Record your values below.

apical count ______________________ beats/min

radial count ______________________ pulses/min

pulse deficit ______________________ /min

BLOOD PRESSURE DETERMINATIONS

Blood pressure is defined as the pressure the blood exerts against any unit area of the blood vessel walls, and it is generally measured in the arteries. Because the heart alternately contracts and relaxes, the resulting rhythmic flow of blood into the arteries causes the blood pressure to rise and fall during each beat. Thus you must take two blood pressure readings: the **systolic pressure,** which is the pressure in the arteries at the peak of ventricular ejection, and the **diastolic pressure,** which reflects the pressure during ventricular relaxation. Blood pressures are reported in millimeters of mercury (mm Hg), with the systolic pressure appearing first; 120/80 translates to 120 over 80, or a systolic pressure of 120 mm Hg and a diastolic pressure of 80 mm Hg. Normal blood pressure varies considerably from one person to another.

In this procedure, you will measure arterial and venous pressures by indirect means and under various conditions. You will investigate and demonstrate factors affecting blood pressure, the rapidity of blood pressure changes, and the large differences between arterial and venous pressures.

Indirect Measurement of Arterial Blood Pressure

The **sphygmomanometer,** commonly called a *blood pressure cuff,* is an instrument used to obtain blood pressure readings by the auscultatory method (Figure 33.4). It consists of an inflatable cuff with an attached pressure gauge. The cuff is placed around the arm and inflated to a pressure higher than systolic pressure to occlude circulation to the forearm. As cuff pressure is gradually released, the examiner listens with a stethoscope for characteristic sounds called the **sounds of Korotkoff,** which indicate the resumption of blood flow into the forearm. The pressure at which the first soft tapping sounds can be detected is recorded as the systolic pressure. As the pressure is reduced further, blood flow becomes more turbulent, and the sounds become louder. As the pressure is reduced still further, below the diastolic pressure, the artery is no longer compressed; and blood flows freely and without turbulence. At this point, the sounds of Korotkoff can no longer be detected. The pressure at which the sounds disappear is recorded as the diastolic pressure.

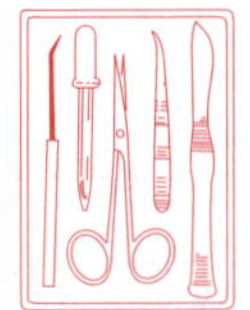

1. Work in pairs to obtain radial artery blood pressure readings. Obtain a stethoscope, alcohol swabs, and a sphygmomanometer. Clean the earpieces of the stethoscope with the alcohol swabs, and check the cuff for the presence of trapped air by compressing it against the laboratory table. (A partially inflated cuff will cause erroneous measurements.)

2. The subject should sit in a comfortable position with one arm resting on the laboratory table (approxi-

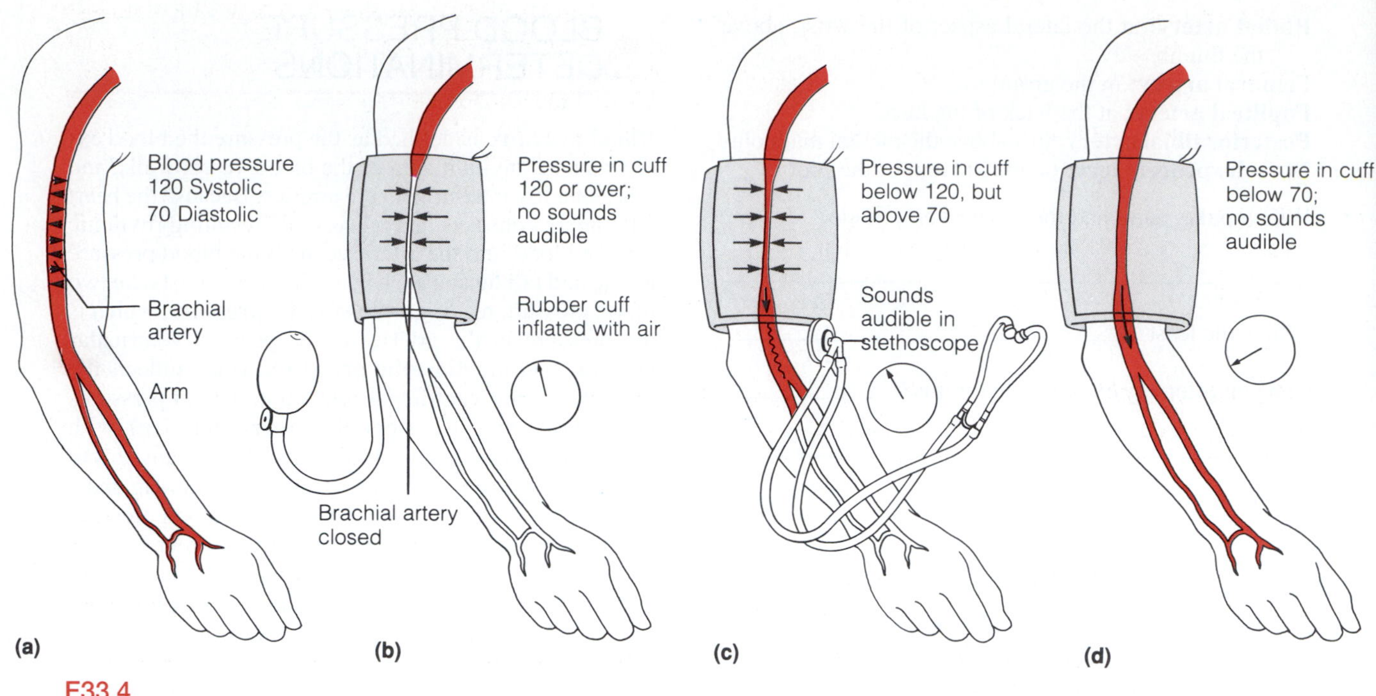

F33.4

Procedure for measurement of blood pressure. (a) The course of the brachial artery of the arm. Assume a blood pressure of 120/70. **(b)** The blood pressure cuff is wrapped snugly around the arm just above the elbow and inflated until blood flow into the forearm is stopped and a brachial pulse cannot be felt or heard. **(c)** The pressure in the cuff is gradually reduced while the examiner listens (auscultates) carefully for sounds in the brachial artery with a stethoscope. The pressure read as the first soft tapping sounds are heard (the first point at which a small amount of blood is spurting through the constricted artery) is recorded as the systolic pressure. **(d)** As the pressure is reduced still further, the sounds become louder and more distinct, but when the artery is no longer restricted and blood flows freely, the sounds can no longer be heard. The pressure at which the heart sounds disappear is routinely recorded as the diastolic pressure.

mately at heart level if possible). Wrap the cuff around the subject's arm, just above the elbow, with the inflatable area on the medial arm surface. The cuff may be marked with an arrow; if so, the arrow should be positioned over the brachial artery (Figure 33.4). Secure the cuff by tucking the distal end under the wrapped portion or by bringing the Velcro areas together.

3. Palpate the brachial pulse, and lightly mark its position with a felt pen. Don the stethoscope, and place its diaphragm over the pulse point.

⚠ The cuff should not be kept inflated for more than 1 minute. If you have any trouble obtaining a reading within this time, deflate the cuff, wait 1 or 2 minutes, and try again. (A prolonged interference with BP homeostasis can lead to fainting.)

4. Inflate the cuff to approximately 160 mm Hg pressure, and slowly release the pressure valve. Watch the pressure gauge as you listen carefully for the first soft thudding sounds of the blood spurting through the partially occluded artery. Mentally note this pressure (systolic pressure), and continue to release the cuff pressure. You will notice first an increase, then a muffling, of the sound. Note, as the diastolic pressure, the pressure at which the sound becomes muffled or disappears. Controversy exists over which of the two points should be recorded as the diastolic pressure; so in some cases you may see readings such as 120/80/78, which indicates the systolic pressure followed by the *first* and *second diastolic end points.* The first diastolic end point is the pressure at which the sound muffles; the second is the pressure at which the sound disappears. It makes little difference here which of the two diastolic pressures is recorded, but be consistent. Make two blood pressure determinations, and record your results below.

First trial:

systolic pressure ______ diastolic pressure ______

Second trial:

systolic pressure ______ diastolic pressure ______

5. Compute the **pulse pressure** for each trial. The pulse pressure is the difference between the systolic and diastolic pressures, and indicates the amount of blood forced from the heart during systole, or the actual "working" pressure.

Pulse pressure:

first trial _______ second trial _______

6. Compute the **mean arterial pressure (MAP)** for each trial using the following equation:

$$\text{MAP} = \text{diastolic pressure} + \frac{\text{pulse pressure}}{3}$$

first trial _______ second trial _______

Estimation of Venous Pressure

It is not possible to measure venous pressure with the sphygmomanometer. The methods available for measuring it produce estimates at best, because venous pressures are so much lower than arterial pressures. The difference in pressure becomes obvious when these vessels are cut. If a vein is cut, the blood flows evenly from the cut. A lacerated artery produces rapid spurts of blood.

1. Ask your lab partner to stand with his or her right side toward the blackboard, arms hanging freely at the sides. On the board, mark the approximate level of the right atrium. (This will be just slightly higher than the point at which you auscultated the apical pulse.)

2. Observe the superficial veins on the dorsum of the right hand as the subject alternately raises and lowers it. Notice the collapsing and filling of the veins as internal pressures change. Have the subject repeat this action until you can determine the point at which the veins have just collapsed. Then measure, in millimeters, the distance in the vertical plane from this point to the level of the right atrium (previously marked). Record this value. Distance of right arm from right atrium at point of venous collapse:

_______________ mm

3. Compute the venous pressure (P_V), in millimeters of mercury, with the following formula:

$$P_V = \frac{1.056 \text{ (specific gravity of blood)} \times \text{mm (measured)}}{13.6 \text{ (specific gravity of Hg)}}$$

Venous pressure computed: _______________ mm Hg

Normal venous pressure varies from approximately 30 to 90 mm Hg. That of the hand ranges between 30 and 40 mm Hg. How does your computed value compare?

____

4. Because venous walls are so thin, pressure within them is readily affected by external factors such as muscle activity, deep pressure, and pressure changes occurring in the thorax during breathing. The Valsalva maneuver, which increases intrathoracic pressure, is used to demonstrate the effect of thoracic pressure changes on venous pressure.

To perform this maneuver take a deep breath, and then mimic the motions of exhaling forcibly, but without actually exhaling. In reaction to this, the glottis will close; and intrathoracic pressure will increase. (Most of us have performed this maneuver unknowingly in acts of defecation in which there is "straining at stool.") Measure and record below the distance of the right arm from right atrium at the point of venous collapse while the subject is performing the Valsalva maneuver. Compute the venous pressure and record it below.

_______ mm Venous pressure: _______ mm Hg

How does this value compare with the venous pressure measurement computed for the relaxed state?

____

Explain: _______________________________________

____

____

EFFECT OF VARIOUS FACTORS ON BLOOD PRESSURE AND HEART RATE

Arterial blood pressure is directly proportional to cardiac output (amount of blood pumped out of the left ventricle per unit time) and peripheral resistance to blood flow, that is

$$\text{BP} = \text{CO} \times \text{PR}$$

Peripheral resistance is increased by constriction of blood vessels (most importantly the arterioles), by an increase in blood viscosity or volume, and by a loss of elasticity of the arteries (seen in arteriosclerosis). Any factor that increases either the cardiac output or the peripheral resistance causes an almost immediate reflex rise in blood pressure. A close examination of these relationships reveals that many factors—age, weight, time of day, exercise, body position, emotional state, and various drugs, for example—alter blood pressure. The influence of a few of these factors is investigated here.

The following tests are done most efficiently if one student acts as the subject; two are examiners (one taking the radial pulse and the other auscultating the brachial blood pressure); and a fourth student collects and records data. The sphygmomanometer cuff should be left on the subject's arm throughout the experiments (in a deflated state, of course) so that, at the proper times, the blood pressure can be taken quickly. In each case, take the measurements at least twice.

Posture

To monitor circulatory adjustments to changes in position, take blood pressure and pulse measurements under the conditions noted in Chart 1 on p. 312. Also record your results on that chart.

CHART 1 Posture

	Trial 1 BP	Trial 1 Pulse	Trial 2 BP	Trial 2 Pulse
Sitting quietly	______	______	______	______
Reclining (after 2 to 3 min)	______	______	______	______
Immediately on standing from the reclining position ("at attention" stance)	______	______	______	______
After standing for 3 min	______	______	______	______

Exercise

Blood pressure and pulse changes occurring during and after exercise provide a good yardstick for measuring one's overall cardiovascular fitness. Although there are more sophisticated and more accurate tests that evaluate fitness according to a specific point system, the *Harvard step test* described here is a quick way to compare the relative fitness level of a group of people.

You will be working in groups of four, duties assigned as indicated above, except that student 4, in addition to recording the data, will act as the timer and call the cadence.

⚠ Any student with a known heart problem should refuse to participate as the subject.

All four students may participate as the subject in turn, if desired, but the bench stepping is to be performed *at least twice* in each group—once with a well-conditioned person acting as the subject, and once with a poorly conditioned subject.

Bench stepping is the following series of movements repeated sequentially:

1. Place one foot on the step.
2. Step up with the other foot so that both feet are on the platform. Straighten the legs and the back.
3. Step down with one foot.
4. Bring the other foot down.

The pace for the stepping will be set by the "timer" (student 4), who will repeat "Up-2-3-4, up-2-3-4" at such a pace that each "up-2-3-4" sequence takes 2 sec (i.e., 30 cycles/min).

1. Student 4 should obtain the step (20-in. height for male subject, or 16 in. for a female subject) while baseline measurements are being obtained on the subject.
2. Once the baseline pulse and blood pressure measurements have been recorded on Chart 2, the subject is to stand quietly at attention for 2 min to allow his or her blood pressure to stabilize before beginning to step.
3. The subject is to perform the bench stepping for as long as possible, up to a maximum of 5 min, according to the cadence called by the timer. The subject is to be watched for and warned against crouching (posture must remain erect). If he or she is unable to keep the pace for a span of 15 sec, the test is to be terminated.
4. When the subject is stopped by the pacer, stops voluntarily because he or she is unable to continue, or has completed 5 min of bench stepping, he or she is to sit down. The duration of exercise (in seconds) is to be recorded, and the blood pressure and pulse are to be measured immediately and thereafter at 1-min intervals for 4 min post-exercise.
5. The subject's *index of physical fitness* is to be calculated using the formula given below:

$$\text{Index} = \frac{\text{duration of exercise in seconds} \times 100}{2 \times \text{sum of the 3 pulse counts in recovery}}$$

   Scores are interpreted according to the following scale:

below 55	poor physical condition
55 to 62	low average
63 to 71	average
72 to 79	high average
80 to 89	good
90 and over	excellent

6. Record the test values on Chart 2, and repeat the testing and recording procedure with the second subject.

When did you notice a greater elevation of blood pressure and pulse?

____

Explain: ______________________________

____

Was there a sizeable difference between the after-exercise values for well-conditioned and poorly conditioned individuals?

________ Explain: ________________________

____

Did the diastolic pressure also increase? __________

Explain: ___

___

___

Nicotine (Optional)

To investigate the effects of nicotine on blood pressure and pulse, take a baseline blood pressure and pulse of a smoker who is in a standing position. Ask the subject to light up and smoke in the usual manner. After 2 min, take blood pressure and pulse readings. Repeat the measurements at 1-min intervals for the next 2 min, and record your values on Chart 3.

How do the effects of nicotine bring about the changes noted? (Hint: nicotine has vasoconstrictor effects initially.)

___

___

___

A Noxious Sensory Stimulus (Cold)

There is little question that blood pressure is affected by emotions and pain. This lability of blood pressure will be investigated through use of the **cold pressor test,** in which one hand will be immersed in unpleasantly (even painfully) cold water.

Measure the blood pressure and pulse of the subject as he or she sits quietly. Obtain a basin and thermometer, fill the basin with ice cubes, and add water. When the temperature of the ice bath has reached 5°C, immerse the subject's other hand (the noncuffed limb) in the ice water. With the hand still immersed, take blood pressure and pulse readings at 1-min intervals for a period of 3 min, and record the values on Chart 4.

How did the blood pressure change during cold exposure?

___

___

Was there any change in pulse? ___

CHART 2 Exercise

Harvard step test for 5 min at 30/min	Baseline		Interval Following Test							
			Immediately		1 min		2 min		3 min	
	BP	P	BP	P	BP	P	BP	P	BP	P
Well-conditioned individual	___	___	___	___	___	___	___	___	___	___
Poorly conditioned individual	___	___	___	___	___	___	___	___	___	___

CHART 3 Nicotine

Baseline		After smoking for 2 min		After smoking for 3 min		After smoking for 4 min	
BP	P	BP	P	BP	P	BP	P
___	___	___	___	___	___	___	___

CHART 4 A Noxious Sensory Stimulus (Cold)

Baseline		1 min		2 min		3 min	
BP	P	BP	P	BP	P	BP	P
___	___	___	___	___	___	___	___

Subtract the respective baseline readings of systolic and diastolic blood pressure from the highest single reading of systolic and diastolic pressure obtained during cold immersion. (For example, if the highest experimental reading is 140/88 and the baseline reading is 120/70, then the differences in blood pressure would be systolic pressure, 20 mm Hg, and diastolic pressure, 18 mm Hg.) These differences are called the index of response. According to their index of response, subjects can be classified as follows:

Hyporeactors: (stable blood pressure)—exhibit a rise of diastolic and/or systolic pressure ranging from 0 to 22 mm Hg; or a drop in pressures

Hyperreactors: (labile blood pressure)—exhibit a rise of 23 mm Hg or more in the diastolic and/or systolic blood pressure

Is the subject tested a hypo- or hyperreactor?

____

SKIN COLOR AS AN INDICATOR OF LOCAL CIRCULATORY DYNAMICS

Skin color reveals with surprising accuracy the state of the local circulation, and allows inferences concerning the larger blood vessels and the circulation as a whole. The experiments on local circulation outlined below illustrate a number of factors that affect blood flow to the tissues.

Clinical expertise often depends upon good observation skills, accurate recording of data, and logical interpretation of the findings. A single example will be given to demonstrate this statement: A massive hemorrhage may be internal and hidden (thus, not obvious), but will still threaten the blood delivery to the brain and other vital organs. One of the earliest compensatory responses of the body to such a threat is constriction of cutaneous blood vessels, which reduces blood flow to the skin and diverts it into the circulatory mainstream to serve other, more vital tissues. As a result, the skin of the face and particularly of the extremities becomes pale, cold, and eventually moist with perspiration. Therefore, pale, cold, clammy skin should immediately lead the careful diagnostician to suspect that the circulation is dangerously inefficient. Other conditions, such as local arterial obstruction and venous congestion, as well as certain pathologies of the heart and lungs, also alter skin texture, color, and circulation in characteristic ways.

The local blood supply to the skin (indeed, to any tissue) is influenced by (1) local metabolites, (2) oxygen supply, (3) local temperature, (4) autonomic nervous system impulses, (5) local vascular reflexes, (6) certain hormones, and (7) substances released by injured tissues. A number of these factors are examined in the simple experiments that follow. Each experiment should be conducted by students in groups of three or four. One student will act as the subject; the others will conduct the tests, and make and record observations.

Vasodilation and Flushing of the Skin Due to Local Metabolites

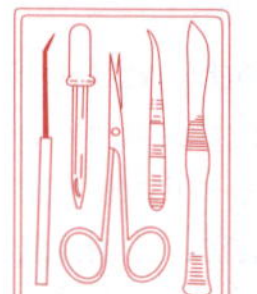

1. Obtain a blood pressure cuff (sphygmomanometer) and stethoscope. You will also need a watch with a second hand.

2. The subject should bare both arms by rolling up the sleeves as high as possible and then lay the forearms side by side on the bench top.

3. Observe the general color of the subject's forearm skin, and the normal contour and size of the veins. Notice whether skin color is bilaterally similar. Record your observations:

____

____

____

4. Apply the blood pressure cuff to one arm, and inflate it to 250 mm Hg. Keep it inflated for 1 min. During this period, repeat the observations made above and record the results:

____

____

5. Release the pressure in the cuff (leaving the deflated cuff in position), and again record the forearm skin color and the condition of the forearm veins. Make this observation immediately after deflation and then again 30 sec later.

Immediately after deflation ______________________________

____

30 sec after deflation ______________________________

____

The above observations constitute your baseline information. Now conduct the following tests.

6. Instruct the subject to raise the cuffed arm above his or her head and to clench the fist as tightly as possible. While the hand and forearm muscles are tightly contracted, rapidly inflate the cuff to 240 mm Hg or more. This maneuver partially empties the hand and forearm of blood, and stops most blood flow to the hand and forearm. Once the cuff has been inflated, the subject is to relax the fist and return the forearm to the bench top so that it can easily be compared to the other forearm.

7. Leave the cuff inflated for exactly 1 min. During this interval, compare the skin color in the "ischemic" (blood-deprived) hand to that of the "normal" (non-cuffed-limb) hand. Quickly release the pressure immediately after the 1-min count.

What are the subjective effects* of stopping blood flow to the arm and hand for 1 min?

___

___

What are the objective effects (color of skin and venous condition)?

___

___

How long does it take for the subject's ischemic hand to regain its normal color?

___

Effects of Venous Congestion

1. Again, but with a different subject, observe and record the appearance of the skin and veins on the forelimbs resting on the bench top. This time, pay particular attention to the color of the fingers, particularly the distal phalanges, and the nail beds. Record this information:

___

___

2. Wrap the blood pressure cuff around one of the subject's arms, and inflate it to 40 mm Hg. Maintain this pressure for 5 min. Make a record of the subjective and objective findings just before the 5 min are up, and then again immediately after release of the pressure at the end of 5 min.

Subjective (arm cuffed) ___

___

Objective (arm cuffed) ___

___

Subjective (pressure released) ___

___

*Subjective effects are sensations—such as pain, coldness, warmth, tingling, and weakness—experienced by the subject. They are "symptoms" of a change in function.

Objective (pressure released) ___

___

3. With still another subject, conduct the following simple experiment: Raise one arm above the head, and let the other hang by the side for 1 min. After 1 min, quickly lay both arms on the bench top, and compare their color.

Color of raised arm ___

Color of dependent arm ___

From this and the two preceding observations, analyze the factors that determine tint of color (pink or blue) and intensity of skin color (deep pink or blue as opposed to light pink or blue). Record your conclusions.

___

___

___

___

___

Collateral Blood Flow

In some diseases, blood flow to an organ through one or more arteries may be completely and irreversibly obstructed. Fortunately, in most cases a given body area is supplied both by one main artery and by anastomosing channels connecting the main artery with one or more neighboring blood vessels. Consequently, an organ may remain viable even though its main arterial supply is occluded, as long as the **collateral vessels** are still functional.

The effectiveness of collateral blood flow in preventing ischemia can be easily demonstrated.

1. Check the subject's hands to be sure they are *warm* to the touch. If not, choose another, "warm-handed" subject, or warm the subject's hands in 35°C water for 10 min before beginning.

2. Palpate the subject's radial and ulnar arteries approximately 1 in. above the wrist flexure, and mark their locations with a felt marker.

3. Instruct the subject to supinate one forearm and to hold it in a partially flexed (about a 30° angle) position, with the elbow resting on the bench top.

4. Face the subject and grasp his or her forearm with both of your hands, the thumb and fingers of one hand compressing the marked radial artery and the thumb and fingers of the other hand compressing the ulnar artery.

Maintain the pressure for 5 min, noticing the progression of the subject's hand to total ischemia.

5. At the end of 5 min, release the pressure abruptly. Record the subject's sensations, as well as the intensity and duration of the flush in the previously occluded hand. (Use the other hand as a baseline for comparison.)

____

____

____

6. Allow the subject to relax for 5 min; then repeat the maneuver, but this time *compress only the radial artery.* Record your observations.

____

____

How do the results of the first test differ from those of the second test with respect to color changes during compression, and the intensity and duration of reactive hyperemia (redness of the skin)?

____

____

____

____

7. Once again allow the subject to relax for 5 min. Then repeat the maneuver, *with only the ulnar artery compressed.* Record your observations:

____

____

What can you conclude about the relative sizes of, and hand areas served by, the radial and ulnar arteries?

____

____

Effect of Mechanical Stimulation of Blood Vessels of the Skin

With moderate pressure, draw the blunt end of your pen across the skin of a subject's forearm. Wait 3 min to observe the effects, and then repeat with firmer pressure.

What changes in skin color do you observe with light-to-moderate pressure?

____

With heavy pressure? ______________________________

The redness, or *flare,* observed after mechanical stimulation of the skin results from a local inflammatory response promoted by chemical mediators released by injured tissues. These mediators stimulate increased blood flow into the area and leaking of fluid (from the capillaries) into the local tissues. (Note: People differ considerably in skin sensitivity. Those most sensitive will show **dermographism,** a condition in which the direct line of stimulation will swell quite obviously. This excessively swollen area is called a *wheal.*)

EXERCISE 34

Frog Cardiovascular Physiology

OBJECTIVES

1. To list the properties of cardiac muscle as automaticity and rhythmicity and define each.
2. To explain the statement "Cardiac muscle has an intrinsic ability to beat."
3. To compare the intrinsic rate of contraction of the "pacemaker" of the frog heart (sinus venosus) to that of the atria and ventricle.
4. To compare the relative length of the refractory period of cardiac muscle with that of skeletal muscle, and explain why it is not possible to tetanize cardiac muscle.
5. To define *extrasystole* and to explain at what point in the cardiac cycle (and on an ECG tracing) an extrasystole can be induced.
6. To describe the effect of the following on heart rate: cold, heat, vagal stimulation, pilocarpine, digitalis, atropine sulfate, epinephrine, and potassium, sodium, and calcium ions.
7. To define *vagal escape* and discuss its value.
8. To define *ectopic pacemaker.*
9. To define *partial* and *total heart block.*
10. To describe how heart block was induced in the laboratory and explain the results, and to explain how heart block might occur in the body.
11. To list the components of the microcirculatory system.
12. To identify an arteriole, venule, and capillaries in a frog's web, and to cite differences between relative size, rate of blood flow, and regulation of blood flow in these vessels.
13. To describe the effect of heat, cold, local irritation, and histamine on the rate of blood flow in the microcirculation, and to explain how these responses help maintain homeostasis.

MATERIALS

Frogs*
Dissecting pan and instruments
Disposable plastic gloves
Petri dishes
Medicine dropper
Millimeter ruler
Frog Ringer's solutions (at room temperature, 5°C, and 32°C)
Thread, large rubber bands, fine common pins
Frog board
Laboratory ring stand, or transducer stand; right-angle clamps (two)
Cotton balls
Freshly prepared solutions (using frog Ringer's solution as the solvent) of the following in dropper bottles:
- 2.5% pilocarpine
- 2% digitalis
- 5% atropine sulfate
- 1% epinephrine
- 5% potassium chloride (KCl)
- 2% calcium chloride ($CaCl_2$)
- 0.7% sodium chloride (NaCl)
- 0.01% histamine
- 0.01*N* HCl

Apparatus A or B:
- A: Kymograph, kymograph paper (smoking stand, burner and glazing fluid if using a smoke-writing apparatus), heart lever, signal magnet, electronic stimulator, platinum electrodes
- B: Physiograph (polygraph), polygraph paper and ink, myograph transducer, transducer coupler, stimulator output extension cable, electrodes

See Appendix E, Exercise 34 for links to *Anatomy and PhysioShow: The Videodisc.*

*Instructor will double-pith frogs as required for student experimentation.

Investigation of human cardiovascular physiology is most interesting, but many areas obviously do not lend themselves to experimentation. It would be tantamount to murder to inject a human subject with various drugs to observe their effects on heart activity, and to expose the human heart in order to study the length of its refractory period. However, this type of investigation can be done on frogs or small laboratory animals and provides valuable data, because the physiologic mechanisms in these animals are similar if not identical to those in humans.

In this exercise, you will conduct the cardiac investigations just mentioned and others. In addition, you will observe the microcirculation in a frog's web and subject it to various chemical and thermal agents to demonstrate their influence on blood flow.

AUTOMATICITY AND RHYTHMICITY OF CARDIAC MUSCLE

Cardiac muscle differs from skeletal muscle both functionally and in its fine structure. Skeletal muscle must be electrically stimulated to contract. In contrast, heart muscle can and does depolarize spontaneously in the absence of external stimulation. This property, called **automaticity,** is due to plasma membranes that have reduced permeability to potassium ions, but still allow sodium ions to slowly leak into the cells. This leakage causes the muscle cells to slowly depolarize until the action potential threshold is reached and *fast calcium channels* open, allowing Ca^{2+} entry from the extracellular fluid. Shortly thereafter, contraction occurs.

Additionally, the spontaneous depolarization-repolarization events occur in a regular and continuous manner in cardiac muscle, a property referred to as **rhythmicity.**

In the following experiment, you will observe these properties of cardiac muscle in an *in vitro* (outside, or removed from the body) situation. Work together in groups of three or four. (The instructor may choose to demonstrate this procedure if time or frogs are at a premium.)

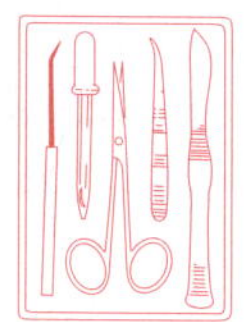

1. Obtain a dissecting pan and instruments, disposable plastic gloves, two petri dishes, frog Ringer's solution, and a medicine dropper, and bring them to your laboratory bench.

2. Don the gloves, and then request and obtain a doubly pithed frog from your instructor. Quickly open the thoracic cavity and observe the heart rate *in situ* (at the site or within the body).

Record the heart rate: ____________________ beats/min

3. Dissect out the heart and gastrocnemius muscle of the calf and place the removed organs in separate petri dishes containing frog Ringer's solution. (Note: Just in case you don't remember, the procedure for removing the gastrocnemius muscle is provided on pp. 145-146 in Exercise 16A. The extreme care used in that procedure for the removal of the gastrocnemius muscle need not be exercised here.)

4. Observe the activity of the two organs for a few seconds.

Which is contracting? ____________________

At what rate? ______________ beats/min

Is the contraction rhythmic? ____________________

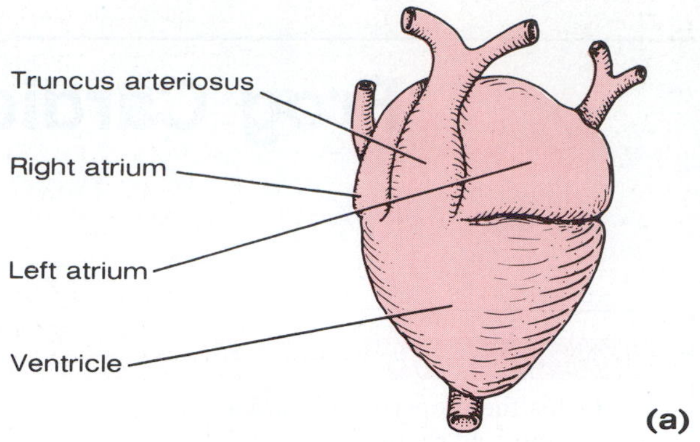

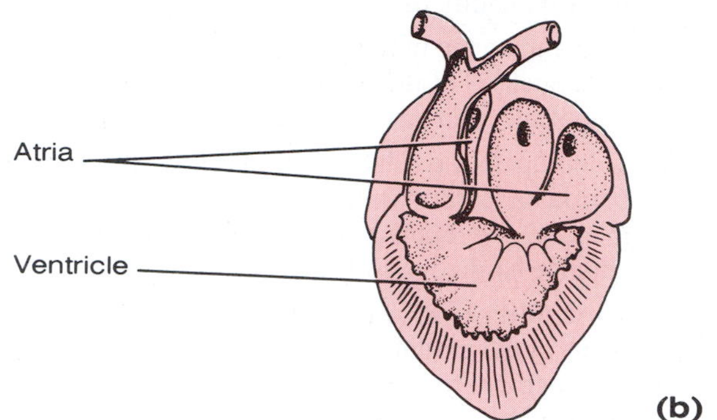

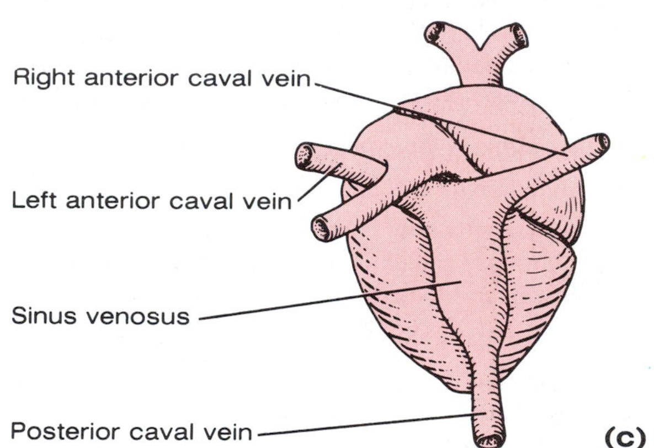

F34.1

Anatomy of the frog heart. (a) Ventral view showing the single truncus arteriosus leaving the undivided ventricle. **(b)** Longitudinal section showing the two atrial and single ventricular chambers. **(c)** Dorsal view showing the sinus venosus (pacemaker).

5. Sever the sinus venosus from the heart (see Figure 34.1). The sinus venosus of the frog's heart corresponds to the SA node of the human heart. Does the sinus venosus continue to beat?

____

If not, lightly touch it with a probe to stimulate it. Record its rate of contraction.

Rate: ______________ beats/min

6. Sever the right atrium from the heart; then the left atrium. Does each atrium continue to beat?

______________ Rate: ________________ beats/min

Does the ventricle continue to beat? ______________

Rate: ______________ beats/min

7. Continue to fragment the ventricle to determine how small the ventricular fragments must be before the automaticity of ventricular muscle is abolished. Measure these fragments and record their approximate size.

_______ mm × _______ mm × _______ mm

Which portion of the heart exhibited the most marked automaticity? ______________________________

Which the least? ______________________________

8. Properly dispose of the frog and heart fragments before continuing with the next procedure.

RECORDING BASELINE FROG HEART ACTIVITY

The heart's effectiveness as a pump is dependent both on intrinsic (within the heart) and extrinsic (external to the heart) controls. In this experiment, you will investigate some of these factors.

The nodal system, in which the "pacemaker" imposes its depolarization rate on the rest of the heart, is one intrinsic factor that influences the heart's pumping action. If its impulses fail to reach the ventricles (as in heart block), the ventricles continue to beat but at their own inherent rate, which is much slower than that usually imposed on them. Although heart contraction does not depend on nerve impulses, its rate can be modified by extrinsic impulses reaching it through the autonomic nerves. Additionally, cardiac activity is modified by various chemicals, hormones, ions, and metabolites. The effects of several of these chemical factors are examined in the next experimental series.

The frog heart has two atria and a single, incompletely divided ventricle (see Figure 34.1). The pacemaker is located in the sinus venosus, an enlarged region between the venae cavae and the right atrium. The SA node of mammals may have evolved from the sinus venosus.

To record frog heart activity, work in groups of four—two students handling the equipment setup and two preparing the frog for experimentation. Two sets of instructions are provided for apparatus setup—one for the kymograph (Figure 34.2), the other for the physiograph (Figure 34.3). Follow the procedure outlined for the apparatus you will be using.

Kymograph Apparatus Setup

1. Obtain a kymograph, recording paper, ring stand, frog board, two right-angle clamps, signal magnet, stimulator, platinum electrodes, and a heart lever.
2. If you are not using an ink-writing kymograph apparatus, smoke the drum lightly; it is important that the recording be made with minimum pressure against the drum.
3. As in the setup depicted in Figure 34.2, connect the electrodes to the two output terminals on the stimulator, and attach the signal magnet to the two binding posts on the back of the stimulator. Connect the signal magnet to the ring stand with a clamp so it will record close to the bottom of the drum. Attach the heart lever to the stand above and in line with the signal magnet.

Physiograph Apparatus Setup

1. Obtain a myograph transducer, transducer cable, and stand, and bring them to the recording site.
2. Attach the myograph transducer to the transducer stand as shown in Figure 34.3.
3. Then attach the transducer cable to the transducer coupler (input) on the amplifier of the physiograph and to the myograph transducer.
4. Attach the stimulator output extension cable to output on the stimulator panel (red to red, black to black).

Preparation of the Frog

1. Obtain room-temperature frog Ringer's solution, a medicine dropper, dissecting instruments and pan, disposable gloves, fine common pins, and some thread, and bring them to your bench.

2. Don the gloves and obtain a doubly pithed frog from your instructor.

3. Make a longitudinal incision through the abdominal and thoracic walls with scissors, and then cut through the sternum to expose the heart.

4. Grasp the pericardial sac with forceps, and cut it open so that the beating heart can be observed.

Is the sequence an atrial-ventricular one? ______________

5. Locate the vagus nerve, which runs down the lateral aspect of the neck and parallels the trachea and carotid artery. (In good light, it appears to be striated.) Slip an 18-inch length of thread under the vagus nerve

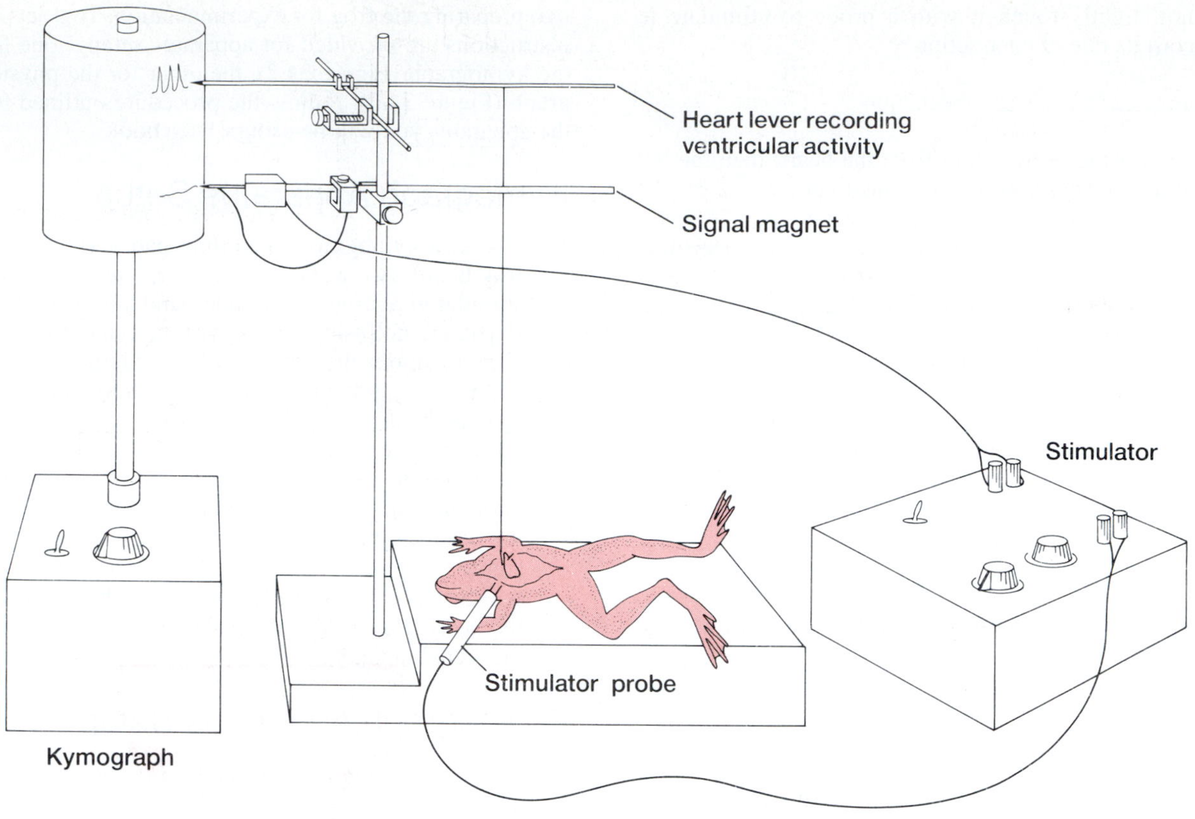

F34.2

Kymograph apparatus setup for recording the activity of the frog heart.

so that it can later be lifted away from the surrounding tissues by the thread. Then place a saline-soaked cotton ball over the nerve to keep it moistened until you are ready to stimulate it later in the procedure.

6. Flush the heart with the saline (Ringer's) solution. *From this point on the heart must be kept continually moistened with room temperature Ringer's solution unless other solutions are being used for the experimentation.*

7. Attach the frog to the frog board.

8. Bend a common pin to a 90° angle, and tie an 18- to 20-inch length of thread to its head. Force the pin through the apex of the heart (but do not penetrate the ventricular chamber) until the apex is well secured in the angle of the pin.

9. Tie the thread to the end of the muscle lever directly over the heart if using the kymograph, or to the hook on the myograph transducer if using the physiograph apparatus. Do not pull the thread too tightly. It should be taut enough to lift the heart apex upward, away from the thorax, but should *not* stretch the heart. Adjust the muscle lever or myograph transducer as necessary.

10. If using a kymograph, adjust the writing tip of the heart lever so that it is directly in line with the stylus of the signal magnet. Leave enough room between the two writing tips to give each adequate working distance. When properly set up, the heart lever should be in a horizontal position.

Recording Baseline Activity

1. Kymograph users should set the drum speed at 7, and briefly turn on the kymograph to ensure that the styluses are recording properly. If using the physiograph, turn the amplifier on and balance the apparatus according to instructions provided by your instructor. Set the paper speed at 0.5 cm/sec. Press the record and paper advance buttons.

2. Set the signal magnet or time marker at 1/sec.

3. Record several normal heartbeats (12 to 15). Be sure you can distinguish atrial and ventricular contractions (see Figure 34.4). Then adjust the paper speed so that the peaks of ventricular contractions are approximately 2 cm apart. (Peaks indicate systole; troughs in-

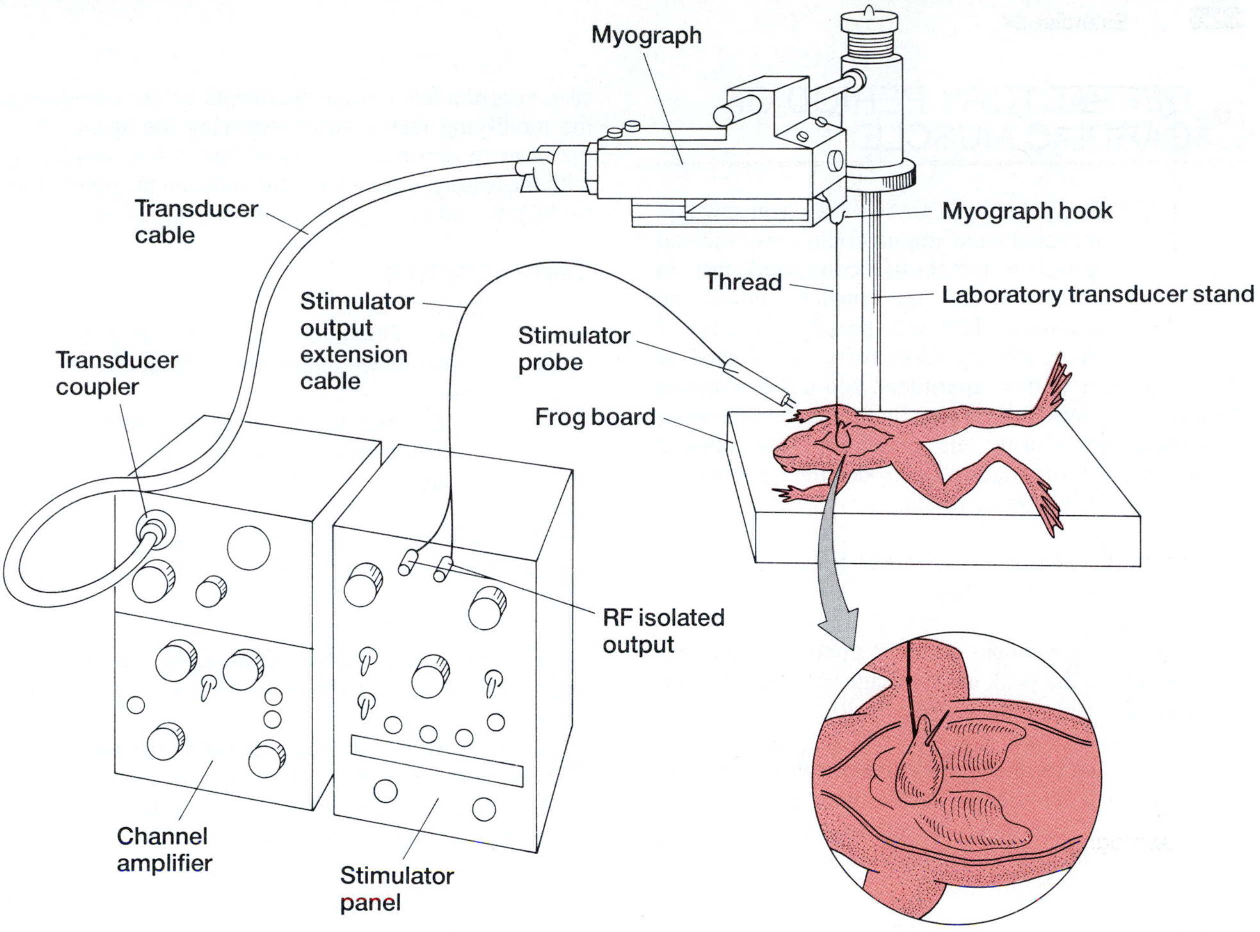

Myograph
Transducer cable
Myograph hook
Thread
Laboratory transducer stand
Stimulator output extension cable
Stimulator probe
Transducer coupler
Frog board
RF isolated output
Channel amplifier
Stimulator panel

F34.3

Physiograph setup for recording the activity of the frog heart.

dicate diastole.) Pay attention to the relative force of heart contractions while recording.

4. Count the number of ventricular contractions per minute and record:

_______________ beats/min

5. Compute the A-V interval (period from the beginning of atrial contraction to the beginning of ventricular contraction).

_______________ sec

How do the two tracings compare in time?

____

____

____

6. Mark the atrial and ventricular systoles on the record. *Remember to keep the heart moistened with Ringer's solution.*

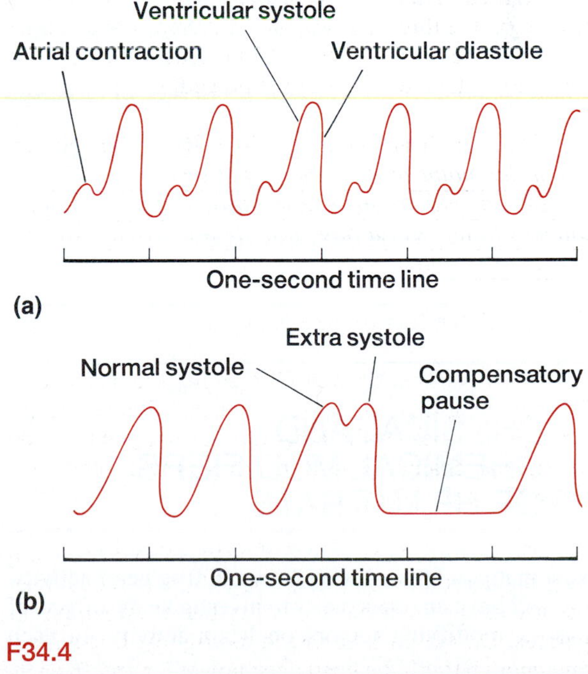

F34.4

Recording of contractile activity of a frog heart. (a) Normal heartbeat. **(b)** Induction of an extrasystole.

REFRACTORY PERIOD OF CARDIAC MUSCLE

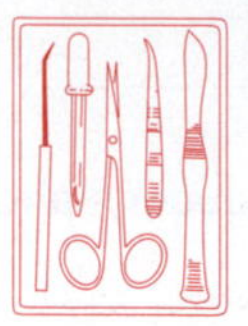

In conducting Exercise 16A, you saw that repeated rapid stimuli could cause skeletal muscle to remain in a contracted state. In other words, the muscle could be tetanized. This was possible because of the relatively short refractory period of skeletal muscle. In this experiment, you will investigate the refractory period of cardiac muscle and its response to stimulation. During the procedure, one student should keep the stimulating electrodes in constant contact with the frog heart ventricle.

1. Set the stimulator to deliver 20-V shocks of 2-ms duration, and begin recording.

2. Deliver single shocks at the beginning of ventricular contraction, the peak of ventricular contraction, and then later and later in the cardiac cycle.

3. Observe the recording for **extrasystoles,** which are extra beats that show up riding on the ventricular contraction peak. Also note the **compensatory pause,** which allows the heart to get back on schedule after an extrasystole. (See Figure 34.4b.)

During which portion of the cardiac cycle was it possible to induce an extrasystole?

____

____

4. Attempt to tetanize the heart by stimulating it at the rate of 20 to 30 impulses per second. What is the result?

____

____

Considering the function of the heart, why is it important that heart muscle cannot be tetanized?

____

____

PHYSICAL AND CHEMICAL MODIFIERS OF HEART RATE

Now that you have observed normal frog heart activity, you will have an opportunity to investigate the effects of various modifying factors on heart activity. In each case, record a few normal heartbeats before introducing the modifying factor. After removing the agent, allow the heart to return to its normal rate before continuing with the testing. On each record, indicate the point of introduction and removal of the modifying agent.

Temperature

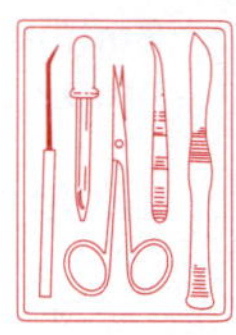

1. Obtain 5°C and 32°C frog Ringer's solutions and medicine droppers.

2. Increase the drum or paper speed so that heartbeats appear as spikes 4 to 5 mm apart.

3. Bathe the heart with 5°C saline, and continue to record until the recording indicates a change in cardiac activity.

4. Stop recording, pipette off the cold Ringer's solution, and flood the heart with room-temperature saline.

5. Start recording again to determine the resumption of the normal heart rate. When this has been achieved, flood the heart with 32°C Ringer's solution and again record until a change is noted.

6. Stop the recording, pipette off the saline, and bathe the heart with room-temperature Ringer's solution once again.

What change occurred with the cold (5°C) Ringer's solution?

____

What change occurred with the warm (32°C) Ringer's solution?

____

7. Count the heart rate at the two temperatures and record below.

______ beats/min at 5°C; __________ beats/min at 32°C

Vagus Nerve Stimulation

The vagus nerve carries parasympathetic impulses to the heart, which modify heart activity.

1. Remove the cotton placed over the vagus nerve. Using the previously tied thread, lift the nerve away from the tissues and place the nerve on the stimulating electrodes.

2. Stimulate the nerve at a rate of 50/sec for 0.5 msec at a voltage of 1 mV. Continue stimulation until the heart stops momentarily and then begins to beat again

(vagal escape). If no effect is observed, increase stimulus intensity and try again. If no effect is observed after a substantial increase in stimulus voltage, reexamine your "vagus nerve" to make sure that it is not simply strands of connective tissue.

3. Discontinue stimulation after you observe vagal escape, and flush the heart with saline until the normal heart rate resumes. What is the effect of vagal stimulation on heart rate?

___

The phenomenon of vagal escape demonstrates that many factors are involved in heart regulation and that any deleterious factor (in this case, excessive vagal stimulation) will be overcome, if possible, by other physiologic mechanisms such as activation of the sympathetic division of the autonomic nervous system (ANS).

Pilocarpine

Flood the heart with a 2.5% solution of pilocarpine. Record until a change in the pattern of the ECG is noticed. Pipette off the excess pilocarpine solution, and proceed immediately to the next test, which uses atropine as the testing solution. What happened when the heart was bathed in the pilocarpine solution?

___

Pilocarpine simulates the effect of parasympathetic nerve (hence, vagal) stimulation by enhancing acetylcholine release; such drugs are called parasympathomimetic drugs.

Atropine Sulfate

Apply a few drops of atropine sulfate to the frog's heart and observe the recording. If no changes are observed within 2 minutes, apply a few more drops. When you observe a response, pipette off the excess atropine sulfate and flood the heart with Ringer's solution. What happens when the atropine sulfate is added?

___

___

Atropine is a drug that blocks the effect of the neurotransmitter acetylcholine, which is liberated by the parasympathetic nerve endings. Do your results accurately reflect this effect of atropine?

___

Are pilocarpine and atropine agonists or antagonists in their effects on heart activity?

___

Epinephrine

Flood the frog heart with epinephrine solution, and continue to record until a change in heart activity is noted.

What are the results? ___

___

Which division of the autonomic nervous system does its effect imitate?

___

Digitalis

Pipette off the excess epinephrine solution, and rinse the heart with room-temperature Ringer's solution. Continue recording, and when the heart rate returns to baseline values, bathe it in digitalis solution. What is the effect of digitalis on the heart?

___

___

___

Digitalis is a drug commonly prescribed for heart patients with congestive heart failure. It slows heart rate, providing more time for venous return and decreasing the load on the weakened heart. These effects are thought to be due to inhibition of the Na^+-K^+ pump and enhancement of Ca^{2+} entry into the myocardial fibers.

Effect of Various Ions

To test the effect of various ions on the heart, apply the designated solution until you observe a change in heart rate or in strength of contraction. Pipette off the solution, flush with Ringer's solution, and allow the heart to resume its normal rate before continuing. *Do not allow the heart to stop.* If the rate decreases dramatically, flood the heart with room-temperature Ringer's solution.

Effect of Ca^{2+} (use 2% $CaCl_2$) ___

___

Effect of Na^+ (use 0.7% NaCl) ___

___

Effect of K^+ (use 5% KCl) ___

___

Potassium ion concentration is normally higher within cells than in the extracellular fluid. *Hyperkalemia* decreases the resting potential of plasma membranes, thus decreasing the force of heart contraction. In some cases, the conduction rate of the heart is so depressed that **ectopic pacemakers** (pacemakers appearing erratically and at abnormal sites in the heart muscle) appear in the ventricles, and fibrillation may occur. Was there any evidence of premature beats in the recording of potassium ion effects?

____

Was arrhythmia produced with any of the ions tested?

________ If so, which? __________________________

Intrinsic Conduction System Disturbance (Heart Block)

1. Moisten a 10-inch length of thread and make a Stannius ligature (loop the thread around the heart at the junction of the atria and ventricle).

2. Decrease the drum or paper speed to achieve intervals of approximately 2 cm between the ventricular contractions, and record a few normal heartbeats.

3. Tighten the ligature in a stepwise manner while observing the atrial and ventricular contraction curves. As heart block occurs, the atria and ventricle will no longer show a 1:1 contraction ratio. Record a few beats each time you observe a different degree of heart block—a 2:1 atria to ventricle, 3:1, 4:1, and so on. As long as you can continue to count a whole number ratio between the two chamber types, the heart is in **partial heart block.** When you can no longer count a whole number ratio, the heart is in **total,** or **complete, heart block.**

4. When total heart block occurs, release the ligature to see if the normal A-V rhythm is reestablished. What is the result?

____

____

5. Attach properly labeled recordings (or copies of the recordings) made during this procedure to the last page of this exercise for future reference.

6. Dispose of the frog remains and gloves in appropriate containers, and dismantle the experimental apparatus before continuing.

MICROCIRCULATION AND FACTORS AFFECTING LOCAL BLOOD FLOW

The thin web of a frog's foot provides an excellent opportunity to observe the flow of blood to, from, and within the capillary beds, where the real business of the circulatory system occurs. The collection of vessels involved in the exchange mechanism is referred to as the **microcirculation;** it consists of arterioles, venules, capillaries, and vascular shunts called *metarterioles-thoroughfare channels.*

The total cross-sectional area of the capillaries in the body is much greater than that of the veins and arteries combined. Thus, the flow through the capillary beds is quite slow. Capillary flow is also intermittent, because if all capillary beds were filled with blood at the same time, there would be no blood at all in the large vessels. The flow of blood into the capillary beds is regulated by the activity of muscular *terminal arterioles,* which feed the beds, and by *precapillary sphincters* at entrances to the true capillaries. The amount of blood flowing into the true capillaries of the bed is regulated most importantly by local chemical controls (local concentrations of carbon dioxide, histamine, pH). Thus a capillary bed may be flooded with blood or almost entirely bypassed depending on what is happening within the body or in a particular body region at any one time. You will investigate some of the local chemical controls in the next group of experiments.

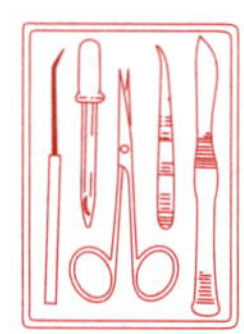

1. Obtain a frog board (with a hole at one end), dissecting pins, disposable gloves, frog Ringer's solution, 0.01N HCl, 0.01% histamine solution, 1% epinephrine solution, a large rubber band, and some paper towels.

2. Put on the gloves and obtain a frog (alive and hopping, *not* pithed). Moisten several paper towels with room-temperature Ringer's solution, and wrap the frog's body securely with them. One hind leg should be left unsecured and extending beyond the paper cocoon.

3. Attach the frog to the frog board (or other supporting structure) with a large rubber band and then carefully spread (but do not stretch) the web of the exposed hindfoot over the hole in the support. Have your partners hold the edges of the web firmly for viewing. Alternatively, secure the toes to the board with dissecting pins.

4. Obtain a compound microscope, and observe the web under low power to find a capillary bed. Focus on the vessels in high power. Keep the web moistened with Ringer's solution as you work. If the circulation seems to stop during your observations, massage the hind leg of the frog gently to restore blood flow.

5. Observe the red blood cells of the frog. Notice that, unlike human RBCs, they are nucleated. Watch their movement through the smallest vessels—the capillaries. Do they move in single file or do they flow through two or three cells abreast?

____

Are they flexible? ________ Explain. ______________

____

Can you see any white blood cells in the capillaries?

_______ If so, which types? _______________

____

6. Notice the relative speed of blood flow through the blood vessels. Differentiate between the arterioles, which feed the capillary bed, and the venules, which drain it. This may be tricky, because images are reversed in the microscope. Thus, the vessel that appears to feed into the capillary bed will actually be draining it. You can distinguish between the vessels, however, if you consider that the flow is more pulsating and turbulent in the arterioles and smoother and steadier in the venules. How does the *rate* of flow in the arterioles compare with that in the venules?

____

In the capillaries? _______________________

What is the relative difference in the diameter of the arterioles and capillaries?

____

Effect of Temperature on Blood Flow

To investigate the effect of temperature on blood flow, flood the web with cold saline two or three times to chill the entire area. Is a change in vessel diameter noticeable?

_______ Which vessels are affected? _______________

How? _______________________________________

Blot the web gently with a paper towel, and then bathe the web with warm saline. Record your observations.

____

Effect of Inflammation on Blood Flow

Pipette 0.01*N* HCl onto the frog's web. Hydrochloric acid will act as an irritant and cause a localized inflammatory response. Is there an increase or decrease in the blood flow into the capillary bed following the application of HCl?

____

What purpose do these local changes serve during a localized inflammatory response?

____

____

____

Flush the web with room-temperature Ringer's solution and blot.

Effect of Histamine on Blood Flow

Histamine, which is released in large amounts during allergic responses, causes extensive vasodilation. Investigate this effect by adding a few drops of histamine solution to the frog web. What happens?

____

____

How does this response compare to that produced by HCl? _______________________________________

Blot the web and flood with warm saline as before. Now add a few drops of 1% epinephrine solution, and observe the web. What are epinephrine's effects on the blood vessels?

____

Epinephrine is used clinically to reverse the vasodilation seen in severe allergic attacks (such as asthma) which are mediated by histamine and other vasoactive molecules.

Return the dropper bottles to the supply area and the frog to the terrarium. Properly clean your work area before leaving the lab.

35

EXERCISE

The Lymphatic System and Immune Response

OBJECTIVES

1. To name the components of the lymphatic system.
2. To relate the function of the lymphatic system to that of the blood vascular system.
3. To describe the formation and composition of lymph, and to describe how it is transported through the lymphatic vessels.
4. To relate immunologic memory, specificity, and differentiation of self from nonself to immune function.
5. To differentiate between the roles of B cells and T cells in the immune response.
6. To describe the structure and function of lymph nodes, and to indicate the localization of T cells, B cells, and macrophages in a typical lymph node.
7. To describe (or draw) the structure of the immunoglobulin monomer, and to name the five immunoglobulin subclasses.
8. To discuss the immunologic "abnormality" evidenced by many rheumatoid arthritis sufferers.

MATERIALS

Large anatomical chart of the human lymphatic system
Prepared microscope slides of lymph nodes
Compound microscope

Microscope slide
Wax marking pencil
Applicator sticks
Arthritis screening test (Wampole)
Plastic (disposable) gloves
Disposable autoclave bag

See Appendix D, Exercise 35 for links to A.D.A.M. Standard.

See Appendix E, Exercise 35 for links to *Anatomy and PhysioShow: The Videodisc.*

THE LYMPHATIC SYSTEM

General Description

The **lymphatic system** consists of a network of lymphatic vessels (lymphatics), lymph nodes, and a number of other lymphoid organs, such as the tonsils, thymus, and spleen (Figure 35.1). We will focus on the lymphatic vessels and lymph nodes in this section. The overall function of the lymphatic system is twofold. It transports tissue fluid (lymph) to the blood vessels. In addition, it protects the body by removing foreign material such as bacteria from the lymphatic stream and by serving as a site for lymphocyte "policing" of body fluids and lymphocyte multiplication. The function of these white blood cells in body immunity is described later in this exercise.

Distribution and Function of Lymphatic Vessels and Lymph Nodes

As blood circulates through the body, the hydrostatic and osmotic pressures operating at the capillary beds result in fluid outflow at the arterial end of the bed and in its return at the venous end. However, not all of the lost fluid is returned to the bloodstream by this mechanism, and the fluid that lags behind in the tissue spaces must eventually return to the blood if the vascular system is to operate properly. (If it does not, fluid accumulates in the tissues, producing a condition called edema.) It is the microscopic, blind-ended **lymphatic capillaries,** which ramify through nearly all the tissues of the body, that pick up this leaked fluid (primarily water and a small amount of dissolved proteins) and carry it through successively larger vessels—**lymphatic collecting vessels to lymphatic trunks**—until the lymph finally returns to the blood vascular system through one of the two large ducts in the thoracic region. The **right lymphatic duct** drains lymph from the right upper extremity, head, and thorax; the large **thoracic duct** receives lymph from the rest of the body (Figure 35.1b). In humans, both ducts empty the lymph into the venous circulation at the junction of the internal jugular vein and the subclavian vein, on their respective sides of the body. Notice that the lymphatic system, lacking both a contractile "heart" and arteries, is a one-way system; it carries lymph only toward the heart.

Like veins of the blood vascular system, the lymphatic collecting vessels have three tunics and are

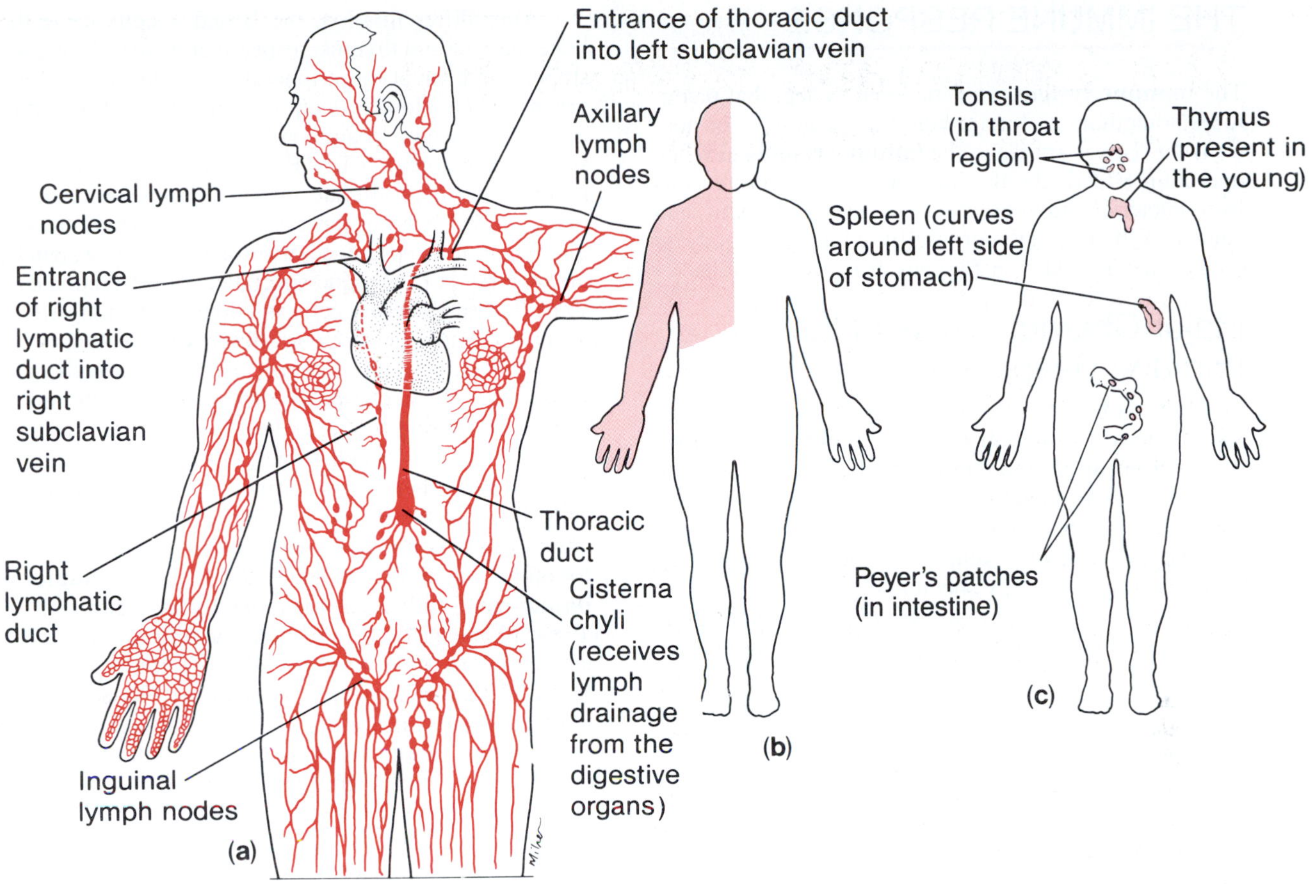

F35.1

Lymphatic system. (a) Distribution of lymphatic vessels and lymph nodes. **(b)** Shaded area represents body area drained by the right lymphatic duct. **(c)** Body location of the tonsils, thymus, spleen, and Peyer's patches.

equipped with valves. However, the lymphatics tend to be thinner-walled, to have *more* valves, and to anastomose more than veins. Since the lymphatic system is a pumpless system, lymph transport depends largely on the milking action of the skeletal muscles and on pressure changes within the thorax during breathing.

As lymph is transported, it filters through bean-shaped **lymph nodes,** which cluster along the lymphatic vessels of the body. There are thousands of lymph nodes; but because they are usually embedded in connective tissue, they are not ordinarily seen. Within the lymph nodes are **macrophages,** phagocytes that destroy bacteria, cancer cells, and other foreign matter in the lymphatic stream, thus rendering many harmful substances or cells harmless before the lymph enters the bloodstream. Particularly large collections of lymph nodes are found in the inguinal, axillary, and cervical regions of the body. Although we are not usually aware of the filtering and protective nature of the lymph nodes, most of us have experienced "swollen glands" during an active infection. This swelling is a manifestation of the trapping function of the nodes.

The tonsils, thymus, and spleen are generally considered to be lymphoid organs because they resemble the lymph nodes histologically, and house similar cell populations (lymphocytes and macrophages).

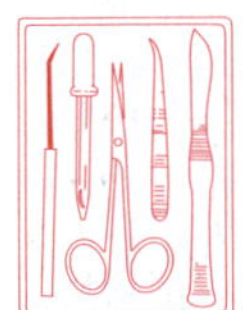

Study the large anatomical chart to observe the general plan of the lymphatic system. Notice the distribution of lymph nodes, various lymphatics, the lymphatic trunks, and the location of the right lymphatic duct and the thoracic duct. Also identify the **cisterna chyli,** the enlarged terminus of the thoracic duct that receives lymph from the digestive viscera.

THE IMMUNE RESPONSE

The **immune system** is a functional system that recognizes something as foreign and acts to destroy or neutralize it. This is known as the **immune response.** When operating effectively, the immune response protects us from bacterial and viral infections, bacterial toxins, and cancer. When it fails or malfunctions, the body is quickly devastated by pathogens or its own assaults.

Major Characteristics of the Immune Response

The most important characteristics of the immune response are its (1) **memory,** (2) **specificity,** and (3) **ability to differentiate self from nonself.** Not only does the immune system have a "memory" for previously encountered antigens (molecules foreign to the body—the chicken pox virus for example), this memory is remarkably accurate and highly specific.

An almost limitless variety of macromolecules is *antigenic*—that is, capable of provoking an immune response and reacting with its products. Nearly all foreign proteins, many polysaccharides, and many small molecules (haptens), when linked to our own body proteins, exhibit this capability. The cells that recognize antigens and initiate the immune response are lymphocytes. (As described in Exercise 29, lymphocytes are the second most numerous members of the leukocyte or white blood cell [WBC] population.) Each immunocompetent lymphocyte is virtually monospecific; that is, it has receptors on its surface allowing it to bind with only one or a few very similar antigens.

As a rule, our own proteins are tolerated, a fact that reflects the ability of the immune system to distinguish our own tissues (self) from foreign antigens (nonself). Nevertheless, an inability to recognize self can and does occasionally happen and our own tissues are attacked by the immune system. This phenomenon is called *autoimmunity.*

Organs, Cells, and Cell Interactions of the Immune Response

The immune system utilizes as part of its arsenal the **lymphoid organs,** including the thymus, lymph nodes, spleen, tonsils, and bone marrow. Of these, the thymus and bone marrow are considered to be the *primary lymphoid organs.* The others are *secondary lymphoid areas.*

The stem cells which give rise to the immune system arise in the bone marrow. Their subsequent differentiation into one of the two populations of immunocompetent lymphocytes occurs in the primary lymphoid organs. The **B cells** differentiate in bone marrow, and the **T cells** differentiate in the thymus. While in their "programming organs," the lymphocytes become *immunocompetent,* an event indicated by the appearance of specific cell-surface proteins that enable the lymphocytes to respond (by binding) to a particular antigen.

After differentiation, the B and T cells leave the bone marrow and thymus, respectively; enter the bloodstream; and travel to peripheral (secondary) lymphoid organs, where clonal selection occurs. **Clonal selection** is triggered when an antigen binds to the specific cell-surface receptors of a T or B cell. This event causes the lymphocyte to proliferate rapidly, forming a clone of like cells, all bearing the same antigen-specific receptors. Then, in the presence of certain regulatory signals, the members of the clone specialize, or differentiate—some forming memory cells and others becoming effector or regulatory cells. Upon subsequent meetings with the same antigen, the immune response proceeds considerably faster because the "troops are already mobilized and awaiting further orders," so to speak.

In the case of B cell clones, some become memory B cells; the others form antibody-producing **plasma cells.** Because the B cells act indirectly through the antibodies that their progeny release into the bloodstream (or other body fluids), they are said to provide **humoral immunity.** T cell clones are more diverse. Although all T cell clones also contain memory cells, some clones contain *killer,* or *cytotoxic,* T cells (effector cells that directly attack virus-infected tissue cells). Others contain regulatory cells such as the *helper cells* (that help activate the B cells and killer T cells) and still others contain *suppressor cells* that can inhibit the immune response. Because the T cells act directly to destroy virus-infected cells, many bacteria, and cancer cells, and to reject foreign grafts, T cells are said to mediate **cellular immunity.**

Absence or failure of thymic differentiation of T lymphocytes results in a marked depression of both antibody and cell-mediated immune functions. Additionally, the observation that the thymus naturally involutes with age has been correlated with the relatively immune-deficient status of elderly individuals. ■

All lymphoid tissues except the thymus and bone marrow contain both T and B cell–dependent regions. The lymph node is used as the example in the following microscopic study.

MICROSCOPIC ANATOMY OF A LYMPH NODE

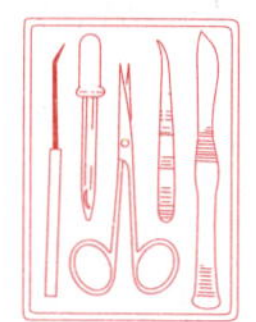

Obtain a prepared slide of a lymph node and a compound microscope. As you examine the slide, notice the following anatomical features, depicted in Figure 35.2. The node is enclosed within a fibrous **capsule,** from which connective tissue septa **(trabeculae)** extend inward to divide the node into several compartments. Very fine strands of reticular connective tissue issue from the trabeculae, forming the stroma of the gland within which cells are found.

In the outer region of the node, the **cortex,** some of the cells are arranged in globular masses, referred to as **germinal centers.** The germinal centers contain rapidly dividing B lymphocytes. The rest of the cortical cells are primarily T lymphocytes that circulate continuously, moving from the blood into the node and then exiting from the node in the lymphatic stream.

In the internal portion of the gland, the **medulla,** the cells are arranged in cordlike fashion. Most of the medullary cells are macrophages. Macrophages are important not only for their phagocytic function, but also because they play an essential role in "presenting" the antigens to the T cells.

Lymph enters the node through a number of afferent vessels, circulates through sinuses within the node, and leaves the node through efferent vessels at the **hilus.** Since each node has fewer efferent than afferent vessels, the lymph flow stagnates somewhat within the node. This allows time for the generation of an immune response and for the macrophages to remove debris from the lymph before it reenters the blood vascular system.

In the space provided here, draw a pie-shaped section of a lymph node, showing the detail of cells in a germinal center, sinusoids, and afferent and efferent vessels. Label all elements.

Before leaving the topic of the lymphoid organs, compare and contrast the structure of the lymph node examined here with the microscopic anatomy of the spleen and tonsils illustrated in Plates 28 and 29 in the Histology Atlas.

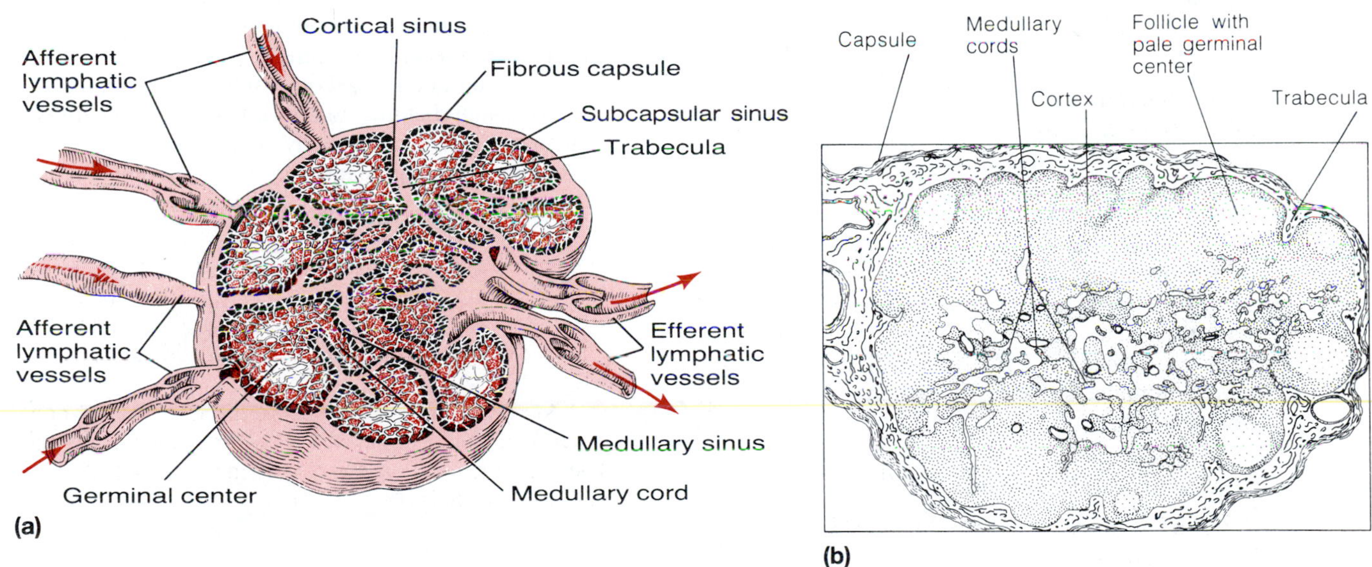

F35.2

Structure of lymph node. (a) Cross section of a lymph node, diagrammatic view. Notice that the afferent vessels outnumber the efferent vessels, which slows the rate of lymph flow. The arrows indicate the direction of the lymph flow. **(b)** Line drawing of a photomicrograph (Plate 27 in the Histology Atlas) showing a part of a lymph node.

ANTIBODIES AND TESTS FOR THEIR PRESENCE

Antibodies, or **immunoglobulins,** produced by sensitized B cells and their plasma cell offspring in response to an antigen, are a heterogenous group of proteins that comprise the general class of plasma proteins called **gamma globulins.** Antibodies are found not only in plasma, but also (to greater or lesser extents) in all body secretions. Five major classes of immunoglobulins (Igs) have been identified: IgM, IgG, IgD, IgA, and IgE. The immunoglobulin classes share a common basic structure, but differ functionally and in their localization in the body.

All Igs are composed of one or more monomers (structural units). A monomer consists of four protein chains bound together by disulfide bridges (Figure 35.3). Two of the chains are quite large and have a high molecular weight; these are the **heavy chains.** The other two chains are only half as long and have a low molecular weight. These are called **light chains.** The two heavy chains have a *constant (C) region,* in which the amino acid sequence is identical in both chains, and a *variable (V) region,* which differs in the Igs formed in response to different antigens. The same is true of the two light chains; each has a constant and a variable region.

The intact Ig molecule has a three-dimensional shape that is generally Y-shaped. Together, the variable regions of the light and heavy chains in each "arm" construct one **antigen-binding site** uniquely shaped to "fit" a specific *antigenic determinant* (portion) of an antigen. Thus, each Ig monomer bears two identical sites that bind to a specific (and the same) antigen. Binding of the immunoglobulins to their complementary antigen(s) effectively immobilizes the antigens until they can be phagocytized or lysed by complement fixation.

Although the role of the immune system is to protect the body, symptoms of certain diseases involve excessively high antibody synthesis (as in multiple myeloma, a cancer of the bone marrow and adjacent bony structures) and/or the production of abnormal antibodies (such as **rheumatoid factor** present in the blood of many rheumatoid arthritis sufferers). The simple experiment described below tests for the presence of the rheumatoid factor in plasma.

Test for Rheumatoid Factor

Although the precise cause of rheumatoid arthritis is not yet known, the disease process itself involves an astonishing number of the body's immune elements—T cells, rheumatoid factor, and antirheumatoid antibodies, to name a few. Rheumatoid arthritis is an autoimmune reaction, and rheumatoid factor (while not fully diagnostic for the disease) is found in the serum of many rheumatoid arthritis victims. Perhaps the initial disease trigger is a virus (or bacterium) bearing molecules similar to some naturally present in the joints, which in turn prompts the clustering, at certain body joints, of T cells, protective antibodies, and other inflammatory cells.

Although rheumatoid arthritis is a systemic disease that may involve many body organs and tissues, it is characterized by the manner in which the joints are af-

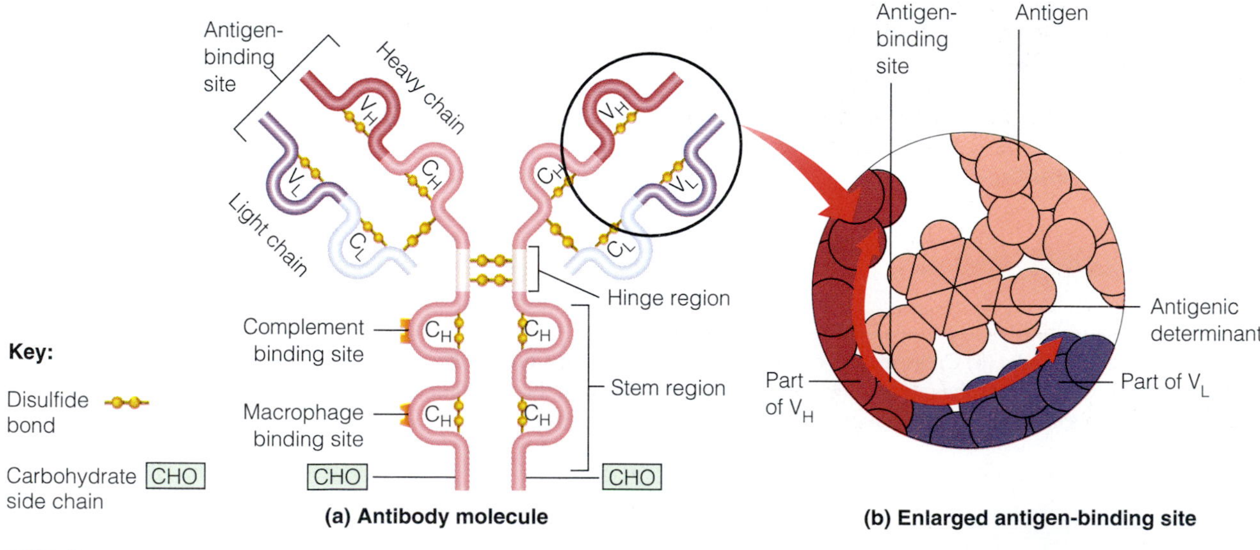

F35.3

Structure of an immunoglobulin monomer. (a) Each monomer is composed of four protein chains (two heavy chains and two light chains) connected by disulfide bonds. Both the heavy and light chains have regions of constant amino acid sequence (C regions) and regions of variable amino acid sequence (V regions). The variable regions differ in each type of antibody and construct the antigen-binding sites. Each immunoglobulin monomer has two such antigen-specific sites. **(b)** Enlargement of an antigen-binding site of an immunoglobulin.

fected. The synovial membranes become inflamed and then thicken as inflammatory cells accumulate. This is followed by the formation of **pannus,** an abnormal tissue that clings to the articular cartilages and eventually destroys them. The final stage involves the elaboration of fibrous tissue that connects the bone ends. Because this fibrous tissue eventually ossifies, the bone ends become firmly fused, and often deformed (Figure 35.4). Not all cases progress to joint immobilization, the severely crippling stage, but all cases do involve some restriction of joint movement and extreme pain.

The test for the rheumatoid factor involves mixing a drop of plasma with serum containing antibodies against the abnormal antibody (that is, antirheumatoid-factor antibodies). If rheumatoid factor is present, antigen-antibody complexes will form, causing agglutination.

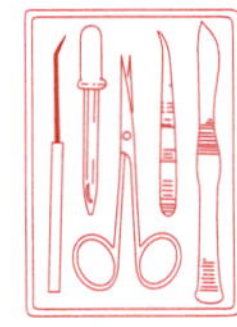

1. Obtain the arthritis screening test materials, a microscope slide, a wax marker, two applicator sticks, and plastic gloves.

2. With the wax marker, label one end of the slide "control" and the other end "test." Don the gloves.

3. Place a drop of test plasma on the end of the slide marked "test" and a drop of the positive control plasma on the end marked "control."

4. Add one drop of reagent 1 to each sample, and mix with an applicator stick. (Use opposite ends of the applicator stick to mix each sample, and then discard the applicator stick in the disposable autoclave bag provided at the supply area.)

5. Add 2 drops of reagent 2 to each sample. Using opposite ends of a fresh applicator stick, mix each sample thoroughly; and then spread the samples over a 1-inch-square area on the slide.

6. Gently rock the slide back and forth for 3 minutes, and then check each sample for the presence of agglutination.

Which sample gave a positive test for the presence of rheumatoid factor?

____

7. Dispose of the second applicator stick and the gloves in the autoclave bag.

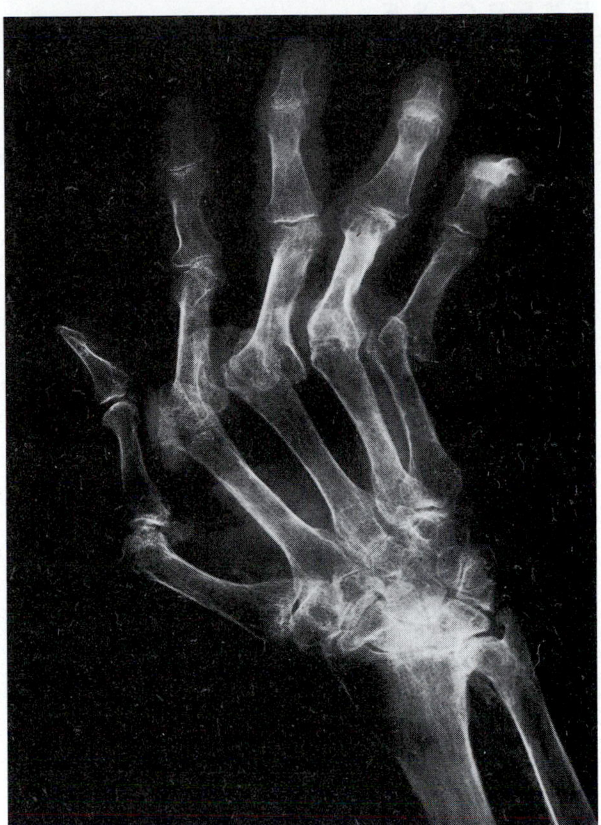

F35.4

X-ray image of a hand with rheumatoid arthritis. Notice the enlarged joints, a product of inflammation of the synovial membrane. Also note evidence of ankylosis (bone fusion) in the ring finger.

36

EXERCISE

Anatomy of the Respiratory System

OBJECTIVES

1. To define the following terms: *respiratory system, pulmonary ventilation, external respiration, and internal respiration.*
2. To label the major respiratory system structures on a diagram (or identify them on a model or dissection) and to describe the function of each.
3. To recognize the histologic structure of the trachea (cross section) and lung tissue on prepared slides, and describe the functions the observed structural modifications serve.

MATERIALS

Human torso model
Respiratory organ system model and/or chart of the respiratory system
Larynx model (if available)
Preserved inflatable lung preparation (obtained from a biological supply house) or sheep pluck fresh from the slaughterhouse
Source of compressed air*
2-foot length of laboratory rubber tubing
Protective skin cream or disposable gloves

Histologic slides of the following (if available): trachea (cross section), lung tissue, both normal and pathologic specimens (e.g., sections taken from lung tissues exhibiting bronchitis, pneumonia, emphysema, or lung cancer)
Compound and dissecting microscopes

See Appendix D, Exercise 36 for links to A.D.A.M. Standard.

See Appendix E, Exercise 36 for links to *Anatomy and PhysioShow: The Videodisc.*

* If a compressed air source is not available, cardboard mouthpieces that fit the cut end of the rubber tubing should be available for student use. Disposable autoclave bags should also be provided for discarding the mouthpiece.

Body cells require an abundant and continuous supply of oxygen. As the cells use oxygen, they release carbon dioxide, a waste product that the body must get rid of. These oxygen-using cellular processes, collectively referred to as *cellular respiration,* are more appropriately described in conjunction with the topic of cellular metabolism. The major role of the **respiratory system,** our focus in this exercise, is to supply the body with oxygen and dispose of carbon dioxide. For it to fulfill this role, at least four distinct processes, collectively referred to as **respiration,** must occur:

Pulmonary ventilation: the tidelike movement of air into and out of the lungs so that the gases in the alveoli are continuously changed and refreshed. Also more simply called *ventilation,* or *breathing.*

External respiration: the gas exchanges to and from the pulmonary circuit blood that occur in the lungs (oxygen loading/carbon dioxide unloading).

Transport of respiratory gases: the transport of respiratory gases between the lungs and tissue cells of the body accomplished by the cardiovascular system, using blood as the transport vehicle.

Internal respiration: exchange of gases to and from the blood capillaries of the systemic circulation (oxygen unloading and carbon dioxide loading).

Only the first two processes are the exclusive province of the respiratory system, but all four must occur for the respiratory system to "do its job." Hence, the respiratory and circulatory system are irreversibly linked. Should either system fail, the cells begin to die from oxygen starvation and the accumulation of carbon dioxide. If uncorrected, this situation soon causes death of the entire organism.

UPPER RESPIRATORY SYSTEM STRUCTURES

The upper respiratory system structures—the nose, pharynx, and larynx—are shown in Figure 36.1 and described below. As you read through the descriptions, identify each structure in the figure.

Air generally passes into the respiratory tract through the **external nares** (nostrils), and enters the **nasal cavity** (divided by the **nasal septum**). It then flows posteriorly over three pairs of lobelike structures, the **inferior, superior,** and **middle nasal conchae,** which increase the air turbulence. As the air passes through the nasal cavity, it is also warmed, moistened, and filtered by the nasal mucosa. The air that flows di-

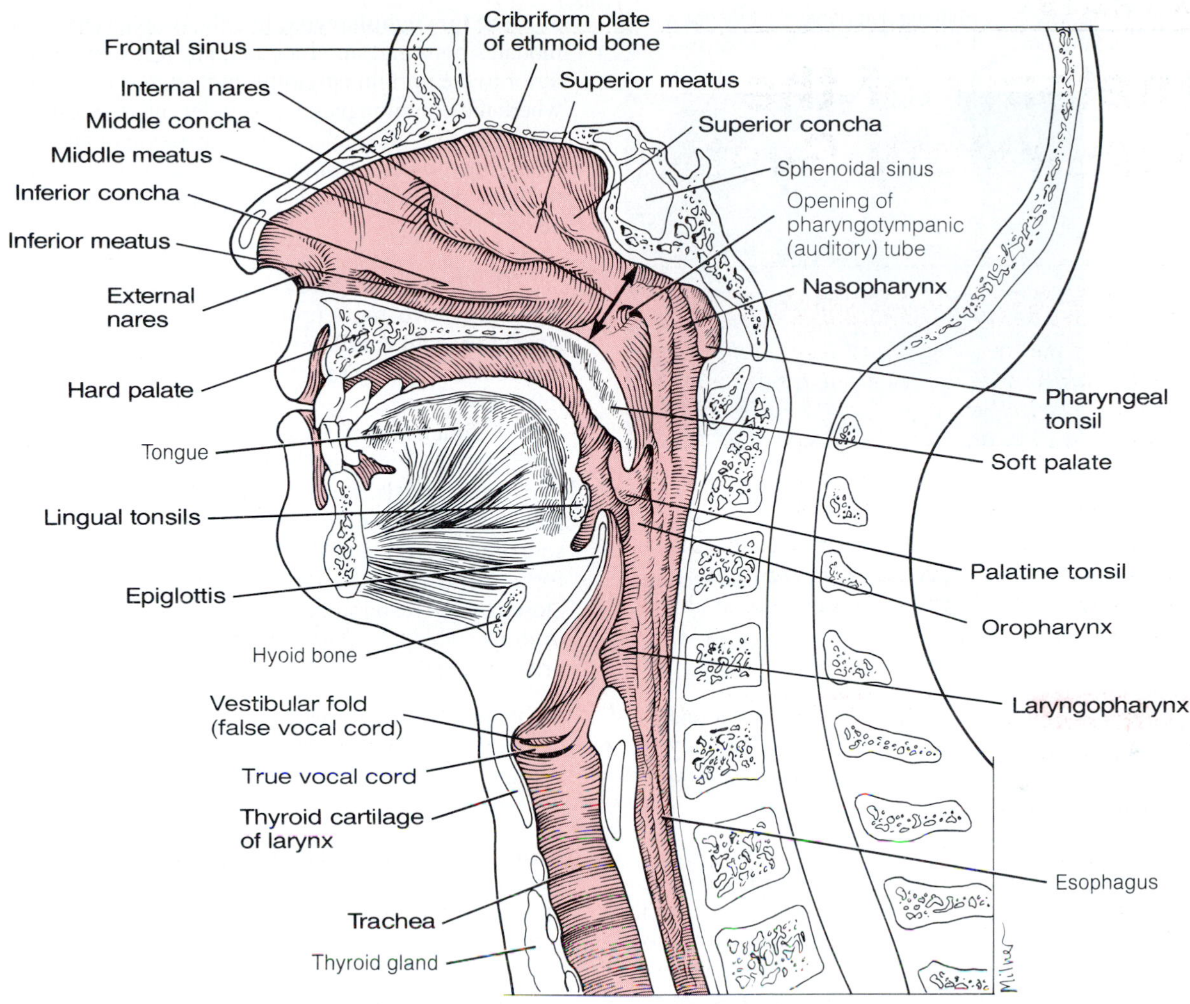

F36.1

Structures of the upper respiratory tract (sagittal section). (a) Diagrammatic view.

rectly beneath the upper part of the nasal cavity may chemically stimulate the olfactory receptors located in the mucosa of that region. The nasal cavity is surrounded by the **paranasal sinuses** in the frontal, sphenoid, ethmoid, and maxillary bones. These sinuses act as resonance chambers in speech and their mucosae, like that of the nasal cavity, warm and moisten the incoming air.

The nasal passages are separated from the oral cavity below by a partition composed anteriorly of the **hard palate** and posteriorly by the **soft palate.**

The genetic defect called **cleft palate** (failure of the palatine bones and/or the palatine processes of the maxillary bones to fuse medially) causes difficulty in breathing and oral cavity functions such as sucking and, later, mastication and speech. ■

Needless to say, air may also enter the body via the mouth, and pass through the oral cavity to move into the pharynx posteriorly, where the oral and nasal cavities are joined temporarily.

Commonly called the *throat,* the funnel-shaped **pharynx** connects the nasal and oral cavities to the larynx and esophagus inferiorly. It has three named parts:

1. The **nasopharynx** lies posterior to the nasal cavity and is continuous with it via the **internal nares.** It lies above the soft palate; hence, it serves only as an air passage. High on its posterior wall are the *pharyngeal tonsils,* paired masses of lymphoid tissue that help to protect the respiratory passages from invading pathogens. The *pharyngotympanic (auditory) tubes,* which allow middle ear pressure to become equalized to atmospheric pressure, drain into the lateral aspects of the nasopharynx.

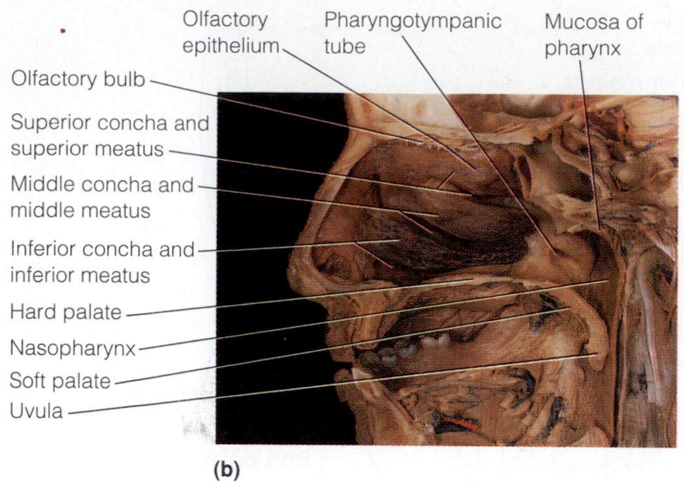

F36.1 (*continued*)

Structures of the upper respiratory tract (sagittal section). (**b**) Photograph.

Because of the continuity of the middle ear and nasopharyngeal mucosae, nasal infections may invade the middle ear cavity causing *otitis media,* which is difficult to treat. ■

2. The **oropharynx** is continuous posteriorly with the oral cavity. Since it extends from the soft palate to the epiglottis of the larynx inferiorly, it serves as a common conduit for food and air. In its lateral walls are the *palatine tonsils.* The *lingual tonsils* cover the base of the tongue.

3. The **laryngopharynx,** like the oropharynx, accommodates both ingested food and air. It lies directly posterior to the upright epiglottis and extends to the larynx, where the common pathway divides into the respiratory and digestive channels. From the laryngopharynx, air enters the lower respiratory passageways by passing through the larynx (voice box) and into the trachea below.

The **larynx** (Figure 36.2) consists of nine cartilages. The two most prominent are the large shield-shaped **thyroid cartilage,** whose anterior medial prominence is commonly referred to as *Adam's apple,* and the inferiorly located, ring-shaped **cricoid cartilage,** whose widest dimension faces posteriorly. All the laryngeal cartilages are composed of hyaline cartilage except the flaplike **epiglottis,** a flexible elastic cartilage located superior to the opening of the larynx. The epiglottis, sometimes referred to as the "guardian of the airways," forms a lid over the larynx when we swallow. This closes off the respiratory passageways to incoming food or drink, which is routed into the posterior esophagus, or food chute.

- Palpate your larynx by placing your hand on the anterior neck surface approximately halfway down its length. Swallow. Can you feel the cartilaginous larynx rising?

If anything other than air enters the larynx, a cough reflex attempts to expel the substance. Note that this reflex operates only when a person is conscious; thus you should never try to feed or pour liquids down the throat of an unconscious person.

The mucous membrane of the larynx is thrown into two pairs of folds—the upper **vestibular folds,** also

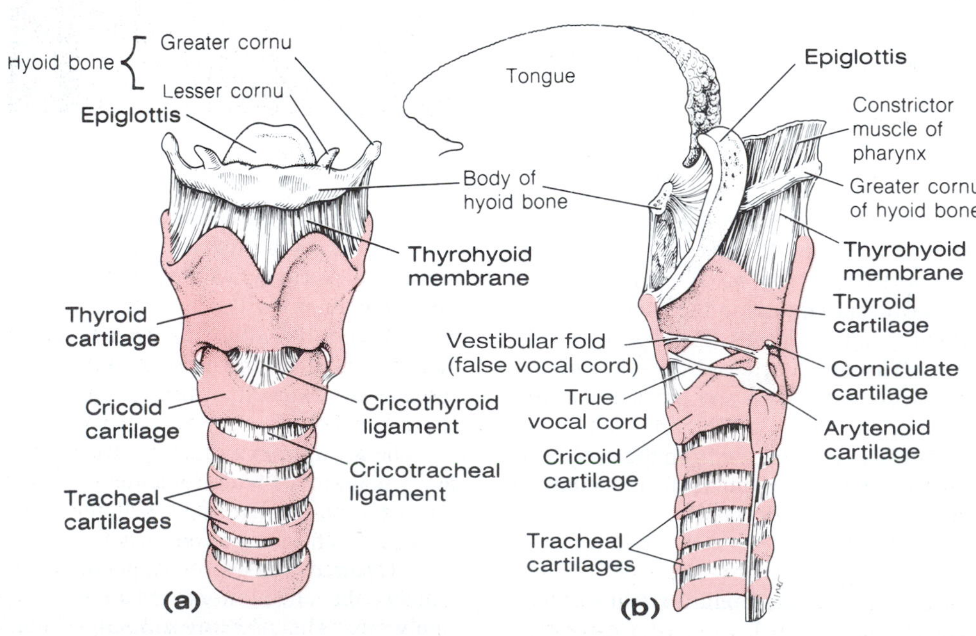

F36.2

Structure of the larynx. (**a**) Anterior view. (**b**) Sagittal section.

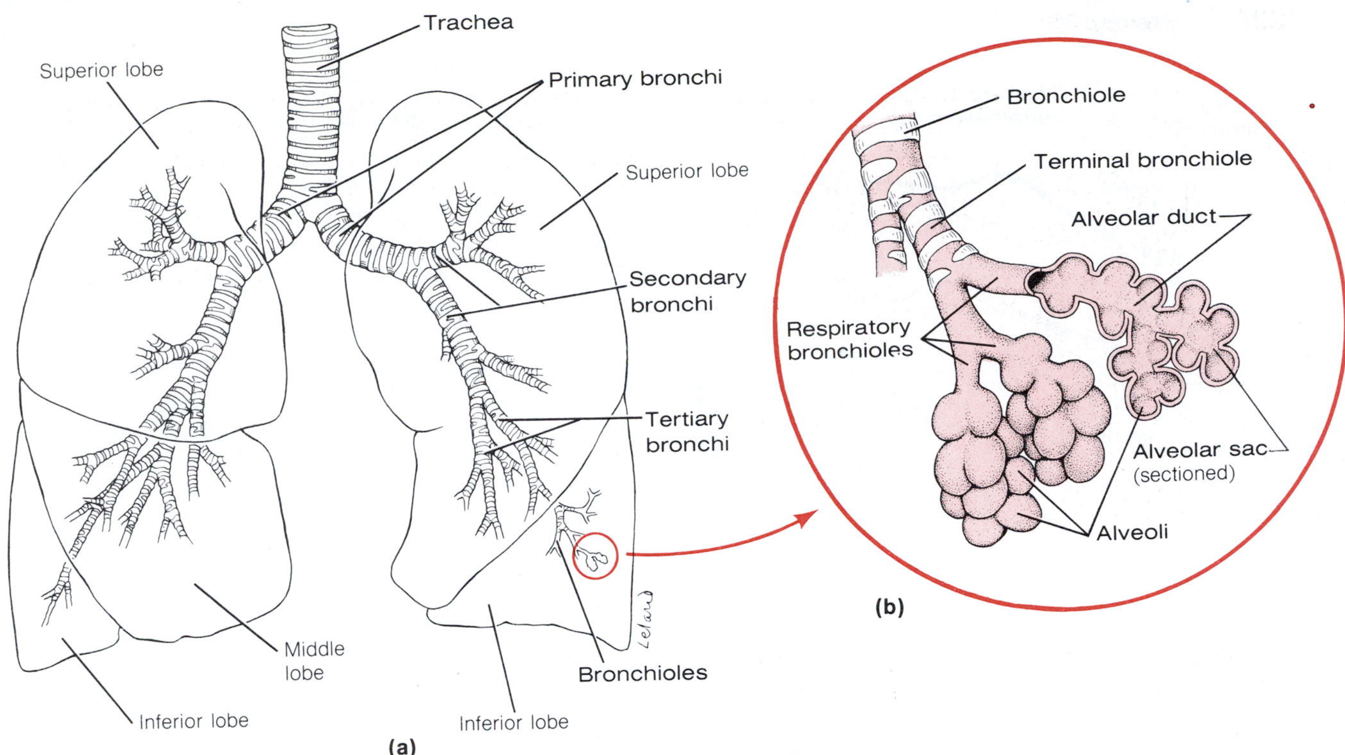

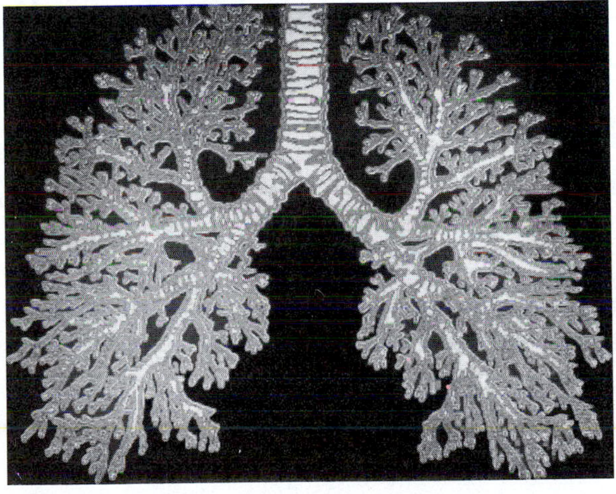

F36.3

Structures of the lower respiratory tract. **(a)** Diagrammatic view. **(b)** Inset shows enlarged view of alveoli. **(c)** Resin cast showing the extensive branching of the respiratory tree.

called the **false vocal cords,** and the lower **vocal folds,** or **true vocal cords,** which vibrate with expelled air for speech. The vocal cords are attached posterolaterally to the small triangular **arytenoid cartilages** by the *vocal ligaments.* The slitlike passageway between the folds is called the **glottis.**

LOWER RESPIRATORY SYSTEM STRUCTURES

Air entering the **trachea,** or windpipe, from the larynx travels down its length (about 11.0 cm) to the level of the *sternal angle* (or the disc between the fourth and fifth thoracic vertebrae). There the passageway divides into the right and left **primary bronchi** (Figure 36.3), which plunge into their respective lungs at an indented area called the **hilus** (see Figure 36.5b). The right primary bronchus is wider, shorter, and more vertical than the left. As a result, foreign objects that enter the respiratory passageways are more likely to become lodged in it.

The trachea is lined with a ciliated mucus-secreting, pseudostratified columnar epithelium as are many of the other respiratory system passageways. The cilia propel mucus (produced by goblet cells) laden with dust particles, bacteria, and other debris away from the lungs and toward the throat, where it can be expectorated or swallowed. The walls of the trachea are reinforced with C-shaped cartilage rings, the incomplete portion located posteriorly. These C-shaped cartilages serve a double function: The incomplete parts allow the esophagus to expand anteriorly when a large food bolus is swallowed. The solid portions reinforce the trachea walls to maintain its open passageway regardless of the pressure changes that occur during breathing.

The primary bronchi further divide into smaller and smaller branches (the secondary, tertiary, on down), finally becoming the **bronchioles,** which have terminal branches called **respiratory bronchioles** (Figure 36.3b). All but the most minute branches have cartilaginous reinforcements in their walls, usually in the form

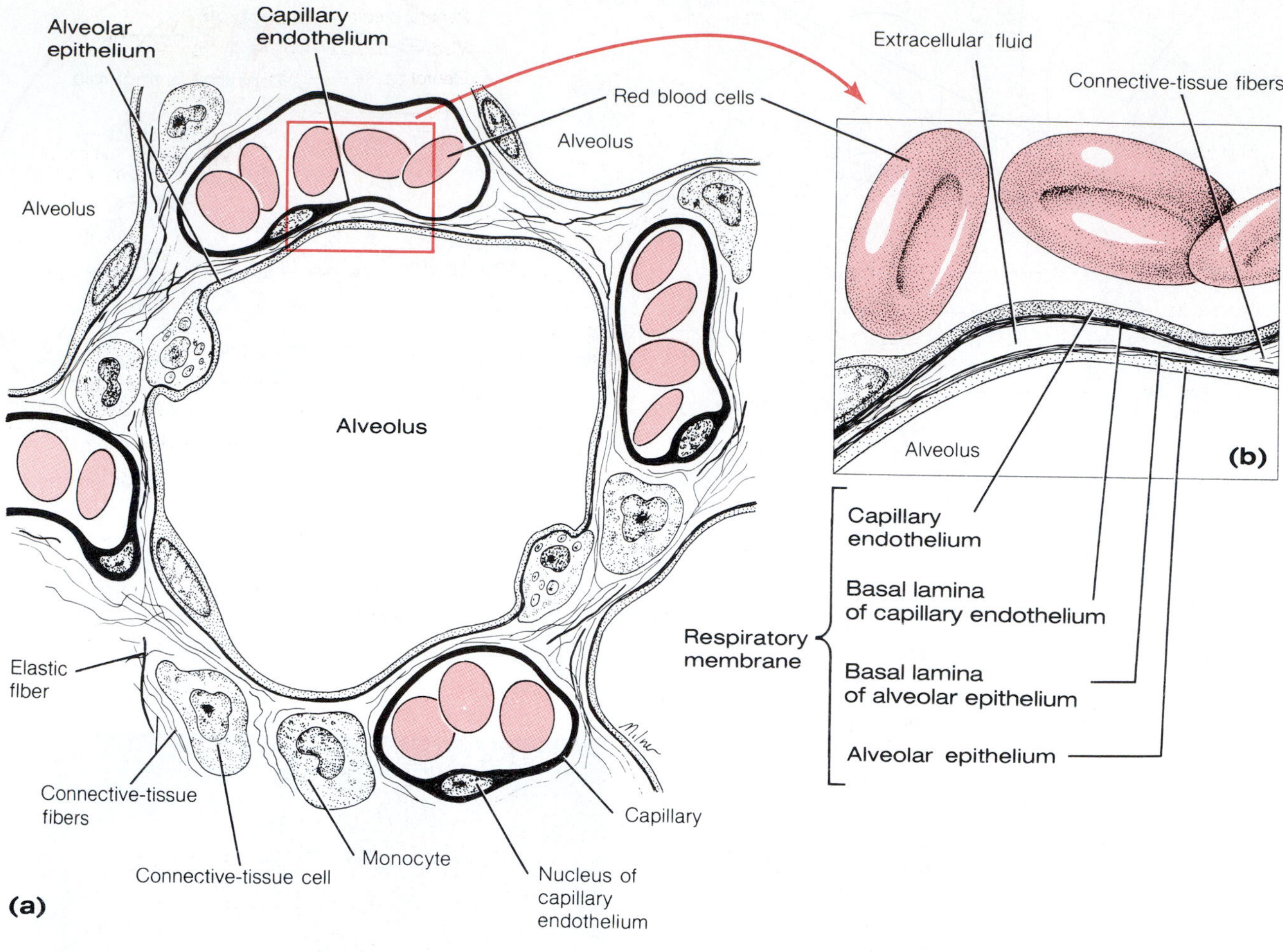

F36.4

Diagrammatic view of the relationship between the alveoli and pulmonary capillaries involved in gas exchange. (a) One alveolus surrounded by capillaries. **(b)** Enlargement of the respiratory membrane.

of small plates of hyaline cartilage rather than cartilaginous rings. As the respiratory tubes get smaller and smaller, the relative amount of smooth muscle in their walls increases as the amount of cartilage declines and finally disappears. The complete layer of smooth muscle present in the bronchioles enables them to provide considerable resistance to air flow under certain conditions (asthma, hay fever, etc.). The continuous branching of the respiratory passageways in the lungs is often referred to as the **respiratory tree.** The comparison becomes much more meaningful if you observe a resin cast of the respiratory passages. (Do so, if one is available for observation in the laboratory; otherwise, refer to Figure 36.3c.)

The respiratory bronchioles in turn subdivide into several **alveolar ducts,** which terminate in alveolar sacs that rather resemble clusters of grapes. **Alveoli,** tiny balloonlike expansions along the alveolar sacs, and occasionally found protruding from the alveolar ducts and respiratory bronchioles, are composed of a single thin layer of squamous epithelium overlying a wispy basal lamina. The external surfaces of the alveoli are densely spiderwebbed with a network of pulmonary capillaries (Figure 36.4). Together, the alveolar and capillary walls and their fused basal laminas form the **respiratory membrane.** Because gas exchanges occur by simple diffusion across the respiratory membrane—oxygen passing from the alveolar air to the capillary blood and carbon dioxide leaving the capillary blood to enter the alveolar air—the alveolar sacs, alveolar ducts, and respiratory bronchioles are referred to collectively as **respiratory zone structures.** All other respiratory passageways (from the nasal cavity to the bronchioles), which simply serve as access or exit routes to and from these gas exchange chambers, are called **conducting zone structures.** Because the conducting zone structures have no exchange function, they are also referred to as *anatomical dead space.*

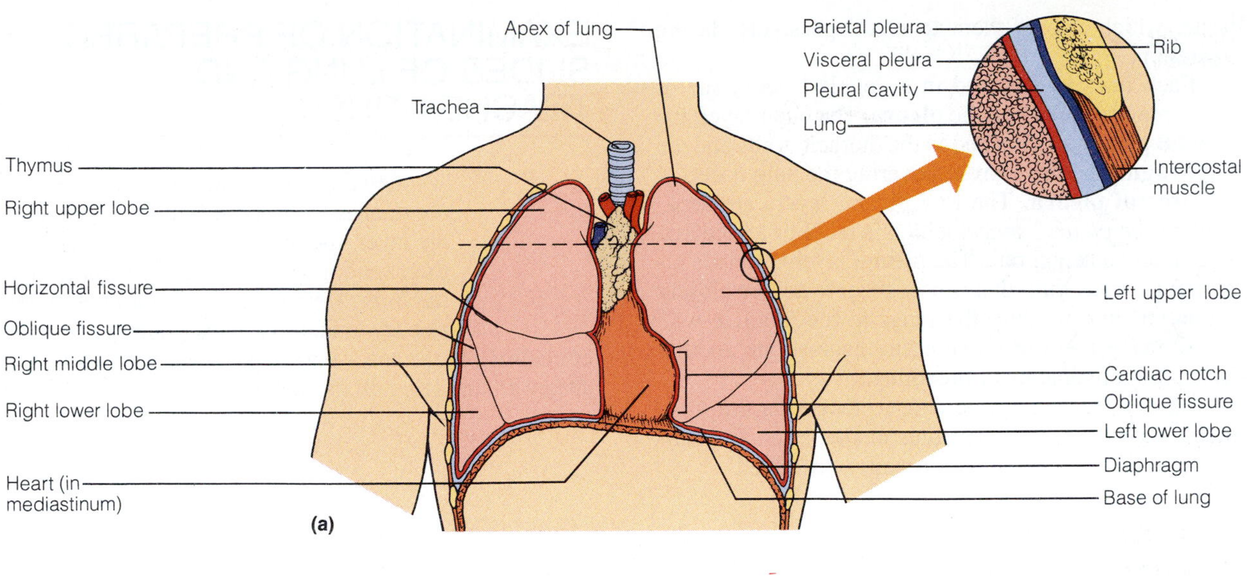

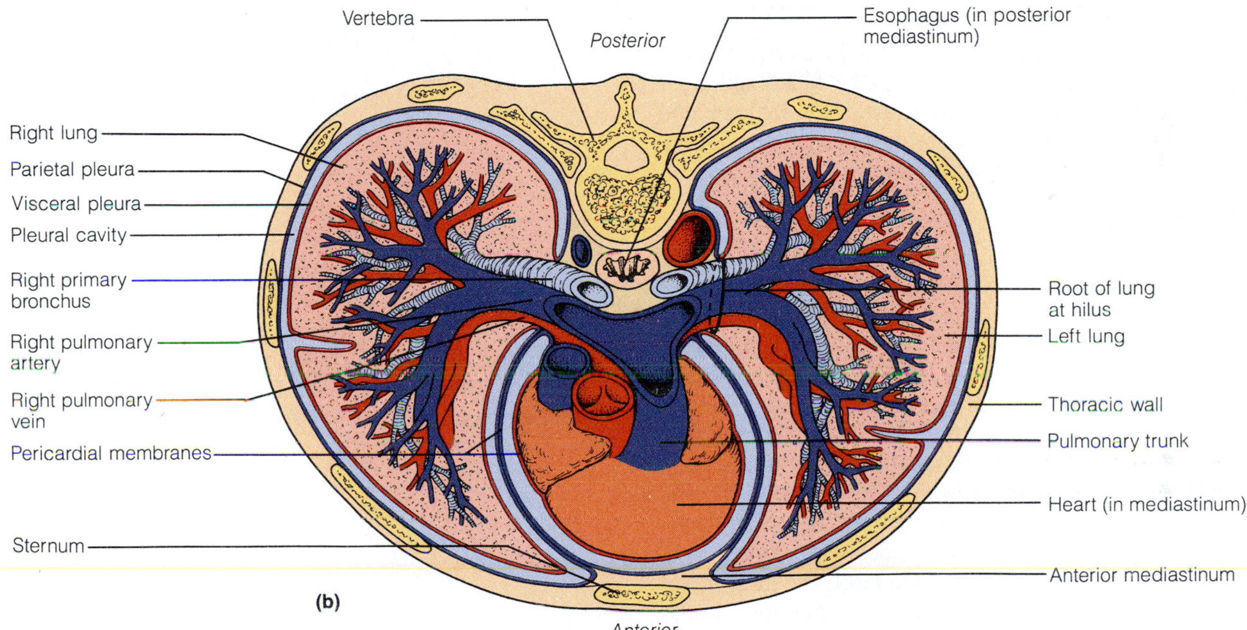

F36.5

Anatomical relationships of organs in the thoracic cavity. (a) Anterior view of the thoracic organs. The lungs flank the central mediastinum. The inset at upper right depicts the pleura and the pleural cavity. **(b)** Transverse section through the superior part of the thorax, showing the lungs and the main organs in the mediastinum. The plane of section is shown by the dotted line in part (a).

The Lungs and Their Pleural Coverings

The paired lungs are soft, spongy organs that occupy the entire thoracic cavity except for the *mediastinum,* which houses the heart, bronchi, esophagus, and other organs (Figure 36.5 and Plate D in the Human Anatomy Atlas). Each lung is connected to the mediastinum by a *root* containing its vascular and bronchial attachments. The structures of the root enter (or leave) the lung via a medial indentation called the *hilus.* All structures distal to the primary bronchi are found within the lung substance. A lung's *apex,* the narrower superior aspect, lies just deep to the clavicle, and its *base,* the inferior concave surface, rests on the diaphragm. Anterior, lateral, and posterior lung surfaces are in close contact with the ribs. The medial surface of the left lung exhibits a concavity called the *cardiac notch* (see Plate E in the Human Anatomy Atlas), which accommodates the heart where it extends left from the body midline. Fissures divide the lungs into a number of lobes—two in the left lung and three in the right. Other than the respiratory passageways and air spaces that make up the bulk of their volume, the lungs are mostly elastic connective tissue, which allows them to recoil passively during expiration.

Each lung is enclosed in a double-layered sac of serous membrane called the **pleura.** The outer layer, the **parietal pleura,** is attached to the thoracic walls and the **diaphragm;** the inner layer, covering the lung tissue, is the **visceral pleura.** The two pleural layers are separated by the *pleural space,* which is more of a potential space than an actual one. The pleural layers produce lubricating serous fluid that causes them to adhere closely to one another, holding the lungs to the thoracic wall and allowing them to move easily against one another during the movements of breathing.

Before proceeding, be sure to locate on the torso model, thoracic cavity structures model, or an anatomical chart, all the respiratory structures described.

SHEEP PLUCK DEMONSTRATION

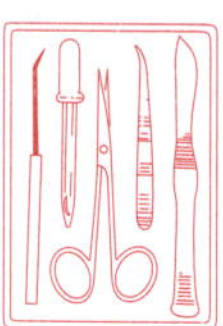

A *sheep pluck* includes the larynx, trachea with attached lungs, the heart and pericardium, and portions of the major blood vessels found in the mediastinum (aorta, pulmonary artery and vein, venae cavae).

Don plastic gloves, obtain a fresh sheep pluck (or a preserved pluck of another animal), and identify the lower respiratory system organs. Once you have completed your observations, insert a hose from an air compressor (vacuum pump) into the trachea and alternately allow air to flow in and out of the lungs. Notice how the lungs inflate. This observation is educational in a preserved pluck but it is a spectacular sight in a fresh one. Another advantage of using a fresh pluck is that the lung pluck changes color (becomes redder) as hemoglobin in trapped RBCs becomes loaded with oxygen.

If air compressors are not available, the same effect may be obtained by using a length of laboratory rubber tubing to blow into the trachea. Obtain a cardboard mouthpiece and fit it into the cut end of the laboratory tubing before attempting to inflate the lungs.

Dispose of the mouthpiece and gloves in the autoclave bag immediately after use.

EXAMINATION OF PREPARED SLIDES OF LUNG AND TRACHEA TISSUE

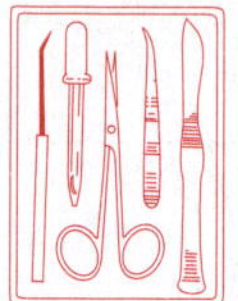

1. Obtain and examine a cross section of the trachea wall. Identify the smooth muscle layer, the hyaline cartilage supporting rings, and the pseudostratified ciliated epithelium. Using Figure 36.6a as a guide, also try to identify a few goblet cells in the epithelium. In the space below, draw a section of the trachea wall and label all tissue layers.

2. Obtain a slide of lung tissue for examination. The alveolus is the main structural and functional unit of the lung and is the actual site of gas exchange. Identify a bronchiole (Figure 36.6b) and the thin squamous epithelium of the alveolar walls (Figure 36.6c). Draw your observations of a small section of the alveolar tissue in the space below. Label the alveoli.

3. Examine slides of pathologic lung tissues, and compare them to the normal lung specimens. Record your observations in the Exercise 36 Review Sheets.

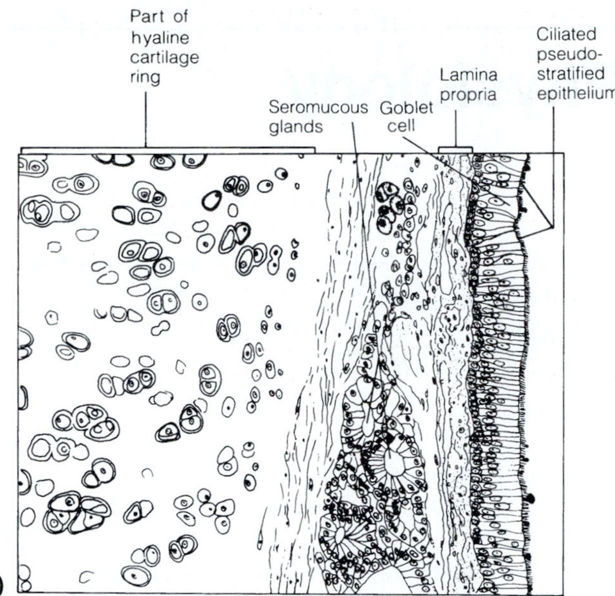

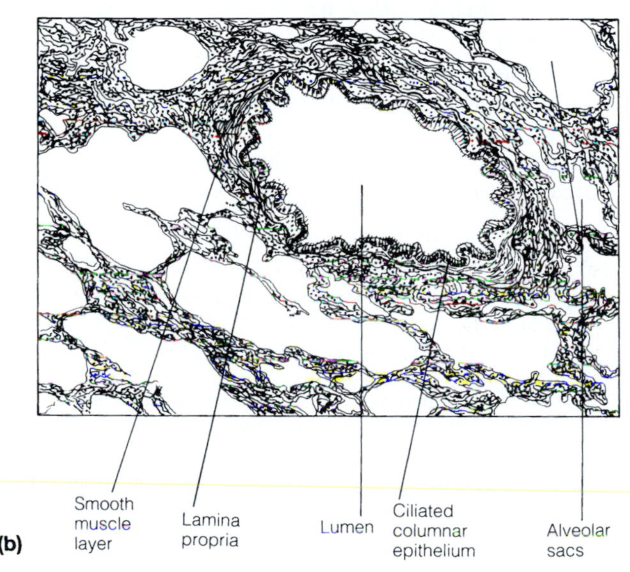

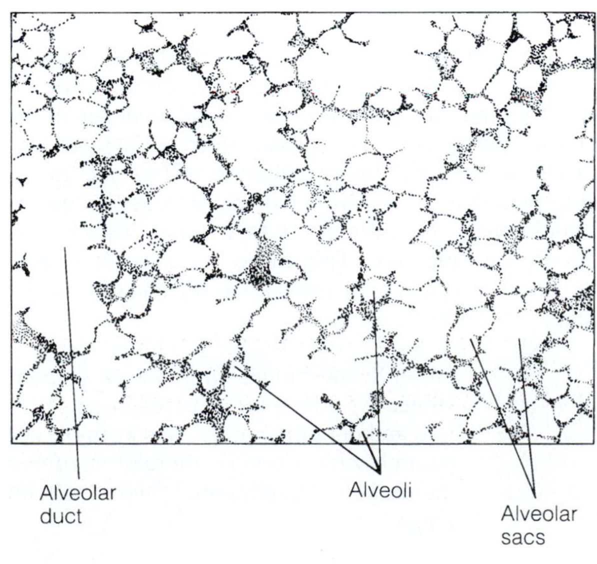

F36.6

Microscopic structure of the trachea, a bronchiole, and alveoli. **(a)** Cross section through the trachea. See corresponding Plate 31 in the Histology Atlas. **(b)** Cross-sectional view of a bronchiole. See corresponding Plate 32 in the Histology Atlas. **(c)** Alveoli. See corresponding Plate 30 in the Histology Atlas.

37 EXERCISE

Respiratory System Physiology

OBJECTIVES

1. To define the following (and be prepared to provide volume figures if applicable):

   inspiration — *expiratory reserve volume*
   expiration — *expiratory end point*
   tidal volume — *inspiratory reserve volume*
   vital capacity — *minute respiratory volume*

2. To explain the role of muscles and volume changes in the mechanical process of breathing.
3. To demonstrate proper usage of the spirometer.
4. To explain the relative importance of various mechanical and chemical factors in producing respiratory variations.
5. To describe bronchial and vesicular breathing sounds.
6. To explain the importance of the carbonic acid–bicarbonate buffer system in maintaining blood pH.

MATERIALS

Model lung (bell jar demonstrator)
Tape measure

Spirometer
Disposable mouthpieces
Nose clips
Alcohol swabs
70% ethanol solution in a battery jar
Disposable autoclave bag

Scotch tape
Paper bag
Pneumograph and recording attachments
Recording apparatus (kymograph or physiograph)*
Stethoscope

Table (on chalkboard) for recording class data

pH meter (standardized with buffer of pH 7)
Buffer solution (pH 7)
Concentrated HCl and NaOH
0.01 *M* HCl
250- and 50-ml beakers
Graduated cylinder (100 ml)
Glass stirring rod
Plastic wash bottles containing distilled water
Animal plasma

See Appendix E, Exercise 37 for links to *Anatomy and PhysioShow: The Videodisc.*

*** Exercise 37i in Appendix C provides instructions for use of Intelitool Spirocomp for this investigation.**

MECHANICS OF RESPIRATION

Pulmonary ventilation, or **breathing,** consists of two phases: **inspiration,** during which air is taken into the lungs, and **expiration,** during which air passes out of the lungs. As the inspiratory muscles (external intercostals and diaphragm) contract during inspiration, the size of the thoracic cavity increases. The diaphragm moves from its relaxed dome shape to a flattened position, increasing the superoinferior volume. The external intercostals lift the rib cage, increasing the anteroposterior and lateral dimensions (Figure 37.1). Since the lungs adhere to the thoracic walls like flypaper because of the cohesive character of the pleurae, the intrapulmonary volume (volume within the lungs) also increases, lowering the air (gas) pressure inside the lungs. The gases then expand to fill the available space, creating a partial vacuum that causes air to flow into the lungs—constituting the act of inspiration. During expiration, the inspiratory muscles relax, and the natural tendency of the elastic lung tissue to recoil acts to decrease the intrathoracic and intrapulmonary volumes. As the gas molecules within the lungs are forced closer together, the intrapulmonary pressure rises to a point higher than atmospheric pressure. This causes gases to flow from the lungs to equalize the pressure inside and outside the lungs—the act of expiration.

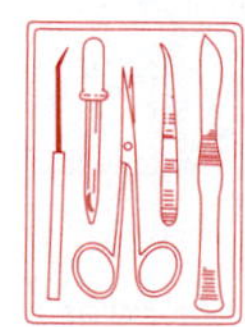

Observe the model lung, which demonstrates the principles involved in gas flows into and out of the lungs. It is a simple apparatus with a bottle "thorax," a rubber membrane "diaphragm," and "balloon lungs."

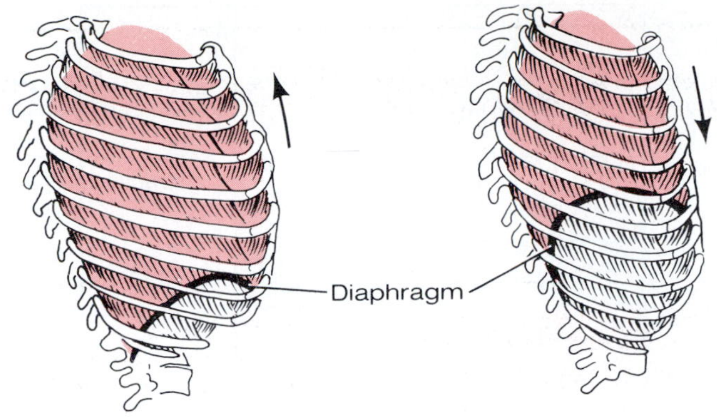

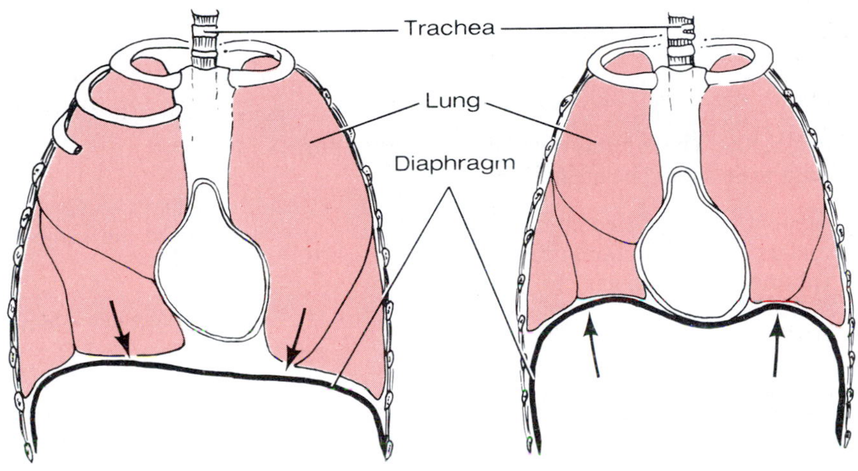

F37.1

Rib cage and diaphragm positions during breathing. (a) At the end of a normal inspiration; chest expanded, diaphragm depressed. **(b)** At the end of a normal expiration; chest depressed, diaphragm elevated.

1. Go to the demonstration area and work the model lung by moving the rubber diaphragm up and down. Notice the changes in balloon (lung) size as the volume of the thoracic cavity is alternately increased and decreased.

2. Check the appropriate columns in the chart concerning these observations in the Exercise 37 Review Sheet on p. RS 159 at the back of this book.

3. After observing the operation of the model lung, conduct the following tests on your lab partner. Use the tape measure to determine his or her chest circumference by placing the tape around the chest as high up under the armpits as possible. Record the measurements in inches in the appropriate space for each of the conditions below.

Quiet breathing:

inspiration ______________ expiration ____________

Forced breathing:

inspiration ______________ expiration ____________

Do the results coincide with what you expected on the basis of what you have learned thus far? ____________

RESPIRATORY VOLUMES AND CAPACITIES—SPIROMETRY

A person's size, sex, age, and physical condition produce variations in respiratory volumes. Normal quiet breathing moves about 500 ml of air in and out of the lungs with each breath. As you have seen in the previous experiment, a person can usually forcibly inhale or exhale much more air than is exchanged in normal quiet breathing. The terms given to the measurable respiratory

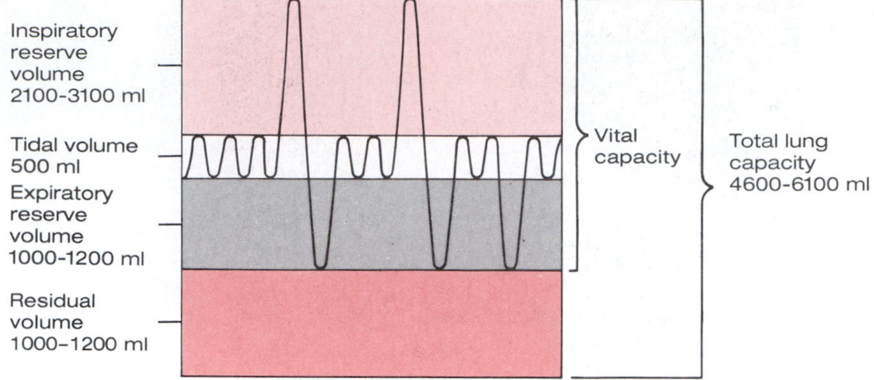

F37.2

Idealized tracing of the various respiratory volumes.

volumes are defined just below. These terms and their normal values for an adult male should be memorized.

Tidal volume (TV): amount of air inhaled or exhaled with each breath under resting conditions (500 ml)
Inspiratory reserve volume (IRV): amount of air that can be forcefully inhaled after a normal tidal volume inhalation (3100 ml)
Expiratory reserve volume (ERV): amount of air that can be forcefully exhaled after a normal tidal volume exhalation (1200 ml)
Vital capacity (VC): maximum amount of air that can be exhaled after a maximal inspiration (4800 ml)

$$VC = TV + IRV + ERV$$

An idealized tracing of the various respiratory volumes and their relationships to each other are shown in Figure 37.2.

Respiratory volumes will be measured with an apparatus called a **spirometer.** There are two major types of spirometers, which give comparable results—the hand-held dry, or wheel, spirometers (such as the Buhl spirometer illustrated in Figure 37.3) and "wet" spirometers, such as the Phipps and Bird spirometer and the Collins spirometer (which is available in both recording and nonrecording varieties). The somewhat more sophisticated wet spirometer consists of a plastic or metal *bell* within a rectangular or cylindrical tank that air can be added to or removed from (Figure 37.4). The outer tank contains water and has a tube running through it to carry air above the water level. The floating bottomless bell is inverted over the water-containing tank and connected to a volume indicator.

In nonrecording spirometers, an indicator moves as air is *exhaled,* and only expired air volumes can be measured directly. By contrast, recording spirometers allow both inspired and expired gas volumes to be measured. Directions for both types of apparatus are provided below. Follow the steps in Procedure A if using a nonrecording spirometer, and those in Procedure B if using a wet recording spirometer.

Procedure A

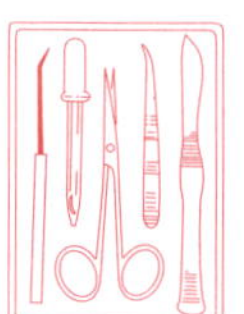

1. Without using the spirometer, count and record the subject's normal respiratory rate.

Respirations per minute ________

2. Identify the parts of the spirometer you will be using by comparing it to the illustration in Figure 37.3 or 37.4a. Examine the spirometer volume indicator *before beginning* to make sure you know how to read the scale.

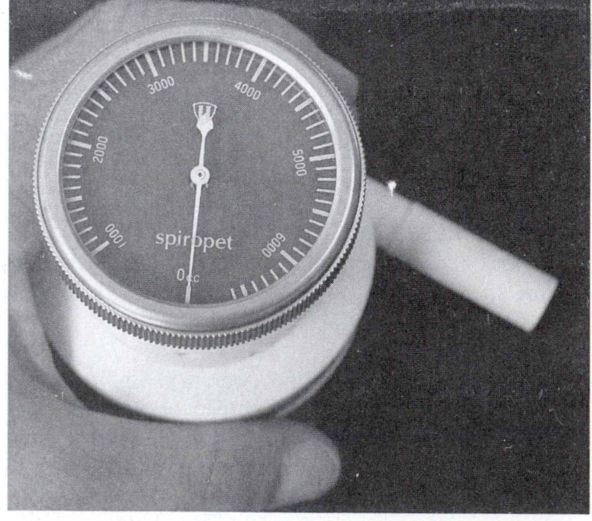

F37.3

The Buhl Spiropet, an example of a hand-held dry spirometer. The dial face of the spirometer is rotated to zero prior to each test.

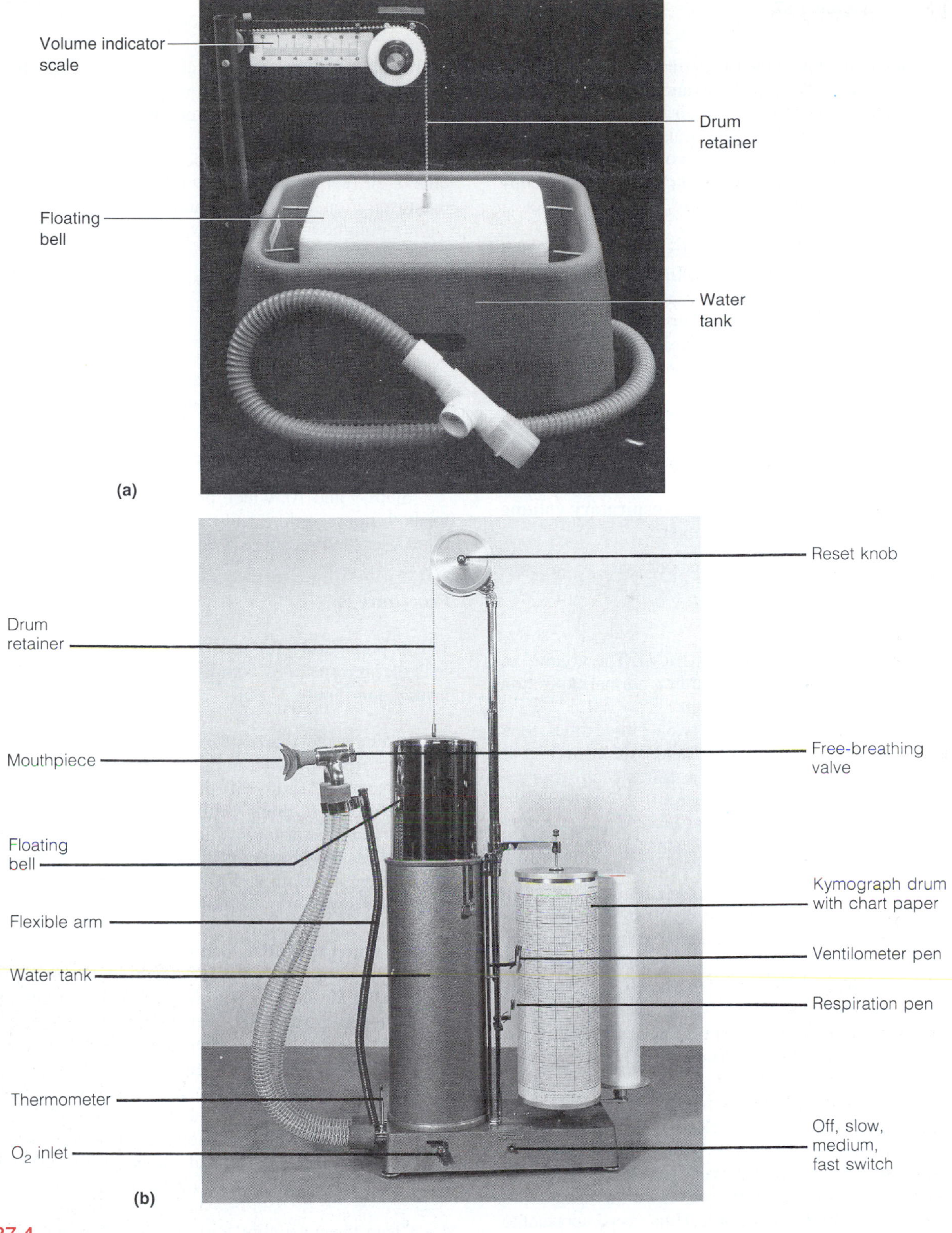

F37.4

Wet spirometers. (a) The Phipps and Bird "wet" spirometer. **(b)** The Collins-9L "wet" recording spirometer.

Work in pairs, with one person acting as the subject while the other records the data of the volume determinations. The subject should stand erect during testing. Reset the indicator to zero before beginning each trial. If you are using the hand-held spirometer, make sure its dial faces upward so that the volumes can be easily read during the tests.

Obtain a disposable cardboard mouthpiece. Insert it in the open end of the valve assembly (attached to the flexible tube) of the wet spirometer or over the fixed stem of the hand-held dry spirometer. Before beginning, the subject should practice exhaling through the mouthpiece without exhaling through the nose, or prepare to use the nose clips (clean them first with an alcohol swab).

3. Conduct the test three times for each required measurement. Record the data here, and then find the average volume figure for that respiratory measurement. After you have completed the trials and computed the averages, enter the average values on the table prepared on the chalkboard for tabulation of class data,* and copy all averaged data onto the Exercise 37 Review Sheet.

4. Tidal volume (TV). The volume of air inhaled and exhaled with each normal respiration is approximately 500 ml. To conduct the test, inhale a normal breath, and then exhale a normal breath of air into the spirometer mouthpiece. (Do not force the expiration!) Record the volume and repeat the test twice.

trial 1 ____________ ml trial 2 ____________ ml

trial 3 ____________ ml average TV __________ ml

5. Compute the subject's **minute respiratory volume (MRV)** using the following formula:

$$\text{MRV} = \text{TV} \times \text{respirations/min}$$

MRV _______ ml/min

6. Expiratory reserve volume (ERV). The volume of air that can be forcibly exhaled after a normal expiration ranges between 1000 and 1200 ml.

Inhale and exhale normally two or three times, then insert the spirometer mouthpiece and exhale forcibly as much of the additional air as you can. Record your results, and repeat the test twice again.

trial 1 ____________ ml trial 2 ______________ ml

trial 3 ____________ ml average ERV _________ ml

The ERV is dramatically reduced in conditions in which the elasticity of the lungs is decreased by a chronic obstructive pulmonary disease (COPD) such as **emphysema.** Since energy must be used to *deflate* the lungs in such conditions, expiration is physically exhausting to individuals suffering from COPD. ■

7. Vital capacity (VC). The total exchangeable air of the lungs (the sum of TV + IRV + ERV) is normally 4500 ml to 4800 ml. Breathe in and out normally two or three times, and then bend over and exhale all the air possible. Then, as you raise yourself to the upright position, inhale as fully as possible. (It is very important to *strain* to inhale the maximum amount of air that you can.) Quickly insert the mouthpiece, and exhale as forcibly as you can. Record your results and repeat the test twice again.

* Note to the Instructor: The format of class data tabulation can be similar to that shown here. However, it would be interesting to divide the class into smokers and nonsmokers and then compare the mean average VC and ERV for each group. Such a comparison might help to determine if smokers are handicapped in any way. It also might be a good opportunity for an informal discussion of the early warning signs of bronchitis and emphysema, which are primarily smokers' diseases.

trial 1 ____________ ml trial 2 ____________ ml

trial 3 ____________ ml average VC __________ ml

8. Inspiratory reserve volume (IRV). The IRV, or volume of air that can be forcibly inhaled following a normal inspiration, can now be computed using the average values obtained for TV, ERV, and VC and plugging them into the equation:

$$\text{IRV} = \text{VC} - (\text{TV} + \text{ERV})$$

Record your average IRV:________________________ ml

The normal IRV is substantial, ranging from 2100 to 3100 ml. How does your computed value compare?

____

Steps 9 and 10, which provide common directions for both nonrecording and recording spirometers, continue after the Procedure B directions.

Procedure B

1. In preparation for recording, familiarize yourself with the spirometer by comparing it to the equipment illustrated in Figure 37.4b.

2. Examine the chart paper, noting that its horizontal lines represent milliliter units. To apply the chart paper to the recording drum, lift the drum retainer and then remove the kymograph drum. Wrap a sheet of chart paper around the drum, making sure that the right edge overlaps the left. Fasten it with tape, and then replace the kymograph drum and lower the drum retainer into its original position in the hole in the top of the drum.

3. Raise and lower the floating bell several times, noting as you do so that the *ventilometer pen* moves up and down on the drum. This pen, which writes in black ink, will be used for recording and should be positioned or adjusted so that it records in the approximate middle of the chart paper. This adjustment is made by repositioning the floating bell using the *reset knob* on the metal pulley at the top of the spirometer apparatus. The other pen, the respirometer pen, which records in red ink, will not be used for these tests and should be moved away from the drum's recording surface.

4. Clean the nose clips with an alcohol swab. While you wait for the alcohol to air dry, count and record your normal respiratory rate.

Respirations per minute _______

After the alcohol has air dried, apply the nose clips to your nose. This will enforce mouth breathing.

5. Open the *free breathing valve.* Insert a disposable cardboard mouthpiece into the end (valve assembly) of the breathing tube, and then insert the mouthpiece into your mouth. Practice breathing for several breaths to get

used to the apparatus. At this time, you are still breathing room air.

6. Set the spirometer switch to SLOW (32 mm/min). Close the free breathing valve, and breathe in a normal manner for 2 minutes to record your tidal volume—the amount of air inspired or expired with each normal respiratory cycle. This recording should show a regular pattern of inspiration-expiration spikes and should gradually move upward on the chart paper. (A downward slope indicates that there is an air leak somewhere in the system—most likely at the mouthpiece.) Notice that on an apparatus using a counterweighted pen, such as the Collins Vitalometer shown in Fig. 37.4b, inspirations are recorded by upstrokes and expirations are recorded by downstrokes.*

7. To record your vital capacity, take the deepest possible inspiration you can and then exhale to the greatest extent possible (really *push* the air out). The recording obtained should resemble that shown in Figure 37.5. Repeat the vital capacity maneuver twice again. Then turn off the spirometer and remove the chart paper from the kymograph drum.

* If a Collins survey spirometer is used, the situation is exactly opposite: Upstrokes are expirations and downstrokes are inspirations.

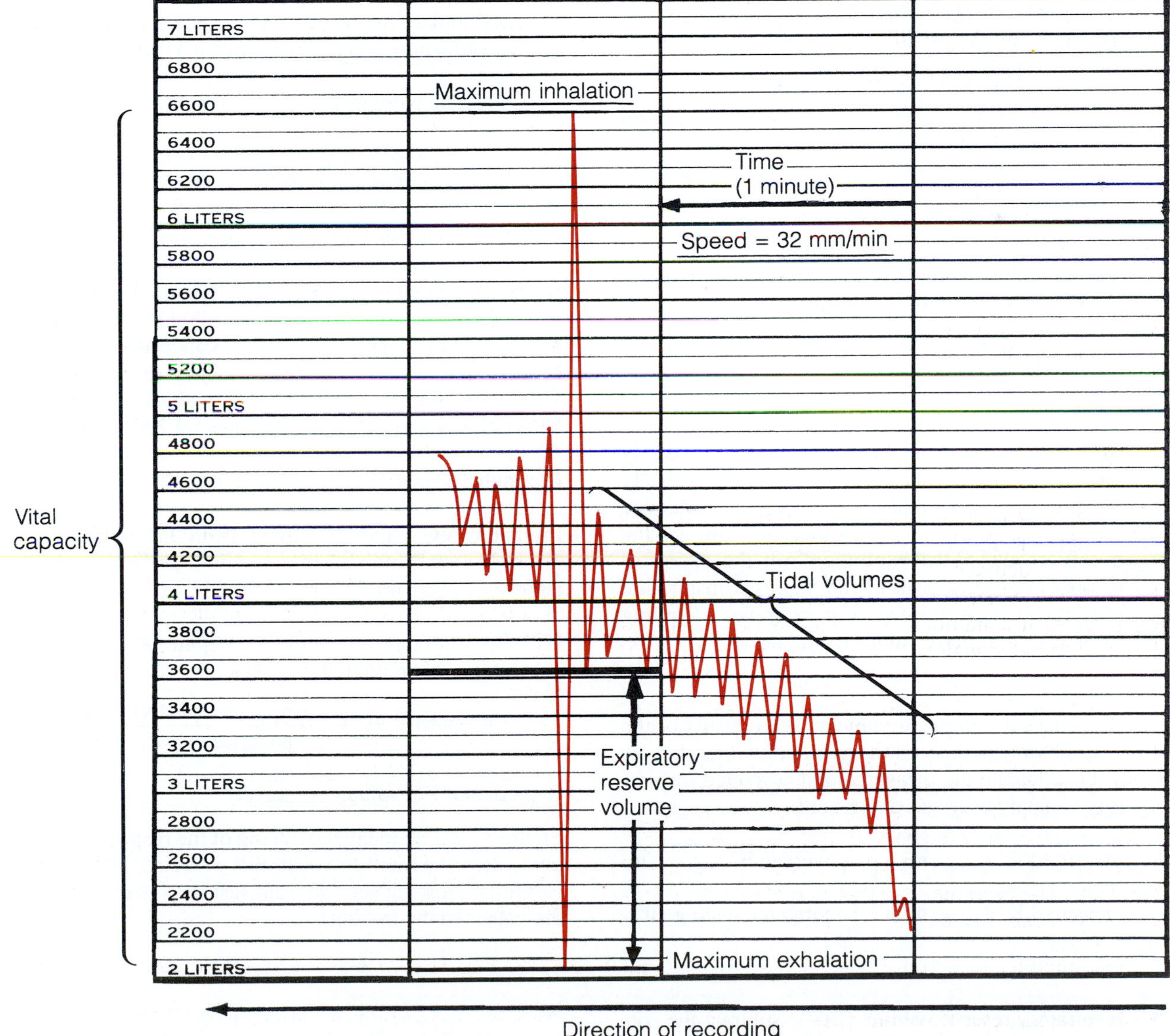

F37.5

A typical spirometry recording of tidal volume, inspiratory capacity, expiratory reserve volume, and vital capacity. At a drum speed of 32 mm/min, each vertical column of the chart represents a time interval of 1 minute.

8. Determine and record your measured, averaged, and corrected respiratory volumes here. Because the pressure and temperature inside the spirometer are influenced by room temperature and differ from those in the body, all measured values are to be multiplied by a **BTPS** (body temperature, atmospheric pressure, and water saturation) **factor.** At room temperature, the BTPS factor is typically 1.1 or very close to that value. Hence, you will multiply your measured values by 1.1 to obtain your corrected respiratory volume values. Copy the averaged and corrected values onto the Exercise 37 Review Sheet.

- Tidal volume (TV). Select a typical resting tidal breath recording. Subtract the millimeter value of the trough (exhalation) from the millimeter value of the peak (inspiration). Record this value below as *measured TV 1.* Select two other TV tracings to determine the TV values for the TV 2 and TV 3 measurements. Then, determine your average TV and multiply it by 1.1 to obtain the BTPS-corrected average TV value.

  measured TV 1 ____ ml   average TV ______ ml

  measured TV 2 ____ ml   corrected average TV

  measured TV 3 ____ ml   ______ ml

  Also compute your **minute respiratory volume (MRV)** using the following formula:

  $$\text{MRV} = \text{TV} \times \text{respirations/min}$$

  MRV _______ ml/min

- Inspiratory capacity (IC). In the first vital capacity recording, find the expiratory trough immediately preceding the maximal inspiratory peak achieved during vital capacity determination. Subtract the milliliter value of that expiration from the value corresponding to the peak of the maximal inspiration that immediately follows. For example, according to Figure 37.5, these values would be

  $$6600 - 3650 = 2950 \text{ ml}$$

  Record your computed value and the results of the two subsequent tests on the appropriate lines below. Then calculate the measured and corrected inspiratory capacity averages and record.

  measured IC 1 ____ ml   average IC ______ ml

  measured IC 2 ____ ml   corrected average IC

  measured IC 3 ____ ml   ______ ml

- Inspiratory reserve volume (IRV). Subtract the corrected average tidal volume from the corrected average for the inspiratory capacity and record below.

  $$\text{IRV} = \text{corrected average IC} - \text{corrected average TV}$$

  corrected average IRV _______ ml

- Expiratory reserve volume (ERV). Subtract the number of milliliters corresponding to the trough of the maximal expiration obtained during the vital capacity maneuver from milliliters corresponding to the last *normal* expiration before the VC maneuver is performed. For example, according to Figure 37.5, these values would be

  $$3650 \text{ ml} - 2050 \text{ ml} = 1650 \text{ ml}$$

  Record your measured and averaged values (three trials) below.

  measured ERV 1 ____ ml   average ERV ______ ml

  measured ERV 2 ____ ml   corrected average ERV

  measured ERV 3 ____ ml   ______ ml

- Vital capacity (VC). Add your corrected values for ERV and IC to obtain the corrected average VC. Record below and on the Exercise 37 Review Sheet.

  corrected average VC ______ ml

[*Now continue with step 9, whether you are following Procedure A or Procedure B.*]

9. Figure out how closely your measured average vital capacity volume compares with the *predicted values* for someone your age, sex, and height. Obtain the predicted figure from Appendix B either from Table 1 (male values) or Table 2 (female values). Notice that you will have to convert your height in inches to centimeters (cm) to find the corresponding value. This is easily done by multiplying your height in inches by 2.54.

Computed height: _______ cm

Predicted VC value (obtained from the appropriate table): _______ ml

Use the following equation to compute your VC as a percentage of the predicted VC value:

$$\% \text{ of predicted VC} = \frac{\text{averaged measured VC}}{\text{predicted value}} \times 100$$

% predicted VC value: _______%

Figure 37.2 is an idealized tracing of the respiratory volumes described and tested in this exercise. Examine it carefully. How closely do your test results compare to the values in the tracing?

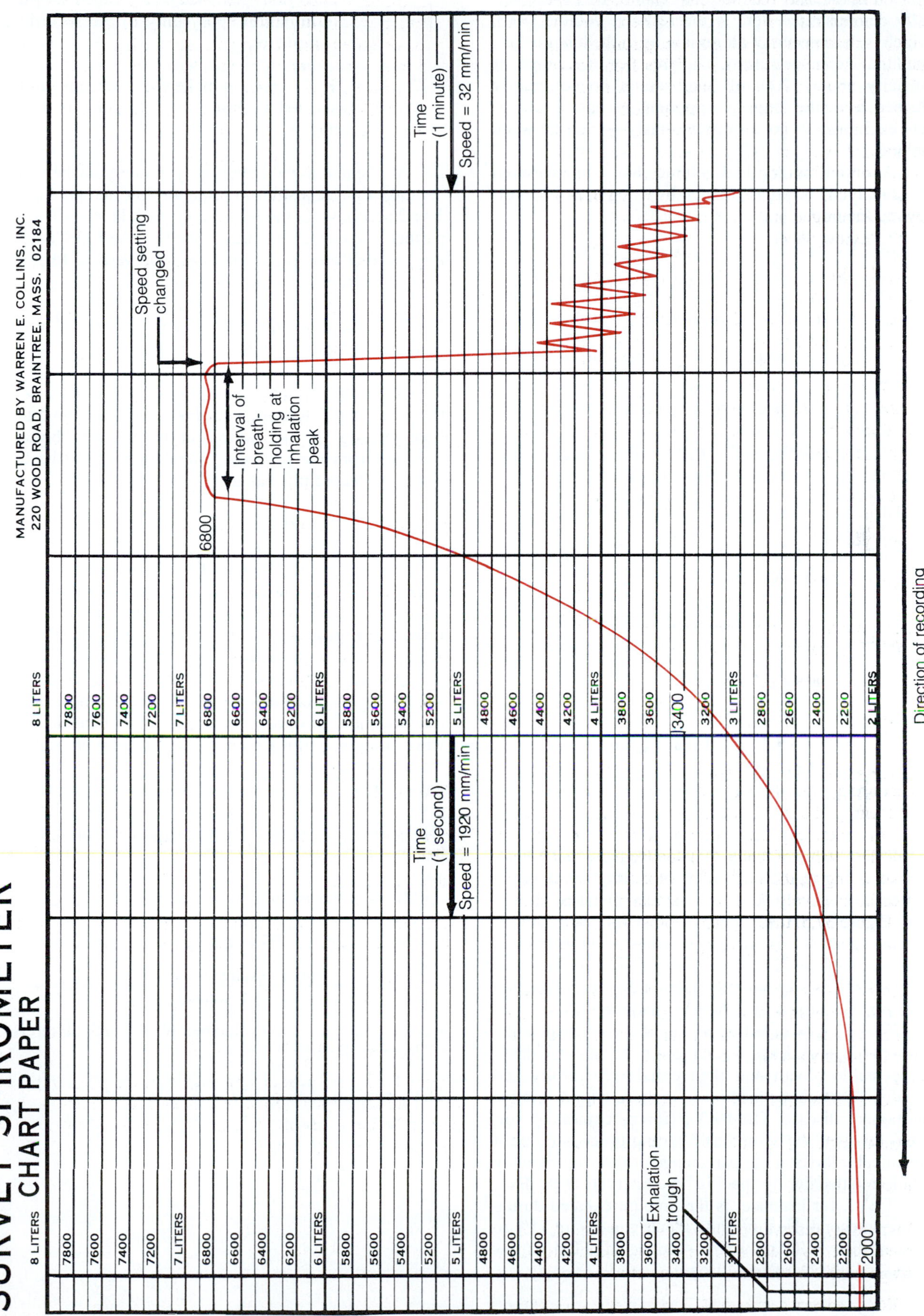

F37.6

A recording of the forced vital capacity (FVC) and forced expiratory volume (FEV) or timed vital capacity test.

10. A respiratory volume that cannot be experimentally demonstrated here is the residual volume (RV), which is the amount of air remaining in the lungs after a maximal expiratory effort. The presence of residual air (usually about 1200 ml) that cannot be voluntarily flushed from the lungs is important because it allows gas exchange to go on continuously—even between breaths.

Although the residual volume cannot be measured directly, it can be approximated by using one of the following factors:

For ages 16–34	Factor = 0.250
For ages 35–49	Factor = 1.305
For ages 50–69	Factor = 1.445

Compute your predicted RV using the following equation:

$$RV = VC \times \text{factor}$$

11. Recording is finished for this subject. Before continuing with the next member of your group:

- Dispose of used cardboard mouthpieces in the autoclave bag.
- Swish the valve assembly (if removable) in the 70% ethanol solution, then rinse with tap water.
- Put a fresh mouthpiece into the valve assembly (or on the stem of the hand-held spirometer) and continue recording all members of your group using the procedures outlined above.

FORCED EXPIRATORY VOLUME (FEV_T) MEASUREMENT

While they are not really diagnostic, pulmonary function tests can help the clinician to distinguish between obstructive and restrictive pulmonary diseases. (In obstructive disorders, like chronic bronchitis and asthma, airway resistance is increased, whereas in restrictive diseases, such as polio and tuberculosis, total lung capacity declines.) Two highly useful pulmonary function tests for this purpose are the FVC and the FEV_T.

The **FVC** (forced vital capacity) measures the amount of gas expelled when the subject takes the deepest possible breath and then exhales forcefully and rapidly. This volume is reduced in those with restrictive pulmonary disease. The **FEV_T** (forced expiratory volume) involves the same basic testing procedure, but it specifically looks at the percentage of the vital capacity that is exhaled during specific time intervals of the FVC test. FEV_1, for instance, is the amount exhaled during the first second. Healthy individuals can expire 75% to 85% of their FVC in the first second. The FEV_1 is low in those with obstructive disease.

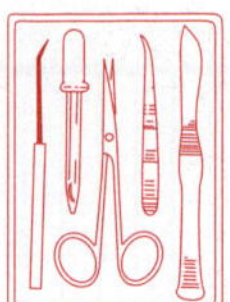

Directions provided here for the FEV_T determination apply only to the recording spirometer.

1. Prepare to make your recording as described in Procedure B, steps 1–6 just above.

2. At a signal agreed upon by you and your lab partner, take the deepest inspiration possible and hold it for 1 to 2 seconds. As the inspiratory peak levels off, your partner is to change the drum speed to FAST (1920 mm/min) so that the distance between the vertical lines on the chart represents a time of 1 second.

3. Once the drum speed has been changed, exhale as much air as rapidly and forcibly as possible.

4. When the tracing plateaus (bottoms out), stop recording and determine your FVC. Subtract the milliliter reading in the expiration trough (the bottom plateau) from the preceding inhalation peak (the top plateau). Record this value.

FVC = _______ ml

5. Prepare to calculate the FEV_1. Draw a vertical line intersecting with the spirogram tracing at the precise point that exhalation began. Identify this line as *line 1.* From line 1, measure 32 mm horizontally to the left, and draw a second vertical line. Label this as *line 2.* The distance between the two lines represents 1 second, and the volume exhaled in the first second is read where line 2 intersects the spirogram tracing. Subtract that milliliter value from the milliliter value of the inhalation peak (at the intersection of line 1), to determine the volume of gas expired in the first second. According to the values given in Figure 37.6, that figure would be 3400 ml (6800 ml − 3400 ml). Record your measured value below.

milliliters of gas expired in second 1:____________ ml

6. To compute the FEV_1 use the following equation:

$$FEV_1 = \frac{\text{volume expired in second 1}}{\text{FVC volume}} \times 100\%$$

Record your calculated value below and on the Exercise 37 Review Sheet.

FEV_1 = _______% of FVC

USE OF THE PNEUMOGRAPH TO DETERMINE FACTORS INFLUENCING RATE AND DEPTH OF RESPIRATION*

The neural centers that control respiratory rhythm and maintain a rate of 12 to 18 respirations/min are located in the medulla and pons. On occasion, input from the stretch receptors in the lungs (via the vagus nerve to the medulla) modifies the respiratory rate, as in cases of extreme overinflation of the lungs (Hering-Breuer reflex).

Death occurs when medullary centers are completely suppressed, as from an overdose of sleeping pills or gross overindulgence in alcohol, and respiration ceases completely. ■

Although the nervous system centers initiate the basic rhythm of breathing, there is no question that physical phenomena such as talking, yawning, coughing, and exercise can modify the rate and depth of respiration. So too can chemical factors such as changes in oxygen or carbon dioxide concentrations in the blood or fluctuations in blood pH. Changes in carbon dioxide blood levels seem to act directly on the medulla control centers, whereas changes in pH and oxygen concentrations are monitored by chemoreceptor regions in the aortic and carotid bodies, which in turn send input to the medulla. The experimental sequence in this section is designed to test the relative importance of various physical and chemical factors in the process of respiration.

The **pneumograph,** an apparatus that records variations in breathing patterns, is the best means of observing respiratory variations resulting from physical and chemical factors. The chest pneumograph is a coiled rubber hose that is attached around the thorax. As the subject breathes, chest movements produce pressure changes within the pneumograph that are transmitted to a recorder.

The instructor will demonstrate the method of setting up the pneumograph and discuss the interpretation of the results. Work in pairs so that one person can mark the record to identify the test for later interpretation. Ideally, the student being tested should face away from the recording apparatus to prevent voluntary modification of the record.

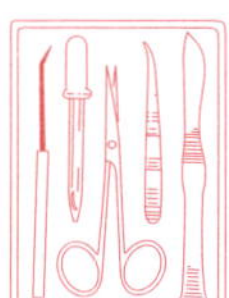

1. Attach the pneumograph tubing firmly, but not restrictively, around the thoracic cage at the level of the sixth rib, leaving room for chest expansion during testing. If the subject is female, position the tubing above the breasts to prevent slippage during testing. Set the pneumograph speed at 1 or 2, and the time signal at 10-second intervals. Record quiet breathing for 1 minute with the subject in a sitting position.

Record breaths per minute. ______________________

2. Make a vital capacity tracing: Record a maximal inhalation followed by a maximal exhalation. This should correlate to the vital capacity measurement obtained earlier and will provide a baseline for comparison during the rest of the pneumograph testing. Stop the recording apparatus and indicate the following on the graph by marking the graph appropriately: tidal volume, expiratory reserve volume, inspiratory reserve volume, and vital capacity (the total of the three measurements). Also mark, with arrows, the direction the recording stylus moves during inspiration and during expiration.

Measure in mm the height of the vital capacity recording. Divide the vital capacity measurement average recorded on the Exercise 37 Review Sheet by the millimeter figure to obtain the volume (in milliliters of air) represented by one mm on the recording. For example, if your vital capacity reading is 4000 ml and the vital capacity tracing occupies a vertical distance of 40 mm on the pneumograph recording, then a vertical distance of 1 mm equals 100 ml of air.

Record your computed value. ____________ ml air/mm

3. Record the subject's breathing as he or she performs activities from the list below. Make sure the record is marked accurately to identify each test conducted. Record your results on the Exercise 37 Review Sheet.

talking	swallowing water
yawning	coughing
laughing	lying down
standing	running in place
doing a math problem (concentrating)	

4. Without recording, have the subject breathe normally for 2 minutes, then inhale deeply and hold his or her breath for as long as he or she can.

Time the breath-holding interval. ________________ sec

As the subject exhales, turn on the recording apparatus and record the recovery period (time to return to normal breathing—usually slightly over 1 minute):

Time of recovery period. ______________________ sec

Did the subject have the urge to inspire *or* expire during breath holding?

____

* Note to the Instructor: This exercise may also be done without using the recording apparatus by simply having the students count the respiratory rate visually.

Without recording, repeat the above experiment, but this time exhale completely and forcefully *after* taking the deep breath. What was observed this time?

____

____

Explain the results. (Hint: the vagus nerve is the sensory nerve of the lungs and plays a role here.)

____

____

____

____

5. Have the subject hyperventilate (breathe deeply and forcefully at the rate of 1 breath/4 sec) for about 30 seconds.* Record both during and after hyperventilation. How does the pattern obtained during hyperventilation compare with that recorded during the vital capacity tracing?

____

Is the respiratory rate after hyperventilation faster *or* slower than during normal quiet breathing?

____

6. Repeat the above test, but do not record until after hyperventilating. After hyperventilation, the subject is to hold his or her breath as long as he or she can. Can the breath be held for a longer or shorter period of time after hyperventilating?

____

7. Without recording, have the subject breathe into a paper bag for 3 minutes, then record his or her breathing movements.

Caution: During the bag-breathing exercise the subject's partner should watch the subject carefully for any untoward reactions.

* A sensation of dizziness may develop. (As the carbon dioxide is washed out of the blood by overventilation, the blood pH increases, leading to a decrease in blood pressure and reduced cerebral circulation.) The subject may experience a lack of desire to breathe after forced breathing is stopped. If the period of breathing cessation—apnea—is extended, cyanosis of the lips may occur.

Is the breathing rate faster *or* slower than that recorded during normal quiet breathing?

____

After hyperventilating? ______________________________

8. Run in place for 2 minutes, and then have your partner determine the length of time that you can hold your breath.

Length of breath-holding. ________ sec

9. To prove that respiration has a marked effect on circulation, conduct the following test. Have your lab partner record the rate and relative force of your radial pulse before beginning.

rate ________ beats/min relative force ________

Inspire forcibly. Immediately close your mouth and nose to retain the inhaled air, and then make a forceful and prolonged expiration. Your lab partner should observe and record the condition of the blood vessels of your neck and face, and again immediately palpate the radial pulse.

Observations ______________________________________

____

Radial pulse _____ beats/min Relative force _____

Explain the changes observed. ______________________

____

____

Dispose of the paper bag in the autoclave bag. Keep the pneumograph records to interpret results and hand them in if requested by the instructor. Observation of the test results should enable you to determine which chemical factor, carbon dioxide or oxygen, has the greatest effect on modifying the respiratory rate and depth.

RESPIRATORY SOUNDS

As air flows in and out of the respiratory tree, it produces two characteristic sounds that can be picked up with a stethoscope. The **bronchial sounds** are produced by air rushing through the large respiratory passageways (the trachea and the bronchi). The second sound type, **vesicular breathing sounds,** apparently results from air filling the alveolar sacs and resembles the sound of a rustling or muffled breeze.

1. Obtain a stethoscope and clean the earpieces with an alcohol swab. Allow the alcohol to dry before donning the stethoscope.

2. Place the diaphragm of the stethoscope on the throat of the test subject just below the larynx. Listen for bronchial sounds on inspiration and expiration. Move the stethoscope downward toward the bronchi until you can no longer hear sounds.

3. Place the stethoscope over the following chest areas and listen for vesicular sounds during respiration (heard primarily during inspiration).

- At various intercostal spaces
- At the *triangle of auscultation* (a small depressed area of the back where the muscles fail to cover the rib cage; located just medial to the inferior part of the scapula)
- Under the clavicle

Diseased respiratory tissue, mucus, or pus can produce abnormal chest sounds such as rales (a rasping sound) and wheezing (a whistling sound). ■

ROLE OF THE RESPIRATORY SYSTEM IN ACID-BASE BALANCE OF BLOOD

As you have already learned, pulmonary ventilation is necessary for continuous oxygenation of the blood and removal of carbon dioxide (a waste product of cellular respiration) from the blood. Blood pH must be relatively constant for the cells of the body to function optimally. Therefore the carbonic acid - bicarbonate buffer system of the blood is extremely important, because it helps stabilize arterial blood pH at 7.4 ± 0.02.

When carbon dioxide diffuses into the blood from the tissue cells, much of it enters the red blood cells, where it combines with water to form carbonic acid:

$$H_2O + CO_2 \xrightarrow[\text{enzyme present in RBC}]{\text{Carbonic anhydrase}} H_2CO_3$$

Some carbonic acid is also formed in the plasma, but its formation there is very slow because of the lack of the carbonic anhydrase enzyme. Shortly after it forms, carbonic acid dissociates to release bicarbonate (HCO_3^-) and hydrogen (H^+) ions. The hydrogen ions that remain in the cells are neutralized, or buffered, when they combine with hemoglobin molecules. If they were not neutralized, the intracellular pH would become very acidic as H^+ ions accumulated. The bicarbonate ions diffuse out of the red blood cells into the plasma, where they become part of the carbonic acid–bicarbonate buffer system. As HCO_3^- follows its concentration gradient into the plasma, an electrical imbalance develops in the RBCs that draws Cl^- into them from the plasma. This exchange phenomenon is called the *chloride shift.*

Acids (more precisely, H^+) released into the blood by the body cells tend to lower the pH of the blood and to cause it to become acidic. On the other hand, basic substances that enter the blood tend to cause the blood to become more alkaline and the pH to rise. Both of these tendencies are resisted in large part by the carbonic acid–bicarbonate buffer system. If H^+ concentration in the blood begins to increase, the H^+ ions combine with bicarbonate ions to form carbonic acid (a weak acid that does not tend to dissociate at physiologic or acid pH) and are thus removed.

$$H^+ + HCO_3^- \longrightarrow H_2CO_3$$

Likewise, as blood H^+ concentration drops below what is desirable and blood pH rises, H_2CO_3 dissociates to release bicarbonate ions and H^+ ions to the blood. The released H^+ lowers the pH again. The bicarbonate ions, being *weak* bases, are poorly functional under alkaline conditions and have little effect on blood pH unless and until blood pH drops toward acid levels.

$$H_2CO_3 \longrightarrow H^+ + HCO_3^-$$

In the case of excessively slow or shallow breathing (hypoventilation) or fast deep breathing (hyperventilation), the amount of carbonic acid in the blood can be greatly modified—increasing dramatically during hypoventilation and decreasing substantially during hyperventilation. In either situation, if the buffering ability of the blood is inadequate, respiratory acidosis or alkalosis can result. Therefore, maintaining the normal rate and depth of breathing is important for proper control of blood pH.

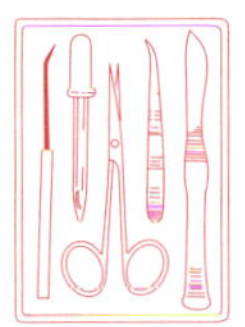

1. To observe the ability of a buffer system to stabilize the pH of a solution, obtain five 250-ml beakers, and a wash bottle containing distilled water. Set up the following experimental samples:

Beaker 1
(150 ml distilled water) pH ______

Beaker 2
(150 ml distilled water and
1 drop concentrated HCl) pH ______

Beaker 3
(150 ml distilled water and
1 drop concentrated NaOH) pH ______

Beaker 4
(150 ml standard buffer solution
[pH 7] and 1 drop concentrated HCl) pH ______

Beaker 5
(150 ml standard buffer solution
[pH 7] and 1 drop concentrated NaOH) pH ______

2. Using a pH meter standardized with a buffer solution of pH 7, determine the pH of the contents of each beaker and record above. After *each and every* pH recording, the pH meter switch should be turned to *standby,* and the electrodes rinsed thoroughly with a stream of distilled water from the wash bottle.

3. Add 3 more drops of concentrated HCl to beaker 4, stir, and record the pH: ______________________

4. Add 3 more drops of concentrated NaOH to beaker 5, stir, and record the pH: ______________________

How successful was the buffer solution in resisting pH changes when a strong acid (HCl) or a strong base (NaOH) was added?

____

To observe the ability of the carbonic acid–bicarbonate buffer system of blood to resist pH changes, perform the following simple experiment.

1. Obtain two small beakers (50 ml), animal plasma, graduated cylinder, glass stirring rod, and a dropper bottle of 0.01 *M* HCl. Using the pH meter standardized with the buffer solution of pH 7.0, measure the pH of the animal plasma. Use only enough plasma to allow immersion of the electrodes and measure the volume used carefully.

pH of the animal plasma: ______________________

2. Add 2 drops of the 0.01 *M* HCl solution to the plasma; stir and measure the pH again.

pH of plasma plus 2 drops of HCl: ______________________

3. Turn the pH meter switch to *standby,* rinse the electrodes, and then immerse them in a quantity of distilled water (pH 7) exactly equal to the amount of animal plasma used. Measure the pH of the distilled water.

pH of distilled water: ______________________

4. Add 2 drops of 0.01 *M* HCl, swirl, and measure the pH again.

pH of distilled water plus the two drops of HCl: ________

Is the plasma a good buffer? ______________________

What component of the plasma carbonic acid–bicarbonate buffer system was acting to counteract a change in pH when HCl was added?

____

Anatomy of the Digestive System

OBJECTIVES

1. To state the overall function of the digestive system.
2. To identify on an appropriate diagram, torso model, or cadaver the organs comprising the alimentary canal; and to name their subdivisions if any.
3. To name and/or identify the accessory digestive organs.
4. To describe the general functions of the digestive system organs or structures.
5. To describe the general histologic structure of the wall of the alimentary canal and/or label a cross-sectional diagram of the wall with the following terms: mucosa, submucosa, muscularis externa, and serosa or adventitia.
6. To list and explain the specializations of anatomical structure of the stomach and small intestine that contribute to their functional roles.
7. To list the major enzymes or enzyme groups produced by each of the following organs: salivary glands, stomach, small intestine, pancreas.
8. To name human deciduous and permanent teeth and describe the anatomy of the generalized tooth.
9. To recognize by microscopic inspection, or by viewing an appropriate diagram or photomicrograph, the histologic structure of the following organs:

small intestine	pancreas	stomach
salivary glands	tooth	liver

MATERIALS

Dissectible torso model
Anatomical chart of the human digestive system
Model of a villus and liver (if available)
Jaw model or human skull
Prepared microscope slides of the liver, pancreas, and mixed salivary glands; of longitudinal sections of the gastroesophageal junction and a tooth; and of cross sections of the stomach, duodenum, and ileum
Hand lens
Compound microscope

See Appendix D, Exercise 38 for links to A.D.A.M. Standard.

See Appendix E, Exercise 38 for links to *Anatomy and PhysioShow: The Videodisc.*

The **digestive system** provides the body with the nutrients, water, and electrolytes essential for health. The organs of this system are responsible for food ingestion, digestion, absorption, and the elimination of the undigested remains as feces.

The digestive system consists of a hollow tube extending from the mouth to the anus, into which various accessory organs or glands empty their secretions (Figure 38.1). Food material within this tube, the *alimentary canal,* is technically outside the body, since it has contact only with the cells lining the tract. For ingested food to become available to the body cells, it must first be broken down *physically* (chewing, churning) and *chemically* (enzymatic hydrolysis) into its smaller diffusible molecules—a process called **digestion.** The digested end products can then pass through the epithelial cells lining the tract into the blood for distribution to the body cells—a process termed **absorption.** In one sense, the digestive tract can be viewed as a disassembly line, in which food is carried from one stage of its digestive processing to the next by muscular activity, and its nutrients are made available to the cells of the body en route.

The organs of the digestive system are traditionally separated into two major groups: the **alimentary canal,** or **gastrointestinal (GI) tract,** and the **accessory digestive organs.** The alimentary canal is approximately 30 feet long in a cadaver but considerably less in a living person. It consists of the mouth, pharynx, esophagus, stomach, small and large intestines, rectum, and anus. The accessory structures include the salivary glands, gallbladder, liver, and pancreas, which secrete their products into the alimentary canal. These individual organs are described shortly.

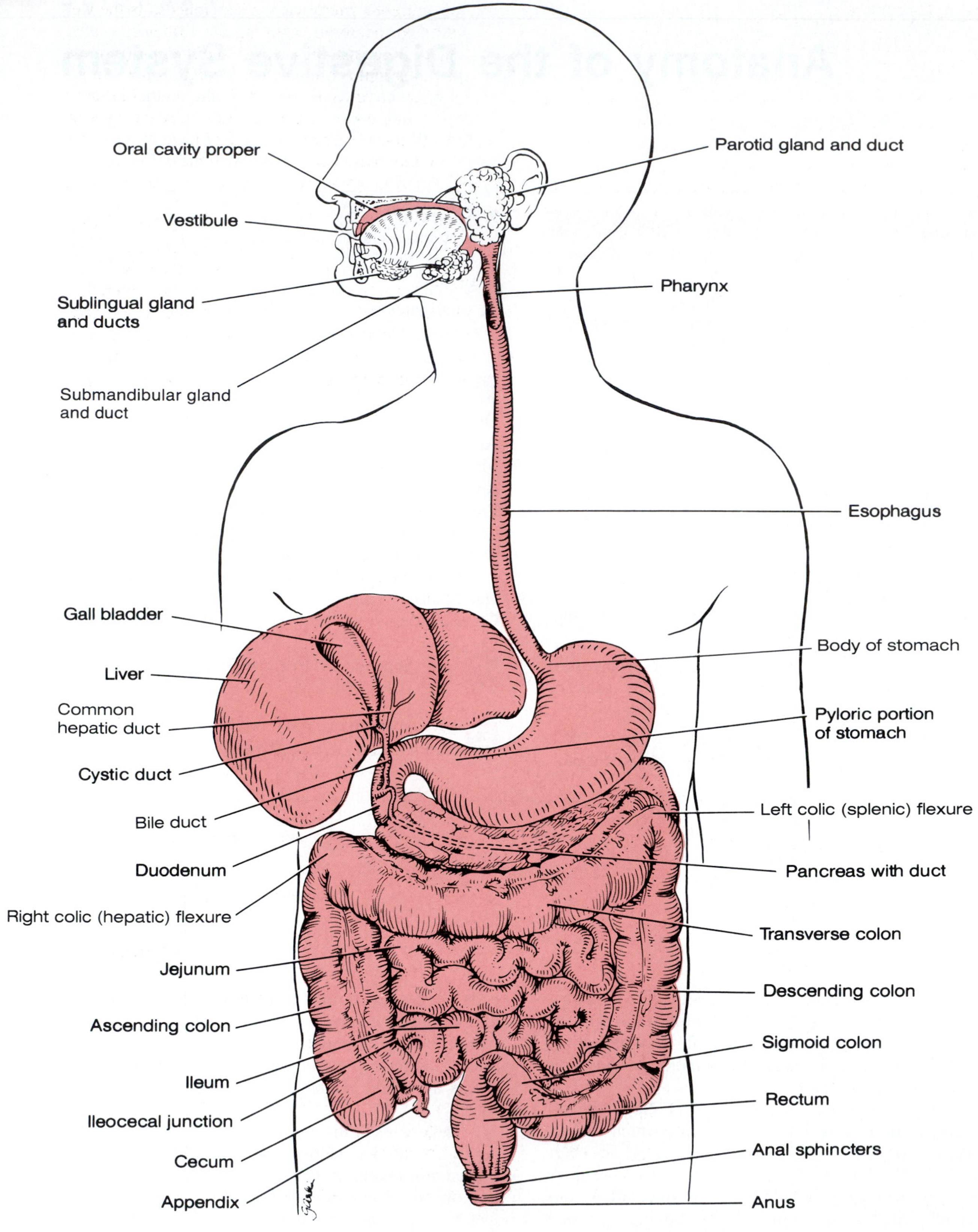

F38.1

The human digestive system: alimentary tube and accessory organs. (Liver and gallbladder are reflected superiorly and to the right.)

GENERAL HISTOLOGICAL PLAN OF THE ALIMENTARY CANAL

Because the alimentary canal has a shared basic structure (particularly from the esophagus to the anus), it makes sense to review that structure as we begin studying this group of organs. Once done, all that need be emphasized as the individual organs are described is their specializations for unique functions in the digestive process.

Essentially the alimentary canal walls have four basic **tunics** (layers). From the lumen outward, these are the *mucosa,* the *submucosa,* the *muscularis externa,* and the *serosa* or *adventitia* (Figure 38.2). Each of these tunics has a predominant tissue type and a specific function in the digestive process.

Mucosa (mucous membrane): The mucosa is the wet epithelial membrane abutting the alimentary canal lumen. It consists of a surface *epithelium* (in most cases, a simple columnar), a *lamina propria* (areolar connective tissue on which the epithelial layer rests), and a *muscularis mucosae* (a scant layer of smooth muscle fibers that enable local movements of the mucosa). The major functions of the mucosa are secretion (of enzymes, mucus, hormones, etc.), absorption of digested foodstuffs, and protection (against bacterial invasion). A particular mucosal region may be involved in one or all three functions.

Submucosa: Superficial to the mucosa, the submucosa is moderately dense connective tissue containing blood and lymphatic vessels, scattered lymph nodules, and nerve fibers. Its intrinsic nerve supply is called the *submucosal plexus.* Its major functions are nutrition and protection.

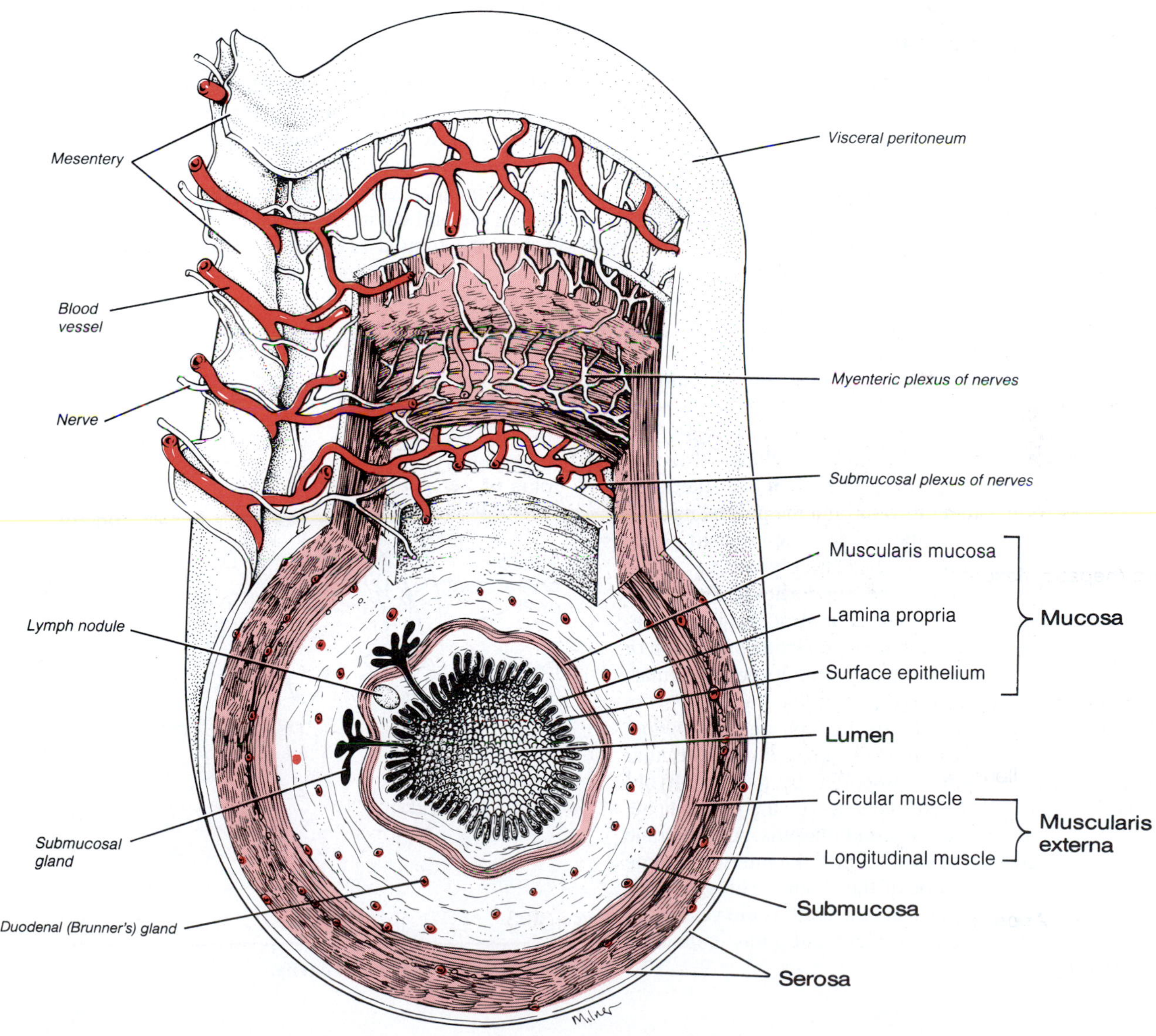

F38.2

Basic structural pattern of the alimentary canal wall.

Muscularis externa: The muscularis externa, also simply called the *muscularis,* typically is a bilayer of smooth muscle, with the deeper layer running circularly and the superficial layer running longitudinally. Another important intrinsic nerve plexus, the *myenteric plexus,* is associated with this tunic. By controlling the smooth muscle of the muscularis, this plexus is the major regulator of GI motility.

Serosa: The outermost serosa is equal to the *visceral peritoneum.* It consists of mesothelium associated with a thin layer of areolar connective tissue. In areas *outside* the abdominopelvic cavity, the serosa is replaced by an **adventitia,** a layer of coarse fibrous connective tissue that binds the organ to surrounding tissues. (This is the case with the esophagus.) The serosa reduces friction as the mobile GI tract organs work and slide across one another and the cavity walls. The adventitia anchors and protects the surrounded GI tract organ.

ORGANS OF THE ALIMENTARY CANAL

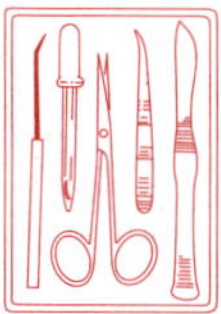

The sequential pathway and fate of food as it passes through the alimentary canal organs are described in the next sections. Identify each structure in Figure 38.1 and on the torso model as you work.

Oral Cavity or Mouth

Food enters the digestive tract through the **oral cavity,** or **mouth** (Figure 38.3). Within this mucous membrane–lined cavity are the gums, teeth, tongue, and openings of the ducts of the salivary glands. The **lips (labia)** protect the opening of the chamber anteriorly, the **cheeks** form its lateral walls, and the **palate,** its roof. The anterior portion of the palate is referred to as the **hard palate,** since bone (the palatine processes of the maxillae and the palatine bones) underlies it. The posterior **soft palate** is a fibromuscular structure that is unsupported by bone. The **uvula,** a fingerlike projection of the soft palate, extends inferiorly from its posterior margin. The soft palate rises to close off the oral cavity from the nasal and pharyngeal passages during swallowing. The floor of the oral cavity is occupied by the muscular **tongue,** which is largely supported by the **mylohyoid muscle** (Figure 38.4) and attaches to the hyoid bone, mandible, styloid processes, and pharynx. A membrane called the **lingual frenulum** secures the inferior midline of the tongue to the floor of the mouth. The space between the lips and cheeks and the teeth is the **vestibule;** the area that lies within the teeth and gums is the **oral cavity** proper.

On each side of the mouth at its posterior end are masses of lymphoid tissue, the **palatine tonsils** (Figure 38.4). Each lies in a concave area bounded anteriorly and posteriorly by membranes, the **palatoglossal arch** (anterior membrane) and the **palatopharyngeal arch** (posterior membrane). Another mass of lymphoid tissue, the **lingual tonsil,** covers the base of the tongue, posterior to the oral cavity proper. The tonsils, in common with other lymphoid tissues, are part of the body's defense system.

Very often in young children, the palatine tonsils become inflamed and enlarge, partially blocking the entrance to the pharynx posteriorly and making swallowing difficult and painful. This condition is called **tonsilitis.** ■

Three pairs of salivary glands duct their secretion, saliva, into the oral cavity. One component of saliva, salivary amylase, begins the digestion of starchy foods within the oral cavity. (The salivary glands are discussed in more detail on pp. 365–366.)

As food enters the mouth, it is mixed with saliva and masticated (chewed). The cheeks and lips help hold the food between the teeth during mastication, and the highly mobile tongue manipulates the food during chewing and initiates swallowing. Thus the mechanical and chemical breakdown of food begins before the food has left the oral cavity. As noted in Exercise 26, the surface of the tongue is covered with papillae, many of which contain taste buds, receptors for taste sensation. So, in addition to its manipulative function, the tongue provides for the enjoyment and appreciation of the food ingested.

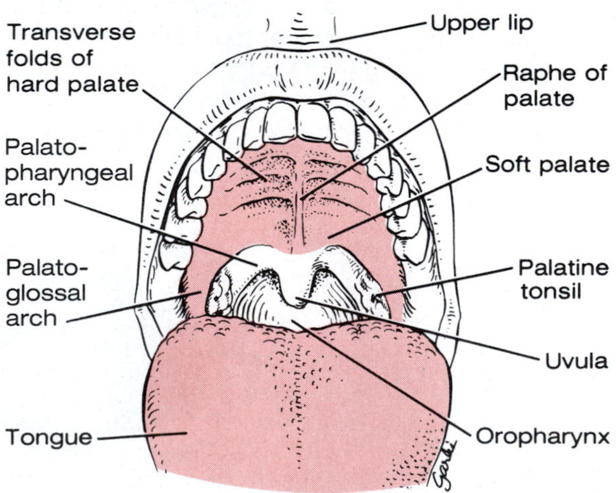

F38.3

Anterior view of the oral cavity.

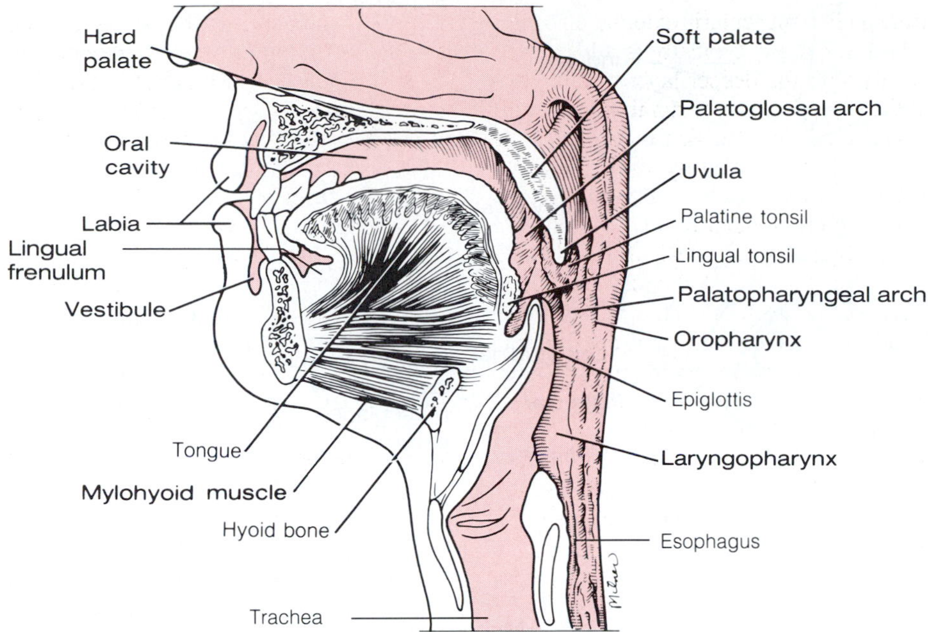

F38.4

Sagittal view of the head showing oral, nasal, and pharyngeal cavities.

Pharynx

When the tongue initiates swallowing, the food passes posteriorly into the pharynx, a common passageway for food, fluid, and air (Figure 38.4). The pharynx is often subdivided anatomically into the **nasopharynx** (behind the nasal cavity), the **oropharynx** (behind the oral cavity extending from the soft palate to the epiglottis overlying the larynx), and the **laryngopharynx** (extending from the epiglottis to the base of the larynx), which is continuous with the esophagus.

The walls of the pharynx consist largely of two layers of skeletal muscles: an inner layer of longitudinal muscle (the levator muscles) and an outer layer of circular constrictor muscles, which initiate wavelike contractions that propel the food inferiorly into the esophagus. Its mucosa, like that of the oral cavity, contains a friction-resistant stratified squamous epithelium.

Esophagus

The **esophagus,** or gullet, extends from the pharynx through the diaphragm to the cardiac sphincter in the superior aspect of the stomach. It is approximately 10 inches long in humans and is essentially a food passageway that conducts food to the stomach in a wavelike peristaltic motion. The esophagus has no digestive or absorptive function. The walls at its superior end contain skeletal muscle, which is replaced by smooth muscle in the area nearing the stomach. Since the esophagus is located in the thoracic rather than the abdominal cavity, its outermost layer is an *adventitia,* rather than the serosa.

Stomach

The **stomach** (Figures 38.1 and 38.5) is on the left side of the abdominal cavity and is hidden by the liver and diaphragm. Different regions of the saclike stomach are the **cardiac region** (the area surrounding the **cardiac sphincter** through which food enters the stomach from the esophagus), the **fundus** (the expanded portion of the stomach, superolateral to the cardiac region), the **body** (midportion of the stomach, inferior to the fundus), and the **pyloric region** (the terminal part of the stomach, which is continuous with the small intestine through the **pyloric sphincter**).

The concave medial surface of the stomach is called the **lesser curvature;** its convex lateral surface is the **greater curvature.** Extending from these curvatures are two mesenteries, called *omenta.* The **lesser omentum** extends from the liver to attach to the lesser curvature of the stomach. The **greater omentum,** a saclike mesentery, extends from the greater curvature of the stomach, reflects downward over the abdominal contents to cover them in an apronlike fashion, and then blends with the **mesocolon** attaching the transverse colon to the posterior body wall. Figure 38.7 (p. 360) illustrates the omenta as well as the other peritoneal attachments of the abdominal organs.

The stomach is a temporary storage region for food as well as a site for mechanical and chemical breakdown of food. It contains a third *obliquely* oriented layer of smooth muscle in its muscularis externa that allows it to churn, mix, and pummel the food, physically reducing it to smaller fragments. Gastric glands of the mucosa

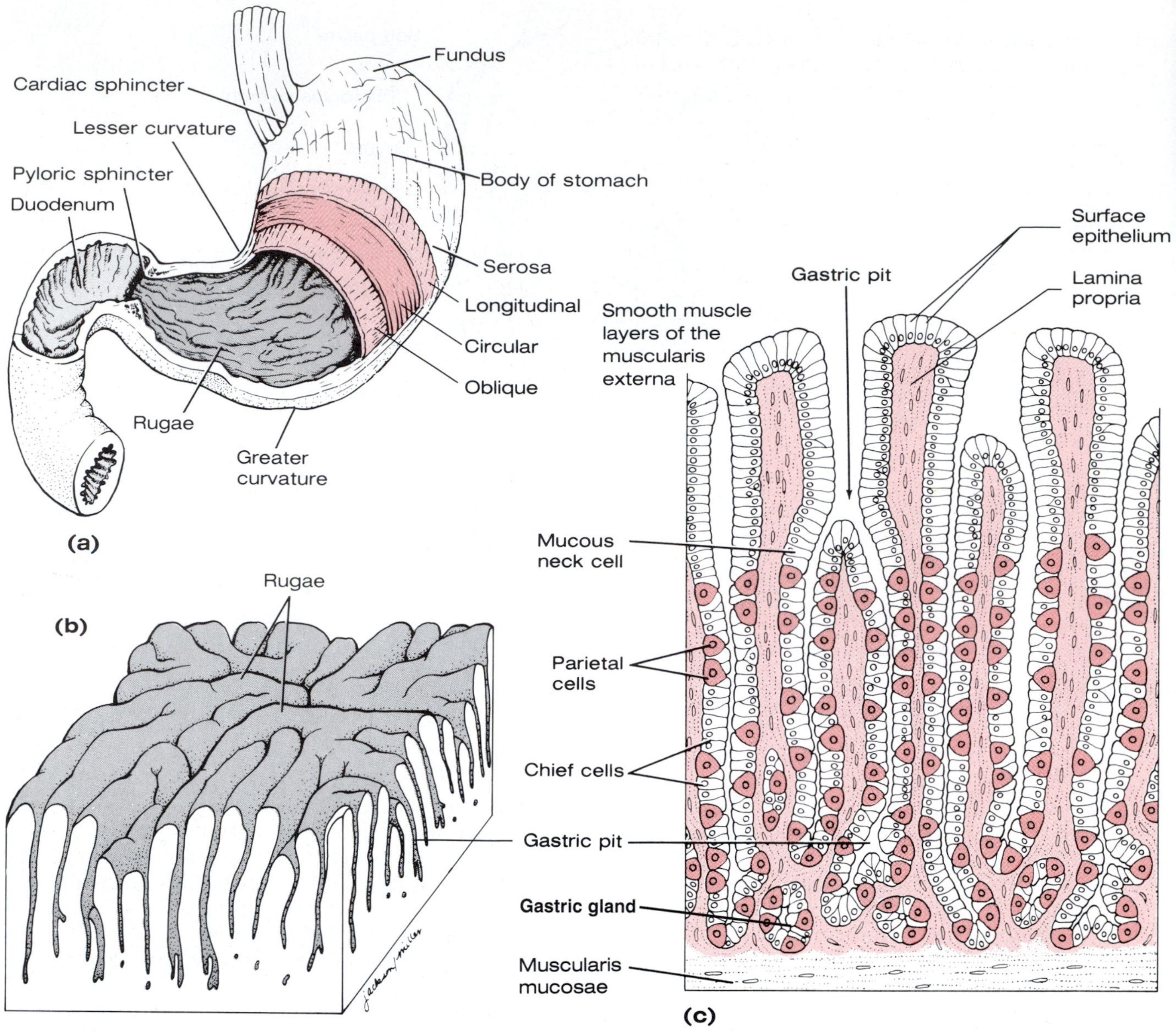

F38.5

Anatomy of the stomach. **(a)** Gross internal and external anatomy. **(b)** Section of the stomach wall showing rugae and gastric pits. **(c)** Enlarged view of gastric pits (longitudinal section).

secrete hydrochloric acid (HCl) and hydrolytic enzymes (primarily pepsinogen, the inactive form of *pepsin,* a protein-digesting enzyme), which begin the enzymatic, or chemical, breakdown of protein foods. The mucosal glands also secrete a viscous mucus that helps prevent the stomach itself from being digested by the proteolytic enzymes. Most digestive activity occurs in the pyloric region of the stomach. After the food has been processed in the stomach, it resembles a creamy mass (chyme), which enters the small intestine through the pyloric sphincter.

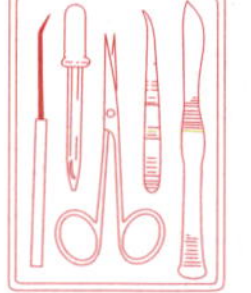

To prepare for the histological study you will be conducting now and later in the lab, obtain a microscope and the following slides: salivary glands (submandibular or sublingual); pancreas; liver; cross sections of the duodenum, ileum, and stomach, and longitudinal sections of a tooth and the gastroesophageal junction.

1. **Stomach:** The stomach slide will be viewed first. Refer to Figure 38.6a as you scan the tissue under low power to locate the muscularis externa; then move to high power to more closely examine this layer. Try to pick out the three smooth muscle layers. How does the extra (oblique) layer of smooth muscle found in the stomach correlate with the stomach's churning movements?

___

___

___

Identify the gastric glands and the gastric pits (see Figures 38.5 and 38.6b). If the section is taken from the stomach fundus and is appropriately stained, you can identify, in the gastric glands, the blue-staining **chief,** or **zymogenic, cells,** which produce pepsinogen, and the red-staining **parietal cells,** which secrete HCl. Draw a small section of the stomach wall and label it appropriately.

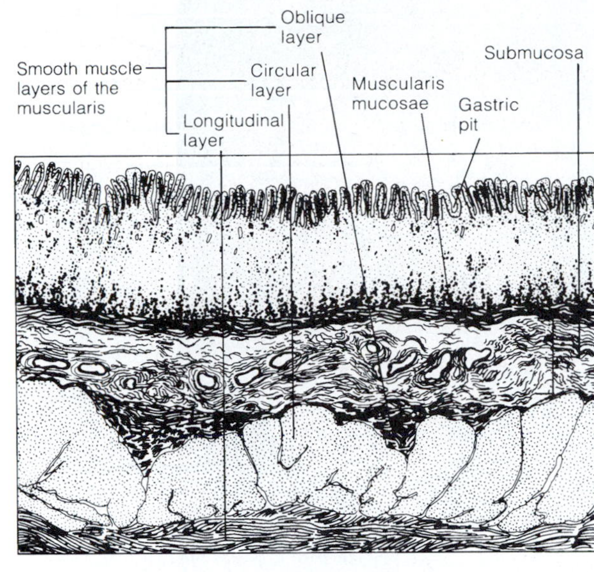

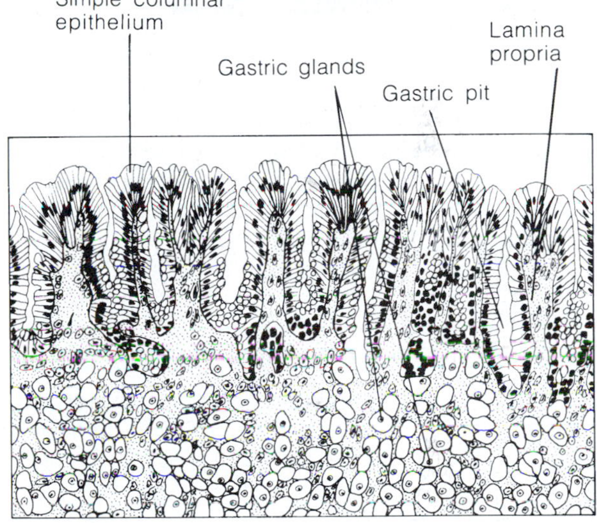

2. **Gastroesophageal junction:** Scan the slide under low power to locate the junction between the end of the esophagus and the beginning of the stomach, the gastroesophageal junction. Compare your observations to Figure 38.6c. What is the functional importance of the epithelial differences seen in the two organs?

___

___

___

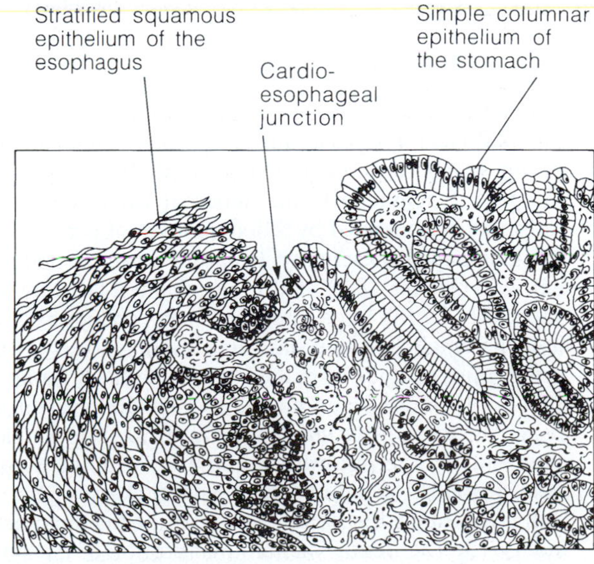

F38.6

Histology of selected regions of the stomach and gastroesophageal junction. (a) Low-power view of the stomach wall. See corresponding Plate 33 in the Histology Atlas. **(b)** High-power view of gastric pits and glands. See corresponding Plate 34 in the Histology Atlas. **(c)** Gastroesophageal junction, longitudinal section. See corresponding Plate 35 in the Histology Atlas.

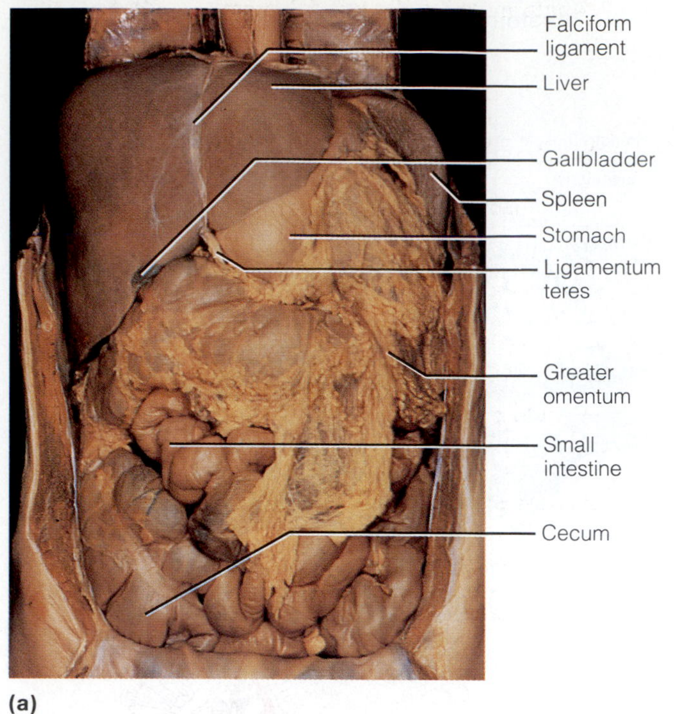

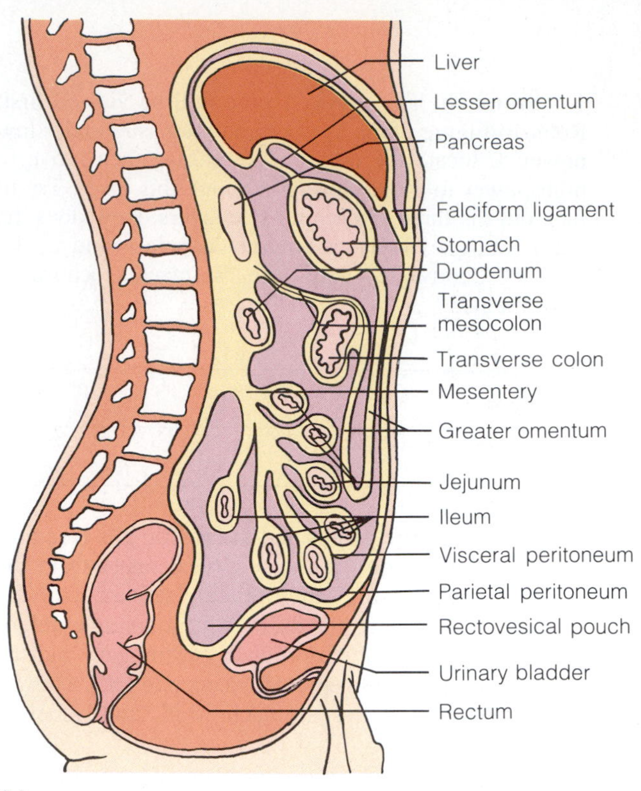

F38.7

Peritoneal attachments of the abdominal organs. **(a)** Superficial anterior view of abdominal cavity with omentum in place. **(b)** Sagittal view of a male torso. **(c)** Mesentery of the small intestine.

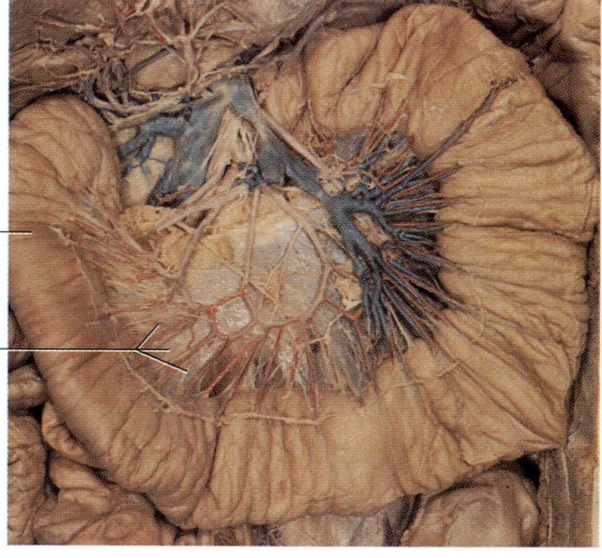

Small Intestine

The **small intestine** is a convoluted tube, 6 to 7 meters (about 20 feet) long in a cadaver but only about 2 m long during life because of its muscle tone. It extends from the pyloric sphincter to the ileocecal valve. The small intestine is suspended by a double layer of peritoneum, the fan-shaped **mesentery,** from the posterior abdominal wall (Figure 38.7), and it lies, framed laterally and superiorly by the large intestine, in the abdominal cavity. The small intestine has three subdivisions: (1) the **duodenum** extends from the pyloric sphincter for about 25 cm (10 inches) and curves around the head of the pancreas; most of the duodenum lies in a retroperitoneal position. (2) The **jejunum,** continuous with the duodenum, extends for 2.5 m (about 8 feet). Most of the jejunum occupies the umbilical region of the abdominal cavity. (3) The **ileum,** the terminal portion of the small intestine, is about 3.6 m (12 feet) long and joins the large intestine at the **ileocecal valve.** It is located inferiorly and somewhat to the right in the abdominal cavity, but its major portion lies in the hypogastric region.

Brush border enzymes, hydrolytic enzymes bound to the microvilli of the columnar epithelial cells, and, more importantly, enzymes produced by the pancreas and ducted into the duodenum via the **pancreatic duct** complete the enzymatic digestion process in the small intestine. Bile (formed in the liver) also enters the duodenum via the **bile duct** in the same area (see Figure 38.1). At the duodenum, the ducts join to form the bulblike **hepatopancreatic ampulla** and empty their products into the duodenal lumen through the **major duodenal papilla,** an orifice controlled by a muscular valve called the **hepatopancreatic sphincter (sphincter of Oddi).**

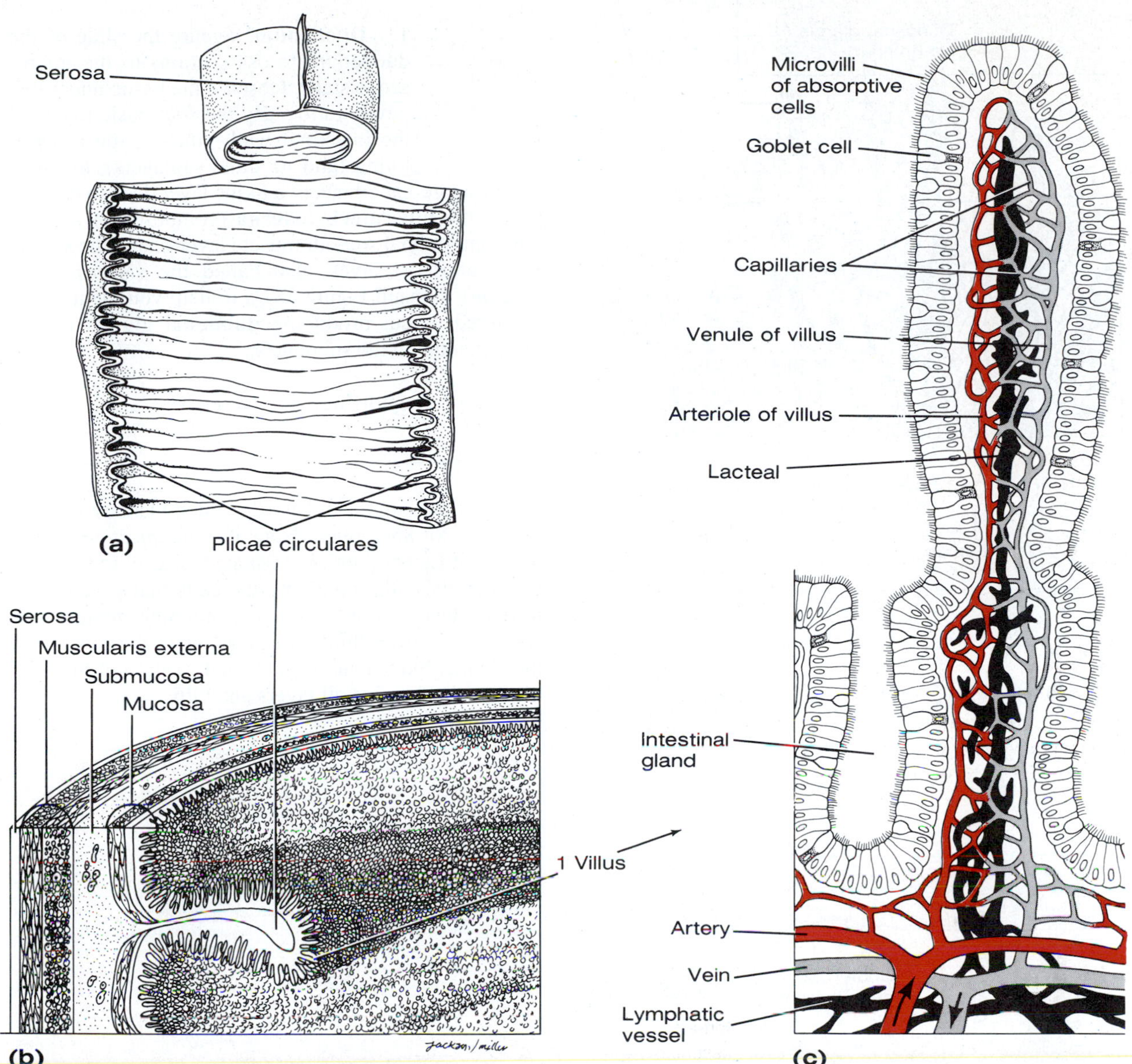

F38.8

Structural modifications of the small intestine. (a) Plicae circulares (circular folds) seen on the internal surface of the small intestine. **(b)** Enlargement of one plica circulare to show villi. **(c)** Detailed anatomy of a villus.

Nearly all nutrient absorption occurs in the small intestine, where three structural modifications that increase the mucosa absorptive area appear—the microvilli, villi, and plicae circulares (Figure 38.8). **Microvilli** are minute projections of the surface plasma membrane of the columnar epithelial lining cells of the mucosa; the **villi** are the fingerlike projections of the mucosa tunic that give it a velvety appearance and texture. The **plicae circulares** are deep folds of the mucosa and submucosa layers that force chyme to spiral through the intestine, mixing it and slowing its progress. These structural modifications, which increase the surface area, decrease in frequency and elaboration toward the end of the small intestine. Any residue remaining undigested and unabsorbed at the terminus of the small intestine enters the large intestine through the ileocecal valve. In contrast, the amount of lymphoid tissue in the submucosa of the small intestine (especially the aggregated lymphoid nodules called **Peyer's patches**) increases along the length of the small intestine and is very apparent in the ileum. This reflects the fact that the remaining undigested food residue contains large numbers of bacteria that must be prevented from entering the bloodstream.

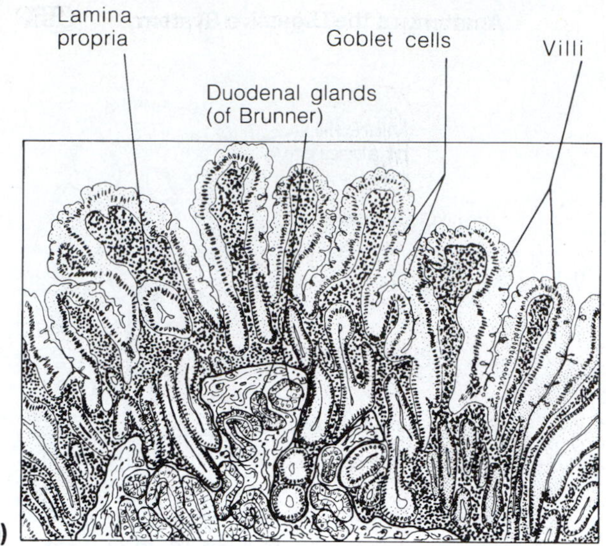

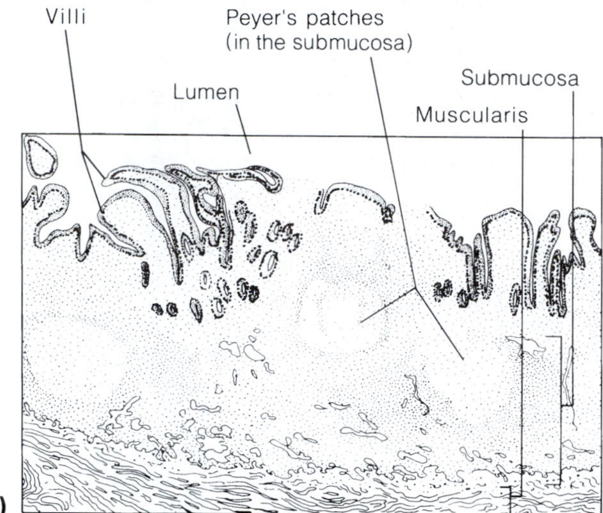

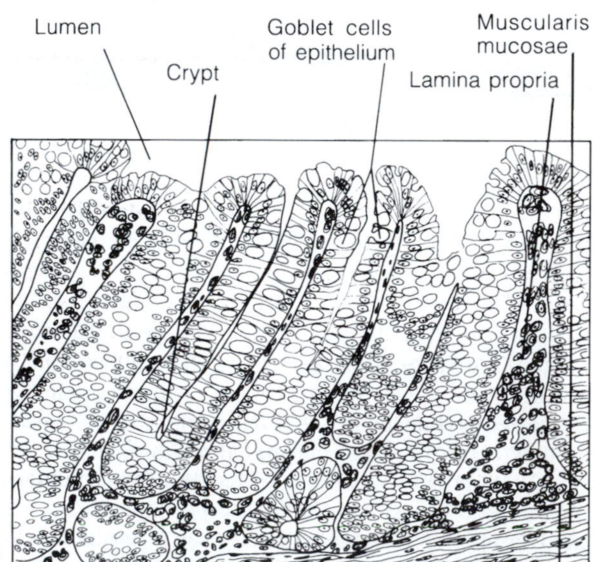

F38.9

Histology of selected regions of the small and large intestines (cross-sectional views). **(a)** Duodenum of the small intestine. See corresponding Plate 36 in the Histology Atlas. **(b)** Ileum of the small intestine. See corresponding Plate 37 in the Histology Atlas. **(c)** Large intestine. See corresponding Plate 38 in the Histology Atlas.

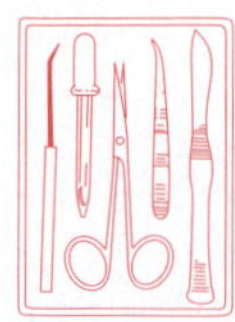

1. **Duodenum:** Secure the slide of the duodenum (cross section) to the microscope stage. Observe the tissue under low power to identify the four basic tunics of the intestinal wall—that is, the **mucosa** lining (and its three sublayers), the **submucosa** (areolar connective tissue layer deep to the mucosa), the **muscularis externa** (composed of circular and longitudinal smooth muscle layers), and the **serosa** (the outermost layer, also called the *visceral peritoneum*). Consult Figure 38.9a to help you identify the scattered mucus-producing **duodenal glands** (Brunner's glands) in the submucosa.

What type of epithelium do you see here? ___________

____

Examine the large leaflike *villi,* which increase the surface area for absorption. Note also the *intestinal crypts* (crypts of Lieberkühn), invaginated areas of the mucosa between the villi containing the cells that produce intestinal juice, a watery mucus-containing mixture that serves as a carrier fluid for absorption of nutrients from the chyme. Sketch and label a small section of the duodenal wall, showing all layers and villi.

2. **Ileum:** The structure of the ileum resembles that of the duodenum, except that the villi are less elaborate (most of the absorption has occurred by the time the ileum is reached). Obtain and secure a slide of the ileum for viewing. Observe the villi, and identify the four layers of the wall and the large generally spherical Peyer's patches (Figure 38.9b). What tissue comprises Peyer's patches?

____

3. If a *villus model* is available, identify the following cells or regions before continuing: absorptive epithelium, goblet cells, lamina propria, slips of the muscularis mucosae, capillary bed, and lacteal. If possible, also identify the intestinal crypts which lie between the villi.

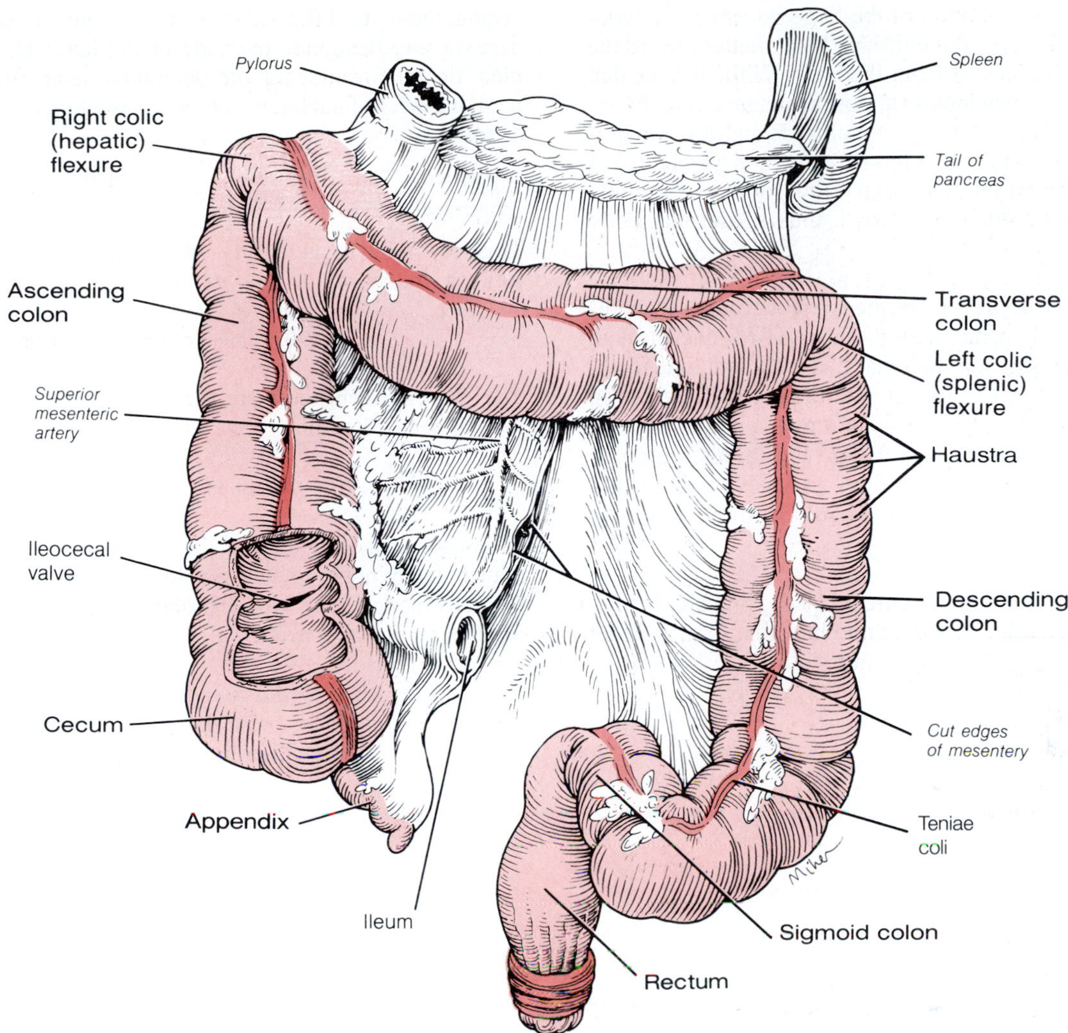

F38.10

The large intestine. (Section of the cecum removed to show the ileocecal valve.)

Large Intestine

The **large intestine** (see Figure 38.10) is about 1.5 m (5 feet) long and extends from the ileocecal valve to the anus. It encircles the small intestine on three sides and consists of the following subdivisions: the **cecum, appendix, colon, rectum,** and **anal canal.**

The blind tubelike appendix, which hangs from the cecum, is a trouble spot in the large intestine. Since it is generally twisted, it provides an ideal location for bacteria to accumulate and multiply. Inflammation of the appendix, or **appendicitis,** is the result. ■

The colon is divided into several distinct regions. The **ascending colon** travels up the right side of the abdominal cavity and makes a right-angle turn at the **right colic (hepatic) flexure** to cross the abdominal cavity as the **transverse colon.** It then turns at the **left colic (splenic) flexure** and continues down the left side of the abdominal cavity as the **descending colon,** where it takes an S-shaped course as the **sigmoid colon.** The sigmoid colon, rectum, and the anal canal lie in the pelvis anterior to the sacrum and thus are not considered abdominal cavity structures. Except for the transverse and sigmoid colons, which are secured to the dorsal body wall by mesocolons (see Figure 38.7), the colon is retroperitoneal.

The anal canal terminates in the **anus,** the opening to the exterior of the body. The anus, which has an external sphincter of skeletal muscle (the voluntary sphincter) and an internal sphincter of smooth muscle (the involuntary sphincter), is normally closed except during defecation when the undigested remains of the food and bacteria are eliminated from the body as feces.

In the large intestine, the longitudinal muscle layer of the muscularis externa is reduced to three longitudinal muscle bands called the **teniae coli.** Since these bands are shorter than the rest of the wall of the large intestine, they cause the wall to pucker into small pocket-like sacs called **haustra.**

The major function of the large intestine is to consolidate and propel the unusable fecal matter toward the anus and eliminate it from the body. While it does that "chore," it (1) provides a site for the manufacture, by intestinal bacteria, of some vitamins (B and K), which it then absorbs into the bloodstream; and (2) reclaims most of the remaining water (and some of the electrolytes) from undigested food, thus conserving body water.

Watery stools, or diarrhea, result from any condition that rushes undigested food residue through the large intestine before it has had sufficient time to absorb the water (as in irritation of the colon by bacteria). Conversely, when food residue remains in the large intestine for extended periods (as with atonic colon or failure of the defecation reflex), excessive water is absorbed and the stool becomes hard and difficult to pass (constipation). ■

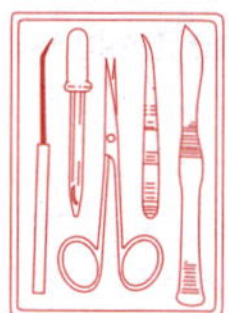

Examine Figure 38.9c to compare the histology of the large intestine to that of the small intestine just studied.

ACCESSORY DIGESTIVE ORGANS

Teeth

By the age of 21, two sets of teeth have developed (Figure 38.11). The initial set, called the **deciduous,** or **milk teeth,** normally appears between the ages of 6 months and 2½ years. The first of these to erupt are the lower central incisors. The child begins to shed the deciduous teeth around the age of 6, and a second set of teeth, the **permanent teeth,** gradually replace them. As the deeper permanent teeth progressively enlarge and develop, the roots of the deciduous teeth are resorbed, leading to their final shedding. During years 6–12, the child has mixed dentition—both permanent and deciduous teeth. Generally, by the age of 12, all of the deciduous teeth have been shed, or exfoliated.

Teeth are classified as **incisors, canines** (eye teeth), **premolars** (bicuspids), and **molars.** Teeth names reflect differences in relative structure and function. The incisors are chisel-shaped and exert a shearing action used in biting. Canines are cone shaped or fanglike, the latter description being much more applicable to the canines of animals whose teeth are used for the tearing of food. Incisors, canines, and premolars typically have single roots, though the first upper premolars may have two. The lower molars have two roots but the upper molars usually have three. The premolars have two cusps (grinding surfaces); the molars have broad crowns with rounded cusps specialized for the fine grinding of food.

Dentition is described by means of a **dental formula,** which designates the numbers, types, and position of the teeth in one side of the jaw. (Since tooth arrangement is bilaterally symmetrical, it is only necessary to designate one side of the jaw.) The complete dental formula for the deciduous teeth from the medial aspect of each jaw and proceeding posteriorly is as follows:

$$\frac{\text{Upper teeth: 2 incisors, 1 canine, 0 premolars, 2 molars}}{\text{Lower teeth: 2 incisors, 1 canine, 0 premolars, 2 molars}} \times 2$$

This formula is generally abbreviated to read as follows:

$$\frac{2,1,0,2}{2,1,0,2} \times 2 = 20 \text{ (number of deciduous teeth)}$$

The 32 permanent teeth are then described by the following dental formula:

$$\frac{2,1,2,3}{2,1,2,3} \times 2 = 32 \text{ (number of deciduous teeth)}$$

Although 32 is designated as the normal number of permanent teeth, not everyone develops a full complement. In many people, the No. 3 molars, commonly called *wisdom teeth,* never erupt.

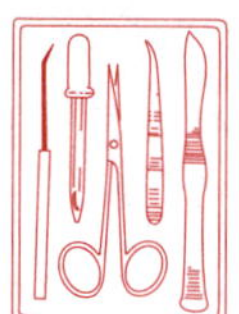

Identify the four types of teeth (incisors, canines, premolars, and molars) on the jaw model or human skull.

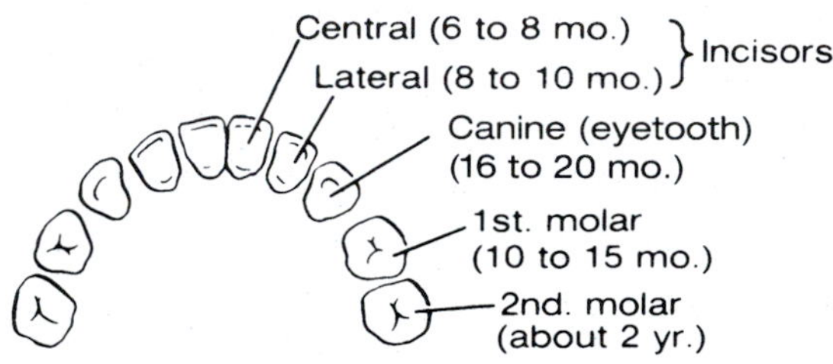

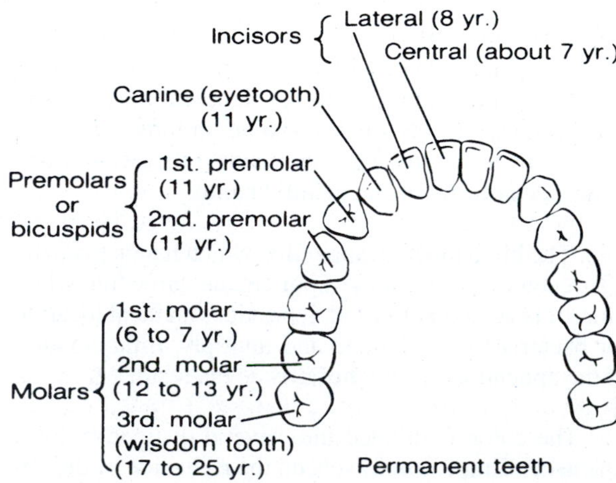

F38.11

Human deciduous teeth and permanent teeth.
(Approximate time of teeth eruption shown in parentheses.)

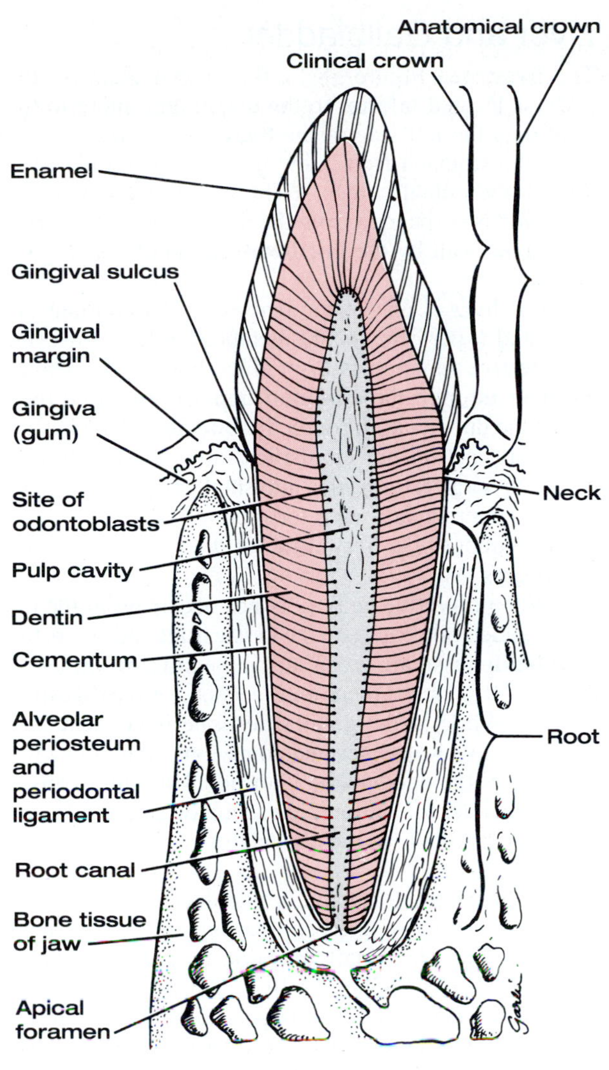

F38.12

Longitudinal section of human canine tooth.

A tooth consists of two major regions, the **crown** and the **root.** A longitudinal section made through a tooth shows the following basic anatomical plan (Figure 38.12). The crown is the superior portion of the tooth. The portion of the crown visible above the **gum,** or **gingiva,** is referred to as the **clinical crown.** The entire area covered by **enamel** is called the **anatomical crown.** Enamel is the hardest substance in the body and is fairly brittle. It consists of 95% to 97% inorganic calcium salts (chiefly $CaPO_4$) and thus is heavily mineralized. The crevice between the end of the anatomical crown and the upper margin of the gingiva is referred to as the **gingival sulcus** and its apical border is the **gingival margin.**

That portion of the tooth embedded in the alveolar portion of the jaw is the root, and the root and crown are connected by a slight constriction, the **neck.** The outermost surface of the root is covered by **cementum,** which is similar to bone in composition and less brittle than enamel. The cementum attaches the tooth to the **periodontal ligament** which holds the tooth in the alveolar socket and exerts a cushioning effect. **Dentin,** which comprises the bulk of the tooth, is the bonelike material medial to the enamel and cementum.

The **pulp cavity** occupies the central portion of the tooth. **Pulp,** connective tissue liberally supplied with blood vessels, nerves, and lymphatics, occupies this cavity and provides for tooth sensation and supplies nutrients to the tooth tissues. Specialized cells, called **odontoblasts,** which reside in the outer margins of the pulp cavity, produce the dentin. The pulp cavity extends into distal portions of the root and becomes the **root canal.** An opening at the root apex, the **apical foramen,** provides a route of entry into the tooth for the blood vessels, nerves, and other structures from the tissues beneath.

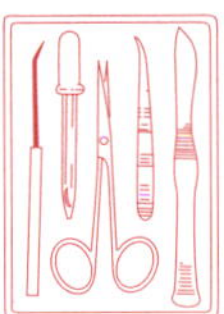

Observe a slide of a longitudinal section of a tooth, and compare your observations with the structures detailed in Figure 38.12. Identify as many of these structures as possible.

Salivary Glands

Three pairs of major **salivary glands** (see Figure 38.1) empty their secretions into the oral cavity.

Parotid glands: large glands located anterior to the ear and ducting into the mouth over the second upper molar through the parotid duct

Submandibular glands: located inside the maxillary arch in the floor of the mouth and ducting under the tongue to the base of the lingual frenulum

Sublingual glands: small glands located most anteriorly in the floor of the mouth and emptying under the tongue via several small ducts

Food in the mouth and mechanical pressure (even chewing rubber bands or wax) stimulate the salivary glands to secrete saliva. Saliva consists primarily of mucin (a viscous glycoprotein), which moistens the food and helps to bind it together into a mass called a **bolus,** and a clear serous fluid containing the enzyme *salivary amylase.* Salivary amylase begins the digestion of starch (a large polysaccharide), breaking it down into disaccharides, or double sugars, and glucose. The secretion of the parotid glands is mainly serous, whereas the submandibular and sublingual glands are mixed glands that produce both mucin and serous components.

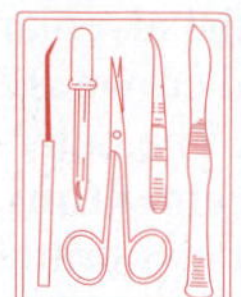

Examine salivary gland tissue under low power and then high power to become familiar with the appearance of a glandular tissue. Notice the clustered arrangement of the cells around their ducts. The cells are basically triangular, with their pointed ends facing the duct orifice. If possible, differentiate between the mucus-producing cells, which look hollow or have a clear cytoplasm, and the serous cells, which produce the clear, enzyme-containing fluid and have granules in their cytoplasm. The serous cells often form *demilunes* ("caps") around the more central mucous cells. Figure 38.13 may be helpful in this task. Draw your version of a small portion of the salivary gland tissue and label it appropriately.

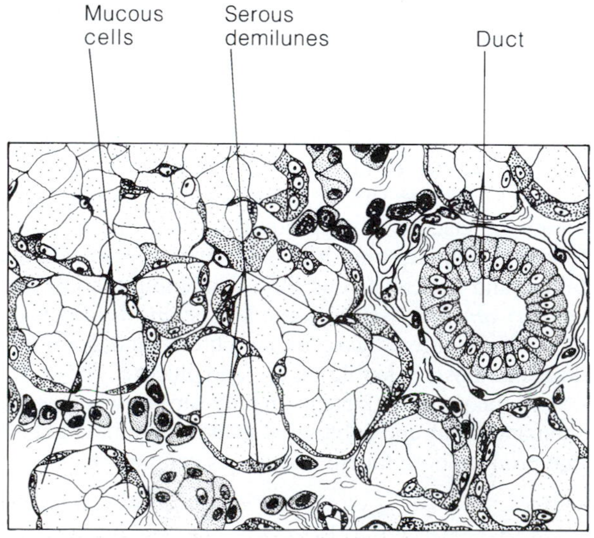

F38.13

A mixed salivary gland. Corresponds to the photomicrograph in Plate 39 in the Histology Atlas.

Liver and Gallbladder

The **liver** (see Figure 38.1), the largest gland in the body, is located inferior to the diaphragm, more to the right than the left side of the body. As noted earlier, it hides the stomach from view in a superficial observation of abdominal contents. The human liver has four lobes and is suspended from the diaphragm and anterior abdominal wall by the **falciform ligament** (see Figure 38.7a).

The liver is one of the body's most important organs, and it performs many metabolic roles. However, its digestive function is to produce bile, which leaves the liver through the **common hepatic duct** and then enters the duodenum through the **bile duct.** Bile has no enzymatic action but emulsifies fats (spreads thin or breaks up large fat particles into smaller ones), thus creating a larger surface area for more efficient lipase activity. Without bile, very little fat digestion or absorption occurs.

When digestive activity is not occurring in the digestive tract, bile backs up the **cystic duct** and enters the **gallbladder,** a small, green sac on the inferior surface of the liver. It is stored there until needed for the digestive process. While in the gallbladder, bile is concentrated by the removal of water and some ions. When fat-rich food enters the duodenum, a hormonal stimulus causes the gallbladder to contract, releasing the stored bile and making it available to the duodenum.

If the common hepatic or bile duct is blocked (for example, by wedged gallstones), bile is prevented from entering the small intestine, accumulates, and eventually backs up into the liver. This exerts pressure on the liver cells, and bile begins to enter the bloodstream. As the bile circulates through the body, the tissues become yellow or jaundiced.

Blockage of the ducts is just one cause of jaundice. More often it results from actual liver problems such as **hepatitis** (an inflammation of the liver) or **cirrhosis,** a condition in which the liver is severely damaged, becoming hard and fibrous. Cirrhosis is almost guaranteed in those who drink excessive alcohol for many years. ■

As demonstrated by its highly organized anatomy, the liver (Figure 38.14) is very important in the initial processing of the nutrient-rich blood draining the digestive organs. Its structural and functional units are called **lobules.** Each lobule is a basically cylindrical structure consisting of cordlike arrays of parenchyma cells, which radiate outward from a central vein running upward in the longitudinal axis of the lobule. At each of the six corners of the lobule is a **portal triad,** so named because three basic structures are always present there: a branch of the *hepatic artery* (the functional blood supply of the liver), a branch of the *hepatic portal vein* (carrying nutrient-rich blood from the digestive viscera), and a *bile duct.* Between the liver parenchyma cells are blood-filled spaces, or **sinusoids,** through which blood from the hepatic portal vein and hepatic artery percolates past the parenchyma cells. Special phagocytic cells, **Kupffer cells,** line the sinusoids and remove debris such as bacteria from the blood as it flows past, while the paren-

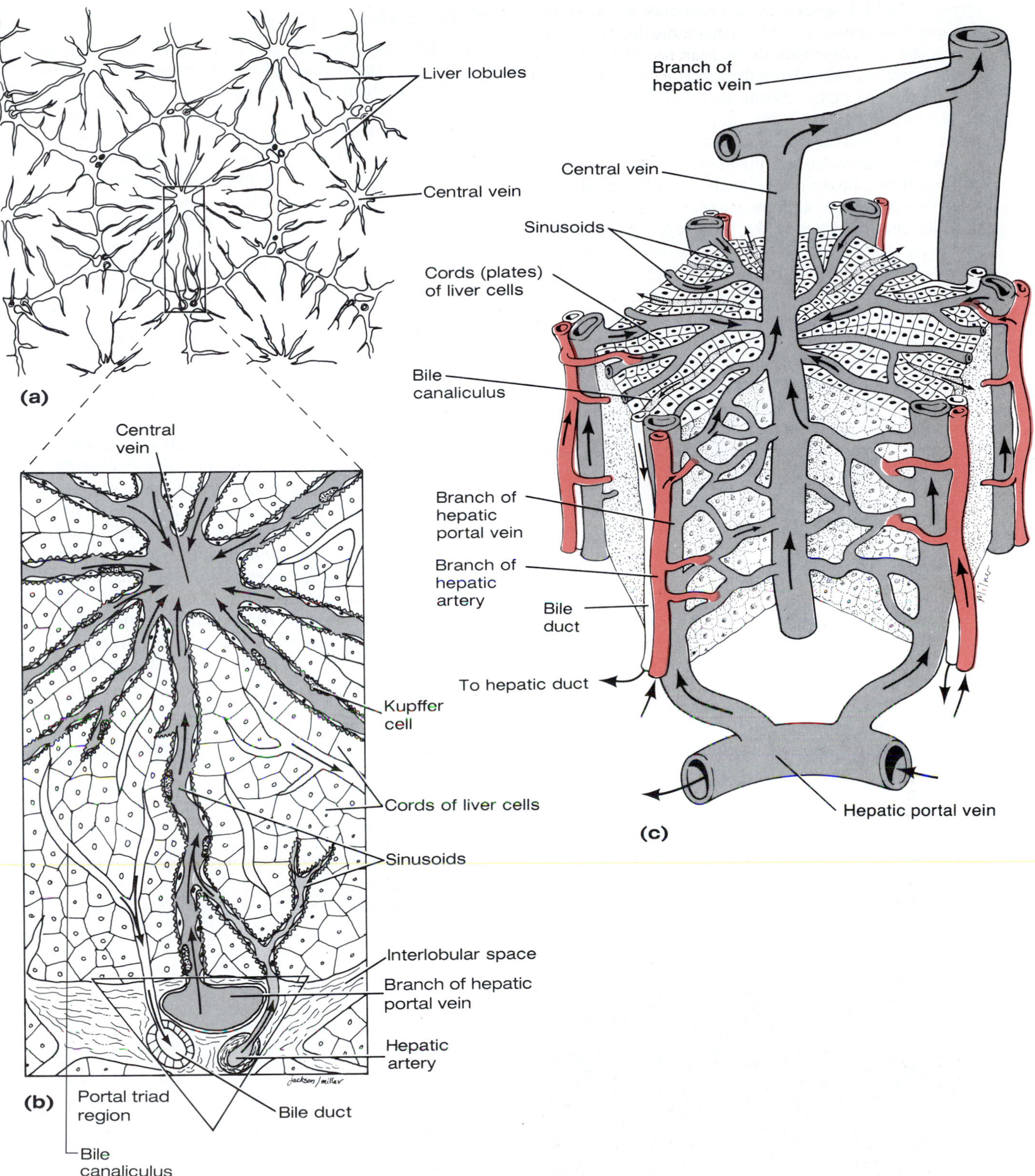

F38.14

Microscopic anatomy of the liver, diagrammatic view. (a) Several liver lobules (cross section). **(b)** Enlarged view of a portion of one liver lobule (cross section). **(c)** Portion of one liver lobule (three-dimensional representation). Arrows show direction of bile and blood flow.

chyma cells pick up oxygen and nutrients. Much of the glucose transported to the liver from the digestive system is stored as glycogen in the liver for later use, and amino acids are taken from the blood by the liver cells and utilized to make plasma proteins. The sinusoids empty into the central vein, and the blood ultimately drains from the liver via the *hepatic vein.*

Bile is continuously being made by the parenchyma cells. It flows through tiny canals, the **bile canaliculi,** which run between adjacent parenchyma cells toward the bile duct branches in the triad regions, where the bile eventually leaves the liver. Notice that the direction of blood and bile flow in the liver lobule is exactly opposite.

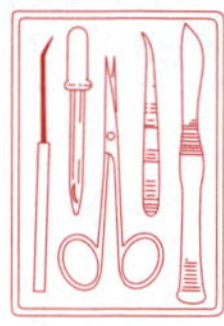

Examine a slide of liver tissue and identify as many of the structural features illustrated in Figure 38.14 and Histology Atlas Plates 41 and 42 as possible. Also examine a three-dimensional model of the liver if this is available. Draw your observations below.

Pancreas

The **pancreas** is a soft, triangular gland that extends horizontally across the posterior abdominal wall from the spleen to the duodenum (see Figure 38.1). Like the duodenum, it is a retroperitoneal organ (see Figure 38.7). As noted in Exercise 27, the pancreas has both an endocrine function (it produces the hormones insulin and glucagon) and an exocrine (enzyme-producing) function. It produces a whole spectrum of hydrolytic enzymes, which it secretes in an alkaline fluid into the duodenum through the pancreatic duct. Pancreatic juice is very alkaline. Its high concentration of bicarbonate ion (HCO_3^-) neutralizes the acidic chyme entering the duodenum from the stomach, enabling the pancreatic and intestinal enzymes to operate at their optimal pH. (Optimal pH for digestive activity to occur in the stomach is very acidic and results from the presence of HCl; that for the small intestine is slightly alkaline.)

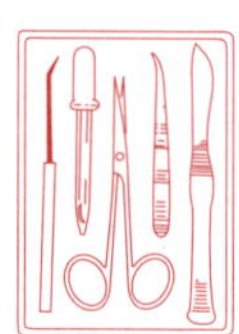

Observe pancreas tissue under low power and then high power to distinguish between the lighter-staining, endocrine-producing clusters of cells (pancreatic islets) and the deeper-staining acinar cells, which produce the hydrolytic enzymes and form the major portion of the pancreatic tissue (see Figure 38.15). Notice the arrangement of the exocrine cells around their central ducts. Draw a small portion of the enzyme-producing pancreatic tissue and appropriately label your drawing (i.e., indicate acinar cells and ducts).

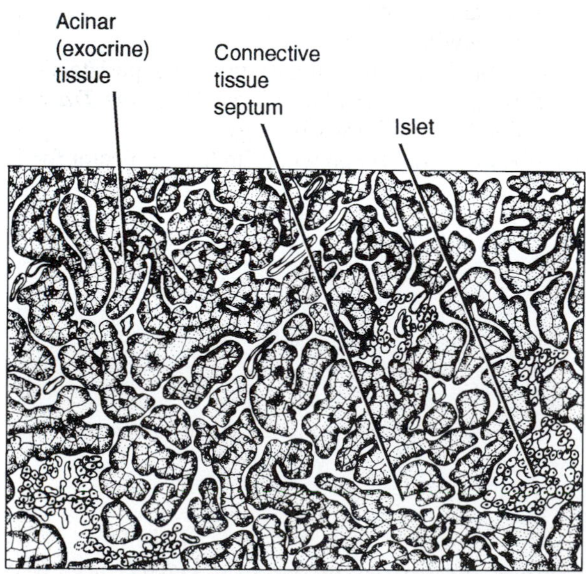

F38.15

Histology of the pancreas. The pancreatic islet cells produce insulin and glucagon (hormones). The acinar cells synthesize digestive enzymes for "export" to the duodenum. See also Plate 40 in the Histology Atlas.

EXERCISE 39

Chemical and Physical Processes of Digestion

OBJECTIVES

1. To list the digestive system enzymes involved in the digestion of proteins, fats, and carbohydrates; to state their site of origin; and to summarize the environmental conditions promoting their optimal functioning.
2. To recognize the variation between different types of enzyme assays.
3. To name the end products of digestion of proteins, fats, and carbohydrates.
4. To perform the appropriate chemical tests to determine if digestion of a particular foodstuff has occurred.
5. To cite the function(s) of bile in the digestive process.
6. To discuss the possible role of temperature and pH in the regulation of enzyme activity.
7. To define *enzyme, catalyst, control, substrate,* and *hydrolase.*
8. To explain why swallowing is both a voluntary and a reflex activity.
9. To discuss the role of the tongue, larynx, and gastroesophageal sphincter in swallowing.
10. To compare and contrast segmentation and peristalsis as mechanisms of propulsion.

MATERIALS

PART I: Enzyme Action

General supply area:
Test tubes and test tube rack
Wax markers
Large beakers
Water bath set at 37°C (if not available, incubate at room temperature and double the time)
Ice water bath
Hot plates
Chart on chalkboard for recording class results

Supply area 1:
Spot plates
Dropper bottles of distilled water
250-ml beakers
Boiling chips
Dropper bottles of
- 1% alpha-amylase solution*
- 1% boiled starch solution, freshly prepared†
- 1% maltose solution
- Lugol's IKI (Lugol's iodine)
- Benedict's solution

Supply area 2:
Dropper bottles of
- 1% trypsin
- 1% BAPNA solution

Supply area 3:
Dropper bottles of
- 1% pancreatin solution
- Litmus cream (fresh cream to which powdered litmus is added to achieve a blue color)
- 0.1 *N* HCl
- Vegetable oil

Bile salts (sodium taurocholate)
Parafilm (small squares to cover the test tubes)

PART II: Physical Processes

Supply area 4:
Water pitcher
Paper cups
Stethoscope
Alcohol swabs
Disposable autoclave bag
Film loop viewing station
- Film loop illustrating deglutition and peristalsis (*Passage of Food Through the Digestive Tract,* available from Ward's Biology)
- Cardboard carton set up with film loop projector for independent viewing of film loop by students.

* The alpha-amylase must be a low-maltose preparation for good results.

† Prepare by adding 1 g starch to 100 ml distilled water; boil and cool; add a pinch of salt (NaCl). Prepare fresh daily.

See Appendix E, Exercise 39 for links to *Anatomy and PhysioShow: The Videodisc.*

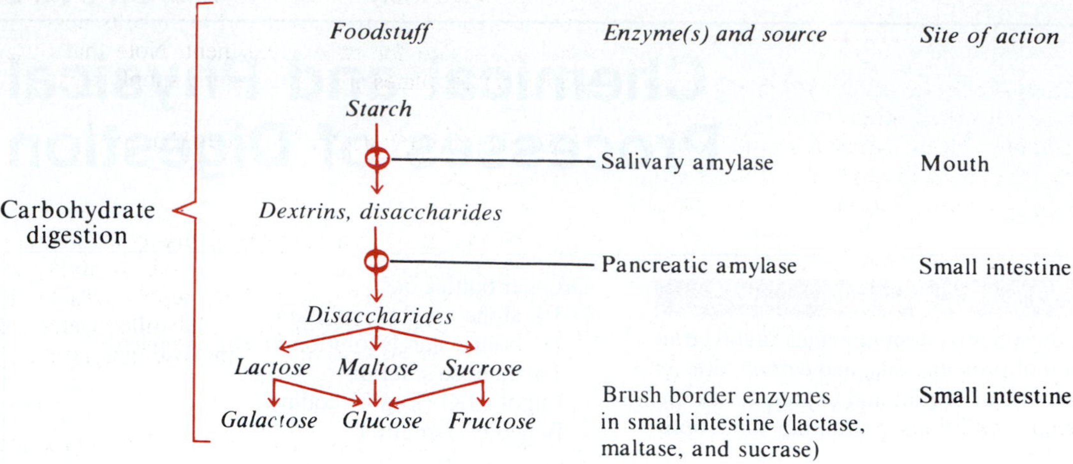

Absorption: The monosaccharides (glucose, galactose, and fructose) are absorbed into the capillary blood in the villi and transported to the liver via the hepatic portal vein

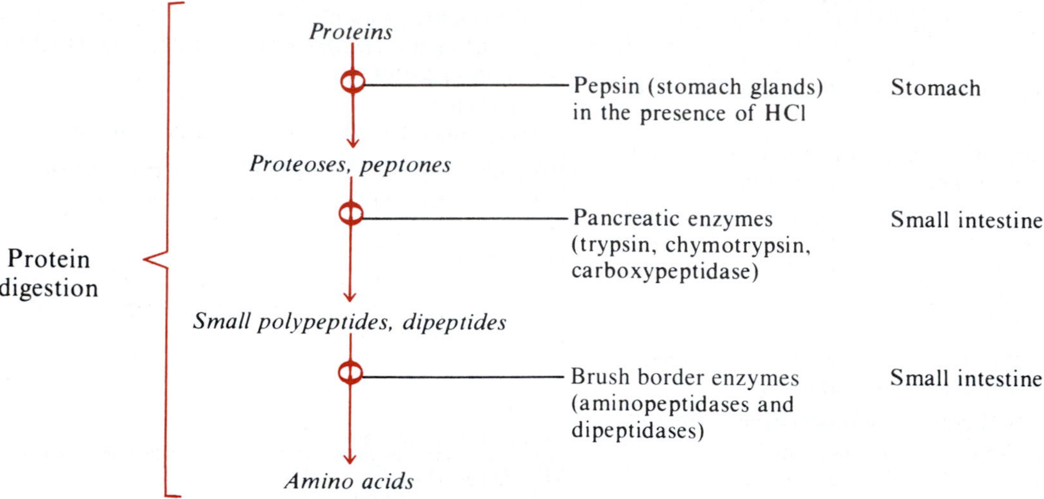

Absorption: The amino acids are absorbed into the capillary blood in the villi and transported to the liver via the hepatic portal vein

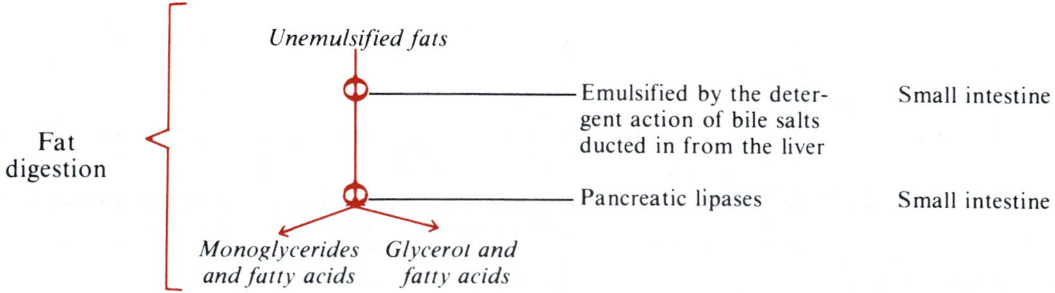

Absorption: Absorbed primarily into the lacteals of the villi and transported to the systemic circulation via the lymph in the thoracic duct. (Glycerol and short-chain fatty acids are absorbed into the capillary blood in the villi and transported to the liver via the hepatic portal vein.)

F39.1

Flow chart of digestion and absorption of foodstuffs.

CHEMICAL DIGESTION OF FOODSTUFFS: ENZYMATIC ACTION

Because nutrients can only be absorbed when broken down to their monomers, food digestion is a prerequisite to food absorption. You have already studied mechanisms of passive and active absorption in Exercise 5. Before proceeding, review that material on pages 33–39.

Enzymes are large protein molecules produced by body cells. They are biologic catalysts that increase the rate of a chemical reaction without themselves becoming part of the product. The digestive enzymes are hydrolytic enzymes, or *hydrolases,* which break down organic food molecules by adding water to the molecular bonds, thus cleaving the bonds between the subunits or monomers.

The various hydrolytic enzymes are highly specific in their action. Each enzyme hydrolyzes only one or a small group of substrate molecules, and specific environmental conditions are necessary for it to function optimally. Since digestive enzymes actually function outside the body cells in the digestive tract, their hydrolytic activity can also be studied in a test tube. Such an *in vitro* study provides a convenient laboratory environment for investigating the effect of such variations on enzymatic activity.

Figure 39.1 is a flow chart of the progressive digestion of proteins, fats, and carbohydrates. It indicates specific enzymes involved, their site of formation, and their site of action. Acquaint yourself with the flow chart before beginning this experiment, and refer to it as necessary during the laboratory session.

Work in groups of 3 or 4, with each group taking responsibility for setting up and conducting one of the following experiments. Each group should then communicate its results to the rest of the class by recording them in a chart on the chalkboard. All members of the class should observe the controls as well as the positive and negative examples of all experimental results. Additionally, all members of the class should be able to explain the tests used and the results observed and anticipated for each experiment. Note that water baths and hot plates are at the general supply area.

Starch Digestion by Salivary Amylase

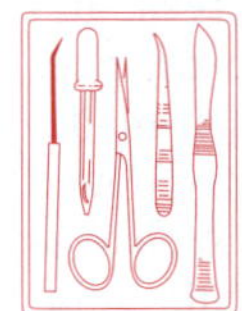

1. From the general supply area, obtain a test tube rack, 10 test tubes, and a wax marking pencil. From supply area 1, obtain a dropper bottle of distilled water and dropper bottles of maltose, amylase, and starch solutions.

2. Since in this experiment you will investigate the hydrolysis of starch to maltose by **salivary amylase** (the enzyme produced by the salivary glands and secreted into the mouth), it is important to be able to identify the presence of these substances to determine to what extent the enzymatic activity has occurred. Thus controls must be prepared to provide a known standard against which comparisons can be made. Starch decreases and sugar increases as digestion occurs, according to the following equation:

$$\text{Starch} + \text{water} \xrightarrow{\text{Amylase}} \text{X* maltose}$$

Two students should prepare the controls (tubes 1A to 3A) while the other two prepare the experimental samples (tubes 4A to 6A).

- Mark each tube with a wax pencil and load the tubes as indicated in Chart 1, using 3 drops (gtt) of each indicated substance.

* X = a variable number (of maltose molecules) depending on the specific starch molecule digested.

CHART 1 Salivary Amylase Digestion of Starch

Tube no.	1A	2A	3A	4A	5A	6A
Additives (3 gtt ea)	Starch, Water	Amylase, Water	Maltose, Water	Amylase, Starch	Amylase, Starch	Amylase, Starch
Incubation condition	37°C	37°C	37°C	Boil 4 min first, then incubate at 37°C	37°C	0°C
IKI test						
Benedict's test						

Additive key: Amylase, Starch, Maltose, Water

- Place all tubes in a rack in the appropriate water bath for approximately 1 hour.
- While these tubes are incubating, proceed to the next section.

Protein Digestion by Trypsin

Trypsin, an enzyme produced by the pancreas, hydrolyzes proteins to small fragments (proteoses, peptones, and peptides). BAPNA is a synthetic (man-made) protein substrate consisting of a dye covalently bound to an amino acid. Trypsin hydrolysis of BAPNA cleaves the dye molecule from the amino acid, causing the solution to change from colorless to bright yellow. Since the covalent bond between the dye molecule and the amino acid is the same as the peptide bonds that link amino acids together, the appearance of a yellow color indicates the presence and activity of an enzyme that is capable of peptide bond hydrolysis. Because the color change from clear to yellow is direct evidence of hydrolysis, additional tests are not required when determining trypsin activity using BAPNA.

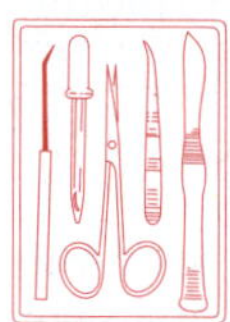

1. From the general supply area, obtain 5 test tubes and a test tube rack, and from supply area 2 get a dropper bottle of trypsin and one of BAPNA and bring them to your bench.

2. Two students should prepare the controls (tubes 1T and 2T) while the other two prepare the experimental samples (tubes 3T to 5T).

- Mark each tube with a wax pencil and load the tubes as indicated in Chart 2, using 3 drops (gtt) of each indicated substance.
- Place all tubes in a rack in the appropriate water bath for approximately 1 hour.
- While these tubes are incubating, proceed to the next section.

Pancreatic Lipase Digestion of Fats and the Action of Bile

The treatment that fats and oils go through during digestion in the small intestine is a bit more complicated than that of carbohydrates or proteins—pretreatment with bile to physically emulsify the fats is required. Hence, two sets of reactions occur.

First: Fats/oils $\xrightarrow[\text{(emulsification)}]{\text{Bile}}$ minute fat/oil droplets

Then: Fat/oil droplets $\xrightarrow{\text{Lipase}}$ monoglycerides and fatty acids

The term **pancreatin** describes the enzymatic product of the pancreas, which includes protein, carbohydrate, nucleic acid, and fat-digesting enzymes. It is used here to investigate the properties of **pancreatic lipase,** which hydrolyzes fats and oils to their component monoglycerides and two fatty acids (and occasionally to glycerol and three fatty acids).

The fact that some of the end products of fat digestion (fatty acids) are organic acids that decrease the pH provides an easy way to recognize that digestion is ongoing or completed. You will be using a pH indicator called *litmus blue* to follow these changes; it changes from *blue* to *pink* as the test tube contents become acid.

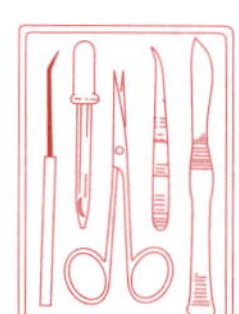

1. From the general supply area, obtain 9 test tubes and a test tube rack, plus one dropper bottle of each of the solutions in supply area 3.

CHART 2 Trypsin Digestion of Protein

Tube no.	1T	2T	3T	4T	5T
Additives (3 gtt ea)	Trypsin, Water	BAPNA, Water	Trypsin, BAPNA	Trypsin, BAPNA	Trypsin, BAPNA
Incubation condition	37°C	37°C	Boil 4 min first, then incubate at 37°C	37°C	0°C
Color change					

Additive key: Trypsin, BAPNA, Water

2. Although *bile,* a secretory product of the liver, is not an enzyme, it is important to fat digestion because of its emulsifying action (the physical breakdown of larger particles into smaller ones) on fats. Emulsified fats provide a larger surface area for enzymatic activity. To demonstrate the action of bile on fats, prepare two test tubes and mark them 1E and 2E.

- To tube 1E add 10 drops of water and 2 drops of vegetable oil.
- To tube 2E add 10 drops of water, 2 drops of vegetable oil, and a pinch of bile salts.
- Cover each tube with a small square of Parafilm, shake vigorously, and allow the tubes to stand at room temperature.

After 10–15 minutes, observe both tubes. If emulsification has not occurred, the oil will be floating on the surface of the water. If emulsification has occurred, the fat droplets will be suspended throughout the water, forming an emulsion.

In which tube has emulsification occurred? __________

3. Two students should prepare the controls (1L and 2L), while the other two students in the group set up the experimental samples (3L to 5L, 4B, and 5B) as indicated in Chart 3.

- Mark each tube with a wax pencil and load the tubes using 5 drops (gtt) of each indicated solution.
- A pinch of bile salts should be placed in tubes 4B and 5B.
- Cover each tube with a small square of Parafilm and shake to mix the contents of the tube.
- Remove the Parafilm and place all tubes in a rack in the appropriate water bath for approximately 1 hour.

ENZYME ASSAYS: DETERMINING RESULTS

Amylase Assay

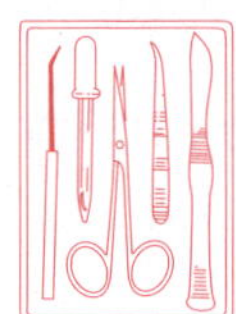

1. From supply area 1 obtain a 250-ml beaker, boiling chips, a spot plate, and dropper bottles of IKI and Benedict's solution. Use a hot plate as necessary from the general supply area.

2. Place a few boiling chips and about 125 ml water into the beaker and bring to a boil. While the water is heating, mark the spot plate *A* (for amylase) and number six of its depressions 1–6 for sample identification.

3. Pour about a drop of the sample from each of the tubes into the appropriately numbered spot. Into each sample droplet, place a drop of IKI solution. A blue-black color indicates the presence of starch and is referred to as a **positive starch test.** If starch is not present, the mixture will not turn blue, which is referred to as a **negative starch test.** Record your results (+ for positive, − for negative) in Chart 1 and on the chalkboard.

4. Into the remaining mixture in each tube, place 3 drops of Benedict's solution. Put each tube into the beaker of boiling water for about 5 minutes. If a green-to-orange precipitate forms, maltose is present; this is a **positive sugar test.** A **negative sugar test** is indicated by no color change. Record your results in Chart 1 and on the chalkboard.

CHART 3 Pancreatic Lipase Digestion of Fats

Tube no.	1L	2L	3L	4L	5L	4B	5B
Additives (5 gtt ea)							
Incubation condition	37°C	37°C	Boil 4 min first, then incubate at 37°C	37°C	0°C	37°C	0°C
Color change							

Additive key: Lipase, Litmus cream, Water, Pinch bile salts

Trypsin Assay

Since BAPNA is a synthetic colorigenic (color-producing) substrate, the presence of yellow color indicates a **positive hydrolysis test;** the dye molecule has been cleaved from the amino acid. If the sample mixture remains clear, no detectable hydrolysis has occurred.

1. Record the color of the experimental tubes in Chart 2 and on the chalkboard.

Lipase Assay

The basis of this assay is a pH change that is detected by a litmus powder indicator. Alkaline or neutral solutions containing litmus are blue but will turn reddish in the presence of acid. Since fats are digested to fatty acids (organic acids) during hydrolysis, they lower the pH of the sample they are in. Litmus cream (fresh cream providing the fat substrate to which litmus powder was added) will turn from a bluish color to pink if the solution is acid. Because the effect of hydrolysis is directly seen, additional assay reagents are not necessary.

1. To prepare a color control, add 0.1 *N* HCl drop by drop to tubes 1L and 2L (covering the tubes with a square of Parafilm after each addition and shaking to mix) until the cream turns pink.

2. Record the color of the tubes in Chart 3 and on the chalkboard.

PHYSICAL PROCESSES: MECHANISMS OF FOOD PROPULSION AND MIXING

Although enzyme activity is a very important part of the overall digestion process, foods must also be processed physically (churning and chewing), and moved by mechanical means along the tract if digestion and absorption are to be completed. Just about any time organs exhibit mobility, muscles are involved, and movements of and in the gastrointestinal tract are no exception. Although we tend to think only of smooth muscles when visceral activities are involved, both skeletal and smooth muscles are involved in digestion. This fact is amply demonstrated by the simple demonstrations that follow.

Deglutition (Swallowing)

Swallowing, or **deglutition,** which is largely the result of skeletal muscle activity, occurs in two phases: *buccal* (mouth) and *pharyngeal-esophageal.* The initial phase —the buccal—is voluntarily controlled and initiated by the tongue. Once begun, the process continues involuntarily in the pharynx and esophagus, through peristalsis, resulting in the delivery of the swallowed contents to the stomach.

1. Obtain a pitcher of water, a stethoscope, a paper cup, an alcohol swab, and an autoclave bag in preparation for making the following observations.

2. While swallowing a mouthful of water, consciously note the movement of your tongue during the process. Record your observations.

____

____

____

3. Repeat the swallowing process while your laboratory partner watches the externally visible movements of your larynx. (This movement is more obvious in a male, who has a larger Adam's apple.) Record your observations.

____

____

What do these movements accomplish? ____________

____

____

4. Before donning the stethoscope your lab partner should clean the earpieces with an alcohol swab. Then, he or she should place the diaphragm of the stethoscope over your abdominal wall, approximately 1 inch below the xiphoid process and slightly to the left, to listen for sounds as you again take two or three swallows of water. There should be two audible sounds—one when the water splashes against the gastroesophageal sphincter and the second when the peristaltic wave of the esophagus arrives at the sphincter and the sphincter opens, allowing water to gurgle into the stomach. Determine, as accurately as possible, the time interval between these two sounds and record it below.

Interval between arrival of water at the sphincter and the

opening of the sphincter: ________ sec

This interval gives a fair indication of the time it takes for the peristaltic wave to travel down the 10-inch-long esophagus. (Actually the time interval is slightly less than it seems, because pressure causes the sphincter to relax before the peristaltic wave reaches it.)

Dispose of the used paper cup in the autoclave bag.

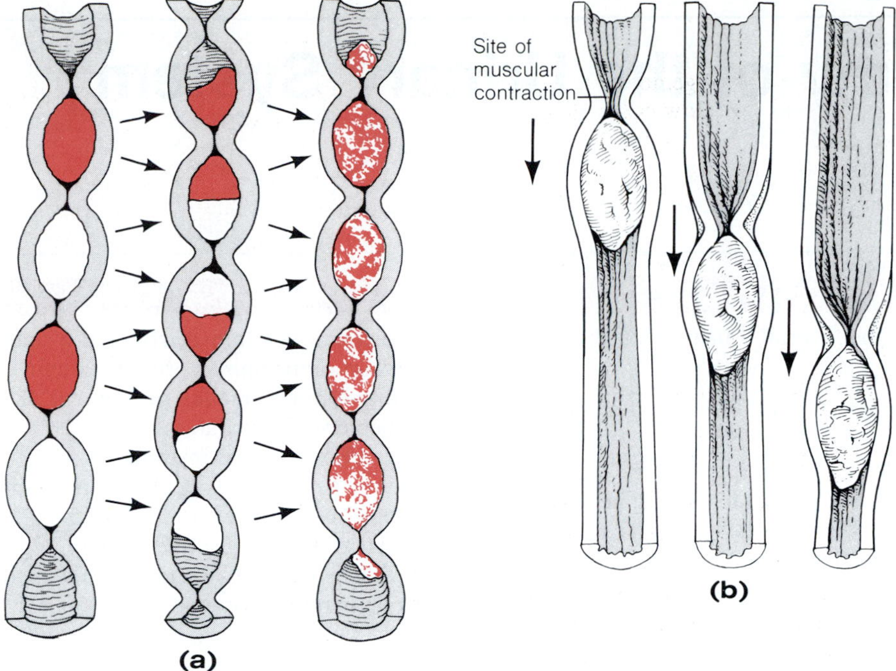

F39.2

Segmental and peristaltic movements of the digestive tract. (a) Segmentation: single segments of intestine alternately contract and relax. Because inactive segments exist between active segments, food mixing occurs to a greater degree than food movement. **(b)** Peristalsis: superimposed on segmentation. Neighboring segments of the intestine alternately contract and relax, moving food along the tract.

Segmentation and Peristalsis

Although several types of movements occur in the digestive tract organs, segmentation and peristalsis are most important as mixing and propulsive mechanisms (Figure 39.2).

Segmental movements are local constrictions of the organ wall that occur rhythmically. They serve mainly to mix the foodstuffs with digestive juices and to increase the rate of absorption by continually moving different portions of the chyme over adjacent regions of the intestinal wall. However, segmentation is also an important means of food propulsion in the small intestine, and slow segmenting movements called haustral contractions are frequently seen in the large intestine.

Peristaltic movements are the major means of propelling food through most of the digestive viscera. Essentially they are waves of contraction followed by waves of relaxation that squeeze foodstuffs through the alimentary canal, and they are superimposed on segmental movements.

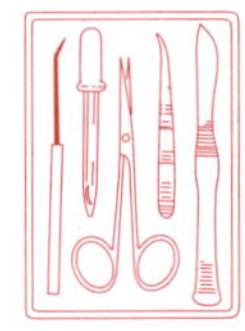

If a film loop showing some of the propulsive movements is available, go to a viewing station to view it before leaving the laboratory.

40

EXERCISE

Anatomy of the Urinary System

OBJECTIVES

1. To describe the overall function of the urinary system.
2. To identify, on an appropriate diagram, torso model, or cadaver, the urinary system organs and to describe the general function of each.
3. To define *micturition,* and to explain pertinent differences in the control of the two bladder sphincters (internal and external).
4. To compare the course and length of the urethra in males and females.
5. To identify the following regions of the dissected kidney (longitudinal section): hilus, cortex, medulla, medullary pyramids, major and minor calyces, pelvis, renal columns, and capsule layers.
6. To trace the blood supply of the kidney from the renal artery to the renal vein.
7. To define the nephron as the physiological unit of the kidney and to describe its anatomy.
8. To define *glomerular filtration, tubular reabsorption,* and *tubular secretion,* and to indicate the nephron areas involved in these processes.
9. To recognize microscopic or diagrammatic views of the histologic structure of the kidney and bladder.

MATERIALS

Human dissectible torso model and/or anatomical chart of the human urinary system
3-dimensional model of the cut kidney and of a nephron (if available)
Dissection tray and instruments
Pig or sheep kidney, doubly or triply injected
Protective skin cream or disposable gloves
Prepared histologic slides of a longitudinal section of kidney and cross sections of the bladder
Compound microscope

See Appendix D, Exercise 40 for links to A.D.A.M. Standard.

See Appendix E, Exercise 40 for links to *Anatomy and PhysioShow: The Videodisc.*

Metabolism of nutrients by the body produces wastes (carbon dioxide, nitrogenous wastes, ammonia, and so on) that must be eliminated from the body if normal function is to continue. Although excretory processes involve several organ systems (the lungs excrete carbon dioxide and skin glands excrete salts and water), it is the **urinary system** that is primarily concerned with the removal of nitrogenous wastes from the body. In addition to this purely excretory function, the kidney maintains the electrolyte, acid-base, and fluid balances of the blood and is thus a major, if not *the* major, homeostatic organ of the body.

To perform its functions, the kidney acts first as a blood "filter," and then as a blood "processor." It allows toxins, metabolic wastes, and excess ions to leave the body in the urine, while simultaneously retaining needed substances and returning them to the blood. Malfunction of the urinary system, particularly of the kidneys, leads to a failure in homeostasis which, unless corrected, is fatal.

GROSS ANATOMY OF THE HUMAN URINARY SYSTEM

The urinary system (Figure 40.1) consists of the paired kidneys and ureters and the single urinary bladder and urethra. The kidneys perform the functions described above and manufacture urine in the process. The remaining organs of the system provide temporary storage reservoirs or transportation channels for urine.

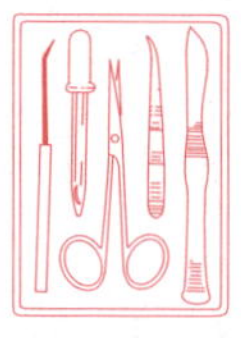

Examine the human torso model, a large anatomical chart, or a three-dimensional model of the urinary system to locate and study the anatomy and relationships of the urinary organs.

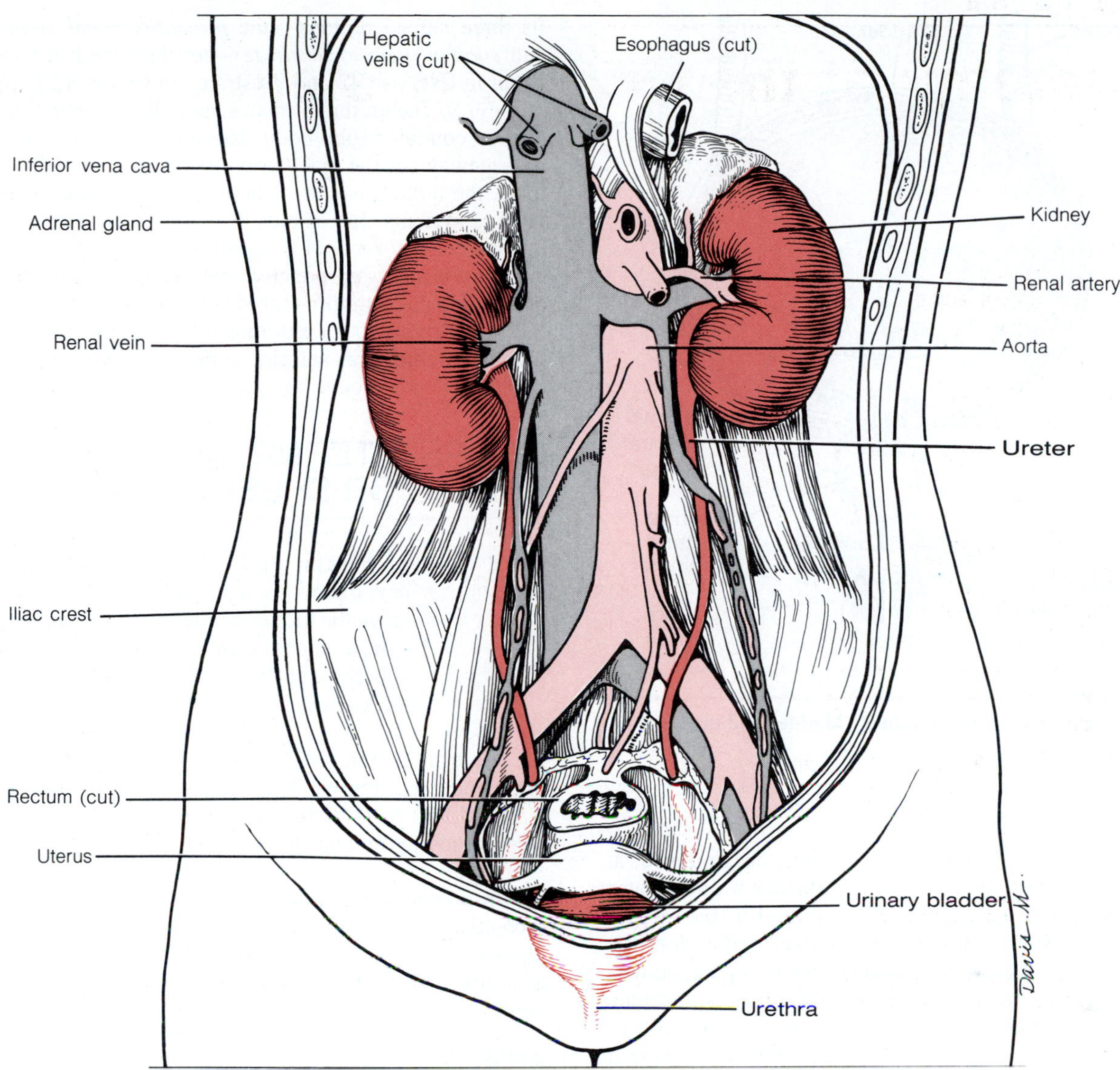

F40.1

Anterior view of the urinary organs of a human female. Most unrelated abdominal organs have been removed. (See also Plates F and G in the Human Anatomy Atlas.)

1. Locate the paired **kidneys** on the dorsal body wall in the superior lumbar region. Note that they are not positioned at exactly the same level. Because it is "crowded" by the liver, the right kidney is slightly lower than the left kidney. In a living person, fat deposits (the *adipose capsules*) hold the kidneys in place in a retroperitoneal position.

When the fatty material surrounding the kidneys is reduced or too meager in amount (in cases of rapid weight loss or in very thin individuals), the kidneys are less securely anchored to the body wall and may drop to a lower or more inferior position in the abdominal cavity. This phenomenon is called **ptosis.** ■

2. Observe the **renal arteries** as they diverge from the descending aorta and plunge into the indented medial region **(hilus)** of each kidney. Note also the **renal veins,** which drain the kidneys (circulatory drainage) and the two **ureters,** which drain urine from the kidneys and conduct it by peristalsis to the bladder for temporary storage.

3. Locate the **urinary bladder,** and observe the point of entry of the two ureters into this organ. Also locate the single **urethra,** which drains the bladder. The triangular region of the bladder, which is delineated by these three openings (two ureteral and one urethral orifice), is referred to as the **trigone** (Figure 40.2). Although urine formation by the kidney is a continuous process, urine is usually removed from the body when voiding is convenient. In the meantime the bladder stores it temporarily.

Voiding, or **micturition,** is the process in which urine empties from the bladder. Two sphincter muscles or valves, the **internal urethral sphincter** (more superiorly located) and the **external urethral sphincter** (more inferiorly located) control the outflow of urine from the bladder. Ordinarily, the bladder continues to

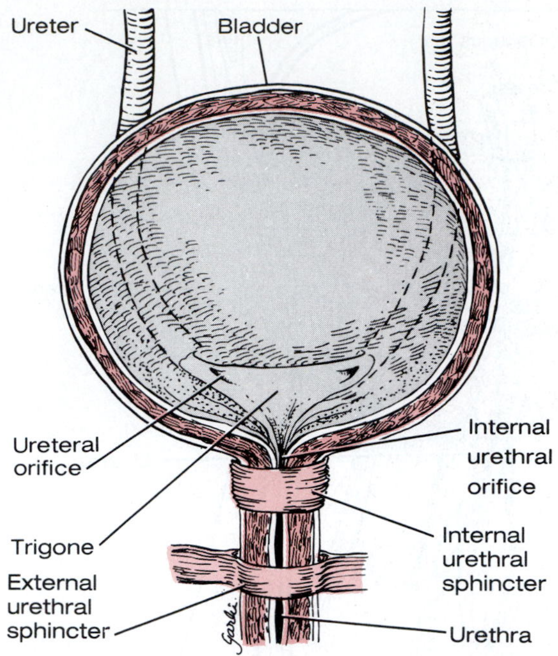

F40.2

Detailed structure of the urinary bladder and urethral sphincters.

collect urine until about 200 ml have accumulated, at which time the stretching of the bladder wall activates stretch receptors. Impulses transmitted to the central nervous system subsequently produce reflex contractions of the bladder wall through parasympathetic nervous system pathways (i.e., via the pelvic splanchnic nerves). As the contractions increase in force and frequency, the stored urine is forced past the internal sphincter, which is a smooth muscle involuntary sphincter, into the superior part of the urethra. It is then that a person feels the urge to void. The inferior external sphincter consists of skeletal muscle and is voluntarily controlled. If it is not convenient to void, the opening of this sphincter can be inhibited. Conversely, if the time is convenient, the sphincter may be relaxed and the stored urine flushed from the body. If voiding is inhibited, the reflex contractions of the bladder cease temporarily and urine continues to accumulate in the bladder. After another 200 to 300 ml of urine have been collected, the *micturition reflex* will again be initiated.

Lack of voluntary control over the external sphincter is referred to as **incontinence.** Incontinence is normal in children 2 years old or younger, as they have not yet gained control over the voluntary sphincter. In adults and older children, incontinence is generally a result of spinal cord injury, emotional problems, bladder irritability, or some other pathology of the urinary tract. ■

4. Follow the course of the urethra to the body exterior. In the male, it is approximately 20 cm (8 inches) long, travels the length of the **penis,** and opens at its tip. Its three named regions—the *prostatic, membranous,* and *spongy (penile) urethrae*—are described in more detail in Exercise 42 and illustrated in Figure 42.1 (pp. 391–392). The urethra of males has a dual function: it is a urine conduit to the body exterior, and it provides a passageway for the ejaculation of semen. Thus, in the male, the urethra is part of both the urinary and reproductive systems. In females, the urethra is very short, approximately 4 cm (1½ inches) long. There are no common urinary-reproductive pathways in the female, and the female's urethra serves only to transport urine to the body exterior. Its external opening, the **external urethral orifice,** lies anterior to the vaginal opening.

GROSS INTERNAL ANATOMY OF THE PIG OR SHEEP KIDNEY

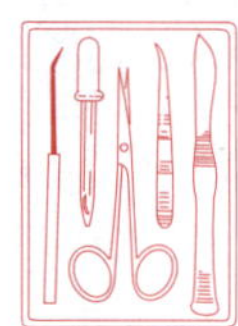

1. Obtain a preserved sheep or pig kidney, dissecting pan, and instruments. Observe the kidney to identify the **renal capsule,** a smooth transparent membrane that adheres tightly to the external aspect of the kidney.

2. Find the ureter, renal vein, and renal artery at the hilus (indented) region. The renal vein has the thinnest wall and will be collapsed. The ureter is the largest of these structures and has the thickest wall.

3. In preparation for dissection, don gloves or apply protective skin cream. Make a cut through the longitudinal axis (frontal section) of the kidney and locate the anatomical areas described below and depicted in Figure 40.3.

Kidney cortex: the superficial kidney region, which is lighter in color. If the kidney is doubly injected with latex, you will see a predominance of red and blue latex specks in this region indicating its rich vascular supply.

Medullary region: deep to the cortex; a darker, reddish-brown color. The medulla is segregated into triangular regions that have a striped, or striated, appearance—the **medullary (renal) pyramids.** The base of each pyramid faces toward the cortex. Its more pointed **apex,** or **papilla,** points to the innermost kidney region.

Renal columns: areas of tissue, more like the cortex in appearance, which segregate and dip inward between the pyramids.

Renal pelvis: medial to the hilus; a relatively flat, basinlike cavity that is continuous with the **ureter,** which exits from the hilus region. Fingerlike extensions of the pelvis should be visible. The larger, or primary, extensions are called the **major calyces;** subdivisions of the major calyces are the **minor calyces.** Notice that the minor calyces terminate in cuplike areas that enclose the apexes of the medullary pyramids and collect urine draining from the pyramidal tips into the pelvis.

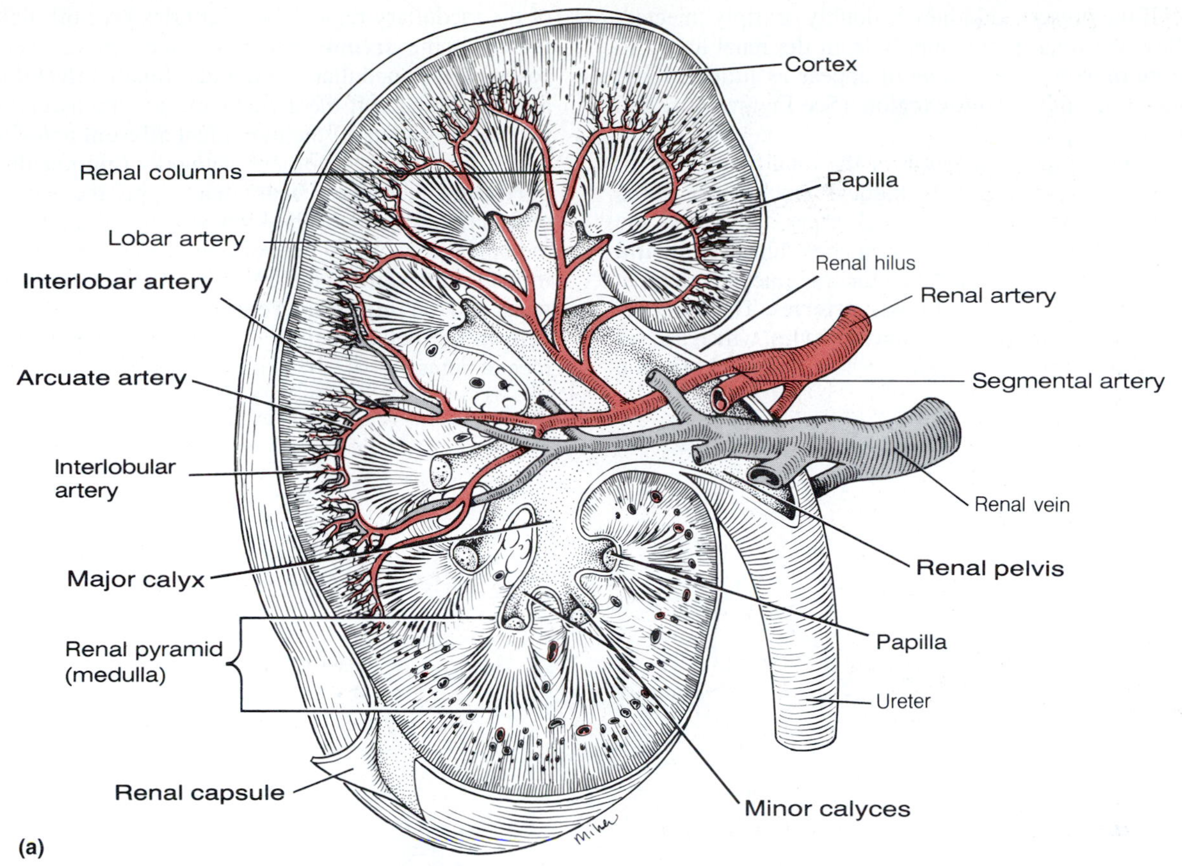

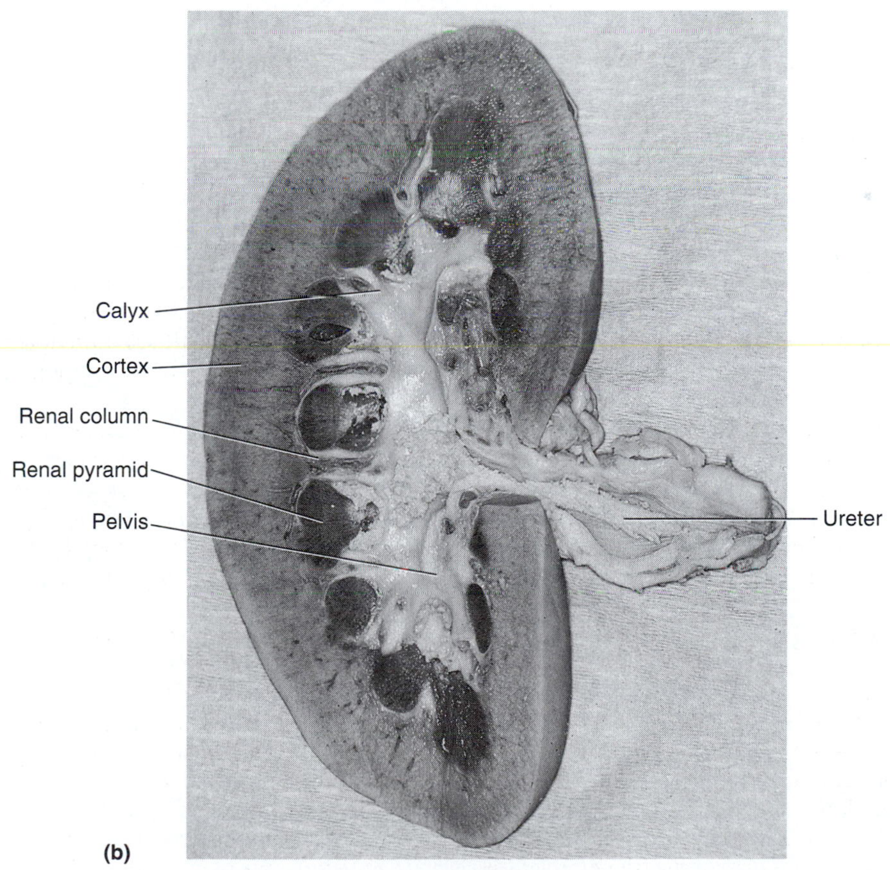

F40.3

Frontal section of a kidney. **(a)** Diagrammatic view, showing the larger arteries supplying the kidney tissue. **(b)** Photograph of a triple-injected pig kidney.

4. If the preserved kidney is doubly or triply injected, follow the renal blood supply from the renal artery to the **glomeruli.** The glomeruli appear as little red and blue specks in the cortex region. (See Figures 40.3 and 40.4.)

Approximately a fourth of the total blood flow of the body is delivered to the kidneys each minute by the large **renal arteries.** As a renal artery approaches the kidney, it breaks up into five branches called **segmental arteries** which enter the hilus. Each segmental artery, in turn, divides into several **lobar arteries.** The lobar arteries branch to form **interlobar arteries,** which ascend toward the cortex in the renal column areas. At the top of the medullary region, these arteries give off arching branches, the **arcuate arteries,** which curve over the bases of the medullary pyramids. Small **interlobular arteries** branch off the arcuate arteries and ascend into the cortex, giving off the individual **afferent arterioles,** which provide the capillary networks (**glomeruli** and **peritubular capillary beds**) that supply the nephrons, or functional units, of the kidney. Blood draining from the nephron capillary networks in the cortex enters the **interlobular veins** and then drains through the **arcuate veins** and the **interlobar veins** to finally enter the **renal vein** in the pelvis region. (There are no lobar or segmental veins.)

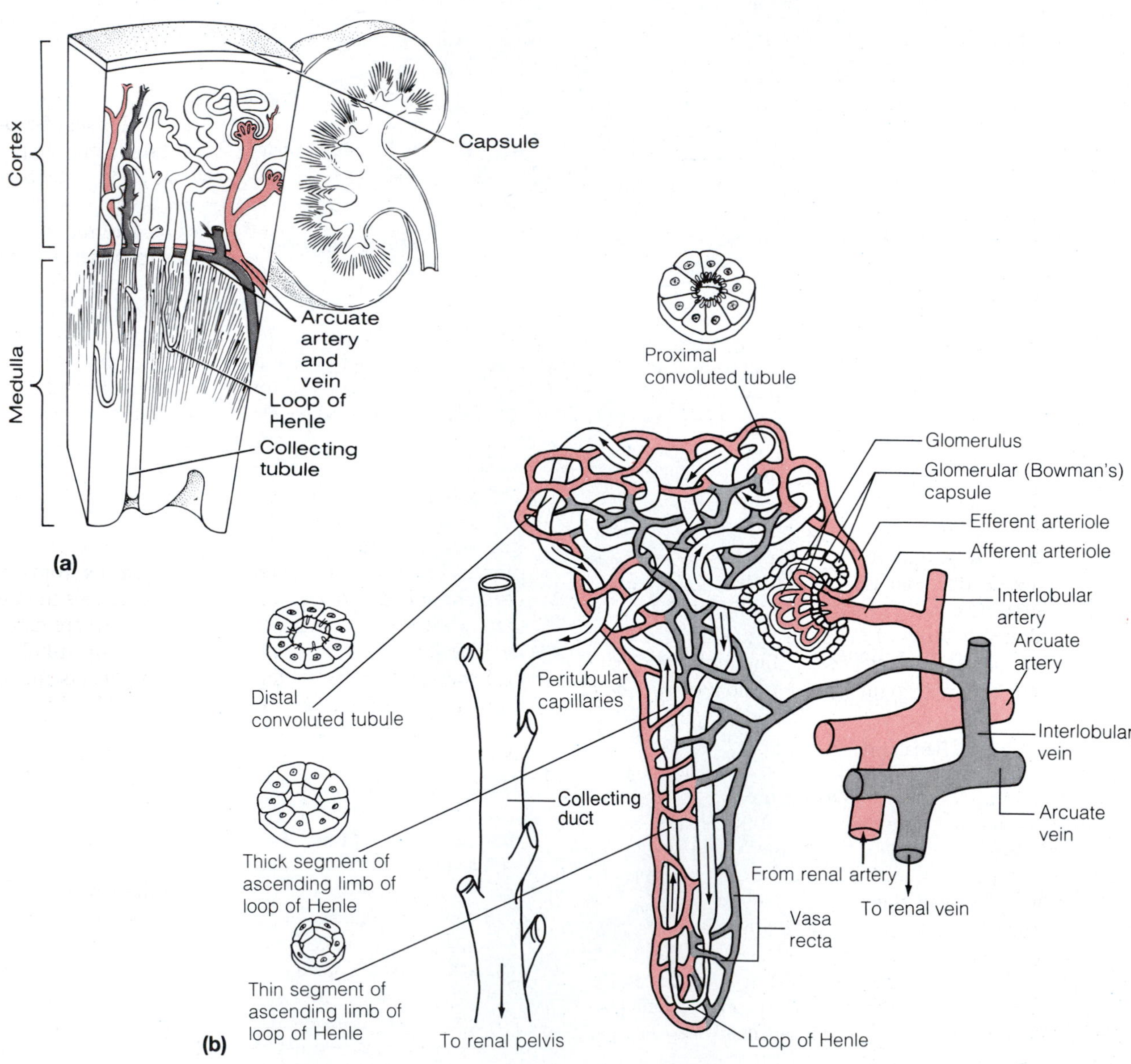

F40.4

Structure of a nephron. (a) Wedge-shaped section of kidney tissue, indicating the position of the nephrons in the kidney. **(b)** Detailed nephron anatomy and associated blood supply.

MICROSCOPIC ANATOMY OF THE KIDNEY AND BLADDER

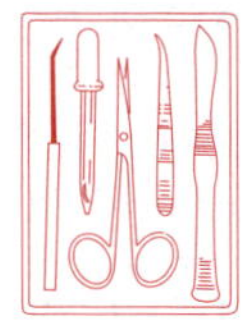

Obtain prepared slides of kidney and bladder tissue, and a compound microscope.

Kidney

Each kidney contains over a million nephrons, which are the anatomical units responsible for forming urine. Figure 40.4 depicts the detailed structure and the relative positioning of the nephrons in the kidney.

Each nephron consists of two major structures: a **glomerulus** (a capillary knot) and a **renal tubule.** During embryologic development, each renal tubule begins as a blind-ended tubule that gradually encloses an adjacent capillary cluster, or glomerulus. The enlarged end of the tubule encasing the glomerulus is the **glomerular (Bowman's) capsule,** and its inner, or visceral, wall consists of highly specialized cells called **podocytes.** Podocytes have long, branching processes (*foot processes*) that interdigitate with those of other podocytes and cling to the endothelial wall of the glomerular capillaries, thus forming a very porous epithelial membrane surrounding the glomerulus. The glomerulus-capsule complex is sometimes called the **renal corpuscle.**

The rest of the tubule is approximately 3 cm (1.25 inches) long. As it emerges from the glomerular capsule, it becomes highly coiled and convoluted, drops down into a long hairpin loop, and then again coils and twists before entering a collecting duct. In order from the glomerular capsule, the anatomical areas of the renal tubule are: the **proximal convoluted tubule, loop of Henle** (descending and ascending limbs), and the **distal convoluted tubule.** The wall of the renal tubule is composed almost entirely of cuboidal epithelial cells, with the exception of part of the descending limb (and sometimes part of the ascending limb) of the loop of Henle, which is simple squamous epithelium. The lumen surfaces of the cuboidal cells in the proximal convoluted tubule have dense microvilli (a cellular modification that greatly increases the surface area exposed to the lumen contents, or filtrate). Microvilli also occur on cells of the distal convoluted tubule but in greatly reduced numbers, revealing its less significant role in reclaiming filtrate contents.

Most nephrons, called **cortical nephrons,** are located entirely within the cortex. However, parts of the loops of Henle of the **juxtamedullary nephrons** (located close to the cortex-medulla junction) penetrate well into the medulla. The **collecting ducts,** each of which receives urine from many nephrons, run downward through the medullary pyramids, giving them their striped appearance. As the collecting ducts approach the renal pelvis, they fuse to form larger *papillary ducts,* which empty the final urinary product into the calyces and pelvis of the kidney.

The function of the nephron depends on several unique features of the renal circulation. The capillary vascular supply consists of two distinct capillary beds, the *glomerulus* and the *peritubular capillary bed.* Vessels leading to and from the **glomerulus,** the first capillary bed, are both arterioles: **the afferent arteriole** feeds the bed while the **efferent arteriole** drains it. The glomerular capillary bed has no parallel elsewhere in the body. It is a high-pressure bed along its entire length. Its high pressure is a result of two major factors: (1) the bed is *fed and drained* by arterioles (arterioles are high-resistance vessels as opposed to venules, which are low-resistance vessels), and (2) the afferent feeder arteriole is larger in diameter than the efferent arteriole draining the bed. The high hydrostatic pressure created by these two anatomical features forces out fluid and blood components smaller than proteins from the glomerulus into the glomerular capsule. That is, it forms the filtrate which is processed by the nephron tubule.

The **peritubular capillary bed** arises from the efferent arteriole draining the glomerulus. This set of capillaries cling intimately to the renal tubule and empty into the interlobular veins that leave the cortex. The peritubular capillaries are *low-pressure* very porous capillaries adapted for absorption rather than filtration and readily take up the solutes and water reabsorbed from the filtrate by the tubule cells. The juxtamedullary nephrons have additional looping vessels, called the **vasa recta** ("straight vessels"), that parallel their long loops of Henle in the medulla (see Figure 40.4). Hence, the two capillary beds of the nephron have very different, but complementary, roles: The glomerulus produces the filtrate and the peritubular capillaries reclaim most of that filtrate.

Urine formation is a result of three processes: *filtration, reabsorption,* and *secretion* (Figure 40.5). **Filtration,** the role of the glomerulus, is largely a passive process in which a portion of the blood passes from the glomerular bed into the glomerular capsule. This filtrate then enters the proximal convoluted tubule where tubular reabsorption and secretion begin. During **tubular reabsorption,** many of the filtrate components move through the tubule cells and return to the blood in the peritubular capillaries. Some of this reabsorption is passive, such as that of water which passes by osmosis, but the reabsorption of most substances depends on active transport processes and is highly selective. Which substances are reabsorbed at a particular time depends on the composition of the blood and needs of the body at that time. Substances that are almost entirely reabsorbed from the filtrate include water, glucose, and amino acids. Various ions are selectively reabsorbed or allowed to go out in the urine according to what is required to maintain appropriate blood pH and electrolyte composition. Waste products (urea, creatinine, uric acid, and drug metabolites) are reabsorbed to a much lesser degree or not at all. Most (75% to 80%) of tubular reabsorption occurs in the proximal convoluted tubule; the balance occurs in other areas, especially the distal convoluted tubules and collecting ducts.

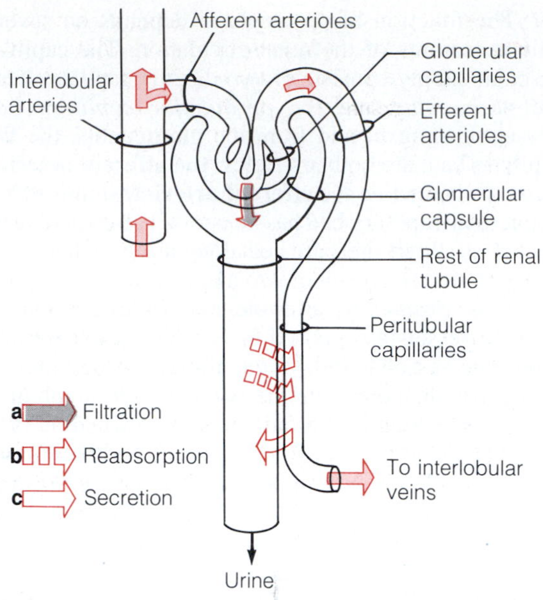

F40.5

The kidney depicted as a single, large nephron. A kidney actually has millions of nephrons acting in parallel. The three major mechanisms by which the kidneys adjust the composition of plasma are **(a)** glomerular filtration, **(b)** tubular reabsorption, and **(c)** tubular secretion. Solid pink arrows show the path of blood flow through the renal microcirculation.

Tubular secretion is essentially the reverse process of tubular reabsorption. Substances such as hydrogen and potassium ions and creatinine move either from the blood of the peritubular capillaries through the tubular cells or from the tubular cells into the filtrate to be disposed of in the urine. This process is particularly important for the disposal of substances not already in the filtrate (such as drug metabolites), and as a device for controlling blood pH.

Observe a model of the nephron before continuing on with the microscope study of the kidney.

1. Hold the longitudinal section of the kidney up to the light to identify cortical and medullary areas. Then secure the slide on the microscope stage and scan the slide under low power.

2. Move the slide so that you can see the cortical area. Identify a glomerulus, which appears as a ball of tightly packed material containing many small nuclei (Figure 40.6). It is usually delineated by a vacant-appearing region (corresponding to the space between the visceral and parietal layers of the glomerular capsule) that surrounds it.

3. Notice that the renal tubules are cut at various angles. Also try to differentiate between the thin-walled loop of Henle portion of the tubules and the cuboidal epithelium of the proximal convoluted tubule, which has dense microvilli.

Bladder

1. Scan the bladder tissue. Identify its three layers—mucosa, muscular layer, and fibrous adventitia.

2. Study the mucosa with its highly specialized transitional epithelium. The plump, transitional epithelial cells have the ability to slide over one another, thus decreasing the thickness of the mucosa layer as the bladder fills and stretches to accommodate the increased urine volume. Depending on the degree of stretching of the bladder, the mucosa may be three to eight cell layers thick. Compare the transitional epithelium of the mucosa to that shown in Figure 6.3h (p. 46).

3. Examine the heavy muscular wall (detrusor muscle), which consists of three irregularly arranged muscular layers. The innermost and outermost muscle layers are arranged longitudinally; the middle layer is arranged circularly. Attempt to differentiate the three muscle layers.

4. Draw a small section of the bladder wall, and label all regions or tissue areas.

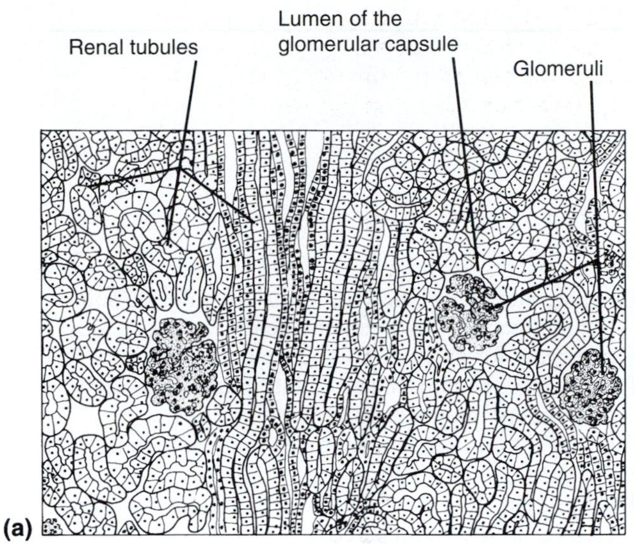

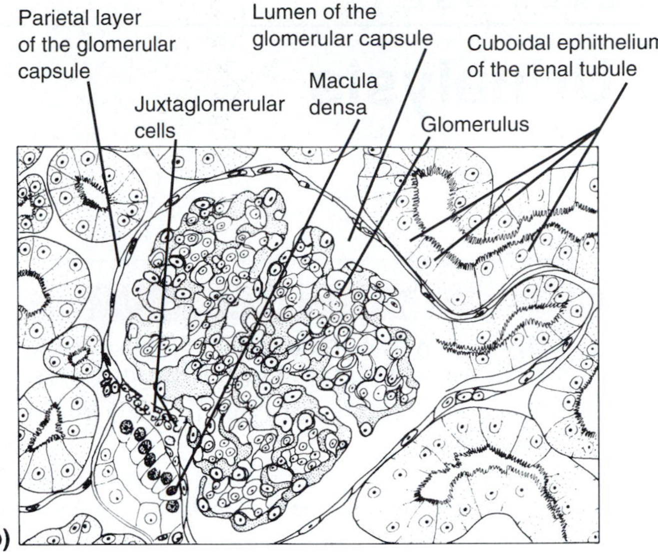

F40.6

Microscopic structure of kidney tissue. (a) Low-power view of the renal cortex. **(b)** Detailed structure of the glomerulus. (See corresponding Plates 43 and 44 in the Histology Atlas.)

5. Compare your sketch of the bladder wall to the structure of the ureter wall shown in Plate 45 in the Histology Atlas. How are the two organs similar histologically?

___

___

___

What is/are the most obvious differences?

___

___

41

EXERCISE

Urinalysis

OBJECTIVES

1. To list the physical characteristics of urine, and to indicate the normal pH and specific gravity ranges.
2. To list substances that are normal urinary constituents.
3. To conduct various urinalysis tests and procedures and use them to determine the substances present in a urine specimen.
4. To define the following urinary conditions:

calculi	*hematuria*
glycosuria	*hemoglobinuria*
albuminuria	*pyuria*
ketonuria	*casts*

5. To explain the implications and possible causes of conditions listed in Objective 4.

MATERIALS

"Normal" artificial urine provided by the instructor*
Numbered "pathologic" urine specimens provided by the instructor*

Urinometer
Hot plate
500-ml beakers
Wide-range pH paper
Dip sticks: individual (Clinistix, Ketostix, Albustix, Hemastix, Bilistix) or combination (Chemstrip or Multistix)
Ictotest reagent
Test reagents for sulfates: 10% barium chloride solution, dilute HCl (hydrochloric acid)
Test reagent for glucose (Clinitest tablets)
Test reagent for phosphates (dilute nitric acid, dilute ammonium molybdate)
Test reagent for urea (concentrated nitric acid in dropper bottle)
Test reagent for Cl: 3.0% silver nitrate solution ($AgNO_3$), freshly prepared
Test tubes, test tube rack, and test tube holders
10-cc graduated cylinders
Glass stirring rods
Medicine droppers
Wax marking pencils

Microscope slides
Coverslips
Compound microscope
Centrifuge and centrifuge tubes
Sedi-stain

Flasks and laboratory buckets containing 10% bleach solution
Disposable plastic gloves
Disposable autoclave bags

Demonstration: Instructor prepared specimen of urine sediment set up for microscopic analysis.

*Directions for making artificial urine are provided in the Instructor's Guide for this manual.

Blood composition depends on three major factors: diet, cellular metabolism, and urinary output. In 24 hours, the kidney's million nephrons filter approximately 150 to 180 liters of blood plasma through their glomeruli into the tubules, where it is selectively processed by tubular reabsorption and secretion. In the same period, urinary output, which contains by-products of metabolism and excess ions, is 1.0 to 1.8 liters. In healthy individuals, the kidneys can maintain blood constancy despite wide variations in diet and metabolic activity. With certain pathologic conditions, urine composition often changes dramatically.

CHARACTERISTICS OF URINE

Freshly voided urine is generally clear and pale yellow to amber in color. This normal yellow color is due to *urochrome,* a pigment metabolite arising from the body's destruction of hemoglobin (via bilirubin or bile pigments). As a rule, color variations from pale yellow to deeper amber indicate the relative concentration of solutes to water in the urine. The greater the solute concentration, the deeper the color. Abnormal urinary color may be due to certain foods, such as beets, various drugs, bile, or blood.

The odor of freshly voided urine is characteristic and slightly aromatic, but bacterial action gives it an ammonialike odor when left standing. Some drugs, vegetables (such as asparagus), and various disease processes (such as diabetes mellitus) alter the characteristic odor of urine. For example, the urine of a person with uncontrolled diabetes mellitus (and elevated levels of ketones) smells fruity or acetonelike.

The pH of urine ranges from 4.5 to 8.0, but its average value, 6.0, is slightly acidic. Diet may markedly influence the pH of the urine. For example, a diet high in protein (meat, eggs, cheese) and whole wheat products increases the acidity of urine. Such foods are called acid *ash foods.* On the other hand, a vegetarian diet (*alkaline ash diet*) increases the alkalinity of the urine. A bacterial infection of the urinary tract may also result in urine with a high pH.

Specific gravity is the relative weight of a specific volume of liquid compared with an equal volume of distilled water. The specific gravity of distilled water is 1.000, because 1 ml weighs 1 g. Since urine contains dissolved solutes, it weighs more than water, and its customary specific gravity ranges from 1.001 to 1.030. Urine with a specific gravity of 1.001 contains few solutes and is considered very dilute. Dilute urine commonly results when a person drinks excessive amounts of water, uses diuretics, or suffers from diabetes insipidus or chronic renal failure. Conditions that produce urine with a high specific gravity include limited fluid intake, fever, and kidney inflammation, called *pyelonephritis.* If urine becomes excessively concentrated, some of the substances normally held in solution begin to precipitate or crystallize, forming **kidney stones,** or **renal calculi.**

Normal constituents of urine (in order of decreasing concentration) include water; urea;* sodium,† potassium, phosphate, and sulfate ions; creatinine,* and uric acid.* Much smaller but highly variable amounts of calcium, magnesium, and bicarbonate ions are also found in the urine. Abnormally high concentrations of any of these urinary constituents may indicate a pathological condition.

ABNORMAL URINARY CONSTITUENTS

Abnormal urinary constituents are substances not normally present in the urine when the body is operating properly.

* Urea, uric acid, and creatinine are the most important nitrogenous wastes found in urine. Urea is an end product of protein breakdown; uric acid is a metabolite of purine breakdown; and creatinine is associated with muscle metabolism of creatine phosphate.

† Sodium ions appear in relatively high concentration in the urine because of reduced urine volume, not because large amounts are being secreted. Sodium is the major positive ion in the plasma; under normal circumstances, most of it is actively reabsorbed.

Glucose

The presence of glucose in the urine, a condition called **glycosuria,** indicates abnormally high blood sugar levels. Normally blood sugar levels are maintained between 80 and 120 mg/100 ml of blood. At this level all glucose in the filtrate is reabsorbed by the tubular cells and returned to the blood. Glycosuria may result from carbohydrate intake so excessive that normal physiologic and hormonal mechanisms cannot clear it from the blood quickly enough. In such cases, the active transport reabsorption mechanisms of the renal tubules for glucose are exceeded, but only temporarily.

Pathologic glycosuria occurs in conditions such as uncontrolled diabetes mellitus, in which the body cells are unable to absorb glucose from the blood because the pancreatic islet cells produce inadequate amounts of the hormone insulin, or there is some abnormality of the insulin receptors. Under such circumstances, the body cells increase their metabolism of fats, and the excess and unusable glucose spills out in the urine.

Albumin

Albuminuria, or the presence of albumin in urine, is an abnormal finding. Albumin is the single most abundant blood protein and is very important in maintaining the osmotic pressure of the blood. Albumin, like other blood proteins, normally is too large to pass through the glomerular filtration membrane; thus albuminuria is generally indicative of an abnormally increased permeability of the glomerular membrane. Certain nonpathologic conditions, such as excessive exertion, pregnancy, or overabundant protein intake, can temporarily increase the membrane permeability, leading to **physiologic albuminuria.** Pathologic conditions resulting in albuminuria include events that damage the glomerular membrane, such as kidney trauma due to blows, the ingestion of heavy metals, bacterial toxins, glomerulonephritis, and hypertension.

Ketone Bodies

Ketone bodies (acetoacetic acid, beta-hydroxybutyric acid, and acetone) normally appear in the urine in very small amounts. **Ketonuria,** the presence of these intermediate products of fat metabolism in excessive amounts, usually indicates that abnormal metabolic processes are occurring. The result may be *acidosis* and its complications. Ketonuria is an expected finding during starvation, when inadequate food intake forces the body to use its fat stores. Ketonuria coupled with a finding of glycosuria is generally diagnostic for diabetes mellitus.

Red Blood Cells

Hematuria, the appearance of red blood cells, or erythrocytes, in the urine, almost always indicates pathology of the urinary tract, because erythrocytes are too large to pass through the glomerular pores. Possible causes include irritation of the urinary tract organs by calculi (kidney stones), which produces frank bleeding; infections; or physical trauma to the urinary organs. In healthy menstruating females, it may reflect accidental contamination of the urine sample with the menstrual flow.

Hemoglobin

Hemoglobinuria, the presence of hemoglobin in the urine, is a result of the fragmentation, or hemolysis, of red blood cells. As a result, hemoglobin is liberated into the plasma and subsequently appears in the kidney filtrate. Hemoglobinuria indicates various pathologic conditions including hemolytic anemias, transfusion reactions, burns, or renal disease.

Bile Pigments

Bilirubinuria, the appearance of bilirubin (bile pigments) in urine, is an abnormal finding and usually indicates liver pathology, such as hepatitis or cirrhosis. Bilirubinuria is signaled by a yellow foam that forms when the urine sample is shaken.

White Blood Cells

Pyuria is the presence of white blood cells or other pus constituents in the urine. It indicates inflammation of the urinary tract.

Casts

Any complete discussion of the varieties and implications of casts is beyond the scope of this exercise. However, because they always represent a pathologic condition of the kidney or urinary tract, they should at least be mentioned. **Casts** are hardened cell fragments, usually cylindrical, which are flushed out of the urinary tract. *White blood cell casts* are a typical finding with pyelonephritis, *red blood cell casts* are commonly seen with glomerulonephritis, and *fatty casts* indicate severe renal damage.

ANALYSIS OF URINE SAMPLES

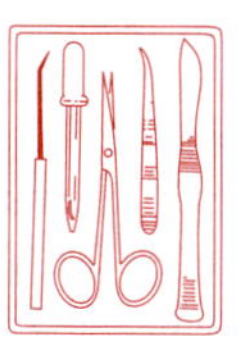

In this part of the exercise, you will use prepared dip sticks and perform chemical tests to determine the characteristics of normal urine as well as to identify abnormal urinary components. You will investigate both "normal" urine *and* an unknown specimen of urine provided by your instructor. Make the following determinations on both samples and record your results by circling the appropriate item or description or by adding data to complete Table 41.1. If you have more than one unknown sample, accurately identify each sample used by number.

Obtain and wear plastic disposable gloves throughout this laboratory session. Although the instructor-provided urine samples are actually artifical urine (concocted in the laboratory to resemble real urine), the techniques of safe handling of body fluids should still be observed as part of your learning process. When you have completed the laboratory procedures: (1) dispose of the gloves and used pH paper strips in the autoclave bag; (2) put used glassware in the bleach-containing laboratory bucket; (3) wash the lab bench down with 10% bleach solution.

Determination of the Physical Characteristics of Urine

1. Determine the color, transparency, and odor of your "normal" sample and one of the numbered pathologic samples, and circle the appropriate descriptions in Table 41.1.

2. Obtain a roll of wide-range pH paper to determine the pH of each sample. Use a fresh piece of paper for each test, and dip the strip into the urine to be tested two or three times before comparing the color obtained with the chart on the dispenser. Record your results in Table 41.1. (If you will be using one of the combination dip sticks—Chemstrip or Multistix—this pH determination can be done later.)

3. To determine specific gravity, obtain a urinometer cylinder and float. Mix the urine well, and fill the urinometer cylinder about two-thirds full with urine.

4. Examine the urinometer float to determine how to read its markings. In most cases, the scale has numbered lines separated by a series of unnumbered lines. The numbered lines give the reading for the first two decimal places. You must determine the third decimal place by reading the lower edge of the meniscus—the curved surface representing the urine-air junction—on the stem of the float.

5. Carefully lower the urinometer float into the urine. Make sure it is floating freely before attempting to take the reading. Record the specific gravity of both samples in the table. *Do not dispose of this urine if the samples that you have are less than 200 ml in volume* because you will need to make several more determinations.

Determination of Inorganic Constituents in Urine

SULFATES Add 5 ml of urine to a test tube, and then add a few drops of dilute hydrochloric acid and 2 ml of 10% barium chloride solution. The appearance of a white precipitate (barium sulfate) indicates the presence of sulfates in the sample. Clean the test tubes well after use. Record your results. Are sulfates a *normal* constituent of urine?

PHOSPHATES Obtain a hot plate and a 500 ml beaker. To prepare the hot water bath, half fill the beaker with tap water and heat it on the hot plate. Add 5 ml of urine to a test tube, and then add three or four drops of dilute nitric acid and 3 ml of ammonium molybdate. Mix well with a glass stirring rod, and then heat gently in a hot water bath. Formation of a yellow precipitate indicates the presence of phosphates in the sample.

TABLE 41.1 Urinalysis Results*

Observation or test	Normal values	Student urine specimen	Unknown specimen (#)
Physical characteristics			
Color	Pale yellow	Yellow: pale medium dark other ______	Yellow: pale medium dark other ______
Transparency	Transparent	Clear slightly cloudy cloudy	Clear slightly cloudy cloudy
Odor	Characteristic	Describe ______	Describe ______
pH	4.5–8.0	______	______
Specific gravity	1.001–1.030	______	______
Inorganic components			
Sulfates	Present	Present Absent	Present Absent
Phosphates	Present	Present Absent	Present Absent
Chlorides	Present	Present Absent	Present Absent
Organic components			
Urea	Present	Present Absent	Present Absent
Glucose Dip stick: ______	Negative	Record results: ______	Record results: ______
Clinitest ______	Negative	Record results: ______	Record results: ______
Albumin Dip stick: ______	Negative	Record results: ______	Record results: ______
Ketone bodies Dip stick: ______	Negative	Record results: ______	Record results: ______
RBCs/hemoglobin Dip stick: ______	Negative	Record results: ______	Record results: ______
WBCs (leukocytes) Dip stick: ______	Negative	Record results: ______	Record results: ______
Bilirubin Dip stick: ______	Negative	Record results: ______	Record results: ______
Ictotest	Negative (no color change)	Negative Positive (purple)	Negative Positive (purple)

* In recording urinalysis data, circle the appropriate description if provided; otherwise record the results you observed. Identify dip sticks used.

CHLORIDES Place 5 ml of urine in a test tube, and add several drops of silver nitrate ($AgNO_3$). The appearance of a white precipitate (silver chloride) is a positive test for chlorides. Record your results.

Determination of Organic Constituents in Urine

Individual dip sticks or combination dip sticks (Chemstrip or Multistix) may be used for many of the tests in this section. If combination dip sticks are used, be prepared to take the readings on several factors (pH, protein [albumin], glucose, ketones, blood/hemoglobin, and leukocytes [pus]) at the same time. Generally speaking, results for all of these tests may be read *during* the second minute after immersion, but readings taken after 2 minutes have passed should be considered invalid. Pay careful attention to the directions for method and time of immersion and disposal of excess urine from the strip, regardless of the dip stick used.

UREA Put two drops of urine on a clean microscope slide and *carefully* add one drop of concentrated nitric acid to the urine. Slowly warm the mixture on a hot plate until it begins to dry at the edges, but do not allow it to boil or to evaporate to dryness. When the slide has cooled, examine the edges of the preparation under low power to identify the rhombic or hexagonal crystals of urea nitrate, which form when urea and nitric acid react chemically. Keep the light low for best contrast. Record your results.

GLUCOSE Use a combination dip stick or obtain a vial of Clinistix, and conduct the dip stick test according to the instructions on the vial. Record the results in Table 41.1.

Because the Clinitest reagent is routinely used in clinical agencies for glucose determinations in pediatric patients (children), it is worthwhile to conduct this test as well. Obtain the Clinitest tablets and the associated color chart. Using a medicine dropper, put 5 drops of urine into a test tube; then rinse the dropper and add 10 drops of water to the tube. Add a Clinitest tablet. Wait 15 seconds and then compare the color obtained to the color chart. Record the results.

ALBUMIN Use a combination dip stick or obtain the Albustix dip sticks, and conduct the determinations as indicated on the vial. Record your results.

BLOOD/HEMOGLOBIN Test your urine samples for the presence of hemoglobin by using a Hemastix dip stick or a combination dip stick according to the directions on the vial. Usually a short drying period is required before making the reading, so read the directions carefully. Record your results.

BILIRUBIN Using a Bilistix dip stick, determine if there is any bilirubin in your urine samples. Record your results.

Also conduct the Ictotest for the presence of bilirubin. Using a medicine dropper, place one drop of urine in the center of one of the special test mats provided with the Ictotest reagent tablets. Place one of the reagent tablets over the drop of urine, and then add two drops of water directly to the tablet. If the mixture turns purple when you add water, bilirubin is present. Record your results.

MICROSCOPIC ANALYSIS OF URINE SEDIMENT (OPTIONAL)

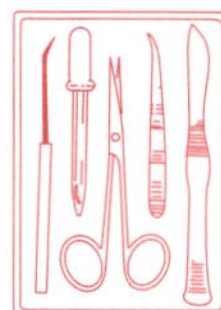

If desired by your instructor, a microscopic analysis of urine sediment (of "real" urine) can be done. The urine sample to be analyzed microscopically has been centrifuged to spin the more dense urine components to the bottom of a tube, and some of the sediment (mounted on a slide) has been stained with Sedi-stain to make the components more visible.

Go to the demonstration microscope to conduct this study. Using the lowest light source possible, examine the slide under low power to determine if any of the sediments illustrated in Figures 41.1 and 41.2 (and described here) can be seen.

Unorganized sediments (Figure 41.1): chemical substances that form crystals or precipitate from solution; for example, calcium oxalates, carbonates, and phosphates; uric acid; ammonium ureates; and cholesterol. Also, if one has been taking antibiotics or certain drugs such as sulfa drugs, these may be detectable in the urine in crystalline form. Normal urine contains very small amounts of crystals, but conditions such as urinary retention or urinary tract infection may result in the appearance of much larger amounts (and their possible consolidation into calculi). The high-power lens may be needed to view the various crystals, which tend to be much more minute than the organized (cellular) sediments.

Organized sediments (Figure 41.2): include epithelial cells (rarely of any pathologic significance), pus cells (white blood cells), red blood cells, and casts. Urine is normally negative for organized sediments, and the presence of the last three categories mentioned, other than trace amounts, always indicates kidney pathology.

Draw a few of your observations below and attempt to identify them by comparing them with Figures 41.1 and 41.2.

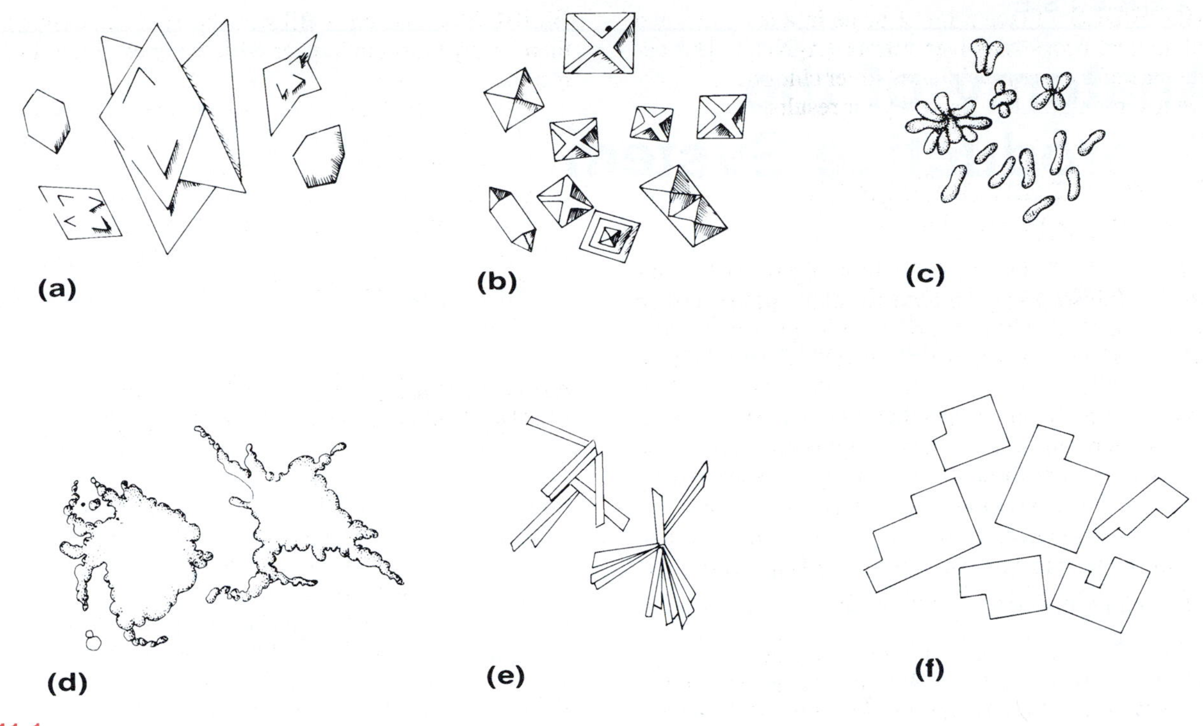

F41.1

Examples of unorganized sediments. (**a**) Uric acid crystals, (**b**) calcium oxalate crystals, (**c**) calcium carbonate crystals, (**d**) ammonium ureate crystals, (**e**) calcium phosphate crystals, (**f**) cholesterol crystals.

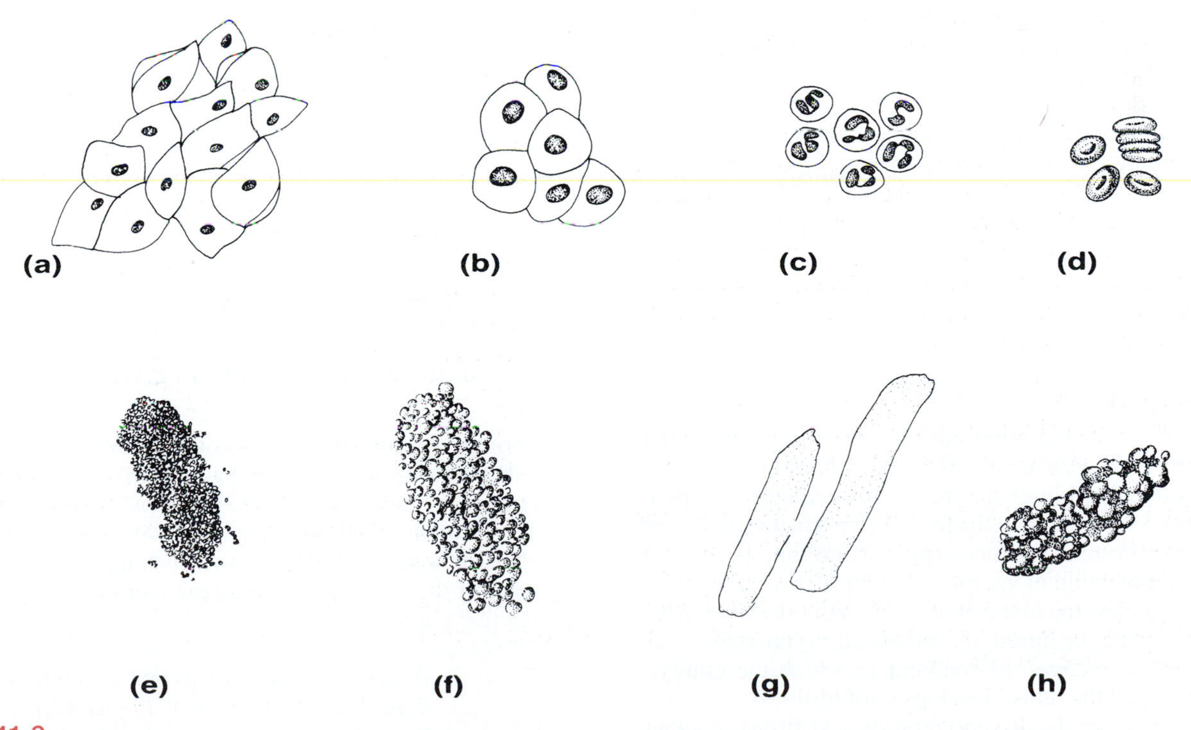

F41.2

Examples of organized sediments. (**a**) Squamous epithelial cells, (**b**) transitional epithelial cells, (**c**) white blood cells, largely neutrophils (pus), (**d**) red blood cells, (**e**) granular casts, (**f**) red blood cell casts, (**g**) hyaline casts, (**h**) fatty casts.

42 EXERCISE

Anatomy of the Reproductive System

OBJECTIVES

1. To discuss the general function of the reproductive system.
2. To identify and name the structures of the male and female reproductive systems when provided with an appropriate model or diagram, and to discuss the general function of each.
3. To define *semen,* discuss its composition, and name the organs involved in its production.
4. To trace the pathway followed by a sperm from its site of formation to the external environment.
5. To name the exocrine and endocrine products of the testes and ovaries, indicating the cell types or structures responsible for the production of each.
6. To identify homologous structures of the male and female systems.
7. To discuss the microscopic structure of the penis, epididymis, uterine tube, and uterus, and to relate structure to function.
8. To define *ejaculation, erection,* and *gonad.*
9. To discuss the function of the fimbriae and ciliated epithelium of the uterine (fallopian) tubes.
10. To identify the fundus, body, and cervical regions of the uterus.
11. To define *endometrium, myometrium,* and *ovulation.*

MATERIALS

Models or large laboratory charts of the male and female reproductive tracts
Prepared microscope slides of cross sections of the penis, epididymis, uterine tube, and uterus showing endometrium (proliferative phase)
Compound microscope
Protective skin cream or plastic gloves
Small metric ruler

See Appendix D, Exercise 42 for links to A.D.A.M. Standard.

See Appendix E, Exercise 42 for links to *Anatomy and PhysioShow: The Videodisc.*

Most simply stated, the biologic function of the **reproductive system** is to perpetuate the species. Thus the reproductive system is unique, since the other organ systems of the body function primarily to sustain the existing individual.

The essential organs of reproduction are those that produce the germ cells—the testes and the ovaries. The reproductive role of the male is to manufacture sperm and to deliver them to the female reproductive tract. The female, in turn, produces eggs. If the time is suitable, the combination of sperm and egg produces a fertilized egg, which is the first cell of a new individual. Once fertilization has occurred, the female uterus provides a nurturing, protective environment in which the embryo, later called the fetus, develops until birth.

Although the drive to reproduce is strong in all animals, in humans this drive is also intricately related to nonbiologic factors. Emotions and social considerations often enhance or thwart its expression.

GROSS ANATOMY OF THE HUMAN MALE REPRODUCTIVE SYSTEM

The primary reproductive organs of the male are the **testes,** the male *gonads* which have both an exocrine (sperm production) and an endocrine (testosterone production) function. All other reproductive structures are conduits or sources of secretions, which aid in the safe delivery of the sperm to the body exterior or female reproductive tract.

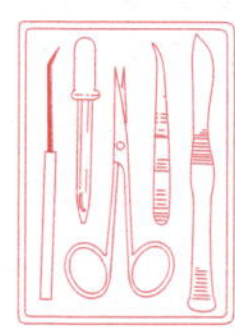

As the following organs and structures are described, locate them on Figure 42.1, and then identify them on a three-dimensional model of the male reproductive system or on a large laboratory chart.

The paired oval testes lie in the **scrotal sac** outside the abdominopelvic cavity. The temperature there (approximately 94°F, or 34°C) is slightly lower than body temperature, a requirement for producing viable sperm.

The accessory structures forming the *duct system* are the epididymis, the ductus deferens, the ejaculatory duct, and the urethra. The **epididymis** is an elongated structure running up the posterolateral aspect of the testis and capping its superior aspect. The epididymis forms the first portion of the duct system and provides a site for immature sperm that enter it from the testis to complete their maturation process. The **ductus deferens** (sperm duct) arches superiorly from the epididymis, passes through the inguinal canal into the pelvic cavity, and courses over the superior aspect of the urinary bladder. In life, the ductus deferens (also called the *vas deferens*) is enclosed along with blood vessels and nerves in a connective tissue sheath called the **spermatic cord.** The terminus of the ductus deferens enlarges to form the region called the **ampulla,** which empties into the **ejaculatory duct.** Contraction of the ejaculatory duct propels the sperm through the prostate gland to the **prostatic urethra,** which in turn empties into the **membranous urethra** and then into the **penile urethra,** which runs through the length of the penis to the body exterior.

The spermatic cord is easily palpated through the skin of the scrotum. When a *vasectomy* is performed, a small incision is made in each side of the scrotum, and each ductus deferens is cut through or cauterized. Although sperm are still produced, they can no longer reach the body exterior; thus a man is sterile after this procedure (and 12 to 15 ejaculations to clear the conducting tubules).

The *accessory glands* include the prostate gland, the paired seminal vesicles, and bulbourethral glands. These glands produce **seminal fluid,** the liquid medium in which sperm leave the body. The **seminal vesicles,** which produce about 60% of seminal fluid, lie at the posterior wall of the urinary bladder close to the terminus of the ductus deferens. They produce a viscous alkaline secretion containing fructose (a simple sugar) and other substances that nourish the sperm passing through the tract or promote the fertilizing capability of sperm in some way. The duct of each seminal vesicle merges with a ductus deferens to form the ejaculatory duct (mentioned above); thus sperm and seminal fluid enter the urethra together.

The **prostate gland** encircles the urethra just inferior to the bladder. It secretes a milky fluid into the urethra, which plays a role in activating the sperm.

Hypertrophy of the prostate gland, a troublesome condition commonly seen in elderly men, constricts the urethra so that urination is difficult. ■

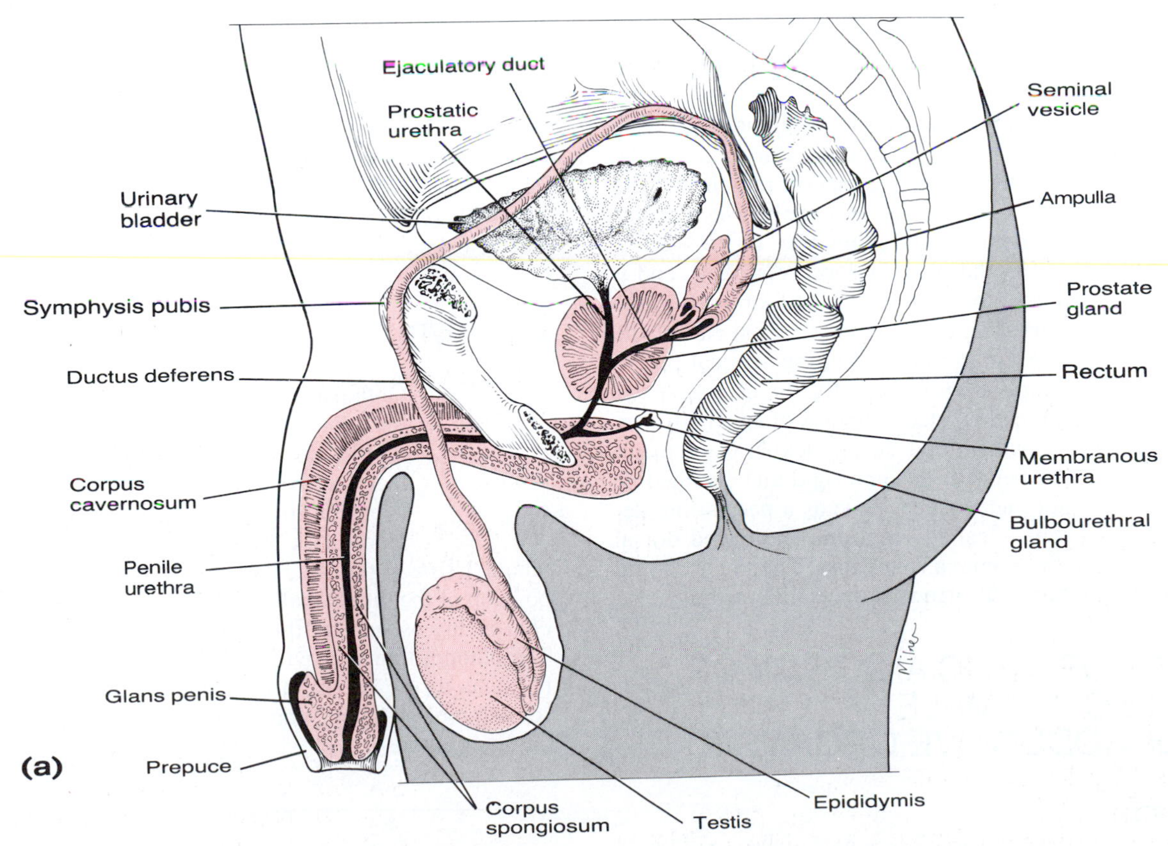

F42.1

Reproductive system of the human male. (a) Midsagittal section.

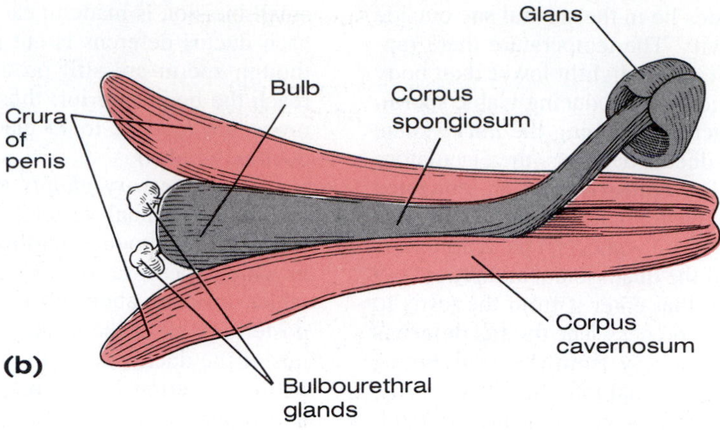

F42.1 *(continued)*

Reproductive system of the human male. (b) Inferior view of the penis with corpus spongiosum partially reflected.

The **bulbourethral glands** are tiny, pea-shaped glands inferior to the prostate. They produce a thick, clear, alkaline mucus that drains into the membranous urethra. This secretion acts to wash residual urine out of the urethra when ejaculation of **semen** (sperm plus seminal fluid) occurs. The relative alkalinity of seminal fluid also buffers the sperm against the acidity of the female reproductive tract.

The **penis,** part of the external genitalia of the male along with the scrotal sac, is the copulatory organ of the male and is designed to deliver sperm into the female reproductive tract. It consists of a shaft, which terminates in an enlarged tip, the **glans** (see Figure 42.1b). The skin covering the penis is loosely applied, and it reflects downward to form a circular fold of skin, the **prepuce,** or **foreskin,** around the proximal end of the glans. (The foreskin is removed in the surgical procedure called *circumcision.*) Internally, the penis consists primarily of three elongated cylinders of erectile tissue, which engorge with blood during sexual excitement. This causes the penis to become rigid and enlarged so that it may more adequately serve as a penetrating device. This event is called **erection.** The paired dorsal cylinders are the **corpora cavernosa.** The single ventral **corpus spongiosum** surrounds the penile urethra.

MICROSCOPIC ANATOMY OF SELECTED MALE REPRODUCTIVE ORGANS

Testis

Each testis is covered by a dense connective tissue capsule called the **tunica albuginea** (literally, "white tunic"). Extensions of this sheath enter the testis, dividing it into a number of lobes, each of which houses one to four highly coiled **seminiferous tubules,** the sperm-forming factories (Figure 42.2). The seminiferous tubules of each lobe converge to empty the sperm into another set of tubules, the **rete testis,** at the mediastinum of the testis. Sperm traveling through the rete testis then enter the epididymis, located on the exterior aspect of the testis, as previously described. Lying between the seminiferous tubules and softly padded with connective tissue are the **interstitial cells,** which produce testosterone, the hormonal product of the testis. You will conduct a microscopic study of testes tissue in Exercise 43.

Epididymis

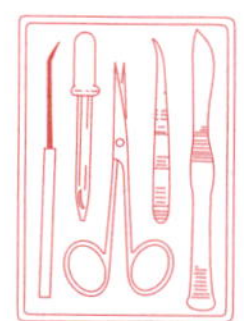

Obtain a microscope and a cross section of the epididymis. Notice the abundant tubule cross sections resulting from the fact that the coiling epididymis tubule has been cut through many times in the specimen. Look for sperm in the lumen of the tubule. Using Figure 42.3a as a guide, examine the composition of the tubule wall carefully. Identify the *stereocilia* of the pseudostratified columnar epithelial lining. These nonmotile microvilli absorb excess fluid and pass nutrients to the sperm in the lumen. Now identify the smooth muscle layer. What do you think the function of the smooth muscle is?

____

____

____

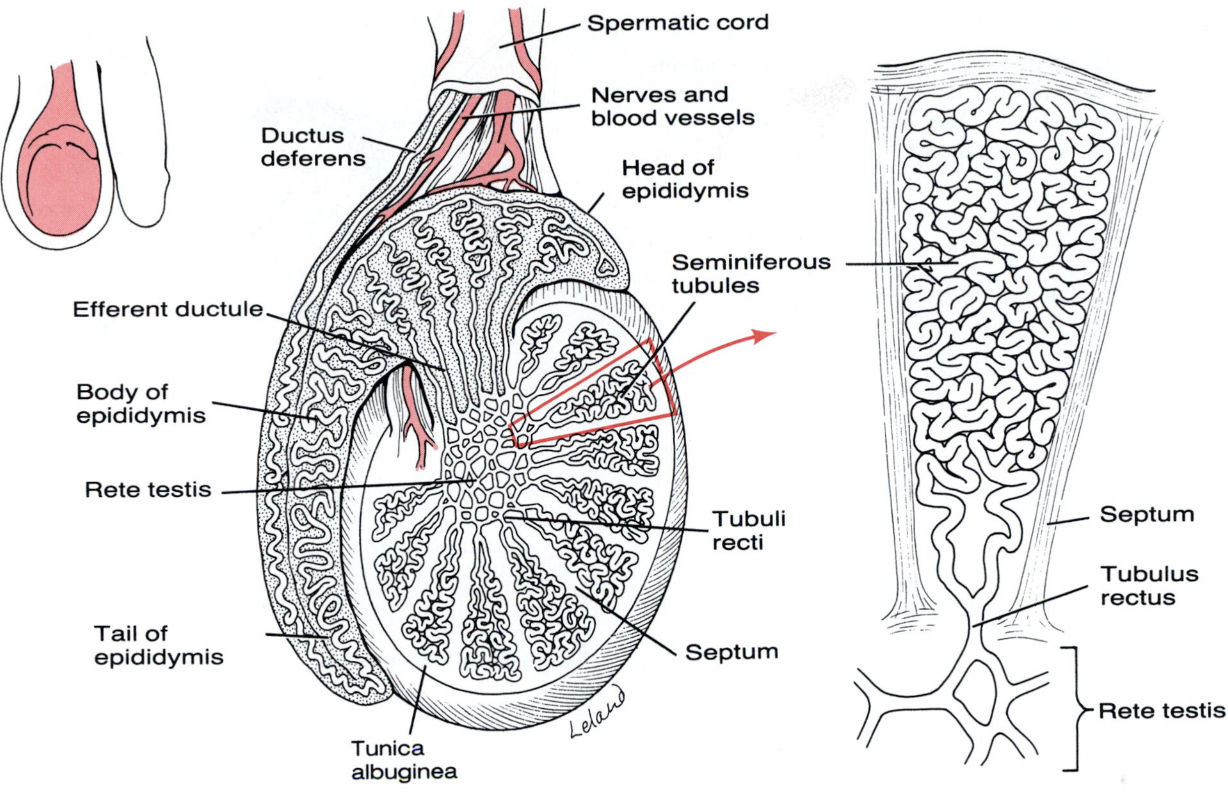

F42.2

Longitudinal-section view of the testis showing seminiferous tubules. (Epididymis and part of the ductus deferens also shown.)

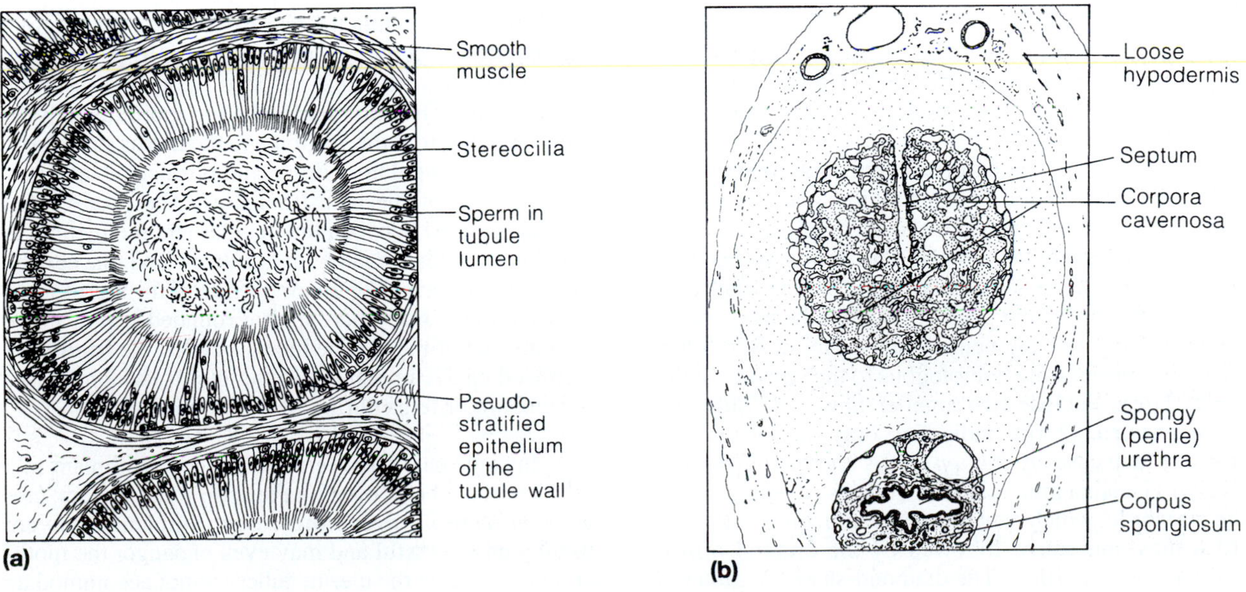

F42.3

Microscopic anatomy of selected organs of the male reproductive duct system. **(a)** Epididymis (see corresponding Plate 46 in the Histology Atlas). **(b)** Cross-sectional view of the penis (see also Plate 47).

Penis

Obtain a cross section of the penis. Scan the tissue under low power to identify the urethra and the cavernous bodies. Compare your observations to Figure 42.3b. Observe the lumen of the urethra carefully. What type of epithelium do you see?

___

Explain the function of this type of epithelium.

___

___

___

___

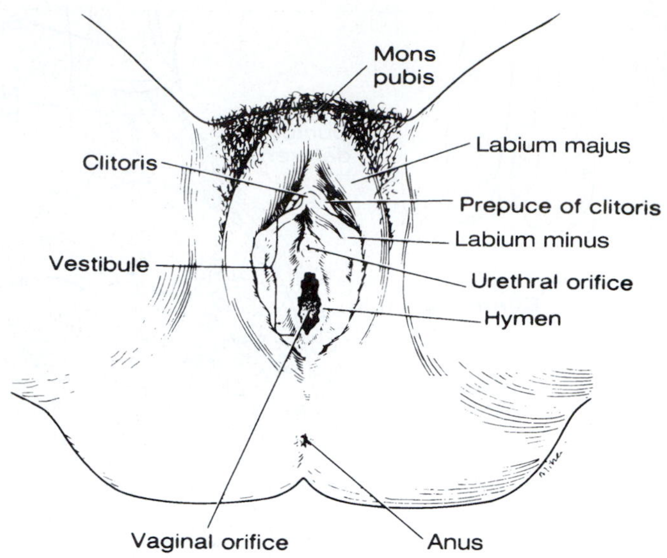

F42.4

External genitalia of the human female.

GROSS ANATOMY OF THE HUMAN FEMALE REPRODUCTIVE SYSTEM

The **ovaries** (female gonads) are the primary reproductive organs of the female. Like the testes of the male, the ovaries produce both an exocrine product (the eggs, or ova) and endocrine products (estrogens and progesterone). The other accessory structures of the female reproductive system transport, house, nurture, or otherwise serve the needs of the reproductive cells and/or the developing fetus.

The reproductive structures of the female are generally considered in terms of internal organs and external organs, or external genitalia.

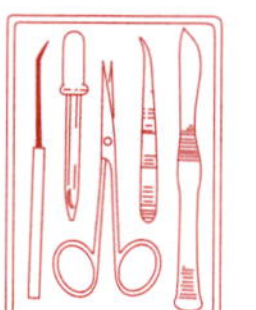

As you read the descriptions of these structures, locate them on Figures 42.4 and 42.5 and then on the female reproductive system model or large laboratory chart.

The **external genitalia (vulva)** consist of the mons pubis, the labia majora and minora, the clitoris, the urethral and vaginal orifices, the hymen, and the greater vestibular glands. The **mons pubis** is a rounded fatty eminence overlying the pubic symphysis. Running inferiorly and posteriorly from the mons pubis are two elongated, pigmented, hair-covered skin folds. The **labia majora** are homologous to the scrotum of the male. These enclose two smaller hair-free folds, the **labia minora.** (Terms indicating only one of the two folds in each case are *labium majus* and *minus,* respectively.) The labia minora, in turn, enclose a region called the **vestibule,** which contains many structures—the clitoris, most anteriorly, followed by the urethral orifice and the vaginal orifice. The diamond-shaped region between the anterior end of the labial folds, the ischial tuberosities laterally, and the anus posteriorly is called the **perineum.**

The **clitoris** is a small protruding structure, homologous to the male penis. Like its counterpart, it is composed of highly sensitive, erectile tissue. It is hooded by skin folds of the anterior labia minora, referred to as the **prepuce of the clitoris.** The urethral orifice, which lies posterior to the clitoris, is the outlet for the urinary system and has no reproductive function in the female. The vaginal opening is partially closed by a thin fold of mucous membrane called the **hymen** and is flanked by the pea-sized, mucus-secreting **greater vestibular glands.** These glands (not depicted in the illustrations) lubricate the distal end of the vagina during coitus.

The internal female organs include the vagina, uterus, uterine tubes, ovaries, and the ligaments and supporting structures that suspend these organs in the pelvic cavity. The **vagina** extends for approximately 10 cm (4 inches) from the vestibule to the uterus superiorly. It serves as a copulatory organ and birth canal, and permits passage of the menstrual flow. The pear-shaped **uterus,** situated between the bladder and the rectum, is a highly muscular organ with its narrow end, the **cervix,** directed inferiorly. The major portion of the uterus is referred to as the **body;** its superior rounded region above the entrance of the uterine tubes is called the **fundus.** A fertilized egg is implanted in the uterus, which houses the embryo or fetus during its development.

In some cases, the fertilized egg may implant in a uterine tube or even on the abdominal viscera, creating an **ectopic pregnancy.** Such implantations are usually unsuccessful and may even endanger the mother's life, because the uterine tubes cannot accommodate the increasing size of the fetus. ■

The apical region of the **endometrium,** the thick mucosal lining of the uterus, sloughs off periodically (about every 28 days) in response to cyclic changes in

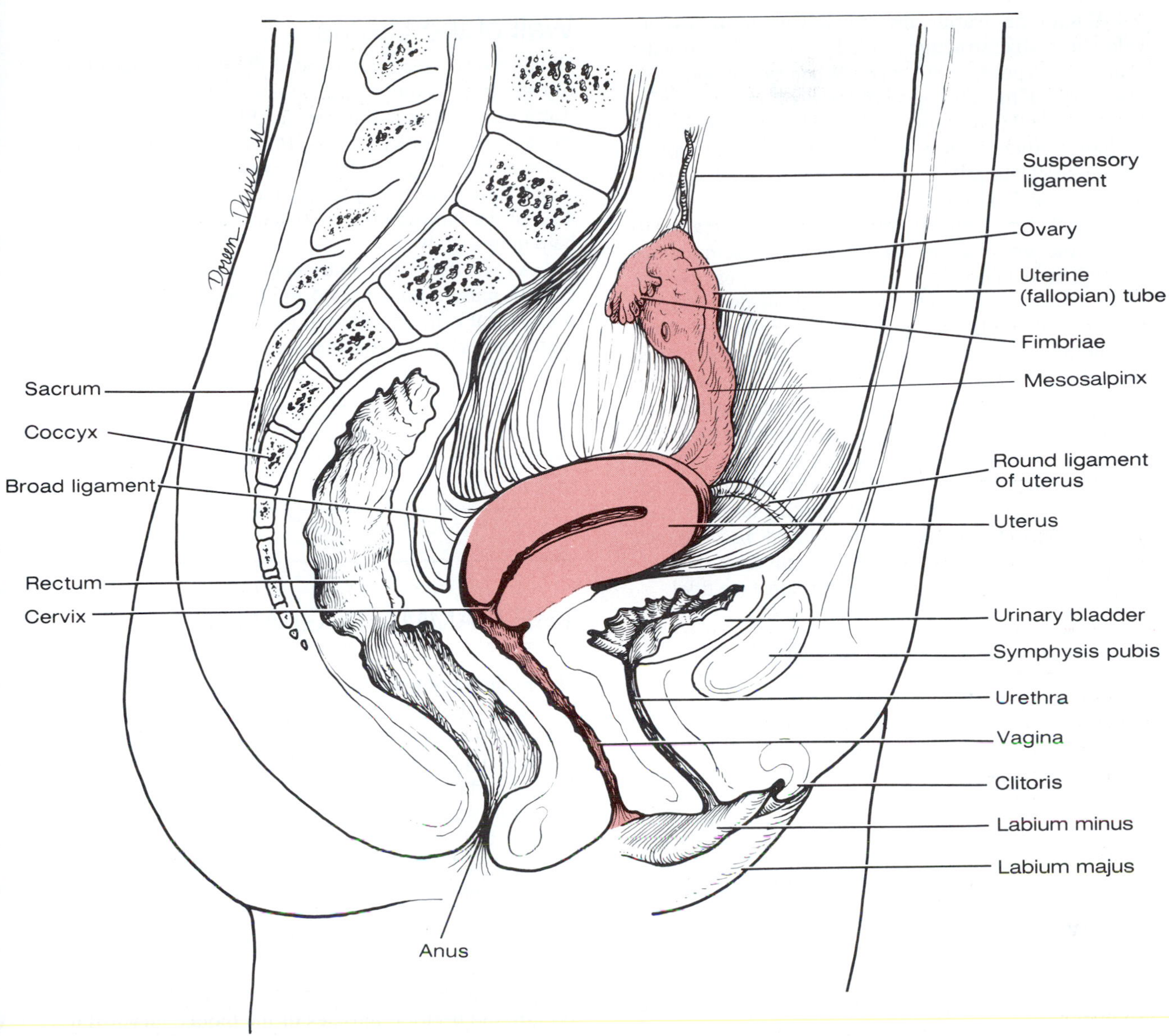

F42.5

Midsagittal section of the human female reproductive system.

the levels of ovarian hormones in the woman's blood. This sloughing-off process, which is accompanied by bleeding, is referred to as **menstruation,** or **menses.**

The **uterine,** or **fallopian, tubes** enter the superior region of the uterus and extend laterally for about 10 cm (4 inches) toward the **ovaries** in the peritoneal cavity. The distal ends of the tubes are funnel-shaped and have fingerlike projections called **fimbriae.** Unlike the male duct system, there is no actual contact between the female gonad and the initial part of the female duct system—the uterine tube.

Because of this open passageway between the female reproductive organs and the peritoneal cavity, reproductive system infections, such as gonorrhea, can spread to cause widespread inflammations of the pelvic viscera, a condition called **pelvic inflammatory disease** or **PID.** ■

The internal female organs are all retroperitoneal, except the ovaries. They are supported and suspended somewhat freely by ligamentous folds of peritoneum. The peritoneum takes an undulating course. From the pelvic cavity floor it moves superiorly over the top of the bladder, reflects over the anterior and posterior surfaces of the uterus, and then over the rectum, and up the posterior body wall. The fold that encloses the uterine tubes and uterus and secures them to the lateral body walls is the **broad ligament.** The part of the broad ligament specifically anchoring the uterus is called the **mesometrium** and that anchoring the uterine tubes, the **mesosalpinx.** The **round ligaments,** fibrous cords that run from the uterus to the labia majora, also help attach the uterus to the body wall. The ovaries are supported medially by the **ovarian ligament** (extending from the uterus to the ovary), laterally by the **suspensory ligaments,** and posteriorly by a fold of the broad ligament, the **mesovarium.**

Within the ovaries, the female gametes (eggs) develop in sac-like structures called *follicles.* The growing follicles also produce *estrogens.* When a developing egg has reached the appropriate stage of maturity, it is ejected from the ovary in an event called **ovulation.** The ruptured follicle is then converted to a second type of endocrine gland, called a *corpus luteum,* which secretes progesterone (and some estrogens).

The flattened almond-shaped ovaries lie adjacent to the uterine tubes but are not connected to them; consequently, an ovulated egg* enters the pelvic cavity. The waving fimbriae of the uterine tubes create fluid currents that, if successful, draw the egg into the lumen of the uterine tube, where it begins its passage to the uterus, propelled by the cilia of the tubule walls. The usual and most desirable site of fertilization is the uterine tube, because the journey to the uterus takes about 3 to 4 days and an egg is viable for up to 24 hours after it is expelled from the ovary. Thus, sperm must swim upward through the vagina and uterus, and into the uterine tubes to reach the egg. This must be an arduous journey, because they must swim against the downward current created by ciliary action—rather like swimming against the tide!

MICROSCOPIC ANATOMY OF SELECTED FEMALE REPRODUCTIVE ORGANS†

Uterine Tube

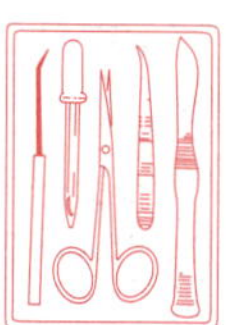

Obtain a prepared slide of a cross-sectional view of a uterine tube for examination. Notice the highly folded mucosa (the folds nearly fill the tubule lumen) as illustrated in Figure 42.6. Then switch to high power to examine the ciliated secretory epithelium. Draw your observations of the tubule mucosa below.

* To simplify this discussion, the ovulated cell is called an egg. What is actually expelled from the ovary is an earlier stage of development called a secondary oocyte. These matters are explained in more detail in Exercise 43.

† A microscopic study of the ovary is described in Exercise 43, pp. 403–404. If that exercise is not to be conducted in its entirety, the instructor might want to include the ovary study in the scope of this exercise.

Wall of the Uterus

Obtain a cross-sectional view of the uterine wall. Identify the three layers of the uterine wall—the endometrium, myometrium, and serosa. Also identify the two strata of the endometrium: the **functional layer** (stratum functionalis) and the **basal layer** (stratum basalis), which forms a new functional layer each month. Figure 42.7 of the proliferative endometrium may be of some help in this study.

As you study the slide, notice that the bundles of smooth muscle are oriented in several different directions. What is the function of the myometrium (smooth muscle layer) during the birth process?

____

____

____

____

THE MAMMARY GLANDS

The **mammary glands** or breasts exist, of course, in both sexes, but they have a reproduction-related function only in females. Since the function of the mammary glands is to produce milk to nourish the newborn infant, their importance is more closely associated with events that occur when reproduction has already been accomplished. Periodic stimulation by the female sex hormones, especially estrogens, increases the size of the female mammary glands at puberty. During this period, the duct system becomes more elaborate, and fat is deposited—fat deposition being the more important contributor to increased breast size.

The rounded, skin-covered mammary glands lie anterior to the pectoral muscles of the thorax, attached to them by connective tissue. Slightly below the center of each breast is a pigmented area, the **areola,** which surrounds a centrally protruding **nipple** (Figure 42.8).

Internally each mammary gland consists of 15 to 20 **lobes** which radiate around the nipple and are separated by fibrous connective tissue and adipose, or fatty, tissue. Within each lobe are smaller chambers called **lobules,** containing the glandular **alveoli** that produce milk during lactation. The alveoli of each lobule pass the milk into a number of **lactiferous ducts,** which join to form an expanded storage chamber, the **lactiferous sinus,** as they approach the nipple. The sinuses open to the outside at the nipple.

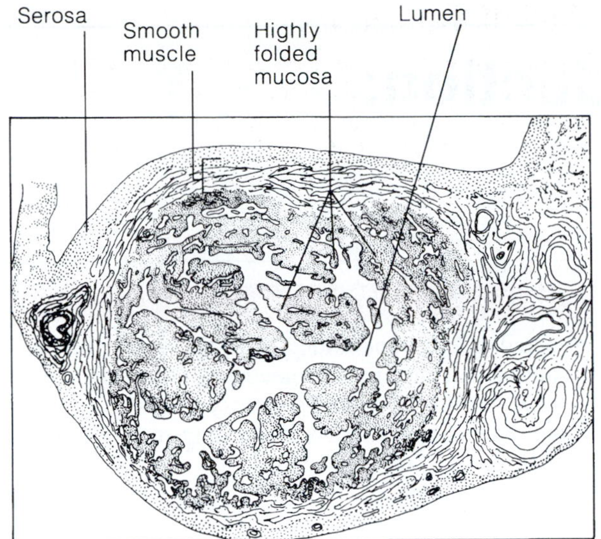

F42.6

Cross-sectional view of a uterine tube. Notice its highly folded mucosa. (See also corresponding Plate 48 in the Histology Atlas.)

F42.7

Structure of the uterine endometrium (proliferative phase). Corresponds to Plate 52 in the Histology Atlas.

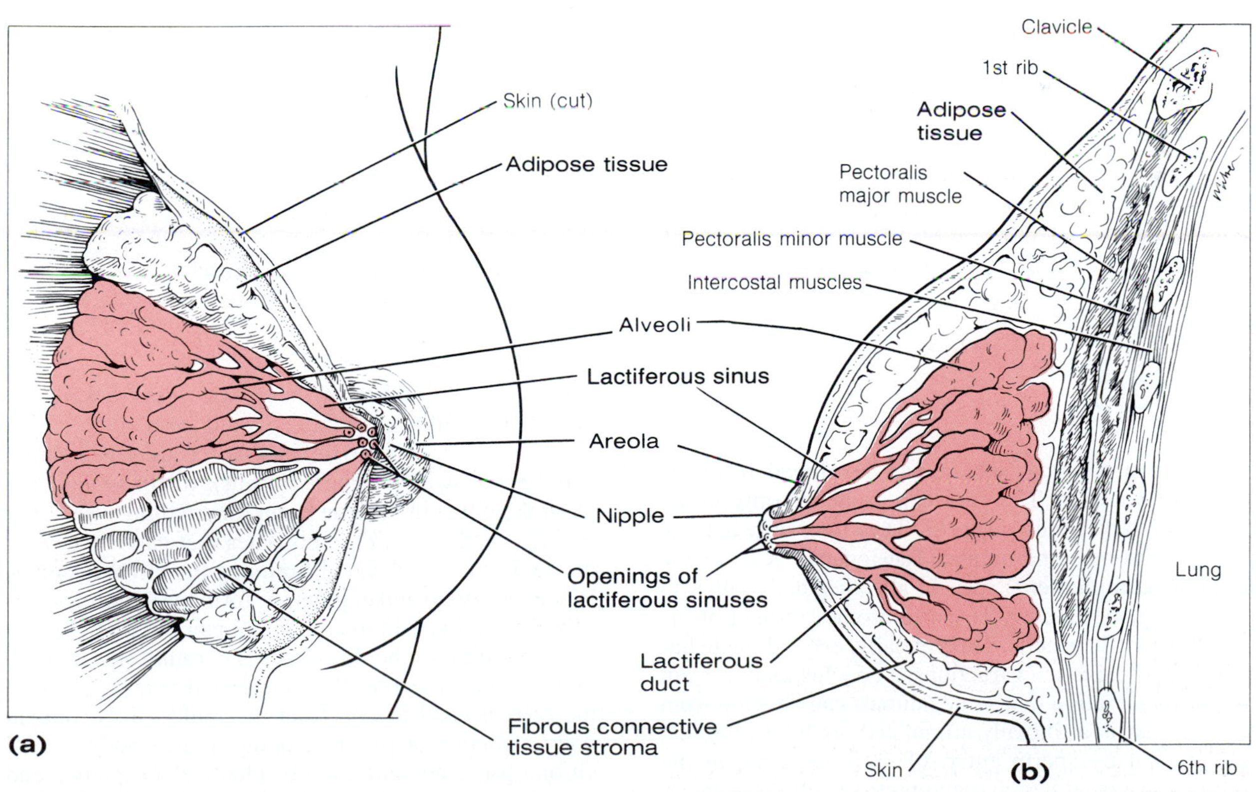

F42.8

Female mammary gland. **(a)** Anterior view. **(b)** Sagittal section.

43

EXERCISE

Physiology of Reproduction: Gametogenesis and the Female Cycles

OBJECTIVES

1. To define *meiosis, gametogenesis, oogenesis, spermatogenesis, spermiogenesis, synapsis, haploid, diploid,* and *menses.*
2. To relate the stages of spermatogenesis to the cross-sectional structure of the seminiferous tubule.
3. To discuss the microscopic structure of the ovary and to be prepared to identify primary, secondary, and vesicular follicles and the corpus luteum, and to state the hormonal products of the last two structures.
4. To relate the stages of oogenesis to follicle development in the ovary.
5. To cite similarities and differences between mitosis and meiosis, and between spermatogenesis and oogenesis.
6. To describe the anatomical structure of the sperm and relate it to function.
7. To discuss the phases and control of the menstrual cycle.
8. To discuss the effect of FSH and LH on the ovary and to describe the feedback relationship between anterior pituitary gonadotropins and ovarian hormones.
9. To describe the effect of FSH and LH (ICSH) on testicular function.

MATERIALS

Prepared microscope slides of testis, ovary, human sperm, and uterine endometrium (showing menstrual, proliferative, and secretory stages)

Three-dimensional models illustrating meiosis, spermatogenesis, and oogenesis

Compound microscope

Demonstration Area: microscopes set up to demonstrate the following stages of oogenesis in *Ascaris megalocephala:*

1. Primary oocyte with fertilization membrane, sperm nucleus, and aligned tetrads apparent
2. Formation of the first polar body
3. Secondary oocyte with dyads aligned
4. Formation of the ovum and second polar body
5. Fusion of the male and female pronuclei to form the fertilized egg

See Appendix E, Exercise 43 for links to *Anatomy and PhysioShow: The Videodisc.*

MEIOSIS

Every human being, so far, has developed from the union of gametes. The gametes, produced only in the testis or ovary, are unique cells. They have only half the normal chromosome number (designated as ***n,*** or the **haploid complement**) seen in all other body cells. In humans, gametes have 23 chromosomes instead of the 46 of other tissue cells. Theoretically, every gamete has a full set of genetic instructions, a conclusion borne out by the observation that some animals can develop from an egg that is artificially stimulated, as by a pinprick, rather than by sperm entry. (This is less true of the sperm, because it has an incomplete sex chromosome.)

Gametogenesis, the process of gamete formation, involves reduction of the chromosome number by half. This event is important to maintain the characteristic chromosomal number of the species generation after generation. Otherwise there would be a doubling of chromosome number with each succeeding generation and the cells would become so chock-full of genetic material there would be little room for anything else.

Egg and sperm chromosomes that carry genes for the same traits are called **homologous chromosomes.** When the sperm and egg fuse to form the **zygote,** or fertilized egg, it is said to contain 23 pairs of homologous chromosomes, or the **diploid (2*n*)** chromosome number of 46. The zygote, once formed, then divides to produce the cells needed to construct the multicellular human body. All cells of the developing human body have a chromosome content exactly identical in quality and quantity to that of the fertilized egg; this is assured by the nuclear division process called **mitosis.** (Mitosis was considered in depth in Exercise 4. You may want to review it at this time.)

To produce gametes with the reduced (haploid) chromosomal number, **meiosis,** a specialized type of nuclear division, occurs in the ovaries and testes during gametogenesis. Before meiosis begins, the chromosomes are replicated in the *mother cell* or stem cell just as they are before mitosis. As a result, the mother cell briefly has double the normal diploid genetic complement. The stem cell then undergoes two consecutive nuclear divisions, termed *meiosis I* and *II,* or the first and second maturation divisions, without replicating the chromosomes before the second division. The result is that four haploid daughter cells are produced, rather than the two diploid daughter cells produced by mitotic division.

The entire process of meiosis is complex and is dealt with here only to the extent necessary to reveal important differences between this type of nuclear division and mitosis. Essentially, each meiotic division involves the same phases and events seen in mitosis (prophase, metaphase, anaphase, and telophase), but during the first maturation division (meiosis I) an event not seen in mitosis occurs during prophase. The homologous chromosomes, each now a duplicated structure, begin to pair so that they become closely aligned along their entire length. This pairing is called **synapsis.** As a result, 23 **tetrads** (groupings of four chromatids) form, become attached to the spindle fibers, and begin to align themselves on the spindle equator. While in synapsis, two of the four strands in each tetrad (one from each homologue) coil around each other, forming many points of **crossover,** or **chiasmata.** (Perhaps this could be called the conjugal bed of the cell!) When anaphase of meiosis I begins, the homologues separate from one another, breaking and exchanging parts at points of crossover, and move apart toward opposite poles of the cell. The centromeres holding the "sister" chromatids or **dyads** together do not break at this point (Figure 43.1).

During the second maturation division, events parallel those in mitosis, except that the daughter cells do *not* replicate their chromosomes before this division, and each daughter cell has only half of the homologous chromosomes rather than a complete set. The crossover events and the way in which the homologues align on the spindle equator during the first maturation division introduce an immense variability in the resulting gametes, which explains why we are all unique.

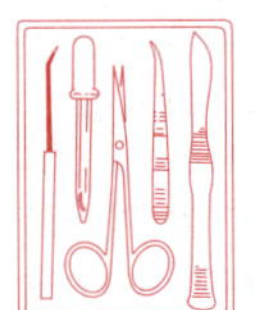

Obtain a model depicting the events of meiosis, and follow the sequence of events during the first and second maturation divisions. Identify prophase, metaphase, anaphase, and telophase in each. Also identify tetrads and chiasmata during the first maturation division and dyads (groupings of two chromatids connected by centromeres) in the second maturation division. Note ways in which the daughter cells resulting from meiosis I differ from the mother cell and how the gametes differ from both cell populations. (Use the key on the model, your textbook, or an appropriate reference as necessary to aid you in these observations.)

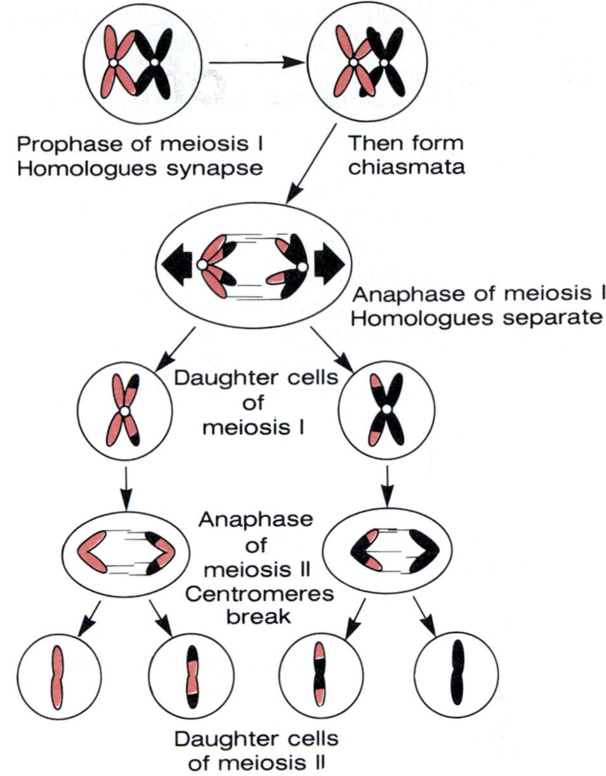

F43.1

Events of meiosis involving one pair of homologous chromosomes. (Male homologue is black; female homologue is red.)

If your instructor wishes you to observe meiosis in *Ascaris* to provide cellular material for comparison, continue with the microscopic study described next. Otherwise, skip to the study of spermatogenesis.

DEMONSTRATION OF OOGENESIS IN *ASCARIS* (OPTIONAL)

Generally speaking, oogenesis (the process of gametogenesis resulting in egg production) in mammals is difficult to demonstrate in a laboratory situation. However, the process of oogenesis and mechanics of meiosis* may be studied rather easily in the transparent eggs of *Ascaris megalocephala,* an invertebrate roundworm parasite found in the intestine of mammals. Since its diploid chromosome number is 4, the chromosomes are easily counted.

* In *Ascaris,* meiosis does not begin until the sperm has penetrated the primary oocyte, whereas in humans, meiosis I occurs before sperm penetration.

Go to the demonstration area where the slides are set up and make the following observations:

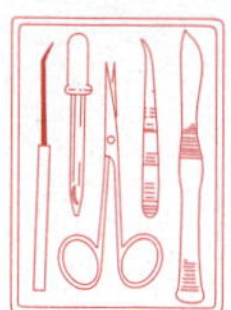

1. Scan the first demonstration slide to identify a *primary oocyte,* the cell type that begins the meiotic process. It will have what appears to be a relatively thick cell membrane; this is the *fertilization membrane* that the oocyte produces after sperm penetration. Find and study a primary oocyte that is undergoing the first maturation division. Look for a barrel-shaped spindle with two tetrads (two groups of four beadlike chromosomes) in it. Most often the spindle is located at the periphery of the cell. (The sperm nucleus may or may not be seen, depending on how the cell was cut.)

2. Observe slide 2. Locate a cell in which half of each tetrad (a dyad) is being extruded from the cell surface into a smaller cell called the *first polar body.*

3. On slide 3, attempt to locate a *secondary oocyte* (a daughter cell produced during meiosis I) undergoing the second maturation division. In this view, two dyads (two groups of two beadlike chromosomes) will be seen on the spindle.

4. On slide 4, locate a cell in which the *second polar body* is being formed. In this case, both it and the ovum will now contain two chromosomes, the haploid number for *Ascaris.*

5. On the fifth slide, identify a *fertilized egg* or a cell in which the sperm and ovum nuclei (actually *pronuclei*) are fusing to form a single nucleus containing four chromosomes.

SPERMATOGENESIS

Human sperm production or **spermatogenesis** begins at puberty and continues without interruption throughout life. The average male ejaculation contains about a quarter-billion sperm. Since only one sperm fertilizes an ovum, it seems that nature has tried to assure that the perpetuation of the species will not be endangered for lack of sperm.

As explained in Exercise 42, spermatogenesis (the process of gametogenesis in males) occurs in the seminiferous tubules of the testes. The primitive stem cells or **spermatogonia,** found at the tubule periphery, divide extensively to build up the stem cell line. Before puberty, all divisions are mitotic divisions that produce more spermatogonia. At puberty, however, under the influence of FSH (follicle-stimulating hormone) secreted by the anterior pituitary gland, each mitotic division of a spermatogonium produces one spermatogonium and one **primary spermatocyte,** which is destined to undergo meiosis. As meiosis occurs, the dividing cells approach the lumen of the tubule. Thus the progression of meiotic events can be followed from the tubule periphery to the lumen. It is important to recognize that **spermatids,** haploid cells that are the actual product of meiosis, are not functional gametes. They are nonmotile cells and have too much excess baggage to function well in a reproductive capacity. Another process, called **spermiogenesis,** which follows meiosis, strips away the extraneous cytoplasm from the spermatid, converting it to a motile, streamlined sperm.

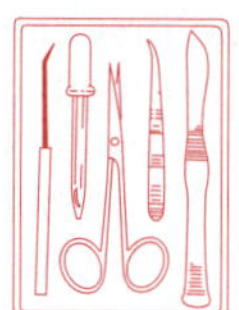

1. Obtain a slide of the testis and a microscope. Examine the slide under low power to identify the cross-sectional views of the cut seminiferous tubules. Then rotate the high power lens into position and observe the wall of one of the cut tubules. As you work, refer to Figure 43.2 and to Plate 49 in the Histology Atlas to make the following identifications.

2. Scrutinize the cells at the periphery of the tubule. The cells in this area are the spermatogonia. About half of their offspring form primary spermatocytes, which begin meiosis. The remaining daughter cells resulting from mitotic divisions of spermatogonia remain at the tubule periphery to maintain the germ cell line.

3. Observe the cells in the middle of the tubule wall. There you should see a large number of cells (spermatocytes) that are obviously undergoing a nuclear division process. Look for the chromosomes, visible only during nuclear division, that have the appearance of coiled springs. Attempt to differentiate between the larger primary spermatocytes and the somewhat smaller secondary spermatocytes.

Can you see tetrads? ______________________

Evidence of crossover? ______________________

Where would you expect to see the tetrads, closer to the spermatogonia or closer to the lumen?

____

In the primary or secondary spermatocytes? __________

____

4. Examine the cells at the tubule lumen. Identify the small round-nucleated spermatids, many of which may appear lopsided and look as though they are starting to lose their cytoplasm. See if you can find a spermatid embedded in an elongated cell type, a **sustentacular (Sertoli) cell,** which extends inward from the periphery of the tubule. The sustentacular cells nourish the spermatids as they begin their transformation into sperm. Also in the adluminal area, locate sperm, which can be identified by their tails. The sperm develop directly from the spermatids by the loss of extraneous cytoplasm and the development of a propulsive tail.

5. Identify the **interstitial cells** lying external to and between the seminiferous tubules. LH (luteinizing hormone), also called *interstitial cell-stimulating hormone (ICSH)* in males, prompts these cells to produce testos-

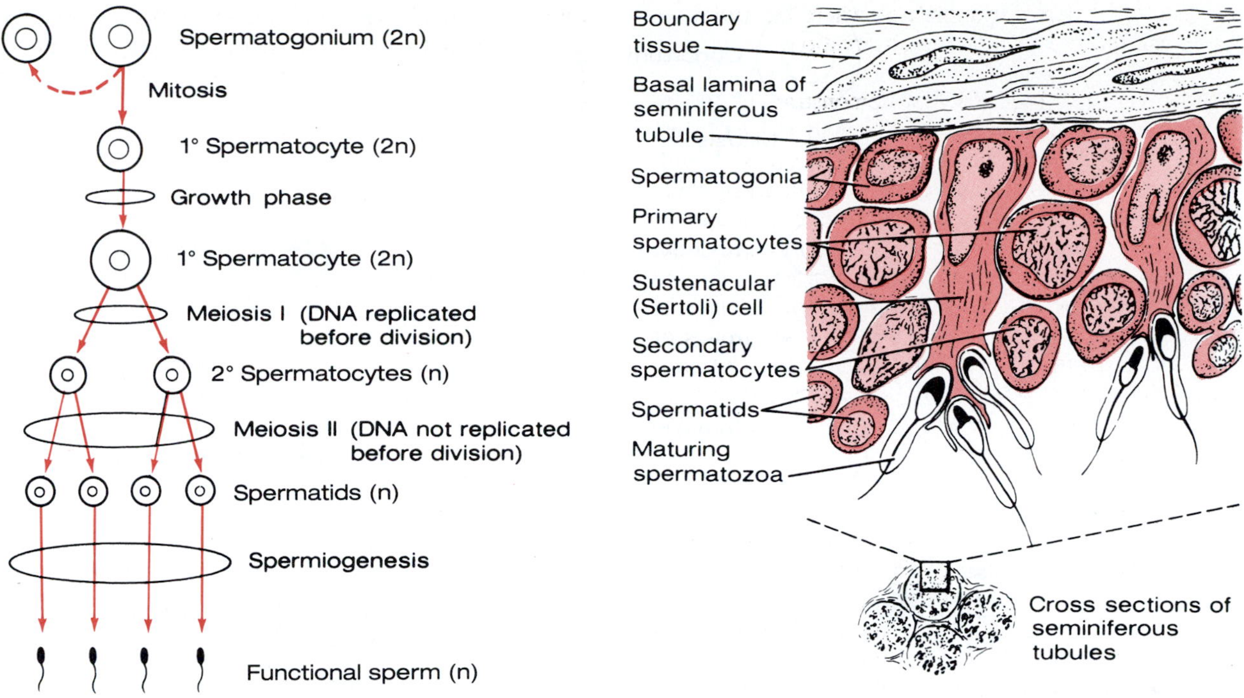

F43.2

Spermatogenesis. Left, flowchart of meiotic and spermiogenesis events. Right, diagrammatic view of seminiferous tubule. Redrawn with permission from C. R. Leeson and T. S. Leeson, *Histology,* 4th ed. (Philadelphia: W. B. Saunders, 1981).

terone which acts synergistically with FSH to stimulate sperm production.

In the next stage of sperm development, spermiogenesis, all the superficial cytoplasm is sloughed off, and the remaining cell organelles are compacted into the three regions of the mature sperm. At the risk of oversimplifying, these anatomical regions are the *head,* the *midpiece,* and the *tail,* which correspond roughly to the activating and genetic region, the metabolic region, and the locomotor region respectively. The mature sperm is a streamlined cell equipped with an organ of locomotion and a high rate of metabolism that enables it to move long distances in jig time to get to the egg. It is a prime example of the correlation of form and function.

The sperm head contains the DNA, or genetic material, of the chromosomes. Essentially it is the nucleus of the spermatid. Anterior to the nucleus is the **acrosome,** which contains enzymes involved in sperm penetration of the egg.

In the midpiece of the sperm is a centriole which gives rise to the filaments that structure the sperm tail. Wrapped tightly around the centriole are mitochondria, which provide the ATP needed for contractile activity of the tail filaments.

The tail filaments are constructed of contractile proteins, much like those in muscle. When powered by ATP, these filaments propel the sperm.

6. Obtain a prepared slide of human sperm and view it with the oil immersion lens. Compare what you see to the photograph of sperm in Plate 50 in the Histology Atlas. Identify the head, acrosome, and tail regions. Draw and appropriately label two or three sperm in the space below.

Deformed sperm, for example sperm with multiple heads or tails, are sometimes present in such preparations. Did you observe any?

_______ If so, describe them. ____________________

____

____

7. Examine the model of spermatogenesis to identify the spermatogonia, the primary and secondary spermatocytes, the spermatids, and the functional sperm.

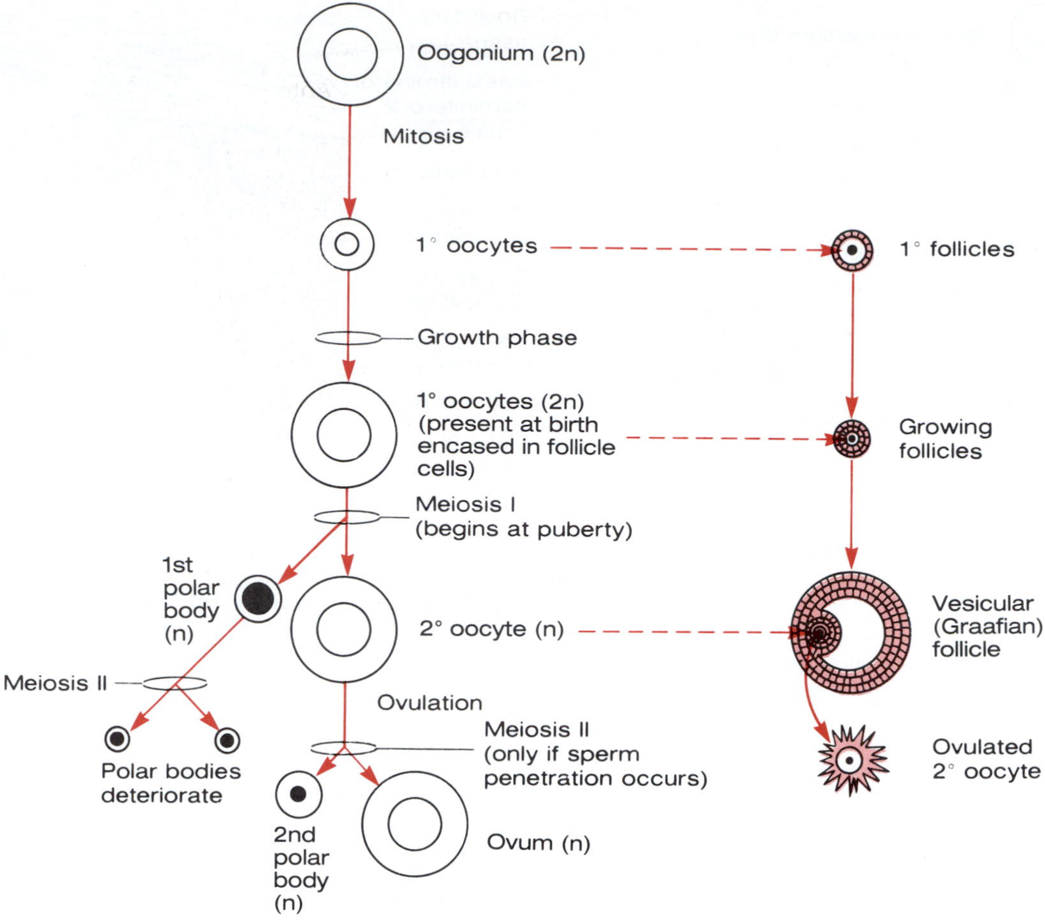

F43.3

Oogenesis. Left, flowchart of meiotic events. Right, correlation with follicular development and ovulation in the ovary.

OOGENESIS AND THE OVARIAN CYCLE

Gonadotropic hormones produced by the anterior pituitary influence the development of ova in the ovaries and their cyclic production of female sex hormones. Within an ovary, each immature ovum develops within a saclike structure called a *follicle,* where it is encased by one or more layers of smaller cells called **follicle cells** (when one layer is present) or **granulosa cells** (when there is more than one layer).

The process of **oogenesis,** or female gamete formation, which occurs in the ovary, is similar to spermatogenesis occurring in the testis, but there are some important differences. The process, schematically outlined in Figure 43.3, begins with primitive stem cells called **oogonia,** located in the ovarian cortices of the developing female fetus. During fetal development, the oogonia undergo mitosis thousands of times until their number reaches 700,000 or more. They then become encapsulated by a single layer of squamouslike follicle cells and form the **primordial follicles** of the ovary. By the time the female child is born, most of her oogonia have increased in size and have become **primary oocytes,** which are in the prophase stage of meiosis I. Thus at birth, the total potential for producing germ cells in the female is already determined; the primitive stem-cell line no longer exists or will exist for only a brief period after birth.

From birth until puberty, the primary oocytes are quiescent. Then, under the influence of FSH, one or sometimes more of the follicles begin to undergo maturation approximately every 28 days.

As a follicle grows, its epithelium changes from squamous to cuboidal cells and it comes to be called a **primary follicle** (see also Figure 43.4). The primary follicle begins to produce estrogens, and the primary oocyte completes its first maturation division, producing two haploid daughter cells that are very disproportionate in size. One of these is the **secondary oocyte,** which contains nearly all of the cytoplasm in the primary oocyte. The other is the tiny **first polar body.** The first polar body then completes the second maturation division, producing two more polar bodies. These eventually disintegrate for lack of sustaining cytoplasm.

As the follicle containing the secondary oocyte continues to enlarge, blood levels of estrogens rise. Initially, estrogen exerts a negative feedback influence on the release of gonadotropins by the anterior pituitary.

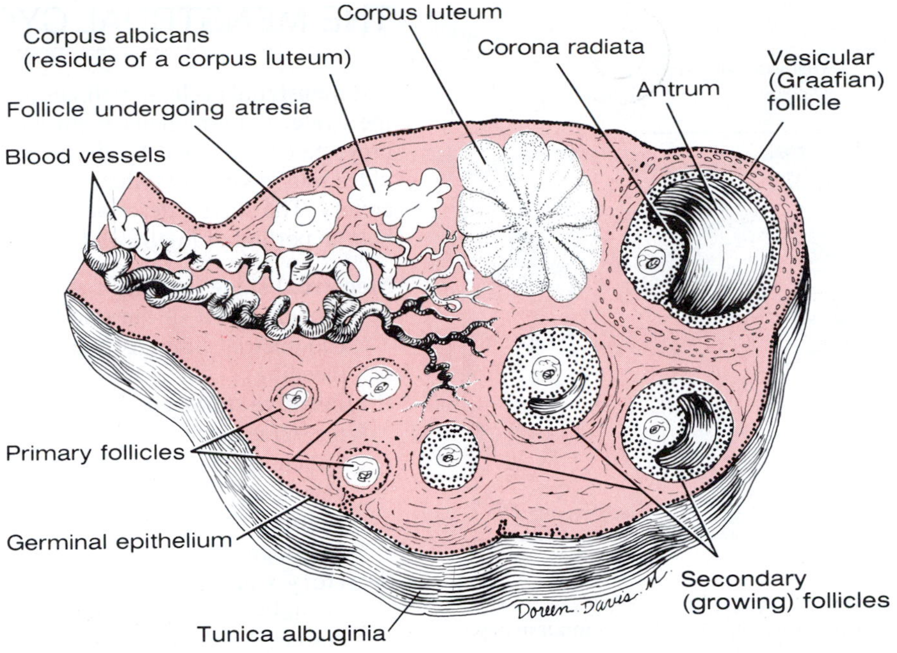

F43.4

Anatomy of the human ovary.

However, approximately in the middle of the 28-day cycle, as the follicle reaches the mature **vesicular,** or **Graafian, follicle** stage, rising estrogen levels become highly stimulatory and a sudden burstlike release of LH (and, to a lesser extent, FSH) by the anterior pituitary triggers ovulation. The secondary oocyte is extruded and begins its journey down the uterine tube to the uterus. If penetrated en route by a sperm, the secondary oocyte will undergo meiosis II, producing one large **ovum** and a tiny **second polar body.** When the second maturation division is complete, the chromosomes of the egg and sperm combine to form the diploid nucleus of the fertilized egg. If sperm penetration does not occur, the secondary oocyte simply disintegrates without ever producing the female gamete in human females.

Thus in the female, meiosis produces only one functional gamete, in contrast to the four produced in the male. Another major difference is in the relative size and structure of the functional gametes. Sperm are tiny and equipped with tails for locomotion. They have few organelles and virtually no nutrient-containing cytoplasm; hence the nutrients contained in semen are essential to their survival. In contrast, the egg is a relatively large nonmotile cell, well stocked with cytoplasmic reserves that nourish the developing embryo until implantation can be accomplished. Essentially all the zygote's organelles are "delivered" by the egg.

Once the secondary oocyte has been extruded from the ovary, LH transforms the ruptured follicle into the **corpus luteum,** which begins producing progesterone and estrogen. Rising blood levels of the two ovarian hormones inhibit FSH release by the anterior pituitary. As FSH declines, its stimulatory effect on follicular production of estrogens ends, and estrogen blood levels begin to decline. Since rising estrogen levels triggered LH release by the anterior pituitary, falling estrogen levels result in declining levels of LH in the blood. Corpus luteum secretory function is maintained by high blood levels of LH. Thus as LH blood levels begin to drop toward the end of the 28-day cycle, progesterone production ends and the corpus luteum begins to degenerate and is replaced by scar tissue **(corpus albicans).** The graphs in Figure 43.6 depict the hormone relationships described here.

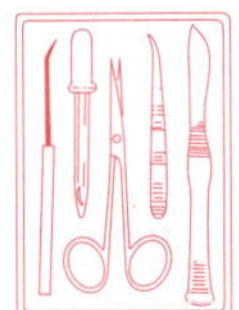

Because many different stages of ovarian development exist within the ovary at any one time, a single microscopic preparation will contain follicles at many different stages of development. Obtain a cross section of ovary tissue, and identify the following structures. Refer to Figures 43.4 and 43.5 as you work.

Germinal epithelium: outermost layer of the ovary.

Primary follicle: one or a few layers of cuboidal follicle cells surrounding the larger central developing ovum.

Secondary (growing) follicles: follicles consisting of several layers of follicle (granulosa) cells surrounding the central developing ovum, and beginning to show evidence of fluid accumulation and **antrum** (central cavity) formation.

Vesicular (Graafian) follicle: at this stage of development, the follicle has a large antrum containing

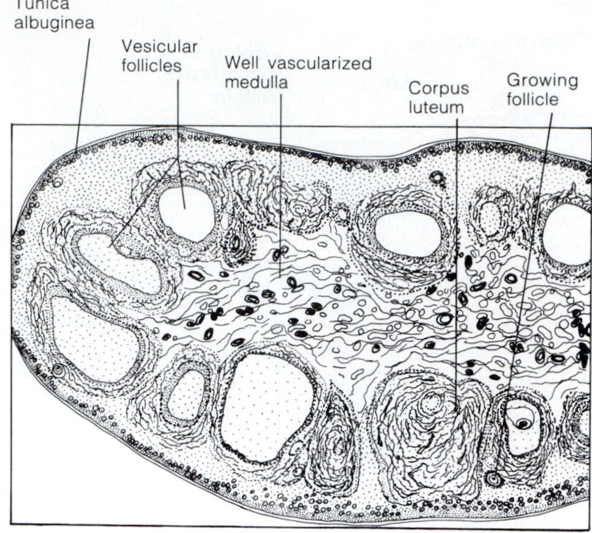

F43.5

Line drawing of a photomicrograph of the human ovary. (See corresponding Plate 51 in the Histology Atlas.)

fluid produced by the granulosa cells. The developing ovum is pushed to one side of the follicle and is surrounded by a capsule of several layers of granulosa cells called the **corona radiata** (radiating crown). When the immature ovum (secondary oocyte) is released, it enters the uterine tubes with its corona radiata intact. The connective tissue stroma (background tissue) adjacent to the mature follicle forms a capsule, called the **theca,** that encloses the follicle. (See also Plate 24 in the Histology Atlas.)

Corpus luteum: a solid glandular structure or a structure containing a scalloped lumen that develops from the ovulated follicle. (See Plate 25 in the Histology Atlas.)

Examine the model of oogenesis and compare it with the spermatogenesis model. Note differences in the size and structure of the functional gametes.

THE MENSTRUAL CYCLE

The **menstrual cycle,** sometimes referred to as the **uterine cycle,** is hormonally controlled by estrogens and progesterone secreted by the ovary. It is normally divided into three stages: menstrual, proliferative, and secretory. The stages, described below, are shown in the bottom portion of Figure 43.6.

Menstrual stage (menses): approximately days 1 to 5. Sloughing off of the thick functional layer of the endometrial lining of the uterus, accompanied by bleeding.

Proliferative stage: approximately days 6 to 14. Under the influence of estrogens produced by the growing follicle of the ovary, the endometrium is repaired, glands and blood vessels proliferate, and the endometrium thickens. Ovulation occurs at the end of this stage.

Secretory stage: approximately days 15 to 28. Under the influence of progesterone produced by the corpus luteum, the vascular supply to the endometrium increases further. The glands increase in size and begin to secrete nutrient substances to sustain a developing embryo, if present, until implantation can occur. If fertilization has occurred, the embryo will produce a hormone much like LH, which will maintain the function of the corpus luteum. Otherwise, as the corpus luteum begins to deteriorate, lack of ovarian hormones in the blood causes blood vessels supplying the endometrium to kink and become spastic, setting the stage for menses to begin by the 28th day.

Although the foregoing explanation assumes a classic 28-day cycle, the length of the menstrual cycle is highly variable, sometimes as short as 21 days or as long as 38. Only one interval is relatively constant in all females: the time from ovulation to the onset of menstruation is almost always 14 days.

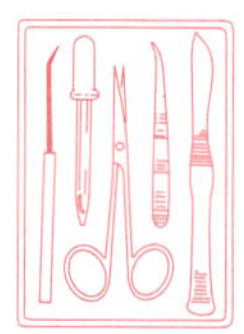

Obtain slides showing the menstrual, secretory, and proliferative phases of the uterine endometrium. Observe each carefully, comparing their relative thicknesses and vascularity. As you work, refer to the corresponding photomicrographs (Plates 52 through 54) in the Histology Atlas.

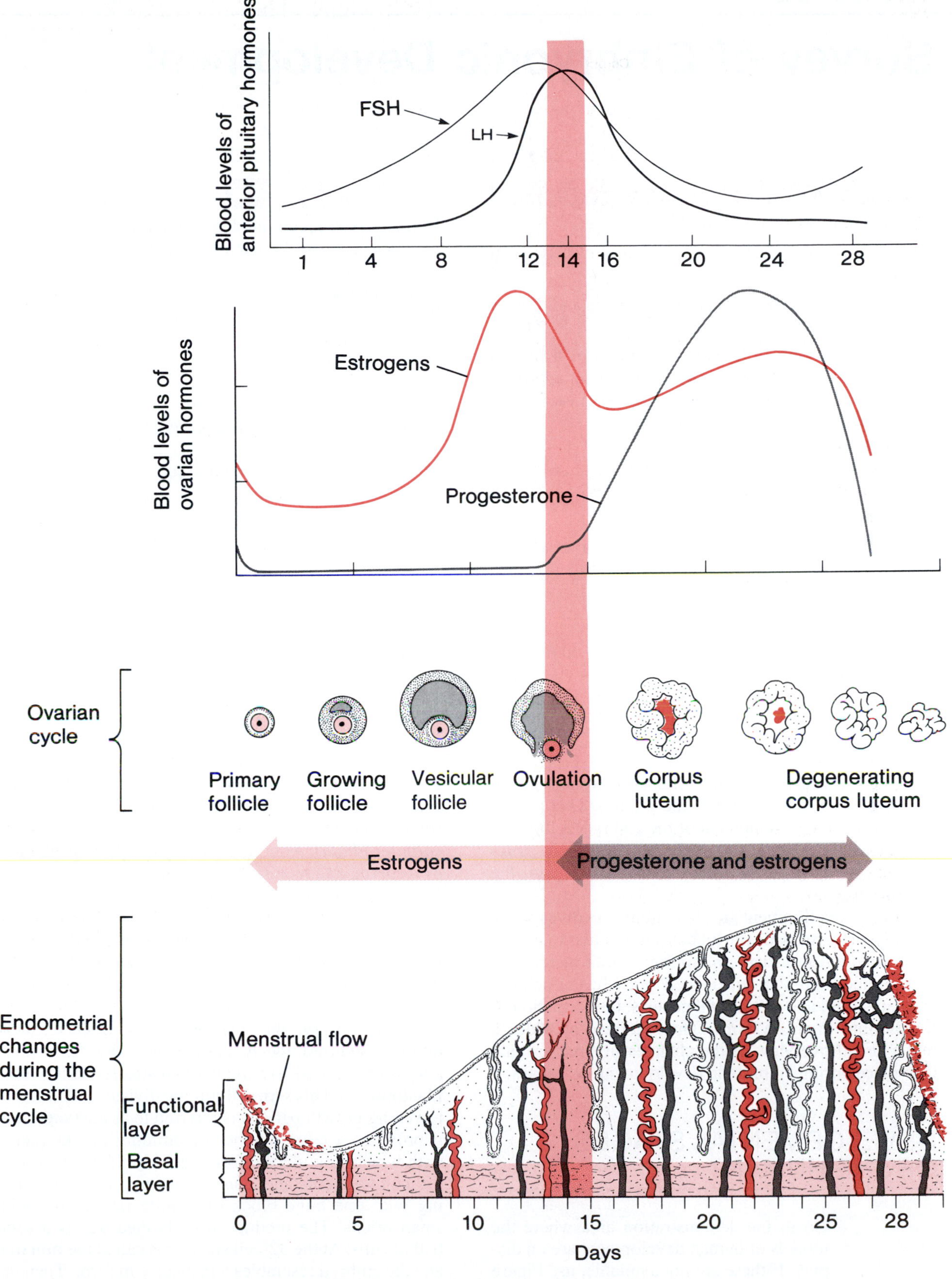

F43.6

Hormonal interactions of the female cycles. Relative level of anterior pituitary hormones correlated with follicular and hormonal changes in the ovary. The menstrual cycle is also depicted.

44 EXERCISE

Survey of Embryonic Development

OBJECTIVES

1. To define *fertilization* and *zygote.*
2. To define and discuss the function of *cleavage* and *gastrulation.*
3. To name the three primary germ layers and discuss the importance of each.
4. To identify the following structures of a human chorionic vesicle when provided with an appropriate diagram, and to state the function of each.

inner cell mass	trophoblast
amnion	allantois
yolk sac	chorionic villi

5. To describe the process and timing of implantation in the human.
6. To define *decidua basalis* and *decidua capsularis.*
7. To state the germ-layer origin of several body organs and organ systems of the human.
8. To describe developmental direction.
9. To describe the gross anatomy and general function of the human placenta.

MATERIALS

Human development models or plaques (if available)
Life Before Birth, Educational Reprint #27, available from Time-Life Educational Materials, Box 834, Radio City P.O., New York, N.Y. 10019
Pregnant cat, rat, or pig uterus (one per laboratory session) with uterine wall dissected to allow student examination
Dissecting instruments
Model of pregnant human torso
Fresh or formalin-preserved placenta (obtained from a clinical agency)
Microscope slide of placenta tissue
Compound microscope

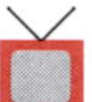
See Appendix E, Exercise 44 for links to *Anatomy and PhysioShow: The Videodisc.*

Because reproduction is such a familiar event, we tend to lose sight of the wonder of the process. One part of that process, the development of the embryo, is the concern of embryologists who study the changes in structure that occur from the time of fertilization until the time of birth.

Early development in all animals involves three basic types of activities, which are integrated to ensure the formation of a viable offspring: (1) an increase in cell number and subsequent cell growth; (2) cellular specialization; and (3) morphogenesis, the formation of functioning organ systems. This exercise provides a rather broad overview of the changes in structure that take place during embryonic development in humans.

DEVELOPMENTAL STAGES OF THE HUMAN

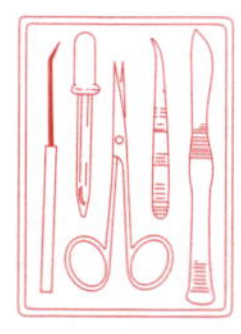

Go to the demonstration area where the models of human development are on display. If these are not available, use Figure 44.1 for this study. Observe the models to identify the various stages of human development as they are described, and respond to the questions posed below.

1. Observe the fertilized egg, or zygote, which appears as a single cell immediately surrounded by a jellylike *zona pellucida* and then a crown of granulosa cells (the *corona radiata*). After a secondary oocyte is penetrated by a sperm, and meiosis is completed to yield the ovum (egg) nucleus, the egg and the sperm nuclei fuse to form a single nucleus. This process is called **fertilization.** Shortly after sperm penetration, a granular barrier forms beneath the zona pellucida to prevent the entry of additional sperm.

2. Next, observe the cleavage stages. Once fertilization has occurred, the zygote begins to divide, forming a mass of successively smaller and smaller cells, called **blastomeres.** This series of mitotic divisions without intervening growth periods is referred to as **cleavage,** and it provides a large number of building blocks (cells) with which to build the forming body. If this is a little difficult to understand, consider trying to erect a building with one huge block of granite rather than with small bricks. The product of early cleavage is a solid ball of cells. At the 32-cell stage, it is called the **morula,** and the embryo resembles a raspberry in form. Then the cell mass hollows out to become the embryonic form called the **blastula,** which is a ball of cells surrounding a central cavity. The blastula, more commonly called the **blastocyst** in humans, is the final product of cleavage.

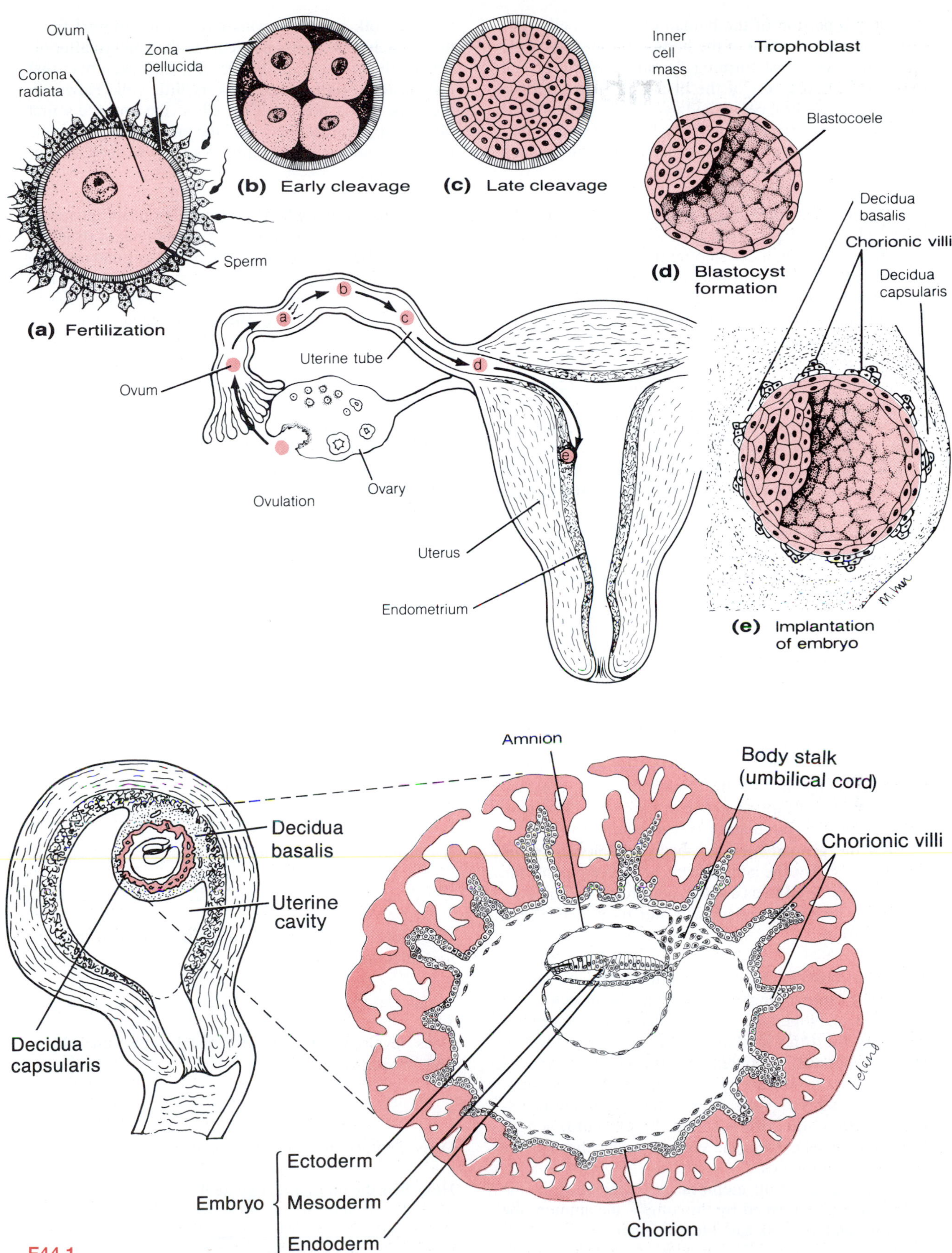

F44.1

Early embryonic development of the human. Top **(a–e),** from fertilization to blastocyst implantation in the uterus. Below, embryo of approximately 22 days. Embryonic membranes and germ layers present.

Only a portion of the human blastocyst cells contribute to the formation of the body—those seen on the top of the blastocyst forming the so-called **inner cell mass (ICM).** The rest of the blastocyst—that portion enclosing the central cavity and overriding the ICM—is referred to as the **trophoblast.** The trophoblast becomes an extraembryonic membrane called the **chorion,** which forms the fetal portion of the *placenta.*

3. Observe the *implanting* blastocyst shown on the model or in the figure. By approximately the seventh day after ovulation, a developing human embryo (blastocyst) is floating free in the uterine cavity. About that time, it adheres to the uterine wall over the ICM area, and implantation begins. The trophoblast cells secrete enzymes that erode the uterine mucosa at the point of attachment to reach the vascular supply in the submucosa. By the fourteenth day after ovulation, implantation is completed and the uterine mucosa has grown over the burrowed-in embryo. The portion of the uterine wall beneath the ICM, destined to take part in placenta formation, is called the **decidua basalis** and that surrounding the rest of the blastocyst is called the **decidua capsularis.** Identify these regions.

By the time implantation is complete, the blastocyst has undergone **gastrulation.** As a result of gastrulation, a three-layered embryo called a **gastrula** forms. Each of the gastrula's three layers corresponds to a **primary germ layer** from which specific body tissues develop. Within the next 6 weeks, virtually all of the body organ systems will have been laid down at least in rudimentary form by the germ layers. The outermost layer, **ectoderm,** gives rise to the epidermis of the skin and the nervous system. The deepest layer, the **endoderm,** forms the mucosa of the digestive and respiratory tracts and associated glands. Mesoderm, the middle layer, forms virtually everything lying between the two (skeleton, walls of the digestive organs, urinary system, skeletal muscles, circulatory system, and others).

All the groundwork has been completed by the eighth week. By the ninth week of development, the embryo is referred to as a **fetus,** and from this point on, the major activities are growth and tissue and organ specialization.

4. Again observe the blastocyst to follow the formation of the embryonic membranes and the placenta (see Figure 44.1). Notice the villus extensions of the trophoblast. By the time implantation is complete, the trophoblast has differentiated into the **chorion,** and its large elaborate villi are lying in the blood-filled sinusoids in the uterine tissue. This composite of uterine tissue and **chorionic villi** is called the **placenta,** and all exchanges to and from the embryo occur through the chorionic membranes.

Three embryonic membranes (originating in the ICM) have also formed by this time—the amnion, the allantois, and the yolk sac. Identify each.

- The **amnion** encases the young embryonic body in a fluid-filled chamber that protects the embryo against mechanical trauma and temperature extremes.

- The **yolk sac** in humans has lost its original function, which was to pass nutrients to the embryo after digesting the yolk mass. The placenta has taken over that task; also, the human egg has very little yolk. However, the yolk sac is not totally useless; the embryo's first blood cells originate here, and the primordial germ cells migrate from it into the embryo's body to seed the gonadal tissue.

- The **allantois,** which protrudes from the posterior end of the yolk sac, is also a largely redundant structure in humans because of the placenta. In birds and reptiles, it is a repository for embryonic wastes. In humans, it is the structural basis on which the mesoderm migrates to form the body stalk, or **umbilical cord,** which attaches the embryo to the placenta. (Refer ahead to p. 444 to refresh your memory of the structure of the umbilical cord if necessary.)

5. Go to the demonstration area to view the photographic series, *Life Before Birth.* This series, by Lennart Nilsson, illustrates human development in a way you will long remember. After viewing it, respond to the following questions.

In your own words, what do the chorionic villi look like?

____

____

____

What organs or organ systems appear *very* early in embryonic development?

____

____

____

Does development occur in a rostral to caudal (head to toe) direction, or vice versa?

____

Does development occur in a distal to proximal direction, or vice versa?

____

Does spontaneous movement occur *in utero?* __________

How does the mother recognize this? ________________

____

The very young embryo has been described as resembling "an astronaut suspended and floating in space." Do you think this definition is appropriate?

_______ Why or why not? ____________________

What is vernix caseosa? ____________________

What is lanugo? ____________________

IN UTERO DEVELOPMENT

1. Go to the appropriate demonstration area and observe the fetuses in the Y-shaped animal uterus. Identify the following fetal or fetal-related structures:

- **Placenta.** (a composite structure formed from the uterine mucosa and the fetal chorion).

Describe its appearance. ____________________

- **Umbilical cord.** Describe its relationship to the placenta and fetus.

- **Amniotic sac.** Identify the transparent amnion surrounding a fetus. Open one amniotic sac and note the amount, color, and consistency of the fluid.

Remove a fetus and observe the degree of development of the head, body, and extremities. Is the skin thick or thin?

What is the basis for your response? ____________________

2. Observe the model of a pregnant human torso. Identify the placenta. How does it differ in shape from the animal placenta observed?

Identify the umbilical cord. In what region of the uterus does implantation usually occur, as indicated by the position of the placenta?

What might be the consequence if it occurred lower?

Why would a feet-first position (breech presentation) be less desirable than the positioning of the model?

GROSS AND MICROSCOPIC ANATOMY OF THE PLACENTA

The placenta is a remarkable temporary organ. Composed of maternal and fetal tissues, it is responsible for providing nutrients and oxygen to the embryo and fetus while removing carbon dioxide and metabolic wastes.

1. Notice that the placenta on display has two very different-appearing surfaces—one smooth and the other spongy, roughened, and torn-looking.

Which is the fetal side? ____________________

Basis of your conclusion? ____________________

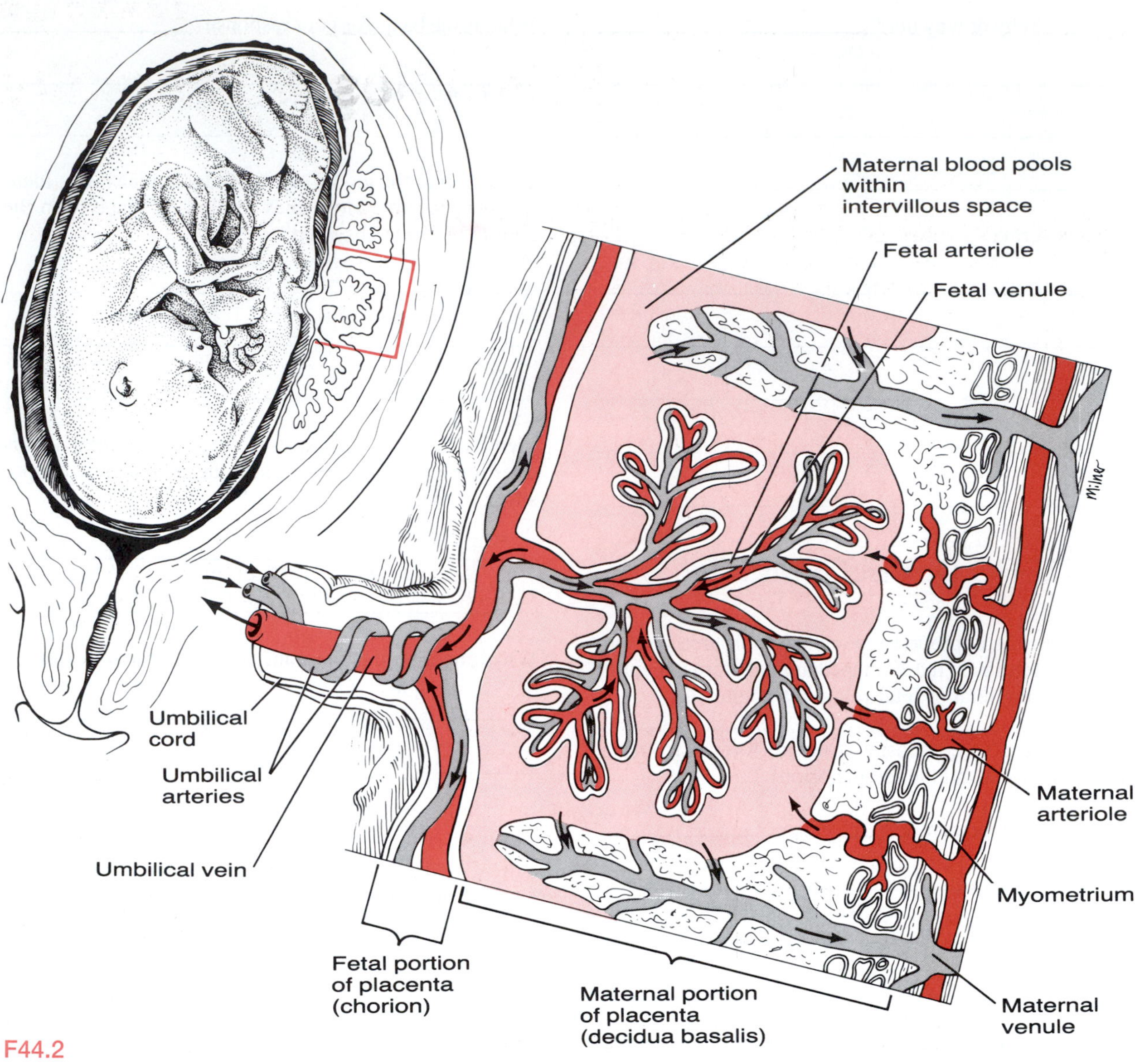

F44.2

Diagrammatic representation of the structure of the placenta.

Identify the umbilical cord. Within the cord, identify the umbilical vein and two umbilical arteries. What is the function of the umbilical vein?

____

____

The umbilical arteries? ______________________________

____

Are any of the fetal membranes still attached?

____

If so, which? ________________________________

2. Obtain a microscope slide of placental tissue. Observe the tissue carefully, comparing it to Figure 44.2. Identify the *intervillous spaces* (maternal sinusoids), which are blood-filled in life. Identify the villi, and notice their rich vascular supply. Draw a small representative diagram of your observations below and label it appropriately.

Principles of Heredity

OBJECTIVES

1. To define *allele, dominance, genotype, heterozygous, homozygous, incomplete dominance, phenotype,* and *recessiveness.*
2. To gain practice working out simple genetics problems, using a Punnett square.
3. To become familiar with basic laws of probability.
4. To observe selected human phenotypes and determine their genotype basis.

MATERIALS

Pennies (for coin tossing)
PTC (phenylthiocarbamide) taste strips
Sodium benzoate taste strips
Chart drawn on chalkboard for tabulation of class results of human phenotype/genotype determinations

Blood typing supplies:
Anti-A and Anti-B sera, slides, toothpicks, wax pencils, sterile lancets, alcohol swabs
Beaker containing 10% bleach solution
Disposable autoclave bag

See Appendix E, Exercise 45 for links to *Anatomy and PhysioShow: The Videodisc.*

The field of genetics currently is bristling with excitement. Complex gene-splicing techniques have allowed researchers to precisely isolate genes coding for specific proteins and then to use those genes to harvest large amounts of particular proteins and even to cure some dreaded human diseases. At present, growth hormone, insulin, erythropoietin, and interferon produced by these genetic engineering techniques are available for clinical use, and the list is growing daily.

Comprehending genetics relative to such studies requires arduous training. However, a basic understanding and appreciation of how genes regulate our various traits (dimples and hair color, for example) can be gained by anyone. The thrust of this exercise is to provide a "genetics sampler" or relatively simple introduction to the principles of heredity.

INTRODUCTION TO THE LANGUAGE OF GENETICS

In humans all cells, except eggs and sperm, contain 46 chromosomes, that is, the diploid number. This number is established when fertilization occurs and the egg and sperm fuse, combining the 23 chromosomes (or haploid complement) each is carrying. The diploid chromosomal number is maintained throughout life in nearly all cells of the body by the precise process of mitosis. As explained in Exercise 43, the diploid chromosomal number actually represents two complete (or nearly complete) sets of genetic instructions—one from the egg and the other from the sperm—or 23 pairs of *homologous chromosomes.*

Genes coding for the same traits on each pair of homologous chromosomes are called **alleles.** The alleles may be identical or different in their influence. For example, the members of the gene pair, or alleles, coding for hairline shape on your forehead may specify either straight across or widow's peak. When both alleles in a homologous chromosome pair have the same expression, the individual is said to be **homozygous** for that trait. When the alleles differ in their expression, the individual is **heterozygous** for the given trait; and typically only one of the alleles, called the **dominant gene,** will exert its effects. The allele with less potency, the **recessive gene,** will be present but masked. Whereas dominant genes, or alleles, exert their effects in both homozygous and heterozygous conditions, as a rule recessive alleles *must* be present in double dose (homozygosity) to exert their influence. An individual's actual genetic makeup, that is, whether he is homozygous or heterozygous for the various alleles, is called **genotype.** The manner in which genotype is expressed (for example, the presence of a widow's peak or not, blue vs. brown eyes) is referred to as **phenotype.**

The complete story of heredity is much more complex than just outlined, and in actuality the expression of many traits (for example, eye color) is determined by the interaction of many allele pairs. However, our emphasis here will be to investigate only the less complex aspects of genetics.

DOMINANT-RECESSIVE INHERITANCE

One of the best ways to master the terminology and learn the principles of heredity is to work out the solutions to some genetic crosses in much the same manner Gregor Mendel did in his classic experiments on pea plants. (Mendel, an Austrian monk of the mid-1800s, found evidence in these experiments that each gamete contributes just one allele to each pair in the zygote.)

To work out the various simple monohybrid (one pair of alleles) crosses in this exercise, you will be given the genotype of the parents. You will then determine the possible genotypes of their offspring by using a grid called the *Punnett square,* and you will record both genotype and phenotype percentages. To illustrate the procedure, an example of one of Mendel's pea plant crosses is outlined next.

Alleles: T (determines *tallness;* dominant)
t (determines *dwarfness;* recessive)
Genotypes of parents: TT(♂) × tt (♀)
Phenotypes of parents: Tall × dwarf

To use the Punnett, or checkerboard, square, write the alleles (actually gametes) of one parent across the top and the gametes of the other parent down the left side. Then combine the gametes across and down to achieve all possible combinations as shown below:

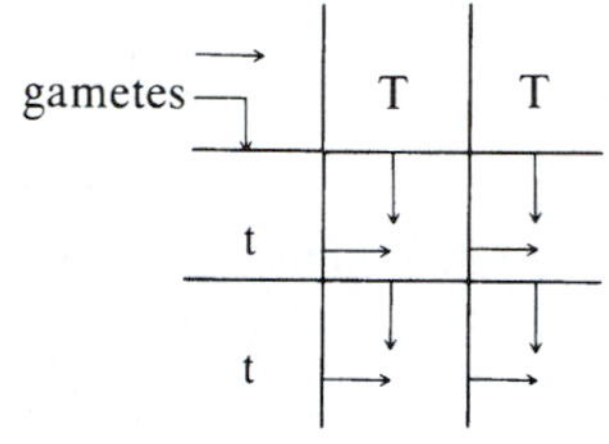

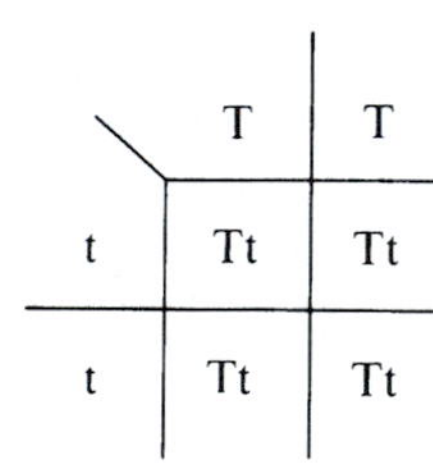

Results: Genotypes 100% Tt (all heterozygous)
Phenotypes 100% Tall (since T, which determines tallness, is dominant and all contain the T allele)

1. Using the technique outlined above, determine the genotypes and phenotypes of the offspring of the following crosses:

a. Genotypes of parents: Tt (♂) × tt (♀)

% of each genotype: ____________________

% of each phenotype: ________ % tall

________ % dwarf

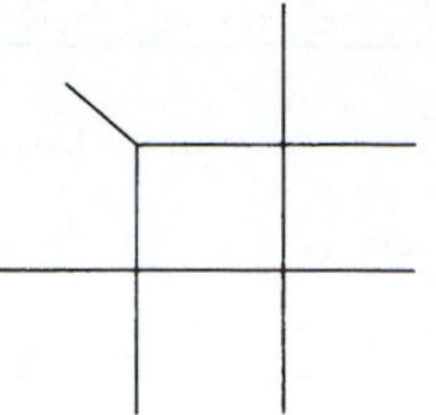

b. Genotypes of parents: Tt (♂) × Tt (♀)

% of each genotype: ____________________

% of each phenotype: ________ % tall

________ % dwarf

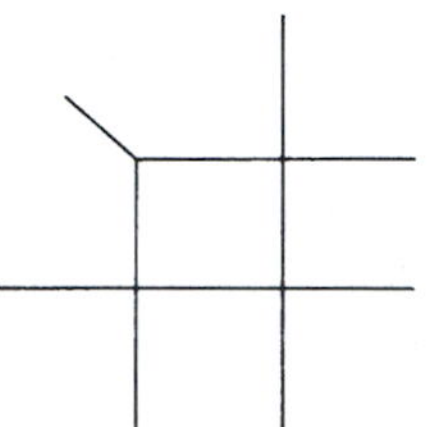

2. In guinea pigs, rough coat (R) is dominant over smooth coat (r). What will be the genotypes and phenotypes of the following monohybrid crosses?

a. Genotypes of the parents: RR × rR

% of each genotype: ____________________

% of each phenotype: ________ % rough

________ % smooth

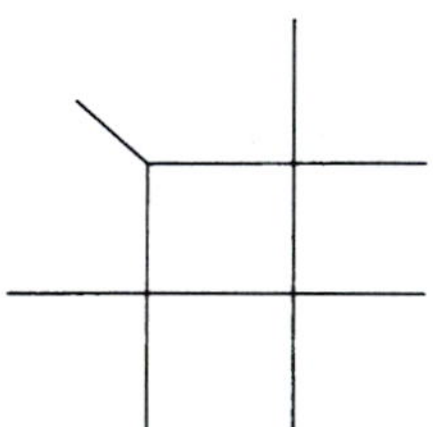

b. Genotypes of the parents: Rr × rr

% of each genotype: ____________________

% of each phenotype: ________ % rough

________ % smooth

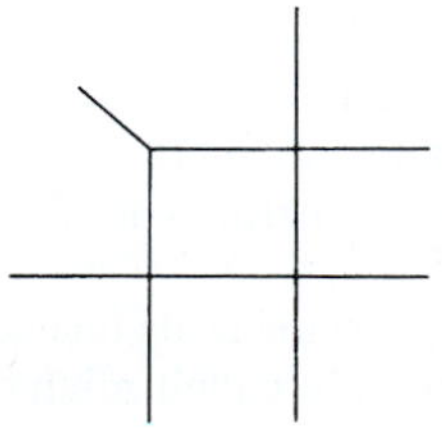

c. Genotypes of the parents: RR × rr

% of each genotype: ____________________

% of each phenotype: ________ % rough

________ % smooth

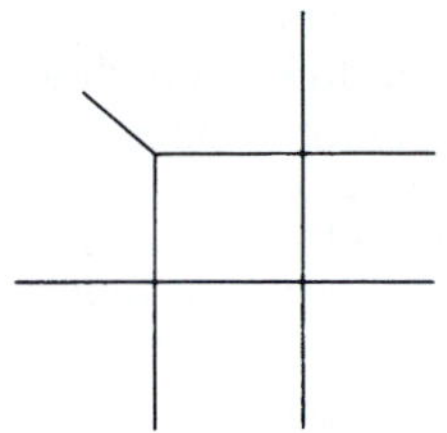

INCOMPLETE DOMINANCE

The concepts of dominance and recessiveness are somewhat arbitrary and artificial in some instances, because so-called dominant genes may be expressed differently in homozygous and heterozygous individuals. This gives rise to a condition called *incomplete dominance* or *intermediate inheritance.* In such cases, both alleles express themselves in the offspring. The crosses are worked out in the same manner as indicated previously, but heterozygous offspring exhibit a phenotype intermediate between that of the homozygous individuals. Some examples follow.

1. The inheritance of flower color in snapdragons illustrates the principle of incomplete dominance. The genotype RR is expressed as a red flower, Rr yields pink flowers, and rr produces white flowers. Work out the following crosses to determine phenotypes seen and both genotype and phenotype percentages.

a. Genotypes of parents: RR × rr

Genotypes and %: ____________________

Phenotypes and %: ____________________

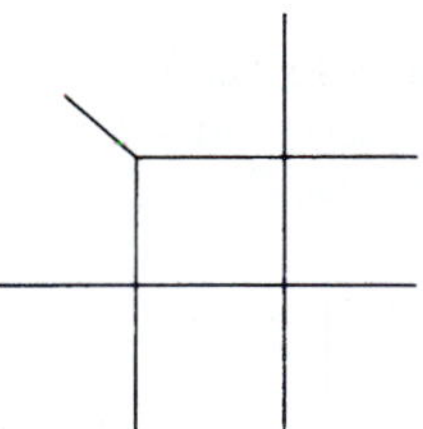

b. Genotypes of parents: Rr × rr

Genotypes and %: ____________________

Phenotypes and %: ____________________

____

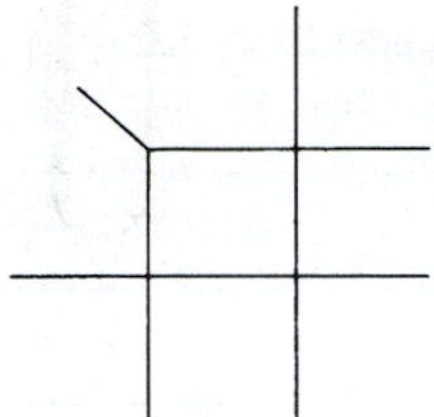

c. Genotypes of parents Rr × Rr

Genotypes and %: ____________________

Phenotypes and %: ____________________

____

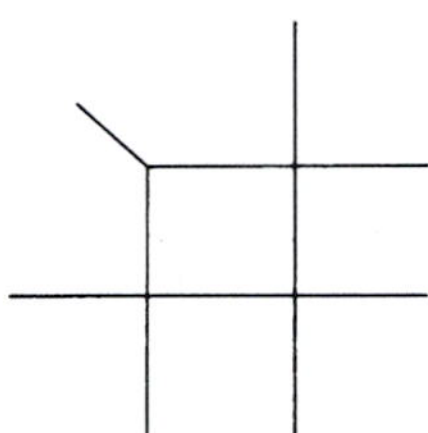

2. In humans, the inheritance of sickle-cell anemia/trait is determined by a single pair of alleles that exhibit incomplete dominance. Individuals homozygous for the sickling gene (s) have *sickle-cell anemia.* In double dose (ss) the sickling gene causes production of a very abnormal hemoglobin, which crystallizes and becomes sharp and spiky under conditions of oxygen deficit. This, in turn, leads to clumping and hemolysis of red blood cells in the circulation, which causes a great deal of pain and can be fatal. Heterozygous individuals (Ss) have the *sickle-cell trait,* which is much less severe. However, they are carriers for the abnormal gene and may pass it on to their offspring. Individuals with the genotype SS form normal hemoglobin. Work out the following crosses:

a. Parental genotypes: SS × ss

Genotypes and %: ____________________

Phenotypes and %: ____________________

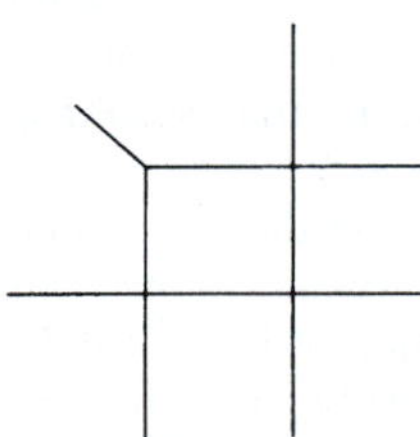

b. Parental genotypes: Ss × Ss

Genotypes and %: ______________________

Phenotypes and %: ______________________

____

c. Parental genotypes: ss × Ss

Genotypes and %: ______________________

Phenotypes and %: ______________________

____

SEX-LINKED INHERITANCE

Of the 23 pairs of homologous chromosomes, 22 pairs are referred to as **autosomes.** Autosomes contain genes that determine most body (somatic) characteristics. The 23rd pair, the **sex chromosomes,** determine the sex of an individual, that is, whether an individual will be male or female. Normal females possess two sex chromosomes that look alike, the X chromosomes. Males possess two dissimilar sex chromosomes, referred to as X and Y. Possession of the Y chromosome determines maleness. A photomicrograph of a male's chromosome complement (male karyotype) is shown in Figure 45.1. The Y sex chromosome is only about a third as large as the X sex chromosome, and lacks many of the genes (directing characteristics other than sex) that are found on the X.

Genes present *only* on the X sex chromosome are called *sex-linked* (or X-linked) genes. Some examples of X-linked genes include those that determine normal color vision (or, conversely, color blindness), and normal clotting ability (as opposed to hemophilia, or bleeder's disease). The alleles that determine color blindness and hemophilia are recessive alleles. In females, *both* X chromosomes must carry the recessive alleles for a woman to express either of these conditions, and thus they tend to be infrequently seen. However, should a male receive even one sex-linked recessive allele for these conditions, he will exhibit the recessive phenotype because his Y chromosome lacks any genes that might dominate or mask the recessive allele.

The critical understanding of X-linked inheritance is the *absence* of male to male (that is, father to son) transmission of X-linked genes. The X of the father *will* pass to each of his daughters but to none of his sons. Males always inherit sex-linked conditions from their mothers (through the X chromosome).

1. A heterozygous woman carrying the recessive gene for color blindness marries a man who is color-blind. Assume the dominant gene is X^C (allele for normal color vision) and the recessive gene is X^c (determines color blindness). The mother's genotype is X^CX^c and the father's X^cY. Do a Punnett square to determine the answers to the following questions.

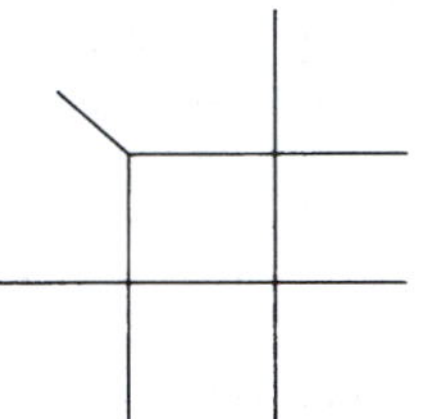

According to the laws of probability, what percent of their children will be color-blind?

_______ %

What is the proportion of color-blind individuals by sex? _______ male; _______ female

What percentage will be carriers? ___________ %

What is the sex of the carriers? ______________

2. A heterozygous woman carrying the recessive gene for hemophilia marries a man who is not a hemophiliac. Assume the dominant gene is X^H and the recessive gene is X^h. The woman's genotype is X^HX^h and her husband's genotype is X^HY. What is the potential percentage and sex of their offspring that will be hemophiliacs?

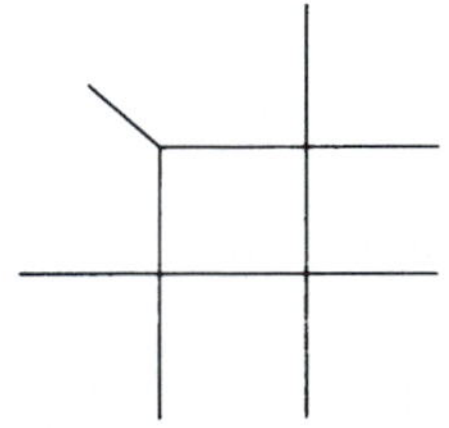

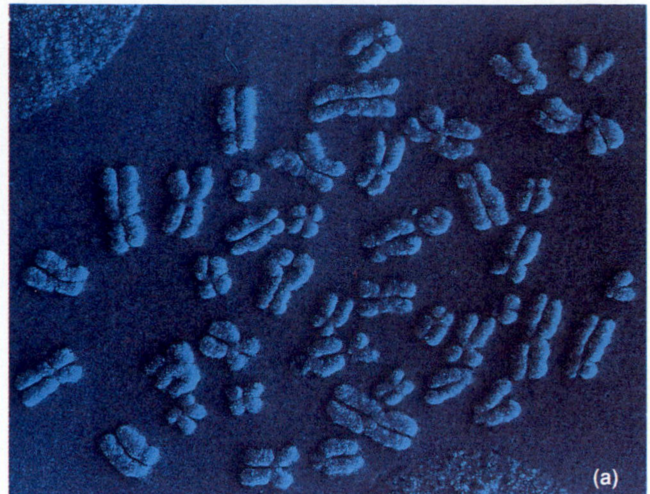

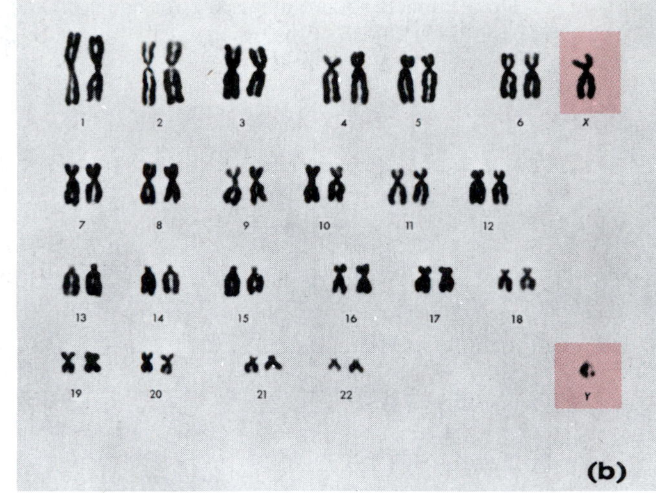

F45.1

Karyotype (chromosomal complement) of human male. Each pair of homologous chromosomes is numbered (1–22) except the sex chromosomes, which are identified by their letters, X and Y.

_______ % males; _______ % females

What percentage can be expected to neither exhibit nor carry the allele for hemophilia?

_______ %

What is the anticipated sex and percentage of individuals that will be carriers for hemophilia?

_______ %; _______ sex

PROBABILITY

Segregation (or parceling out) of chromosomes to daughter cells (gametes) during meiosis and the combination of egg and sperm are random or chance events. Hence, the possibility that certain genomes will arise and be expressed is based on the laws of probability. The randomness of gene recombination from each parent determines individual uniqueness and explains why siblings, however similar, never have totally corresponding traits (unless, of course, they are identical twins). The Punnett square method that you have been using to work out the genetics problems actually provides information on the *probability* of appearance of certain genotypes considering all possible events. Probability (P) is defined as:

$$P = \frac{\text{number of specific events/cases}}{\text{total number of events/cases}}$$

If an event is certain to happen, its probability is 1. If it happens one out of every two times, its probability is $\frac{1}{2}$; if one out of 4 times, its probability is $\frac{1}{4}$, and so on.

When figuring the probability of separate events occurring together (or consecutively), the probability of each event must be multiplied together to get the final probability figure. For example, the probability of a penny coming up "heads" in each toss is $\frac{1}{2}$ (because it has two sides—heads and tails). But the probability of a tossed penny coming up heads 4 times in a row is: $\frac{1}{2} \times \frac{1}{2} \times \frac{1}{2} \times \frac{1}{2} = \frac{1}{16}$.

1. Obtain two pennies and perform the following simple experiment to explore the laws of probability.

   a. Toss one penny into the air 10 times, and record the number of heads/tails observed.

   _______________ heads _______________ tails

   Probability: _____ /10ths tails; _____ /10ths heads

   b. Now simultaneously toss two pennies into the air for 24 tosses, and record the results of each toss below. In each case, report the probability in the lowest fractional terms.

   #HH__________ Probability __________

   #HT__________ Probability __________

   #TT__________ Probability __________

   Does the first toss have any influence on the second?

   ____

   Does the third toss have any influence on the fourth?

   ____

c. Do a Punnett square using HT for one coin and HT for the "alleles" of the other.

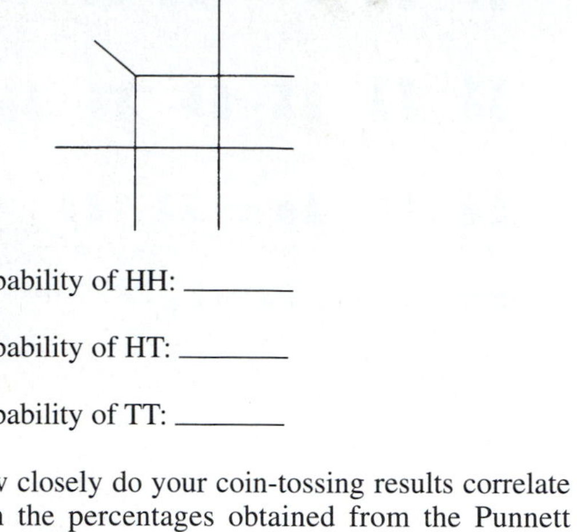

Probability of HH: _______

Probability of HT: _______

Probability of TT: _______

How closely do your coin-tossing results correlate with the percentages obtained from the Punnett square results?

____

____

2. Determine the probability of having a boy or girl offspring for each conception.

Parental genotypes: XY × XX

Probability of males: _______ %

Probability of females: _______ %

3. Dad wants a baseball team! What are the chances of his having nine sons in a row?

_______ (Sorry, Dad!)

GENETIC DETERMINATION OF SELECTED HUMAN CHARACTERISTICS

Many human traits are determined by a single pair of alleles easily identified by simple observation. For each of the characteristics described here, determine (as best you can) both your own phenotype and genotype, and record this information on Table 45.1. Since it is impossible to know whether you are homozygous or heterozygous for a trait when you exhibit its dominant expression, you are to record your genotype as A—(or B—, and so on, depending on the letter used to indicate the alleles) in such cases. If you exhibit the recessive trait, you are homozygous for the recessive allele and

TABLE 45.1 Record of Human Genotypes/Phenotypes

Characteristic	Phenotype	Genotype
Tongue rolling (T,t)		
Attached earlobes (E,e)		
Interlocking fingers (I,i)		
PTC taste (P,p)		
Sodium benzoate taste (S,s)		
Sex (X,Y)		
Dimples (D,d)		
Widow's peak (W,w)		
Double-jointed thumb (J,j)		
Bent little finger (L,l)		
Middigital hair (H,h)		
Freckles (F,f)		
Blaze (B,b)		
ABO blood type (I^A,I^B,i)		

should record it accordingly as aa (bb, cc, and so on). When you have completed your observations, also record your data on the chart for tabulation of class results, on the chalkboard.

Tongue rolling: Extend your tongue and attempt to roll it into a U shape longitudinally. People with this ability have the dominant allele for this trait. Use T for the dominant allele, and t for the recessive allele.

Attached earlobes: Have your lab partner examine your earlobes. If no portion of the lobe hangs free inferior to its point of attachment to the head, you are homozygous recessive (ee) for attached earlobes. If part of the lobe hangs free below the point of attachment, you possess at least one dominant gene (E) (see Figure 45.2).

Interlocking fingers: Clasp your hands together by interlocking your fingers. Observe your clasped hands. Which thumb is uppermost? If the left thumb is uppermost, you possess a dominant allele (I) for this trait. If you clasped your right over your left thumb, you are illustrating the homozygous recessive (ii) phenotype.

PTC taste: Obtain a PTC taste strip. PTC or phenylthiocarbamide is a harmless chemical that some people can taste and others find tasteless. Chew the strip. If it tastes slightly bitter, you are a "taster" and possess the dominant gene (P) for this trait. If you cannot taste anything, you are a nontaster and are homozygous recessive (pp) for the trait. Approximately 70% of the people in the United States are tasters.

Sodium benzoate taste: Obtain a sodium benzoate taste strip and chew it. A different pair of alleles (from that determining PTC taste) determines the ability to taste sodium benzoate. If you can taste it, you have at least one of the dominant alleles (S). If not, you are homozygous recessive (ss) for the trait. Also record whether sodium benzoate tastes salty, bitter, or sweet to you (if a taster). Even though PTC and sodium benzoate taste are inherited independently, they interact to determine a person's taste sensations. Individuals who find PTC bitter and sodium benzoate salty tend to be devotees of sauerkraut, buttermilk, spinach, and other slightly bitter or salty foods.

Sex: The genotype XX determines the female phenotype, whereas XY determines the male phenotype.

Dimpled cheeks: The presence of dimples in one or both cheeks is due to a dominant gene (D). Absence of dimples indicates the homozygous recessive condition (dd).

Widow's peak: A distinct downward V-shaped hairline at the middle of the forehead is referred to as a widow's peak. It is determined by a dominant allele (W), whereas the straight or continuous forehead hairline is determined by the homozygous recessive condition (ww) (see Figure 45.2).

Double-jointed thumb: A dominant gene determines a condition of loose ligaments that allows one to throw the thumb out of joint. The homozygous recessive condition determines tight joints. Use J for the dominant allele and j for the recessive allele (see Figure 45.2).

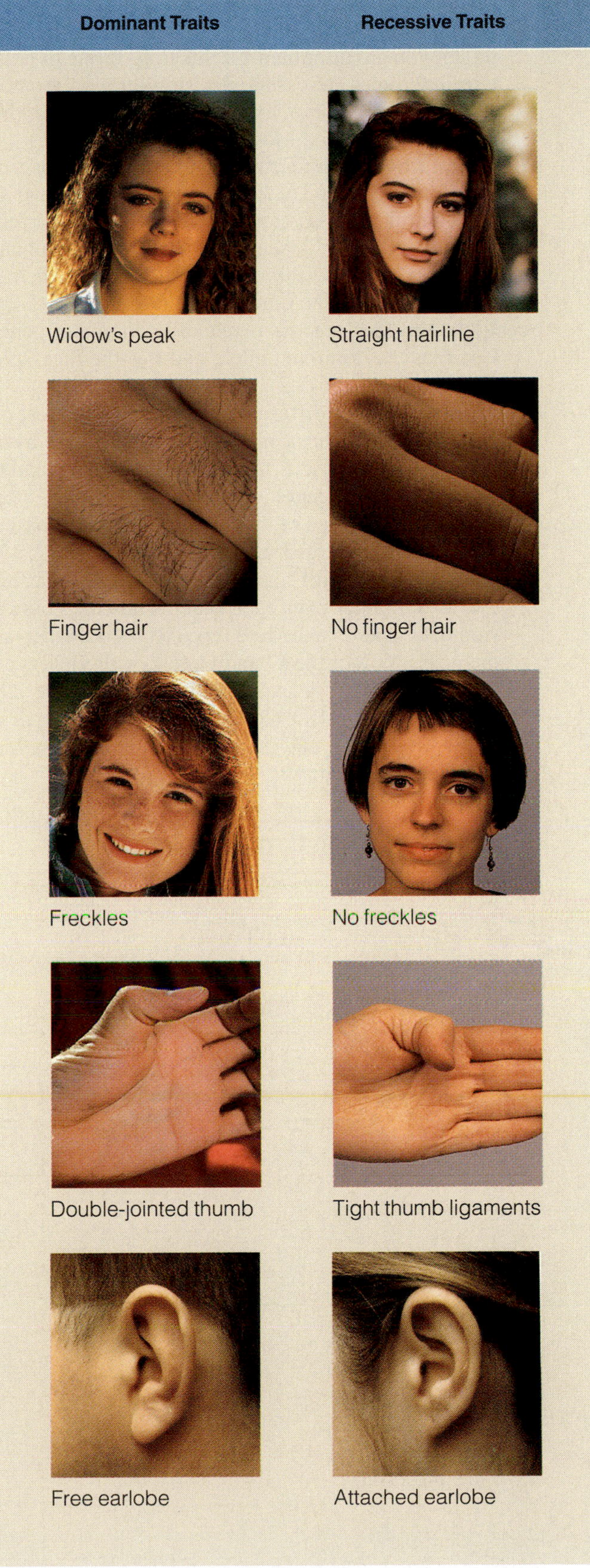

F45.2

Selected examples of human phenotypes.

Bent little finger: Examine your little finger on each hand. If its terminal phalanx angles toward the ring finger, you are dominant for this trait. If one or both terminal digits are essentially straight, you are homozygous recessive for the trait. Use L for the dominant allele and l for the recessive allele.

Middigital finger hair: Critically examine the dorsum of the middle segment (phalanx) of fingers 3 and 4. If no hair is obvious, you are recessive (hh) for this condition. If hair is seen, you have the dominant gene (H) for this trait (which, however, is determined by multigene inheritance) (see Figure 45.2).

Freckles: Freckles are the result of a dominant gene. Use F as the dominant allele and f as the recessive allele (see Figure 45.2).

Blaze: A lock of hair different in color from the rest of scalp hair is called a blaze; it is determined by a dominant gene. Use B for the dominant gene and b for the recessive gene.

Blood type: Inheritance of the ABO blood type is based on the existence of 3 alleles designated as I^A, I^B, and i. Both I^A and I^B are dominant over i, but neither is dominant over each other. Thus the possession of I^A and I^B will yield type AB blood, whereas the possession of the I^A and i alleles will yield type A blood, and so on as explained in Exercise 29. There are four ABO blood groups or phenotypes, A, B, AB, and O, and their correlation to genotype is indicated as follows:

ABO blood group	Genotype
A	I^AI^A or I^Ai
B	I^BI^B or I^Bi
AB	I^AI^B
O	ii

Assuming you have previously typed your blood, record your phenotype and genotype in Table 45.1. If not, type your blood following the instructions on p. 278, and then enter your results in the table.

Dispose of any blood-soiled supplies by placing the glassware in the bleach-containing beaker and all other items in the autoclave bag.

Once class data have been tabulated, scrutinize the results. Is there a single trait that is expressed in an identical manner by all members of the class?

____

Because all human beings have 23 pairs of homologues and each pair segregates independently at meiosis, the number of possible combinations at segregation is more than 8 million! On the basis of this information, what would you guess are the chances of any two individuals in the class having identical phenotypes for all 14 traits investigated?

____

Surface Anatomy Roundup

OBJECTIVES

1. Define *surface anatomy,* and explain why it is an important field of study. Define *palpation.*
2. Describe and palpate the major surface features of the cranium, face, and neck.
3. Describe the easily palpated bony and muscular landmarks of the back. Locate the vertebral spines on the living body.
4. List the bony surface landmarks of the thoracic cage, and explain how they relate to the major soft organs of the thorax. Explain how to find any rib (second to eleventh).
5. Name and palpate the important surface features on the anterior abdominal wall, and explain how to palpate a full bladder.
6. Define and explain the following: *linea alba, umbilical hernia,* examination for an inguinal hernia, *linea semilunaris,* and *McBurney's point.*
7. Locate and palpate the main surface features of the upper limb.
8. Explain the significance of the cubital fossa, pulse points in the distal forearm, and the anatomical snuff box.
9. Describe and palpate the surface landmarks of the lower limb.
10. Explain exactly where to administer an injection in the gluteal region and in the other major sites of intramuscular injection.

MATERIALS

Articulated skeletons
Charts or models of the skeletal muscles of the body
Washable markers
Hand mirror
Stethoscope
Alcohol swabs

See Appendix E, Exercise 46 for links to *Anatomy and PhysioShow: The Videodisc.*

Surface anatomy is an extremely valuable branch of anatomical and medical science. True to its name, surface anatomy does indeed study the *external surface* of the body, but more importantly, it also studies *internal* organs as they relate to external surface landmarks and as they are seen and felt through the skin. Feeling internal structures through the skin with the fingers is called **palpation** (literally, "touching").

Surface anatomy is living anatomy, better studied in live people than in cadavers. It can provide a great deal of information about the living skeleton (almost all bones can be palpated) and about the muscles and blood vessels that lie near the body surface. Furthermore, a skilled examiner can learn a good deal about the heart, lungs, and other deep organs by performing a surface assessment. Thus, surface anatomy serves as the basis of the standard physical examination. For those planning a career in the health sciences or physical education, a study of surface anatomy will show you where to take pulses, where to insert tubes and needles, where to locate broken bones and inflamed muscles, and where to listen for the sounds of the lungs, heart, and intestines.

We will take a regional approach to surface anatomy, exploring the head first and proceeding to the trunk and the limbs. You will be observing and palpating your own body as you work through the exercise, because your body is the best learning tool of all. To aid your exploration of living anatomy, skeletons and muscle models are provided around the lab so that you can review the bones and muscles you will encounter. Where you are asked to mark with a washable marker areas of skin that are personally unreachable, it probably would be best to choose a male student as a subject.

THE HEAD

The head (Figures 46.1 and 46.2) is divided into the cranium and the face.

Cranium

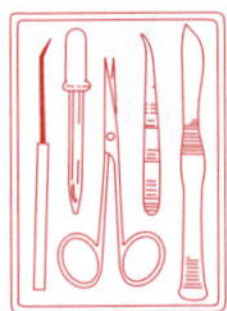

1. Run your fingers over the superior surface of your head. Notice that the underlying cranial bones lie very near the surface. Proceed to your forehead and palpate the *superciliary arches* ("brow ridges") directly superior to your orbits (Figure 46.1).

2. Move your hand to the posterior surface of your skull, where you can feel the knoblike *external occipital protuberance.* Run your finger directly laterally from this protuberance to feel the ridgelike *superior nuchal*

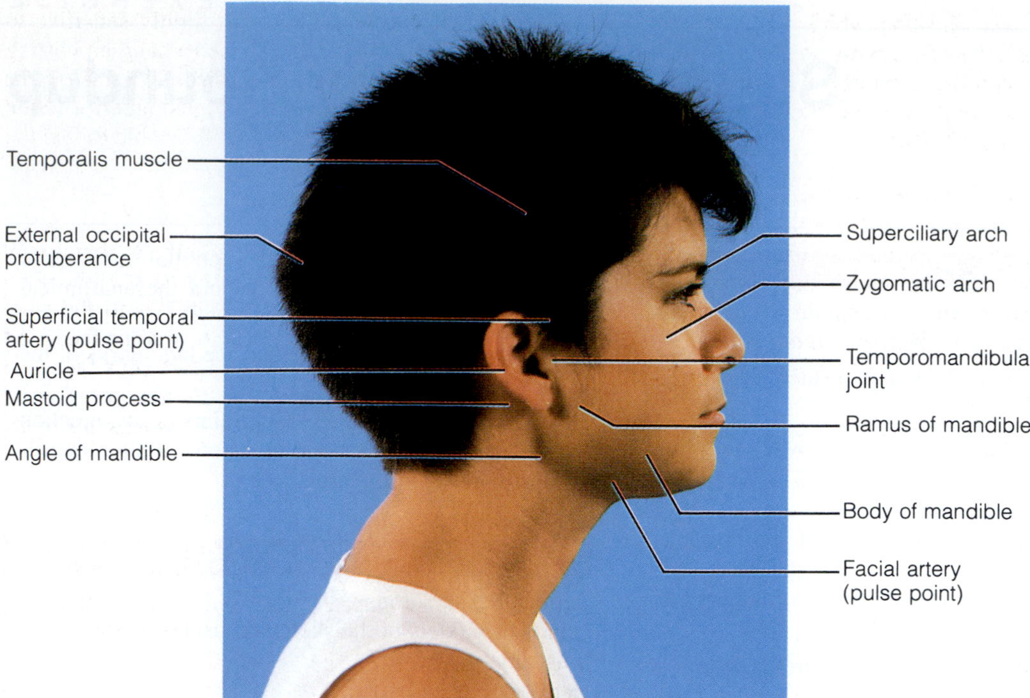

F46.1

Surface anatomy of the lateral aspect of the head.

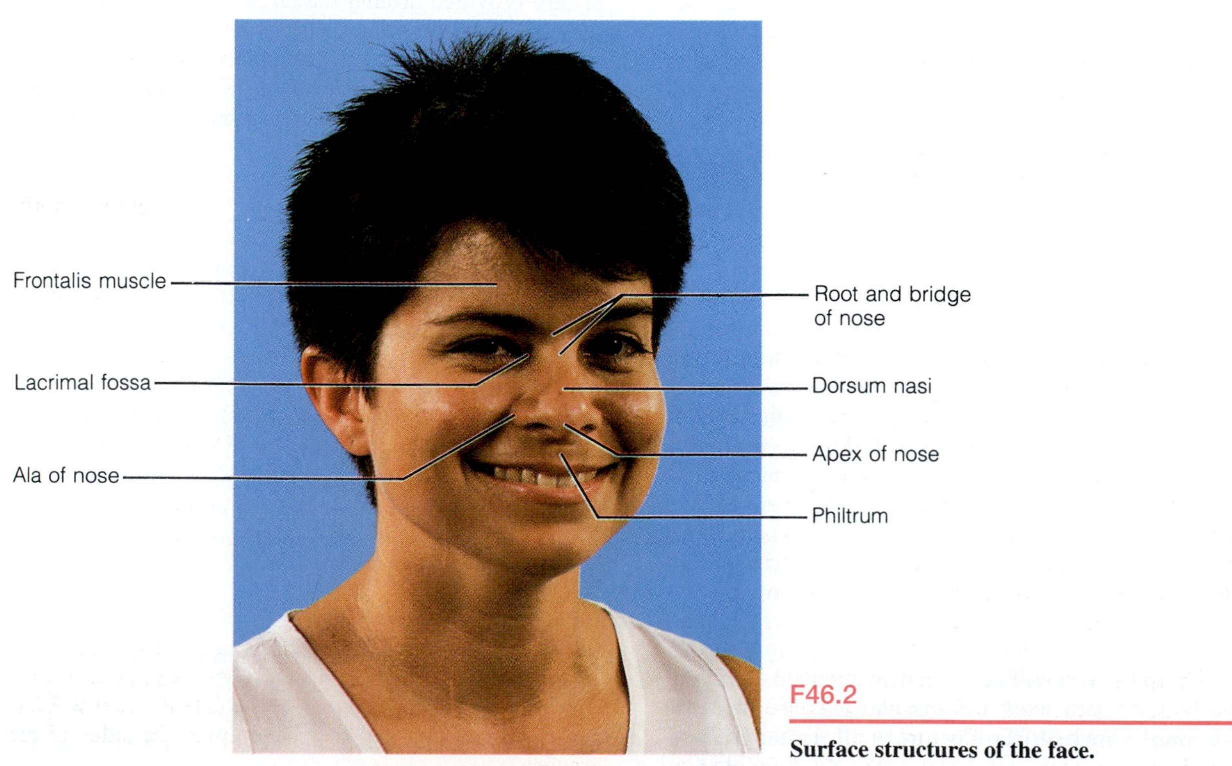

F46.2

Surface structures of the face.

line on the occipital bone. This line, which marks the superior extent of the muscles of the posterior neck, serves as the boundary between the head and the neck. Now feel the prominent *mastoid process* on each side of the cranium just posterior to your ear.

3. Place a hand on your temple, and clench your teeth together in a biting action. You should be able to feel the *temporalis muscle* bulge as it contracts. Next, raise your eyebrows, and feel your forehead wrinkle, an action produced by the *frontalis muscle.* The frontalis inserts superiorly onto a broad aponeurosis called the *galea aponeurotica* (Table 15.1, p. 115), which covers the superior surface of the cranium. This aponeurosis binds tightly to the overlying subcutaneous tissue and skin to form the true **scalp.** Push on your scalp, and confirm that it slides freely over the underlying cranial bones. Because the scalp is only loosely bound to the skull, people can easily be "scalped" (in industrial accidents, for example). The scalp is richly vascularized by a large number of arteries running through its subcutaneous tissue. Most arteries of the body constrict and close after they are cut or torn, but those in the scalp are unable to do so because they are held open by the dense connective tissue surrounding them.

What do these facts suggest about the amount of bleeding that accompanies scalp wounds?

____

____

____

Face

The surface of the face is divided into many different regions, including the *orbital, nasal, oral* (mouth), and *auricular* (ear) areas (Figure 46.2).

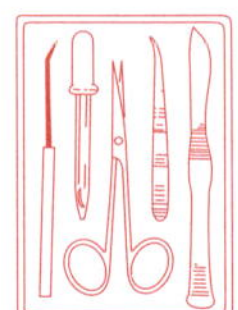

1. Trace a finger around the entire margin of the bony orbit. The *lacrimal fossa,* which contains the tear-gathering lacrimal sac, may be felt on the medial side of the eye socket.

2. Touch the most superior part of your nose, its **root,** which lies between the eyebrows (Figure 46.2). Just inferior to this, between your eyes, is the **bridge** of the nose formed by the nasal bones. Continue your finger's progress inferiorly along the nose's anterior margin, the **dorsum nasi,** to its tip, the **apex.** Place one finger in a nostril and another finger on the flared wing, the **ala,** that defines the nostril's lateral border. Then feel the **philtrum,** the shallow vertical groove on the upper lip below the nose.

3. Grasp your *auricle,* the shell-like part of the external ear that surrounds the opening of the *external auditory canal* (Figure 46.1). Now trace the ear's outer rim, or *helix,* to the *lobule* (earlobe) inferiorly. The lobule is easily pierced, and since it is not highly sensitive to pain, it provides a convenient place to hang an earring or obtain a drop of blood for clinical blood analysis. Next, place a finger on your temple just anterior to the auricle (Figure 46.1). There, you will be able to feel the pulsations of the *superficial temporal artery,* which ascends to supply the scalp.

4. Run your hand anteriorly from your ear toward the orbit, and feel the *zygomatic arch* just deep to the skin. This bony arch is easily broken by blows to the face. Next, place your fingers on the skin of your face, and feel it bunch and stretch as you contort your face into smiles, frowns, and grimaces. You are now monitoring the action of several of the subcutaneous *muscles of facial expression* (Table 15.1, p. 115).

5. On your lower jaw, palpate the parts of the bony *mandible:* its anterior body and its posterior ascending *ramus.* Press on the skin over the mandibular ramus, and feel the *masseter muscle* bulge when you clench your teeth. Palpate the anterior border of the masseter, and trace it to the mandible's inferior margin. At this point, you will be able to detect the pulse of your *facial artery* (Figure 46.1). Finally, to feel the *temporomandibular joint,* place a finger directly anterior to the external auditory canal of your ear, and open and close your mouth several times. The bony structure you feel moving is the *head of the mandible.*

THE NECK

Bony Landmarks

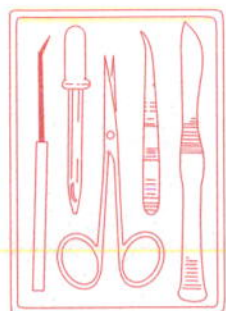

1. Run your fingers inferiorly along the back of your neck, in the posterior midline, to feel the *spinous processes* of the cervical vertebrae. The spine of C_7, the *vertebra prominens,* is especially prominent.

2. Now, beginning at your chin, run a finger inferiorly along the anterior midline of your neck (Figure 46.3). The first hard structure you encounter will be the U-shaped *hyoid bone,* which lies in the angle between the floor of the mouth and the vertical part of the neck (Figure 46.4). Directly inferior to this, you will feel the *laryngeal prominence* (Adam's apple) of the thyroid cartilage. Just inferior to the laryngeal prominence, your finger will sink into a soft depression (formed by the *cricothyroid ligament*) before proceeding onto the rounded surface of the *cricoid cartilage.* Now swallow several times, and feel the whole larynx move up and down.

3. Continue inferiorly to the trachea. Attempt to palpate the *isthmus of the thyroid gland,* which feels like a spongy cushion over the second to fourth tracheal rings (see Figure 46.3a). Then, try to palpate the two soft lateral *lobes* of your thyroid gland along the sides of the trachea.

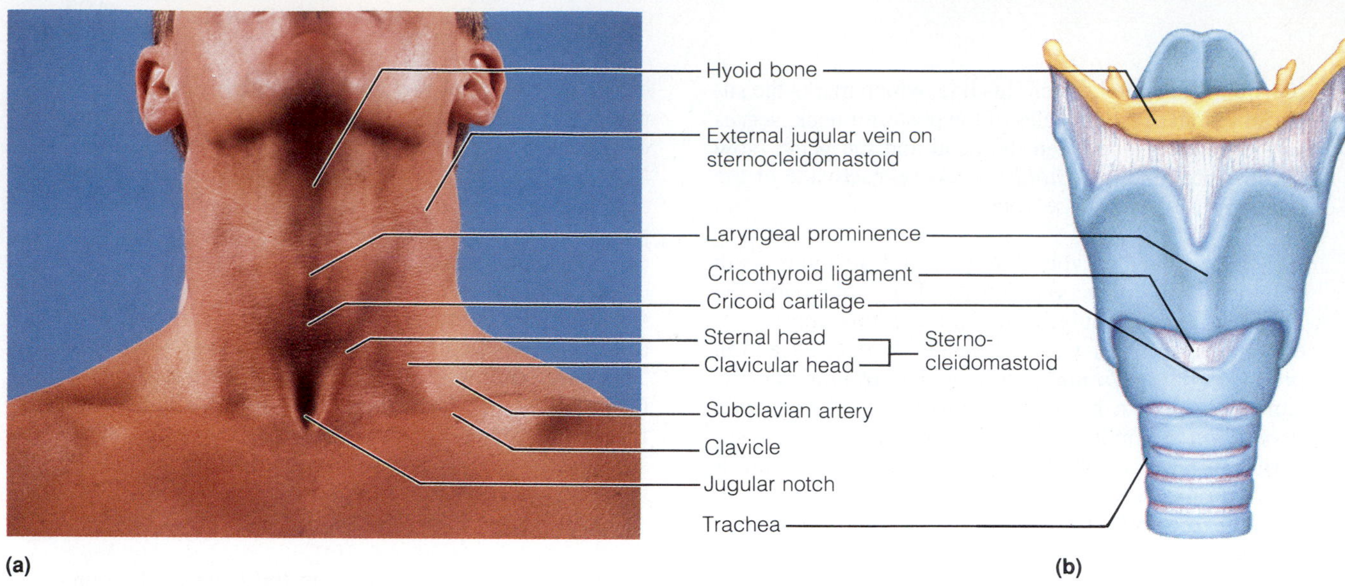

F46.3

Anterior surface of the neck. (a) Photograph. **(b)** Diagram of the underlying skeleton of the larynx.

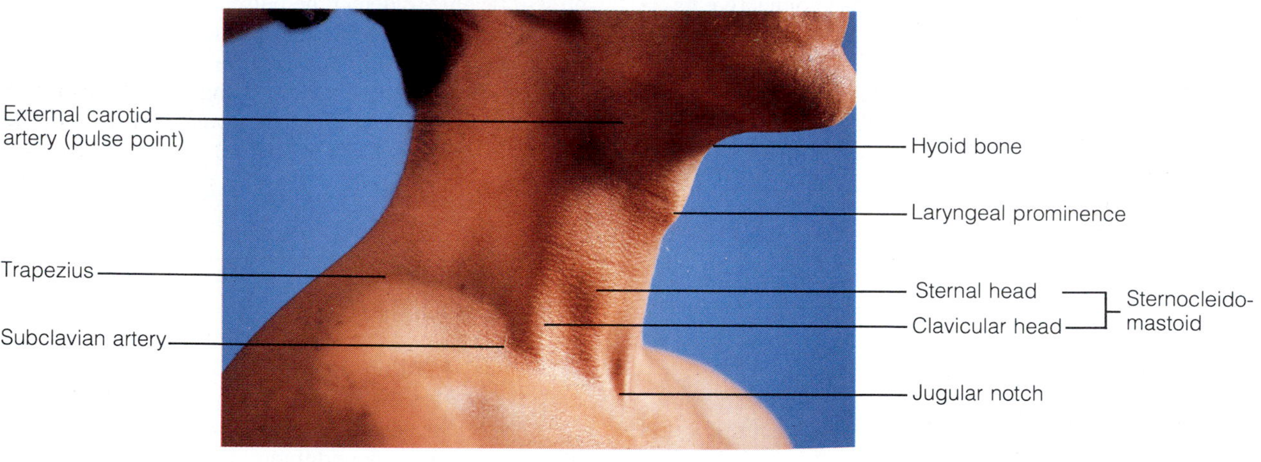

F46.4

Lateral surface of the neck.

4. Move your finger all the way inferiorly to the root of the neck, and rest it in the *jugular notch,* the depression in the superior part of the sternum between the two clavicles. By pushing deeply at this point, you can feel the cartilage rings of the trachea.

Muscles

The *sternocleidomastoid* is the most prominent muscle in the neck and the neck's most important surface landmark. You can best see and feel it when you turn your head to the side.

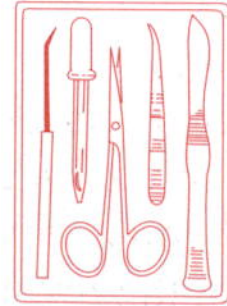

1. Obtain a hand mirror, hold it in front of your face, and turn your head sharply from right to left several times. You will be able to see both heads of this muscle, the *sternal head* medially and the *clavicular head* laterally (Figures 46.3 and 46.4).

Several important structures lie beside or beneath the sternocleidomastoid:

- The *cervical lymph nodes* lie both superficial and deep to this muscle. (Swollen cervical nodes provide evidence of infections or cancer of the head and neck.)
- The *common carotid artery* and *internal jugular vein* lie just deep to the sternocleidomastoid, a relatively superficial location that exposes these vessels to danger in slashing wounds to the neck.
- Just lateral to the inferior part of the sternocleidomastoid is the large *subclavian artery* on its way to supply the upper limb. By pushing on the subclavian artery at this point, one can stop the bleeding from a wound anywhere in the associated limb.
- Just anterior to the sternocleidomastoid, superior to the level of your larynx, you can feel a carotid

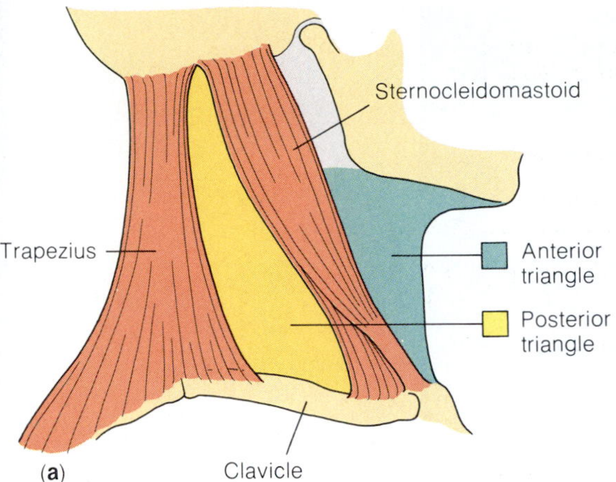

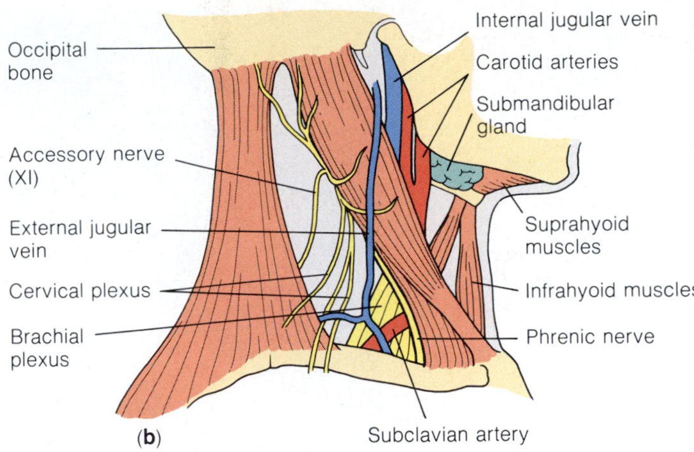

F46.5

Anterior and posterior triangles of the neck. (a) Boundaries of the triangles. **(b)** Some contents of the triangles.

pulse—the pulsations of the *external carotid artery* (Figure 46.4).

- The *external jugular vein* descends vertically, just superficial to the sternocleidomastoid and deep to the skin (Figure 46.3). To make this vein "appear" on your neck, stand before the mirror, and gently compress the skin superior to your clavicle with your fingers.

2. Another large muscle in the neck, on the posterior aspect, is the *trapezius* (Figure 46.4). You can feel this muscle contract just deep to the skin as you shrug your shoulders.

Triangles of the Neck

The sternocleidomastoid muscles divide each side of the neck into the posterior and anterior triangles (Figure 46.5a).

1. The **posterior triangle** is defined by the sternocleidomastoid anteriorly, the trapezius posteriorly, and the clavicle inferiorly. Palpate the muscular borders of the posterior triangle.

The **anterior triangle** is defined by the inferior margin of the mandible superiorly, the midline of the neck anteriorly, and the sternocleidomastoid posteriorly.

2. The contents of these two triangles are shown in Figure 46.5b. The posterior triangle contains many important nerves and blood vessels, including the *accessory nerve* (cranial nerve XI), most of the *cervical plexus,* and the *phrenic nerve.* In the inferior part of the triangle are the external jugular vein, the trunks of the *brachial plexus,* and the subclavian artery. These structures are relatively superficial and are easily cut or injured by wounds to the neck. In the neck's anterior triangle, important structures include the *submandibular gland,* the *suprahyoid* and *infrahyoid muscles,* and parts of the carotid arteries and jugular veins that lie superior to the sternocleidomastoid.

- Palpate your carotid pulse.

A wound to the posterior triangle of the neck can lead to long-term loss of sensation in the skin of the neck and shoulder, as well as partial paralysis of the sternocleidomastoid and trapezius muscles. Can you explain these effects?

____

____

______________________________ ■

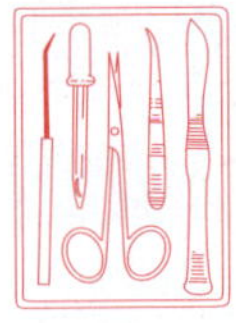

THE TRUNK

The trunk of the body consists of the thorax, abdomen, pelvis, and perineum. The *back* includes parts of all of these regions, but for convenience it is treated separately.

The Back

BONES

1. The vertical groove in the center of the back is called the **posterior median furrow** (Figure 46.6a). The *spinous processes* of the vertebrae are visible in the furrow when the spinal column is flexed. Palpate a few of these processes on your partner's back (C_7 and T_1 are the most prominent and the easiest to find). Also palpate the posterior parts of some ribs, as well as the prominent *spine of the scapula* and the scapula's long *medial border.* The scapula lies superficial to ribs 2 to 7; its *inferior angle* is at the level

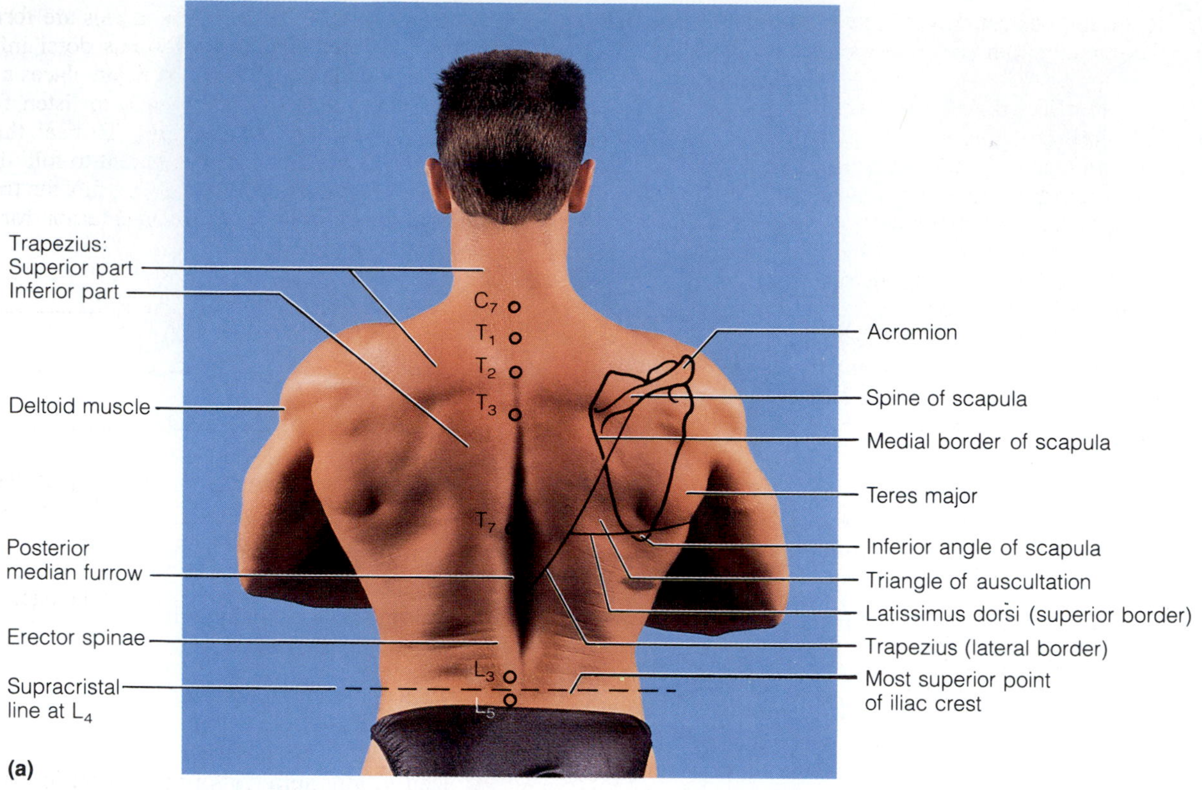

F46.6

Surface anatomy of the back. Two different poses, **(a)** and **(b)**.

of the spinous process of vertebra T_7. The medial end of the scapular spine lies opposite the T_3 spinous process.

2. Now feel the *iliac crests* (superior margins of the iliac bones) in your own lower back. You can find these crests effortlessly by resting your hands on your hips. Locate the most superior point of each crest, a point that lies roughly halfway between the posterior median furrow and the lateral side of the body (Figure 46.6a). A horizontal line through these two superior points, the **supracristal line,** intersects L_4, providing a simple way to locate that vertebra. The ability to locate L_4 is essential for performing a *lumbar puncture,* a procedure in which the clinician inserts a needle into the vertebral canal of the spinal column directly superior or inferior to L_4 and withdraws cerebrospinal fluid.

3. The *sacrum* is easy to palpate just superior to the cleft in the buttocks. You can feel the *coccyx* in the extreme inferior part of that cleft, just posterior to the anus.

MUSCLES The largest superficial muscles of the back are the *trapezius* superiorly and *latissimus dorsi* inferiorly (Figure 46.6b). Furthermore, the deeper *erector spinae* muscles are very evident in the lower back, flanking the vertebral column like thick vertical cords.

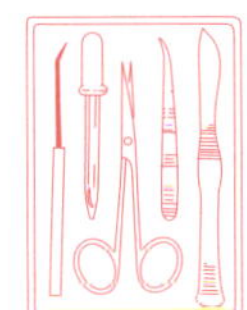

Feel your partner's erector spinae muscles contract and bulge as he extends his spine from a slightly bent-over position.

The superficial muscles of the back fail to cover a small area of the rib cage called the **triangle of auscultation** (Figure 46.6a). This triangle lies just medial to the inferior part of the scapula. Its three boundaries are formed by the trapezius medially, the latissimus dorsi inferiorly, and the scapula laterally. The physician places a stethoscope over the skin of this triangle to listen for lung sounds (*auscultation* = listening). To hear the lungs clearly, the doctor first asks the patient to fold the arms together in front of the chest and then flex the trunk.

What do you think is the precise reason for having the patient take this action?

____

____

____

Have your partner assume the position just described. After cleaning the earpieces with an alcohol swab, use the stethoscope to auscultate his lung sounds. Compare the clarity of the lung sounds heard over the triangle of auscultation to that over other areas of the back.

The Thorax

BONES

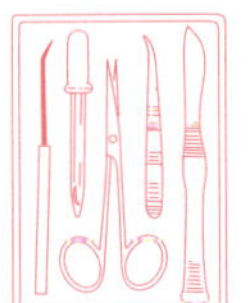

1. Start exploring the anterior surface of your partner's bony *thoracic cage* (Figures 46.7 and 46.8) by defining the extent of the *sternum.* Use a finger to trace the sternum's triangular *manubrium* inferior to the jugular notch, its flat *body,* and the tongue-shaped *xiphoid process.* Now palpate the ridge-like *sternal angle,* where the manubrium meets the body

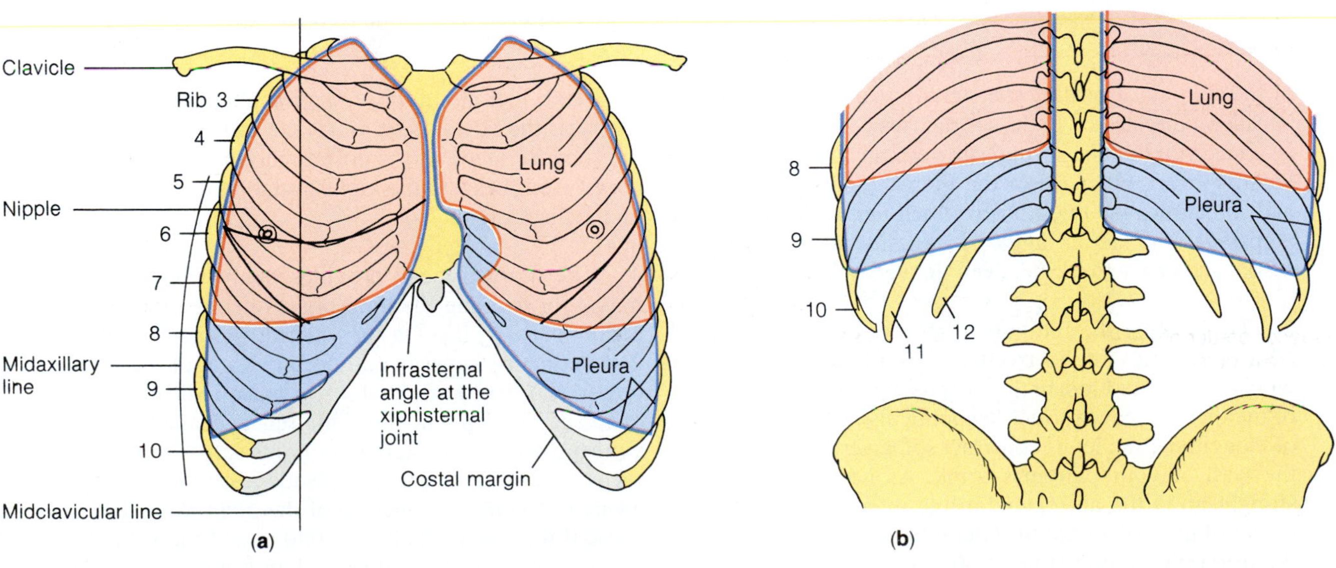

F46.7

The bony rib cage as it relates to the underlying lungs and pleural cavities. Both the pleural cavities (blue) and the lungs (red) are outlined. **(a)** Anterior view. **(b)** Posterior view.

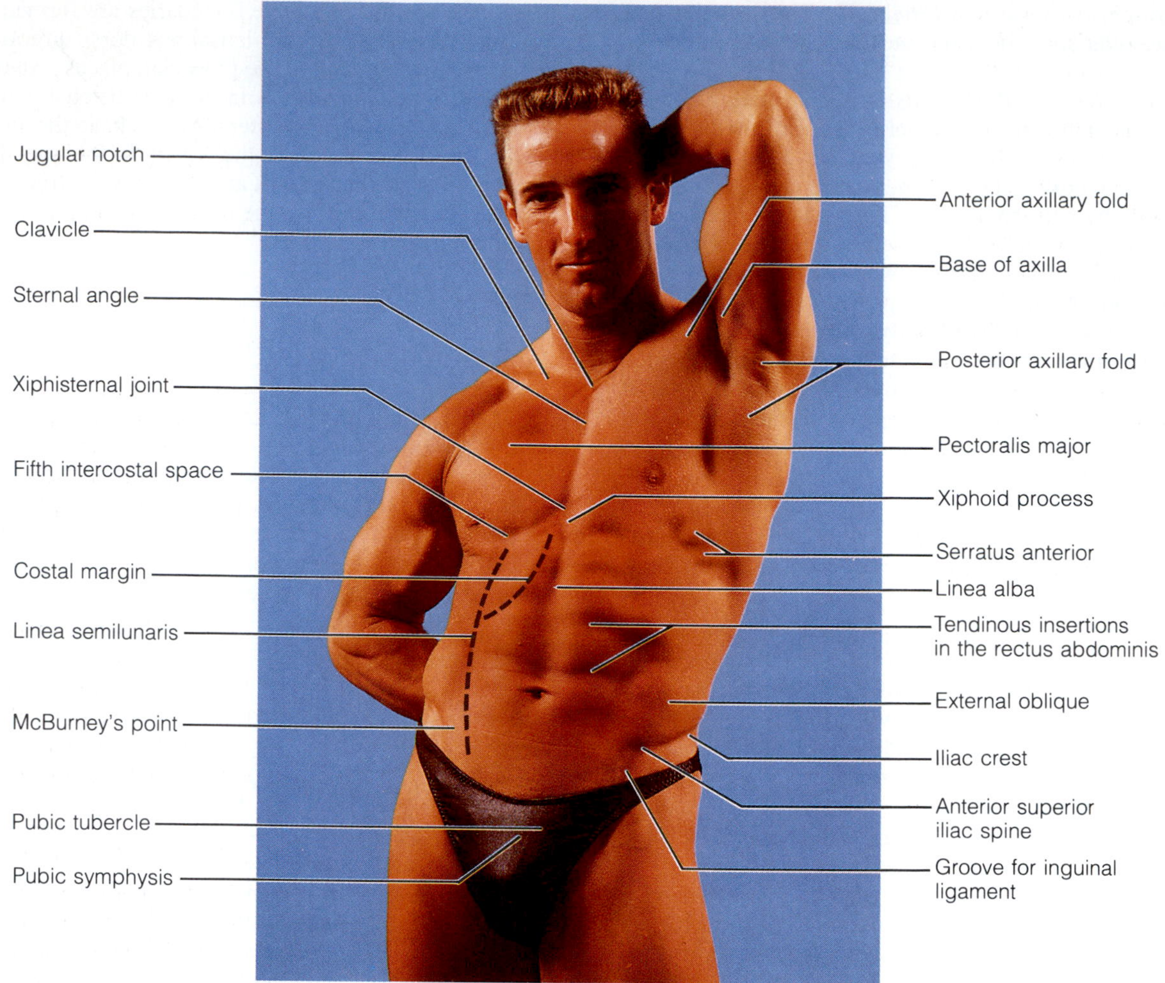

F46.8

The anterior thorax and abdomen.

of the sternum. Locating the sternal angle is important because it directs you to the second ribs (which attach to it). Once you find the second rib, you can count down to identify every other rib in the thorax (except the first and sometimes the twelfth rib, which lie too deep to be palpated). The sternal angle is a highly reliable landmark—it is easy to locate, even in overweight people.

2. By locating the individual ribs, you can mentally "draw" a series of horizontal lines of "latitude" by which to map and locate the underlying visceral organs of the thoracic cavity. Such mapping also requires lines of "longitude," so let us construct some vertical lines on the wall of your partner's trunk. As he lifts an arm straight up in the air, extend a line inferiorly from the center of the axilla onto his lateral thoracic wall. This is the **midaxillary line** (Figure 46.7a). Now estimate the midpoint of his *clavicle,* and run a vertical line inferiorly from that point toward the groin. This is the **midclavicular line,** and it will pass about 1 cm medial to the nipple.

3. Next, feel along the V-shaped inferior edge of the rib cage, the *costal margin.* At the **infrasternal angle,** the superior angle of the costal margin, lies the *xiphisternal joint.* Deep to the xiphisternal joint, the heart lies on the diaphragm.

4. The thoracic cage provides many valuable landmarks for locating the vital organs of the thoracic and abdominal cavities. On the anterior thoracic wall, ribs 2–6 define the superior-to-inferior extent of the female breast, and the fourth intercostal space indicates the location of the *nipple* in men, children, and small-breasted women. The right costal margin runs across the anterior surface of the liver and gallbladder. Surgeons must be aware of the inferior margin of the *pleural cavities* because if they accidentally cut into one of these cavities, a lung collapses. The inferior pleural margin lies adjacent to vertebra T_{12} near the posterior midline (Figure 46.7b) and runs horizontally across the back to reach rib 10 at the midaxillary line. From there, the pleural margin ascends to rib 8 in the midclavicular line (Figure

46.7a) and to the level of the xiphisternal joint near the anterior midline. The *lungs* do not fill the inferior region of the pleural cavity. Instead, their inferior borders run at a level that is two ribs superior to the pleural margin, until they meet that margin near the xiphisternal joint.

5. The relationship of the *heart* to the thoracic cage is considered in Exercise 30 (p. 281). We will review that information here. In essence, the superior right corner of the heart lies at the junction of the third rib and the sternum; the superior left corner lies at the second rib, near the sternum; the inferior left corner lies in the fifth intercostal space in the midclavicular line; and the inferior right corner lies at the sternal border of the sixth rib. You may wish to outline the heart on your chest or that of your lab partner by connecting the four corner points with a washable marker.

MUSCLES The main superficial muscles of the anterior thoracic wall are the *pectoralis major* and the anterior slips of the *serratus anterior.*

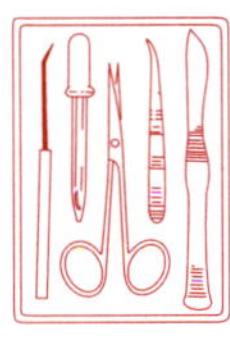

Using Figure 46.8 as a guide, try to palpate these two muscles on your chest. They both contract during push-ups, and you can confirm this by pushing yourself up from your desk with one arm while palpating the muscles with your opposite hand.

The Abdomen

BONY LANDMARKS

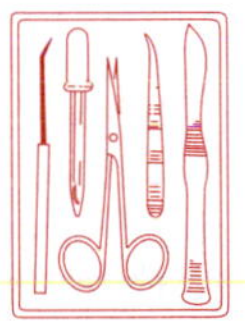

The anterior abdominal wall (Figure 46.8) extends inferiorly from the costal margin to an inferior boundary that is defined by several landmarks. Palpate these landmarks as they are described below.

1. **Iliac crest.** Recall that the iliac crests are the superior margins of the iliac bones, and you can locate them by resting your hands on your hips.

2. **Anterior superior iliac spine.** Representing the most anterior point of the iliac crest, this spine is a prominent landmark. It can be palpated in everyone, even those who are overweight. Run your fingers anteriorly along the iliac crest to its end.

3. **Inguinal ligament.** The inguinal ligament, indicated by a groove on the skin of the groin, runs medially from the anterior superior iliac spine to the pubic tubercle of the pubic bone.

4. **Pubic crest.** You will have to press deeply to feel this crest on the pubic bone near the median *pubic symphysis.* The *pubic tubercle,* the most lateral point of the pubic crest, is easier to palpate, but you will still have to push deeply.

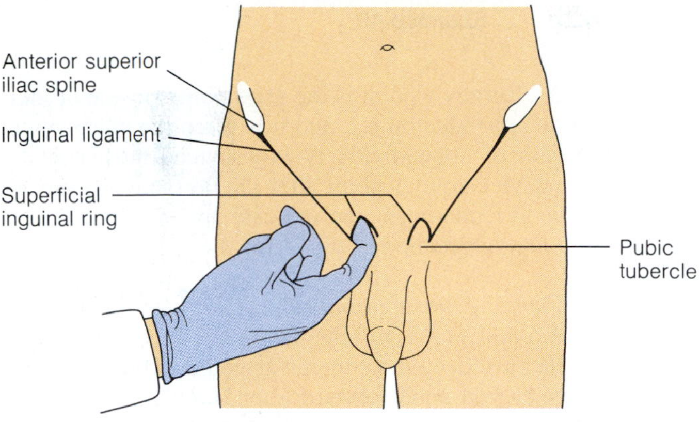

F46.9

Clinical examination for an inguinal hernia in a male. The examiner palpates the patient's pubic tubercle, pushes superiorly to invaginate the scrotal skin into the superficial inguinal ring, and asks the patient to cough. If an inguinal hernia exists, it will push inferiorly and touch the examiner's fingertip.

Inguinal hernias occur immediately superior to the inguinal ligament and may exit from a medial opening called the *superficial inguinal ring.* To locate this ring, one would palpate the pubic tubercle (Figure 46.9). The procedure used by the physician to test whether a male has an inguinal hernia is depicted in Figure 46.9. ■

MUSCLES AND OTHER SURFACE FEATURES

The central landmark of the anterior abdominal wall is the *umbilicus* (navel). Running superiorly and inferiorly from the umbilicus is the *linea alba* ("white line"), represented in the skin of lean people by a vertical groove (Figure 46.8). The linea alba is a tendinous seam that extends from the xiphoid process to the pubic symphysis, just medial to the rectus abdominis muscles (Table 15.3, p. 120). The linea alba is a favored site for surgical entry into the abdominal cavity because the surgeon can make a long cut through this line with no muscle damage and a minimum of bleeding.

Several kinds of hernias involve the umbilicus and the linea alba. In an **acquired umbilical hernia,** the linea alba weakens until intestinal coils push through it just superior to the navel. The herniated coils form a bulge just deep to the skin.

Another type of umbilical hernia is a **congenital umbilical hernia,** present in some infants: The umbilical hernia is seen as a cherry-sized bulge deep to the skin of the navel that enlarges whenever the baby cries. Congenital umbilical hernias are usually harmless, and most correct themselves automatically before the child's second birthday. ■

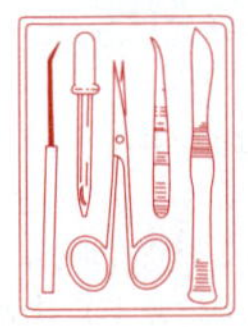

1. **McBurney's point** is the spot on the anterior abdominal skin that lies directly superficial to the base of the appendix (Figure 46.8). It is located one-third of the way along a line between the right anterior superior iliac spine and the umbilicus. Try to find it on your body.

McBurney's point is the most common site of incision in appendectomies, and it is often the place where the pain of appendicitis is experienced most acutely. Pain at McBurney's point after the pressure is removed (rebound tenderness) can indicate appendicitis. This is not a *precise* method of diagnosis, however.

2. Flanking the linea alba are the vertical straplike *rectus abdominis* muscles (Figure 46.8). Feel these muscles contract just deep to your skin as you do a bent-knee situp (or as you bend forward after leaning back in your chair). In the skin of lean people, the lateral margin of each rectus muscle makes a groove known as the **linea semilunaris** ("half-moon line"). On your right side, estimate where your linea semilunaris crosses the costal margin of the rib cage. The *gallbladder* lies just deep to this spot, so this is the standard point of incision for gallbladder surgery. In muscular people, three horizontal grooves can be seen in the skin covering the rectus abdominis. These grooves represent the *tendinous insertions* (or *intersections* or *inscriptions*), fibrous bands that subdivide the rectus muscle. Because of these subdivisions, each rectus abdominis muscle presents four distinct bulges. Try to identify these insertions on yourself or your partner.

3. The only other major muscles that can be seen or felt through the anterior abdominal wall are the lateral *external obliques.* Feel these muscles contract as you cough, strain, or raise your intra-abdominal pressure in some other way.

4. Recall that the anterior abdominal wall can be divided into four quadrants (Figure 46.10). A clinician listening to a patient's **bowel sounds** places the stethoscope over each of the four abdominal quadrants, one after another. Normal bowel sounds, which result as peristalsis moves air and fluid through the intestine, are high-pitched gurgles that occur every 5 to 15 seconds.

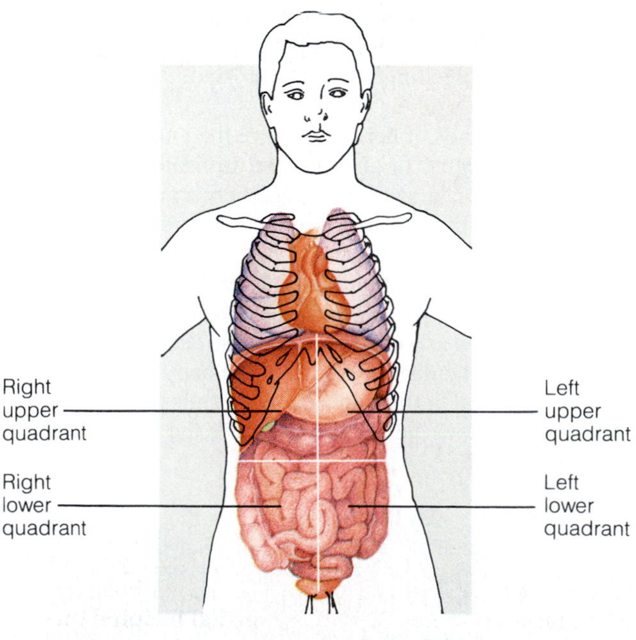

F46.10

The four abdominopelvic quadrants.

- Use the stethoscope to listen to your own or your partner's bowel sounds.

Abnormal bowel sounds can indicate intestinal disorders. Absence of bowel sounds indicates a halt in intestinal activity, which follows long-term obstruction of the intestine, surgical handling of the intestine, peritonitis, or other conditions. Loud tinkling or splashing sounds, by contrast, indicate an increase in intestinal activity. Such loud sounds may accompany gastroenteritis (inflammation and upset of the GI tract) or a partly obstructed intestine. ■

The Pelvis and Perineum

The bony surface features of the *pelvis* are best studied with the abdomen or the lower limb, so they are considered later. Most *internal* pelvic organs are not palpable through the skin of the body surface. A full *bladder,* however, becomes firm and can be felt through the abdominal wall just superior to the pubic symphysis. A bladder that can be palpated more than a few centimeters above this symphysis is retaining urine and dangerously full, and it should be drained by catheterization.

UPPER LIMB

Axilla

The **base of the axilla** is the groove in which the underarm hair grows (Figure 46.8). Deep to this base lie the axillary *lymph nodes* (which swell and can be palpated in breast cancer), the large *axillary vessels* serving the upper limb, and much of the brachial plexus. The base of the axilla forms a "valley" between two thick, rounded ridges, the **axillary folds.** Just anterior to the base, clutch your **anterior axillary fold,** formed by the pectoralis major muscle. Then grasp your **posterior axillary fold.** This fold is formed by the latissimus dorsi and teres major muscles of the back as they course toward their insertions on the humerus.

Shoulder

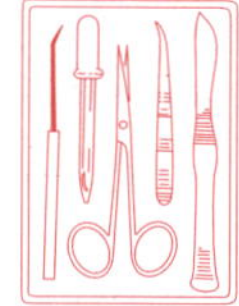

1. Relocate the prominent spine of the scapula posteriorly (Figure 46.11). Follow the spine to its lateral end, the flattened *acromion* on the shoulder's summit. Then, palpate the *clavicle* anteriorly, tracing this bone from the sternum to the shoulder. Notice the clavicle's curved shape.

2. Now locate the junction between the clavicle and the acromion on the superolateral surface of your shoulder, at the *acromioclavicular joint.* To find this joint, thrust your arm anteriorly repeatedly until you can palpate the precise point of pivoting action.

3. Next, place your fingers on the *greater tubercle* of the humerus. This is the most lateral bony landmark on

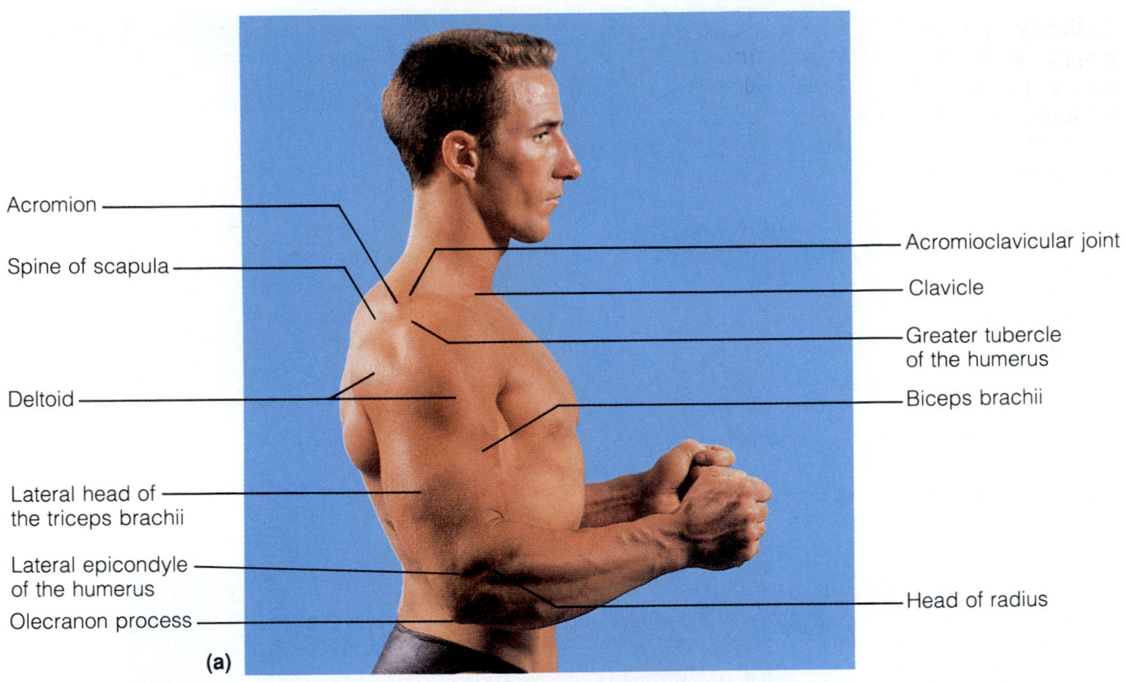

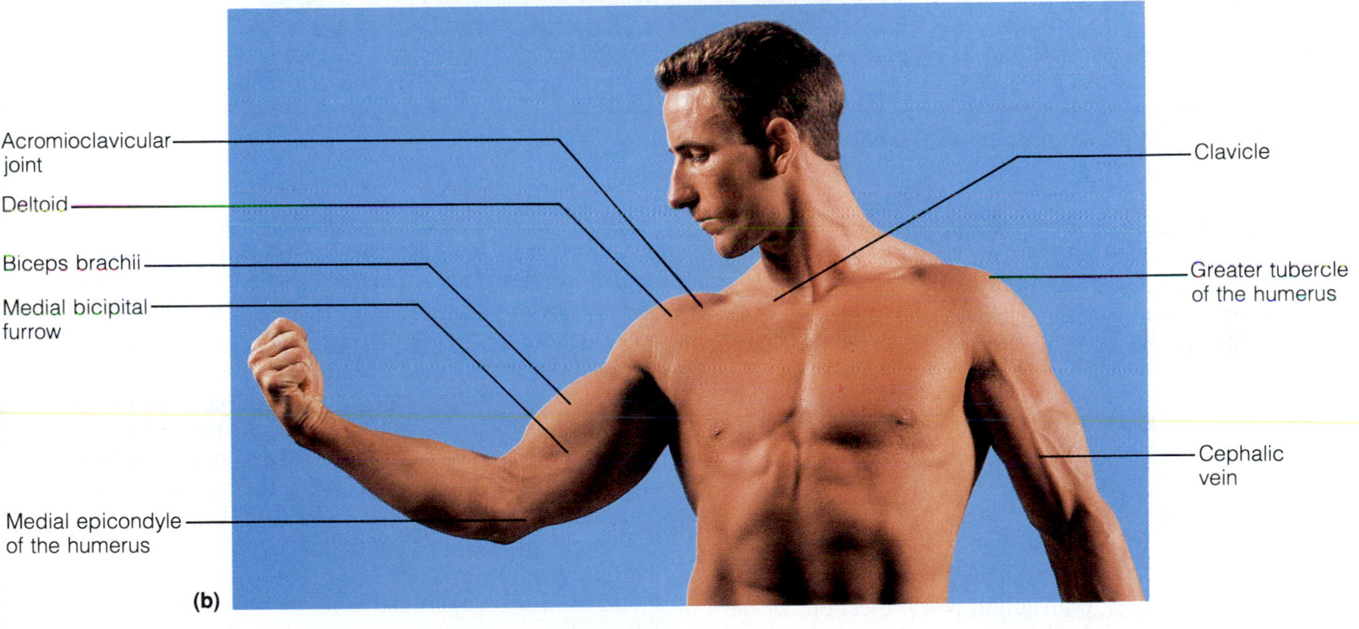

F46.11

Shoulder and arm. (a) Lateral view. **(b)** Anterior and medial view.

the superior surface of the shoulder. It is covered by the thick *deltoid muscle,* which forms the rounded superior part of the shoulder. Intramuscular injections are often given into the deltoid, about 5 cm (2 inches) inferior to the greater tubercle (refer to Figure 46.19a, p. 434).

Arm

Remember, according to anatomists, the arm runs only from the shoulder to the elbow, and not beyond.

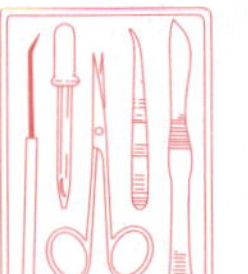

1. In the arm, palpate the humerus along its entire length, especially along its medial and lateral sides.

2. Feel the *biceps brachii* muscle contract on your anterior arm when you flex your forearm against resistance. The medial boundary of the biceps is represented by the **medial bicipital furrow** (Figure 46.11b). This groove contains the large *brachial artery,* and by pressing on it with your finger-

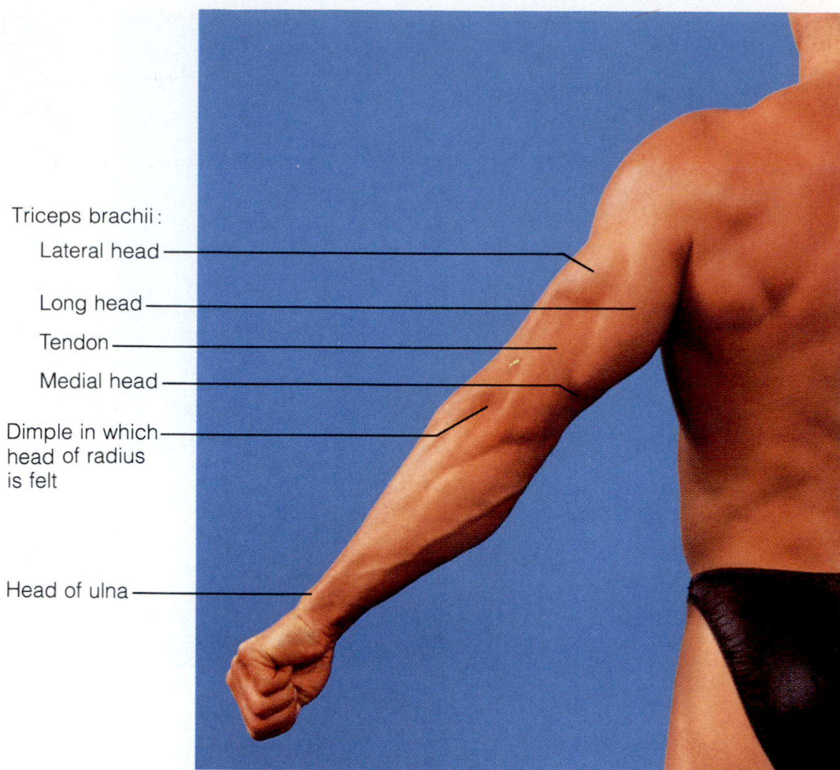

F46.12

Surface anatomy of the upper limb, posterior view.

tips you can feel your *brachial pulse.* Recall that the brachial artery is the artery routinely used in measuring blood pressure with a sphygmomanometer.

3. Extend your forearm against resistance, and feel your *triceps brachii* muscle bulge in the posterior arm. All three heads of the triceps (lateral, long, and medial) are visible through the skin of a muscular person (Figure 46.12).

Elbow Region

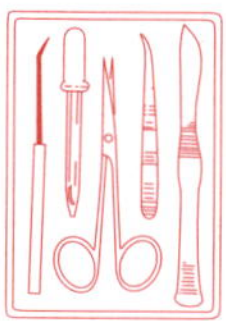

1. In the distal part of your arm, near the elbow, palpate the two projections of the humerus, the *lateral* and *medial epicondyles* (Figure 46.11). Midway between the epicondyles, on the posterior side, feel the *olecranon process* of the ulna, which forms the point of the elbow.

2. Confirm that the two epicondyles and the olecranon all lie in the same horizontal line when the elbow is extended. If these three bony processes do not line up, the elbow is dislocated.

3. Now feel along the posterior surface of the medial epicondyle. You are palpating your ulnar nerve.

4. On the anterior surface of the elbow is a triangular depression called the **cubital fossa** or **antecubital fossa** (Figure 46.13). The triangle's superior *base* is formed by a horizontal line between the humeral epicondyles; its two inferior sides are defined by the *brachioradialis* and *pronator teres* muscles (Figure 46.13). Try to define these boundaries on your own limb. To find the brachioradialis muscle, flex your forearm against resistance, and watch this muscle bulge through the skin of your lateral forearm. To feel your pronator teres contract, palpate the cubital fossa as you pronate your forearm against resistance. (Have your partner provide the resistance.)

Superficially, the cubital fossa contains the *median cubital* vein (Figure 46.13a). Clinicians often draw blood from this superficial vein and insert intravenous (IV) catheters into it to administer medications, transfused blood, and nutrient fluids. The large *brachial artery* lies just deep to the median cubital vein (Figure 46.13b), so a needle must be inserted into the vein from a shallow angle (almost parallel to the skin) to avoid puncturing the artery. Other structures that lie deep in the fossa are also shown in Figure 46.13b.

5. The median cubital vein interconnects the larger cephalic and basilic veins of the upper limb. Recall from Exercise 32 that the *cephalic vein* ascends along the lateral side of the forearm and arm, whereas the *basilic vein* ascends through the limb's medial side. These veins are visible through the skin of lean people (Figure 46.13a). Examine your arm to see if your cephalic and basilic veins are visible.

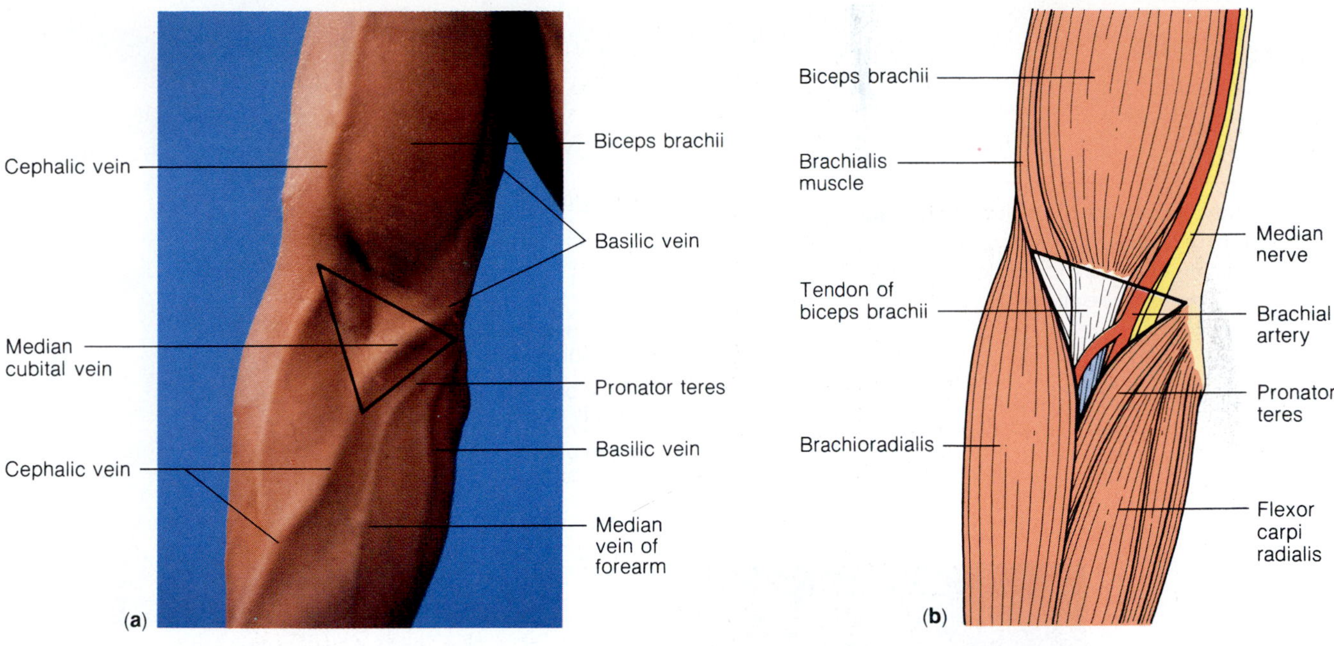

F46.13

The cubital (antecubital) fossa on the anterior surface of the right elbow (outlined by the triangle). (a) Photograph. **(b)** Diagram of deeper structures in the fossa.

Forearm and Hand

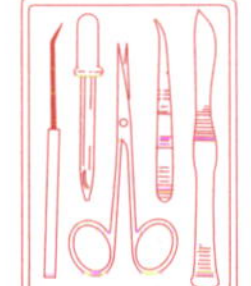

The two parallel bones of the forearm are the medial *ulna* and the lateral *radius.*

1. Feel the ulna along its entire length as a sharp ridge on the posterior forearm (confirm that this ridge runs inferiorly from the olecranon process). As for the radius, you can feel its distal half, but most of its proximal half is covered by muscle. You can, however, feel the rotating *head* of the radius. To do this, extend your forearm, and note that a dimple forms on the posterior lateral surface of the elbow region (Figure 46.12). Press three fingers into this dimple, and rotate your free hand as if you were turning a doorknob. You will feel the head of the radius rotate as you perform this action.

2. Both the radius and ulna have a knoblike *styloid process* at their distal end. Figure 46.14 shows a way to locate these processes. Do not confuse the ulna's styloid process with the conspicuous *head of the ulna,* from which the styloid process stems. Confirm that the styloid process of the radius lies about 1 cm (0.4 inch) distal to that of the ulna.

Colles' fracture of the wrist is an impacted fracture in which the distal end of the radius is pushed proximally into the shaft of the radius. This sometimes occurs when someone falls on outstretched hands, and it most often happens to elderly women with osteoporosis. Colles' fracture bends the wrist into curves that resemble those on a fork. ■

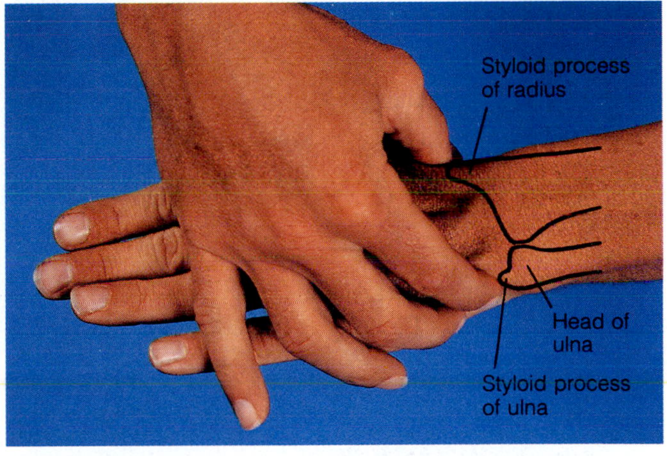

F46.14

A way to locate the styloid processes of the ulna and radius. The right hand is palpating the left hand in this picture. Note that the head of the ulna is not the same as its styloid process. The styloid process of the radius lies about 1 cm distal to the styloid process of the ulna.

Can you deduce how physicians use palpation to diagnose a Colles' fracture?

____

____

3. Next, feel the major groups of muscles within your forearm. Flex your hand and fingers against resistance, and feel the anterior *flexor muscles* contract. Then ex-

F46.15

The anterior surface of the forearm and fist. (a) The entire forearm. **(b)** Enlarged view of the distal forearm and hand. The tendons of the flexor muscles guide the clinician to several sites for pulse taking.

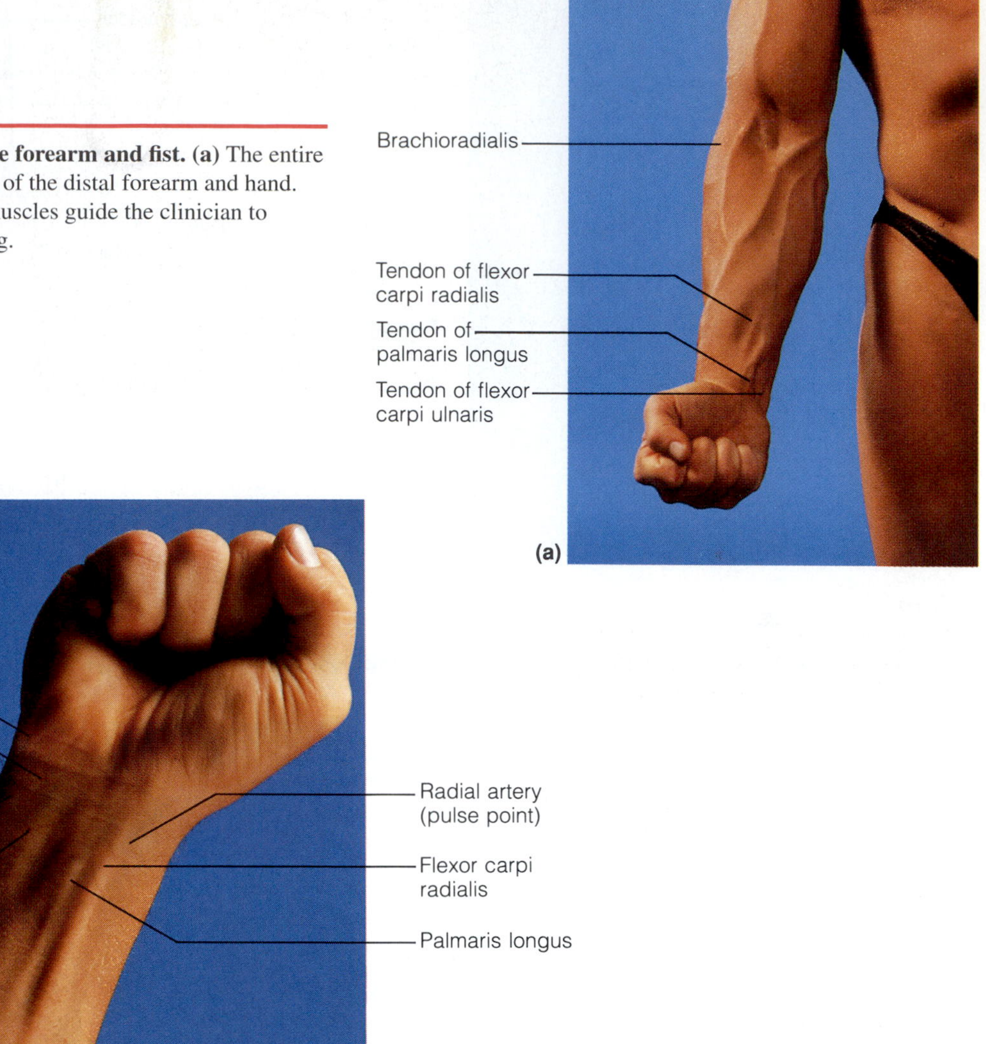

tend your hand at the wrist, and feel the tightening of the posterior *extensor muscles.*

4. Near the wrist, the anterior surface of the forearm reveals many significant features (Figure 46.15). Flex your fist against resistance; the tendons of the main wrist flexors will bulge the skin of the distal forearm. The tendons of the *flexor carpi radialis* and *palmaris longus* muscles are most obvious. (The palmaris longus, however, is absent from at least one arm in 30% of all people, so your forearm may exhibit just one prominent tendon instead of two.) The *radial artery* lies just lateral to (on the thumb side of) the flexor carpi radialis tendon, where the pulse is easily detected (Figure 46.15b). Feel your radial pulse here. The *median nerve* (which innervates the thumb) lies deep to the palmaris longus tendon. Finally, the *ulnar artery* lies on the medial side of the forearm, just lateral to the tendon of the *flexor carpi ulnaris.* Using Figure 46.15b as a guide, locate and feel your ulnar arterial pulse.

5. Extend your thumb and point it posteriorly to form a triangular depression in the base of the thumb on the back of your hand. This is the **anatomical snuff box** (Figure 46.16). Its two elevated borders are defined by the tendons of the thumb extensor muscles, *extensor pollicis brevis* and *extensor pollicis longus.* The radial artery runs within the snuff box, so this is another site for taking a radial pulse. The main bone on the floor of the snuff box is the scaphoid bone of the wrist, but the styloid process of the radius is also present here. (If displaced by a bone fracture, the radial styloid process will be felt outside of the snuff box rather than within it.) The "snuff box" took its name from the fact that people once put snuff (tobacco for sniffing) in this hollow before lifting it up to the nose.

6. On the dorsum of your hand, observe the superficial veins just deep to the skin. This is the *dorsal venous network,* which drains superiorly into the cephalic vein. This venous network provides a site for drawing blood

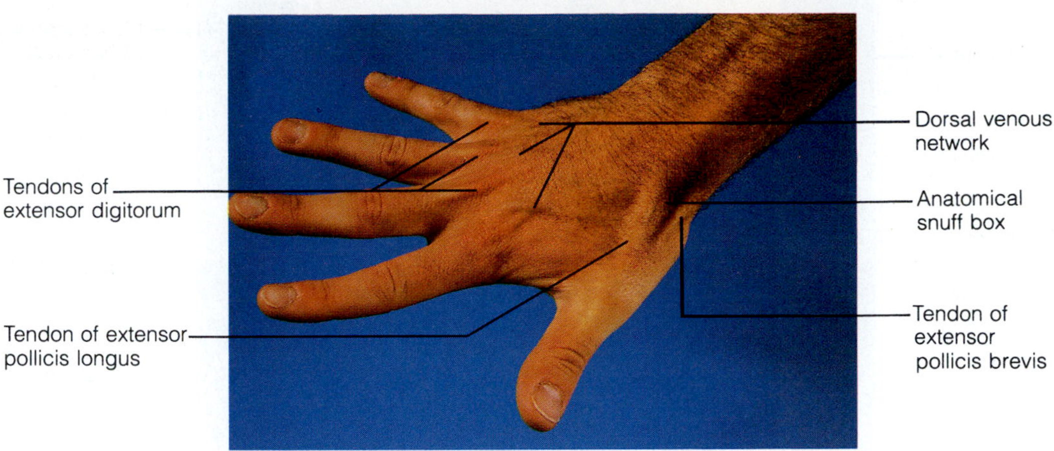

F46.16

The dorsum of the hand. Note especially the anatomical snuff box and dorsal venous network.

and inserting intravenous catheters and is preferred over the median cubital vein for these purposes. Next, extend your hand and fingers, and observe the tendons of the *extensor digitorum* muscle.

7. The anterior surface of the hand also contains some features of interest (Figure 46.17). These features include the *epidermal ridges* ("fingerprints") and many *flexion creases* in the skin. Grasp your *thenar eminence* (the bulge on the palm that contains the thumb muscles) and your *hypothenar eminence* (the bulge on the medial palm that contains muscles that move the little finger).

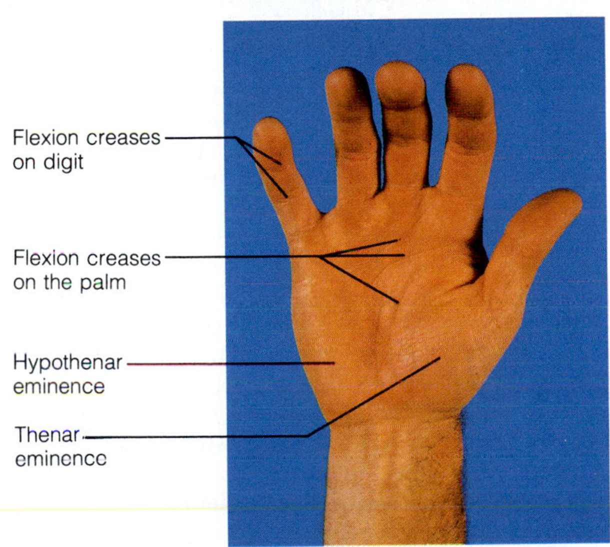

F46.17

The palmar surface of the hand.

LOWER LIMB

Gluteal Region

Dominating the gluteal region are the two *prominences* ("cheeks") of the buttocks. These are formed by subcutaneous fat and by the thick *gluteus maximus* muscles. The midline groove between the two prominences is called the **natal cleft** (*natal*=rump) or *gluteal cleft*. The inferior margin of each prominence is the horizontal **gluteal fold,** which roughly corresponds to the inferior margin of the gluteus maximus.

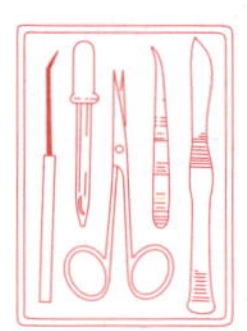

1. Try to palpate your *ischial tuberosity* just above the medial side of each gluteal fold (it will be easier to feel if you sit down or flex your thigh first). The ischial tuberosities are the robust inferior parts of the ischial bones, and they support the body's weight during sitting.

2. Next, palpate the *greater trochanter* of the femur on the lateral side of your hip (Figures 46.18 and 46.20). This trochanter lies just anterior to a hollow and about 10 cm (one hand's breadth) inferior to the iliac crest. To confirm that you have found the greater trochanter, alternately flex and extend your thigh. Because this trochanter is the most superior point on the lateral femur, it moves with the femur as you perform this movement.

3. To palpate the sharp *posterior superior iliac spine* (Figure 46.18), locate your iliac crests again, and trace each to its most posterior point. You may have difficulty feeling this spine, but it is indicated by a distinct dimple in the skin that is easy to find. This dimple lies two to three fingers' breadths lateral to the midline of the back. The dimple also indicates the position of the *sacroiliac joint,* where the hip bone attaches to the sacrum of the spinal column. (You can check *your* "dimples" out in the privacy of your home.)

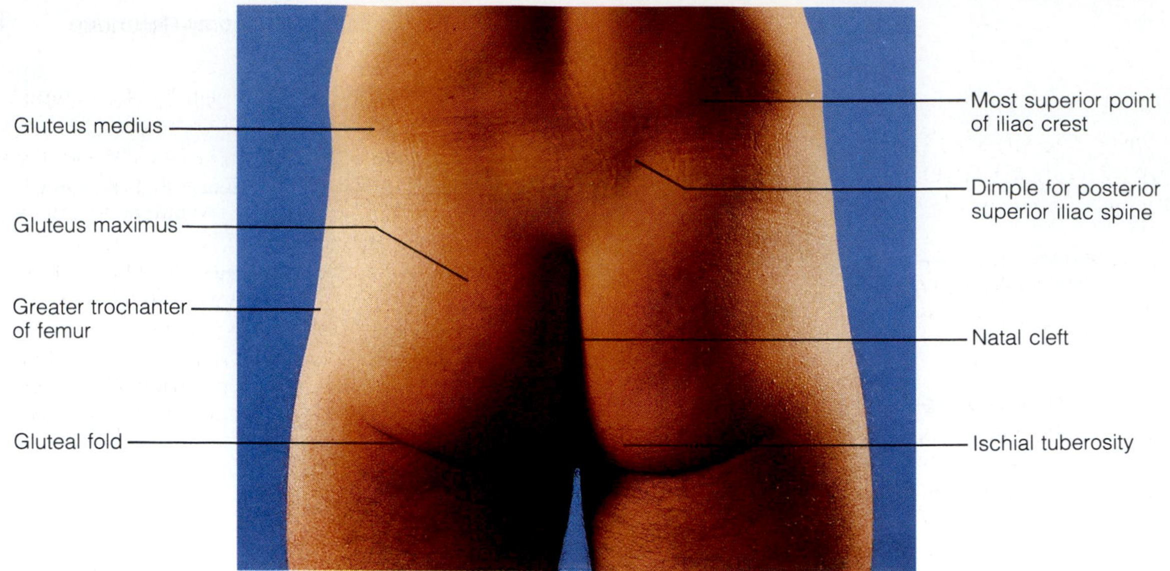

F46.18

The gluteal region. The region extends from the iliac crests superiorly to the gluteal folds inferiorly. Therefore, it includes more than just the prominences of the buttock.

The gluteal region is a major site for administering intramuscular injections. When giving such injections, extreme care must be taken to avoid piercing a major nerve that lies just deep to the gluteus maximus muscle. Can you guess what nerve this is?

It is the thick *sciatic nerve,* which innervates much of the lower limb. Furthermore, the needle must avoid the gluteal nerves and gluteal blood vessels, which also lie deep to the gluteus maximus.

To avoid harming these structures, the injections are most often applied to the gluteus *medius* (not maximus) muscle superior to the cheeks of the buttocks, in a safe area called the **ventral gluteal site** (Figure 46.19b). To locate this site, mentally draw a line laterally from the posterior superior iliac spine (dimple) to the greater trochanter; the injection would be given 5 cm (2 inches) superior to the midpoint of that line. Another safe way to locate the ventral gluteal site is to approach the lateral side of the patient's left hip with your extended right hand (or the right hip with your left hand), then place your thumb on the anterior superior iliac spine and your index finger as far posteriorly on the iliac crest as it can reach. The heel of your hand comes to lie on the greater trochanter, and the needle is inserted in the angle of the V formed between your thumb and index finger about 4 cm (1.5 inches) inferior to the iliac crest.

Gluteal injections are not given to small children because their "safe area" is too small to locate with certainty and because the gluteal muscles are thin at this age. Instead, infants and toddlers receive intramuscular shots in the prominent vastus lateralis muscle of the thigh.

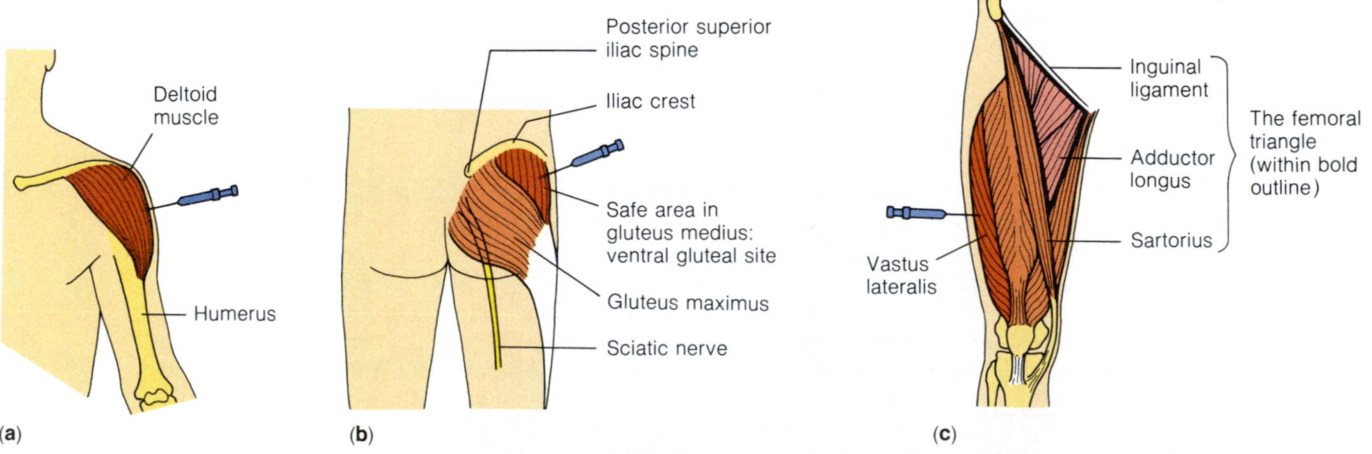

F46.19

Three major sites of intramuscular injections. **(a)** Deltoid muscle of the arm. **(b)** Ventral gluteal site (gluteus medius). **(c)** Vastus lateralis in the lateral thigh. The femoral triangle is also shown.

Thigh

The thigh is pictured in Figures 46.20, 46.21, and 46.22. Much of the femur is clothed by thick muscles, so the thigh has few palpable bony landmarks.

1. Distally, feel the *medial* and *lateral condyles of the femur* and the *patella* anterior to the condyles (see Figure 46.21c and a).

2. Next, palpate your three groups of thigh muscles (Figures 46.20, 46.21a, and 46.21b)—the *quadriceps femoris muscles* anteriorly, the *adductor muscles* medially, and the *hamstrings* posteriorly. The *vastus lateralis,* the lateral muscle of the quadriceps group, is a site for intramuscular injections. Such injections are administered about halfway down the length of this muscle (see Figure 46.19c).

3. The anterosuperior surface of the thigh exhibits a three-sided depression called the **femoral triangle** (Figure 46.21a). As shown in Figure 46.19c, the superior border of this triangle is formed by the inguinal ligament, and its two inferior borders are defined by the *sartorius* and *adductor longus* muscles. The large *femoral artery* and *vein* descend vertically through the center of the femoral triangle. To feel the pulse of your femoral artery, press inward just inferior to your midinguinal point (halfway between the anterior superior iliac spine and the pubic tubercle). Be sure to push hard, because the artery lies somewhat deep. By pressing very hard on this point, one can stop the bleeding from a hemorrhage in the lower limb. The femoral triangle also contains most of the *inguinal lymph nodes* (which are easily palpated if swollen).

Leg and Foot

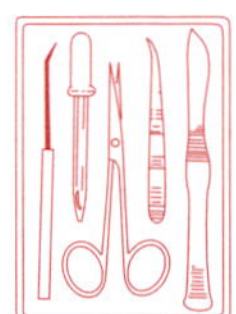

1. Locate your patella again, then follow the thick *patellar ligament* inferiorly from the patella to its insertion on the superior tibia (Figure 46.21c). Here you can feel a rough projection, the *tibial tuberosity.* Continue running your fingers inferiorly along the tibia's sharp *anterior crest* and its flat *medial*

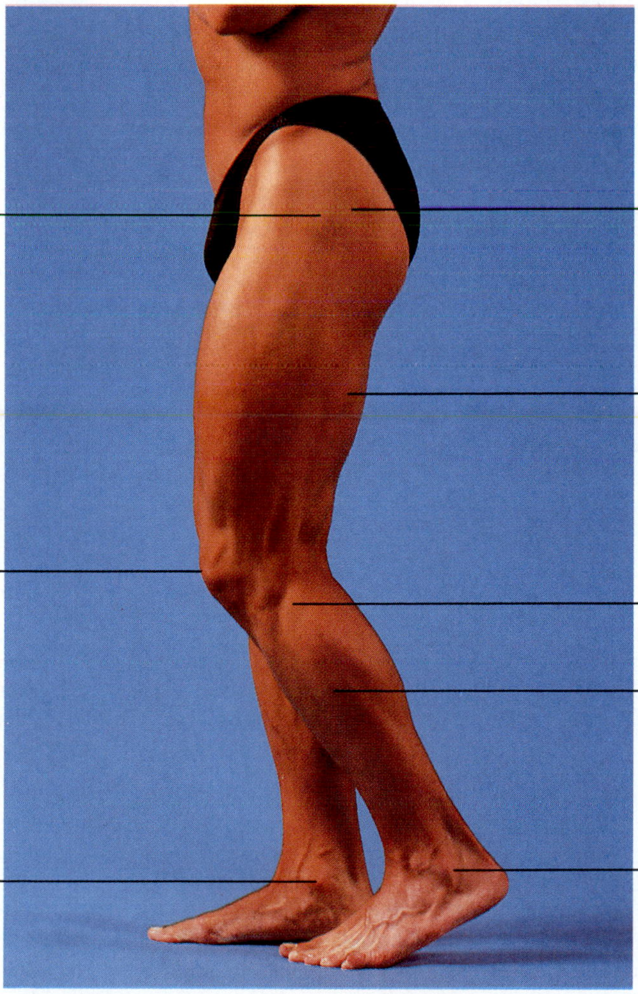

Greater trochanter of femur
Hollow posterior to the greater trochanter
Hamstring muscles
Patella
Head of fibula
Peroneal muscles
Medial malleolus
Lateral malleolus

F46.20

Lateral surface of the lower limb.

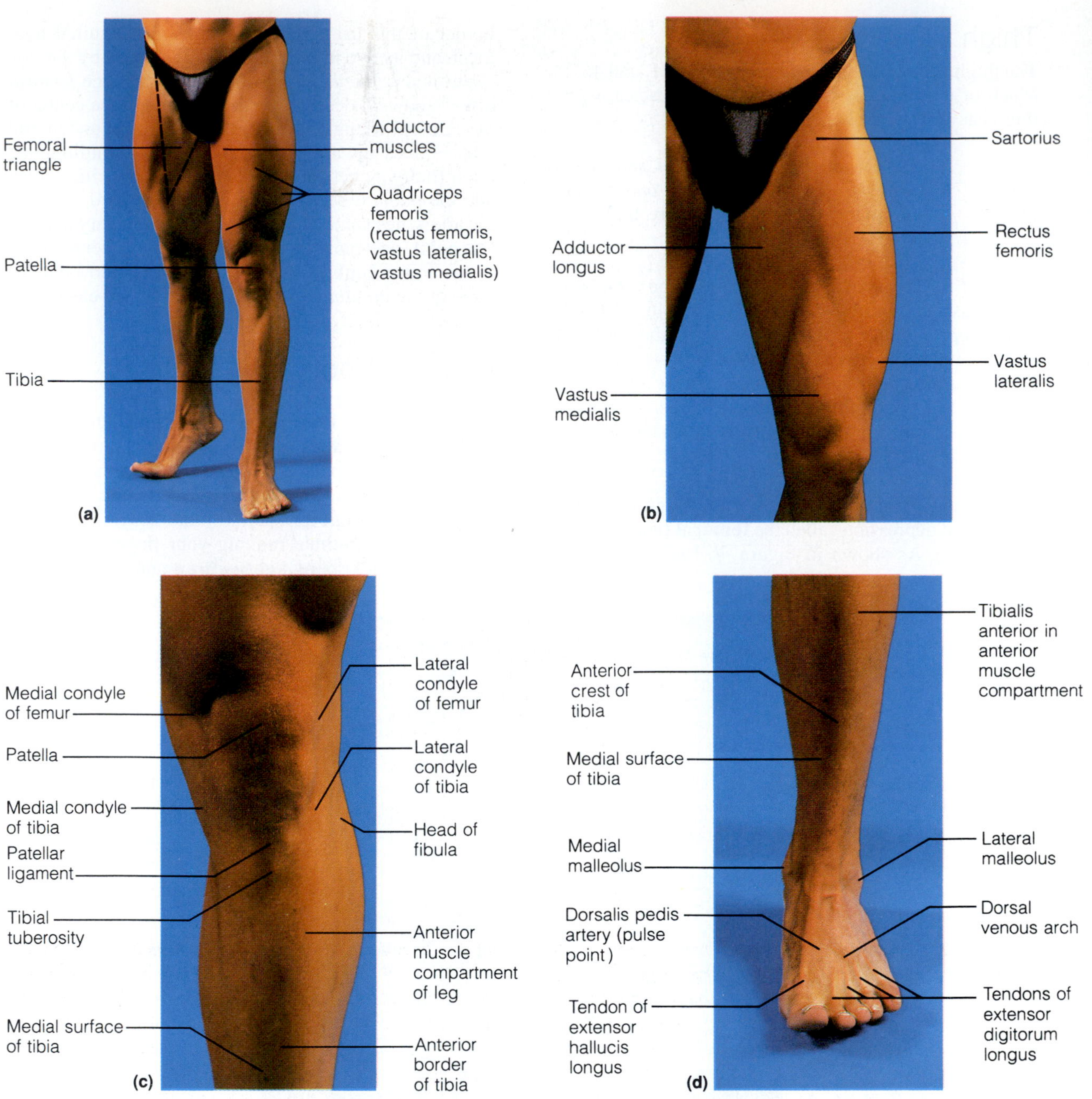

F46.21

Anterior surface of the lower limb. (a) Both limbs, with the right limb revealing its medial aspect. The femoral triangle is outlined on the right limb. **(b)** Enlarged view of the left thigh. **(c)** The left knee region. **(d)** The dorsum of the left foot.

surface—bony landmarks that lie very near the surface throughout their length.

2. Now, return to the superior part of your leg, and palpate the expanded *lateral* and *medial condyles of the tibia* just inferior to the knee. (You can distinguish the tibial condyles from the femoral condyles because you can feel the tibial condyles move with the tibia during knee flexion.) Feel the bulbous *head of the fibula* in the superolateral region of the leg (Figures 46.20 and 46.21c). Try to feel the *common peroneal nerve* (nerve to the anterior leg and foot) where it wraps around the fibula's *neck* just inferior to its head. This nerve is often bumped against the bone here and damaged.

3. In the most distal part of the leg, feel the *lateral malleolus* of the fibula as the lateral prominence of the ankle (Figure 46.21d). Notice that this lies slightly inferior to the *medial malleolus* of the tibia, which forms the ankle's medial prominence. Place your finger just posterior to the medial malleolus to feel the pulse of your *posterior tibial artery.*

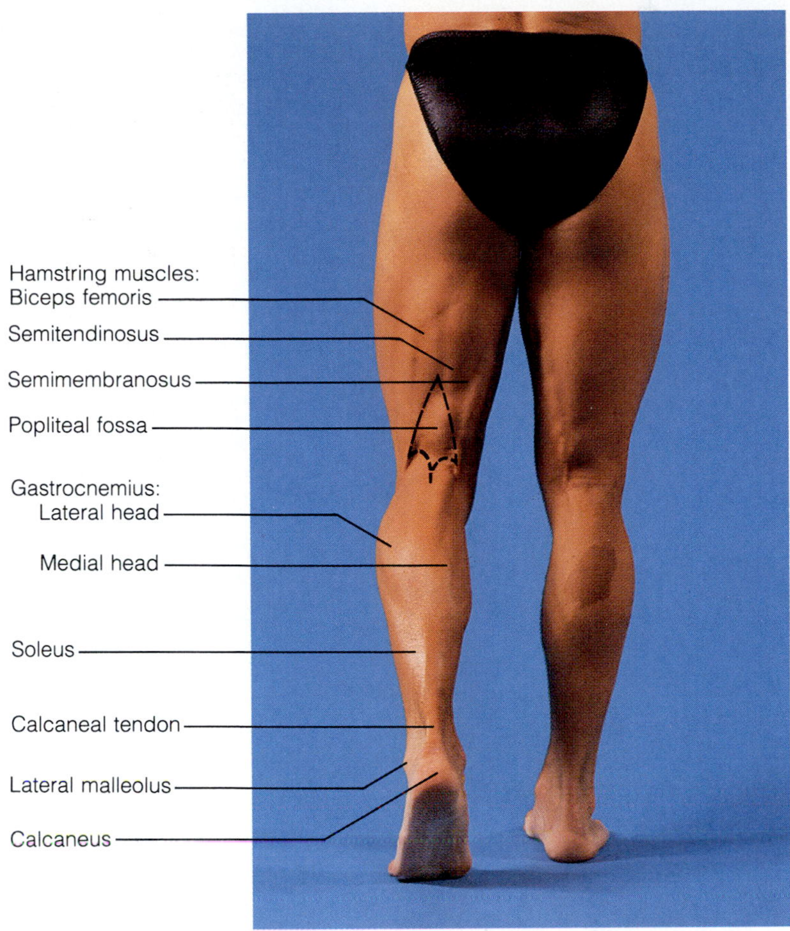

F46.22

Posterior surface of the lower limb. Notice the diamond-shaped popliteal fossa posterior to the knee.

4. On the posterior aspect of the knee is a diamond-shaped hollow called the **popliteal fossa** (Figure 46.22). Palpate the large muscles that define the four borders of this fossa: The *biceps femoris* forming the superolateral border, the *semitendinosus* and *semimembranosus* defining the superomedial border, and the two heads of the *gastrocnemius* forming the inferior border. The *popliteal artery* and *vein* (main vessels to the leg) lie deep within this fossa. To feel a popliteal pulse, flex your leg at the knee and push your fingers firmly into the popliteal fossa. If a physician is unable to feel a patient's popliteal pulse, the femoral artery may be narrowed by atherosclerosis.

5. Next, palpate the main muscle groups of your leg, starting with the calf muscles posteriorly (Figure 46.22). Standing on tiptoes will help you feel the *lateral* and *medial heads of the gastrocnemius* and, inferior to these, the broad *soleus* muscle. Also feel the tension in your *calcaneal* (*Achilles*) *tendon* and at the point of insertion of this tendon onto the calcaneus bone of the foot.

6. Return to the anterior surface of the leg, and palpate the *anterior muscle compartment* (Figure 46.21c and d) while alternately dorsiflexing then plantar flexing your foot. You will feel the tibialis anterior and extensor digitorum muscles contracting then relaxing. Then, palpate the *peroneal muscles* that cover most of the fibula laterally (Figure 46.20). The tendons of these muscles pass posterior to the lateral malleolus and can be felt at a point posterior and slightly superior to that malleolus.

7. Observe the dorsum (superior surface) of your foot. You may see the superficial *dorsal venous arch* overlying the proximal part of the metatarsal bones (Figure 46.21d). This arch gives rise to both saphenous veins (the main superficial veins of the lower limb). Visible in lean people, the *great saphenous vein* ascends along the medial side of the entire limb (see Figure 32.3, p. 298), and the *small saphenous vein* ascends through the center of the calf.

As you extend your toes, observe the tendons of the *extensor digitorum longus* and *extensor hallucis longus* muscles on the dorsum of the foot. Finally, place a finger on the extreme proximal part of the space between the first and second metatarsal bones. Here you should be able to feel the pulse of the *dorsalis pedis artery.*

STUDENT NAME ____________________

LAB TIME/DATE ____________________

Review Sheet

EXERCISE 1 The Language of Anatomy

Surface Anatomy

1. Match each of the following descriptions with a key equivalent, and record the key letter or term in front of the description.

Key: a. buccal c. deltoid e. patellar
b. calcaneal d. digital f. scapular

________ 1. cheek

________ 2. pertaining to the fingers

________ 3. shoulder blade region

________ 4. anterior aspect of knee

________ 5. heel of foot

________ 6. curve of shoulder

2. Indicate the following body areas on the accompanying diagram by placing the correct key letter at the end of each line.

Key:

a. abdominal
b. antecubital
c. axillary
d. brachial
e. cervical
f. femoral
g. gluteal
h. inguinal
i. lumbar
j. occipital
k. oral
l. popliteal
m. pubic
n. sural
o. thoracic
p. umbilical

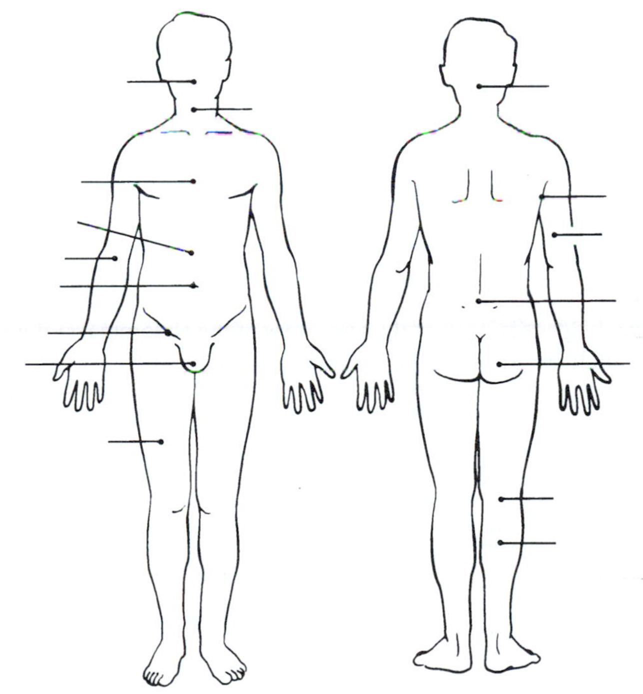

3. Classify each of the surface anatomy terms in the key of question 2 above, into one of the body regions indicated below. Insert the appropriate key letters on the answer blanks.

________ 1. Appendicular

________ 2. Axial

Body Orientation, Direction, Planes, and Sections

1. Describe completely the standard human anatomical position. ______________________________

____

2. Define *section:* ____

3. Several incomplete statements are listed below. Correctly complete each statement by choosing the appropriate anatomical term from the key. Record the key letters and/or terms on the correspondingly numbered blanks below.

Key:			
	a. anterior	e. lateral	i. sagittal
	b. distal	f. medial	j. superior
	c. frontal	g. posterior	k. transverse
	d. inferior	h. proximal	

In the anatomical position, the face and palms are on the __1__ body surface; the buttocks and shoulder blades are on the __2__ body surface; and the top of the head is the most __3__ part of the body. The ears are __4__ and __4__ to the shoulders and __5__ to the nose. The heart is __6__ to the vertebral column (spine) and __7__ to the lungs. The elbow is __8__ to the fingers but __9__ to the shoulder. The abdominopelvic cavity is __10__ to the thoracic cavity and __11__ to the spinal cavity. In humans, the dorsal surface can also be called the __12__ surface; however, in quadruped animals, the dorsal surface is the __13__ surface.

If an incision cuts the heart into right and left parts, the section is a __14__ section; but if the heart is cut so that superior and inferior portions result, the section is a __15__ section. You are told to cut a dissection animal along two planes so that the kidneys are observable in both sections. The two sections that meet this requirement are the __16__ and __17__ sections.

1. ____________	7. ____________	13. ____________
2. ____________	8. ____________	14. ____________
3. ____________	9. ____________	15. ____________
4. ____________	10. ____________	16. ____________
5. ____________	11. ____________	17. ____________
6. ____________	12. ____________	

4. Correctly identify each of the nine areas of the abdominal surface by inserting the appropriate term for each of the letters indicated in the drawing on the next page.

a. ____________________	d. ____________________
b. ____________________	e. ____________________
c. ____________________	f. ____________________

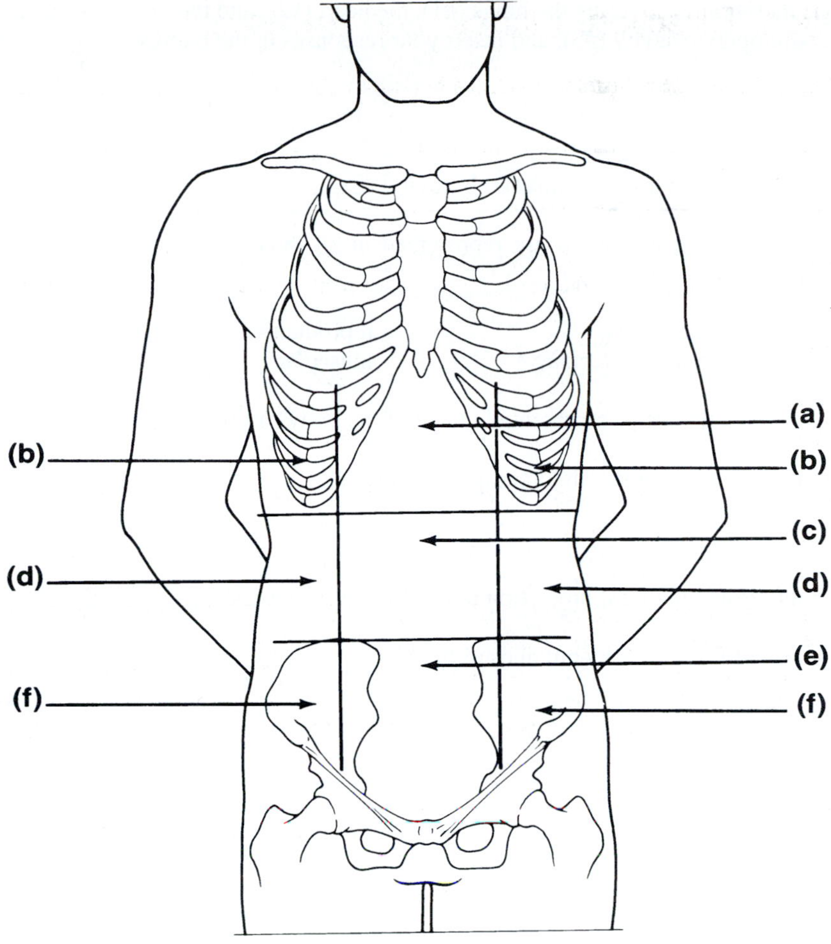

Body Cavities

1. Which body cavity would have to be opened for the following types of surgery? (Insert letter of key choice in same-numbered blank.)

Key: a. abdominopelvic c. dorsal e. thoracic
b. cranial d. spinal f. ventral

1. surgery to remove a cancerous lung lobe
2. removal of the uterus or womb
3. removal of a brain tumor
4. appendectomy
5. stomach ulcer operation

The abdominopelvic and thoracic cavities are subdivisions of the __6__ body cavity, while the cranial and spinal cavities are subdivisions of the __7__ body cavity. The __8__ body cavity is totally surrounded by bone, and thus affords its contained structures very good protection.

1. __________
2. __________
3. __________
4. __________
5. __________
6. __________
7. __________
8. __________

2. Name the serous membranes covering the lungs (#1), the heart (#2), and the organs of the abdominopelvic cavity (#3), and insert your responses in the blanks on the right.

1. ____________________

2. ____________________

3. ____________________

3. Name the muscle that subdivides the ventral body cavity. ____________________

4. Which of the following organ systems are represented in all three subdivisions of the ventral body cavity? (Circle all appropriate responses.)

respiratory	circulatory	reproductive	lymphatic
nervous	excretory (urinary)	muscular	integumentary

5. Which organ system would not be represented in any of the body cavities? ____________________

6. What are the bony landmarks of the abdominopelvic cavity? ____________________

____

7. Which body cavity affords the least protection to its internal structures? ____________________

8. What is the function of the serous membranes of the body? ____________________

____

____

9. A nurse informs you that she is about to take blood from the antecubital region. What portion of your body should you present to her? ____________________

10. What do peritonitis, pleurisy, and pericarditis (pathologic conditions) have in common? ____________________

____

11. Why are these conditions accompanied by a great deal of pain? ____________________

____

12. The mouth, or buccal cavity, and its extension, which stretches through the body inside the digestive system, is not listed as an internal body cavity. Why is this so? ____________________

____

____

STUDENT NAME ______________________

LAB TIME/DATE ______________________

Review Sheet

EXERCISE 2

Organ Systems Overview

1. Use the key below to indicate the body systems that perform the following functions for the body.

Key:
a. cardiovascular
b. digestive
c. endocrine
d. integumentary
e. lymphatic
f. muscular
g. nervous
h. reproductive
i. respiratory
j. skeletal
k. urinary

__________ 1. rids the body of nitrogen-containing wastes

__________ 2. is affected by removal of the thyroid gland

__________ 3. provides support and levers on which the muscular system acts

__________ 4. includes the heart

__________ 5. causes the onset of the menstrual cycle

__________ 6. protects underlying organs from drying out and from mechanical damage

__________ 7. protects the body; destroys bacteria and tumor cells

__________ 8. breaks down ingested food into its building blocks

__________ 9. removes carbon dioxide from the blood

__________ 10. delivers oxygen and nutrients to the tissues

__________ 11. moves the limbs; facilitates facial expression

__________ 12. conserves body water or eliminates excesses

__________ and __________ 13. facilitate conception and childbearing

__________ 14. controls the body by means of chemical molecules called hormones

__________ 15. is damaged when you cut your finger or get a severe sunburn

2. Using the above key, choose the *organ system* to which each of the following sets of organs or body structures belong:

__________ 1. thymus, spleen, lymphatic vessels

__________ 2. bones, cartilages, tendons

__________ 3. pancreas, pituitary, adrenals

__________ 4. trachea, bronchi, alveoli

__________ 5. kidneys, bladder, ureters

__________ 6. testis, vas deferens, urethra

__________ 7. esophagus, large intestine, rectum

__________ 8. arteries, veins, heart

3. Using the key below, place the following organs in their proper body cavity.

Key: a. abdominopelvic b. cranial c. spinal d. thoracic

________	1. stomach	________	6. urinary bladder
________	2. esophagus	________	7. heart
________	3. large intestine	________	8. trachea
________	4. liver	________	9. brain
________	5. spinal cord	________	10. rectum

4. Using the organs listed in item 3 above, record, by number, which would be found in the abdominal regions listed below:

________	1. hypogastric region	________	4. epigastric region
________	2. right lumbar region	________	5. left iliac region
________	3. umbilical region	________	6. left hypochondriac region

5. The five levels of organization of a living body are cell, ________________,

________________, ________________, and organism.

6. Define *organ.* ____

____

____

7. During the course of this laboratory exercise, a rat was dissected. What is the *value* of observing the anatomy of a rat (or any other small mammal) when *human anatomy* is the actual topic of study?

____

____

____

STUDENT NAME ______________________________

LAB TIME/DATE ______________________________

Review Sheet

EXERCISE 3

The Microscope

Care and Structure of the Compound Microscope

1. Label all indicated parts of the microscope.

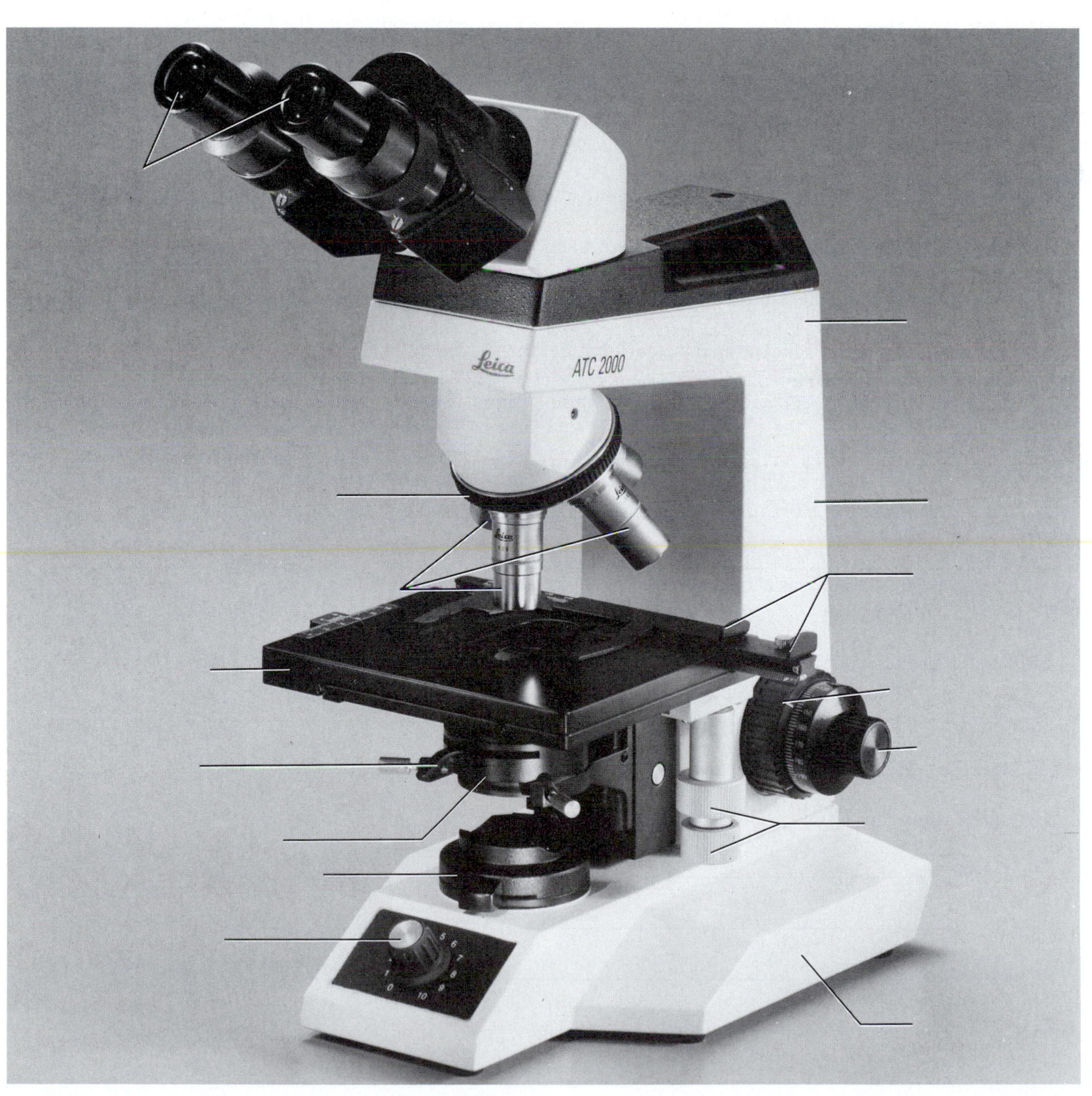

2. The following statements are true or false. If true, write *T* on the answer blank. If false, correct the statement by writing on the blank the proper word or phrase to replace the one that is underlined.

________________ 1. The microscope lens may be cleaned with any soft tissue.

________________ 2. The coarse adjustment knob may be used in focusing with all three objectives.

________________ 3. The microscope should be stored with the oil immersion lens in position over the stage.

________________ 4. When beginning to focus, the low-power lens should be used.

________________ 5. In low power, always focus toward the specimen.

________________ 6. A coverslip should always be used with the high-power and oil lenses.

________________ 7. The greater the amount of light delivered to the objective lens, the less the resolution.

3. Match the microscope structures given in column B with the statements in column A that identify or describe them:

Column A	Column B
______ 1. platform on which the slide rests for viewing	a. coarse adjustment knob
______ 2. lens located at the superior end of the body tube	b. condenser
______ 3. secure(s) the slide to the stage	c. fine adjustment knob
______ 4. delivers a concentrated beam of light to the specimen	d. iris diaphragm
______ 5. used for precise focusing once initial focusing has been done	e. mechanical stage or spring clips
______ 6. carries the objective lenses; rotates so that the different objective lenses can be brought into position over the specimen	f. movable nosepiece
______ 7. used to increase the amount of light passing through the specimen	g. objective lenses
	h. ocular
	i. stage

4. Explain the proper technique for transporting the microscope.

____

____

5. Define the following terms.

real image: ____

____

virtual image: ____

____

6. Define *total magnification:* ____

____

7. Define *resolution:* ____________________

Viewing Objects Through the Microscope

1. Complete, or respond to, the following statements:

1. ____________________ The distance from the bottom of the objective lens in use to the specimen is called the ______.

2. ____________________ The resolution of the human eye is approximately ______ μm.

3. ____________________ The area of the specimen seen when looking through the microscope is the ______.

4. ____________________ If a microscope has a 10× ocular and the total magnification at a particular time is 950×, the objective lens in use at that time is ______ ×.

5. ____________________ Why should the light be dimmed when looking at living (nearly transparent) cells?

6. ____________________ If, after focusing in low power, only the fine adjustment need be used to focus the specimen at the higher powers, the microscope is said to be ______.

7. ____________________ If, when using a 10× ocular and a 15× objective, the field size is 1.5 mm, the approximate field size with a 30× objective is ______ mm.

8. ____________________ If the size of the high-power field is 1.2 mm, an object that occupies approximately a third of that field has an estimated diameter of ______ mm.

9. ____________________ Assume there is an object on the left side of the field that you want to bring to the center (that is, toward the apparent right). In what direction would you move your slide?

10. ____________________ If the object is in the top of the field and you want to move it downward to the center, you would move the slide ______.

2. You have been asked to prepare a slide with letter *k* on it (as below). In the circle below, draw the *k* as seen in the low-power field.

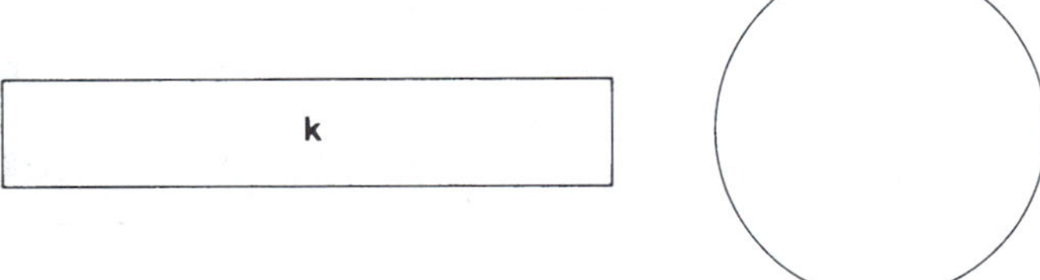

3. The numbers for the field sizes below are too large to represent the typical compound microscope lens system, but the relationships depicted are accurate. Figure out the magnification of fields 1 and 3, and the field size of 2. (*Hint:* Use your ruler.)

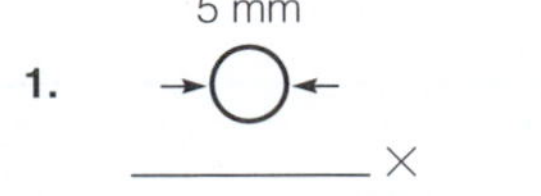

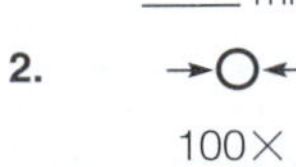

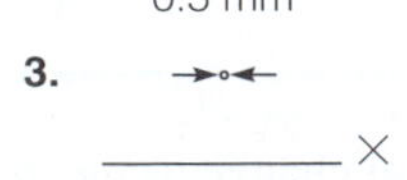

4. Say you are observing an object in the low-power field. When you switch to high-power, it is no longer in your field of view. Why might this occur? ______________________________

____

____

What should be done initially to prevent this from happening? ______________________________

____

5. Do the following factors increase or decrease as one moves to higher magnifications with the microscope?

resolution ______________ amount of light needed ______________

working distance ______________ depth of field ______________

6. A student has the high-power lens in position and appears to be intently observing the specimen. The instructor, noting a working distance of about 1 cm, knows the student isn't actually seeing the specimen.

How so? ______________________________

7. Why is it important to be able to use your microscope to perceive depth when studying slides?

____

____

8. If you are observing a slide of tissue two cell layers thick, how can you determine which layer is superior?

____

____

____

9. Describe the proper procedure for preparing a wet mount.

____

____

____

10. Give two reasons why the light should be dimmed when viewing living or unstained material.

____

____

11. Indicate the probable cause of the following situations arising during use of a microscope.

a. Only half of field illuminated: ______________________________

____

b. Field does not change as mechanical stage is moved: ______________________________

____

STUDENT NAME ______________________

LAB TIME/DATE ______________________

Review Sheet

EXERCISE 4

The Cell—Anatomy and Division

Anatomy of the Composite Cell

1. Define the following:

   organelle: ______________________

   ____

   cell: ______________________

2. Although cells have differences that reflect their specific functions in the body, what functional capabilities do all cells exhibit? ______________________

3. Identify the following cell parts:

______________ 1. external boundary of cell; regulates flow of materials into and out of the cell; site of cell signaling

______________ 2. contains digestive enzymes of many varieties; "suicide sac" of the cell

______________ 3. scattered throughout the cell; major site of ATP synthesis

______________ 4. slender extensions of the plasma membrane that increase its surface area

______________ 5. stored glycogen granules, crystals, pigments, and so on

______________ 6. membranous system consisting of flattened sacs and vesicles; packages proteins for export

______________ 7. control center of the cell; necessary for cell division and cell life

______________ 8. two rod-shaped bodies near the nucleus; direct formation of the mitotic spindle

______________ 9. dense, darkly staining nuclear body; packaging site for ribosomes

______________ 10. contractile elements of the cytoskeleton

______________ 11. membranous system; involved in intracellular transport of proteins and synthesis of membrane lipids

______________ 12. attached to membrane systems or scattered in the cytoplasm; synthesize proteins

______________ 13. threadlike structures in the nucleus; contain genetic material (DNA)

______________ 14. site of free radical detoxification

4. In the following diagram, label all parts provided with a leader line.

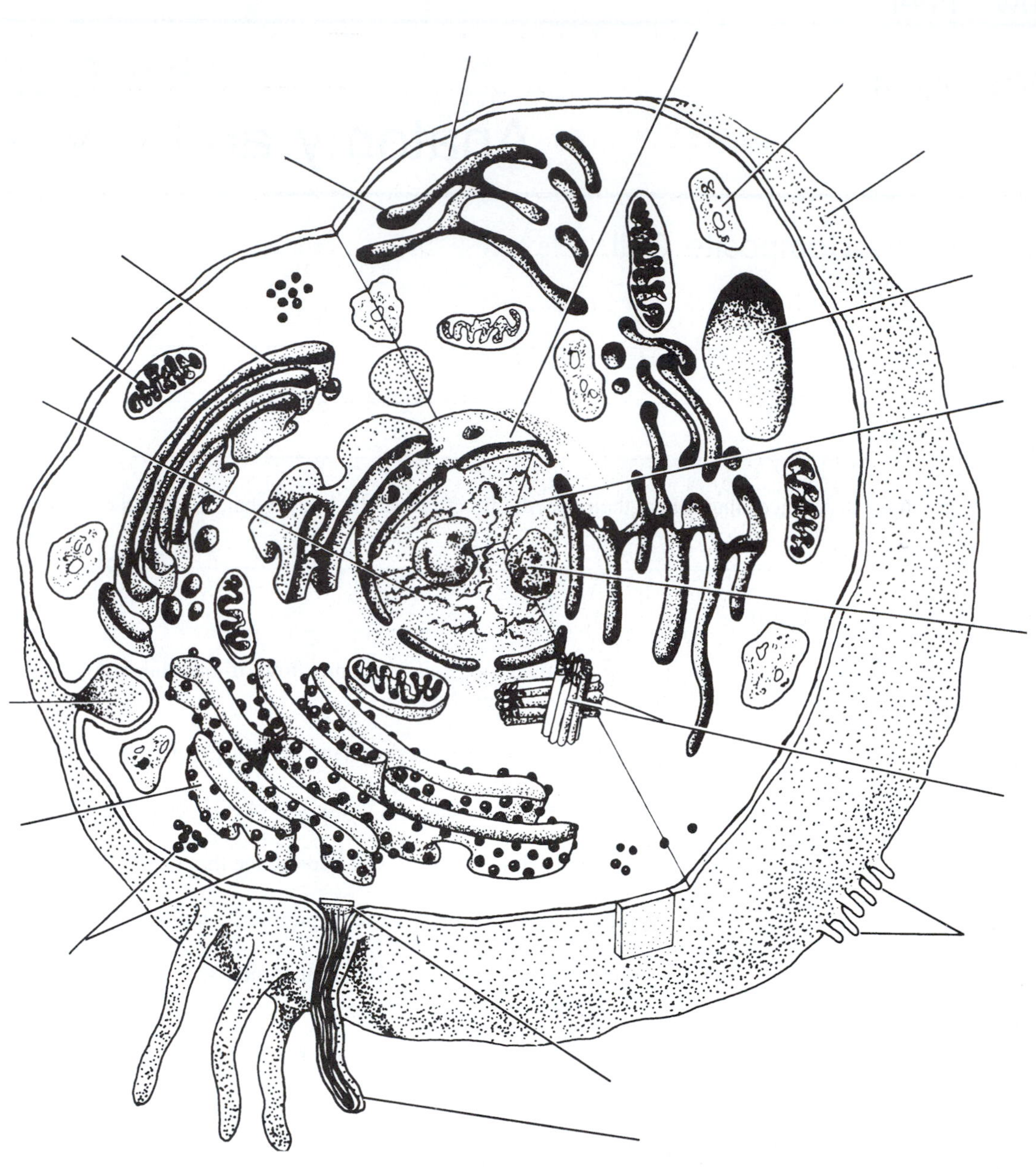

Observing Differences and Similarities in Cell Structure

1. For each of the following cell types, list (a) *one* important structural characteristic observed in the laboratory, and (b) the function that structure complements or ensures.

squamous epithelium a. ______________________________

b. ______________________________

sperm a. ______________________________

b. ______________________________

smooth muscle a. ______________________

b. ______________________

red blood cells a. ______________________

b. ______________________

2. What is the significance of the red blood cell being anucleate (without a nucleus)? ______________________

____

Did it ever have a nucleus? ________ When? ______________________

Cell Division: Mitosis and Cytokinesis

1. Identify the three phases of mitosis in the following photomicrographs.

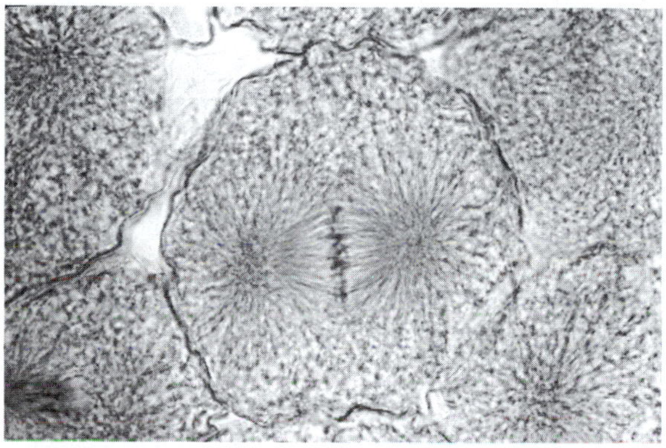

1. ______________________

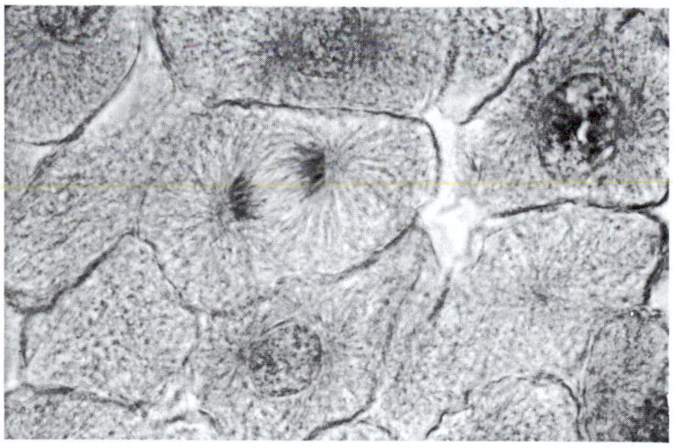

2. ______________________

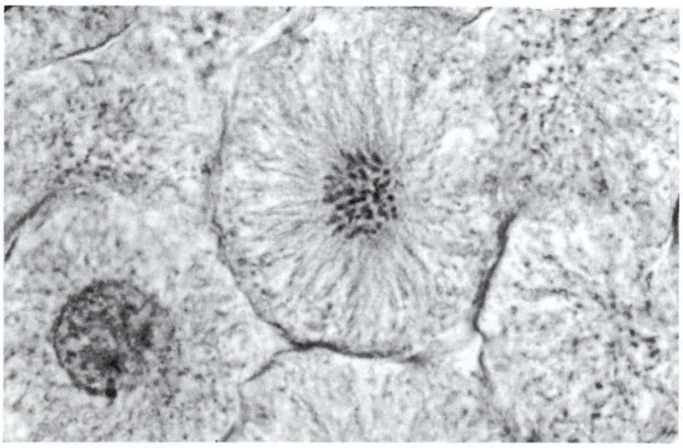

3. ______________________

2. What is the importance of mitotic cell division? ____

____

3. Complete or respond to the following statements:

Division of the __1__ is referred to as mitosis. Cytokinesis is division of the __2__ . The major structural difference between chromatin and chromosomes is that the latter is __3__ . Chromosomes attach to the spindle fibers by undivided structures called __4__ . If a cell undergoes mitosis but not cytokinesis, the product is __5__ . The structure that acts as a scaffolding for chromosomal attachment and movement is called the __6__ . __7__ is the period of cell life when the cell is not involved in division. Two cell populations in the body that do not undergo cell division are __8__ and __9__ . The implication of an inability of a cell population to divide is that when some of its members die, they are replaced by __10__ .

1. ____________________
2. ____________________
3. ____________________
4. ____________________
5. ____________________
6. ____________________
7. ____________________
8. ____________________
9. ____________________
10. ____________________

4. Using the key, categorize each of the events described below according to the phase in which it occurs.

Key: a. prophase b. anaphase c. telophase d. metaphase e. none of these

____________ 1. Chromatin coils and condenses, forming chromosomes.

____________ 2. The chromosomes (chromatids) are V-shaped.

____________ 3. The nuclear membrane re-forms.

____________ 4. Chromosomes stop moving toward the poles.

____________ 5. Chromosomes line up in the center of the cell.

____________ 6. The nuclear membrane fragments.

____________ 7. The mitotic spindle forms.

____________ 8. DNA synthesis occurs.

____________ 9. Centrioles replicate.

____________ 10. Chromosomes first appear to be duplex structures.

____________ 11. Chromosomal centromeres are attached to the kinetochore fibers.

____________ 12. Cleavage furrow forms.

____________ 13. The nuclear membrane(s) is absent.

5. What is the physical advantage of the chromatin coiling and condensing to form short chromosomes at the onset of mitosis? ____

____

STUDENT NAME ____________________

LAB TIME/DATE ____________________

Review Sheet

EXERCISE 5

The Cell—Transport Mechanisms and Cell Permeability

Choose all answers that apply to items 1 and 2, and place their letters on the response blanks to the right.

1. Brownian motion ____________________

   a. reflects the kinetic energy of the smaller solvent molecules
   b. reflects the kinetic energy of the larger solute molecules
   c. is ordered and predictable
   d. is random and erratic

2. Kinetic energy ____________________

   a. is higher in larger molecules
   b. is lower in larger molecules
   c. increases with increasing temperature
   d. decreases with increasing temperature
   e. is reflected in the speed of molecular movement

3. The following refer to the laboratory experiment using dialysis sacs 1 through 4 to study diffusion through nonliving membranes:

   Sac 1: 40% glucose suspended in distilled water

   Did glucose pass out of the sac? ____________________

   Test used to determine presence of glucose: ____________________

   Did the sac weight change? ____________________

   If so, explain the reason for its weight change: ____________________

   ____

   ____

   Sac 2: 40% glucose suspended in 40% glucose

   Was there net movement of glucose in either direction? ____________________

   Explanation: ____________________

   Did the sac weight change? ____________________ Explanation: ____________________

   ____

Sac 3: 10% NaCl in distilled water

Was there net movement of NaCl out of the sac? ______________________________

Test used to determine the presence of NaCl: ______________________________

____

Direction of net osmosis: ______________________________

Sac 4: Boiled starch in distilled water

Was there net movement of starch out of the sac? ______________________________

Test used to determine presence of starch: ______________________________

Direction of net osmosis: ______________________________

4. What single characteristic of the semipermeable membranes used in the laboratory determines the substances that can pass through them? ______________________________

In addition to this characteristic, what other factors influence the passage of substances through living membranes? ______________________________

____

5. A semipermeable sac containing 4% NaCl, 9% glucose, and 10% albumin is suspended in a solution with the following composition: 10% NaCl, 10% glucose, and 40% albumin. Assume that the sac is permeable to all substances except albumin. State whether each of the following will (a) move into the sac, (b) move out of the sac, or (c) not move.

glucose ______________________________ albumin ______________________________

water ______________________________ NaCl ______________________________

6. The diagrams below represent three microscope fields containing red blood cells. Arrows show the direction of net osmosis. Which field contains a hypertonic solution? ________ The cells in this field are said to be ______________________________. Which field contains an isotonic bathing solution? ________ Which field contains a hypotonic solution? ________ What is happening to the cells in this field? ______________________________

____

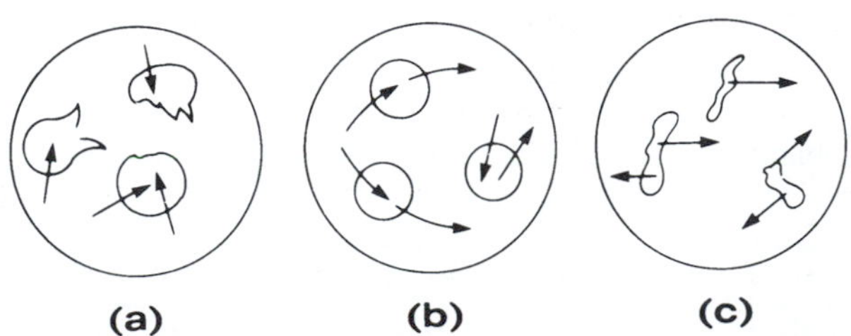

7. Assume you are conducting the experiment illustrated below. Both hydrochloric acid (HCl) with a molecular weight of about 36.5 and ammonium hydroxide (NH_4OH) with a molecular weight of 35 are volatile and easily enter the gaseous state. When they meet, the following reaction will occur:

$$HCl + NH_4OH \rightarrow H_2O + NH_4Cl$$

Ammonium chloride (NH_4Cl) will be deposited on the glass tubing as a smoky precipitate where the two gases meet. Predict which gas will diffuse more quickly and indicate to which end of the tube the smoky precipitate will be closer.

a. The faster diffusing gas is ____________________.

b. The precipitate forms closer to the ____________________ end.

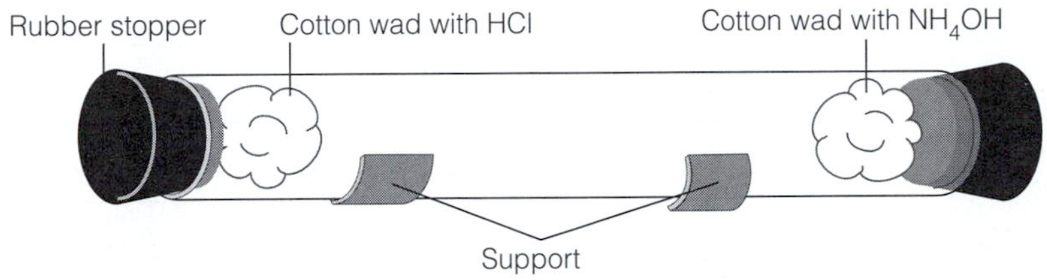

Rubber stopper
Cotton wad with HCl
Cotton wad with NH_4OH
Support

8. When food is pickled for human consumption, as much water as possible is removed from the food. What method is used to achieve this dehydrating effect? ____________________

____

9. What determines whether a transport process is active or passive? ____________________

____

10. Characterize passive and active transport as fully as possible by choosing all the phrases that apply and inserting their letters on the answer blanks.

Passive transport: ______________ Active transport: ______________

a. accounts for the movement of fats and respiratory gases through the plasma membrane
b. explains solute pumping, phagocytosis, and pinocytosis
c. includes osmosis, simple diffusion, and filtration
d. may occur against concentration and/or electrical gradients
e. uses hydrostatic pressure or molecular energy as the driving force
f. moves ions, amino acids, and some sugars across the plasma membrane

11. For the osmometer demonstration (#2), explain why the level of the water column rose during the laboratory session. ____________________

____

____

____

____

12. Define the following terms:

Brownian motion: ______________________________

____

____

diffusion: ______________________________

____

____

osmosis: ______________________________

____

____

simple diffusion: ______________________________

____

____

filtration: ______________________________

____

____

active transport: ______________________________

____

____

phagocytosis: ______________________________

____

____

pinocytosis: ______________________________

____

____

STUDENT NAME ____________________

LAB TIME/DATE ____________________

Review Sheet

EXERCISE 6

Classification of Tissues

Tissue Structure and Function—General Review

1. Define *tissue:* ____________________

____

2. Use the key choices to identify the major tissue types described below.

Key: a. connective tissue b. epithelium c. muscle d. nervous tissue

__________ 1. lines body cavities and covers the body's external surface

__________ 2. pumps blood, flushes urine out of the body, allows one to swing a bat

__________ 3. transmits electrochemical impulses

__________ 4. anchors, packages, and supports body organs

__________ 5. cells may absorb, secrete, and protect

__________ 6. most involved in regulating and controlling body functions

__________ 7. major function is to contract

__________ 8. synthesizes hormones

__________ 9. the most durable tissue type

__________ 10. abundant nonliving extracellular matrix

__________ 11. most widespread tissue in the body

__________ 12. forms nerves and the brain

Epithelial Tissue

1. Describe the general characteristics of epithelial tissue. ____________________

____

____

____

2. On what bases are epithelial tissues classified? ____________________

3. What are the major functions of epithelium in the body? (Give examples.) ____________________

____

____

____

4. How is the function of epithelium reflected in its arrangement? ____________________

____

____

5. Where is ciliated epithelium found? ____________________

____

What role does it play? ____________________

____

____

6. Transitional epithelium is actually stratified squamous epithelium, but there is something special about it.

How does it differ structurally from other stratified squamous epithelia? ____________________

____

How does this reflect its function in the body? ____________________

____

____

7. How do the endocrine and exocrine glands differ in structure and function? ____________________

____

____

____

8. Respond to the following with the key choices.

Key: a. pseudostratified ciliated columnar c. simple cuboidal e. stratified squamous
b. simple columnar d. simple squamous f. transitional

____________________ 1. lining of the esophagus

____________________ 2. lining of the stomach

____________________ 3. lung tissue, alveolar sacs

____________________ 4. tubules of the kidney

____________________ 5. epidermis of the skin

____________________ 6. lining of bladder; peculiar cells that have the ability to slide over each other

____________________ 7. forms the thin serous membranes; a single layer of flattened cells

Connective Tissue

1. What are the general characteristics of connective tissues? ______________________________

____

____

____

____

2. What functions are performed by connective tissue? ______________________________

____

3. How are the functions of connective tissue reflected in its structure? ______________________________

____

____

4. Using the key, choose the best response to identify the connective tissues described below.

Key:
a. adipose connective tissue
b. areolar connective tissue
c. dense connective tissue
d. elastic cartilage
e. fibrocartilage
f. hematopoietic tissue
g. hyaline cartilage
h. osseous tissue

______________ 1. attaches bones to bones and muscles to bones

______________ 2. acts as a storage depot for fat

______________ 3. the dermis of the skin

______________ 4. makes up the intervertebral discs

______________ 5. forms your hip bone

______________ 6. composes basement membranes; a soft packaging tissue with a jellylike matrix

______________ 7. forms the larynx, the costal cartilages of the ribs, and the embryonic skeleton

______________ 8. provides a flexible framework for the external ear

______________ 9. firm, structurally amorphous matrix heavily invaded with fibers; appears glassy and smooth

______________ 10. matrix hard owing to calcium salts; provides levers for muscles to act on

______________ 11. insulates against heat loss

5. Why are adipose cells called "signet ring" cells? ______________________________

____

Muscle Tissue

The three types of muscle tissue exhibit similarities as well as differences. Check the appropriate space in the chart below to indicate which muscle types exhibit each characteristic.

Characteristic	Skeletal	Cardiac	Smooth
Voluntarily controlled			
Involuntarily controlled			
Striated			
Has a single nucleus in each cell			
Has several nuclei per cell			
Found attached to bones			
Allows you to direct your eyeballs			
Found in the walls of the stomach, uterus, and arteries			
Contains spindle-shaped cells			
Contains branching cylindrical cells			
Contains long, nonbranching cylindrical cells			
Has intercalated discs			
Concerned with locomotion of the body as a whole			
Changes the internal volume of an organ as it contracts			
Tissue of the heart			

Nervous Tissue

1. What two physiologic characteristics are highly developed in nervous tissue? ______________________

____

2. In what ways are nerve cells similar to other cells? ______________________

____

How are they different? ______________________

3. Sketch a neuron, recalling in your diagram the most important aspects of its structure. Below the diagram, describe how its particular structure relates to its function in the body.

____

____

For Review

Label the following tissue types here and on the next pages, and identify all major structures.

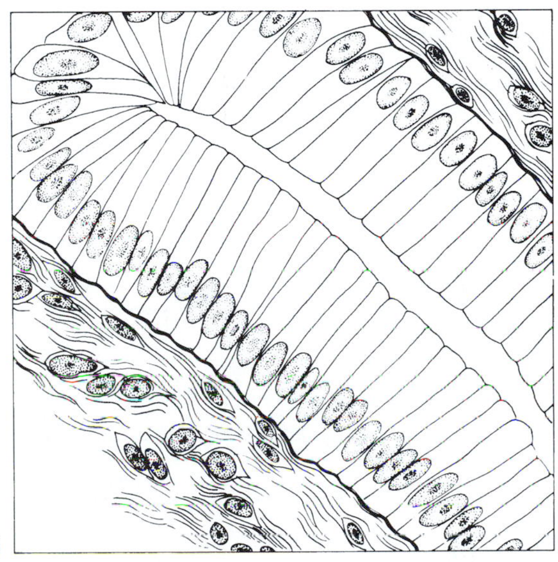

(a) ______________________

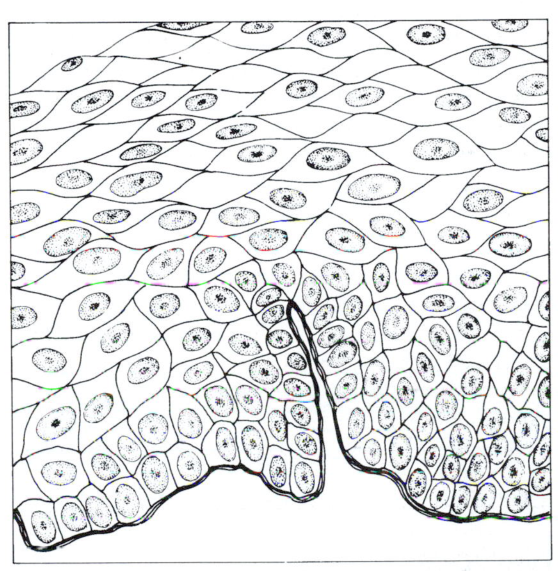

(c) ______________________

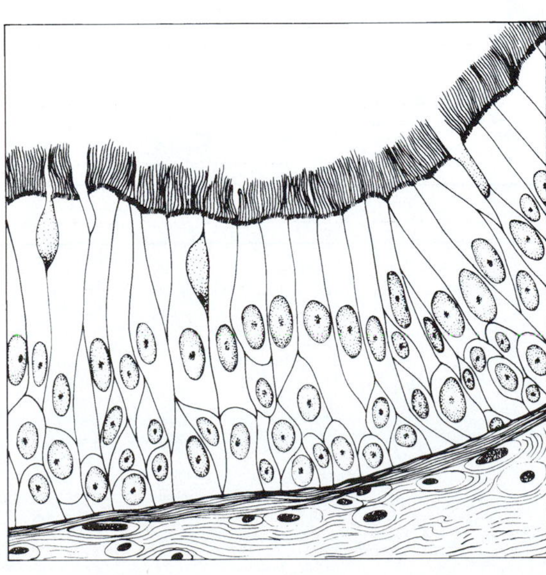

(b) ______________________

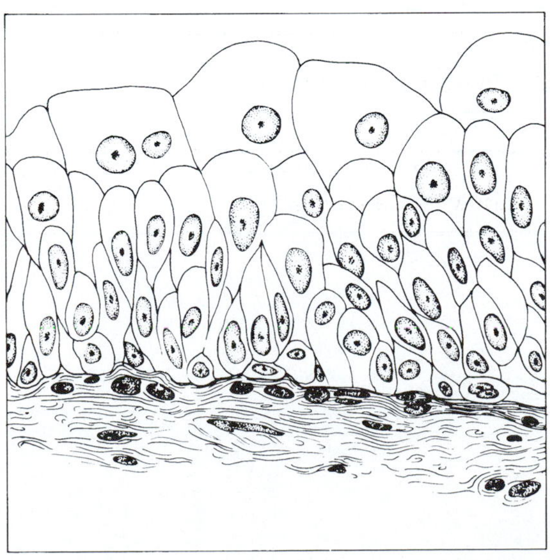

(d) ______________________

(e)

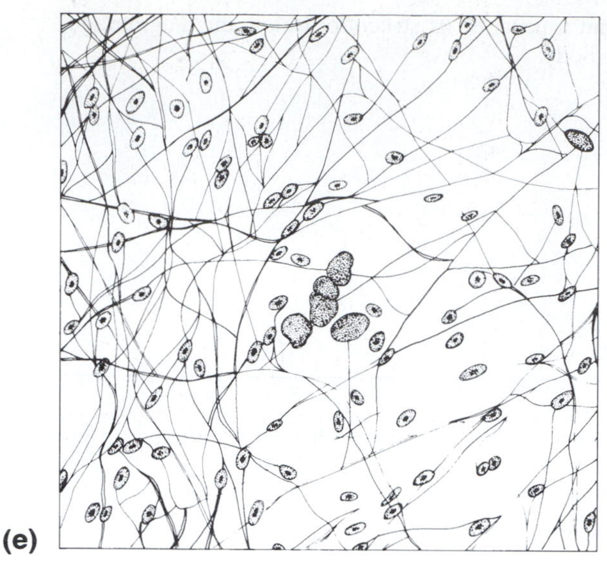

(g)

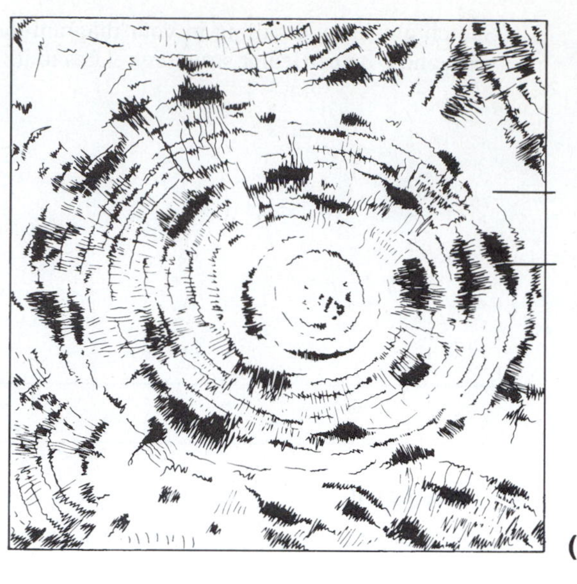

(f)

(h)

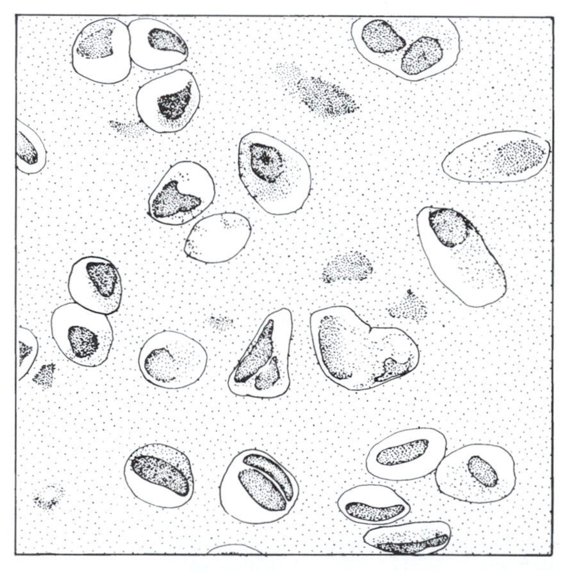

(i)

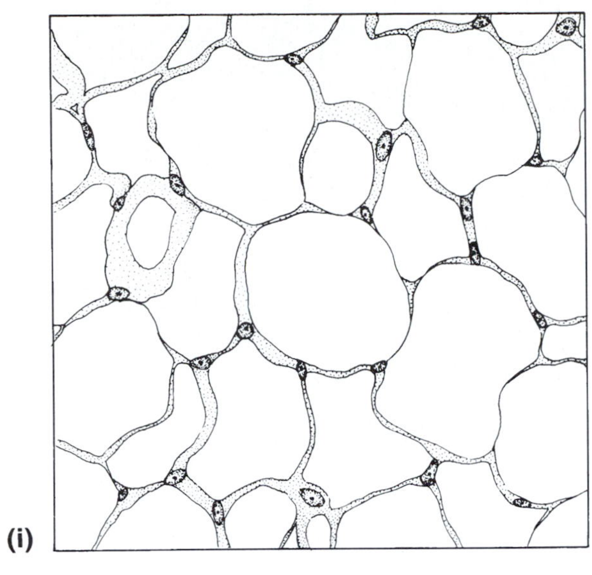

(k)

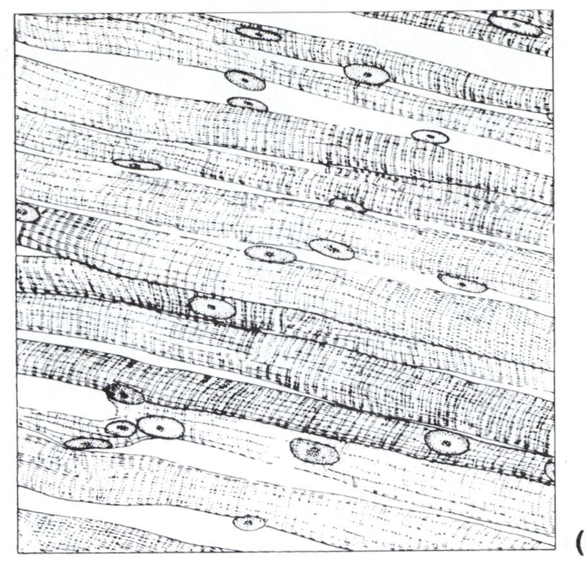

(j)

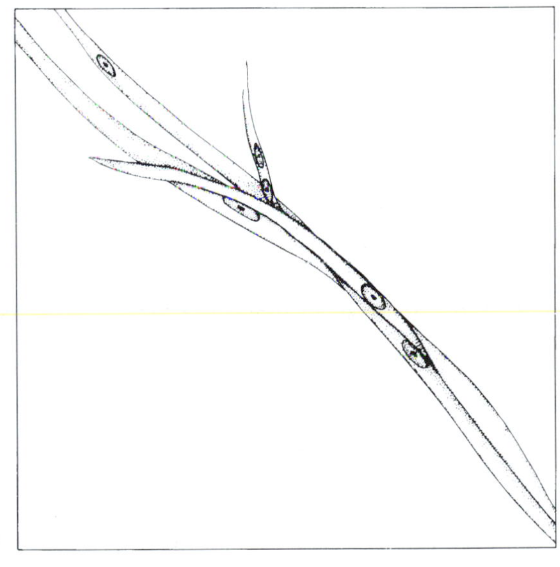

(l)

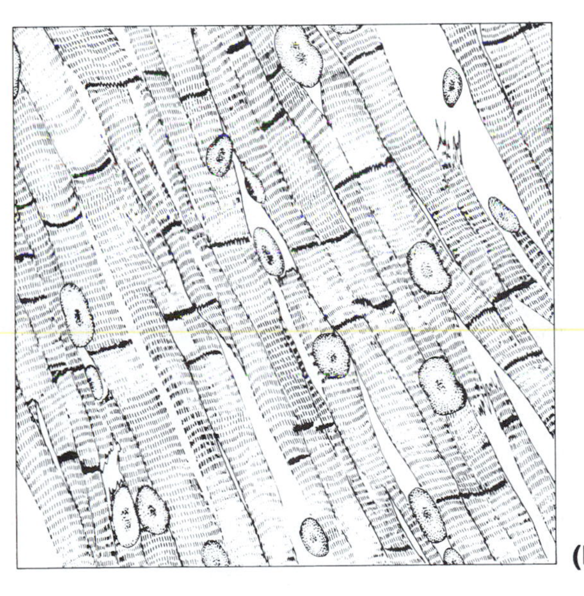

STUDENT NAME ______________________________

LAB TIME/DATE ______________________________

Review Sheet

EXERCISE 7 The Integumentary System

Basic Structure of the Skin

1. Complete the following statements by writing the appropriate word or phrase on the correspondingly numbered blanks:

The two basic tissues of which the skin is composed are dense irregular connective tissue, which makes up the dermis, and __1__ , which forms the epidermis. The tough water-repellent protein found in the epidermal cells is called __2__ . The pigments, melanin and __3__ , contribute to skin color. A localized concentration of melanin is referred to as a __4__ .

1. ______________________________
2. ______________________________
3. ______________________________
4. ______________________________

2. Four protective functions of the skin are ______________________________

______________________________.

3. Using the key choices, choose all responses that apply to the following descriptions.

Key:
a. stratum basale
b. stratum corneum
c. stratum granulosum
d. stratum lucidum
e. stratum spinosum
f. papillary layer
g. reticular layer
h. epidermis as a whole
i. dermis as a whole

______________________ 1. translucent cells containing keratin fibrils

______________________ 2. dead cells

______________________ 3. dermis layer responsible for fingerprints

______________________ 4. vascular region

______________________ 5. major skin area where derivatives (nails and hair) arise

______________________ 6. epidermal region exhibiting rapid cell division

______________________ 7. scalelike dead cells, full of keratin, that constantly slough off

______________________ 8. also called the stratum germinativum

______________________ 9. has abundant elastic and collagenic fibers

______________________ 10. site of melanin formation

______________________ 11. area where tonofilaments first appear

4. Label the skin structures and areas indicated in the accompanying diagram of skin. Then, complete the statements that follow.

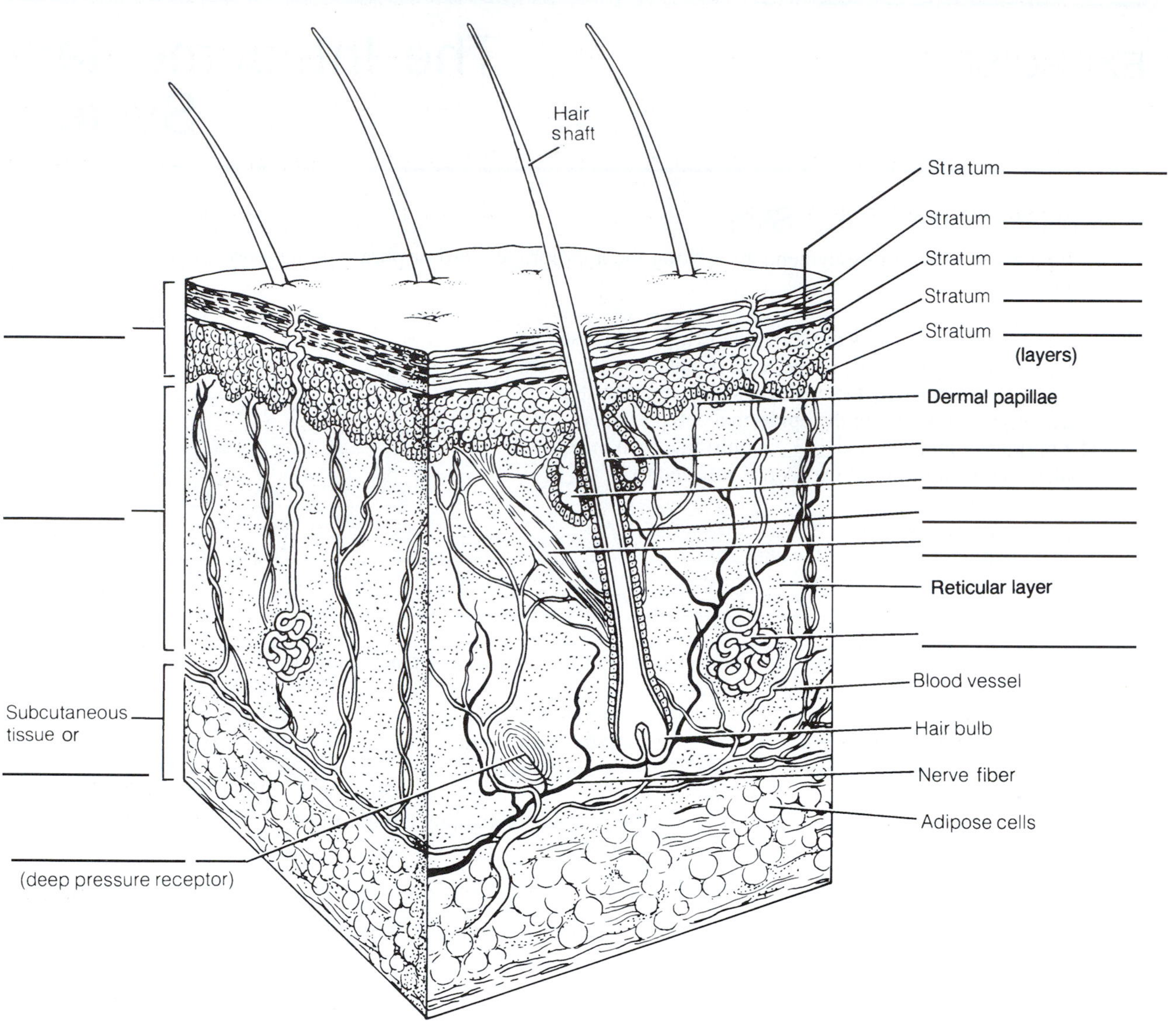

1. ______________________ granules extruded from the keratinocytes prevent water loss by diffusion through the epidermis.

2. The origin of the fibers in the dermis is ______________________.

3. Glands that respond to rising androgen levels are the ______________________ glands.

4. Phagocytic cells that occupy the epidermis are called ______________________.

5. A unique touch receptor formed from a stratum basale cell and a nerve fiber is a ______________________.

5. What substance is manufactured in the skin (but is not a secretion) to play a role elsewhere in the body?

____

6. What sensory receptors are found in the skin? ____

____

____

7. A nurse tells a doctor that a patient is cyanotic. What is cyanosis? ______________________

What does its presence imply? ____

8. What is the mechanism of a suntan? ____

____

9. What is a decubitus ulcer? ____

Why does it occur? ____

____

10. Some injections hurt more than others. On the basis of what you have learned about skin structure, can you determine why this is so? ____

____

Appendages of the Skin

1. Using key choices, respond to the following descriptions.

Key: a. arrector pili
b. cutaneous receptors
c. hair
d. hair follicle
e. nail
f. sebaceous glands
g. sweat gland—apocrine
h. sweat gland—eccrine

______________ 1. A blackhead is an accumulation of oily material that is produced by ________.

______________ 2. Tiny muscles, attached to hair follicles, that pull the hair upright during fright or cold.

______________ 3. More numerous variety of perspiration gland.

______________ 4. Sheath formed of both epithelial and connective tissues.

______________ 5. Less numerous type of perspiration-producing gland; found mainly in the pubic and axillary regions.

______________ 6. Found everywhere on body except palms of hands, and soles of feet.

______________ 7. Primarily dead/keratinized cells.

______________ 8. Specialized nerve endings that respond to temperature, touch, etc.

______________ 9. Its secretion contains cell fragments.

______________ 10. "Sports" a lunula and a cuticle.

2. How does the skin help in regulating body temperature? (Describe two different mechanisms.) ______________

____

____

____

____

Plotting the Distribution of Sweat Glands

1. With what substance in the bond paper does the iodine painted on the skin react? ______________

2. Which skin area—the forearm or palm of hand—has more sweat glands? ______________

   Which other body areas would, if tested, prove to have a high density of sweat glands? ______________

   ____

3. What organ system controls the activity of the eccrine sweat glands? ______________

STUDENT NAME ______________________

LAB TIME/DATE ______________________

Review Sheet

EXERCISE 8

Classification of Body Membranes

Classification of Body Membranes

1. Complete the following chart:

Membrane	Tissue type (epithelial/connective)	Common locations	Functions
cutaneous			
mucous			
serous			
synovial			

2. Respond to the following statements by choosing an answer from the key.

Key: a. cutaneous b. mucous c. serous d. synovial

________________ 1. membrane type associated with skeletal system structures

________________ 2. always formed of simple squamous epithelium

________________, ________________ 3. membrane types *not* found in the ventral body cavity

________________ 4. the only membrane type in which goblet cells are found

________________ 5. the only *external* membrane

________________ 6. "wet" membranes

________________ 7. adapted for absorption and secretion

________________ 8. has parietal and visceral layers

STUDENT NAME ______________________

LAB TIME/DATE ______________________

Review Sheet

EXERCISE 9 Overview of the Skeleton: Classification and Structure of Bones and Cartilages

Bone Markings

1. Match the terms in column B with the appropriate description in column A:

	Column A	Column B
______	1. sharp, slender process*	a. condyle
______	2. small rounded projection*	b. crest
______	3. narrow ridge of bone*	c. epicondyle
______	4. large rounded projection*	d. fissure
______	5. structure supported on neck†	e. foramen
______	6. armlike projection†	f. fossa
______	7. rounded, convex projection†	g. head
______	8. narrow depression or opening‡	h. meatus
______	9. canal-like structure‡	i. ramus
______	10. opening through a bone‡	j. sinus
______	11. shallow depression†	k. spine
______	12. air-filled cavity	l. trochanter
______	13. large, irregularly shaped projection*	m. tubercle
______	14. raised area of a condyle*	n. tuberosity

* A site of muscle attachment.

† Takes part in joint formation.

‡ A passageway for nerves or blood vessels.

Classification of Bones

1. The four major anatomical classifications of bones are long, short, flat, and irregular. Which category has the least amount of spongy bone relative to its total volume? ______________________

2. Classify each of the bones in the next chart into one of the four major categories by checking the appropriate column. Use appropriate references as necessary.

	Long	Short	Flat	Irregular
humerus				
phalanx				
parietal				
calcaneus				
rib				
vertebra				
ulna				

Gross Anatomy of the Typical Long Bone

1. Use the terms below to identify the structures marked by leader lines and braces in the diagrams (some terms are used more than once). After labeling the diagrams, use the listed terms to characterize the statements following the diagrams.

Key:
a. articular cartilage
b. compact bone
c. diaphysis
d. endosteum
e. epiphyseal line
f. epiphysis
g. medullary cavity
h. periosteum
i. red marrow cavity
j. trabeculae of spongy bone
k. yellow marrow

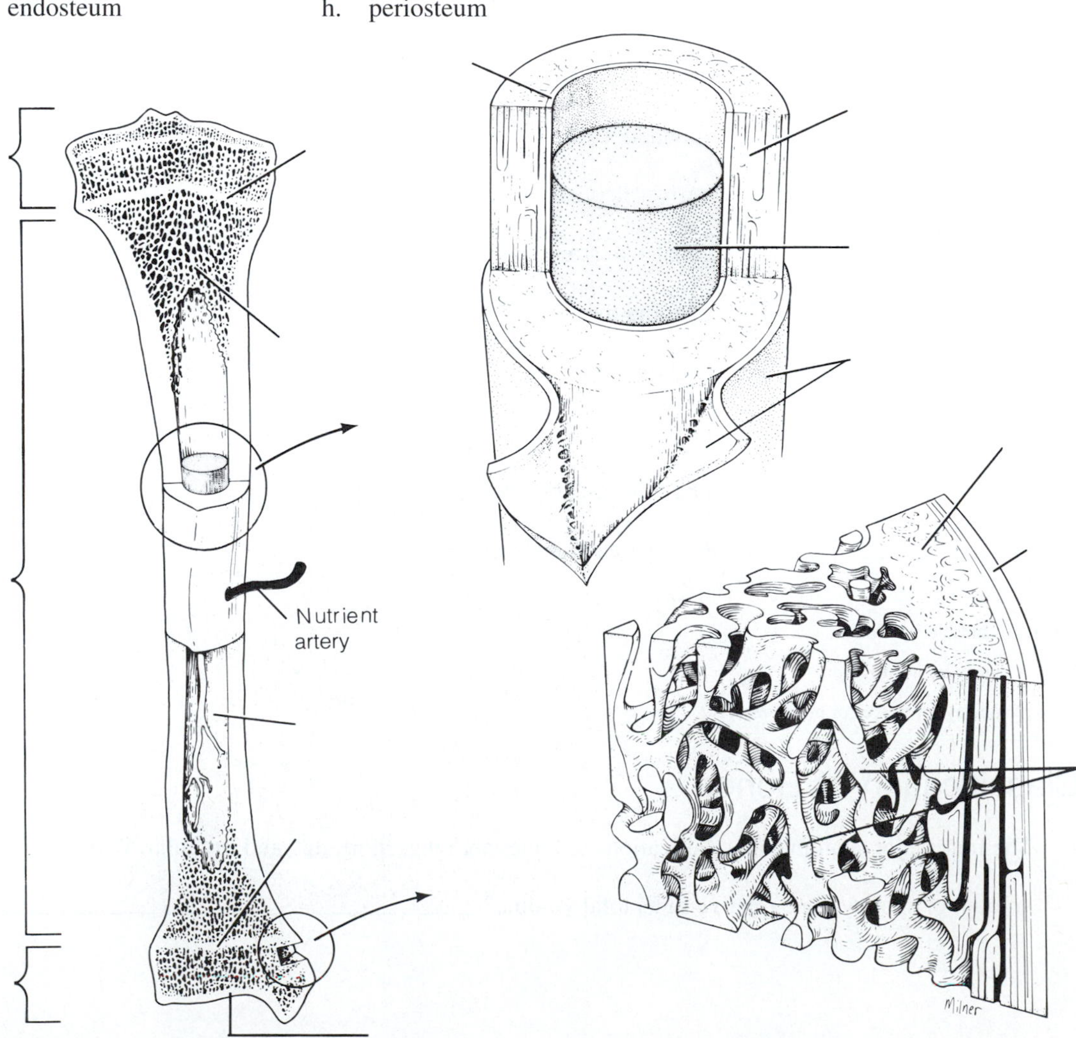

_______ 1. contains spongy bone in adults

_______ 2. made of compact bone

_______ 3. site of blood cell formation

_______ 4. major submembranous site of osteoclasts

_______ 5. scientific term for bone shaft

_______ 6. contains fat in adult bones

_______ 7. growth plate remnant

_______ 8. major submembranous site of osteoblasts

2. What differences between compact and spongy bone can be seen with the naked eye? _______________

____

3. What is the function of the periosteum? _______________

____

Microscopic Structure of Compact Bone

1. Trace the route taken by nutrients through a bone, starting with the periosteum and ending with an osteocyte in a lacuna. Periosteum → _______________ → _______________

_______________ → _______________ → osteocyte

2. Several descriptions of bone structure are given in column B. Identify the structure involved by choosing the appropriate term from column A and placing its letter in the blank.

Column A		Column B
a. canaliculi	_______	1. layers of bony matrix around a central canal
b. central canal	_______	2. site of osteocytes
c. concentric lamellae	_______	3. longitudinal canal carrying blood vessels, lymphatics, and nerves
d. lacunae	_______	4. minute canals connecting osteocytes of an osteon

3. On the photomicrograph of bone below, identify all structures named in column A in question 2 above.

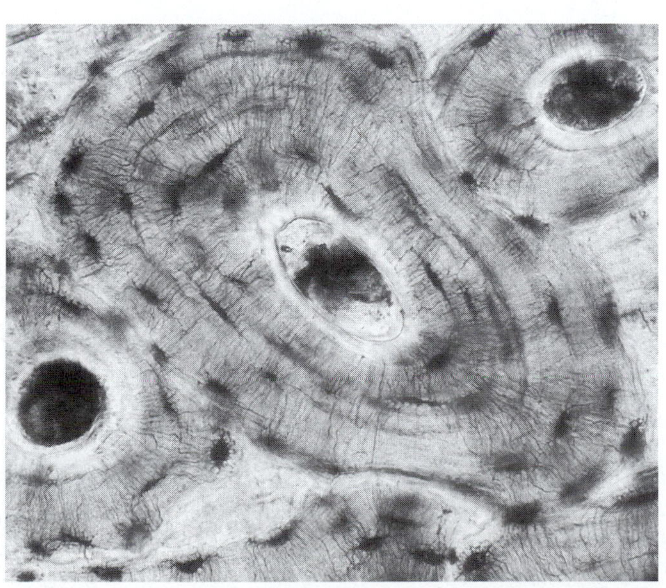

Chemical Composition of Bone

1. What is the function of the organic matrix in bone? ____________________

2. Name the important organic bone components. ____________________

3. Calcium salts form the bulk of the inorganic material in bone. What is the function of the calcium salts?

____

4. Which is responsible for bone structure? (circle the appropriate response)

inorganic portion   organic portion   both contribute

Cartilages of the Skeleton

Using key choices, identify each type of cartilage described (in terms of its body location or function) below.

Key: a. elastic   b. fibrocartilage   c. hyaline

____________ 1. supports the external ear

____________ 2. between the vertebrae

____________ 3. forms the walls of the voice box (larynx)

____________ 4. the epiglottis

____________ 5. articular cartilages

____________ 6. meniscus in a knee joint

____________ 7. connects the ribs to the sternum

____________ 8. most effective at resisting compression

____________ 9. most springy and flexible

____________ 10. most abundant

STUDENT NAME ____________________

LAB TIME/DATE ____________________

Review Sheet

EXERCISE 10

The Axial Skeleton

The Skull

1. The skull is one of the major components of the axial skeleton. Name the other two:

   ____________________ and ____________________

   What structures do each of these areas protect? ____________________

   ____

   ____

2. Define *suture:* ____________________

3. With one exception, the skull bones are joined by sutures. Name the exception. ____________________

   ____

4. What are the four major sutures of the skull, and what bones do they connect? ____________________

   ____

   ____

   ____

   ____

5. Name the eight bones composing the cranium.

   ____________ ____________ ____________ ____________

   ____________ ____________ ____________ ____________

6. Give two possible functions of the sinuses. ____________________

   ____

7. What is the orbit? ____________________

   What bones contribute to the formation of the orbit? ____________________

   ____

8. Why can the sphenoid bone be called the keystone of the cranial floor? ____________________

   ____

   ____

9. What is a cleft palate? ____

____

10. Match the bone names in column B with the descriptions in column A.

	Column A		Column B
________	1. forehead bone	a.	ethmoid
________	2. cheekbone	b.	frontal
________	3. lower jaw	c.	hyoid
________	4. bridge of nose	d.	lacrimal
________	5. posterior bones of the hard palate	e.	mandible
________	6. much of the lateral and superior cranium	f.	maxilla
________	7. most posterior part of cranium	g.	nasal
________	8. single, irregular, bat-shaped bone forming part of the cranial floor	h.	occipital
		i.	palatine
________	9. tiny bones bearing tear ducts	j.	parietal
________	10. anterior part of hard palate	k.	sphenoid
________	11. superior and medial nasal conchae formed from its projections	l.	temporal
________	12. site of mastoid process	m.	vomer
________	13. site of sella turcica	n.	zygomatic

________ 14. site of cribriform plate

________ 15. site of mental foramen

________ 16. site of styloid processes

________, ________, ________, and

________ 17. four bones containing paranasal sinuses

________ 18. condyles here articulate with the atlas

________ 19. foramen magnum contained here

________ 20. small U-shaped bone in neck, where many tongue muscles attach

________ 21. middle ear found here

________ 22. nasal septum

________ 23. bears an upward protrusion, the “cock’s comb,” or crista galli

________, ________ 24. contain alveoli bearing teeth

11. Using choices from column B in question 10 and from the numbered key to the right, identify all bones and bone markings provided with leader lines in the two diagrams below.

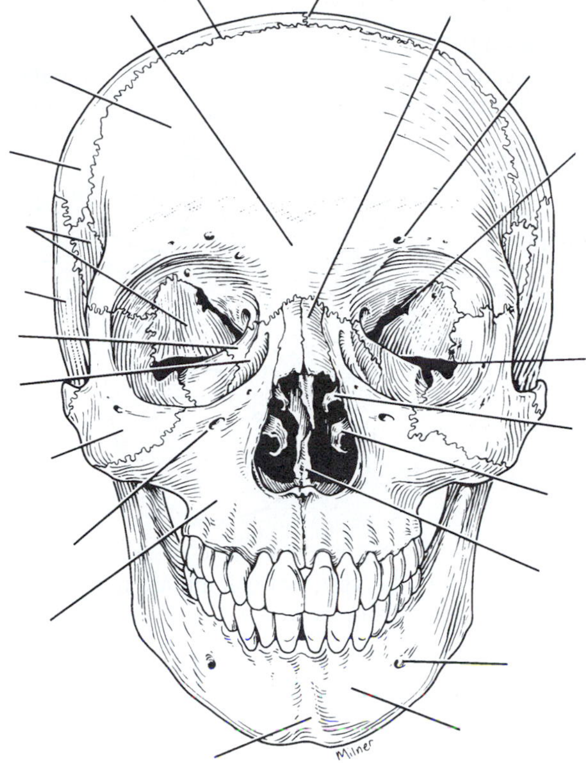

1. carotid canal
2. coronal suture
3. external occipital protuberance
4. foramen lacerum
5. foramen magnum
6. foramen ovale
7. glabella
8. incisive foramen
9. inferior nasal concha
10. inferior orbital fissure
11. infraorbital foramen
12. jugular foramen
13. mandibular fossa
14. mandibular symphysis
15. mastoid process
16. mental foramen
17. middle nasal concha of ethmoid
18. occipital condyle
19. palatine process of maxilla
20. sagittal suture
21. styloid process
22. stylomastoid foramen
23. superior orbital fissure
24. supraorbital foramen
25. zygomatic process of temporal bone

The Vertebral Column

1. Using the key, correctly identify the vertebral parts/areas described below. (More than one choice may apply in some cases.) Also use the key letters to correctly identify the vertebral areas in the diagram.

Key:
a. body
b. intervertebral foramina
c. lamina
d. pedicle
e. spinous process
f. superior articular process
g. transverse process
h. vertebral arch
i. vertebral foramen

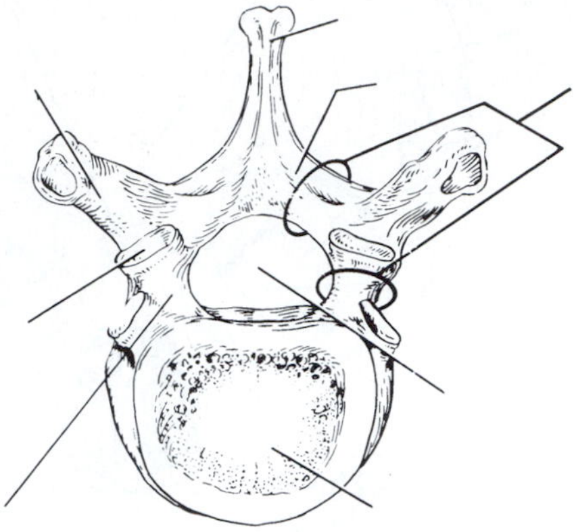

_______ 1. cavity enclosing the nerve cord

_______ 2. weight-bearing portion of the vertebra

_______ 3. provide levers against which muscles pull

_______ 4. provide an articulation point for the ribs

_______ 5. openings providing for exit of spinal nerves

_______ 6. structures that form an enclosure for the spinal cord

2. The distinguishing characteristics of the vertebrae composing the vertebral column are noted below. Correctly identify each described structure/region by choosing a response from the key.

Key:
a. atlas
b. axis
c. cervical vertebra—typical
d. coccyx
e. lumbar vertebra
f. sacrum
g. thoracic vertebra

_______________ 1. vertebral type containing foramina in the transverse processes, through which the vertebral arteries ascend to reach the brain

_______________ 2. dens here provides a pivot for rotation of the first cervical vertebra (C_1)

_______________ 3. transverse processes faceted for articulation with ribs; spinous process pointing sharply downward

_______________ 4. composite bone; articulates with the hip bone laterally

_______________ 5. massive vertebrae; weight-sustaining

_______________ 6. "tail bone"; vestigial fused vertebrae

_______________ 7. supports the head; allows a rocking motion in conjunction with the occipital condyles

_______________ 8. seven components; unfused

_______________ 9. twelve components; unfused

3. Name two factors/structures that allow for flexibility of the vertebral column.

_______________________________ and _______________________________

4. Describe how a spinal nerve exits from the vertebral column. ______________________________

____

5. Which two spinal curvatures are obvious at birth? ______________ and ______________

Under what conditions do the secondary curvatures develop? ______________________________

____

____

6. Diagram the normal spinal curvatures and the abnormal spinal curvatures named below. (Use posterior or lateral views as necessary.)

Normal curvature Lordosis Scoliosis Kyphosis

7. What kind of tissue comprises the intervertebral discs? ______________________________

8. What is a herniated disc? ______________________________

What problems might it cause? ______________________________

____

The Bony Thorax

1. The major components of the thorax (excluding the vertebral column) are the ______________

and the ______________ .

2. Differentiate between a true rib and a false rib. ______________________________

____

____

Is a floating rib a true or a false rib? ______________

3. What is the general shape of the thoracic cage? ______________________________

4. Using the terms at the right, identify the regions and landmarks of the bony thorax.

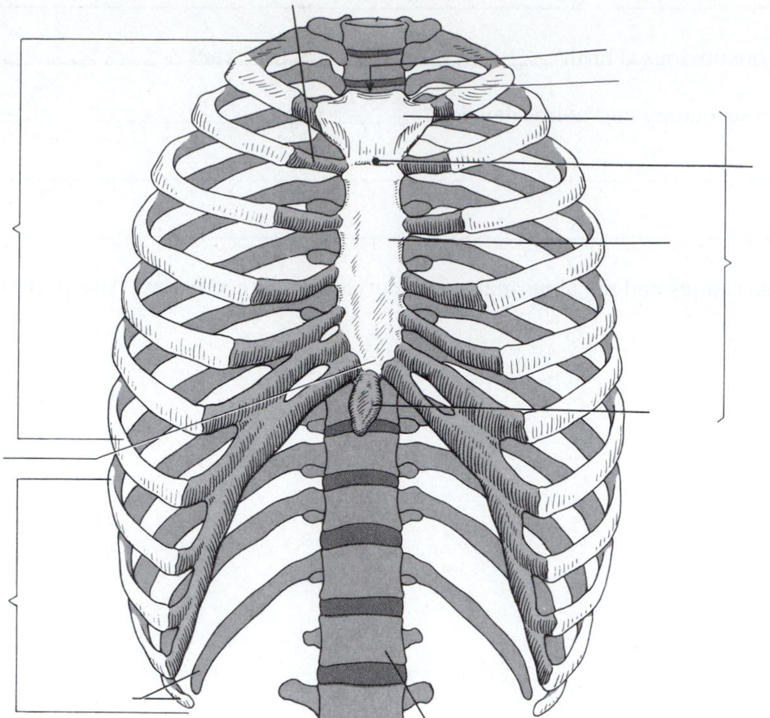

a. body

b. clavicular notch

c. costal cartilage

d. false ribs

e. floating ribs

f. jugular notch

g. manubrium

h. sternal angle

i. sternum

j. true ribs

k. xiphisternal joint

l. xiphoid process

5. Provide the more scientific name for the following rib types.

a. True ribs ____________________

b. False ribs (not including c) ____________________

c. Floating ribs ____________________

STUDENT NAME ____________________

LAB TIME/DATE ____________________

Review Sheet

EXERCISE 11 — The Appendicular Skeleton

Bones of the Shoulder Girdle and Upper Limb

1. Match the bone names or markings in column B with the descriptions in column A.

Column A

_______ 1. raised area on lateral surface of humerus to which deltoid muscle attaches

_______ 2. arm bone

_______, _______ 3. bones of the shoulder girdle

_______, _______ 4. forearm bones

_______ 5. scapular region to which the clavicle connects

_______ 6. shoulder girdle bone that is unattached to the axial skeleton

_______ 7. shoulder girdle bone that transmits forces from the upper limb to the bony thorax

_______ 8. depression in the scapula that articulates with the humerus

_______ 9. process above the glenoid cavity that permits muscle attachment

_______ 10. the "collarbone"

_______ 11. distal condyle of the humerus that articulates with the ulna

_______ 12. medial bone of forearm in anatomical position

_______ 13. rounded knob on the humerus; adjoins the radius

_______ 14. anterior depression, superior to the trochlea, which receives part of the ulna when the forearm is flexed

_______ 15. forearm bone involved in formation of the elbow joint

_______ 16. wrist bones

_______ 17. finger bones

_______ 18. heads of these bones form the knuckles

_______, _______ 19. bones that articulate with the clavicle

Column B

a. acromion
b. capitulum
c. carpals
d. clavicle
e. coracoid process
f. coronoid fossa
g. deltoid tuberosity
h. glenoid cavity
i. humerus
j. metacarpals
k. olecranon fossa
l. olecranon process
m. phalanges
n. radial tuberosity
o. radius
p. scapula
q. sternum
r. styloid process
s. trochlea
t. ulna

2. Why is the clavicle at risk to fracture when a person falls on his or her shoulder? ______________________________

____

3. Why is there generally no problem in the arm clearing the widest dimension of the thoracic cage?

____

____

4. What is the total number of phalanges in the hand? ______________

5. What is the total number of carpals in the wrist? ______________

   In the proximal row, the carpals are (medial to lateral) ______________________________

   In the distal row, they are (medial to lateral) ______________________________

6. Using items from the list at the right, identify the anatomical landmarks and regions of the scapula.

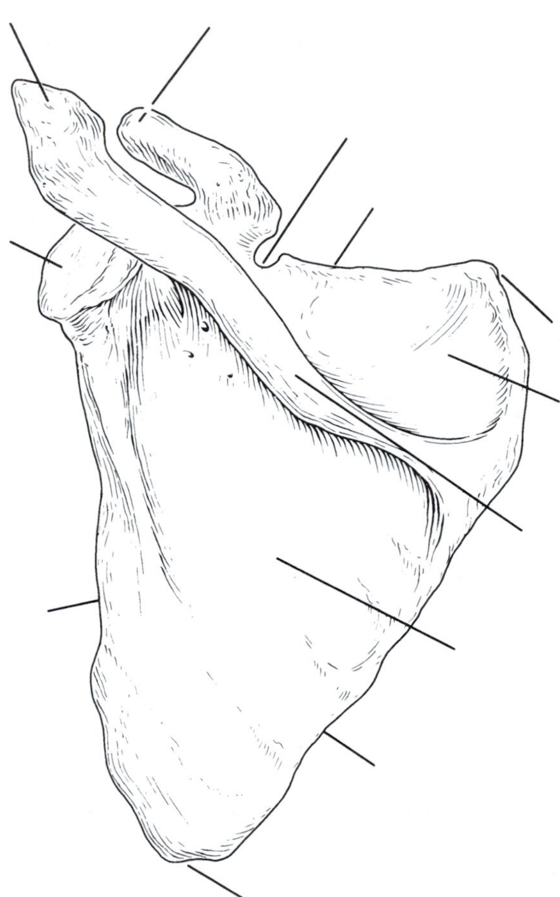

a. acromion

b. coracoid process

c. glenoid cavity

d. inferior angle

e. infraspinous fossa

f. lateral border

g. medial border

h. spine

i. superior angle

j. superior border

k. suprascapular notch

l. supraspinous fossa

Bones of the Pelvic Girdle and Lower Limb

1. Compare the pectoral and pelvic girdles by choosing appropriate descriptive terms from the key.

   Key:
   a. flexibility most important
   b. massive
   c. lightweight
   d. insecure axial and limb attachments
   e. secure axial and limb attachments
   f. weightbearing most important

   Pectoral: ________, ________, ________

   Pelvic: ________, ________, ________

2. What organs are protected, at least in part, by the pelvic girdle? ________________________________

   ____

3. Distinguish between the true pelvis and the false pelvis. ________________________________

   ____

   ____

4. Name five differences between the male and female pelves. ________________________________

   ____

   ____

   ____

   ____

5. Deduce why the pelvic bones of a four-legged animal such as the cat or pig are much less massive than those of the human. ________________________________

   ____

6. A person instinctively curls over his abdominal area in times of danger. Why? ________________________________

   ____

7. For what anatomical reason do many women appear to be slightly knock-kneed? ________________________________

   ____

   ____

   ____

8. What does *fallen arches* mean? ________________________________

   ____

9. Match the bone names and markings in column B with the descriptions in column A.

Column A	Column B
________, ________, and	a. acetabulum
________ 1. fuse to form the coxal bone	b. calcaneus
________ 2. inferoposterior "bone" of the coxal bone	c. femur
________ 3. point where the coxal bones join anteriorly	d. fibula
________ 4. superiormost margin of the coxal bone	e. gluteal tuberosity
________ 5. deep socket in the coxal bone that receives the head of the thigh bone	f. greater sciatic notch
	g. greater and lesser trochanters
________ 6. joint between the axial skeleton and the pelvic girdle	h. iliac crest
________ 7. longest, strongest bone in body	i. ilium
________ 8. thin lateral leg bone	j. ischial tuberosity
________ 9. heavy medial leg bone	k. ischium
________, ________ 10. bones forming the knee joint	l. lateral malleolus
	m. lesser sciatic notch
________ 11. point where the patellar ligament attaches	n. linea aspera
________ 12. kneecap	o. medial malleolus
________ 13. shin bone	p. obturator foramen
________ 14. medial ankle projection	q. metatarsals
________ 15. lateral ankle projection	r. patella
________ 16. largest tarsal bone	s. pubic symphysis
________ 17. ankle bones	t. pubis
________ 18. bones forming the instep of the foot	u. sacroiliac joint
________ 19. opening in hip bone formed by the pubic and ischial rami	v. talus
________ and ________ 20. sites of muscle attachment on the proximal femur	w. tarsals
	x. tibia
________ 21. tarsal bone that "sits" on the calcaneus	y. tibial tuberosity

Summary of Skeleton

1. Identify all indicated bones (or groups of bones) in the diagram of the articulated skeleton.

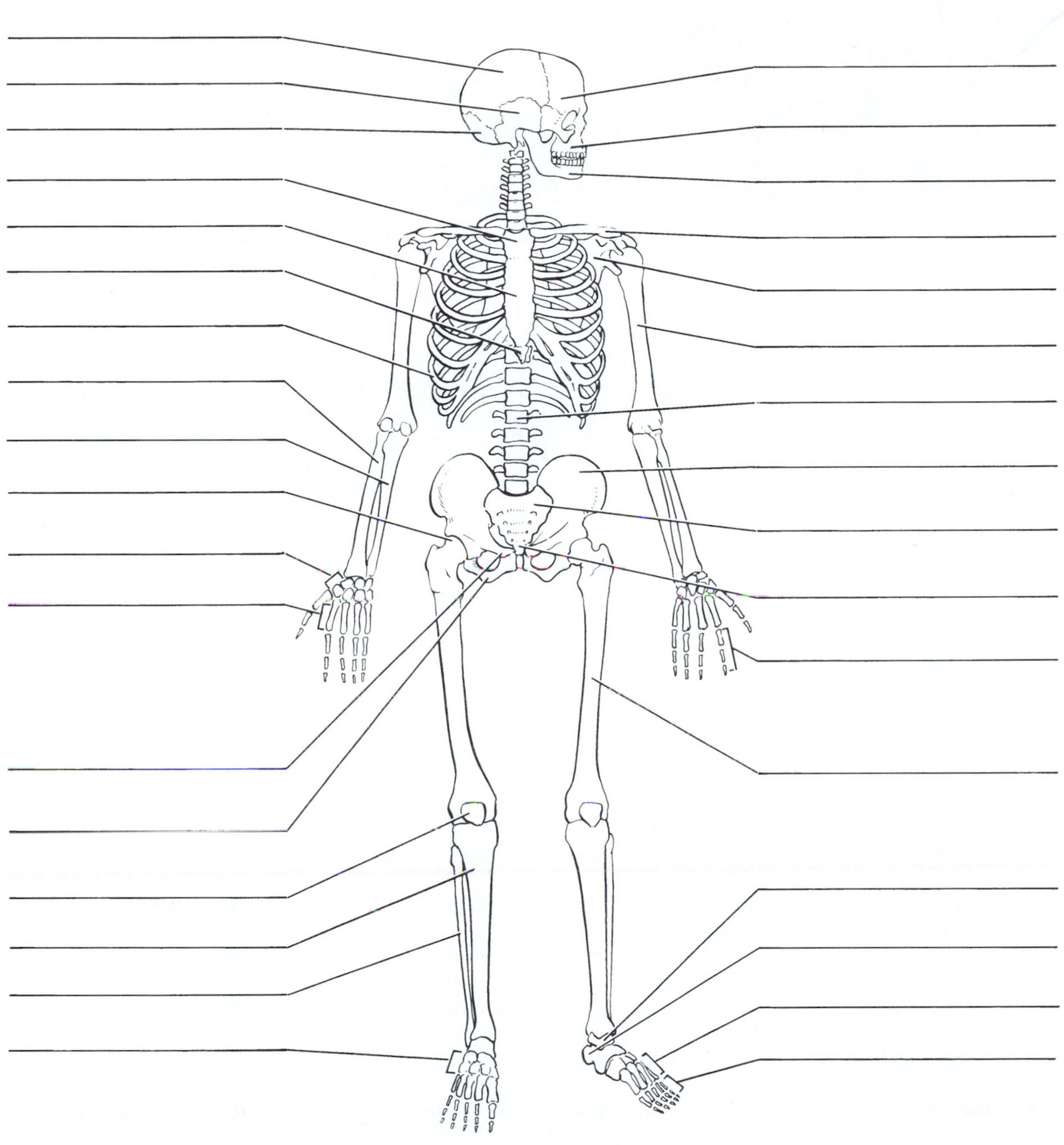

2. Three bones of the appendicular skeleton are diagrammed below. Using the choices from below, correctly identify all bone markings provided with leader lines.

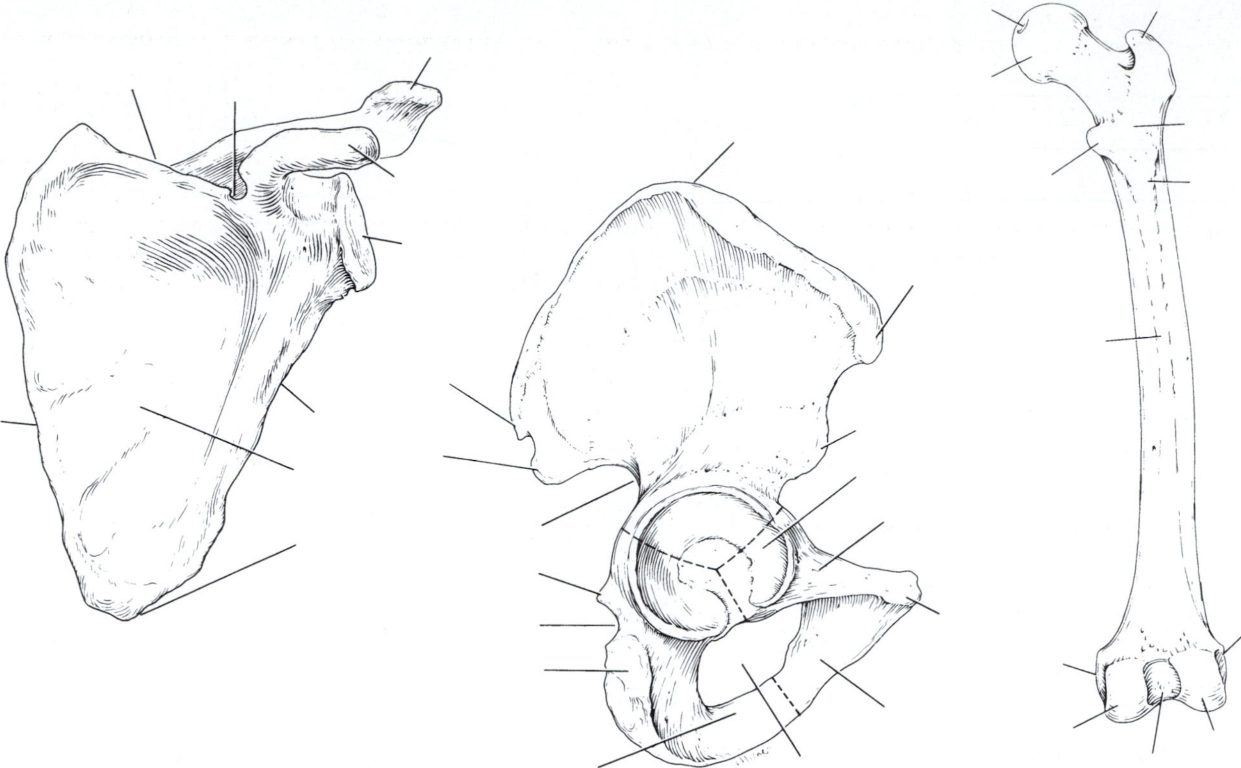

Choices for girdle bones

a. acetabulum
b. acromion
c. anterior inferior iliac spine
d. anterior superior iliac spine
e. coracoid process
f. glenoid cavity
g. greater sciatic notch
h. iliac crest
i. inferior angle of scapula
j. ischial ramus
k. ischial spine
l. ischial tuberosity
m. lateral border of scapula
n. lesser sciatic notch
o. medial border of scapula
p. obturator foramen
q. posterior inferior iliac spine
r. posterior superior iliac spine
s. pubic tubercle
t. pubis—inferior ramus
u. pubis—superior ramus
v. subscapular fossa
w. superior border of scapula
x. suprascapular notch

Choices for limb bone

1. fovea capitis
2. gluteal tuberosity
3. greater trochanter
4. head
5. intercondylar fossa
6. intertrochanteric crest
7. lateral condyle
8. lateral epicondyle
9. lesser trochanter
10. linea aspera
11. medial condyle
12. medial epicondyle

STUDENT NAME ______________________

LAB TIME/DATE ______________________

Review Sheet

EXERCISE 12

The Fetal Skeleton

1. Are the same skull bones seen in the adult found in the fetal skull? ______________________

2. How does the size of the fetal face compare to its cranium? ______________________

____

How does this compare to the adult skull? ______________________

____

____

3. What are the outward conical projections in some of the fetal cranial bones? ______________________

____

4. What is a fontanel? ______________________

What is its fate? ______________________

What is the function of the fontanels in the fetal skull? ______________________

____

5. Describe how the fetal skeleton compares with the adult skeleton in the following areas:

vertebrae ______________________

____

os coxae ______________________

____

carpals and tarsals ______________________

sternum ______________________

frontal bone ______________________

patella ______________________

rib cage ______________________

____

6. How does the size of the fetus's head compare to the size of its body? ______________________

____

7. Using the terms listed, identify each of the fontanels shown on the fetal skull below.

a. anterior fontanel b. mastoid fontanel c. posterior fontanel d. sphenoidal fontanel

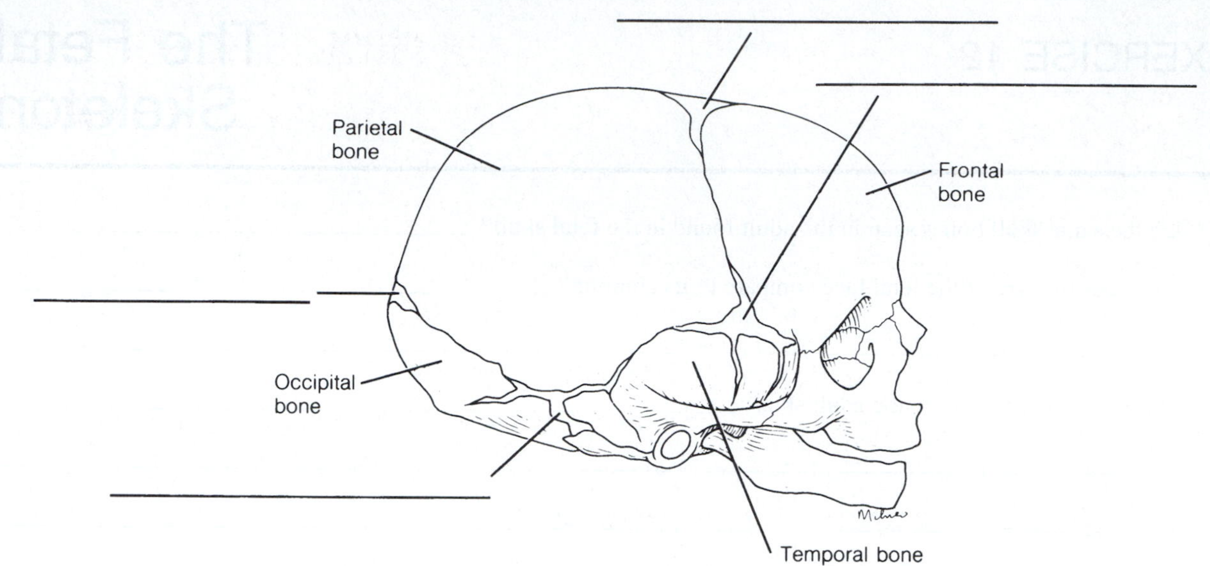

STUDENT NAME ___________________________

LAB TIME/DATE ___________________________

Review Sheet

EXERCISE 13

Articulations and Body Movements

Types of Joints

1. Use key responses to identify the joint types described below.

   Key: a. cartilaginous b. fibrous c. synovial

   ____________ 1. typically allows a slight degree of movement

   ____________ 2. includes joints between the vertebral bodies and the pubic symphysis

   ____________ 3. essentially immovable joints

   ____________ 4. sutures are the most remembered examples

   ____________ 5. characterized by cartilage connecting the bony portions

   ____________ 6. all characterized by a fibrous articular capsule lined with a synovial membrane surrounding a joint cavity

   ____________ 7. all are freely movable or diarthrotic

   ____________ 8. bone regions are united by fibrous connective tissue

   ____________ 9. include the hip, knee, and elbow joints

2. Match the joint subcategories in column B with their descriptions in column A, and place an asterisk (*) beside all choices that are examples of synovial joints.

	Column A	Column B
____________	1. joint between skull bones	a. ball and socket
____________	2. joint between the axis and atlas	b. condyloid
____________	3. hip joint	c. gliding
____________	4. intervertebral joints (between articular processes)	d. hinge
____________	5. joint between forearm bones and wrist	e. pivot
____________	6. elbow	f. saddle
____________	7. interphalangeal joints	g. suture
____________	8. intercarpal joints	h. symphysis
____________	9. joint between tarsus and tibia/fibula	i. synchondrosis
____________	10. joint between skull and vertebral column	j. syndesmosis
____________	11. joint between jaw and skull	
____________	12. joints between proximal phalanges and metacarpal bones	
____________	13. epiphyseal plate of a child's long bone	
____________	14. a multiaxial joint	
____________, ____________	15. biaxial joints	
____________, ____________	16. uniaxial joints	

3. What characteristics do all joints have in common? ______________________________

____

4. Describe the structure and function of the following structures or tissues in relation to a synovial joint and label the structures indicated by leader lines in the diagram.

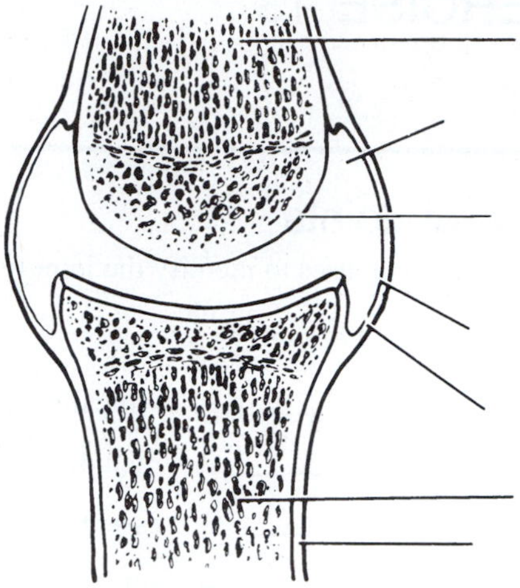

ligament ______________________________

____

tendon ______________________________

____

hyaline cartilage ______________________________

____

synovial membrane ______________________________

____

____

bursa ____

5. Which joint, the hip or the knee, is more stable? ______________________________

Name two important factors that contribute to the stability of the hip joint.

______________________________ and ______________________________

Name two important factors that contribute to the stability of the knee.

______________________________ and ______________________________

Body Movements

Complete the following statements:

The movable attachment of a muscle is called its __1__ , and its stationary attachment is called the __2__ . Winding up for a pitch (as in baseball) can properly be called __3__ . To keep your seat when riding a horse, the tendency is to __4__ your thighs. In running, the action at the hip joint is __5__ in reference to the leg moving forward and __6__ in reference to the leg in the posterior position. In kicking a football, the action at the knee is __7__ . In climbing stairs, the hip and knee of the forward leg are both __8__ . You have just touched your chin to your chest. This is __9__ of the neck. Using a screwdriver with your arm straight requires __10__ of the arm. Consider all the movements of which the arm is capable. One often used for strengthening all the upper arm and shoulder muscles is __11__ . Movement of the head that signifies "no" is __12__ . Standing on your toes, as in ballet, requires __13__ of the foot. Action that moves the distal end of the radius across the ulna is __14__ . Raising the arms laterally away from the body is called __15__ of the arms. Walking on one's heels is __16__ .

1. ______________________
2. ______________________
3. ______________________
4. ______________________
5. ______________________
6. ______________________
7. ______________________
8. ______________________
9. ______________________
10. ______________________
11. ______________________
12. ______________________
13. ______________________
14. ______________________
15. ______________________
16. ______________________

Joint Disorders

1. What structural joint changes are common to the elderly? ______________________

____

____

2. Define:

sprain ______________________

____

dislocation ______________________

____

STUDENT NAME ______________________________

LAB TIME/DATE ______________________________

Review Sheet

EXERCISE 14 Microscopic Anatomy, Organization, and Classification of Skeletal Muscle

Skeletal Muscle Cells and Their Packaging into Muscles

1. What capability is most highly expressed in muscle tissue? ______________________________

2. Use the items on the right to correctly identify the structures described on the left.

		Key
______________	1. connective tissue ensheathing a bundle of muscle cells	a. endomysium
______________	2. bundle of muscle cells	b. epimysium
______________	3. contractile unit of muscle	c. fascicle
______________	4. a muscle cell	d. fiber
______________	5. thin reticular connective tissue investing each muscle cell	e. myofilament
______________	6. plasma membrane of the muscle fiber	f. myofibril
______________	7. a long filamentous organelle with a banded appearance found within muscle cells	g. perimysium
		h. sarcolemma
______________	8. actin- or myosin-containing structure	i. sarcomere
______________	9. cord of collagen fibers that attaches a muscle to a bone	j. sarcoplasm
		k. tendon

3. Why are the connective tissue wrappings of skeletal muscle important? (Give at least three reasons.)

____

____

____

____

4. Why are indirect—that is, tendinous—muscle attachments to bone seen more often than direct attachments?

____

____

____

5. How does an aponeurosis differ from a tendon? ______________________________

____

____

6. The diagram illustrates a small portion of a muscle myofibril. Using letters from the key, correctly identify each structure indicated by a leader line or a bracket. Below the diagram make a sketch of how this segment of the myofibril would look if contracted.

Key:
a. actin filament
b. A band
c. I band
d. myosin filament
e. sarcomere
f. Z line

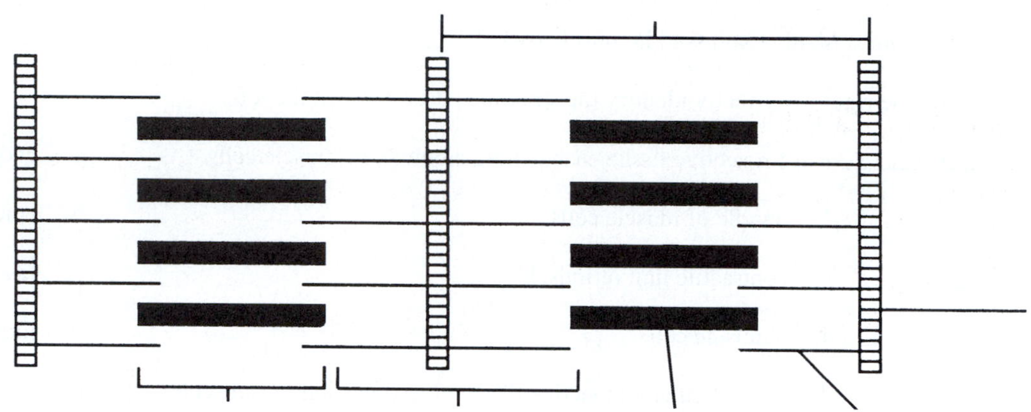

The Neuromuscular Junction

Complete the following statements:

The junction between a motor neuron's axon and the muscle cell membrane is called a neuromuscular junction or a __1__ junction. A motor neuron and all of the skeletal muscle cells it stimulates is called a __2__ . The axonal terminals of each motor axon have numerous projections called __3__ . The actual gap between the axonal terminal and the muscle cell is called a __4__ . Within the axonal terminal are many small vesicles containing a neurotransmitter substance called __5__ . When the __6__ reaches the ends of the axon, the neurotransmitter is released and diffuses to the muscle cell membrane to combine with receptors there. The combining of the neurotransmitter with the muscle membrane receptors causes the membrane to become permeable to sodium, which results in the influx of sodium ions and __7__ of the membrane. Then contraction of the muscle cell occurs. Before a muscle cell can be stimulated to contract again, __8__ must occur.

1. ______________________
2. ______________________
3. ______________________
4. ______________________
5. ______________________
6. ______________________
7. ______________________
8. ______________________

Classification of Skeletal Muscles

1. Several criteria were given relative to the naming of muscles. Match the criteria (column B) to the muscle cell names (column A). Note that more than one criterion may apply in some cases.

Column A	Column B
________ 1. gluteus maximus	a. action of the muscle
________ 2. adductor magnus	b. shape of the muscle
________ 3. biceps femoris	c. location of the origin and/or insertion of the muscle
________ 4. abdominis transversus	d. number of origins
________ 5. extensor carpi ulnaris	e. location of muscle relative to a bone or body region
________ 6. trapezius	f. direction in which the muscle fibers run relative to some imaginary line
________ 7. rectus femoris	g. relative size of the muscle
________ 8. external oblique	

2. When muscles are discussed relative to the manner in which they interact with other muscles, the terms shown in the key are often used. Match the key terms with the appropriate definitions.

Key: a. antagonist b. fixator c. prime mover d. synergist

________________ 1. agonist

________________ 2. postural muscles, for the most part

________________ 3. reverses and/or opposes the action of a prime mover

________________ 4. stabilizes a joint so that the prime mover may act at more distal joints

________________ 5. performs the same movement as the prime mover

________________ 6. immobilizes the origin of a prime mover

STUDENT NAME ________________________________

LAB TIME/DATE ________________________________

Review Sheet

EXERCISE 15 Gross Anatomy of the Muscular System

Muscles of the Head and Neck

1. Using choices from the list at the right, correctly identify muscles provided with leader lines on the diagram.

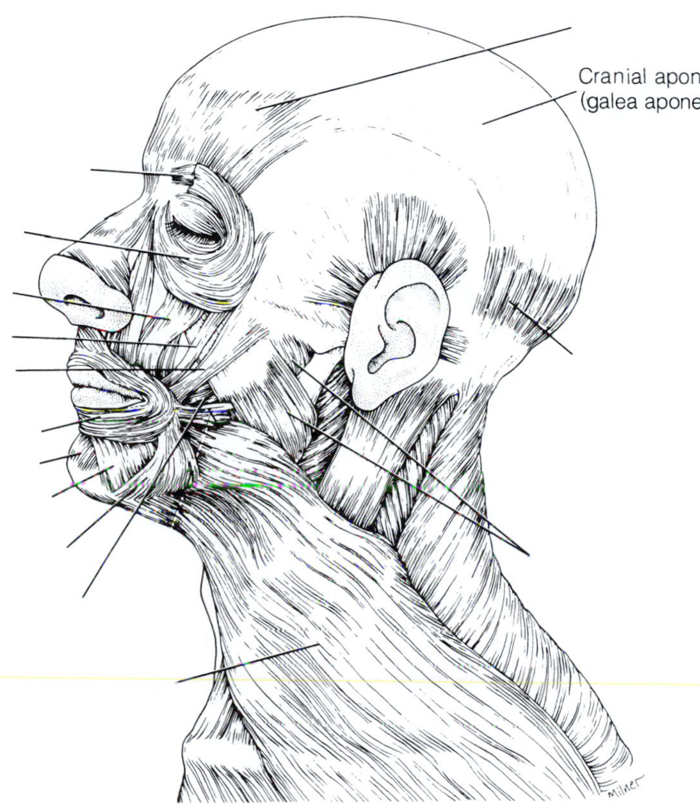

a. buccinator

b. corrugator supercilii

c. depressor anguli oris

d. depressor labii inferioris

e. epicranius frontalis

f. epicranius occipitalis

g. levator labii superioris

h. masseter

i. mentalis

j. platysma

k. orbicularis oculi

l. orbicularis oris

m. zygomaticus

2. Using the terms provided above, identify the muscles described next.

________ 1. used in smiling

________ 2. used to suck in your cheeks

________ 3. used in blinking and squinting

________ 4. used to pout (pulls the corners of the mouth downward)

________ 5. raises your eyebrows for a questioning expression

________ 6. used to form the vertical frown crease on the forehead

________ 7. your "kisser"

________ 8. prime mover to raise the lower jawbone

________ 9. tenses skin of the neck during shaving

Muscles of the Trunk

1. Correctly identify both intact and transected (cut) muscles depicted in the diagram, using the terms given at the right. (Not all terms will be used in this identification.)

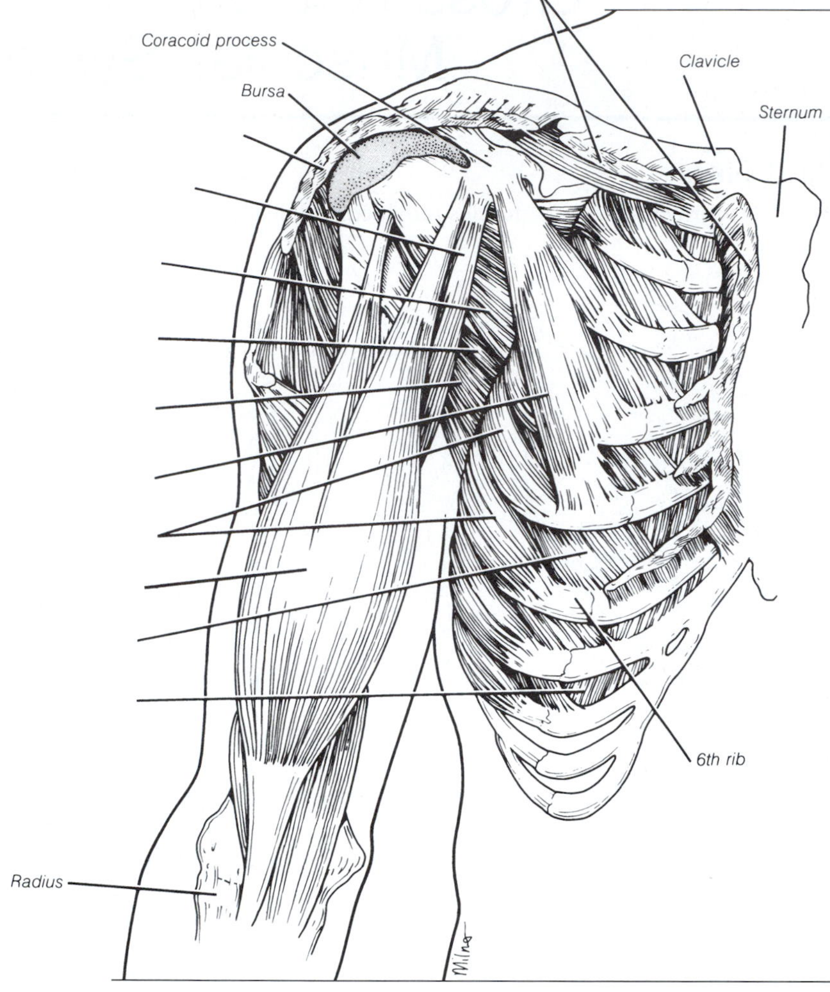

a. biceps brachii
b. coracobrachialis
c. deltoid (cut)
d. external intercostals
e. external oblique
f. internal intercostals
g. internal oblique
h. latissimus dorsi
i. pectoralis major (cut)
j. pectoralis minor
k. rectus abdominis
l. rhomboids
m. serratus anterior
n. subscapularis
o. teres major
p. teres minor
q. transversus abdominis
r. trapezius

2. Using choices from the terms provided in question 1 above, identify the major muscles described next:

_______ 1. a major spine flexor

_______ 2. prime mover for pulling the arm posteriorly

_______ 3. prime mover for shoulder flexion

_______ 4. assume major responsibility for forming the abdominal girdle (three pairs of muscles)

_______ 5. pulls the shoulder backward and downward

_______ 6. prime mover of shoulder abduction

_______ 7. important in shoulder adduction; antagonists of the shoulder abductor (two muscles)

_______ 8. moves the scapula forward and downward

_______ 9. small, inspiratory muscles between the ribs; elevate the ribs

_______ 10. extends the head

_______ 11. pull the scapulae medially

Muscles of the Upper Limb

1. Using terms from the list on the right, correctly identify all muscles provided with leader lines in the diagram. Note that not all the listed terms will be used in this exercise.

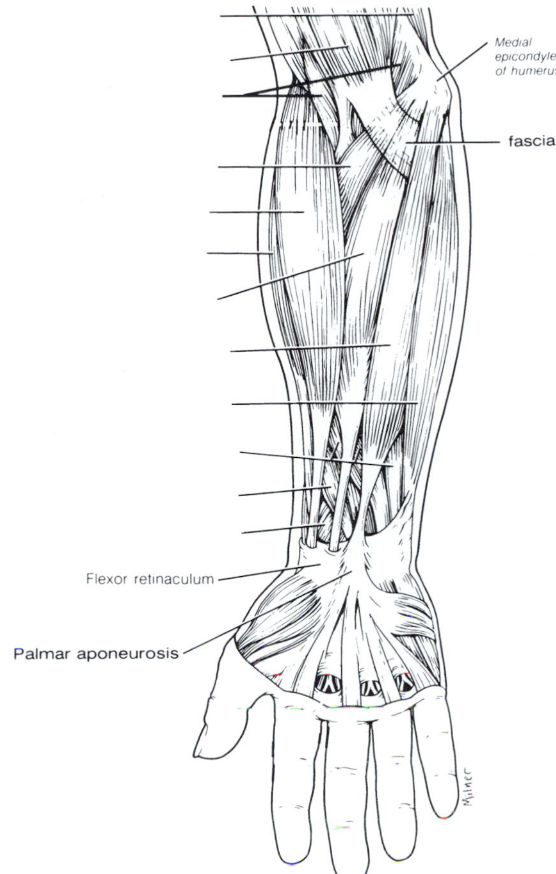

a. biceps brachii
b. brachialis
c. brachioradialis
d. extensor carpi radialis longus
e. extensor digitorum
f. flexor carpi radialis
g. flexor carpi ulnaris
h. flexor digitorum superficialis
i. flexor pollicis longus
j. palmaris longus
k. pronator quadratus
l. pronator teres
m. supinator
n. triceps brachii

2. Use the terms provided in question 1 to identify the muscles described next.

_______ 1. places the palm upward (two muscles)

_______ 2. flexes the forearm and supinates the hand

_______ 3. forearm flexors; no role in supination (two muscles)

_______ 4. elbow extensor

_______ 5. power wrist flexor and abductor

_______ 6. flexes wrist and distal phalanges

_______ 7. pronate the hand (two muscles)

_______ 8. flexes the thumb

_______ 9. extends and abducts the wrist

_______ 10. extends the wrist and digits

_______ 11. flat muscle that is a weak wrist flexor

Muscles of the Lower Limb

1. Using the terms listed to the right, correctly identify all muscles provided with leader lines in the diagram below. Not all listed terms will be used in this exercise.

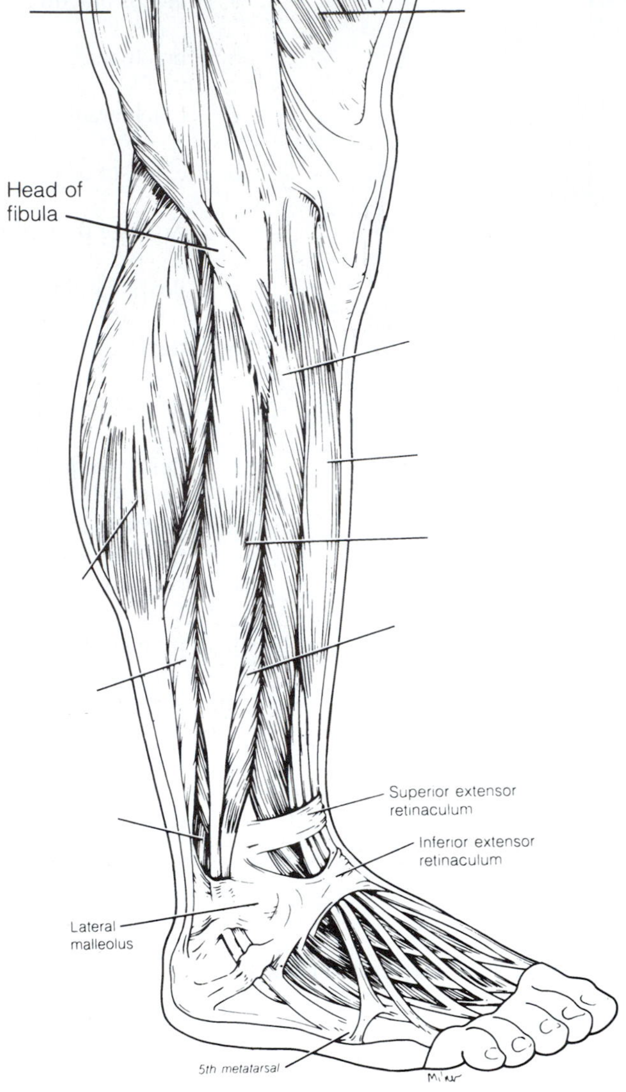

a. adductor group
b. biceps femoris
c. extensor digitorum longus
d. flexor hallucis longus
e. gastrocnemius
f. gluteus maximus
g. gluteus medius
h. peroneus brevis
i. peroneus longus
j. rectus femoris
k. semimembranosus
l. semitendinosus
m. soleus
n. tensor fasciae latae
o. tibialis anterior
p. tibialis posterior
q. vastus muscles

2. Use the key terms in exercise 1 to respond to the descriptions below.

________ 1. flexes the great toe and inverts the ankle

________ 2. lateral compartment muscles that plantar flex and evert the ankle (two muscles)

________ 3. move the thigh laterally to take the "at ease" stance (two muscles)

________ 4. used to extend the hip when climbing stairs

________ 5. prime movers of ankle plantar flexion (two muscles)

________ 6. major foot inverter

________ 7. prime mover of ankle dorsiflexion

________ 8. allow you to draw your legs to the midline of your body, as when standing at attention

________ 9. extends the toes

________ 10. extend thigh and flex knee (three muscles)

________ 11. extends knee and flexes thigh

Muscle Recognition: General Review

1. Identify the lettered muscles in the diagram of the human anterior superficial musculature by matching the letter with one of the following muscle names:

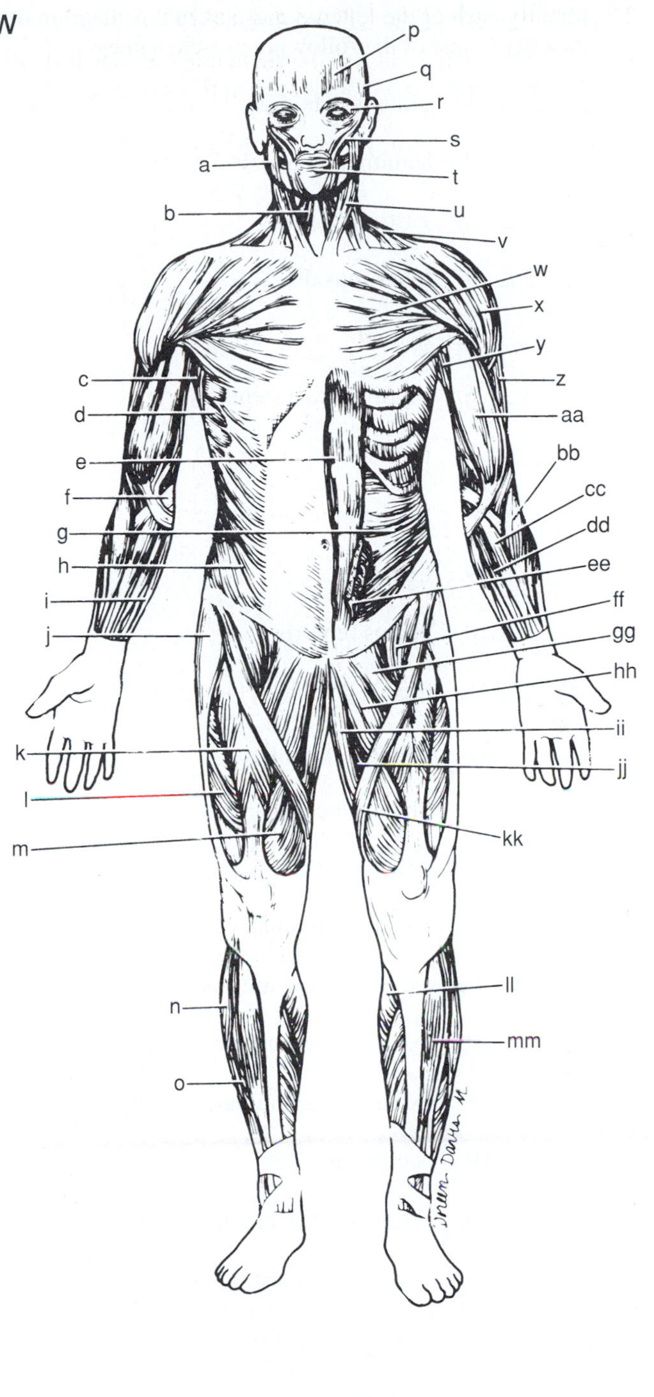

_____ 1. orbicularis oris

_____ 2. pectoralis major

_____ 3. external oblique

_____ 4. sternocleidomastoid

_____ 5. biceps brachii

_____ 6. deltoid

_____ 7. vastus lateralis

_____ 8. brachioradialis

_____ 9. frontalis

_____ 10. rectus femoris

_____ 11. pronator teres

_____ 12. rectus abdominis

_____ 13. sartorius

_____ 14. gracilis

_____ 15. flexor carpi ulnaris

_____ 16. adductor longus

_____ 17. palmaris longus

_____ 18. flexor carpi radialis

_____ 19. latissimus dorsi

_____ 20. orbicularis oculi

_____ 21. gastrocnemius

_____ 22. masseter

_____ 23. trapezius

_____ 24. tibialis anterior

_____ 25. extensor digitorum longus

_____ 26. tensor fasciae latae

_____ 27. pectineus

_____ 28. sternohyoid

_____ 29. serratus anterior

_____ 30. adductor magnus

_____ 31. vastus medialis

_____ 32. transversus abdominis

_____ 33. peroneus longus

_____ 34. iliopsoas

_____ 35. temporalis

_____ 36. zygomaticus

_____ 37. coracobrachialis

_____ 38. triceps brachii

_____ 39. internal oblique

2. Identify each of the lettered muscles in this diagram of the human posterior superficial musculature by matching its letter to one of the following muscle names:

_______ 1. gluteus maximus

_______ 2. semimembranosus

_______ 3. gastrocnemius

_______ 4. latissimus dorsi

_______ 5. deltoid

_______ 6. iliotibial tract (tendon)

_______ 7. teres major

_______ 8. semitendinosus

_______ 9. trapezius

_______ 10. biceps femoris

_______ 11. triceps brachii

_______ 12. external oblique

_______ 13. gluteus medius

_______ 14. gracilis

_______ 15. flexor carpi ulnaris

_______ 16. extensor carpi ulnaris

_______ 17. extensor digitorum communis

_______ 18. extensor carpi radialis longus

_______ 19. occipitalis

_______ 20. extensor carpi radialis brevis

_______ 21. sternocleidomastoid

_______ 22. adductor magnus

_______ 23. anconeus

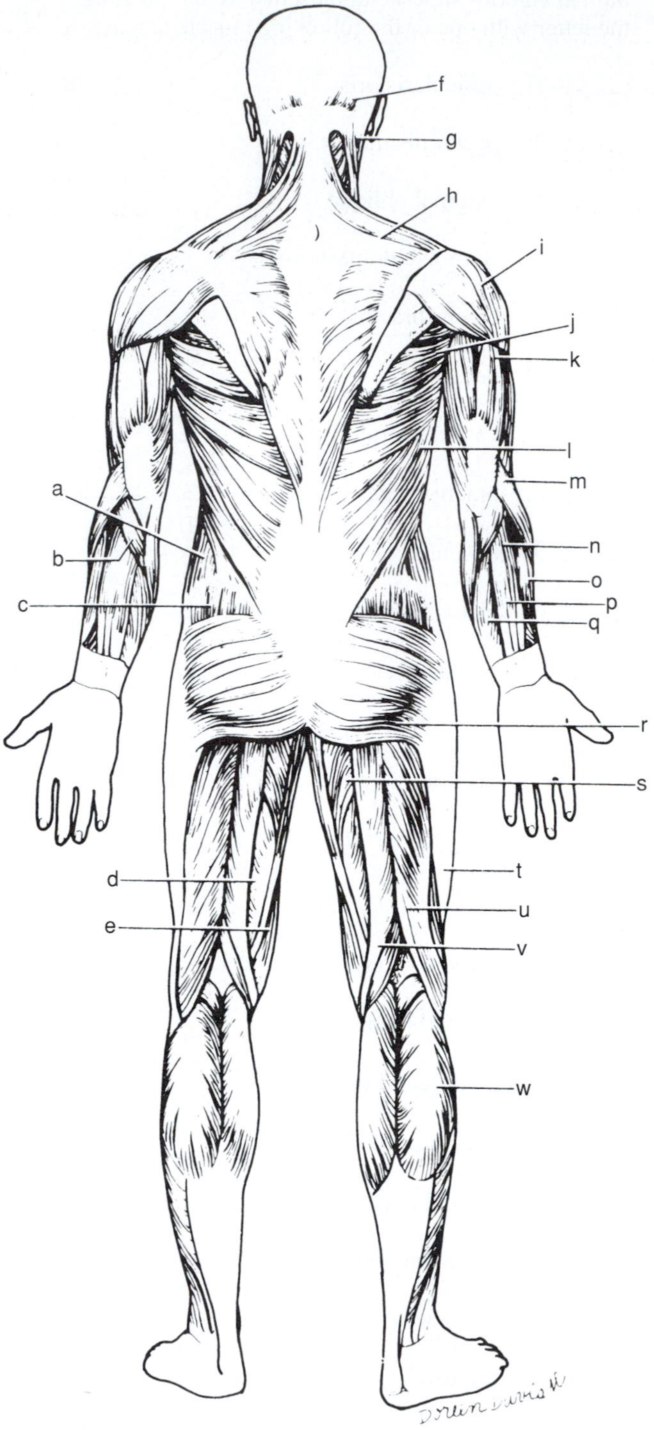

Muscle Descriptions: General Review

Identify the muscles described below by completing the statements:

____________________, ____________________, and ____________________ are commonly used for intramuscular injections (three muscles).

The insertion tendon of the ____________________ group contains a large sesamoid bone, the patella.

The triceps surae insert in common into the ____________________ tendon.

The bulk of the tissue of a muscle tends to lie ____________________ to the part of the body it causes to move.

The extrinsic muscles of the hand originate on the ____________________.

Most flexor muscles are located on the ____________________ aspect of the body; most extensors are located ____________________. An exception to this generalization is the extensor-flexor musculature of the ____________________.

STUDENT NAME ____________________

LAB TIME/DATE ____________________

Review Sheet

EXERCISE 16A

Muscle Physiology (Frog Experimentation)

Muscle Activity

1. The following group of incomplete statements refers to a muscle cell in the resting or polarized state just before stimulation. Complete each statement by choosing the correct response from the key items below.

Key:
a. Na^+ diffuses out of the cell
b. K^+ diffuses out of the cell
c. Na^+ diffuses into the cell
d. K^+ diffuses into the cell
e. inside the cell
f. outside the cell
g. relative ionic concentrations on the two sides of the membrane
h. electrical conditions
i. activation of the sodium-potassium pump, which moves K^+ into the cell and Na^+ out of the cell
j. activation of the sodium-potassium pump, which moves Na^+ into the cell and K^+ out of the cell

There is a greater concentration of Na^+ ________; there is a greater concentration of K^+ ________. When the stimulus is delivered, the permeability of the membrane at that point is changed; and ________ initiating the depolarization of the membrane. Almost as soon as the depolarization wave has begun, a repolarization wave follows it across the membrane. This occurs as ________. Repolarization restores the ________ of the resting cell membrane. The ________ is (are) reestablished by ________.

2. Number the following statements in the proper sequence to describe the contraction mechanism in a skeletal muscle cell. Number 1 has already been designated.

________ Acetylcholine is released into the neuromuscular junction by the axonal terminal.

________ The action potential, carried deep into the cell by the T system, triggers the release of calcium ions from the sarcoplasmic reticulum.

________ The muscle cell relaxes and lengthens.

________ Acetylcholine diffuses across the neuromuscular junction and binds to receptors on the sarcolemma.

________ The calcium ion concentrations at the myofilaments increase; the myofilaments slide past one another, and the cell shortens.

________ Depolarization occurs, and the action potential is generated.

________ The concentration of the calcium ions at the myofilaments decreases as they are actively transported into the sarcoplasmic reticulum.

3. Muscle contraction is commonly explained by the sliding filament hypothesis. What are the essential points of this hypothesis? ____________________

____

____

4. Relative to your observations of muscle fiber contraction (p. 142):

   a. What percentage of contraction was observed with the solution containing ATP, K^+, and Mg^{2+}? ________%

      With *just* ATP? ________% With *just* Mg^{2+} and K^+? ________%

   b. *Explain* your observations fully. ____________________

      ____

   c. What zones or bands disappear when the muscle cell contracts? ____________________

   d. *Draw* a relaxed and a contracted sarcomere below.

Relaxed　　　　　　　　　　Contracted

Induction of Contraction in the Frog Gastrocnemius Muscle

1. Why is it important to destroy the brain and spinal cord of a frog before conducting physiologic experiments on muscle contraction? ____________________

   ____

2. What sources of stimuli, other than electrical shocks, cause a muscle to contract? ____________________

   ____

3. What is the most common stimulus for muscle contraction in the body? ____________________

4. Name the three phases of the muscle twitch, and describe what is happening during each phase:

   (1) ____________, ____________________

   ____

   (2) ____________, ____________________

   (3) ____________, ____________________

5. Use the terms given on the right to identify the conditions described on the left:

______________ 1.	sustained contraction without any evidence of relaxation	a.	maximal stimulus
______________ 2.	stimulus that results in no perceptible contraction	b.	multiple motor unit summation
______________ 3.	stimulus at which the muscle first contracts perceptibly	c.	subthreshold stimulus
______________ 4.	increasingly stronger contractions in the absence of increased stimulus intensity	d.	tetanus
______________ 5.	increasingly stronger contractions owing to stimulation at a rapid rate	e.	threshold stimulus
______________ 6.	increasingly stronger contractions owing to increased stimulus strength	f.	treppe
______________ 7.	weakest stimulus at which all muscle cells in the muscle are contracting	g.	wave summation

6. With brackets and labels, identify the portions of the tracing below that best correspond to three of the phenomena listed in the preceding key. Assume that only the timing of the stimulus has changed.

7. Complete the following statements by writing the appropriate words on the correspondingly numbered blanks at the right.

The "all or none" law applies to skeletal muscle functions at the __1__ level. When a weak but smooth muscle contraction is desired, a few motor units are stimulated at a __2__ rate. Treppe is referred to as the "warming up" process. It is believed that muscles contract more strongly after the first few contractions because the __3__ become more efficient. If blue litmus paper is pressed to the cut surface of a fatigued muscle, the paper color changes to red, indicating low pH. This situation is caused by the accumulation of __4__ in the muscle. Within limits, as the load on a muscle is increased, the muscle contracts __5__ strongly. The relative refractory period is the time when the muscle cell will not respond to a stimulus because __6__ is occurring.

1. ______________
2. ______________
3. ______________
4. ______________
5. ______________
6. ______________

8. During the experiment on muscle fatigue, how did the muscle contraction pattern change as the muscle began to fatigue? ____

____

How long was stimulation continued before fatigue was apparent? ______________________

If the sciatic nerve that stimulates the living frog's gastrocnemius muscle had been left attached to the muscle and the stimulus had been applied to the nerve rather than the muscle, would fatigue have become apparent sooner or later? ____

Explain your answer. ____

____

9. Explain how the weak but sustained (smooth) muscle contractions of precision movements are produced.

____

10. What do you think happens to a muscle in the body when its nerve supply is destroyed or badly damaged?

____

____

11. Explain the relationship between the load on a muscle and its strength of contraction. ______________________

____

____

12. The skeletal muscles are maintained in a slightly stretched condition for optimal contraction. How is this accomplished? ____

Why does overstretching a muscle drastically reduce its ability to contract? (Include an explanation of the events at the level of the myofilaments.) ____

____

____

13. If the length but not the tension of a muscle is changed, the contraction is called an isotonic contraction. In an isometric contraction the tension is increased but the muscle does not shorten. Which type of contraction did you observe most often during the laboratory experiments? ____

What is the role of isometric contractions in normal body functioning? ______________________

____

____

STUDENT NAME ____________________

LAB TIME/DATE ____________________

Review Sheet

EXERCISE 16B

Muscle Physiology (Computerized Simulations)

Electrical Stimulation

1. Complete the following statements.

A motor unit consists of a __1__ and all the __2__ it innervates. If a single motor unit is stimulated, it will respond in a(n) __3__ fashion, whereas whole muscle contraction is a(n) __4__ response. In order for muscles to work in a practical sense, __5__ is the method used to produce a slow steady increase in muscle force.

When we see the slightest evidence of force production on a tracing, the stimulus applied must have reached __6__ .

The weakest stimulus that will elicit the strongest contraction that a muscle is capable of is called the __7__ , and that level of contraction is called the __8__ .

When the __9__ of stimulation is so high that the muscle tracing shows fused twitch peaks, __10__ has been achieved.

1. ____________________
2. ____________________
3. ____________________
4. ____________________
5. ____________________
6. ____________________
7. ____________________
8. ____________________
9. ____________________
10. ____________________

2. Name and describe what is happening in each phase of the typical muscle twitch.

(1) ____________________, ____________________

____

(2) ____________________, ____________________

(3) ____________________, ____________________

____

3. What are the two ways in which mode of stimulation can affect the force a muscle produces?

____________________ and ____________________

Explain. ____________________

____

____

____

Isometric Contraction

1. Identify the following conditions by choosing one of the key terms listed on the right.

   _______ is generated by muscle tissue when it is being stretched

   _______ requires the input of energy

   _______ is measured by recording instrumentation *during* contraction

   Key:

   a. total force

   b. resting force

   c. active force

2. Using the letters of the following statements, correctly label the area on the given Force-Length curves.

   a. an increase in resting length produced an *increase* in the active force generated

   b. an increase in resting length produced a *decrease* in active force generated

   c. an increase in resting length produced an *increase* in resting force

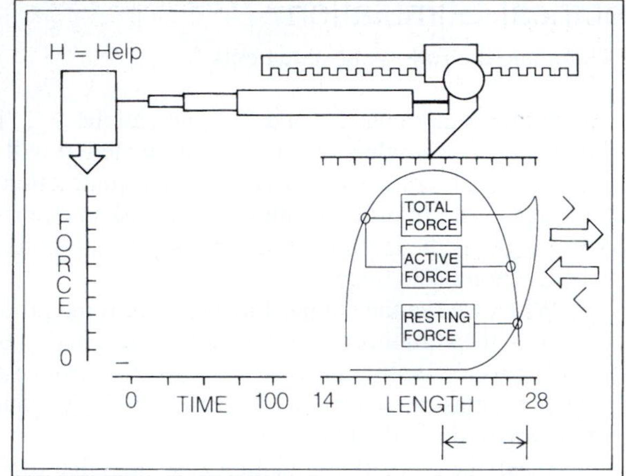

3. Explain what happens to muscle force production at extremes of length (too short or too long). (Relate to sarcomere structure.)

   Muscle too short: ____

   ____

   Muscle too long: ____

   ____

Isotonic Contraction

1. Assuming a fixed starting length, describe the effect afterload has on the initial velocity of shortening, and explain why.

____

____

____

____

2. A muscle has just been stimulated under conditions that will allow both isometric and isotonic contractions. Describe what is happening in terms of length and force.

Isometric: ____

____

Isotonic: ____

____

Terms

Select the condition from column B that most correctly identifies the term in column A.

	Column A	Column B
_______	1. muscle twitch	a. response is all or none
_______	2. wave summation	b. affects the force a muscle can generate
_______	3. multiple motor unit summation	c. a single contraction of intact muscle
_______	4. resting length	d. recruitment
_______	5. afterload	e. increasing force produced by increasing stimulus frequency
_______	6. initial velocity of shortening	f. muscle length changing due to relaxation
_______	7. isotonic shortening	g. caused by application of maximal stimulus
_______	8. isotonic lengthening	h. weight
_______	9. motor unit	i. exhibits graded response
_______	10. whole muscle	j. high values with low afterloads
_______	11. tetanus	k. changing muscle length due to active forces
_______	12. maximal response	l. recording shows no evidence of muscle relaxation

Attach any tracings required by your instructor to the reverse side of this page.

STUDENT NAME ______________________

LAB TIME/DATE ______________________

Review Sheet

EXERCISE 17

Histology of Nervous Tissue

1. The cellular unit of the nervous system is the neuron. What is the major function of this cell type?

____

2. Name four types of neuroglia and list at least four functions of these cells. (You will need to consult your textbook for this.)

Types	**Functions**
____________	____________
	____________
____________	____________
	____________
____________	____________
	____________
____________	____________
	____________

3. Match each statement with a response chosen from the key.

Key:

a.	afferent neuron	e.	ganglion	i.	nuclei
b.	association neuron	f.	neuroglia	j.	peripheral nervous system
c.	central nervous system	g.	neurotransmitters	k.	synapse
d.	efferent neuron	h.	nerve	l.	tract

_______ 1. the brain and spinal cord collectively

_______ 2. specialized supporting cells in the CNS

_______ 3. junction or point of close contact between neurons

_______ 4. a bundle of nerve processes inside the central nervous system

_______ 5. neuron serving as part of the conduction pathway between sensory and motor neurons

_______ 6. spinal and cranial nerves and ganglia

_______ 7. collection of nerve cell bodies found outside the CNS

_______ 8. neuron that conducts impulses away from the CNS to muscles and glands

_______ 9. neuron that conducts impulses toward the CNS from the body periphery

_______ 10. chemicals released by neurons that stimulate or inhibit other neurons or effectors

Neuron Anatomy

1. Match the following anatomical terms (column B) with the appropriate description or function (column A).

	Column A	Column B
________	1. region of the cell body from which the axon originates	a. axon
________	2. secretes neurotransmitters	b. axonal terminal
________	3. receptive region of a neuron	c. axon hillock
________	4. insulates the nerve fibers	d. dendrite
________	5. is site of the nucleus and the most important metabolic area	e. myelin sheath
________	6. may be involved in the transport of substances within the neuron	f. neuronal cell body
________	7. essentially rough endoplasmic reticulum, important metabolically	g. neurofibril
________	8. impulse generator and transmitter	h. Nissl bodies

2. Draw a "typical" neuron in the space below. Include and label the following structures on your diagram: cell body, nucleus, Nissl bodies, dendrites, axon, axon collaterals, myelin sheath, and nodes of Ranvier.

3. How is one-way conduction at synapses assured? ____

____

4. What anatomical characteristic determines whether a particular neuron is classified as unipolar, bipolar, or multipolar? ____

Make a simple line drawing of each type here.

Unipolar neuron   Bipolar neuron   Multipolar neuron

5. Describe how the Schwann cells form the myelin sheath and the neurilemma encasing the nerve processes. (You may want to diagram the process.) ______________________________

____

____

____

Structure of a Nerve

1. What is a nerve? ______________________________

____

2. State the location of each of the following connective tissue coverings:

endoneurium ______________________________

perineurium ______________________________

epineurium ______________________________

3. What is the value of the connective tissue wrappings found in a nerve? ______________________________

____

4. Define *mixed nerve:* ______________________________

____

5. Identify all indicated parts of the nerve section.

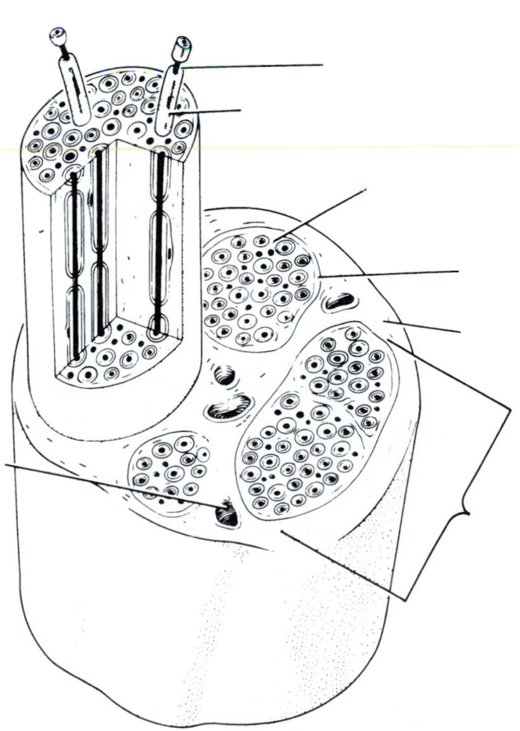

STUDENT NAME ______________________________

LAB TIME/DATE ______________________________

Review Sheet

EXERCISE 18

Neurophysiology of Nerve Impulses

The Nerve Impulse

1. Match each of the terms in column B to the appropriate definition in column A.

Column A

_______ 1. period of depolarization of the neuron membrane during which it cannot respond to a second stimulus

_______ 2. reversal of the resting potential owing to an influx of sodium ions

_______ 3. period during which potassium ions diffuse out of the neuron owing to a change in membrane permeability

_______ 4. self-propagated transmission of the depolarization wave along the neuronal membrane

_______ 5. mechanism in which ATP is used to move sodium out of the cell and potassium into the cell; restores the resting membrane voltage and intracellular ionic concentrations

Column B

a. action potential

b. absolute refractory period

c. depolarization

d. relative refractory period

e. repolarization

f. sodium-potassium pump

2. Respond appropriately to each statement below either by completing the statement or by answering the question raised. Insert your responses in the corresponding numbered blanks on the right.

1. The cellular unit of the nervous system is the neuron. What is the major function of this cell type?

2 and 3. What characteristics are highly developed to allow the neuron to perform this function?

4. Would a substance that decreases membrane permeability to sodium increase *or* decrease the probability of generating a nerve impulse?

1. ______________________________

2. ______________________________

3. ______________________________

4. ______________________________

3. Why don't the terms *depolarization* and *action potential* mean the same thing? (*Hint:* under which conditions will a local depolarization *not* lead to the action potential?) ______________________________

____

____

4. A nerve generally contains many thickly myelinated fibers that typically exhibit nodes of Ranvier. An action potential is generated along these fibers by "saltatory conduction." Use an appropriate reference to explain how saltatory conduction differs from conduction along unmyelinated fibers.

____

____

____

Physiology of Nerve Fibers: Eliciting and Inhibiting the Nerve Impulse

1. Respond appropriately to each question posed below. Insert your responses in the corresponding numbered blanks to the right.

   1–3. Name three types of stimuli that resulted in action potential generation in the sciatic nerve of the frog during the laboratory experiments.

   4. Which of the stimuli resulted in the most effective nerve stimulation?

   5. Which of the stimuli employed in that experiment might represent types of stimuli to which nerves in the human body are subjected?

   6. What is the usual mode of stimulus transfer in neuron-to-neuron interactions?

   7. Since the action potentials themselves were not visualized with an oscilloscope during this initial set of experiments, how did you recognize that impulses were being transmitted?

   1. ______________________________

   2. ______________________________

   3. ______________________________

   4. ______________________________

   5. ______________________________

   6. ______________________________

   7. ______________________________

2. Describe the observed effects of ether on nerve-muscle interaction. ______________________________

   ____

   Does ether exert its blocking effects on nerve *or* muscle? ____________________ What observations made during the experiment support your conclusion? ______________________________

   ____

   ____

3. At what site did the tubocurarine block the impulse transmission? ____________________ Provide evidence from the experiment to substantiate your conclusion. ______________________________

   ____

   ____

   Why was one of the frog's legs ligated in this experiment? ______________________________

   ____

   ____

Visualizing the Action Potential with an Oscilloscope

1. What is a stimulus artifact? ________________________________

____

2. Explain why the amplitude of the action potential recorded from the frog sciatic nerve increased when the voltage of the stimulus was increased above the threshold value. ________________________________

____

3. What was the effect of cold temperature (flooding the nerve with iced Ringer's solution) on the functioning of the sciatic nerve tested? ________________________________

____

4. When the nerve was reversed in position, was the impulse conducted in the opposite direction? ____________

   How can this result be reconciled with the concept of one-way conduction in neurons? ____________

____

____

____

____

____

STUDENT NAME ____________________

LAB TIME/DATE ____________________

Review Sheet

EXERCISE 19

Gross Anatomy of the Brain and Cranial Nerves

The Human Brain

1. Match the letters on the diagram of the human brain (right lateral view) to the appropriate terms listed at the left:

_______ 1. frontal lobe

_______ 2. parietal lobe

_______ 3. temporal lobe

_______ 4. precentral gyrus

_______ 5. parieto-occipital sulcus

_______ 6. postcentral gyrus

_______ 7. lateral sulcus

_______ 8. central sulcus

_______ 9. cerebellum

_______ 10. medulla

_______ 11. occipital lobe

_______ 12. pons

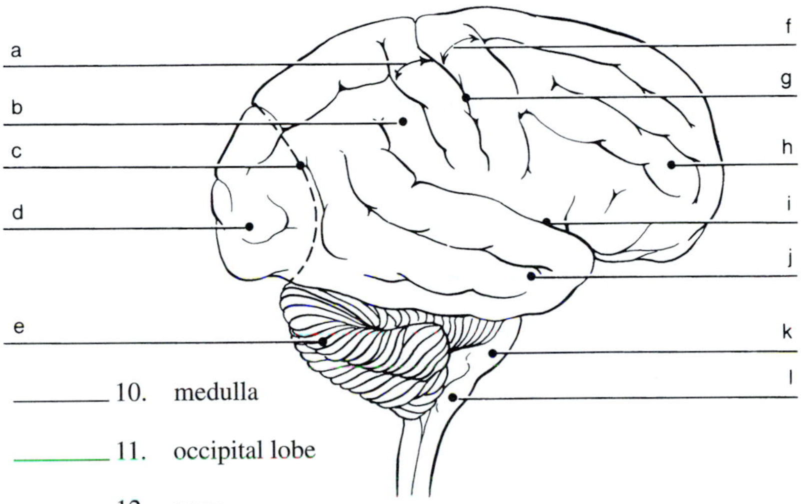

a b c d e f g h i j k l

2. In which of the cerebral lobes would the following functional areas be found?

auditory area ____________________ olfactory area ____________________

primary motor area ____________________ visual area ____________________

primary sensory area ____________________ Broca's area ____________________

3. Which of the following structures are not part of the brain stem? (Circle the appropriate response or responses.)

cerebral hemispheres pons midbrain cerebellum medulla diencephalon

4. Complete the following statements by writing the proper word or phrase on the corresponding blanks at the right.

A(n) __1__ is an elevated ridge of cerebral tissue. The convolutions seen in the cerebrum are important because they increase the __2__ . Gray matter is composed of __3__ . White matter is composed of __4__ . A fiber tract that provides for communication between different parts of the same cerebral hemisphere is called a(n) __5__ , whereas one that carries impulses to and from the cerebrum from and to lower CNS areas is called a(n) __6__ tract. The lentiform nucleus along with the amygdaloid and caudate nuclei are collectively called the __7__ .

1. ____________________

2. ____________________

3. ____________________

4. ____________________

5. ____________________

6. ____________________

7. ____________________

5. Identify the structures on the following sagittal view of the human brain by matching the lettered areas to the proper terms at the left:

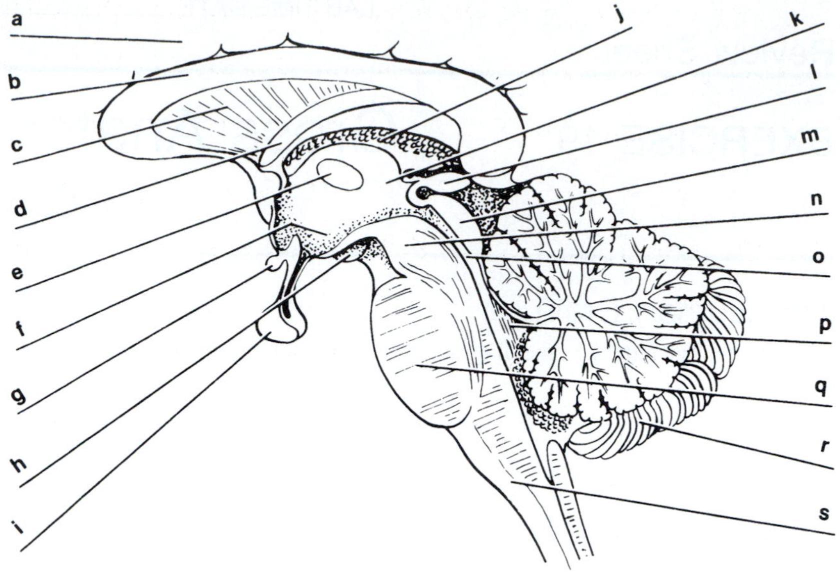

a b c d e f g h i j k l m n o p q r s

_______ 1. cerebellum

_______ 2. cerebral aqueduct

_______ 3. cerebral hemisphere

_______ 4. cerebral peduncle

_______ 5. choroid plexus

_______ 6. corpora quadrigemina

_______ 7. corpus callosum

_______ 8. fornix

_______ 9. fourth ventricle

_______ 10. hypothalamus

_______ 11. mammillary bodies

_______ 12. massa intermedia

_______ 13. medulla oblongata

_______ 14. optic chiasma

_______ 15. pineal body

_______ 16. pituitary gland

_______ 17. pons

_______ 18. septum pellucidum

_______ 19. thalamus

6. Using the letters from the diagram in item 5, match the appropriate structures with the descriptions given below:

_______ 1. site of regulation of body temperature and water balance; most important autonomic center

_______ 2. consciousness depends on the function of this part of the brain

_______ 3. located in the midbrain; contains reflex centers for vision and audition

_______ 4. responsible for regulation of posture and coordination of complex muscular movements

_______ 5. important synapse site for afferent fibers traveling to the sensory cortex

_______ 6. contains autonomic centers regulating blood pressure, heart rate, and respiratory rhythm, as well as coughing, sneezing, and swallowing centers

_______ 7. large commissure connecting the cerebral hemispheres

_______ 8. fiber tract involved with olfaction

_______ 9. connects the third and fourth ventricles

_______ 10. encloses the third ventricle

7. Embryologically, the brain arises from the rostral end of a tubelike structure that quickly becomes divided into three major regions. Groups of structures that develop from the embryonic brain are listed below. Designate the embryonic origin of each group as the hindbrain, midbrain, or forebrain.

_______________ 1. the diencephalon, including the thalamus, optic chiasma, and hypothalamus

_______________ 2. the medulla, pons, and cerebellum

_______________ 3. the cerebral hemispheres

8. What is the function of the basal nuclei? _______________

____

9. What is the corpus striatum, and how is it related to the fibers of the internal capsule? _______________

____

____

10. A brain hemorrhage within the region of the right internal capsule results in paralysis of the left side of the body.

Explain why the left side (rather than the right side) is affected. _______________

____

11. Explain why trauma to the base of the brain is often much more dangerous than trauma to the frontal lobes. (*Hint:* Think about the relative functioning of the cerebral hemispheres and the brain stem structures. Which contain centers more vital to life?)

____

____

____

12. In "split brain" experiments, the main commissure connecting the cerebral hemispheres is cut. First, name this

commissure: _______________

Then, describe what results (in terms of behavior) can be anticipated in such experiments. (Use an appropriate reference if you need help with this one!)

____

____

____

____

____

Meninges of the Brain

Identify the meningeal (or associated) structures described below:

________________________ 1. outermost meninx covering the brain; composed of tough fibrous connective tissue

________________________ 2. innermost meninx covering the brain; delicate and highly vascular

________________________ 3. structures instrumental in returning cerebrospinal fluid to the venous blood in the dural sinuses

________________________ 4. structure that forms the cerebrospinal fluid

________________________ 5. middle meninx; like a cobweb in structure

________________________ 6. its outer layer forms the periosteum of the skull

________________________ 7. a dural fold that attaches the cerebrum to the crista galli of the skull

________________________ 8. a dural fold separating the cerebrum from the cerebellum

Cerebrospinal Fluid

Fill in the following flowchart by delineating the circulation of cerebrospinal fluid from its formation site (assume that this is one of the lateral ventricles) to the site of its reabsorption into the venous blood:

Lateral ventricle --------------> ________________ --------------> Third ventricle --->

-------------------------------> ________________ --------------> ________________

________________ --------------> ____________
--------------> ____________ --->

________________ surrounding the brain and cord --------------> Arachnoid villi --------->
(and central canal of the cord)

-------------------------------> ________________ containing venous blood

Cranial Nerves

1. Using the terms below, correctly identify all structures indicated by leader lines on the diagram below.

a. abducens nerve (I)

b. accessory nerve (XI)

c. cerebellum

d. cerebral peduncle

e. decussation of the pyramids

f. facial nerve (VII)

g. frontal lobe of cerebral hemisphere

h. glossopharyngeal nerve (IX)

i. hypoglossal nerve (XII)

j. longitudinal fissure

k. mammillary body

l. medulla oblongata

m. oculomotor nerve (III)

n. olfactory bulb

o. olfactory tract

p. optic chiasma

q. optic nerve (II)

r. optic tract

s. pituitary gland

t. pons

u. spinal cord

v. temporal lobe of cerebral hemisphere

w. trigeminal nerve (V)

x. trochlear nerve (IV)

y. vagus nerve (X)

z. vestibulocochlear nerve (VIII)

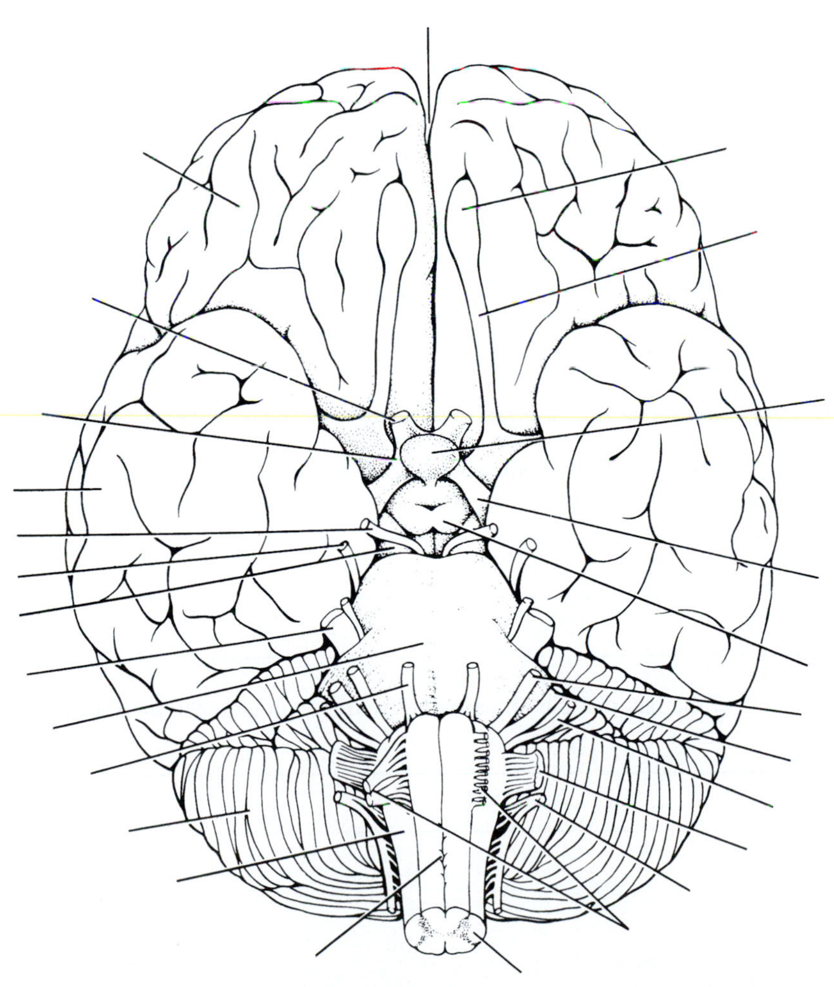

2. Provide the name and number of the cranial nerves involved in each of the following activities, sensations, or disorders:

________________ 1. shrugging the shoulders

________________ 2. smelling a flower

________________ 3. raising the eyelids; focusing the lens of the eye for accommodation; and pupillary constriction

________________ 4. slows the heart; increases the mobility of the digestive tract

________________ 5. involved in Bell's palsy (facial paralysis)

________________ 6. chewing food

________________ 7. listening to music; seasickness

________________ 8. secretion of saliva; tasting well-seasoned food

________________ 9. involved in "rolling" the eyes (three nerves—provide numbers only)

________________ 10. feeling a toothache

________________ 11. reading *Playgirl* or *Playboy* magazine

________________ 12. purely sensory in function (three nerves—provide numbers only)

Dissection of the Sheep Brain

1. In your own words, describe the relative hardness of the sheep brain tissue as observed when cutting into it.

____

Because formalin hardens all tissue, what conclusions might you draw about the relative hardness and texture of living brain tissue? ________________

2. How does the relative size of the cerebral hemispheres compare in sheep and human brains? ________________

____

What is the significance? ________________

____

3. What is the significance of the fact that the olfactory bulbs are much larger in the sheep brain than in the human brain? ________________

____

STUDENT NAME ___________________________

LAB TIME/DATE ___________________________

Review Sheet

EXERCISE 20 Electroencephalography

Brain Wave Patterns and the Electroencephalogram

1. Define *EEG.* ___________________________

2. What are the four major types of brain wave patterns? ___________________________

   Match each statement below to a type of brain wave pattern:

   __________ below 4 Hz; slow, large waves; normally seen during deep sleep

   __________ rhythm generally apparent when an individual is in a relaxed, nonattentive state with the eyes closed

   __________ correlated to the alert state; usually about 15 to 25 Hz

3. What is meant by the term *alpha block?* ___________________________

   ____

   ____

4. List at least four types of brain lesions that may be determined by EEG studies. ___________________________

   ____

5. What is the common result of hypoactivity or hyperactivity of the brain neurons? ___________________________

Observing Brain Wave Patterns

1. How was alpha block demonstrated in the laboratory experiment? ___________________________

   ____

2. What was the effect of mental concentration on the brain wave pattern? ___________________________

   ____

   ____

3. What effect on the brain wave pattern did hyperventilation have? ___________________________

   Why? ___________________________

   ____

STUDENT NAME ____________________

LAB TIME/DATE ____________________

Review Sheet

EXERCISE 21

Spinal Cord Spinal Nerves, and the Autonomic Nervous System

Anatomy of the Spinal Cord

1. Match the descriptions given below to the proper anatomical term:

a. cauda equina b. conus medullaris c. filium terminale d. foramen magnum

_______ 1. most superior boundary of the spinal cord

_______ 2. meningeal extension beyond the spinal cord terminus

_______ 3. spinal cord terminus

_______ 4. collection of spinal nerves traveling in the vertebral canal below the terminus of the spinal cord

2. Using the terms below, correctly identify on the diagram all structures provided with leader lines.

a. anterior (ventral) horn
b. arachnoid mater
c. central canal
d. dorsal ramus of spinal nerve
e. dorsal root ganglion
f. dorsal root of spinal nerve
g. dura mater
h. gray commissure
i. lateral horn
j. pia mater
k. posterior (dorsal) horn
l. spinal nerve
m. ventral ramus of spinal nerve
n. ventral root of spinal nerve
o. white matter

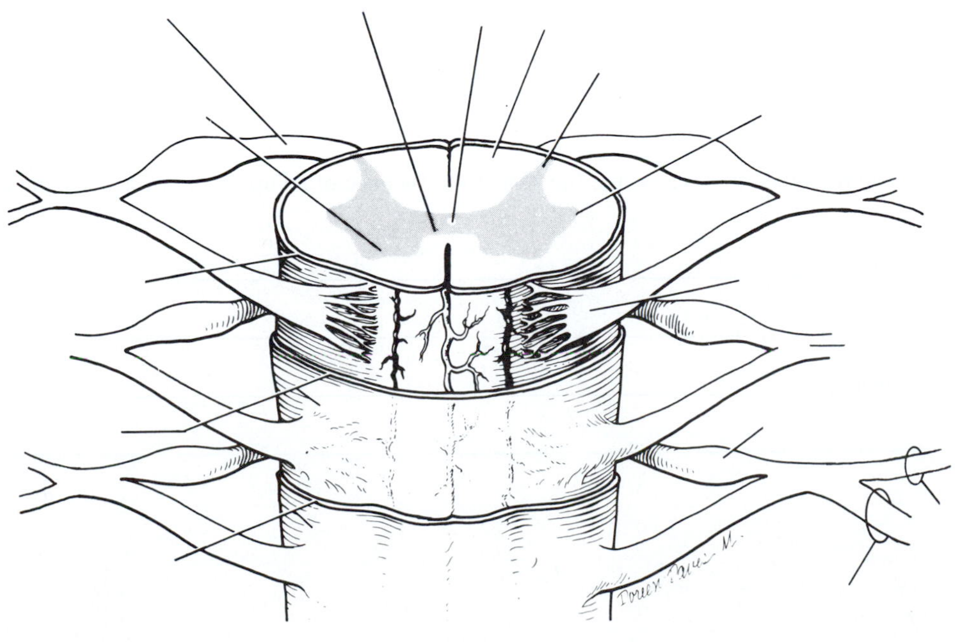

3. Choose the proper answer from the following key to respond to the descriptions relating to spinal cord anatomy.

Key: a. afferent b. efferent c. both afferent and efferent d. association

_______ 1. neuron type found in dorsal horn

_______ 2. neuron type found in ventral horn

_______ 3. neuron type in dorsal root ganglion

_______ 4. fiber type in ventral root

_______ 5. fiber type in dorsal root

_______ 6. fiber type in spinal nerve

4. Where in the vertebral column is a lumbar puncture generally done? _______________

Why is this the site of choice? _______________

____

5. The spinal cord is enlarged in two regions, the _______________ and the _______________ regions. What is the significance of these enlargements? _______________

____

6. How does the position of the gray and white matter differ in the spinal cord and the cerebral hemispheres?

____

____

7. Choose the name of the tract, from the following key, that might be damaged when the following conditions are observed. (More than one choice may apply.)

_______________ 1. uncoordinated movement

_______________ 2. lack of voluntary movement

_______________ 3. tremors, jerky movements

_______________ 4. diminished pain perception

_______________ 5. diminished sense of touch

Key:

a. fasciculus gracilis
b. fasciculus cuneatus
c. lateral corticospinal tract
d. ventral corticospinal tract
e. tectospinal tract
f. rubrospinal tract
g. lateral spinothalamic tract
h. ventral spinothalamic tract
i. dorsal spinocerebellar tract
j. vestibulospinal tract
k. olivospinal tract
l. ventral spinocerebellar tract

8. Use an appropriate reference to describe the functional significance of an upper motor neuron and a lower motor neuron:

upper motor neuron _______________

____

lower motor neuron _______________

____

Will contraction of a muscle occur if the lower motor neurons serving it have been destroyed? ________ If the upper motor neurons serving it have been destroyed? ________ Using an appropriate reference, differentiate between flaccid and spastic paralysis and note the possible causes of each. ________________________________

____

____

____

____

Spinal Nerves and Nerve Plexuses

1. In the human, there are 31 pairs of spinal nerves named according to the region of the vertebral column from which they issue. The spinal nerves are named below; note, by number, the vertebral level at which they emerge:

   cervical nerves ________________________ sacral nerves ________________________

   lumbar nerves ________________________ thoracic nerves ________________________

2. The ventral rami of spinal nerves C_1 through T_1 and T_{12} through S_4 take part in forming ________________, which serve the ________________________ of the body. The ventral rami of T_2 through T_{12} run between the ribs to serve the ________________________. The dorsal rami of the spinal nerves serve ________________________.

3. What would happen if the following structures were damaged or transected? (Use key choices for responses.)

   Key: a. loss of motor function b. loss of sensory function c. loss of both motor and sensory function

   ________ 1. dorsal root of a spinal nerve

   ________ 2. ventral root of a spinal nerve

   ________ 3. anterior ramus of a spinal nerve

4. Define *plexus:* ____

   ____

5. Name the major nerves that serve the following body areas:

   ________________________ 1. head, neck, shoulders (name plexus only)

   ________________________ 2. diaphragm

   ________________________ 3. posterior thigh

   ________________________ 4. leg and foot (name two)

   ________________________ 5. most anterior forearm muscles

   ________________________ 6. arm muscles (name two)

   ________________________ 7. abdominal wall (name plexus only)

   ________________________ 8. anterior thigh

   ________________________ 9. medial side of the hand

The Autonomic Nervous System

1. For the most part, sympathetic and parasympathetic fibers serve the same organs and structures. How can they exert antagonistic effects? (After all, nerve impulses are nerve impulses—aren't they?)

____

2. Name three structures that receive sympathetic but not parasympathetic innervation.

____

____

3. A pelvic splanchnic nerve contains (circle one):

   (a) preganglionic sympathetic fibers
   (b) postganglionic sympathetic fibers
   (c) preganglionic parasympathetic fibers
   (d) postganglionic parasympathetic fibers

4. The following chart states a number of conditions. Use a check mark to show which division of the autonomic nervous system is involved in each.

Sympathetic division	Condition	Parasympathetic division
	Secretes norepinephrine; adrenergic fibers	
	Secretes acetylcholine; cholinergic fibers	
	Long preganglionic axon; short postganglionic axon	
	Short preganglionic axon; long postganglionic axon	
	Arises from cranial and sacral nerves	
	Arises from spinal nerves T_1 through L_3	
	Normally in control	
	"Fight or flight" system	
	Has more specific control (Look it up!)	

5. You are alone in your home late in the evening, and you hear an unfamiliar sound in your backyard. List four physiologic events promoted by the sympathetic nervous system that would aid you in coping with this rather frightening situation:

____

____

____

____

____

6. Often after surgery, people are temporarily unable to urinate, and bowel sounds are absent. What division of the ANS is affected by the anesthesia? _______________

STUDENT NAME ____________________

LAB TIME/DATE ____________________

Review Sheet

EXERCISE 22

Human Reflex Physiology

The Reflex Arc

1. Define *reflex:* ____________________

____

2. Name five essential components of a reflex arc: ____________, ____________,

____________, ____________, and ____________

3. In general, what is the importance of reflex testing in a routine physical examination? ____________

____

____

Somatic and Autonomic Reflexes

1. Use the key terms to complete the statements given below.

Key:
a. abdominal reflex
b. Achilles jerk
c. ciliospinal reflex
d. corneal reflex
e. crossed extensor reflex
f. gag reflex
g. patellar reflex
h. plantar reflex
i. pupillary light reflex

Reflexes classified as somatic reflexes include a ______, ______, ______, ______, ______, ______,

and ______. Of these, the simple stretch reflexes are ______ and ______, and the superficial cord reflexes

are ______ and ______. Reflexes classified as autonomic reflexes include ______ and ______.

2. In what way do cord-mediated reflexes differ from those involving higher brain centers? ____________

____

Name two cord-mediated reflexes: ____________ and ____________

____

Name two somatic reflexes in which the higher brain centers participate: ____________

and ____________________

3. Can the stretch reflex be elicited in a pithed animal? ____________

Explain your answer. ____________________

____

4. Trace the reflex arc, naming efferent and afferent nerves, receptors, effectors, and integration centers, for the following reflexes:

   patellar reflex ______________________________

   ____

   ____

   Achilles reflex ______________________________

   ____

   ____

5. Three factors that influence the rapidity and effectiveness of reflex arcs were investigated in conjunction with patellar reflex testing—mental distraction, effect of simultaneous muscle activity in another body area, and fatigue.

   Which of these factors increases the excitatory level of the spinal cord? ______________________________

   Which factor decreases the excitatory level of the muscles? ______________________________

   When the subject was concentrating on an arithmetic problem, did the change noted in the patellar reflex indicate that brain activity is necessary for the patellar reflex or only that it may modify it? ______________________________

   ____

6. Name the division of the autonomic nervous system responsible for each of the following reflexes:

   ciliospinal reflex ______________________________ salivary reflex ______________________________

   pupillary light reflex ______________________________

7. The pupillary light reflex, the crossed extensor reflex, and the corneal reflex illustrate the purposeful nature of reflex activity. Describe the protective aspect of each:

   pupillary light reflex ______________________________

   corneal reflex ______________________________

   crossed extensor reflex ______________________________

   ____

   ____

   ____

   ____

8. Was the pupillary consensual response contralateral or ipsilateral? ______________________________

   Why would such a response be of significant value in this particular reflex? ______________________________

   ____

9. Differentiate between the types of activities accomplished by somatic and autonomic reflexes. ______________

____

____

____

10. Several types of reflex activity were not investigated in this exercise. The most important of these are autonomic reflexes, which are difficult to illustrate in a laboratory situation. To rectify this omission, complete the following chart, using references as necessary.

Reflex	Organ involved	Receptors stimulated	Action
Micturition (urination)			
Hering-Breuer			
Defecation			
Carotid sinus			

Reaction Time of Unlearned Responses

1. Name at least three factors that may modify reaction time to a stimulus. ______________________________

   ____

2. In general, how did the response time for the unlearned activity performed in the laboratory compare to that for the simple patellar reflex? ______________________________

3. Did the response time without verbal stimuli decrease with practice? ________ Explain the reason for this.

   ____

4. Explain, in detail, why response time increased when the subject had to react to a word stimulus.

   ____

   ____

STUDENT NAME ______________________

LAB TIME/DATE ______________________

Review Sheet

EXERCISE 23

General Sensation

Structure of Sensory Receptors

1. Differentiate between interoceptors and exteroceptors relative to location and stimulus source:

   Interoceptor: ______________________

   Exteroceptor: ______________________

2. A number of activities and sensations are listed in the chart below. For each, check whether the receptors would be exteroceptors or interoceptors; and then name the specific receptor types. (Because visceral receptors were not described in detail in this exercise, you need only indicate that the receptor is a visceral receptor if it falls into that category.)

Activity or sensation	Exteroceptor	Interoceptor	Specific receptor type
Backing into a sun-heated iron railing			
Someone steps on your foot			
Reading a book			
Leaning on your elbows			
Doing sit-ups			
The "too full" sensation			
Seasickness			

Receptor Physiology

1. Explain how the sensory receptors act as transducers: ______________________

____

2. Define *stimulus:* ______________________

____

3. What was demonstrated by the two-point discrimination test? ______________________

____

   How did the accuracy of the subject's tactile localization correlate with the results of the two-point discrimination test? ______________________

____

4. Define *punctate distribution:* ______________________________

____

5. Several questions regarding general sensation are posed below. Answer each by placing your response in the appropriately numbered blanks to the right.

1.	Which cutaneous receptors are the most numerous?	1. ______________
2–3.	Which two body areas tested were most sensitive to touch?	2. ______________
		3. ______________
4–5.	Which two body areas tested were least sensitive to touch?	4. ______________
		5. ______________
6.	Which appear to be more numerous—receptors that respond to cold or to heat?	6. ______________
7–9.	Where would referred pain appear if the following organs were receiving painful stimuli—(7) gallbladder, (8) kidneys, and (9) appendix? (Use your textbook if necessary.)	7. ______________
		8. ______________
		9. ______________
10.	Where was referred pain felt when the elbow was immersed in ice water during the laboratory experiment?	10. ______________
11.	What region of the cerebrum interprets the kind and intensity of stimuli that cause cutaneous sensations?	11. ______________

6. Define *adaptation:* ______________________________

____

7. Why is it advantageous to have pain receptors that are sensitive to all vigorous stimuli, whether heat, cold, or pressure? ______________________________

Why is the nonadaptability of pain receptors important? ______________________________

____

8. Imagine yourself without any cutaneous sense organs. Why might this be very dangerous? ______________

____

____

9. Define *referred pain:* ______________________________

____

What is the probable explanation for referred pain? (Consult your textbook or an appropriate reference if necessary.) ______________________________

____

____

STUDENT NAME ______________________________

LAB TIME/DATE ______________________________

Review Sheet

EXERCISE 24

Special Senses: Vision

Anatomy of the Eye

1. Three accessory eye structures contribute to the formation of tears and/or aid in lubrication of the eyeball. Name each and then name its major secretory product. Indicate which has antibacterial properties by circling the correct secretory product.

Accessory structures	Product

2. The eyeball is wrapped in adipose tissue within the orbit. What is the function of the adipose tissue?

____

What seven bones form the bony orbit? (Think! If you can't remember, check a skull or your text.)

____________________ ____________________ ____________________

____________________ ____________________

____________________ ____________________

3. Why does one often have to blow one's nose after having a good cry? ______________________________

____

4. Identify the extrinsic eye muscle predominantly responsible for the actions described below.

____________________ 1. turns the eye laterally

____________________ 2. turns the eye medially

____________________ 3. turns the eye up and laterally

____________________ 4. turns the eye inferiorly

____________________ 5. turns the eye superiorly

____________________ 6. turns the eye down and laterally

5. What is a sty? ____

____

Conjunctivitis? ____

6. Using the terms listed on the right, correctly identify all structures provided with leader lines in the diagram.

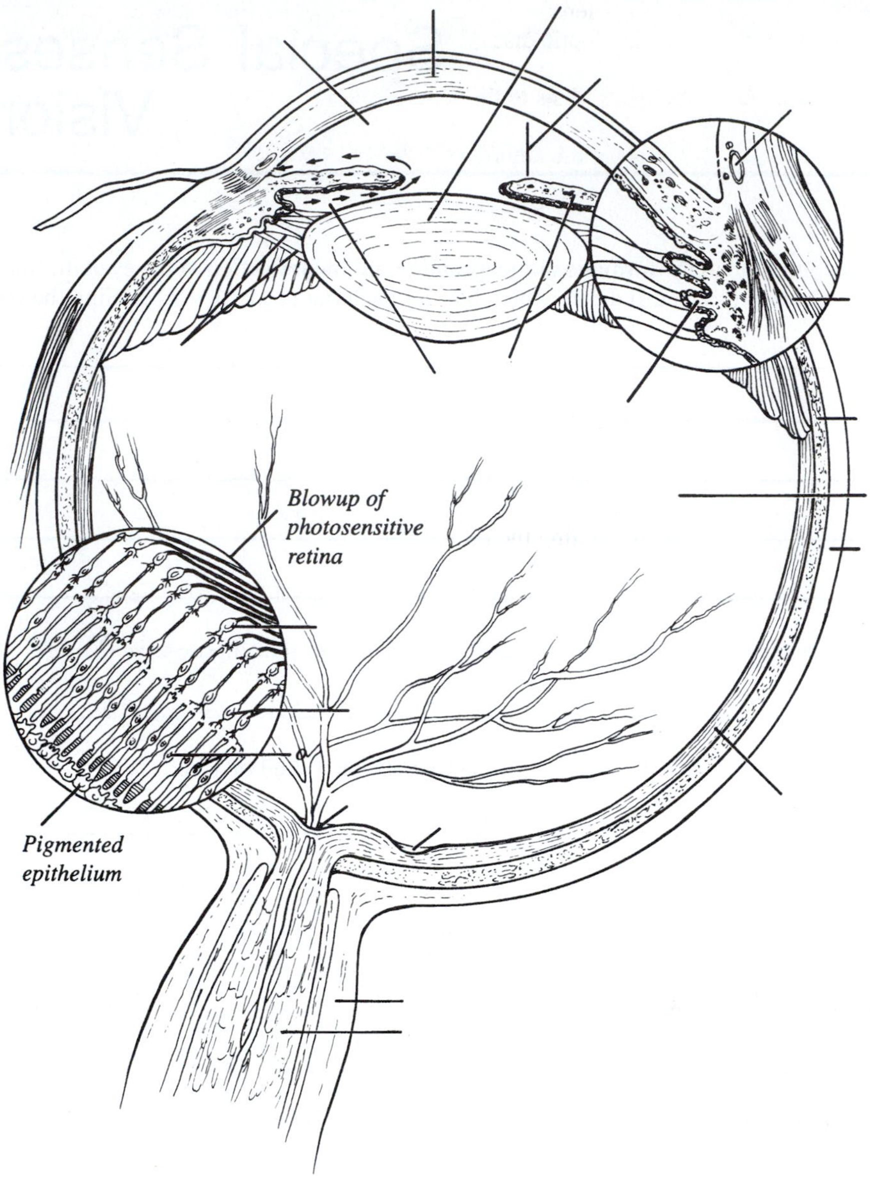

Blowup of photosensitive retina

Pigmented epithelium

a. anterior chamber

b. anterior segment containing aqueous humor

c. bipolar neurons

d. ciliary body and processes

e. ciliary muscle

f. choroid

g. cornea

h. dura mater

i. fovea centralis

j. ganglion cells

k. iris

l. lens

m. optic disc

n. optic nerve

o. photoreceptors

p. posterior chamber

q. retina

r. sclera

s. scleral venous sinus

t. suspensory ligaments

u. vitreous body in posterior segment

Notice the arrows drawn close to the left side of the iris in the diagram above. What do they indicate?

7. Match the key responses with the descriptive statements that follow.

Key:
a. aqueous humor
b. choroid
c. ciliary body
d. ciliary processes of the ciliary body
e. cornea
f. fovea centralis
g. iris
h. lens
i. optic disc
j. retina
k. sclera
l. scleral venous sinus
m. suspensory ligament
n. vitreous humor

__________________ 1. attaches the lens to the ciliary body

__________________ 2. fluid filling the anterior segment of the eye

__________________ 3. the "white" of the eye

__________________ 4. part of the retina that lacks photoreceptors

__________________ 5. modification of the choroid that controls the shape of the crystalline lens

__________________ 6. contains the ciliary muscle

__________________ 7. drains the aqueous humor from the eye

__________________ 8. tunic containing the rods and cones

__________________ 9. substance occupying the posterior segment of the eyeball

__________________ 10. forms the bulk of the heavily pigmented vascular tunic

__________________, __________________ 11. smooth muscle structures

__________________ 12. area of critical focusing and discriminatory vision

__________________ 13. form (by filtration) the aqueous humor

__________________, __________________, __________________,

__________________ 14. light-bending media of the eye

__________________ 15. anterior continuation of the sclera—your "window on the world"

__________________ 16. composed of tough, white, opaque, fibrous connective tissue

8. The iris is composed primarily of two smooth muscle layers, one arranged radially and the other circularly.

Which of these dilates the pupil? __________________

9. You would expect the pupil to be dilated in which of the following circumstances? Circle the correct response(s).

a. in brightly lit surroundings

b. in dimly lit surroundings

c. during focusing for near vision

d. in observing distant objects

10. The intrinsic eye muscles are under the control of which of the following? (Circle the correct response.)

autonomic nervous system somatic nervous system

Dissection of the Cow (Sheep) Eye

1. What modification of the choroid that is not present in humans is found in the cow eye? ______________________

____

What is its function? ____

____

____

2. What is the anatomical appearance of the retina? ____

____

At what point is it attached to the posterior aspect of the eyeball? ______________________

Microscopic Anatomy of the Retina

1. The two major layers of the retina are the epithelial and nervous layers. In the nervous layer, the neuron populations are arranged as follows from the epithelial layer to the vitreous humor. (Circle all proper responses.)

bipolar cells, ganglion cells, photoreceptors

photoreceptors, ganglion cells, bipolar cells

ganglion cells, bipolar cells, photoreceptors

photoreceptors, bipolar cells, ganglion cells

2. The axons of the ______________________ cells form the optic nerve, which exits from the eyeball.

3. Complete the following statements by writing either *rods* or *cones* on each blank:

The dim light receptors are the ______________________. Only ______________________ are found in the fovea centralis, whereas mostly ______________________ are found in the periphery of the retina. ______________________ are the photoreceptors that operate best in bright light and allow for color vision.

Visual Pathways to the Brain

1. The visual pathway to the occipital lobe of the brain consists most simply of a chain of five neurons. Beginning with the photoreceptor cell of the retina, name them and note their location in the pathway.

(1) ______________________ (4) ______________________

(2) ______________________ (5) ______________________

(3) ______________________ ______________________

2. Visual field tests are done to reveal destruction along the visual pathway from the retina to the optic region of the brain. Note where the lesion is likely to be in the following cases:

Normal vision in left eye visual field; absence of vision in right eye visual field: ______________________

Normal vision in both eyes for right half of the visual field; absence of vision in both eyes for left half of the visual field: ____

3. How is the right optic *tract* anatomically different from the right optic *nerve*? ______________________________

____

____

____

Visual Tests and Experiments

1. Match the terms in column B with the descriptions in column A:

	Column A	Column B
__________	1. light bending	a. accommodation
__________	2. ability to focus for close (under 20 ft) vision	b. astigmatism
__________	3. normal vision	c. convergence
__________	4. inability to focus well on close objects (farsightedness)	d. emmetropia
__________	5. nearsightedness	e. hyperopia
__________	6. blurred vision due to unequal curvatures of the lens or cornea	f. myopia
__________	7. medial movement of the eyes during focusing on close objects	g. refraction

2. Complete the following statements:

In farsightedness, the light is focused __1__ the retina. The lens required to treat myopia is a __2__ lens. The "near point" increases with age because the __3__ of the lens decreases as we get older. A convex lens, like that of the eye, produces an image that is upside down and reversed from left to right. Such an image is called a __4__ image.

1. ______________________

2. ______________________

3. ______________________

4. ______________________

3. Use terms from the key to complete the statements concerning near and distance vision.

Key: a. contracted b. decreased c. increased d. relaxed e. taut

During distance vision: The ciliary muscle is ____, the suspensory ligament is ____, the convexity of the lens is ____, and light refraction is ____. During close vision: The ciliary muscle is ____, the suspensory ligament is ____, lens convexity is ____, and light refraction is ____.

4. Explain why vision is lost when light hits the blind spot. ______________________________

5. What is meant by the term *negative afterimage* and what does this phenomenon indicate? ______________________________

____

____

____

6. Record your Snellen eye test results below:

Left eye (without glasses) ______________________ (with glasses)______________________

Right eye (without glasses) ______________________ (with glasses) ______________________

Is your visual acuity normal, less than normal, or better than normal? ______________________

Explain. ____

____

Explain why each eye is tested separately when using the Snellen eye chart. ______________________

____

Explain 20/40 vision. ____

____

Explain 20/10 vision. ____

____

7. Define *astigmatism:* ____

____

How can it be corrected? ____

8. Record the distance of your near point of accommodation as tested in the laboratory:

right eye ______________________ left eye ______________________

Is your near point within the normal range for your age? ______________________

9. Define *presbyopia:* ____

____

What causes it? ____

10. To which wavelengths of light do the three cone types of the retina respond maximally?

______________________, ______________________, and ______________________

11. How can you explain the fact that we see a great range of colors even though only three cone types exist?

____

____

12. From what condition does color blindness result? ____

____

13. Record the results of the demonstration of the relative positioning of rods and cones in the circle below (use appropriately colored pencils).

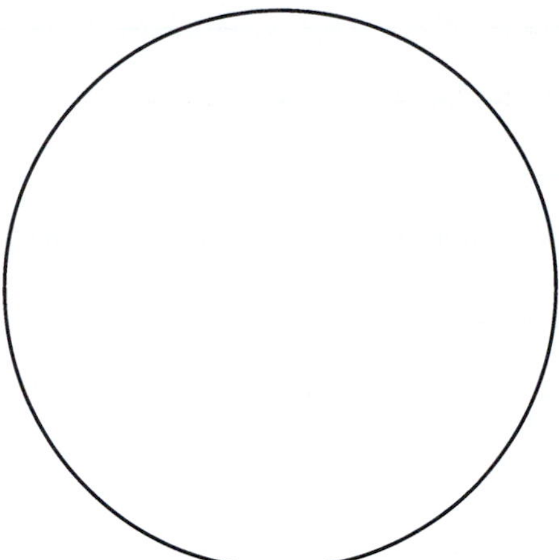

14. Explain the difference between binocular and panoramic vision. ______________________________

____

____

____

____

What is the advantage of binocular vision? ______________________________

What factor(s) are responsible for binocular vision? ______________________________

____

____

15. In the experiment on the convergence reflex, what happened to the position of the eyeballs as the object was moved closer to the subject's eyes? ______________________________

What extrinsic eye muscles control the movement of the eyes during this reflex? ______________________________

What is the value of this reflex? ______________________________

____

What would be the visual result of an inability of these muscles to function? ______________________________

____

16. In the experiment on the photopupillary reflex, what happened to the pupil of the eye exposed to light?

____________________ What happened to the pupil of the nonilluminated eye? ____________________

Explanation? ____________________

17. Why is the ophthalmoscopic examination an important diagnostic tool? ____________________

18. Many college students struggling through mountainous reading assignments are told that they need glasses for "eyestrain." Why is it more of a strain on the extrinsic and intrinsic eye muscles to look at close objects than at far objects? ____________________

STUDENT NAME ______________________________

LAB TIME/DATE ______________________________

Review Sheet

EXERCISE 25

Special Senses: Hearing and Equilibrium

Anatomy of the Ear

1. Select the terms from column B that apply to the column A descriptions. Some terms are used more than once.

Column A	Column B
______________, ______________, ______________ 1. structures comprising the outer or external ear	a. auditory (pharyngo-tympanic) tube
______________, ______________, ______________ 2. structures composing the inner ear	b. anvil (incus)
______________, ______________, ______________ 3. collectively called the ossicles	c. cochlea
______________, ______________ 4. ear structures not involved with audition	d. endolymph
______________ 5. involved in equalizing the pressure in the middle ear with atmospheric pressure	e. external auditory canal
______________ 6. vibrates at the same frequency as sound waves hitting it; transmits the vibrations to the ossicles	f. hammer (malleus)
______________, ______________ 7. contain receptors for the sense of balance	g. oval window
______________ 8. transmits the vibratory motion of the stirrup to the fluid in the scala vestibuli of the inner ear	h. perilymph
______________ 9. acts as a pressure relief valve for the increased fluid pressure in the scala tympani; bulges into the tympanic cavity	i. pinna
______________ 10. passage between the throat and the tympanic cavity	j. round window
______________ 11. fluid contained within the membranous labyrinth	k. semicircular canals
______________ 12. fluid contained within the osseous labyrinth and bathing the membranous labyrinth	l. stirrup (stapes)
	m. tympanic membrane
	n. vestibule

2. Sound waves hitting the eardrum initiate its vibratory motion. Trace the pathway through which vibrations and fluid currents are transmitted to finally stimulate the hair cells in the organ of Corti. (Name the appropriate ear structures in their correct sequence.) Eardrum → ______________________________

____

____

3. Identify all indicated structures and ear regions in the following diagram.

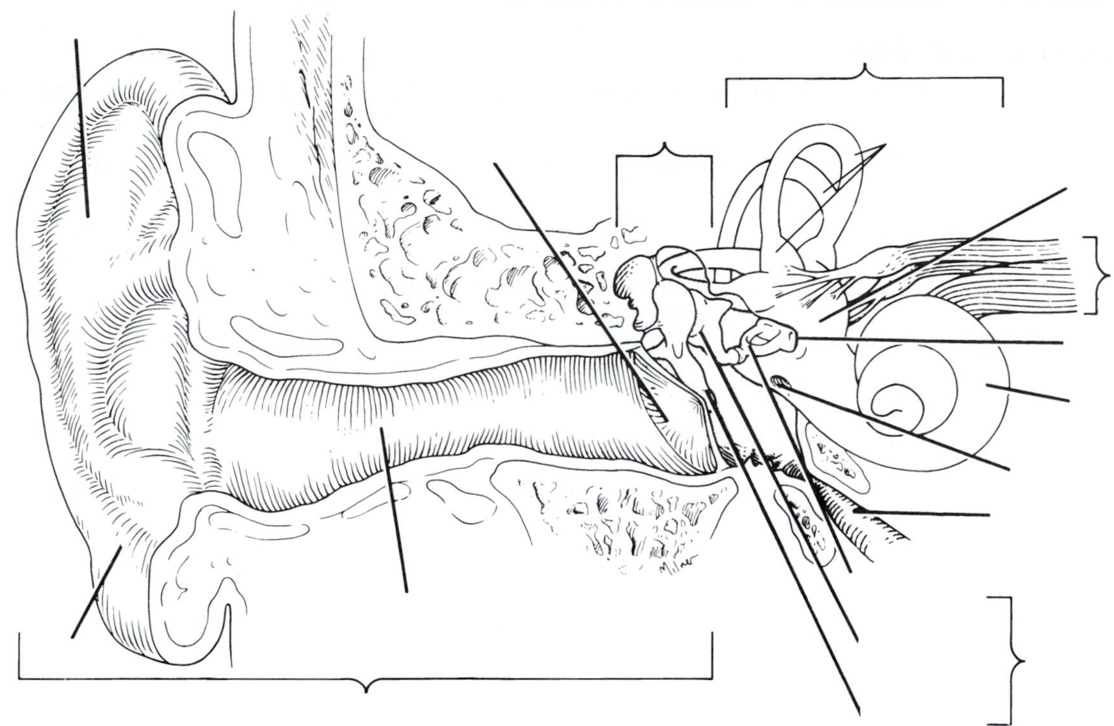

4. Match the membranous labyrinth structures listed in column B with the descriptive statements in column A:

Column A	Column B
________, ________ 1. sacs found within the vestibule	a. ampulla
________ 2. contains the organ of Corti	b. basilar membrane
________, ________ 3. sites of the maculae	c. cochlear duct
________ 4. positioned in all spatial planes	d. cochlear nerve
________ 5. hair cells of organ of Corti rest on this membrane	e. cupula
________ 6. gelatinous membrane overlying the hair cells of the organ of Corti	f. otoliths
________ 7. contains the crista ampullaris	g. saccule
________, ________, ________, ________ 8. function in static equilibrium	h. semicircular ducts
________, ________, ________, ________ 9. function in dynamic equilibrium	i. tectorial membrane
________ 10. carries auditory information to the brain	j. utricle
________ 11. gelatinous cap overlying hair cells of the crista ampullaris	k. vestibular nerve
________ 12. grains of calcium carbonate in the maculae	

5. Describe how sounds of different frequency (pitch) are differentiated in the cochlea. ____________

____

____

____

6. Explain the role of the endolymph of the semicircular canals in activating the receptors during angular motion.

____

____

____

7. Explain the role of the otoliths in perception of static equilibrium (head position). ____________

____

____

____

Laboratory Tests

1. Was the auditory acuity measurement made during the experiment on page 238 the same or different for both ears? ____________ What factors might account for a difference in the acuity of the two ears?

   ____

   ____

2. During the sound localization experiment on page 238, in which position(s) was the sound least easily located?

   ____

   How can this phenomenon be explained? ____________

   ____

3. In the experiment on page 239, which tuning fork was the most difficult to hear? ____________ Hz

   What conclusion can you draw? ____________

   ____

4. When the tuning fork handle was pressed to your forehead during the Weber test, where did the sound seem to originate? ____________

   Where did it seem to originate when one ear was plugged with cotton? ____________

   How do sound waves reach the cochlea when conduction deafness is present? ____________

   ____

5. Indicate whether the following conditions relate to conduction deafness (C) or sensorineural (central) deafness (S):

_______ 1. can result from the fusion of the ossicles

_______ 2. can result from a lesion on the cochlear nerve

_______ 3. sound heard in one ear but not in the other during bone and air conduction

_______ 4. can result from otitis media

_______ 5. can result from impacted cerumen or a perforated eardrum

_______ 6. can result from a blood clot in the auditory cortex

6. The Rinne test evaluates an individual's ability to hear sounds conducted by air or bone. Which is more indicative of normal hearing? _______________

7. Define *nystagmus:* _______________

____

Define *vertigo:* _______________

8. The Barany test investigated the effect that rotatory acceleration had on the semicircular canals. Explain *why* the subject still had the sensation of rotation immediately after being stopped. _______________

____

9. What is the usual reason for conducting the Romberg test? _______________

____

Was the degree of sway greater with the eyes open or closed? _______________

Why? _______________

10. Normal balance, or equilibrium, depends on input from a number of sensory receptors. Name them.

____

____

STUDENT NAME ______________________

LAB TIME/DATE ______________________

Review Sheet

EXERCISE 26

Special Senses: Taste and Olfaction

Localization and Anatomy of Taste Buds

1. Name three sites where receptors for taste are found, and circle the predominant site:

   ______________________, ______________________, and

   ____

2. Describe the cellular makeup and arrangement of a taste bud. (Use a diagram, if helpful.) ______________________

   ____

   ____

Localization and Anatomy of the Olfactory Receptors

1. Describe the cellular composition and the location of the olfactory epithelium. ______________________

   ____

   ____

2. How and why does sniffing improve your sense of smell? ______________________

   ____

Laboratory Experiments

1. Taste and smell receptors are both classified as ______________________, because they both respond to ______________________

2. Why is it impossible to taste substances with a dry tongue? ______________________

   ____

3. State the most important sites of the taste-specific receptors, as determined during the plotting exercise in the laboratory:

   salt ______________________ sour ______________________

   bitter ______________________ sweet ______________________

4. The basic taste sensations are elicited by specific chemical substances or groups. Name them:

   salt ______________________ sour ______________________

   bitter ______________________ sweet ______________________

5. Name three factors that influence our appreciation of foods. Substantiate each choice with an example from the laboratory experience.

________________ Substantiation ________________

________________ Substantiation ________________

________________ Substantiation ________________

Which of the factors chosen is most important? ________________

Substantiate your choice with an example from everyday life. ________________

Expand on your explanation and choices by explaining why a cold, greasy hamburger is unappetizing to most people. ________________

6. Babies tend to favor bland foods, whereas adults tend to like highly seasoned foods. What is the basis for this phenomenon? ________________

7. How palatable is food when you have a cold? ________________

Explain. ________________

8. What is the mechanism of olfactory adaptation? ________________

In your opinion, is olfactory adaptation desirable? ________________ Explain your answer.

STUDENT NAME ______________________

LAB TIME/DATE ______________________

Review Sheet

EXERCISE 27 Anatomy and Basic Function of the Endocrine Glands

Gross Anatomy and Basic Function of the Endocrine Glands

1. Both the endocrine and nervous systems are major regulating systems of the body; however, the nervous system has been compared to an airmail delivery system and the endocrine system to the pony express. Briefly explain this comparison.

____

____

____

2. Define *hormone:* ______________________

____

3. Chemically, hormones belong chiefly to two molecular groups, the ______________________ and the ______________________.

4. What do all hormones have in common? ______________________

____

5. Define *target organ:* ______________________

____

6. Why don't all tissues respond to all hormones? ______________________

____

7. Identify the endocrine organ described by the following statements:

______________________ 1. located in the throat; bilobed gland connected by an isthmus

______________________ 2. found close to the kidney

______________________ 3. a mixed gland, located close to the stomach and small intestine

______________________ 4. paired glands suspended in the scrotum

______________________ 5. ride "horseback" on the thyroid gland

______________________ 6. found in the pelvic cavity of the female, concerned with ova and female hormone production

______________________ 7. found in the upper thorax overlying the heart; large during youth

______________________ 8. found in the roof of the third ventricle

8. For each statement describing hormonal effects, identify the hormone(s) involved by choosing a number from key A, and note the hormone's site of production with a letter from key B. More than one hormone may be involved in some cases.

Key A:

1. ACTH
2. ADH
3. aldosterone
4. cortisone
5. epinephrine
6. estrogens
7. FSH
8. glucagon
9. GH
10. insulin
11. LH
12. melatonin
13. MSH
14. oxytocin
15. progesterone
16. prolactin
17. PTH
18. serotonin
19. testosterone
20. thymosin
21. thyrocalcitonin/calcitonin
22. T_4 / T_3
23. TSH

Key B:

a. adrenal cortex
b. adrenal medulla
c. anterior pituitary
d. hypothalamus
e. ovaries
f. pancreas
g. parathyroid glands
h. pineal gland
i. posterior pituitary
j. testes
k. thymus gland
l. thyroid gland

_______, _______ 1. basal metabolism hormone

_______, _______ 2. programming of T lymphocytes

_______, _______ and _______, _______ 3. regulate blood calcium levels

_______, _______ and _______, _______ 4. released in response to stressors

_______, _______ and _______, _______ 5. drives development of secondary sexual characteristics

_______, _______; _______, _______; _______, _______; and _______, _______ 6. regulate the function of another endocrine gland

_______, _______ 7. mimics the sympathetic nervous system

_______, _______ and _______, _______ 8. regulate blood glucose levels; produced by the same "mixed" gland

_______, _______ and _______, _______ 9. directly responsible for regulation of the menstrual cycle

_______, _______ and _______, _______ 10. regulate the ovarian cycle

_______, _______ and _______, _______ 11. maintenance of salt and water balance in the ECF

_______, _______ and _______, _______ 12. directly involved in milk production and ejection

_______, _______ 13. questionable function; may stimulate the melanocytes of the skin

9. Although the pituitary gland is often referred to as the master gland of the body, the hypothalamus exerts some control over the pituitary gland. How does the hypothalamus control both anterior and posterior pituitary functioning?

____

____

____

____

____

10. Indicate whether the release of the hormones listed below is stimulated by (A), another hormone; (B), the nervous system (neurotransmitters, or releasing factors); or (C), humoral factors (the concentration of specific non-hormonal substances in the blood or extracellular fluid):

_______ 1. T_4 / T_3 _______ 4. parathyroid hormone _______ 7. ADH

_______ 2. insulin _______ 5. testosterone _______ 8. TSH, FSH

_______ 3. estrogens _______ 6. norepinephrine _______ 9. aldosterone

11. Name the hormone that would be produced in *inadequate* amounts under the following conditions. (Use your textbook as necessary.)

_______________________________ 1. sexual immaturity

_______________________________ 2. tetany

_______________________________ 3. excessive diuresis without high blood glucose levels

_______________________________ 4. polyurea, polyphagia, and polydipsia

_______________________________ 5. abnormally small stature, normal proportions

_______________________________ 6. miscarriage

_______________________________ 7. lethargy, hair loss, low BMR, obesity

12. Name the hormone that is produced in *excessive* amounts in the following conditions. (Use your textbook as necessary.)

_______________________________ 1. lantern jaw and large hands and feet in the adult

_______________________________ 2. bulging eyeballs, nervousness, increased pulse rate

_______________________________ 3. demineralization of bones, spontaneous fractures

Microscopic Anatomy of Selected Endocrine Glands (optional)

1. Choose a response from the key below to name the hormone(s) produced by the cell types listed:

Key:
a. insulin
b. GH, prolactin
c. T_4 / T_3
d. calcitonin
e. TSH, ACTH, FSH, LH
f. mineralocorticoids
g. glucagon
h. PTH
i. glucocorticoids

_______ 1. parafollicular cells of the thyroid

_______ 2. follicular epithelial cells of the thyroid

_______ 3. beta cells of the islets of Langerhans

_______ 4. alpha cells of the islets of Langerhans

_______ 5. basophil cells of the anterior pituitary

_______ 6. zona fasciculata cells

_______ 7. zona glomerulosa cells

_______ 8. chief cells

_______ 9. acidophil cells of the anterior pituitary

2. Five diagrams of the microscopic structures of the endocrine glands are presented here. Identify each and name all indicated structures.

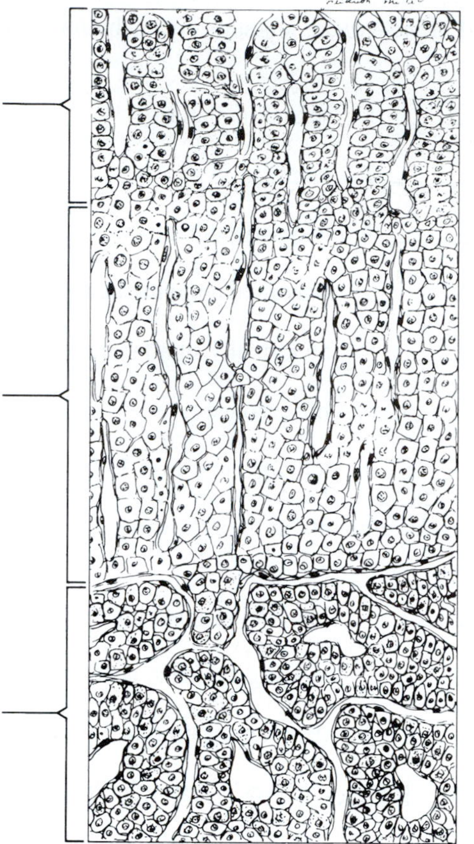

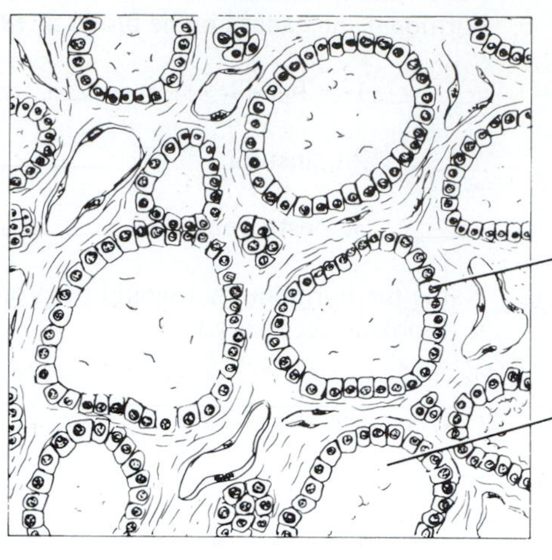

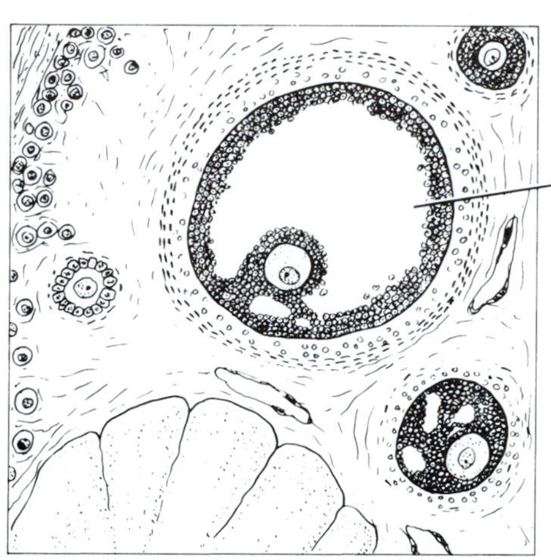

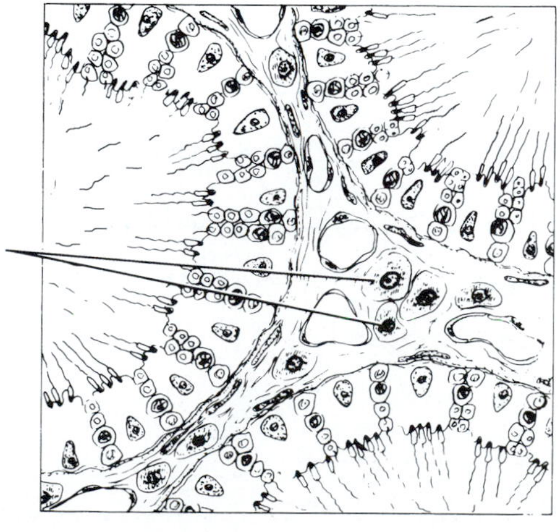

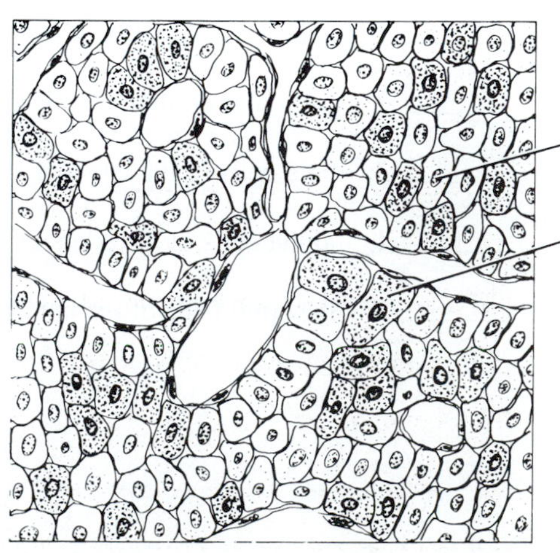

STUDENT NAME ______________________

LAB TIME/DATE ______________________

Review Sheet

EXERCISE 28

Experiments on Hormonal Action

Effect of Thyroid Hormone on Metabolic Rate

1. Relative to the measurement of oxygen consumption in rats, which group had the highest metabolic rate?

   ______________________ Which group had the lowest metabolic rate? ______________________

   Correlate these observations with the pretreatment these animals received. ______________________

   Which group of rats was hyperthyroid? ______________________

   Which euthyroid? ______________________ Which hypothyroid? ______________________

2. Since oxygen used = carbon dioxide evolved, how were you able to measure the oxygen consumption in the experiments? ______________________

3. What did changes in the fluid levels in the manometer arms indicate? ______________________

4. The techniques used in this set of laboratory experiments probably allowed for several inaccuracies. One was the inability to control the activity of the rats. How would changes in their activity levels affect the results observed?

   ____

   Another possible source of error was the lack of control over the amount of food consumed by the rats in the 14-day period preceding the laboratory session. If each of the rats had been force-fed equivalent amounts of food in that 14-day period, which group (do you think) would have gained the most weight?

   ______________________ Which the least? ______________________ Explain your answers. ______________________

5. TSH, produced by the anterior pituitary, prods the thyroid gland to release thyroid hormone to the blood. Which group of rats can be assumed to have the *highest* blood levels of TSH? ______________________

   Which the lowest? ______________________ Explain your reasoning. ______________________

6. Use an appropriate reference to determine how each of the following factors modifies metabolic rate. Indicate increase by ↑ and decrease by ↓.

increased exercise _______ aging _______ infection/fever _______

small/slight stature _______ obesity _______ sex (♂ or ♀) _______

Effect of Pituitary Hormones on the Ovary

1. In the experiment on the effects of pituitary hormones, two anterior pituitary hormones caused ovulation to occur in the experimental animal. Which of these actually triggered ovulation or egg expulsion?

_______________ The normal function of the second hormone involved, _______________,

is to _______________

2. Why was a second frog injected with saline? _______________

Effects of Hyperinsulinism

1. Briefly explain what was happening within the fish's system when the fish was immersed in the insulin solution.

____

____

2. What is the mechanism of the recovery process observed? _______________

____

3. What would you do to help a friend who had inadvertently taken an overdose of insulin? _______________

_______________ Why? _______________

____

4. What is a glucose tolerance test? (Use an appropriate reference, as necessary, to answer this question.)

____

____

____

5. How does diabetic coma differ from insulin shock? _______________

____

____

____

____

Effect of Epinephrine on the Heart

1. Based on your observations, what is the effect of epinephrine on the force and rate of the heartbeat?

____

2. What is the role of this effect in the "fight or flight" response?

____

STUDENT NAME ______________________

LAB TIME/DATE ______________________

Review Sheet

EXERCISE 29 Blood

Composition of Blood

1. What is the blood volume of an average-size adult? __________ liters

2. What determines whether blood is bright red or a dull brick-red? ______________________

____

3. Use the key to identify the cell type(s) or blood elements that fit the following descriptive statements.

Key:
a. red blood cell
b. megakaryocyte
c. eosinophil
d. basophil
e. monocyte
f. neutrophil
g. lymphocyte
h. formed elements
i. plasma

__________ 1. most numerous leukocyte

__________, __________, and

__________ 2. granulocytes

__________ 3. also called an erythrocyte; anucleate formed element

__________, __________ 4. actively phagocytic leukocytes

__________, __________ 5. agranulocytes

__________ 6. ancestral cell of platelets

__________ 7. (a) through (g) are all examples of these

__________ 8. number rises during parasite infections

__________ 9. releases histamine; promotes inflammation

__________ 10. many formed in lymphoid tissue

__________ 11. transports oxygen

__________ 12. primarily water, noncellular; the fluid matrix of blood

__________ 13. increases in number during prolonged infections

__________, __________, __________,

__________, __________ 14. also called white blood cells

4. List three classes of nutrients normally found in plasma: ______________________________

______________________, and ______________________

Name two gases. ______________________ and ______________________

Name three ions. ______________________, ______________________, and ______________________

5. Describe the consistency and color of the plasma you observed in the laboratory. ______________________

____

6. What is the average life span of a red blood cell? How does its anucleate condition affect this life span?

____

____

____

7. From memory, describe the structural characteristics of each of the following blood cell types as accurately as possible, and note the percentage of each in the total white blood cell population.

eosinophils ____

____

neutrophils ____

____

lymphocytes ____

____

basophils ____

____

monocytes ____

____

8. Correctly identify the blood pathologies described in column A by matching them with selections from column B:

	Column A	Column B
____________	1. abnormal increase in the number of WBCs	a. anemia
____________	2. abnormal increase in the number of RBCs	b. leukocytosis
____________	3. condition of too few RBCs or of RBCs with hemoglobin deficiencies	c. leukopenia
____________	4. abnormal decrease in the number of WBCs	d. polycythemia

Hematologic Tests

1. Broadly speaking, why are hematologic studies of blood so important in the diagnosis of disease?

____

____

2. In the chart below, record information from the blood tests you conducted. Complete the chart by recording values for healthy male adults and indicating the significance of high or low values for each test.

Test	Student test results	Normal values (healthy male adults)	Significance High values	Significance Low values
Total WBC count				
Total RBC count				
Hematocrit				
Hemoglobin determination				
Sedimentation rate				
Coagulation time				

3. Why is a differential WBC count more valuable than a total WBC count when trying to pin down the specific source of pathology? ____

____

____

4. Explain the reasons for using different diluents for the RBC and WBC counts. ____

____

____

____

5. What name is given to the process of RBC production? ____________________

What acts as a stimulus for this process? ____________________

What organ provides this stimulus and under what conditions? ____________________

____

____

6. Discuss the effect of each of the following factors on RBC count. Consult an appropriate reference as necessary, and explain your reasoning.

athletic training (for example, running 4 to 5 miles per day over a period of 6 to 9 months) ____________________

____

____

a permanent move from sea level to a high-altitude area ____________________

____

____

7. Define *hematocrit:* ____________________

8. If you had a high hematocrit, would you expect your hemoglobin determination to be high or low?

____________________ Why? ____________________

____

9. What is an anticoagulant? ____________________

Name two anticoagulants used in conducting the hematologic tests. ____________________

and ____________________

What is the body's natural anticoagulant? ____________________

10. If your blood clumped with both anti-A and anti-B sera, your ABO blood type would be ____________________

To what ABO blood groups could you give blood? ____________________

From which ABO donor types could you receive blood? ____________________

Which ABO blood type is most common? ____________________ Least common? ____________________

11. Explain why an Rh-negative person does not have a transfusion reaction on the first exposure to Rh-positive blood but *does* have a reaction on the second exposure. ____________________

____

What happens when an ABO blood type is mismatched for the first time? ____________________

____

12. Record your observations of the five demonstration slides viewed.

a. Macrocytic hypochromic anemia: ____________________

b. Microcytic hypochromic anemia: ____________________

____

c. Sickle-cell anemia: ____________________

d. Lymphocytic leukemia (chronic): ____________________

____

e. Eosinophilia: ____________________

Which of slides a through e above corresponds with the following conditions?

_______ 1. iron-deficient diet

_______ 2. a type of bone marrow cancer

_______ 3. genetic defect that causes hemoglobin to become sharp/spiky

_______ 4. lack of vitamin B_{12}

_______ 5. a tapeworm infestation in the body

_______ 6. a bleeding ulcer

13. Provide the normal, or at least "desirable," range for plasma cholesterol concentration:

____________________ mg/100 ml

14. Describe the relationship between high blood cholesterol levels and cardiovascular diseases such as hypertension, heart attacks, and strokes.

____

____

____

____

____

STUDENT NAME ____________________

LAB TIME/DATE ____________________

Review Sheet

EXERCISE 30

Anatomy of the Heart

Gross Anatomy of the Human Heart

1. An anterior view of the heart is shown here. Identify each numbered structure by writing its name on the correspondingly numbered line:

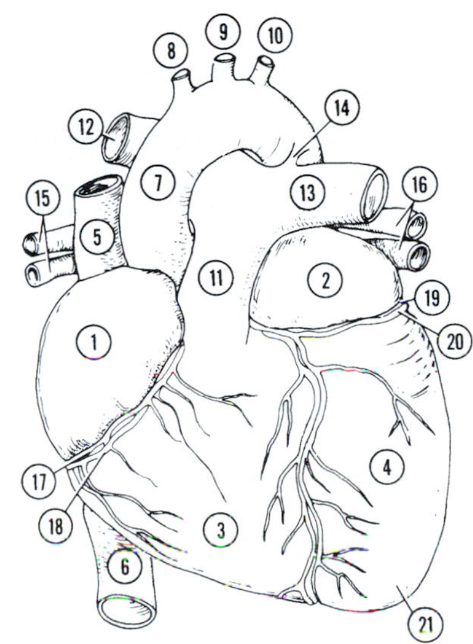

1. ____________________
2. ____________________
3. ____________________
4. ____________________
5. ____________________
6. ____________________
7. ____________________
8. ____________________
9. ____________________
10. ____________________
11. ____________________
12. ____________________
13. ____________________
14. ____________________
15. ____________________
16. ____________________
17. ____________________
18. ____________________
19. ____________________
20. ____________________
21. ____________________

2. What is the function of the fluid that fills the pericardial sac? ____________________

3. Match the terms in the key to the descriptions provided below.

Key:

a. atria
b. coronary arteries
c. coronary sinus
d. endocardium
e. epicardium
f. mediastinum
g. myocardium
h. ventricles

_______ 1. location of the heart in the thorax

_______ 2. superior heart chambers

_______ 3. inferior heart chambers

_______ 4. visceral pericardium

_______ 5. “anterooms” of the heart

_______ 6. equals cardiac muscle

_______ 7. provide nutrient blood to the heart muscle

_______ 8. lining of the heart chambers

_______ 9. actual “pumps” of the heart

_______ 10. drains blood into the right atrium

4. What is the function of the valves found in the heart? ____

____

5. Can the heart function with leaky valves? (Think! Can a water pump function with leaky valves?) ___________

6. What is the role of the chordae tendineae? ____

____

7. Define:

angina pectoris ____

____

pericarditis ____

Pulmonary, Systemic, and Cardiac Circulations

1. A simple schematic of a so-called general circulation is shown below. What part of the circulation is missing from this diagram? ____

Add to the diagram as best you can to make it depict a complete systemic/pulmonary circulation and reidentify “general circulation” as the correct subcirculation.

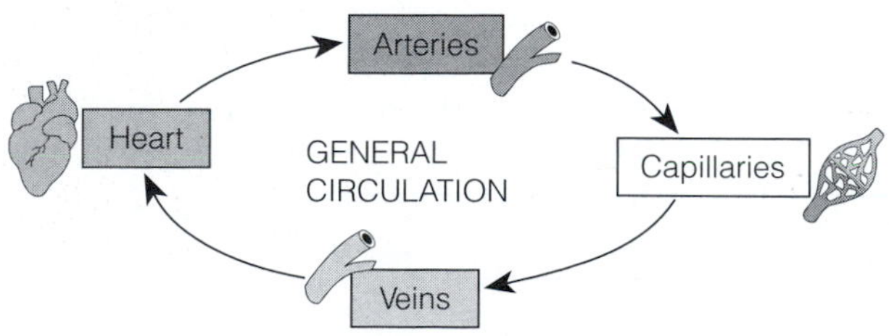

2. Differentiate clearly between the roles of the pulmonary and systemic circulations. ______

______

______

3. Complete the following scheme of circulation in the human body:

Right atrium through the tricuspid valve to the ______ through the ______ valve to the pulmonary trunk to the ______ to the capillary beds of the lungs to the ______ to the ______ of the heart through the ______ valve to the ______ through the ______ valve to the ______ to the systemic arteries to the ______ of the tissues to the systemic veins to the ______ and ______ entering the right atrium of the heart.

4. If the mitral valve does not close properly, which circulation is affected? ______

5. Why might a thrombus (blood clot) in the anterior descending branch of the left coronary artery cause sudden death? ______

______

Microscopic Anatomy of Cardiac Muscle

1. How would you distinguish cardiac muscle from skeletal muscle? ______

______

______

______

______

2. Add the following terms to the diagram of cardiac muscle at the right:

   a. intercalated disc

   b. nucleus of cardiac fiber

   c. striations

   d. cardiac muscle fiber

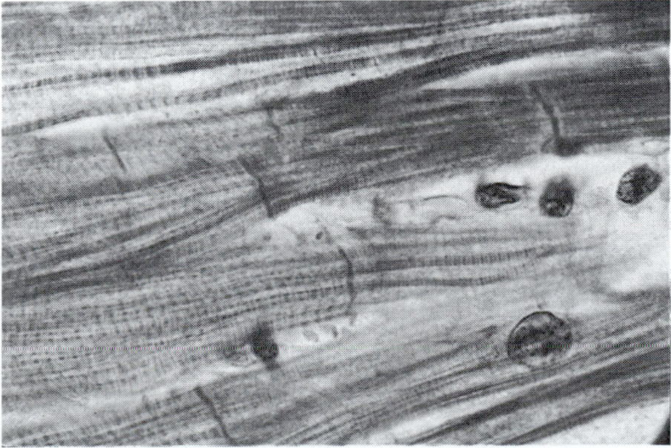

3. What role does the unique structure of cardiac muscle play in its function? (Note: before attempting a response, *describe* the unique anatomy.) ____________________

____

____

____

Dissection of the Sheep Heart

1. During the sheep heart dissection, you were asked initially to identify the right and left ventricles without cutting into the heart. During this procedure, what differences did you observe between the two chambers?

____

____

Knowing that structure and function are related, how would you say this structural difference reflects the relative functions of these two heart chambers? ____________________

____

____

____

2. Semilunar valves prevent backflow into the ____________________; AV valves prevent backflow into the ____________________. Using your own observations, explain how the operation of the semilunar valves differs from that of the AV valves. ____________________

____

____

____

____

3. Differentiate clearly between the location and appearance of pectinate muscle and trabeculae carneae. ____________________

____

____

4. Two remnants of fetal structures are observable in the heart—the ligamentum arteriosum and the fossa ovalis. What were they called in the fetal heart, where was each located, and what common purpose did they serve as functioning fetal structures?

____

____

____

____

STUDENT NAME ______________________________

LAB TIME/DATE ______________________________

Review Sheet

EXERCISE 31 Conduction System of the Heart and Electrocardiography

The Intrinsic Conduction System

1. List the elements of the intrinsic conduction system in order starting from the SA node.

   SA node ⟶ ______________________ ⟶ ______________________ ⟶

   ______________________ ⟶ ______________________

   Which of those structures is replaced when an artificial pacemaker is installed? ______________________

   At what structure in the transmission sequence is the impulse temporarily delayed? ______________________

   Why? ______________________

   ____

2. Even though cardiac muscle has an inherent ability to beat, the nodal system plays a critical role in heart physiology. What is that role? ______________________

   ____

   ____

3. How does the "all-or-none" law apply to normal heart operation? ______________________

   ____

   ____

Electrocardiography

1. Define *ECG.* ______________________

   ____

2. Draw an ECG wave form representing one heart beat. Label the P, QRS, and T waves; the P-R interval; the S-T segment, and the Q-T interval.

3. What changes from *baseline* were noted in the ECG recorded during running? ______

Explain. ______

In that recorded during breath holding? ______

Explain. ______

4. Describe what happens in the cardiac cycle in the following situations:
   1. during the P wave ______
   2. immediately before the P wave ______
   3. immediately after the P wave ______
   4. during the QRS wave ______
   5. immediately after the QRS wave (S-T interval) ______
   6. during the T wave ______
5. Define the following terms:
   1. *tachycardia* ______
   2. *bradycardia* ______
   3. *flutter* ______
   4. *fibrillation* ______
   5. *myocardial infarction* ______
6. Which would be more serious, atrial or ventricular fibrillation? ______

   Why? ______
7. Abnormalities of heart valves can be detected more accurately by auscultation than by electrocardiography. Why is this so? ______

STUDENT NAME ______________________________

LAB TIME/DATE ______________________________

Review Sheet

EXERCISE 32

Anatomy of Blood Vessels

Microscopic Structure of the Blood Vessels

1. Use key choices to identify the blood vessel tunic described.

   Key: a. tunica intima b. tunica media c. tunica externa

   ______________ 1. most internal tunic

   ______________ 2. bulky middle tunic contains smooth muscle and elastin

   ______________ 3. its smooth surface decreases resistance to blood flow

   ______________ 4. tunic(s) of capillaries

   ______________, ______________, ______________ 5. tunic(s) of arteries and veins

   ______________ 6. is especially thick in elastic arteries

   ______________ 7. most superficial tunic

2. Servicing the capillaries is the essential function of the organs of the circulatory system. Explain this statement.

   ____

   ____

   ____

3. Cross-sectional views of an artery and of a vein are shown here. Identify each; and on the lines beneath, note the structural details that enabled you to make these identifications:

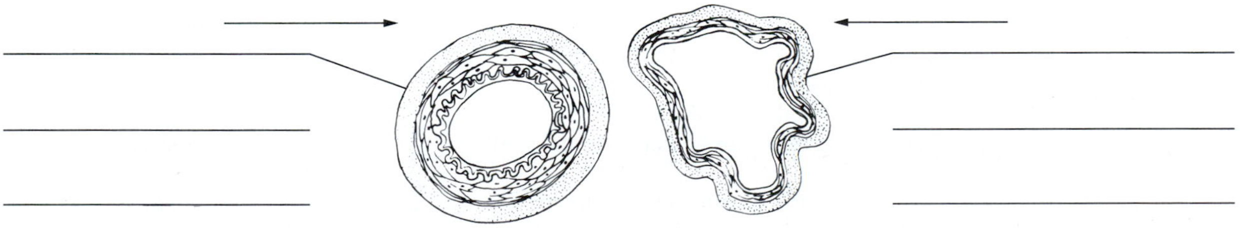

   ____

4. Why are valves present in veins but not in arteries? ______________________________

   ____

   ____

5. Name two events *occurring within the body* that aid in venous return:

   ____

   and ______________________________

6. Why are the walls of arteries proportionately thicker than those of the corresponding veins? ________________

____

____

Major Systemic Arteries and Veins of the Body

1. Use the key on the right to identify the arteries or veins described on the left.

_______ 1. the arterial system has one of these; the venous system has two

_______ 2. these arteries supply the myocardium

_______, _______ 3. two paired arteries serving the brain

_______ 4. longest vein in the lower limb

_______ 5. artery on the dorsum of the foot checked after leg surgery

_______ 6. serves the posterior thigh

_______ 7. supplies the diaphragm

_______ 8. formed by the union of the radial and ulnar veins

_______, _______ 9. two superficial veins of the arm

_______ 10. artery serving the kidney

_______ 11. veins draining the liver

_______ 12. artery that supplies the distal half of the large intestine

_______ 13. drains the pelvic organs

_______ 14. what the external iliac artery becomes on entry into the thigh

_______ 15. major artery serving the arm

_______ 16. supplies most of the small intestine

_______ 17. join to form the inferior vena cava

_______ 18. an arterial trunk that has three major branches, which run to the liver, spleen, and stomach

_______ 19. major artery serving the tissues external to the skull

_______, _______, _______ 20. three veins serving the leg

_______ 21. artery generally used to take the pulse at the wrist

Key:
a. anterior tibial
b. basilic
c. brachial
d. brachiocephalic
e. celiac trunk
f. cephalic
g. common carotid
h. common iliac
i. coronary
j. deep femoral
k. dorsalis pedis
l. external carotid
m. femoral
n. greater saphenous
o. hepatic
p. inferior mesenteric
q. internal carotid
r. internal iliac
s. peroneal
t. phrenic
u. posterior tibial
v. radial
w. renal
x. subclavian
y. superior mesenteric
z. vertebral

2. The human arterial and venous systems are diagrammed on the next two pages. Identify all indicated blood vessels.

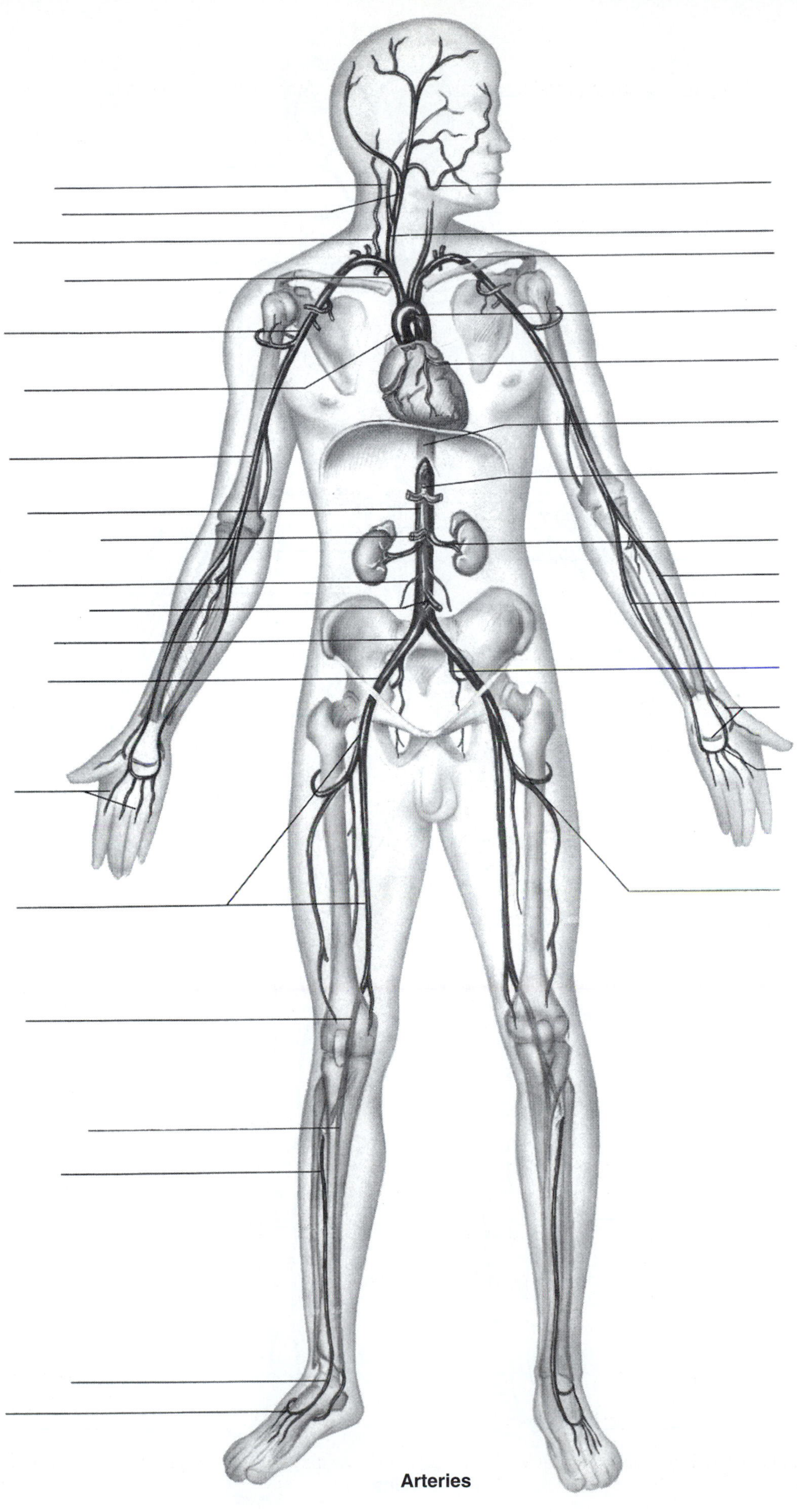

Arteries

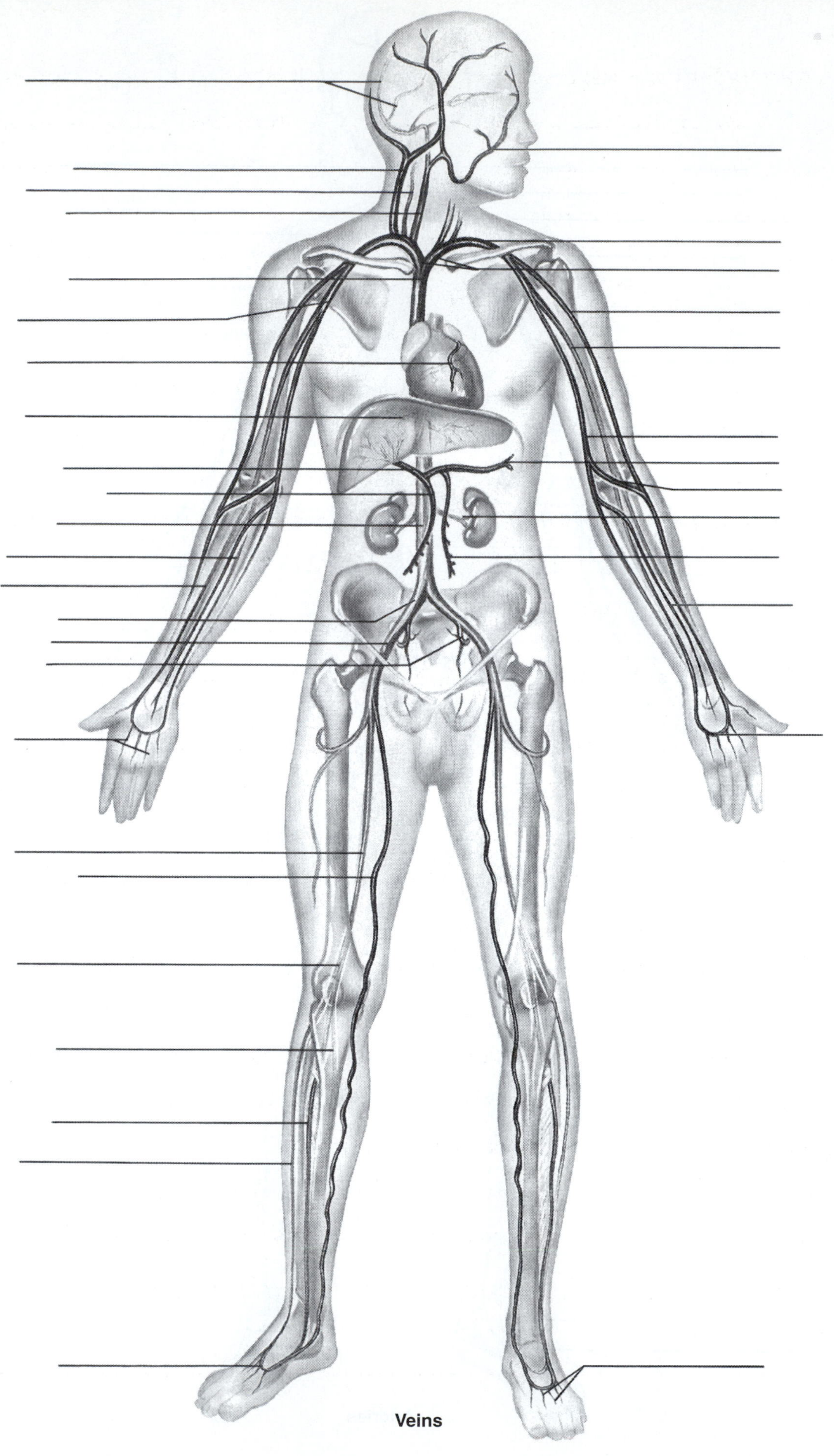
Veins

3. Trace the blood flow for the following situations:

   a. From the capillary beds of the left thumb to the capillary beds of the right thumb ______________________

   b. From the bicuspid valve to the tricuspid valve by way of the great toe ______________________

   c. From the pulmonary vein to the pulmonary artery by way of the right side of the brain ______________________

Special Circulations

Pulmonary circulation:

1. Trace the pathway of a carbon dioxide gas molecule in the blood from the inferior vena cava until it leaves the bloodstream. Name all structures (vessels, heart chambers, and others) passed through en route.

2. Trace the pathway of an oxygen gas molecule from an alveolus of the lung to the right atrium of the heart. Name all structures through which it passes. ______________________________

____

____

3. Most arteries of the adult body carry oxygen-rich blood, and the veins carry oxygen-depleted, carbon dioxide–rich blood. What is different about the pulmonary arteries and veins? ______________________________

____

____

4. How do the arteries of the pulmonary circulation differ structurally from the systemic arteries? What condition is indicated by this anatomical difference? ______________________________

____

____

Arterial supply of the brain and the circle of Willis:

1. What two paired arteries enter the skull to supply the brain?

   ______________________________ and ______________________________

2. Branches of the paired arteries just named cooperate to form a ring of blood vessels encircling the pituitary gland, at the base of the brain. What name is given to this communication network? ______________________________

   What is its function? ______________________________

   ____

3. What portion of the brain is served by the anterior and middle cerebral arteries? ______________________________

   ____

   Both the anterior and middle cerebral arteries arise from the ______________________________ arteries.

4. Trace the pathway of a drop of blood from the aorta to the left occipital lobe of the brain, noting all structures through which it flows. ______________________________

   ____

Hepatic portal circulation:

1. Complete the labeling in the diagram of vessels of the hepatic portal system.

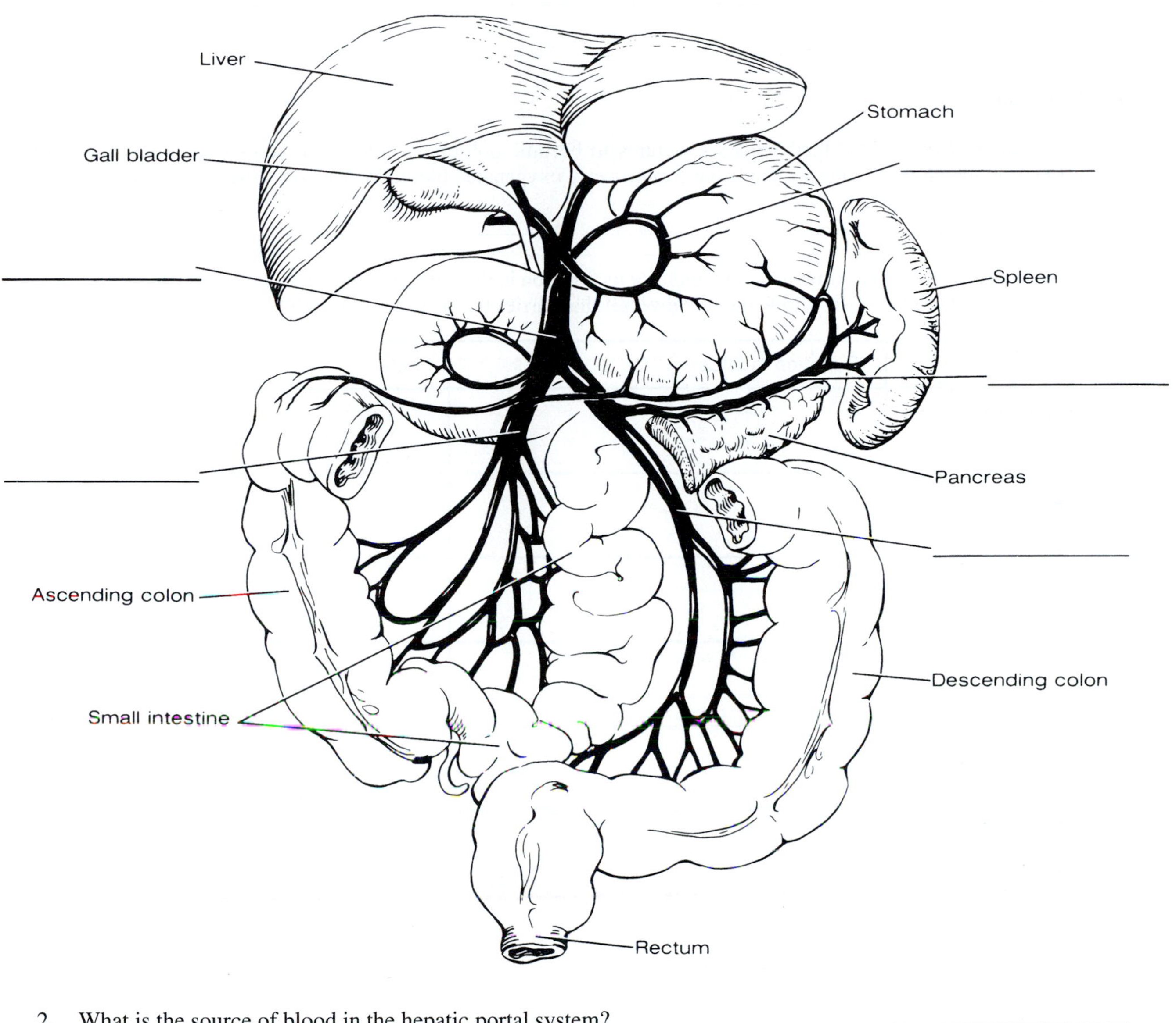

2. What is the source of blood in the hepatic portal system? ______________________________

____

3. Why is this blood carried to the liver before it enters the systemic circulation? ______________________________

____

____

4. The hepatic portal vein is formed by the union of the ______________________________, which drains the ______________________________, ______________________________, ______________________________, and the ______________________________, which drains the ______________________________ and ______________________________. The ______________________________ vein, which drains the lesser curvature of the stomach, empties directly into the hepatic portal vein.

5. Trace the flow of a drop of blood from the small intestine to the right atrium of the heart, noting all structures encountered or passed through on the way. ______________________________

____

____

Fetal circulation:

1. The failure of two of the fetal bypass structures to become obliterated after birth can cause congenital heart disease, in which the youngster would have improperly oxygenated blood. Which two structures are these?

______________________ and ______________________

2. For each of the following structures, first indicate its function in the fetus; and then note what happens to it or what it is converted to after birth. Circle the blood vessel that carries the most oxygen-rich blood.

Structure	Function in fetus	Fate
Umbilical artery		
Umbilical vein		
Ductus venosus		
Ductus arteriosus		
Foramen ovale		

3. What organ serves as a respiratory/digestive/excretory organ for the fetus? ______________________

STUDENT NAME ____________________

LAB TIME/DATE ____________________

Review Sheet

EXERCISE 33

Human Cardiovascular Physiology—Blood Pressure and Pulse Determinations

Cardiac Cycle

1. Define the following terms:

   systole ____________________

   diastole ____________________

   cardiac cycle ____________________

2. Using the key below, indicate the time interval occupied by the following events of the cardiac cycle.

   Key: a. 0.4 sec b. 0.3 sec c. 0.1 sec d. 0.8 sec

   ________ 1. the length of the normal cardiac cycle

   ________ 2. the time interval of atrial systole

   ________ 3. the quiescent period, or pause

   ________ 4. the ventricular contraction period

3. If an individual's heart rate is 80 beats/min, what is the length of the cardiac cycle? ________ What portion of the cardiac cycle is shortened by this more rapid heart rate? ____________________

   ____

4. Answer the following questions, which concern events of the cardiac cycle:

   When are the AV valves closed? ____________________

   Open? ____________________

   What event within the heart causes the AV valves to open? ____________________

   What causes them to close? ____________________

   When are the semilunar valves closed? ____________________

   ____

   Open? ____________________

   What event causes the semilunar valves to open? ____________________

   ____

   To close? ____________________

   Are both sets of valves closed during any part of the cycle? ____________________

If so, when? ______

Are both sets of valves open during any part of the cycle? ______

At what point in the cardiac cycle is the pressure in the heart highest? ______

Lowest? ______

What event results in the pressure deflection called the dicrotic notch? ______

______

5. What two factors promote the movement of blood through the heart? ______

______ and ______

______

Auscultation of Heart Sounds

1. Complete the following statements:

The monosyllables describing the heart sounds are __1__ . The first heart sound is a result of closure of the __2__ valves, whereas the second is a result of closure of the __3__ valves. The heart chambers that have just been filled when you hear the first heart sound are the __4__ , and the chambers that have just emptied are the __5__ . Immediately after the second heart sound, the __6__ are filling with blood, and the __7__ are empty.

1. ______
2. ______
3. ______
4. ______
5. ______
6. ______
7. ______

2. As you listened to the heart sounds during the laboratory session, what differences in pitch, length, and amplitude (loudness) of the two sounds did you observe? ______

______

______

3. Indicate where you would place your stethoscope to auscultate most accurately the following:

closure of the tricuspid valve ______

closure of the aortic semilunar valve ______

apical heartbeat ______

Which valve is heard most clearly when the apical heartbeat is auscultated? ______

4. No one expects you to be a full-fledged physician on such short notice; but on the basis of what you have learned about heart sounds, how might abnormal sounds be used to diagnose heart problems?

______

______

______

Palpation of the Pulse

1. Define *pulse.* ______________________________

____

2. Describe the procedure used to take the pulse. ______________________________

____

____

3. Identify the artery palpated at each of the following pressure points:

at the wrist ______________________ on the dorsum of the foot ______________________

in the front of the ear ______________________ at the side of the neck ______________________

4. When you were palpating the various pulse or pressure points, which appeared to have the greatest amplitude or tension? ______________________ Why do you think this was so? ______________________

____

5. Assume someone has been injured in an auto accident and is hemorrhaging badly. What pressure point would you compress to help stop bleeding from each of the following areas?

the thigh ______________________ the calf ______________________

the forearm ______________________ the thumb ______________________

6. How could you tell by simple observation whether bleeding is arterial or venous? ______________________

____

7. You may sometimes observe a slight difference between the value obtained from an apical pulse (beats/min) and that from an arterial pulse taken elsewhere on the body. What is this difference called?

____

Blood Pressure Determinations

1. Define *blood pressure.* ______________________________

____

2. Identify the phase of the cardiac cycle to which each of the following apply:

systolic pressure ______________________ diastolic pressure ______________________

3. What is the name of the instrument used to compress the artery and record pressures in the auscultatory method of determining blood pressure? ______________________

4. What are the sounds of Korotkoff? ______________________________

____

What causes the systolic sound? ______________________________

____

The disappearance of sound? ______________________________

5. Interpret 145/85/82. ______________________________

____

____

6. Assume the following BP measurement was recorded for an elderly patient with severe arteriosclerosis: 170/110/–. Explain the inability to obtain the third reading.

____

____

7. Define *pulse pressure:* ______________________________

Why is this measurement important? ______________________________

____

8. How do venous pressures compare to arterial pressures? ______________________________

Why? ______________________________

9. What maneuver to increase the thoracic pressure illustrates the effect of external factors on venous pressure?

______________ How was it performed? ______________________________

____

____

10. What might an abnormal increase in venous pressure indicate? (Think!) ______________________________

____

Effect of Various Factors on Blood Pressure and Heart Rate

1. What effect do the following have on blood pressure? (Indicate increase by I and decrease by D.)

_______ 1. increased diameter of the arterioles

_______ 2. increased blood viscosity

_______ 3. increased cardiac output

_______ 4. hemorrhage

_______ 5. arteriosclerosis

_______ 6. increased pulse rate

2. In which position (sitting, reclining, or standing) is the blood pressure normally the highest?

______________ The lowest? ______________

What immediate changes in blood pressure did you observe when the subject stood up after being in the sitting or reclining position? ______________________________

____

____

What changes in the blood vessels might account for the change? ______________________________

____

____

After the subject stood for 3 minutes, what changes in blood pressure were observed? ______

______

How do you account for this change? ______

3. What was the effect of exercise on blood pressure? ______

On pulse? ______ Do you think these effects reflect changes in cardiac output *or* in peripheral resistance? ______

______

Why are there normally no significant increases in diastolic pressure after exercise? ______

______

______

4. What effects of the following did you observe on blood pressure in the laboratory?

nicotine ______ cold temperature ______

What do you think the effect of heat would be? ______

Why? ______

5. Differentiate between a hypo- and a hyperreactor relative to the cold pressor test. ______

______

______

Skin Color as an Indicator of Local Circulatory Dynamics

1. Describe normal skin color and the appearance of the veins in the subject's forearm before any testing was conducted. ______

2. What changes occurred when the subject emptied his forearm of blood (by raising his arm and making a fist) and the flow was occluded with the cuff? ______

______

What changes occurred during venous congestion? ______

______

3. What is the importance of collateral blood supplies? ______

______

4. Explain the mechanism by which mechanical stimulation of the skin produced a flare. ______

______

______

STUDENT NAME ______________________

LAB TIME/DATE ______________________

Review Sheet

EXERCISE 34 Frog Cardiovascular Physiology

Intrinsic Properties of Cardiac Muscle: Automaticity and Rhythmicity

1. Define the following terms:

   automaticity ______________________

   rhythmicity ______________________

2. Discuss the anatomical differences you observed between frog and human hearts. ______________________

   ____

   ____

3. Which region of the dissected frog heart had the highest intrinsic rate of contraction? ______________________

   The greatest automaticity? ______________________

   The greatest regularity or rhythmicity? ____________ How do these properties correlate with the duties of a pacemaker? ______________________

   ____

   Is this region the pacemaker of the frog heart? ______________________

   Which region had the lowest intrinsic rate of contraction? ______________________

Refractory Period of Cardiac Muscle

1. Define *extrasystole:* ______________________

   ____

2. In responding to the following questions, refer to the recordings you made during this exercise:

   What was the effect of stimulation of the heart during ventricular contraction? ______________________

   During ventricular relaxation (first portion)? ______________________

   During the pause interval? ______________________

   What does this indicate about the refractory period of cardiac muscle? ______________________

   ____

   Can cardiac muscle be tetanized? ________ Why or why not? ______________________

Physical and Chemical Factors Modifying Heart Rate

1. Describe the effect of thermal factors on the frog heart:

   cold ________________ heat ________________

2. What was the effect of vagal stimulation on heart rate? ________________

   Which of the following factors cause the same (or very similar) heart rate–reducing effects? Epinephrine, acetylcholine, atropine sulfate, pilocarpine, sympathetic nervous system activity, digitalis, potassium ions?

   ____

   Which of the factors listed above would reverse or antagonize vagal effects? ________________

   ____

3. What is vagal escape? ________________

   ____

   Why is vagal escape valuable in maintaining homeostasis? ________________

   ____

4. Once again refer to your recordings. Did the administration of the following produce any changes in force of contraction (shown by peaks of increasing or decreasing height)? If so, explain the mechanism.

   epinephrine ________________

   ____

   acetylcholine ________________

   ____

   ____

   calcium ions ________________

   ____

   ____

5. Excessive amounts of each of the following ions would most likely interfere with normal heart activity. Note the type of changes caused in each case.

   K^+ ________________

   Ca^{2+} ________________

   Na^+ ________________

6. How does the Stannius ligature used in the laboratory produce heart block? ________________

   ____

   ____

7. Define *partial heart block,* and describe how it was recognized in the laboratory. ____________________

____

____

8. Define *total heart block,* and describe how it was recognized in the laboratory. ____________________

____

____

____

9. What do your heart block experiment results indicate about the spread of impulses from the atria to the ventricles?

____

____

Observation of the Microcirculation and Investigation of Factors Affecting Local Blood Flow

1. In what way are the red blood cells of the frog different from those of the human? ____________________

____

On the basis of this one factor, would you expect their life spans to be longer or shorter? ____________________

2. The following statements refer to your observation of one or more of the vessel types observed in the microcirculation in the frog's web. Characterize each statement by choosing one or more responses from the key.

Key: a. arteriole b. venule c. capillary

________ 1. smallest vessels observed

________ 2. vessel within which the blood flow is rapid, pulsating

________ 3. vessel in which blood flow is least rapid

________ 4. red blood cells pass through these vessels in single file

________ 5. blood flow smooth and steady

________ 6. most numerous vessels

________ 7. vessels that deliver blood to the capillary bed

________ 8. vessels that serve the needs of the tissues via exchanges

________ 9. vessels that drain the capillary beds

3. Which of the vessel diameters changed most? ____________________

What division of the nervous system controls the vessels? ____________________

4. Discuss the effects of the following on blood vessel diameter (state specifically the blood vessels involved) and rate of blood flow. Then explain the importance of the reaction observed to the general well-being of the body.

local application of cold ____________________

____

____

local application of heat ____________________

____

____

inflammation (or application of HCl) ____________________

____

____

histamine ____________________

____

____

STUDENT NAME ____________________

LAB TIME/DATE ____________________

Review Sheet

EXERCISE 35 The Lymphatic System and Immune Response

The Lymphatic System

1. Identify all structures that have leader lines on the diagram below. Use the key for your selection.

Key:

a. bone marrow
b. cisterna chyli
c. lymph nodes
d. lymphatic vessels
e. right lymphatic duct
f. spleen
g. thoracic duct
h. thymus
i. tonsils

1. ____________________
2. ____________________
3. ____________________
4. ____________________
5. ____________________
6. ____________________
7. ____________________
8. ____________________
9. ____________________

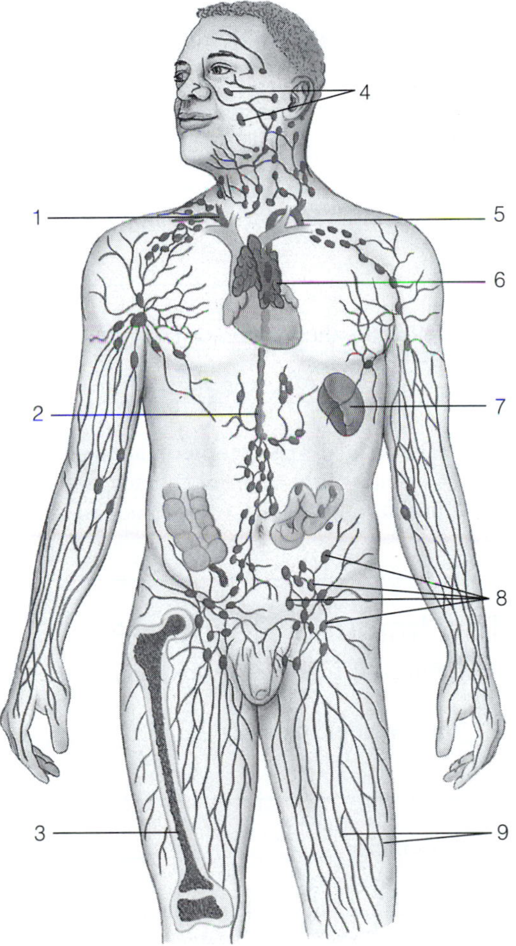

2. Explain why the lymphatic system is a one-way system, whereas the blood vascular system is a two-way system.

____

____

3. How do lymphatic vessels resemble veins? ______________________________

   How do lymphatic capillaries differ from blood capillaries? ______________________________

   ____

4. What is the function of the lymphatic vessels? ______________________________

   ____

5. What is lymph? ______________________________

6. What factors are involved in the flow of lymphatic fluid? ______________________________

   ____

7. What name is given to the terminal duct draining most of the body? ______________________________

8. What is the cisterna chyli? ______________________________

   ____

   How does the composition of lymph in the cisterna chyli differ from that in the general lymphatic stream?

   ____

9. Which portion of the body is drained by the right lymphatic duct? ______________________________

10. Note three areas where lymph nodes are densely clustered: ______________________________,

    ______________________________, and ______________________________

11. What are the two major functions of the lymph nodes? ______________________________

    ____

12. The radical mastectomy is an operation in which a cancerous breast, surrounding tissues, and the underlying muscles of the anterior thoracic wall, plus the axillary lymph nodes, are removed. After such an operation, the arm usually swells, or becomes edematous, and is very uncomfortable—sometimes for months. Why?

    ____

The Immune Response

1. Define the following terms which relate to the operation of the immune system:

   Immunologic memory ______________________________

   ____

   Specificity ______________________________

   ____

   Recognition of self from nonself ______________________________

   ____

2. What is the function of B cells in the immune response? ______________________________

   ____

3. What is the role of T cells? ____

____

____

Microscopic Anatomy of a Lymph Node

1. In the space below, make a rough drawing of the structure of a lymph node. Identify the cortex area, germinal centers, and medulla. For each identified area, note the cell type (T cell, B cell, or macrophage) most likely to be found there.

2. What structural characteristic ensures a *slow* flow of lymph through a lymph node? ______________________________

____

Why is this desirable? ____

____

Antibodies and Tests for Their Presence

1. Describe the structure of the immunoglobulin monomer. ______________________________

____

2. Are the genes coding for one antibody entirely different from those coding for a different antibody?

________ Explain. ____

____

____

3. Rheumatoid arthritis is said to involve an autoimmune response. What is an autoimmune response?

____

4. In the test to identify the presence of rheumatoid factor in plasma, a clumping, or agglutination, indicated a positive response (that is, the presence of the abnormal antibody). What event did the clumping reflect?

____

STUDENT NAME ____________________

LAB TIME/DATE ____________________

Review Sheet

EXERCISE 36 Anatomy of the Respiratory System

Upper and Lower Respiratory Structures

1. Complete the labeling of the diagram of the upper respiratory structures (sagittal section).

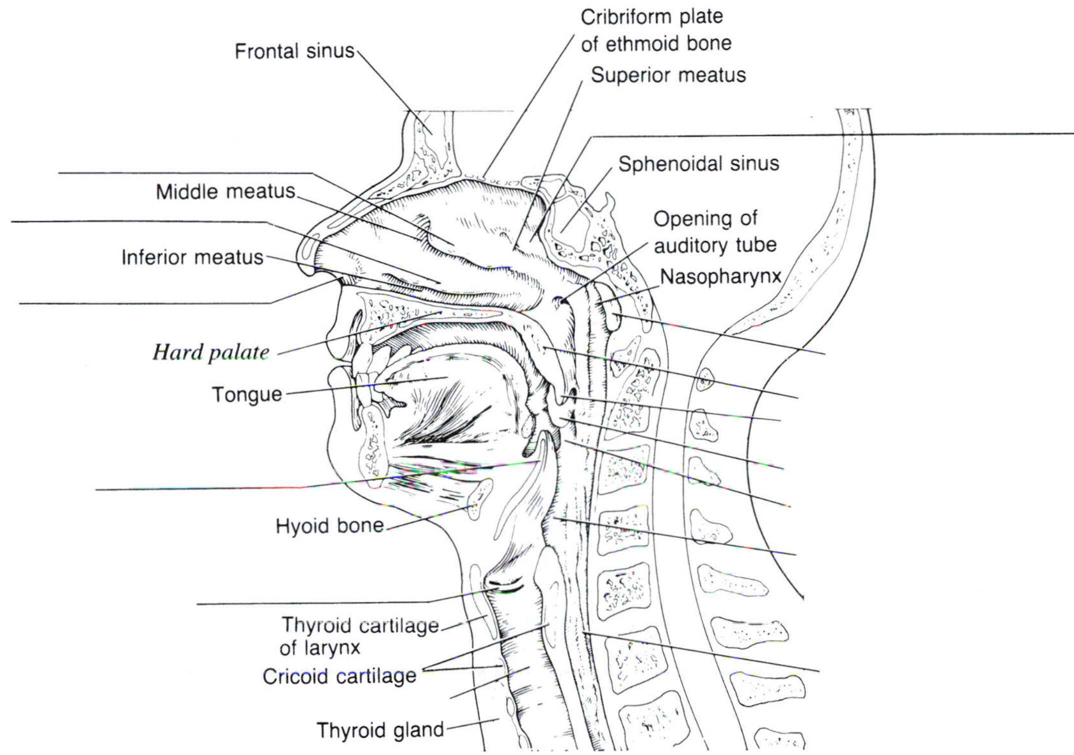

2. Two pairs of vocal folds are found in the larynx. Which pair are the true vocal cords (superior or inferior)?

____

3. What is the significance of the fact that the human trachea is reinforced with cartilage rings?

____

Of the fact that the rings are incomplete posteriorly? ____________________

____

4. Name the specific cartilages in the larynx that correspond to the following descriptions:

1. forms the Adam's apple ____________________
2. a "lid" for the larynx ____________________
3. shaped like a signet ring ____________________
4. vocal cord attachment ____________________

5. Trace a molecule of oxygen from the external nares to the pulmonary capillaries of the lungs: External nares →

____

____

____

6. What is the function of the pleural membranes? ____

____

____

7. Name two functions of the nasal cavity mucosa: ____

8. The following questions refer to the primary bronchi:

Which is longer? ____________ Larger in diameter? ____________ More horizontal? ____________

The more common site for lodging of a foreign object that had entered the respiratory passageways? ________

9. Match the terms in column B to those in column A.

	Column A	Column B
________	1. nerve that activates the diaphragm during inspiration	a. alveolus
________	2. "floor" of the nasal cavity	b. bronchiole
________	3. food passageway posterior to the trachea	c. conchae
________	4. flaps over the glottis during swallowing of food	d. epiglottis
________	5. contains the vocal cords	e. esophagus
________	6. part of the conducting pathway between the larynx and the primary bronchi	f. glottis
________	7. pleural layer lining the walls of the thorax	g. larynx
________	8. site from which oxygen enters the pulmonary blood	h. palate
________	9. autonomic nervous system nerve serving the thoracic region	i. parietal pleura
________	10. opening between the vocal folds	j. phrenic nerve
________	11. increases air turbulence in the nasal cavity	k. primary bronchi
		l. trachea
		m. vagus nerve
		n. visceral pleura

10. What portions of the respiratory system are referred to as anatomical dead space? ____

____

____

Why? ____

11. Define *external respiration:* ______________________________

____

internal respiration: ______________________________

____

12. On the diagram below identify alveolar epithelium, capillary endothelium, alveoli, and red blood cells, and bracket the respiratory membrane.

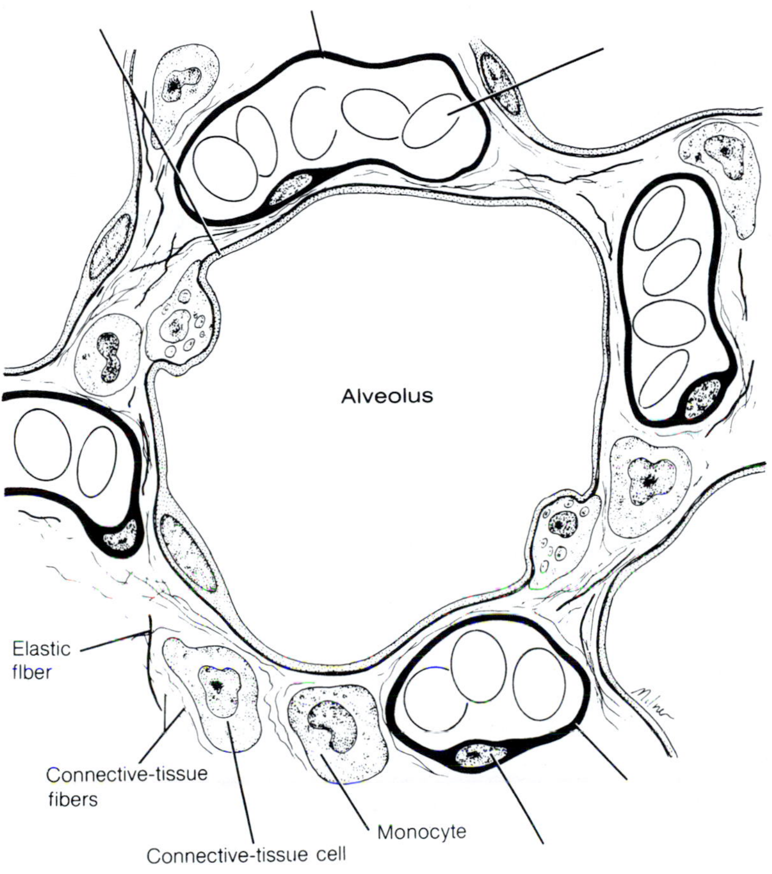

Sheep Pluck Demonstration

1. Does the lung inflate part by part or as a whole, like a balloon? ______________________________

What happened when the pressure was released? ______________________________

What type of tissue ensures this phenomenon? ______________________________

Examination of Prepared Slides of Lung and Trachea Tissue

1. The tracheal epithelium is ciliated and has goblet cells. What is the function of each of these modifications?

Cilia? ______________________________

Goblet cells? ______________________________

2. The tracheal epithelium is said to be pseudostratified. Why? ______________________________

____

____

3. What structural characteristics of the alveoli make them an ideal site for the diffusion of gases?

____

Why does oxygen move from the alveoli into the pulmonary capillary blood? ______________________

____

____

4. If you observed pathologic lung sections, what were the responsible conditions and how did the tissue differ from normal lung tissue?

Slide Type	Observations

STUDENT NAME ____________________

LAB TIME/DATE ____________________

Review Sheet

EXERCISE 37

Respiratory System Physiology

Mechanics of Respiration

1. For each of the following cases, check the column appropriate to your observations on the operation of the model lung.

	Diaphragm pushed up		Diaphragm pulled down	
Change	**Increased**	**Decreased**	**Increased**	**Decreased**
In internal volume of the bell jar (thoracic cage)				
In internal pressure				
In the size of the balloons (lungs)				
In direction of air flow	Into lungs	Out of lungs	Into lungs	Out of lungs

2. Base your answers to the following on your observations in question 1.

Under what internal conditions does air tend to flow into the lungs? ____________________

____

Under what internal conditions does air tend to flow out of the lungs? Explain. ____________________

____

3. Activation of the diaphragm and the external intercostal muscles begins the inspiratory process. What results from the contraction of these muscles, and how is this accomplished? ____________________

____

____

____

4. What was the approximate increase in diameter of chest circumference during a quiet inspiration?

_______________ inches During forced inspiration? _______________ inches

What temporary physiologic advantage does the substantial increase in chest circumference during forced inspiration create? _______________

____

____

5. The presence of a partial vacuum between the pleural membranes is integral to normal breathing movements. What would happen if an opening were made into the chest cavity, as with a puncture wound?

____

How is this condition treated medically? _______________

Respiratory Volumes and Capacities—Spirometry

1. Write the respiratory volume term and the normal value that is described by the following statements:

volume of air present in the lungs after a forceful expiration _______________

volume of air that can be expired forcibly after a normal expiration _______________

volume of air that is breathed in and out during a normal respiration _______________

volume of air that can be inspired forcibly after a normal inspiration _______________

volume of air corresponding to TV + IRV + ERV _______________

2. Record experimental respiratory volumes as determined in the laboratory.

Average tidal volume _______________ ml

Corrected value for TV _______________ ml

Average IRV _______________ ml

Corrected value for IRV _______________ ml

Minute respiratory volume _______________ ml/min

Average ERV _______________ ml

Corrected value for ERV _______________ ml

Average VC _______________ ml

Corrected value for VC _______________ ml

% predicted VC _______________ %

FEV, _______________ % FVC

3. Would your vital-capacity measurement differ if you performed the test while standing? _______ While lying down? _______ Explain. _______________

____

4. Which respiratory ailments can respiratory volume tests be used to detect?

____

____

____

5. Using an appropriate reference, complete the chart below:

		O_2	CO_2	N_2
% of composition of air	Inspired			
	Expired			

Use of the Pneumograph to Determine Factors Influencing Rate and Depth of Respiration

1. Where are the neural control centers of respiratory rhythm? ______________________ and _____________

2. Based on pneumograph reading of respiratory variation, what was the rate of quiet breathing?

   Initial testing ______________________ breaths/min

   Record observations of how the initial pneumograph recording was modified during the various testing procedures described below. Indicate the respiratory rate, and include comments on the relative depth of the respiratory peaks observed.

Test performed	Observations
Talking	
Yawning	
Laughing	
Standing	
Concentrating	
Swallowing water	
Coughing	
Lying down	
Running in place	

3. Student data:

   breath-holding interval after a deep inhalation _______ sec   length of recovery period _______ sec

   breath-holding interval after a forceful expiration _______ sec   length of recovery period _______ sec

   After breathing quietly and taking a deep breath (which you held), was your urge to inspire or expire?

   ____

   After exhaling and then holding one's breath, was the desire for inspiration or expiration? _______________

   Explain these results. (*Hint:* what reflex is involved here?) ______________________________________

   ____

   ____

4. Observations after hyperventilation: ____________________

____

5. Length of breath holding after hyperventilation: ________ sec

Why does hyperventilation produce apnea or a reduced respiratory rate? ____________________

____

____

6. Observations for rebreathing breathed air: ____________________

____

Why does rebreathing breathed air produce an increased respiratory rate? ____________________

____

7. What was the effect of running in place (exercise) on the duration of breath holding? ____________________

Explain: ____________________

____

8. Relative to the test illustrating the effect of respiration on circulation: *(student data)*

Radial pulse before beginning test ________ /min   Radial pulse after testing ________ /min

Relative pulse force before beginning test ______   Relative force of radial pulse after testing ______

Condition of neck and facial veins after testing ____________________

Explain: ____________________

____

9. Do the following factors generally increase (indicate with I) or decrease (indicate with D) the respiratory rate and depth?

1. increase in blood CO_2 ________
2. decrease in blood O_2 ________
3. increase in blood pH ________
4. decrease in blood pH ________

Did it appear that CO_2 or O_2 had a more marked effect on modifying the respiratory rate? ____________________

10. Sensory receptors sensitive to changes in blood pressure are located where? ____________________

____

11. Where are sensory receptors sensitive to changes in O_2 levels in the blood located? ____________________

____

12. What is the primary factor that initiates breathing in a newborn infant? ____________________

____

13. Blood CO_2 levels and blood pH are related. When blood CO_2 levels increase, does the pH increase or decrease?

______________________ Explain why. ______________________

____

____

Respiratory Sounds

1. Which of the respiratory sounds is heard during both inspiration and expiration? ______________________

   Which is heard primarily during inspiration? ______________________

2. Where did you best hear the vesicular respiratory sounds? ______________________

Role of the Respiratory System in Acid-Base Balance of Blood

1. Define *buffer.* ______________________

2. How successful was the laboratory buffer (pH 7) in resisting changes in pH when the acid was added?

   ____

   When the base was added? ______________________

   How successful was the buffer in resisting changes in pH when the additional aliquots (3 more drops) of the acid and base were added to the original samples? ______________________

3. What buffer system operates in blood plasma? ______________________

   Which of its species resists a *drop* in pH? ______________________

   Which resists a *rise* in pH? ______________________

4. Explain how the carbonic acid–bicarbonate buffer system of the blood operates. ______________________

   ____

   ____

   ____

   ____

   ____

STUDENT NAME ____________________

LAB TIME/DATE ____________________

Review Sheet

EXERCISE 38

Anatomy of the Digestive System

General Histological Plan of the Alimentary Canal

1. The general anatomical features of the digestive tube have been presented. Fill in the table below to complete the information listed.

Wall layer	Subdivisions of the layer	Major functions
mucosa		
submucosa		
muscularis externa		
serosa or adventitia		

Organs of the Alimentary Canal

1. The tubelike digestive system canal that extends from the mouth to the anus is the ____________________ canal.

2. How is the muscularis externa of the stomach modified? ____________________

____

How does this modification relate to the function of the stomach? ____________________

____

3. Match the items in column B with the descriptive statements in column A.

Column A

_______ 1. structure that suspends the small intestine from the posterior body wall

_______ 2. fingerlike extensions of the intestinal mucosa that increase the surface area for absorption

_______ 3. large collections of lymphoid tissue found in the submucosa of the small intestine

_______ 4. deep folds of the mucosa and submucosa that extend completely or partially around the circumference of the small intestine

_______, _______ 5. regions that break down foodstuffs mechanically

_______ 6. mobile organ that manipulates food in the mouth and initiates swallowing

_______ 7. conduit for both air and food

_______, _______, _______ 8. three structures continuous with and representing modifications of the peritoneum

_______ 9. the "gullet"; no digestive/absorptive function

_______ 10. folds of the gastric mucosa

_______ 11. sacculations of the large intestine

_______ 12. projections of the plasma membrane of a mucosal epithelial cell

_______ 13. valve at the junction of the small and large intestines

_______ 14. primary region of food and water absorption

_______ 15. membrane securing the tongue to the floor of the mouth

_______ 16. absorbs water and forms feces

_______ 17. area between the teeth and lips/cheeks

_______ 18. wormlike sac that outpockets from the cecum

_______ 19. initiates protein digestion

_______ 20. structure attached to the lesser curvature of the stomach

_______ 21. organ distal to the stomach

_______ 22. valve controlling food movement from the stomach into the duodenum

_______ 23. posterosuperior boundary of the oral cavity

_______ 24. location of the hepatopancreatic sphincter through which pancreatic secretions and bile pass

_______ 25. serous lining of the abdominal cavity wall

_______ 26. principal site for the synthesis of vitamin K by microorganisms

_______ 27. region containing two sphincters through which feces are expelled from the body

_______ 28. bone-supported anterosuperior boundary of the oral cavity

Column B

a. anus

b. appendix

c. esophagus

d. frenulum

e. greater omentum

f. hard palate

g. haustra

h. ileocecal valve

i. large intestine

j. lesser omentum

k. mesentery

l. microvilli

m. oral cavity

n. parietal peritoneum

o. Peyer's patches

p. pharynx

q. plicae circulares

r. pyloric valve

s. rugae

t. small intestine

u. soft palate

v. stomach

w. tongue

x. vestibule

y. villi

z. visceral peritoneum

4. Correctly identify all organs depicted in the diagram below.

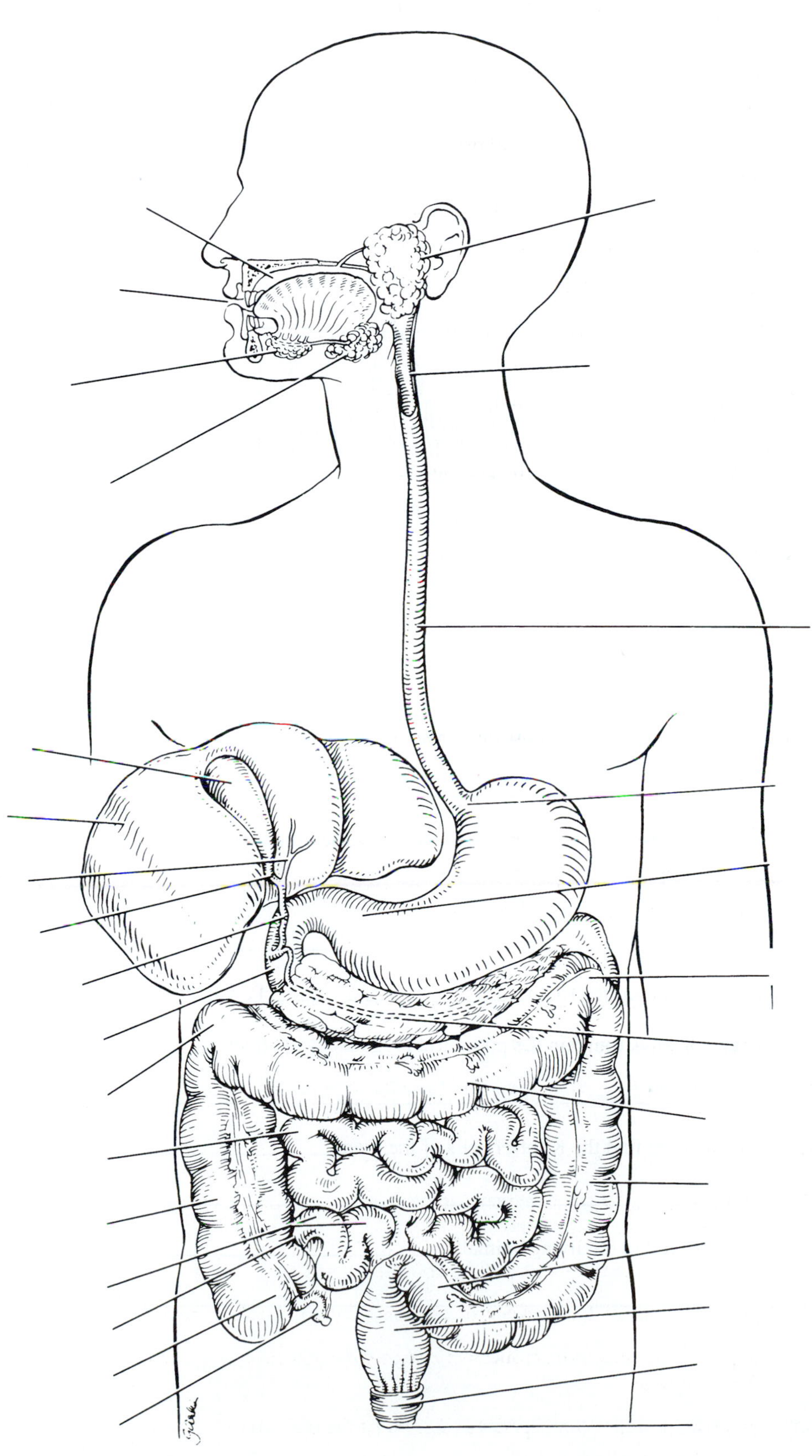

RS167

5. You have studied the histologic structure of a number of organs in this laboratory. Three of these are diagrammed below. Identify each.

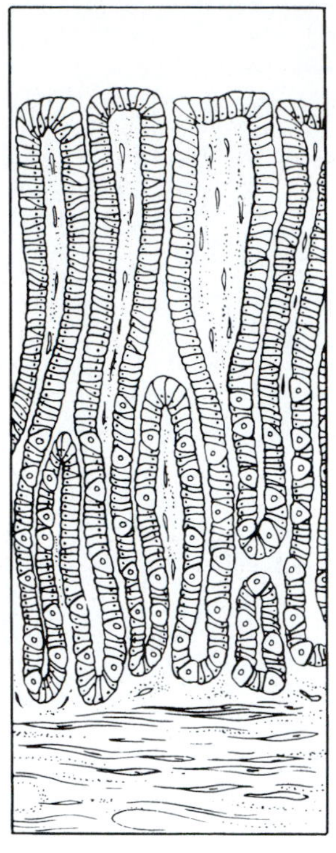

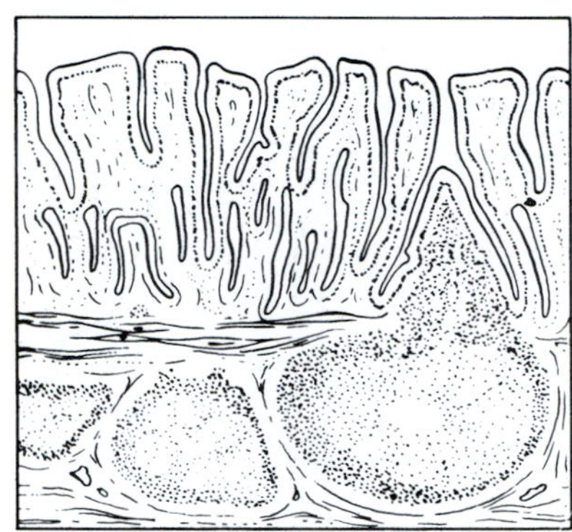

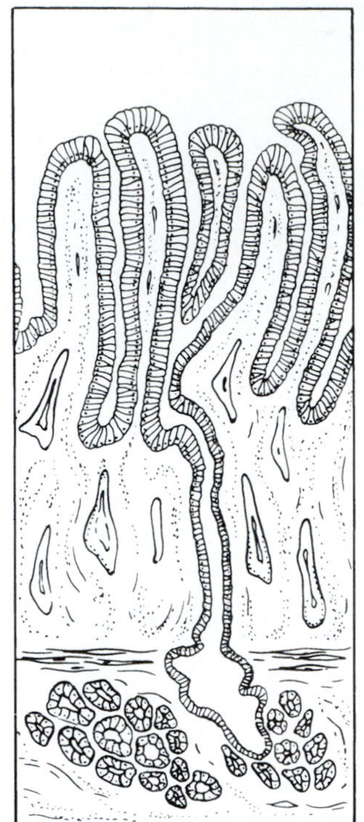

______________________ ______________________ ______________________

6. What transition in epithelium type exists at the cardio-esophageal junction? ______________________

____

How do the epithelia of these two organs relate to their specific functions? ______________________

____

____

7. What cells of the stomach produce HCl? ______________________ Pepsinogen? ______________________

8. Name three structures always found in the portal triad regions of the liver. ______________________,

______________________, and ______________________.

9. Where would you expect to find the Kupffer cells of the liver? ______________________

What is their function? ______________________

10. Why is the liver so dark red in the living animal? ______________________

____

Accessory Digestive Organs

1. Various types of glands form a part of the alimentary tube wall or duct their secretions into it. Match the glands listed in column B with the function/locations described in column A.

Column A	Column B
________ 1. produce(s) mucus; found in the submucosa of the small intestine	a. duodenal glands
________ 2. produce(s) a product containing amylase that begins starch breakdown in the mouth	b. gastric glands
________ 3. produce(s) a whole spectrum of enzymes and an alkaline fluid that is secreted into the duodenum	c. intestinal crypts
________ 4. produce(s) bile that it secretes into the duodenum via the bile duct	d. liver
________ 5. produce(s) HCl and pepsinogen	e. pancreas
________ 6. found in the mucosa of the small intestine; produce(s) intestinal juice	f. salivary glands

2. Which of the salivary glands produces a secretion that is mainly serous? ________________________

3. What is the role of the gallbladder? ________________________

4. Use the key to identify each tooth area described below.

	Key:
________ 1. visible portion of the tooth *in situ*	a. anatomical crown
________ 2. material covering the tooth root	b. cementum
________ 3. hardest substance in the body	c. clinical crown
________ 4. attaches the tooth to bone and surrounding alveolar structures	d. dentin
________ 5. portion of the tooth embedded in bone	e. enamel
________ 6. forms the major portion of tooth structure; similar to bone	f. gingiva
________ 7. form the dentin	g. odontoblasts
________ 8. site of blood vessels, nerves, and lymphatics	h. periodontal ligament
________ 9. entire portion of the tooth covered with enamel	i. pulp
	j. root

5. In the human, the number of deciduous teeth is ________; the number of permanent teeth is ________.

6. The dental formula for permanent teeth is $\frac{2, 1, 2, 3}{2, 1, 2, 3}$

   Explain what this means: ________________________

   ____

   What is the dental formula for the deciduous teeth?

7. What teeth are the "wisdom teeth"? ________________________

STUDENT NAME ______________________

LAB TIME/DATE ______________________

Review Sheet

EXERCISE 39

Chemical and Physical Processes of Digestion

Part 1: Chemical Breakdown of Foodstuffs: Enzymatic Action

1. Match the following definitions with the proper choices from the key.

Key: a. catalyst b. control c. enzyme d. substrate

__________ 1. increases the rate of a chemical reaction without becoming part of the product

__________ 2. provides a standard of comparison for test results

__________ 3. biologic catalyst: protein in nature

__________ 4. substance on which a catalyst works

2. List the three characteristics of enzymes. ______________________

____

____

3. The enzymes of the digestive system are classified as hydrolases. What does this mean?

____

____

4. Fill in the following chart about the various digestive system enzymes encountered in this exercise.

Enzyme	Organ producing it	Site of action	Substrate(s)	Optimal pH
Salivary amylase				
Trypsin				
Lipase (pancreatic)				

5. Name the end products of digestion for the following types of foods:

proteins: __________ carbohydrates: __________

fats: __________ and __________

6. You used several indicators or tests in the laboratory to determine the presence or absence of certain substances. Choose the correct test or indicator from the key to correspond to the condition described below.

Key: a. IKI (Lugol's iodine) b. Benedict's solution c. litmus d. BAPNA

_______ 1. used to test for protein hydrolysis, which was indicated by a yellow color

_______ 2. used to test for the presence of starch, which was indicated by blue-black color

_______ 3. used to test for the presence of fatty acids, which was evidenced by a color change from blue to pink

_______ 4. used to test for the presence of reducing sugars (maltose, sucrose, glucose) as indicated by a blue to green color change

7. What conclusions can you draw when an experimental sample gives both a positive starch test and a positive maltose test after incubation? _______________

Why was 37°C the optimal incubation temperature? _______________

Why did very little, if any, starch digestion occur in test tube 4A? _______________

____

Why did very little, if any, starch digestion occur in test tube 6A? _______________

____

Assume you have made the statement to a group of your peers that amylase is capable of starch hydrolysis to maltose. If you had not done control tube 1A, what could be a possible objection raised to your statement?

____

What if you had not done tube 2A? _______________

8. In the exercise concerning trypsin function, why was an enzyme assay like Benedict's or Lugol's IKI (which test for the presence of a reaction product) not necessary? _______________

Why was tube 1T necessary? _______________

Why was tube 2T necessary? _______________

Trypsin is a protease similar to pepsin, the protein-digesting enzyme in the stomach. Would trypsin work well in the stomach? __________ Why? _______________

9. In the procedure concerning pancreatic lipase digestion of fats and the action of bile salts, how did the appearance of tubes 1E and 2E differ? _______________

____

Can you explain the difference? _______________

____

Why did the litmus indicator change from blue to pink during fat hydrolysis? _______________

____

Why is bile not considered an enzyme? ____

How did the tubes containing bile compare with those not containing bile? ____________________

What role does bile play in fat digestion? ____

____

10. The three-dimensional structure of a functional protein is altered by intense heat or nonphysiological pH even though peptide bonds may not break. Such inactivation is called denaturation, and denatured enzymes are nonfunctional. Explain why. ____

____

____

What specific experimental conditions resulted in denatured enzymes? ________________________

____

11. Pancreatic and intestinal enzymes operate optimally at a pH that is slightly alkaline, yet the chyme entering the duodenum from the stomach is very acid. How is the proper pH for the functioning of the pancreatic-intestinal enzymes assured? ____

____

12. Assume you have been chewing a piece of bread for 5 or 6 minutes. How would you expect its taste to change during this interval? ____

Why? ____

13. Note the mechanism of absorption (passive or active transport) of the following food breakdown products, and indicate by a check mark (✔) whether the absorption would result in their movement into the blood capillaries or the lymph capillaries (lacteals).

Substance	Mechanism of absorption	Blood	Lymph
Monosaccharides			
Fatty acids and glycerol			
Amino acids			
Water			
Na^+, Cl^-, Ca^{2+}			

14. People on a strict diet to lose weight begin to metabolize stored fats at an accelerated rate. How does this condition affect blood pH? ____

15. Trace the pathway of a ham sandwich (ham=protein and fat; bread=starch) from the mouth to the site of absorption of its breakdown products, noting where digestion occurs and what specific enzymes are involved.

16. Some of the digestive organs have groups of secretory cells that liberate hormones (parahormones) into the blood. These exert an effect on the digestive process by acting on other cells or structures and causing them to release digestive enzymes, expel bile, or increase the mobility of the digestive tract. For each hormone below, note the organ producing the hormone and its effects on the digestive process. Include the target organs affected.

Hormone	Target organ(s) and effects
Secretin	
Gastrin	
Cholecystokinin	

Part II: Food Propulsion Mechanisms

Complete the following statements.

Swallowing, or __1__ , occurs in two phases—the __2__ and __3__ . One of these phases, the __4__ phase, is voluntary. During the voluntary phase, the __5__ is used to push the food into the back of the throat. During swallowing, the __6__ rises to ensure that its passageway is covered by the epiglottis so that the ingested substances don't enter the respiratory passageways. It is possible to swallow water while standing on your head because the water is carried along the esophagus involuntarily by the process of __7__ . The pressure exerted by the foodstuffs on the __8__ sphincter causes it to open, allowing the foodstuffs to enter the stomach.

The two major types of propulsive movements that occur in the small intestine are __9__ and __10__ . One of these movements, the __11__ , acts to continually mix the foods and to increase the absorption rate by moving different parts of the chyme mass over the intestinal mucosa, but has less of a role in moving foods along the digestive tract.

1. ______________________________

2. ______________________________

3. ______________________________

4. ______________________________

5. ______________________________

6. ______________________________

7. ______________________________

8. ______________________________

9. ______________________________

10. ______________________________

11. ______________________________

STUDENT NAME ____________________

LAB TIME/DATE ____________________

Review Sheet

EXERCISE 40

Anatomy of the Urinary System

Gross Anatomy of the Human Urinary System

1. Complete the following statements:

The kidney is referred to as an excretory organ because it excretes __1__ wastes. It is also a major homeostatic organ because it maintains the electrolyte, __2__ , and __3__ balance of the blood.

Urine is continuously formed by the __4__ and is routed down the __5__ by the mechanism of __6__ to a storage organ called the __7__ . Eventually, the urine is conducted to the body __8__ by the urethra. In the male, the urethra is __9__ inches long and transports both urine and __10__ . The female urethra is __11__ inches long and transports only urine.

Voiding or emptying the bladder is called __12__ . Voiding has both voluntary and involuntary components. The voluntary sphincter is the __13__ sphincter. An inability to control this sphincter is referred to as __14__ .

1. ____________________
2. ____________________
3. ____________________
4. ____________________
5. ____________________
6. ____________________
7. ____________________
8. ____________________
9. ____________________
10. ____________________
11. ____________________
12. ____________________
13. ____________________
14. ____________________

2. What is the function of the fat cushion that surrounds the kidneys in life? ____________________

____

3. Define *ptosis*. ____________________

____

4. Why is incontinence a normal phenomenon in the child under 1½ to 2 years old? ____________________

____

What events may lead to its occurrence in the adult? ____________________

____

____

5. Complete the labeling of the diagram to correctly identify the urinary system organs.

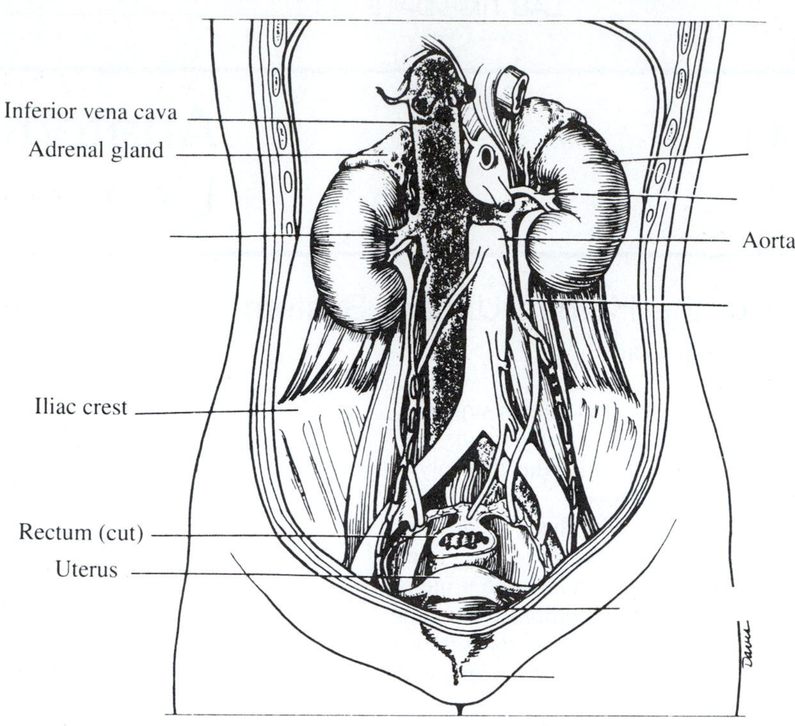

Gross Internal Anatomy of the Pig or Sheep Kidney

Match the appropriate structure in column B to its description in column A.

	Column A	Column B
________	1. smooth membrane, tightly adherent to the kidney surface	a. cortex
________	2. portion of the kidney containing mostly collecting ducts	b. medulla
________	3. portion of the kidney containing the bulk of the nephron structures	c. minor calyx
________	4. superficial region of kidney tissue	d. renal capsule
________	5. basinlike area of the kidney, continuous with the ureter	e. renal column
________	6. a cup-shaped extension of the pelvis that encircles the apex of a pyramid	f. renal pelvis
________	7. area of cortical tissue running between the medullary pyramids	

Microscopic Anatomy of the Kidney and Bladder

1. Match each of the lettered structures on the diagram of the nephron (and associated renal blood supply) on the left with the terms on the right:

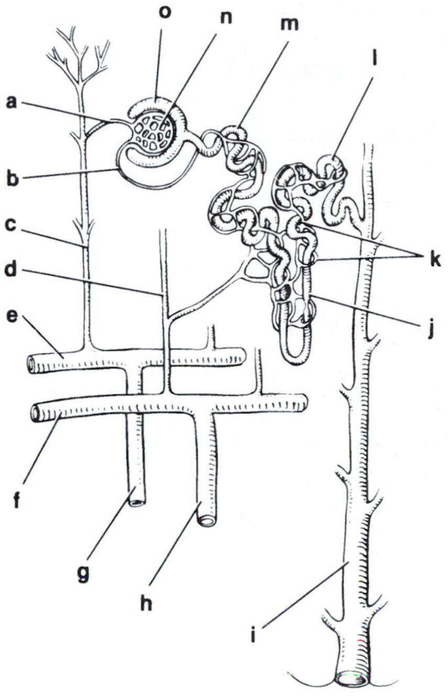

a
b
c
d
e
f
g
h
i
j
k
l
m
n
o

_______ 1. collecting duct

_______ 2. glomerulus

_______ 3. peritubular capillaries

_______ 4. distal convoluted tubule

_______ 5. proximal convoluted tubule

_______ 6. interlobar artery

_______ 7. interlobular artery

_______ 8. arcuate artery

_______ 9. interlobular vein

_______ 10. efferent arteriole

_______ 11. arcuate vein

_______ 12. loop of Henle

_______ 13. afferent arteriole

_______ 14. interlobar vein

_______ 15. glomerular capsule

2. Using the terms provided in item 1, identify the following:

_______________ 1. site of filtrate formation

_______________ 2. primary site of tubular reabsorption

_______________ 3. secondarily important site of tubular reabsorption

_______________ 4. structure that conveys the processed filtrate (urine) to the renal pelvis

_______________ 5. blood supply that directly receives substances from the tubular cells

_______________ 6. its inner (visceral) membrane forms part of the filtration membrane

3. Explain *why* the glomerulus is such a high-pressure capillary bed. _______________

____

How does its high pressure condition aid its function of filtrate formation? _______________

____

4. What structural modification of certain tubule cells enhances their ability to reabsorb substances from the filtrate?

____

5. Explain the mechanism of tubular secretion and explain its importance in the urine formation process. ________

____

____

____

6. Compare and contrast the composition of blood plasma and glomerular filtrate. ________

____

7. Trace a drop of blood from the time it enters the kidney in the renal artery until it leaves the kidney through the renal vein. Renal artery → ________

____

________ → renal vein

8. Trace the anatomical pathway of a molecule of creatinine (metabolic waste) from the glomerular capsule to the urethra. Note each microscopic and/or gross structure it passes through in its travels. Name the subdivisions of the renal tubule. Glomerular capsule → ________

____

________ → urethra

9. What is important functionally about the specialized epithelium (transitional epithelium) in the bladder?

____

____

STUDENT NAME ____________________

LAB TIME/DATE ____________________

Review Sheet

EXERCISE 41 Urinalysis

1. What is the normal volume of urine excreted in a 24-hour period? ______

   a. 0.1–0.5 liters   b. 0.5–1.2 liters   c. 1.0–1.8 liters

2. Assuming normal conditions, note whether each of the following substances would be (a) in greater relative concentration in the urine than in the glomerular filtrate, (b) in lesser concentration in the urine than in the glomerular filtrate, or (c) absent in both the urine and the glomerular filtrate.

______ 1. water	______ 6. amino acids	______ 10. urea
______ 2. phosphate ions	______ 7. glucose	______ 11. uric acid
______ 3. sulfate ions	______ 8. albumin	______ 12. creatinine
______ 4. potassium ions	______ 9. red blood cells	______ 13. pus (WBC)
______ 5. sodium ions		

3. Explain why urinalysis is a routine part of any good physical examination. ____________________

   ____

4. What substance is responsible for the normal yellow color of urine? ____________________

5. Which has a greater specific gravity: 1 ml of urine or 1 ml of distilled water? ____________________

   Explain. ____________________

   ____

6. Explain the relationship between the color, specific gravity, and volume of urine. ____________________

   ____

   ____

   ____

7. A microscopic examination of urine may reveal the presence of certain abnormal urinary constituents.

   Name three constituents that might be present if a urinary tract infection exists. ____________________,

   ____________________, and ____________________

8. How does a urinary tract infection influence urine pH? ____________________

   How does starvation influence urine pH? ____________________

9. Several specific terms have been used to indicate the presence of abnormal urine constituents. Identify each of the abnormalities described below by inserting a term from the list at the right that names the condition.

______ 1.	presence of erythrocytes in the urine	a. albuminuria
______ 2.	presence of hemoglobin in the urine	b. glycosuria
______ 3.	presence of glucose in the urine	c. hematuria
______ 4.	presence of albumin in the urine	d. hemoglobinuria
______ 5.	presence of ketone bodies (acetone and others) in the urine	e. ketonuria
		f. pyuria
______ 6.	presence of pus (white blood cells) in the urine	

10. What are renal calculi and what conditions favor their formation? ______

______

11. All urine specimens become alkaline and cloudy on standing at room temperature. Explain. ______

______

12. Glucose and albumin are both normally absent in the urine, but the reason for their exclusion differs. Explain the reason for the absence of glucose. ______

______

Explain the reason for the absence of albumin. ______

______

13. Several conditions (both pathologic and nonpathologic) are named below. Using the key provided, characterize the probable abnormal constituents or conditions of the urinary product of each. More than one choice may be necessary to fully characterize the condition in most cases.

______ 1. glomerulonephritis	______ 7. starvation	Key:	
______ 2. diabetes mellitus	______ 8. diabetes insipidus	a.	albumin
		b.	hemoglobin
______ 3. pregnancy, exertion	______ 9. kidney stones	c.	blood cells
		d.	glucose
______ 4. hepatitis, cirrhosis of the liver	______ 10. eating a 5-lb box of candy at one sitting	e.	ketone bodies
		f.	bilirubin
		g.	pus
______ 5. pyelonephritis	______ 11. hemolytic anemias	h.	high specific gravity
		i.	low specific gravity
______ 6. gonorrhea	______ 12. cystitis (inflammation of the bladder)	j.	casts

14. Name the three major nitrogenous wastes found in the urine. ______,

______, and ______

15. Explain the difference between organized and unorganized sediments. ______

______

______

______

STUDENT NAME ___________________

LAB TIME/DATE ___________________

Review Sheet

EXERCISE 42 Anatomy of the Reproductive System

Gross Anatomy of the Human Male Reproductive System

1. List the two principal functions of the testis: ___________________

2. Identify all indicated structures or portions of structures on the diagrammatic view of the male reproductive system below.

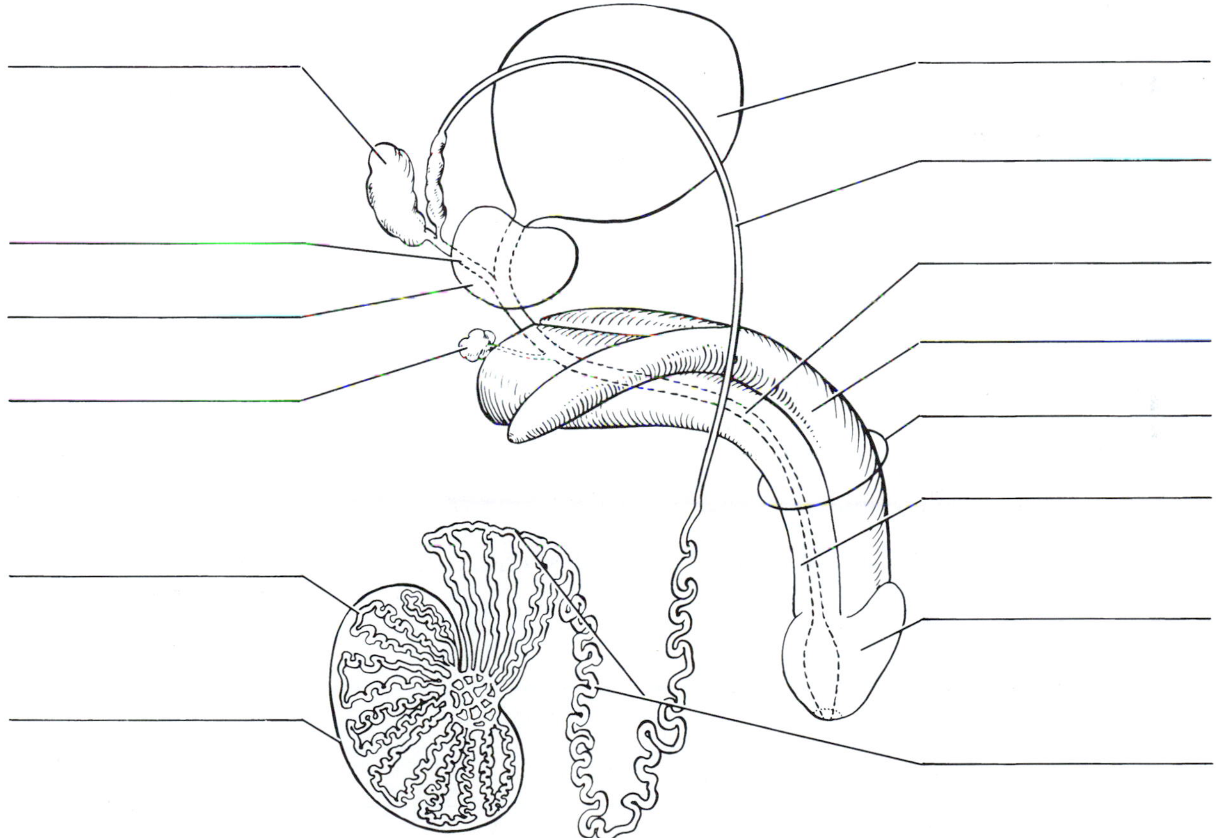

3. A common part of any physical examination of the male is palpation of the prostate gland. How is this accomplished? (Think!) ___________________

4. How might enlargement of the prostate gland interfere with urination or the reproductive ability of the male?

____

____

5. Match the terms in column B to the descriptive statements in column A.

	Column A	Column B
________	1. copulatory organ/penetrating device	a. bulbourethral glands
________	2. site of sperm/androgen production	b. epididymis
________	3. muscular passageway conveying sperm to the ejaculatory duct; in the spermatic cord	c. glans penis
		d. membranous urethra
________	4. transports both sperm and urine	e. penile urethra
________	5. sperm maturation site	f. penis
________	6. location of the testis in adult males	g. prepuce
________	7. loose fold of skin encircling the glans penis	h. prostate gland
________	8. portion of the urethra between the prostate gland and the penis	i. prostatic urethra
________	9. empties a secretion into the prostatic urethra	j. seminal vesicles
________	10. empties a secretion into the membranous urethra	k. scrotum
		l. testes
		m. vas (ductus) deferens

6. Why are the testes located in the scrotum? ________________________________

____

7. Describe the composition of semen and name all structures contributing to its formation. ________________

____

____

8. Of what importance is the fact that seminal fluid is alkaline? ________________________________

____

9. What structures comprise the spermatic cord? ________________________________

____

Where is it located? ________________________________

____

10. Using the following terms, trace the pathway of sperm from the testes to the urethra: rete testis, epididymis, seminiferous tubule, ductus deferens.

________________ → ________________ → ________________ → ________________

11. Using an appropriate reference, define *cryptorchidism* and discuss its significance.

____

____

Gross Anatomy of the Human Female Reproductive System

1. On the diagram of a frontal section of a portion of the female reproductive system seen below, identify all indicated structures.

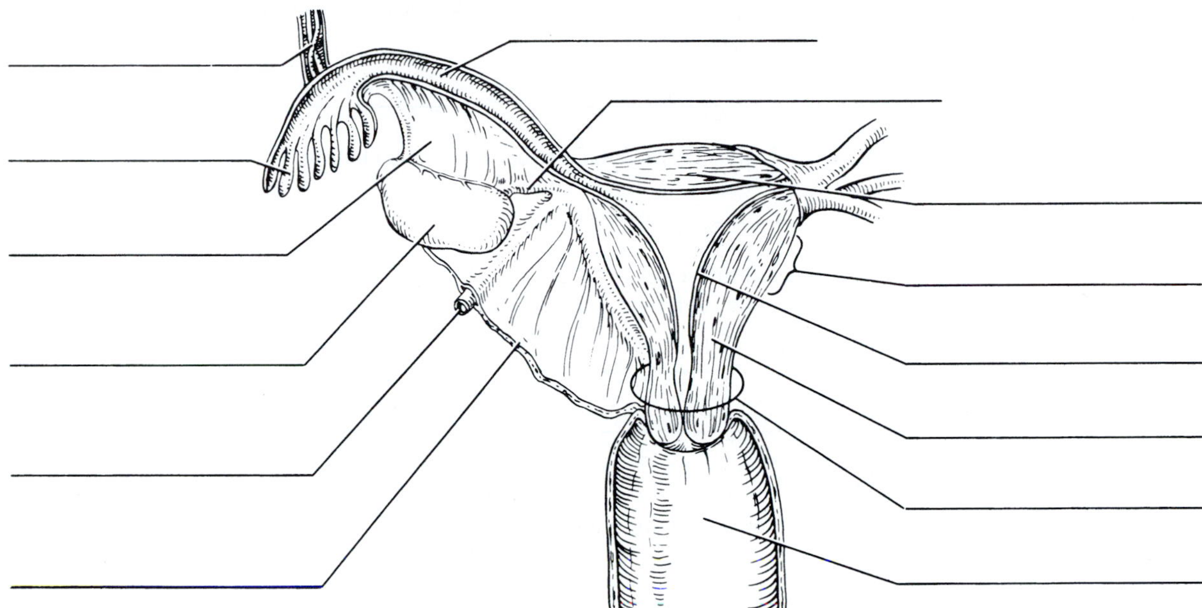

2. Identify the female reproductive system structures described below:

_______________ 1. site of fetal development

_______________ 2. copulatory canal

_______________ 3. "fertilized egg" typically formed here

_______________ 4. becomes erectile during sexual excitement

_______________ 5. duct extending superolaterally from the uterus

_______________ 6. partially closes the vaginal canal; a membrane

_______________ 7. produces eggs, estrogens, and progesterone

_______________ 8. fingerlike ends of the fallopian tube

3. Do any sperm enter the pelvic cavity of the female? Why or why not? _______________

____

4. What is an ectopic pregnancy, and how can it happen? _______________

____

____

____

5. Name the structures composing the external genitalia, or vulva, of the female. ____________________

____

____

6. Put the following vestibular-perineal structures in their proper order from the anterior to the posterior aspect: vaginal orifice, anus, urethral opening, and clitoris.

Anterior limit: ______________ → ______________ → ______________ → ______________

7. Name the male structure that is homologous to the female structures named below.

labia majora ____________________ clitoris ____________________

8. Assume a couple has just consummated the sex act and the male's sperm have been deposited in the woman's vagina. Trace the pathway of the sperm through the female reproductive tract.

____

9. Define *ovulation:* ____

The Mammary Glands

1. To describe breast function, complete the following sentences:

Milk is formed by ____________________ within the ____________________ of the breast. Milk is then excreted into enlarged storage regions called ____________________ and then finally through the ____.

2. Describe the procedure for self-examination of the breasts. (Men are not exempt from breast cancer, you know!)

____

____

____

____

____

____

Microscopic Anatomy of Selected Male and Female Reproductive Organs

1. The testis is divided into a number of lobes by connective tissue. Each of these lobes contains one to four

   ______________________________, which converge on a tubular region at the testis hilus called the

   ______________________.

2. What is the function of the cavernous bodies seen in the male penis? ______________________________

   ____

   ____

3. Name the three layers of the uterine wall from the inside out.

   ____________________, ____________________, ____________________

   Which of these is sloughed during menses? ______________________________

   Which contracts during childbirth? ______________________________

4. What is the function of the stereocilia exhibited by the epithelial cells of the mucosa of the epididymis? ________

   ____

5. On the diagram showing the sagittal section of the human testis, correctly identify all structures provided with leader lines.

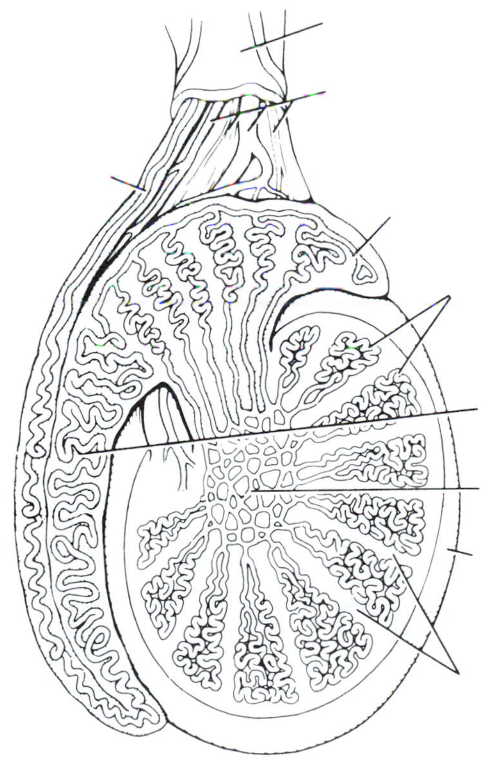

STUDENT NAME ____________________

LAB TIME/DATE ____________________

Review Sheet

EXERCISE 43 Physiology of Reproduction: Gametogenesis and the Female Cycles

Meiosis

1. The following statements refer to events occurring during mitosis and/or meiosis. For each statement, decide if the event occurs in (a) mitosis only, (b) meiosis only, or (c) both mitosis and meiosis.

____ 1. dyads are visible

____ 2. tetrads are visible

____ 3. product is two diploid daughter cells

____ 4. product is four haploid daughter cells

____ 5. involves the phases prophase, metaphase, anaphase, and telophase

____ 6. occurs throughout the body

____ 7. occurs only in the ovaries and testes

____ 8. provides cells for growth and repair

____ 9. homologues synapse and chiasmata are seen

____ 10. daughter cells are quantitatively and qualitatively different from the mother cell

____ 11. daughter cells are genetically identical to the mother cell

____ 12. chromosomes are replicated before the division process begins

____ 13. provides cells for replication of the species

____ 14. consists of two consecutive nuclear divisions, without chromosomal replication occurring before the second division

2. Describe the process of synapsis. ____________________

____

3. How does crossover introduce variability in the daughter cells? ____________________

____

____

4. Define *homologous chromosomes:* ____________________

____

Spermatogenesis

1. The cell types seen in the seminiferous tubules are listed in the key. Match the correct cell type(s) with the descriptions given below.

Key: a. primary spermatocyte
b. secondary spermatocyte
c. spermatogonium
d. sustentacular cell
e. spermatid
f. sperm

_______________ 1. primitive stem cell

_______________ 2. haploid

_______________ 3. provides nutrients to developing sperm

_______________ 4. product of meiosis II

_______________ 5. product of spermiogenesis

_______________ 6. product of meiosis I

2. Why are spermatids not considered functional gametes? _______________

____

3. Define *spermatogenesis:* _______________

Define *spermiogenesis:* _______________

____

4. Draw a sperm below and identify the *acrosome, head, midpiece,* and *tail.* Then beside each label, note the composition and function of each of these sperm structures.

5. The life span of a sperm is very short. What anatomical characteristics might lead you to suspect this even if you didn't know its life span? _______________

Oogenesis, the Ovarian Cycle, and the Menstrual Cycle

1. The sequence of events leading to germ cell formation in the female begins during fetal development. By the time the child is born, all the oogonia have been converted to _______________.

In view of this fact, how does the total germ cell potential of the female compare to that of the male?

____

2. The female gametes develop in structures called *follicles*. What is a follicle? ______________________

____

How are primary and vesicular follicles anatomically different? ______________________

____

____

____

What is a corpus luteum? ______________________

____

3. What hormone is produced by the vesicular follicle? ______________________

By the corpus luteum? ______________________

4. Use the key to identify the cell type you would expect to find in the following structures.

Key: a. oogonium b. primary oocyte c. secondary oocyte d. ovum

________ 1. forming part of the primary follicle in the ovary

________ 2. in the uterine tube before fertilization

________ 3. in the mature vesicular follicle of the ovary

________ 4. in the uterine tube shortly after sperm penetration

5. The cellular product of spermatogenesis is four ______________; the final product of oogenesis is one ______________ and three ______________. What is the function of this unequal cytoplasmic division seen during oogenesis in the female? ______________________

____

____

What is the fate of the three tiny cells produced during oogenesis? ______________________

Why? ______________________

6. The following statements deal with anterior pituitary and ovarian hormonal interrelationships. Name the hormone(s) described in each statement.

______________________ 1. stimulates ovarian follicles to grow and to produce estrogens

______________________ 2. ovulation occurs after its burstlike release

______________________ and ______________________ 3. exert negative feedback on the anterior pituitary relative to FSH secretion

______________________ 4. stimulates LH release by the anterior pituitary

______________________ 5. stimulates the corpus luteum to produce progesterone and estrogen

______________________ 6. maintains the hormonal production of the corpus luteum in a nonpregnant woman

7. Why does the corpus luteum deteriorate toward the end of the ovarian cycle? ______________________________

____

8. For each statement below dealing with hormonal blood levels during the female ovarian and menstrual cycles, decide whether the condition in column A is usually (a) greater than, (b) less than, or (c) essentially equal to the condition in column B.

	Column A		Column B
________	1. amount of estrogen in the blood during menses	↔	amount of estrogen in the blood at ovulation
________	2. amount of progesterone in the blood on the fourteenth day	↔	amount of progesterone in the blood on the twenty-third day
________	3. amount of LH in the blood during menses	↔	amount of LH in the blood at ovulation
________	4. amount of FSH in the blood on day 6 of the cycle	↔	amount of FSH in the blood on day 20 of the cycle
________	5. amount of estrogen in the blood on the tenth day	↔	amount of progesterone in the blood on the tenth day

9. Ovulation and menstruation usually cease by the age of ____________.

10. What uterine tissue undergoes dramatic changes during the menstrual cycle? ______________________________

11. When during the female menstrual cycle would fertilization be unlikely? ______________________________

____

12. Assume that a woman could be an "on demand" ovulator like the rabbit, in which copulation stimulates the hypothalamic-anterior pituitary axis and causes LH release, and an oocyte was ovulated and fertilized on day 26 of her 28-day cycle. Why would a successful pregnancy be unlikely at this time?

____

____

13. The menstrual cycle depends on events within the female ovary. The stages of the menstrual cycle are listed below. For each, note its approximate time span and the related events in the uterus; and then to the right, record the ovarian events occurring simultaneously. Pay particular attention to hormonal events.

Menstrual cycle stage	Uterine events	Ovarian events
Menstruation		
Proliferative		
Secretory		

STUDENT NAME ____________________

LAB TIME/DATE ____________________

Review Sheet

EXERCISE 44

Survey of Embryonic Development

Developmental Stages of the Human

1. Use the key choices to identify the embryonic stage or process described below.

   Key: a. cleavage c. zygote e. blastula
   b. morula d. fertilization f. gastrulation

   ________ 1. the event most immediately following sperm penetration

   ________ 2. solid ball of embryonic cells

   ________ 3. process of rapid mitotic cell division without intervening growth periods

   ________ 4. combination of egg and sperm

   ________ 5. process involving cell rearrangements to form the three primary germ layers

   ________ 6. embryonic stage in which the embryo consists of a hollow ball of cells

2. What is the importance of cleavage in embryonic development? ____________________

   ____

   How is cleavage different from mitotic cell division, which occurs late in life? ____________________

   ____

   ____

3. Explain the importance of gastrulation. ____________________

   ____

   ____

4. Name the primary germ layers, and describe their relative positions in the embryo.

   ____________, ____________________

   ____________, ____________________

   ____________, ____________________

5. The cells of the human blastula (more commonly called the blastocyst or chorionic vesicle) have various fates. Which blastocyst structures have the following fates?

______________ 1. produces the embryonic body

______________ 2. becomes the chorion and cooperates with uterine tissues to form the placenta

______________ 3. produces the amnion, yolk sac, and allantois

______________ 4. produces the primordial germ cells (an embryonic membrane)

______________ 5. an embryonic membrane that provides the structural basis for the body stalk or umbilical cord

6. What is the function of the amnion and the amniotic fluid? ______________

____

____

7. Describe the process of implantation, noting the role of the trophoblast cells. ______________

____

____

8. How many days after fertilization is implantation generally completed? ________ What event in the female menstrual cycle ordinarily occurs just about this time if implantation does not occur? ______________

____

9. What name is given to the part of the uterine wall directly under the implanting embryo? ______________

That surrounding the rest of the embryonic structure? ______________

10. Using an appropriate reference, find out what *decidua* means and state the definition. ______________

____

How is this terminology applicable to the deciduas of pregnancy? ______________

____

11. Referring to the illustrations and text of *Life Before Birth,* answer the following:

Which two organ systems are extensively developed in the *very young* embryo?

______________ and ______________

Describe the direction of development by circling the correct descriptions below:

proximal-distal   distal-proximal   caudal-rostral   rostral-caudal

Does bodily control during infancy develop in the same directions? Think! Can an infant pick up a common pin (pincer grasp) or wave his arms earlier? Is arm-hand or leg-foot control achieved earlier?

____

12. Note whether each of the following organs or organ systems develop from the (a) ectoderm, (b) endoderm, or (c) mesoderm. Use an appropriate reference as necessary.

________ 1. skeletal muscle	________ 4. respiratory mucosa	________ 7. nervous system
________ 2. skeleton	________ 5. circulatory system	________ 8. serosa membrane
________ 3. lining of gut	________ 6. epidermis of skin	________ 9. liver, pancreas

In Utero Development

1. Make the following comparisons between a human and the pregnant dissected animal structures.

Comparison object	Human	Dissected animal
Shape of the placenta		
Shape of the uterus		

2. Where in the human uterus do implantation and placentation ordinarily occur? ____________________

____

3. Describe the function(s) of the placenta. ____________________

____

____

What embryonic membranes has it more or less "put out of business"? ____________________

____

4. When does the human embryo come to be called a fetus? ____________________

5. What is the usual and most desirable fetal position in utero? ____________________

Why is this the most desirable position? ____________________

____

____

____

Gross and Microscopic Anatomy of the Placenta

1. Describe fully the gross structure of the human placenta as observed in the laboratory. ____________________

____

____

2. What is the tissue origin of the placenta: fetal, maternal, or both? ____________________

3. What are the placental barriers that must be crossed to exchange materials? ____________________

____

STUDENT NAME ________________________________

LAB TIME/DATE ________________________________

Review Sheet

EXERCISE 45

Principles of Heredity

Introduction to the Language of Genetics

1. Match the key choices with the definitions given below.

Key:
a. alleles
b. autosomes
c. dominant
d. genotype
e. heterozygous
f. homozygous
g. phenotype
h. recessive
i. sex chromosomes

____________ 1. actual genetic makeup

____________ 2. chromosomes determining maleness/femaleness

____________ 3. situation in which an individual has identical alleles for a particular trait

____________ 4. genes not expressed unless they are present in homozygous condition

____________ 5. expression of a genetic trait

____________ 6. situation in which an individual has different alleles making up his genotype for a particular trait

____________ 7. genes for the same trait that may have different expressions

____________ 8. chromosomes regulating most body characteristics

____________ 9. the more-potent gene allele; masks the expression of the less-potent allele

Dominant-Recessive Inheritance

1. In humans, farsightedness is inherited by possession of a dominant gene. If a man who is homozygous for normal vision (aa) marries a woman who is heterozygous for farsightedness, what proportion of their children would be expected to be farsighted? ________ %

2. A metabolic disorder called PKU is due to an abnormal recessive gene (p). Only homozygous recessive individuals exhibit this disorder. What percentage of the offspring will be anticipated to have PKU if the parents are Pp and pp? ________ %

3. A man obtained 32 spotted and 10 solid-color rabbits from a mating of two spotted rabbits.

   Which trait is dominant? ____________________ Recessive? ____________________

   What is the probable genotype of the rabbit parents? ________ × ________

4. Assume that the allele controlling brown eyes (B) is dominant over that controlling blue eyes (b) in human beings. (In actuality, eye color in humans is an example of multigene inheritance, which is much more complex than this.) A blue-eyed man marries a brown-eyed woman; and they have six children, all brown-eyed. What is the most likely genotype of the father? ______ Of the mother? ______ If the seventh child had *blue* eyes, what could you conclude about the parents' genotypes? ______________________________

____

Incomplete Dominance

1. Tail length on a bobcat is controlled by incomplete dominance. The alleles are T for normal tail length and t for tail-less. What name could/would you give to the tails of heterozygous (Tt) cats? ______________________

   How would their tail length compare with that of TT or tt bobcats? ______________________

2. If curly-haired individuals are genotypically CC, straight-haired individuals are cc, and wavy-haired individuals are heterozygotes (Cc), what percentage of the various phenotypes would be anticipated from a cross between a CC woman and a cc man?

   ________ % curly ________ % wavy ________ % straight

Sex-Linked Inheritance

1. What does it mean when someone says a particular characteristic is sex-linked? ______________________

   ____

2. You are a male, and you have been told that hemophilia "runs in your genes." Whose ancestors, your mother's or your father's, should you investigate? ______________ Why? ______________________

   ____

   ____

3. An X^CX^c female marries an X^CY man. Do a Punnett square for this match.

   What is the probability of producing a color-blind son? ____________

   A color-blind daughter? ____________

   A daughter that is a carrier for the color-blind gene? ____________

4. Why are consanguineous marriages (marriages between blood relatives) prohibited in most cultures?

   ____

   ____

Probability

1. What is the probability of having three daughters in a row? ______

2. A man and a woman, each of seemingly normal intellect, marry. Although neither is aware of the fact, each is a heterozygote for the allele for feeblemindedness. Is the allele for feeblemindedness dominant or recessive?

______

What are the chances of their having one feebleminded child? ______

What are the chances that all of their children (they plan a family of four) will be feebleminded?

______

Genetic Determination of Selected Human Characteristics

1. Look back at your data to complete this section. For each of the situations described here, determine if an offspring with the characteristics noted is possible with the parental genotypes listed. Check (√) the appropriate column.

		Possibility	
Parental genotypes	**Phenotype of child**	**Yes**	**No**
Jj × jj	Double-jointed thumbs		
FF × Ff	Straight little finger		
EE × ee	Detached ear lobes		
HH × Hh	Mid-digital hair		
$I^A i \times I^B i$	Type O blood		
$I^A I^B \times ii$	Type B blood		

2. You have dimples, and you would like to know if you are homozygous or heterozygous for this trait. You have six brothers and sisters. By observing your siblings, how could you tell, with some degree of certainty, that you are a heterozygote?

______

______

______

______

STUDENT NAME ________________________________

LAB TIME/DATE ________________________________

Review Sheet

EXERCISE 46

Surface Anatomy Roundup

________ 1. A blow to the cheek is most likely to break what superficial bone or bone part? (a) superciliary arches, (b) the philtrum, (c) zygomatic arch, (d) the tragus.

________ 2. Rebound tenderness (a) occurs in appendicitis, (b) is whiplash of the neck, (c) is a sore foot from playing basketball, (d) occurs when the larynx falls back into place after swallowing.

________ 3. The anatomical snuff box (a) is in the nose, (b) contains the styloid process of the radius, (c) is defined by tendons of the flexor carpi radialis and palmaris longus, (d) cannot really hold snuff.

________ 4. Some landmarks on the body surface can be seen or felt, but others are abstractions that you must construct by drawing imaginary lines. Which of the following pairs of structures is abstract and invisible? (a) umbilicus and costal margin, (b) anterior superior iliac spine and natal cleft, (c) linea alba and linea semilunaris, (d) McBurney's point and midaxillary line, (e) philtrum and sternocleidomastoid.

________ 5. Many pelvic organs can be palpated by placing a finger in the rectum or vagina, but only one pelvic organ is readily palpated through the skin. This is the (a) nonpregnant uterus, (b) prostate gland, (c) full bladder, (d) ovaries, (e) rectum.

________ 6. A muscle that contributes to the posterior axillary fold is the (a) pectoralis major, (b) latissimus dorsi, (c) trapezius, (d) infraspinatus, (e) pectoralis minor, (f) a and e.

________ 7. Which of the following is not a pulse point? (a) anatomical snuff box, (b) inferior margin of mandible anterior to masseter muscle, (c) center of distal forearm at palmaris longus tendon, (d) medial bicipital furrow on arm, (e) dorsum of foot between the first two metatarsals.

________ 8. Which pair of ribs inserts on the sternum at the sternal angle? (a) first, (b) second, (c) third, (d) fourth, (e) fifth.

________ 9. The inferior angle of the scapula is at the same level as the spinous process of this vertebra: (a) C_5, (b) C_7, (c) T_3, (d) T_7, (e) L_4.

________ 10. An important bony landmark that can be recognized by a distinct dimple in the skin is the (a) posterior superior iliac spine, (b) styloid process of the ulna, (c) shaft of the radius, (d) acromion.

________ 11. A nurse missed a patient's median cubital vein while trying to withdraw blood and then inserted the needle far too deeply into the cubital fossa. This error could cause any of the following problems, except this one: (a) paralysis of the ulnar nerve, (b) paralysis of the median nerve, (c) bruising the insertion tendon of the biceps brachii muscle, (d) blood spurting from the brachial artery.

________ 12. Which of these organs is almost impossible to study with the techniques of surface anatomy? (a) heart, (b) lungs, (c) brain, (d) nose.

________ 13. A preferred site for inserting an intravenous medication line into a blood vessel is the (a) medial bicipital furrow on arm, (b) external carotid artery, (c) dorsal venous arch of hand, (d) popliteal fossa.

________ 14. One listens for bowel sounds with a stethoscope that is placed (a) on the four quadrants of the abdominal wall; (b) in the triangle of auscultation; (c) in the right and left midaxillary line, just superior to the iliac crests; (d) inside the patient's bowels (intestines), on the tip of an endoscope.

15. Define *palpation:* ______

16. Explain how one locates the proper site for intramuscular injections into:

 (a) Deltoid muscle. ______

 (b) Ventral gluteal site: ______

______

______

17. Ashley, a pre-physical therapy student, was trying to locate the vertebral spinous processes on the flexed back of her friend Amber, but she kept losing count. Amber told her to check her count against several reliable "guideposts" along the way: the spinous processes of C_7, T_3, T_7, and L_4. Can you describe how to find each of these four particular vertebrae without having to count any vertebrae?

 C_7: ______

______

 T_3: ______

______

 T_7: ______

______

 L_4: ______

______

18. How does one find the midinguinal point? ______

______

19. Locate the standard points of surgical incision for reaching both the appendix and the gallbladder. ______

______

______

______

20. Gregory hit his funny bone. What nerve was hit against what bony process? ______

______

21. An athletic trainer was helping a college basketball player find the site of a pulled muscle. The trainer asked the athlete to extend her thigh at the hip forcefully, but she found this action too painful to perform. Then the trainer palpated her posterior thigh and felt some swelling of the muscles there. In simplest terms, which basic muscle group was injured?

______

22. Walking to her car after her sixty-fifth birthday party, Mrs. Schultz tripped on ice and fell forward on her outstretched palms. When she arrived at the emergency room, her right wrist and hand were bent like a fork handle. Dr. Jefferson felt that the styloid process of her radius was outside the anatomical snuff box and slightly proximal to the styloid process of the ulna. When he checked her elbow, he found that the olecranon process lay 2 cm proximal to the two epicondyles of the right humerus. Explain all these observations, and describe what had happened to Mrs. Schultz's limb.

____

____

____

____

____

A APPENDIX

The Metric System

Measurement	Unit and abbreviation	Metric equivalent	Metric to English conversion factor	English to metric conversion factor
Length	1 kilometer (km)	= 1000 (10^3) meters	1 km = 0.62 mile	1 mile = 1.61 km
	1 meter (m)	= 100 (10^2) centimeters = 1000 millimeters	1 m = 1.09 yards 1 m = 3.28 feet 1 m = 39.37 inches	1 yard = 0.914 m 1 foot = 0.305 m
	1 centimeter (cm)	= 0.01 (10^{-2}) meter	1 cm = 0.394 inch	1 foot = 30.5 cm 1 inch = 2.54 cm
	1 millimeter (mm)	= 0.001 (10^{-3}) meter	1 mm = 0.039 inch	
	1 micrometer (μm) [formerly micron (μ)]	= 0.000001 (10^{-6}) meter		
	1 nanometer (nm) [formerly millimicron (mμ)]	= 0.000000001 (10^{-9}) meter		
	1 angstrom (Å)	= 0.0000000001 (10^{-10}) meter		
Area	1 square meter (m^2)	= 10,000 square centimeters	1 m^2 = 1.1960 square yards 1 m^2 = 10.764 square feet	1 square yard = 0.8361 m^2 1 square foot = 0.0929 m^2
	1 square centimeter (cm^2)	= 100 square millimeters	1 cm^2 = 0.155 square inch	1 square inch = 6.4516 cm^2
Mass	1 metric ton (t)	= 1000 kilograms	1 t = 1.103 ton	1 ton = 0.907 t
	1 kilogram (kg)	= 1000 grams	1 kg = 2.205 pounds	1 pound = 0.4536 kg
	1 gram (g)	= 1000 milligrams	1 g = 0.0353 ounce 1 g = 15.432 grains	1 ounce = 28.35 g
	1 milligram (mg)	= 0.001 gram	1 mg = approx. 0.015 grain	
	1 microgram (μg)	= 0.000001 gram		
Volume (solids)	1 cubic meter (m^3)	= 1,000,000 cubic centimeters	1 m^3 = 1.3080 cubic yards 1 m^3 = 35.315 cubic feet	1 cubic yard = 0.7646 m^3 1 cubic foot = 0.0283 m^3
	1 cubic centimeter (cm^3 or cc)	= 0.000001 cubic meter = 1 milliliter	1 cm^3 = 0.0610 cubic inch	1 cubic inch = 16.387 cm^3
	1 cubic millimeter (mm^3)	= 0.000000001 cubic meter		
Volume (liquids and gases)	1 kiloliter (kl or kL)	= 1000 liters	1 kL = 264.17 gallons	1 gallon = 3.785 L 1 quart = 0.946 L
	1 liter (l or L)	= 1000 milliliters	1 L = 0.264 gallons 1 L = 1.057 quarts	
	1 milliliter (ml or mL)	= 0.001 liter = 1 cubic centimeter	1 ml = 0.034 fluid ounce 1 ml = approx. $\frac{1}{4}$ teaspoon 1 ml = approx. 15–16 drops (gtt.)	1 quart = 946 ml 1 pint = 473 ml 1 fluid ounce = 29.57 ml 1 teaspoon = approx. 5 ml
	1 microliter (μl or μL)	= 0.000001 liter		
Time	1 second (s)	= $\frac{1}{60}$ minute		
	1 millisecond (ms)	= 0.001 second		
Temperature	Degrees Celsius (°C)		$°F = \frac{9}{5}°C + 32$	$°C = \frac{5}{9}(°F - 32)$

APPENDIX B Predicted Vital Capacities for Males

Age	146	148	150	152	154	156	158	160	162	164	166	168	170	172	174	176	178	180	182	184	186	188	190	192	194
	Height in centimeters																								
16	3765	3820	3870	3920	3975	4025	4075	4130	4180	4230	4285	4335	4385	4440	4490	4540	4590	4645	4695	4745	4800	4850	4900	4955	5005
18	3740	3790	3840	3890	3940	3995	4045	4095	4145	4200	4250	4300	4350	4405	4455	4505	4555	4610	4660	4710	4760	4815	4865	4915	4965
20	3710	3760	3810	3860	3910	3960	4015	4065	4115	4165	4215	4265	4320	4370	4420	4470	4520	4570	4625	4675	4725	4775	4825	4875	4930
22	3680	3730	3780	3830	3880	3930	3980	4030	4080	4135	4185	4235	4285	4335	4385	4435	4485	4535	4585	4635	4685	4735	4790	4840	4890
24	3635	3685	3735	3785	3835	3885	3935	3985	4035	4085	4135	4185	4235	4285	4330	4380	4430	4480	4530	4580	4630	4680	4730	4780	4830
26	3605	3655	3705	3755	3805	3855	3905	3955	4000	4050	4100	4150	4200	4250	4300	4350	4395	4445	4495	4545	4595	4645	4695	4740	4790
28	3575	3625	3675	3725	3775	3820	3870	3920	3970	4020	4070	4115	4165	4215	4265	4310	4360	4410	4460	4510	4555	4605	4655	4705	4755
30	3550	3595	3645	3695	3740	3790	3840	3890	3935	3985	4035	4080	4130	4180	4230	4275	4325	4375	4425	4470	4520	4570	4615	4665	4715
32	3520	3565	3615	3665	3710	3760	3810	3855	3905	3950	4000	4050	4095	4145	4195	4240	4290	4340	4385	4435	4485	4530	4580	4625	4675
34	3475	3525	3570	3620	3665	3715	3760	3810	3855	3905	3950	4000	4045	4095	4140	4190	4225	4285	4330	4380	4425	4475	4520	4570	4615
36	3445	3495	3540	3585	3635	3680	3730	3775	3825	3870	3920	3965	4010	4060	4105	4155	4200	4250	4295	4340	4390	4435	4485	4530	4580
38	3415	3465	3510	3555	3605	3650	3695	3745	3790	3840	3885	3930	3980	4025	4070	4120	4165	4210	4260	4305	4350	4400	4445	4495	4540
40	3385	3435	3480	3525	3575	3620	3665	3710	3760	3805	3850	3900	3945	3990	4035	4085	4130	4175	4220	4270	4315	4360	4410	4455	4500
42	3360	3405	3450	3495	3540	3590	3635	3680	3725	3770	3820	3865	3910	3955	4000	4050	4095	4140	4185	4230	4280	4325	4370	4415	4460
44	3315	3360	3405	3450	3495	3540	3585	3630	3675	3725	3770	3815	3860	3905	3950	3995	4040	4085	4130	4175	4220	4270	4315	4360	4405
46	3285	3330	3375	3420	3465	3510	3555	3600	3645	3690	3735	3780	3825	3870	3915	3960	4005	4050	4095	4140	4185	4230	4275	4320	4365
48	3255	3300	3345	3390	3435	3480	3525	3570	3615	3655	3700	3745	3790	3835	3880	3925	3970	4015	4060	4105	4150	4190	4235	4280	4325
50	3210	3255	3300	3345	3390	3430	3475	3520	3565	3610	3650	3695	3740	3785	3830	3870	3915	3960	4005	4050	4090	4135	4180	4225	4270
52	3185	3225	3270	3315	3355	3400	3445	3490	3530	3575	3620	3660	3705	3750	3795	3835	3880	3925	3970	4010	4055	4100	4140	4185	4230
54	3155	3195	3240	3285	3325	3370	3415	3455	3500	3540	3585	3630	3670	3715	3760	3800	3845	3890	3930	3975	4020	4060	4105	4145	4190
56	3125	3165	3210	3255	3295	3340	3380	3425	3465	3510	3550	3595	3640	3680	3725	3765	3810	3850	3895	3940	3980	4025	4065	4110	4150
58	3080	3125	3165	3210	3250	3290	3335	3375	3420	3460	3500	3545	3585	3630	3670	3715	3755	3800	3840	3880	3925	3965	4010	4050	4095
60	3050	3095	3135	3175	3220	3260	3300	3345	3385	3430	3470	3500	3555	3595	3635	3680	3720	3760	3805	3845	3885	3930	3970	4015	4055
62	3020	3060	3110	3150	3190	3230	3270	3310	3350	3390	3440	3480	3520	3560	3600	3640	3680	3730	3770	3810	3850	3890	3930	3970	4020
64	2990	3030	3080	3120	3160	3200	3240	3280	3320	3360	3400	3440	3490	3530	3570	3610	3650	3690	3730	3770	3810	3850	3900	3940	3980
66	2950	2990	3030	3070	3110	3150	3190	3230	3270	3310	3350	3390	3430	3470	3510	3550	3600	3640	3680	3720	3760	3800	3840	3880	3920
68	2920	2960	3000	3040	3080	3120	3160	3200	3240	3280	3320	3360	3400	3440	3480	3520	3560	3600	3640	3680	3720	3760	3800	3840	3880
70	2890	2930	2970	3010	3050	3090	3130	3170	3210	3250	3290	3330	3370	3410	3450	3480	3520	3560	3600	3640	3680	3720	3760	3800	3840
72	2860	2900	2940	2980	3020	3060	3100	3140	3180	3210	3250	3290	3330	3370	3410	3450	3490	3530	3570	3610	3650	3680	3720	3760	3800
74	2820	2860	2900	2930	2970	3010	3050	3090	3130	3170	3200	3240	3280	3320	3360	3400	3440	3470	3510	3550	3590	3630	3670	3710	3740

Courtesy of Warren E. Collins, Inc., Braintree, Mass.

APPENDIX B Predicted Vital Capacities for Females

	Height in centimeters																								
Age	146	148	150	152	154	156	158	160	162	164	166	168	170	172	174	176	178	180	182	184	186	188	190	192	194
16	2950	2990	3030	3070	3110	3150	3190	3230	3270	3310	3350	3390	3430	3470	3510	3550	3590	3630	3670	3715	3755	3800	3840	3880	3920
17	2935	2975	3015	3055	3095	3135	3175	3215	3255	3295	3335	3375	3415	3455	3495	3535	3575	3615	3655	3695	3740	3780	3820	3860	3900
18	2920	2960	3000	3040	3080	3120	3160	3200	3240	3280	3320	3360	3400	3440	3480	3520	3560	3600	3640	3680	3720	3760	3800	3840	3880
20	2890	2930	2970	3010	3050	3090	3130	3170	3210	3250	3290	3330	3370	3410	3450	3490	3525	3565	3605	3645	3695	3720	3760	3800	3840
22	2860	2900	2940	2980	3020	3060	3095	3135	3175	3215	3255	3290	3330	3370	3410	3450	3490	3530	3570	3610	3650	3685	3725	3765	3800
24	2830	2870	2910	2950	2985	3025	3065	3100	3140	3180	3220	3260	3300	3335	3375	3415	3455	3490	3530	3570	3610	3650	3685	3725	3765
26	2800	2840	2880	2920	2960	3000	3035	3070	3110	3150	3190	3230	3265	3300	3340	3380	3420	3455	3495	3530	3570	3610	3650	3685	3725
28	2775	2810	2850	2890	2930	2965	3000	3040	3070	3115	3155	3190	3230	3270	3305	3345	3380	3420	3460	3495	3535	3570	3610	3650	3685
30	2745	2780	2820	2860	2895	2935	2970	3010	3045	3085	3120	3160	3195	3235	3270	3310	3345	3385	3420	3460	3495	3535	3570	3610	3645
32	2715	2750	2790	2825	2865	2900	2940	2975	3015	3050	3090	3125	3160	3200	3235	3275	3310	3350	3385	3425	3460	3495	3535	3570	3610
34	2685	2725	2760	2795	2835	2870	2910	2945	2980	3020	3055	3090	3130	3165	3200	3240	3275	3310	3350	3385	3425	3460	3495	3535	3570
36	2655	2695	2730	2765	2805	2840	2875	2910	2950	2985	3020	3060	3095	3130	3165	3205	3240	3275	3310	3350	3385	3420	3460	3495	3530
38	2630	2665	2700	2735	2770	2810	2845	2880	2915	2950	2990	3025	3060	3095	3130	3170	3205	3240	3275	3310	3350	3385	3420	3455	3490
40	2600	2635	2670	2705	2740	2775	2810	2850	2885	2920	2955	2990	3025	3060	3095	3135	3170	3205	3240	3275	3310	3345	3380	3420	3455
42	2570	2605	2640	2675	2710	2745	2780	2815	2850	2885	2920	2955	2990	3025	3060	3100	3135	3170	3205	3240	3275	3310	3345	3380	3415
44	2540	2575	2610	2645	2680	2715	2750	2785	2820	2855	2890	2925	2960	2995	3030	3060	3095	3130	3165	3200	3235	3270	3305	3340	3375
46	2510	2545	2580	2615	2650	2685	2715	2750	2785	2820	2855	2890	2925	2960	2995	3030	3060	3095	3130	3165	3200	3235	3270	3305	3340
48	2480	2515	2550	2585	2620	2650	2685	2715	2750	2785	2820	2855	2890	2925	2960	2995	3030	3060	3095	3130	3160	3195	3230	3265	3300
50	2455	2485	2520	2555	2590	2625	2655	2690	2720	2755	2785	2820	2855	2890	2925	2955	2990	3025	3060	3090	3125	3155	3190	3225	3260
52	2425	2455	2490	2525	2555	2590	2625	2655	2690	2720	2755	2790	2820	2855	2890	2925	2955	2990	3020	3055	3090	3125	3155	3190	3220
54	2395	2425	2460	2495	2530	2560	2590	2625	2655	2690	2720	2755	2790	2820	2855	2885	2920	2950	2985	3020	3050	3085	3115	3150	3180
56	2365	2400	2430	2460	2495	2525	2560	2590	2625	2655	2690	2720	2755	2790	2820	2855	2885	2920	2950	2980	3015	3045	3080	3110	3145
58	2335	2370	2400	2430	2460	2495	2525	2560	2590	2625	2655	2690	2720	2750	2785	2815	2850	2880	2920	2945	2975	3010	3040	3075	3105
60	2305	2340	2370	2400	2430	2460	2495	2525	2560	2590	2625	2655	2685	2720	2750	2780	2810	2845	2875	2915	2940	2970	3000	3035	3065
62	2280	2310	2340	2370	2405	2435	2465	2495	2525	2560	2590	2620	2655	2685	2715	2745	2775	2810	2840	2870	2900	2935	2965	2995	3025
64	2250	2280	2310	2340	2370	2400	2430	2465	2495	2525	2555	2585	2620	2650	2680	2710	2740	2770	2805	2835	2865	2895	2925	2955	2990
66	2220	2250	2280	2310	2340	2370	2400	2430	2460	2495	2525	2555	2585	2615	2645	2675	2705	2735	2765	2800	2825	2860	2890	2920	2950
68	2190	2220	2250	2280	2310	2340	2370	2400	2430	2460	2490	2520	2550	2580	2610	2640	2670	2700	2730	2760	2795	2820	2850	2880	2910
70	2160	2190	2220	2250	2280	2310	2340	2370	2400	2425	2455	2485	2515	2545	2575	2605	2635	2665	2695	2725	2755	2780	2810	2840	2870
72	2130	2160	2190	2220	2250	2280	2310	2335	2365	2395	2425	2455	2480	2510	2540	2570	2600	2630	2660	2685	2715	2745	2775	2805	2830
74	2100	2130	2160	2190	2220	2245	2275	2305	2335	2360	2390	2420	2450	2475	2505	2535	2565	2590	2620	2650	2680	2710	2740	2765	2795

Courtesy of Warren E. Collins, Inc., Braintree, Mass.

Intelitool Computerized Exercises

This appendix contains four lab exercises that use Intelitool equipment and students as subjects. These exercises can be used in place of this manual's traditional exercises on the same topics. For example, reflex physiology is presented as both a traditional lab exercise (Exercise 22) and an Intelitool exercise (Exercise 22i in Appendix C). While each Intelitool exercise is numbered to correspond with the traditional exercise on the same topic, the Intelitool exercise numbers all end in ***i.*** The following table lists the four lab topics that appear as both traditional and Intelitool exercises.

For Intelitool Exercises 22i, 31i, and 37i, use the review sheets that correspond with traditional lab Exercises 22, 31, and 37. For Intelitool Exercise 16i, use the review sheet on page C-33 at the end of this appendix.

The Intelitool exercises are written for the MS-DOS version of Intelitool. Please consult the Intelitool documentation for specific procedures if you are using the Macintosh or Apple II version of Intelitool.

Exercise topic	Traditional lab exercise	Intelitool lab exercise
Muscle Physiology	Ex. 16A, p. 140	Ex. 16i, p. C-2
Human Reflex Physiology	Ex. 22, p. 207	Ex. 22i, p. C-7
Conduction System of the Heart and Electrocardiography	Ex. 31, p. 289	Ex. 31i, p. C-17
Respiratory System Physiology	Ex. 37, p. 340	Ex. 37i, p. C-24

16i EXERCISE

Muscle Physiology

OBJECTIVES

1. To define:

agonist	*antagonist*
muscle tension	*load*
muscle tone	*isotonic contraction*
isometric contraction	*electromyogram*

2. To distinguish between muscle tone and muscle contractions that cause movement.
3. To demonstrate the difference between isometric and isotonic muscle contraction.
4. To realize that the electrical activity of internal organs can be measured at the surface of the body.
5. To recognize that whole muscle contraction is a result of electrical activity of many muscle cells.
6. To realize the difference between intracellular electrical recordings and those using surface electrodes.
7. To demonstrate that variations in recorded electrical activity are a result of changing muscle activity.
8. To understand the relationship between agonists and antagonists.
9. To explain results obtained in terms of muscle structure.

MATERIALS

5-lb exercise weights (a large book provides a good substitute)
Cot (if available)
Tape measures
Hand dynamometer
Alcohol swabs
Lead-Lok self-adhesive electrode pads
Minimum computer equipment required:
- IBM PC/XT/AT or compatible
- 640K RAM—single disk (3.5 in.) drive
- Graphics adapter and compatible graphics monitor
- Intelitool data acquisition hardware (Cardiocomp 1)

Software:
- Intelitool ECG software

See Appendix E, Exercise 16i for links to *Anatomy and PhysioShow: The Videodisc.*

MUSCLE ACTIVITY

Because muscles are capable only of shortening in length (they always pull and never push), muscle shortening (contraction) is required to move a body part. A muscle that causes a desired movement is called a **prime mover** or **agonist.** For example, the biceps brachii is a muscle that flexes (bends) the forearm, and is called the agonist of forearm flexion. Muscles that oppose or reverse the action of the agonist are called **antagonists.** Because the triceps brachii causes extension (straightening) of the forearm, it is the antagonist in forearm flexion. On the other hand, when the desired action is forearm **extension,** the triceps is the agonist and the biceps is the antagonist.

Muscle Tone

Although skeletal muscles are called voluntary muscles, many of their responses are controlled reflexively. For example, even relaxed muscles are in a slight state of contraction called *muscle tone* (a state of slight but sustained contraction), due to spinal reflexes which activate different groups of muscle cells in sequence. Muscle tone helps keep muscle cells healthy and ready to respond quickly. Skeletal muscle tone also helps maintain body posture; indeed, we would collapse without the continuous, even if slight, muscle contractions.

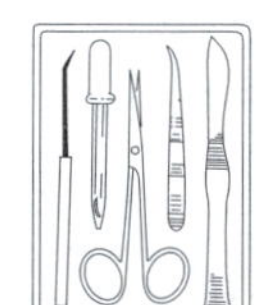

1. Obtain a 5-lb exercise weight from the supply area.

2. As you perform the activities listed below, your partner is to record the activity of your right biceps brachii and triceps brachii by placing a check mark in the appropriate column of Chart 1 when the muscle contracts (hardens or bulges).

- Stand comfortably with arms relaxed at the sides.
- Stand comfortably while holding the weight at the right side of the body.
- Sit comfortably with forearm relaxed on the lab bench, and in a supinated position (palm facing up).
- Sit comfortably with forearm relaxed on the lab bench, and in a pronated position (palm facing down).
- Sit comfortably while lifting and holding the weight a few inches above the lab bench.

CHART 1

Activity	Muscle(s) contracted Biceps brachii	 Triceps brachii
Standing, arms at sides		
Standing, holding weight at side		
Sitting, forearm supinated		
Sitting, forearm pronated		
Sitting, lifting the weight		

Which activity produced the greatest contraction (evidenced by hardness of the muscle)? ______________

3. Again working with your lab partner, carry out a similar evaluation on each gastrocnemius (calf) muscle under the following conditions and record results in Chart 2.

- Standing comfortably, but with legs relaxed
- Standing on your toes, reaching overhead toward the ceiling
- Standing erect with knees locked
- Sitting comfortably on the edge of the lab bench, with legs relaxed and not touching the floor
- Sitting comfortably on a chair with both feet on the floor
- Lying prone (face down) on a cot if available, or on the bench top

CHART 2

Activity	Gastrocnemius contracted Right leg	 Left leg
Standing, with legs relaxed		
Standing, reaching toward the ceiling		
Standing, knees locked		
Sitting on lab bench		
Sitting on chair, feet touching floor		
Lying prone		

During which activities was muscle tone exhibited?

____

____

Isometric and Isotonic Contractions

The force exerted by the muscle during contraction is referred to as *muscle tension,* and the weight (resistance to movement) exerted by the object being moved is called the *load.* However, muscles do not always shorten during contraction even though they are producing force. For example, when an object is too heavy to lift, the muscle exerts maximum force, but nothing moves. The condition in which a muscle develops tension but does not shorten is called *isometric* (*iso* = same; *metric* = measure) contraction. More typically, we are able to move most of the objects we attempt to lift, a form of muscle shortening called *isotonic* (*tonic* = tension) contraction.

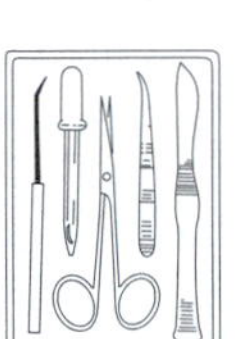

1. Obtain two tape measures from the supply area and use the same weight from the previous exercise on muscle tone.

2. Measure the circumference of the subject's arm around the largest part of the muscle bulge while he or she carries out each of the following muscle activities. Record the results in Chart 3.

- Extend both arms directly in front of the body.
- With forearms supinated, flex both forearms five times and then remeasure.
- Extend the right arm while holding the weight.
- Extend the left arm while holding the weight.
- Flex and then extend the right forearm five times while holding the weight and then remeasure.
- Flex and then extend the left forearm five times while holding the weight and then remeasure.

CHART 3

Activity	Measurements (cm) Right arm	 Left arm
Upper limbs extended		
Forearms flexed, supinated		
Right arm extended, holding weight		
Left arm extended, holding weight		
Right arm flexed, holding weight		
Left arm flexed, holding weight		

If there was any difference in the measurements between right and left arms, explain why. ______________

____

____

3. Evaluate gastrocnemius muscle activity under the following conditions and record results in Chart 4:

- Stand erect, comfortably relaxed.
- Stand on your toes, reaching toward the ceiling.
- Stand on your toes with arms at your sides.
- Sit on the lab bench with legs dangling relaxed.
- While you sit on the lab bench, your lab partner holds your legs in place while you try to move your feet forward and upward.
- Lie prone on a cot or lab bench with the gastrocnemius relaxed.
- Still lying prone, contract the gastrocnemius.

CHART 4

Activity	Measurements (cm) Right leg	Left leg
Standing erect		
Standing on toes, reaching toward ceiling		
Standing on toes, arms at sides		
Sitting with legs relaxed		
Sitting, legs restrained		
Lying prone, gastrocnemius relaxed		
Lying prone, gastrocnemius contracted		

If there was any difference in the measurements between right and left gastrocnemius, explain why. ______

____

____

Electromyography

Single muscle cells (and motor units) respond to action potentials in an all-or-none way, but a whole muscle contracts with varying degrees of force. Since muscle contraction is a direct result of cell membrane depolarization, we can detect muscle activity by measuring the electrical potential associated with the muscle cells. Placing the tip of a needle recording electrode *inside* a muscle cell produces a typical action potential tracing when that single cell is stimulated, whereas a flat plate recording electrode placed on the skin surface measures the combined electrical activity of all *active* muscle cells in the area near the electrode. Thus, when a whole skeletal muscle contracts, and increasing numbers of motor units are called into play, we should be able to measure an increase in electrical activity.

An electromyogram (EMG) is a recording of the electrical activity associated with muscle contraction. Muscle cell depolarizations (which result in muscle cell contraction) and repolarizations (which allow relaxation) are electrical events resulting in the numerous irregular, overlapping spikes seen on an EMG tracing. Each spike indicates the electrical activity arising from an individual motor unit. Clinically, the EMG is used as a noninvasive method to assess muscle cell damage.

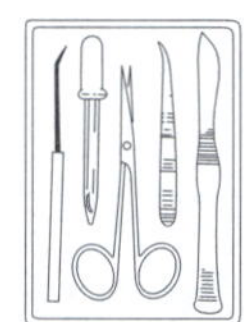

Work in groups of 2 to 4 students. During the experiment, group members should strive to gain experience operating the computer. Since the computer displays the keys that control the program, prior computer experience is not necessary. Each group will be expected to explain all operations to their experimental subject.

Preparing the Subject

Because most common problems encountered are due to incorrect electrode application, the following instructions should be carefully followed to establish proper electrical connection between the subject and the computer.

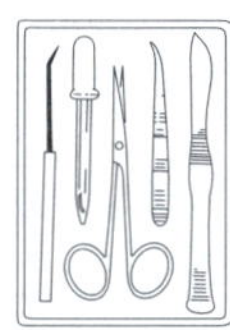

One at a time, remove the Lead-Lok electrode pads from the protective backing and apply to the skin surface at each of the sites described below *after first scrubbing the designated area with an alcohol swab:*

- One over the proximal end of the biceps brachii
- Another at the distal end of the same biceps brachii
- The third over the *wrist* of the same limb

Recording the EMG

The Cardiocomp software used in this exercise is a dual-purpose program designed to record an electrocardiogram (ECG). Although the term *EMG* is not specifically listed in the menus, you may think of EMG in place of ECG in all references.

1. Have the subject sit comfortably, holding a hand dynamometer in a relaxed state on the countertop. The elbow should be slightly bent, with the back of the hand against the countertop.

2. Attach the wires to the Lead-Lok pad positions as follows:

- Connect the *black* and *white* wires to the electrode pads placed at the proximal and distal ends of the muscle (it does not matter which wire is connected to which electrode pad).

- Attach the *green* wire to the wrist electrode pad.

3. Start the software. Type **ITOOL** at the DOS prompt. Select Cardiocomp by making sure the highlight (moved with the arrow keys) is on the selection and pressing (ENTER). Then carry out the following sequence of steps:

- Select Experiment menu from the Cardiocomp Main menu.
- Select Real-Time ECG to display the data collection screen (Figure 16i.1).

The data collection screen should be displaying a typical EMG tracing (looking like random electrical spikes) sweeping from left to right.

- The top of the screen displays keys that control data collection.
- Experimental data is recorded in the computer's memory as individual screens (frames), the numbers of which are displayed in the Frame # box. Cardiocomp software can record a maximum of 20 frames, and then starts to overwrite the data first obtained. Because each frame is a single "snapshot" of a few seconds of muscle electrical activity, the complete EMG tracing consists of a series of frames.
- See Table 16i.1 for the keys that control data collection:

TABLE 16i.1 Control Keys for Real-Time ECG

Key	Function
(ESC)	Return to the Experiment menu
S	Start/stop toggle to alternately start or stop the tracing
C	Clear and restart data collection
(↑) (↓)	Change (increase or decrease) the voltage scale
P	Print screen
1–V₆	Change lead employed (not used in the EMG experiment)

SPONTANEOUS ACTIVITY The electrical activity recorded in an EMG when a muscle is at rest is called **spontaneous activity,** and represents muscle tone.

1. Press **S** to start the tracing and then immediately press **C**. This action clears the computer's memory of all practice frames collected to this point, and the prompt "All data will be lost. Continue? Y/N" will be displayed. (This is normal; do not panic!)

2. Press **Y** (yes) to start EMG recording of spontaneous activity in frame 1.

3. In the third frame of data collection, press **S** to pause the recording. (Notice that the tracing continues to the end of the screen after you press **S**.) Record the frame numbers for spontaneous activity:

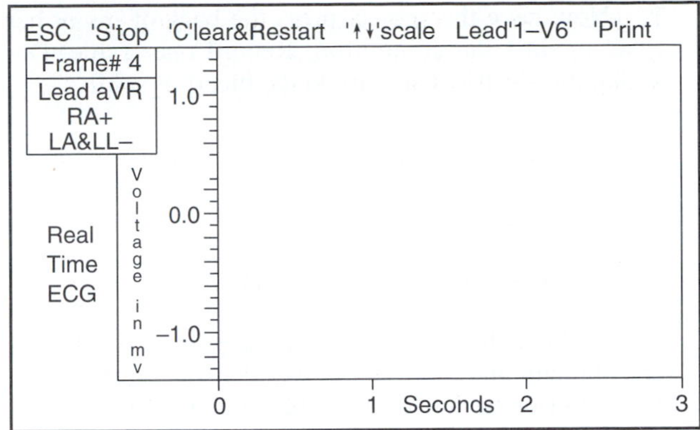

F16i.1

Data collection screen.

Frames ________ to ________

If you are not satisfied with the overall results, you can repeat the EMG recording by selecting **C** (clear and restart). All data collected thus far will be discarded.

Describe the EMG of a resting muscle. (Be sure to indicate the relative amplitude [height] of the spikes seen.)

____

____

RECRUITMENT When we try to lift a heavy object, voluntary control of the involved skeletal muscles activates increasing amounts of muscle mass until enough force is achieved to move the object. The process by which increasing numbers of motor units are activated in whole muscle is called *recruitment* or *multiple motor unit summation.*

1. To see the effect of recruitment on the EMG, have the subject squeeze the hand dynamometer to 5 kg for one frame (screen) of recording. Then, without relaxing the hand between frames, the subject is to increase the force by 5 kg with each subsequent frame until reaching his or her maximum force. Record the frame numbers for the recruitment series:

Frames ________ to ________

How does the electrical activity recorded change with each incremental increase in force?

____

Explain the observed increase in electrical activity in terms of motor units.

____

____

____

2. Now have the subject press the back of his or her hand against the countertop. Record one frame. Describe the electrical activity in the biceps brachii.

____

____

Record the frame number: _______

3. Peel the electrodes from their positions over the biceps brachii and place them over the triceps brachii on the posterior surface of the arm. Again have the subject press the back of his or her hand against the countertop. Record one frame.

Record the frame number: _______

How does this electrical activity compare with that seen during the biceps activity previously recorded?

____

The biceps brachii and triceps brachii have an agonist/antagonist relationship. When the subject is pressing the back of the hand against the countertop, which muscle is acting as the agonist? _______________

Explain your conclusion in terms of the EMG results.

____

____

____

DETERMINING MUSCLE FATIGUE Have the subject vigorously exercise the hand by alternately squeezing and releasing the grip of the hand dynamometer until he or she is fatigued (*really* tired) and then relax the hand on the countertop. Immediately start recording again (by pressing **S**) and stop when frame 19 is reached. Record the frame numbers:

Frames _______ to _______

How does the EMG tracing of fatigue differ from normal spontaneous activity after vigorous exercise?

____

____

____

What happens to the EMG as resting time progresses?

____

____

____

When finished with the exercise:

1. Press (ESC) to return to the Experiment submenu.
2. Press (ESC) again to return to the Main menu.
3. Use the arrow keys to move to the Review/Analyze submenu.
4. Select Review/Time Analysis to see the frames recorded during the session.

The Review/Time Analysis screen is similar to the Experiment screen with the exception of the controls. Table 16i.2 explains the program control keys.

TABLE 16i.2 Control Keys for Review/Time Analysis

Key	Function
(ESC)	Return to the submenu
S	Stop the current tracing
B	Move the screen back one frame
F	Move the screen forward one frame
(↑) (↓)	Change the voltage scale

Although exact measurements cannot be made from your EMG, the amplitude of the spikes should give you a perception of how much electrical activity occurred in the muscle that was recorded. Compare your results from frame to frame while thinking about what electrical activity should have occurred during the experiment.

Save your data.

1. Select Data File/Disk from the Main menu
2. Select Save.
3. You should see a yellow blinking cursor near the top of the screen. Press (BACKSPACE) to clear the default save location, and type a file name according to your instructor's directions.
4. Once your data is saved to disk, it may be retrieved anytime for viewing.

EXERCISE **22i**

Human Reflex Physiology

OBJECTIVES

1. To define *reflex* and *reflex arc.*
2. To name, identify, and describe the function of each element of a reflex arc.
3. To state why reflex testing is an important part of every physical examination.
4. To describe and discuss several types of reflex activities as observed in the laboratory; to indicate the functional or clinical importance of each; and to categorize each as a somatic or autonomic reflex action.
5. To explain why cord-mediated reflexes are generally much faster than those involving input from the higher brain centers.
6. To investigate the differences in reaction time of reflexes and unlearned responses.

MATERIALS

Sharp pencils
Cot (if available)
Absorbent cotton (sterile)
Tongue depressor
Flashlight
100- or 250-ml beaker
10- or 25-ml graduated cylinder
Lemon juice
Wide-range pH paper
Large laboratory bucket containing freshly prepared 10% household bleach solution (for saliva-soiled glassware)
Disposable autoclave bag
Wash bottle containing 10% bleach solution
Minimum computer equipment required:
 IBM PC/XT/AT or compatible computer.
 640K RAM—single disk (3.5 in.) drive
 Graphics adapter and monitor
 Intelitool Flexicomp equipment
Software:
 Flexicomp software

See Appendix E, Exercise 22i for links to *Anatomy and PhysioShow: The Videodisc.*

THE REFLEX ARC

Reflexes are rapid, predictable, involuntary motor responses to stimuli; they are mediated over neural pathways called **reflex arcs.** All reflex arcs have five essential parts (Figure 22i.1a):

1. The *receptor,* which reacts to a stimulus
2. The *sensory neuron,* which conducts the afferent impulses to the CNS
3. The *integration center,* consisting of one or more synapses in the CNS
4. The *motor neuron,* which conducts the efferent impulses from the integration center to the effector
5. The *effector,* the muscle fibers or glands that respond to the efferent impulses by contracting or secreting a product, respectively

The simple patellar or knee-jerk reflex shown in Figure 22i.1b is an example of a simple, two-neuron, *monosynaptic* (literally, "one synapse") reflex arc. It will be demonstrated in the laboratory. However, most reflexes are more complex and *polysynaptic,* involving the participation of one or more association neurons in the reflex arc pathway. A three-neuron reflex arc (flexor reflex) is diagrammed in Figure 22i.1c. Since delay or inhibition of the reflex may occur at the synapses, the more synapses encountered in a reflex pathway, the more time is required to effect the reflex.

Reflexes of many types may be considered programmed into the neural anatomy. For example, many *spinal reflexes* (reflexes that are initiated and completed at the spinal cord level, such as the flexor reflex) occur without the involvement of higher brain centers. These reflexes work equally well in decerebrate animals (those in which the brain has been destroyed), as long as the spinal cord is functional. Conversely, other reflexes require the involvement of functional brain tissue, since many different types of inputs must be evaluated before the appropriate reflex is determined. Superficial cord reflexes and pupillary responses to light are in this category. In addition, although many spinal cord reflexes do not require the involvement of higher centers, the brain is "advised" of spinal cord reflex activity and may alter it by facilitating or inhibiting the reflexes.

Reflex testing is an important diagnostic tool for assessing the condition of the nervous system. Distorted, exaggerated, or absent responses may indicate degeneration or pathology of portions of the nervous system, often before other signs are apparent.

If the spinal cord is damaged, the easily performed

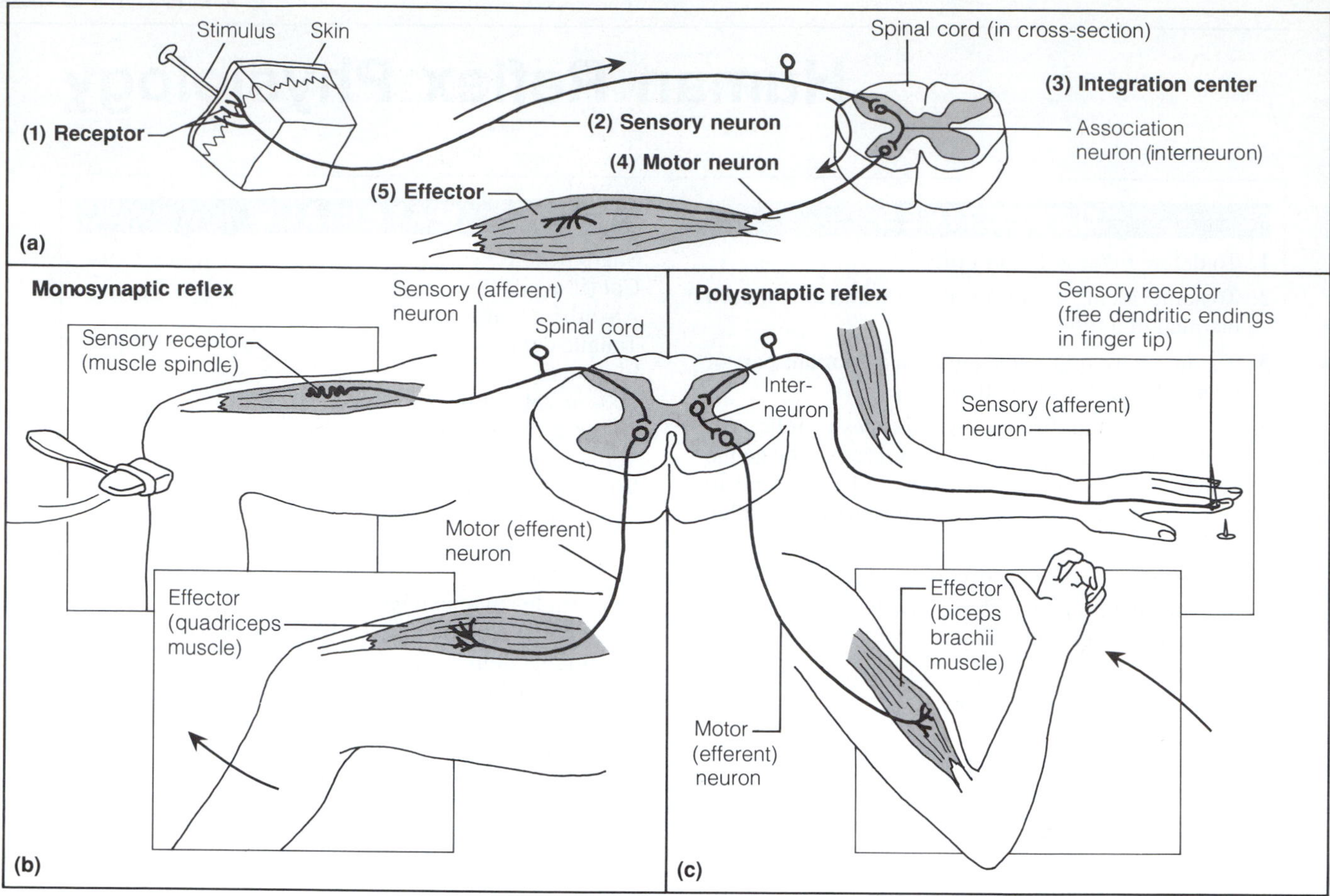

F22i.1

Simple reflex arcs. (a) Components of all human reflex arcs: receptor, sensory neuron, integration center (one or more synapses in the CNS), motor neuron, and effector. **(b)** Monosynaptic reflex arc. **(c)** Polysynaptic reflex arc. The integration center is in the spinal cord, and in each example the receptor and effector are in the same limb.

reflex tests can help pinpoint the area (level) of spinal cord injury. Motor nerves above the injured area may be unaffected, whereas those at or below the lesion site may be unable to participate in normal reflex activity. ■

Reflexes can be categorized into one of two large groups: somatic reflexes and autonomic reflexes. **Autonomic** (or visceral) **reflexes** are mediated through the autonomic nervous system and are not subject to conscious control. These reflexes activate smooth muscles, cardiac muscle, and the glands of the body, and they regulate body functions such as digestion, elimination, blood pressure, salivation, and sweating. **Somatic reflexes** include all those reflexes that involve stimulation of skeletal muscles by the somatic division of the nervous system. An example of such a reflex is the rapid withdrawal of the hand from a hot object.

SOMATIC REFLEXES

There are several types of somatic reflexes, including several that you will be eliciting during this laboratory session—the stretch, crossed extensor, superficial cord, corneal, and gag reflexes. Some require only spinal cord activity; others require brain involvement as well.

Spinal Reflexes

STRETCH REFLEXES **Stretch reflexes** are important postural reflexes, normally acting to maintain posture, balance, and locomotion. Stretch reflexes are initiated by tapping a tendon, which stretches the muscle the tendon is attached to. This stimulates the muscle spindles and causes reflex contraction of the stretched muscle or muscles, which resists further stretching. Even as the primary stretch reflex is occurring, impulses are being sent to other destinations as well. For example, branches of the afferent fibers (from the muscle spindles) also synapse with interneurons (association neurons) controlling the antagonistic muscles (Figure 22i.2). The inhibition of the antagonistic muscles that follows, called *reciprocal inhibition,* causes them to relax and prevents them from resisting (or reversing) the contraction of the stretched muscle caused by the main reflex arc. Additionally, impulses are relayed to higher brain centers (largely via the dorsal white columns) to advise of muscle length, speed of shortening, and the

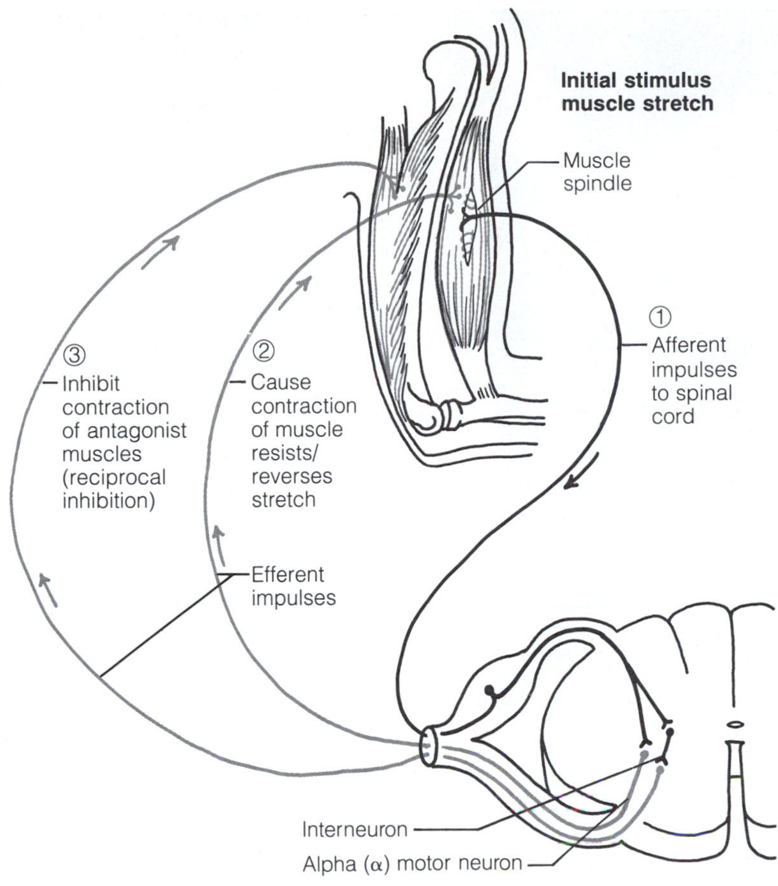

F22i.2

Events of the stretch reflex by which muscle stretch is damped. The events are shown in circular fashion. **(1)** Stretching of the muscle activates a muscle spindle. **(2)** Impulses transmitted by afferent fibers from muscle spindle to alpha motor neurons in the spinal cord result in activation of the stretched muscle, causing it to contract. **(3)** Impulses transmitted by afferent fibers from muscle spindle to interneurons in the spinal cord result in reciprocal inhibition of the antagonist muscle.

like—information needed to maintain muscle tone and posture. Stretch reflexes tend to be hypoactive or absent in cases of peripheral nerve damage or ventral horn disease, and hyperactive in corticospinal tract lesions. They are absent in deep sedation and coma.

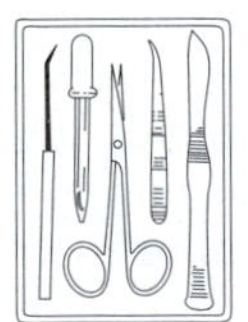

We will use computer-driven Intelitool equipment to demonstrate the stretch reflex in this section. The **patellar reflex,** which assesses the L_2–L_4 level of the spinal cord, will be used as the main example of the stretch reflex. Besides simply eliciting the reflex, the effects of a number of factors (mental distraction, exercise, and others) on its expression will be investigated, and the ankle jerk reflex will be elicited for comparison.

Work in groups of 2 to 4 students. During the experiment, all group members should strive to gain experience operating the computer. Because the computer displays the keys that control the program, prior computer experience is not necessary. Each group member will be expected to explain all operations to his or her experimental subject.

1. Prepare the subject. Do not select a subject who has a history of poor reflex responses (that is, in whom a physician has been able to demonstrate reflexes only after repeated tries). Have the subject sit on the edge of the counter or laboratory bench so that the legs are relaxed with feet well above the floor.

⚠ You will attach a transducer, but first loosen its thumbscrew locking tab. *Always loosen the thumbscrew locking tab when the transducer is not being read,* for example, when strapping it on, removing it, or storing it. This will lessen the chance of damage when it is not in actual use. *Also, never force the transducer if it has hit an internal stop!*

Carefully strap the transducer to the subject's right thigh and leg with the displacement transducer box facing outward (Figure 22i.3). The hinge of the transducer should be aligned with the hinge of the knee. Once connected, the subject must remain seated.

2. Start the software. Type **ITOOL** at the DOS prompt. Select Flexicomp by making sure the highlight (moved with the arrow keys) is on the selection and

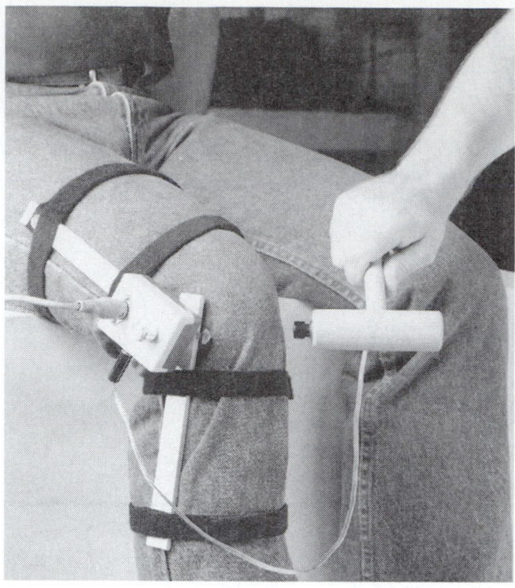

F22i.3

Alignment of the force transducer.

pressing (ENTER). Then carry out the following sequence of steps.

- Select Experiment menu.
- Select Real-Time Experiment. This selection will display the screen prompting you to "initialize" (calibrate) the force transducer. Perform the calibration as follows:

  1. Use the paper template provided in the Flexicomp box to align the transducer to 90°.
  2. Turn the aluminum knob on the side of the transducer until the calibration number on the computer screen reads close to zero (±100).
  3. Tighten the locking thumbscrew. If this causes the transducer arms to move from 90° or the screen reading to move from the zero (±100) mark, loosen and readjust the aluminum knob. Press the spacebar when ready.
  4. Using the template again, align the transducer arms to 150°, and press the spacebar when the alignment is correct. Now, loosen the thumbscrew and return the paper template to the Flexicomp box.
  5. Plug the connector of the provided mallet (reflex hammer) into the force transducer to complete the calibration process.

  The data collection screen should be visible once you have correctly calibrated the transducer (Figure 22i.4).

3. Examine the data collection screen to get oriented. The top of the screen displays the keys that control data collection.

- The boxes at the upper left of the screen contain information about collected data.
- Experimental data is recorded in the computer's memory as individual screens (frames), the numbers of which are displayed in the Frame # box. Flexicomp software will record a maximum of 20 frames, after which the first data obtained is overwritten by the new information.
- The Maximum Excursion box reports information about the force of the "kick" (the largest rotation of the force transducer generated during a reflex demonstration) measured in degrees, and the time period (seconds) from the hammer strike to the maximum excursion.
- The Latent Period box displays the response time from the hammer strike to the first positive deflection of the leg. The latent period is the time interval from stretching the sensory receptors to effector (muscle) movement.
- Study Table 22i.1 for the keys that control the operation of the program.

TABLE 22i.1 Control Keys for Real-Time Experiment

Key	Function
(ESC)	Return to the Experiment menu
C	Clear and reset data collection
P	Print the current tracing

4. Practice eliciting the patellar reflex. First, make some adjustments to ensure that the scan line (trace) will remain on the screen. Press the button on the end of the mallet. As the trace moves across the screen, turn the aluminum knob so that the trace is just below the *center* of the screen, and then tighten the thumbscrew. It may take a few tries to get it right.

Now, tap the patellar ligament sharply with the button on the end of the reflex hammer just below the knee to elicit the knee-jerk response (Figure 22i.5). Watch the screen and the subject's reflex response to see the relationship between the screen tracing and the actual leg kick. It may take a few tries to find the correct technique and strike point.

5. Begin data recording. Once you develop the proper method, press **C** to reset the computer's memory to begin data collection. Test the right knee and record* the maximum degree of excursion, the time it took to get to maximum, the latent period, and the frame number.

Maximum excursion ________°

Time to maximum ________ sec

Latent period ________ sec

Frame number ________

* You may read indicated values directly off the screen or return to recorded frames later to perform more accurate measurements in the Displacement/Time Analysis part of the program.

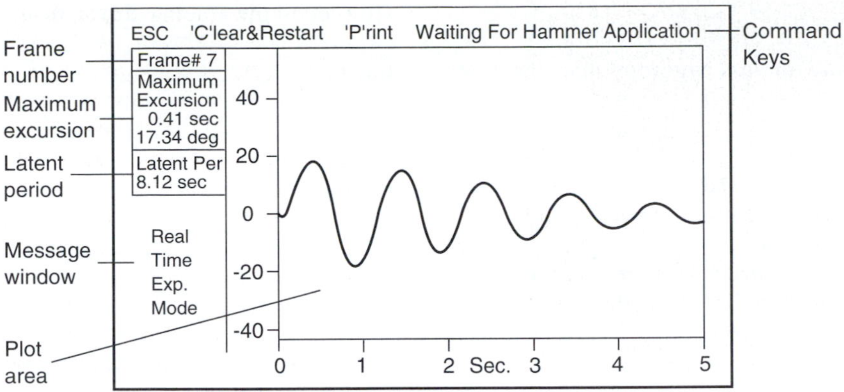

F22i.4

Data collection screen.

Which muscles contracted? ____________________

____

What nerve is carrying the afferent and efferent impulses?

____

6. Test the effect of mental distraction on the patellar reflex by having the subject add a column of three-digit numbers while you test the reflex again.

Maximum excursion ________ °

Time to maximum ________ sec

Latent period ________ sec

Frame number ________

Is the response greater than or less than the first response? ____________________

What are your conclusions about the effect of mental distraction on reflex activity? ____________________

____

____

____

7. Now test the effect of muscular activity occurring simultaneously in other areas of the body. Have the subject clasp the edge of the laboratory bench and vigorously attempt to pull it upward with both hands. At the same time, test the patellar reflex again.

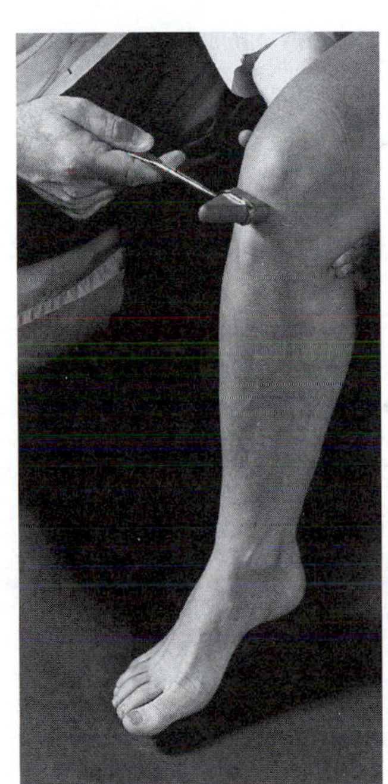

F22i.5

Testing the patellar reflex. The examiner supports the subject's knee so that the subject's muscles are relaxed, and then strikes the patellar ligament with the reflex hammer. The proper location may be ascertained by palpation of the patella.

Maximum excursion ________ °

Time to maximum ________ sec

Latent period ________ sec

Frame number _______

Is the response more or less vigorous than the first response? ________________________________

8. Fatigue also influences the reflex response. After removing the force transducer, the subject should jog in position until she or he is very fatigued (*really* fatigued—no slackers). Attach the transducer and test the patellar reflex again and record whether it is more or less vigorous.

Maximum excursion _______°

Time to maximum _______ sec

Latent period _______ sec

Frame number _______

More or less vigorous? ________________________________

Would you say that the nervous system activity *or* muscle function is responsible for the changes you have just observed?

____

____

Explain your reasoning. ________________________________

____

____

9. Now you are ready to compare the *ankle jerk reflex* (another stretch reflex) to the patellar reflex. To demonstrate the ankle jerk, or Achilles, reflex, loosen the thumbscrew on the transducer, remove it, and kneel on a chair with the feet dangling (relaxed) over the seat edge. Strap the transducer on the leg and foot. Dorsiflex the foot slightly to increase the tension on the gastrocnemius muscle and then tighten the thumbscrew. Have your partner sharply tap the Achilles (calcaneal) tendon with the reflex hammer.

Maximum excursion _______°

Time to maximum _______ sec

Latent period _______ sec

Frame number _______

What is the result? ________________________________

____

Does the contraction of the gastrocnemius normally result in the activity you have observed? ______________

How does the tracing differ from the patellar reflex tracing? ________________________________

____

Explain: ________________________________

____

____

Loosen the thumbscrew and remove the force transducer from the subject.

10. Save your data:*

- Press (ESC) for the Main menu and select Data File/Disk.
- Select Save.
- You should see a yellow blinking cursor near the top of the screen. Press (BACKSPACE) to erase the current location, and type a file name according to your instructor's directions.
- Once your data is saved to disk, it may be retrieved later for analysis.

So far, you have tested the femoral and sciatic nerves. If time allows, and with your instructor's permission, you might want to demonstrate other stretch reflexes. Other possible pathways to test include the following:

- Musculocutaneous nerve (biceps jerk): Attach the transducer to the arm at the elbow. Strike the biceps tendon found in the hollow of the elbow (antecubital fossa).
- Radial nerve (triceps jerk): Keep the transducer attached to the elbow. Strike the triceps tendon about 2 inches above the elbow. The arm should be slightly bent and held in front of the subject (supported by the other arm).

Be sure to use a different file name if you want to save the data for any of these additional reflexes.

Do not exit the program at this time because you will be using the apparatus to record data for the experiment Reaction Time of Unlearned Responses shortly.

CROSSED EXTENSOR REFLEX The **crossed extensor reflex** is more complex than the stretch reflex. It consists of a flexor, or withdrawal, reflex followed by extension of the opposite limb.

This reflex is quite obvious when, for example, a stranger suddenly and strongly grips one's arm. The immediate response is to withdraw the clutched arm and push the intruder away with the other arm. The reflex is more difficult to demonstrate in a laboratory because it is anticipated, and under these conditions the extensor part of the reflex may be inhibited.

* Note to the Instructor: Students will be saving two sets of data under separate file names. Assign specific file names so one file does not overwrite the other's data.

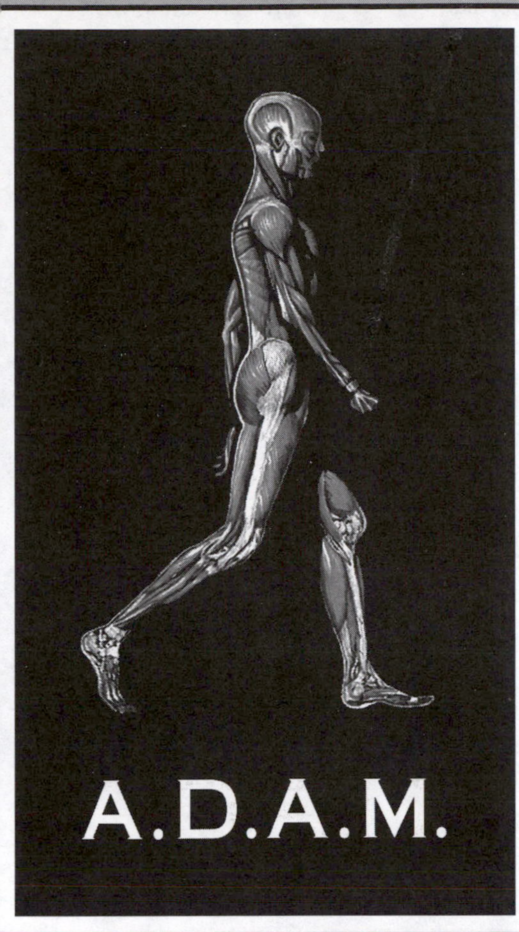

The A.D.A.M. Reference Guide For

Human Anatomy and Physiology, Third Edition

by Elaine N. Marieb

LINKING TEXT AND SOFTWARE TO GIVE YOU EXCITING NEW LEARNING OPTIONS.

With **A.D.A.M. Standard 2.0** and ***Human Anatomy and Physiology, Third Edition*** by Elaine N. Marieb, you can connect software, art, and text to create a powerful learning experience that helps you understand complex, three-dimensional anatomical relationships in a way never before possible.

This handy reference guide, organized by chapter, shows you where the links occur between A.D.A.M. Standard 2.0 and *Human Anatomy and Physiology, Third Edition.*

The guide includes:

- Selected **figures** from *Human Anatomy and Physiology, Third Edition*—identified by figure number and key body structure — that are listed with the appropriate view and layer from A.D.A.M. Standard 2.0.

Good luck and enjoy your study of human anatomy!

Figures From *Human Anatomy and Physiology, Third Edition* by Elaine N. Marieb That Correlate With A.D.A.M Standard 2.0

Figure	Key Structure	A.D.A.M. View and Layer
1.2 a	Integumentary system	Using "View" on the Menu Bar, choose Integumentary system
1.2 b	Skeletal system	Using "View" on the Menu Bar, choose Skeletal system
1.2 c	Muscular system	Using "View" on the Menu Bar, choose Muscular system
1.2 d	Nervous system	Using "View" on the Menu Bar, choose Nervous system
1.2 e	Endocrine system	Using "View" on the Menu Bar, choose Endocrine system
1.2 f	Cardiovascular system	Using "View" on the Menu Bar, choose Cardiovascular system
1.2 g	Lymphatic system	Using "View" on the Menu Bar, choose Lymphatic system
1.2 h	Respiratory system	Using "View" on the Menu Bar, choose Respiratory system
1.2 i	Digestive system	Using "View" on the Menu Bar, choose Digestive system
1.2 j	Male urinary system	Using "Options", choose Male; using "View" on the Menu Bar, choose Urinary system
1.2 k	Male reproductive system	Using "Options", choose Male; using "View" on the Menu Bar, choose Reproductive system

Figure	Key Structure	A.D.A.M. View and Layer
1.2 l	Female reproductive system	Using "Options", choose Female; using "View" on the Menu Bar, choose Reproductive system
4.9 c	Peritoneum	Go to "Full Anatomy" View: Medial Layer: Mesentery
4.9 c	Pericardium	Go to "Full Anatomy" View: Anterior Layer: Heart-cut section
4.9 c	Pleura	Go to "Full Anatomy" View: Anterior Layer: Lungs
6.1	Costal cartilage	Go to "Full Anatomy" View: Anterior Layer: Ribs
6.1	Articular cartilage	Go to "Full Anatomy" View: Anterior Layer: Articular cartilage
6.1	Tracheal cartilage	Go to "Full Anatomy" View: Anterior Layer: Trachea
6.3 a	Humerus–coronal section	Go to "Full Anatomy" View: Anterior Layer: Bones-coronal section
7.1 a	Skeletal system	Using "View" on the Menu Bar, choose Skeletal system

Figure Number	Key Structure	A.D.A.M. View and Layer
7.1 b	Skeletal system	Go to "Full Anatomy" View: Posterior Layer: Posterior vertebral column
7.2 a	Skull bones	Go to "Full Anatomy" View: Anterior Layer: Skull
7.2 b	Skull bones	Go to "Full Anatomy" View: Posterior Layer: Articular cartilage
7.3 a	Skull bones	Go to "Full Anatomy" View: Lateral Layer: Skull
7.5	Temporal bone	Go to "Full Anatomy" View: Lateral Layer: Skull
7.8 a	Mandible	Go to "Full Anatomy" View: Lateral Layer: Nasal cartilage
7.8 b	Maxilla	Go to "Full Anatomy" View: Lateral Layer: Skull
7.9	Bones of the orbit	Go to "Full Anatomy" View: Anterior Layer: Skull
7.10 a	Bones of the nasal cavity	Go to "Full Anatomy" View: Medial Layer: Nasal conchae
7.11 a	Paranasal air sinuses	Go to "Full Anatomy" View: Anterior Layer: Anterior jugular vein
7.13	Vertebral column	Go to "Full Anatomy" View: Anterior Layer: Bones–coronal section
7.13	Vertebral column	Go to "Full Anatomy" View: Medial Layer: Skin
7.18 a	Sacrum and coccyx	Go to "Full Anatomy" View: Anterior Layer: Femur
7.18 b	Sacrum and coccyx	Go to "Full Anatomy" View: Posterior Layer: Posterior vertebral column
7.19 a	Thorax	Go to "Full Anatomy" View: Anterior Layers: Ribs; Bones–coronal section
7.21 b	Clavicle	Go to "Full Anatomy" View: Anterior Layer: Clavicle
7.21 e	Scapula	Go to "Full Anatomy" View: Posterior Layer: Scapula
7.22 a	Humerus	Go to "Full Anatomy" View: Anterior Layer: Humerus
7.22 b	Humerus	Go to "Full Anatomy" View: Posterior Layer: Humerus
7.23 a	Radius and ulna	Go to "Full Anatomy" View: Anterior Layer: Radius
7.23 b	Radius and ulna	Go to "Full Anatomy" View: Posterior Layer: Radius
7.24	Bones of the hand	Go to "Full Anatomy" View: Anterior Layer: Bones–coronal section
7.25 a	Pelvic girdle	Go to "Full Anatomy" View: Anterior Layer: Femur
7.25 b	Coxal bone	Go to "Full Anatomy" View: Lateral Layer: Femur
7.27 a	Patella	Go to "Full Anatomy" View: Anterior Layer: Patella
7.27 b	Femur	Go to "Full Anatomy" View: Anterior Layer: Femur
7.27 b	Femur	Go to "Full Anatomy" View: Posterior Layer: Femur
7.28 a	Tibia and fibula	Go to "Full Anatomy" View: Anterior Layer: Tibia
7.28 b	Tibia and fibula	Go to "Full Anatomy" View: Posterior Layer: Tibia
7.29 a	Bones of the foot	Go to "Full Anatomy" View: Anterior Layer: Bones–coronal section
8.1 b	Syndesmosis	Go to "Full Anatomy" View: Anterior Layer: Tibia
8.2 b	Synchondrosis	Go to "Full Anatomy" View: Anterior Layer: Ribs
8.2 c	Symphysis	Go to "Full Anatomy" View: Anterior Layer: Articular cartilage
8.3 a	Synovial joint	Go to "Full Anatomy" View: Anterior Layers: Articular cartilage; Bones–coronal section
8.4 a	Bursae and tendon sheaths	Go to "Full Anatomy" View: Anterior Layer: Ligaments of the upper limb
8.8 a-f	Types of synovial joints	Using "View" on the Menu Bar, choose Skeletal system

Figure Number	Key Structure	A.D.A.M. View and Layer	
24.14 a	Internal anatomy of the stomach	Go to View: Layer:	"Full Anatomy" Anterior Stomach–coronal section
24.20 a	Gallbladder and associated ducts, pancreas and duodenum	Go to View: Layers:	"Full Anatomy" Anterior Gallbladder; Pancreas
24.23 b,c	Greater omentum and lesser omentum	Go to View: Layer:	"Full Anatomy" Anterior Greater omentum
24.23 d	Mesentery	Go to View: Layer:	"Full Anatomy" Medial Mesentery
24.24	Liver	Go to View: Layer:	"Full Anatomy" Anterior Liver
24.30 a	Large intestine	Go to View: Layer:	"Full Anatomy" Anterior Transverse colon
26.1 a	Female urinary system	Using "Options", choose Female; using "View" on the Menu Bar, choose Urinary system	
26.2	Cross section through kidney	Select the "Cross Sections" book from the Library and click on the 8th "Eye" button from the top on the left side	
26.3 b	Internal kidney	Go to View: Layer:	"Full Anatomy" Anterior Ureter
28.1	Male reproductive system	Go to View: Layer:	"Full Anatomy" Using "Options", choose Male Medial Urinary bladder and prostate gland
28.2	Testes	Go to View: Layers:	"Full Anatomy" Using "Options", choose Male Anterior Internal abdominal oblique muscle; Testes
28.11 a	Female reproductive system	Go to View: Layer:	"Full Anatomy" Using "Options", choose Female Anterior Uterus
28.11 b	Female reproductive system	Go to View: Layer:	"Full Anatomy" Using "Options", choose Female Medial Uterus
28.17 a	Mammary gland	Go to View: Layer:	"Full Anatomy" Using "Options", choose Female Anterior Female Breast

End-of-Chapter Questions from *Human Anatomy and Physiology, Third Edition*, by Elaine N. Marieb that correlate with A.D.A.M. Standard 2.0:

Chapter	Review Questions
1	6, 9, 18
7	2, 3, 4
8	5
10	2, 4, 6, 8, 12, 16, 18
12	12
13	5
17	16
19	10, 13
20	11, 12, 22
21	2, 5
23	17, 18
24	11, 17, 26
26	11
28	4,25

Chapter	Critical Thinking and Clinical Application Questions
7	1
13	5
16	6
20	1
21	2
23	3

THE BENJAMIN/CUMMINGS PUBLISHING COMPANY, INC.
2725 Sand Hill Road
Menlo Park, CA 94025

ISBN 0-8053-4797-6

IGURE NUMBER	KEY STRUCTURE	A.D.A.M. VIEW AND LAYER
.9 a	Shoulder joint	Go to "Full Anatomy" View: Anterior Layer: Fibrous joint capsule of the shoulder and elbow
.9 b	Shoulder joint	Go to "Full Anatomy" View: Lateral Layer: Right shoulder joint
.10 c	Hip joint	Go to "Full Anatomy" View: Anterior Layer: Obturator nerve
.10 d	Hip joint	Go to "Full Anatomy" View: Posterior Layer: Ligaments of the hip
.11 b,d	Elbow joint	Go to "Full Anatomy" View: Anterior Layer: Fibrous joint capsule of the shoulder and elbow
.12 b	Knee joint	Go to "Full Anatomy" View: Anterior Layer: Anterior cruciate ligament
.12 d	Knee joint	Go to "Full Anatomy" View: Posterior Layer: Adductor hallucis muscle
0.4 b	Superficial muscles	Using "View" on the Menu Bar, choose Muscular system
0.5 b	Superficial muscles	Go to "Full Anatomy" View: Posterior Layer: Trapezius muscle
0.6	Muscles of the scalp, face, and neck	Go to "Full Anatomy" View: Lateral Layer: Orbicularis oculi muscle
0.7 a	Muscles of mastication	Go to "Full Anatomy" View: Lateral Layer: Masseter muscle
0.8 a	Muscles of the anterior neck	Go to "Full Anatomy" View: Anterior Layers: Sternocleidomastoid muscle; Sternohyoid muscle
0.8 b	Muscles of the throat	Go to "Full Anatomy" View: Lateral Layer: Esophagus
0.9 a	Muscles of the neck and vertebral column	Go to "Full Anatomy" View: Anterior Layers: Sternocleidomastoid muscle; Anterior scalene muscle
0.9 b	Muscles of the neck and vertebral column	Go to "Full Anatomy" View: Posterior Layer: Splenius muscle
0.9 d	Deep muscles of the back	Go to "Full Anatomy" View: Posterior Layers: Iliocostalis lumborum muscle; Semispinalis capitis muscle

FIGURE NUMBER	KEY STRUCTURE	A.D.A.M. VIEW AND LAYER
10.10 a	Deep muscles of the thorax	Go to "Full Anatomy" View: Anterior Layers: External intercostal muscle; Internal intercostal muscle
10.10 b	Diaphragm	Go to "Full Anatomy" View: Anterior Layer: Diaphragm
10.11 a	Muscles of the abdominal wall	Go to "Full Anatomy" View: Anterior Layers: External abdominal oblique muscle; Internal abdominal oblique muscle; Rectus abdominis muscle
10.11 b	Muscles of the abdominal wall	Go to "Full Anatomy" View: Lateral Layer: External abdominal oblique muscle
10.13 a	Muscles of the anterior thorax and shoulder	Go to "Full Anatomy" View: Anterior Layers: Deltoid muscle; Pectoralis minor muscle
10.13 b	Muscles of the back and shoulder	Go to "Full Anatomy" View: Posterior Layers: Deltoid muscle; Rhomboideus muscle
10.14 a	Muscles crossing the shoulder and elbow joints	Go to "Full Anatomy" View: Anterior Layer: Deltoid muscle
10.14 d	Muscles crossing the shoulder and elbow joints	Go to "Full Anatomy" View: Posterior Layers: Latissimus dorsi muscle; Rhomboideus muscle
10.15 a	Muscles of the anterior forearm	Go to "Full Anatomy" View: Anterior Layer: Biceps brachii muscle
10.15 b	Muscles of the anterior forearm	Go to "Full Anatomy" View: Anterior Layer: Flexor digitorum superficialis muscle
10.15 c	Muscles of the anterior forearm	Go to "Full Anatomy" View: Anterior Layer: Flexor digitorum profundus muscle
10.16 a	Muscles of the posterior forearm	Go to "Full Anatomy" View: Posterior Layer: Extensor digitorum muscle
10.16 b	Muscles of the posterior forearm	Go to "Full Anatomy" View: Posterior Layers: Supinator muscle; Abductor pollicis longus muscle
10. 18 a	Muscles of the anterior and medial thigh	Go to "Full Anatomy" View: Anterior Layer: Tensor fasciae latae muscle

FIGURE NUMBER	KEY STRUCTURE	A.D.A.M. VIEW AND LAYER
10.18 b,c	Muscles of the anterior and medial thigh	Go to "Full Anatomy" View: Anterior Layer: Vastus intermedius
10.19 a	Muscles of the posterior hip and thigh	Go to "Full Anatomy" View: Posterior Layer: Gluteus maximus muscle
10.19 b,c	Deep muscles of the gluteal region	Go to "Full Anatomy" View: Posterior Layers: Gluteus minimus muscle; Obturator externus muscle
10.20 a	Muscles of the anterior leg	Go to "Full Anatomy" View: Anterior Layer: Tibialis anterior muscle
10.20 c,d	Muscles of the anterior leg	Go to "Full Anatomy" View: Anterior Layer: Extensor hallucis longus muscle
10.21 a	Muscles of the lateral leg	Go to "Full Anatomy" View: Lateral Layer: Peroneus longus muscle
10.22 a	Muscles of the posterior leg	Go to "Full Anatomy" View: Posterior Layer: Gastrocnemius muscle
10.22 b	Muscles of the posterior leg	Go to "Full Anatomy" View: Posterior Layer: Soleus muscle
10.22 c	Muscles of the posterior leg	Go to "Full Anatomy" View: Posterior Layer: Flexor digitorum longus muscle
12.8 a	Brain	Go to "Full Anatomy" View: Lateral Layer: Brain
12.8 b	Medial surface of brain	Go to "Full Anatomy" View: Medial Layer: Central nervous system and sacral plexus
12.9	Internal structure of the brain	Go to "Full Anatomy" View: Anterior Layer: Atlas
12.14 a	Diencephalon and brain stem	Go to "Full Anatomy" View: Medial Layer: Central nervous system and sacral plexus
12.21	Meninges of the brain	Go to "Full Anatomy" View: Anterior Layer: Atlas
12.22	Falx cerebri	Go to "Full Anatomy" View: Lateral Layer: Falx cerebri
12.23 b	Midsagittal brain and spinal cord	Go to "Full Anatomy" View: Medial Layer: Central nervous system and sacral plexus
12.24 a	Spinal cord	Go to "Full Anatomy" View: Posterior Layer: Spinal cord

FIGURE NUMBER	KEY STRUCTURE	A.D.A.M. VIEW AND LAYER
13.7	Cervical plexus	Go to "Full Anatomy" View: Anterior Layer: Inferior root of ansa cervicalis
13.7	Cervical plexus	Go to "Full Anatomy" View: Lateral Layer: Ansa cervicalis
13.8 a,c	Brachial plexus	Using "View" on the Menu Bar, choose Nervous system
13.8 a,c	Brachial plexus	Go to "Full Anatomy" View: Anterior Layer: Brachial plexus and branches
13.9 a,b	Lumbar plexus	Using "View" on the Menu Bar, choose Nervous system
13.9 a,b	Lumbar plexus	Go to "Full Anatomy" View: Anterior Layer: Deep arteries
13.10 a,b	Sacral plexus	Using "View" on the Menu Bar, choose Nervous system
13.10 a,b	Sacral plexus	Go to "Full Anatomy" View: Posterior Layers: Posterior femoral cutaneous nerve; Sciatic nerve and branches
14.4 a	Sympathetic trunk	Go to "Full Anatomy" View: Anterior Layer: Sympathetic trunk
16.6 a	Extrinsic muscles of the eye	Go to "Full Anatomy" View: Lateral Layer: Eye muscles–lateral
17.1	Male endocrine system	Using "Options", choose Male; using "View" on the Menu Bar, choose Endocrine system
17.1	Female endocrine system	Using "Options", choose Female; using "View" on the Menu Bar, choose Endocrine system
17.5	Pituitary gland	Go to "Full Anatomy" View: Lateral Layer: Falx cerebri
17.8 a	Thyroid gland	Go to "Full Anatomy" View: Anterior Layer: Thyroid gland
17.12	Adrenal gland	Go to "Full Anatomy" View: Anterior Layer: Suprarenal gland
17.15 a	Pancreas	Go to "Full Anatomy" View: Anterior Layer: Pancreas
19.1c	Heart and great vessels in mediastinum	Go to "Full Anatomy" View: Anterior Layer: Coronary arteries
19.2	Pericardium and heart wall	Go to "Full Anatomy" View: Anterior Layer: Heart–cut section

FIGURE NUMBER	KEY STRUCTURE	A.D.A.M. VIEW AND LAYER
19.4 a	External anatomy of the heart	Go to "Full Anatomy" View: Anterior Layer: Coronary arteries
19.4 c	Internal anatomy of the heart	Go to "Full Anatomy" View: Anterior Layer: Heart–cut section
19.10 a	Coronary arteries	Go to "Full Anatomy" View: Anterior Layer: Coronary arteries
19.10 b	Cardiac veins	Go to "Full Anatomy" View: Anterior Layer: Cardiac veins
20.15 b	Pulmonary circulation	Go to "Full Anatomy" View: Anterior Layers: Heart–cut section; Pulmonary arteries
20.17 b	Arteries of the systemic circulation	Using "View" on the Menu Bar, choose Cardiovascular system
20.18 b	Arteries of the head and neck	Go to "Full Anatomy" View: Lateral Layer: Common carotid artery and branches
20.19 b	Arteries of the upper limb and thorax	Using "View" on the Menu Bar, choose Cardiovascular system
20.20 b	Celiac trunk and branches	Go to "Full Anatomy" View: Anterior Layer: Celiac trunk and branches
20.20 c	Abdominal aorta and branches	Go to "Full Anatomy" View: Anterior Layer: Abdominal aorta and branches
20.20 d	Superior and inferior mesenteric arteries	Go to "Full Anatomy" View: Anterior Layers: Superior mesenteric artery; Inferior mesenteric and testicular arteries (Male); Ovarian and inferior mesenteric arteries (Female)
20.21 b	Arteries of the pelvis and lower limb	Using "View" on the Menu Bar, choose Cardiovascular system
20.21 b	Arteries of the pelvis and lower limb	Go to "Full Anatomy" View: Anterior Layer: Abdominal aorta and branches
20.22 b	Veins of the systemic circulation	Using "View" on the Menu Bar, choose Cardiovascular system
20.23 b	Veins of the head and neck	Go to "Full Anatomy" View: Lateral Layers: External jugular vein and tributaries; Tributaries of right brachiocephalic vein
20.23 c	Dural sinuses of the brain	Go to "Full Anatomy" View: Lateral Layer: Falx cerebri
20.24 b	Veins of the upper limb and thorax	Using "View" on the Menu Bar, choose Cardiovascular system
20.24 b	Veins of the deep thorax	Go to "Full Anatomy" View: Anterior Layer: Azygos vein and tributaries
20.25 b	Tributaries of the inferior vena cava	Go to "Full Anatomy" View: Anterior Layer: Inferior vena cava
20.25 c	Hepatic portal circulation	Go to "Full Anatomy" View: Anterior Layer: Portal vein and tributaries
20.26 b	Veins of the lower limb	Using "View" on the Menu Bar, choose Cardiovascular system
20.26 c	Veins of the lower limb	Go to "Full Anatomy" View: Posterior Layer: Posterior tibial vein
21.2 a	Lymphatic vessels and nodes	Using "View" on the Menu Bar, choose Lymphatic system
21.2 b	Thoracic duct	Go to "Full Anatomy" View: Anterior Layer: Thoracic duct
21.4	Remnant of thymus gland	Go to "Full Anatomy" View: Anterior Layer: Remnant of thymus gland
21.4	Spleen	Go to "Full Anatomy" View: Anterior Layer: Spleen
23.1	Respiratory system	Using "View" on the Menu Bar, choose Respiratory system
23.2 a	External nose	Go to "Full Anatomy" View: Anterior Layer: Nasal cartilage
23.3 a	Upper respiratory tract	Go to "Full Anatomy" View: Medial Layer: Nasal conchae
23.4 a	Larynx	Go to "Full Anatomy" View: Anterior Layer: Tracheal lymph nodes
23.7 a	Lower respiratory tract	Go to "Full Anatomy" View: Anterior Layer: Trachea
23.10 a	Thoracic organs	Go to "Full Anatomy" View: Anterior Layers: Lungs; Phrenic nerve
24.1	Digestive system	Using "View" on the Menu Bar, choose Digestive system
24.7 a	Oral cavity	Go to "Full Anatomy" View: Medial Layer: Tongue
24.7 b	Oral cavity	Go to "Full Anatomy" View: Anterior Layer: Palatoglossus muscle
24.9 a	Salivary glands	Go to "Full Anatomy" View: Lateral Layers: Parotid gland and duct; Deep salivary glands

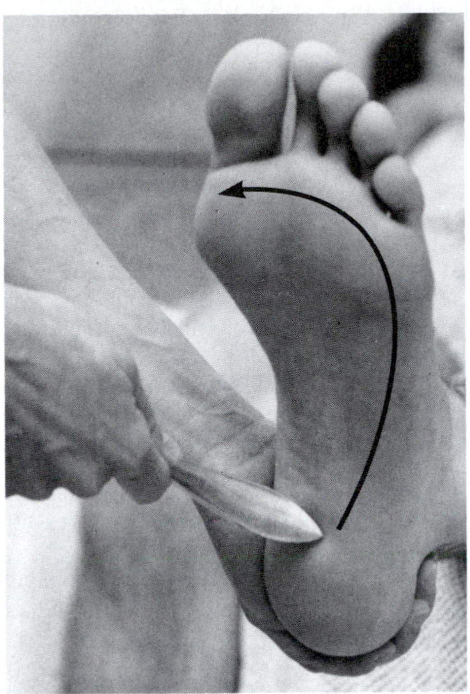

F22i.6

Demonstrating the plantar reflex.

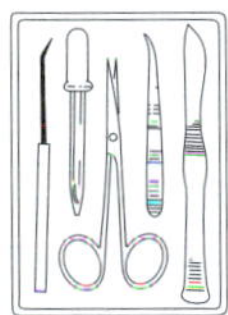

The subject should sit with eyes closed and with the dorsum of one hand resting on the laboratory bench. Obtain a sharp pencil and suddenly prick the subject's index finger. What are the results?

___

___

Did the extensor part of this reflex seem to be slow compared to other reflexes you have observed? ___

What are the reasons for this? ___

___

SUPERFICIAL CORD REFLEXES The **superficial cord reflexes** (abdominal, cremaster, and plantar reflexes) result from pain and temperature changes. They are initiated by stimulation of receptors in the skin and mucosae. The superficial cord reflexes depend *both* on functional upper-motor pathways and on the cord-level reflex arc. Since only the plantar reflex can be tested conveniently in a laboratory setting, we will use this as our example.

The plantar reflex, an important neurological test, is elicited by stimulating the cutaneous receptors in the sole of the foot. In adults, stimulation of these receptors causes the toes to flex and move closer together. Damage to the pyramidal (or corticospinal) tract, however, produces *Babinski's sign,* an abnormal response in which the toes flare and the great toe moves in an upward direction. (In newborn infants, Babinski's sign is due to incomplete myelination of the nervous system.)

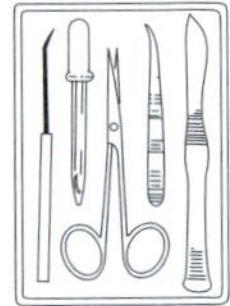

Have the subject remove a shoe and lie on the cot or laboratory bench with the knees slightly bent and thighs rotated so that the lateral side of the foot rests on the cot. Alternatively, the subject may sit up and rest the lateral surface of the foot on a chair. Draw the handle of the reflex hammer firmly down the lateral side of the exposed sole from the heel to the base of the great toe (Figure 22i.6).

What is the response? ___

___

Is this a normal plantar reflex or Babinski's sign?

___

REACTION TIME OF UNLEARNED RESPONSES

The time required for reaction to a stimulus depends on many factors—sensitivity of the receptors, velocity of nerve conduction, the number of neurons and synapses involved, and the speed of effector activation, just to name a few. The type of response to be elicited is also important. If the response involves a reflex arc, the synapses are facilitated and response time will be short. If, on the other hand, the response can be categorized as an unlearned response, then a far larger number of neural pathways and many types of higher intellectual activities—including choice and decision making—will be involved, and the time for response will be considerably lengthened.

There are various ways of testing reaction time of unlearned responses. The tests range in difficulty from simple to ultrasophisticated. Since the objective here is to demonstrate the major time difference between reflexes and unlearned responses, an intermediate approach will suffice.

1. Once again, have the subject sit on the edge of the counter so that the legs are relaxed with feet well above the floor. Carefully strap the force transducer again to the right knee. To enter the data collection screen:

- Select Experiment menu from the Main menu.
- Select Real-Time Experiment Mode.

2. Now test the reaction time for unlearned responses. Have the subject face away and randomly tap the button on the hammer gently on the tabletop. Instruct the subject to respond to the sound of the tap as quickly as possible by gently kicking the right leg away from the table. Record the information for five trials.

Trial	Max. excur.°	Latent (sec)	Frame #
1			
2			
3			
4			
5			

3. Perform the test again, but this time say a simple word each time you tap the tabletop. Designate a specific word as a signal for the subject to respond. On all other words, the subject is to remain motionless. Trials in which the subject erroneously kicks out their leg are to be disregarded. Record the information for five *successful* trials.

Trial	Max. excur.°	Latent (sec)	Frame #
1			
2			
3			
4			
5			

4. Perform the test once again to investigate the subject's reaction to word association. As you tap the tabletop, say a word—for example, *hot.* The subject is to respond with a word he or she associates with the stimulus word—for example, *cold*—kicking as he or she responds. If the subject is unable to make a word association, he or she must not kick out the leg. Record the information for five successful trials, as well as the number of times the subject erroneously kicks.

Trial	Max. excur.°	Latent (sec)	Frame #
1			
2			
3			
4			
5			

The number of times the subject erroneously kicked out.

____

You should have noticed quite a large variation in reaction time in this series of trials. Why is this so?

____

____

____

5. Save your data:

- Select Data File/Disk from the Main menu.
- Select Save.
- Again type a file name according to your instructor's directions.
- Once your data is saved to disk, it may be retrieved later for a more detailed analysis.

Cranial Nerve Reflex Tests

In these experiments, you will be working with your lab partner to illustrate two somatic reflexes mediated by cranial nerves.

CORNEAL REFLEX The **corneal reflex** is mediated through the trigeminal nerve (cranial nerve V). The absence of this reflex is an ominous sign, because it often indicates damage to the brain stem, resulting from compression of the brain or other trauma.

Stand to one side of the subject; the subject should look away from you toward the opposite wall. Wait a few seconds and then quickly, but gently, touch the subject's cornea (on the side toward you) with a wisp of absorbent cotton. What is the reaction?

____

What is the function of this reflex? ______________

____

Was the sensation that of touch or pain? ______________

Why? ______________

____

GAG REFLEX The **gag reflex** tests the somatic motor responses of cranial nerves IX and X. When the oral mucosa on the side of the uvula is stroked, each side of the mucosa should rise, and the amount of elevation should be equal.*

For this experiment, select a subject who does not have a queasy stomach, because regurgitation is a possibility. Stroke the subject's oral mucosa on each side of the uvula with a tongue depressor. What happens?

____

____

* The uvula is the fleshy tab hanging from the roof of the mouth just above the root of the tongue.

Discard the used tongue depressor in the disposable autoclave bag before continuing. Do *not* lay it on the laboratory bench at any time.

AUTONOMIC REFLEXES

The autonomic reflexes include the pupillary, ciliospinal, and salivary reflexes, as well as a multitude of other reflexes. Work with your lab partner to demonstrate the four autonomic reflexes described next.

Pupillary Reflexes

There are several types of pupillary reflexes. The **pupillary light reflex** and the **consensual reflex** will be examined here. In both of these pupillary reflexes, the retina of the eye is the receptor, the optic nerve (cranial nerve II) contains the afferent fibers, the occulomotor nerve (cranial nerve III) is responsible for conducting efferent impulses to the eye, and the smooth muscle of the iris is the effector. Many central nervous system centers are involved in the integration of these responses. Absence of the normal pupillary reflexes is generally a late indication of severe trauma or deterioration of the vital brain stem tissue due to metabolic imbalance.

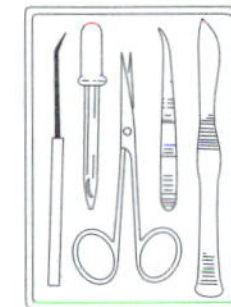

1. Conduct the reflex testing in an area that is relatively dim. Before beginning, obtain a metric ruler to measure and record the size of the subject's pupils as best you can.

Right pupil: _______ mm

Left pupil: _______ mm

2. Stand to the left of the subject to conduct the testing. The subject should shield the right eye by holding a hand vertically between the eye and the right side of the nose.

3. Shine a flashlight into the subject's left eye. What is the pupillary response? ______________________________

Measure the size of the left pupil: _______ mm

4. Observe the right pupil. Has the same type of change (called a consensual response) occurred in the right eye?

____

Measure the size of the right pupil: _______ mm

The consensual response, or any reflex observed on one side of the body when the other side has been stimulated, is called a **contralateral response.** The pupillary light response, or any reflex occurring on the side that was stimulated, is referred to as an **ipsilateral response.**

When a contralateral response occurs, what does this indicate about the pathways involved?

____

____

____

Was the sympathetic or parasympathetic division of the autonomic nervous system active during the testing of these reflexes?

____

What is the function of these pupillary responses?

____

____

____

Ciliospinal Reflex

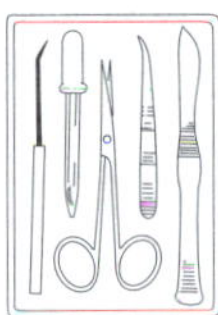

The **ciliospinal reflex** is another example of reflex activity in which pupillary responses can be observed. This response may initially seem a little bizarre, especially in view of the consensual reflex just demonstrated. While observing the subject's eyes, gently stroke the skin (or just the hairs) on the left side of the back of the subject's neck, close to the hairline.

What is the reaction of the left pupil? ______________

The reaction of the right pupil? ______________

If you see no reaction, repeat the test using a gentle pinch in the same area.

The response you should have noted—pupillary dilation—is consistent with the pupillary changes occurring when the sympathetic nervous system is stimulated. Such a response may also be elicited in a single pupil when more impulses from the sympathetic nervous system reach it for any reason. For example, when the left side of the subject's neck was stimulated, sympathetic impulses to the left eye increased, resulting in the ipsilateral reaction of the left pupil.

On the basis of your observations, would you say that the sympathetic innervation of the two irises is closely integrated?

____

Why or why not? ______________________________

____

Salivary Reflex

Unlike the other reflexes, in which the effectors were smooth or skeletal muscles, the effectors of the **salivary reflex** are glands. The salivary glands secrete varying amounts of saliva in response to reflex activation.

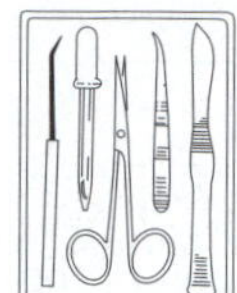

1. Obtain a small beaker, a graduated cylinder, lemon juice, and wide-range pH paper. After refraining from swallowing for 2 minutes, the subject is to expectorate (spit) the accumulated saliva into a small beaker. Using the graduated cylinder, measure the volume of the expectorated saliva and determine its pH.

Volume: _______ cc pH: ____________________

2. Now place 2 or 3 drops of lemon juice on the subject's tongue. Allow the lemon juice to mix with the saliva for 5 to 10 seconds, and then determine the pH of the subject's saliva by touching a piece of pH paper to the tip of the tongue.

pH: _______

As before, the subject is to refrain from swallowing for 2 minutes. After the 2 minutes is up, again collect and measure the volume of the saliva and measure its pH.

Volume: _______ cc pH: ____________________

3. How does the volume of saliva collected after the application of the lemon juice compare with the volume of the first saliva sample? ____________________

How does the final saliva pH reading compare to the initial reading? ____________________

To that obtained 10 seconds after the application of the lemon juice? ____________________

What division of the autonomic nervous system mediates the reflex release of saliva? ____________________

Dispose of the saliva-containing beakers and the graduated cylinders in the laboratory bucket that contains bleach and put the used pH paper into the disposable autoclave bags. Wash the bench down with 10% bleach solution before continuing.

Conduction System of the Heart and Electrocardiography

OBJECTIVES

1. To list and localize the elements of the intrinsic conduction, or nodal, system of the heart; and to describe the initiation and conduction of impulses through this system and the myocardium.
2. To recognize that heart contraction results from electrical activity of many muscle cells.
3. To realize that the electrical activity of the heart can be measured at the surface of the body, and that such measurements give different results than intracellular recording.
4. To operate computerized recording equipment.
5. To describe the different parts of the ECG wave form.
6. To realize that ECG patterns can reveal possible heart dysfunction.
7. To explain the results of an ECG recording.

MATERIALS

Minimum computer equipment required:
- IBM PC/XT/AT or compatible computer
- 640K RAM—single disk (3.5 in.) drive
- Graphics adapter and monitor
- Intelitool Cardiocomp 1, 7, or 12 equipment

Software:
- Intelitool ECG software

Alcohol swabs

Lead-Lok adhesive electrode pads or other surface electrodes

See Appendix E, Exercise 31i for links to *Anatomy and PhysioShow: The Videodisc.*

THE INTRINSIC CONDUCTION SYSTEM

Heart contraction results from a series of electrical potential changes (depolarization waves) that travel through the heart preliminary to each beat. The ability of cardiac muscle to beat is intrinsic. The heart does not depend on nerve impulses to initiate its contraction and will continue to contract rhythmically even if all nerve connections are severed. However, two controlling systems act on the heart. One of these involves nerves of the autonomic nervous system, which accelerate or decrease the heartbeat rate depending on which division is activated. The second system is the **intrinsic conduction system,** or **nodal system,** of the heart, consisting of specialized noncontractile myocardial tissue. The intrinsic conduction system ensures that heart muscle depolarizes in an orderly and sequential manner (from atria to ventricles) and that the heart beats as a coordinated unit.

The components of the intrinsic conduction system include the **SA (sinoatrial) node,** located in the upper right atrium just inferior to the entrance to the superior vena cava; the **AV (atrioventricular) node** in the lower atrial septum at the junction of the atria and ventricles; the **AV bundle (bundle of His)** and the right and left **bundle branches,** located in the interventricular septum; and the **Purkinje fibers,** which ramify within the muscle bundles of the ventricular walls. The Purkinje fiber network is much denser and more elaborate in the left ventricle because of the larger size of this chamber (Figure 31i.1).

The SA node, which has the highest rate of discharge, provides the stimulus for contraction. Because it sets the rate of depolarization for the heart as a whole, the SA node is often referred to as the *pacemaker.* From the SA node, the impulse spreads throughout the atria and to the AV node. The electrical wave is immediately followed by atrial contraction. At the AV node, the impulse is momentarily delayed (approximately 0.1 sec), allowing the atria to complete their contraction. It then passes through the AV bundle, the right and left bundle branches, and the Purkinje fibers, finally resulting in ventricular contraction. Note that the atria and the ventricles are separated from one another by a region of electrically inert tissue, so the depolarization wave can be transmitted to the ventricles only via the tract between the AV node and AV bundle. Thus, any damage to the AV node-bundle pathway partially or totally insulates the ventricles from the influence of the SA node. Although autorhythmic cells are found throughout the heart, their rates of spontaneous depolarization differ. The nodal system increases the rate of heart depolarization and synchronizes heart activity.

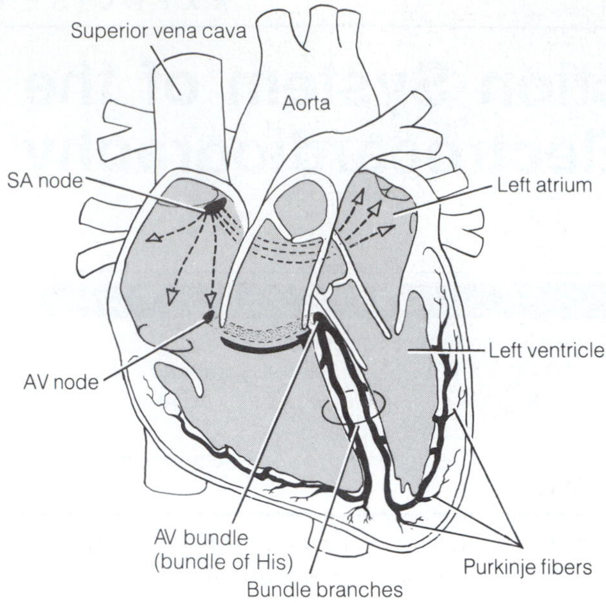

F31i.1

The intrinsic conduction system of the heart. Dashed-line arrows indicate transmission of the impulse from the SA node through the atria. Solid arrow indicates transmission of the impulse from the AV node to the AV bundle.

ELECTROCARDIOGRAPHY

The conduction of impulses through the heart generates electrical currents that eventually spread throughout the body. These impulses can be detected on the body surface and recorded with an instrument called an *electrocardiograph.* The graphic recording of the electrical changes (depolarization and repolarization) occurring in the cardiac cycle is called an **electrocardiogram (ECG)** (Figure 31i.2). The typical ECG consists of a series of three recognizable waves called *deflection waves.* The first wave, the **P wave,** is a small wave that indicates the depolarization of the atria immediately before atrial contraction. The large **QRS complex,** resulting from ventricular depolarization, has a complicated shape (primarily because of the variability in size of the two ventricles and the time differences required for these chambers to depolarize). It precedes ventricular contraction. The **T wave** results from currents propagated during ventricular repolarization. The repolarization of the atria, which occurs during the QRS interval, is generally obscured by the large QRS complex.

It is important to understand what an ECG does and does not show: An ECG is a record of voltage and time—nothing else. Although we can and do infer that muscle contraction follows its excitation, sometimes it does not. Secondly, an ECG records electrical events occurring in relatively large amounts of muscle tissue (i.e., the bulk of the heart muscle), *not* the electrical activity of nodal tissue which, like muscle contraction, can only be inferred. Nonetheless, abnormalities of the deflection waves and changes in the time intervals of the ECG are useful in detecting myocardial infarcts or problems with the conduction system of the heart. The P-R (P-Q) interval represents the time between the beginning of atrial depolarization and ventricular depolarization. Thus, it typically includes the period during which the depolarization wave passes to the AV node, atrial systole, and the passage of the excitation wave to the balance of the conducting system. Generally, the P-R interval is about 0.16 to 0.18 sec. A longer interval may suggest a partial AV block caused by damage to the AV node. In total heart block, no impulses are transmitted through the AV node, and the atria and ventricles beat independently of one another—the atria at the SA node rate and the ventricles at their intrinsic rate, which is considerably slower.

If the QRS interval (normally 0.06 to 0.10 sec) is prolonged, it may indicate a right or left bundle branch block in which one ventricle is contracting later than the other. The Q-T interval is the period from the beginning of ventricular depolarization through repolarization and includes the time of ventricular contraction (the S-T segment). With a heart rate of 70 beats/min, this interval is normally 0.31 to 0.41 sec. As the rate increases, this interval becomes shorter; conversely, when the heart rate drops, the interval is longer.

A heart rate over 100 beats/min is referred to as **tachycardia;** a rate below 60 beats/min is **bradycardia.** Although neither condition is pathological, prolonged tachycardia may progress to **fibrillation,** a condition of rapid uncoordinated heart contractions which makes the heart useless as a pump. Bradycardia in athletes is a positive finding; that is, it indicates increased efficiency of cardiac functioning. Because *stroke volume* (the amount of blood ejected by a ventricle with each contraction) increases with physical conditioning, the heart can contract more slowly and still meet circulatory demands.

Twelve standard leads are used to record an ECG for diagnostic purposes. Three of these are bipolar leads that measure the voltage difference between the arms, or an arm and a leg, and nine are unipolar leads. Together the 12 leads provide a fairly comprehensive picture of the electrical activity of the heart.

An important distinction is the difference between the definitions of the terms *electrode* and *lead.*

Electrode: An electrode is a physical piece of wire. The polarity of each wire (electrode) of the cable is labeled by color:

- White—negative
- Black—positive
- Green—ground

Note: Cardiocomp 7 and 12 have labeled interface boxes to allow simultaneous connection of all electrodes.

Lead: A lead measures the electrical potential difference between electrodes.

We will use Intelitool Cardiocomp ECG recording equipment in this exercise. For this investigation, a total

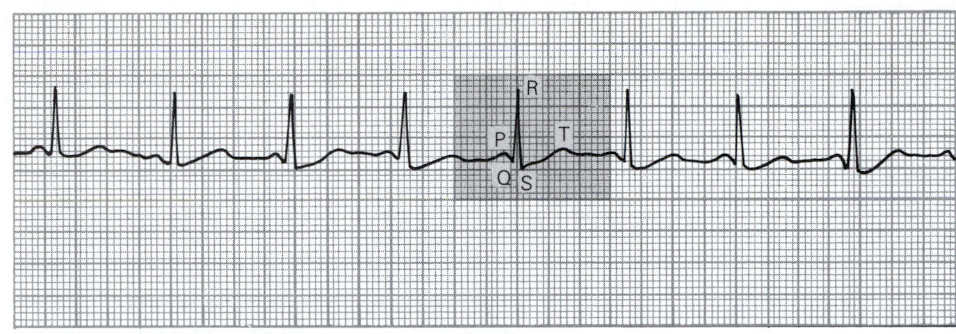

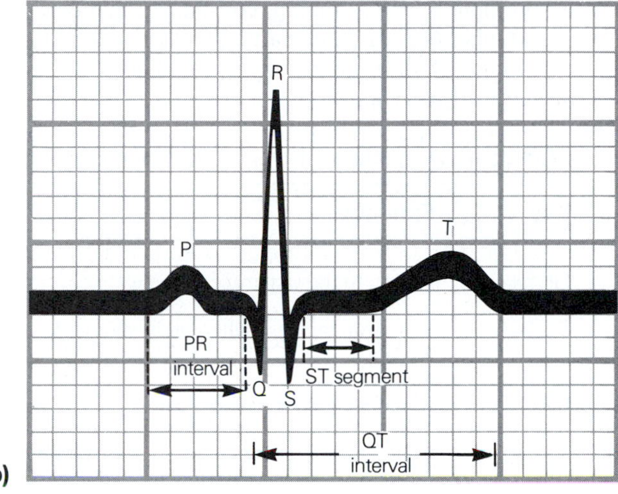

Time: small squares = 0.04 sec
1 large square = 0.20 sec
5 large squares = 1.00 sec

F31i.2

The normal electrocardiogram. (a) Regular sinus rhythm. **(b)** Waves, segments, and intervals of a normal ECG.

of 4 electrodes are used and results are obtained from the three bipolar (limb) leads (Figure 31i.3). With each of the leads, the electrical changes between two of the electrodes are determined. As shown in the figure, the potential difference between the right and left arms (AL-AR) is determined with lead I, between the right arm and left leg (RA-LL) with lead II, and between the left arm and left leg (LA-LL) with lead III.

Understanding the Standard Limb Leads

As you might expect, electrical activity recorded by any lead depends on the location and orientation of the recording electrodes. Clinically, it is assumed that the heart lies in the center of a triangle with sides of equal length (*Einthoven's triangle*) and that the recording connections are made at the vertices (corners) of that triangle. But, in practice, the electrodes connected to each arm and to the left leg are considered to connect to the triangle vertices. The standard limb leads record the voltages generated in the extracellular fluids surrounding the heart by the ion flows occurring simultaneously in many cells between any two of the connections. A recording using lead I (AL-AR), which connects the right arm (RA) and the left arm (LA), is most sensitive to electrical activity spreading horizontally across the heart. Lead II (RA-LL) and lead III (LA-LL) record activity along the vertical axis (from the base of the heart to its apex), but from different orientations. The significance of Einthoven's triangle is that the sum of the volt-

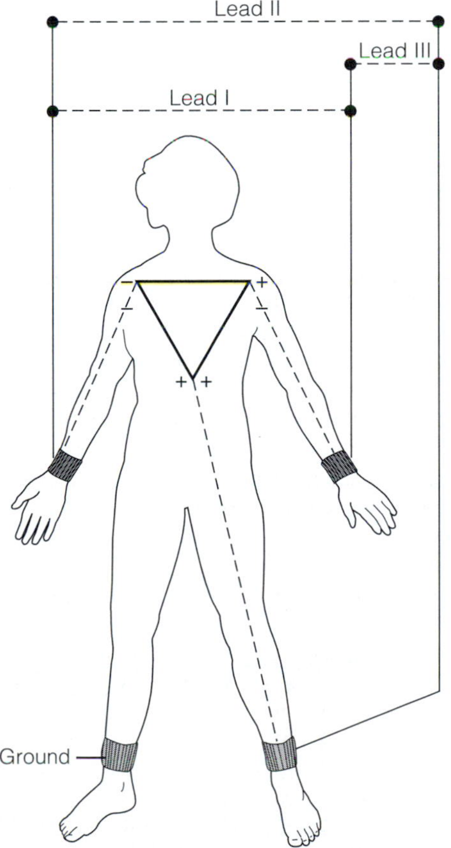

F31i.3

ECG recording positions for the standard limb leads.

ages of leads I and III equals that in lead II (Einthoven's law). Hence, if the voltages of two of the standard leads are recorded, that of the third lead can be determined mathematically.

Preparing the Subject

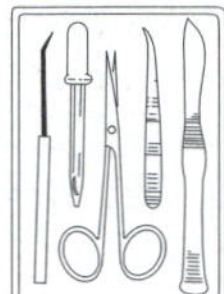

Because most common problems encountered with Intelitool ECG equipment are due to incorrect electrode application, the following instructions should be carefully followed to establish the proper electrical connection.

One at a time, remove the Lead-Lok electrode pads from the protective backing and apply to the skin surface at each of the sites described below after first scrubbing the designated area with an alcohol swab:

- One over the anterior part of each forearm, about 2–3 inches above the wrist
- One approximately 2–3 inches above each medial malleolus (inner aspect of the ankle)

(Cardiocomp 7 and Cardiocomp 12 users should place electrodes at the locations described on the Cardiocomp interface box.)

Recording the ECG

1. Start the software. Type **ITOOL** at the DOS prompt. Select Cardiocomp by making sure the highlight (moved with the arrow keys) is on the selection and pressing (ENTER). Then carry out the following sequence of steps.

- Select Experiment menu from the Cardiocomp Main menu.
- Select Real-Time ECG, which automatically displays the data collection screen (Figure 31i.4).

The data collection screen should be displaying a typical ECG trace sweeping from left to right if all the electrode connections to the subject are satisfactory. If random-looking electrical spikes are evident in the tracing, remove the electrodes, clean the skin again with an alcohol scrub, then reapply the electrode pads to establish good electrical contact.

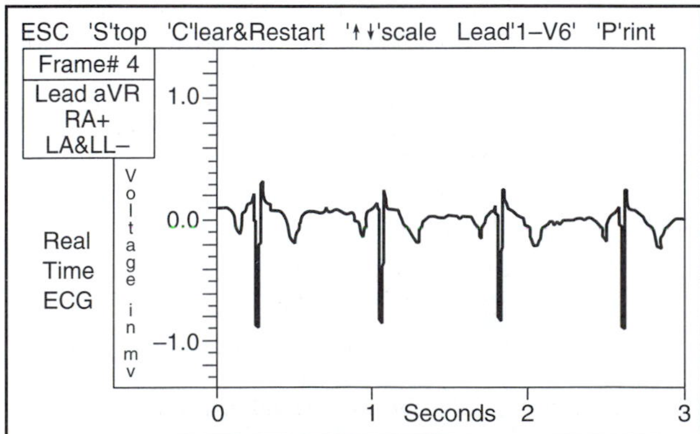

F31i.4

The real-time ECG display screen.

- The top screen edge displays keys that control data collection.
- Experimental data is recorded in the computer's memory as individual screens (frames), the numbers of which are displayed in the Frame # box. Cardiocomp software can record a maximum of 20 frames, after which the data first obtained is overwritten by the new information. Because each frame is a single "snapshot" of a few heartbeats, the complete ECG tracing consists of a continuing number of frames, one after another.
- Table 31i.1 explains the keys that control data collection.

TABLE 31i.1 Control Keys for Real-Time ECG

Key	Function
(ESC)	Return to the Experiment menu
S	Start/stop toggle to alternately start or stop the tracing
C	Clear and restart data collection
↑ ↓	Change (increase or decrease) the voltage scale
P	Print screen
1–V6	Change lead number*

2. Press **1** (number one) to set the computer to collect lead I data. The computer's screen will prompt you to change electrode positions to record lead I.

3. For lead I recording, attach the appropriate electrode to the Lead-Lok pad at the following locations:

- Black—left wrist (positive)
- White—right wrist (negative)
- Green—right ankle (ground)

4. When you are satisfied that you see a clean, stable tracing, press **S** to stop the tracing at the end of the screen. Before you start actual data collection, carefully read steps 1 through 7 below.

BASELINE RECORDING

1. Position the subject comfortably in a supine position on a cot (if available) or sitting relaxed on a laboratory chair. If a cot is not available, the subject must remain still to limit electrical interference from skeletal muscle activity.

2. Press **S** to start the ECG tracing, and then immediately press **C**. This action clears the computer's memory

* Lead numbers differ depending on the Cardiocomp version used.

of all practice frames collected to this point, and the prompt: "All data will be lost. Continue? Y/N" will be displayed. (This is normal; do not panic!)

- Press **Y** (yes) and **1** to start ECG recording in frame 1.
- In the *third* frame of lead I data collection, press **2**.

3. Pressing **2** sets the computer for lead II ECG recording. It also causes the tracing to stop automatically at the end of the frame, giving you time and prompting you to change the electrode positions to record lead II as follows:

- White—right wrist (negative)
- Black—left ankle (positive)
- Green—right ankle (ground)

(Cardiocomp 7 and 12 users need not change electrode cable positions because the equipment automatically switches to the appropriate lead configuration.)

4. Press **S** to start lead II data collection. Press **3** in the third frame of lead II collection. As before, the tracing will stop at the end of the frame.

5. Change the electrode positions to record lead III as follows:

- White—left wrist (negative)
- Black—left ankle (positive)
- Green—right ankle (ground)

6. Press **S** to collect lead III data. Press **S** again in the third frame of data collection and as before, the tracing will automatically stop at the end of the frame.

7. You should now have three frames of data for each lead in the computer's memory. If you are not satisfied with the overall results, you can repeat the exercise by selecting **C** to clear and restart, which removes all frames from the computer's memory and restarts ECG recording at frame 1.

When you are satisfied that you have a good overall baseline ECG recording:

- Press (ESC) to return to the Experiment menu.
- Press (ESC) again to return to the Main menu.

Do not remove the electrodes from the subject at this time because you will make additional recordings shortly in the Running in Place Recording and Breath-Holding Recording experiments.

Computerized Data Analysis

From the Main menu, select Review/Time Analysis. The Review/Time Analysis screen is similar in appearance to the experiment screen, and automatically displays your recorded ECG tracing, frame by frame. Carry out the following steps to acquaint yourself with the data analysis functions.

1. Stop the tracing for analysis by pressing **S**, which "freezes" the frame being viewed on the screen. You can also navigate through your recorded ECG tracing by pressing **B** to move back one frame or **F** to go forward one frame.

2. Press (→) to move a vertical line (the data analysis line) across the screen to the right.

3. Watch the Difference box in the left part of the screen as you move the data analysis line. The numbers represent the time (sec) and voltage (mV) values of the tracing at the intersection point between the tracing and the data analysis line.

4. Using (→) or (←), move the data analysis line to the beginning of a P wave and press **Z** (zero). This action "zeroes" the numbers in the difference box. When the data analysis line is next moved away from this position, the time and voltage values in the Difference box will be relative to this "zero" point.

5. Now, move the data analysis line to the end of the P wave to see the duration (in seconds) of the P wave read directly in the Difference box.

Refer to Table 31i.2 for a summary of the data analysis control keys.

TABLE 31i.2 Control Keys for Data Analysis

Key	Function
(ESC)	Return to the submenu.
S	Start/stop toggle.
B	Move the screen back one frame.
F	Move the screen forward one frame.
P	Print screen.
(↑) (↓)	Change the voltage scale.
(←)	Move the data analysis line left.
(→)	Move the data analysis line right.
Z	Make the current location of the data analysis line the zero point. That is, the voltage and time displays in the Difference box reset to zero. When (→) or (←) is pressed, the data analysis line moves, and the program keeps track of the difference in time and voltage between the zero point and the new, current location of the data analysis line.
J	Allow the data analysis line to move in large jumps. Pressing it again toggles "jump" off.

Now complete the following analysis steps.

1. Each student should choose a representative frame of one of the lead recordings and measure the following time values using the data analysis line.

Name of subject ______________________________

Lead used: _______

2. Compute the heart rate. Place the cursor at the beginning of one QRS complex by using the arrow keys, press **Z** to zero all displays, and move the data analysis line to the beginning of the next QRS complex. Record the time for one heartbeat:

_______ sec/beat.

3. Now find the beats per minute, or heart rate, by plugging the figure obtained above into the following equation:

Beats/min = 1/ _______ sec/beat × 60 sec/min

Beats/min = _______

Is the value obtained within normal limits? _______

4. Measure and record the following durations using the data analysis line.

P-R interval _______ sec

QRS interval _______ sec

Q-T interval _______ sec

Are the computed values within normal limits?

____

Which of the leads indicates the greatest R-wave deflection (overall height of the wave)? ________________

Because the QRS complex represents ventricular depolarization, a greater R-wave magnitude in one of the three bipolar leads means that the greater mass of ventricular tissue is oriented in the axis of that lead. The greatest R-wave deflection is usually recorded by lead II, which indicates that the largest electrical potential (and therefore muscle mass) is in the axis leading from the right shoulder to the left foot. (Notice that the electrode placement for lead II is right arm to left leg.) If a greater R-wave deflection is seen in another lead, then the heart must be oriented less typically, i.e., in the direction indicated by that lead.

How could we determine that the heart is positioned more horizontally than normal by looking at a three-lead ECG tracing?

____

____

Compared to the isoelectric line (baseline), we can immediately see if an electrical deflection is positive or negative. What would happen if we switched the polarity of the leads (e.g., positive for negative)?

____

____

Describe the differences seen in the tracings for each lead.

____

____

____

____

____

5. If your instructor requires you to attach a representative part of your ECG tracing, press **P** to print the frame being viewed.

6. Save your data:

- Select Data File/Disk from the Main menu.
- Select Save.
- You should see a yellow blinking cursor near the top of the screen. Press (BACKSPACE) to erase the current location and type a file name according to your instructor's directions.
- Once your data is saved to disk, it may be retrieved anytime for analysis.

7. Press (ESC) to return to the Main menu.

RUNNING IN PLACE RECORDING

1. Select the Experiment menu from the Main menu.

2. Make sure the electrodes are securely attached to the subject to prevent electrode movement while recording the ECG.

3. As before in the Baseline Recording, prepare the subject and the computer for a lead I recording.

4. With the electrodes in place for lead I recording, have the subject run in place for 3 min.

- In the last 30 sec of running, press **S** to start recording ECG data.
- In the third frame of data collection, press **S** again to stop data collection.
- Instruct the subject to sit down and relax.
- Using **S** as a start/stop toggle, record three frames of ECG data at 1 minute after exercise, and again at 4 minutes after exercise.
- Record the frame numbers for later data analysis:

Frames _______ to _______ : running in place

Frames _______ to _______ : 1 minute after exercise

Frames _______ to _______ : 4 minutes after exercise

Do not exit the experiment; continue with the Breath-Holding Recording experiment.

BREATH-HOLDING RECORDING

1. Position the subject comfortably in the sitting position.

2. After the subject has fully rested and relaxed, press **S** to record a few frames and then stop the tracing by pressing **S** again. Record frame numbers:

Frames _______ to _______

3. After approximately 10 seconds, instruct the subject to begin breath-holding and note the time. After another 50 seconds have passed, record for a total of 10 seconds by pressing **S** to start and then to stop the tracing. Record frame numbers:

Frames _______ to _______

At this time, the subjects have been holding their breath for 1 minute; stop the recording and remind the subject to breathe.

4. Save your data as before, using a different file name.

Computerized Data Analysis

1. Return to the Review/Time Analysis screen.

2. Compute the beats/min during the third minute of running, at 1 minute after exercise, and at 4 minutes after exercise. Record below:

_______ beats/min while running in place

_______ beats/min at 1 minute after exercise

_______ beats/min at 4 minutes after exercise

3. Compare these recordings with the previous (baseline) recording from lead I. Which intervals are shorter

in the "running" recording? _______________________

Does the subject's heart rate return to baseline levels by

4 minutes after exercise? _______________________

4. Compute the beats/min during the 1-minute experimental (breath-holding) period.

_______ beats/min during breath-holding

5. Compare this recording with the lead I recording obtained under baseline conditions. What differences are seen?

____

____

6. Attempt to *explain* the physiological reason for the differences you have seen. (Hint: a good place to start might be to check "hypoventilation" or the role of the *respiratory* system in acid-base balance of the blood.)

____

____

____

37i EXERCISE

Respiratory System Physiology

OBJECTIVES

1. To define the following (and be prepared to provide volume figures if applicable):

inspiration	*expiratory reserve volume*
expiration	*expiratory end point*
tidal volume	*inspiratory reserve volume*
vital capacity	*minute respiratory volume*

2. To explain the role of muscles and volume changes in the mechanical process of breathing.
3. To demonstrate the proper usage of the spirometer.
4. To explain the relative importance of various mechanical and chemical factors in producing respiratory variations.
5. To describe bronchial and vesicular breathing sounds.
6. To explain the importance of the carbonic acid–bicarbonate buffer system in maintaining blood pH.

MATERIALS

Model lung (bell jar demonstration)
Tape measure
Spirometer
Disposable mouthpieces
Nose clips
Alcohol swabs
Minimum computer equipment required:
IBM PC/XT/AT or compatible computer
640K RAM—single disk (3.5 in.) drive
Graphics adapter and monitor
Intelitool equipment (Spirocomp)
Software:
Intelitool Spirocomp software
Surface electrodes
75% ethanol solution in battery jar
Pneumograph and recording attachments
Recording apparatus (kymograph or physiograph)
Stethoscope
Paper bag
pH meter (standardized with buffer of pH 7)
Buffer solution (pH 7)
Concentrated HCl and NaOH
0.01 *M* HCl
250- and 50-ml beakers
Graduated cylinder (100 ml)
Glass stirring rod
Plastic wash bottles containing distilled water
Animal plasma
Disposable autoclave bag

See Appendix E, Exercise 37i for links to *Anatomy and PhysioShow: The Videodisc.*

THE MECHANICS OF RESPIRATION

Pulmonary ventilation, or **breathing,** consists of two phases: **inspiration,** during which air is taken into the lungs, and **expiration,** during which air passes out of the lungs. As the inspiratory muscles (external intercostals and diaphragm) contract during inspiration, the size of the thoracic cavity increases. The diaphragm moves from its relaxed dome shape to a flattened position, increasing the superoinferior volume. The external intercostals lift the rib cage, increasing the anteroposterior and lateral dimensions (Figure 37i.1). Since the lungs adhere to the thoracic walls like flypaper because of the cohesive character of the pleurae, the intrapulmonary volume (volume within the lungs) also increases, lowering the air (gas) pressure inside the lungs. The gases then expand to fill the available space, creating a partial vacuum that causes air to flow into the lungs—constituting the act of inspiration. During expiration, the inspiratory muscles relax, and the natural tendency of the elastic lung tissue to recoil acts to decrease intrathoracic and intrapulmonary volumes. As the gas molecules within the lungs are forced closer together, the intrapulmonary pressure rises to a point higher than atmospheric pressure. This causes gases to flow from the lungs to equalize the pressure inside and outside the lungs—the act of expiration.

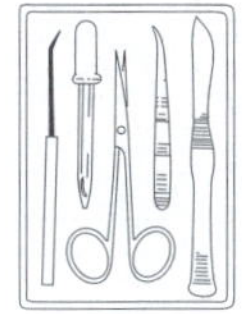

You will be observing the model lung, which demonstrates the principles involved in gas flows into and out of the lungs. It is a simple apparatus with a bottle "thorax," a rubber membrane "diaphragm," and "balloon lungs."

1. Go to the demonstration area and work the model lung by moving the rubber diaphragm up and down. Notice the changes in balloon (lung) size as the volume of the thoracic cavity is alternately increased and decreased.

2. Check the appropriate columns in the chart concerning these observations in the Exercise 37 Review Sheet on page RS 161 (the review sheets precede the appendixes).

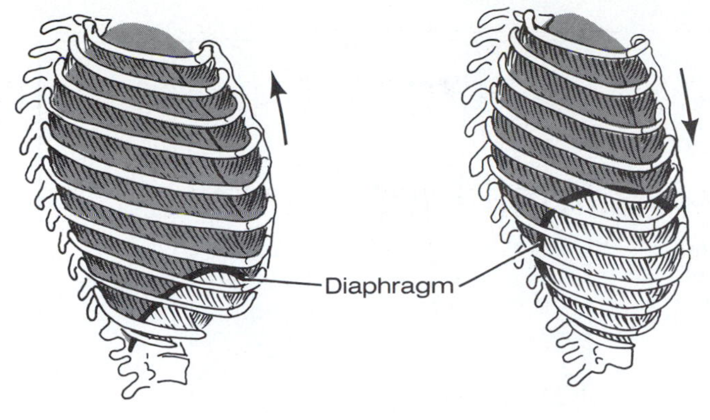

(a) Inspiration (b) Expiration

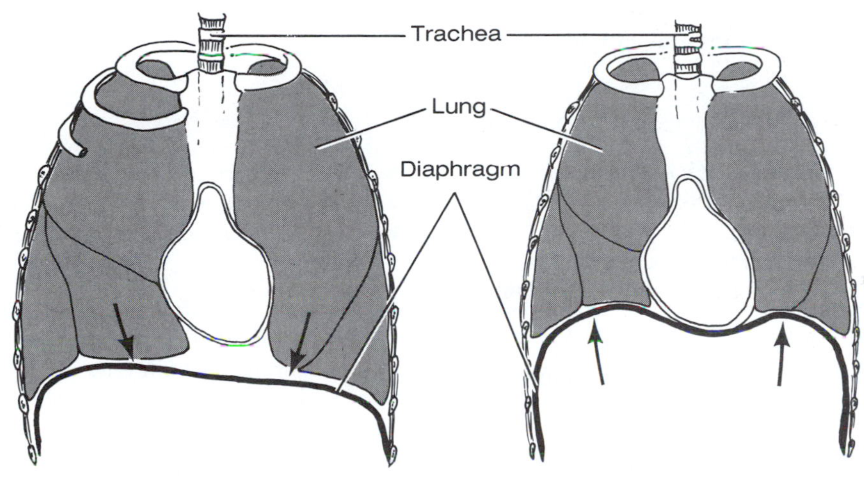

F37i.1

Rib cage and diaphragm positions during breathing. (a) At the end of a normal inspiration; chest expanded, diaphragm depressed. **(b)** At the end of a normal expiration; chest depressed, diaphragm elevated.

3. After observing the operation of the model lung, conduct the following tests on your lab partner. Use the tape measure to determine his or her chest circumference by placing the tape around the chest as high up under the armpits as possible. Record the measurements in inches in the appropriate space for each of the conditions below.

Quiet breathing:

inspiration ______________ expiration ____________

Forced breathing:

inspiration ______________ expiration ____________

Do the results coincide with what you expected on the basis of what you have learned thus far? ____________

RESPIRATORY VOLUMES AND CAPACITIES—SPIROMETRY

A person's size, sex, and physical condition produce variations in respiratory volumes. Normal quiet breathing moves about 500 ml of air in and out of the lungs with each breath. As you have seen in the previous experiment, a person can usually forcibly inhale or exhale much more air than is exchanged in normal quiet breathing. The terms given to the measurable respiratory volumes are defined just below. These terms and their normal values for an adult male should be memorized.

Tidal volume (TV): amount of air inhaled or exhaled with each breath under resting conditions (500 ml)

Inspiratory reserve volume (IRV): amount of air that can be forcefully inhaled after a normal tidal volume inhalation (3100 ml)

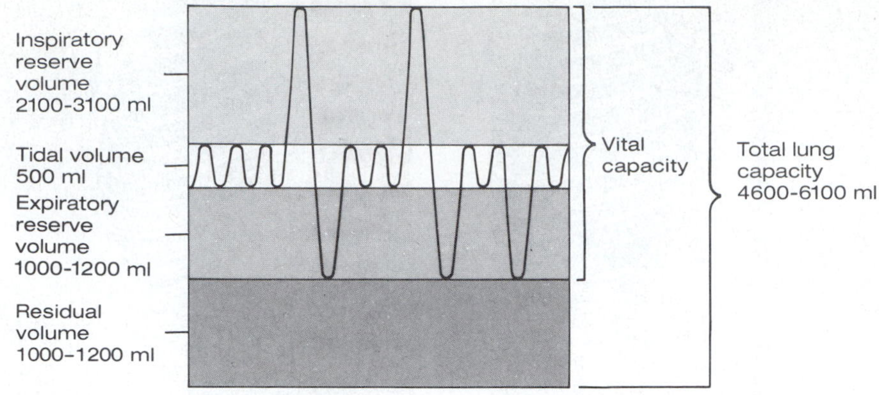

F37i.2

Idealized tracing of the various respiratory volumes.

Expiratory reserve volume (ERV): amount of air that can be forcefully exhaled after a normal tidal volume exhalation (1200 ml)

Vital capacity (VC): maximum amount of air that can be exhaled after a maximal inspiration (4800 ml)

$$VC = TV + IRV + ERV$$

An ideal tracing of the various respiratory volumes and their relationships to each other are shown in Figure 37i.2.

Respiratory volumes will be measured with an apparatus called a spirometer. There are two major types of spirometers, which give comparable results—the hand-held dry, or wheel, spirometers (such as the Buhl spirometer illustrated in Figure 37i.3) and "wet" spirometers such as the Phipps and Bird spirometer. The somewhat more sophisticated wet spirometer consists of a plastic float within a rectangular base; air can be added to or removed from the float (Figure 37i.4). The rectangular base contains water and has a tube running through it to carry air above the water level. The plastic float is a "tub" inverted over the water-containing tank and connected to a volume indicator. The Intelitool equipment connects the wet spirometer to a computer so lung volume recordings can be recorded and analyzed using the included software.

Conducting Spirometer Testing

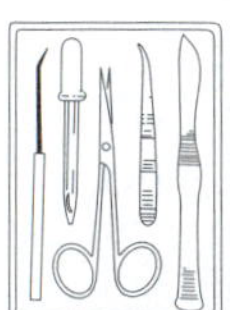

1. Clean the nose clips with an alcohol swab. While you wait for the alcohol to air dry, count and record *your* normal respiratory rate.

Respirations per minute ______

2. Start the software. Type **ITOOL** at the DOS prompt. Select Spirocomp by making sure the highlight (moved with the arrow keys) is on the selection and pressing (ENTER). Then carry out the following sequence of steps.

- Select Experiment menu from the Main menu.
- Select Start a New Group Record.
- Calibrate the transducer for the *first* subject only, as follows.

1. Carefully lift the chain *off* the pulley.
2. While holding the chain, turn the knob on the spirometer pulley until the calibration number on the computer's screen reads close to zero (±100).
3. Lay the chain back on the pulley, being careful that the pulley does not turn. *If the pulley does turn, start the calibration process over.* Note: if the chain slips off the pulley during the experiment, you must recalibrate and redo all records.
4. Press any key to continue with calibration if the computer's screen still reads close to zero (±100).
5. For the last step in calibration, place a disposable mouthpiece in the valve assembly. After the first subject is comfortably seated in a laboratory chair, instruct him or her to inflate the spirometer to *exactly* 5.0 liters by reading the value directly off the spirometer's sliding scale. The subject may use multiple breaths to achieve the 5.0-liter mark. Then take the valve assembly apart to deflate the spirometer.
6. Press any key to begin recording for the first subject. You will be prompted for your subject's initials, height in cm,* age, and gender.
7. The data collection screen is automatically displayed (Figure 37i.5).

* If the subject's height in cm is not known, enter his or her height in inches, followed by a single quote mark, and the program will perform the conversion.

F37i.3

The Buhl Spiropet, an example of a hand-held dry spirometer. The dial face of the spirometer is rotated to zero prior to each test.

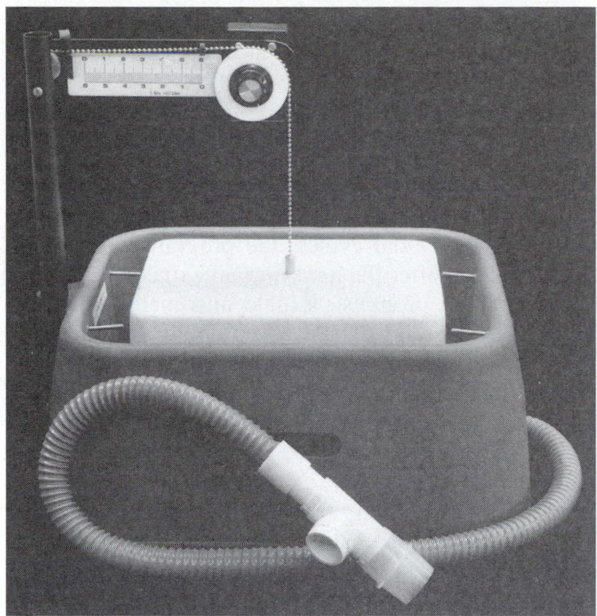

F37i.4

Phipps and Bird "wet" recording spirometer.

- For accurate results:

  1. Instruct the subject to use the nose clip and keep her or his lips sealed tightly around the disposable mouthpiece.
  2. Have the subject look away from lab partners and the computer screen during recording because subconscious alterations in breathing patterns may occur if the subject watches her or his "progress."

3. Since the pressure and temperature inside the spirometer are influenced by room temperature and differ from those in the body, all measured values are to be multiplied by a **BTPS** (body temperature, atmospheric pressure, and water saturation) **factor.** At room temperature, the BTPS factor is typically 1.1 or very close to that value. Hence, measured values are multiplied by the BTPS factor to obtain the corrected respiratory volume values. The Spirocomp software automatically factors in the BTPS factor, so additional calculations are not necessary with computerized spirometry.

4. Repeat the following tests with the first subject until you achieve consistent results.

- Tidal volume (TV). The volume of air inhaled and exhaled with each normal respiration is approximately 500 ml. To conduct the test:

  1. Have the subject inhale a normal breath and exhale a normal breath into the spirometer mouthpiece for several cycles. (Do not force the expiration!)

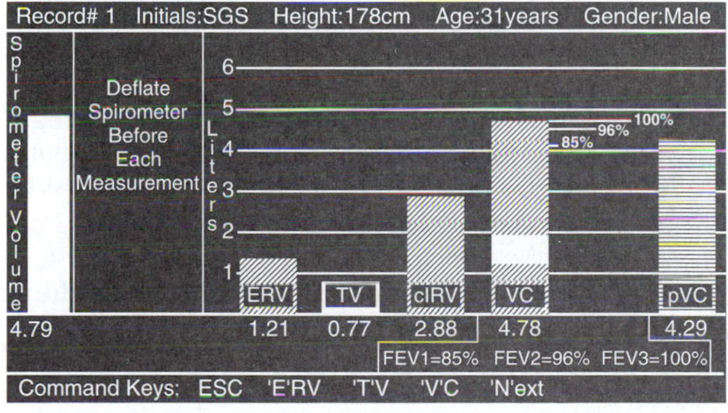

F37i.5

Data collection screen.

  2. Press **T** during an inhalation and follow the instructions on the screen.
  3. When the computer automatically displays a bar graph of the tidal volume, go to the next measurement. If the results are not satisfactory, you may repeat the measurement by pressing **T** again. A common source of error is the psychological effect of "competition," so have the subject breathe normally, as if she or he were on the couch watching television. Empty the spirometer between tests by disconnecting the valve assembly from the spirometer hose.

- Expiratory reserve volume (ERV). The volume of air that can be forcibly exhaled after normal expiration ranges between 1000 and 1200 ml.
  1. Have the subject breathe normally for a cycle or two, then press **E** during an inhalation.
  2. Instruct the subject to follow the directions on the screen. Here is a summary: After measuring several tidal cycles, the program will prompt "Stop after the next normal exhale." When the screen prompts "Exhale maximally," the subject should exhale as completely as possible *without prior inhalation;* this is what you can exhale *above* what you normally exhale. You may repeat the measurement by pressing **E** again. Empty the spirometer for the next reading.

The ERV is dramatically reduced in conditions in which the elasticity of the lungs is decreased by a chronic obstructive pulmonary disease (COPD) such as emphysema. Since energy must be used to deflate the lungs in such conditions, expiration is physically exhausting to individuals suffering from COPD. ■

- Vital capacity (VC). The total exchangeable air of the lungs (the sum of TV + IRV + ERV) is normally 4500 ml to 4800 ml.
  1. Press **V** and follow the screen directions. When prompted, take the deepest possible inspiration you can and then exhale to the greatest extent possible (really *push* the air out).
  2. You may repeat the measurement by pressing **V** again. The program automatically computes and displays the predicted vital capacity (pVC) for your age, height, and sex.

5. Also compute your **minute respiratory volume (MRV)** using the following formula.

$$MRV = TV \times \text{respirations/min}$$

MRV = _______ liters/min

Forced Expiratory Volume (FEV_T) Measurement

While they are not really diagnostic, pulmonary function tests can help the clinician to distinguish between obstructive and restrictive pulmonary diseases. (In obstructive disorders, like chronic bronchitis and asthma, airway resistance is increased, whereas in restrictive diseases such as polio and tuberculosis, total lung capacity declines.) Two highly useful pulmonary function tests for this purpose are the FVC and FEV_T.

The FVC (forced vital capacity) measures the amount of gas expelled when the subject takes the deepest possible breath and then exhales forcefully and rapidly. This volume is reduced in those with restrictive pulmonary disease. The FEV_T (forced expiratory volume) involves the same basic testing procedure, but it looks at the percentage of vital capacity that is exhaled during specific time intervals of the FVC test. FEV_1, for instance, is the amount exhaled during the first second. Healthy individuals can expire about 80% of their FVC in the first second; the FEV_1 is low in those with obstructive disease. FEV_1, FEV_2, FEV_3 may be read directly off the computer screen after completing the vital capacity reading. Normal FEV values are as follows:

FEV_1 75% to 85% of FVC
FEV_2 about 94% of FVC
FEV_3 about 97% of FVC

The recording is finished for this subject. Before continuing with the next member of your group:

1. Dispose of used cardboard mouthpieces in the autoclave bag.
2. Swish the valve assembly in the ethanol solution, then rinse with water.
3. Put a fresh mouthpiece into the valve assembly.

When ready, press **N** for the next subject, and continue recording all members of your group using the procedure above.

Computerized Data Analysis

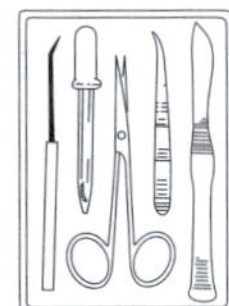

1. From the Main menu, select the Review/Analyze menu, and then Graph Averages.

The Graph Averages screen is similar in appearance to the experiment screen. Refer to Table 37i.1 for how to use the control keys.

TABLE 37i.1 Control Keys for Graph Averages

Key	Function
ESC	Return to the submenu
A	Display the average values for all members of the group
F	Display the average values for all females in the group
M	Display the average values for all males in the group
S	Display a single record
D	Delete record
P	Print screen

2. Compare the average readings for the entire group (press **A**), for females (press **F**), and for males (press **M**) to normal (published) average values. If there are any significant variations, try to explain why (think about smoking, illness, asthma, etc.).

____

____

3. Female values tend to be about 20% to 25% less than male values. Why? ______________________

____

____

4. Bring up your own record by pressing **S** (single record), and typing your initials. Record the values below:

Tidal volume (TV) ________ ml

Expiratory reserve volume (ERV) ________ ml

Calculated inspiratory reserve vol. (cIRV) ________ ml

Vital capacity (VC) ________ ml

Predicted vital capacity (pVC) ________ ml

Forced expiratory volume$_{1second}$ (FEV_1) ________ %

Forced expiratory volume$_{2seconds}$ (FEV_2) ________ %

Forced expiratory volume$_{3seconds}$ (FEV_3) ________ %

5. The normal IRV is substantial, ranging from 2100 ml to 3100 ml. The program calculates the IRV based on the average values obtained for TV, ERV, and VC. How does your cIRV compare to the normal value?

____

6. Compare your vital capacity reading to the predicted vital capacity displayed on your data screen. If there is any significant difference, explain why:

____

____

If you wish to look at the data in a spreadsheet format rather than graphically, press (ESC) to return to the Review/Analyze menu and then select View Spreadsheet.

7. Record respiratory values as requested in question 2 on p. RS 162 for later reference.

8. Save your data:

- Select Data File/Disk from the Main menu.
- Select Save.
- You should see a yellow blinking cursor near the top of the screen. Press (BACKSPACE) to clear the default save location, and type a file name according to your instructor's directions.
- Once your data is saved to disk, it may be retrieved anytime for analysis.

USE OF THE PNEUMOGRAPH TO DETERMINE FACTORS INFLUENCING RATE AND DEPTH OF RESPIRATION*

The neural centers that control respiratory rhythm and maintain a rate of 12 to 18 respirations per minute are located in the medulla and pons. On occasion, input from the stretch receptors in the lungs (via the vagus nerve to the medulla) modifies the respiratory rate, as in the case of extreme overinflation of the lungs (Hering-Breuer reflex).

Death occurs when medullary centers are completely suppressed, as from an overdose of sleeping pills or gross overindulgence in alcohol, and respiration ceases completely. ■

Although these nervous system centers initiate the basic rhythm of breathing, there is no question that physical phenomena such as talking, yawning, coughing, and exercise can modify the rate and depth of respiration. So too can chemical factors, such as changes in oxygen or carbon dioxide concentrations in the blood, or fluctuations in blood pH. Changes in carbon dioxide blood levels seem to act directly on the medullary control centers, whereas changes in pH and oxygen concentrations are monitored by chemoreceptor regions in the aortic and carotid bodies, which in turn send input to the medulla. The experimental sequence in this section is designed to test the relative importance of various physical and chemical factors in the process of respiration.

The pneumograph, an apparatus that records variations in breathing patterns, is the best means of observing respiratory variations resulting from physical and chemical factors. The chest pneumograph is a coiled rubber hose that is attached around the thorax. As the subject breathes, chest movements produce pressure changes within the pneumograph that are transmitted to a recorder.

The instructor will demonstrate the method for setting up the pneumograph and discuss the interpretation of the results. Work in pairs so that one person can mark the record to identify the test for later interpretation. Ideally, the student being tested should face away from the recording apparatus to prevent voluntary modification of the record.

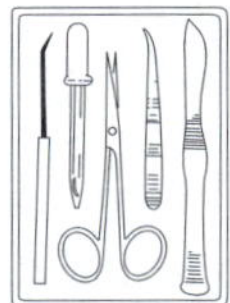

1. Attach the pneumograph tubing firmly, but not restrictively, around the thoracic cage at the level of the sixth rib, leaving room for chest expansion during testing. If the subject is female, position the tubing above the breasts to prevent slippage during testing. Set the pneumograph speed at 1 or 2, and the time signal at 10-second intervals. Record

* Note to the instructor: this exercise may also be done without using the recording apparatus by simply having the students count the respiratory rate visually.

quiet breathing for 1 minute with the subject in a sitting position.

Record breaths per minute: ______________________

2. Make a vital capacity tracing: Record a maximal inhalation followed by a maximal exhalation. This should correlate to the vital capacity measurement obtained earlier and will provide a baseline for comparison during the rest of the pneumograph testing. Stop the recording apparatus testing and indicate the following on the graph by marking the graph appropriately: tidal volume, expiratory reserve volume, inspiratory reserve volume, and vital capacity (the total of the three measurements). Also mark, with arrows, the direction the recording stylus moves during inspiration and expiration.

Measure in millimeters the height of the vital capacity recording. Divide the vital capacity measurement recorded on p. C-29 by the millimeter figure to obtain the volume (in milliliters of air) represented by 1 mm on the recording. For example, if your vital capacity reading is 4000 ml and the vital capacity tracing occupies a vertical distance of 40 mm on the pneumograph recording, then a vertical distance of 1 mm equals 100 ml of air.

Record your computed value: ________ ml air/mm

3. Record the subject's breathing as she or he performs activities from the list below. Make sure the record is marked accurately to identify each test conducted. Record your results on the Exercise 37 Review Sheet.

talking	swallowing water
yawning	coughing
laughing	lying down
standing	running in place

doing a math problem (concentrating)

4. Without recording, have the subject breathe normally for 2 minutes, then inhale deeply and hold her or his breath for as long as possible.

Time the breath-holding interval: ________ sec

As the subject exhales, turn on the recording apparatus and record the recovery period (time to return to normal breathing—usually slightly over 1 minute):

Time the recovery period: ________ sec

Did the subject have the urge to inspire or expire during the breath holding? ______________

Without recording, repeat the above experiment, but this time have the subject exhale completely and forcefully *after* taking the deep breath. What was observed this time?

____

____

Explain the results. (*Hint:* the vagus nerve is the sensory nerve of the lungs and plays a role here.)

____

____

5. Have the subject hyperventilate (breathe deeply and forcefully at the rate of 1 breath/4 sec) for about 30 seconds.* Record both during and after hyperventilation. How does the pattern obtained during hyperventilation compare with that recorded during the vital capacity tracing?

____

Is the respiratory rate recorded after hyperventilation faster *or* slower than during normal quiet breathing?

____

6. Repeat the above test, but do not record until after the hyperventilation. After hyperventilating, the subject is to hold his or her breath as long as possible. Can the breath be held for a longer or shorter period of time after hyperventilation?

____

7. Without recording, have the subject breathe into a paper bag for 3 minutes; then record breathing movements.

⚠ *Caution:* during the bag-breathing exercise the subject's partner should watch the subject carefully for any untoward reactions.

Is the breathing rate faster or slower than the rate recorded during normal quiet breathing?

____

After hyperventilating? ______________________

8. Run in place for 2 minutes: then have your partner determine the length of your breath-holding.

________ sec

9. To prove that respiration has a marked effect on circulation, conduct the following test. Have your lab part-

* A sensation of dizziness may develop. (As the carbon dioxide is washed out of the blood by overventilation, the blood pH increases, leading to a decrease in blood pressure and reduced cerebral circulation.) The subject may experience a lack of desire to breathe after forced breathing is stopped. (If the period of breathing cessation—**apnea**—is extended, cyanosis of the lips may occur.)

ner record the rate and relative force of your radial pulse before you begin.

Rate ________ beats/min Relative force ________

Inspire forcibly. Immediately close your mouth and nose to retain the inhaled air, and then make a forceful and prolonged expiration. Your lab partner should observe and record the condition of the blood vessels of your neck and face, and again immediately palpate the radial pulse.

Observations ____

____

Radial pulse ________ beats/min Relative force ________

Explain the changes observed.

____

____

Dispose of the paper bag in the autoclave bag. Keep the pneumograph records to interpret results and hand them in if requested by the instructor. Observation of the test results should enable you to determine which chemical factor, carbon dioxide or oxygen, has the greatest effect on modifying the respiratory rate and depth.

RESPIRATORY SOUNDS

As air flows in and out of the respiratory tree, it produces two characteristic sounds that can be picked up with a stethoscope. One is **bronchial sounds,** produced by air rushing through the large respiratory passageways (the trachea and bronchi). The second type, **vesicular breathing sounds,** apparently results from air filling the alveolar sacs and resembles the sound of a rustling or muffled breeze.

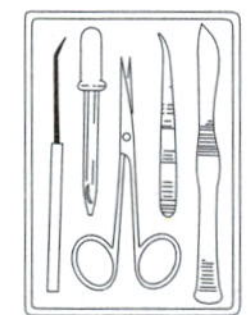

1. Obtain a stethoscope and clean the earpieces with an alcohol swab. Allow the alcohol to dry before donning the stethoscope. Place the diaphragm of the stethoscope on the throat of the test subject just below the larynx. Listen for bronchial sounds on inspiration and expiration.

2. Move the stethoscope over the following chest areas and listen for vesicular sounds during respiration (heard primarily during inspiration).

- At various intercostal spaces
- At the triangle of auscultation (a small depressed area of the back where the muscles fail to cover the rib cage; located just medial to the inferior part of the scapula)
- Under the clavicle

Diseased respiratory tissue, mucus, or pus can produce abnormal chest sounds such as rales (a rasping sound) and wheezing (a whistling sound). ■

ROLE OF THE RESPIRATORY SYSTEM IN ACID-BASE BALANCE OF THE BLOOD

As you have already learned, pulmonary ventilation is necessary for continuous oxygenation of the blood and removal of carbon dioxide (a waste product of cellular respiration) from the blood. Blood pH must be relatively constant for the cells of the body to function optimally. Therefore the carbonic acid–bicarbonate buffer system of the blood is extremely important, because it helps stabilize arterial blood pH at 7.4 ± 0.02.

When carbon dioxide diffuses into the blood from the tissue cells, much of it enters the red blood cells, where it combines with water to form carbonic acid:

$$H_2O + CO_2 \longrightarrow H_2CO_3$$

Some carbonic acid is also formed in the plasma, but its formation there is very slow because of the lack of the carbonic anhydrase enzyme. Shortly after it forms, carbonic acid dissociates to release bicarbonate (HCO_3^-) and hydrogen (H^+) ions. The hydrogen ions that remain in the cells are neutralized, or buffered, when they combine with hemoglobin molecules. If they were not neutralized, the intracellular pH would become very acidic as H^+ ions accumulated. Once formed, the bicarbonate ions diffuse out of the red blood cells into the plasma where they become part of the carbonic acid–bicarbonate buffer system. As HCO_3^- follows its concentration gradient into the plasma, an electrical imbalance develops in the RBCs that draws Cl^- into them from the plasma. This exchange phenomenon is called the *chloride shift.*

Acids (more precisely, H^+) released into the blood by the body cells tend to lower the pH of the blood and cause it to become acidic. On the other hand, basic substances that enter the blood tend to cause the blood to become more alkaline and the pH to rise. Both of these tendencies are resisted in large part by the carbonic acid–bicarbonate buffer system. If H^+ concentration in the blood begins to increase, the H^+ ions combine with bicarbonate ions to form carbonic acid (a weak acid that does not tend to dissociate at physiological or acid pH) and is thus removed.

$$H^+ + HCO_3^- \longrightarrow H_2CO_3$$

Likewise, as blood H^+ concentration drops below what is desirable, and blood pH rises, H_2CO_3 dissociates to release bicarbonate ions and H^+ ions to the blood. The released H^+ lowers the pH again. Bicarbonate ions, being weak bases, are poorly functional under alkaline conditions and have little effect on blood pH unless and until blood pH drops toward acid levels.

$$H_2CO_3 \longrightarrow H^+ + HCO_3$$

In the case of excessively slow or shallow breathing (hypoventilation) or fast deep breathing (hyperventilation), the amount of carbonic acid in the blood can be greatly modified—increasing dramatically during hypoventilation and decreasing substantially during hyperventilation. In either situation, the buffering ability of the blood may be inadequate, and respiratory acidosis or alkalosis can result. Therefore, maintaining the normal rate and depth of breathing is important for proper control of blood pH.

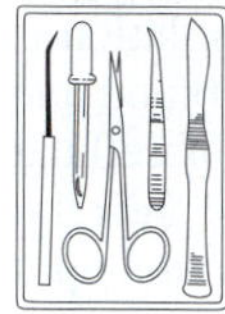

1. To observe the ability of a buffer system to stabilize the pH of a solution, obtain five 250-ml beakers, and a wash bottle containing distilled water. Set up the following experimental samples:

Beaker 1
(150 ml distilled water) pH ______

Beaker 2
(150 ml distilled water and
1 drop concentrated HCl) pH ______

Beaker 3
(150 ml distilled water and
1 drop concentrated NaOH) pH ______

Beaker 4
(150 ml standard buffer solution
[pH 7] and 1 drop concentrated HCl) pH ______

Beaker 5
(150 ml standard buffer solution
[pH 7] and 1 drop concentrated NaOH) pH ______

2. Using a pH meter standardized with a buffer solution of pH 7, determine the pH of the contents of each beaker and record above. After *each and every* pH recording, the pH meter switch should be turned to *standby,* and the electrodes rinsed thoroughly with a stream of distilled water from the wash bottle.

3. Add 3 more drops of concentrated HCl to beaker 4, stir, and record the pH: ______________________

4. Add 3 more drops of concentrated NaOH to beaker 5, stir, and record the pH: ______________________

How successful was the buffer solution in resisting pH changes when a strong acid (HCl) or a strong base (NaOH) was added? ______________________

To observe the ability of the carbonic acid–bicarbonate buffer system of blood to resist pH changes, perform the following simple experiment.

1. Obtain two small beakers (50 ml), animal plasma, graduated cylinder, glass stirring rod, and a dropper bottle of 0.01 *M* HCl. Using the pH meter standardized with the buffer solution of pH 7.0, measure the pH of the animal plasma. Use only enough plasma to allow immersion of the electrodes, and measure the volume used carefully.

pH of the animal plasma: ______________________

2. Add 2 drops of the 0.01 *M* HCl solution to the plasma, stir, and measure pH again.

pH of the plasma plus 2 drops of HCl: ______________________

3. Turn the pH meter switch to standby, rinse the electrodes, and then immerse them in a quantity of distilled water (pH 7) exactly equal to the amount of animal plasma used. Measure the pH of the distilled water.

pH of distilled water: ______________________

4. Add 2 drops of the 0.01 *M* HCl; swirl, and measure pH again.

pH of distilled water plus 2 drops of HCl: ______________________

Is the plasma a good buffer? ______________________

What component of the plasma carbonic acid–bicarbonate buffer system was acting to counteract a change in pH when HCl was added?

____

STUDENT NAME ____________________

LAB TIME/DATE ____________________

Review Sheet

EXERCISE 16i

Muscle Physiology

Muscle Activity

1. Match the following definitions with the proper key letters.

Key: a. agonist b. load c. muscle tone d. isotonic e. isometric f. EMG

__________ 1. state of partial but sustained contraction

__________ 2. recording of electrical activity

__________ 3. causes desired muscle action

__________ 4. resistance to movement

__________ 5. force changes during contraction

__________ 6. length changes during contraction

2. Why do we need a prime mover (agonist) and an antagonist for normal limb movement?

3. Explain the difference between muscle tone and muscle contractions that cause movement.

4. What is muscle tension?

5. Explain how a muscle can produce force but not cause any movement.

Activity	Agonist	Antagonist
Flexing forearm		
Extending forearm		
Flexing the knee		
Extending the knee		
Dorsiflexion		
Plantar flexion		

6. Fill in the chart with the name of the muscle that causes the indicated action.

7. Why do we not see a typical action potential when recording an EMG?

____

____

8. Even at rest, a typical EMG displays a certain amount of muscle electrical activity. Why?

____

____

9. Choose the correct term from the key to correspond to the statement below:

Key: a. recruitment b. spontaneous activity c. action potential

_______ 1. recorded by using an intracellular electrode

_______ 2. increased EMG amplitude (height of "spikes") corresponding to increased force production

_______ 3. low-level muscle electrical activity even at rest

10. Complete the following statements:

Muscles are able to __1__ their length, which explains why we need an __2__ to cause a desired movement at a joint and an __3__ that opposes that action. For example, if we possessed only the biceps brachii muscle, we would only be able to __4__ the forearm. When the prime mover is active, the __5__ is relaxed and often stretched.

Even though we usually think of muscles moving body parts, __6__ occurs without our even realizing that slight contraction is always present. __7__ is generated even when the muscle does not shorten, an activity called __8__ contraction. On the other hand, if we are able to cause movement, we are performing an __9__ contraction.

1. _______________
2. _______________
3. _______________
4. _______________
5. _______________
6. _______________
7. _______________
8. _______________
9. _______________

A.D.A.M. Correlations

Appendix D lists correlations to *A.D.A.M. Standard* or *A.D.A.M. Comprehensive* for appropriate exercises in this manual. When you choose a system from the "View" menu of the main menu bar, the selected system will be highlighted and the rest of the anatomy will be dimmed. You will also see a text overview of the selected system in the right corner. When in "Full Anatomy" in *A.D.A.M. Standard* or *A.D.A.M. Comprehensive,* "View" refers to the "Anterior," "Lateral," "Posterior," or "Medial" view in the primary window. "Layer" refers to one or more of the layers listed in the lower right corner of the computer screen. Note that the layer name in *A.D.A.M. Standard* or *A.D.A.M. Comprehensive* may be different from the key structure you are studying, but using the indicated layer will bring the structure into full view.

Exercise 1
The Language of Anatomy

Go to "Full Anatomy" to illustrate a coronal plane
View: Anterior
Layer: Skull—coronal section

Go to "Full Anatomy" to illustrate a midsagittal plane
View: Medial
Layer: Skin

Go to the "Cross Sections" book to illustrate a transverse plane
Choose the eye button for the location of the cross-sectional illustration or MRI image of your choice*

Go to "Full Anatomy"
View: Medial
Layer: Central nervous system and sacral plexus
Key Structure: Dorsal body cavity

Go to "Full Anatomy"
View: Anterior
Layer: Diaphragm
Key Structure: Ventral body cavity

Go to "Full Anatomy"
View: Anterior
Layers: Pericardial sac, Parietal peritoneum, Peritoneum
Key Structures: Serous membranes of the ventral body cavity

Go to "Full Anatomy"
View: Medial
Layer: Parietal pleura
Key Structure: Parietal pleura

Exercise 2
Organ Systems Overview

Go to "View" on the Menu Bar
Choose each body system and review text and illustrations.

Exercise 8
Classification of Body Membranes

Go to "Full Anatomy"
View: Anterior
Layer: Lungs
Key Structures: Parietal pleural membranes

Go to "Full Anatomy"
View: Medial
Layer: Parietal pleura
Key Structures: Peritoneal membranes

Exercise 9
Overview of the Skeleton

Go to "View" on Menu Bar
Choose "Skeletal System"

Exercise 10
The Axial Skeleton

Go to "View," choose "Skeletal System"

Go to "Full Anatomy"
View: Anterior
Layer: Skull
Key Structures: Bones of the cranium

Go to "Full Anatomy"
View: Posterior
Layer: Bones—coronal section
Key Structures: Bones of the cranium

Go to "Full Anatomy"
View: Lateral
Layer: Skull
Key Structures: Bones of the cranium and face

Go to "Full Anatomy"
View: Anterior
Layer: Skull
Key Structures: Bones of the face

Go to "Full Anatomy"
View: Lateral
Layer: Hyoid bone
Key Structure: Hyoid bone

Go to "Full Anatomy"
View: Anterior
Layer: Skull—coronal section
Key Structures: Paranasal sinuses

Go to "Full Anatomy"
View: Anterior
Layer: Atlas
Key Structure: Vertebral column

Go to "Full Anatomy"
View: Posterior
Layer: Posterior vertebral column
Key Structure: Posterior vertebral column

Go to "Full Anatomy"
View: Medial
Layer: Central nervous system and sacral plexus
Key Structures: Vertebral column and intervertebral discs

Go to "Full Anatomy"
View: Posterior
Layer: Bones—coronal section
Key Structures: Sacrum and lumbar vertebrae

Go to "Full Anatomy"
View: Anterior
Layer: Ribs
Key Structure: Bony thorax

*Open the Cross Sections book in the Library window, and then choose the eye button as requested. "Level 1" refers to the cross section at the top of the image and "Level 7" refers to the bottom cross section.

Exercise 11 The Appendicular Skeleton

Go to "View," choose "Skeletal System"

Go to "Full Anatomy"
View: Anterior
Layer: Clavicle
Key Structure: Pectoral girdle

Go to "Full Anatomy"
View: Posterior
Layer: Scapula
Key Structure: Pectoral girdle

Go to "Full Anatomy"
View: Anterior
Layer: Humerus
Key Structure: Humerus

Go to "Full Anatomy"
View: Anterior
Layer: Ulna
Key Structure: Humerus

Go to "Full Anatomy"
View: Anterior
Layer: Bones—coronal section
Key Structures: Bones of the wrist and hand

Go to "Full Anatomy"
View: Anterior
Layer: Femur
Key Structures: Bones of the pelvic girdle

Using "Options," choose "Female"
Go to "Full Anatomy"
View: Anterior
Layer: Bones—coronal section
Key Structure: Female pelvis

Using "Options," choose "Male"
Go to "Full Anatomy"
View: Anterior
Layer: Bones—coronal section
Key Structure: Male pelvis

Go to "Full Anatomy"
View: Anterior
Layer: Femur
Key Structure: Femur

Go to "Full Anatomy"
View: Anterior
Layer: Tibia
Key Structure: Tibia

Go to "Full Anatomy"
View: Anterior
Layer: Bones of the foot
Key Structures: Bones of the foot

Exercise 13 Articulations and Body Movements

Go to "Full Anatomy"
View: Lateral
Layer: Synovial joint of the knee
Key Structure: Synovial joint of the knee

Go to "Full Anatomy"
View: Anterior
Layer: Synovial joint capsule of the hip
Key Structure: Synovial joint capsule of the hip

Exercise 15 Gross Anatomy of the Muscular System

Go to "Full Anatomy"
View: Anterior
Layer: Orbicularis oculi
Key Structures: Muscles of the face

Go to "Full Anatomy"
View: Anterior
Layer: Masseter muscle
Key Structures: Muscles of mastication

Go to "Full Anatomy"
View: Lateral
Layers: Platysma muscle, Sternocleidomastoid muscle
Key Structures: Superficial muscles of the neck

Go to "Full Anatomy"
View: Lateral
Layer: Strap muscles
Key Structures: Deep muscles of the neck

Go to "Full Anatomy"
View: Lateral
Layers: Pectoralis major muscle, Serratus anterior muscle
Key Structures: Thorax and shoulder muscles

Go to "Full Anatomy"
View: Anterior
Layers: External intercostal muscles, Internal intercostal muscle
Key Structures: Thorax muscles

Go to "Full Anatomy"
View: Anterior
Layers: Rectus abdominis muscle, External abdominal oblique muscle, Internal abdominal oblique muscle
Key Structures: Abdominal wall muscles

Go to "Full Anatomy"
View: Lateral
Layer: Latissimus dorsi muscle
Key Structures: Thorax muscles

Go to "Full Anatomy"
View: Posterior
Layer: Rhomboideus muscle
Key Structures: Posterior muscles of the trunk

Go to "Full Anatomy"
View: Posterior
Layers: Semispinalis muscle, Splenius muscle
Key Structures: Muscles associated with the vertebral column

Go to "Full Anatomy"
View: Posterior
Layer: Brachialis muscle
Key Structures: Muscles of the humerus that act on the forearm

Go to "Full Anatomy"
View: Anterior
Layers: Flexor carpi radialis muscle, Palmaris longus muscle, Flexor carpi ulnaris muscle, Flexor digitorum superficialis muscle
Key Structures: Anterior muscles of the forearm that act on the hand and fingers

Go to "Full Anatomy"
View: Anterior
Layer: Abductor pollicis longus muscle
Key Structures: Deep muscles of the forearm that act on the hand and fingers

Go to "Full Anatomy"
View: Anterior
Layers: Sartorius muscle
Key Structures: Muscles acting on the thigh

Go to "Full Anatomy"
View: Anterior
Layers: Rectus femoris muscle, Vastus lateralis muscle, Vastus medialis muscle, Vastus intermedius muscle, Tensor fasciae latae
Key Structures: Quadriceps

Go to "Full Anatomy"
View: Posterior
Layers: Gluteus maximus muscle, Gluteus medius muscle, Gluteus minimus muscle
Key Structures: Muscles acting on the thigh and originating on the pelvis

Go to "Full Anatomy"
View: Posterior
Layers: Long head of the biceps femoris muscle, Semitendinosus muscle, Semimembranosus muscle
Key Structures: Hamstrings

Go to "Full Anatomy"
View: Posterior
Layers: Gastrocnemius muscle, Soleus muscle, Popliteus muscle, Tibialis posterior
Key Structures: Superficial muscles acting on the foot and ankle

Go to "Full Anatomy"
View: Anterior
Layers: Tibialis anterior muscle, Extensor digitorum longus muscle, Extensor hallucis longus muscle
Key Structures: Muscles acting on the foot and ankle

Exercise 19 Gross Anatomy of the Brain and Cranial Nerves

Go to "View," choose "Nervous System"

Go to "Full Anatomy"
View: Anterior
Layer: Skull—coronal section
Key Structure: Cerebrum

Go to "Full Anatomy"
View: Lateral
Layer: Brain
Key Structure: Cerebrum

Go to the "Cross Sections" book, choose Level 1 (in the cerebral region)*
Key Structure: Cerebrum

Go to "Full Anatomy"
View: Medial
Layer: Central nervous system and sacral plexus
Key Structures: Diencephalon and pituitary gland

Go to the "Cross Sections" book, choose Level 2*
Key Structures: Cerebellum, cerebrum, and cranial nerves V and VI

Go to "Full Anatomy"
View: Lateral
Layer: Falx cerebri
Key Structures: Internal anatomy of the cerebrum and meningeal extensions

Go to the "Cross Sections" book, choose Level 1 (in the cerebral region)
Key Structures: Internal anatomy of the cerebrum

Go to "Full Anatomy"
View: Lateral
Layer: Cranial nerves
Key Structures: Cranial nerves

Go to "Full Anatomy"
View: Lateral
Layer: Right vagus nerve
Key Structure: Vagus nerve

Go to "Full Anatomy"
View: Anterior
Layer: Phrenic and vagus nerves
Key Structures: Phrenic and vagus nerves

Exercise 21 Spinal Cord, Spinal Nerves, and the Autonomic Nervous System

Go to "View," choose "Nervous System"

Go to "Full Anatomy"
View: Posterior
Layer: Spinal Cord
Key Structure: Spinal cord

Go to the "Cross Sections" book, choose Level 3 (in the cervical region)
Key Structure: Spinal cord

Go to "Full Anatomy"
View: Anterior
Layer: Deep nerve plexuses and intercostal nerves
Key Structures: Deep nerve plexuses and intercostal nerves

Go to "Full Anatomy"
View: Anterior
Layer: Brachial plexus and branches
Key Structures: Brachial plexus and branches

Go to "Full Anatomy"
View: Medial
Layer: Central nervous system and sacral plexus
Key Structures: Central nervous system and sacral plexus

Go to "Full Anatomy"
View: Medial
Layer: Sciatic nerve and branches
Key Structures: Sciatic nerve and branches

Go to "Full Anatomy"
View: Posterior
Layer: Spinal cord
Key Structure: Spinal cord

Go to "Full Anatomy"
View: Posterior
Layer: Tibial nerve and branches
Key Structure: Tibial nerve

Go to "Full Anatomy"
View: Medial
Layer: Autonomic nerve plexuses
Key Structures: Autonomic nerve plexuses

Go to "Full Anatomy"
View: Anterior
Layer: Sympathetic trunk
Key Structure: Sympathetic trunk

Exercise 24 Special Senses: Vision

Go to "Full Anatomy"
View: Anterior
Layer: Muscles of the eye
Key Structures: Muscles and external anatomy of the eye

Go to "Full Anatomy"
View: Lateral
Layer: Eye muscles
Key Structures: Eye muscles

Go to the "Cross Sections" book, choose the eye button at Level 2 (in the cephalic region)*
Key Structure: Eye

*Open the Cross Sections book in the Library window, and then choose the eye button as requested. "Level 1" refers to the cross section at the top of the image and "Level 7" refers to the bottom cross section.

Exercise 27 Anatomy and Basic Function of the Endocrine Glands

Go to "Options," choose "Female"
Go to "View," choose "Endocrine System"

Go to "Options," choose "Male"
Go to "View," choose "Endocrine System"

Go to "Full Anatomy"
View: Lateral
Layer: Tongue
Key Structure: Pituitary gland

Go to "Full Anatomy"
View: Anterior
Layer: Thyroid gland
Key Structure: Thyroid gland

Go to "Full Anatomy"
View: Anterior
Layer: Suprarenal gland
Key Structure: Suprarenal gland

Go to "Full Anatomy"
View: Anterior
Layer: Pancreas
Key Structure: Pancreas

Go to "Options," choose "Female"
View: Medial
Layer: Ovarian artery
Key Structure: Ovaries

Go to "Options," choose "Male"
View: Anterior
Layer: Testes
Key Structure: Testes

Go to "Full Anatomy"
View: Anterior
Layer: Remnant of thymus gland
Key Structure: Remnant of thymus gland

Go to "Full Anatomy"
View: Anterior
Layer: Testes
Key Structure: Testes

Exercise 30 Anatomy of the Heart

Go to "Full Anatomy"
View: Anterior
Layer: Heart
Key Structure: Heart

Go to "Full Anatomy"
View: Anterior
Layer: Pericardiacophrenic vein
Key Structure: Fibrous pericardium

Go to "Full Anatomy"
View: Anterior
Layer: Epicardium
Key Structure: Visceral pericardium

Go to "Full Anatomy"
View: Anterior
Layer: Heart—cut section
Key Structures: Heart chambers and heart valves

Go to "Full Anatomy"
View: Anterior
Layer: Coronary arteries
Key Structure: Coronary arteries

Exercise 32 Anatomy of Blood Vessels

Go to "View," choose "Cardiovascular System"

Go to "Full Anatomy"
View: Anterior
Layer: Aortic arch and branches
Key Structures: Aortic arch and branches

Go to "Full Anatomy"
View: Anterior
Layer: Common carotid artery and branches
Key Structures: Branches of the aortic arch

Go to "Full Anatomy"
View: Anterior
Layer: Celiac trunk and branches
Key Structures: Major branches of the descending aorta

Go to "Full Anatomy"
View: Anterior
Layer: Descending thoracic aorta
Key Structure: Descending thoracic aorta

Go to "Full Anatomy"
View: Anterior
Layer: Abdominal aorta and branches
Key Structures: Abdominal aorta and branches

Go to "Options," choose "Male"
View: Anterior
Layer: Gonadal veins
Key Structures: Abdominal aorta and branches

Go to "Full Anatomy"
View: Lateral
Layer: Aortic arch and branches
Key Structures: Lateral of the aorta and vena cava

Go to "Full Anatomy"
View: Anterior
Layer: Superior vena cava and tributaries
Key Structures: Veins draining into the vena cava

Go to "Full Anatomy"
View: Anterior
Layer: Pulmonary arteries
Key Structure: Pulmonary circulation

Go to "Full Anatomy"
View: Lateral
Layer: Aortic arch and branches
Key Structure: Arterial supply of the brain

Go to "Full Anatomy"
View: Anterior
Layer: Portal vein and tributaries
Key Structure: Hepatic portal circulation

Exercise 33 Human Cardiovascular Physiology—Blood Pressure and Pulse Determination

Go to "Full Anatomy"
View: Anterior
Layers: Ribs, Cardiac veins, Heart—cut section
Key Structure: Position of the heart and valves in the thoracic cavity

Go to "Full Anatomy"
View: Anterior
Layers: Radial artery, Deep femoral artery
Key Structure: Pulse points

Exercise 35 The Lymphatic System and Immune Response

Go to "View," choose "Immune System"
Key Structures: Lymphatic vessels and lymphoid organs

Go to "View," choose "Lymphatic System"
Key Structures: Lymphatic vessels and lymph nodes

Go to "Full Anatomy"
View: Anterior
Layer: Thoracic duct
Key Structure: Thoracic duct

Go to "Full Anatomy"
View: Anterior
Layer: Axillary and cervical lymph nodes
Key Structures: Axillary and cervical lymph nodes

Go to "Full Anatomy"
View: Anterior
Layer: Remnant of thymus gland
Key Structure: Remnant of thymus gland
(View histology also)*

Go to "Full Anatomy"
View: Anterior
Layer: Spleen
Key Structure: Spleen
(View histology also)*

Go to "Full Anatomy"
View: Anterior
Layer: Palatine glands and tonsils
Key Structures: Palatine glands and tonsils
(View histology also)*

Go to "Full Anatomy"
View: Anterior
Layer: Axillary and cervical lymph nodes
Key Structures: Axillary and cervical lymph nodes
(View histology also)*

Exercise 36 Anatomy of the Respiratory System

Go to "View," choose "Respiratory System"

Go to "Full Anatomy"
View: Medial
Layer: Nasal conchae
Key Structure: Upper respiratory tract

Go to "Full Anatomy"
View: Lateral
Layer: Mediastinal parietal pleura of left pleural cavity
Key Structure: Upper respiratory tract

Go to "Full Anatomy"
View: Anterior
Layer: Thyroid gland
Key Structure: Larynx

Go to "Full Anatomy"
View: Anterior
Layer: Epiglottis
Key Structure: Epiglottis

Go to "Full Anatomy"
View: Medial
Layer: Structures in the hilum of the right lung
Key Structure: Lower respiratory structures

Go to "Full Anatomy"
View: Anterior
Layer: Trachea
Key Structure: Lower respiratory structures

Go to "Full Anatomy"
View: Anterior
Layer: Lungs
Key Structures: Lungs

Go to "Full Anatomy"
View: Anterior
Layer: Lungs—coronal section
Key Structure: Lungs—coronal section

Go to "Full Anatomy"
View: Lateral
Layer: Right lung
Key Structure: Right lung

Go to "Full Anatomy"
View: Lateral
Layer: Parietal pleura
Key Structure: Parietal pleura

Go to "Full Anatomy"
View: Anterior
Layer: Diaphragm
Key Structure: Diaphragm

Go to the "Cross sections" book, choose Level 5 and Level 6
Key Structure: Lungs

Go to "Full Anatomy"
View: Lateral
Layer: Trachea
Key Structure: Trachea
(View histology also)*

Exercise 38 Anatomy of the Digestive System

Go to "View," choose "Digestive System"

Go to "Full Anatomy"
View: Lateral
Layer: Esophagus
Key Structure: Oral cavity

Go to "Full Anatomy"
View: Anterior
Layer: Oral mucosa
Key Structures: Tonsils

Go to "Full Anatomy"
View: Lateral
Layers: Parotid gland and duct, Deep salivary glands
Key Structures: Salivary glands

Go to "Full Anatomy"
View: Lateral
Layer: Esophagus
Key Structure: Esophagus
(View histology also)*

Go to "Full Anatomy"
View: Anterior
Layers: Stomach, Stomach—coronal section
Key Structure: Stomach

Go to "Full Anatomy"
View: Medial
Layer: Stomach
Key Structure: Stomach

Go to "Full Anatomy"
View: Lateral
Layer: Esophagus
Key Structure: Esophagus

*In "Full Anatomy", select the structure with the Identify tool, and then select "Histology" from the pop-up submenu.

Go to "Full Anatomy"
View: Anterior
Layer: Small intestine
Key Structure: Ileum

Go to "Full Anatomy"
View: Anterior
Layer: Duodenum
Key Structure: Duodenum

Go to "Full Anatomy"
View: Anterior
Layer: Transverse colon
Key Structure: Transverse colon

Go to "Full Anatomy"
View: Anterior
Layer: Ascending and descending colon
Key Structure: Ascending and descending colon

Go to "Full Anatomy"
View: Medial
Layer: Colon
Key Structure: Colon

Go to "Full Anatomy"
View: Anterior
Layer: Skull
Key Structures: Teeth

Go to "Full Anatomy"
View: Anterior
Layers: Common bile duct, Liver, Liver—coronal section, Gallbladder
Key Structures: Common bile duct, liver, gallbladder

Go to "Full Anatomy"
View: Anterior
Layer: Pancreas
Key Structure: Pancreas
(View histology also)*

Exercise 40 Anatomy of the Urinary System

Go to "Options," choose "Female"
Go to "View," choose "Urinary System"

Go to "Options," choose "Male"
Go to "View" on the Menu Bar
Choose Urinary System

Go to "Full Anatomy"
View: Anterior
Layers: Kidney, Kidney—longitudinal section
Key Structure: Kidneys

Go to "Full Anatomy"
View: Anterior
Layer: Ureter
Key Structure: Ureter

Go to "Full Anatomy"
View: Anterior
Layer: Urinary bladder
Key Structure: Urinary bladder

Go to "Options," choose "Female"
Go to "Full Anatomy"
View: Medial
Layer: Urinary bladder
Key Structure: Urethra

Go to "Options," choose "Male"
View: Medial
Layer: Urinary bladder and prostate gland
Key Structure: Urethra

Exercise 42 Anatomy of the Reproductive System

Go to "Options," choose "Male"
Go to "View," choose "Reproductive System"

Go to "Options," choose "Male"
View: Medial
Layer: Urinary bladder and prostate gland
Key Structure: Urethra

Go to "Options," choose "Male"
View: Anterior
Layer: Testes
Key Structure: Testes
(View histology also)*

Go to "Options," choose "Male"
View: Anterior
Layer: Penis
Key Structure: Penis

Go to "Options," choose "Female"
Go to "View," choose "Reproductive System"

Go to "Options," choose "Female"
View: Medial
Layer: Uterus
Key Structure: Uterus

Go to "Options," choose "Female"
View: Anterior
Layer: Uterus
Key Structure: Uterus
(View histology also)*

Go to "Options," choose "Female"
View: Anterior
Layer: Breast
Key Structure: Mammary glands

*In "Full Anatomy", select the structure with the Identify tool, and then select "Histology" from the pop-up submenu.

APPENDIX E

Videodisc Correlations

Appendix E includes barcodes for images from *Anatomy and PhysioShow: The Videodisc* for *Human Anatomy and Physiology,* third edition by Elaine Marieb for appropriate exercises in this lab manual. The barcodes are listed in the order that topics are presented in each exercise.

Exercise 1 The Language of Anatomy

Planes of the body

Search frame 37–39

Dorsal and ventral body cavities and their subdivisions

Search frame 40–42

Serous membrane relationships

Search frame 4428–4431

Exercise 4 The Cell: Anatomy and Division

Structure of the generalized cell

Search frame 5970

The endoplasmic reticulum

Search frame 6310

Role of the Golgi apparatus in packaging proteins for cellular use and for secretion

Search frame 6315

Cytoskeleton

Search frame 6319

Cilia structure and function

Search frame 7810–7811

The cell cycle

Search frame 8983

The stages of mitosis

Search frame 9444

Human blood smear

Search frame 53018

Simple squamous epithelium

Search frame 53024

Exercise 5 The Cell: Transport Mechanisms and Cell Permeability

Simple diffusion

Search frame 5978

The effect of solutions of varying tonicities on living red blood cells

Search frame 6299–6300

Three types of endocytosis

Search frame 6304–6306

Exercise 6 Classification of Tissues

Epithelial tissue: simple squamous epithelium

Search frame 14022–14043

Areolar connective tissue: a prototype (model) connective tissue

Search frame 14047

Connective tissues: embryonic connective tissue

Search frame 14049–14069

Muscle tissues: skeletal muscle

Search frame 14074–14078

Nervous tissue

Search frame 14080

Alveoli, alveolar ducts and sacs

Search frame 53000

Cross section of the trachea

Search frame 53001

Simple squamous epithelium

Search frame 53024

Simple cuboidal epithelium

Search frame 53025

Simple columnar epithelium

Search frame 53026

Pseudostratified epithelium

Search frame 53027

Transitional epithelium

Search frame 53028

Loose areolar connective tissue

Search frame 53029

Reticular tissue of the spleen

Search frame 53030

Adipose tissue

Search frame 53031

Dense regular connective tissue

Search frame 53032

Dense irregular connective tissue

Search frame 53033

Hyaline cartilage

Search frame 53034

Elastic cartilage

Search frame 53035

Fibrocartilage

Search frame 53036

Skeletal muscle

Search frame 53037

Visceral muscle

Search frame 53038

Thin section of cardiac muscle

Search frame 53039

Skeletal muscle

Search frame 52977

Multipolar neuron

Search frame 52979

Exercise 7 The Integumentary System

Skin structure

Search frame 14089

Diagram showing the main features—layers and relative distribution of the different cell types—in epidermis of thin skin

Search frame 14092

Structure of a hair and hair follicle

Search frame 14093

Scalp tissue

Search frame 53040

Exercise 8 Classification of Body Membranes

Classes of epithelial membranes

Search frame 14071–14073

Goblet cells, example of unicellular exocrine glands

Search frame 14046

General structure of a synovial joint

Search frame 14240

Exercise 9 Overview of the Skeleton

The human skeleton

Search frame 14128–14132

The structure of a long bone (humerus of arm)

Search frame 14104

Microscopic structure of compact bone

Search frame 14107

Connective tissues: others

Search frame 14067

Cartilage in the adult skeleton and body

Search frame 14101

Connective tissues: cartilage

Search frame 14063

Epiphyseal plate of a bone

Search frame 53041

Hyaline cartilage

Search frame 53034

Elastic cartilage

Search frame 53035

Fibrocartilage

Search frame 53036

Exercise 10 The Axial Skeleton

Anatomy of the anterior and posterior aspects of the skull

Search frame 14134–14137

Anatomy of the lateral aspects of the skull

Search frame 14139–14142

Anatomy of the inferior portion of the skull

Search frame 14145–14151

Detailed anatomy of some isolated facial bones

Search frame 14152–14156

Special anatomical characteristics of the orbits

Search frame 14158

Special anatomical characteristics of the nasal cavity

Search frame 14163–14164

Paranasal sinuses

Search frame 14165–14166

The vertebral column

Search frame 14167

Structure of a typical vertebra

Search frame 14170

The first and second cervical vertebrae

Search frame14171–14173

Posterolateral views of articulated vertebrae

Search frame 14174–14176

The sacrum and coccyx

Search frame 14177–14178

The bony thorax

Search frame 14179–14181

Exercise 11 The Appendicular Skeleton

The human skeleton

Search frame 14128–14132

Bones of the pectoral girdle

Search frame 14182

The humerus of the arm

Search frame 14188–14191

Bones of the forearm

Search frame 14194–14197

Bones of the hand

Search frame 14200–14201

Bones of the pelvic girdle

Search frame 14202–14204

Bones of the right thigh and knee

Search frame 14205–14210

The tibia and fibula of the right leg

Search frame 14211–14216

Bones of the right foot

Search frame 14217–14218

Exercise 12 Fetal Skeleton

Differences in the growth rates for some parts of the body compared to others determine body proportions

Search frame 14221

Exercise 13 Articulations and Body Movements

Fibrous joints

Search frame 14235–14236

Cartilaginous joints

Search frame 14237–14239

General structure of a synovial joint

Search frame 14240

Types of synovial joints

Search frame 14242–14245

Knee joint relationships

Search frame 14248–14251

Photograph of a hand deformed by rheumatoid arthritis

Search frame 14253

Exercise 14 Microscopic Anatomy, Organization, and Classification of Skeletal Muscle

Connective tissue wrappings of skeletal muscle

Search frame 14256

Microscopic anatomy of a skeletal muscle fiber

Search frame 14257

Myofilament composition in skeletal muscle

Search frame 14261–14264

Relationship of the sarcoplasmic reticulum and T tubules to the myofibrils of skeletal muscle

Search frame 14265

The neuromuscular junction

Search frame 16254

Skeletal muscle

Search frame 52977

Part of a motor unit

Search frame 52978

Skeletal muscle

Search frame 53037

Exercise 15 Gross Anatomy of the Muscular System

Lateral view of muscles of the scalp, face, and neck

Search frame 21315

Muscles promoting mastication and tongue movement

Search frame 21322–21323

Muscles of the anterior neck and throat that promote swallowing

Search frame 21326–21329

Muscles of the neck and vertebral column causing movements of the head and trunk

Search frame 21331–21335

Superficial muscles of the thorax and shoulder acting on the scapula and arm

Search frame 21347–21349

Muscles crossing the shoulder and elbow joint, respectively causing movements of the arm and forearm

Search frame 21351–21355

Muscles of the abdominal wall

Search frame 21336–21340

Muscles of the anterior fascial compartment of the forearm acting on the wrist and fingers

Search frame 21356

Muscles of the posterior fascial compartment of the forearm acting on the wrist and fingers

Search frame 21362

Anterior and medial muscles promoting movements of the thigh and leg

Search frame 21366–21370

Posterior muscles of the right hip and thigh

Search frame 21371–21374

Muscles of the anterior compartment of the right leg

Search frame 21375–21379

Muscles of the lateral compartment of the right leg

Search frame 21380–21383

Muscles of the posterior compartment of the right leg

Search frame 21384–21392

Exercise 16A or 16i Muscle Physiology

Summary of events in the generation and propagation of an action potential in a skeletal muscle fiber

Search frame 16257–16260

Relationship of the sarcoplasmic reticulum and T tubules to the myofibrils of skeletal muscle

Search frame 14265

Skeletal muscle

Search frame 52977

Exercise 17 Histology of Nervous Tissue

Supporting cells of nervous tissue

Search frame 21395–21399

Structure of a motor neuron

Search frame 22301

The neuromuscular junction

Search frame 16254

Relationship of Schwann cells to axons in the peripheral nervous system

Search frame 22303–22306

Structure of a nerve

Search frame 27597

Teased myelinated axons

Search frame 52981

Multipolar neuron

Search frame 52979

Neurons differentially stained

Search frame 52980

Exercise 18 Neurophysiology of Nerve Impulses

The mechanisms of a graded potential

Search frame 25097–25098

The phases of the action potential and the role of gated ion channels in those phases

Search frame 25099

Exercise 19 Gross Anatomy of the Brain and Cranial Nerves

Embryonic development of the human brain

Search frame 27536

Three-dimensional views of the ventricles of the brain

Search frame 27540–27541

Regions of the brain

Search frame 27539

Lobes and fissures of the cerebral hemispheres

Search frame 27542

Functional areas of the left cerebral cortex

Search frame 27549

Diencephalon and brain stem structures

Search frame 27554

Basal nuclei

Search frame 27552–27553

Meninges of the brain

Search frame 27560

Formation, location, and circulation of CSF

Search frame 27563–27564

Summary of location and function of the cranial nerves

Search frame 27602

Exercise 21 Spinal Cord, Spinal Nerves, and the Autonomic Nervous System

Anatomy of the spinal cord

Search frame 27567–27569

Major ascending (sensory) and descending (motor) tracts of the spinal cord

Search frame 27573

Formation and branches of spinal nerves

Search frame 27606–27607

The cervical plexus

Search frame 27610

The brachial plexus

Search frame 27612–27616

The lumbar plexus

Search frame 27619–27622

The sacral plexus

Search frame 27625–27627

Comparison of the somatic and autonomic nervous systems

Search frame 27646

Sympathetic pathways

Search frame 27649–27651

Exercise 22 or 22i Human Reflex Physiology

The basic components of all human reflex arcs

Search frame 27631

Exercise 23 General Sensation

General sensory receptors classified by structure and function

Search frame 27588

Exercise 24 Special Senses: Vision

Accessory structures of the eye

Search frame 27668–27670

Extrinsic muscles of the eye

Search frame 27671–27672

Internal structure of the eye

Search frame 27673–27677

Schematic view of the major three neuronal populations composing the neural layer of the retina

Search frame 27678

Photoreceptors of the retina

Search frame 27682

Visual fields of the eyes and visual pathway to the brain

Search frame 27686

Focusing for distant and close vision

Search frame 27680–27681

Retina of the eye

Search frame 52988

Exercise 25
Special Senses: Hearing and Equilibrium

Structure of the ear

Search frame 27687–27690

Anatomy of the cochlea

Search frame 28262–28264

The effect of gravitational pull on a macula receptor cell in the utricle

Search frame 28265

Localization and structure of a crista ampullaris

Search frame 28269–28271

One turn of the cochlea

Search frame 52989

The organ of Corti

Search frame 52990

Exercise 26
Special Senses: Taste and Olfaction

Location and structure of taste buds

Search frame 27662–27664

Olfactory receptors

Search frame 27665

Taste buds

Search frame 52991

Olfactory epithelium

Search frame 52992

Exercise 27
Anatomy and Basic Function of the Endocrine Glands

Location of the major endocrine organs of the body

Search frame 28276

Structural and functional relationships of the pituitary and the hypothalamus

Search frame 29556–29559

Gross and microscopic anatomy of the thyroid gland

Search frame 29560

Summary of parathyroid hormone's effects on bone, the intestines, and the kidneys

Search frame 29568

Microscopic structure of the adrenal gland

Search frame 29571

Regulation of blood sugar levels by insulin and glucagon

Search frame 29576

The thyroid gland

Search frame 52993

Pancreatic islet

Search frame 52994

Posterior pituitary

Search frame 52995

Adrenal gland

Search frame 52996

Exercise 29 Blood

Major components of whole blood

Search frame 29583

Life cycle of red blood cells

Search frame 30216

Structure of hemoglobin

Search frame 30211–30212

Human blood smear

Search frame 53018

Monocyte

Search frame 53019

Neutrophils

Search frame 53020

Eosinophil

Search frame 53021

Lymphocyte

Search frame 53022

Basophil

Search frame 53023

Exercise 30 Anatomy of the Heart

Location of the heart in the mediastinum of the thorax

Search frame 30228–30230

The pericardial layers and layers of the heart wall

Search frame 30232

Gross anatomy of the heart

Search frame 30235–30241

Beating heart

Search frame 30245

Heart valves

Search frame 34828

Operation of the atrioventricular valves of the heart

Search frame 38381–38383

Operation of the semilunar valves

Search frame 39433–39436

Microscopic anatomy of cardiac muscle

Search frame 39437–39438

Thin section of cardiac muscle

Search frame 53039

Exercise 31 or 31i Conduction System of the Heart and Electrocardiography

The sequence of excitation of the heart related to the deflection waves of an ECG tracing

Search frame 40232

Echocardiogram (video)

Search frame 38385

Exercise 32 Anatomy of Blood Vessels

Structure of arteries, veins, and capillaries

Search frame 40246–40249

Anatomy of a capillary bed

Search frame 40250

Arteries of the abdomen

Search frame 43333–43340

Venous drainage of the head, neck, and brain

Search frame 43344–43347

Pulmonary circulation

Search frame 43324

Arteries of the head, neck, and brain

Search frame 43327–43330

Veins of the abdomen

Search frame 43348–43350

Circulation in fetus and newborn

Search frame 52952–52955

Exercise 33 Human Cardiovascular Physiology: Blood Pressure and Pulse Determination

Systemic blood pressure

Search frame 41867

Exercise 34 Frog Cardiovascular Physiology

Microscopic anatomy of cardiac muscle

Search frame 39437–39438

Changes in membrane potential and membrane permeability during action potentials of contractile cardiac muscle cells

Search frame 40228–40229

Pacemaker and action potentials of autorhythmic cells of the heart

Search frame 40230

Anatomy of a capillary bed

Search frame 40250

Exercise 35 The Lymphatic System and Immune Response

The lymphatic system

Search frame 43359–43361

Lymphoid organs

Search frame 43364

Structure of the spleen

Search frame 43367–43368

Simplified version of clonal selection of a B cell by antigen binding

Search frame 43381

The central role of helper T cells

Search frame 43392–43394

Lymphocyte traffic

Search frame 43378

Basic antibody structure

Search frame 43385–43387

Neutrophils

Search frame 53020

Spleen

Search frame 52998

Palatine tonsil

Search frame 52999

Lymphocyte

Search frame 53022

Exercise 36 Anatomy of the Respiratory System

Basic anatomy of the upper respiratory tract

Search frame 45725

Anatomical relationships of organs in the thoracic cavity

Search frame 46428

The major respiratory organs shown in relation to surrounding structures

Search frame 45724

Respiratory zone structures

Search frame 46423

Anatomy of the respiratory membrane

Search frame 46425–46426

Tissue composition of the tracheal wall

Search frame 46421

Alveoli, alveolar ducts and sacs

Search frame 53000

Cross section of the trachea

Search frame 53001

Exercise 37 or 37i Respiratory System Physiology

Changes in thoracic volume that lead to the flow of air (gases) during breathing

Search frame 46430

Exercise 38 Anatomy of the Digestive System

Organs of the alimentary canal and related accessory organs

Search frame 46449

The basic structure of the wall of the alimentary canal consists of four layers: the mucosa, submucosa, muscularis externa, and serosa

Search frame 47808

Anatomy of the oral cavity (mouth)

Search frame 47811–47813

Major salivary glands

Search frame 47815–47816

Microscopic structure of the wall of the esophagus

Search frame 47820

Anatomy of the stomach

Search frame 47826–47828

Microscopic anatomy of the stomach

Search frame 47829–47831

The duodenum of the small intestine, and related organs

Search frame 47832

Structural modifications of the small intestine that increase its surface area for digestion and absorption

Search frame 47834

Scanning electron micrographs of the villi and microvilli of the small intestine

Search frame 47836

Gross anatomy of the large intestine

Search frame 47857–47859

Longitudinal section of a canine tooth within its bony alveolus

Search frame 47817

Major salivary glands

Search frame 47815–47816

Microscopic anatomy of the liver

Search frame 47840

Mechanisms promoting the secretion of bile and its entry into the duodenum

Search frame 47844

Stomach wall showing three tunics

Search frame 53002

Palatine tonsil

Search frame 52999

Gastroesophageal junction

Search frame 53004

Gastric glands and pits

Search frame 53003

Villi and duodenal glands

Search frame 53005

Ileum showing Peyer's patches

Search frame 53006

Large intestine mucosa

Search frame 53007

Pig liver

Search frame 53009

Pancreas

Search frame 53008

Pancreatic islet

Search frame 52994

Exercise 39 Chemical and Physical Processes of Digestion

Schematic summary of gastrointestinal tract activities

Search frame 46452

Exercise 40 Anatomy of the Urinary System

Organs of the urinary system

Search frame 50752

Internal anatomy of the kidney

Search frame 50753

Location and structure of nephrons

Search frame 50755–50759

Detailed anatomy of nephrons and their vasculature

Search frame 50760–50763

The kidney depicted schematically as a single, large, uncoiled nephron

Search frame 50764

The filtration membrane

Search frame 52822–52824

Directional movements of reabsorbed substances

Search frame 52825

Summary of nephron functions

Search frame 52828

Structure of the urinary bladder and urethra

Search frame 52834–52836

Glomerulus

Search frame 53046

Urinary bladder wall

Search frame 53047

Exercise 42
Anatomy of the Reproductive System

Reproductive organs of the male

Search frame 52870

Structure of the penis

Search frame 52875

Internal structure of the testis

Search frame 52873–52874

Internal organs of the female reproductive system

Search frame 52903–52907

Structure of an ovary

Search frame 52910–52913

Anterior view of the internal reproductive organs from a female cadaver

Search frame 52914

Epididymis

Search frame 53010

The penis (transverse section)

Search frame 53011

Uterus: proliferative phase

Search frame 53015

Exercise 43
Physiology of Reproduction: Gametogenesis and the Female Cycles

Meiotic cell division

Search frame 52882

Spermatogenesis

Search frame 52891–52894

Structure of an ovary

Search frame 52910–52913

The female menstrual cycle: structural and hormonal changes

Search frame 52927

Feedback interaction involved in the regulation of ovarian function

Search frame 52924

The ovary

Search frame 53014

Uterus: proliferative phase

Search frame 53015

Uterus: secretory phase

Search frame 53016

Uterus: menstrual phase

Search frame 53017

Exercise 44
Survey of Embryonic Development

Sperm penetration and the cortical reaction

Search frame 52933

Cleavage is a rapid series of mitotic divisions that begins with the zygote and ends with the blastocyst

Search frame 52939–52943

Events of placentation, early embryonic development, and formation of the embryonic membranes

Search frame 52944–52950

Exercise 45 Principles of Heredity

ABO blood groups

Search frame 52975

Exercise 46 Surface Anatomy Roundup

Lateral view of muscles of the scalp, face, and neck

Search frame 21315

Muscles of the anterior neck and throat that promote swallowing

Search frame 21326

Location of the heart in the mediastinum of the thorax

Search frame 30228

Posterior view of superficial muscles of the body

Search frame 21308

The bony thorax

Search frame 14181

Anterior view of superficial muscles of the body

Search frame 21296

Muscles of the abdominal wall

Search frame 21336–21340

CREDITS

Photograph Credits

Plates 1–60: © The Benjamin/Cummings Publishing Company. Photos by Victor Eroschenko, University of Idaho, except as noted.

Plate 10: Courtesy of Churchill-Livingstone.

Plate 20: © Carolina Biological Supply Company.

Plate 30: Courtesy of Marian Rice.

Beauchene skull photos: © Somso Modelle.

Plates A-G, 10.5c, 15.3c, 15.4c,15.5c, 15.6a, 15.10b, 15.11b, 15.12c, 19.2c, 19.4b, 19.6a,b, 19.7c, 21.1b, 36.1b, 38.7: From *A Stereoscopic Atlas of Human Anatomy* by David L. Bassett, M.D.

3.5, 6.5k, 7.6, page RS13: © The Benjamin/Cummings Publishing Company. Photo by Victor Eroschenko.

6.3, 6.5a,c–j, 6.6a–c, 30.6, page RS35: © The Benjamin/Cummings Publishing Company. Photo by Allen Bell, University of New England.

2.1, 2.2, 2.3b, 24.6, 29.6, 29.7, 29.9: © Benjamin/Cummings Publishing Company. Photo by Jack Scanlon, Holyoke Community College.

22.4, 22.5, 22i.6: B. Kozier and G. Erb, *Techniques in Clinical Nursing,* 4th ed. (Redwood City, CA: Addison-Wesley Nursing, a division of Benjamin/Cummings, 1993). © 1993 Addison-Wesley Publishing Company.

19.11, 30.7, 30.8, 40.3b: © The Benjamin/Cummings Publishing Company. Photo by Wally Cash, Kansas State University.

16i.1, 22i.4, 31i.4, 37i.5: Reprinted with permission of Intelitool, Inc.

6.5b, 9.3b, 14.2b, page RS129: Courtesy of Marian Rice.

1.1, Chapter 46: © Paul Waring/BioMed Arts Associates, Inc.

2.5: © Carolina Biological Supply Company/Phototake.

3.1: Courtesy of Leica Inc., Deerfield, IL.

4.1b: © Photo Researchers, Inc.

4.3: © Ed Reschke.

5.2: © Richard Megna/Fundamental Photographs.

5.3a: © K. R. Porter/Photo Researchers, Inc.

5.3b: © David M. Phillips/The Population Council, Science Source/Photo Researchers.

5.3c: Courtesy of Dr. Mohandas Narla, Lawrence Berkeley Laboratory.

6.7c: © Ed Reschke.

7.3: B. Kozier and G. Erb, *Fundamentals of Nursing,* 5th ed. (Redwood City, CA: Addison-Wesley Nursing, a division of Benjamin/Cummings, 1995). © 1995 Addison-Wesley Publishing Comany.

10.8c: A. P. Spence, E. B. Mason, *Human Anatomy and Physiology,* 3rd ed. (Redwood City, CA: Benjamin/Cummings, 1987). © 1987 The Benjamin/Cummings Publishing Company.

11.5b: Courtesy of the Department of Anatomy and Histology, University of California, San Francisco.

11.6d: © The Benjamin/Cummings Publishing Company. Photo by San Francisco State University.

11.8c, 11.9b, 11.10b: Courtesy of Richard Hambert, Stanford University.

12.2: © The Benjamin/Cummings Publishing Company. Photo by San Francisco State University.

13.2: Courtesy of University of California, San Francisco, Department of Anatomy and Histology.

15.8f: Courtesy of University of Kansas Medical Center. Photo by Stephen Spector.

16B.2, 16B.3, and 16B.4: Computer screens from *Mechanical Properties of Active Muscle* by Richard A. Meiss, © 1987 COM Press, a division of Queue, Inc., 1-800-232-2224.

17.2a: © Manfred Kage/Peter Arnold, Inc.

17.3g: Courtesy of G. L. Scott, J. A. Feilbach, T. A. Duff.

19.10c, 19.12b, 19.13: © The Benjamin/Cummings Publishing Company. Photo by Dr. Sharon Cummings, University of California, Davis.

20.1a: © Alexander Tsiaras, Photo Researchers.

22.3: Courtesy of William Thompson.

24.12a : Courtesy of Dr. William Spencer.

24.12b : © Elena Dorfman

24.13: Courtesy of Christopher S. Sherman, Swedish Hospital Medical Center.

29.4: Courtesy of Becton Dickinson Vacutainer Systems.

29.8b: © Boehringer Ingelheim International GmbH. Photo by Lennart Nilsson.

30.3b: © Lennart Nilsson, *Behold Man,* Little, Brown and Company.

35.4: © Carroll H. Weiss, Camera M.D. Studios.

36.3c: © SPL, Custom Medical Stock Photo.

37.4a: © The Benjamin/Cummings Publishing Company. Photo by San Francisco State University.

37.4b: Courtesy of the Warren E. Collins Company.

45.1a: © Lennart Nilsson, *The Body Victorious,* Dell Publishing Company.

45.1b: Christine J. Harrison et al., "Scanning Electron Microscopy of the G-Banded Human Karyotype," *Exp Cell Res* 134(1981): 141–153: © 1981 Academic Press, Inc.

45.2a: © Marc Romanelli/The Image Bank.

45.2b,e : © Eliane Sulle/The Image Bank.

45.2c,d,g: © Ogust/The Image Works.

45.2f,h–j: © The Benjamin/Cummings Publishing Company. Photo by Anthony Loveday.

Page RS7: Courtesy of Leica, Inc., Deerfield, IL.

22i.3: Courtesy of Intelitool, Inc.

22i.5: Courtesy of William Thompson.

37i.4: Courtesy of San Francisco State University.

Trademark Acknowledgements

Adrenalin is a registered trademark of Warner-Lambert Company.

Albustix, Clinitest, Hernastix, Icotest, Ketostik, and Multistix are registered trademarks of Miles Laboratories, Inc.

Betadine is a registered trademark of The Purdue Frederick Company.

Cardiocomp and Flexicomp are registered trademarks of Hewitt Associates.

Chemstrip is a registered trademark of Boehringer Mannheim Corporation.

Clinistix is a registered trademark of Ames Company, Inc.

Harleco is a registered trademark of Hartman-Leddon Company, Inc.

Hefty is a registered trademark of Mobil Oil Company.

IBM is a registered trademark of International Business Machines Corp.

Intelitool is a registered trademark of Intelitool, Inc.

Landau is a registered trademark of Landau Uniform.

Lead-Loc is a registered trademark of Lead-Loc, Inc.

Mechanical Properties of Active Muscle is a registered trademark of Queue, Inc.

Novocain is a registered trademark of Sterling Drug, Inc.

Parafilm is a registered trademark of American National Can Company.

Seal-Ease, Sedi-Stain, and Unopette are registered trademarks of Becton, Dickinson, and Company.

Spirocomp is a registered trademark of Cyber Diagnostics, Inc.

Velcro is a registered trademark of Velcro, U.S.A., Inc.

Wampole is a registered trademark of Henry K. Wampole and Company.

Index

Note: *Italicized* page numbers locate figures; pages with a *-t* suffix locate tables; and pages with a "C" locate entries in Appendix C.